# Molecular Biology of B Cells

# Molecular Biology of B Cells

SECOND EDITION

*Edited by*

**Frederick W. Alt**
**Tasuku Honjo**
**Andreas Radbruch**
**Michael Reth**

AMSTERDAM • BOSTON • HEIDELBERG • LONDON • NEW YORK • OXFORD • PARIS
SAN DIEGO • SAN FRANCISCO • SINGAPORE • SYDNEY • TOKYO

Academic Press is an imprint of Elsevier

Academic Press is an imprint of Elsevier
32 Jamestown Road, London NW1 7BY, UK
525 B Street, Suite 1800, San Diego, CA 92101-4495, USA
225 Wyman Street, Waltham, MA 02451, USA
The Boulevard, Langford Lane, Kidlington, Oxford OX5 1GB, UK

First edition 2003

Reprinted 2015



**Notices**

Knowledge and best practice in this field are constantly changing. As new research and experience broaden our understanding, changes in research methods, professional practices, or medical treatment may become necessary.

Practitioners and researchers must always rely on their own experience and knowledge in evaluating and using any information, methods, compounds, or experiments described herein. In using such information or methods they should be mindful of their own safety and the safety of others, including parties for whom they have a professional responsibility.

To the fullest extent of the law, neither the Publisher nor the authors, contributors, or editors, assume any liability for any injury and/or damage to persons or property as a matter of products liability, negligence or otherwise, or from any use or operation of any methods, products, instructions, or ideas contained in the material herein.

ISBN : 978-0-12-397933-9

**British Library Cataloguing-in-Publication Data**

A catalogue record for this book is available from the British Library

**Library of Congress Cataloging-in-Publication Data**

A catalog record for this book is available from the Library of Congress

For Information on all Academic Press publications visit our website at http://store.elsevier.com/

Typeset by TNQ Books and Journals
www.tnq.co.in

Printed and bound in the United Kingdom

# Dedication

## To Michael Neuberger (1953–2013)

We wish to dedicate this volume to the memory of Michael Neuberger, who was a coauthor of the prior volume, wonderful colleague, and truly outstanding B-cell biologist. In his lab at the MRC laboratory of Molecular Biology in Cambridge England he made groundbreaking contributions to elucidate the mechanisms of antibody gene regulation and antibody repertoire diversification. His work was characterized by his careful but very innovative approaches and his remarkable scientific insight.

Neuberger's earlier studies led to his discovery of major enhancer elements downstream of the IgH and Igκ loci, which have been shown to play critical roles in the secondary diversification of antibody genes. The enhancers that Neuberger first found downstream of IgH are now known to be important both for regulation of IgH class switch recombination (CSR), which generates different antibody classes through a DNA breakage and joining mechanism, as well as for IgH variable region somatic hypermutation (SHM) which provides the basis for antibody affinity maturation. In another line of research, Neuberger, together with Greg Winter, was the first to use recombinant DNA techniques to generate human antibodies with desired antigen binding specificities. He also collaborated with Marianne Bruggemann to pioneer the development of human antibody production in transgenic mice. His work in these areas helped lay a foundation for the development therapeutic human antibodies, which now have proven so successful in the clinic.

Neuberger's more recent work on how antibody genes are further diversified by a DNA deamination mechanism has been among his most important and influential. One of the coauthors of this volume, namely Tasuku Honjo discovered the activation-induced cytidine deaminase (AID) and showed that it is required for CSR as well as for SHM. A mystery at the time of the discovery was how the small AID protein could initiate DNA breaks during CSR and mutations during SHM. Neuberger, along with Mathew Scharf and others, proposed that AID could serve this dual role by deaminating deoxycytidines in Ig loci to trigger a cascade of events that lead to DNA double-strand brakes during CSR and mutations during SHM. Neuberger's model was based on his careful, several decade-long studies of the SHM process of antibody genes. Following up on his model, Neuberger performed a beautiful series of genetic studies that provided compelling evidence for his DNA deamination model of AID function in SHM and IgH class switching, a model, which is now widely accepted by nearly all workers in the field and that has formed the basis for their ongoing studies.

Beyond science, Michael was a warm and enthusiastic individual with quite diverse set of interests in history and literature, as well as in various sports. Interacting with him both scientifically and personally was always a great pleasure. It was a privilege to have him as a friend. Michael also was well known as a great mentor and leader. His outstanding scientific contributions and leadership have been acknowledged by many prestigious awards including his recent election as a Foreign Associate of the US National Academy of Sciences which acknowledged both his leadership and discoveries outlined above on antibody diversification.

# Contents

## 10. Structure and Signaling Function of the B-Cell Antigen Receptor and Its Coreceptors

*Michael Reth, Lars Nitschke, Masaki Hikida and Tomohiro Kurosaki*



## 11. Fc and Complement Receptors

*Jeffrey V. Ravetch, Falk Nimmerjahn and Michael C. Carroll*



## 12. B Cell Localization and Migration in Health and Disease

*Anja E. Hauser and Uta E. Höpken*



## 13. B Cells as Regulators

*Ping Shen, Ellen Hilgenberg, Andreia C. Lino, Van Duc Dang, Stefanie Ries, Imme Sakwa and Simon Fillatreau*

## 17. Gut Microbiota and Their Regulation

*Frederick W. Alt and Duane R. Wesemann*



## 18. Molecular Mechanisms of AID Function

*Nasim A. Begum, Hitoshi Nagaoka, Maki Kobayashi and Tasuku Honjo*



## 19. The Mechanism of IgH Class Switch Recombination

*Fei-Long Meng, Frederick W. Alt and Ming Tian*



## 20. Somatic Hypermutation: The Molecular Mechanisms Underlying the Production of Effective High-Affinity Antibodies

*Alberto Martin, Richard Chahwan, Jahan Yar Parsa and Matthew D. Scharff*

## 27. Anti-interleukin-6 Receptor Antibody Therapy Against Autoimmune Inflammatory Diseases

*Toshio Tanaka, Masashi Narazaki and Tadamitsu Kishimoto*



## 28. Targeting the IL-17/IL-23 Axis in Chronic Inflammatory Immune-Mediated Diseases

*Alexander Tsoukas, Carl S. Goodyear and Iain B. McInnes*



## 29. Discovery and Development of Anti-TNF Therapy: Pillar of a Therapeutic Revolution

*Marc Feldmann*

# Preface

B lymphocytes are an essential part of the humoral immune system and produce the specific antibodies generated during infection or vaccination. Soon after their discovery as antitoxins more than a hundred years ago by Emil Behring and Shibasaburo Kitasato it became clear that antibodies are enormously diverse and can be generated against many different structures. The elucidation of the mechanism underlying the generation of diversity (GOD) of the antibody system kept immunologists busy over the last hundred years. With the discovery of the V(D)J recombination mechanism of the immunoglobulin (Ig) genes and the somatic hypermutation (SHM) processes many aspects of this problem were solved. The discovery of the V(D)J recombinase machinery consisting of the lymphocyte lineage-specific proteins recombination activating gene (RAG) 1 and RAG2 which provide the specific endonuclease function and the generally expressed nonhomologous DNA end joining factors that join RAG-generated DNA breaks was a fascinating development as was the discovery of AID and its associated protein machinery mediating class switch recombination (CSR) as well as SHM and gene conversion. The progress in the B-cell field in this area is well documented by the different editions of this book series which started in 1989 under the title "Immunoglobulin genes" and in 2004 became "Molecular Biology of B cells."

This new edition of the "Molecular Biology of B cells" brings the reader up to date with these Ig diversifying processes. However, this book has now a much broader scope as it covers many different aspects of B cell Biology. B cells are a well-characterized and easily assessable cellular system that can be used to answer many issues of general cell biology. Furthermore due to the availability of several different B cell-specific Cre deleter mice the B cell system is also useful for general gene function studies. These are reasons why many scientists worldwide start to work on B cells although they are not regarded as classical 'B cell immunologists' and for these this book is particularly useful. In fact B cells have served as excellent model systems to study many common biological questions including transcriptional regulation of differentiation, signal transduction, and tumorigenesis. Furthermore, the mechanism of GOD has been considered to be unique to lymphocytes especially B cells as the expression of AID is specific to B cells. Subsequently, however, the DNA repair phases of GOD are shown to be common between B cells and other types of cells, indicating that GOD employs factors that normally are entrusted with preserving genome stability. More recently, IgA synthesis has been shown to influence the whole body metabolic regulation through symbiotic interaction with the gut microbiota. These striking developments convince us that molecular biology of B cells does not only mean molecular biology 'for' B cells but also molecular biology 'through' B cells to understand the whole biology.

The chapters of this new edition of "Molecular Biology of the B cells" have been written by an international faculty of experts in their fields and the readers can expect to get a comprehensive overview about the current status and the future development of the B-cell field. In particular the reader will learn how a B-cell-specific transcriptional network drives differentiation of hematopoietic stem cells (HSC) through B lymphopoiesis and how during their development B cells repeatedly switch between molecular programs promoting proliferation and those involved in differentiation. B-cell research in the age of genomics also means that we now know many more details about the gene expression program of different B-cell developmental stages. This is due to the joint efforts of many scientists, for example those working in the ImmGen Consortium. We also learned much more about the signaling mechanisms controlling the development and activation of B cells. For example, it was found that tumor necrosis factor (TNF) superfamily members such as lymphotoxin and BAFF play an important role in the homeostasis and survival of B cells in the periphery.

New topics that were not at all covered in the last edition of this book are, for example, the role of evolutionarily conserved microRNAs during B-cell development, function, and transformation and the role of IL-10- and IL-35-producing regulatory B cells. Furthermore, jawless vertebrates (lampreys and hagfish) were recently shown to have B cells employing the leucine-rich repeat (LRR)-based variable lymphocyte receptors (VLR), a type of antigen receptor completely different from their mammalian counterpart.

Apart from being a useful tool for basic cell biological and signaling studies B cells are also playing an increasingly important role in the clinic. Genetic defects of the B-cell system are the cause of important human immunodeficiency diseases. A dysregulation of this system causes autoimmunity or tumor diseases, topics which are both well covered in this new edition. Furthermore, not only the B cells but also their product, namely antibodies made a remarkable career in the recent years. As the reader will learn by reading this book, antibodies are highly versatile tools that are not only used in clinical diagnostics but also play an important role as therapeutic agents. Thus most drug companies have antibody departments and factories that produce antibodies and bring them into the clinic for the treatment of various diseases.

In spite of the importance of B cells in basic research and the clinic, there are still many questions that B biology needs to address in the future. We still have to learn more about how the B cell system distinguishes between self and foreign antigens, in particular after the finding that many newly generated B cells have a certain level of autoreactivity. What is the role of B cells in specialized compartments such as mucosal tissues? The maintenance of B-cell memory and antibody production over a long period of time is also a topic of active research where major breakthroughs are to be expected.

**Frederick W. Alt**
**Tasuku Honjo**
**Radbruch Andreas**
**Michael Reth**
May, 2014.

# Contributors

**Frederick W. Alt** Howard Hughes Medical Institute, Program in Cellular and Molecular Medicine, Boston Children's Hospital, Boston, MA, USA; Department of Genetics, Harvard Medical School, Boston, MA, USA

**Radbruch Andreas** German Rheumatism Research Center Berlin, a Leibniz Institute, Berlin, Germany

**Nasim A. Begum** Department of Immunology and Genomic Medicine, Kyoto University Graduate School of Medicine, Kyoto, Japan

**Michael C. Carroll** Program in Cell and Molecular Medicine, Boston Childrens Hospital, Harvard Medical School, Boston, Massachusetts, USA

**Andrea Cerutti** Institut Hospital del Mar d'Investigacions Mèdiques, Barcelona, Spain; Immunology Institute, Department of Medicine, Icahn School of Medicine at Mount Sinai, New York, NY, USA

**Richard Chahwan** Department of Biosciences, University of Exeter, Exeter, UK

**Tsutomu Chiba** Department of Gastroenterology and Hepatology, Graduate School of Medicine, Kyoto University, Kyoto, Japan

**Max D. Cooper** Department of Pathology and Laboratory Medicine, Emory Vaccine Center, Emory University School of Medicine, Atlanta, GA, USA

**Riccardo Dalla-Favera** Institute for Cancer Genetics, Herbert Irving Comprehensive Cancer Center, Columbia University, New York, NY, USA; Department of Pathology and Cell Biology, Columbia University, New York, NY, USA; Department of Genetics and Development, Columbia University, New York, NY, USA; Department of Microbiology and Immunology, Columbia University, New York, NY, USA

**Van Duc Dang** Deutsches Rheuma-Forschungszentrum, Leibniz Institute, Berlin, Germany

**Sabyasachi Das** Department of Pathology and Laboratory Medicine, Emory Vaccine Center, Emory University School of Medicine, Atlanta, GA, USA

**Betty Diamond** Center for Autoimmune and Musculoskeletal Disease, The Feinstein Institute for Medical Research, Manhasset, NY, USA

**Anne Durandy** National Institute of Health and Medical Research INSERM U768, Necker Children's Hospital, Paris, France; Descartes-Sorbonne Paris Cité University of Paris, Imagine Institute, France; Department of Immunology and Hematology, Assistance Publique-Hopitaux de Paris, Necker Children's Hospital, Paris, France

**Sidonia Fagarasan** Laboratory for Mucosal Immunity, RIKEN Center for Integrative Medical Sciences, RIKEN Yokohama, Tsurumi, Yokohama, Japan

**Ann J. Feeney** Department of Immunology and Microbial Science, The Scripps Research Institute, CA, USA

**Marc Feldmann** Nuffield Department of Orthopaedics, Rheumatology, and Musculoskeletal Sciences, Kennedy Institute of Rheumatology, University of Oxford, Headington, Oxford, UK

**Simon Fillatreau** Deutsches Rheuma-Forschungszentrum, Leibniz Institute, Berlin, Germany

**Alain Fischer** National Institute of Health and Medical Research INSERM U768, Necker Children's Hospital, Paris, France; Descartes-Sorbonne Paris Cité University of Paris, Imagine Institute, France; Department of Immunology and Hematology, Assistance Publique-Hopitaux de Paris, Necker Children's Hospital, Paris, France

**Jennifer L. Gommerman** Department of Immunology, University of Toronto, Toronto, ON, Canada

**Carl S. Goodyear** Institute of Infection, Immunity and Inflammation, Glasgow Biomedical Research Centre, University of Glasgow, Glasgow, UK

**James Hagman** National Jewish Health, Denver, CO, USA

**Richard R. Hardy** Fox Chase Cancer Center, Philadelphia, PA, USA

**Anja E. Hauser** Immune Dynamics, Deutsches Rheumaforschungszentrum, Berlin, Germany; Charité Universitätsmedizin, Berlin, Germany

**Kyoko Hayakawa** Fox Chase Cancer Center, Philadelphia, PA, USA

**Barton F. Haynes** Departments of Medicine and Immunology, Duke Human Vaccine Institute, Duke University School of Medicine, Durham, NC, USA

**Brantley R. Herrin** Department of Pathology and Laboratory Medicine, Emory Vaccine Center, Emory University School of Medicine, Atlanta, GA, USA

**Falk Hiepe** Department of Rheumatology and Clinical Immunology, Charité University Hospital, Berlin, Germany

**Masaki Hikida** Center for Innovation in Immunoregulative Technology and Therapeutics, Graduate School of Medicine, Kyoto University, Kyoto, Japan

**Ellen Hilgenberg** Deutsches Rheuma-Forschungszentrum, Leibniz Institute, Berlin, Germany

**Masayuki Hirano** Department of Pathology and Laboratory Medicine, Emory Vaccine Center, Emory University School of Medicine, Atlanta, GA, USA

**Tasuku Honjo** Department of Immunology and Genomic Medicine, Kyoto University Graduate School of Medicine, Kyoto, Japan

**Uta E. Höpken** Department of Tumor- and, Immunogenetics, Max-Delbrück Center for Molecular Medicine, Berlin, Germany

**Ellen Hsu** Department of Physiology and Pharmacology, The State University of New York Health Science Center at Brooklyn, Brooklyn, NY, USA

**Hassan Jumaa** Institute of Immunology, University Clinics Ulm, Ulm, Germany; Faculty of Biology, Albert-Ludwigs University of Freiburg, Freiburg, Germany; Centre for Biological Signaling Studies BIOSS, Albert-Ludwigs University of Freiburg, Freiburg, Germany

**John F. Kearney** Department of Microbiology, University of Alabama at Birmingham, Birmingham, AL, USA

**Garnett Kelsoe** Department of Immunology, Duke Human Vaccine Institute, Duke University School of Medicine, Durham, NC, USA

**Tadamitsu Kishimoto** Laboratory of Immune Regulation, World Premier International Immunology Frontier Research Center, Osaka University, Osaka, Japan

**Maki Kobayashi** Department of Immunology and Genomic Medicine, Kyoto University Graduate School of Medicine, Kyoto, Japan

**Sven Kracker** National Institute of Health and Medical Research INSERM U768, Necker Children's Hospital, Paris, France; Descartes-Sorbonne Paris Cité University of Paris, Imagine Institute, France

**Tomohiro Kurosaki** Laboratory for Lymphocyte Differentiation, Center for Integrative Medical Sciences (IMS), RIKEN, Yokohama, Japan; Laboratory for Lymphocyte Differentiation, WPI Immunology Frontier Research Center, Osaka University, Osaka, Japan

**Marie-Paule Lefranc** IMGT®, The International ImMunoGeneTics Information System®, Université Montpellier 2, CNRS, Institut de Génétique Humaine, Montpellier, France

**Susanna M. Lewis** Department of Pathology and Laboratory Medicine, Perelman School of Medicine, University of Pennsylvania, Philadelphia, PA, USA

**Jianxu Li** Department of Pathology and Laboratory Medicine, Emory Vaccine Center, Emory University School of Medicine, Atlanta, GA, USA

**Andreia C. Lino** Deutsches Rheuma-Forschungszentrum, Leibniz Institute, Berlin, Germany

**Alicia J. Little** Department of Immunobiology, Yale University School of Medicine, New Haven, CT, USA

**Joseph S. Lucas** Division of Biological Sciences, Department of Molecular Biology, University of California, San Diego, CA, USA

**Fabienne Mackay** Department of Immunology, Monash University, Clayton, VIC, Australia

**Giuliana Magri** Institut Hospital del Mar d'Investigacions Mèdiques, Barcelona, Spain

**Alberto Martin** Department of Immunology, University of Toronto, Toronto, ON, Canada

**Hiroyuki Marusawa** Department of Gastroenterology and Hepatology, Graduate School of Medicine, Kyoto University, Kyoto, Japan

**John R. Mascola** Vaccine Research Center, National Institutes of Allergy and Infectious Diseases, National Institutes of Health, Bethesda, MD, USA

**Adam Matthews** Department of Molecular Biology, Massachusetts General Hospital, Boston, MA, USA; Department of Genetics, Harvard Medical School, Boston, MA, USA; Department of Biological Sciences, Wellesley College, Wellesley, MA, USA

**Iain B. McInnes** Institute of Infection, Immunity and Inflammation, Glasgow Biomedical Research Centre, University of Glasgow, Glasgow, UK

**Fei-Long Meng** Howard Hughes Medical Institute, Program in Cellular and Molecular Medicine, Boston Children's Hospital, Boston, MA, USA; Department of Genetics, Harvard Medical School, Boston, MA, USA

**Byoung-Gon Moon** Fox Chase Cancer Center, Philadelphia, PA, USA

**Stefan A. Muljo** Laboratory of Immunology, National Institute of Allergy and Infectious Diseases, National Institutes of Health, Bethesda, Maryland, USA

**Cornelis Murre** Division of Biological Sciences, Department of Molecular Biology, University of California, San Diego, CA, USA

**Gary J. Nabel** Sanofi, Cambridge, MA, USA

**Hitoshi Nagaoka** Department of Immunology and Genomic Medicine, Kyoto University Graduate School of Medicine, Kyoto, Japan

**Masashi Narazaki** Department of Respiratory Medicine, Allergy and Rheumatic Diseases, Osaka University Graduate School of Medicine, Osaka University, Osaka, Japan

**Falk Nimmerjahn** Institute of Genetics, University of Erlangen-Nuernberg, Germany

**Lars Nitschke** Division of Genetics, Department of Biology, University of Erlangen-Nürnberg, Erlangen, Germany

**Alberto Nobrega** Department of Immunology, Paulo de Goes Institute of Microbiology, Federal University of Rio de Janeiro, Rio de Janeiro, Brazil

**Marjorie Oettinger** Department of Molecular Biology, Massachusetts General Hospital, Boston, MA, USA

**Jahan Yar Parsa** Department of Immunology, University of Toronto, Toronto, ON, Canada

**Laura Pasqualucci** Institute for Cancer Genetics, Herbert Irving Comprehensive Cancer Center, Columbia University, New York, NY, USA; Department of Pathology and Cell Biology, Columbia University, New York, NY, USA

**Klaus Rajewsky** Max Delbrück Center for Molecular Medicine, Berlin, Germany

**Jeffrey V. Ravetch** Laboratory of Molecular Genetics and Immunology, The Rockefeller University, New York, USA

**Michael Reth** BIOSS Centre of Biological Signaling Studies, University of Freiburg, Freiburg im Breisgau, Germany; Faculty of Biology, Institute of Biology III, University of Freiburg, Freiburg im Breisgau, Germany; MPI for Immunobiology and Epigenetics, Freiburg im Breisgau, Germany

**Roy Riblet** Torrey Pines Institute for Molecular Studies, San Diego, CA, USA

**Stefanie Ries** Deutsches Rheuma-Forschungszentrum, Leibniz Institute, Berlin, Germany

**David B. Roth** Department of Pathology and Laboratory Medicine and the Center for Personalized Diagnostics, Perelman School of Medicine of the University of Pennsylvania, PA, USA

**Imme Sakwa** Deutsches Rheuma-Forschungszentrum, Leibniz Institute, Berlin, Germany

**Kevin O. Saunders** Vaccine Research Center, National Institutes of Allergy and Infectious Diseases, National Institutes of Health, Bethesda, MD, USA

**Matthew D. Scharff** Department of Cell Biology, Albert Einstein College of Medicine, Bronx, NY, USA

**David G. Schatz** Department of Immunobiology, Yale University School of Medicine, New Haven, CT, USA; Department of Molecular Biophysics and Biochemistry, Yale University School of Medicine, New Haven, CT, USA; Howard Hughes Medical Institute, New Haven, CT, USA

**Harry W. Schroeder** Department of Microbiology, University of Alabama at Birmingham, Birmingham, AL, USA; Department of Medicine, University of Alabama at Birmingham, Birmingham, AL, USA; Department of Genetics, University of Alabama at Birmingham, Birmingham, AL, USA

**Ping Shen** Deutsches Rheuma-Forschungszentrum, Leibniz Institute, Berlin, Germany

**Susan A. Shinton** Fox Chase Cancer Center, Philadelphia, PA, USA

**Yoichi Sutoh** Department of Pathology and Laboratory Medicine, Emory Vaccine Center, Emory University School of Medicine, Atlanta, GA, USA

**Atsushi Takai** Department of Gastroenterology and Hepatology, Graduate School of Medicine, Kyoto University, Kyoto, Japan

**Toshitada Takemori** RIKEN Center for Integrative Medical Sciences, Yokohama, Japan

**Toshio Tanaka** Department of Clinical Application of Biologics, Osaka University Graduate School of Medicine, Osaka University, Osaka, Japan; Department of Immunopathology, World Premier International Immunology Frontier Research Center, Osaka University, Osaka, Japan

**David Tarlinton** The Walter and Eliza Hall Institute of Medical Research, Parkville, VIC, Australia

**Ming Tian** Howard Hughes Medical Institute, Program in Cellular and Molecular Medicine, Boston Children's Hospital, Boston, MA, USA; Department of Genetics, Harvard Medical School, Boston, MA, USA

**Alexander Tsoukas** Institute of Infection, Immunity and Inflammation, Glasgow Biomedical Research Centre, University of Glasgow, Glasgow, UK

**Andre M. Vale** Program in Immunobiology, Carlos Chagas Filho Institute of Biophysics, Federal University of Rio de Janeiro, Rio de Janeiro, Brazil

**Markus Werner** Institute of Immunology, University Clinics Ulm, Ulm, Germany

**Duane R. Wesemann** Howard Hughes Medical Institute, Program in Cellular and Molecular Medicine, Boston Children's Hospital, Boston, MA, USA; Department of Genetics, Harvard Medical School, Boston, MA, USA

**Yan Zhou** Fox Chase Cancer Center, Philadelphia, PA, USA

**Yong-Rui Zou** Center for Autoimmune and Musculoskeletal Disease, The Feinstein Institute for Medical Research, Manhasset, NY, USA

Chapter 1

# The Structure and Regulation of the Immunoglobulin Loci

Joseph S. Lucas[1], Cornelis Murre[1], Ann J. Feeney[2] and Roy Riblet[3]

[1]*Division of Biological Sciences, Department of Molecular Biology, University of California, San Diego, CA, USA,* [2]*Department of Immunology and Microbial Science, The Scripps Research Institute, CA, USA,* [3]*Torrey Pines Institute for Molecular Studies, San Diego, CA, USA*

## 1. INTRODUCTION

Our adaptive immune system relies heavily on the use of antibodies, produced by B cells, to eliminate foreign pathogens and toxins. It is estimated that mammalian organisms have the ability to generate on the order of $10^{11}$ different antibodies [1]. This staggering number is made possible by an elaborate mechanism that assembles complete antigen binding site sequences from arrays of gene segments encoding portions of the complete antibody proteins. This assembly involves combinatorial selection of each type of gene segment and creation of short novel sequences at the junctions of these DNA segments, thus generating a vast variety of assembled gene sequences and antigen binding specificities. Sequence diversity is further increased by the somatic hypermutation process that is active in these sequences.

Antibodies are proteins made up of two identical heavy chains and two identical light chains. All heavy chains are expressed from the Igh locus, whereas light chains are expressed from one of two loci, Igκ or Igλ [2]. The DNA segments that rearrange to create antibody heavy and light chain genes in developing B cells include variable (V) and joining (J) elements, and heavy chain genes include a third, diversity (Dh), gene segment [3]. Single light chain V and J and heavy chain V, D, and J segments are selected and joined in a seemingly random process out of the many such genomic segments that span as much as 3 Mbp of DNA

Over the past 30 years, a large number of studies have been performed to describe the process of V(D)J recombination in molecular terms, including how it is regulated during development; how it is controlled by cell signaling; how it is modulated by transcription; and how chromatin modifications, nuclear positioning and three-dimensional (3D) topology affect rearrangement. Collectively, these experiments aimed to address a number of critical questions. Some of the prominent questions follow: How do DNA sequences separated by millions of base pairs interact to allow recombination? How are gene segments selected randomly to allow for the generation of antibody diversity?

## 2. GENOMIC ORGANIZATION OF THE MOUSE IMMUNOGLOBULIN HEAVY CHAIN LOCUS

The Igh locus is made up of multiple variable (Vh), Dh, joining (Jh), and constant (Ch) segments, arrayed in adjacent regions of the Igh locus. In mice the locus spans 2.75 million base pairs, and there are about 100 Vh segments that have seemingly functional coding sequences and nearly as many nonfunctional Vh segments, 10–15 Dh segments depending on mouse strain, and four Jh segments. Eight constant regions encode the Igh isotypes [4,5]. Vh genes encode most of the heavy chain variable region including the first two hypervariable loops that form the binding site for antigen [6,7]. Dh segments are very short, 10–15 nucleotides of coding sequence, but they are critically important in the generation of antibody diversity. Dh segments determine most of the heavy chain's third hypervariable region or complementarity determining region 3 (CDR3) [8–10]. In the folded protein's antigen binding site, this region makes major, sometimes dominant, contacts with antigen.

Each set of gene segments has evolved, and continues to evolve, via gene addition by duplication, divergence through mutation, and loss by nonsense or frameshift mutations or by partial or total deletion. A prominent example is the ancestral duplication of the constant region IgG2a gene and divergence and alternate gene deletion leaving the present IgG2a gene in BALB/c and many other mouse strains and the IgG2c gene in C57BL/6 and related strains [11,12]. A larger scale case in the Igh-V (Vh) array is the duplication of the entire proximal region of the Vh array comprising the

Molecular Biology of B Cells. http://dx.doi.org/10.1016/B978-0-12-397933-9.00001-1

interspersed Vh7183 and VhQ52 genes (Ighv5 and Ighv2 families in Figure 1) from 400 kb in C57BL/6 to 800 kb in 129 and BALB/c [5].

Alignment of Vh gene sequences and grouping them by similarity yields an evolutionary structure of the gene population, revealing three major branches or groups, splitting further into 15 families of genes that share 80% sequence identity. The three groups of Vh genes are an ancient evolutionary division; Vh genes from any vertebrate species fall into several or all three of these groups. Vh gene family structure is apparently not generally shared beyond closely related species [13], although the Vh7183 or Ighv5 family is found widely [14]. In mice, at least, a family of Vh genes occupies a restricted region of the Vh array and are interspersed there with other families; the proximal, 3′ 300 kb region of the Vh locus contains only the interspersed Vh7183 and VhQ52 (Ighv5 and Ighv2) families, the distal, 5′ 1.5 Mb portion contains only VhJ558 and Vh3609 (Ighv1 and Ighv8) genes, and the central 680 kb contains the remaining families.

In developing B cells, Vh, Dh, and Jh segments recombine to encode the antigen-binding domain of the antibody heavy chain [3]. Each Vh, Dh, and Jh segment is flanked by a short DNA sequence called the recombination signal sequence (RSS). This sequence is recognized by recombination activating gene enzymes RAG1 and RAG2. RAG1 and RAG2 form a complex with additional proteins and act to create a loop between the two RSSs, bringing them into proximity. The RAG complex then induces double-strand DNA breaks and promotes ligation of appropriate coding segments via the nonhomologous end-joining machinery [15–17].

In the Igh locus, V, D, J, and C gene segments are in separate clusters, and all in the same 5′→3′ polarity. VDJ rearrangements proceed via serial excisions of intervening sequences, D to J, then V to DJ, and loss of the excised

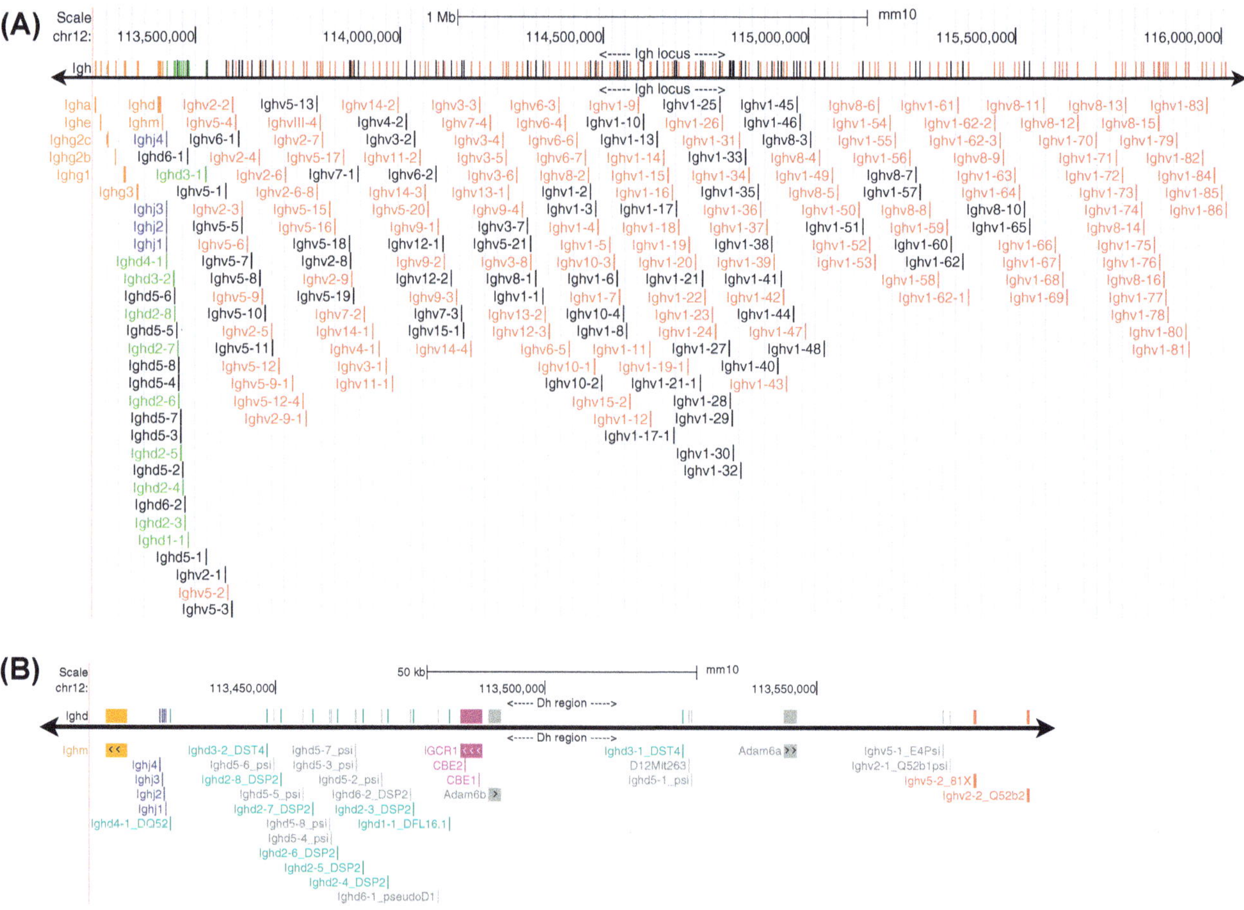


FIGURE 1 **The genomic structure of the immunoglobulin heavy chain locus.** The 2.75 Mb Igh locus is located on the distal region of mouse chromosome 12. (A) The various types of gene segments are represented in different colors: functional Vh segments are red, functional Dh segments are green, the four Jh segments are blue, and the eight Ch loci are orange. Nonfunctional Vh and Dh pseudogenes are black. The sequence is from mm10, the Dec. 2011 *Mus musculus* assembly (Genome Reference Consortium Mouse Build 38), documented in NCBI Gene ID: 111507 [http://ncbi.nlm.nih.gov/gene/?term=111507]. (B) Expanded view of the Dh region; the IgM constant region at the left is orange; the Jh regions are blue; functional Dh segments are cyan; nonfunctional Dh, Vh, and Adam6 pseudogenes are gray; the intergenic control region 1 (88) with its two CCCTC-binding factor binding elements, CBE1 and CBE2, are magenta; and functional Vh genes at the right are red.

sequences as free closed DNA circles. VDJ recombination is irreversible, although a secondary V–V replacement mechanism occurs at low frequency [18].

## 3. GENOMIC ORGANIZATION OF THE MOUSE IMMUNOGLOBULIN KAPPA LIGHT CHAIN LOCUS

The kappa light chain locus (Figure 2) is even larger than the heavy chain locus, extending 3.17 Mb; it too contains approximately 100 functional and 60 nonfunctional V segments and four functional (plus one nonfunctional) J segments [19]. Light chain genes do not use D segments. Interestingly, half of all Vk genes are in the opposite polarity to the Jk and Ck loci and must rearrange through a nondestructive inversion of the portion of the Vk gene array between the joining Vk and Jk segments. As a result, many Vk segments are retained rather than excised, for potential use in a secondary rearrangement with any remaining downstream Jk segments. This conserves a larger set of Vk genes for "B cell editing," i.e., replacement of an initial Vk-Jk choice that yields an antiself antibody or is otherwise a poor partner for the rearranged heavy chain. There seems to be no bias between the excision and inversion mechanisms because both sets of Vk genes contain similar frequencies of both functional genes and highly expressed genes [19].

## 4. GENOMIC ORGANIZATION OF THE MOUSE IMMUNOGLOBULIN LAMBDA LIGHT CHAIN LOCUS

In mice the lambda light chain locus is quite small, 200 kb, and of limited diversity (Figure 3). There are two adjacent VJC clusters, 5′-V2-V3 (originally termed VλX)-J2C2 followed by V1-J3C3-J1C1. The major rearranged combinations expressed in serum antibodies are λ1, V1J1C1, λ2, V2J2C2, and λ3, V1J3C3. Rarely, λx, V3J2C2, and V2J3C3 and V2J1C1 products have been noted [20]. The large difference in germline-encoded diversity between kappa and lambda light chains is reflected in the population of serum antibodies that is 95% kappa [20].

## 5. B CELL DEVELOPMENT AND REGULATION OF V(D)J RECOMBINATION

Recombination steps define the stages of development in B cells. Before V(D)J recombination, the cells are called pre-pro-B cells and are not fully committed to the B cell lineage [21–23]. However, at the Igh locus Dh to Jh recombination can occur at this stage [24]. Vh to DJh rearrangement occurs in the pro-B cell stage, and the successful production of a heavy chain protein is necessary to permit the transition from the pro-B stage to the pre-B cell stage [25,26]. Upon productive Igh locus V(D)J gene rearrangement, cycling large pre-B cells express the pre-B cell receptor that, in turn, functions to suppress RAG1/2 activity, preventing continued Igh locus rearrangement [27]. Large pre-B cells divide several times to increase the population of cells with functional heavy chains. Large pre-B cells differentiate into resting small pre-B cells [26]. In small pre-B cells, the RAG genes are expressed again to allow for Ig light chain gene rearrangement [27]. At this point, assuming successful recombination, the cells will express a B cell receptor. Before the cells are released from the bone marrow, they undergo negative selection. Self-antigens are presented by antigen-presenting cells such as dendritic cells. B-cells expressing receptors with high affinity toward self-antigens will undergo apoptosis or additional recombination of the light chains [28,29]. Surviving B cells migrate to the periphery as naïve cells, waiting to be activated by their specific antigens [30].

V(D)J recombination is a highly regulated and ordered event. Dh to Jh recombination always precedes Vh to DhJh recombination, and recombination of the heavy chain locus always precedes recombination of the light chain loci [31,32]. Furthermore, Vh to DhJh recombination occurs on only one allele at a time, although both alleles have previously undergone Dh to Jh rearrangement [33]. Such a strict order of events is crucial to ensure that each B cell produces antibodies with affinity toward a single antigen. Although the expression of RAG genes is tightly regulated, the fact that they catalyze all recombination steps means that additional mechanisms need to be in place to ensure the ordered progression of rearrangement [27]. Hence only at the correct developmental stage does an antigen receptor locus become accessible to recombination.

The Igh locus is, by default, nonpermissive to V(D)J recombination. This state is achieved by several mechanisms, many of which resemble general mechanisms used in transcriptional repression. First, the locus is tethered at the transcriptionally repressive nuclear periphery [34]. The Igh locus is highly enriched for nucleosomes that are dimethylated at lysine nine of histone 3, correlating with transcriptionally inactive genomic regions [35]. Furthermore, the RSS sequences may act as weak nucleosome-positioning sequences, causing the recombination sequences to be buried within the histones and inaccessible to the RAG complex [36]. In addition, the Igh locus is decontracted, positioning many of the Vh gene segments away from the DhJh elements [34,37]. Recombination of the Igh locus requires a reversal of the above-mentioned trends. The mechanisms by which this occurs vary for different regions of the locus.

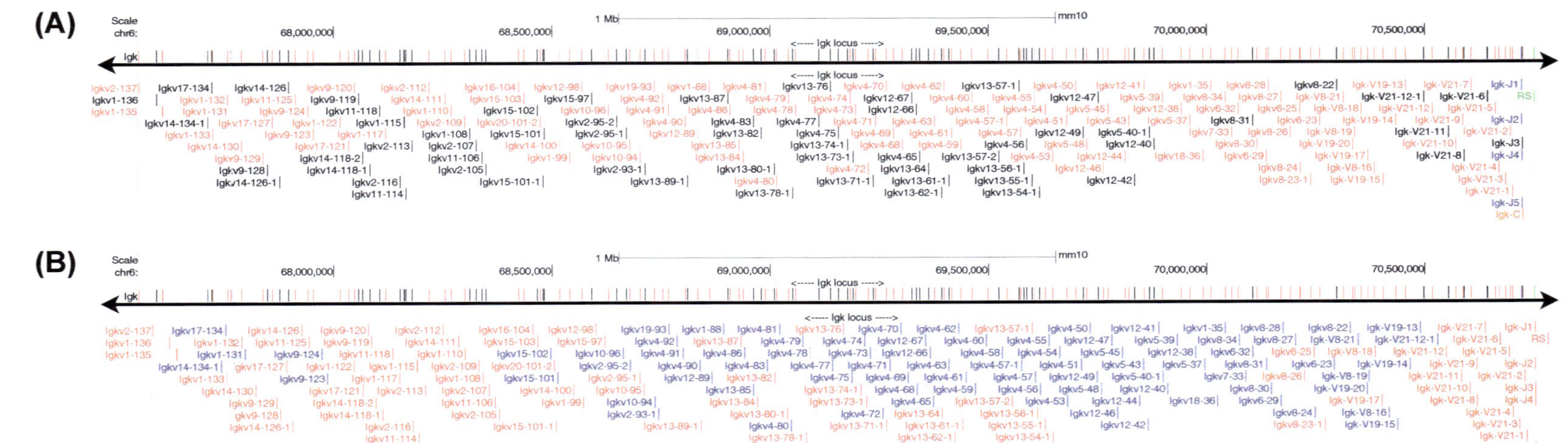


**FIGURE 2** **The genomic structure of the immunoglobulin kappa light chain locus.** The 3.2 Mb Igk locus is located centrally on mouse chromosome 6. (A) The various types of gene segments are represented in different colors: functional Vk segments are red, the four functional Jk segments are blue, and the single Ck locus is orange. Nonfunctional Vk and Jk pseudogenes are black. The downstream "recombining sequence" (RS) is green. (B) On this map gene segments on the plus strand are colored red, and those Vk segments on the minus strand that rearrange by inversion are blue. The sequence is from mm10, the Dec. 2011 *Mus musculus* assembly (Genome Reference Consortium Mouse Build 38), documented in NCBI Gene ID: 243469 [http://ncbi.nlm.nih.gov/gene/?term=243469].

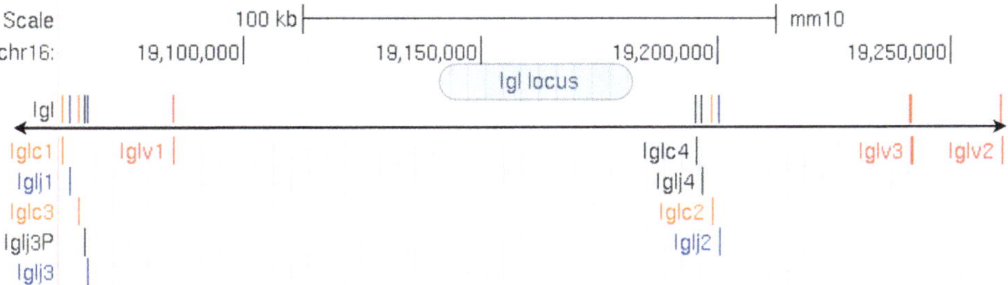


FIGURE 3 **The genomic structure of the immunoglobulin lambda light chain locus.** The small 200kb Igl locus is located in the proximal region of mouse chromosome 16. The various types of gene segments are represented in different colors: the three Vl segments are red, the three functional Jl segments are blue, and the three functional Cl loci are orange. Nonfunctional Jl and Cl pseudogenes are black. The sequence is from mm10, the Dec. 2011 *Mus musculus* assembly (Genome Reference Consortium Mouse Build 38), documented in NCBI Gene ID: 111519 [http://ncbi.nlm.nih.gov/gene/?term=111519].

## 6. JUNCTIONAL DIVERSITY

The primary antibody repertoire derives its diversity from two sources: combinatorial diversity (mix-and-match usage of V, D, and J gene segments) and junctional diversity [38]. Junctional diversity occurs because the RAG1/2 recombinase makes a covalent hairpin at the coding end. Because the hairpin can be opened at any location near its end, this can result in a random loss of a small number of nucleotides at the coding end, or the gain of a few base pairs from the opposite strand that is now covalently linked to the coding end. These latter nucleotides are called "P" nucleotides because they are palindromic to the coding end. In addition, tremendous diversity is added to the junctions by the nontemplated addition of nucleotides to the coding ends by terminal deoxynucleotidyl transferase [38–41]. This "N" region diversity provides extensive diversity to all junctions of Ig and of T cell receptors, with the exception of light chains. In mice, there are no N nucleotides at the light chain V–J junctions, and in humans there is less N addition than in the heavy chain. N region diversity is also absent early in ontogeny, greatly restricting the fetal repertoire of antibodies [42,43].

## 7. COMBINATORIAL DIVERSITY

Initial estimates of the diversity of the antibody repertoire assumed that all V, D, and J genes recombined with equal frequencies, thus making combinatorial association of the large number of V, D, and J gene segments a major aspect of creating a diverse repertoire. However, the advent of polymerase chain reaction (PCR) technology provided the means to determine that this was not the case, but rather Vh and Vk genes were shown to rearrange at very different intrinsic frequencies [44–48]. D and J genes also were used at different frequencies, but the differences were much more modest. Early repertoire studies consisted of PCR, cloning, and sequencing, but after 30 years of this approach, next generation sequencing now is producing vastly more detailed studies [49]. Nonetheless, early studies demonstrating nonrandom V gene usage on a small scale are now substantiated and extended by the deep sequencing approach [19,50]. The reasons for this unequal V gene usage are myriad.

The quality of the flanking RSS is one factor in the nonrandom usage of V genes in the primary repertoire [51]. The RSS consists of a conserved heptamer and nonamer separated by a spacer of approximately 12 or 23 bp. Although there is a consensus, very few RSSs actually have the full consensus heptamer and nonamer sequence. In vitro plasmid-based recombination substrate assays have shown the effect of variation from the consensus at each position [52,53]. These studies demonstrated that the first 3 bp of the heptamer are essential. The remaining 4 bp of the heptamer can tolerate variation to a greater extent, with reduction in recombination frequency of 2- to 50-fold. The nonamer has more flexibility and should have a core of 3–5 A's, flanked by non-A's. Again, recombination substrates have demonstrated the importance of each site. The 12 and 23 bp spacers are one or two turns of the DNA helix, respectively. Hence, they can tolerate only a slight variation in length, and such variation in length will reduce the recombination efficiency. The sequence of the spacer is much less conserved than the heptamer and nonamer, but some spacer sequences support more recombination than others [54,55]. Because the RSS is the binding site for the recombinase, one might expect that variation from the consensus sequence might support less efficient RAG binding, and thus less efficient recombination of that gene. Indeed, this has been shown for allelic variants of a V gene [56]. However, it has also been demonstrated that V genes with identical RSS can rearrange at very different frequencies, thus demonstrating that other factors play important roles in V, D, and J gene use [47].

## 8. NONCODING TRANSCRIPTION AND IMMUNOGLOBULIN LOCUS REARRANGEMENT

It was demonstrated many years ago that J gene and constant region genes undergo robust "sterile" or "germline" transcription preceding V(D)J rearrangement at that locus [39,57,58]. For many years it was not clear whether this made those regions accessible for rearrangement, or if the transcription was a byproduct of the fact that the regions had been made accessible. Key studies using transcriptional terminators inserted into the genome demonstrated that the J genes downstream of the terminator rearranged less well, providing evidence that the germline transcription directly facilitated the accessibility of those downstream gene segments [59]. We know now that noncoding RNA plays many regulatory roles, and these "germline transcripts" through the J/C regions at the Igh and Igk loci were among the first noncoding RNAs to be identified [60].

In addition to the high abundance of transcripts through the J/C region, it was also shown many years ago that the V genes underwent germline transcription [61]. These transcripts started at the V gene promoter, proceeded through the coding region of the V gene segment, and ended at variable sites a little further downstream of the gene. These "sense" V gene transcripts were easily observed at the VhJ558 genes, but they were much more difficult to observe in other V genes. Much more recently, it was demonstrated that there is antisense transcription also in the Igh locus [62]. The D genes undergo antisense transcription precisely at the stage of D to J rearrangement [63], whereas the V gene part of the locus is transcribed only after DJ rearrangement has completed [62]. The precise timing of the antisense transcripts to the stage at which those segments are rearranging led to the proposal that the transcription was making the region accessible for rearrangement, although other hypotheses were also proposed [62–64]. Recently, RNA-sequencing (RNA-seq) analysis of the Igh locus revealed that there were three regions of extensive antisense transcription in the Vh locus, all at the distal end of the locus [65]. These originated at three of the "PAIR" elements, sites that bound Pax5, E2A, and CCCTC-binding factor (CTCF) [66]. RNA-seq also demonstrated that there is extensive, but low level, sense transcription of V genes in the distal half of the locus, with lower transcription in the proximal half of the locus. Nonfunctional pseudogenes had significantly lower level of sense transcripts than functional V genes.

## 9. THE PROCESS OF DH–JH RECOMBINATION

Dh to Jh recombination of the Igh locus is perhaps the least regulated step of V(D)J recombination as it is found on both alleles in a majority of B cells [31]. DhJh joints can also be detected in the common lymphoid progenitor compartment and in a fraction of developing T-lineage cells [67–69]. At the time of Dh to Jh recombination, the locus is still tethered at the nuclear periphery on its 5′ side, allowing the 3′ end, containing the Dh and Jh elements, to extend toward the cell's interior, perhaps allowing it to interact with transcription factories [34,70]. The intronic enhancer (Eμ) located just downstream of the Jh segments is involved in switching the Dh and Jh regions of the locus from a repressive to an active state, allowing for germline transcription across the region [71]. Several transcription factors are involved, including E2A, EBF, and Pax5 [72–74]. Transcription, along with recruitment of the Swi/Snf chromatin-remodeling complex is thought to expose RSS sequences on the 5′ end of Jh segments and the 3′ end of Dh segments, allowing for their recombination [75]. A successful recombination exposes the 5′ RSS of the used Dh segment, targeting it for recombination with an upstream Vh segment [35].

## 10. EPIGENETICS AND IMMUNOGLOBULIN LOCUS REARRANGEMENT

The epigenetic status of V, D, and J genes can also impact rearrangement. Histone proteins can be posttranslationally modified by acetylation and methylation, and indeed there is significant variation in the local epigenetic environment of each V, D, and J gene. Actively transcribed regions acquire trimethylation of lysine 4 on histone 3 (H3K4me3) because the H3K4 methyltransferase travels with the RNA polymerase II complex [76]. Not surprisingly, therefore, the highly transcribed J and C genes have the highest amount of H3K4me3 [77,78]. This is of great importance for repertoire development, and for ordered rearrangement of DJ genes before V to DJ gene segments, because it was demonstrated that RAG2 specifically binds to H3K4me3 [79,80]. Thus, RAG2 will initially be bound at the J genes. However, there are also regions of H3K4me3 throughout the V regions at sites of higher transcription (65). In addition, acetylation is found more generally throughout actively rearranging loci [81]. Indeed, V genes have been demonstrated to have more acetylation when that locus is undergoing rearrangement than when the locus is inactive [82]. Histone posttranslational modifications such as acetylation and methylation can act as docking sites for other chromatin-modifying enzymes, e.g., ATP-dependent chromatin-remodeling complexes. These could be important because the RSS needs to be exposed to bind the RAG complex, and thus nucleosomes may need to be moved to achieve this goal [83,84]. Bioinformatic analysis of the epigenetic status of each V gene demonstrated that there was a correlation of the extent of histone acetylation or methylation and rearrangement frequency for the distal half of the Igh locus [50].

## 11. INSULATORS AND IMMUNOGLOBULIN LOCUS REARRANGEMENT

The transition from a repressive to active chromatin state that occurs during Dh to Jh recombination is highly localized and does not extend past the 5′ most Dh segment. Recent studies have identified an insulator element located 5′ of the DhJh cluster that interferes with rearrangement of Vh gene elements before DhJh joining [85,86]. The insulator element is characterized by the presence of two CTCF binding sites, also named the CTCF-binding element (CBE). In pre-pro-B cells, these CTCF sites interact with CTCF sites downstream of the 3′ regulatory regions, creating a looped domain that includes all of the Dh and Jh genes, but excludes all Vh genes, thus facilitating DhJh rearrangement before Vh to DhJh [87]. Deletion of the CBE results in a loss of ordered and lineage-specific Igh locus recombination [88]. This leads to the cardinal conclusion that the process of ordered rearrangement is controlled, at least in part, by an insulator element, acting to suppress VhDhJh gene rearrangement before the formation of a DhJh joint. We suggest that after Dh to Jh recombination, this insulator is inactivated, plausibly through the removal of cohesin, a binding partner of CTCF [85,87,88].

## 12. 3D STRUCTURE AND COMPACTION OF THE IMMUNOGLOBULIN HEAVY CHAIN LOCUS

It is well established that the folding of the chromatin fiber plays a critical role in genome function. For more than a century, the organization of the chromatin fiber has been extensively analyzed using light microscopy. Each chromosome is folded into its own territory [89–91]. The DNA from each chromosome can intermingle with other chromosomes at the boundaries, but in general it is kept within its own territory [92]. In eukaryotes, DNA is restricted to the nucleus throughout most of the cell cycle where it is highly compacted. The first level of compaction is achieved by wrapping of the DNA around core histones (H2A, H2B, H3, and H4) to form nucleosomes [93,94]. Addition of histone H1 brings the nucleosomes together in a zigzag pattern to form a fiber approximately 30 nm in diameter [95]. The exact architecture of the chromatin past the 30 nm fiber is still a topic of intense study. It has been well established that the level of DNA compaction varies throughout the nucleus. Gene poor regions and repressed genes tend to be located in highly compacted heterochromatin, whereas highly expressed genes are present in less compacted euchromatin [96,97]. Furthermore, heterochromatin tends to aggregate at the nuclear periphery. Highly expressed genes are often found near the center of the nucleus where they are thought to interact with complexes of polymerases and other factors, termed transcription factories [98,99]. Other highly expressed genes, though, are known to associate with the nuclear pores during transcription to expedite nuclear export of transcripts [100].

Electron microscopic studies have demonstrated the presence of ~90 kb loops in mitotic cells that assemble into rosettes that are comprised of ~18 loops [101–103]. Similar bundles of loops have been seen in interphase nuclei [104]. Despite the progress in studying the genomic organization of the Igh locus, the topology of the Igh locus in developing B cells remains unresolved. As a first approach to address this critical issue, recent studies have used 3D fluorescence in situ hybridization to measure end–to–end physical distances separating multiple markers across the Igh locus. These studies showed that in developing B cells, the distance distributions are compatible with a multiloop-subcompartment (MLS) model [37]. The MLS model predicts that the chromatin fiber is structured as rosettes of small loops, folding into 1 Mbp domains, connected by flexible linkers [105,106]. Furthermore, using computational geometry, this study revealed the striking conformational compaction of the Igh topology during the transition from the pre-pro-B to the pro-B cell stage [37].

Recent observations have identified CTCF binding located directly adjacent to many of the proximal Vh segments. This has raised the interesting possibility that CTCF helps to position Vh segments at the center of a rosette. Such a configuration would permit DhJh segments to have equal access to any Vh segment [85,107,108]. The comparison of experimental and simulated measurements suggests that the distal Vh segments also adopt a rosette-like structure and are repositioned during B-lineage progression to merge with the proximal Vh region cluster [37,73,109]. In theory, this should allow the DhJh regions equal access to nearly any Vh segment regardless of genomic separation. Although the exact mechanism remains to be determined, it has been shown that interleukin-7 signaling as well as Pax5, YY1, and Ezh2 are required to promote the merging of proximal and distal Vh segments [73,110,111].

## 13. CONCLUSION

During the past two decades, much has been learned as to how the immunoglobulin loci are organized in terms of their genomic structure. Recent studies also have revealed as to how the immunoglobulin heavy chain locus is spatially structured. However, the majority of these studies have used populations of studies to generate statistical topologies. What is needed now is to reveal the distribution of structures in single cells and to determine the mechanics that permit V, D, and J elements to encounter each other with the appropriate frequencies.

## REFERENCES


[1] Nussenzweig MC, Alt FW. Antibody diversity: one enzyme to rule them all. Nat Med 2004;10(12):1304–5.

[2] Jung D, Giallourakis C, Mostoslavsky R, Alt FW. Mechanism and control of V(D)J recombination at the immunoglobulin heavy chain locus. Annu Rev Immunol 2006;24:541–70.

[3] Early P, Huang H, Davis M, Calame K, Hood L. An immunoglobulin heavy chain variable region gene is generated from three segments of DNA: VH, D and JH. Cell 1980;19(4):981–92.

[4] Brodeur PH, Riblet R. The immunoglobulin heavy chain variable region (Igh-V) locus in the mouse I One hundred Igh-V genes comprise seven families of homologous genes. Eur J Immunol 1984;14(10):922–30.

[5] Retter I, Chevillard C, Scharfe M, Conrad A, Hafner M, Im TH, et al. Sequence and characterization of the Ig heavy chain constant and partial variable region of the mouse strain 129S1. J Immunol 2007;179(4):2419–27.

[6] Chothia C, Lesk AM. Canonical structures for the hypervariable regions of immunoglobulins. J Mol Biol 1987;196:901–17.

[7] Al-Lazikani B, Lesk AM, Chothia C. Standard conformations for the canonical structures of immunoglobulins. J Mol Biol 1997;273:927–48.

[8] Wilson IA, Stanfield RL. Antibody-antigen interactions: new structures and new conformational changes. Curr Opin Struct Biol 1994;4:857–67.

[9] Morea V, Tramontano A, Rustici M, Chothia C, Lesk AM. Conformations of the third hypervariable region in the VH domain of immunoglobulins. J Mol Biol 1998;275:269–94.

[10] Larimore K, McCormick MW, Robins HS, Greenberg PD. Shaping of human germline IgH repertoires revealed by deep sequencing. J Immunol 2012;189(6):3221–30.

[11] Fukui K, Hamaguchi Y, Shimizu A, Nakai S, Moriwaki K, Wang CH, et al. Duplicated immunoglobulin gamma 2a genes in wild mice. J Mol Cell Immunol 1984;1(5):321–30.

[12] Morgado MG, Cam P, Gris-Liebe C, Cazenave PA, Jouvin-Marche E. Further evidence that BALB/c and C57BL/6 gamma 2a genes originate from two distinct isotypes. EMBO J 1989;8(11):3245–51.

[13] Tutter A, Riblet R. Evolution of the immunoglobulin heavy chain variable region (Igh-V) locus in the genus Mus. Immunogenetics 1989;30(5):315–29.

[14] Tutter A, Riblet R. Conservation of an immunoglobulin variable-region gene family indicates a specific, noncoding function. Proc Natl Acad Sci USA 1989;86(19):7460–4.

[15] Oettinger MA, Schatz DG, Gorka C, Baltimore D. RAG-1 and RAG-2, adjacent genes that synergistically activate V(D)J recombination. Science 1990;248(4962):1517–23.

[16] van Gent DC, McBlane JF, Ramsden DA, Sadofsky MJ, Hesse JE, Gellert M. Initiation of V(D)J recombination in a cell-free system. Cell 1995;81(6):925–34.

[17] Hendrickson EA, Qin XQ, Bump EA, Schatz DG, Oettinger M, Weaver DT. A link between double-strand break-related repair and V(D)J recombination: the scid mutation. Proc Natl Acad Sci USA 1991;88(10):4061–5.

[18] Kleinfield R, Hardy RR, Tarlinton D, Dangl J, Herzenberg LA, Weigert M. Recombination between an expressed immunoglobulin heavy-chain gene and a germline variable gene segment in a Ly 1+ B-cell lymphoma. Nature 1986;322(6082):843–6.

[19] Aoki-Ota M, Torkamani A, Ota T, Schork N, Nemazee D. Skewed primary Igk repertoire and V-J joining in C57BL/6 Mice: Implications for recombination accessibility and receptor editing. J Immunol 2012;188(5):2305–15.

[20] Eisen HN, Reilly EB. Lambda chains and genes in inbred mice. Annu Rev Immunol 1985;3:337–65.

[21] Ikawa T, Kawamoto H, Wright LY, Murre C. Long-term cultured E2A-deficient hematopoietic progenitor cells are pluripotent. Immunity 2004;20(3):349–60.

[22] Osmond DG, Rolink A, Melchers F. Murine B lymphopoiesis: towards a unified model. Immunol Today 1998;19(2):65–8.

[23] Hardy RR, Carmack CE, Shinton SA, Kemp JD, Hayakawa K. Resolution and characterization of pro-B and pre-pro-B cell stages in normal mouse bone marrow. J Exp Med 1991;173(5):1213–25.

[24] Li YS, Hayakawa K, Hardy RR. The regulated expression of B lineage associated genes during B cell differentiation in bone marrow and fetal liver. J Exp Med 1993;178(3):951–60.

[25] Opstelten D, Osmond DG. Pre-B cells in mouse bone marrow: immunofluorescence stathmokinetic studies of the proliferation of cytoplasmic mu-chain-bearing cells in normal mice. J Immunol 1983;131(6):2635–40.

[26] Osmond DG, Owen JJ. Pre-B cells in bone marrow: size distribution profile, proliferative capacity and peanut agglutinin binding of cytoplasmic mu chain-bearing cell populations in normal and regenerating bone marrow. Immunology 1984;51(2):333–42.

[27] Grawunder U, Leu TM, Schatz DG, Werner A, Rolink AG, Melchers F, et al. Down-regulation of RAG1 and RAG2 gene expression in preB cells after functional immunoglobulin heavy chain rearrangement. Immunity 1995;3(5):601–8.

[28] Goodnow CC, Crosbie J, Adelstein S, Lavoie TB, Smith-Gill SJ, Brink RA, et al. Altered immunoglobulin expression and functional silencing of self-reactive B lymphocytes in transgenic mice. Nature 1988;334(6184):676–82.

[29] Hartley SB, Crosbie J, Brink R, Kantor AB, Basten A, Goodnow CC. Elimination from peripheral lymphoid tissues of self-reactive B lymphocytes recognizing membrane-bound antigens. Nature 1991;353(6346):765–9.

[30] Allman DM, Ferguson SE, Lentz VM, Cancro MP. Peripheral B cell maturation. II. Heat-stable antigen(hi) splenic B cells are an immature developmental intermediate in the production of long-lived marrow-derived B cells. J Immunol 1993;151(9):4431–44.

[31] Alt FW, Yancopoulos GD, Blackwell TK, Wood C, Thomas E, Boss M, et al. Ordered rearrangement of immunoglobulin heavy chain variable region segments. Embo J 1984;3(6):1209–19.

[32] Maki R, Kearney J, Paige C, Tonegawa S. Immunoglobulin gene rearrangement in immature B cells. Science 1980;209:1366–9.

[33] Chowdhury D, Sen R. Mechanisms for feedback inhibition of the immunoglobulin heavy chain locus. Curr Opin Immunol 2004;16(2):235–40.

[34] Kosak ST, Skok JA, Medina KL, Riblet R, Le Beau MM, Fisher AG, et al. Subnuclear compartmentalization of immunoglobulin loci during lymphocyte development. Science 2002;296(5565):158–62.

[35] Subrahmanyam R, Du H, Ivanova I, Chakraborty T, Ji Y, Zhang Y, et al. Localized epigenetic changes induced by DH recombination restricts recombinase to DJH junctions. Nat Immunol 2012;13(12):1205–12.

[36] Baumann M, Mamais A, McBlane F, Xiao H, Boyes J. Regulation of V(D)J recombination by nucleosome positioning at recombination signal sequences. Embo J 2003;22(19):5197–207.

[37] Jhunjhunwala S, van Zelm MC, Peak MM, Cutchin S, Riblet R, van Dongen JJ, et al. The 3D structure of the immunoglobulin heavy-chain locus: implications for long-range genomic interactions. Cell 2008;133(2):265–79.

[38] Desiderio SV, Yancopoulos GD, Paskind M, Thomas E, Boss MA, Landau N, et al. Insertion of N regions into heavy-chain genes is correlated with expression of terminal deoxytransferase in B cells. Nature 1984;311(5988):752–5.

[39] Yancopoulos GD, Alt FW. Regulation of the assembly and expression of variable-region genes. Annu Rev Immunol 1986;4:339–68.

[40] Komori T, Okada A, Stewart V, Alt FW. Lack of N regions in antigen receptor variable region genes of TdT-deficient lymphocytes. Science 1993;261(5125):1171–5.

[41] Gilfillan S, Dierich A, Lemeur M, Benoist C, Mathis D. Mice lacking TdT: mature animals with an immature lymphocyte repertoire. Science 1993;261(5125):1175–8.

[42] Feeney AJ. Lack of N regions in fetal and neonatal mouse immunoglobulin V-D-J junctional sequences. J Exp Med 1990;172(5):1377–90.

[43] Gu H, Förster I, Rajewsky K. Sequence homologies, N sequence insertion and JH gene utilization in VHDJH joining: implications for the joining mechanism and the ontogenetic timing of Ly1 B cell and B-CLL progenitor generation. EMBO J 1990;9(7):2133–40.

[44] Yancopoulos GD, Desiderio SV, Paskind M, Kearney JF, Baltimore D, Alt FW. Preferential utilization of the most JH-proximal VH gene segments in pre-B-cell lines. Nature 1984;311(5988):727–33.

[45] Perlmutter RM, Kearney JF, Chang SP, Hood LE. Developmentally controlled expression of immunoglobulin VH genes. Science 1985;227(4694):1597–601.

[46] Feeney AJ, Lugo G, Escuro G. Human cord blood kappa repertoire. J Immunol 1997;158(8):3761–8.

[47] Williams GS, Martinez A, Montalbano A, Tang A, Mauhar A, Ogwaro KM, et al. Unequal VH gene rearrangement frequency within the large VH7183 gene family is not due to recombination signal sequence variation, and mapping of the genes shows a bias of rearrangement based on chromosomal location. J Immunol 2001;167(1):257–63.

[48] Feeney AJ, Goebel P, Espinoza CR. Many levels of control of V gene rearrangement frequency. Immunol Rev 2004;200:44–56.

[49] Boyd SD, Marshall EL, Merker JD, Maniar JM, Zhang LN, Sahaf B, et al. Measurement and clinical monitoring of human lymphocyte clonality by massively parallel VDJ pyrosequencing. Sci Transl Med 2009;1(12):12–23.

[50] Choi NM, Loguercio S, Verma-Gaur J, Degner SC, Torkamani A, Su AI, et al. Deep sequencing of the murine IgH repertoire reveals complex regulation of nonrandom v gene rearrangement frequencies. J Immunol 2013;191(5):2393–402.

[51] Feeney AJ. Genetic and epigenetic control of V gene rearrangement frequency. Adv Exp Med Biol 2009;650:73–81.

[52] Hesse JE, Lieber MR, Mizuuchi K, Gellert M. V(D)J recombination: a functional definition of the joining signals. Genes Dev 1989;3(7):1053–61.

[53] Akamatsu Y, Tsurushita N, Nagawa F, Matsuoka M, Okazaki K, Imai M, et al. Essential residues in V(D)J recombination signals. J Immunol 1994;153(10):4520–9.

[54] Montalbano A, Ogwaro KM, Tang A, Matthews AG, Larijani M, Oettinger MA, et al. V(D)J recombination frequencies can be profoundly affected by changes in the spacer sequence. J Immunol 2003;171(10):5296–304.

[55] Lee AI, Fugmann SD, Cowell LG, Ptaszek LM, Kelsoe G, Schatz DG. A functional analysis of the spacer of V(D)J recombination signal sequences. PLoS Biol 2003;1(1). E1.

[56] Feeney AJ, Atkinson MJ, Cowan MJ, Escuro G, Lugo G. A defective Vkappa A2 allele in Navajos which may play a role in increased susceptibility to haemophilus influenzae type b disease. J Clin Invest 1996;97(10):2277–82.

[57] Van Ness BG, Weigert M, Coleclough C, Mather EL, Kelley DE, Perry RP. Transcription of the unrearranged mouse C kappa locus: sequence of the initiation region and comparison of activity with a rearranged V kappa-C kappa gene. Cell 1981;27(3 Pt 2):593–602.

[58] Lennon GG, Perry RP. C mu-containing transcripts initiate heterogeneously within the IgH enhancer region and contain a novel 5'-nontranslatable exon. Nature 1985;318(6045):475–8.

[59] Abarrategui I, Krangel MS. Regulation of T cell receptor-alpha gene recombination by transcription. Nat Immunol 2006;7(10):1109–15.

[60] Perry RP, Kelley DE, Coleclough C, Seidman JG, Leder P, Tonegawa S, et al. Transcription of mouse kappa chain genes: implications for allelic exclusion. Proc Natl Acad Sci USA 1980;77(4):1937–41.

[61] Yancopoulos G, Alt FW. Developmentally controlled and tissue-specific expression of unrearranged VH gene segments. Cell 1985;40:271–81.

[62] Bolland DJ, Wood AL, Johnston CM, Bunting SF, Morgan G, Chakalova L, et al. Antisense intergenic transcription in V(D)J recombination. Nat Immunol 2004;5(6):630–7.

[63] Bolland DJ, Wood AL, Afshar R, Featherstone K, Oltz EM, Corcoran AE. Antisense intergenic transcription precedes Igh D-to-J recombination and is controlled by the intronic enhancer Emu. Mol Cell Biol 2007;27(15):5523–33.

[64] Chakraborty T, Chowdhury D, Keyes A, Jani A, Subrahmanyam R, Ivanova I, et al. Repeat organization and epigenetic regulation of the DH-Cmu domain of the immunoglobulin heavy-chain gene locus. Mol Cell 2007;27(5):842–50.

[65] Verma-Gaur J, Torkamani A, Schaffer L, Head SR, Schork NJ, Feeney AJ. Noncoding transcription within the Igh distal V(H) region at PAIR elements affects the 3D structure of the Igh locus in pro-B cells. Proc Natl Acad Sci USA 2012;109(42):17004–9.

[66] Ebert A, McManus S, Tagoh H, Medvedovic J, Salvagiotto G, Novatchkova M, et al. The distal V(H) gene cluster of the Igh locus contains distinct regulatory elements with Pax5 transcription factor-dependent activity in pro-B cells. Immunity 2011;34(2):175–87.

[67] Kurosawa Y, von Boehmer H, Haas W, Sakano H, Trauneker A, Tonegawa S. Identification of D segments of immunoglobulin heavy-chain genes and their rearrangement in T lymphocytes. Nature 1981;290(5807):565–70.

[68] Allman D, Sambandam A, Kim S, Miller JP, Pagan A, Well D, et al. Thymopoiesis independent of common lymphoid progenitors. Nat Immunol 2003;4(2):168–74.

[69] Borghesi L, Hsu LY, Miller JP, Anderson M, Herzenberg L, Herzenberg L, et al. B lineage-specific regulation of V(D)J recombinase activity is established in common lymphoid progenitors. J Exp Med 2004;199(4):491–502.

[70] Yang Q, Riblet R, Schildkraut CL. Sites that direct nuclear compartmentalization are near the 5' end of the mouse immunoglobulin heavy-chain locus. Mol Cell Biol 2005;25(14):6021–30.

[71] Perlot T, Alt FW, Bassing CH, Suh H, Pinaud E. Elucidation of IgH intronic enhancer functions via germ-line deletion. Proc Natl Acad Sci USA 2005;102(40):14362–7.

[72] Romanow WJ, Langerak AW, Goebel P, Wolvers-Tettero IL, van Dongen JJ, Feeney AJ, et al. E2A and EBF act in synergy with the V(D)J recombinase to generate a diverse immuno- globulin repertoire in nonlymphoid cells. Mol Cell 2000;5:343–53.

[73] Fuxa M, Skok J, Souabni A, Salvagiotto G, Roldan E, Busslinger M. Pax5 induces V-to-DJ rearrangements and locus contraction of the immunoglobulin heavy-chain gene. Genes Dev 2004;18(4):411–22.

[74] Seet CS, Brumbaugh RL, Kee BL. Early B cell factor promotes B lymphopoiesis with reduced Interleukin 7 responsiveness in the absence of E2A. J Exp Med 2004;199:1689–700.

[75] Osipovich OA, Subrahmanyam R, Pierce S, Sen R, Oltz EM. Cutting edge: SWI/SNF mediates antisense Igh transcription and locus-wide accessibility in B cell precursors. J Immunol 2009;183(3):1509–13.

[76] Milne TA, Dou Y, Martin ME, Brock HW, Roeder RG, Hess JL. MLL associates specifically with a subset of transcriptionally active target genes. Proc Natl Acad Sci USA 2005;102(41):14765–70.

[77] Xu CR, Feeney AJ. The epigenetic profile of Ig genes is dynamically regulated during B cell differentiation and is modulated by pre-B cell receptor signaling. J Immunol 2009;182(3):1362–9.

[78] Ji Y, Resch W, Corbett E, Yamane A, Casellas R, Schatz DG. The in vivo pattern of binding of RAG1 and RAG2 to antigen receptor loci. Cell 2010;141(3):419–31.

[79] Matthews AG, Kuo AJ, Ramón-Maiques S, Han S, Champagne KS, Ivanov D, et al. RAG2 PHD finger couples histone H3 lysine 4 trimethylation with V(D)J recombination. Nature 2007;450(7172):1106–10.

[80] Liu Y, Subrahmanyam R, Chakraborty T, Sen R, Desiderio S. A plant homeodomain in RAG-2 that binds Hypermethylated lysine 4 of histone H3 is necessary for efficient antigen-receptor-gene rearrangement. Immunity 2007;27(4):561–71.

[81] McMurry MT, Krangel MS. A role for histone acetylation in the developmental regulation of VDJ recombination. Science 2000;287(5452):495–8.

[82] Chowdhury D, Sen R. Stepwise activation of the immunoglobulin mu heavy chain gene locus. EMBO J 2001;20(22):6394–403.

[83] Kwon J, Imbalzano AN, Matthews A, Oettinger MA. Accessibility of nucleosomal DNA to V(D)J cleavage is modulated by RSS positioning and HMG1. Mol Cell 1998;2(6):829–39.

[84] Golding A, Chandler S, Ballestar E, Wolffe AP, Schlissel MS. Nucleosome structure completely inhibits in vitro cleavage by the V(D)J recombinase. EMBO J 1999;18(13):3712–23.

[85] Degner SC, Wong TP, Jankevicius G, Feeney AJ. Cutting edge: developmental stage-specific recruitment of cohesin to CTCF sites throughout immunoglobulin loci during B lymphocyte development. J Immunol 2009;182(1):44–8.

[86] Featherstone K, Wood AL, Bowen AJ, Corcoran AE. The mouse immunoglobulin heavy chain V-D intergenic sequence contains insulators that may regulate ordered V(D)J recombination. J Biol Chem 2010;285(13):9327–38.

[87] Degner SC, Verma-Gaur J, Wong TP, Bossen C, Iverson GM, Torkamani A, et al. CCCTC-binding factor (CTCF) and cohesin influence the genomic architecture of the Igh locus and antisense transcription in pro-B cells. Proc Natl Acad Sci USA 2011;108(23):9566–71.

[88] Guo C, Yoon HS, Franklin A, Jain S, Ebert A, Cheng HL, et al. CTCF-binding elements mediate control of V(D)J recombination. Nature 2011;477(7365):424–30.

[89] Cremer T, Lichter P, Borden J, Ward DC, Manuelidis L. Detection of chromosome aberrations in metaphase and interphase tumor cells by in situ hybridization using chromosome-specific library probes. Hum Genet 1988;80(3):235–46.

[90] Lichter P, Cremer T, Borden J, Manuelidis L, Ward DC. Delineation of individual human chromosomes in metaphase and interphase cells by in situ suppression hybridization using recombinant DNA libraries. Hum Genet 1988;80(3):224–34.

[91] Pinkel D, Landegent J, Collins C, Fuscoe J, Segraves R, Lucas J, et al. Fluorescence in situ hybridization with human chromosome-specific libraries: detection of trisomy 21 and translocations of chromosome 4. Proc Natl Acad Sci USA 1988;85(23):9138–42.

[92] Branco MR, Pombo A. Intermingling of chromosome territories in interphase suggests role in translocations and transcription-dependent associations. PLoS Biol 2006;4(5):e138.

[93] Kornberg RD, Thomas JO. Chromatin structure; oligomers of the histones. Science 1974;184(4139):865–8.

[94] Oudet P, Gross-Bellard M, Chambon P. Electron microscopic and biochemical evidence that chromatin structure is a repeating unit. Cell 1975;4(4):281–300.

[95] Bednar J, Horowitz RA, Grigoryev SA, Carruthers LM, Hansen JC, Koster AJ, et al. Nucleosomes, linker DNA, and linker histone form a unique structural motif that directs the higher-order folding and compaction of chromatin. Proc Natl Acad Sci USA 1998;95(24):14173–8.

[96] Hsu TC. Differential rate in RNA synthesis between euchromatin and heterochromatin. Exp Cell Res 1962;27:332–4.

[97] Frenster JH, Allfrey VG, Mirsky AE. Repressed and active chromatin Isolated from interphase lymphocytes. Proc Natl Acad Sci USA 1963;50:1026–32.

[98] Boyle S, Gilchrist S, Bridger JM, Mahy NL, Ellis JA, Bickmore WA. The spatial organization of human chromosomes within the nuclei of normal and emerin-mutant cells. Hum Mol Genet 2001;10(3):211–9.

[99] Sadoni N, Langer S, Fauth C, Bernardi G, Cremer T, Turner BM, et al. Nuclear organization of mammalian genomes. Polar chromosome territories build up functionally distinct higher order compartments. J Cell Biol 1999;146(6):1211–26.

[100] Chan JK, Park PC, De Boni U. Association of DNAse sensitive chromatin domains with the nuclear periphery in 3T3 cells in vitro. Biochem Cell Biol 2000;78(2):67–78.

[101] Paulson JR, Laemmli UK. The structure of histone-depleted metaphase chromosomes. Cell 1977;12(3):817–28.

[102] Paulson JR. Scaffolding and radial loops: the structural organization of metaphase chromosomes. In: Adolph KW, editor. Chromosomes and chromatin, vol. III. Boca Raton, Florida: CRC Press; 1988. pp. 3–36.

[103] Pienta KJ, Coffey DS. A structural analysis of the role of the nuclear matrix and DNA loops in the organization of the nucleus and chromosome. J Cell Sci Suppl 1984;1:123–35.

[104] Okada TA, Comings DE. Higher order structure of chromosomes. Chromosoma 1979;72(1):1–14.

[105] Münkel C, Langowski J. Chromosome structure described by a polymer model. Phys Rev E Stat Phys Plasmas Fluids Relat Interdiscip Top 1998;57(5B):5888–96.

[106] Knoch TA. Approaching the three-dimensional organization of the human genome: Structural-, scaling- and Dynamic-Properties in the simulation of interphase chromosomes and cell nuclei, Long-Range correlations in complete genomes, in vivo quantification of the chromatin distribution, Construct Conversions in Simultaneous Co-Transfections. Mannheim, Germany: TAK Press; 2002.

[107] Lin YC, Jhunjhunwala S, Benner C, Heinz S, Welinder E, Mansson R, et al. A global network of transcription factors, involving E2A, EBF1 and Foxo1, that orchestrates B cell fate. Nat Immunol 2010;11:635–43.

[108] Lucas JS, Bossen C, Murre C. Transcription and recombination factories: common features? Curr Opin Cell Biol 2011;23(3):318–24.

[109] Sayegh CE, Jhunjhunwala S, Riblet R, Murre C. Visualization of looping involving the immunoglobulin heavy-chain locus in developing B cells. Genes Dev 2005;19(3):322–7.

[110] Su IH, Basavaraj A, Krutchinsky AN, Hobert O, Ullrich A, Chait BT, et al. Ezh2 controls B cell development through histone H3 methylation and Igh rearrangement. Nat Immunol 2003;4:124–31.

[111] Liu H, Schmidt-Supprian M, Shi Y, Hobeika E, Barteneva N, Jumaa H, et al. Yin Yang 1 is a critical regulator of B-cell development. Genes Dev 2007;21(10):1179–89.

Chapter 2

# The Mechanism of V(D)J Recombination

Alicia J. Little[1], Adam Matthews[2,3,4], Marjorie Oettinger[2], David B. Roth[5] and David G. Schatz[1,6,7]

[1]*Department of Immunobiology, Yale University School of Medicine, New Haven, CT, USA;* [2]*Department of Molecular Biology, Massachusetts General Hospital, Boston, MA, USA;* [3]*Department of Genetics, Harvard Medical School, Boston, MA, USA;* [4]*Department of Biological Sciences, Wellesley College, Wellesley, MA, USA;* [5]*Department of Pathology and Laboratory Medicine and the Center for Personalized Diagnostics, Perelman School of Medicine of the University of Pennsylvania, PA, USA;* [6]*Department of Molecular Biophysics and Biochemistry, Yale University School of Medicine, New Haven, CT, USA;* [7]*Howard Hughes Medical Institute, New Haven, CT, USA*

## 1. OVERVIEW

In 1976, the revolutionary discovery was made that the DNA in lymphoid cells encoding the antigen receptors is altered from that of other somatic tissues and germ-line cells [1]. This DNA rearrangement is at the heart of the ability of B and T cells to generate a highly diverse array of antigen receptor molecules, allowing a virtually unlimited set of antigen molecules to be recognized with a high degree of specificity. A series of site-specific recombination events, collectively termed V(D)J recombination, serves to assemble antigen receptor genes from arrays of gene segments (for additional reviews on this topic see Refs [2–5]). Defects in the V(D)J recombination mechanism can result in immunodeficiencies, such as severe combined immunodeficiency and Omenn syndrome.

The process of V(D)J recombination can be divided into two highly coordinated stages: DNA cleavage and repair. In broad terms, rearrangement is initiated by the lymphoid-specific proteins recombination-activating gene (RAG) 1 and RAG2 [6,7]. In the first phase of this process, RAG1 and RAG2, aided by either of the DNA-binding/bending proteins high mobility group box protein (HMGB) 1 or HMGB2 [8], bind to the recombination signal sequences (RSSs) that flank each gene segment and introduce a double-strand break (DSB) between the RSS and the flanking coding DNA (Figure 1(A)). The DNA ends generated by cleavage are asymmetrical, with the coding end covalently sealed into a hairpin and the signal end present as a 5′-phosphorylated blunt DNA end [9–11]. This first stage of V(D)J rearrangement is the point at which most of the regulation of the recombination reaction is imposed.

The second phase of V(D)J recombination, end processing and end joining, involves RAG1 and 2 as well as the DNA-repair proteins of the nonhomologous end joining (NHEJ) pathway [12]. The broken DNA generated by RAG cleavage can be resolved through several different mechanisms. The first is standard V(D)J joining, in which the hairpins are opened and joined imprecisely to each other to form a coding joint (CJ) and the two signal ends are ligated together, to generate a signal joint (SJ) (Figure 1(A)). Joining is mediated by components of the NHEJ DNA repair pathway (reviewed in Refs [2,12]). Alternatively, two nonstandard products of V(D)J recombination can be generated by the joining of signal ends to coding ends [13]. Rejoining of a signal to its original coding flank yields an open and shut joint, whereas joining of a signal to its reaction partner's coding flank generates a hybrid joint. Finally, RAG1/2 bound to the signal ends can catalyze the transpositional attack of the signal ends on unrelated target DNA [14,15]. The mechanisms and factors involved in each of these reactions are considered separately later in this chapter.

This chapter considers in detail what is known about the V(D)J recombination reaction including the biochemistry of the cleavage reactions, the activities of the RAG proteins, and the regulation of recombination.

## 2. ANTIGEN RECEPTOR GENE ASSEMBLY

Immunoglobulin (Ig) genes and T cell receptor (TCR) genes exist in the germ line as linear arrays of clustered gene segments. During V(D)J recombination, the many germ-line V (variable), in some cases D (diversity), and J (joining) segments undergo stochastic rearrangement, with one V, one D in some cases, and one J segment being brought together by several highly coordinated DNA cleavage and repair steps, resulting in extensive combinatorial diversity (Figure 2). Lymphocyte antigen receptor diversity is enhanced by the imperfect joining of the V, D, and J segments, resulting in

Molecular Biology of B Cells. http://dx.doi.org/10.1016/B978-0-12-397933-9.00002-3

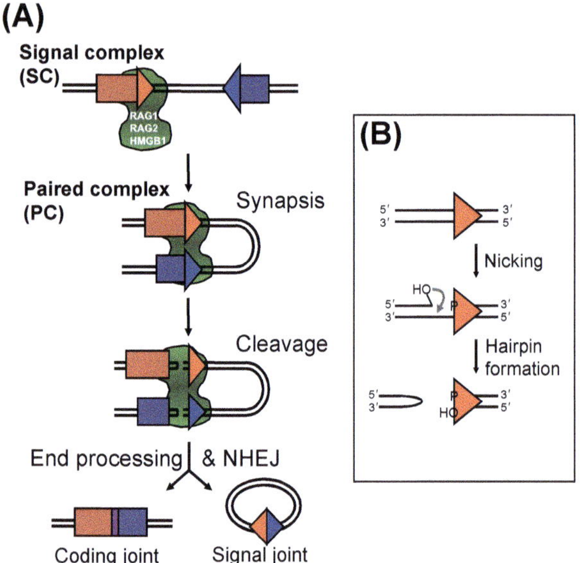


FIGURE 1 **Schematic of V(D)J recombination.** (A) A RAG1/RAG2/HMGB1 complex can bind a single RSS, forming a signal complex (SC), which then captures another RSS to form the paired complex (PC). Under physiological conditions, nicking can occur in the SC or PC, but hairpin formation can occur only in the context of the PC. After cleavage, hairpin coding ends and blunt signal ends are processed by the nonhomologous end-joining pathway, creating an imperfectly joined coding joint (imperfect joining indicated by purple section between coding gene segments) and a perfectly joined signal joint. (B) RAG-mediated DNA cleavage occurs in two steps, nicking and hairpin formation. Triangle depicts RSS, flanking coding gene segment is omitted for clarity. HO, hydroxyl group; HMGB1, high mobility group box protein 1; RSS, recombination signal sequence; RAG, recombination-activating gene; NHEJ, nonhomologous end joining.

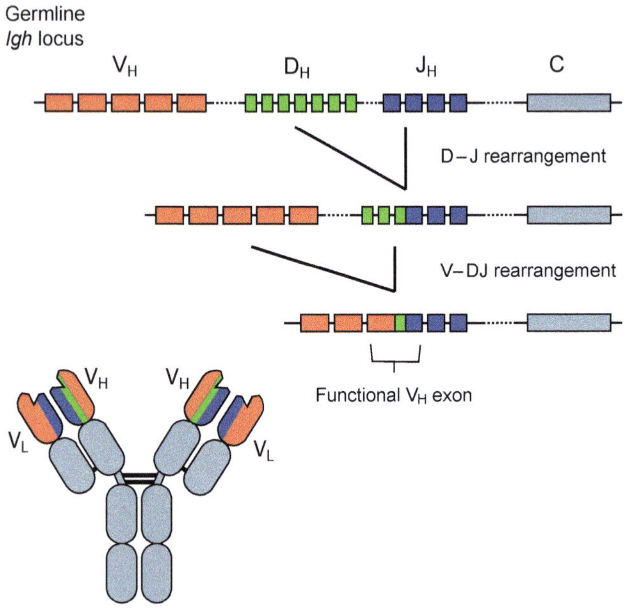


FIGURE 2 **Schematic of V(D)J recombination in the immunoglobulin heavy chain locus.** The germ-line immunoglobulin heavy chain (IgH) locus is composed of many V (variable), D (diversity), and J (joining) gene segments upstream of the constant (C) regions. For simplicity, only one C region is depicted. During the stochastic gene rearrangement process of V(D)J recombination, one D gene segment is rearranged to join one J gene segment (D–J rearrangement). A second recombination event rearranges one V gene segment to join the previously rearranged DJ gene segment (V–DJ rearrangement), creating the functional $V_H$ exon that encodes the variable domain of the immunoglobulin heavy chain. Not shown: the immunoglobulin light chain loci Igλ and Igκ are composed of many V and J gene segments, and a separate V(D)J recombination event in either the Igλ or the Igκ locus results in V to J recombination, yielding the functional light-chain variable domain. The immunoglobulin schematic depicts the contribution of the V, D, and J gene segments to the variable domains of the IgH and IgL chains. Disulfide bonds are shown as black lines.

junctional diversity. Each Ig molecule is made up of two identical heavy chains (IgH) and two identical light chains (IgL), and the many different possible combinations of individual IgH and IgL polypeptide chains that pair up to form a functional lymphocyte antigen receptor add an additional layer of combinatorial diversity [16–18].

Seven antigen receptor loci exist: TCR α, β, γ, and δ and Ig H, κ, and λ. Ig κ and λ are the two IgL loci. All loci contain V and J segments, whereas three (TCR β and δ and IgH) also contain D segments between the V and the J clusters. The heterodimeric immune receptors are always composed of one polypeptide derived from a locus containing V, D, and J elements and one from a locus with just V and J elements. At each locus, the variable region exon consisting of VJ or VDJ elements created during V(D)J recombination is then fused to a C (constant) region through RNA splicing. There are several IgH constant regions, each of which corresponds to a different Ig isotype. The incorporation of some IgH constant regions is determined by a separate somatic DNA rearrangement process, class-switch recombination, which is discussed in later chapters.

In the germ-line configuration, gene segments to be recombined are flanked by RSSs, each of which contains two highly conserved sequence elements, a heptamer (consensus sequence 5′-CACAGTG-3′) and a nonamer (consensus sequence 5′-ACAAAAACC-3′), separated by a spacer of conserved length (12 or 23 bp±1) but poorly conserved sequence. These are referred to as the 12RSS and the 23RSS, respectively (Figure 3). Efficient recombination occurs only between a 12RSS and a 23RSS (the "12/23 rule"), helping to ensure that the correct types of gene segments are joined. All segments of a particular type (e.g., $V_\kappa$ segments) are flanked by one type of signal, and all the segments to which they could be joined ($J_\kappa$) are flanked by the opposite type. In the IgH locus, the 12/23 rule ensures that V segments recombine with D segments, but not directly to J segments (Figure 3) [5,17].

The 12/23 rule is not sufficient, however, to account for the rearrangement patterns observed at the TCRβ locus. At this locus, the $V_\beta$ segments are flanked by 23RSSs and $J_\beta$ segments by 12RSSs, whereas the $D_\beta$ segments have a 5′ 12RSS and a 3′ 23RSS. Direct $V_\beta$ to $J_\beta$ joining is rarely observed in vivo, although it would be in accordance with

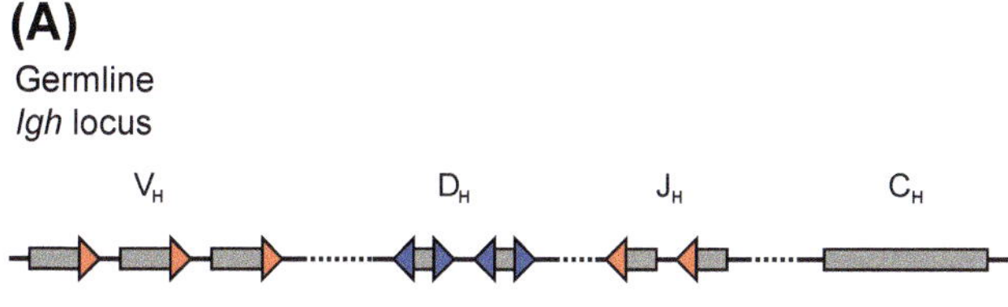


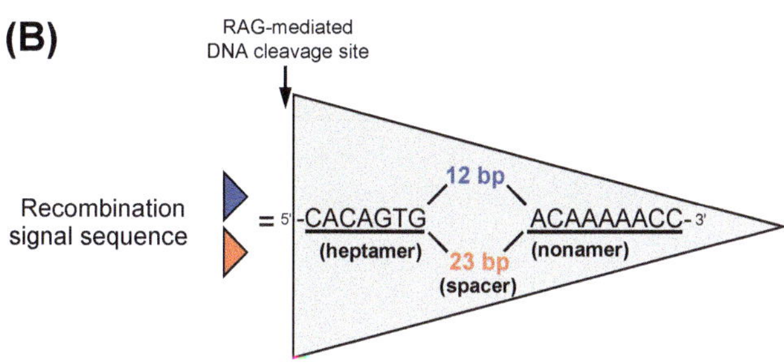


FIGURE 3 **Diagram of the IgH locus and recombination signal sequence.** (A) The germ-line immunoglobulin heavy chain locus (IgH) contains many V, D, and J gene segments (for simplicity, only two or three segments are shown for each), each of which is flanked by a recombination signal sequence (RSS) (triangles). (B) The consensus RSS is depicted as a gray triangle containing the consensus heptamer and nonamer sequences. For simplicity, only top-strand sequences are shown. The heptamer and nonamer sequences are separated by a poorly conserved 12 (blue) or 23 (red)-bp spacer. Recombination activating gene (RAG)-mediated DNA cleavage occurs at the 5′ border of the heptamer, cleaving between the heptamer and the gene segment (coding flank).

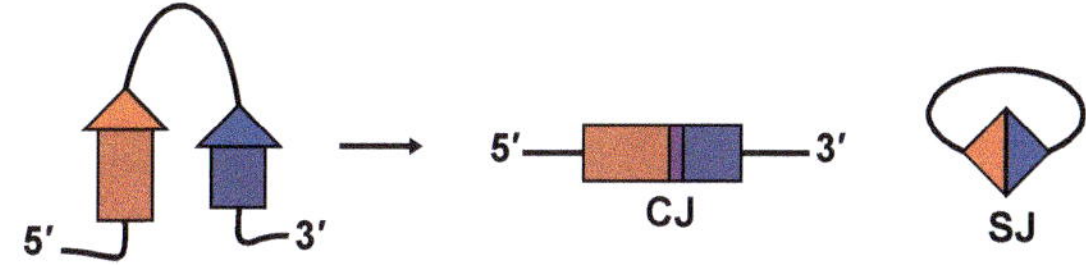


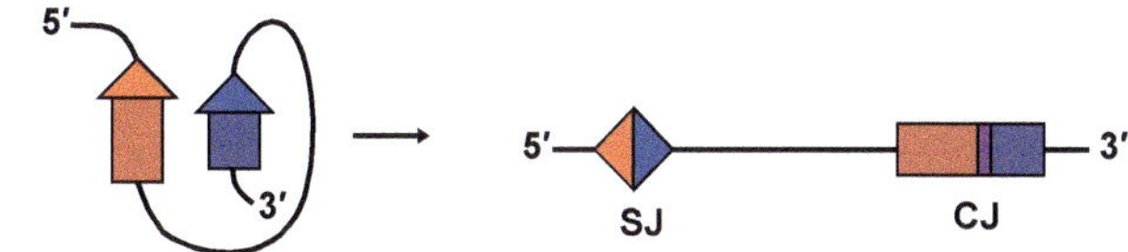


FIGURE 4 **Deletional and inversional V(D)J recombination.** (A) The intervening sequences between the recombining coding segments are deleted when the 12 and 23 RSSs are oriented as depicted, forming a CJ on the chromosome and an SJ on an extrachromosomal circle. (B) RSSs oriented in the same direction along the chromosome, as depicted, lead to inversional V(D)J recombination, in which both CJ and SJ remain in the chromosome. SJs are typically joined precisely end to end, and CJs are imperfectly joined, as indicated by the purple section between coding gene segments. RSS, recombination signal sequence; CJ, coding joint; SJ, signal joint.

the 12/23 rule. In vivo, $V_\beta$ gene segments preferentially recombine with the 12RSS 5′ of a $D_\beta$ gene segment over any of the 12RSSs flanking the $J_\beta$ gene segments, a restriction termed the "beyond 12/23 (B12/23) rule" [19,20]. The B12/23 restriction is largely enforced by the intrinsic properties of interactions within the functional cleavage complexes (discussed later) containing the TCRβ DNA substrates, the RAG proteins, and HMGB1 or 2 [21–23].

The RSSs are all that is required to render a piece of DNA a substrate for V(D)J recombination. The orientation of the RSSs with respect to each other determines the outcome of the reaction. Deletional V(D)J recombination occurs when the flanking RSSs are oriented in opposite directions, resulting in retention of the CJ in the chromosome and deletion of a circular molecule containing the SJ. Alternatively, recombination can lead to inversion of the DNA between the RSSs with retention of both the SJ and the CJ in the chromosome (Figure 4). Both of these arrangements are found in vivo [24–26].

## 3. RECOMBINATION SIGNAL SEQUENCES

Although the consensus RSS is cleaved and recombined most efficiently, and thus most in vitro experiments have used consensus sequences, virtually all endogenous RSSs vary from the consensus in some positions [27]. Only the first three nucleotides of the heptamer (CAC) are 99–100% conserved and are absolutely required for cleavage. The remaining positions exhibit 77–91% conservation [27,28]. Similarly, not all positions of the nonamer are equally conserved. In the nonamer, the cytosine at the second position and several adenines in the A tract are over 85% conserved. The remaining nonamer positions show 56–77% conservation [27]. This distribution of sequence conservation has functional correlates, with the most highly conserved positions of the nonamer being most important for initial protein binding and the conserved heptamer positions being required for efficient RAG-mediated cleavage [28,29]. The spacer sequences are much less conserved than the heptamer or nonamer sequences, with conservation at individual positions ranging from 29% to 68%. Nonetheless, the sequence of the spacer can have a substantial effect on RSS recombination efficiency [30].

Although both a heptamer and a nonamer sequence are required for efficient DNA cleavage, in the complete absence of a heptamer, the nonamer alone can still direct some nicking (but no hairpinning) under permissive conditions in vitro. This somewhat imprecise and very inefficient nicking occurs at the boundary of where the heptamer would have been in a 12RSS or 23RSS (that is, 19 or 30 nucleotides from the 5′ end of the nonamer), as if the proteins reach out a defined distance from the bound nonamer. Under the same permissive in vitro conditions, substrates containing a heptamer alone also work, although cleavage is reduced severalfold [29,31].

The length of the spacer sequence is also important. The length difference between the 12RSS and the 23RSS is almost precisely one helical turn. Adding an extra one-half helical turn (spacer length of 18 or 29 bp) substantially inhibits cleavage, below that seen with an isolated heptamer, whereas adding one full turn (spacer length of 33 or 34 bp) permits a substantial level of cleavage [29,31]. Taken together these results suggest that the heptamer and nonamer can each function individually and can act together synergistically with the proper spacing, but they conflict when the spacing is wrong.

In addition to the RSS, the first two or three nucleotides of the coding flank immediately abutting the heptamer of the RSS can have considerable effects on the efficiency of the recombination reaction. Although most sequences are essentially neutral, certain nucleotide combinations can be favorable or unfavorable to the reaction. The effect of the coding flank sequence appears to be primarily at the cleavage phase of V(D)J recombination [29,31,32]. This may be related to the structural distortion of the coding flank and heptamer regions during cleavage, as discussed later.

The variation in endogenous RSS sequences and coding flanks has an effect on their cleavage efficiency, and this has an impact on gene segment usage in the antibody or TCR repertoire. In addition, other factors, including RSS accessibility within chromatin, nuclear localization, and antigen receptor locus architecture, all play roles in the control of V(D)J recombination targeting and the tightly regulated order of recombination events in vivo [4], though their specific contributions are still under investigation.

## 4. BIOCHEMISTRY OF V(D)J CLEAVAGE

V(D)J cleavage requires only RAG1 and RAG2, a divalent metal ion, and a DNA substrate containing the RSS [33]. In addition, the nonspecific DNA-bending protein HMGB1 (or HMGB2) plays an important role in augmenting the reaction. No external source of energy is needed [33,34]. DNA cleavage in V(D)J recombination occurs in two separable steps. First, the top strand is nicked 5′ of the RSS precisely at the heptamer-coding flank junction, generating a 3′ hydroxyl group. In the second step, this 3′ hydroxyl attacks the phosphate on the bottom strand of the DNA, performing direct transesterification and resulting in a double-stranded DNA (dsDNA) break with a hairpin coding end and a blunt signal end [33,35] (Figure 1(B)) During hairpin formation, the energy required for the formation of the new bond is derived from the breakage of the old one. Stereochemical studies have shown that a covalent bond between the RAG proteins and the DNA is not formed [35]. This distinguishes the RAG cleavage reaction from that of a number of other site-specific recombinases, such as Cre, Flp, and Lambda-Int, which rely on a covalent intermediate. Instead, the mechanism used by RAG1/RAG2 is similar to that of bacterial transposases and HIV integrase [35–37].

The V(D)J recombinase machinery can first assemble on a single 12 or 23RSS, creating a signal complex (SC) in which nicking can occur. Most evidence suggests that the paired complex (PC), which contains both a 12RSS and a 23RSS, is then formed by the "capture model," in which one SC captures the appropriate partner RSS to form the PC, maintaining the same protein content in both the SC and the PC [38–41]. However, there is not universal agreement with the capture model. Some studies favor the "association model," in which two preformed SCs associate to form the PC [42,43]. Under physiological conditions, dsDNA breaks are generated only by "coupled cleavage" in the PC [44] (Figure 1(A)). These conditions are reproduced in vitro by the use of magnesium ($Mg^{2+}$) as the divalent metal cation in the reaction.

Alteration of the structure of the substrate DNA suggests that DNA unpairing and structural distortion play a role in V(D)J cleavage [29,31]. More recent studies suggest that base-flipping of a nucleotide in the coding flank may play an important role. Use of abasic substrates identified a nucleotide in the bottom strand of the coding flank as a candidate for base-flipping during V(D)J recombination. An RSS substrate with an abasic site at that location was capable of hairpin formation in the absence of a partner RSS, permitting uncoupled cleavage [45]. Potassium permanganate footprinting has also been used to examine distortion at the coding flank due to its selective oxidation of single-stranded or extrahelical thymines in DNA. The same nucleotide in the coding flank identified by the abasic substrate experiments was found to exhibit permanganate hypersensitivity when a pair of prenicked 12/23RSS oligonucleotide substrates is incubated in the presence of RAG1, RAG2, and HMGB1 [46,47]. This permanganate sensitivity is similar to that found in the DNA substrates of the Tn5 and Tn10 transposases, which flip out a thymine base during their cleavage reaction. Like RAG1/2, Tn5 and Tn10 also perform DNA recombination via a nicked and subsequently hairpinned DNA intermediate [48,49].

After coupled cleavage, the recombinase proteins and both the coding and the signal ends are contained in a postcleavage complex [50,51], from which the coding ends are rapidly shepherded to the NHEJ proteins for end processing and joining (discussed in Section 18.6). The hairpin-sealed coding ends are first opened by a complex of the proteins Artemis and DNA-dependent protein kinase catalytic subunit (DNA-$PK_{CS}$), generally resulting in P (palindromic) nucleotides due to hairpin opening occurring two nucleotides 3′ of the tip of the hairpin [52]. The ends are then processed by other ubiquitous proteins of the NHEJ pathway, resulting in addition of N (nontemplated) nucleotides and deletion of nucleotides by nucleolytic trimming, before being joined to form a CJ. This coding-end processing is responsible for the imprecise joining of the V, D, and J segments, which creates junctional diversity. The signal ends are generally precisely joined to each other by the NHEJ pathway, forming an SJ that is in many cases lost from the genome [12].

## 5. RAG1 AND RAG2

The *Rag1* and *Rag2* genes are not homologous to each other, but highly conserved homologs of both genes have been identified in all jawed vertebrates studied. In all species examined thus far *Rag1* and *Rag2* are tightly linked, and in most species, the entire open reading frame of each gene is contained within a single exon, both features that are more commonly found in prokaryotic than in eukaryotic organisms. These findings, along with mechanistic similarities between RAG-mediated DNA cleavage and that of some transposases, suggest that the RAG genes evolved from an ancient transposable element, though conclusive evidence of this evolutionary origin has not yet been identified [53,54]. The amino acid sequences of murine and human RAG1 and RAG2 are 90% and 88% identical, respectively [6,55], and most studies of RAG protein biochemistry have used the mouse proteins. In addition, because the minimal functional truncation mutants, core RAG1 and RAG2, are more soluble in vitro than the full-length proteins, most in vitro biochemical experiments to date have used the core proteins. However, the noncore regions have been found to play essential roles in ensuring physiological V(D)J recombination in vivo. Mice expressing core RAG1 or RAG2 in the absence of the wild-type protein exhibit reduced V(D)J recombination efficiency and fidelity as well as defective ordering of gene segment rearrangement [56–60].

Full-length murine RAG1 is a 1040-amino-acid polypeptide, with the minimal core protein capable of recombination encoded by residues 384–1008 [61,62]. Distinct domains within core RAG1 include the nonamer-binding domain (NBD), which forms a tight dimer and binds to the RSS nonamer [63,64]; the central domain, which interacts with RAG2 and preferentially binds the heptamer sequence in single-stranded DNA (ssDNA) [65]; and the C-terminal domain (CTD), which dimerizes and can bind dsDNA nonspecifically [66,67] (Figure 5(A)). The Asp and Glu residues that comprise the catalytic triad, also known as the DDE motif, are located within the central and C-terminal domains [68–70]. The catalytic triad in the recombinase active site is thought to coordinate $Mg^{2+}$ ions that can activate a water molecule for nucleophilic attack on the DNA [71]. The CTD is also implicated in the interaction with the nuclear importin karyopherin α (KPNA) 2, which has been shown to regulate RAG1 nuclear localization [72,73]. Under most in vitro conditions, core RAG1 forms a tight dimer in solution [74–77]. Higher-order oligomers (tetramer and octamer) may be favored under conditions of low salt concentration (0.2M NaCl) or higher temperature (25–37°C) [78,79], but the physiologic relevance of these higher-order RAG1 oligomers is unclear.

The noncore regions of RAG1 include the central noncore domain (CND), which contains several motifs involved in coordinating zinc ions and binds dsDNA weakly and nonspecifically [80]. The CND also interacts with nuclear importin KPNA1, but this interaction does not mediate RAG1 nuclear transport [72,81]. The zinc dimerization domain, located just N-terminal to the NBD of

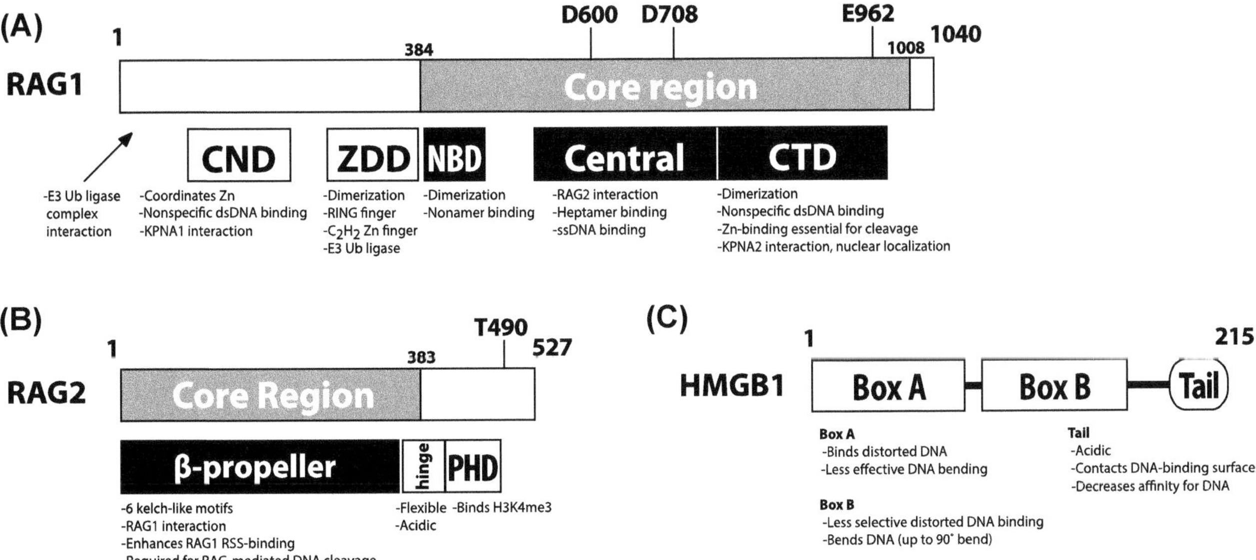


FIGURE 5 **Schematic of RAG1, RAG2, and HMGB1 proteins.** Murine RAG1 (A) and RAG2 (B) contain minimal core regions (gray) sufficient for RAG-mediated cleavage in vitro. RAG1 core contains the DDE active-site residues. Noncore regions (white) are important for V(D)J recombination in vivo and include RAG2 Thr490, which is phosphorylated in a cell-cycle-dependent manner, leading to RAG2 ubiquitylation and degradation. Core (black) and noncore (white) domains are indicated below the full protein diagram. (C) HMGB1 contains DNA-binding HMG box domains A and B (rectangles) and C-terminal acidic tail (rounded rectangle). Basic linkers between domains are depicted as thick black lines. CND, central noncore domain; ZDD, zinc dimerization domain; NBD, nonamer-binding domain; CTD, C-terminal domain; RSS, recombination signal sequence; KPNA, karyopherin α; RING, really interesting new gene; Zn, zinc cation; PHD, plant homeodomain; H3K4me3, trimethylated lysine 4 of histone H3.

core RAG1, dimerizes and contains a RING (really interesting new gene) finger and a $C_2H_2$ zinc finger [82,83]. The RING finger has been identified as an E3 ubiquitin ligase that can ubiquitylate RAG1 itself, KPNA1, and histone H3 [84–86], although the significance of this E3 ubiquitin ligase activity in vivo has not been determined. The extreme N-terminal region of RAG1 interacts with an E3 ubiquitin ligase complex containing VprBP, DDB1, Cul4A, and Roc1 [87].

Murine RAG2 is a 527-amino-acid polypeptide. Like RAG1, RAG2 has a minimal core region necessary for catalysis (amino acids 1–383) [88,89]. Less is known about the structure and function of RAG2 than those of RAG1. Based on amino acid sequence analyses, core RAG2 is predicted to form six Kelch-like motifs that assemble into a six-bladed β-propeller structure [90]. This structure is attached via a flexible acidic hinge to a structurally characterized plant homeodomain that specifically binds trimethylated lysine 4 on histone H3 [91,92] (Figure 5(B)). The far C-terminal region of RAG2 contains Thr490, which is phosphorylated in a cell-cycle-dependent manner, resulting in ubiquitylation and proteasomal degradation [93–96]. The far C-terminal region also contains a nuclear localization signal, which interacts with importin 5 and can mediate RAG2 nuclear transport [97]. Recent studies using RAG2 core knock-in mice have shown that the RAG2 C-terminus also plays an important role in maintaining genome integrity during V(D)J recombination in vivo [98,99]. RAG1 and RAG2 have been found to associate by coimmunoprecipitation, copurification, and pulldown assays, and this interaction has been localized primarily to the central domain of RAG1 and to core RAG2, specifically the sixth predicted Kelch repeat motif [67,100–102].

## 6. A ROLE FOR HMGB1 IN V(D)J RECOMBINATION

The high mobility group box proteins HMGB1 and HMGB2 were identified as a component of the V(D)J recombinase through an effort to isolate factors in whole-cell extract that stimulate RSS cleavage [8,103]. In vitro, the RAG1 and RAG2 proteins alone are necessary and sufficient to catalyze the cleavage reaction. They can efficiently assemble on a 12RSS where nicking and, in the presence of a 23RSS, low levels of hairpin formation can occur. However, the RAG proteins alone bind and nick the 23RSS at very low levels, and coupled hairpin cleavage at the 23RSS is nearly undetectable [33,104]. HMGB1 and 2 enhance 23SC formation, incorporating stably into the complex and stimulating nicking while suppressing aberrant nicking at inappropriate sites within the 23RSS. With the addition of a partner 12RSS, HMGB1 stimulates coupled cleavage of both the 12RSS and the 23RSS by more than 100-fold in vitro, achieving a level of coupled cleavage that would be physiologically relevant for V(D)J recombination in vivo [8].

HMGB1 and 2 are sequence-nonspecific DNA-binding proteins found only in vertebrates. HMGB1 and 2 are highly homologous and each contains two DNA-binding HMG-box domains, which bind in the minor groove, connected via basic linkers to each other and to a highly acidic C-terminal tail domain (Figure 5(C)) [105]. HMGB1 and HMGB2 are structurally and functionally very similar, and as a cofactor in V(D)J cleavage, they have been found to be functionally redundant in vitro [8,106,107]. For simplicity, we refer only to HMGB1 hereafter.

HMGB1 binds DNA without sequence specificity but binds B-form DNA weakly, preferring noncanonical DNA structures such as ssDNA and four-way junctions [108]. HMGB1 binding can also introduce DNA bends via partial intercalation of hydrophobic residues [109,110]. Within HMGB1, box A favors binding to distorted DNA structures, whereas box B can introduce up to a 90° bend in the DNA structure. The acidic tail domain makes extensive contacts with the DNA-binding surface of both HMG boxes, which regulates HMGB1 DNA binding by acting as a DNA mimetic, resulting in a "closed" HMGB1 conformation incapable of binding DNA [111].

Studies have identified the components of the HMGB1 protein essential for stimulation of RSS cleavage [106,112–114]. Although either HMG box, A or B, alone can be incorporated into the RAG1–RAG2–DNA complex, enhancement of V(D)J cleavage is much more efficient with the addition of the tandem box protein. In addition, the intercalating residues play a role in stimulating RAG-mediated RSS binding and cleavage [112–114]. The HMGB1 acidic tail is not required for enhancing 23RSS cleavage. Its deletion increases nonspecific DNA-binding activity and rescues some HMGB1 mutants unable to promote 23SC formation when linked to the acidic tail [112]. HMGB1 demonstrates both association with the RAG1 NBD in the absence of RSS [107] and RSS binding in the absence of RAG1 [106], but HMGB1 is recruited into a ternary HMGB1–RAG1–RSS complex much more efficiently than to either RAG1 or RSS alone [115]. Circular permutation analyses and fluorescence resonance energy transfer experiments using fluorophore-labeled 23RSS have shown that addition of HMGB1 enhances the DNA bend induced by the RAG1–RAG2 complex [107,116].

Genetic evidence for the role of HMGB1 in V(D)J recombination in vivo has been difficult to obtain because of the functional redundancy and the diverse but essential roles of the HMGB1 and HMGB2 proteins in a variety of nuclear processes. Single-knockout mice do not display obvious defects in V(D)J recombination or lymphocyte development [117,118], and *Hmgb1*$^{-/-}$:*Hmgb2*$^{-/-}$ double knockouts are lethal [119]. Evidence for the importance of HMGB1 in V(D)J recombination has been generated by one-hybrid and extrachromosomal recombination assays in cell culture, demonstrating that HMGB1 overexpression increases RSS-binding by RAG1 alone or RAG1 and RAG2, and increases recombination of a V(D)J plasmid substrate [107]. These data indicate that, despite its relative abundance in the nucleus, HMGB1 is limiting for V(D)J recombination, suggesting its importance for V(D)J recombination in vivo.

## 7. RECOMBINATION COMPLEXES: ANALYSIS OF STOICHIOMETRY AND ORGANIZATION

The stoichiometry and organization of the SC and PC have been the subject of extensive study and debate, and many questions remain unanswered. Most studies have shown that the SC on either the 12RSS or the 23RSS contains a very tight dimer of RAG1 that cannot reassort in solution and either one or two molecules of RAG2 that can reassort in solution [41,74,76,120]. The addition of HMGB1 to the binding reaction supershifts the RSS-bound complex without altering the RAG1/RAG2 protein stoichiometry [41], but the number of HMGB1 molecules within the RSS-bound precleavage complex has not been addressed. The RAG1/RAG2 stoichiometry of the PC is less clear, with suggestions that it contains the same proteins as the SC [40,41], consistent with the "capture model" of PC assembly that is supported by additional evidence [38,39]. However, two studies suggest that the PC contains twice the RAG protein content of the SC, supporting the "association model" of assembly [42,43]. The postcleavage signal end complex, composed of RAG1 and RAG2 plus both a 12RSS and a 23RSS that are precleaved and thus lack a coding flank, was found to have a RAG1:RAG2:HMGB1 stoichiometry of 2:2:1 [121].

DNA footprinting and interference as well as ultraviolet (UV) crosslinking experiments have been used to identify locations on the RSS bound by the RAG proteins. DNase I footprinting, as well as ethylation, methylation, and potassium permanganate interference, has shown that RAG1 alone makes extensive phosphate backbone contacts in the nonamer and nonamer-proximal spacer on one side of the DNA helix on both the 12RSS and the 23RSS [122–124]. Base-specific interactions were observed only in the nonamer [122,124], consistent with surface plasmon resonance experiments and the recent crystal structure of the nonamer bound by the RAG1 NBD [63,64]. The addition of RAG2 extends both phosphate-backbone and base-specific interactions to the 3′ end of the heptamer and the heptamer-proximal spacer on the 12RSS, which are on the same side of the DNA helix as the nonamer and nonamer-proximal spacer contacts present for both RAG1 alone and RAG1 plus RAG2. In the absence of HMGB1, however, neither protection nor interference is observed at the heptamer of the 23RSS [122–124]. UV crosslinking experiments of RSS incubated with both RAG1 and RAG2 demonstrate robust RAG1 and weak RAG2 crosslinking to the coding flank and heptamer [125]. Overall, these studies show that RAG1 alone can bind to the nonamer with some sequence specificity, and they suggest that the addition of RAG2 induces a change that allows RAG1 to bind the RSS at the heptamer in addition to the nonamer; this could then allow RAG1 to assemble the enzymatic active site at the heptamer-coding flank border.

Similar footprinting, interference, and UV crosslinking experiments with complexes containing HMGB1 have not been as extensive but have indicated a role for HMGB1 in DNA binding at the RSS. One study of HMGB1 in the SC using prenicked substrates demonstrated UV crosslinking of HMGB1 in the heptamer and heptamer-proximal spacer regions of the 12RSS and predominately in the center of the spacer of the 23RSS. These sites of HMGB1 crosslinking on the 23RSS are on the side of the DNA opposite those in contact with the RAG proteins in previous interference studies. In contrast to these UV crosslinking results, interference footprinting experiments on intact RSS suggest that HMGB1 is located at the nonamer-proximal end of the spacer [106]. Addition of HMGB1 to the 23SC extends the nonspecific protein-phosphate backbone interactions into the spacer immediately 5′ of the nonamer as well as to the heptamer-spacer junction of the 23RSS, where there are new base-specific contacts as well, creating a DNA interference footprint in the heptamer-spacer junction that is very similar to that of the RAG proteins in the 12SC. As HMGB1 has two DNA-binding HMG boxes with slightly different DNA binding and bending properties, it has been speculated that one HMG box may bind near the heptamer to stabilize the RAG-induced DNA bend, whereas the other binds near the nonamer, facilitating DNA bending [126].

The results of investigations into RAG1 and RAG2 biochemistry as well as SC and PC stoichiometry and organization have been combined to create a model for the binding of RAG1 to a 12/23RSS pair, with much more tentative suggestions for RAG2 and HMGB1 locations as well (Figure 6) [126]. The finding that the RAG1 DDE active-site motif is contributed by one RAG1 protomer (of a RAG1 dimer) and the NBD is contributed by the other RAG1 protomer has led to a *trans* cleavage model [120] in which each RSS is in contact with both RAG1 protomers in a dimer. This model is consistent with electron microscopy studies [121], the crystal structure of the NBD dimer bound to DNA [64], and most functional studies. Investigation into the structure and orientation of the 12 and 23RSS in the PC suggests that the RSSs are not planar but rather cross over each other, each bent in a U-shaped configuration that may be organized around a RAG protein core [116,127]. Despite the fairly detailed understanding of the interaction between core RAG1 and RSS within the paired complex, there remain many unanswered questions about the locations and roles of RAG2, HMGB1, and the noncore regions of RAG1.

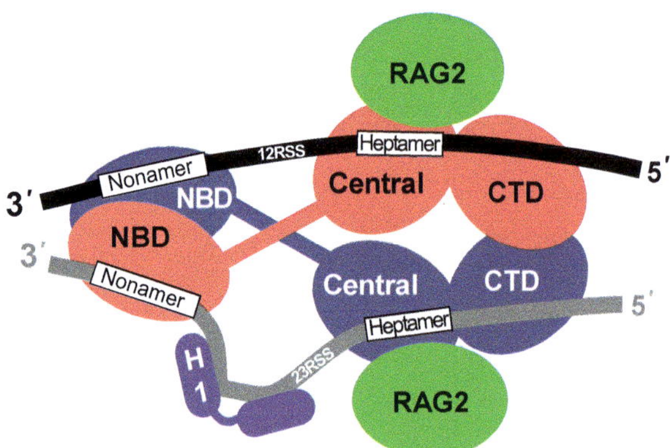


FIGURE 6 **Model of paired complex protein organization.** A core RAG1 dimer plus two monomers of RAG2 and one HMGB1 molecule are depicted binding a 12RSS (black) and a 23RSS (gray) to form a paired complex. Each RAG1 protomer (red, blue) is depicted as primarily binding the nonamer (white rectangles) of one RSS and the heptamer (white rectangles) of the opposite RSS. Although the majority of the RAG1–nonamer contacts are made by the nonamer-binding domain (NBD) as depicted, the two NBDs are known to form a tight dimer such that each subunit makes some contact with both RSSs. The active site residues are contributed by the central and C-terminal domains (CTD), and the DNA is cleaved at the border of the heptamer and coding flank. RSSs are arranged in a parallel orientation, such that the coding flanks are bound by the CTDs of RAG1. RAG1 core dimerization is mediated by the NBD and CTD, and the central domain interacts with the heptamer and RAG2. RAG2 (green ovals) is shown binding the central domain of RAG1 near the site of cleavage. HMGB1 (purple rounded rectangles, H1) is shown binding in the 23RSS spacer near the NBD of RAG1. HMGB1 stoichiometry is unknown and it is possible that it also makes contact with the 12RSS. For simplicity, only one HMGB1 molecule is modeled. The locations of RAG2 and HMGB1 are speculative. RSS, recombination signal sequence; HMGB1, high mobility group box protein 1; RAG, recombination-activating gene. *Based on Ref. [126].*

## 8. V(D)J RECOMBINATION IS TIGHTLY REGULATED DURING LYMPHOCYTE DEVELOPMENT

Given the importance of generating functional B and T cell antigen receptors and the need to protect the integrity of the genome from inappropriate RAG-catalyzed DNA double-strand breaks, it is essential that V(D)J recombination is tightly regulated to ensure that DNA double-strand breaks are generated at the right place (i.e., the appropriate antigen receptor locus) and the right time (i.e., the appropriate stage of lymphocyte development). Indeed, the process of V(D)J recombination is strictly regulated at multiple levels: tissue specificity (rearrangement occurs only in lymphocytes); lineage specificity (complete Ig gene assembly occurs only in B cells; complete TCR gene assembly occurs only in T cells); developmental-stage specificity (IgH genes are assembled in pro-B cells while Igκ/λ genes are assembled in pre-B cells; TCRβ genes are assembled in $CD4^-CD8^-$ (DN) pro-T cells while TCRα genes are assembled in $CD4^+CD8^+$ (DP) pre-T cells); temporal ordering (D–J joining occurs on both alleles before V–DJ joining occurs on one allele); allelic exclusion (productive rearrangement of a given antigen receptor gene on one allele prevents rearrangement on the other allele); isotypic exclusion (productive rearrangement of the κ isotype of the immunoglobulin light chain prevents rearrangement of the λ isotype and vice versa); and cell cycle specificity (rearrangement occurs only during G0/G1) (Figure 7).

Certain aspects of this regulation are RAG-intrinsic and are easily explained. For instance, the tissue specificity of V(D)J recombination can be explained by the lymphoid-restricted expression pattern of RAG1 and RAG2 [7]. Similarly, the cell cycle specificity can be explained by RAG2's accumulation in G0/G1 cells, followed by its regulated degradation at the G1–S transition [94]. This cell cycle-regulated degradation of RAG2 results from cyclin

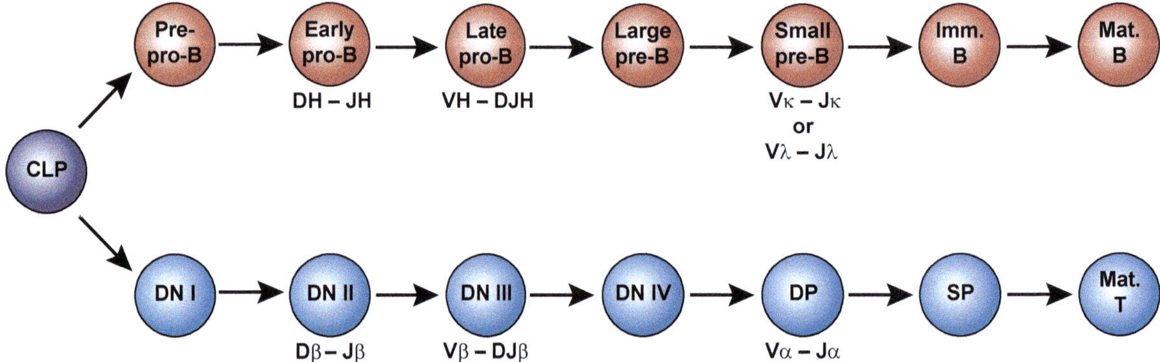


FIGURE 7 **Regulation of V(D)J recombination.** V(D)J recombination exhibits tissue specificity, lineage specificity, developmental-stage specificity, temporal ordering, and isotypic exclusion (pictured here). V(D)J recombination also exhibits cell cycle specificity and allelic exclusion (not depicted here). CLP, common lymphoid progenitor; DN, double-negative ($CD4^-CD8^-$) thymocyte; DP, double-positive ($CD4^+CD8^+$) thymocyte; SP, single-positive ($CD4^+$ or $CD8^+$) thymocyte; Imm., immature; Mat., mature.

A/CDK2 phosphorylation of T490 in RAG2 at the G1–S boundary, which causes RAG2 to be exported to the cytoplasm where it is destroyed by the ubiquitin/proteasome system [95,96,128,129]. Whereas the mechanisms underlying V(D)J recombination's tissue specificity and cell cycle specificity are fairly well characterized, the mechanisms underlying lineage specificity, temporal ordering, allelic exclusion, and isotypic exclusion have remained elusive.

## 9. ACCESSIBILITY MODEL OF REGULATION

Because the same recombinase (RAG1/2) catalyzes all V(D)J recombination events, and yet only a subset of all recombinationally competent gene segments are actually rearranged in a given cell type at a particular developmental stage, antigen receptor gene assembly must be regulated by modulating the accessibility of the V, D, and J gene segments to the V(D)J recombinase. This "accessibility hypothesis" [130] is attractive because it provides a simple conceptual framework within which to understand the lineage specificity, temporal ordering, allelic exclusion, and isotypic exclusion of V(D)J recombination.

Early studies provided evidence supporting the principle of the accessibility hypothesis. One study showed that when fibroblasts—which do not normally perform V(D)J recombination—were transfected with RAG1 and RAG2 expression vectors, these cells would rearrange extrachromosomal plasmid substrates, but failed to recombine their endogenous, chromosomal antigen receptor loci [7]. Therefore, artificial, extrachromosomal RSSs are more accessible to the RAG1/2 recombinase than are endogenous, chromosomal RSSs. Moreover, when RAG1 and RAG2 expression vectors were transfected into fibroblasts that harbored an artificial recombination substrate that had stably integrated into a random location in the genome, these cells could rearrange this artificial, chromosomal recombination substrate, but still failed to recombine their endogenous, chromosomal antigen receptor loci [6,7,131]. Thus, even though some regions of the genome—such as the integration sites of these artificial recombination substrates—are accessible to the V(D)J recombinase in fibroblast cells, other regions of the genome—such as the immunoglobulin and T cell receptor loci—are inaccessible to the V(D)J recombinase in these same cells. Taken together, these observations demonstrate that the RAG1/2 recombinase is necessary, but insufficient for chromosomal V(D)J recombination; before the RAG1/2 recombinase can initiate V(D)J recombination at the endogenous antigen receptor loci, the antigen receptor loci must be made accessible to the RAG proteins. However, the mechanisms by which the antigen receptor loci are made accessible or inaccessible were not addressed by these early studies.

## 10. OVERVIEW OF CHROMATIN STRUCTURE

Since the 1990s, increasing evidence has pointed to chromatin structure as a key determinant of antigen receptor gene accessibility. Chromatin, the physiological template of genetic information in all eukaryotic cells, is a dynamic, regulatable polymer of nucleosomes. Each nucleosome is composed of 146 base pairs of DNA (approximately two superhelical turns) wrapped around an octamer of the core histone proteins—H2A, H2B, H3, and H4 [132–134]. Additional proteins, such as histone H1, that bind to the linker DNA connecting nucleosomes, allow nucleosomes to be packaged into 30 nm fibers [135,136]. These 30 nm fibers can be compacted further, but the architecture of these higher order structures remains poorly understood.

Studies investigating the role of chromatin in transcriptional regulation have revealed two types of enzymatic activity that can regulate the accessibility of chromatinized DNA. The first family consists of ATP-dependent chromatin-remodeling complexes that can noncovalently alter nucleosome position, spacing, and conformation [137,138]. The second family consists of an increasingly complex group of protein complexes that covalently modify the histone proteins, primarily on their surface-exposed N-terminal tails (e.g., lysine acetylation, serine/threonine phosphorylation, and lysine/arginine methylation) [139]. Together, these two families of enzymes play key roles in regulating the accessibility of the eukaryotic genome to nucleic acid enzymes, including the RAG1/2 recombinase.

## 11. REGULATION BY NUCLEOSOME STRUCTURE AND HISTONE ACETYLATION

After early studies provided proof-of-principle for the accessibility hypothesis, the next question became, what specific aspects of chromatin structure are important for targeting the V(D)J recombinase to the appropriate antigen receptor locus in a particular cell at a specific stage of lymphoid development? The simplest aspect of chromatin structure that could regulate V(D)J recombination is the nucleosome itself. Indeed, initial studies showed that incorporating an RSS into a mononucleosome can inhibit RAG1/2-catalyzed V(D)J cleavage in vitro up to 100-fold [140–142]. Notably, this inhibition can be largely relieved by histone acetylation and ATP-dependent chromatin remodeling [143], thereby providing a plausible mechanism for controlling the accessibility of antigen receptor gene segments. Likewise, positioning an RSS within an array of nucleosomes can also strongly inhibit V(D)J cleavage in vitro, but ATP-dependent chromatin remodeling can greatly stimulate RAG1/2-catalyzed cleavage of these nucleosome arrays [144]. Thus, when an RSS is incorporated into a hypoacetylated nucleosome in vitro, it is effectively inaccessible to the RAG1/2 complex. However, when this nucleosome is hyperacetylated and mobilized by chromatin remodeling, the RSS becomes accessible to the RAG1/2 recombinase.

To determine whether histone acetylation and chromatin remodeling also modulate antigen receptor gene segment accessibility in vivo, chromatin immunoprecipitation was used to map histone acetylation [145–150] and Brg1 (a chromatin remodeling factor) [146] to the antigen receptor loci in a variety of cell types. Consistent with the in vitro studies, Brg1 and histone acetylation were enriched at recombinationally active antigen receptor loci and depleted at recombinationally inactive antigen receptor loci, suggesting that histone acetylation and chromatin remodeling play a role in increasing the accessibility of the RSSs in these loci during B cell and T cell development. Furthermore, it was observed that treating developing lymphocytes with drugs that globally increase histone acetylation in the cell can, either directly or indirectly, increase the efficiency of V(D)J recombination at the endogenous antigen receptor loci [141,151]. Taken together, these studies suggest a model in which the antigen receptor loci normally exist in a repressed state that is inaccessible to the RAG1/2 recombinase. However, after localized ATP-dependent chromatin remodeling and histone acetylation, the nucleosomes can adopt a permissive state that allows the RAG1/2 recombinase to access and recombine the gene segments within that particular region of the antigen receptor locus.

## 12. REGULATION BY HISTONE METHYLATION

Although histone acetylation and chromatin remodeling seem to play important roles in modulating the accessibility of gene segments within the antigen receptor loci, it has become clear that they constitute just two layers of a multilayered regulatory process. For instance, when key transcription factors such as OcaB [152] or Pax5 [153] are removed from the cell, V(D)J recombination of particular antigen receptor loci is impaired, even though the levels of histone acetylation at these loci are unaffected. Therefore, histone acetylation alone is not sufficient to activate V(D)J recombination at the antigen receptor loci during lymphoid development. To better understand which other aspects of chromatin structure contribute to the regulation of V(D)J recombination, chromatin immunoprecipitation was utilized to map other posttranslational histone modifications—for example, dimethylated H3K9 [146,154], dimethylated H3K79 [155], dimethylated H3K4 [146,156], and trimethylated H3K4 [92,156]—to the antigen receptor loci in various cell types.

Based on these studies, a surprising mechanistic link between chromatin structure and V(D)J recombination emerged. Using a variety of biochemical assays, it was demonstrated that a domain—termed the plant homeodomain (PHD) finger—present in the C-terminal region of RAG2 specifically recognizes the N-terminal tail of histone H3 when it is trimethylated at lysine 4 (H3K4me3) [91,92]. Intriguingly, not only is H3K4me3 enriched at gene segments that are poised to undergo V(D)J recombination, but its pattern of enrichment is consistent with the tissue specificity, lineage specificity, and temporal ordering of antigen receptor gene rearrangement [92,156]. With the aid of a high-resolution crystal structure of the RAG2 PHD finger bound to H3K4me3, the molecular basis of H3K4me3 recognition was determined, leading to a functional analysis of this interaction's significance [92]. Strikingly, mutations that disrupted RAG2 binding to H3K4me3 severely impaired V(D)J recombination in vivo [91,92]. Likewise, reducing the abundance of H3K4me3 also leads to a dramatic decrease in V(D)J recombination [92]. Perhaps most surprisingly, a conserved tryptophan residue (W453) that constitutes a key structural component of the H3K4me3-binding pocket and is essential for RAG2's recognition of H3K4me3 is mutated in patients with a specific primary immunodeficiency, termed Omenn syndrome. Overall, these studies demonstrated that recognition of H3K4me3 by RAG2 is crucial for V(D)J recombination in vivo, thereby providing a mechanism by which chromatin structure can directly regulate the efficiency of V(D)J recombination.

Although it is clear that recognition of histone H3 lysine 4 trimethylation by RAG2 is essential for endogenous V(D)J recombination, the mechanism by which this interaction promotes V(D)J recombination is still somewhat unclear. Some studies suggest that H3K4me3 functions to allosterically stimulate the catalytic activity the RAG1/2 recombinase [157,158] by relieving the autoinhibition mediated by the C-terminal regions of RAG1 and RAG2 [158]. However, other studies suggest that H3K4me3 serves to recruit RAG2 to recombinationally active antigen receptor gene segments [159]. Regardless of whether the RAG2–H3K4me3 interaction is required for RAG1/2 recruitment to active antigen receptor loci, retention at these loci, or stimulation of RAG1/2-catalyzed DNA double-strand breaks, it is clear that this interaction represents an important, direct link between chromatin structure and V(D)J recombination.

## 13. HOW IS THE CHROMATIN STRUCTURE OF ANTIGEN RECEPTOR LOCI DEVELOPMENTALLY REGULATED?

Although it is increasingly apparent that alterations in chromatin structure regulate V(D)J recombination by modulating the accessibility of the various antigen receptor loci to the RAG1/2 recombinase (Figure 8), it is still not clear how the appropriate combination of ATP-dependent chromatin remodelers and histone-modifying enzymes are recruited to a particular region within a specific antigen receptor locus at

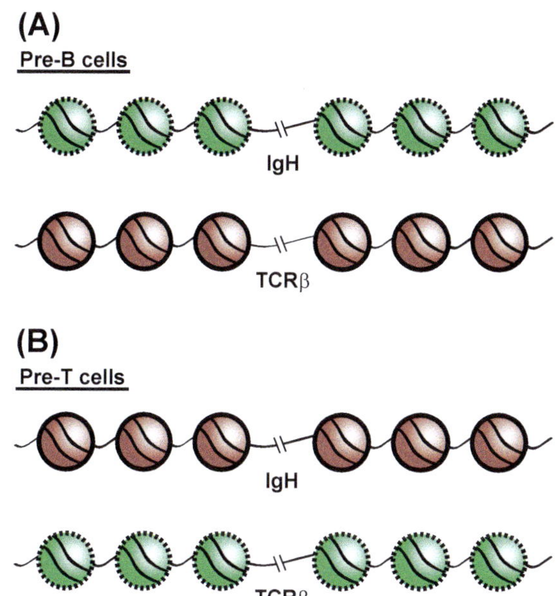


FIGURE 8 **Accessibility control of V(D)J recombination.** (A) In pre-B cells, the IgH locus is accessible to the RAG1/2 recombinase. This accessibility is due, in part, to nucleosome mobilization (indicated by dashed outlines), as well as H3/H4 acetylation and H3K4 trimethylation of nucleosomes in the locus (indicated by green spheres). In these same cells, the TCRβ locus is inaccessible to the RAG1/2 recombinase. This inaccessibility is due, in part, to a lack of nucleosome mobility (indicated by solid outlines), as well as a lack of H3/H4 acetylation and H3K4 trimethylation (indicated by red spheres). (B) In pre-T cells, the opposite pattern is observed: the IgH locus is inaccessible, whereas the TCRβ locus is accessible to the RAG1/2 recombinase. IgH, immunoglobulin heavy chain; TCRβ, T cell receptor β.

the right stage of lymphocyte development. Because each of the antigen receptor loci contains at least one enhancer that is essential for efficient V(D)J recombination at that locus [160–168], early studies investigated whether these enhancers play a role in regulating the accessibility of the antigen receptor loci. These studies demonstrated that enhancers play an important role in regulating the efficiency of RAG1/2-catalyzed double-strand breaks at antigen receptor gene segments [169], as well as the posttranslational modification status [170] and mobility [171] of nucleosomes within the antigen receptor loci. Subsequent studies demonstrated that promoter positioning near antigen receptor gene segments plays a key role in modulating the levels of histone acetylation [172,173] and methylation [172], nucleosome mobility [171], and overall accessibility of antigen receptor gene segments [174] to the RAG1/2 recombinase.

Taken together, these studies suggest a model in which locus-specific enhancers and developmentally regulated promoters within the antigen receptor loci alter local chromatin structure—by modulating both nucleosome mobility and nucleosome modifications—to developmentally regulate the ability of the RAG1/2 recombinase to bind and cleave antigen receptor gene segments. However, the mechanisms by which these accessibility control elements specifically recruit particular chromatin remodeling complexes and histone modifying enzymes to particular regions of the antigen receptor loci at various stages of lymphocyte development remain unclear.

## 14. ADDITIONAL LAYERS OF REGULATION

Although the developmental regulation of antigen receptor loci accessibility is best understood at the level of the nucleosome, accessibility control of V(D)J recombination by chromatin also extends to larger scales of chromosome structure. For instance, $V_H$ gene segments located in the $D_H$-distal portion of the murine $V_H$ domain are differentially accessible compared to $V_H$ gene segments located in the $D_H$-proximal portion of the $V_H$ domain, and the accessibility of the distal gene segments is regulated by a distinct set of proteins [153,175]. Thus, the accessibility of an entire cluster of antigen receptor gene segments can be regulated at the scale of several hundred kilobases of DNA. Furthermore, it has also been shown that the antigen receptor loci undergo large-scale locus contraction during lymphocyte development [176–178]. This locus contraction, which probably occurs via chromosomal looping [177–179] and seems to be regulated by transcription factors such as Pax5 [180], YY1 [175], Ikaros [181], and CTCF [182], brings recombinationally active gene segments into proximity with one another, thereby facilitating the joining of otherwise distant V, D, and J gene segments. In addition to long-range intrachromosomal contacts, the antigen receptor loci also exhibit chromosomal pairing, which may play a role in mediating allelic exclusion [183,184]. Finally, V(D)J recombination also seems to be regulated at the largest scale of chromosome movement within the nucleus. That is, during different stages of lymphoid development, the chromosomes that harbor the antigen receptor loci exhibit altered subnuclear localization and are positioned at the nuclear periphery when they are not poised to undergo recombination, but are positioned centrally when they are poised to undergo recombination [176].

## 15. REGULATION OF V(D)J RECOMBINATION: SUMMARY

Based on these overlapping layers of regulation, it has been proposed that antigen receptor gene assembly occurs within a specific, subnuclear compartment—which can be referred to as either a V(D)J recombination factory [3] or a recombination center [4]—within which V(D)J recombination is tightly regulated and carefully orchestrated. These specialized, subnuclear compartments would bring together the RAG1/2 recombinase, the NHEJ machinery, and the many proteins required for chromosomal pairing, chromosomal looping, chromatin remodeling, and posttranslational histone modification of the antigen receptor loci. By colocalizing the various factors that participate in chromosomal V(D)J recombination, the cell would be able to functionally coordinate the execution and regulation of antigen receptor gene assembly.

## 16. ONCOGENIC LESIONS IN LYMPHOID NEOPLASMS: THE PRICE OF A DIVERSE ANTIGEN RECEPTOR REPERTOIRE

The discovery that B and T cell antigen receptor diversity is created through V(D)J recombination provided an elegant solution to one of the most intriguing mysteries facing immunologists of that era [1]. Solving this problem, however, raised another important question: might this system, which involves breakage and rejoining of chromosomal DNA, allow occasional errors to undermine genomic stability, perhaps leading to oncogenic transformation? Shortly after the discovery of V(D)J recombination, characteristic chromosome rearrangements involving antigen receptor loci were indeed reported in lymphoid neoplasms [185,186]. In the following decades, a great deal of evidence accumulated to support the hypothesis that errors made by this mechanism are responsible for oncogenic DNA rearrangements in early B and T cell neoplasms, as well as in some follicular lymphomas [187,188]. Our understanding of the mechanisms responsible for generating recombination errors and the safeguards employed to limit them, however, remains incomplete.

Gross genomic alterations detected in lymphoid neoplasms include megabase-sized deletions and inversions, chromosomal translocations, and reinsertion of DNA segments (excised during normal V(D)J recombination) elsewhere in the genome. These events can activate oncogenes, inactivate tumor suppressor genes, and cause inappropriate expression of developmentally regulated genes. Biochemical and cell-based studies of V(D)J recombination have suggested a variety of possible mechanisms that might underlie such oncogenic rearrangements, as described below. In many cases, however, it has been difficult to attribute definitively a particular oncogenic lesion to a specific mechanism. Furthermore, our appreciation of the variety of mechanisms that can lead to oncogenic DNA rearrangements has been limited, until very recently, by the techniques used to detect aberrant events, which have been targeted to individual lesions that could be observed cytogenetically (e.g., chromosome translocations) or detected by PCR targeted to specific candidate locations (e.g., activating *Notch1* deletions). The deployment of large-scale tumor genome sequencing will facilitate more comprehensive surveys of the types, locations, and relative frequencies of lesions generated in neoplastic B and T cells. Such studies are beginning to provide a much richer understanding of the contributions of V(D)J recombination errors to neoplastic transformation.

## 17. PROPOSED MECHANISMS UNDERLYING RAG-MEDIATED GENOMIC LESIONS

V(D)J recombination errors can be classified into two broad categories: those that occur before cleavage (recognition errors) and those occurring after DNA strand breakage (joining errors and transposition) [189] (Figure 9). Both categories can, in theory, generate deletions, inversions, and chromosome translocations. Misrepair of broken ends can also fuel gene amplification, probably through a breakage–fusion–bridge mechanism [190]. Another type of joining error involves insertion of DNA segments flanked by signal ends, which are normally excised during V(D)J recombination, into the genome. Such insertions could, in principle, be mediated by joining errors (joining of the signal ends to a genomic double-strand break) or by RAG-mediated transposition (in which the RAG complex catalyzes integration of the signal ends into the new genomic location via a transesterification mechanism) [14,15].

### 17.1 Recognition Errors: Ectopic Recombination between cRSS

Ectopic V(D)J recombination between pairs of DNA sequences bearing a fortuitous resemblance to authentic RSS, termed cryptic RSSs (cRSSs), have been described in oncogenic lesions isolated from lymphoid neoplasms (Figure 9(A)). Such events have been well documented in experimental systems (see, for example, Refs [191–193]). Indeed, detailed analysis of ectopic recombination events in extrachromosomal substrates revealed that RSS-like sequences capable of supporting low levels of recombination are found roughly once per kilobase of random DNA sequence [191]. Such poor discrimination between authentic RSSs and cRSSs, if borne out in developing lymphocytes, would signify tremendous potential for RAG-mediated genomic instability. Given this, it is not surprising that lesions that could be attributed to ectopic recombination have been identified in oncogenic rearrangements. These include an activating Tal1–Sil fusion in T cell acute lymphoblastic leukemia (T-ALL) [194], oncogenic chromosome translocations in pre-B cell neoplasms [195], activating deletions in the *Notch1* gene in T-ALL [196] and in murine T lymphomas [197], and deletions in the *Izkf1* gene that create dominant negative forms of the Ikaros protein in B-ALL [198]. Ectopic recombination involving cRSS pairs can also inactivate tumor suppressor genes [199].

Two main criteria are used to determine whether a given rearrangement is likely to have been mediated by ectopic V(D)J recombination: (1) identification of sequences resembling authentic RSSs in the vicinity of both breakpoints and (2) the presence of sequence features reminiscent of normal V(D)J recombination at the junction (such as characteristic

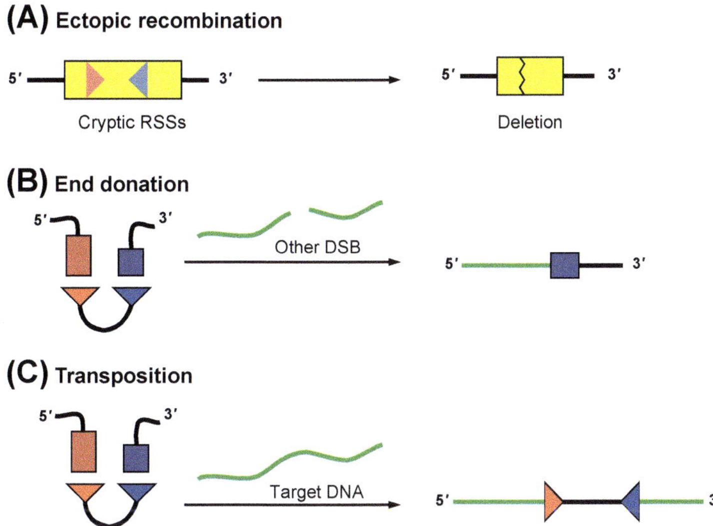


FIGURE 9 **V(D)J recombination errors.** (A) Ectopic recombination. RAG recognizes and cleaves at sequences in the genome that accidentally resemble a 12RSS and a 23RSS (cryptic RSSs), leading to deletion of a portion of the genome or, if the RSSs are oriented differently, to inversion (not shown). (B) End donation. After RAG cleaves in a normal manner at a 12RSS–23RSS pair in an antigen receptor locus, the repair step of the reaction fails and the ends are released. One or more of the ends generated by RAG can then be joined to an end generated by a fortuitous break generated elsewhere in the genome, typically leading to a chromosomal translocation. (C) Transposition. After performing cleavage normally, the RAG proteins remain bound to the signal ends, capture a target DNA molecule, and then insert the signal ends into the target, resulting in movement of the sequences bounded by the RSSs from one region in the genome to another. Such events are rare and have yet to be reported in primary human lymphocytes or lymphoid tumors. DSB, double-strand DNA break.

patterns of nucleotide loss and nucleotide addition). Several factors complicate this determination. First, the putative cRSSs may be poor matches to the consensus RSS, raising the possibility that they are nonfunctional DNA sequences present simply by chance. Indeed, when tested in vitro, some putative cRSSs isolated from oncogenic rearrangements yield low (or even undetectable) recombination activity [192,200]. These data suggest that these sequences do not function as cRSSs (the caveat being that even very rare recombination events can lead to oncogenic transformation by conferring a selective advantage). Another complicating factor is that some candidate cRSSs identified in putative ectopic recombination events can be located far from the breakpoints, or in the wrong orientation with respect to the putative cleavage site, both of which argue against involvement of the V(D)J recombinase. These observations raise an important question: how does one reliably distinguish an authentic, RAG-mediated ectopic recombination event from other rearrangements that, by chance, happen to lie within some (unspecified) distance from a sequence that bears some resemblance to an authentic RSS? One approach is to determine whether the events are recurrent in multiple tumors (such as the *Notch1* and *Ikaros* deletions noted earlier) and whether the putative rearrangements follow the 12/23 rule (again, observed in some of the recurrent rearrangements cited above). Other criteria derive from close examination of the junction sequences to see whether they bear characteristics of authentic V(D)J recombination. These include loss of small numbers of nucleotides from the presumptive coding ends, the presence of characteristic nucleotide insertions (N nucleotides, which could be inserted by TdT, and P nucleotides, which can be derived from opening of hairpin coding ends). Again, these criteria are not absolute, as non-RAG-mediated rearrangements may bear extra nucleotides and harbor short deletions from the broken DNA ends [201].

Another way to approach this problem is to analyze the quality of the putative cRSS. One example is provided by the "RIC score," a method designed to calculate the "recombination information content" of a particular sequence by determining its similarity to authentic RSSs [202]. This model, however, does not consistently identify cRSSs involved in RAG-induced rearrangements isolated from sequenced lymphoid neoplasms [203]. Perhaps the most diagnostic criterion is the presence of a reciprocal junction in DNA harvested from the tumor, although these have rarely, if ever, been reported. Further work needs to be done to develop more reliable methods to identify RAG-mediated genomic rearrangements in sequenced tumor genomes.

## 17.2 Recognition Errors: RAG-Mediated Cleavage at Non-B-Form DNA Structures

Apart from the (understandable) ability of the RAG nuclease to mistakenly recognize DNA sequences that resemble authentic RSSs, can the recombinase cleave at sites that bear no sequence similarity to an RSS? This possibility is suggested by biochemical studies that showed that purified RAG proteins can (under certain conditions) generate single-strand nicks at certain unusual DNA structures, such as hairpins, 3′ single-stranded tails, and non-B-form DNA structures [204–207]. Analysis of oncogenic rearrangements in B cell lymphomas involving particular hot spots for breakage (termed "major breakpoint regions" or MBRs) in the vicinity of oncogenes (*Bcl1* and *Bcl2*) failed to reveal cRSSs. Studies by Lieber's group, however, have indicated that the RAG proteins can cleave at a non-B DNA structure in the *Bcl2* MBR [206].

### 17.3 Joining Errors: End Donation

The second broad category of mistakes made during V(D)J recombination involves errors made after strand cleavage. Errors made in repair of broken DNA ends, in which an end created by RAG cleavage is joined to an end created through some other process, have been termed "end donation" [189,208], or "strand donation" [209] (Figure 9(B)). Such rearrangements can, for example, place normally silent oncogenes in the proximity of strong Ig or TCR enhancers. Examples include IgH–Myc translocations in Burkitt's lymphomas [186] and TCRα–TCL3 translocations involved in human B and T cell neoplasms [210]. The DNA ends may be joined either by classical NHEJ (cNHEJ), the high-fidelity end-joining mechanism normally responsible for repairing RAG-mediated DNA breaks, or by the poorly characterized, error-prone mechanism (or set of mechanisms) termed alternative NHEJ (aNHEJ) [201]. Junction sequences from chromosome translocations classifiable as end-donation events often show sequence features considered characteristic of aNHEJ, which is increasingly recognized as a genotoxic process in other systems [201]. Work has revealed, however, that junctions created by aNHEJ in developing lymphocytes or in T cell lymphomas need not show such features [211].

### 17.4 Joining Errors: Transposition

RAG-mediated transposition, joining an excised DNA fragment bounded by signal ends to an unrelated DNA sequence (Figure 9(C)), was suggested as a potential genotoxic mechanism by in vitro observations using purified, truncated RAG proteins [14,15]. Such events are expected to generate characteristic short duplications of DNA sequences at the site of integration (target site duplications) and should lack insertion of extra nucleotides at the junctions. Transposition events that meet these strict criteria have been observed in cell culture systems, but these events are uncommon [212,213] and have not yet been identified in oncogenic DNA rearrangements, including in lymphomas from mice lacking the C-terminus of RAG2 (which increases the frequency of transposition in vitro) [211,214]. Reinsertion of excised DNA fragments terminating in signal ends by end joining is more common than transposition in these systems [212,213] and has been observed in follicular lymphomas [215] and in mouse thymocytes [216]. Overall, the available evidence suggests that, although end-joining events that superficially resemble transposition are readily detected, true transposition events are uncommon. One caveat, however, is that some transposition events need not leave distinguishable footprints [217].

## 18. REGULATORY CONTROLS PROPOSED TO SUPPRESS RAG-MEDIATED GENOMIC INSTABILITY

Given the diversity of potential mechanisms by which the recombinase may commit the errors described above, it is reasonable to consider the possibility that the entire reaction is carefully regulated at multiple steps. Several potential regulatory controls have been examined, although we do not yet know which of these are the most critical to prevent oncogenic rearrangements, or how they might work together to suppress aberrant events.

### 18.1 Control Cell Type and Developmental-Stage-Specific Expression of Recombination Activity

To minimize opportunities for errors, it seems logical to limit RAG expression to the right cells and to the proper developmental stage, terminating expression once V(D)J recombination activity is no longer needed. Support for this idea is provided by the observation that constitutive expression of RAG1/RAG2 transgenes in mice severely reduced life span, with features reminiscent of DSB repair deficiency [218]. Transcriptional control elements responsible for regulating expression of RAG1/2 have been identified. Other mechanisms may also control the duration of RAG activity in lymphocyte precursors undergoing active V(D)J recombination, such as an ATM-dependent process described in 2013 [98].

The importance of minimizing the window of time during which RAG activity is expressed is emphasized by accelerated lymphomagenesis in engineered mice lacking p53 in which V(D)J recombination persists, without receiving an "off" signal from a completed antigen receptor [219]. This acceleration presumably occurs by increasing the time window in which errors can occur. This is supported by the identification of SLP65 deficiencies in human B cell neoplasms, resulting in "perpetual activation" of RAG activity [220].

### 18.2 Restrict RAG Activity to an Appropriate Stage of the Cell Cycle

V(D)J recombination normally occurs during the G0 phase of the cell cycle. This restriction requires the presence of a degradation motif located in the noncore C-terminus of

RAG2, ablation of which results in abnormal persistence of RAG-induced DSBs throughout the cell cycle [129]. Mice bearing a point mutation that disables this function experience accelerated lymphomagenesis in a p53-deficient background, with chromosome translocations providing evidence of genomic instability [221]. The mechanism(s) underlying the accelerated lymphomagenesis remains unclear. Persistence of signal ends throughout the cell cycle was also observed in mice lacking the C-terminus of RAG2 [60], with aberrant integration of signal end- or signal joint-containing DNA fragments into chromosomal DNA [216]. Indeed, mice lacking the entire dispensable C-terminus of RAG2 experience a much stronger acceleration of lymphomagenesis, suggesting that suppressive mechanisms in addition to cell cycle regulation may be disabled [99,211].

## 18.3 Suppress Recognition of Inappropriate, Ectopic Target Sequences

Recognition of cRSSs at ectopic genomic locations plays a clear role in V(D)J recombinase-mediated oncogenic DNA rearrangements. Reducing the length of time RAGs are expressed as well as maintaining cell cycle and cell-type-specific controls probably plays some role in minimizing opportunities for off-target recombination activity. Another layer of control that is likely to be important is targeting RAG1/2 to appropriate chromosomal locations. The RAG proteins are preferentially localized to appropriate antigen receptor loci [159]. This localization may involve the PHD finger in RAG2's C-terminus, which recognizes a chromatin mark associated with active chromatin, trimethylated histone H3K4 [91,92], consistent with the chromatin immunoprecipitation studies noted above [159]. Another role for histone modifications in regulating RAG activity is the autoinhibition of cleavage mediated by noncore portions of RAG1 and RAG2 that is relieved by binding to epigenetic marks characteristic of active chromatin [158].

## 18.4 Downregulate RAG's Inherent Transposition Activity

Expression of robust transposition activity would clearly have mutagenic consequences. This appears to be regulated, at least in part, by the noncore C-terminus of RAG2, which suppresses transposition in vitro [222–224]. Noncore regions of RAG1 and RAG2 also appear to limit another aberrant outcome of V(D)J recombination, hybrid joint formation [225].

## 18.5 Suppress Interchromosomal V(D)J Recombination

RAG2's noncore C-terminus also plays a role in suppressing interchromosomal rearrangements between antigen receptor loci (the TCRδ and TCRβ loci) [60]. The underlying mechanism has not yet been clarified, but could reflect persistent RAG activity through the cell cycle or elevated overall levels of recombinase activity.

## 18.6 Regulate Joining: Ensure High-Fidelity cNHEJ Machinery Is Used and Discourage End Donation

Postcleavage roles for the RAG proteins were noted upon analysis of certain mutant RAG1 and RAG2 proteins in cell-based assays, in which a defect in joining was observed [226–228], leading to the suggestion that a RAG1/RAG2 postcleavage DNA–protein complex might play an architectural role in orchestrating the joining reaction. By coordinating the joining of both pairs of ends, such a complex could serve to limit opportunities for end donation. Subsequent work revealed a role for the RAG proteins in shepherding the broken ends to the high-fidelity cNHEJ pathway, minimizing opportunities for the broken ends to engage in inappropriate repair pathways such as homologous recombination and aNHEJ [229]. Indeed, the dispensable C-terminus of RAG2 is critical for preventing access of the ends to aNHEJ [230], leading to the proposal that RAG2 is important for enforcing repair pathway choice. Interestingly, the RAG mutants that affect postcleavage steps appear to perturb the RAG postcleavage complex [228–232]. These observations led to the hypothesis that the RAG postcleavage complex might function to couple the cleavage reaction to the high-fidelity cNHEJ pathway, thereby limiting opportunities for the ends to engage in aNHEJ [230,232], which is known to be translocation-prone [233].

The above hypothesis predicts that mice bearing such RAG mutants should experience an elevated frequency of aberrant V(D)J recombination events, perhaps leading to an elevated incidence of lymphoid neoplasms. This prediction has, at least in part, been borne out. A RAG1 mutant that disrupts the postcleavage complex in vitro results in accelerated lymphomagenesis in knock-in mice lacking p53 [234]. Mice lacking both p53 and the dispensable RAG2 C-terminus experience a much greater acceleration of lymphomagenesis, again with chromosome translocations involving antigen receptor loci not observed in $p53^{-/-}$ littermates bearing wild-type RAG2 [99,211]. These observations suggested that an identifiable type of V(D)J recombination error may be enhanced in RAG2-mutant animals, perhaps of the end-donation type, and perhaps involving aNHEJ. Very recent studies tested these hypotheses by performing whole genome sequencing of lymphomas derived from two different RAG2 C-terminal-deletion mutant mice. Although some chromosome translocations were observed, the predominant genomic lesions, and all of those that cause known or suspected driver lesions, are ectopic recombination events between RSS pairs [211,214]. These data, although incomplete (only a few tumors have been studied to date), suggest that the C-terminus of RAG2 may play additional roles in discouraging ectopic recombination events.

These events, which are mostly deletions in the size range of normal V(D)J recombination events, would be largely undetectable by traditional cytogenetics-based approaches.

## 19. V(D)J RECOMBINATION ERRORS AS PATHOGENIC LESIONS IN LYMPHOID NEOPLASMS: SUMMARY

The discovery of V(D)J recombinase-induced mutations in lymphoid neoplasms has been largely restricted to events involving recurrent rearrangements that could be detected either cytogenetically or by targeted analysis of candidate genes. The advent of new techniques for genome-wide identification of structural variants will fuel the rapid development of comprehensive information about the types and locations of aberrant, RAG-mediated genome rearrangements. This information should illuminate both the nature of oncogenic V(D)J recombination errors and the mechanisms employed to prevent them.

Portions of this chapter were previously published in The Mechanism of V(D)J Recombination by Joann Sekiguchi, Frederick W. Alt, and Marjorie Oettinger © 2004.

## REFERENCES


[1] Hozumi N, Tonegawa S. Evidence for somatic rearrangement of immunoglobulin genes coding for variable and constant regions. Proc Natl Acad Sci USA 1976;73:3628–32.

[2] Bassing CH, Swat W, Alt FW. The mechanism and regulation of chromosomal V(D)J recombination. Cell 2002;109(Suppl.): S45–55.

[3] Matthews AG, Oettinger MA. RAG: a recombinase diversified. Nat Immunol 2009;10:817–21.

[4] Schatz DG, Ji Y. Recombination centres and the orchestration of V(D)J recombination. Nat Rev Immunol 2011;11:251–63.

[5] Schatz DG, Swanson PC. V(D)J recombination: mechanisms of initiation. Annu Rev Genet 2011;45:167–202.

[6] Schatz DG, Oettinger MA, Baltimore D. The V(D)J recombination activating gene, RAG-1. Cell 1989;59:1035–48.

[7] Oettinger MA, Schatz DG, Gorka C, Baltimore D. RAG-1 and RAG-2, adjacent genes that synergistically activate V(D)J recombination. Science 1990;248:1517–23.

[8] van Gent DC, Hiom K, Paull TT, Gellert M. Stimulation of V(D)J cleavage by high mobility group proteins. Embo J 1997;16: 2665–70.

[9] Roth DB, Menetski JP, Nakajima PB, Bosma MJ, Gellert M. V(D)J recombination: broken DNA molecules with covalently sealed (hairpin) coding ends in scid mouse thymocytes. Cell 1992;70: 983–91.

[10] Roth DB, Zhu C, Gellert M. Characterization of broken DNA molecules associated with V(D)J recombination. Proc Natl Acad Sci USA 1993;90:10788–92.

[11] Schlissel M, Constantinescu A, Morrow T, Baxter M, Peng A. Double-strand signal sequence breaks in V(D)J recombination are blunt, 5'-phosphorylated, RAG-dependent, and cell cycle regulated. Genes Dev 1993;7:2520–32.

[12] Lieber MR. The mechanism of double-strand DNA break repair by the nonhomologous DNA end-joining pathway. Annu Rev Biochem 2010;79:181–211.

[13] Lewis SM, Hesse JE, Mizuuchi K, Gellert M. Novel strand exchanges in V(D)J recombination. Cell 1988;55:1099–107.

[14] Agrawal A, Eastman QM, Schatz DG. Transposition mediated by RAG1 and RAG2 and its implications for the evolution of the immune system. Nature 1998;394:744–51.

[15] Hiom K, Melek M, Gellert M. DNA transposition by the RAG1 and RAG2 proteins: a possible source of oncogenic translocations. Cell 1998;94:463–70.

[16] Early P, Huang H, Davis M, Calame K, Hood L. An immunoglobulin heavy chain variable region gene is generated from three segments of DNA: VH, D and JH. Cell 1980;19:981–92.

[17] Tonegawa S. Somatic generation of antibody diversity. Nature 1983;302:575–81.

[18] Davis MM, Bjorkman PJ. T-cell antigen receptor genes and T-cell recognition. Nature 1988;334:395–402.

[19] Bassing CH, Alt FW, Hughes MM, D'Auteuil M, Wehrly TD, Woodman BB, et al. Recombination signal sequences restrict chromosomal V(D)J recombination beyond the 12/23 rule. Nature 2000;405:583–6.

[20] Sleckman BP, Bassing CH, Hughes MM, Okada A, D'Auteuil M, Wehrly TD, et al. Mechanisms that direct ordered assembly of T cell receptor beta locus V, D, and J gene segments. Proc Natl Acad Sci USA 2000;97:7975–80.

[21] Jung D, Bassing CH, Fugmann SD, Cheng HL, Schatz DG, Alt FW. Extrachromosomal recombination substrates recapitulate beyond 12/23 restricted VDJ recombination in nonlymphoid cells. Immunity 2003;18:65–74.

[22] Tillman RE, Wooley AL, Khor B, Wehrly TD, Little CA, Sleckman BP. Cutting edge: targeting of V beta to D beta rearrangement by RSSs can be mediated by the V(D)J recombinase in the absence of additional lymphoid-specific factors. J Immunol 2003;170:5–9.

[23] Drejer-Teel AH, Fugmann SD, Schatz DG. The beyond 12/23 restriction is imposed at the nicking and pairing steps of DNA cleavage during V(D)J recombination. Mol Cell Biol 2007;27:6288–99.

[24] Fujimoto S, Yamagishi H. Isolation of an excision product of T-cell receptor alpha-chain gene rearrangements. Nature 1987;327:242–3.

[25] Okazaki K, Davis DD, Sakano H. T cell receptor beta gene sequences in the circular DNA of thymocyte nuclei: direct evidence for intramolecular DNA deletion in V-D-J joining. Cell 1987;49:477–85.

[26] Zachau HG. The immunoglobulin kappa locus - or - what has been learned from looking closely at one-tenth of a percent of the human genome. Gene 1993;135:167–73.

[27] Ramsden DA, Baetz K, Wu GE. Conservation of sequence in recombination signal sequence spacers. Nucleic Acids Res 1994;22: 1785–96.

[28] Hesse JE, Lieber MR, Mizuuchi K, Gellert M. V(D)J recombination: a functional definition of the joining signals. Genes Dev 1989;3:1053–61.

[29] Ramsden DA, McBlane JF, van Gent DC, Gellert M. Distinct DNA sequence and structure requirements for the two steps of V(D)J recombination signal cleavage. Embo J 1996;15:3197–206.

[30] Feeney AJ, Goebel P, Espinoza CR. Many levels of control of V gene rearrangement frequency. Immunol Rev 2004;200:44–56.
[31] Cuomo CA, Mundy CL, Oettinger MA. DNA sequence and structure requirements for cleavage of V(D)J recombination signal sequences. Mol Cell Biol 1996;16:5683–90.
[32] Yu K, Lieber MR. Mechanistic basis for coding end sequence effects in the initiation of V(D)J recombination. Mol Cell Biol 1999;19:8094–102.
[33] McBlane JF, van Gent DC, Ramsden DA, Romeo C, Cuomo CA, Gellert M, et al. Cleavage at a V(D)J recombination signal requires only RAG1 and RAG2 proteins and occurs in two steps. Cell 1995;83:387–95.
[34] van Gent DC, McBlane JF, Ramsden DA, Sadofsky MJ, Hesse JE, Gellert M. Initiation of V(D)J recombination in a cell-free system. Cell 1995;81:925–34.
[35] van Gent DC, Mizuuchi K, Gellert M. Similarities between initiation of V(D)J recombination and retroviral integration. Science 1996;271:1592–4.
[36] Mizuuchi K, Adzuma K. Inversion of the phosphate chirality at the target site of Mu DNA strand transfer: evidence for a one-step transesterification mechanism. Cell 1991;66:129–40.
[37] Engelman A, Mizuuchi K, Craigie R. HIV-1 DNA integration: mechanism of viral DNA cleavage and DNA strand transfer. Cell 1991;67:1211–21.
[38] Curry JD, Geier JK, Schlissel MS. Single-strand recombination signal sequence nicks in vivo: evidence for a capture model of synapsis. Nat Immunol 2005;6:1272–9.
[39] Jones JM, Gellert M. Ordered assembly of the V(D)J synaptic complex ensures accurate recombination. Embo J 2002;21:4162–71.
[40] Mundy CL, Patenge N, Matthews AG, Oettinger MA. Assembly of the RAG1/RAG2 synaptic complex. Mol Cell Biol 2002;22:69–77.
[41] Swanson PC. A RAG-1/RAG-2 tetramer supports 12/23-regulated synapsis, cleavage, and transposition of V(D)J recombination signals. Mol Cell Biol 2002;22:7790–801.
[42] Shlyakhtenko LS, Gilmore J, Kriatchko AN, Kumar S, Swanson PC, Lyubchenko YL. Molecular mechanism underlying RAG1/RAG2 synaptic complex formation. J Biol Chem 2009;284:20956–65.
[43] Landree MA, Kale SB, Roth DB. Functional organization of single and paired V(D)J cleavage complexes. Mol Cell Biol 2001;21:4256–64.
[44] Eastman QM, Leu TM, Schatz DG. Initiation of V(D)J recombination in vitro obeying the 12/23 rule. Nature 1996;380:85–8.
[45] Grundy GJ, Hesse JE, Gellert M. Requirements for DNA hairpin formation by RAG1/2. Proc Natl Acad Sci USA 2007;104:3078–83.
[46] Nishihara T, Nagawa F, Imai T, Sakano H. RAG-heptamer interaction in the synaptic complex is a crucial biochemical checkpoint for the 12/23 recombination rule. J Biol Chem 2008;283:4877–85.
[47] Bischerour J, Lu C, Roth DB, Chalmers R. Base flipping in V(D)J recombination: insights into the mechanism of hairpin formation, the 12/23 rule, and the coordination of double-strand breaks. Mol Cell Biol 2009;29:5889–99.
[48] Bischerour J, Chalmers R. Base-flipping dynamics in a DNA hairpin processing reaction. Nucleic Acids Res 2007;35:2584–95.
[49] Bischerour J, Chalmers R. Base flipping in tn10 transposition: an active flip and capture mechanism. PLoS One 2009;4:e6201.
[50] Agrawal A, Schatz DG. RAG1 and RAG2 form a stable postcleavage synaptic complex with DNA containing signal ends in V(D)J recombination. Cell 1997;89:43–53.
[51] Hiom K, Gellert M. Assembly of a 12/23 paired signal complex: a critical control point in V(D)J recombination. Mol Cell 1998;1:1011–9.
[52] Ma Y, Pannicke U, Schwarz K, Lieber MR. Hairpin opening and overhang processing by an Artemis/DNA-dependent protein kinase complex in nonhomologous end joining and V(D)J recombination. Cell 2002;108:781–94.
[53] Schatz DG. Antigen receptor genes and the evolution of a recombinase. Semin Immunol 2004;16:245–56.
[54] Fugmann SD. The origins of the Rag genes–from transposition to V(D)J recombination. Semin Immunol 2010;22:10–6.
[55] Ichihara Y, Hirai M, Kurosawa Y. Sequence and chromosome assignment to 11p13-p12 of human RAG genes. Immunol Lett 1992;33:277–84.
[56] Liang HE, Hsu LY, Cado D, Cowell LG, Kelsoe G, Schlissel MS. The "dispensable" portion of RAG2 is necessary for efficient V-to-DJ rearrangement during B and T cell development. Immunity 2002;17:639–51.
[57] Akamatsu Y, Monroe R, Dudley DD, Elkin SK, Gartner F, Talukder SR, et al. Deletion of the RAG2 C terminus leads to impaired lymphoid development in mice. Proc Natl Acad Sci USA 2003;100:1209–14.
[58] Dudley DD, Sekiguchi J, Zhu C, Sadofsky MJ, Whitlow S, DeVido J, et al. Impaired V(D)J recombination and lymphocyte development in core RAG1-expressing mice. J Exp Med 2003;198:1439–50.
[59] Talukder SR, Dudley DD, Alt FW, Takahama Y, Akamatsu Y. Increased frequency of aberrant V(D)J recombination products in core RAG-expressing mice. Nucleic Acids Res 2004;32:4539–49.
[60] Curry JD, Schlissel MS. RAG2's non-core domain contributes to the ordered regulation of V(D)J recombination. Nucleic Acids Res 2008;36:5750–62.
[61] Sadofsky MJ, Hesse JE, McBlane JF, Gellert M. Expression and V(D)J recombination activity of mutated RAG-1 proteins. Nucleic Acids Res 1993;21:5644–50.
[62] Silver DP, Spanopoulou E, Mulligan RC, Baltimore D. Dispensable sequence motifs in the RAG-1 and RAG-2 genes for plasmid V(D)J recombination. Proc Natl Acad Sci USA 1993;90:6100–4.
[63] Spanopoulou E, Zaitseva F, Wang FH, Santagata S, Baltimore D, Panayotou G. The homeodomain region of Rag-1 reveals the parallel mechanisms of bacterial and V(D)J recombination. Cell 1996;87:263–76.
[64] Yin FF, Bailey S, Innis CA, Ciubotaru M, Kamtekar S, Steitz TA, et al. Structure of the RAG1 nonamer binding domain with DNA reveals a dimer that mediates DNA synapsis. Nat Struct Mol Biol 2009;16:499–508.
[65] Peak MM, Arbuckle JL, Rodgers KK. The central domain of core RAG1 preferentially recognizes single-stranded recombination signal sequence heptamer. J Biol Chem 2003;278:18235–40.
[66] Mo X, Bailin T, Sadofsky MJ. A C-terminal region of RAG1 contacts the coding DNA during V(D)J recombination. Mol Cell Biol 2001;21:2038–47.
[67] Arbuckle JL, Fauss LA, Simpson R, Ptaszek LM, Rodgers KK. Identification of two topologically independent domains in RAG1 and their role in macromolecular interactions relevant to V(D)J recombination. J Biol Chem 2001;276:37093–101.
[68] Kim DR, Dai Y, Mundy CL, Yang W, Oettinger MA. Mutations of acidic residues in RAG1 define the active site of the V(D)J recombinase. Genes Dev 1999;13:3070–80.

[69] Landree MA, Wibbenmeyer JA, Roth DB. Mutational analysis of RAG1 and RAG2 identifies three catalytic amino acids in RAG1 critical for both cleavage steps of V(D)J recombination. Genes Dev 1999;13:3059–69.

[70] Fugmann SD, Villey IJ, Ptaszek LM, Schatz DG. Identification of two catalytic residues in RAG1 that define a single active site within the RAG1/RAG2 protein complex. Mol Cell 2000;5:97–107.

[71] De P, Rodgers KK. Putting the pieces together: identification and characterization of structural domains in the V(D)J recombination protein RAG1. Immunol Rev 2004;200:70–82.

[72] Spanopoulou E, Cortes P, Shih C, Huang CM, Silver DP, Svec P, et al. Localization, interaction, and RNA binding properties of the V(D)J recombination-activating proteins RAG1 and RAG2. Immunity 1995;3:715–26.

[73] Cuomo CA, Kirch SA, Gyuris J, Brent R, Oettinger MA. Rch1, a protein that specifically interacts with the RAG-1 recombination-activating protein. Proc Natl Acad Sci USA 1994;91:6156–60.

[74] Bailin T, Mo X, Sadofsky MJ. A RAG1 and RAG2 tetramer complex is active in cleavage in V(D)J recombination. Mol Cell Biol 1999;19:4664–71.

[75] Ciubotaru M, Ptaszek LM, Baker GA, Baker SN, Bright FV, Schatz DG. RAG1-DNA binding in V(D)J recombination. Specificity and DNA-induced conformational changes revealed by fluorescence and CD spectroscopy. J Biol Chem 2003;278:5584–96.

[76] Swanson PC, Desiderio S. RAG-2 promotes heptamer occupancy by RAG-1 in the assembly of a V(D)J initiation complex. Mol Cell Biol 1999;19:3674–83.

[77] Rodgers KK, Villey IJ, Ptaszek L, Corbett E, Schatz DG, Coleman JE. A dimer of the lymphoid protein RAG1 recognizes the recombination signal sequence and the complex stably incorporates the high mobility group protein HMG2. Nucleic Acids Res 1999;27: 2938–46.

[78] De P, Zhao S, Gwyn LM, Godderz LJ, Peak MM, Rodgers KK. Thermal dependency of RAG1 self-association properties. BMC Biochem 2008;9:5.

[79] Godderz LJ, Rahman NS, Risinger GM, Arbuckle JL, Rodgers KK. Self-association and conformational properties of RAG1: implications for formation of the V(D)J recombinase. Nucleic Acids Res 2003;31:2014–23.

[80] Arbuckle JL, Rahman NS, Zhao S, Rodgers W, Rodgers KK. Elucidating the domain architecture and functions of non-core RAG1: the capacity of a non-core zinc-binding domain to function in nuclear import and nucleic acid binding. BMC Biochem 2011;12:23.

[81] Cortes P, Ye ZS, Baltimore D. RAG-1 interacts with the repeated amino acid motif of the human homologue of the yeast protein SRP1. Proc Natl Acad Sci USA 1994;91:7633–7.

[82] Rodgers KK, Bu Z, Fleming KG, Schatz DG, Engelman DM, Coleman JE. A zinc-binding domain involved in the dimerization of RAG1. J Mol Biol 1996;260:70–84.

[83] Bellon SF, Rodgers KK, Schatz DG, Coleman JE, Steitz TA. Crystal structure of the RAG1 dimerization domain reveals multiple zinc-binding motifs including a novel zinc binuclear cluster. Nat Struct Biol 1997;4:586–91.

[84] Yurchenko V, Xue Z, Sadofsky M. The RAG1 N-terminal domain is an E3 ubiquitin ligase. Genes Dev 2003;17:581–5.

[85] Jones JM, Gellert M. Autoubiquitylation of the V(D)J recombinase protein RAG1. Proc Natl Acad Sci USA 2003;100:15446–51.

[86] Grazini U, Zanardi F, Citterio E, Casola S, Goding CR, McBlane F. The RING domain of RAG1 ubiquitylates histone H3: a novel activity in chromatin-mediated regulation of V(D)J joining. Mol Cell 2010;37:282–93.

[87] Kassmeier MD, Mondal K, Palmer VL, Raval P, Kumar S, Perry GA, et al. VprBP binds full-length RAG1 and is required for B-cell development and V(D)J recombination fidelity. Embo J 2012;31:945–58.

[88] Cuomo CA, Oettinger MA. Analysis of regions of RAG-2 important for V(D)J recombination. Nucleic Acids Res 1994;22:1810–4.

[89] Sadofsky MJ, Hesse JE, Gellert M. Definition of a core region of RAG-2 that is functional in V(D)J recombination. Nucleic Acids Res 1994;22:1805–9.

[90] Callebaut I, Mornon JP. The V(D)J recombination activating protein RAG2 consists of a six-bladed propeller and a PHD finger-like domain, as revealed by sequence analysis. Cell Mol Life Sci 1998;54:880–91.

[91] Liu Y, Subrahmanyam R, Chakraborty T, Sen R, Desiderio S. A plant homeodomain in RAG-2 that binds Hypermethylated lysine 4 of histone H3 is necessary for efficient antigen-receptor-gene rearrangement. Immunity 2007;27:561–71.

[92] Matthews AG, Kuo AJ, Ramon-Maiques S, Han S, Champagne KS, Ivanov D, et al. RAG2 PHD finger couples histone H3 lysine 4 trimethylation with V(D)J recombination. Nature 2007;450: 1106–10.

[93] Jiang H, Chang FC, Ross AE, Lee J, Nakayama K, Nakayama K, et al. Ubiquitylation of RAG-2 by Skp2-SCF links destruction of the V(D)J recombinase to the cell cycle. Mol Cell 2005;18:699–709.

[94] Lin WC, Desiderio S. Cell cycle regulation of V(D)J recombination-activating protein RAG-2. Proc Natl Acad Sci USA 1994;91: 2733–7.

[95] Lee J, Desiderio S. Cyclin A/CDK2 regulates V(D)J recombination by coordinating RAG-2 accumulation and DNA repair. Immunity 1999;11:771–81.

[96] Mizuta R, Mizuta M, Araki S, Kitamura D. RAG2 is down-regulated by cytoplasmic sequestration and ubiquitin-dependent degradation. J Biol Chem 2002;277:41423–7.

[97] Ross AE, Vuica M, Desiderio S. Overlapping signals for protein degradation and nuclear localization define a role for intrinsic RAG-2 nuclear uptake in dividing cells. Mol Cell Biol 2003;23:5308–19.

[98] Chaumeil J, Micsinai M, Ntziachristos P, Roth DB, Aifantis I, Kluger Y, et al. The RAG2 C-terminus and ATM protect genome integrity by controlling antigen receptor gene cleavage. Nat Commun 2013;4:2231.

[99] Deriano L, Chaumeil J, Coussens M, Multani A, Chou Y, Alekseyenko AV, et al. The RAG2 C terminus suppresses genomic instability and lymphomagenesis. Nature 2011;471:119–23.

[100] Leu TM, Schatz DG. rag-1 and rag-2 are components of a high-molecular-weight complex, and association of rag-2 with this complex is rag-1 dependent. Mol Cell Biol 1995;15:5657–70.

[101] McMahan CJ, Sadofsky MJ, Schatz DG. Definition of a large region of RAG1 that is important for coimmunoprecipitation of RAG2. J Immunol 1997;158:2202–10.

[102] Aidinis V, Dias DC, Gomez CA, Bhattacharyya D, Spanopoulou E, Santagata S. Definition of minimal domains of interaction within the recombination-activating genes 1 and 2 recombinase complex. J Immunol 2000;164:5826–32.

[103] Sawchuk DJ, Weis-Garcia F, Malik S, Besmer E, Bustin M, Nussenzweig MC, et al. V(D)J recombination: modulation of RAG1 and RAG2 cleavage activity on 12/23 substrates by whole cell extract and DNA-bending proteins. J Exp Med 1997;185:2025–32.
[104] Hiom K, Gellert M. A stable RAG1-RAG2-DNA complex that is active in V(D)J cleavage. Cell 1997;88:65–72.
[105] Thomas JO, Travers AA. HMG1 and 2, and related "architectural" DNA-binding proteins. Trends Biochem Sci 2001;26:167–74.
[106] Swanson PC. Fine structure and activity of discrete RAG-HMG complexes on V(D)J recombination signals. Mol Cell Biol 2002;22:1340–51.
[107] Aidinis V, Bonaldi T, Beltrame M, Santagata S, Bianchi ME, Spanopoulou E. The RAG1 homeodomain recruits HMG1 and HMG2 to facilitate recombination signal sequence binding and to enhance the intrinsic DNA-bending activity of RAG1-RAG2. Mol Cell Biol 1999;19:6532–42.
[108] Stros M. HMGB proteins: interactions with DNA and chromatin. Biochim Biophys Acta 2010;1799:101–13.
[109] Klass J, Murphy FV, Fouts S, Serenil M, Changela A, Siple J, et al. The role of intercalating residues in chromosomal high-mobility-group protein DNA binding, bending and specificity. Nucleic Acids Res 2003;31:2852–64.
[110] Saito K, Kikuchi T, Yoshida M. The mechanism of sequence non-specific DNA binding of HMG1/2-box B in HMG1 with DNA. Protein Eng 1999;12:235–42.
[111] Stott K, Watson M, Howe FS, Grossmann JG, Thomas JO. Tail-mediated collapse of HMGB1 is dynamic and occurs via differential binding of the acidic tail to the A and B domains. J Mol Biol 2010;403:706–22.
[112] Bergeron S, Madathiparambil T, Swanson PC. Both high mobility group (HMG)-boxes and the acidic tail of HMGB1 regulate recombination-activating gene (RAG)-mediated recombination signal synapsis and cleavage in vitro. J Biol Chem 2005;280:31314–24.
[113] Dai Y, Wong B, Yen YM, Oettinger MA, Kwon J, Johnson RC. Determinants of HMGB proteins required to promote RAG1/2-recombination signal sequence complex assembly and catalysis during V(D)J recombination. Mol Cell Biol 2005;25:4413–25.
[114] Kriatchko AN, Bergeron S, Swanson PC. HMG-box domain stimulation of RAG1/2 cleavage activity is metal ion dependent. BMC Mol Biol 2008;9:32.
[115] Little AJ, Corbett E, Ortega F, Schatz DG. Cooperative recruitment of HMGB1 during V(D)J recombination through interactions with RAG1 and DNA. Nucleic Acids Res 2013;41:3289–301.
[116] Ciubotaru M, Trexler AJ, Spiridon LN, Surleac MD, Rhoades E, Petrescu AJ, et al. RAG and HMGB1 create a large bend in the 23RSS in the V(D)J recombination synaptic complexes. Nucleic Acids Res 2013;41:2437–54.
[117] Calogero S, Grassi F, Aguzzi A, Voigtlander T, Ferrier P, Ferrari S, et al. The lack of chromosomal protein Hmg1 does not disrupt cell growth but causes lethal hypoglycemia in newborn mice. Nat Genet 1999;22:276–80.
[118] Ronfani L, Ferraguti M, Croci L, Ovitt CE, Scholer HR, Consalez GG, et al. Reduced fertility and spermatogenesis defects in mice lacking chromosomal protein Hmgb2. Development 2001;128:1265–73.
[119] Bianchi ME, Agresti A. HMG proteins: dynamic players in gene regulation and differentiation. Curr Opin Genet Dev 2005;15:496–506.
[120] Swanson PC. The DDE motif in RAG-1 is contributed in trans to a single active site that catalyzes the nicking and transesterification steps of V(D)J recombination. Mol Cell Biol 2001;21:449–58.
[121] Grundy GJ, Ramon-Maiques S, Dimitriadis EK, Kotova S, Biertumpfel C, Heymann JB, et al. Initial stages of V(D)J recombination: the organization of RAG1/2 and RSS DNA in the postcleavage complex. Mol Cell 2009;35:217–27.
[122] Swanson PC, Desiderio S. V(D)J recombination signal recognition: distinct, overlapping DNA-protein contacts in complexes containing RAG1 with and without RAG2. Immunity 1998;9:115–25.
[123] Akamatsu Y, Oettinger MA. Distinct roles of RAG1 and RAG2 in binding the V(D)J recombination signal sequences. Mol Cell Biol 1998;18:4670–8.
[124] Nagawa F, Ishiguro K, Tsuboi A, Yoshida T, Ishikawa A, Takemori T, et al. Footprint analysis of the RAG protein recombination signal sequence complex for V(D)J type recombination. Mol Cell Biol 1998;18:655–63.
[125] Eastman QM, Villey IJ, Schatz DG. Detection of RAG protein-V(D)J recombination signal interactions near the site of DNA cleavage by UV cross-linking. Mol Cell Biol 1999;19:3788–97.
[126] Swanson PC. The bounty of RAGs: recombination signal complexes and reaction outcomes. Immunol Rev 2004;200:90–114.
[127] Ciubotaru M, Kriatchko AN, Swanson PC, Bright FV, Schatz DG. Fluorescence resonance energy transfer analysis of recombination signal sequence configuration in the RAG1/2 synaptic complex. Mol Cell Biol 2007;27:4745–58.
[128] Lin WC, Desiderio S. Regulation of V(D)J recombination activator protein RAG-2 by phosphorylation. Science 1993;260:953–9.
[129] Li Z, Dordai DI, Lee J, Desiderio S. A conserved degradation signal regulates RAG-2 accumulation during cell division and links V(D)J recombination to the cell cycle. Immunity 1996;5:575–89.
[130] Yancopoulos GD, Blackwell TK, Suh H, Hood L, Alt FW. Introduced T cell receptor variable region gene segments recombine in pre-B cells: evidence that B and T cells use a common recombinase. Cell 1986;44:251–9.
[131] Schatz DG, Baltimore D. Stable expression of immunoglobulin gene V(D)J recombinase activity by gene transfer into 3T3 fibroblasts. Cell 1988;53:107–15.
[132] Kornberg RD, Thomas JO. Chromatin structure; oligomers of the histones. Science 1974;184:865–8.
[133] Luger K, Mader AW, Richmond RK, Sargent DF, Richmond TJ. Crystal structure of the nucleosome core particle at 2.8 A resolution. Nature 1997;389:251–60.
[134] Richmond TJ, Davey CA. The structure of DNA in the nucleosome core. Nature 2003;423:145–50.
[135] Robinson PJ, Fairall L, Huynh VA, Rhodes D. EM measurements define the dimensions of the "30-nm" chromatin fiber: evidence for a compact, interdigitated structure. Proc Natl Acad Sci USA 2006;103:6506–11.
[136] Dorigo B, Schalch T, Kulangara A, Duda S, Schroeder RR, Richmond TJ. Nucleosome arrays reveal the two-start organization of the chromatin fiber. Science 2004;306:1571–3.
[137] Kingston RE, Narlikar GJ. ATP-dependent remodeling and acetylation as regulators of chromatin fluidity. Genes Dev 1999;13:2339–52.
[138] Narlikar GJ, Fan HY, Kingston RE. Cooperation between complexes that regulate chromatin structure and transcription. Cell 2002;108:475–87.

[139] Kouzarides T. Chromatin modifications and their function. Cell 2007;128:693–705.

[140] Kwon J, Imbalzano AN, Matthews A, Oettinger MA. Accessibility of nucleosomal DNA to V(D)J cleavage is modulated by RSS positioning and HMG1. Mol Cell 1998;2:829–39.

[141] McBlane F, Boyes J. Stimulation of V(D)J recombination by histone acetylation. Curr Biol 2000;10:483–6.

[142] Golding A, Chandler S, Ballestar E, Wolffe AP, Schlissel MS. Nucleosome structure completely inhibits in vitro cleavage by the V(D)J recombinase. Embo J 1999;18:3712–23.

[143] Kwon J, Morshead KB, Guyon JR, Kingston RE, Oettinger MA. Histone acetylation and hSWI/SNF remodeling act in concert to stimulate V(D)J cleavage of nucleosomal DNA. Mol Cell 2000;6:1037–48.

[144] Patenge N, Elkin SK, Oettinger MA. ATP-dependent remodeling by SWI/SNF and ISWI proteins stimulates V(D)J cleavage of 5 S arrays. J Biol Chem 2004;279:35360–7.

[145] McMurry MT, Krangel MS. A role for histone acetylation in the developmental regulation of VDJ recombination. Science 2000;287:495–8.

[146] Morshead KB, Ciccone DN, Taverna SD, Allis CD, Oettinger MA. Antigen receptor loci poised for V(D)J rearrangement are broadly associated with BRG1 and flanked by peaks of histone H3 dimethylated at lysine 4. Proc Natl Acad Sci USA 2003;100:11577–82.

[147] Agata Y, Katakai T, Ye SK, Sugai M, Gonda H, Honjo T, et al. Histone acetylation determines the developmentally regulated accessibility for T cell receptor gamma gene recombination. J Exp Med 2001;193:873–80.

[148] Chowdhury D, Sen R. Stepwise activation of the immunoglobulin mu heavy chain gene locus. EMBO J 2001;20:6394–403.

[149] Maes J, O'Neill LP, Cavelier P, Turner BM, Rougeon F, Goodhardt M. Chromatin remodeling at the Ig loci prior to V(D)J recombination. J Immunol 2001;167:866–74.

[150] Johnson K, Angelin-Duclos C, Park S, Calame KL. Changes in histone acetylation are associated with differences in accessibility of V(H) gene segments to V-DJ recombination during B-cell ontogeny and development. Mol Cell Biol 2003;23:2438–50.

[151] Mathieu N, Hempel WM, Spicuglia S, Verthuy C, Ferrier P. Chromatin remodeling by the T cell receptor (TCR)-beta gene enhancer during early T cell development: Implications for the control of TCR-beta locus recombination. J Exp Med 2000;192:625–36.

[152] Casellas R, Jankovic M, Meyer G, Gazumyan A, Luo Y, Roeder R, et al. OcaB is required for normal transcription and V(D)J recombination of a subset of immunoglobulin kappa genes. Cell 2002;110:575–85.

[153] Hesslein DG, Pflugh DL, Chowdhury D, Bothwell AL, Sen R, Schatz DG. Pax5 is required for recombination of transcribed, acetylated, 5' IgH V gene segments. Genes Dev 2003;17:37–42.

[154] Johnson K, Pflugh DL, Yu D, Hesslein DG, Lin KI, Bothwell AL, et al. B cell-specific loss of histone 3 lysine 9 methylation in the V(H) locus depends on Pax5. Nat Immunol 2004;5:853–61.

[155] Ng HH, Ciccone DN, Morshead KB, Oettinger MA, Struhl K. Lysine-79 of histone H3 is hypomethylated at silenced loci in yeast and mammalian cells: a potential mechanism for position-effect variegation. Proc Natl Acad Sci USA 2003;100:1820–5.

[156] Perkins EJ, Kee BL, Ramsden DA. Histone 3 lysine 4 methylation during the pre-B to immature B-cell transition. Nucleic Acids Res 2004;32:1942–7.

[157] Shimazaki N, Tsai AG, Lieber MR. H3K4me3 stimulates the V(D)J RAG complex for both nicking and hairpinning in trans in addition to tethering in cis: implications for translocations. Mol Cell 2009;34:535–44.

[158] Grundy GJ, Yang W, Gellert M. Autoinhibition of DNA cleavage mediated by RAG1 and RAG2 is overcome by an epigenetic signal in V(D)J recombination. Proc Natl Acad Sci USA 2010;107:22487–92.

[159] Ji Y, Resch W, Corbett E, Yamane A, Casellas R, Schatz DG. The in vivo pattern of binding of RAG1 and RAG2 to antigen receptor loci. Cell 2010;141:419–31.

[160] Monroe RJ, Sleckman BP, Monroe BC, Khor B, Claypool S, Ferrini R, et al. Developmental regulation of TCR delta locus accessibility and expression by the TCR delta enhancer. Immunity 1999;10:503–13.

[161] Sleckman BP, Bardon CG, Ferrini R, Davidson L, Alt FW. Function of the TCR alpha enhancer in alphabeta and gammadelta T cells. Immunity 1997;7:505–15.

[162] Gorman JR, van der Stoep N, Monroe R, Cogne M, Davidson L, Alt FW. The Ig(kappa) enhancer influences the ratio of Ig(kappa) versus Ig(lambda) B lymphocytes. Immunity 1996;5:241–52.

[163] Sakai E, Bottaro A, Davidson L, Sleckman BP, Alt FW. Recombination and transcription of the endogenous Ig heavy chain locus is effected by the Ig heavy chain intronic enhancer core region in the absence of the matrix attachment regions. Proc Natl Acad Sci USA 1999;96:1526–31.

[164] Inlay M, Alt FW, Baltimore D, Xu Y. Essential roles of the kappa light chain intronic enhancer and 3' enhancer in kappa rearrangement and demethylation. Nat Immunol 2002;3:463–8.

[165] Xu Y, Davidson L, Alt FW, Baltimore D. Deletion of the Ig kappa light chain intronic enhancer/matrix attachment region impairs but does not abolish V kappa J kappa rearrangement. Immunity 1996;4:377–85.

[166] Bories JC, Demengeot J, Davidson L, Alt FW. Gene-targeted deletion and replacement mutations of the T-cell receptor beta-chain enhancer: the role of enhancer elements in controlling V(D)J recombination accessibility. Proc Natl Acad Sci USA 1996;93:7871–6.

[167] Bouvier G, Watrin F, Naspetti M, Verthuy C, Naquet P, Ferrier P. Deletion of the mouse T-cell receptor beta gene enhancer blocks alphabeta T-cell development. Proc Natl Acad Sci USA 1996;93:7877–81.

[168] Lauzurica P, Krangel MS. Temporal and lineage-specific control of T cell receptor alpha/delta gene rearrangement by T cell receptor alpha and delta enhancers. J Exp Med 1994;179:1913–21.

[169] McMurry MT, Hernandez-Munain C, Lauzurica P, Krangel MS. Enhancer control of local accessibility to V(D)J recombinase. Mol Cell Biol 1997;17:4553–61.

[170] Hawwari A, Krangel MS. Regulation of TCR delta and alpha repertoires by local and long-distance control of variable gene segment chromatin structure. J Exp Med 2005;202:467–72.

[171] Kondilis-Mangum HD, Cobb RM, Osipovich O, Srivatsan S, Oltz EM, Krangel MS. Transcription-dependent mobilization of nucleosomes at accessible TCR gene segments in vivo. J Immunol 2010;184:6970–7.

[172] Abarrategui I, Krangel MS. Noncoding transcription controls downstream promoters to regulate T-cell receptor alpha recombination. EMBO J 2007;26:4380–90.

[173] Hawwari A, Bock C, Krangel MS. Regulation of T cell receptor alpha gene assembly by a complex hierarchy of germline Jalpha promoters. Nat Immunol 2005;6:481–9.

[174] Sikes ML, Meade A, Tripathi R, Krangel MS, Oltz EM. Regulation of V(D)J recombination: a dominant role for promoter positioning in gene segment accessibility. Proc Natl Acad Sci USA 2002;99:12309–14.

[175] Liu H, Schmidt-Supprian M, Shi Y, Hobeika E, Barteneva N, Jumaa H, et al. Yin Yang 1 is a critical regulator of B-cell development. Genes Dev 2007;21:1179–89.

[176] Kosak ST, Skok JA, Medina KL, Riblet R, Le Beau MM, Fisher AG, et al. Subnuclear compartmentalization of immunoglobulin loci during lymphocyte development. Science 2002;296:158–62.

[177] Roldan E, Fuxa M, Chong W, Martinez D, Novatchkova M, Busslinger M, et al. Locus "decontraction" and centromeric recruitment contribute to allelic exclusion of the immunoglobulin heavy-chain gene. Nat Immunol 2005;6:31–41.

[178] Skok JA, Gisler R, Novatchkova M, Farmer D, de Laat W, Busslinger M. Reversible contraction by looping of the TCRα and TCRβ loci in rearranging thymocytes. Nat Immunol 2007;8: 378–87.

[179] Sayegh CE, Jhunjhunwala S, Riblet R, Murre C. Visualization of looping involving the immunoglobulin heavy-chain locus in developing B cells. Genes Dev 2005;19:322–7.

[180] Fuxa M, Skok J, Souabni A, Salvagiotto G, Roldan E, Busslinger M. Pax5 induces V-to-DJ rearrangements and locus contraction of the immunoglobulin heavy-chain gene. Genes Dev 2004;18:411–22.

[181] Reynaud D, Demarco IA, Reddy KL, Schjerven H, Bertolino E, Chen Z, et al. Regulation of B cell fate commitment and immunoglobulin heavy-chain gene rearrangements by Ikaros. Nat Immunol 2008;9:927–36.

[182] Degner SC, Verma-Gaur J, Wong TP, Bossen C, Iverson GM, Torkamani A, et al. CCCTC-binding factor (CTCF) and cohesin influence the genomic architecture of the Igh locus and antisense transcription in pro-B cells. Proc Natl Acad Sci USA 2011;108:9566–71.

[183] Hewitt SL, Farmer D, Marszalek K, Cadera E, Liang HE, Xu Y, et al. Association between the Igk and Igh immunoglobulin loci mediated by the 3' Igk enhancer induces "decontraction" of the Igh locus in pre-B cells. Nat Immunol 2008;9:396–404.

[184] Hewitt SL, Yin B, Ji Y, Chaumeil J, Marszalek K, Tenthorey J, et al. RAG-1 and ATM coordinate monoallelic recombination and nuclear positioning of immunoglobulin loci. Nat Immunol 2009;10:655–64.

[185] Tsujimoto Y, Gorham J, Cossman J, Jaffe E, Croce CM. The t(14;18) chromosome translocations involved in B-cell neoplasms result from mistakes in VDJ joining. Science 1985;229:1390–3.

[186] Haluska FG, Finver S, Tsujimoto Y, Croce CM. The t(8; 14) chromosomal translocation occurring in B-cell malignancies results from mistakes in V-D-J joining. Nature 1986;324:158–61.

[187] Alt FW, Zhang Y, Meng FL, Guo C, Schwer B. Mechanisms of programmed DNA lesions and genomic instability in the immune system. Cell 2013;152:417–29.

[188] Papaemmanuil E, Rapado I, Li Y, Potter NE, Wedge DC, Tubio J, et al. RAG-mediated recombination is the predominant driver of oncogenic rearrangement in ETV6-RUNX1 acute lymphoblastic leukemia. Nat Genet 2014;46:116–25.

[189] Lewis SM. The mechanism of V(D)J joining: lessons from molecular, immunological, and comparative analyses. Adv Immunol 1994;56:27–150.

[190] Zhu C, Mills KD, Ferguson DO, Lee C, Manis J, Fleming J, et al. Unrepaired DNA breaks in p53-deficient cells lead to oncogenic gene amplification subsequent to translocations. Cell 2002;109: 811–21.

[191] Lewis SM, Agard E, Suh S, Czyzyk L. Cryptic signals and the fidelity of V(D)J joining. Mol Cell Biol 1997;17:3125–36.

[192] Raghavan SC, Kirsch IR, Lieber MR. Analysis of the V(D)J recombination efficiency at lymphoid chromosomal translocation breakpoints. J Biol Chem 2001;276:29126–33.

[193] Arnal SM, Holub AJ, Salus SS, Roth DB. Non-consensus heptamer sequences destabilize the RAG post-cleavage complex, making ends available to alternative DNA repair pathways. Nucleic Acids Res 2010;38:2944–54.

[194] Aplan PD, Lombardi DP, Ginsberg AM, Cossman J, Bertness VL, Kirsch IR. Disruption of the human SCL locus by "illegitimate" V-(D)-J recombinase activity. Science 1990;250:1426–9.

[195] Dik WA, Nadel B, Przybylski GK, Asnafi V, Grabarczyk P, Navarro JM, et al. Different chromosomal breakpoints impact the level of LMO2 expression in T-ALL. Blood 2007;110:388–92.

[196] Haydu JE, De Keersmaecker K, Duff MK, Paietta E, Racevskis J, Wiernik PH, et al. An activating intragenic deletion in NOTCH1 in human T-ALL. Blood 2012;119:5211–4.

[197] Ashworth TD, Pear WS, Chiang MY, Blacklow SC, Mastio J, Xu L, et al. Deletion-based mechanisms of Notch1 activation in T-ALL: key roles for RAG recombinase and a conserved internal translational start site in Notch1. Blood 2010;116:5455–64.

[198] Mullighan CG, Miller CB, Radtke I, Phillips LA, Dalton J, Ma J, et al. BCR-ABL1 lymphoblastic leukaemia is characterized by the deletion of Ikaros. Nature 2008;453:110–4.

[199] Sakata J, Inoue J, Ohi H, Kosugi-Okano H, Mishima Y, Hatakeyama K, et al. Involvement of V(D)J recombinase in the generation of intragenic deletions in the Rit1/Bcl11b tumor suppressor gene in gamma-ray-induced thymic lymphomas and in normal thymus of the mouse. Carcinogenesis 2004;25:1069–75.

[200] Kitagawa Y, Inoue K, Sasaki S, Hayashi Y, Matsuo Y, Lieber MR, et al. Prevalent involvement of illegitimate V(D)J recombination in chromosome 9p21 deletions in lymphoid leukemia. J Biol Chem 2002;277:46289–97.

[201] Deriano L, Roth DB. Modernizing the nonhomologous end-joining repertoire: alternative and classical NHEJ share the stage. Annu Rev Genet 2013;47:433–55.

[202] Murray JM, O'Neill JP, Messier T, Rivers J, Walker VE, McGonagle B, et al. V(D)J recombinase-mediated processing of coding junctions at cryptic recombination signal sequences in peripheral T cells during human development. J Immunol 2006;177:5393–404.

[203] Zhang J, Ding L, Holmfeldt L, Wu G, Heatley SL, Payne-Turner D, et al. The genetic basis of early T-cell precursor acute lymphoblastic leukaemia. Nature 2012;481:157–63.

[204] Besmer E, Mansilla-Soto J, Cassard S, Sawchuk DJ, Brown G, Sadofsky M, et al. Hairpin coding end opening is mediated by RAG1 and RAG2 proteins. Mol Cell 1998;2:817–28.

[205] Santagata S, Besmer E, Villa A, Bozzi F, Allingham JS, Sobacchi C, et al. The RAG1/RAG2 complex constitutes a 3' flap endonuclease: implications for junctional diversity in V(D)J and transpositional recombination. Mol Cell 1999;4:935–47.

[206] Raghavan SC, Swanson PC, Wu X, Hsieh CL, Lieber MR. A non-B-DNA structure at the Bcl-2 major breakpoint region is cleaved by the RAG complex. Nature 2004;428:88–93.

[207] Raghavan SC, Gu J, Swanson PC, Lieber MR. The structure-specific nicking of small heteroduplexes by the RAG complex: implications for lymphoid chromosomal translocations. DNA Repair (Amst) 2007;6:751–9.

[208] Roth DB. Restraining the V(D)J recombinase. Nat Rev Immunol 2003;3:656–66.
[209] Marculescu R, Le T, Simon P, Jaeger U, Nadel B. V(D)J-mediated translocations in lymphoid neoplasms: a functional assessment of genomic instability by cryptic sites. J Exp Med 2002;195:85–98.
[210] Kagan J, Finan J, Letofsky J, Besa EC, Nowell PC, Croce CM. Alpha-chain locus of the T-cell antigen receptor is involved in the t(10;14) chromosome translocation of T-cell acute lymphocytic leukemia. Proc Natl Acad Sci USA 1987;84:4543–6.
[211] Gigi V, Lewis SM, Shestova M, Mijuskovic M, Deriano L, Meng W, et al. RAG2 mutants alter DSB repair pathway choice in vivo and illuminate the nature of "alternative NHEJ". Nucleic Acids Res, 2014;42:6352–64.
[212] Chatterji M, Tsai CL, Schatz DG. Mobilization of RAG-generated signal ends by transposition and insertion in vivo. Mol Cell Biol 2006;26:1558–68.
[213] Reddy YV, Perkins EJ, Ramsden DA. Genomic instability due to V(D)J recombination-associated transposition. Genes Dev 2006;20:1575–82.
[214] Mijuskovic M, Brown SM, Tang Z, Lindsay CR, Efstathiadis E, Deriano L, et al. A streamlined method for detecting structural variants in cancer genomes by short read paired-end sequencing. PLoS One 2012;7:e48314.
[215] Vanura K, Montpellier B, Le T, Spicuglia S, Navarro JM, Cabaud O, et al. In vivo reinsertion of excised episomes by the V(D)J recombinase: a potential threat to genomic stability. PLoS Biol 2007;5:e43.
[216] Curry JD, Schulz D, Guidos CJ, Danska JS, Nutter L, Nussenzweig A, et al. Chromosomal reinsertion of broken RSS ends during T cell development. J Exp Med 2007;204:2293–303.
[217] Roth DB, Craig NL. VDJ recombination: a transposase goes to work. Cell 1998;94:411–4.
[218] Barreto V, Marques R, Demengeot J. Early death and severe lymphopenia caused by ubiquitous expression of the Rag1 and Rag2 genes in mice. Eur J Immunol 2001;31:3763–72.
[219] Haines BB, Ryu CJ, Chang S, Protopopov A, Luch A, Kang YH, et al. Block of T cell development in P53-deficient mice accelerates development of lymphomas with characteristic RAG-dependent cytogenetic alterations. Cancer Cell 2006;9:109–20.
[220] Sprangers M, Feldhahn N, Liedtke S, Jumaa H, Siebert R, Muschen M. SLP65 deficiency results in perpetual V(D)J recombinase activity in pre-B-lymphoblastic leukemia and B-cell lymphoma cells. Oncogene 2006;25:5180–6.
[221] Zhang L, Reynolds TL, Shan X, Desiderio S. Coupling of V(D)J recombination to the cell cycle suppresses genomic instability and lymphoid tumorigenesis. Immunity 2011;34:163–74.
[222] Elkin SK, Matthews AG, Oettinger MA. The C-terminal portion of RAG2 protects against transposition in vitro. EMBO J 2003;22:1931–8.
[223] Tsai CL, Schatz DG. Regulation of RAG1/RAG2-mediated transposition by GTP and the C-terminal region of RAG2. EMBO J 2003;22:1922–30.
[224] Swanson PC, Volkmer D, Wang L. Full-length RAG-2, and not full-length RAG-1, specifically suppresses RAG-mediated transposition but not hybrid joint formation or disintegration. J Biol Chem 2004;279:4034–44.
[225] Sekiguchi JA, Whitlow S, Alt FW. Increased accumulation of hybrid V(D)J joins in cells expressing truncated versus full-length RAGs. Mol Cell 2001;8:1383–90.
[226] Qiu JX, Kale SB, Yarnell Schultz H, Roth DB. Separation-of-function mutants reveal critical roles for RAG2 in both the cleavage and joining steps of V(D)J recombination. Mol Cell 2001;7:77–87.
[227] Yarnell Schultz H, Landree MA, Qiu JX, Kale SB, Roth DB. Joining-deficient RAG1 mutants block V(D)J recombination in vivo and hairpin opening in vitro. Mol Cell 2001;7:65–75.
[228] Tsai CL, Drejer AH, Schatz DG. Evidence of a critical architectural function for the RAG proteins in end processing, protection, and joining in V(D)J recombination. Genes Dev 2002;16:1934–49.
[229] Lee GS, Neiditch MB, Salus SS, Roth DB. RAG proteins shepherd double-strand breaks to a specific pathway, suppressing error-prone repair, but RAG nicking initiates homologous recombination. Cell 2004;117:171–84.
[230] Corneo B, Wendland RL, Deriano L, Cui X, Klein IA, Wong SY, et al. Rag mutations reveal robust alternative end joining. Nature 2007;449:483–6.
[231] Wang G, Dhar K, Swanson PC, Levitus M, Chang Y. Real-time monitoring of RAG-catalyzed DNA cleavage unveils dynamic changes in coding end association with the coding end complex. Nucleic Acids Res 2012;40:6082–96.
[232] Coussens MA, Wendland RL, Deriano L, Lindsay CR, Arnal SM, Roth DB. RAG2's acidic hinge restricts repair-pathway choice and promotes genomic stability. Cell Rep 2013;4:870–8.
[233] Simsek D, Brunet E, Wong SY, Katyal S, Gao Y, McKinnon PJ, et al. DNA ligase III promotes alternative nonhomologous end-joining during chromosomal translocation formation. PLoS Genet 2011;7:e1002080.
[234] Giblin W, Chatterji M, Westfield G, Masud T, Theisen B, Cheng HL, et al. Leaky severe combined immunodeficiency and aberrant DNA rearrangements due to a hypomorphic RAG1 mutation. Blood 2009;113:2965–75.

# Transcriptional Regulation of Early B Cell Development

James Hagman
*National Jewish Health, Denver, CO, USA*

Sequential progression through a series of developmental stages results in the generation of antibody-producing B lymphocytes (Figure 1). In the bone marrow, hematopoietic stem cells (HSCs) provide a multipotent, self-renewing reservoir of progenitors for all lineages of the blood. Differentiation of HSCs involves the loss of potential and gain of commitment to specific cell types. These stages are largely achieved by the sequential activation of lineage-specific transcription factors.

Although many transcription factors participate in B lymphopoiesis, several are crucial for this process. The E26 transforming sequence (ETS) factor purine box transcription factor 1 (PU.1) is required for the development of all myeloid and lymphoid cells. Ikaros synergizes with PU.1, and through its control of lineage priming sets the stage for lymphoid cell development and reduced myeloid potential. E2A proteins perform critical functions in HSCs through B cell specification and commitment. E2A modifies the epigenetic landscape and activates key downstream factors including Early B cell Factor (EBF1), which drives the expression of the B cell program. Early B cell Factor synergizes with Foxo1 to activate and repress genes in early B cell progenitors. Collectively, E2A, EBF1 and Foxo1 constitute a regulatory network that drives the B cell fate. Commitment to this fate requires activation of Pax5 by these factors. In turn, Pax5 and EBF1 repress genes of other hematopoietic lineages. This chapter reviews past and current literature that defines the global network of transcriptional control in B lymphopoiesis in murine bone marrow.

## 1. PU.1 SETS THE STAGE FOR LYMPHOID AND MYELOID DEVELOPMENT

Purine box transcription factor 1 was first identified as a member of the ETS gene family [4], which is composed of a large group of transcription factors in B lineage cells and other cell types (reviewed in Ref. [5]). Purine box transcription factor 1 was named for its DNA-binding specificity, which is a purine-rich sequence encompassing a GGA-containing core (Table 1). This sequence is recognized by the protein's ETS domain, a winged helix-turn-helix motif (Figure 2) [9]. Other domains in PU.1 activate or repress transcription in different contexts. The gene encoding PU.1 was confirmed to coincide with sites of proviral integration in Friend leukemias; consequently, the gene was renamed *Spi1* (or *Sfpi1*) [17]. Expression of PU.1 is limited to cells of the hematopoietic lineages (Figure 1) [1,3]. Purine box transcription factor 1 is essential for early progenitors in both fetal and adult hematopoiesis, including thymic progenitors, as well as for generating differentiated cells including macrophages, granulocytes, and B cells. Effects of the lack of PU.1 on the development of many hematopoietic cell types confirmed its importance as a key director of cell differentiation in the blood.

Although candidate target genes including *Igκ light chain* (*L*) genes [18] were identified biochemically in B cells, experiments using gene-targeted mice were crucial for determining the functions of PU.1 in early hematopoiesis and B cell development. Purine box transcription factor 1 is expressed in fetal mice as early as the yolk sac. Fetal mice lacking PU.1 fail to generate most blood cell types or to complete development in utero [19,20]. Experiments in PU.1-deficient mice suggested the existence of a multipotent PU.1-dependent progenitor with lymphoid and myeloid potential. Other studies demonstrated PU.1's importance for generating the lymphoid/myeloid branch of hematopoiesis, and particularly for the generation of terminally differentiated myeloid cells [21,22]. With the identification of multipotent progenitor cells (MPP) that depend on PU.1 for their differentiation [23], elegant experiments confirmed that the myeloid versus B lymphoid lineage decision depends on high versus low expression of PU.1 in these cells [24]. These data provided one of the earliest examples of the

Molecular Biology of B Cells. http://dx.doi.org/10.1016/B978-0-12-397933-9.00003-5

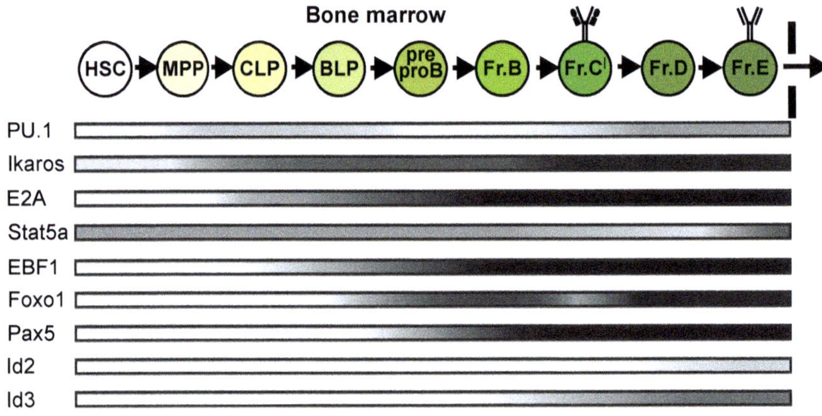


The intensity of shading represents amounts of mRNA of the indicated protein.

FIGURE 1 **Progression of B cell development in the bone marrow and expression of key transcription factors**. Stages shown include those analyzed in Seita et al. [1]. Pluripotent hematopoietic stem cells (HSC) differentiate in the bone marrow to generate multipotent progenitors (MPP), common lymphoid progenitors (CLP), B lymphoid progenitors (BLP), pre–pro-B cells, and B cell fractions as defined previously by Hardy and colleagues [2]. Although not indicated, lymphoid-primed multipotent progenitors (LMPP) would be present between MPP and CLP. Cell surface expression of the pre-BCR on Fr.C′ and BCR on Fr.E are indicated. The arrow and dashed line indicate transit to the periphery. For each transcription factor indicated, relative amounts of mRNAs were estimated using data at Gene Commons [1] and the Immunological Genome Project (ImmGen) [3].

**TABLE 1** Regulation Transcription Factors in B Cell Development

| Common Name | Gene Name Mouse/Human | Other Names | Transcription Factor Family | DNA Binding Specificity[a] | References |
|---|---|---|---|---|---|
| PU.1 | *Spi1/SPI1* | Sfpil; Dis1; Tcfpu1 | ETS | | [6,7] |
| Ikaros | *Ikzf1/IKZF1* | LYF1; ZNFNA1 | Zinc finger; IK family | | [7,8] |
| E2A | *Tcf3/TCF3* | E12; E47:Tcfe2a; Pan1; bHLHb21; A1 | bHLH | | [6,7] |
| STAT5 | *Stat5a/STAT5A* | MGF | STAT | | [6,7] |
| EBF1 | *Ebf1/EBF1* | EBF; OE-1; Olf1;O/E-1; COE1 | EBF; COE | | [6,7] |
| Foxo1 | *Foxo1/FOXO1* | FOXO1A; FKH1; FKHR | Forkhead | | [6,7] |
| Pax5 | *Pax5/PAX5* | BSAP; ALL3; KLP; EBB-1 | PAX; Paired box | | [6,7] |

[a]*DNA binding specificity is represented by sequence logo motifs. These motifs are graphical representations generated by multiple sequence alignments of experimentally determined binding sites. The relative height of each letter represents the frequency of nucleotide binding at a given position.*

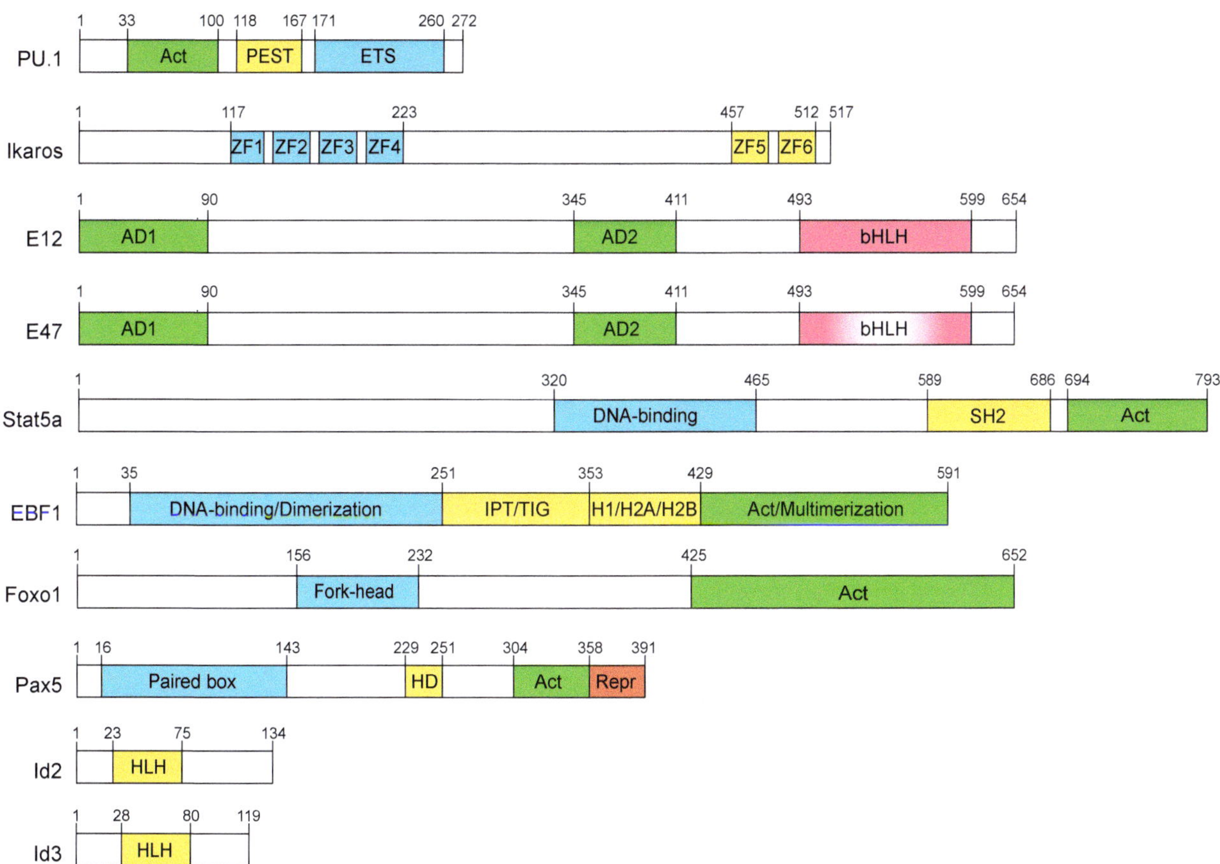


FIGURE 2 **Schematic structure of transcription factors required for B cell development in the bone marrow.** Diagrams include transcription factors discussed in the text. Functional domains are color coded as blue (DNA-binding), yellow (dimerization/multimerization and stability), green (activation), and red (repression). The basic-helix-loop-helix domains of E12 and E47, which combine DNA-binding and dimerization functions, are shown in magenta. These two domains are indicated with different shading to emphasize alternative splicing. PU.1 features an acidic and glutamine-rich activation domain and the proline, glutamic acid, serine and threonine domain (PEST) that governs its stability. Ikaros is shown as the alternatively spliced Ik-1 isoform. Stat5a: SH2- Src-homology domain. EBF1: IPT/TIG-Ig, plexin, transcription factor-like/transcription factor-Ig domain. Pax5: HD-partial homeodomain. Id2 and Id3 have only HLH domains and do not bind DNA [10–16].

control of differentiation outcomes by graded expression of a transcription factor in mammals.

Functions of PU.1 in differentiated B cells have been controversial; however, development in mice with a conditionally targeted allele may have settled this issue [25]. Similar to previous knockout mice, total deletion of *floxed Spi1* genes in early hematopoiesis resulted in the expected loss of B cells and other lineages. In contrast, conditional deletion of these genes in differentiated B cells using *Cd19-Cre* had little or no effect on B cell–specific gene expression. This may be partly due of functional redundancies between PU.1 and the related ETS protein Spi-B. However, PU.1, but not Spi-B, is required by early lymphoid progenitors [26]. These data suggest that the two proteins have overlapping but distinct functions during B lymphopoiesis.

How does PU.1 activity in early MPPs direct B lymphopoiesis at later stages of differentiation? Purine box transcription factor 1 is required for the expression of factors that shepherd B cells through key developmental checkpoints. In the absence of PU.1, cells fail to express interleukin-7 receptor-α (IL-7Rα/Cd127; *Il7r*) [27], which is crucial for B cell lineage specification and development [28–31]. Reconstitution of PU.1-deficient MPPs with IL-7Rα mimicked reconstitution with PU.1 itself [27]. PU.1 is also required for expression of EBF1 (see section below) and the protein tyrosine phosphatase CD45R/B220 [32]. Moreover, PU.1 acts upstream of other transcription factors necessary for B cell development.

As mentioned above, the expression levels of PU.1 orchestrate different lineage fate decisions during hematopoiesis. Low amounts of PU.1 promote the development of B cells over myeloid cells [24]. Levels of PU.1 are influenced by the following two mechanisms: (1) growth factor–independent 1 (Gfi1), which represses transcription of *Spi1* (PU.1) [33], and (2) relative to B cells, slower cell cycle progression in myeloid cells that allows PU.1 to accumulate [34]. Together, PU.1, Gfi1, and a recurrent network of repressors control lineage decisions for the generation of cells necessary for adaptive (B and T) and innate (myeloid) immunity.

In early progenitors, entry into lymphoid lineages is specified by PU.1 and other transcription factors including Runx1 [35], Ikaros [36], and E2A (see the following section). This mechanism of cell fate decision, termed lineage priming, activates or represses different lineage programs [37]. The direct function of PU.1 in this circuit is again dosage-dependent. It likely involves PU.1 acting as a pioneer factor that induces changes in chromatin structure. Indeed, PU.1 initiates nucleosome remodeling, which precedes deposition of histone H3 lysine 4 monomethylation (H3K4me1), a mark of poised transcriptional enhancers [38]. Thus, PU.1 leaves gene loci with functional adaptations for later activation by other temporally regulated factors (such as EBF1; see the following section).

## 2. LINEAGE PRIMING IN LYMPHOID PROGENITORS BY IKAROS

Ikaros (*Ikzf1*; also known as Lyf-1) was identified first as a T and B lineage–specific zinc finger protein responsible for the regulation of genes related to antigen receptor assembly (*Cd3δ* and terminal *deoxynucleotidyl transferase*) [39,40]. Ikaros and its relatives, including Aiolos (*Ikzf3*), each possess DNA binding domains containing Krüppel-like zinc fingers ($Cys_2$-$His_2$) and a pair of carboxy-terminal zinc fingers that mediate multimerization (Figure 2). Multimerization is required for Ikaros to bind its DNA consensus site (Table 1). Ikaros is essential for the generation of all lymphoid lineages [41]. It is expressed in mouse fetal liver progenitors, in thymus, and in the bone marrow of adult mice. Of particular note, studies identified at least eight alternatively spliced isoforms of Ikaros. Conservation or omission of amino-terminal zinc fingers 1–4 (ZF1–4) in these isoforms contributes to potentially different DNA-binding specificities, whereas all isoforms possess the two carboxy-terminal zinc fingers (ZF5–6) (Figure 2) [42]. Isoforms containing ZF5–6, but not ZF1–4, act as dominant negative regulators of Ikaros function. This was revealed in the first *Ikzf1* gene–targeted mice in which deletion of the DNA-binding domain resulted in a dominant negative phenotype [41].

Mice lacking functional Ikaros proteins (Ikaros-null mice) fail to produce B, T, and NK cells, which suggests that Ikaros has key functions in early lymphoid progenitors [43]. Similar to proposed activities of PU.1, Ikaros has been linked with transcriptional priming required for lineage potential and cell fate decisions [36,44]. Ikaros is necessary for the development of functional lymphoid-primed multipotent progenitors (LMPP), which have enhanced lymphoid potential, but retain potential for differentiating into myeloid cells. In LMPPs, Ikaros is necessary for expression of the following lymphoid lineage transcripts: nuclear regulators (e.g., *Sox4*), *Ig* genes, cytokine receptors, and signaling proteins (e.g., *Notch1*), in addition to the repression of genes that contribute to stem cell, myeloid, and erythroid lineages. Ikaros may begin this process as early as HSCs. In the absence of Ikaros, LMPPs do not differentiate into common lymphoid progenitors (CLPs) (Figure 1). Instead, LMPPs exhibit extensive changes in gene expression that promote myeloid cell development. In connection with the earlier discussion of PU.1, Ikaros promotes expression of Gfi1. In turn, Gfi1 limits PU.1 expression to low levels to promote B cell development, as opposed to high levels that promote myeloid cell development [33]. Ikaros may be regulated by E2A proteins (see subsequent discussion), which suggests that Ikaros and E2A act together in a feedforward loop to promote lymphoid development [45].

Recent studies using conditionally deleted or partially modified *Ikzf1* alleles revealed important roles of Ikaros in B cell development in the bone marrow. After conditional expression of Cre recombinase in mice, *floxed* sequences encompassing ZF5–6 of Ikaros were deleted [8]. B cell–specific deletion was achieved using *mb-1(Cd79a)-Cre,* such that Cre expression was initiated in pro-B cells. In this system, loss of Ikaros expression resulted in impairment of pro-B cells transitioning to pre-B cells. In addition, defective proliferation of the developmentally arrested cells was associated with defective pre-BCR signaling. A second study using similar mice also detected developmental arrest at the early pre-B (Fraction C′) stage after conditional deletion of *Ikzf1* genes [46]. Use of endogenously tagged, biotinylated Ikaros in chromatin interaction analysis sequencing (ChIP-seq) assays identified > 8000 target genes in pro-B cells [8], including activated genes involved in signal transduction, surface receptors, and transcriptional regulation. Repressed genes included many of the same categories, including genes required at later stages of B cell differentiation (*Ikzf3*, *Bach2*, and *Irf4*). Repression of transcription by Ikaros is mediated by its recruitment of the Mi-2/nucleosome remodeling and deacetylase (Mi-2/NuRD) complex. Mi-2/NuRD is a multiprotein assemblage of ATPase/helicase, histone deacetylase, and lysine demethylase proteins. Mi-2/nucleosome remodeling and deacetylase localizes genes with Ikaros within pericentromeric heterochromatin in developing lymphocytes [47–50].

An interesting new development concerns the functions of various zinc fingers in Ikaros. The alternatively spliced isoforms of Ikaros contain various combinations of the amino-terminal zinc fingers [51]. A recent study demonstrated that mRNA splicing necessary to generate functional isoforms of Ikaros requires Foxo1 [52]. To address the differential activities of these isoforms, ZF1 or ZF4 was deleted using gene targeting in embryonic stem cells. Analysis of the mice demonstrated that ZF1 is required for the maturation of late pro-B/early pre-BI (fractions B/C) to large pre-BII cells (C′). However, ZF4 is selectively required for tumor suppression. A second study generated similar results by conditional deletion in pro-B cells of the entire Ikaros DNA-binding domain [53]. Together, these studies further support roles of Ikaros as a gatekeeper of B cell checkpoints during early differentiation.

## 3. E2A REGULATES THE CHROMATIN LANDSCAPE TO PROMOTE GENE EXPRESSION IN B CELL DEVELOPMENT

In early studies of the immunoglobulin gene enhancers, evidence was sought for a transcription factor(s) binding to a motif they shared in common: an inverted repeat sequence termed the E box (5′-CANNTG-3′) (Table 1) [54,55]. Two closely regulated but distinct proteins that bound in vitro to the κE2 enhancer motif were identified (Figure 2) [56,57]. These proteins, E12 and E47, were identified as alternatively spliced isoforms of the *E2A* gene (also referred to as *Tcf3/Tcfe2a* in mice and *TCF3/Pan1/2* in humans; Table 1). E12 and E47 possess different basic-helix-loop-helix (bHLH) domains (related to the daughterless transcription factor of *Drosophila*) that mediate their homo- or heterodimerization. Each protein's basic region binds one half-site of the E box. This model was confirmed by X-ray crystallographic analysis of E47 homodimers bound to E box DNA [58]. E12 and E47 also include domains (AD1 and AD2) that function in transcriptional activation [10]. Because of the strict recognition of E boxes by E12 and E47 and related family members (encoded by the *E2-2/TCF4* and *HEB/TCF12* genes), these proteins are referred to as E proteins. The following section will focus on functions of E2A in early B cell development.

The generation of mice lacking functional *E2A* genes confirmed their importance for B cell development and function. These gene-targeted mice produce neither E12 nor E47 proteins. The mutant mice completely lacked B cells [59,60]. Indeed, the absence of cells expressing the pan-B cell marker B220 and a lack of immunoglobulins, with no evidence of V(D)J recombination, suggested that B cell development was arrested at a very early stage. Expression of key B cell–specific genes, including *mb-1(Cd79a)*, the surrogate light chain (*ψL*) genes *λ5* (*Igll1*) and *VpreB* (*Vpreb1*), *Cd19*, and *Recombination activating gene 1* (*Rag1*) were not detected in *E2A* gene targeted mice. Also, the *Pax5* gene, which regulates much of the early B cell transcriptome, was not transcribed in these mice.

Initial observations of the lack of B cells in E2A-deficient mice suggested that functions of E2A proteins are restricted to B cells. However, these factors are required for the normal development of both B and T cells. Experiments using conditional knockout mice demonstrated that E2A proteins are required for the development and function of thymocytes, mature CD4$^+$ cells, invariant natural killer T (iNKT) cells, and Th17 cells as well [61–66]. Deficiencies in B and T cell development in *Tcf3* (E2A) knockout mice likely result from reduced development of LMPPs (see below), which require E2A for the expression of lymphoid-specific genes [67]. Even reduced expression of E2A proteins in *Tcf3* haplo-insufficient mice severely impairs the generation of lymphoid progenitors [68]. Interestingly, E2A-deficient progenitors exhibit pluripotency and inappropriate expression of regulators of other lineages including the Gata1 (erythroid lineage-specific) and Tcf1 (T cell–specific) factors [69].

In B cells, E2A proteins have long been associated with the process of antigen receptor gene assembly via V(D)J recombination. In E2A-deficient B cells, V(D)J recombination is not detected [59,70]. Importantly, expression of E2A proteins in non-lymphoid cells activated *Ig* germline transcription and expression of proteins (terminal deoxynucleotidyl transferase) involved in *Ig heavy chain* (*IgH*) gene rearrangements [71]. Moreover, in the presence of co-expressed RAG1/2 proteins, E2A proteins allowed for completion of *Igκ light (L) chain* gene rearrangements [72,73]. Although some studies suggested similar capabilities of E12 and E47 [72,74], other experiments highlight distinct functional differences between the two isoforms [71]. For example, E47 and E12 are required differentially in a temporally controlled fashion for *Ig* gene recombination during B cell development [75]. E47 is required in pre–pro-B cells, which undergo $D_H$ to $J_H$, followed by $V_H$ to $DJ_H$ rearrangements. Although E12 is not required in pre–pro-B cells (Figure 1), both E12 and E47 are increased in pre-B cells coincidently with *IgL* gene rearrangements [76–78].

Recent evidence suggests that a complex set of mechanisms controlled by E2A proteins and their partners EBF1 and Pax5 regulate *Ig* locus accessibility, nuclear positioning, and long-range interactions necessary for successful recombination (reviewed in Ref. [79]). E2A and cofactors that mediate epigenetic modifications control chromatin accessibility necessary for V(D)J recombination at both *IgH* and *IgL* genes [80–82]. Accessibility and use of Vκ segments may be controlled by E2A, perhaps by increasing looping between germline segments before rearrangements [83]. E12 may be more important for germline transcription, rearrangement, and expression of *Igλ L chain* genes [75,84,85]. Furthermore, expression of the *Rag1/2* genes depends on E2A, which drives their promoters via the *Erag* enhancer in CLPs [86,87].

How do E2A proteins control gene transcription? At least two levels of interactions are important: (1) DNA binding together with other transcription factors and (2) recruitment of auxiliary proteins that bridge between E2A proteins and the transcription machinery. In either case, associations with proteins that mediate epigenetic modifications of chromatin are an important consequence of E protein DNA binding.

E2A cooperates with other transcription factors bound at adjacent sites. Among the first identified functional partners of E2A was EBF1, which binds adjacent sites with E2A in promoters of the φ*L* genes λ5 (*Igll1*), VpreB1 (*Vpreb1*), and BCR accessory protein mb-1/Ig-α (*Cd79a*) [88–90]. Chromatin interaction analysis sequencing studies identified >4500 sites in pre–pro-B cells and nearly 12,000 sites in pro-B cells occupied by E2A proteins [91]. Binding of E2A proteins correlated strongly with H3K4 monomethylation (which marks promoters and enhancers), but less so with H3 acetylation (which marks actively transcribed genes). Many of these sites were already occupied in pre–pro-B cells in the absence of EBF1, which suggests a hierarchy of DNA binding. However, co-occupancy (binding within 150 bp of each other) of regulatory modules by E2A and EBF1 was detected at half of the EBF1-binding sites identified in pro-B cells. These findings suggest that synergy between E2A and EBF1 is essential for early B cell specification and commitment. Mutual positive feedback circuits in B cell differentiation were confirmed by the binding of E2A to regulatory elements of *Ebf1* genes and to the *Foxo1* gene in collaboration with EBF1. Genes regulated by E2A, EBF1, and Foxo1 binding include those encoding components of the pre-BCR and BCR, as well as signaling and survival proteins from distinct pathways (Figure 3) [91]. An independent study reported many of the same target genes and confirmed requirements for E2A proteins in the maintenance of target gene expression in early B cell progenitors [92]. Chromatin interaction analysis sequencing also detected co-occupancy of promoters by E2A proteins and Runx proteins (likely Runx1) in pre–pro-B cells. It has also been proposed that E2A proteins function cooperatively with NF-κB as well [93].

E2A proteins function by recruiting proteins that drive epigenetic remodeling that precedes transcription. The recruitment of the archetypal coactivators, CREB-binding protein (CBP)/p300, by E2A proteins is well described [81]. Recruitment of CBP/p300 brings their histone acetyltransferase activities to target genes. E2A proteins also recruit the lysyl-acetyltransferase GCN5/PCAF subunit of the Spt-Ada-Gcn5-Acetyltransferase (SAGA) coactivator complex [94]. More recently, biochemical analysis identified additional interactions between E2A and epigenetic regulators including lysine-specific histone demethylase 1 (LSD1), protein arginine methyltransferase 5 (PRMT5), and the co-repressor CoREST in association with E47 [95]. The diverse

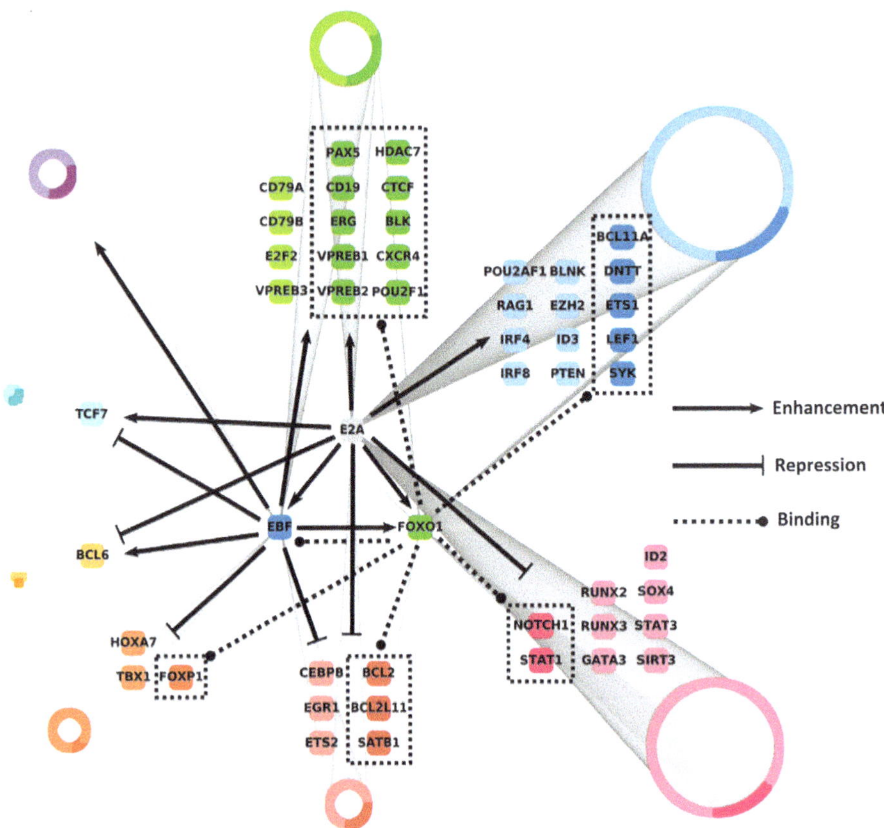


FIGURE 3 **Regulatory network linking transcriptional activators, signaling proteins, and survival factors in developing B cells.** Transcription factor binding sites were identified using ChIP-seq and integrative genome-wide bioinformatic analysis [91]. Transcriptional regulatory targets of E2A and EBF1, or that are bound by Foxo1, are indicated in groups based on their activation or repression in pro-B cells (relative to E2A- or EBF1-deficient pre–pro-B cells). Genes that are critical for B cell specification and commitment are shown. Circles reflect the number of genes identified within each group. Genes with confirmed binding sites for Foxo1 are indicated by darker colors. *Provided by Cornelis Murre.*

functions of these partner proteins suggest their participation in context-dependent activation or repression by E2A proteins. E2A proteins also interact directly with the basal transcription machinery. E47 binds to core promoters cooperatively with TFIID through its TAF4 subunit [96].

As described above, E2A facilitates the activation of downstream regulators commencing in LMPPs and CLPs. However, it is now apparent that E2A primes lymphoid cell development during the earliest stages of hematopoiesis. E2A proteins may prepare genes for activation or silencing as early as HSCs. In these cells, it is proposed that E2A proteins bind and pre-mark enhancers for activation [67,97]. This is evidenced by the deposition of histone marks including H3K4me1, which are predictive of both transcriptional enhancers and multilineage priming for transcription [98]. Although the mechanisms necessary for the deposition of H3K4me1 are incompletely understood, restoration of E47 expression in E2A-deficient pre–pro-B cells was sufficient for genome-wide deposition of H3K4me1 [91]. Before lymphopoiesis, it is likely that E2A proteins and other collaborating factors (e.g., PU.1 and Ikaros), directly or indirectly recruit histone methyltransferases, including those of the mixed-lineage leukemia (MLL) protein family [99]. Other data suggest that E2A positions genes within loops of chromatin associated with the nuclear lamina [100]. These topological structures are associated with poised genes that change location within subcompartments of the nucleus during B cell development. In this regard, E2A may synergize with PU.1 to anchor interactions between enhancers of coregulated genes. This mechanism may preclude premature gene activation, resulting in appropriate temporal regulation of downstream regulatory genes (such as *Ebf1*).

## 4. E2A IS INHIBITED BY ID PROTEINS

Functional amounts of E2A proteins are controlled at the 1) transcriptional level and 2) posttranslationally by the Inhibitor of DNA binding (Id) family of helix-loop-helix (HLH) proteins. In the bone marrow, developing B cells express *Id2* and *Id3* at discrete phases of lymphopoiesis (Figure 1). Neither Id2 nor Id3 contains an independent DNA-binding domain; however, each protein can heterodimerize with E2A proteins [101–103]. Heterodimerization dampens E2A activity because E2A:Id complexes are unable to bind DNA stably. Repression of *Id2* and *Id3* genes, first by Gfi1 and then by EBF1, aids in establishment and maintenance of sufficient levels of E2A proteins for B lineage specification and commitment [104,105]. Id3 induces growth arrest and controls survival signals in B cell progenitors [106]. Upregulation of Id3 in immature B cells may be important for controlling allelic exclusion and secondary rearrangements associated with receptor editing [76]. Id2 is also upregulated in immature B cells [107]. Importantly, inhibition of E47 by Id2 in early hematopoiesis limits lineage priming and the generation of B cells [108].

## 5. INTERLEUKIN-7/STAT5 SIGNALING PROVIDES AN EARLY SIGNAL FOR B CELL LINEAGE SPECIFICATION

Interleukin-7 (IL-7) provides key signals for early B cell–specific proliferation, differentiation, and lineage progression. In developing B cells, IL-7 engages receptors including the α subunit encoded by the *Il7r* (also known as *Cd127/Il7ra*) gene, together with the cytokine receptor common γ chain (γc) to activate Janus kinases 1 (Jak1) and 3 (Jak3). Subsequently, Jak1 phosphorylates Stat5A/B on specific tyrosines (Tyr694/699, respectively) necessary for homodimerization, translocation to the nuclear compartment where it functions as a transcription factor (reviewed in Ref. [109]). DNA binding by Stat5 homodimers results in interactions with other proteins that modify chromatin. These interactions govern key checkpoints during bone marrow B cell development, including the transition between pro-B and pre-B cells that is regulated by the pre-BCR (reviewed in Ref. [110]). An important feature of this mechanism is the control of *Rag1* gene expression by IL-7/Stat5. The basis of control of this developmental progression by IL-7 is poorly understood, but new findings shed much light on this mechanism.

The onset of early B lymphopoiesis critically depends on IL-7 signaling [28]. In the absence of IL-7 or its receptor, B cell development is arrested at an early progenitor stage [111,112]. Because the *Il7r* gene is activated by PU.1, PU.1-deficient MPPs are developmentally arrested. Differentiation is restored by enforced expression of the IL-7 receptor [27]. The lack of IL-7 receptors is complemented by constitutively active Stat5a, which restores *IgH* gene rearrangements and lineage progression in IL-7R–deficient mice [113].

Synergy between IL-7 and Flt3L/Flt3 signaling in hematopoietic progenitors activates MAPK/Erk and PI3K/Akt pathways. In turn, this increases the survival and proliferation of pre–pro-B cells [114]. Many of the effects of IL-7 signaling result from its activation of EBF1 expression, which is necessary for the B cell transcriptional program [30–32]. IL-7 signaling drives EBF1 expression directly. IL-7/Stat5, E2A, and EBF1 itself activate the distal promoter of the *Ebf1* gene [115–117]. Because it results in EBF1 expression, IL-7 signaling is considered to have an instructive role in activating the early B cell–specific transcriptome [112]. Furthermore, B cell development in the thymus is increased by enforced expression of constitutively active Stat5 [118].

Assembly of the pre-BCR is an important checkpoint in early B cell development. In pro-B cells, Stat5 activates *IgH* rearrangements and represses *IgL* rearrangements.

Interleukin-7/Stat5 is necessary for germline transcription, histone acetylation, and DNA recombination of distal $V_H$ to $DJ_H$ gene segments in pro-B cells [119]. Then, loss of the IL-7/Stat5 signal, together with pre-BCR signaling, drives pre-B cells to rearrange *Igκ* genes. In Stat5-deficient pro-B cells, *Igκ* rearrangements are detected prematurely [120]. The molecular basis of Stat5's temporal regulation of *Igκ* rearrangements has been dissected. Stat5 binds as a tetramer at the κ intronic enhancer ($E\kappa_i$), which inhibits *Igκ* rearrangements. Tetrameric Stat5 recruits Polycomb Repressive Complex 2 (PRC2) and its core subunit Enhancer of zeste homolog 2 (Ezh2), a histone lysine methyltransferase [121]. Consequently, Ezh2 adds the repressive H3K27me3 mark to nucleosomes. This reduces accessibility to the $E\kappa_i$ enhancer, the *Jκ* cluster, and *Cκ*.

After expression of the pre-BCR, developing B cells traffic to microenvironments with low IL-7. With reduced IL-7 signals, pre-B cells exit the cell cycle in preparation for *Igκ* rearrangements [122]. Ikaros facilitates the transition of large pre-B (Fr.C′) to small non-cycling pre-B cells, which rearrange *Igκ* genes [8,46]. E2A opens *Igκ* loci to facilitate binding of recombination signal sequences by Rag proteins [73]. Together, Ikaros and E2A synergize with other factors including Pax5 and Interferon regulatory factor 4 (Irf4). In preparation for *Igκ* gene rearrangements, these proteins (1) promote increased local accessibility by the removal of H3K27me3 and the addition of histone acetylation and (2) upregulate *Rag1* in pre-B cells [8,46,122].

## 6. EARLY B CELL FACTOR: CENTRAL COORDINATOR OF B CELL DEVELOPMENT

The onset of B cell differentiation is initiated by Early B cell Factor 1 (EBF1, also known as EBF/COE1/O/E−1). Although many important transcription factors, including PU.1, Ikaros, and E2A, are expressed before EBF1 and prepare cells for the B cell fate, it is EBF1 that drives B cell specification and commitment (see section on Regulation of B Cell Lineage Commitment). EBF1 functions by interacting and cooperating with partner proteins, including many that are controlled by EBF1 itself (e.g., *Pax5*). Recent studies, including elucidation of its three-dimensional structure, identification of its target genes in developing B cells, and its orchestration of epigenetic modifications, have greatly increased understanding of how EBF1 accomplishes specification and commitment to the B lineage.

Early B cell Factor 1 was first identified as a B cell–specific protein that binds the *mb-1/Cd79a* promoter [123,124]. As expected for an early coordinator of B cell differentiation, EBF1 was implicated in the regulation of multiple genes (e.g., *B29/Cd79b*, *Vpreb1*, *λ5/Igll1*) and functions cooperatively with other known regulators of early B cell–specific transcription, including E2A [86–89,125,126]. Cloning of cDNAs encoding EBF1 and subsequent functional studies identified its atypical helix-loop-helix-loop-helix (HLHLH) domain (Figure 2), which mediates its homodimerization [125,127]. However, these studies also revealed a novel DNA-binding domain that binds conserved inverted repeat sequences (Table 1). Structural studies of EBF1 bound to DNA or of its individual domains confirmed the similarity between its HLHLH motif and the HLH motif of E2A [126,128]; however, heterodimerization between HLH domains of E2A and EBF1 has not been reported. The structure of the DNA-binding domain of EBF1 and its mode of DNA recognition has several remarkable features. DNA binding is mediated by recognition of each of the inverted repeats by a Rel-like (with similarities to Ig-like folds) core, similar to those of NFAT and NF-κB subunits, including loop modules that contact the major groove of DNA [128]. The structure confirmed previous studies suggesting the presence of an unusual zinc-binding domain, the zinc knuckle, which participates in DNA recognition [125,129]. Both the very short length (14 residues) and configuration of residues coordinating zinc (His–$Cys_3$) distinguish the zinc knuckle from Krüppel-like ($Cys_2$-$His_2$) and other types of zinc fingers. The structure of a second Ig-like fold of unknown function, termed the Ig plexin transcription factor-like (IPT)/transcription factor-Ig (TIG) domain, confirmed its Ig-like fold as well.

The absence of EBF1 in gene-targeted mice results in developmental arrest at an early pre–pro-B, or CLP-like stage of differentiation [130]. Within the hematopoietic system, only B cell development is affected. Early B cell Factor 1–deficient mice fail to express sterile transcripts of the *IgH* locus (*Iμ* and *μ0*) that precede V(D)J recombination and exhibit a total lack of *Ig* gene rearrangements. Many early B cell–specific genes, including *Cd19* and *mb-1/Cd79a*, are not expressed in the absence of EBF1. Importantly, this B cell–specific phenotype differentiates EBF1-deficient mice from mice lacking PU.1, Ikaros, or E2A, which lack cells of multiple hematopoietic lineages.

Early studies demonstrated the ability of EBF1 to drive HSCs to the B cell fate at the expense of other lineages [131]. Early B cell Factor 1 also has abilities to activate the B cell program in progenitors in the absence of upstream factors, including PU.1 and E2A [32,132]. In normal mice, B lineage restriction in $Ly6d^{+}Il7r^{+}$ CLPs correlates with the earliest onset of *Ebf1* transcripts [112,133]. In these cells, expression of *Ebf1* precedes expression of critical B cell–specific markers, including *ψL* genes. Activation of B cell–specific genes occurs together with the silencing of genes of T and NK cells that are expressed in early CLPs. Expression of *Ebf1* in CLPs and pre–pro-B cells is sufficient for activation of other important members of the B cell–specific transcriptional network, including *Foxo1, Pax5,* and *Pou2af1* (the coactivator Oca-B) [134]. In addition, the histone H2A deubiquitinase Myb-like, SWIRM, and MPN domains 1 (*MYSM1*) is necessary for efficient activation of *Ebf1* gene expression in CLPs [135].

Two promoters drive *Ebf1* transcription at different stages of development. *Ebf1* transcripts increase in pro-B cells [1,117]. Then, maximal expression is reached in pre-B cells (Hardy Frs. C′ and D). At early stages, the distal (−4.4 kb) *Ebf1α* promoter is activated by IL-7 signaling, E2A, and EBF1 itself [115–117]. In contrast, the *Ebf1β* proximal promoter is activated by Pax5 and ETS proteins. The proximal protein is also activated by Runx1 and its partner CBFβ [203]. Together, the two promoters finely tune EBF1 expression. The importance of *Ebf1* gene dosage is evidenced further by critical requirements for normal amounts of EBF1 at different stages of differentiation. For example, *Ebf1* haplo-insufficient mice exhibit reduced abilities to rearrange *Ig* genes and impaired B cell lineage commitment and progression [68,136,137].

In addition to transcriptional mechanisms, levels of EBF1 in B cells are also controlled by posttranslational mechanisms. The carboxy-terminal domain of zinc finger protein 521 (Evi3/Zfp521/ZNF521) interacts with EBF1 and inhibits its DNA binding in vitro and in transfected cells [138,139]. *Zfp521* is expressed together with *Ebf1* in early B cell progenitors (from CLPs to pro-B cells) [1,3] and in non-lymphoid cells including osteoblasts and adipocytes [140,141]. These observations suggest that the two factors may have coevolved in a regulatory partnership.

Gene expression profiling and ChIP-seq identified 565 genes that are occupied and transcriptionally regulated by EBF1 [91,142]. Binding sites for EBF1 were largely localized at promoters and enhancers. In pro-B cells, EBF1 preferentially regulates genes involved in signal transduction, including the expected φL chains (*Vpreb1*, *Vpreb2*, *Vpreb3*, and *Igll1*), *Ceacam1*, and *Notch3*. Transcription factor genes were also activated, such as *Hes1,* a mediator of Notch signaling. Other classes of activated genes include kinases (*Blk*), phosphatases (*Sbk1*), metabolic enzymes (*Neu1*), membrane transporters, and genes associated with the cytoskeleton and motility (*Sema7a*). A number of important genes were repressed directly by EBF1, including *Icosl*, *Dusp7,* and *Itgb7.*

The ability of EBF1 to both activate and repress gene targets suggests that it directs changes in the structure of surrounding chromatin to facilitate activation or repression depending on the local context. In this regard, active genes bound by EBF1 in pro-B cells feature enriched local densities of H3K4me3 and H3Ac epigenetic marks, whereas genes that shut off in these cells after EBF1 binding are enriched for repressive H3K27me3. Chromatin interaction analysis sequencing studies also suggested that EBF1 is associated with poised chromatin (associated with the H3K4me2 mark) at many targets, including genes that are inactive in pro-B cells but activated at later stages of development.

Similar to E2A, EBF1 interacts with other DNA binding proteins and cofactors to activate or repress transcription. Early B cell Factor 1 binds genes together with Runx1, including the *mb-1/Cd79a promoter* [143]. Chromatin interaction analysis sequencing and subsequent bioinformatic analysis suggested that approximately one-fourth of the regulatory modules bound by EBF1 include or are near Runx1-binding sites [91]. Early B cell Factor 1 is also known to recruit p300, which may be involved in activation or repression in different contexts [144]. Importantly, EBF1 recruits SWI/SNF complexes and initiates chromatin remodeling through its interactions with the Brahma-related gene 1 (Brg1) ATPase [145]. In relation to EBF1's ability to initiate demethylation of DNA (see subsequent discussion) [145], a recent report suggests that EBF1 interacts with the Ten-Eleven Translocase 2 (Tet2) protein [146], a hydroxylase that oxidizes 5-methylcytosine to 5-hydroxymethylcytosine [147]. This chemical modification is the first step in converting methylated to unmethylated cytosine necessary for removal of this epigenetic mark (see following sections).

The importance of EBF1 for activation of B cell–specific developmental programs was evidenced further by studies using *Ebf1* conditional knockout mice [148,149]. As expected from results of previous studies using cell lines and knockout mice, conditional mutagenesis of *Ebf1* genes confirmed requirements for EBF1 for lineage commitment in early progenitors (see subsequent discussion), proliferative expansion, and survival of pro-B cells and their transition to pre-B cells in the bone marrow. Genes and mechanisms affected by the loss of EBF1 include regulators of cell cycle progression and signaling by the pre-BCR and BCR, including the PI3K/Akt and BAFF receptor pathways in the bone marrow. Interestingly, although EBF1 is required for activating hundreds of genes in pro-B cells, it is not uniformly required for the maintenance of their expression as development proceeds. Moreover, EBF1 activates different sets of genes at different stages of B cell development. Conditional deletion of EBF1 at later stages of development did not result in significant changes in the expression of a number of EBF1 target genes (e.g., *Cd79a, Pax5*, and *Blnk/SLP-65*). This suggests that EBF1 functions by establishing epigenetic modifications that are propagated in its absence. These results explain why other genes dependent on EBF1, such as *Cd79b/Ig-β*, remain transcriptionally active after the silencing of EBF1 in terminally differentiated plasma cells [123,127].

## 7. COLLABORATION BETWEEN EBF1 AND FOXO1

The *Foxo1* gene encodes Forkhead box O1, a key effector of the PI3K-Akt/PKB signaling pathways in B cells and a host of other tissues [150]. Foxo1's subcellular localization is regulated by phosphorylation of its nuclear localization or nuclear export signals. Foxo1 is required at multiple stages of B cell development [151]. In CLPs, E2A and the bHLH protein HEB collaborate to activate Foxo1 expression [152]. Ablation of *floxed Foxo1* genes at the pro-B cell stage accumulated additional pro-B cells, but also a 90% reduction in numbers of B cells in the bone marrow relative to wild-type mice [151]. Moreover, Foxo1-deficient B cells that progressed to the periphery lacked surface expression of IgM and IgD. The increase in pro-B cells suggests that Foxo1 deficiency creates a partial block in the transition of these cells to pre-B cells. In part, the phenotype of Foxo1-deficient pro-B cells may result from low responsiveness to IL-7 owing to reduced amounts of IL-7 receptors. Reduced expression of *Rag1* and *Rag2* in pre-B cells lacking Foxo1 also contributes to decreased developmental progression at this stage. In the periphery, Foxo1 also regulates genes such as *Aicda* that are required for *Ig* gene maturation in germinal center B cells.

Foxo1 DNA binding was observed coordinately with E2A and EBF1 in pro-B cells [91]. Genetic studies revealed synergistic activities of E2A and Foxo1. Together, these studies assembled a global network consisting of E2A, EBF1, Foxo1, downstream regulators, signaling, and survival genes that control B cell specification and commitment (Figure 3). Genes bound by Foxo1 included genes that are important for B cell survival, including the previously identified target, *Bcl2l11* (Bim); [153,154]. Together with E2A, Foxo1 binds the *Bcl11a* gene, which encodes another transcription factor that contributes to lymphocyte development [155]. Interestingly, Foxo1 and EBF1 crossregulate each other in a positive feedback circuit, because *Ebf1* transcripts are absent from Ly6d$^+$ CLPs in the absence of Foxo1 [156]. Gene profiling and ChIP-seq revealed that EBF1 and Foxo1 regulate similar sets of genes, which suggests that the two factors coregulate B cell lineage specification. However, EBF1 and Foxo1 do not necessarily regulate transcription to the same outcome. Examples were noted of genes that are repressed by EBF1 alone (e.g., *Cd5*), Foxo1 alone (*Thy1*), or EBF1 and Foxo1 together (*Gfib*), or positively regulated by EBF1 alone (*Cd79b*) or Foxo1 alone (*Blnk/SLP-65*), or activated by EBF1 and Foxo1 together (*Blk*). Together, these proteins collaborate in different ways, involving multiple feedback circuits, to establish B cell identity.

## 8. REGULATION OF THE B CELL–SPECIFIC PROGRAM BY PAX5

*Paired box gene 5* (*Pax5*) is a member of the paired domain, or PAX, family of DNA-binding proteins, which includes important regulators of development in humans and mice [157]. Originally identified in a wide array of contexts as the B cell–specific activator protein (BSAP), Pax5 is the only PAX protein expressed in hematopoietic cells (reviewed in Ref. [158]). Early studies suggested regulation by Pax5 of components of the pre-BCR and BCR (i.e., *mb-1/Cd79a*), *Cd19*, and *IgH* and *IgL* genes. As this section will detail, Pax5 is among the most important regulators of the B cell lineage.

Noteworthy features of Pax5 begin with its paired domain, which includes two helix-turn-helix (HTH) motifs separated by a short flexible spacer (Figures 2, 4) [159]. Both HTH motifs are capable of binding DNA and are used together or separately in different contexts. Differential use of these domains and the numerous possible contacts made by Pax5 with DNA results in its complex DNA-binding specificity, which is considerably more degenerate than those of other proteins described here (Table 1). Pax5 also comprises other domains necessary for protein–protein interactions, transcriptional activation, and repression. In turn, Pax5 activates the B cell–specific transcriptome and represses the transcriptional programs of other lineages.

Pax5's quintessential roles in B lymphopoiesis were confirmed in gene-targeted mice [160–163]. In the absence of Pax5, B cell development is arrested at a stage between normal pre–pro-B and pro-B cells. Arrested cells fail to express many B cell–specific genes including *mb-1/Cd79a* and *Cd19*. Unlike EBF1- and E2A-deficient mice that exhibit a complete lack of *Ig* gene rearrangements, Pax5-deficient mice express *Rag1/2* and are partially defective in V(D)J recombination. *IgH* gene $D$-$J_H$ rearrangements are detected in normal amounts, but $V_H$ to $D$-$J_H$ rearrangements are rarely detected. Moreover, in the absence of Pax5, only the most $D_H$-proximal $V_H$ segment ($V_H7183$) is assembled into complete *IgH* genes. Distal $V_H$ segments ($V_HJ558$), which rearrange frequently in wild-type mice, are not used.

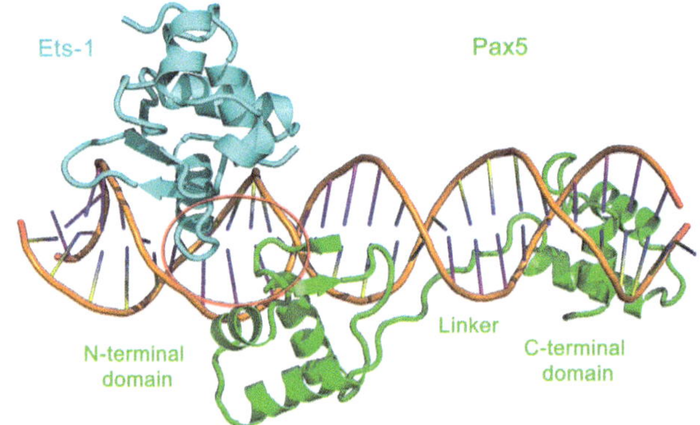


FIGURE 4 One context of Pax5 binding to DNA: overview of the crystal structure of the paired domain of Pax5 and ETS domain of Ets-1 bound to the *mb-1/Cd79a* promoter. This view shows the crystal structure of the paired domain of Pax5 (green) in complex with the ETS domain of Ets-1 (turquoise) on the 25 bp *mb-1/Cd79a* promoter binding site. The structure was solved at a resolution of 2.25 Å [159]. The Pax5 paired domain includes amino- and carboxy-terminal subdomains connected by a flexible spacer. The region marked by a red circle includes multiple side chains that mediate direct contacts between the two proteins and with DNA. *The figure was prepared from the PDB file IK78 using PyMOL (www.pymol.org)* (provided by Mair Churchill).

The lack of Pax5 also affected pre-BCR signaling on many levels [77]. In particular, *Blnk/SLP-65* was not expressed in Pax5-deficient pro-B cells. Together, these data demonstrate critical roles for Pax5 in the generation of immunoglobulins and expression of proteins necessary for signaling and function of B cells. Perhaps of most importance, pro-B cells lacking Pax5 exhibited latent potential for differentiation to other hematopoietic lineages [164,165]. In the absence of Pax5, cytokine receptors and other factors that can steer lineage choices to non-B cell fates are expressed inappropriately. These observations are discussed in detail in the next section.

Gene expression profiling of conditionally deleted, Pax5-deficient pro-B cells identified 170 genes activated by Pax5 [166]. This extensive profile included genes encoding receptors and adhesion molecules (e.g., *Vpreb3* and *Enpep/BP-1*), B cell signaling proteins (*Bcar3* and *Nedd9*), transcription factors (*Ikzf3*, *Spib*, *Irf4*, *Irf8*, and *Id3*), and other proteins involved in cell-cycle control. Pax5 particularly affected genes involved in B lymphocyte adhesion, migration, and trafficking (e.g., *Cd44*, *Cd97*, and *Tnfrsf19*). Analysis of DNA binding in wild-type pro-B cells using ChIP-seq detected enrichment of Pax5 binding to promoters and enhancers and its association with positive histone marks H3K9ac, H3K4me2, and H3K4me3 on active genes. In light of these data, Pax5's ability to repress over 110 genes in other contexts is striking [167].

Whether Pax5 activates or represses transcription is usually attributed to the context of its binding site. The target sequence may be permissive for interactions with other DNA-binding proteins or coactivators/corepressors that modulate transcriptional activity. In this regard, Pax5 interacts directly with the coactivators Daxx and CBP [168], retinoblastoma (Rb1) [169], and the SAGA histone acetyltransferase complex [170]. Previous studies suggest Pax5's direct recruitment of the TFIID subunit TATA binding protein (TBP) and other general transcription factors [169]. Recent studies using endogenously biotinylated Pax5 as bait for interacting proteins detected its associations with additional subunits of TFIID and components (BAF57 and BAF170) of BAF chromatin remodeling complexes [171]. Pax5 also bound PAX-interacting protein 1 (Paxip1/PTIP), which recruits MLL-containing H3K4 methyltransferase complexes. Together, these proteins add positive H3K4 methylation after DNA binding by Pax5.

Concerning how Pax5 mediates repression, a specialized domain in its carboxy-terminus recruits four corepressors of the Groucho (Grg)/transducin-like enhancer of split (Tle) family [11]. In this regard, context may include binding of Pax5 together with PU.1, which recruits Grg4 to an enhancer at the *IgH* locus [172]. Pax5 also recruits the NCoR1 repressor. Genes repressed by Pax5 featured a high density of H3K27me3 marks that are likely added by PRC2 complexes [171].

One context of Pax5 binding that is clearly defined is its recruitment of ETS family proteins to activate the *mb-1/Cd79a* gene [173]. Activation of *mb-1/Cd79a* promoter transcription is a stepwise process. In early B cell progenitors, including cells derived from E2A- or EBF1-deficient fetal livers, *mb-1/Cd79a* promoters are hypermethylated on CpG dinucleotides. Moreover, the local chromatin structure is closed and inaccessible. Before the expression of Pax5, a complex of EBF1, Runx1 (and its partner CBFβ) and E2A is assembled at the distal end (−178 to −148) of the promoter [143]. These factors begin the process of DNA demethylation at distal CpGs and facilitate chromatin accessibility necessary for the binding of Pax5 [145]. These early events require recruitment of SWI/SNF and Tet2 (implicated, but not yet proven) by EBF1 and its partners [145,146]. Bound factors collaborate to deplete *mb-1/Cd79a* promoters of Mi-2/NuRD complexes, which help maintain the hypermethylated state of the promoter in the absence of activators [145,174]. After the local opening of chromatin and loss of DNA methylation (which blocks ETS protein DNA binding), Pax5 recruits Ets-1 to bind the promoter [143,145,175]. Interactions with Pax5 enhance Ets-1 DNA binding by >800-fold [176]. Thus, DNA binding by Pax5 and Ets-1 is highly cooperative. The basis for Pax5-Ets-1 interactions was revealed by the X-ray crystallographic structure of the two proteins' DNA-binding domains assembled with *mb-1/Cd79a* promoter DNA (Figure 4) [159]. Highly conserved residues in Pax5 and Ets1 interact in a network of contacts that allows Ets-1 to bind a non-consensus site. Together, these data describe a mechanism in which E2A, EBF1, and Pax5 synergistically activate a model B cell–specific gene. Recently, long-range analysis of interactions at the genomic level identified two enhancers that may drive *mb-1/Cd79a* promoter activation [177]. A similar stepwise epigenetic process activates *Cd19* expression in B cells, which crucially depends on E2A, EBF1, and Pax5 (but not Pax5–Ets complexes) [178].

*Pax5* expression itself depends on a host of upstream factors, including other transcription factors detailed in this chapter. As described earlier, Pax5 was not expressed in gene-targeted mice lacking E2A, EBF1, or Foxo1. This suggests that the four factors coordinate early B cell development as part of a functional hierarchy of transcription factors [60,89,130,156]. Once active, *Pax5* transcription is detected at all stages of bone marrow B cell development and in peripheral B cells. However, Pax5 expression must be shut off to allow the onset of plasma cell differentiation [179]. During development, *Pax5* and *Ebf1* transcripts concomitantly increase at the stage of B cell–committed CLPs [134]. Not surprisingly, *Pax5* genes are activated rapidly after ectopic expression of EBF1 in EBF1-deficient progenitors [145].

Chromatin interaction analysis sequencing confirmed binding of EBF1 to a novel regulatory module upstream of exon 1B of the *Pax5* gene [91,180]. In addition, a potent enhancer in intron 5 drives B cell–specific expression of *Pax5*. The enhancer binds the following transcription factors that are all important for B cell development: PU.1, Irf4, Irf8, and NF-κB. Irf4 and Irf8 may crossregulate *Pax5* because Pax5 is required for their efficient expression [166]. Foxo1 binds additional elements of the *Pax5* gene and is essential for Pax5 expression [91,156]. Interestingly, a recent examination of the *Pax5* locus (and the entire B cell genome) using Chromatin Interaction Analysis by Paired-End-Tag (ChIA-PET) sequencing identified two additional enhancers ~250 kb upstream of the transcription start site [177]. Functional consequences of these interactions are currently unknown, but this observation reveals the power of the ChIA-PET technology to identify new regulatory modules on a genome-wide scale.

One of the more intriguing questions concerning Pax5 is its ability to facilitate V(D)J recombination between rearranged $DJ_H$ gene segments and remote $V_H$ segments [162]. In the absence of Pax5, only the most proximal $V_H$ segment ($V_H$*7183*) is used. Ectopic expression of Pax5 in Pax5-deficient pro-B cells restored the use of distal $V_H$ segments during V(D)J recombination [181]. As expected [182], the restoration of normal recombination frequencies was accompanied by repositioning of the *IgH* locus away from the nuclear periphery. Moreover, Pax5 induced contraction of the *IgH* locus. The murine *IgH* locus spans ~3 Mb of DNA Thus, contraction is necessary to synapse and recombine distant $V_H$ and $DJ_H$ gene segments. This process involves localized epigenetic modifications (i.e., acetylation of histones at accessible $V_H$ segments [183]. Rearrangements of distal $V_H$ segments are controlled by other transcription factors, including Ikaros, Yin-yang 1 (YY1), and Ezh2 [52,184–187]. How these other proteins function together with Pax5 is currently an intense area of study.

*IgH* locus recombination has been proposed to involve the reorganization of chromatin topology into loops, or rosettes, that facilitate long-range interactions between $V_H$ segments and rearranged $DJ_H$ genes in V(D)J recombination [188]. Similar to how transcription factors mediate the association of distant enhancers and promoters by looping the intervening DNA, DNA-binding proteins are necessary for DNA compaction via loop formation within *Ig* loci. Pax5 facilitates looping of intervening DNA in *IgH* recombination by binding Pax5-activated intergenic repeat (PAIR) elements, which are interspersed with $V_H$ gene segments in the distal $V_H$*J558* gene cluster [82,189]. Pax5, E2A, CCCTC-binding factor (CTCF), and cohesin bind PAIR elements [186]. Three-dimensional structural studies (chromosome conformation capture, or 3C) suggest that loops generated by PAIR elements interact with other loops including rearranged $D$-$J_H$ and the Eμ enhancer of the *IgH* locus during V(D)J recombination in pro-B cells [190]. Pax5-activated intergenic repeat elements give rise to long non-coding antisense transcripts in pro-B cells that regulate distal rearrangements, possibly by influencing locus contraction [82,186,190]. Pax5 also binds elements within the 3′ regulatory region (3′RR) at the far end of the *IgH* locus [191], which also interacts with upstream regions in pro-B cells [192]. Interestingly, other elements associated with DNA looping, but not Pax5-dependent PAIR elements, were identified in more proximal $V_H$ segment clusters [192]. This model helps explain the requirement for Pax5 during distal, but not proximal, $V_H$ gene rearrangements.

## 9. REGULATION OF B LINEAGE COMMITMENT

Lineage commitment has been defined as "the decision made by a cell to enter, or generate progeny that enter a particular maturation lineage" [193]. Lineage commitment during B lymphopoiesis has been elegantly shown to require sequential activities of most of the transcription factors described previously, including PU.1, Ikaros, EBF1 (and Foxo1), and Pax5. Activities of Pax5 and EBF1 were revealed in studies using gene-targeted mice that have lost B lineage commitment, and these experiments are the focus of this section.

As described earlier, mice lacking Pax5 do not generate functional B cells and exhibit developmental arrest at a pre–pro-B/pro-B–like stage (referred to hereafter as pro-B cells) [160,163]. B cell progenitors from these mice exhibit a number of special properties [194]. The cells grow indefinitely in the presence of IL-7 and do not differentiate in vitro as long as this cytokine is maintained. However, after removal of IL-7, Pax5-deficient pro-B cells demonstrate expanded lymphomyeloid potential but cannot differentiate into erythroid cells. Coculture with instructive cytokines drives Pax5-deficient pro-B cells to differentiate into all of the following types of cells: functional, phagocytic macrophages (M-CSF), granulocytes expressing Gr-1 (IL-3, IL-6, SCF, and G-CSF), dendritic cells that present antigen (GM-CSF), and NK cells that kill tumor cell targets (IL-2). Strikingly, Pax5-deficient pro-B cells grown in the presence of RANKL (Trance/Tnfsf11) developed into osteoclasts. Furthermore, after transfer into RAG-deficient mice, Pax5-deficient pro-B cells were transcriptionally reprogrammed in the thymus and generated functional T cells [165]. The molecular basis of these effects includes the inability of Pax5-deficient cells to repress transcription of 110 genes [167]. The inappropriate expression of these genes, including cytokine receptors (e.g., *c-fms/M-CSFR*) and transcription factors (e.g., GATA1) promote expression of transcriptional programs of non–B cell lineages and promiscuous differentiation. Restoration of Pax5 expression blocks this differentiation and reinstates commitment to the B lineage via the mechanisms described in the previous section.

Linkage of EBF1 with lineage commitment was evidenced by its ability to drive the differentiation of progenitors toward B cell development and the absence of B cells in mice lacking EBF1 [130,131]. Indeed, ectopic expression of EBF1 promoted B cell development of MPPs more potently than did Pax5 [195]. Further studies of EBF1-deficient B cell progenitors revealed a role of EBF1 that is similar to that of Pax5 in B lineage commitment. After transfer into RAG-deficient mice, EBF1-deficient cells were able to differentiate into cells including Mac-1$^+$ and Gr-1$^+$ myeloid cells, dendritic cells, NK cells, and T cells [195]. Enforced expression of EBF1 in EBF1-deficient cells activated Pax5 and antagonized transcription of genes that contribute to alternative potential, including C/EBPα, PU.1 (which collaborate to drive myeloid differentiation), and Id2 (which promotes NK cell development). EBF1 also mediates commitment by blocking expression of GATA3, which is essential for T cell development [196]. Ectopic expression of EBF1 restored normal patterns of B cell–specific transcription, including its repression of non-B cell genes. Importantly, Pax5 was not required for the repression of alternative lineage genes by EBF1. Another study demonstrated altered fate choices and lineage conversion of EBF1-deficient pro-B cells, which differentiated into innate lymphoid cells and T cells in vivo [197]. Although Pax5$^{+/-}$ mice are phenotypically normal [68,194], the full dosage of EBF1 is necessary for lineage commitment. Pro-B cells from *Ebf1*$^{+/-}$ mice are functionally haplo-insufficient and express genes of alternative lineages, including macrophage- and NK cell–specific genes [137,195]. These results are supported by the partial loss of many EBF1-dependent functions (e.g., IL-7 responsiveness) in *Ebf1* heterozygous knockout mice [68].

## 10. CONCLUSION

This chapter has focused on transcription factors that are essential for the development and commitment of B cells from progenitors in the bone marrow. By no means has this discussion been totally inclusive, because many other factors are required for the regulation of B cell–specific transcription. As a goal for the field, the elucidation of regulatory networks in B cell development and function has just begun. Similar to data shown in Figure 3 for E2A, EBF1, and Foxo1 in pro-B cells, extension of our current understanding of these networks has been achieved at only a few stages of development and for a small number of transcription factors. However, with additional, more inclusive ChIP-seq studies, it should be possible to expand gene networks to later stages of development and a wider range of transcription factors. To some degree, data already available as part of the Encyclopedia of DNA Elements Consortium at the University of California, Santa Cruz (an international collaboration of research groups funded by the National Human Genome Research Institute; http://genome.ucsc.edu) complements studies described here, but its predictive value is currently limited by a small number of contributed datasets relevant to B cells.

Other types of studies are moving the field ahead in different ways. Genomic analysis of long-range interactions using multiplexed chromosome conformation capture sequencing (4C, 5C, and HiC) can predict locations of promoters and enhancers for bioinformatic analysis of their composition (reviewed in Ref. [198]). Measurements of chromatin connectivity using ChIA-PET have revealed the differential use of enhancers between embryonic stem cells and B cells [177]. Other methods that survey the B cell–specific interactome can reveal new interactions between transcription factors [199].

The relationships between the various transcription factors and cancer were also not discussed. It is notable that human leukemias, including B cell (and T cell) acute-lymphoblastic leukemia, have been linked with mutations in *Ikzf1*, *E2A*, *Ebf1*, and *Pax5* loci [200–202]. In many ways, functional analyses of transcription factors in B cells have been instructive about cancer development and prognosis, and vice versa.

## ACKNOWLEDGMENT

The author thanks the NIH for support via Grants AI081878 and AI098417, Seth Welsh and Kara Lukin for assistance with figures, and Kara Lukin, Mikael Sigvardsson and Peter Nielsen for critically reading the manuscript. Grateful thanks are made to Cornelis Murre and Mair Churchill, who provided figures.

## REFERENCES

[1] Seita J, Sahoo D, Rossi DJ, Bhattacharya D, Serwold T, Inlay MA, et al. Gene expression commons: an open platform for absolute gene expression profiling. PLoS One 2012;7:e40321.

[2] Hardy RR, Kincade PW, Dorshkind K. The protean nature of cells in the B lymphocyte lineage. Immunity 2007;26:703–14.

[3] Heng TS, Painter MW. The immunological genome project: networks of gene expression in immune cells. Nat Immunol 2008;9:1091–4.

[4] Klemsz MJ, McKercher SR, Celada A, Beveren C v, Maki RA. The macrophage and B-cell-specific transcription factor PU.1 is related to the ets oncogene. Cell 1990;61:113–24.

[5] Hollenhorst PC, McIntosh LP, Graves BJ. Genomic and biochemical insights into the specificity of ETS transcription factors. Annu Rev Biochem 2011;80:437–71.

[6] Mathelier A, Zhao X, Zhang AW, Parcy F, Worsley-Hunt R, et al. JASPAR 2014: an extensively expanded and updated open-access database of transcription factor binding profiles. Nucleic Acids Research 2013;42:D142–147.

[7] Maglott D, Ostell J, Pruitt KD, Tatusova T. Entrez gene: gene-centered information at NCBI. Nucleic Acids Research 2005;33:D54–8.

[8] Schwickert TA, Tagoh H, Gultekin S, Dakic A, Axelsson E, Minnich M, et al. Stage-specific control of early B cell development by the transcription factor Ikaros. Nat Immunol 2014;15:283–93.

[9] Pio F, Kodandapani R, Ni CZ, Shepard W, Klemsz M, McKercher SR, et al. New insights on DNA recognition by ets proteins from the crystal structure of the PU.1 ETS domain-DNA complex. J Biol Chem 1996;271:23329–37.

[10] Massari M, Jennings P, Murre C. The AD1 transactivation domain contains a highly conserved helix which is required for its activity in both *Saccharomyces cerevisiae* and mammalian cells. Mol Cell Biol 1996;16:121–9.

[11] Eberhard D, Jimenez G, Heavey B, Busslinger M. Transcriptional repression by Pax5 (BSAP) through interaction with corepressors of the Groucho family. EMBO J 2000;19:2292–303.

[12] Fisher RC, Olson MC, Pongubala JM, Perkel JM, Atchison ML, Scott EW, et al. Normal myeloid development requires both the glutamine-rich transactivation domain and the PEST region of transcription factor PU.1 but not the potent acidic transactivation domain. Mol Cell Biol 1998;18:4347–57.

[13] Hagman J, Ramirez J, Lukin K. B lymphocyte lineage specification, commitment and epigenetic control of transcription by early B cell factor 1. Curr Top Microbiol Immunol 2011.

[14] Friedbichler K, Hoelbl A, Li G, Bunting KD, Sexl V, Gouilleux F, et al. Serine phosphorylation of the Stat5a C-terminus is a driving force for transformation. Front Biosci (Landmark Ed) 2011;16:3043–56.

[15] Consortium, TU. Activities at the universal protein resource (UniProt). Nucleic Acids Res 2014;42:D191–8.

[16] Kerdiles YM, Beisner DR, Tinoco R, Dejean AS, Castrillon DH, DePinho RA, et al. Foxo1 links homing and survival of naive T cells by regulating L-selectin, CCR7 and interleukin 7 receptor. Nat Immunol 2009;10:176–84.

[17] Paul R, Schuetze S, Kozak SL, Kozak CA, Kabat D. The Sfpi-1 proviral integration site of Friend erythroleukemia encodes the ets-related transcription factor Pu.1. J Virol 1991;65:464–7.

[18] Pongubala JMR, Nagulapalli S, Klemsz MJ, McKercher SR, Maki RA, Atchinson ML. PU.1 recruits a second nuclear factor to a site important for immunoglobulin k 3' enhancer activity. Mol Cell Biol 1992;12:368–78.

[19] Scott EW, Simon MC, Anastasi J, Singh H. Requirement of transcription factor PU.1 in the development of multiple hematopoietic lineages. Science 1994;265:1573–7.

[20] McKercher SR, Torbett BE, Anderson KL, Henkel GW, Vestal DJ, Baribault H, et al. Targeted disruption of the *PU.1* gene results in multiple hematopoietic abnormalities. EMBO J 1996;15:5647–58.

[21] Olson MC, Scott EW, Hack AA, Su GH, Tenen DG, Singh H, et al. PU. 1 is not essential for early myeloid gene expression but is required for terminal myeloid differentiation. Immunity 1995;3:703–14.

[22] Henkel GW, McKercher SR, Yamamoto H, Anderson KL, Oshima RG, Maki RA. PU.1 but not ets-2 is essential for macrophage development from embryonic stem cells. Blood 1996;88:2917–26.

[23] Scott EW, Fisher RC, Olson MC, Kehrli EW, Simon MC, Singh H. PU.1 functions in a cell-autonomous manner to control the differentiation of multipotential lymphoid-myeloid progenitors. Immunity 1997;6:437–47.

[24] DeKoter RP, Singh H. Regulation of B lymphocyte and macrophage development by graded expression of PU.1. Science 2000;288:1439–41.

[25] Polli M, Dakic A, Light A, Tarlinton DM, Nutt SL. The development of functional B lymphocytes in conditional PU.1 knock-out mice. Blood 2005;106:2083–90.

[26] Dahl R, Ramirez-Bergeron DL, Rao S, Simon MC. Spi-B can functionally replace PU.1 in myeloid but not lymphoid development. EMBO J 2002;21:2220–30.

[27] DeKoter RP, Lee HJ, Singh H. PU.1 regulates expression of the interleukin-7 receptor in lymphoid progenitors. Immunity 2002;16:297–309.

[28] Miller JP, Izon D, DeMuth W, Gerstein R, Bhandoola A, Allman D. The earliest step in B lineage differentiation from common lymphoid progenitors is critically dependent upon interleukin 7. J Exp Med 2002;196:705–11.

[29] Kee BL, Bain G, Murre C. IL-7Ralpha and E47: independent pathways required for development of multipotent lymphoid progenitors. EMBO J 2002;21:103–13.

[30] Kikuchi K, Kasai H, Watanabe A, Lai AY, Kondo M. IL-7 specifies B cell fate at the common lymphoid progenitor to pre-proB transition stage by maintaining early B cell factor expression. J Immunol 2008;181:383–92.

[31] Purohit SJ, Stephan RP, Kim HG, Herrin BR, Gartland L, Klug CA. Determination of lymphoid cell fate is dependent on the expression status of the IL-7 receptor. EMBO J 2003;22:5511–21.

[32] Medina KL, Pongubala JMR, Reddy KL, Lancki DW, DeKoter RP, Kieslinger M, et al. Defining a regulatory network for specification of the B cell fate. Dev Cell 2004;7:607–17.

[33] Spooner CJ, Cheng JX, Pujadas E, Laslo P, Singh H. A recurrent network involving the transcription factors PU.1 and Gfi1 orchestrates innate and adaptive immune cell fates. Immunity 2009;31:576–86.

[34] Kueh HY, Champhekar A, Nutt SL, Elowitz MB, Rothenberg EV. Positive feedback between PU.1 and the cell cycle controls myeloid differentiation. Science 2013;341:670–3.

[35] Lichtinger M, Ingram R, Hannah R, Muller D, Clarke D, Assi SA, et al. RUNX1 reshapes the epigenetic landscape at the onset of haematopoiesis. EMBO J 2013;31:4318–33.

[36] Yoshida T, Ng SY, Zuniga-Pflucker JC, Georgopoulos K. Early hematopoietic lineage restrictions directed by Ikaros. Nat Immunol 2006;7:382–91.

[37] Laslo P, Spooner CJ, Warmflash A, Lancki DW, Lee HJ, Sciammas R, et al. Multilineage transcriptional priming and determination of alternate hematopoietic cell fates. Cell 2006;126:755–66.

[38] Heinz S, Benner C, Spann N, Bertolino E, Lin YC, Laslo P, et al. Simple combinations of lineage-determining transcription factors prime cis-regulatory elements required for macrophage and B cell identities. Mol Cell 2010;38:576–89.

[39] Georgopoulos K, Moore DD, Derfler B. Ikaros, an early lymphoid-specific transcription factor and a putative mediator for T cell commitment. Science 1992;258:808–12.

[40] Hahm K, Ernst P, Lo K, Kim GS, Turck C, Smale ST. The lymphoid transcription factor LyF-1 is encoded by specific, alternatively spliced mRNAs derived from the *Ikaros* gene. Mol Cell Biol 1994;14:7111–23.

[41] Georgopoulos K, Bigby M, Wang JH, Molnar A, Wu P, Winandy S, et al. The *Ikaros* gene is required for the development of all lymphoid lineages. Cell 1994;79:143–56.

[42] Molnár A, Wu P, Largespada DA, Vortkamp A, Scherer S, Copeland NG, et al. The *Ikaros* gene encodes a family of lymphocyte-restricted zinc finger DNA binding proteins, highly conserved in human and mouse. J Immunol 1996;156:585–92.

[43] Wang JH, Nichogiannopoulou A, Wu L, Sun L, Sharpe AH, Bigby M, et al. Selective defects in the development of the fetal and adult lymphoid system in mice with an Ikaros null mutation. Immunity 1996;5:537–49.

[44] Ng SY, Yoshida T, Zhang J, Georgopoulos K. Genome-wide lineage-specific transcriptional networks underscore Ikaros-dependent lymphoid priming in hematopoietic stem cells. Immunity 2009;30:493–507.
[45] Yoshida T, Landhuis E, Dose M, Hazan I, Zhang J, Naito T, et al. Transcriptional regulation of the Ikzf1 locus. Blood 2013;122:3149–59.
[46] Heizmann B, Kastner P, Chan S. Ikaros is absolutely required for pre-B cell differentiation by attenuating IL-7 signals. J Exp Med 2013;210:2823–32.
[47] Koipally J, Renold A, Kim J, Georgopoulos K. Repression by Ikaros and Aiolos is mediated through histone deacetylase complexes. EMBO J 1999;18:3090–100.
[48] Kim J, Sif S, Jones B, Jackson A, Koipally J, Heller E, et al. Ikaros DNA-binding proteins direct formation of chromatin remodeling complexes in lymphocytes. Immunity 1999;10:345–55.
[49] Koipally J, Heller EJ, Seavitt JR, Georgopoulos K. Unconventional potentiation of gene expression by Ikaros. J Biol Chem 2002;277:13007–15.
[50] Sridharan R, Smale ST. Predominant interaction of both Ikaros and Helios with the NuRD complex in immature thymocytes. J Biol Chem 2007;282:30227–38.
[51] Schjerven H, McLaughlin J, Arenzana TL, Frietze S, Cheng D, Wadsworth SE, et al. Selective regulation of lymphopoiesis and leukemogenesis by individual zinc fingers of Ikaros. Nat Immunol 2013;14:1073–83.
[52] Alkhatib A, Werner M, Hug E, Herzog S, Eschbach C, Faraidun H, et al. FoxO1 induces Ikaros splicing to promote immunoglobulin gene recombination. J Exp Med 2012;209:395–406.
[53] Joshi I, Yoshida T, Jena N, Qi X, Zhang J, Van Etten RA, et al. Loss of Ikaros DNA-binding function confers integrin-dependent survival on pre-B cells and progression to acute lymphoblastic leukemia. Nat Immunol 2014;15:294–304.
[54] Ephrussi A, Church GM, Tonegawa S, Gilbert W. B lineage–specific interactions of an immunoglobulin enhancer with cellular factors in vivo. Science 1985;227:134–40.
[55] Church GM, Ephrussi A, Gilbert W, Tonegawa S. Cell-type-specific contacts to immunoglobulin enhancers in nuclei. Nature 1985;313:798–801.
[56] Murre C, McCaw PS, Baltimore D. A new DNA binding and dimerization motif in immunoglobulin enhancer binding- *daughterless, MyoD*, and *myc* proteins. Cell 1989;56:777–83.
[57] Murre C, McCaw PS, Vaessin H, Caudy M, Jan LY, Jan YN, et al. Interactions between heterologous helix-loop-helix proteins generate complexes that bind specifically to a common DNA sequence. Cell 1989;58:537–44.
[58] Ellenberger T, Fass D, Arnaud M, Harrison SC. Crystal structure of transcription factor E47: E box recognition by a basic region helix-loop-helix dimer. Genes Dev 1994;8:970–80.
[59] Zhuang Y, Soriano P, Weintraub H. The helix-loop-helix gene E2A is required for B cell formation. Cell 1994;79:875–84.
[60] Bain G, Maandag EC, Izon DJ, Amsen D, Kruisbeek AM, Weintraub BC, et al. E2A proteins are required for proper B cell development and initiation of immunoglobulin gene rearrangements. Cell 1994;79:885–92.
[61] Bain G, Quong MW, Soloff RS, Hedrick SM, Murre C. Thymocyte maturation is regulated by the activity of the helix-loop-helix protein, E47. J Exp Med 1999;190:1605–16.
[62] Schwartz R, Engel I, Fallahi-Sichani M, Petrie HT, Murre C. Gene expression patterns define novel roles for E47 in cell cycle progression, cytokine-mediated signaling, and T lineage development. Proc Natl Acad Sci USA 2006;103:9976–81.
[63] Wojciechowski J, Lai A, Kondo M, Zhuang Y. E2A and HEB are required to block thymocyte proliferation prior to pre-TCR expression. J Immunol 2007;178:5717–26.
[64] Jones-Mason ME, Zhao X, Kappes D, Lasorella A, Iavarone A, Zhuang Y. E protein transcription factors are required for the development of CD4+ lineage T cells. Immunity 2012;36:348–61.
[65] Stradner MH, D'Cruz LM, Yang CY, Goldrath AW. A8.23 E and Id proteins influence invariant natural killer T cell sublineage differentiation and proliferation. Ann Rheum Dis 2014;73(Suppl. 1):A85.
[66] Zhang F, Fuss IJ, Yang Z, Strober W. Transcription of RORgammat in developing Th17 cells is regulated by E-proteins. Mucosal Immunol 2013.
[67] Dias S, Mansson R, Gurbuxani S, Sigvardsson M, Kee BL. E2A proteins promote development of lymphoid-primed multipotent progenitors. Immunity 2008;29:217–27.
[68] Ahsberg J, Ungerback J, Strid T, Welinder E, Stjernberg J, Larsson M, et al. Early B-cell factor 1 regulates the expansion of B-cell progenitors in a dose-dependent manner. J Biol Chem 2013;288: 33449–61.
[69] Ikawa T, Kawamoto H, Wright LY, Murre C. Long-term cultured E2A-deficient hematopoietic progenitor cells are pluripotent. Immunity 2004;20:349–60.
[70] Bain G, Romanow WJ, Albers K, Havran WL, Murre C. Positive and negative regulation of V(D)J recombination by the E2A proteins. J Exp Med 1999;189:289–300.
[71] Choi JK, Shen C-P, Radomska HS, Eckhardt LA, Kadesch T. E47 activates the Ig-heavy chain and Tdt loci in non-B cells. EMBO J 1996;15:5014–21.
[72] Romanow WJ, Langerak AW, Goebel P, Wolvers-Tettero ILM, van Dongen JJM, Feeney AJ, et al. E2A and EBF act in synergy with the V(D)J recombinase to generate a diverse immunoglobulin repertoire in nonlymphoid cells. Mol Cell 2000;5:343–53.
[73] Goebel P, Janney N, Valenzuela JR, Romanow WJ, Murre C, Feeney AJ. Localized gene-specific induction of accessibility to V(D)J recombination induced by E2A and early B cell factor in nonlymphoid cells. J Exp Med 2001;194:645–56.
[74] Bain G, Maandag EC, te Riele HP, Feeney AJ, Sheehy A, Schlissel M, et al. Both E12 and E47 allow commitment to the B cell lineage. Immunity 1997;6:145–54.
[75] Beck K, Peak MM, Ota T, Nemazee D, Murre C. Distinct roles for E12 and E47 in B cell specification and the sequential rearrangement of immunoglobulin light chain loci. J Exp Med 2009;206:2271–84.
[76] Quong MW, Martensson A, Langerak AW, Rivera RR, Nemazee D, Murre C. Receptor editing and marginal zone B cell development are regulated by the helix-loop-helix protein, E2A. J Exp Med 2004;199:1101–12.
[77] Schebesta M, Pfeffer PL, Busslinger M. Control of pre-BCR signaling by Pax5-dependent activation of the *BLNK* gene. Immunity 2002;17:473–85.
[78] Lazorchak AS, Schlissel MS, Zhuang Y. E2A and IRF-4/Pip promote chromatin modification and transcription of the immunoglobulin kappa locus in pre-B cells. Mol Cell Biol 2006;26:810–21.
[79] Hewitt SL, Chaumeil J, Skok JA. Chromosome dynamics and the regulation of V(D)J recombination. Immunol Rev 2010;237:43–54.
[80] Feeney AJ. Genetic and epigenetic control of V gene rearrangement frequency. Adv Exp Med Biol 2009;650:73–81.
[81] Sakamoto S, Wakae K, Anzai Y, Murai K, Tamaki N, Miyazaki M, et al. E2A and CBP/p300 act in synergy to promote chromatin accessibility of the immunoglobulin kappa locus. J Immunol 2012;188:5547–60.

[82] Ebert A, McManus S, Tagoh H, Medvedovic J, Salvagiotto G, Novatchkova M, et al. The distal $V_H$ gene cluster of the Igh locus contains distinct regulatory elements with Pax5 transcription factor-dependent activity in pro-B cells. Immunity 2011;34: 175–87.

[83] Stadhouders R, de Bruijn MJ, Rother MB, Yuvaraj S, Ribeiro de Almeida C, et al. Pre-b cell receptor signaling induces immunoglobulin kappa locus accessibility by functional redistribution of enhancer-mediated chromatin interactions. PLoS Biol 2014;12:e1001791.

[84] Yabuki M, Ordinario EC, Cummings WJ, Fujii MM, Maizels N. E2A acts in cis in G1 phase of cell cycle to promote Ig gene diversification. J Immunol 2009;182:408–15.

[85] Conlon TM, Meyer KB. The chicken Ig light chain 3'-enhancer is essential for gene expression and regulates gene conversion via the transcription factor E2A. Eur J Immunol 2006;36:139–48.

[86] Hsu L-Y, Lauring J, Liang H-E, Greenbaum S, Cado D, Zhuang Y, et al. A conserved transcriptional enhancer regulates RAG gene expression in developing B cells. Immunity 2003;19:105–17.

[87] Borghesi L, Hsu LY, Miller JP, Anderson M, Herzenberg L, Schlissel MS, et al. B lineage-specific regulation of V(D)J recombinase activity is established in common lymphoid progenitors. J Exp Med 2004;199:491–502.

[88] Sigvardsson M, O'Riordan M, Grosschedl R. EBF and E47 collaborate to induce expression of the endogenous immunoglobulin surrogate light chain genes. Immunity 1997;7:25–36.

[89] O'Riordan M, Grosschedl R. Coordinate regulation of B cell differentiation by the transcription factors EBF and E2A. Immunity 1999;11:21–31.

[90] Sigvardsson M, Clark DR, Fitzsimmons D, Doyle M, Åkerblad P, Breslin T, et al. Early B cell Factor, E2A, and Pax-5 cooperate to activate the early B cell-specific *mb-1* promoter. Mol Cell Biol 2002;22:8539–51.

[91] Lin YC, Jhunjhunwala S, Benner C, Heinz S, Welinder E, Mansson R, et al. A global network of transcription factors, involving E2A, EBF1 and Foxo1, that orchestrates B cell fate. Nat Immunol 2010;11:635–43.

[92] Kwon K, Hutter C, Sun Q, Bilic I, Cobaleda C, Malin S, et al. Instructive role of the transcription factor E2A in early B lymphopoiesis and germinal center B cell development. Immunity 2008;28:751–62.

[93] Karczewski KJ, Tatonetti NP, Landt SG, Yang X, Slifer T, Altman RB, et al. Cooperative transcription factor associations discovered using regulatory variation. Proc Natl Acad Sci USA 2011;108:13353–8.

[94] Massari ME, Grant PA, Pray-Grant MG, Berger SL, Workman JL, Murre C. A conserved motif present in a class of helix-loop-helix proteins activates transcription by direct recruitment of the SAGA complex. Mol Cell 1999;4:63–73.

[95] Teachenor R, Beck K, Wright LY, Shen Z, Briggs SP, Murre C. Biochemical and phosphoproteomic analysis of the helix-loop-helix protein E47. Mol Cell Biol 2012;32:1671–82.

[96] Chen WY, Zhang J, Geng H, Du Z, Nakadai T, Roeder RG. A TAF4 coactivator function for E proteins that involves enhanced TFIID binding. Genes Dev 2013;27:1596–609.

[97] Mercer EM, Lin YC, Benner C, Jhunjhunwala S, Dutkowski J, Flores M, et al. Multilineage priming of enhancer repertoires precedes commitment to the B and myeloid cell lineages in hematopoietic progenitors. Immunity 2011;35:413–25.

[98] Heintzman ND, Stuart RK, Hon G, Fu Y, Ching CW, Hawkins RD, et al. Distinct and predictive chromatin signatures of transcriptional promoters and enhancers in the human genome. Nat Genet 2007;39:311–8.

[99] Kaikkonen MU, Spann NJ, Heinz S, Romanoski CE, Allison KA, Stender JD, et al. Remodeling of the enhancer landscape during macrophage activation is coupled to enhancer transcription. Mol Cell 2013;51:310–25.

[100] Lin YC, Benner C, Mansson R, Heinz S, Miyazaki K, Miyazaki M, et al. Global changes in the nuclear positioning of genes and intra- and interdomain genomic interactions that orchestrate B cell fate. Nat Immunol 2012;13:1196–204.

[101] Lassar AB, Davis RL, Wright WE, Kadesch T, Murre C, Voronova A, et al. Functional activity of myogenic HLH proteins requires hetero-oligomerization with E12/E47-like proteins in vivo. Cell 1991;66:305–15.

[102] Wilson RB, Kiledjian M, Shen C-P, Benezra R, Zwollo P, Dymecki SM, et al. Repression of immunoglobulin enhancers by the helix-loop-helix protein Id: implications for B-lymphoid-cell development. Mol Cell Biol 1991;11:6185–91.

[103] Kreider BL, Benezra R, Rovera G, Kadesch T. Inhibition of myeloid differentiation by the helix-loop-helix protein Id. Science 1992;255:1700–2.

[104] Thal M, Carvalho TL, He T, Kim HG, Gao H, Hagman J, et al. Ebf1-mediated down-regulation of Id2 and Id3 is essential for specification of the B cell lineage. Proc Natl Acad Sci USA 2009;106:552–7.

[105] Li H, Ji M, Klarmann KD, Keller JR. Repression of Id2 expression by Gfi-1 is required for B-cell and myeloid development. Blood 2010;116:1060–9.

[106] Kee BL. Id3 induces growth arrest and caspase-2-dependent apoptosis in B lymphocyte progenitors. J Immunol 2005;175:4518–27.

[107] Becker-Herman S, Lantner F, Shachar I. Id2 negatively regulates B cell differentiation in the spleen. J Immunol 2002;168:5507–13.

[108] van Galen P, Kreso A, Wienholds E, Laurenti E, Eppert K, Lechman ER, et al. Reduced lymphoid lineage priming promotes human hematopoietic stem cell expansion. Cell Stem Cell 2014;14:94–106.

[109] Corfe SA, Paige CJ. The many roles of IL-7 in B cell development; mediator of survival, proliferation and differentiation. Semin Immunol 2012;24:198–208.

[110] Clark MR, Mandal M, Ochiai K, Singh H. Orchestrating B cell lymphopoiesis through interplay of IL-7 receptor and pre-B cell receptor signalling. Nat Rev Immunol 2014;14:69–80.

[111] Kikuchi K, Lai AY, Hsu C-L, Kondo M. IL-7 receptor signaling is necessary for stage transition in adult B cell development through up-regulation of EBF. J Exp Med 2005;201:1197–203.

[112] Tsapogas P, Zandi S, Ahsberg J, Zetterblad J, Welinder E, Jonsson JI, et al. IL-7 mediates Ebf-1-dependent lineage restriction in early lymphoid progenitors. Blood 2011;118:1283–90.

[113] Goetz CA, Harmon IR, O'Neil JJ, Burchill MA, Farrar MA. STAT5 activation underlies IL7 receptor-dependent B cell development. J Immunol 2004;172:4770–8.

[114] Ahsberg J, Tsapogas P, Qian H, Zetterblad J, Zandi S, Mansson R, et al. Interleukin-7-induced Stat-5 acts in synergy with Flt-3 signaling to stimulate expansion of hematopoietic progenitor cells. J Biol Chem 2010;285:36275–84.

[115] Hirokawa S, Sato H, Kato I, Kudo A. EBF-regulating Pax5 transcription is enhanced by STAT5 in the early stage of B cells. Eur J Immunol 2003;33:1824–9.

[116] Smith EMK, Gisler R, Sigvardsson M. Cloning and characterization of a promoter flanking the early B cell factor (EBF) gene indicates roles for E-proteins and autoregulation in the control of EBF expression. J Immunol 2002;169:261–70.
[117] Roessler S, Gyory I, Imhof S, Spivakov M, Williams RR, Busslinger M, et al. Distinct promoters mediate the regulation of Ebf1 gene expression by interleukin-7 and Pax5. Mol Cell Biol 2007;27: 579–94.
[118] Goetz CA, Harmon IR, O'Neil JJ, Burchill MA, Johanns TM, Farrar MA. Restricted STAT5 activation dictates appropriate thymic B versus T cell lineage commitment. J Immunol 2005;174:7753–63.
[119] Bertolino E, Reddy K, Medina KL, Parganas E, Ihle J, Singh H. Regulation of interleukin 7-dependent immunoglobulin heavy-chain variable gene rearrangements by transcription factor STAT5. Nat Immunol 2005;6:836–43.
[120] Malin S, McManus S, Cobaleda C, Novatchkova M, Delogu A, Bouillet P, et al. Role of STAT5 in controlling cell survival and immunoglobulin gene recombination during pro-B cell development. Nat Immunol 2010;11:171–9.
[121] Mandal M, Powers SE, Maienschein-Cline M, Bartom ET, Hamel KM, Kee BL, et al. Epigenetic repression of the Igk locus by STAT5-mediated recruitment of the histone methyltransferase Ezh2. Nat Immunol 2011;12:1212–20.
[122] Ochiai K, Maienschein-Cline M, Mandal M, Triggs JR, Bertolino E, Sciammas R, et al. A self-reinforcing regulatory network triggered by limiting IL-7 activates pre-BCR signaling and differentiation. Nat Immunol 2012;13:300–7.
[123] Hagman J, Travis A, Grosschedl R. A novel lineage-specific nuclear factor regulates mb-1 gene transcription at the early stages of B cell differentiation. EMBO J 1991;10:3409–17.
[124] Feldhaus A, Mbangkollo D, Arvin K, Klug C, Singh H. BlyF, a novel cell-type- and stage-specific regulator of the B-lymphocyte gene *mb*-1. Mol Cell Biol 1992;12:1126–33.
[125] Hagman J, Gutch MJ, Lin H, Grosschedl R. EBF contains a novel zinc coordination motif and multiple dimerization and transcriptional activation domains. EMBO J 1995;14:2907–16.
[126] Siponen MI, Wisniewska M, Lehtio L, Johansson I, Svensson L, Raszewski G, et al. Structural determination of functional domains in early B-cell factor (EBF) family of transcription factors reveals similarities to Rel DNA-binding proteins and a novel dimerization motif. J Biol Chem 2010;285:25875–9.
[127] Hagman J, Belanger C, Travis A, Turck CW, Grosschedl R. Cloning and functional characterization of early B-cell factor, a regulator of lymphocyte-specific gene expression. Genes Dev 1993;7:760–73.
[128] Treiber N, Treiber T, Zocher G, Grosschedl R. Structure of an Ebf1:DNA complex reveals unusual DNA recognition and structural homology with Rel proteins. Genes Dev 2010;24:2270–5.
[129] Fields S, Ternyak K, Gao H, Ostraat R, Akerlund J, Hagman J. The "zinc knuckle" motif of early B cell factor is required for transcriptional activation of B cell-specific genes. Mol Immunol 2008;45:3786–96.
[130] Lin H, Grosschedl R. Failure of B-cell differentiation in mice lacking the transcription factor EBF. Nature 1995;376:263–7.
[131] Zhang Z, Cotta CV, Stephan RP, deGuzman CG, Klug CA. Enforced expression of EBF in hematopoietic stem cells restricts lymphopoiesis to the B cell lineage. EMBO J 2003;22:4759–69.
[132] Seet CS, Brumbaugh RL, Kee BL. Early B cell factor promotes B lymphopoiesis with reduced interleukin-7 responsiveness in the absence of E2A. J Exp Med 2004;199:1689–700.
[133] Zandi S, Ahsberg J, Tsapogas P, Stjernberg J, Qian H, Sigvardsson M. Single-cell analysis of early B-lymphocyte development suggests independent regulation of lineage specification and commitment in vivo. Proc Natl Acad Sci USA 2012;109:15871–6.
[134] Mansson R, Zandi S, Welinder E, Tsapogas P, Sakaguchi N, Bryder D, et al. Single-cell analysis of the common lymphoid progenitor compartment reveals functional and molecular heterogeneity. Blood 2010;115:2601–9.
[135] Jiang X-X, Nguyen Q, Chou Y, Wang T, Nandakumar V, Yates P, et al. Control of B cell development by the histone H2A deubiquitinase MYSM1. Immunity 2012;35:883–96.
[136] Lukin K, Fields S, Lopez D, Cherrier M, Ternyak K, Ramirez J, et al. Compound haploinsufficiencies of Ebf1 and Runx1 genes impede B cell lineage progression. Proc Natl Acad Sci USA 2010;107: 7869–74.
[137] Lukin K, Fields S, Guerrettaz L, Straign D, Rodriguez V, Zandi S, et al. A dose-dependent role for EBF1 in repressing non-B-cell-specific genes. Eur J Immunol 2011;41:1787–93.
[138] Warming S, Liu P, Suzuki T, Akagi K, Lindtner S, Pavlakis GN, et al. Evi3, a common retroviral integration site in murine B-cell lymphoma, encodes an EBFAZ-related kruppel-like zinc finger protein. Blood 2003;101:1934–40.
[139] Mega T, Lupia M, Amodio N, Horton SJ, Mesuraca M, Pelaggi D, et al. Zinc finger protein 521 antagonizes early B-cell factor 1 and modulates the B-lymphoid differentiation of primary hematopoietic progenitors. Cell Cycle 2011;10:2129–39.
[140] Kiviranta R, Yamana K, Saito H, Ho DK, Laine J, Tarkkonen K, et al. Coordinated transcriptional regulation of bone homeostasis by Ebf1 and Zfp521 in both mesenchymal and hematopoietic lineages. J Exp Med 2013;210:969–85.
[141] Kang S, Akerblad P, Kiviranta R, Gupta RK, Kajimura S, Griffin MJ, et al. Regulation of early adipose commitment by Zfp521. PLoS Biol 2012;10:e1001433.
[142] Treiber T, Mandel EM, Pott S, Gyory I, Firner S, Liu ET, et al. Early B cell factor 1 regulates B cell gene networks by activation, repression, and transcription- independent poising of chromatin. Immunity 2010;32:714–25.
[143] Maier H, Ostraat R, Gao H, Fields S, Shinton SA, Medina KL, et al. Early B cell factor cooperates with Runx1 and mediates epigenetic changes associated with mb-1 transcription. Nat Immunol 2004;5:1069–77.
[144] Zhao F, McCarrick-Walmsley R, Akerblad P, Sigvardsson M, Kadesch T. Inhibition of p300/CBP by early B-cell factor. Mol Cell Biol 2003;23:3837–46.
[145] Gao H, Lukin K, Ramirez J, Fields S, Lopez D, Hagman J. Opposing effects of SWI/SNF and Mi-2/NuRD chromatin remodeling complexes on epigenetic reprogramming by EBF and Pax5. Proc Natl Acad Sci USA 2009;106:11258–63.
[146] Guilhamon P, Eskandarpour M, Halai D, Wilson GA, Feber A, Teschendorff AE, et al. Meta-analysis of IDH-mutant cancers identifies EBF1 as an interaction partner for TET2. Nat Commun 2013;4:2166.
[147] Ito S, D'Alessio AC, Taranova OV, Hong K, Sowers LC, Zhang Y. Role of Tet proteins in 5mC to 5hmC conversion, ES-cell self-renewal and inner cell mass specification. Nature 2010;466:1129–33.
[148] Gyory I, Boller S, Nechanitzky R, Mandel E, Pott S, Liu E, et al. Transcription factor Ebf1 regulates differentiation stage-specific signaling, proliferation, and survival of B cells. Genes Dev 2012;26:668–82.

[149] Vilagos B, Hoffmann M, Souabni A, Sun Q, Werner B, Medvedovic J, et al. Essential role of EBF1 in the generation and function of distinct mature B cell types. J Exp Med 2012;209:775–92.

[150] Szydlowski M, Jablonska E, Juszczynski P. FOXO1 transcription factor: a critical effector of the PI3K-AKT Axis in b-cell development. Int Rev Immunol 2014;33:146–57.

[151] Dengler HS, Baracho GV, Omori SA, Bruckner S, Arden KC, Castrillon DH, et al. Distinct functions for the transcription factor Foxo1 at various stages of B cell differentiation. Nat Immunol 2008;9:1388–98.

[152] Welinder E, Mansson R, Mercer EM, Bryder D, Sigvardsson M, Murre C. The transcription factors E2A and HEB act in concert to induce the expression of FOXO1 in the common lymphoid progenitor. Proc Natl Acad Sci USA 2011;108:17402–7.

[153] Dijkers PF, Birkenkamp KU, Lam EW, Thomas NS, Lammers JW, Koenderman L, et al. FKHR-L1 can act as a critical effector of cell death induced by cytokine withdrawal: protein kinase B-enhanced cell survival through maintenance of mitochondrial integrity. J Cell Biol 2002;156:531–42.

[154] Stahl M, Dijkers PF, Kops GJ, Lens SM, Coffer PJ, Burgering BM, et al. The forkhead transcription factor FoxO regulates transcription of p27Kip1 and Bim in response to IL-2. J Immunol 2002;168:5024–31.

[155] Liu P, Keller JR, Ortiz M, Tessarollo L, Rachel RA, Nakamura T, et al. Bcl11a is essential for normal lymphoid development. Nat Immunol 2003;4:525–32.

[156] Mansson R, Welinder E, Ahsberg J, Lin YC, Benner C, Glass CK, et al. Positive intergenic feedback circuitry, involving EBF1 and FOXO1, orchestrates B-cell fate. Proc Natl Acad Sci USA 2012;109:21028–33.

[157] Blake JA, Ziman MR. Pax genes: regulators of lineage specification and progenitor cell maintenance. Development 2014;141:737–51.

[158] Medvedovic J, Ebert A, Tagoh H, Busslinger M. Pax5: a master regulator of B cell development and leukemogenesis. Adv Immunol 2011;111:179–206.

[159] Garvie CW, Hagman J, Wolberger C. Structural studies of Ets-1/Pax-5 complex formation on DNA. Mol Cell 2001;8:1267–76.

[160] Urbanek P, Wang ZQ, Fetka I, Wagner EF, Busslinger M. Complete block of early B cell differentiation and altered patterning of the posterior midbrain in mice lacking Pax5/BSAP. Cell 1994;79:901–12.

[161] Thevenin C, Nutt SL, Busslinger M. Early function of Pax5 (BSAP) before the pre-B cell receptor stage of B lymphopoiesis. J Exp Med 1998;188:735–44.

[162] Nutt SL, Urbanek P, Rolink A, Busslinger M. Essential functions of Pax5 (BSAP) in pro-B cell development: difference between fetal and adult B lymphopoiesis and reduced V-to-DJ recombination at the IgH locus. Genes Dev 1997;11:476–91.

[163] Nutt SL, Morrison AM, Dorfler P, Rolink A, Busslinger M. Identification of BSAP (Pax-5) target genes in early B-cell development by loss- and gain-of-function experiments. EMBO J 1998;17:2319–33.

[164] Nutt SL, Heavey B, Rolink AG, Busslinger M. Commitment to the B-lymphoid lineage depends on the transcription factor Pax5. Nature 1999;401:556–62.

[165] Rolink AG, Nutt SL, Melchers F, Busslinger M. Long-term *in vivo* reconstitution of T-cell development by Pax5-deficient B-cell progenitors. Nature 1999;401:603–6.

[166] Schebesta A, McManus S, Salvagiotto G, Delogu A, Busslinger GA, Busslinger M. Transcription factor Pax5 activates the chromatin of key genes involved in B cell signaling, adhesion, migration, and immune function. Immunity 2007;27:49–63.

[167] Delogu A, Schebesta A, Sun Q, Aschenbrenner K, Perlot T, Busslinger M. Gene repression by Pax5 in B cells is essential for blood cell homeostasis and is reversed in plasma cells. Immunity 2006;24:269–81.

[168] Emelyanov AV, Kovac CR, Sepulveda MA, Birshtein BK. The interaction of Pax5 (BSAP) with Daxx can result in transcriptional activation in B cells. J Biol Chem 2002;277:11156–64.

[169] Eberhard D, Busslinger M. The partial homeodomain of the transcription factor Pax-5(BSAP) is an interaction motif for the retinoblastoma and TATA-binding proteins. Cancer Res 1999;59(Suppl. 7):1716s–24s.

[170] Barlev NA, Emelyanov AV, Castagnino P, Zegerman P, Bannister AJ, Sepulveda MA, et al. A novel human Ada2 homologue functions with Gcn5 or Brg1 to coactivate transcription. Mol Cell Biol 2003;23:6944–57.

[171] McManus S, Ebert A, Salvagiotto G, Medvedovic J, Sun Q, Tamir I, et al. The transcription factor Pax5 regulates its target genes by recruiting chromatin-modifying proteins in committed B cells. EMBO J 2011;30:2388–404.

[172] Linderson Y, Eberhard D, Malin S, Johansson A, Busslinger M, Pettersson S. Corecruitment of the Grg4 repressor by PU.1 is critical for Pax5-mediated repression of B-cell-specific genes. EMBO Rep 2004;5:291–6.

[173] Fitzsimmons D, Hodsdon W, Wheat W, Maira S-M, Wasylyk B, Hagman J. Pax-5 (BSAP) recruits Ets proto-oncogene family proteins to form functional ternary complexes on a B-cell-specific promoter. Genes Dev 1996;10:2198–211.

[174] Ramirez J, Dege C, Kutateladze TG, Hagman J. MBD2 and multiple domains of CHD4 are required for transcriptional repression by Mi-2/NuRD complexes. Mol Cell Biol 2013;32:5078–88.

[175] Maier H, Colbert J, Fitzsimmons D, Clark DR, Hagman J. Activation of the early B cell-specific mb-1 (Ig-a) gene by Pax-5 is dependent on an unmethylated Ets binding site. Mol Cell Biol 2003;23:1946–60.

[176] Fitzsimmons D, Lukin K, Lutz R, Garvie CW, Wolberger C, Hagman J. Highly cooperative recruitment of Ets-1 and release of autoinhibition by Pax5. J Mol Biol 2009;392:452–64.

[177] Kieffer-Kwon KR, Tang Z, Mathe E, Qian J, Sung MH, Li G, et al. Interactome maps of mouse gene regulatory domains reveal basic principles of transcriptional regulation. Cell 2013;155:1507–20.

[178] Walter K, Bonifer C, Tagoh H. Stem cell specific epigenetic priming and B cell specific transcriptional activation at the mouse Cd19 locus. Blood 2008;112:1673–82.

[179] Nutt SL, Taubenheim N, Hasbold J, Corcoran LM, Hodgkin PD. The genetic network controlling plasma cell differentiation. Semin Immunol 2011;23:341–9.

[180] Decker T, Pasca di Magliano M, McManus S, Sun Q, Bonifer C, Tagoh H, et al. Stepwise activation of enhancer and promoter regions of the B cell commitment gene Pax5 in early lymphopoiesis. Immunity 2009;30:508–20.

[181] Fuxa M, Skok J, Souabni A, Salvagiotto G, Roldan E, Busslinger M. Pax5 induces V-to-DJ rearrangements and locus contraction of the immunoglobulin heavy-chain gene. Genes Dev 2004;18:411–22.

[182] Kosak ST, Skok JA, Medina KL, Riblet R, Le Beau MM, Fisher AG, et al. Subnuclear compartmentalization of immunoglobulin loci during lymphocyte development. Science 2002;296:158–62.

[183] Espinoza CR, Feeney AJ. The extent of histone acetylation correlates with the differential rearrangement frequency of individual VH genes in pro-B cells. J Immunol 2005;175:6668–75.

[184] Su IH, Basavaraj A, Krutchinsky AN, Hobert O, Ullrich A, Chait BT, et al. Ezh2 controls B cell development through histone H3 methylation and Igh rearrangement. Nat Immunol 2003;4:124–31.

[185] Liu H, Schmidt-Supprian M, Shi Y, Hobeika E, Barteneva N, Jumaa H, et al. Yin Yang 1 is a critical regulator of B-cell development. Genes Dev 2007;21:1179–89.

[186] Medvedovic J, Ebert A, Tagoh H, Tamir IM, Schwickert TA, Novatchkova M, et al. Flexible long-range loops in the VH gene region of the Igh locus facilitate the generation of a diverse antibody repertoire. Immunity 2013;39:229–44.

[187] Pan X, Papasani M, Hao Y, Calamito M, Wei F, Quinn Iii WJ, et al. YY1 controls Igkappa repertoire and B-cell development, and localizes with condensin on the Igkappa locus. EMBO J 2013;32:1168–82.

[188] Jhunjhunwala S, van Zelm MC, Peak MM, Cutchin S, Riblet R, van Dongen JJ, et al. The 3D structure of the immunoglobulin heavy-chain locus: implications for long-range genomic interactions. Cell 2008;133:265–79.

[189] Roldan E, Fuxa M, Chong W, Martinez D, Novatchkova M, Busslinger M, et al. Locus "decontraction" and centromeric recruitment contribute to allelic exclusion of the immunoglobulin heavy-chain gene. Nat Immunol 2005;6:31–41.

[190] Verma-Gaur J, Torkamani A, Schaffer L, Head SR, Schork NJ, Feeney AJ. Noncoding transcription within the Igh distal $V_H$ region at PAIR elements affects the 3D structure of the Igh locus in pro-B cells. Proc Natl Acad Sci USA 2012;109:17004–9.

[191] Michaelson JS, Singh M, Snapper CM, Sha WC, Baltimore D, Birshtein BK. Regulation of 3' IgH enhancers by a common set of factors, including kB-binding proteins. J Immunol 1996;156:2828–39.

[192] Degner SC, Verma-Gaur J, Wong TP, Bossen C, Iverson GM, Torkamani A, et al. CCCTC-binding factor (CTCF) and cohesin influence the genomic architecture of the Igh locus and antisense transcription in pro-B cells. Proc Natl Acad Sci USA 2011;108:9566–71.

[193] Metcalf D. Lineage commitment and maturation in hematopoietic cells: the case for extrinsic regulation. Blood 1998;92:345–7. discussion 352.

[194] Nutt SL, Rolink AG, Busslinger M. The molecular basis of B-cell lineage commitment. Cold Spring Harb Symp Quant Biol 1999;64:51–9.

[195] Pongubala JM, Northrup DL, Lancki DW, Medina KL, Treiber T, Bertolino E, et al. Transcription factor EBF restricts alternative lineage options and promotes B cell fate commitment independently of Pax5. Nat Immunol 2008;9:203–15.

[196] Banerjee A, Northrup D, Boukarabila H, Jacobsen SE, Allman D. Transcriptional repression of Gata3 is essential for early B cell commitment. Immunity 2013;38:930–42.

[197] Nechanitzky R, Akbas D, Scherer S, Gyory I, Hoyler T, Ramamoorthy S, et al. Transcription factor EBF1 is essential for the maintenance of B cell identity and prevention of alternative fates in committed cells. Nat Immunol 2013;14:867–75.

[198] Dekker J, Marti-Renom MA, Mirny LA. Exploring the three-dimensional organization of genomes: interpreting chromatin interaction data. Nat Rev Genet 2013;14:390–403.

[199] Lefebvre C, Rajbhandari P, Alvarez MJ, Bandaru P, Lim WK, Sato M, et al. A human B-cell interactome identifies MYB and FOXM1 as master regulators of proliferation in germinal centers. Mol Syst Biol 2010;6:377.

[200] Mullighan CG, Goorha S, Radtke I, Miller CB, Coustan-Smith E, Dalton JD, et al. Genome-wide analysis of genetic alterations in acute lymphoblastic leukaemia. Nature 2007;446:758–64.

[201] Tijchon E, Havinga J, van Leeuwen FN, Scheijen B. B-lineage transcription factors and cooperating gene lesions required for leukemia development. Leukemia 2013;27:541–52.

[202] Holmfeldt L, Wei L, Diaz-Flores E, Walsh M, Zhang J, Ding L, et al. The genomic landscape of hypodiploid acute lymphoblastic leukemia. Nat Genet 2013;45:242–52.

[203] Seo W, Ikawa T, Kawamoto H, Taniuchi I. Runx-1Cbfβ facilitates early B lymphocyte development by regulating expression of Ebf1. J Exp Med 2012;209:1255–62.

Chapter 4

# Relationships among B Cell Populations Revealed by Global Gene Analysis

Yan Zhou, Byoung-Gon Moon, Susan A. Shinton, Kyoko Hayakawa and Richard R. Hardy
*Fox Chase Cancer Center, Philadelphia, PA, USA*

## 1. INTRODUCTION

We have focused our analysis on gene expression in B cells, specifically differences in expression that distinguish maturing and mature B cell populations. B cells form one arm of the immune system and are responsible for mediating humoral immunity [1]. They are generated through a tightly regulated sequence of developmental stages in which immunoglobulin heavy chain and light chain genes are rearranged and their products are selected, yielding a diverse antigen receptor repertoire that is largely purged of high-affinity pathogenic self-reactivity [2]. Over the past 20 years considerable progress has been made in mapping out the differentiation stages from committed B cell progenitors (Pro-B cells) through transitional populations in spleen to mature B cells. Yet some controversy persists regarding the nature of some transitional stages and the precise origins of certain functional B cell subsets. Thus, for example, transitional stage 3 (T3) cells were initially described as intermediate in maturation to follicular (Fo) B cells, but later work described this pool of cells as an anergic self-reactive subset [3]. Furthermore, whereas most functional B cells are considered to have been purged of autoreactive B cell antigen receptors (BCRs), certain functionally distinct subsets, including the marginal zone and CD5$^+$B cell ("B1a") subsets, show some degree of self-recognition [4,5].

As members of the ImmGen project, we generated a wide panel of gene expression datasets that follow the full range of B cell differentiation. These data, in relation to profiles from other lineages, allowed us to: (1) characterize the relationship of the T3 subset to other transitional cell fractions; (2) compare gene expression in three major mature functionally distinct B cell subsets; and (3) assess tissue specific specialization of Fo-like B cells.

## 2. GENE PROFILE CHANGES WITH B CELL MATURATION SUGGEST AN ORDERING OF TRANSITIONAL STAGES

In adult mice, B cells are produced in the bone marrow and then migrate to the spleen where they mature into functional cells. Before the mature stage, B cells express CD94 (AA4.1) and these maturing cells can be subdivided based on changes in cell surface proteins. Previously we proposed a subdivision of these cells into three transitional stages, using expression of CD94, CD23, and changes in the immunoglobulin (Ig)M/IgD ratio [6]. Flow cytometry gating for isolation of these T1, T2, T3, and Fo (follicular, mature stage) cells is shown in Figure 1(A). However, studies with autoreactive B cell antigen receptor (BCR) transgenic models have shown that self-reactive B cells may become arrested at a CD94$^+$ stage, even as such cells upregulate CD23 and downregulate IgM expression [7]. Based on such observations, one report proposed that T3 cells found in normal (nontransgenic) mice consisted of a pool of self-reactive functionally inactivated ("anergic") B cells [3].

To further test this idea, we purified B cells representing these four stages from the spleen and carried out a series of cell transfer experiments. In one set of experiments, we directly purified B cell fractions from B6 mouse spleen according to the gating strategy in Figure 1(A) and injected cells intravenously into Ly5 congenic B6 recipients. Mice were analyzed 6 days later, gating on the Ly5 allele of the donor mice. Results presented for CD19$^+$IgM$^+$ cells in Figure 1(B) showed a gain of CD23 by T1 cells and progressive loss of CD93/AA4.1 from T1 to T2 and T3. We observed no distinctive failure to engraft by T3 cells, unlike the previous report. Next, we used an irradiation/self-repopulation approach to generate larger numbers of transitional B cells in BALB/c spleen, sorted the transitional and follicular fractions as before, and transferred these into allotype congenic

**Molecular Biology of B Cells. http://dx.doi.org/10.1016/B978-0-12-397933-9.00004-7**

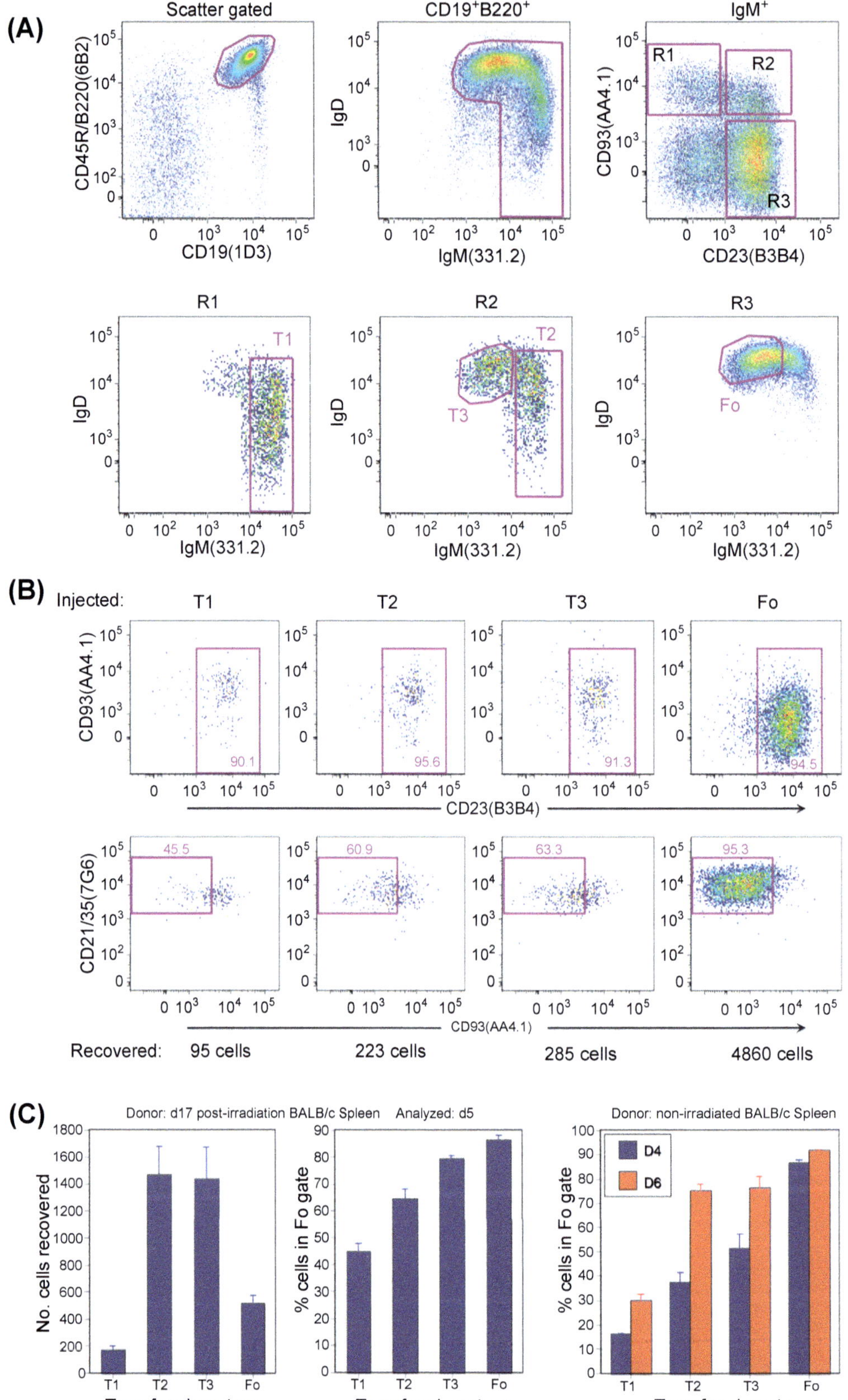


FIGURE 1 **Progression of transitional B cells in intact hosts to B cells with a mature follicular (Fo) phenotype.** (A) Flow cytometry resolution of transitional B cell stages in spleen. Gates shown were used for isolation of transitional cells in spleen, for preparation of RNA and for cell transfer experiments. (B) Data from B6 Ly5 congenic transfers. Spleen cells from 4 B6 mice were sorted as shown in Figure 1A and transferred into Ly5 congenic B6 recipients. Recipients were sacrificed 6 days later and spleen cells were analyzed for the presence of B6 B cells using Ly5 allele-specific reagents. CD19+IgM+ cells are shown in the panels. (C) Data from BALB/c B cell transfers. For left-most and center panels, spleen cells from 10 irradiated BALB/c mice were stained and sorted as in Figure 1A, then transferred to C.B-17 recipients by i.v. injection. Recipients were sacrificed 5 days later and spleen cells were analyzed for the presence of BALB/c B cells using IgH-a allotype IgM and IgD reagents. For right-most panel, spleen cells from BALB/c mice were sorted as shown in Figure 1A, then transferred into C.B-17 recipients by i.v. injection. Four or six days later, recipients were sacrificed and spleen cells were analyzed for engrafted BALB/c B cells.

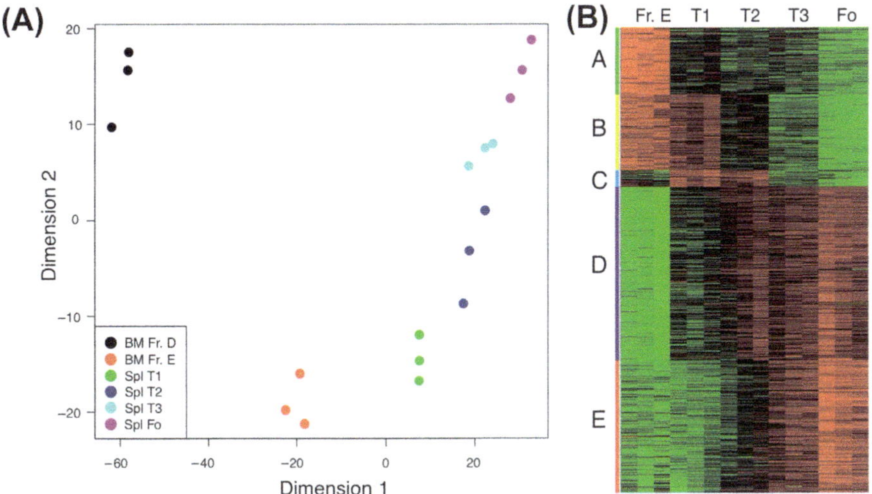


FIGURE 2 **Changes in gene expression during maturation from newly formed B cells in bone marrow to mature follicular B cells in speen.** (A) Multidimentional scaling analysis shows progressive gene changes from T1 to T2 to T3 to Fo. MDS using Euclidean distance, with genes filtered by coefficient of variation (cv) >0.2 and maximum expression >100 (4329 gene probes passed this filter). Bone marrow samples identified as CD19+CD43–AA4+, IgM– (Fr. D) or IgM+ (Fr. E) were also used in this analysis. Genes selected based on criteria provided in Methods. (B) Heatmap display of genes changing during B cell maturation. Three samples for each cell type show extent of variability. Genes were clustered into five sets using k-means and Pearson distance metric. A 1.5-fold cutoff was used to filter the gene set.

C.B-17 recipients. After 5 days, we analyzed the spleens of recipient mice for the presence of donor-allotype B cells. As shown in the left-most and center plots of Figure 1(C), we found comparable engraftment by T2 and T3 fractions and progressive increasing generation of Fo-type B cells from T1 to T2 to T3. Finally, we directly sorted these B cell fractions from un-irradiated BALB/c mice and transferred the cells by intravenous injection into C.B-17 recipients. As shown in the right-most panel of Figure 1(C), we found similar results analyzing recipient spleen cells at day 4 or 6 post-transfer, with an increasing proportion of B cells showing an Fo phenotype from T1 to T2 to T3. Again, we found no distinctive deficit in repopulation with T3 type B cells. In summary, different from the previous report, we found no distinctive deficit in engraftment of T3 stage cells or any block in their progression to cells with an Fo phenotype. Therefore, we searched for changes in global gene expression in these B cells.

First, to understand the similarity and relationships of different cell populations, we used multidimensional scaling (MDS) analysis [8] to capture the dissimilarity measured by the Euclidean distance among groups using genes that passed nonspecific filtering (coefficient of variation >0.2 and maximum normalized expression across samples >100), and then projected the distance/difference in a two-dimensional space. The resulting plot (Figure 2(A)) shows a linear progression from T1 to T2 to T3 to Fo (mature) B cells, consistent with identifying T3 cells as an intermediate stage in the progression from T2 to Fo cells. However, one could argue that this approach might not be sensitive enough to identify distinctive features in T3, so we also analyzed genes with significant expression whose level differed among five B cell fractions, newly formed B cells in bone marrow (Fr. E), and the splenic transitional (T1, T2 and T3) and mature Fo B cells. Genes with at least a 1.5-fold difference were included, identifying a set of 1101 members that were then subject to cluster analysis, to group those with similar patterns of expression. The k-means cluster analysis identifies five groups of genes with distinctive patterns of expression across the five B cell fractions analyzed (Figure 2(B)). Of note, we did not identify a distinctive gene cluster restricted to the T3 or T2/T3 stages.

We next used Ingenuity Systems IPA software to examine the five gene clusters shown in the heat map. Considering that newly formed B cells have been recently generated from cells that were proliferating, it is not surprising that the first two clusters (A and B), high in the earliest fractions, include genes in pathways related to regulation of the cell cycle. The small cluster C, abundant in T1 and T2, is harder to analyze by this approach, but it contains a Notch pathway gene, *Hes1*, possibly indicating the branch point for marginal zone B cells that require Notch signaling for their generation [9]. Cluster D contains genes increasing in expression through the Fo stage and genes from pathways important in mature B cell function, including interleukin-4 signaling, antigen presentation, CD40 signaling, phosphatidylinositol 3-kinase signaling, B cell receptor signaling, and B cell activating factor (BAFF) signaling [10]. Also, a number of genes involved in Toll-like receptor signaling are present in this cluster [11]. Cluster E, genes that show late expression in this sequence, also include genes in pathways crucial to mature B cell function, including BAFF, chemokine, and apoptosis signaling.

Because this analysis did not identify a cluster of genes in the T3 cell stage considered to reflect anergy, we took

**TABLE 1** Genes with Altered Expression in T3 B Cells Relative to T2 and Fo B Cells

| Gene | T1 | T2 | T3 | Fo | T3/Fo | T3/T2 | Fo/T3 | T2/T3 | Citation |
|---|---|---|---|---|---|---|---|---|---|
| *Myb* | 282 | 221 | **401** | 68 | 5.9 | 1.8 | | | |
| *Lef1* | 192 | 241 | **428** | 92 | 4.7 | 1.8 | | | [17] |
| *Sox4* | 198 | 147 | **300** | 74 | 4.1 | 2.0 | | | [18] |
| *Pcp4* | 66 | 104 | **190** | 73 | 2.6 | 1.8 | | | [12] |
| *ATP6* | 3386 | 1334 | **2291** | 907 | 2.5 | 1.7 | | | |
| *Rgl1* | 117 | 98 | **275** | 118 | 2.3 | 2.8 | | | |
| *Rgs2* | 305 | 161 | **220** | 96 | 2.3 | 1.4 | | | [20] |
| *Ephx1* | 663 | 726 | **1666** | 838 | 2.0 | 2.3 | | | |
| *Ndrg1* | 156 | 120 | **253** | 127 | 2.0 | 2.1 | | | |
| *Egr2* | 102 | 121 | **167** | 89 | 1.9 | 1.4 | | | [12,21] |
| *Egr3* | 271 | 358 | **539** | 305 | 1.8 | 1.5 | | | |
| *Nr4a1* | 182 | 367 | **520** | 307 | 1.7 | 1.4 | | | [15] |
| *Lck* | 109 | 113 | **286** | 170 | 1.7 | 2.5 | | | |
| *Dexi* | 400 | 371 | **532** | 323 | 1.7 | 1.4 | | | |
| *Lass6* | 111 | 86 | **116** | 72 | 1.6 | 1.4 | | | |
| *Spn* | 112 | 95 | **152** | 99 | 1.5 | 1.6 | | | |
| *Bcl2* | 101 | 160 | **222** | 151 | 1.5 | 1.4 | | | |
| *Faim3* | 3230 | 3630 | **6128** | 4212 | 1.5 | 1.7 | | | [23] |
| *Grap2* | 677 | 888 | **1263** | 875 | 1.4 | 1.4 | | | [24] |
| *Rnf157* | 119 | 177 | **259** | 183 | 1.4 | 1.5 | | | |
| *Nab2* | 240 | 288 | **491** | 353 | 1.4 | 1.7 | | | [12,21] |
| *Cd83* | 356 | 674 | **935** | 675 | 1.4 | 1.4 | | | [25] |
| *Fhl3* | 104 | 119 | **221** | 160 | 1.4 | 1.9 | | | |
| *Snora44* | 156 | 201 | **145** | 296 | 0.5 | 0.7 | 2.0 | 1.4 | |
| *Cacna1i* | 478 | 504 | **351** | 518 | 0.7 | 0.7 | 1.5 | 1.4 | |
| *Slfn5* | 400 | 372 | **243** | 328 | 0.7 | 0.7 | 1.4 | 1.5 | |

a more direct approach, examining genes that were altered in T3 compared with both adjacent stages, T2 and Fo. We searched for genes in the list of the 1101 members that showed a selective increase or decrease in T3 relative to these adjacent stages, and relaxed our selection criteria to require a 1.4× change, because more stringent analyses identified few genes and we are searching for evidence of a biological process that may not be regulated by sharp changes in mRNA levels. This less stringent filter generated a list of 26 genes, shown in Table 1. Significantly, a number of these genes have been described previously as indicating an anergic state [21], including two of the three genes mentioned in a report from the Cambier laboratory [3].

## 3. DISTINCTIONS IN GENE NETWORKS ACTIVATED IN MATURE B CELL POPULATIONS

We can identify three functionally distinct mature populations of B cells: the splenic follicular pool that is responsible for mediating T cell–dependent immune responses, the spleen marginal zone B cell population that makes T-independent responses to blood-borne pathogens, and the B1a subset, localized in the peritoneal cavity, responsible for production of natural autoantibody. Based on experiments with BCR transgenic mice, we proposed that these cells show significant differences in the level of self-reactivity, with B1a the highest, MZ B intermediate, and Fo B cells the lowest [22]. Thus, we would expect significant differences in global gene expression when comparing these three mature B cell populations. Figure 3(A) shows how these cells differ in surface phenotype and presents an approach to separating these cells from spleen and peritoneal cavity. Our analysis of gene expression is shown in the heat map in Figure 3(B), in which genes expressed at significant levels and differing among these subsets were clustered based on comparable expression, identifying six patterns. Thus, we found gene clusters with distinctive low- and high-level expression for each B cell subset.

As before, we applied MDS to investigate the similarities and differences among these three cell populations using the Euclidean distance measure. The MDS

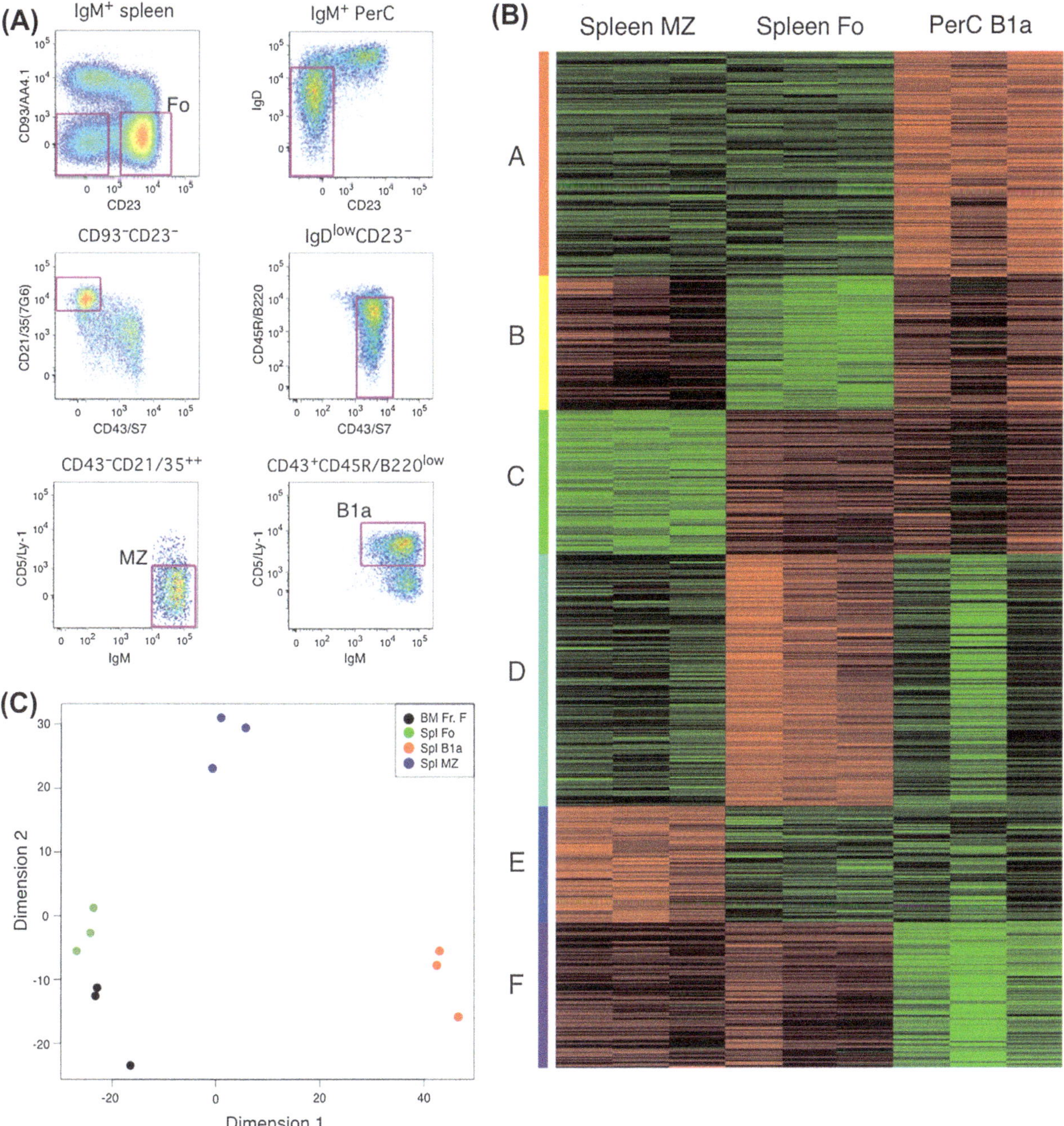


FIGURE 3 **Differences in genes expression among three mature B cell populations.** (A) Flow cytometry identification of follicular (Fo) and marginal zone (MZ) B cells in spleen, and CD5+ (B1a) B cells in peritoneal cavity (PerC). Samples were first gated as CD19+IgM+, then as shown in the figure for preparation of RNA. (B) Heatmap display of genes that differ among MZ, Fo, and B1a B cells. Analysis as in Figure 1C. Genes clustered into six sets, identifying those that are up- and down-regulated in each of the three types of B cells. (C) MDS analysis illustrating overall population distance between MZ, Fo, and B1a B cells using Euclidean metric. Analysis as Figure 1B; 3610 gene probe passed the filtering criteria.

**TABLE 2** B Cell Antigen Receptor Signaling Genes Distinguishing Mature B Cell Populations

| Cluster | Gene Symbol | Fo | MZ | B1a |
|---|---|---|---|---|
| A | Akt1 | 0.85 | 0.87 | 1.28 |
| A | Inppl1 | 0.65 | 0.81 | 1.54 |
| A | Nfatc1 | 0.78 | 0.95 | 1.26 |
| A | Pik3c2b | 0.87 | 0.83 | 1.29 |
| A | Pik3cd | 0.88 | 0.62 | 1.50 |
| A | Pik3r1 | 0.85 | 0.70 | 1.45 |
| A | Pik3r5 | 0.59 | 0.58 | 1.83 |
| A | Pou2f2 | 0.89 | 0.82 | 1.29 |
| A | Sos1 | 0.81 | 0.74 | 1.46 |
| A | Jun | 0.76 | 0.46 | 1.78 |
| F | Inpp5f | 1.01 | 1.52 | 0.47 |
| F | Pax5 | 1.18 | 1.09 | 0.73 |
| F | Pik3c2a | 1.23 | 1.07 | 0.71 |
| F | Ptpn6 | 1.21 | 1.08 | 0.71 |
| F | Rras2 | 1.38 | 1.24 | 0.38 |
| F | Ebf1 | 1.20 | 1.11 | 0.69 |
| | | | | |
| D | Akt3 | 1.74 | 0.69 | 0.57 |
| D | Bcl6 | 2.17 | 0.52 | 0.31 |
| D | Foxo1 | 1.40 | 0.94 | 0.66 |
| B | Fcrl5 | 0.15 | 1.77 | 1.08 |
| B | Cd300lf | 0.15 | 0.84 | 2.00 |
| B | Ptpn22 | 0.21 | 0.81 | 1.98 |
| | | | | |
| E | Bcl2a1a | 0.82 | 1.41 | 0.77 |
| E | Blnk | 0.92 | 1.33 | 0.75 |
| E | Fcgr2b | 0.86 | 1.32 | 0.81 |
| C | Ap1m1 | 1.20 | 0.67 | 1.13 |
| C | Mapk11 | 1.64 | 0.30 | 1.06 |
| C | Mapk12 | 1.03 | 0.49 | 1.49 |

Clusters A and F showed the strongest differential expression in B1a B cells; clusters D and B showed the strongest differential expression in Fo B cells, and clusters E and C in MZ B cells. Red indicates increased expression in the specified cluster; and green indicates decreased expression.

plot (Figure 3(C)) shows the distinct separation among these cell types, consistent with their functional distinctions. Next, we used IPA network analysis to investigate the major pathways in the gene clusters shown in the heat map. Using this approach, we found that some pathways were represented in most of the clusters: for example B cell receptor signaling, PI3K signaling in B lymphocytes, and apoptosis signaling. In contrast, some pathways were restricted to a particular subset, such as low CD40 signaling in B1a B cells, high Notch signaling in MZ B cells, and high B cell activating factor signaling in Fo B cells.

Finally, we carried out a focused analysis of differences in the BCR signaling network, because this is likely to underlie functional differences among these three B cell subsets. We computed a mean-normalized signal level among the three cell populations for genes in the IPA BCR signaling pathway (Table 2). This analysis revealed differences in the BCR signaling network among these three types of B cells. Thus, MZ B cells show higher levels of BCR-proximal signaling components, such as BLNK, whereas Fo B cells show a higher level of Bcl6, a transcription factor that functions to repress transcription, potentially mediating the quiescent state of this naive B cell pool and having a key role in their progression to germinal center B cells. Curiously, B1a B cells do not show distinctively high levels of BCR-proximal genes. Instead, they have high levels of Akt1, Pou2f2 (OCT2) and several members of the PI3-kinase family. Pik3cd is important in Akt1 signaling and Pik3r1 and Pik3r5 are both regulatory subunits that can associate with Pik3cd, which suggests the importance of these genes and this pathway in the survival and/or proliferation of B1a B cells.

## 4. Fo/B2 B CELLS IN DIFFERENT TISSUES ARE SIMILAR, BUT SPECIALIZATION WITH LOCATION EMERGES

Mature B cells that resemble follicular B cells in spleen can be identified in peritoneal cavity (where they are referred to as B2), in bone marrow (where they may represent cells recirculating in blood that contaminate bone marrow cell preparations), and in lymph nodes (Figure 4(A)). An analysis of the distance between follicular-like/B2 B cells in different tissues using MDS, comparing bone marrow, peritoneal cavity, and lymph nodes, showed the general similarity of global gene expression, compared with immature B cells in bone marrow, marginal zone B cells in spleen, and B1 B cells in spleen or peritoneal cavity (Figure 4(B)).

However, when we focus on genes that differ among the Fo-like/B2-type mature B cells from these different tissue sites, distinctions emerge, as revealed by cluster analysis of the samples shown in the heat map (Figure 4(C)). For example, the B2 cells in peritoneal cavity can be distinguished from the follicular-like B cells in all other sites, whereas Fo/B2 cells in bone marrow (Fr. F) are most similar to the Fo subset in spleen. The Fo/B2 cells in peripheral and mesenteric lymph nodes are similar and are distinct from all other Fo/B2-type B cells.

We applied gene ontology to examine the gene clusters generated in our analysis and found several interesting features (Table 3). The Fo/B2 cells in peritoneal cavity expressed at higher levels genes related to programmed cell death, cell adhesion, and the response to granulocyte macrophage–cerebrospinal fluid (GM-CSF). This latter enrichment likely reflects the novel environment of the peritoneal cavity, where lymphoid cells may be triggered to produce GM-CSF [23]. Adhesion signaling may indicate selective retention in this novel anatomical site. Elevation expression of genes related to apoptosis suggests that these cells, more than other Fo/B2 B cells, are particularly sensitive to extracellular signals inducing death rather than proliferation or Ig secretion. The cluster of

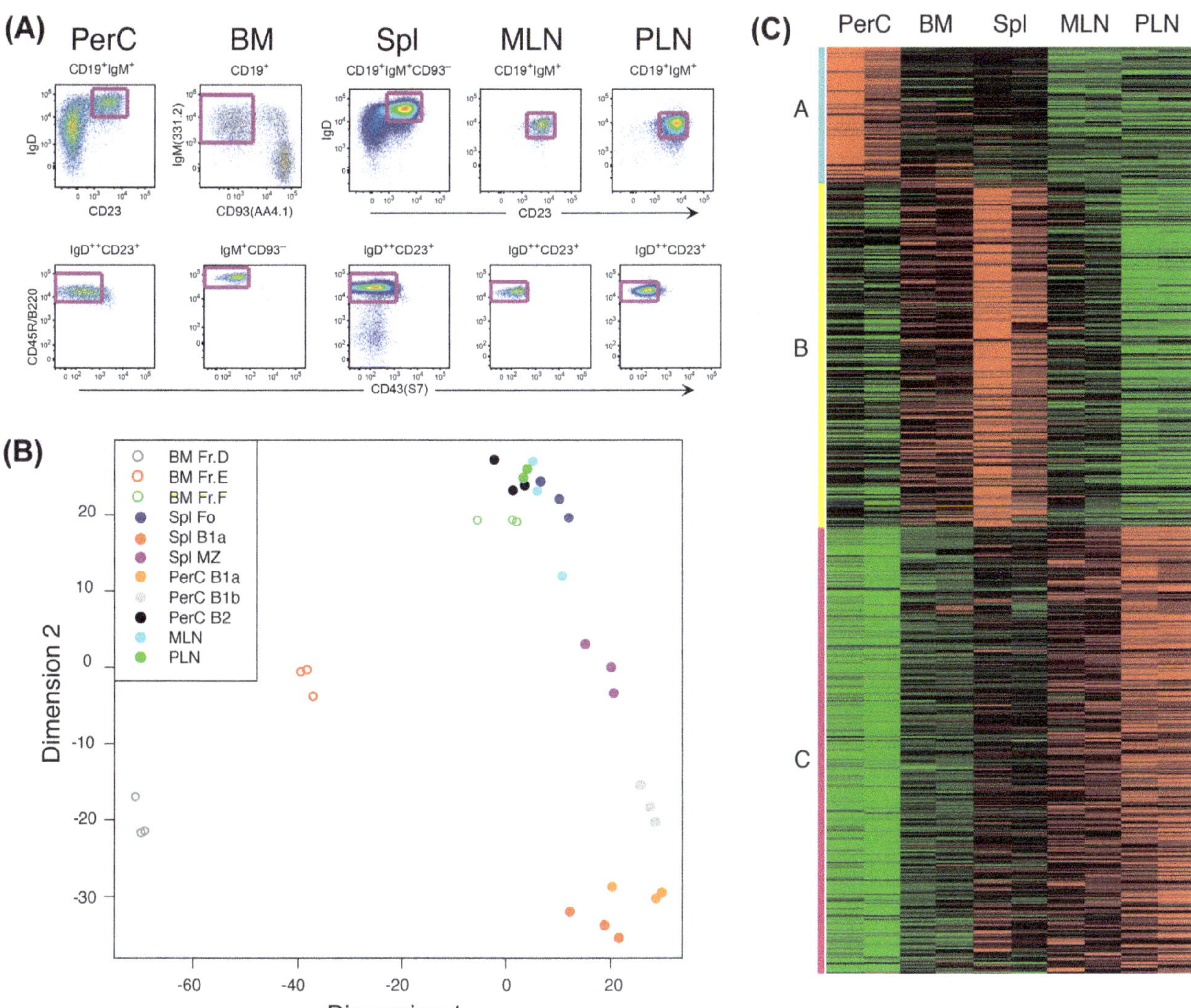


FIGURE 4 **Comparison of gene expression in follicular-like (Fo/B2) B cells from different tissues.** (A) Flow cytometry identification of Fo/B2 B cells in peritoneal cavity, bone marrow, spleen, mesenteric lymph node and peripheral lymph node. Cells gated as shown were purified for RNA preparation. All samples resemble follicular B cells in spleen. (B) MDS analysis illustrating overall population distances among Fo/B2 type B cells determined by the Euclidean metric. Pre-B cells (Fr. D) and newly-formed B cells (Fr. E) in bone marrow, MZ B cells in spleen, and B1 B cells in spleen and peritoneal cavity are also included to illustrate distances found with different types of B cells. Analysis as in Figure 1B; 7639 gene probes passed the filtering criteria. (C) Heatmap display of genes that differ among Fo/B2 B cells present in different tissues. Analysis as in Figure 1C. Optimal cluster separation was found using three gene sets.

genes that is most expressed in spleen Fo and BM (Fr. F) B cells is related to cell growth and may reflect the recent generation of these cells from transitional and precursor stages where cells were not quiescent. The cluster of genes that is most expressed in lymph nodes identifies B cells with a mature functional capacity, including the ability to present antigen and respond to antigen receptor signaling. Interestingly, we also identified a set of genes detected at the highest expression levels that are classified as mediating innate immunity. This suggests that both adaptive and innate responses can derive from the cells we identified as Fo/B2-like, where we specifically excluded cells resembling MZ or B1 B cells.

## 5. CONCLUSIONS

Analyses of global gene expression in the ImmGen data sets reveals interesting aspects of B cell function and development. First, looking at global gene expression in cell fractions representing putative maturation stages, we find support for this model, because MDS analysis shows a progressive change in gene expression from transitional stages (T1 to T2 to T3) to mature follicular (Fo) B cells in spleen. Furthermore, our cluster analysis does not identify any stage as a uniquely distinctive dead end or branch population. However, because a prior publication suggested that the T3 stage represented a pool of anergic B cells [3], we focused

**TABLE 3** Gene Differences in Fo/B2-Type B Cells with Tissue Location

| Cluster | Biological Cell Function | Genes |
|---|---|---|
| A | Programmed cell death | *Nod1, Rhob, Casp4, cd44, Hipk1, Anxa1, Sik1, Tnfrsf8, Dyrk2, Jak2, Amigo2, Lmna, Pim1, Rara* |
| A | Cell adhesion | *Rhob, Cd44, Nid1, Jak2, Amigo2, Cd2, Cd9, Emb, Itgam, Pde3b, Sele, Cd97* |
| A | GM-CSF response | *Jak2, Pde2a* |
| B | Cell growth | *B430203M17Rik, Ccna2, Wapal, Tipin, Klhl9, Cetn3, Kif2a, Pafah1b1, Cep63, Rrs1, Lrrcc1, Cdkn1b, Rny1, Top2a, Itgb1, Taok1, Cspp1* |
| C | Antigen presentation | *H2-dma, H2-Oa, H2-Eb2, Iifi30, Psme1* |
| C | Lymphocyte activation | *Fgr, Tlr7, Gfr, Tnf, H2-Dma, Lck, Lyl1, Igbp1, Psmb10, Gadd46g, Chd7, H2-Oa, Cr2, Slamf1, Tnfrsf13c, Tnfrsf13b, Cd83, Samsn1* |
| C | Innate immunity | *Ciita, Crcp, Fgr, Ptpn6, Tlr7, Mif\|Gm16379, Mx1, Zbp1, Ifitm3, Irgm1, Clec2d, Irgm2* |
| C | Antiviral activity | *Tlr7, Irf9, Irf7, Tnf, Crcp, Mx1, Zbp1, Ifitm3, Rps15a, Ifit2, Ifit27l2a* |

Cluster A—Fo/B2 type B cells in peritoneal cavity; Cluster B—Fo/B2 type B cells in spleen and bone marrow; Cluster C—Fo/B2 type B cells in mesenteric and peripheral lymph nodes.

our attention on identifying any gene with a discontinuous expression in this progression, searching for genes whose expression was either higher or lower than both T2 and Fo stages. This analysis yielded a set of 26 genes, 10 of which had previously been described as having some relevance to the anergic state. However, the difference in gene expression compared with T2 or Fo in most cases is not large, particularly for Egr2 and Nab2, mentioned by Goodnow and Cambier, both less than twofold upregulated in T3. Considering a recent report [24] that monitored one of these 10, Nur77(Nr4a1), at the individual cell level using a BAC-GFP reporter, it seems reasonable to speculate that only a fraction of cells in the T2, T3, and Fo stages may have self-reactivity sufficient to render them unresponsive (anergic). However, because most Fo B cells are clearly capable of mediating T cell–dependent responses, even while expressing similar or higher levels of the Nur77 reporter than most T3 stage cells, it seems likely that most B cells in the T3 stage constitute an intermediate transitional stage cells that will progress to fully functional cells, given that space is available (in the competition for BAFF). This conclusion is also supported by cell transfer experiments that showed no deficit in generation of cells with an Fo phenotype when T3 cells were used as a donor pool.

A comparison of the three mature functionally distinct B cell subpopulations, B1a, MZ, and Fo, highlighted both similarities and differences. Analyzing clusters with genes distinctively higher or lower in each subset showed the expected presence of Notch signaling in MZ B cells, because these cells require Notch-2 signaling for generation [9]. We also found enrichment for BAFF signaling genes in the Fo subset, as expected because this subset of B cells absolutely depends on this cytokine for maintenance [10]. Another finding is the downregulation of CD40 signaling in B1a B cells, which reflects their relative independence from T cell cognate help. A comparison focused on the BCR signaling network in these three mature B cell types showed a different emphasis in the network for each, possibly reflecting the balance between BCR activation and its outcome, whether positive or negative, that varies among these cell types.

We also asked if B cells with a similar cell surface phenotype and generally considered to function similarly, Fo/B2 B cells, exhibited distinctive gene signatures. The idea of anatomically specialized B2 cells has a precedent because a report showed that B2 cells in peritoneal cavity could more efficiently reenter this site compared with Fo/B2 cells from the spleen [25]. This analysis revealed two clear-cut distinctions: (1) Fo/B2 cells in the peritoneal cavity expressed a set of genes different from Fo/B2 B cells in other sites analyzed; and (2) lymph node Fo/B2 B cells could be distinguished from spleen Fo/B2 B cells. Fo/B2 B cells in the peritoneal cavity had a higher expression of genes related to cell death, cell adhesion, and the GM-CSF response. Elevated apoptotic pathway genes may indicate a greater sensitivity to activation-induced cell death, rather than a productive response, upon antigen encounter. The increase in genes related to cell adhesion likely indicates the specialization of these B cells to the peritoneal cavity, and the GM-CSF response may indicate cell conditioning by the peritoneal microenvironment. In contrast to the peritoneal cavity Fo/B2 B cells, Fo B cells in the lymph nodes were enriched for genes related to immune responses in general. For example, pathways such as antigen presentation, lymphocyte activation, innate response, and antiviral activity were enriched in the cluster with elevated expression. Thus, it appears that the broadly defined Fo population in the spleen is not fully mature compared with cells recirculating through the lymphatics.

In summary, we addressed three issues related to B cells and B cell development using the ImmGen Consortium Project datasets. We examined transitional B cells undergoing maturation to a functional population, which suggested that although certain genes related to anergy are increased in the T3 subset, this signal likely arises from only a fraction of the cells, with most representing intermediates between the T2 and Fo stages. We identified sets of genes distinctively up- and downregulated in spleen Fo and MZ B cells and in the peritoneal cavity B1a B cells. This analysis highlighted similarities among these subsets, but also revealed differences such as the Notch and BAFF dependence of MZ and Fo B cells. Our analysis also revealed differences between Fo/B2 B cells in peritoneal cavity and lymph nodes compared with cells with that phenotype in spleen. The B2 cells in peritoneal cavity are most distinct, likely owing to the novel microenvironment, but we also found that lymph node B cells appear more functional compared with the follicular pool in spleen. Such studies and results reveal the power and richness of information present in the ImmGen Consortium database.

## ACKNOWLEDGMENT

This work used key facilities, DNA sequencing, flow cytometry, laboratory animals, and biostatistics and bioinformatics supported by Fox Chase Cancer Center Core Grant P30CA006927. RRH was supported by a Grant R01AI026782 from the NIH and by the PA Department of Health Tobacco Settlement funds. KH was supported by Grant R01AI49335 from the NIH. Array expression analysis was generated by the Immunological Genome Project Consortium.

## REFERENCES

[1] Murphy K. The humoral immune response. In: Janeway's immunobiology. 8th ed. New York: Garland Science; 2011.

[2] Hardy RR, Hayakawa K. B cell development pathways. Annu Rev Immunol 2001;19:595–621.

[3] Merrell KT, Benschop RJ, Gauld SB, Aviszus K, Decote-Ricardo D, Wysocki LJ, et al. Identification of anergic B cells within a wild-type repertoire. Immunity 2006;25:953–62.

[4] Martin F, Oliver AM, Kearney JF. Marginal zone and B1 B cells unite in the early response against T-independent blood-borne particulate antigens. Immunity 2001;14:617–29.

[5] Hardy RR. B-1 B cell development. J Immunol 2006;177:2749–54.

[6] Allman D, Lindsley RC, DeMuth W, Rudd K, Shinton SA, Hardy RR. Resolution of three nonproliferative immature splenic B cell subsets reveals multiple selection points during peripheral B cell maturation. J Immunol 2001;167:6834–40.

[7] Hayakawa K, Asano M, Shinton SA, Gui M, Wen LJ, Dashoff J, et al. Positive selection of anti-thy-1 autoreactive B-1 cells and natural serum autoantibody production independent from bone marrow B cell development. J Exp Med 2003;197:87–99.

[8] Cox TF, Cox MAA. Multidimensional scaling. 2nd ed. London: Chapman and Hall; 2001.

[9] Tanigaki K, Han H, Yamamoto N, Tashiro K, Ikegawa M, Kuroda K, et al. Notch-RBP-J signaling is involved in cell fate determination of marginal zone B cells. Nat Immunol 2002;3:443–50.

[10] Schiemann B, Gommerman JL, Vora K, Cachero TG, Shulga-Morskaya S, Dobles M, et al. An essential role for BAFF in the normal development of B cells through a BCMA-independent pathway. Science 2001;293:2111–4.

[11] Leadbetter EA, Rifkin IR, Hohlbaum AM, Beaudette BC, Shlomchik MJ, Marshak-Rothstein A. Chromatin-IgG complexes activate B cells by dual engagement of IgM and toll-like receptors. Nature 2002;416:603–7.

[12] Glynne R, Ghandour G, Rayner J, Mack DH, Goodnow CC. B-lymphocyte quiescence, tolerance and activation as viewed by global gene expression profiling on microarrays. Immunol Rev 2000;176:216–46.

[13] Wen L, Brill-Dashoff J, Shinton SA, Asano M, Hardy RR, Hayakawa K. Evidence of marginal-zone B cell-positive selection in spleen. Immunity 2005;23:297–308.

[14] Rauch PJ, Chudnovskiy A, Robbins CS, Weber GF, Etzrodt M, Hilgendorf I, et al. Innate response activator B cells protect against microbial sepsis. Science 2012;335:597–601.

[15] Zikherman J, Parameswaran R, Weiss A. Endogenous antigen tunes the responsiveness of naive B cells but not T cells. Nature 2012;489:160–4.

[16] Berberich S, Forster R, Pabst O. The peritoneal micromilieu commits B cells to home to body cavities and the small intestine. Blood 2007;109:4627–34.

[17] Suryani S, Fulcher DA, Santner-Nanan B, Nanan R, Wong M, Shaw PJ, et al. Differential expression of CD21 identifies developmentally and functionally distinct subsets of human transitional B cells. Blood 2010;115:519–29.

[18] Isnardi I, Ng YS, Menard L, Meyers G, Saadoun D, Srdanovic I, et al. Complement receptor 2/CD21- human naive B cells contain mostly autoreactive unresponsive clones. Blood 2010;115:5026–36.

[19] Glynne R, Ghandour G, Rayner J, Mack DH, Goodnow CC. B-lymphocyte quiescence, tolerance and activation as viewed by global gene expression profiling on microarrays. Immunol Rev 2000;176:216–46.

[20] Reif K, Cyster JG. RGS molecule expression in murine B lymphocytes and ability to down-regulate chemotaxis to lymphoid chemokines. J Immunol 2000;164:4720–9.

[21] Merrell KT, Benschop RJ, Gauld SB, Aviszus K, Decote-Ricardo D, Wysocki LJ, et al. Identification of anergic B cells within a wild-type repertoire. Immunity 2006;25:953–62.

[22] Zikherman J, Parameswaran R, Weiss A. Endogenous antigen tunes the responsiveness of naive B cells but not T cells. Nature 2012;489:160–4.

[23] Shima H, Takatsu H, Fukuda S, Ohmae M, Hase K, Kubagawa H, et al. Identification of TOSO/FAIM3 as an Fc receptor for IgM. Int Immunol 2010;22:149–56.

[24] Garrett-Sinha LA, Hou P, Wang D, Grabiner B, Araujo E, Rao S, et al. Spi-1 and Spi-B control the expression of the Grap2 gene in B cells. Gene 2005;353:134–46.

[25] Kretschmer B, Luthje K, Schneider S, Fleischer B, Breloer M. Engagement of CD83 on B cells modulates B cell function in vivo. J Immunol 2009;182:2827–34.

Chapter 5

# Roles of MicroRNAs in B Lymphocyte Physiology and Oncogenesis

Stefan A. Muljo[1] and Klaus Rajewsky[2]

[1]*Laboratory of Immunology, National Institute of Allergy and Infectious Diseases, National Institutes of Health, Bethesda, Maryland, USA;*

[2]*Max Delbrück Center for Molecular Medicine, Berlin, Germany*

## 1. INTRODUCTION

The first microRNAs (miRNAs) lin-4 and let-7 were discovered upon careful characterization of heritable genetic mutations in *Caenorhabditis elegans* that affect the timing of larval development [1,2]. Autosomal dominant mutations in the 3′ untranslated region (UTR) of the *lin-14* gene phenocopied *lin-4* loss-of-function mutants [3]. As will be discussed later, in retrospect this is surprising because genetics suggests that the lin-4 miRNA may only have one functional target, whereas biochemical, genomic, and computational analyses suggest that miRNA targeting is more extensive by several orders of magnitude. Nevertheless, this work proposed the prescient model that the lin-4 miRNA binds to complementary sequences in the 3′ UTR of *lin-14* mRNA to repress expression of the latter post-transcriptionally [1,3]. In principle, similar mutations or (epi)genetic variations in miRNA genes and their targets may occur in humans to mediate quantitative traits or even disease such as cancer, immunodeficiency, or autoimmunity.

Mature miRNAs are tiny regulatory RNAs, ~21 nucleotides long, generated by an evolutionarily conserved pathway. In the human genome, there are ~1000 miRNA genes [4]. Many miRNAs are evolutionarily conserved to different degrees, but some are species specific [5]. Most miRNA genes are transcribed by RNA polymerase II into long primary miRNA (pri-miRNA) transcripts that are processed in the nucleus by Drosha and its partner Dgcr8 that together recognize and cleave the base of stem-loop RNA to release a precursor miRNA (pre-miRNA). The short hairpin pre-miRNA, ~65–70 nucleotides long, is exported to the cytoplasm by exportin-5 [6,7] and further processed by Dicer to generate a double-stranded RNA ~21 nucleotides long [8]. Generally, one of the two mature strands is preferentially loaded into the RNA-binding protein Argonaute (Ago).

These miRNAs act as sequence-specific guides for Ago to bind cognate mRNAs and mediate post-transcriptional gene silencing. In mammals, Ago proteins are encoded by four paralogous genes (commonly referred to as Ago1-4, but officially named *Eif2c1-4*), and they interact with miRNAs to form a multi-component ribonucleoprotein complex called the miRNA-induced silencing complex (miRISC) that is programmed to silence targets with sequence complementarity to the 5′ end of the mature miRNA, the so-called "seed" region (nucleotides 2–8). The crystal structure of human Ago2 complexed with a miRNA provides a molecular basis for target recognition by the seed sequence [9]. The most recent systematic analyses of the evolutionary conservation of putative miRNA-binding sites in the human 3′ UTRome predict that miRNAs may collectively target ~60% of the protein-coding transcriptome, identifying >45,000 potential sites [10].

Elucidating the molecular mechanisms of miRNA-mediated silencing is an area of intense research. Of the mammalian Ago proteins, only Ago2 harbors an enzymatically active RNaseH-like PIWI domain called "slicer," which is responsible for mediating RNA interference (RNAi) [11] and Dicer independent biogenesis of mature miR-451 [12,13]. Although Ago2 is required for early B cell development, this particular function does not seem to require its catalytic activity [14]. In mammals, Ago recruits Tnrc6a, -b, and -c, also known as the GW182 (glycine–tryptophan (GW) repeat-containing protein of 182 kDa) family of proteins. GW182 proteins are essential for miRNA-mediated silencing because they recruit additional proteins including the CCR4-NOT deadenylase complex. Indeed, mRNA decay initiated by deadenylation of poly-A tails is one mechanism by which miRNAs can silence cognate targets [15]. In addition, Ago2 and miRNAs have been found to cosediment with polysomes [16,17]. This may reflect the

**Molecular Biology of B Cells. http://dx.doi.org/10.1016/B978-0-12-397933-9.00005-9**

ability of miRISC to inhibit translation by mechanisms that are still debated and not fully understood. Notably, one recent study suggested that in mammalian somatic cells, miRNAs primarily act by reducing mRNA stability rather than translation [18].

A prominent feature of miRNA control is the dose-dependent modulation of target protein expression over a generally narrow range. This suggests that miRNAs naturally possess drug-like properties, and in turn miRNA targets may serve as potentially interesting drug targets. In B cell physiology, this was exemplified in an early study in which it was shown that miR-150, a miRNA selectively and highly expressed in mature lymphocytes, controls certain aspects of B cell differentiation by regulating the expression level of a key transcription factor, c-Myb, over a roughly twofold range [19]. In this system, the effect of miR-150 overexpression could be largely reproduced by c-Myb hemizygosity. The ensemble of small changes of gene expression, the targeting of hundreds of potential functional players, and the paradoxical functional prominence of just a few of them are what make the study of miRNA control so difficult. It seems likely that miRNAs are tailored to control pathways in which the precise concentrations of specific proteins matter.

## 2. CONTROL OF CELL SURVIVAL AND PROLIFERATION BY MIR-17~92 IN B CELL DEVELOPMENT AND LYMPHOMAGENESIS

A number of miRNAs are expressed in B lineage cells, a few of which display intriguing patterns of developmental stage-specific expression [20–22]. To determine whether miRNAs are required for B cell development, Dicer was ablated genetically in progenitor B (pro-B) cells [23]. This, along with the work of Ventura et al., led to the finding that the polycistronic miR-17~92 cluster is required for the maturation of pro-B cells into precursor B (pre-B) cells by inhibiting apoptosis [23,24]. Thus, a reasonable target was the pro-apoptotic molecule Bim (*Bcl2l11*), which harbors several conserved binding sites for miR-17~92 in its 3′ UTR, and whose expression is elevated in the absence of miR-17~92. Furthermore, deletion of Bim could partially rescue development of Dicer-deficient mature B cells. Ultimately, however, in vivo validations of miRNA-target relationships await mutation of the binding sites in the endogenous 3′ UTR. Thus, Bim upregulation in the absence of miR-17~92 could equally well reflect direct miRNA-seed match interaction as well as a more complex miRNA-inhibited cell death program with accompanying Bim upregulation. Consistent with the latter interpretation, B cell–specific mutation of six putative miR-17~92 binding sites in the 3′ UTR of *Bcl2l11* has no functional consequence in a genetically engineered mouse model, where the same mutations produced a major phenotype in a different cellular context (V. Labi, S. Peng and KR, unpublished data). Thus, although other interpretations are possible, Bim may not be a direct functional target of miR-17~92 in developing B cells.

Before the discovery of miRNAs, the locus encoding the miR-17~92 cluster was designated Chromosome 13 open reading frame 25 (C13orf25), a misnomer because this transcript is most likely untranslated and instead serves as a polycistronic pri-miRNA. The miR-17~92 cluster encodes six miRNAs belonging to four families: miR-17, -18a, -19a, -20a, -19b-1, and -92a-1 (Figure 1). Although the miR-17~92 cluster is essential for B cell development, the paralogous miR-106b~25 and miR-106a~363 clusters seem dispensable for B cell development [24]. Based on seed sequence identity, miR-17 and -20 belong to one family, along with the paralogous miR-93 and -106. A second, closely related family is miR-18. A third family of distinct sequence is miR-19; and finally a fourth family of distinct sequence is composed of miR-92 and the paralogous miR-25 and -363. Within the miR-17~92 cluster, miR-92 is the most ancient evolutionarily being conserved in *Drosophila* [25]. In principle, each of these miRNA families can repress many target mRNAs either on their own or cooperatively. Evolutionarily, it would make sense for individual miRNAs within the miR-17~92 cluster to coevolve to act cooperatively; however, there is currently no evidence that such a general principle applies in this case. Furthermore, it is not known whether a subset or all of the miRNAs within the miR-17~92 cluster are required for B cell development. Interestingly, in the case of mouse pre-B cell transformation by the Eμ-myc transgene (discussed below), only miR-19 appears to be required. Surprisingly, in this tumor model, miR-92 antagonizes miR-19 [26].

Consistent with the idea that the tumor suppressors Bim and Pten are targets of this miRNA cluster, transgenic expression of the miR-17~92 cluster in lymphocytes driven by CD2-cre resulted in lymphoproliferation and reduced apoptosis, with the most dramatic cellular expansion in the $CD4^+$ T cell compartment [27]. In these miR-17~92 transgenic mice, there are more B-1a and fewer marginal zone B cells, but the molecular basis for this lineage skewing was not studied [27]. In addition, there are more germinal center B cells, which could result from increased follicular helper T (Tfh) cells in these mice [28,29]. Thus, some of the in vivo B cell phenotypes in this mouse model may not be B cell autonomous. However, ex vivo transgenic B cells proliferated more upon activation with anti-immunoglobulin (Ig)M or lipopolysaccharide, and were more resistant to Fas-mediated cell death [27]. In any case, the augmented germinal center response resulted in higher serum Ig titers and formation of autoantibodies to DNA that was comparable to the MRL-lpr/lpr autoimmune mouse model [27].

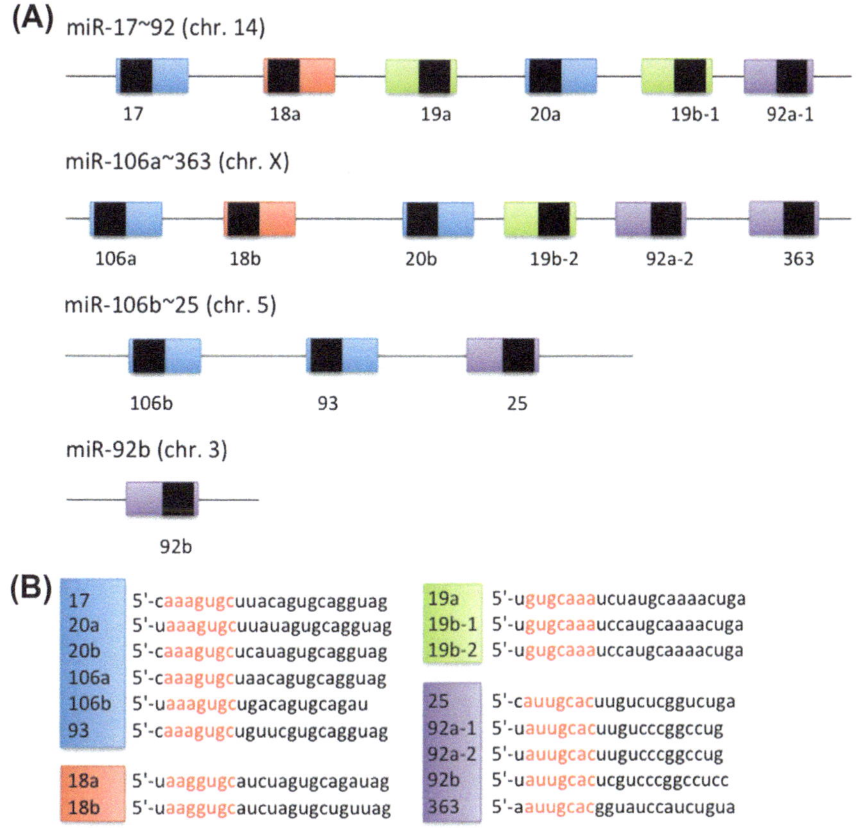


FIGURE 1 (A) Schematic of chromosomal organization of miR-17~92 family members and paralogs in the mouse genome. Pre-miRNAs and mature miRNAs are denoted by colored and black boxes, respectively. Mature miRNAs with the same seed sequence are depicted in the same color. Information was based on the UCSC Genome Browser (http://genome.ucsc.edu) and miRBase (http://www.mirbase.org). (B) Mature miRNA sequences of miR-17~92 family members are aligned and grouped together when they share the same seed sequence (red font). Families have been grouped and color-coded as in (A). Information was obtained from miRBase (http://www.mirbase.org).

The miR-17~92 cluster exhibits genomic amplification (13q31-q32) and high expression in some cases of diffuse large B cell lymphomas (DLBCL) in humans [30], specifically in the germinal center B cell–like (GCB) but not the activated B cell–like (ABC) subtype [31]. Transgenic expression of the miR-17~92 cluster in B cells driven by CD19-cre is sufficient to mediate lymphomagenesis in mice, reminiscent of human DLBCL in 12 of 24 cases [32]. Interestingly, one key transcriptional target of the c-Myc oncogene may be the miR-17~92 cluster, with both acting together to control E2F1 expression and thus cell proliferation [33]. In Burkitt lymphoma, a condition in which Myc is highly expressed owing to its translocation to the IgH enhancer, the miR-17~92 cluster is also highly expressed [34]. A similar correlation was apparent in a mouse model of Myc-induced B cell lymphomas, in which conditional deletion of the miR-17~92 cluster using the CD19-cre knock-in transgene significantly delayed lymphomagenesis [32]. In an attempt to exclude that this was the result of *Cd19* haploinsufficiency and/or Cre toxicity through induction of DNA breaks [35], the authors demonstrated in an unpublished control experiment that CD19-cre alone does not have a significant effect on development of lymphomas or survival of these Myc-transgenic mice, albeit on a different genetic background (M. Lai and C. Xiao, personal communication).

In one Myc-transformed B cell lymphoma line, deletion of the miR-17~92 cluster again seemed to mainly increase apoptosis and did not affect cellular proliferation [36]. Pioneering work has demonstrated that retroviral expression of miR-17~19b-1 (not including miR-92a-1) significantly accelerated pre B cell lymphomagenesis by the Eμ-myc transgene [37]; thus, miR-92 is not required for oncogenesis in this mouse model. In fact, it was later revealed that miR-92 enhances Myc-induced proliferation and apoptosis, which results overall in less efficient tumorigenesis [26]. Follow-up studies found that within the miR-17~92 cluster only miR-19 is required and sufficient for oncogenesis in this model, with PTEN being a critical target of miR-19 [36,38]. Of 46 putative targets that were functionally validated, RNAi-mediated knock-down of PTEN best mimicked enforced expression of miR-19a and -19b [36]. Furthermore, heterozygous germline deletion of Bim or Pten phenocopied miR-17~19b-1 overexpression to different degrees in the Eμ-Myc transgenic mouse model [39,40], which reaffirms the notion that inhibition of apoptosis is critical in this cancer model.

## 3. THE PROBLEM OF MIRNA TARGET IDENTIFICATION AND VALIDATION

Direct identification of miRNA targets has been a challenge for the field. One problem is the incomplete annotation of 3′ UTRs in different cell types. Alternative 3′ UTR use could have an impact on miRNA targeting [41,42]. Thus, comprehensive identification of polyadenylated RNA termini in transcriptomes would be a valuable resource for identifying miRNA binding sites [42,43]. Recently, high-throughput sequencing of RNA isolated by crosslinking immunoprecipitation (HITS-CLIP), and a variant method, photoactivatable-ribonucleoside-enhanced crosslinking and immunoprecipitation (PAR-CLIP) of Ago complexes allowed transcriptome-wide identification of miRNA–mRNA interactions experimentally [44,45]. Further refinements to this technology have been proposed to improve identification of miRNA–mRNA interactions [46,47]. However, these methods require large numbers of cells, and in the case of PAR-CLIP, the cells of interest need to be fed ribonucleotide analogs, which potentially limits the ability to identify mammalian miRNA targets in vivo. These studies found widespread evidence of noncanonical miRISC binding sites lacking perfect seed matches [46,48]; whether these are functionally relevant in vivo remains to be demonstrated.

Photoactivatable-ribonucleoside-enhanced crosslinking and immunoprecipitation of Ago2 was recently performed in human lymphoblastoid B cell lines transformed with Epstein Barr virus [49]. Previously reported targets of miR-17~92 including Bim, E2F3, Phlpp2, and Pten were validated by this PAR-CLIP analysis, whereas others (e.g., E2F1, E2F2, p21, and Rbl2) were not. In addition, these experiments revealed that multiple inhibitors of the NFκB pathway (Cyld, A20, Itch, Rnf11, and Tax1bp1) are directly targeted by miR-17~92, thus contributing to the activation of the canonical NFκB pathway in Myc-driven lymphomas [32]. Using CLIP methods, thousands of miRNA-binding sites can be identified through deep sequencing, but these may not all be functional and will need to be experimentally validated in vivo by site-directed mutagenesis. Frequently, artificial reporter constructs and heterologous transformed cell lines are used in the context of overexpression to validate miRNA–mRNA target interaction. As a case in point, Bim is implicated as a direct target of miR-17~92 by PAR-CLIP and in vitro reporter assays, but as previously discussed, mutation of six of the binding sites in the endogenous 3′ UTR suggests otherwise.

A computational search for miR-17~92 targets in this dataset identified 3744 unique binding sites (considering only 7mer-A1 and 7mer-m8 seed matches) in this transcriptome, 30% of which are conserved between mouse and human [32]. Of 3744, 1699 mapped to 3′ UTRs of protein-coding mRNAs, and nearly half of those sites (772) are conserved in mice but only 397 overlap with miRNA target predictions from TargetScan (http://www.targetscan.org/). Thus, PAR-CLIP may identify miRNA targets that would otherwise be missed. About 80% of these identified miR-17~92 binding sites were located in protein-coding genes, whereas the rest mapped to intervening regions and may correspond to long noncoding RNAs (lncRNAs). Interactions between miRNAs and competing endogenous RNAs have been reported [50–52], but need to be explicitly tested in B cells. In the future, it would be interesting to perform Ago PAR-CLIP or HITS-CLIP in additional lymphomas and primary B cell subsets whenever feasible. Ideally, a matched sample in which the miRNA of interest has been perturbed would be analyzed in parallel. As more datasets become available, it may become apparent why some targets are accessible to miRISC in one cell type but not in another. Presumably, RNA within a cell is not naked but coated with RNA-binding proteins that can compete with each other. In addition, RNA is probably structured and not linear.

Although it is useful and important to demonstrate miRISC binding, it remains a great challenge to prove that these interactions are functional. Often, artificial reporter constructs are tested in a heterologous cell line, but such experiments do not address what truly happens in vivo. Increasingly, there is a desire to mutate the endogenous miRNA-binding sites, but this line of investigation has low throughput. Recent advances in genome editing technologies may facilitate such studies, but they pose significant challenges in practice. It is not trivial to predict which binding sites will result in gene derepression upon mutation. Furthermore, such perturbations may not be sufficient to phenocopy ablation of a miRNA, which causes dysregulation of multiple target genes—potentially thousands of transcripts.

## 4. MIR-155 IN GERMINAL CENTER B CELLS AND LYMPHOMAGENESIS

Another interesting miRNA that has been determined to be required by B cells is miR-155 [53,54]. This miRNA is remarkably well conserved during evolution (Figure 2) and is even present in the primitive sea lamprey (*Petromyzon marinus*) and sea squirts (*Ciona intestinalis* and *savignyi*). Furthermore, a human and chicken virus appears to have an ortholog of cellular miR-155 [55–57]. The *Mir155* gene was first identified as a common retroviral integration site in B cell lymphomas from chickens infected with avian leukosis virus, and named B cell integration cluster before the discovery of miRNAs [58,59]. Since then, miR-155

| Accession | Sequence | Genome |
|---|---|---|
| MI0000681 | 5'-uuaaugcuaaucgugauaggggu | *Homo sapiens* |
| MI0014843 | 5'-uuaaugcuaaucgugauaggggu | *Pongo pygmaeus* |
| MI0020768 | 5'-uuaaugcuaaucgugauagggg | *Gorilla gorilla* |
| MI0008554 | 5'-uuaaugcuaaucgugauaggggu | *Pan troglodytes* |
| MI0007645 | 5'-uuaaugcuaaucgugauaggggu | *Macaca mulatta* |
| MI0012927 | 5'-uuaaugcuaaucgugauaggggu | *Equus caballus* |
| MI0009752 | 5'-uuaaugcuaaucgugauaggggu | *Bos taurus* |
| MI0015907 | 5'-uuaaugcuaauugugauagggg | *Sus scrofa* |
| MI0008078 | 5'-uuaaugcuaaucgugauaggggu | *Canis familiaris* |
| MI0020415 | 5'-uuaaugcuaaucgugauagggg | *Cricetulus griseus* |
| MI0025509 | 5'-uuaaugcuaauugugauaggggu | *Rattus norvegicus* |
| MI0000177 | 5'-uuaaugcuaauugugauaggggu | *Mus musculus* |
| MI0001176 | 5'-uuaaugcuaaucgugauagggg | *Gallus gallus* |
| MI0013842 | 5'-uuaaugcuaaucgugauaggg | *Taeniopygia guttata* |
| MI0006775 | 5'-uuaaugcuaaucgugauaggggu | *Ornithorhynchus anatinus* |
| MI0018764 | 5'-uuaaugcuaaucgugauagggg | *Anolis carolinensis* |
| MI0004848 | 5'-uuaaugcuaaucgugauagggg | *Xenopus tropicalis* |
| MI0002023 | 5'-uuaaugcuaaucgugauagggg | *Danio rerio* |
| MI0023336 | 5'-uuaaugcuaaucgugauagggg | *Cyprinus carpio* |
| MI0024518 | 5'-uuaaugcuaaucgugauaggggguu | *Ictalurus punctatus* |
| MI0017088 | 5'-uuaaugcuaucguugaugggguu | *Petromyzon marinus* |
| MI0007170 | 5'-uuaaugcuaauagugauaggg | *Ciona intestinalis* |
| MI0007198 | 5'-uuaaugcuaauaagugauuuaug | *Ciona savignyi* |
| MI0002474 | 5'-uuaaugcuuagccuguguccga | HHV-8/KSHV |
| MI0005096 | 5'-uuaaugcuguaucggaacccuuc | GaHV-2/MDV |

FIGURE 2 Mature miRNA sequences of miR-155 orthologs from 25 different genomes (two of which are viral) are aligned, and the seed sequences are highlighted in red. Information was obtained from miRBase (http://www.mirbase.org). HHV-8/KSHV, human herpesvirus 8/Kaposi sarcoma–associated herpesvirus; GaHV-2/MDV, Gallid herpesvirus 2/Marek disease virus.

gain-of-function has been implicated in a variety of malignancies [60–63], and it has been shown in a mouse model that B cell–specific overexpression of miR-155 results in B cell malignancy [64]. Interestingly, miR-K11, the miR-155 ortholog in Kaposi sarcoma–associated herpesvirus, has been suggested to contribute to B cell lymphomagenesis [55]. In B cells, miR-155 is highly expressed in germinal center B cells [65]. In vitro, miR-155 expression can be induced upon B cell activation via the B cell receptor, TLR4 or TLR9 in a manner dependent on NFAT and NFκB [53]. Germline deletion of miR-155 results in fewer germinal center B cells and consequently defective Ig isotype switching and affinity maturation, and humoral memory response [53,54] such that autoantibodies are reduced in a mouse model of lupus [65]. Conversely, transgenic expression of miR-155 in mature B cells driven by CD21-Cre promoted development of germinal center B cells [53]. A possible target of miR-155 is *Sfpi1* encoding the transcription factor PU.1, because overexpression of the latter can inhibit development of IgG1-expressing B cells in vitro [66]. In addition, it was found that miR-155 is responsible for fine-tuning activation-induced cytidine deaminase (AID or AICDA) expression in germinal center B cells to prevent excessive genomic translocations [67]. This was partially due to direct interaction of the miRNA with a single seed match in the 3′ UTR of *AICDA*, as demonstrated by targeted mutation of that sequence. However, further elucidation of the mechanisms by which miR-155 promotes the germinal B cell reaction may yet reveal some novel biology.

## 5. DISCOVERY OF AN ELUSIVE TUMOR SUPPRESSOR: MIR-15A~16-1 CLUSTER

A pioneering study found that the miR-15a and miR-16-1 cluster located within an intron of *DLEU2* (Figure 3), a long noncoding RNA gene of unknown function at 13q14, is frequently deleted or repressed in B cell chronic lymphocytic leukemias (CLL) [68]. Consequently, bcl-2 a putative target may be derepressed [69]. However, upregulation of bcl-2 alone is not sufficient for CLL pathogenesis and deletion of the miR-15a and miR-16-1 cluster does not significantly affect cell survival, but instead proliferation [70]. Interestingly and surprisingly, the most related paralogous miR-15b and miR-16-2 cluster on human chromosome 3 seems to perform a distinct function because it is not redundant with the miR-15a and miR-16-1 cluster. Large deletions surrounding the *DLEU2* locus in the mouse have been performed to model CLL [71]. These experiments demonstrated that deletion of only the miR-15a and miR-16-1 cluster produced a milder phenotype than larger deletions, and suggested that there may be uncharacterized element(s) encoded in the *DLEU2* locus that collaborate to mediate tumor suppression.

## 6. LIN28B REGULATES THE FETAL–ADULT B CELL DEVELOPMENT SWITCH

During ontogeny, there is a fascinating switch in B cell development that is akin to the fetal–adult hemoglobin switch occurring during erythropoiesis [72]. In fetal liver, B cell development mainly generates innate-like $CD5^+$ B-1 cells, whereas in adult bone marrow, B cell development primarily generates conventional adaptive B-2 cells [73,74]. Searching for a molecular basis for this phenomenon, Yuan et al. discovered that fetal liver pro-B cells do not express the let-7 family of miRNAs, whereas pro-B cells from adult bone marrow do [75]. Interestingly, fetal liver pro-B cells expressed the RNA-binding protein Lin28b that is known to inhibit let-7 biogenesis, whereas pro-B cells from adults do not harbor this inhibitor [75,76]. Furthermore, ectopic expression of Lin28b in adult bone marrow pro-B cells efficiently blocked let-7 expression and enhanced development of peritoneal B-1a cells and marginal zone B cells [75]. However, it is important to consider that Lin28b may have targets in addition to the let-7 miRNAs. It will therefore be important to perform CLIP of Lin28b in fetal liver hematopoietic stem and progenitor cells if feasible. More generally, identification of RNA-binding proteins that directly interact with miRNAs may be important experimental avenues to consider in the future.

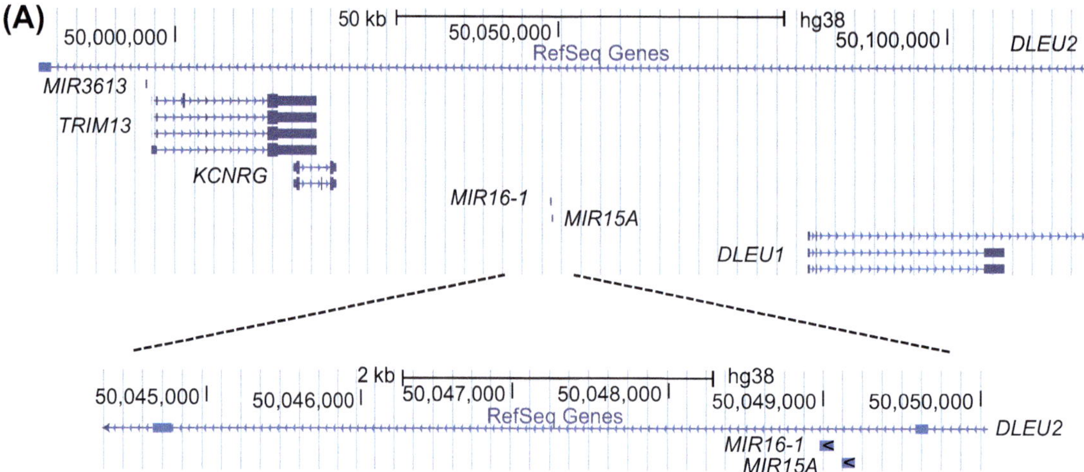


(B)

| Sequence | miRNA | Human | Mouse |
|---|---|---|---|
| 5'- uagcagcacauaaugguuugug | miR-15a | 13 | 14 |
| 5'- uagcagcacguaaauauuggcg | miR-16-1 | | |
| 5'- uagcagcacaucaugguuuaca | miR-15b | 3 | 3 |
| 5'- uagcagcacguaaauauuggcg | miR-16-2 | | |
| 5'- uagcagcacagaaauauuggc | miR-195 | 17 | 11 |
| 5'- cagcagcacacugugguuugu | miR-497 | | |
| 5'- uagcagcacagaaauauuggc | miR-195b† | n/a | 2 |
| 5'- cagcagcaauucauguuuugaa | miR-424 | X | n/a |
| 5'- cagcagcaauucauguuuugga | miR-322 | n/a | X |

FIGURE 3 (A) Screenshot of the UCSC Genome Browser (http://genome.ucsc.edu) depicts chromosomal organization of the human 13q14 locus. The indicated region has been magnified to highlight the miR-15a and miR-16-1 cluster located within an intron of *DLEU2*, a lncRNA gene. Boxes denote exons and arrowheads indicate the direction of transcription of the corresponding genes. (B) Mature miRNA sequences of paralogs related to miR-15a and miR-16-1 are aligned and the seed sequences are highlighted in red. Information was obtained from miRBase (http://www.mirbase.org). Their location on human and mouse chromosomes is indicated. Human miR-424 and mouse miR-322 are orthologs of each other and are syntenic, but have been named differently. Human miR-424 and similarly mouse miR-322 are clustered next to miR-503, which is not listed here but may also be related because their seed sequences differ only by one nucleotide. n/a, not available in genome. † miR-195b is not detected in human genome.

Determining whether the let-7 miRNAs have a role in B cell development is not trivial. Genetic deletion of the let-7 family will be challenging because there are 12 members encoded by eight distinct loci (Figure 4). Overexpression of let-7 in fetal liver pro-B cells is conceivable, but it may not be feasible to match the high aggregate expression level of the let-7 family in adult pro-B cells. Identification of let-7 targets may offer some useful clues. A prominent target of let-7 described in the literature is *Hmga2*, which harbors seven potential binding sites in its 3′ UTR that are evolutionarily conserved [77]. Consistent with the idea that it is a target of let-7, Hmga2 expression is high in fetal hematopoietic stem cells (HSCs) and Lin28-transduced adult HSCs compared with appropriate adult control populations [75,78]. However, unlike Lin28b, enforced expression of Hmga2 in adult HSCs did not result in fetal-like B lymphopoiesis [78]. Thus, additional targets of let-7 and/or Lin28b will need to be uncovered to better understand this regulatory network that controls the switching of fetal and adult B cell development programs.

## 7. TO BE FURTHER DETERMINED

There have been notable gain-of-function studies of miRNAs that require further clarification. For instance, retroviral expression of miR-181 in hematopoietic stem/progenitor cells resulted in a preference for developing B cells at the expense of T cells [20]. A tetracycline-regulated miR-21 construct in mice mediated pre-B cell transformation upon

| Name | Chromosome | Sequence (5'->3') |
|---|---|---|
| let-7a-1 | 13 (intergenic) | ugagguaguagguuguauaguu |
| let-7f-1 | 13 (intergenic) | ugagguaguagauuguauaguu |
| let-7d | 13 (intergenic) | agagguaguagguugcauaguu |
| let-7a-2 | 9 (intergenic [1]) | ugagguaguagguuguauaguu |
| let-7b | 15 (intergenic) | ugagguaguagguugugugguu |
| let-7c-2 | 15 (intergenic) | ugagguaguagguuguaugguu |
| let-7c-1 | 16 (intergenic [2]) | ugagguaguagguuguaugguu |
| let-7e | 17 (intergenic [3]) | ugagguaggagguuguauaguu |
| let-7f-2 | X (intronic) | ugagguaguagauuguauaguu |
| miR-98 | X (intronic) | ugagguaguaaguuguauuguu |
| let-7g | 9 (intronic) | ugagguaguaguuuguacaguu |
| let-7i | 10 (intergenic) | ugagguaguaguuugugcuguu |

FIGURE 4 Twelve Let-7 paralogs are encoded by eight distinct chromosomal loci. Mature miRNA sequences of paralogs belonging to let-7 family are aligned and the seed sequences are highlighted in red. Clusters are grouped together and their location on mouse chromosomes is indicated. Information was obtained from miRBase (http://www.mirbase.org). [1]clustered with miR-100; [2]clustered with miR-99a; [3]clustered with miR-99b and miR-125a.

overexpression [79] although there was no control for a specific action of miR-21 in that experiment. In these cases, the targets of these miRNAs are not definitively known, and it would be interesting to determine the roles of miR-21 and miR-181 in B cell development and their mechanisms of action. We independently observed that conditional deletion of miR-21 does not have an overt effect on mouse B cell development (B. Brady and SM; T. Chakraborty and KR, unpublished data), nor did we observe B cell expansion or transformation upon B cell–specific miR-21 overexpression in vivo (T. Chakraborty and KR, unpublished). In the case of miR-181, a family of six mature miRNAs encoded by three paralog gene clusters, recent work has identified a major role of these miRNAs in NKT and T cell physiology through the targeting of PTEN and thus the phosphatidylinositol 3-kinase (PI3K) signaling pathway [80]. Because PI3K signaling has a key role in B cell development and homeostasis, the miR-181 family may well have a critical role in B cell physiology if PTEN is also targeted. It remains to be seen whether miR-181 has a role in the T versus B cell lineage decision, as initially proposed. Overall, miRNA overexpression experiments should be interpreted with caution until complementary loss-of-function studies have been performed and the relevant miRNA targets have been validated in vivo.

## 8. CONCLUDING REMARKS

Despite their small size and modest effect on the repression of direct target genes, it is clear that particular miRNAs have important roles in B cells, as illustrated by the examples cited previously. The list of such miRNAs will certainly grow beyond those that we have discussed because other miRNAs, such as miR-142 and miR-148, for example, are also highly and stage-specifically expressed in the B cell lineage. Once the issue of synthetic miRNA delivery is solved, it will be possible to use miRNAs themselves or antisense miRNA inhibitors as drugs. The targets of miRNAs may also be interesting candidates for drug development because evolution has already selected them as key nodes in pathways that can and need to be modulated. It will be highly instructive to integrate miRNAs into models of genetic regulatory networks that program B cell development and function. In most cases, we still lack a comprehensive understanding of the pathways upstream and downstream of miRNAs. Such knowledge may bring us one step closer to being able to (re)program B cells at will, with the goal of augmenting desirable immune responses or inhibiting pathogenesis of cancer or autoimmunity.

## ACKNOWLEDGMENTS

We thank C. Kanellopoulou for reading the manuscript. S.A.M. is supported by the NIH Intramural Research Program of NIAID. K.R. is supported by the Helmholtz Association and an Advanced Grant from the European Research Council.

## REFERENCES

[1] Lee RC, Feinbaum RL, Ambros V. The *C. elegans* heterochronic gene lin-4 encodes small RNAs with antisense complementarity to lin-14. Cell 1993;75:843–54.

[2] Reinhart BJ, Slack FJ, Basson M, Pasquinelli AE, Bettinger JC, Rougvie AE, et al. The 21-nucleotide let-7 RNA regulates developmental timing in *Caenorhabditis elegans*. Nature 2000;403:901–6.

[3] Wightman B, Ha I, Ruvkun G. Posttranscriptional regulation of the heterochronic gene lin-14 by lin-4 mediates temporal pattern formation in *C. elegans*. Cell 1993;75:855–62.

[4] Kozomara A, Griffiths-Jones S. miRBase: integrating microRNA annotation and deep-sequencing data. Nucleic Acids Res 2011;39:D152–7.

[5] Bentwich I, Avniel A, Karov Y, Aharonov R, Gilad S, Barad O, et al. Identification of hundreds of conserved and nonconserved human microRNAs. Nat Genetics 2005;37:766–70.

[6] Lund E, Guttinger S, Calado A, Dahlberg JE, Kutay U. Nuclear export of microRNA precursors. Science 2004;303:95–8.

[7] Yi R, Qin Y, Macara IG, Cullen BR. Exportin-5 mediates the nuclear export of pre-microRNAs and short hairpin RNAs. Genes Development 2003;17:3011–6.

[8] Bernstein E, Caudy AA, Hammond SM, Hannon GJ. Role for a bidentate ribonuclease in the initiation step of RNA interference. Nature 2001;409:363–6.

[9] Elkayam E, Kuhn CD, Tocilj A, Haase AD, Greene EM, Hannon GJ, et al. The structure of human argonaute-2 in complex with miR-20a. Cell 2012;150:100–10.

[10] Friedman RC, Farh KK, Burge CB, Bartel DP. Most mammalian mRNAs are conserved targets of microRNAs. Genome Res 2009;19:92–105.

[11] Liu J, Carmell MA, Rivas FV, Marsden CG, Thomson JM, Song JJ, et al. Argonaute2 is the catalytic engine of mammalian RNAi. Science 2004;305:1437–41.

[12] Cheloufi S, Dos Santos CO, Chong MM, Hannon GJ. A dicer-independent miRNA biogenesis pathway that requires Ago catalysis. Nature 2010;465:584–9.

[13] Cifuentes D, Xue H, Taylor DW, Patnode H, Mishima Y, Cheloufi S, et al. A novel miRNA processing pathway independent of dicer requires Argonaute 2 catalytic activity. Science 2010;328:1694–8.

[14] O'Carroll D, Mecklenbrauker I, Das PP, Santana A, Koenig U, Enright AJ, et al. A slicer-independent role for Argonaute 2 in hematopoiesis and the microRNA pathway. Genes Dev 2007;21:1999–2004.

[15] Wu L, Fan J, Belasco JG. MicroRNAs direct rapid deadenylation of mRNA. Proc Natl Acad Sci USA 2006;103:4034–9.

[16] Maroney PA, Yu Y, Fisher J, Nilsen TW. Evidence that microRNAs are associated with translating messenger RNAs in human cells. Nat Struct Mol Biol 2006;13:1102–7.

[17] Nelson PT, Hatzigeorgiou AG, Mourelatos Z. miRNP: mRNA association in polyribosomes in a human neuronal cell line. RNA 2004;10:387–94.

[18] Guo H, Ingolia NT, Weissman JS, Bartel DP. Mammalian microRNAs predominantly act to decrease target mRNA levels. Nature 2010;466:835–40.

[19] Xiao C, Calado DP, Galler G, Thai TH, Patterson HC, Wang J, et al. MiR-150 controls B cell differentiation by targeting the transcription factor c-Myb. Cell 2007;131:146–59.

[20] Chen CZ, Li L, Lodish HF, Bartel DP. MicroRNAs modulate hematopoietic lineage differentiation. Science 2004;303:83–6.

[21] Monticelli S, Ansel KM, Xiao C, Socci ND, Krichevsky AM, Thai TH, et al. MicroRNA profiling of the murine hematopoietic system. Genome Biol 2005;6:R71.

[22] Kuchen S, Resch W, Yamane A, Kuo N, Li Z, Chakraborty T, et al. Regulation of microRNA expression and abundance during lymphopoiesis. Immunity 2010;32:828–39.

[23] Koralov SB, Muljo SA, Galler GR, Krek A, Chakraborty T, Kanellopoulou C, et al. Dicer ablation affects antibody diversity and cell survival in the B lymphocyte lineage. Cell 2008;132:860–74.

[24] Ventura A, Young AG, Winslow MM, Lintault L, Meissner A, Erkeland SJ, et al. Targeted deletion reveals essential and overlapping functions of the miR-17 through 92 family of miRNA clusters. Cell 2008;132:875–86.

[25] Tanzer A, Stadler PF. Molecular evolution of a microRNA cluster. J Mol Biol 2004;339:327–35.

[26] Olive V, Sabio E, Bennett MJ, De Jong CS, Biton A, et al. A component of the mir-17-92 polycistronic oncomir promotes oncogene-dependent apoptosis. eLife 2013;2:e00822.

[27] Xiao C, Srinivasan L, Calado DP, Patterson HC, Zhang B, Wang J, et al. Lymphoproliferative disease and autoimmunity in mice with increased miR-17-92 expression in lymphocytes. Nat Immunol 2008;9:405–14.

[28] Kang SG, Liu WH, Lu P, Jin HY, Lim HW, Shepherd J, et al. MicroRNAs of the miR-17~92 family are critical regulators of T(FH) differentiation. Nat Immunol 2013;14:849–57.

[29] Baumjohann D, Kageyama R, Clingan JM, Morar MM, Patel S, de Kouchkovsky D, et al. The microRNA cluster miR-17~92 promotes TFH cell differentiation and represses subset-inappropriate gene expression. Nat Immunol 2013;14:840–8.

[30] Ota A, Tagawa H, Karnan S, Tsuzuki S, Karpas A, Kira S, et al. Identification and characterization of a novel gene, C13orf25, as a target for 13q31-q32 amplification in malignant lymphoma. Cancer Res 2004;64:3087–95.

[31] Lenz G, Wright GW, Emre NC, Kohlhammer H, Dave SS, Davis RE, et al. Molecular subtypes of diffuse large B-cell lymphoma arise by distinct genetic pathways. Proc Natl Acad Sci USA 2008;105: 13520–5.

[32] Jin HY, Oda H, Lai M, Skalsky RL, Bethel K, Shepherd J, et al. MicroRNA-17~92 plays a causative role in lymphomagenesis by coordinating multiple oncogenic pathways. EMBO J 2013;32: 2377–91.

[33] O'Donnell KA, Wentzel EA, Zeller KI, Dang CV, Mendell JT. c-Myc-regulated microRNAs modulate E2F1 expression. Nature 2005;435:839–43.

[34] Schmitz R, Young RM, Ceribelli M, Jhavar S, Xiao W, Zhang M, et al. Burkitt lymphoma pathogenesis and therapeutic targets from structural and functional genomics. Nature 2012;490:116–20.

[35] Schmidt-Supprian M, Rajewsky K. Vagaries of conditional gene targeting. Nat Immunol 2007;8:665–8.

[36] Mu P, Han YC, Betel D, Yao E, Squatrito M, Ogrodowski P, et al. Genetic dissection of the miR-17~92 cluster of microRNAs in Myc-induced B-cell lymphomas. Genes Dev 2009;23:2806–11.

[37] He L, Thomson JM, Hemann MT, Hernando-Monge E, Mu D, Goodson S, et al. A microRNA polycistron as a potential human oncogene. Nature 2005;435:828–33.

[38] Olive V, Bennett MJ, Walker JC, Ma C, Jiang I, Cordon-Cardo C, et al. miR-19 is a key oncogenic component of mir-17-92. Genes Dev 2009;23:2839–49.

[39] Egle A, Harris AW, Bouillet P, Cory S. Bim is a suppressor of Myc-induced mouse B cell leukemia. Proc Natl Acad Sci USA 2004;101:6164–9.

[40] Wendel HG, Malina A, Zhao Z, Zender L, Kogan SC, Cordon-Cardo C, et al. Determinants of sensitivity and resistance to rapamycin-chemotherapy drug combinations in vivo. Cancer Res 2006;66: 7639–46.

[41] Nam JW, Rissland OS, Koppstein D, Abreu-Goodger C, Jan CH, Agarwal V, et al. Global analyses of the effect of different cellular contexts on MicroRNA targeting. Mol Cell 2014;53:1031–43.

[42] Mangone M, Manoharan AP, Thierry-Mieg D, Thierry-Mieg J, Han T, Mackowiak SD, et al. The landscape of *C. elegans* 3'UTRs. Science 2010;329:432–5.

[43] Jan CH, Friedman RC, Ruby JG, Bartel DP. Formation, regulation and evolution of *Caenorhabditis elegans* 3' UTRs. Nature 2011;469:97–101.

[44] Hafner M, Landthaler M, Burger L, Khorshid M, Hausser J, Berninger P, et al. Transcriptome-wide identification of RNA-binding protein and microRNA target sites by PAR-CLIP. Cell 2010;141:129–41.

[45] Chi SW, Zang JB, Mele A, Darnell RB. Argonaute HITS-CLIP decodes microRNA-mRNA interaction maps. Nature 2009;460: 479–86.

[46] Helwak A, Kudla G, Dudnakova T, Tollervey D. Mapping the human miRNA interactome by CLASH reveals frequent noncanonical binding. Cell 2013;153:654–65.

[47] Grosswendt S, Filipchyk A, Manzano M, Klironomos F, Schilling M, Herzog M, et al. Unambiguous identification of miRNA: target site interactions by different types of ligation reactions. Mol Cell 2014;54:1042–54.

[48] Loeb GB, Khan AA, Canner D, Hiatt JB, Shendure J, Darnell RB, et al. Transcriptome-wide miR-155 binding map reveals widespread noncanonical microRNA targeting. Mol Cell 2012;48:760–70.
[49] Skalsky RL, Corcoran DL, Gottwein E, Frank CL, Kang D, Hafner M, et al. The viral and cellular microRNA targetome in lymphoblastoid cell lines. PLoS Pathog 2012;8:e1002484.
[50] Memczak S, Jens M, Elefsinioti A, Torti F, Krueger J, Rybak A, et al. Circular RNAs are a large class of animal RNAs with regulatory potency. Nature 2013;495:333–8.
[51] Taulli R, Loretelli C, Pandolfi PP. From pseudo-ceRNAs to circ-ceRNAs: a tale of cross-talk and competition. Nat Struct Mol Biol 2013;20:541–3.
[52] Hansen TB, Jensen TI, Clausen BH, Bramsen JB, Finsen B, Damgaard CK, et al. Natural RNA circles function as efficient microRNA sponges. Nature 2013;495:384–8.
[53] Thai TH, Calado DP, Casola S, Ansel KM, Xiao C, Xue Y, et al. Regulation of the germinal center response by microRNA-155. Science 2007;316:604–8.
[54] Rodriguez A, Vigorito E, Clare S, Warren MV, Couttet P, Soond DR, et al. Requirement of bic/microRNA-155 for normal immune function. Science 2007;316:608–11.
[55] Gottwein E, Mukherjee N, Sachse C, Frenzel C, Majoros WH, Chi JT, et al. A viral microRNA functions as an orthologue of cellular miR-155. Nature 2007;450:1096–9.
[56] Skalsky RL, Samols MA, Plaisance KB, Boss IW, Riva A, Lopez MC, et al. Kaposi's sarcoma-associated herpesvirus encodes an ortholog of miR-155. J Virol 2007;81:12836–45.
[57] Morgan R, Anderson A, Bernberg E, Kamboj S, Huang E, Lagasse G, et al. Sequence conservation and differential expression of Marek's disease virus microRNAs. J Virol 2008;82:12213–20.
[58] Tam W, Ben-Yehuda D, Hayward WS. bic, a novel gene activated by proviral insertions in avian leukosis virus-induced lymphomas, is likely to function through its noncoding RNA. Mol Cell Biol 1997;17:1490–502.
[59] Clurman BE, Hayward WS. Multiple proto-oncogene activations in avian leukosis virus-induced lymphomas: evidence for stage-specific events. Mol Cell Biol 1989;9:2657–64.
[60] Eis PS, Tam W, Sun L, Chadburn A, Li Z, Gomez MF, et al. Accumulation of miR-155 and BIC RNA in human B cell lymphomas. Proc Natl Acad Sci USA 2005;102:3627–32.
[61] Kluiver J, Poppema S, de Jong D, Blokzijl T, Harms G, Jacobs S, et al. BIC and miR-155 are highly expressed in Hodgkin, primary mediastinal and diffuse large B cell lymphomas. J Pathol 2005;207:243–9.
[62] Metzler M, Wilda M, Busch K, Viehmann S, Borkhardt A. High expression of precursor microRNA-155/BIC RNA in children with Burkitt lymphoma. Genes Chromosomes Cancer 2004;39:167–9.
[63] Yanaihara N, Caplen N, Bowman E, Seike M, Kumamoto K, Yi M, et al. Unique microRNA molecular profiles in lung cancer diagnosis and prognosis. Cancer Cell 2006;9:189–98.
[64] Costinean S, Zanesi N, Pekarsky Y, Tili E, Volinia S, Heerema N, et al. Pre-B cell proliferation and lymphoblastic leukemia/high-grade lymphoma in E(mu)-miR155 transgenic mice. Proc Natl Acad Sci USA 2006;103:7024–9.
[65] Thai TH, Patterson HC, Pham DH, Kis-Toth K, Kaminski DA, Tsokos GC. Deletion of microRNA-155 reduces autoantibody responses and alleviates lupus-like disease in the Fas(lpr) mouse. Proc Natl Acad Sci USA 2013;110:20194–9.
[66] Vigorito E, Perks KL, Abreu-Goodger C, Bunting S, Xiang Z, Kohlhaas S, et al. microRNA-155 regulates the generation of immunoglobulin class-switched plasma cells. Immunity 2007;27:847–59.
[67] Dorsett Y, McBride KM, Jankovic M, Gazumyan A, Thai TH, Robbiani DF, et al. MicroRNA-155 suppresses activation-induced cytidine deaminase-mediated Myc-Igh translocation. Immunity 2008;28:630–8.
[68] Calin GA, Dumitru CD, Shimizu M, Bichi R, Zupo S, Noch E, et al. Frequent deletions and down-regulation of micro- RNA genes miR15 and miR16 at 13q14 in chronic lymphocytic leukemia. Proc Natl Acad Sci USA 2002;99:15524–9.
[69] Cimmino A, Calin GA, Fabbri M, Iorio MV, Ferracin M, Shimizu M, et al. miR-15 and miR-16 induce apoptosis by targeting BCL2. Proc Natl Acad Sci USA 2005;102:13944–9.
[70] Klein U, Lia M, Crespo M, Siegel R, Shen Q, Mo T, et al. The DLEU2/miR-15a/16-1 cluster controls B cell proliferation and its deletion leads to chronic lymphocytic leukemia. Cancer Cell 2010;17:28–40.
[71] Lia M, Carette A, Tang H, Shen Q, Mo T, Bhagat G, et al. Functional dissection of the chromosome 13q14 tumor-suppressor locus using transgenic mouse lines. Blood 2012;119:2981–90.
[72] Lee YT, de Vasconcellos JF, Yuan J, Byrnes C, Noh SJ, Meier ER, et al. LIN28B-mediated expression of fetal hemoglobin and production of fetal-like erythrocytes from adult human erythroblasts ex vivo. Blood 2013;122:1034–41.
[73] Hardy RR, Hayakawa K. A developmental switch in B lymphopoiesis. Proc Natl Acad Sci USA 1991;88:11550–4.
[74] Kantor AB, Stall AM, Adams S, Herzenberg LA. Differential development of progenitor activity for three B-cell lineages. Proc Natl Acad Sci USA 1992;89:3320–4.
[75] Yuan J, Nguyen CK, Liu X, Kanellopoulou C, Muljo SA. Lin28b reprograms adult bone marrow hematopoietic progenitors to mediate fetal-like lymphopoiesis. Science 2012;335:1195–200.
[76] Yuan J, Muljo SA. Exploring the RNA world in hematopoietic cells through the lens of RNA-binding proteins. Immunol Rev 2013;253:290–303.
[77] Mayr C, Hemann MT, Bartel DP. Disrupting the pairing between let-7 and Hmga2 enhances oncogenic transformation. Science 2007;315:1576–9.
[78] Copley MR, Babovic S, Benz C, Knapp DJ, Beer PA, Kent DG, et al. The Lin28b-let-7-Hmga2 axis determines the higher self-renewal potential of fetal haematopoietic stem cells. Nat Cell Biol 2013;15:916–25.
[79] Medina PP, Nolde M, Slack FJ. OncomiR addiction in an in vivo model of microRNA-21-induced pre-B-cell lymphoma. Nature 2010;467:86–90.
[80] Henao-Mejia J, Williams A, Goff LA, Staron M, Licona-Limon P, Kaech SM, et al. The microRNA miR-181 is a critical cellular metabolic rheostat essential for NKT cell ontogenesis and lymphocyte development and homeostasis. Immunity 2013;38:984–97.

Chapter 6

# Proliferation and Differentiation Programs of Developing B Cells

Markus Werner[1] and Hassan Jumaa[1,2,3]

[1]*Institute of Immunology, University Clinics Ulm, Ulm, Germany,* [2]*Faculty of Biology, Albert-Ludwigs University of Freiburg, Freiburg, Germany,* [3]*Centre for Biological Signaling Studies BIOSS, Albert-Ludwigs University of Freiburg, Freiburg, Germany*

B lymphocytes derive from undifferentiated pluripotent cells, termed hematopoietic stem cells (HSCs), in a highly regulated and dynamic process. This developmental process takes place in the fetal liver before birth and after birth in the bone marrow, and it is driven by soluble and membrane-bound growth factors. To this end, in the fetal liver, and later in the bone marrow, distinct microenvironments exist in which individual sets of these factors are produced in high concentrations to trigger developmental stage-specific proliferation and differentiation programs that cooperatively govern the development of uncommitted progenitors into cells of the B lineage [1].

HSCs are the source not only for B cells but also for all other cells of the immune system, such as T cells, macrophages, dendritic cells, or natural killer (NK) cells [2]. Therefore, the hematopoietic microenvironment has to provide a plethora of factors required to drive the differentiation of uncommitted precursor cells into the distinct hematopoietic lineages. Out of this large set of growth factors, uncommitted precursors have to obtain access to lineage-specific sets of these factors for a critical time period to receive sufficient signals that allow their differentiation into a specific hematopoietic lineage. The large number of cells in the bone marrow suggests, however, that precursor cells compete for these growth factors and that the development of a precursor cell into a certain hematopoietic lineage is not a one-way road but rather a permanent struggle between survival and differentiation signals delivered by the growth factors. Therefore, uncommitted precursors may be able to switch between distinct hematopoietic lineages until one specific set of growth factors becomes dominant and induces a lineage-specific program that promotes the differentiation along this lineage and simultaneously represses the potential to adopt other hematopoietic lineages.

B cell differentiation takes place in distinct steps that can be identified according to the expression of B cell lineage-related markers and cell surface proteins as well as the recombination status of the *immunoglobulin* (*Ig*) *heavy chain* (*IgH*), and *Ig light chain* (*IgL*) genes [3]. From the common lymphoid progenitor (CLP), developing B cells progress in the bone marrow through a pre-pro-B, pro-B, and pre-B cell stage before they become immature B cells. For the generation of B lineage cells from uncommitted precursors, the chemokine stromal cell derived factor 1 (SDF-1, also known as C-X-C motif chemokine CXCL12; encoded by *CXCL12*), and the cytokine interleukin-7 (IL-7; encoded by *IL7*) emerged as important factors [1]. During ontogeny CXCL12 is responsible for the migration of hematopoietic cells from the fetal liver into the bone marrow where cells expressing the SDF-1 receptor, also known as C-X-C chemokine receptor type 4 (CXCR4; encoded by *CXCR4*), are retained [4,5]. Uncommitted precursor cells, as well as pre-pro-B cells, are associated with stromal cells that produce high levels of the chemokine CXCL12 [5], indicating that CXCL12 helps to retain uncommitted precursor cells in those microenvironments where B cell lineage-specific factors are generated [1].

IL-7 induces intracellular signaling cascades that promote survival and proliferation of B cell precursors as well as their further differentiation along the B cell lineage [6]. Interestingly, in contrast to pre-pro-B cells, pro-B cells were shown to interact with IL-7-producing stromal cells that secrete only low amounts of CXCL12 [5]. It is currently not completely clear how the movement of very early B lineage precursors away from the CXCL12$^{hi}$ stromal cells toward the IL-7-producing stromal cells is regulated, but dynamics in CXCR4 surface expression during B cell development may play a role. The necessity of IL-7 receptor (IL-7R)-derived signals for early B cell development is impressively demonstrated by the almost complete absence of mature B cells in mice lacking IL-7R or IL-7 [7,8]. Within the bone marrow, B cell development in IL-7R-deficient mice is

Molecular Biology of B Cells. http://dx.doi.org/10.1016/B978-0-12-397933-9.00006-0

blocked at the early pro-B cell stage with cells unable to differentiate further along the B cell lineage and prone to undergo apoptosis [7]. IL-7R signaling is also essential for recombination of the *IgH* genes, which if successful results in the expression of a precursor B cell antigen receptor (pre-BCR) and transition to the pre-B cell stage [9]. Similar to the IL-7R in pro-B cells, pre-BCR signaling at the pre-B cell stage is essential to induce proliferative expansion and further differentiation into immature B cells expressing for the first time a mature BCR complex on the cell surface. Intriguingly, IL-7R and pre-BCR are thought to cooperate at the pre-B cell stage, as absence of either receptor interferes with B cell development [9].

This chapter focuses on the proliferation and differentiation programs that drive early B lymphopoiesis. Starting from the pro-B cell stage, we describe the surface receptors and downstream signaling events, which guide pro-B cells stepwise into mature B cells. During this maturation, B cell lineage-specific signaling systems are established and culminate in the BCR signaling complex, which provides basal survival signals ensuring long-term persistence of mature naïve B cells. Because of repeated switching between proliferation and differentiation, developing B cells are prone to unrestrained proliferation and cellular transformation, or apoptosis. Therefore, tight regulation of the proliferation and differentiation programs during B cell development is of fundamental importance.

## 1. PROLIFERATION AND DIFFERENTIATION PROGRAMS AT THE PRO-B CELL STAGE

At the pro-B cell stage, crucial processes take place leading to expansion of the B cell precursors, commitment to the B cell lineage, as well as recombination of the *IgH* genes. Interestingly, signaling cascades triggered by the IL-7R regulate most of these processes at the pro-B cell stage.

### 1.1 The IL-7R

As outlined above, the IL-7R complex is an important surface receptor on murine pro-B cells. It consists of the IL-7Rα-chain (CD127) associated with a common gamma chain (γc, also known as interleukin-2 receptor subunit gamma; encoded by *IL2RG*) (Figure 1) [6]. Both receptor chains are not unique to the IL-7R as the γc is also part of the receptors for IL-2, IL-4, IL-9, IL-15 and IL-21, and the IL-7Rα chain also associates with the thymus stroma lymphopoeitin (TSLP) receptor chain to form a receptor complex that is specific for TSLP, a growth factor thought to be important for fetal B cell development [10,11]. Binding of IL-7 to the IL-7R promotes heterodimerization of the two receptor chains and leads to the activation of the receptor-associated Janus kinases JAK1 and JAK3 (encoded by *JAK1* and *JAK3*, respectively), which are associated with the IL-7Rα and the γc, respectively [6]. Activated JAKs phosphorylate specific tyrosine residues located within the cytoplasmic tails of the IL-7Rα chain, thereby providing binding sites for the sarcoma (Src) homology 2 (SH2) domain of signal transducers and activators of transcription (STATs). Subsequently, JAK-mediated phosphorylation of STATs enables them to form homo- and heterodimers and to translocate into the nucleus where they bind to specific promoter motifs and regulate a variety of genes [6,10]. The family of STAT transcription factors comprises, in total, seven members (STAT1, 2, 3, 4, 5a, 5b, and 6) of which only STAT1, STAT3, STAT5a, and STAT5b (hereafter collectively referred to as STAT5) are activated in response to IL-7 [10,12]. Although STAT1 plays only a minor role in B cell development, STAT3 was shown to be involved in inducing proliferation in pro-B cells in response to IL-7 [13]. In addition, numbers of pro-B, pre-B, and the more mature B cell stages were markedly reduced in STAT3-deficient mice [13]. Nevertheless, STAT5 is considered to be the major component downstream of the IL-7R, as introduction of solely a constitutively active form of STAT5 largely restores B cell development in IL-7Rα-deficient mice, and B cell-specific deletion of STAT5 closely resembles the phenotype of IL-7R-deficient mice [14–16].

Signaling from the IL-7R involves in addition to Src-kinases such as the protein tyrosine kinases Fyn or Lyn, signaling pathways activating the mitogen-activated protein kinases (MAPK)/extracellular signal-regulated kinase (Erk) signaling pathways [10,17,18]. The precise role of Src-kinases in IL-7R signaling is not completely clear, but Src-kinases can directly phosphorylate and activate STATs, providing an additional possibility for STAT activation, independent of JAKs [19]. The importance of Erk/MAPK signaling downstream of the IL-7R was impressively demonstrated in the absence of Erk1/2, where B cell development is blocked at the pro-B to pre-B cell transition because of reduced survival and proliferation in response to IL-7 [20]. How the Erk signaling cascade is activated downstream of the IL-7R is to date not completely elucidated. It is however possible that the adaptor protein Src homology and collagen homology (Shc; encoded by *SHC*) plays a role in the IL-7R-dependent activation of MAPK signaling by recruiting the adaptor protein Grb2 and the Ras GTPase SOS [21]. Consistent with this, Shc becomes phosphorylated and activated after IL-7 stimulation [22]. The finding that B cell development is blocked at the pro-B cell stage in mice lacking Shc further supports the idea of an important role of Shc in IL-7R signaling at the pro-B cell stage [22]. Hence it may well be that, in response to IL-7, JAKs become activated and phosphorylate target motifs within the cytoplasmic domains of the IL-7R chains, thereby generating docking sites for Shc, leading to activation of MAPK/Erk downstream signaling.

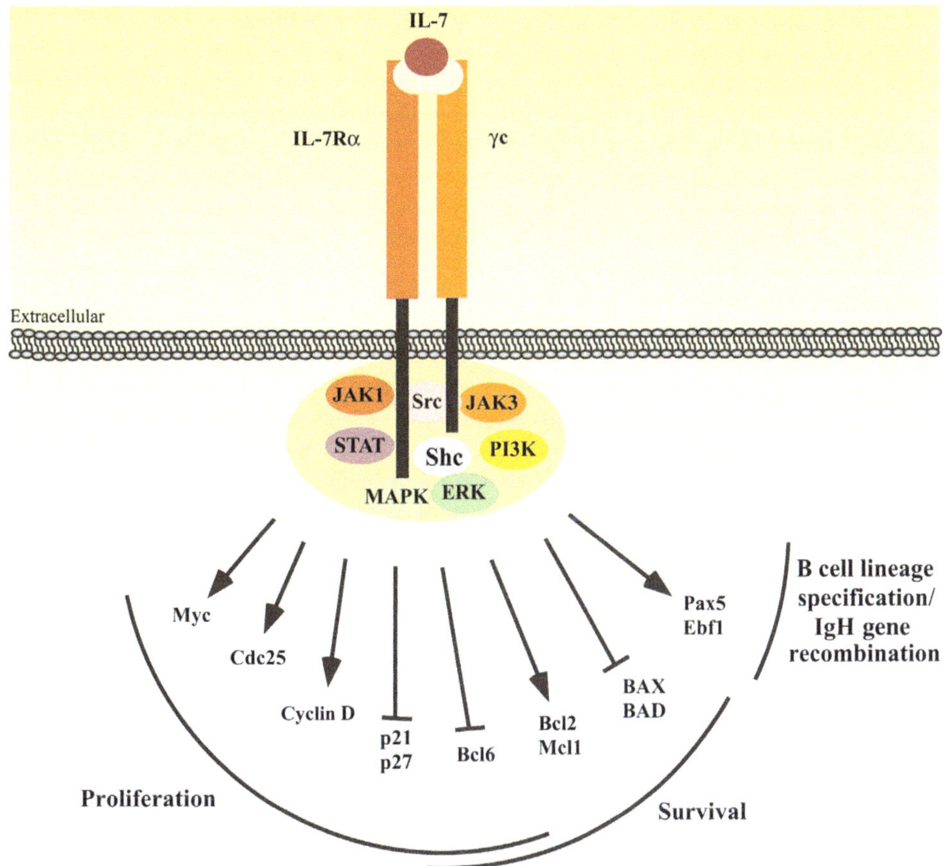


FIGURE 1 **IL-7R signaling at the pro-B cell stage.** IL-7R engagement leads to the activation of numerous signaling cascades, including JAK/STAT, PI3K, and MAPK/Erk. These signaling cascades collectively promote proliferation by inducing Myc, Cdc25, and cyclin D and repressing the cell cycle inhibitors p21 and p27 as well as Bcl6. Furthermore, IL-7R signaling promotes survival of early B lineage cells by inducing the prosurvival factors Bcl2 and Mcl1 and inhibiting the proapoptotic factors BAX and BAD. In addition, IL-7R signaling induces Ebf1 and Pax5 and thereby B cell lineage specification as well as *IgH* gene recombination.

Another important signaling pathway activated by IL-7R engagement is the phosphatidylinositol 3-kinase (PI3K) signaling cascade [23]. How PI3K signaling becomes activated by the IL-7R as well as the molecular outcomes of PI3K signaling are discussed later in this chapter.

## 1.2 SOCS are Negative Regulators of IL-7R Signaling

Because unrestrained IL-7R signaling is frequently observed in leukemic cells, a tight control of IL-7R activity is required. The family of suppressors of cytokine signaling (SOCS), consisting of seven members termed SOCS1 to SOCS7 and cytokine-inducible SH2 protein, evolved as important negative regulators of IL-7R signaling [24]. SOCS can directly bind to and inhibit the activity of JAKs, or become recruited to phosphorylated motifs within the IL-7R cytoplasmic chains, thereby functioning as competitive inhibitors of STAT5 binding and activation [24]. In addition, SOCS can associate with other proteins and become part of an E3-ubiquitin ligase complex that is able to mark JAKs and the IL-7R chains for proteasomal degradation. Interestingly, IL-7R engagement itself was shown to induce SOCS1 expression [10,25], suggesting that the IL-7R establishes a negative feedback to control its own activity. The importance of this negative feedback is evident, as mutations of SOCS1 are frequently observed in primary mediastinal B cell lymphoma and in classic Hodgkin lymphoma and cause nuclear accumulation of activated STAT5 [26].

## 1.3 Human B Cells Develop Independently of the IL-7R

In contrast to murine B cell development, the generation of human B cells seems not to depend on IL-7 or IL-7R signaling. Human individuals harboring mutations in the IL-7Rα chain develop severe combined immunodeficiency disease (SCID), characterized by a lack of T cells but normal numbers of NK and B cells [27]. Similarly, levels of B cells are normal, whereas T and NK cells are reduced in individuals with SCID due to absence of JAK3 [28–32]. It is currently not clear whether other cytokines compensate for the loss of IL-7R derived signals in the above-mentioned human individuals. Recently, FLT-3L was shown to promote human B lymphopoiesis in vitro from neonatal cord blood precursors independent of IL-7 [33,34], suggesting that certain growth factors may be important to drive IL-7-independent human B cell development. However, whether the FLT-3L-driven human B cell generation reflects the in vivo situation is not completely clear and requires further investigation.

## 1.4 IL-7R Signaling and the Proliferation of Pro-B Cells

For cell cycle entry and progression, developing B cells must activate specific G1 cyclins and cyclin-dependent kinases (CDKs). Concomitantly, cell cycle inhibitors of the CDK interaction protein (CIP)/kinase-interacting protein (KIP) family, which impair cyclin or CDK activity, must be either downregulated or inhibited [35].

IL-7R signaling promotes cell cycle entry by several mechanisms (Figure 1). For instance, in response to IL-7R signaling and STAT5 activation, expression of the transcription factors n-myc and c-myc (hereafter collectively referred to as Myc) is induced [36]. Whether STAT5 induces Myc expression directly is controversial. However, STAT5 was shown to regulate Myc gene expression by repressing B cell lymphoma 6 protein (Bcl6; encoded by BCL6), a zinc-finger-containing transcription factor that functions as a sequence specific repressor of Myc gene transcription [37–39]. STAT5-mediated repression of Bcl6 may therefore free Myc from Bcl6-mediated repression, leading to increased Myc expression in response to IL-7R engagement. Myc plays a fundamental role in processes that ensure cell growth and cell cycle initiation, such as ribosome biogenesis, DNA replication, and telomere maintenance [40–42]. In addition, Myc promotes cell cycle entry by repressing the cell cycle inhibitors p21 and p27 (also referred to as CIP1 and KIP1, respectively, and encoded by *CDKN1A* and *CDKN1B*, respectively) and by inducing CDK4, CDK6, and cyclin D gene transcription [43–50]. Cyclin D forms heterodimers with either CDK4 or CDK6, resulting in the activation of CDK4/6 and entry into the G1 phase of the cell cycle. Besides this, CDK4-D-type cyclin and CDK6-D-type cyclin complexes promote further progression of the cell cycle by relieving transcription factors of the E2F family from retinoblastoma protein-mediated repression [51]. E2F transcription factors are essential for G1-to-S phase transition by inducing transcription of the genes for cyclin A (encoded by *CCNA2*) and E-type cyclins (cyclin E1 and cyclin E2; encoded by *CCNE1* and *CCNE2*, respectively), as well as other proteins involved in cell cycle progression and DNA replication [35].

Another mechanism that promotes cell cycle progression at the pro-B cell stage is the IL-7R-mediated stabilization of the dual specific phosphatase Cdc25 (cell division cycle 25) family [52]. Members of the Cdc25 family activate CDKs by removing inhibitory phosphates from the CDK active site, thereby promoting transition through several phases of the cell cycle, even in the presence of cell cycle inhibitors [52]. Notably, the finding that IL-7R signaling stabilizes Cdc25 protein expression was discovered in T cells but may well be applicable for IL-7R signaling in B cells.

Erk signaling also regulates Myc expression by phosphorylating and marking Bcl6 for proteasomal degradation [53,54]. In addition, Erk phosphorylates Myc at serine 62, leading to increased protein stability of Myc [55]. Erk was also shown to interact with Cdc25C in interphase and to activate Cdc25 by phosphorylation at threonine 48 in mitosis [56]. Hence, MAPK/Erk signaling supports the STAT5-mediated Myc induction and the subsequent cell cycle entry downstream of the IL-7R.

## 1.5 IL-7R Signaling and the Survival of Pro-B Cells

Besides inducing pro-B cell proliferation, IL-7R signaling provides survival signals (Figure 1). Pro-apoptotic proteins, such as the Bcl-2 homology domain 3 (BH3)-only proteins interfere with cell survival by directly binding to and inactivating members of the prosurvival Bcl2-family [57]. Hence, the ratio of proapoptotic BH3-only proteins (such as BAX, BAD, BIM) to members of the Bcl2-family (Bcl2, Mcl-1) is a critical determinant of cell survival. Interestingly, IL-7R signaling was shown to maintain an antiapoptotic ratio between Bcl2 and the BH3-only protein Bax, as mice lacking IL-7Rα or Jak3 show strongly increased levels of the BH3-family member Bax and consequently reduced survival [58,59]. Mice expressing an IL-7 transgene instead suppressed precursor B cell apoptosis by producing high Bcl2 and low Bax levels [59]. The importance of Bcl2 synthesis downstream of the IL-7R is further demonstrated by the ability of Bcl2 to rescue survival of IL-7R- or STAT5-deficient precursor B cells when expressed as a transgene [60]. Besides affecting the Bax/Bcl2 ratio, IL-7R signaling promotes survival by inactivating Bad, thereby preventing its movement to the mitochondrial membrane where it binds to and inactivates Bcl2 [61]. Also, the activity of the BH3-only family member BIM is inhibited on the posttranslational level by IL-7 [62]. Interestingly, BIM deficiency partially rescued B cell development in IL-7-deficient mice [63,64], demonstrating the importance of the IL-7-dependent inactivation of BIM. Finally, IL-7R signaling induces transcription of the Mcl1 gene, whose product emerged to be a key molecule for B cell precursor survival. Expression levels of Mcl1 are approximately 8-fold reduced in STAT5-deficient pro-B cells [60], suggesting that STAT5 directly promotes expression of Mcl1. Consistent with an important role of Mcl1 in early B cell development, Mcl1-deficient mice have markedly reduced numbers of mature B cells due to a B cell developmental block at the pro-B cell stage [60].

## 1.6 The Role of IL-7R in Inducing Differentiation Programs

In addition to promoting survival and proliferation of pro-B cells, IL-7R signaling is involved in establishing differentiation programs that lead to transition from the pro-B to the pre-B cell stage (Figure 1). This became apparent by studies dissecting the function of distinct cytoplasmic domains within the IL-7Rα chain [65]. The introduction of an IL-7Rα-mutant chain lacking a specific tyrosine into IL-7R-deficient precursor cells severely interfered with the proliferative capacity of the IL-7R, but not with its ability to promote pro-B cell differentiation to the μHC-positive stage [65]. Replacement of the whole IL-7Rα cytoplasmic domain with corresponding sequences from the IL-2Rβ chain in contrast did not affect the initiation of proliferation, but it markedly impaired differentiation of pro-B cells toward the pre-B cell stage [65]. These data clearly demonstrate that signals from the IL-7R are essential to promote B cell differentiation and that these signals are distinct from those inducing proliferation.

One possibility how IL-7R signaling promotes pro-B cell differentiation stems from the observation that expression of the early B cell factor 1 (Ebf1, encoded by *EBF1*), a crucial transcription factor for B cell lineage specification, is induced by STAT5 in response to IL-7 [66,67]. Similar to IL-7R-deficient mice, Ebf1-deficient mice show a B cell developmental block at the very early pro-B cell stage and already in Ebf1-deficient CLPs, the expression of numerous B cell-specific genes is lacking [68,69]. This is consistent with the ability of Ebf1 to act as a potent B cell lineage specifier by inducing expression of numerous B cell-specific genes that are required for further differentiation [70]. In addition, Ebf1 is an essential component of the transcriptional program that regulates survival and proliferation in B cells. This so far unknown role of Ebf1 was uncovered by studies showing that the deletion of Ebf1 in various B cell subsets results in rapid cell death [71]. Genome-wide analysis of Ebf1-dependent transcriptional targets revealed that Ebf1 regulates the expression of several molecules involved in proliferation and survival [71]. This points to a key role of Ebf1 downstream of the IL-7R by directly connecting gene transcription of survival/proliferation-specific molecules to the expression of B cell-specific genes that promote further differentiation of pro-B cells.

An important gene that is unequivocally required for further B cell development and is induced by Ebf1, is the transcription factor paired box protein 5 (Pax5; encoded by *BSAP*) [72,73]. Notably, downstream of the IL-7R, Pax5 can also be induced independent of Ebf1, as it was shown that ectopic expression of a constitutively active form of STAT5 was able to induce Pax5 expression in T cell progenitors, which usually do not express Ebf1 [74,75]. Consistent with this, functional binding sites for STAT5 have been identified in the Pax5 promoter [74]. Therefore, the IL-7R/STAT5 axis can induce Pax5 expression directly as well as indirectly via Ebf1. Within the hematopoietic system, Pax5 is only expressed in B cells and considered to be a key factor for B cell lineage commitment [76]. This is based on the findings that Pax5 promotes expression of numerous prominent proteins that are essential for B cell development, including CD19, immunoglobulin alpha (Ig-α; encoded by *MB-1*) and SH2 domain-containing leukocyte protein of 65 kDa (SLP-65; encoded by *BLNK*) [76]. Moreover, Pax5 simultaneously represses B cell lineage inappropriate genes such as *csf1r* and *notch1*, whose products (CSF1R and Notch1) are essential for myeloid and T cell development, respectively [76]. In line with this instructive role of Pax5 in early B cell development, Pax5-deficient mice show a B cell developmental block at the pro-B-cell stage with cells retaining the ability to develop into T cells, macrophages, or NK cells [76–78]. This indicates that as soon as B lineage precursors receive continuous input via the IL-7/IL-7R signaling axis leading to expression of Ebf1 and Pax5, B cell-specific gene expression is induced, and the potential of developing B cells to respond to growth factors of other cell lineages becomes reduced. The important role of the IL-7R in B cell lineage specification is further demonstrated by the observation that Pax5-deficient progenitor B cells are only able to differentiate into other hematopoietic lineages upon withdrawal of IL-7 [79]. It is notable that complete deletion of Pax5 seems not to be required to confer alternative lineage potential to early B cell precursors, as it was shown that already slight reductions in Pax5 expression were sufficient to allow conversion of B cell precursors into myeloid cells [80]. It is therefore possible, that during the dynamic process of hematopoiesis, the amount of IL-7 available for precursor B cells is a critical determinant for their choice to differentiate further along the B lineage or to adopt other hematopoietic lineages.

Another process how the IL-7R promotes the differentiation of B cell precursors is by regulating V(D)J recombination at the *IgH* locus. In fact, D to $J_H$ joining is not impaired in developing B cells lacking IL-7 or IL-7Rα, but the rearrangement of the distal $V_H$ genes is severely perturbed [81]. This indicates that IL 7R-derived signals are dispensable for $DJ_H$ recombination but essential for proper $V_H$ to $DJ_H$ joining and for broad $V_H$ gene usage. It appears that IL-7R signaling plays a role in regulating distal $V_H$ gene accessibility, as IL-7 treatment increases histone acetylation and nuclease sensitivity of the distal $V_H$J558 gene segments [82,83]. STAT5 is thought to be the link between $V_H$ gene accessibility and IL-7, as reintroduction of a constitutively active STAT5 in STAT5-deficient pro-B cells increased histone acetylation, germline transcription, as well as recombination of distal $V_H$ genes [84]. The perturbed distal $V_H$ gene recombination in the absence of IL-7R signaling is also in line with the above-mentioned induction of Pax5 by

IL-7R signaling. Pax5 itself was shown to be essential for distal $V_H$ gene rearrangement by inducing compaction of the *IgH* locus, thereby bringing distal $V_H$ genes into proximity to the rearranged $DJ_H$ gene segments [85]. In addition, Pax5 promotes recombination of the $V_H$ genes by recruiting the recombination machinery including the recombination activation genes 1 and 2 (Rag1 and 2, hereafter collectively referred to as Rag1/2; encoded by *RAG1* and *RAG2*, respectively) to the recombination signal sequences located 3′ of every $V_H$ gene [76,86]. Studies suggest that Pax5 requires the help of the transcription factor Yin Yang 1 (YY1) and Ikaros to establish distal $V_H$ gene rearrangement [87,88]. It is therefore tempting to speculate that IL-7R signaling regulates Pax5, and probably also YY1 and Ikaros, leading to efficient recombination of the distal $V_H$ genes.

## 2. PROLIFERATION AND DIFFERENTIATION PROGRAMS AT THE PRE-B CELL STAGE

The successful recombination of the *IgH* genes at the pro-B cell stage leads to the expression of the pre-BCR complex and transition to the pre-B cell stage. The pre-BCR complex contains two identical μHCs each associated with a surrogate light chain (SLC) composed of two germline-encoded proteins, VpreB and λ5, as well as the signaling transducing subunits immunoglobulin (Ig) alpha (Ig-α) and Ig-β [9,89–91]. Each single μHC protein carries one variable (Vh) domain, four constant ($C_h1$–$C_h4$) domains, and a transmembrane region, which anchors the receptor to the plasma membrane [89]. Expression of the pre-BCR is the first important checkpoint in B cell development by establishing, similar to the IL-7R at the pro-B cell stage, molecular programs inducing proliferative expansion and further differentiation of pre-B cells.

### 2.1 Signal Induction from the Pre-BCR

Several mechanisms have been suggested for how pre-BCR downstream signaling is initiated. One mechanism is based on the interaction of the pre-BCR with external ligands such as heparan sulfate and galectin-1, both presented by or derived from stromal cells in the bone marrow [92–94]. In line with this mechanism, secreted pre-BCRs are polyvalent and able to bind to multiple antigens [95]. However, surface expression of the pre-BCR is not inevitably required for triggering the downstream signaling cascades, as pre-BCR signaling can already be initiated upon pre-BCR assembly within the *trans*-Golgi network compartment [96]. In addition, biochemical analysis revealed that the pre-BCR is impaired in binding antigen and rather self-associates in a ligand-independent manner [97]. This is in line with the finding that signaling from the pre-BCR is induced autonomously and that this is dependent on the arginine-rich tail of the SLC component λ5 [95,98,99]. Further studies revealed that an N-linked glycosylation site (N46) in the $C_h1$-domain of the μHC, which is conserved between mouse and human, is critical for pre-BCR formation and function [100]. It was shown that the λ5 tail is able to interact with the $C_h1$-domain of the μHC, but failed to bind the μHC $C_h1$-domain in the absence of the N46 glycosylation site [100]. This interaction might provide the structural basis for receptor assembly by releasing λ5 from its autoinhibited form and enabling the formation of a functional SLC able to associate with another μHC protein, thereby inducing ligand-independent pre-BCR signaling in an autonomous manner.

### 2.2 Signals from the Pre-BCR Induce Both Proliferation and Differentiation of Pre-B Cells

The signaling cascades activated by the pre-BCR induce, on one hand, a phase of intense proliferation that increases the amount of cells having successfully recombined the *IgH* genes. On the other hand, pre-BCR signaling was shown to lead to the downregulation of the pre-BCR, as well as the initiation of *IgL* gene recombination [9]. Hence, the pre-BCR is able to establish both proliferation and differentiation programs of developing pre-B cells. It is therefore not surprising that defects in pre-BCR expression lead to severe developmental defects during the early stages of B cell development. For instance, developing B cells that cannot express a functional μHC protein, due to deletion of the μHC transmembrane region, are blocked at the pro-B cell stage and unable to proceed in differentiation [101]. Similarly, mice lacking either Ig-α or Ig-β have a complete block at the pro-B cells stage, as a signaling competent pre-BCR can only be expressed in the presence of both signaling transducing subunits [102,103]. Consistent with this, also in mice lacking the SLC components λ5 or VpreB, developing B cells are impaired in transitioning from the pro-B to the pre-B cell stage [104,105]. However, the developmental block caused by SLC deficiency is not complete, indicating that pre-B cells are able to compensate for the loss of a normal pre-BCR under certain circumstances. Premature expression of a conventional light chain (LC), together with the μHC protein, or receptors consisting only of the μHC protein, were shown to compensate for the loss of the pre-BCR, possibly because they can generate the signals required to establish the pre-B cell survival and differentiation programs [106–109]. Notably, the partial rescue of B cell development in the absence of SLC only occurs in mice, whereas humans with deficiency in SLC show severe B cell deficiency and agammaglobulinemia [110], indicating that pre-BCR expression and signaling are fundamental for human B cell development.

The signals induced by a normal pre-BCR are transmitted via the signaling transducing subunits Ig-α and Ig-β and involve the activation of Src-kinases (Lyn, Fyn, and Blk) that are continuously associated with the pre-BCR complex [111]. Activation of the Src-kinases leads to phosphorylation of the immunoreceptor tyrosine-based activation motifs (ITAMs) present within the cytoplasmic tails of Ig-α and Ig-β [112,113]. The spleen tyrosine kinase (Syk) is recruited to these phosphorylated ITAMs by its tandem SH2 domain and is subsequently activated by (auto)phosphorylation [114]. Interestingly, the Syk-kinase activity can be further enhanced by phosphorylation of the upstream Src-kinases and by Syk-mediated phosphorylation of ITAMs from neighboring pre-BCRs, thereby creating an amplification loop for signal initiation [114]. Extensive efforts in the recent years revealed that three major signaling pathways are activated by the pre-BCR. The activation of the PI3K and MAPK/Erk signaling cascades is, for instance, linked to the induction of survival and proliferation programs [20,115], whereas Syk-dependent activation of SLP-65 is necessary to initiate differentiation processes that result in developmental progression [9,115]. Consistent with this, B cell development is arrested at the pre-B cell stage in Syk-deficient mice [116].

The extensive phase of proliferation triggered by the pre-BCR is supported by an upregulation of IL-7R expression and thus IL-7R signaling [117–119]. This provides pre-BCR-positive cells with a selective advantage over pre-BCR-negative cells, leading to an enrichment of pre-B cells in the bone marrow. Interestingly, pre-BCR signaling and IL-7R signaling were shown to use similar signaling cascades that share intense crosstalk and lead to activation of a similar set of genes to promote survival and proliferation [120]. For instance, as is the case for the pre-BCR, the IL-7R is able to activate PI3K signaling [23], and therefore PI3K-deficient pre-B cells show defective proliferation in response to IL-7 stimulation [121]. However, the precise role of PI3K downstream of the IL-7R is not completely clear to date, as introduction of a mutation into the IL-7Rα chain, which abrogates activation of PI3K, does not block B cell development [122].

## 2.3 Role of Class IA PI3K in Pre-BCR-Induced Proliferation

As mentioned above, expression of the pre-BCR on large pre-B cells induces a phase of intense proliferation and survival that are largely mediated by PI3K signaling. In mammals, three classes of PI3K are expressed, termed class I, II, and III [123]. The class I of PI3K evolved as especially important for PI3K signaling in lymphocytes and is further subdivided into class IA and class IB. The class IB of PI3K comprises one catalytic subunit (p110γ) that is associated with one of two regulatory subunits (p84 and p101) and becomes predominantly activated downstream of G protein-coupled receptors [123]. The class IA of PI3K instead is preferentially activated upon antigen receptor engagement [123]. In B cells, three class IA PI3K catalytic subunits (p110α, p110β, and p110δ) are expressed, and they form heterodimers with one of five regulatory subunits (p85α, p85β, p55α, p55γ, and p50α) [124]. Because every regulatory subunit can associate with every catalytic subunit, the class IA of PI3K shows great redundancy and studying the specific roles of selective catalytic or regulatory subunits is very difficult because of compensation by the remaining subunits. Downstream of the pre-BCR, signaling through class IA PI3K is initiated by Lyn- or- Syk-mediated phosphorylation of a common tyrosine-motif (YxxM) present in adaptor proteins such as CD19 and B cell adaptor protein (BCAP) (Figure 2) [124,125]. Binding of a class IA PI3K heterodimer to phosphorylated YxxM-motifs within CD19 or BCAP via its SH2 domains results in activation of the catalytic subunit. It is notable that members of the Ras-family that directly bind to the Ras-binding domain present in every catalytic subunit can also activate p110α, p110β, and p110δ independent of a regulatory subunit [123,126]. Because pre-BCR signaling was shown to activate Ras-dependent signaling pathways [127], it is possible that the activation of specific Ras-family members by pre-BCR signaling results in activation of distinct class IA PI3K catalytic subunits, which would add another level of complexity to the class IA PI3K signaling system downstream of the pre-BCR.

After activation, class IA PI3Ks catalyze the generation of the signaling intermediate phosphatidylinositol (3,4,5)-trisphosphate [PI(3,4,5)$P_3$] from phosphatidylinositol (4,5)-bisphosphate [PI(4,5)$P_2$] at the inner leaflet of the plasma membrane. PI(3,4,5)$P_3$ provides a molecular scaffold for signaling molecules possessing a pleckstrin homology (PH) domain, resulting in recruitment of selected signaling proteins to the plasma membrane and establishment of important signaling complexes involved in the regulation of survival, proliferation and differentiation of pre-B cells [128]. Important proteins possessing PH domains are protein kinase B (PKB or AKT), 3-phosphoinositide-dependent protein kinase-1 (PDK1), the Tec-family kinase Bruton's tyrosine kinase (Btk) or phospholipase C gamma 2 (PLC-γ2) [123]. Upon recruitment to the plasma membrane via its PH domain, Btk is rendered active and able to phosphorylate PLC-γ2, an enzyme generating signaling intermediates required to induce calcium flux from intracellular and extracellular stores [129–131]. The importance of class IA PI3K signaling for Btk activation is impressively demonstrated by the observation that a mutation within the PH domain of Btk is as severe as complete deletion of Btk, and similar to p85α or p110δ deficiency [123,129]. In humans, Btk inactivation is responsible for the X-linked agammaglobulinemia, a severe immunodeficiency syndrome characterized by the lack of peripheral B cells and hypogammaglobulinemia [132,133].

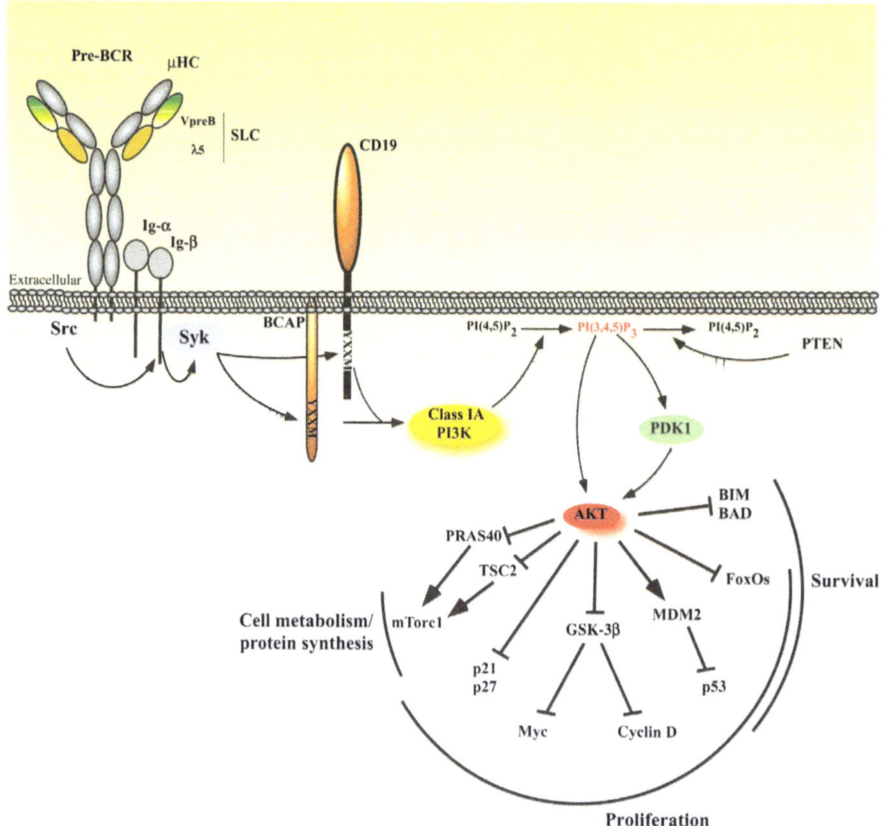


FIGURE 2 **The activation of PI3K signaling by the pre-BCR.** The pre-BCR consists of two identical μHCs, each associated with the invariant SLC, composed of λ5 and VpreB. The pre-BCR signals through the Ig-α and Ig-β heterodimer, resulting in the activation of Src- and Syk-kinases. Phosphorylation of YXXM-motifs within the coreceptors CD19 and BCAP by Src- and Syk-kinases then lead to the recruitment and activation of class IA PI3K, which converts $PI(4,5)P_2$ at the inner leaflet of the plasma membrane to $PI(3,4,5)P_3$. $PI(3,4,5)P_3$ functions as a molecular scaffold for proteins possessing PH domains, such as PDK1 or AKT. Upon full activation by PDK1, AKT promotes the survival and proliferation of pre-B cells by regulating expression and function of numerous proteins.

So far, the investigations of how PI3K signaling regulates survival and proliferation have mainly focused on the AKT signaling axis (Figure 2). In mammals, three highly related AKT isoforms encoded by three different genes are expressed (AKT1, AKT2, and AKT3, hereafter collectively referred to as AKT) [134]. It is thought that all three AKT isoforms share similar and thus overlapping substrate specificities. However, with the help of isoform-specific deletions, specific functions of the different AKT isoforms are becoming apparent, although, identifying the specific roles of the AKT isoforms awaits further analysis. Activation of the serine/threonine kinase AKT depends on sequential phosphorylation of two highly conserved amino acid residues. The phosphorylation of AKT at serine 473 is mediated by the rapamycin-insensitive mammalian target of rapamycin (mTOR) complex 2 (mTORC2, consisting of mTOR, Rictor, Sin1, and MLST8) and seems to be a precondition for the PDK1-mediated phosphorylation on threonine 308, which renders AKT fully active [128].

## 2.4 Regulation of Survival and Cell Cycle by the Class IA PI3K Downstream Target AKT

AKT is responsible for most of the class IA PI3K-mediated effects on survival and cell cycle regulation and controls these processes by phosphorylating its target proteins possessing serine/threonine residues within the consensus motif RXRXXS/T (R represents arginine, X represents any amino acid) (Figure 2) [135]. For instance, AKT-mediated phosphorylation of the proapoptotic protein BAD, which binds directly to and inactivates members of the prosurvival Bcl2-family, results in dissociation of Bad from its target proteins and thereby promotes survival [136,137]. In addition, AKT phosphorylates the E3 ubiquitin-ligase mouse double minute 2 homolog (MDM2; encoded by *MDM2*) at serine 166 and serine 186, which enables MDM2 to translocate into the nucleus [138,139]. An important function of nuclear MDM2 is the marking of tumor protein 53 (p53; encoded by *TP53*) for proteasomal degradation, thereby preventing p53-dependent induction of proapoptotic proteins and cell cycle arrest [139,140].

As mentioned above, for cell cycle entry as well as cell cycle progression, cells have to activate specific G1-cyclins and CDKs. Concomitantly, cell cycle inhibitors, which impair cyclin or CDK activity, have to be either downregulated or inhibited. AKT directly phosphorylates the cell cycle inhibitors p21 and p27. The AKT-mediated phosphorylation of p21 reduces its CDK2 inhibitory activity, and the phosphorylation of p27 renders p27 unable to enter the nucleus and to suppress the cell cycle [141–144]. In addition, class IA PI3K/AKT signaling regulates p27 protein stability as p27 levels are reduced upon activation of class IA PI3K and increased upon downregulation of class IA PI3K activity by the pan-class IA PI3K inhibitor LY294002 or introduction of a kinase-dead version of AKT [145]. Hence, by negatively regulating the activity, localization, and protein stability of the cell cycle inhibitors p21 and p27, AKT directly abrogates the p21/p27-mediated G1 cell cycle arrest and thereby promotes cell cycle progression. Another prominent target of AKT is glycogen synthase kinase 3 beta (GSK-3β), which marks cylin D, cyclin E, and Myc for proteasomal degradation by phosphorylating threonine 286 of cyclin D, tyrosine 380 of cyclin E, and threonine 58 of Myc [146–148]. Intriguingly, the phosphorylation of Myc on threonine 58 by GSK-3β is dependent on prior phosphorylation of serine 62 by Erk [55]. The activity of GSK-3β is negatively regulated by AKT, meaning that an activated class IA PI3K/AKT signaling axis relieves Myc, cyclin D, and cyclin E from GSK-3β-mediated repression.

AKT also regulates the rapamycin-sensitive mTOR complex 1 (mTORC1) that promotes cell cycle and stimulates protein synthesis as well as cell growth [134]. AKT activates mTORC1 by inhibiting the tuberous sclerosis complex 2, a tumor suppressor that effectively counteracts mTORC1 activation by functioning as a GTPase-activating protein for the Ras-related small G protein Rheb [149]. Rheb, when bound to guanosine triphosphate, strongly activates mTORC1 [149]. Another mechanism how AKT can activate mTORC1 involves the proline-rich AKT substrate of 40 kDa (PRAS40), which binds to and inhibits mTORC1 [150]. Active AKT phosphorylates PRAS40 at threonine 246, which marks PRAS40 for 14-3-3 binding and proteasomal degradation [151]. Hence, there exist at least two independent ways how AKT can activate mTORC1. Whether both act redundantly or independently of each other is currently not clear. Treatment of proliferating tumor B cells with the chemical compound rapamycin blocked G1-to-S phase transition by downmodulating expression of cyclins A, D, and E [152], impressively demonstrating the importance of mTORC1 for cell cycle.

It is also noteworthy that signaling through PI3K/AKT promotes cell survival and proliferation by sharing crosstalk with other signaling pathways. A prominent example is the AKT/mTORC1-dependent activation of nuclear factor-κB signaling, which promotes survival and proliferation in various cell types [153]. Signaling through class IA PI3K also inhibits the stress-induced proapoptotic c-Jun $NH_2$-terminal kinase (JNK) and p38 signaling cascades by inactivating the JNK/p38 upstream kinase apoptosis signal-regulated kinase 1 [154]. Thus, although class IA PI3K signaling plays a major role downstream of the pre-BCR in inducing survival and proliferation, it does not act separately but is rather a central component of a complex signaling network within developing B cells.

## 2.5 The Family of FoxO Transcription Factors is a Highly Conserved Downstream Target of PI3K Signaling

Besides modifying the activity and function of survival/proliferation-specific factors on the protein level, class IA PI3K signaling also controls the transcription of genes implicated in the regulation of survival or cell cycle. The family of FoxO transcription factors is a highly conserved downstream target of AKT, and it comprises, in mammals, FoxO1, FoxO3a, FoxO4, and FoxO6, all of which are orthologous to the *Caenorhabditis elegans* transcription factor DAF-16 [155]. FoxO transcription factors are controlled by several posttranslational modifications, including ubiquitination, acetylation, and phosphorylation [156]. Specifically, AKT, and accordingly, class IA PI3K, regulates FoxOs through phosphorylation of three highly conserved serine/threonine residues [157]. This phosphorylation pattern creates binding sites for 14-3-3 chaperones that guide FoxO transcription factors out of the nucleus into the cytoplasm, where FoxOs are degraded in a proteasome-dependent manner [155]. It is generally thought that FoxO transcription factors negatively regulate survival and cell cycle. For instance, FoxOs induce expression of the proapoptotic factors p53, the BH3-only family member Bcl2-like protein 11 (Bim; encoded by *BCL2L11*), and Fas ligand (FasL; encoded by *FASLG*) and repress expression of cyclin D, which is essential for cell cycle entry [158–160]. Another important target of FoxOs involved in the regulation of the cell cycle is the CDK inhibitor p27, which blocks progression from G1 to S phase by binding to CDK2-E-type cyclin complexes [160]. Together, it appears likely that the activation of class IA PI3K signaling by the pre-BCR promotes cell cycle entry and progression as well as cell survival by reducing the expression levels of FoxO transcription factors.

## 2.6 Differentiation of Pre-B Cells Involves Allelic Exclusion of the IgH Gene Locus

Normally, incorporation of the μHC protein into a pre-BCR generates the signal for successful recombination of one of the two *IgH* alleles. That signals from a pre-BCR are involved in establishing allelic exclusion became apparent by studies showing that impaired pre-BCR signaling, either by preventing pre-BCR expression or by introducing mutations within the cytoplasmic tails of Ig-α or Ig-β, perturbs allelic exclusion and blocks pre-B cell development [161,162]. Consistent with these data, the absence of particular signaling components downstream of the pre-BCR interferes with allelic exclusion. For instance, in the absence of Syk and its family member zeta-chain-associated protein kinase 70, pre-B cells show strongly impaired allelic exclusion of the *IgH* gene loci [163]. Also, defective activation of the Syk-downstream signaling target class IA PI3K, by deletion of p110α and concomitant inactivation of p110δ, leads to ongoing recombination of the *IgH* genes and impaired allelic exclusion [121]. Together, these data strongly support a model in which signals by the pre-BCR provide feedback for successful recombination of one of the *IgH* genes by inducing allelic exclusion of the second *IgH* allele and making it inaccessible for the Rag1/2-dependent recombination machinery. According to this model, a nonproductive recombination of the first allele does not result in expression of the pre-BCR complex and the second allele remains accessible for the Rag1/2-dependent recombination machinery. Recombination of the second *IgH* allele may then result in the generation of a functional pre-BCR, which promotes further development. In addition, the pre-BCR-driven activation of class IA PI3K supports the termination of recombination processes at the *IgH* genes by downregulating FoxO1 activity and Rag1/2 expression, thereby resulting in inactivation of the recombination machinery [115,164–166].

## 2.7 Pre-BCR Signals Induce Differentiation by Activating SLP-65-Dependent Signaling Pathways

It is notable that the ubiquitously expressed class IA PI3K signaling system is not only involved in promoting pre-B cell expansion and survival, but also in regulating highly B cell-specific processes such as *IgH* gene recombination or allelic exclusion. It appears therefore likely that class IA PI3K signaling downstream of the pre-BCR is connected to B cell-specific signaling elements to regulate class IA PI3K activity in the B cell context. Recent work revealed that the adaptor protein SLP-65 regulates class IA PI3K activity and is thus a key molecule for pre-B cell differentiation by interconnecting multiple signaling cascades downstream of the pre-BCR or BCR (Figure 3) [167]. Signaling through SLP-65 depends on Syk-mediated phosphorylation of five highly conserved tyrosine residues (tyrosine 72, 84, 96, 178, and 189 in SLP-65) [168,169]. The phosphorylation of these tyrosines results in the relocalization of SLP-65 from the cytoplasm to the plasma membrane, as well as recruitment of various signaling molecules possessing SH2 domains such as Btk (to tyrosine 178) and PLC-γ2 (to tyrosine 189) [167,170]. By these mechanisms, SLP-65 promotes the establishment of signaling complexes near the plasma membrane and brings them into proximity to other signaling molecules such as Src-kinases, Syk, or class IA PI3K. SLP-65 exerts an important role in the activation of calcium-dependent signaling pathways by recruiting PLC-γ2 that cleaves $PI(4,5)P_2$ into the second messengers inositol triphosphate ($IP_3$) and diacylgycerol. Binding of $IP_3$ to its receptor $IP_3R$ located at the endoplasmic reticulum (ER) results in release of calcium ions ($Ca^{2+}$) from the ER into the cytosol. This $Ca^{2+}$ release provides further opening of extracellular $Ca^{2+}$ channels in the plasma membrane and an additional $Ca^{2+}$ influx from the extracellular space into the cytosol [171]. In B cells, the mobilization of $Ca^{2+}$ is for instance important to activate calcineurin-dependent signaling pathways that lead to activation of nuclear factor of activated T cells, a transcription factor important for numerous cellular processes [167,172].

Besides establishing $Ca^{2+}$-dependent signaling pathways, signaling through SLP-65 induces the exit from the proliferative phase and thereby drives pre-B cells into differentiation. This important role of SLP-65 became apparent by studies showing that activation of SLP-65 downmodulates class IA PI3K activity, leading to stabilization of FoxO transcription factors [115]. How SLP-65 signaling interferes with PI3K activity is currently not clear as the regulation of class IA PI3K activity could occur at numerous levels. It is, for instance, possible that SLP-65 interferes with the activity of pre-BCR proximal signaling elements such as Src-kinases, Syk, CD19, or BCAP, or that SLP-65 inhibits class IA PI3K activity through yet not defined mechanisms. Alternatively, SLP-65 may activate lipid phosphatases such as phosphatase and tensin homolog (PTEN; encoded by *PTEN*) or SH2-containing inositol phosphatase 1 (SHIP; encoded by *SHIP*), which dephosphorylate $PI(3,4,5)P_3$ and thereby abrogate membrane recruitment and activation of downstream signaling molecules such as AKT or PDK1 [124]. Finally, SLP-65 may activate the PH domain and leucine-rich repeat protein phosphatases 1 and 2 (PHLPP1 and PHLPP2) that remove the phosphate at serine 473 of Akt [173,174]. It is notable that besides the downregulation of class IA PI3K activity, additional mechanisms through which SLP-65 counteracts proliferation exist. SLP-65-dependent signaling, for instance, was shown to induce expression of Ikaros and Aiolos (encoded by *IKZF1* and *IKZF3*, respectively), transcription factors that directly repress λ5 gene transcription and pre-BCR expression [98,175,176]. The termination of pre-BCR expression and signaling may therefore represent another mechanism for how activation of SLP-65 results in downmodulation of class IA PI3K signaling. Moreover, Ikaros and Aiolos counteract proliferation by inducing expression of p27 and by repressing gene transcription of Myc and cyclin D3 [177]. Based on these

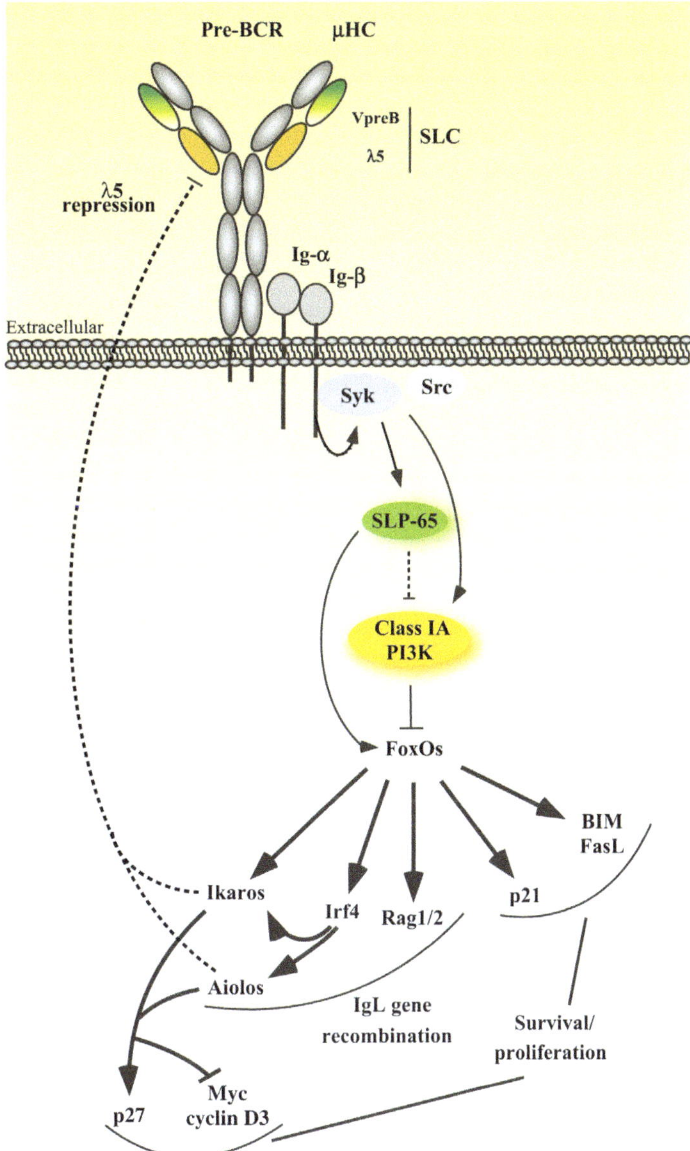


FIGURE 3 **Activation of SLP-65 by the pre-BCR.** Activation of Syk leads to activation of class IA PI3K and SLP-65. The Syk-dependent activation of SLP-65 results in the activation of FoxO transcription factors by inhibiting class IA PI3K activity through as yet unidentified mechanisms. FoxOs exert an important role at the pre-B cell stage by promoting cell cycle arrest (e.g., by inducing p21), as well as by activating *IgL* gene recombination through the induction of Rag1/2 and Ikaros. In addition, Ikaros and Aiolos lead to repression of λ5 and thereby pre-BCR expression.

findings, it is not surprising that defects in the activation or expression of SLP-65 result in a B cell developmental block at the pre-B cell stage with cells showing stable pre-BCR surface expression and mice having increased incidence of pre-B cell leukemia [178,179]. In addition, a recent study showed that the pre-BCR-induced cell cycle arrest in BCR-ABL-positive acute lymphoblastic leukemia (Ph+ALL) critically depends on SLP-65 and Ikaros. This might provide a molecular explanation for the high frequencies of Ikaros deletions in Ph+ALL and suggest that SLP-65 exerts its tumor suppressor function, at least partially, by activating Ikaros [180]. Another important mechanism for how SLP-65 prevents tumorigenesis is based on the ability of SLP-65 to interact with and inhibit the IL-7R downstream kinase JAK3. In line with this, a hyperactive IL-7R/JAK3 signaling axis that results in enhanced STAT5 activity promoting survival and proliferation is characteristic of SLP-65-deficient pre-B cell tumors [181]. In addition, SLP-65-deficient pre-B cell tumors were described to produce IL-7 in an autocrine manner, suggesting that these cells do not depend on external IL-7 and that signaling from the IL-7R is constitutively triggered in these cells, in a cell autonomous manner [181]. Therefore, SLP-65 is also essential to control IL-7-responsiveness in developing B cells.

## 2.8 SLP-65-Dependent Signaling Promotes IgL Gene Recombination and Differentiation

In addition to promoting the exit from the proliferative phase, the characterization of SLP65-deficient cells revealed an important role of SLP-65 in the initiation and regulation of *IgL* gene recombination (Figure 3), which is necessary to replace the SLC of the pre-BCR with a conventional LC and thus to express a mature BCR complex on the cell surface. The downmodulation of class IA PI3K activity and the concomitant stabilization of FoxO transcription factors is key for the initiation of the recombination processes at the *IgL* genes, because FoxOs directly induce expression of Rag1 and 2 [115,182]. This important role of FoxO transcription factors for *IgL* gene recombination became apparent in studies showing that, upon inactivation of FoxO1 at the late pre-B cell stage, the transition of pre-B cells to the immature and mature B cell stages was impaired [183]. Consistent with this, a pre-B cell population lacking *IgL* gene recombination could be detected in the periphery of these mice [183], which supports the crucial function of FoxO1 in *IgL* gene rearrangement. As outlined above, FoxO transcription factors also induce expression of cell cycle inhibitors and repress expression of cyclins. Cell cycle regulation is essential for *Ig* gene recombination because Rag2 becomes rapidly degraded is S phase of the cell cycle due to phosphorylation by cyclin A-CDK2 complexes [184]. Therefore, the Rag recombination complex is stably expressed and able to initiate *Ig* gene recombination only in specific phases of the cell cycle.

In addition to Rag1/2, FoxO1 was recently shown to induce Ikaros [88]. Upon binding to DNA, Ikaros forms chromatin-remodeling complexes in cooperation with other proteins, thereby regulating gene accessibility and transcription of its target genes [185]. FoxO1 does not regulate basal gene transcription of Ikaros but rather seems to be essential for proper Ikaros splicing [88]. How FoxO1 regulates the splicing of Ikaros is not yet completely understood. However, there is to date no evidence available that FoxO1 itself can function as a splicing factor. It appears therefore likely, that the effects of FoxO1 on Ikaros splicing are indirect.

Another factor induced by SLP-65-dependent signaling is the interferon regulatory factor 4 (Irf4), a member of the Irf transcription factor family [98]. Irf4 can bind to DNA either alone or in cooperation with other transcription factors, thereby functioning as a repressor or activator of gene transcription, respectively [186]. It is notable that in adipocytes FoxO1 directly induces Irf4 [187] and that in pre-B cells, the deletion of FoxO1 leads to a loss of Irf4 expression [188]. This strongly suggests that at the late pre-B cell stage, Irf4 becomes induced through the SLP-65-dependent stabilization of FoxO1. A role of Irf4 in *IgL* gene recombination became apparent by analyzing pre-B cells lacking Irf4. As is true for *IgH* gene recombination, germline transcription of the *IgL* gene loci is a precondition for *IgL* gene rearrangement by making the different *IgL* genes accessible to the Rag1/2-dependent recombination machinery. Irf4 has been shown to induce Igκ germline transcription and *Igk* recombination cooperatively with the Ets-family transcription factor Spi-B [189,190]. Accordingly, Igκ gene recombination is severely perturbed in Irf4-deficient pre-B cells [189–192]. In addition, Irf4 induces the expression of Ikaros and its family member Aiolos, thereby also promoting cell cycle exit as well as pre-BCR downregulation through the above-mentioned mechanisms [193].

Together, activation of SLP-65 results in the activation of a transcription factor circuitry that collectively regulates pre-BCR surface expression, cell cycle exit, and *IgL* gene recombination (Figure 3). The action of these transcription factors, switches off proliferation programs and installs differentiation programs that ultimately result in the exchange of the SLC by a conventional LC and the expression of a mature BCR complex on the cell surface.

## 2.9 Interplay of the Proliferation and Differentiation Programs at the Pre-B Cell Stage

The above-described molecular programs downstream of the pre-BCR provide the basis for an ordered and sequential occurrence of proliferation and differentiation at the pre-B cell stage. The successful recombination of one of the two *IgH* alleles at the pro-B cell stage, results in expression of the μHC protein together with an SLC and the signaling transducing subunits Ig-α and Ig-β on the cell surface. The pre-BCR-dependent activation of Src- and Syk-kinases then predominantly activates the class IA PI3K/AKT signaling axis leading to proliferation. In the absence of SLP-65, pre-B cells persist in this proliferative state, possibly leading to cellular transformation and tumorigenesis. However, in the presence of SLP-65, Syk-dependent activation of SLP-65 results in downmodulation of class IA PI3K activity, thereby terminating pre-B cell proliferation and allowing the establishment of the differentiation-inducing program. How pre-BCR signals are integrated to ensure that proliferation precedes differentiation is currently not completely clear. However, it appears likely that to allow several cycles of proliferation before the differentiation processes are initiated, the activation of SLP-65 has to be delayed relative to the activation of class IA PI3K. One possibility is that at the early pre-B cell stage the expression level of SLP-65 is too low to induce differentiation processes and that expression of SLP-65 first has to be induced during the pre-B cell stage. Interestingly, it was shown that a reduction of IL-7R signaling induces the expression of SLP-65 in pre-B cells [188]. In vivo, one could therefore imagine that once pro-B cells, whose survival and proliferation is critically dependent on stromal cells producing IL-7, recombine their *IgH* genes and start to express a pre-BCR, they undergo an intense phase of proliferation. Assuming the amount of IL-7 is limited within this niche, proliferating pre-B cells compete with each other for the available IL-7, resulting in attenuation of IL-7R signaling and induction of SLP-65. Once a critical threshold of SLP-65 is exceeded, the pre-BCR-dependent activation of SLP-65 would then terminate proliferation and induce differentiation programs. Intriguingly, FoxO1 and Irf4, which become, as mentioned, activated downstream of SLP-65, induce the expression of CXCR4 and possibly thereby the movement of differentiating pre-B cells away from the IL-7-producing stromal cells toward CXCL12$^{hi}$-expressing stromal cells, which do not secrete IL-7 [188,191]. As in addition the expression levels of SOCS1 were shown to peak at the small pre-B cell stage [25], this suggests that IL-7 consumption may act in concert with the movement of small pre-B cells away from IL-7-producing stromal cells and the upregulation of negative regulators of IL-7R signaling to augment SLP-65 expression and the activation of the pre-B cell differentiation program.

It is noteworthy that this model is only applicable for murine B cell development, which is clearly dependent on IL-7. As mentioned, human B cell development in contrast seems to occur independent of IL-7, suggesting that other mechanisms have to exist that ensure a delayed activation of SLP-65 in the human context. Possibly, a mechanism similar to the IL-7R-dependent induction of SLP-65 in murine pre-B cells could increase SLP-65 expression in human pre-B cells during their proliferative expansion. Interestingly, the phenotypes of various mice defective in the activation of class IA PI3K suggest that the pre-BCR itself may generate the signals promoting establishment of the differentiation-inducing programs. For instance, the combined deletion of CD19/BCAP severely perturbs activation of class IA PI3K/AKT signaling downstream of the pre-BCR/BCR, and B cell development in CD19/BCAP double-deficient mice is blocked at the large pre-B cell stage, with cells being unable to differentiate into small pre-B cells [125]. Also, the deletion of particular regulatory or catalytic class IA PI3K subunits results in an impaired pro-B to pre-B transition. Absence of p110α combined with mutational inactivation of p110δ, for example, leads to an almost complete block in B cell development at the pre-B cell stage [121], similar to CD19/BCAP-deficient mice. Interference with activation of AKT by deletion of AKT1 together with AKT2, leads to an accumulation pro-B and pre-B cells in the bone marrow, whereas the more mature B cell stages are severely reduced [194]. Collectively, these studies indicate that the class IA of PI3K plays an important role in pre-B cell differentiation. In line with this, PDK1-deficient pre-B cells, which have reduced AKT activity, show reduced levels of SLP-65 [195], supporting the idea that defects in the class IA PI3K/AKT signaling system downstream of the pre-BCR interfere with pre-B cell differentiation. A positive correlation between class IA PI3K and SLP-65 can also be observed in individuals lacking the regulatory p85α subunit and showing almost complete absence of CD19$^{+}$ B cells, as do SLP-65-deficient human individuals [196,197]. It is to date unclear how class IA PI3K/AKT activity is linked on the molecular level to SLP-65 expression. However, the idea that the pre-BCR-induced class IA PI3K/AKT signals, which promote survival and proliferation, are coupled to the basal expression level of SLP-65 provides an elegant mechanism for how activation of SLP-65 would be delayed relative to the activation of class IA PI3K, downstream of the pre-BCR. Most importantly, because the induction of SLP-65 does not depend on signaling through the IL-7R or other cytokine receptors, this model could also be applied to human pre-B cells.

## 2.10 The Conventional BCR Replaces the Pre-BCR in Providing Survival Signals

The process of *IgL* gene recombination allows pre-B cells to replace the SLC by a conventional LC and to develop into immature B cells. However, during the process of *IgL* gene recombination, late pre-B cells are deprived from survival signals, due to the above-mentioned SLP-65-mediated repression of both class IA PI3K and IL-7R activity. This dampening of survival and proliferation signaling is on one hand important to allow activation of Rag1/2 and recombination processes, but on the other hand, it may trigger an apoptotic program if not terminated before a critical time point. It appears therefore likely that late pre-B cells have a fixed time frame in which the recombination processes have to be successfully accomplished to shut off the differentiation program and regain survival signals. There is accumulating evidence that the expression of a conventional BCR is essential for the maintenance of mature B cells. This became apparent by studies taking advantage of mouse models in which BCR surface expression could by inducibly deleted resulting in deletion of mature B cells within 1–2 weeks [198]. Later studies investigating the nature of the basal BCR signal revealed that a resting BCR transmits these survival-inducing signals via the class IA PI3K signaling cascade [199]. In fact, enhanced class IA PI3K signaling activity, either by introducing a constitutively active form of p110α or by deletion of PTEN, could rescue the survival of BCR-depleted B cells [199]. Consistent with the involvement of class IA PI3K/AKT signaling in basal BCR signaling, deletion of FoxO1 was sufficient to allow BCR-deficient B cells to survive for prolonged times [199]. These data suggest that a resting BCR activates class IA PI3K signaling, thereby establishing a B cell-specific survival program. How a resting BCR can trigger the activation of the class IA PI3K signaling cascade is not completely elucidated at present. One recent study suggested that the small GTPase Ras-related protein R-Ras2 (TC21; encoded by *RRAS2*) interacts continuously with the unphosphorylated ITAMs of a resting BCR and recruits the p110δ catalytic subunit, leading to the initiation of class IA PI3K downstream signaling and the generation of $PI(3,4,5)P_3$ [200]. The subsequent activation of AKT would then result in inactivation of FoxO transcription factors and to survival of mature B cells.

Besides the basal BCR signal, survival of mature B cells is dependent on the cytokine B cell-activating factor of the TNF family (BAFF) [201]. BAFF receptor (BAFFR) is expressed at low levels on immature B cells and increases during further development [201]. Interestingly, reminiscent of the pre-BCR-driven upregulation of IL-7R expression, signals generated by a conventional BCR influence the expression of BAFFR [202]. These signals depend on class IA PI3K, as deletion of p110δ or application of the class IA PI3K inhibitor wortmannin impairs the BCR-dependent upregulation of BAFFR expression [203]. This indicates that the generation of a conventional BCR complex and the accompanying basal class IA PI3K signals are a precondition for the establishment of the BCR/BAFFR survival system. Studies in which BCR and BAFFR were deleted revealed that the signaling cascades downstream of both receptors share intense crosstalk and depend on each other [204]. For instance, one study suggested that BAFF stimulation induces phosphorylation of the BCR signaling components Syk and Ig-α, leading to activation of BCR-related downstream signaling cascades such as PI3K [205]. Consistent with this, in the absence of Syk peripheral mature B cells were not able to respond to BAFF and as a consequence underwent massive cell death [205]. This suggests that efficient B cell maintenance relies on both expression of a conventional BCR and BAFF-BAFFR engagement.

## 3. SELECTION MECHANISMS AT THE IMMATURE B CELL STAGE

The variable nature of *IgH* and *IgL* gene recombination during early B cell development generates a mature B cell pool expressing BCRs that are able to recognize an almost unlimited number of foreign substances. However, this random assembly of gene segments bears the inherent risk of generating nonfunctional BCRs or BCRs with specificities directed against self-structures (Figure 4). Indeed, it has been reported that the majority of the early immature B cells express BCRs with marked self-reactivity, including specificities to several nuclear and cytoplasmic self-antigens such as DNA, insulin, lipopolysaccharide, or antinuclear antibodies [206]. It is traditionally thought that self-reactive receptor species have to be eliminated to prevent the destruction of self-structures in the periphery and development of autoimmune diseases. Several concerted mechanisms ensure the generation of a self-tolerant B cell pool and are collectively referred to as central tolerance. A prominent feature of central tolerance is clonal deletion, which results in the apoptotic elimination of self-reactive immature B cells. This was unraveled by studies showing that enforced expression of a self-reactive BCR in the presence of its self-antigen results in a complete block in B cell development at the immature B cell stage [207]. A second mechanism ensuring central tolerance is termed receptor editing, a process in which the original BCR specificity is altered by secondary recombination of the previously rearranged *IgL* genes [208]. It is also possible that during receptor editing, the *IgH* genes undergo secondary recombination, a process named $V_H$ replacement [209]. This process also contributes to altered receptor specificity by recombining the $V_H$ genes located 5′ of the $V_H$ gene used for the original $V_H$ to $DJ_H$ gene recombination. Moreover, self-reactive B cells may be rendered inactive or unresponsive to self-antigen binding, a state that is termed anergy [210]. Together, clonal deletion, receptor editing and anergy are traditionally thought to prevent the development of autoimmune diseases. It appears therefore likely, that defects in central tolerance can result in autoimmune diseases [211–213].

### 3.1 Role of Self-Reactivity in Early B Cell Development

In contrast to the traditional view, however, recent studies suggest that self-reactivity plays a positive role or is even necessary for B cell development in the bone marrow. This is reminiscent of the development of T cells in the thymus where only precursors recognizing self-peptides presented by major histocompatibility complex class I or II (MHC I/II) are allowed to mature further [214]. Interestingly, it was shown that the pre-BCR and a self-reactive BCR show functional similarities and that the pre-BCR represents the "prototype" of a self-reactive BCR [95,206]. As already mentioned above, studies using a secreted form of the pre-BCR revealed that its self-reactivity is based on a unique polypeptide region of λ5 [95]. Interestingly, this part of λ5 comprises several positively charged amino acids, which represents also a hallmark of self-reactive antibodies [99]. In line with this, replacing the unique polypeptide region of λ5 by transferring the CDR3 region of a self-reactive BCR to λ5 retains pre-BCR function [95]. These studies show that the pre-BCR has features similar to a self-reactive BCR and that the germline encoded SLC confers a certain degree of self-reactivity to every newly generated pre-BCR.

The assumption that a pre-BCR functionally resembles a self-reactive BCR predicts that a BCR with self-reactive specificity should be able to promote B cell development in the absence of the pre-BCR. This interesting question was investigated by crossing λ5-deficient mice, which cannot form a pre-BCR, to 3-83 BCR knockin mice, whose

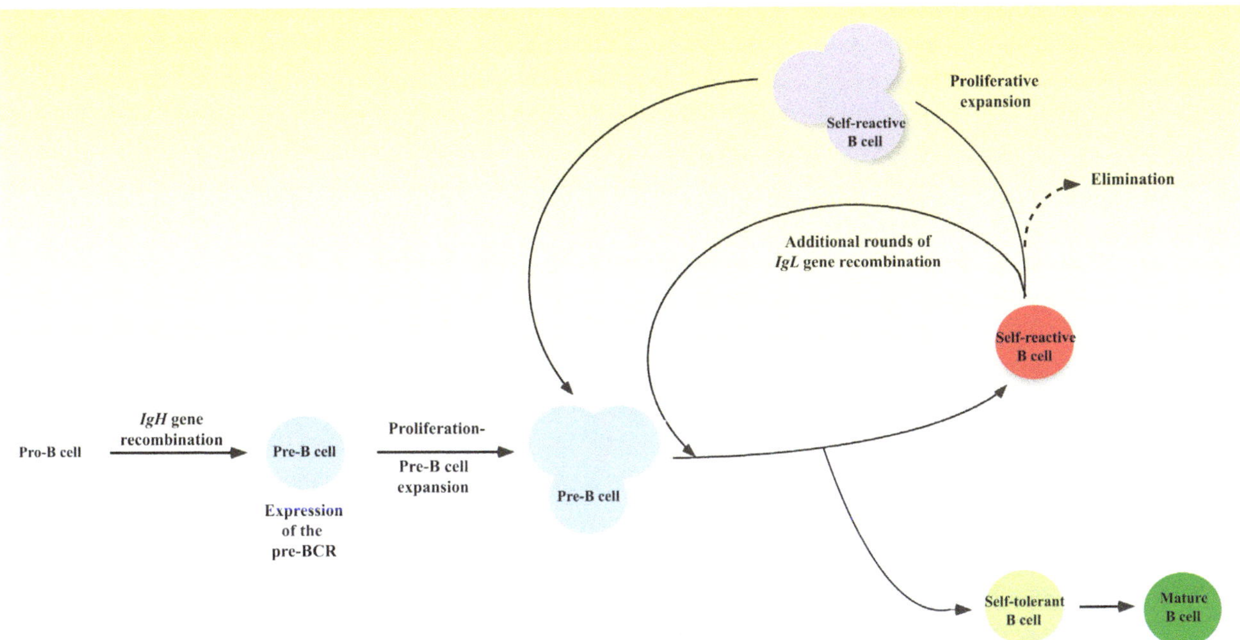


FIGURE 4 **Checkpoints during B cell development.** Upon successful recombination of a μHC, pro-B cells transit to the large pre-B cell stage and express a pre-BCR complex. Signals from the pre-BCR induce an intense phase of proliferation, leading to expansion of pre-B cells having successfully recombined their *IgH* genes. In addition, signals from the pre-BCR promote the recombination of the *IgL* genes, resulting in replacement of the SLC by a conventional LC and expression of a mature BCR complex. Due to the random nature of *Ig* gene recombination, newly generated BCRs can possess either self-tolerant or self-reactive receptor specificities. As only self-tolerant B cells are allowed to mature further, several mechanisms evolved to prevent the generation of self-reactive mature B cells. For instance, self-reactive immature B cells undergo additional rounds of recombination of the *IgL* genes to remove the self-reactive BCR specificity. However, if this is not successful, self-reactive B cells become eliminated.

developing B cells exclusively carry the 3-83 BCR recognizing different MHC I haplotypes with distinct affinities [215]. Interestingly, the 3-83 BCR was able to promote early B cell development in the presence of its cognate self-antigen with efficiencies similar to the pre-BCR in a wild-type situation [215]. In contrast, in the absence of the cognate self-antigen and thereby self-reactivity, the accumulation of precursor B cells was severely compromised. As a consequence, only a few mature B cells accumulated in the periphery [215]. This revealed that a self-reactive BCR is indeed able to rescue a pre-BCR-deficiency and that self-reactivity plays an important role for positive selection and expansion of B cells early in development.

If self-reactivity is essential for an efficient generation of B cells, the question arises how development of self-reactive B cells and autoimmune diseases is prevented. As already described earlier in this chapter, self-reactive immature B cells, which become exposed to self-antigen, reactivate SLP-65, resulting in secondary recombination mainly of the *IgL* genes and removal of the original self-reactive BCR specificity. Indeed, in the above-mentioned study using the λ5 deficiency combined with the 3-83 BCR knockin, mature B cells that developed out of the self-reactive precursor B cells in the presence of their cognate self-antigen in the bone marrow, changed their BCR specificity by receptor editing [215].

Together, a scheme emerges in which self-reactivity provided by the pre-BCR or a self-reactive BCR is important for positive selection of precursor B cells early in development (Figure 4). The recognition of self-structures by the pre-BCR induces proliferation of precursor B cells, which is followed by cell cycle arrest, stabilization of Rag1/2, and rearrangement of the *IgL* gene loci. Successful recombination of the *IgL* genes can have two outcomes. First, upon *IgL* gene recombination a self-tolerant BCR is expressed. This results in basal activation of class IA PI3K signaling, termination of recombination processes at the *Ig* gene loci, and further B cell maturation in the periphery. Second, *IgL* gene recombination results in the generation of a self-reactive BCR that immediately upon expression encounters self-antigen, resulting in the generation of a BCR signal which is stronger relative to the basal signal of a self-tolerant BCR. This then induces receptor editing of the *IgL* and *IgH* genes, or possibly clonal deletion. The data suggest that the signaling cascades activated downstream of a pre-BCR and a self-reactive BCR are similar. However, it is not clear to date whether self-reactive immature B cells immediately reactivate SLP-65 and induce secondary recombination processes, or undergo first a proliferative expansion, similar to pre-B cells. Because the above-mentioned studies revealed that accumulation of B cell precursors and generation of mature B cells was severely affected in the absence of self-reactivity [95,215], it appears, that in the complete absence of self-reactivity, only limited proliferative expansion of precursor B cells with recombined *Ig* genes is possible. One could therefore speculate, that self-reactive immature B cells might undergo also proliferative expansion upon self-antigen encounter but that this expansion phase may be shorter relative to that from pre-B cells, as self-reactive immature B cells have already passed through the pre-B cell stage and may be "poised" to re-activate SLP-65. Once a self-reactive immature B cell undergoes secondary rearrangement of the *Ig* gene segments, this can again have two different outcomes. If developing B cells lose their self-reactivity by expressing a self-tolerant BCR, they are allowed to proceed in development. Conversely, if the B cells still recognize self-antigens upon receptor editing, they may undergo further rearrangement of their *Ig* genes until the self-reactive BCR-specificity is removed. However, if recombination processes take too long or all available gene segments have been used and no more recombination processes are possible, the cells may be deleted due to FoxO-induced apoptotic programs. In line with this, transgenic expression of the pre-BCR, which cannot be removed, was shown to promote ongoing *IgL* gene rearrangement, resulting in a severe developmental block at the immature B cell stage [216,217]. This clearly demonstrates that although self-reactivity is important during early B cell developmental stages, self-reactive BCR specificities have to be removed to allow for proper B cell maturation and development.

## 3.2 The Role of BCR Signaling in the Differentiation of Peripheral B Cells

Immature B cells expressing BCRs of the IgM isotype leave the bone marrow and travel to the spleen. During their maturation, they pass through early and late translational stages before they become selected into the marginal zone (MZ) or follicular B cell (FO) compartment [218]. B1 B cells represent a separate subset of B cells in the periphery. Unlike conventional B (B2) cells of the MZ and FO B cell compartments, which develop throughout life out of transitional B cells, B1 B cells are thought to be mainly generated in the fetal liver and to depend on self-renewal in the periphery for their persistence [218,219]. However, B1 B cells show functional similarities to MZ B cells, as both subsets rapidly secrete so-called natural IgM antibodies of low affinity in response to common bacteria and viruses [219,220]. FO B cells, in contrast, are able to induce specific immune responses by increasing the affinity of their BCR for the invading pathogen and by replacing the isotype of their BCR through processes that are termed somatic hypermutation and class switch recombination, respectively [221].

How the selection of developing B cells into the B1, MZ and FO B cell compartments is regulated is not completely clear to date, but the analysis of numerous gene-targeted mice with reduced or enhanced BCR signaling activity indicated that the strength/intensity of BCR signaling critically influences peripheral B cell development. For instance, a study using the Epstein-Barr virus-encoded signaling protein LMP2A as a constitutively active BCR surrogate revealed that high signal intensities promote B1 B cell development, whereas low signal intensities drive differentiation of developing B cells into the MZ and FO B cell compartments [222]. Similarly, a study, taking advantage of the BCR transgene specific for the T cell-restricted cell surface molecule Thy1.2, found that high levels of the self-antigen promoted the differentiation of developing B cells into the B1 cell compartment at the expense of MZ and FO B cell development [223,224]. Lower doses of this antigen instead favored differentiation into MZ B cells [225]. In combination these studies suggest, that strong signals delivered by the BCR promote development into B1 B cells, whereas weaker signals drive developing B cells into MZ and FO B cell compartments.

An important question in this context is the source that induces this BCR signal and the downstream signaling cascades that drive the differentiation of peripheral B cells into the distinct compartments. As above-mentioned, self-reactivity is important for B cell development in the bone marrow. It has been reported that peripheral B cells constantly encounter endogenous antigen and that B1 and MZ B cell compartments are enriched for polyreactive receptor specificities recognizing microbial and/or self-antigens [219,220,224–227]. However, whether self-reactivity plays a role and how it may regulate peripheral B cell differentiation remain to be determined by future studies.

Several studies suggest that the activation of class IA PI3K signaling downstream of a BCR is a critical determinant for peripheral B cell differentiation. The deletion of specific regulatory or catalytic subunits, for example, results in a reduction of B1 and MZ B cells [228–230]. Similarly, reduced class IA PI3K signaling due to the absence of CD19 also leads to reduced B1 and MZ B cell compartments [231,232]. This skewed peripheral B cell differentiation of CD19-deficient mice could be largely rescued by concomitant deletion of PTEN and increased class IA PI3K activity [233]. Together, these studies point to an important role of class IA PI3K signaling in the decision of developing B cells to differentiate into the distinct mature B cell compartments.

## 3.3 The Role of BCR Signaling in Malignant Transformation of B Cells

As outlined above, expression of conventional BCR complexes is crucial for proper survival of B cells. Interestingly, most B lymphoma cells express functional BCRs even in the presence of additional oncogenic hits [234]. Accordingly, the nonproductively rearranged *Ig* gene loci are preferentially used for translocation events [234,235]. In line with the role of the BCR in malignancies and the function of Syk to activate class IA PI3K signaling and thereby trigger survival and proliferation, several B cell lymphomas show a hyperactive PI3K signaling cascade [236–239]. For instance, transgenic expression of T cell leukemia/lymphoma 1, which binds to AKT and enhances its kinase activity, was used to generate a mouse model representing human B chronic lymphocytic leukemia (CLL) [240]. Consistent with this, Syk inhibitors potently induce apoptosis of CLL B cells and the application of a p110δ-specific inhibitor (GS-1101) led to reduction in lymph node sizes of a large proportion of CLL patients, demonstrating the importance of Syk and class IA PI3K signaling in CLL B cells [241–246]. Therefore, a better understanding of the signaling networks regulating normal B cell proliferation and differentiation is of fundamental importance for the development of strategies to treat and possibly cure B cell-related malignancies in the future.

## REFERENCES


[1] Nagasawa T. Microenvironmental niches in the bone marrow required for B-cell development. Nat Rev Immunol 2006;6:107–16.

[2] Orkin SH, Zon LI. Hematopoiesis: an evolving paradigm for stem cell biology. Cell 2008;132:631–44.

[3] Hardy RR, Hayakawa K. B cell development pathways. Annu Rev Immunol 2001;19:595–621.

[4] Nagasawa T, Hirota S, Tachibana K, Takakura N, Nishikawa S, Kitamura Y, et al. Defects of B-cell lymphopoiesis and bone-marrow myelopoiesis in mice lacking the CXC chemokine PBSF/SDF-1. Nature 1996;382:635–8.

[5] Tokoyoda K, Egawa T, Sugiyama T, Choi BI, Nagasawa T. Cellular niches controlling B lymphocyte behavior within bone marrow during development. Immunity 2004;20:707–18.

[6] Milne CD, Paige CJ. IL-7: a key regulator of B lymphopoiesis. Semin Immunol 2006;18:20–30.

[7] Peschon JJ, Morrissey PJ, Grabstein KH, Ramsdell FJ, Maraskovsky E, Gliniak BC, et al. Early lymphocyte expansion is severely impaired in interleukin 7 receptor-deficient mice. J Exp Med 1994;180:1955–60.

[8] von Freeden-Jeffry U, Vieira P, Lucian LA, McNeil T, Burdach SE, Murray R. Lymphopenia in interleukin (IL)-7 gene-deleted mice identifies IL-7 as a nonredundant cytokine. J Exp Med 1995;181:1519–26.

[9] Herzog S, Reth M, Jumaa H. Regulation of B-cell proliferation and differentiation by pre-B-cell receptor signalling. Nat Rev Immunol 2009;9:195–205.

[10] Corfe SA, Paige CJ. The many roles of IL-7 in B cell development; mediator of survival, proliferation and differentiation. Semin Immunol 2012;24:198–208.

[11] Vosshenrich CA, Cumano A, Muller W, Di Santo JP, Vieira P. Thymic stromal-derived lymphopoietin distinguishes fetal from adult B cell development. Nat Immunol 2003;4:773–9.

[12] van der Plas DC, Smiers F, Pouwels K, Hoefsloot LH, Lowenberg B, Touw IP. Interleukin-7 signaling in human B cell precursor acute lymphoblastic leukemia cells and murine BAF3 cells involves activation of STAT1 and STAT5 mediated via the interleukin-7 receptor alpha chain. Leukemia 1996;10:1317–25.

[13] Chou WC, Levy DE, Lee CK. STAT3 positively regulates an early step in B-cell development. Blood 2006;108:3005–11.

[14] Goetz CA, Harmon IR, O'Neil JJ, Burchill MA, Farrar MA. STAT5 activation underlies IL7 receptor-dependent B cell development. J Immunol 2004;172:4770–8.

[15] Hoelbl A, Kovacic B, Kerenyi MA, Simma O, Warsch W, Cui Y, et al. Clarifying the role of Stat5 in lymphoid development and Abelson-induced transformation. Blood 2006;107:4898–906.

[16] Yao Z, Cui Y, Watford WT, Bream JH, Yamaoka K, Hissong BD, et al. Stat5a/b are essential for normal lymphoid development and differentiation. Proc Natl Acad Sci USA 2006;103:1000–5.

[17] Seckinger P, Fougereau M. Activation of src family kinases in human pre-B cells by IL-7. J Immunol 1994;153:97–109.

[18] Venkitaraman AR, Cowling RJ. Interleukin 7 receptor functions by recruiting the tyrosine kinase p59fyn through a segment of its cytoplasmic tail. Proc Natl Acad Sci USA 1992;89:12083–7.

[19] Ingley E, Klinken SP. Cross-regulation of JAK and Src kinases. Growth Factors 2006;24:89–95.

[20] Yasuda T, Sanjo H, Pages G, Kawano Y, Karasuyama H, Pouyssegur J, et al. Erk kinases link pre-B cell receptor signaling to transcriptional events required for early B cell expansion. Immunity 2008;28:499–508.

[21] Baldari CT, Telford JL. Lymphocyte antigen receptor signal integration and regulation by the SHC adaptor. Biol Chem 1999;380:129–34.

[22] Giles AJ, Bender TP, Ravichandran KS. The adaptor protein Shc plays a key role during early B cell development. J Immunol 2009;183:5468–76.

[23] Venkitaraman AR, Cowling RJ. Interleukin-7 induces the association of phosphatidylinositol 3-kinase with the alpha chain of the interleukin-7 receptor. Eur J Immunol 1994;24:2168–74.

[24] Croker BA, Kiu H, Nicholson SE. SOCS regulation of the JAK/STAT signalling pathway. Semin Cell Dev Biol 2008;19:414–22.

[25] Corfe SA, Rottapel R, Paige CJ. Modulation of IL-7 thresholds by SOCS proteins in developing B lineage cells. J Immunol 2011;187:3499–510.

[26] Weniger MA, Melzner I, Menz CK, Wegener S, Bucur AJ, Dorsch K, et al. Mutations of the tumor suppressor gene SOCS-1 in classical Hodgkin lymphoma are frequent and associated with nuclear phospho-STAT5 accumulation. Oncogene 2006;25:2679–84.

[27] Puel A, Ziegler SF, Buckley RH, Leonard WJ. Defective IL7R expression in T(-)B(+)NK(+) severe combined immunodeficiency. Nat Genet 1998;20:394–7.

[28] Russell SM, Tayebi N, Nakajima H, Riedy MC, Roberts JL, Aman MJ, et al. Mutation of Jak3 in a patient with SCID: essential role of Jak3 in lymphoid development. Science 1995;270:797–800.

[29] Macchi P, Villa A, Giliani S, Sacco MG, Frattini A, Porta F, et al. Mutations of Jak-3 gene in patients with autosomal severe combined immune deficiency (SCID). Nature 1995;377:65–8.

[30] Buckley RH, Schiff RI, Schiff SE, Markert ML, Williams LW, HarvilleTO,etal.Humansevere combinedimmunodeficiency:genetic, phenotypic, and functional diversity in one hundred eight infants. J Pediatr 1997;130:378–87.

[31] Fischer A, Cavazzana-Calvo M, De Saint Basile G, DeVillartay JP, Di Santo JP, et al. Naturally occurring primary deficiencies of the immune system. Annu Rev Immunol 1997;15:93–124.

[32] Leonard WJ. The molecular basis of X-linked severe combined immunodeficiency: defective cytokine receptor signaling. Annu Rev Med 1996;47:229–39.

[33] Parrish YK, Baez I, Milford TA, Benitez A, Galloway N, Rogerio JW, et al. IL-7 Dependence in human B lymphopoiesis increases during progression of ontogeny from cord blood to bone marrow. J Immunol 2009;182:4255–66.
[34] Giliani S, Mori L, de Saint Basile G, Le Deist F, Rodriguez-Perez C, Forino C, et al. Interleukin-7 receptor alpha (IL-7Ralpha) deficiency: cellular and molecular bases. Analysis of clinical, immunological, and molecular features in 16 novel patients. Immunol Rev 2005;203:110–26.
[35] Malumbres M, Barbacid M. Cell cycle, CDKs and cancer: a changing paradigm. Nat Rev Cancer 2009;9:153–66.
[36] Morrow MA, Lee G, Gillis S, Yancopoulos GD, Alt FW. Interleukin-7 induces N-myc and c-myc expression in normal precursor B lymphocytes. Genes Dev 1992;6:61–70.
[37] Walker SR, Nelson EA, Frank DA. STAT5 represses BCL6 expression by binding to a regulatory region frequently mutated in lymphomas. Oncogene 2007;26:224–33.
[38] Basso K, Dalla-Favera R. Roles of BCL6 in normal and transformed germinal center B cells. Immunol Rev 2012;247:172–83.
[39] Nahar R, Ramezani-Rad P, Mossner M, Duy C, Cerchietti L, Geng H, et al. Pre-B cell receptor-mediated activation of BCL6 induces pre-B cell quiescence through transcriptional repression of MYC. Blood 2011;118:4174–8.
[40] Dang CV. MYC on the path to cancer. Cell 2012;149:22–35.
[41] Eilers M, Eisenman RN. Myc's broad reach. Genes Dev 2008;22:2755–66.
[42] Grandori C, Cowley SM, James LP, Eisenman RN. The Myc/Max/Mad network and the transcriptional control of cell behavior. Annu Rev Cell Dev Biol 2000;16:653–99.
[43] Claassen GF, Hann SR. A role for transcriptional repression of p21CIP1 by c-Myc in overcoming transforming growth factor beta-induced cell-cycle arrest. Proc Natl Acad Sci USA 2000;97:9498–503.
[44] Mitchell KO, El-Deiry WS. Overexpression of c-Myc inhibits p21WAF1/CIP1 expression and induces S-phase entry in 12-O-tetradecanoylphorbol-13-acetate (TPA)-sensitive human cancer cells. Cell Growth Differ 1999;10:223–30.
[45] Gartel AL, Ye X, Goufman E, Shianov P, Hay N, Najmabadi F, et al. Myc represses the p21(WAF1/CIP1) promoter and interacts with Sp1/Sp3. Proc Natl Acad Sci USA 2001;98:4510–5.
[46] Yang W, Shen J, Wu M, Arsura M, FitzGerald M, Suldan Z, et al. Repression of transcription of the p27(Kip1) cyclin-dependent kinase inhibitor gene by c-Myc. Oncogene 2001;20:1688–702.
[47] Hermeking H, Rago C, Schuhmacher M, Li Q, Barrett JF, Obaya AJ, et al. Identification of CDK4 as a target of c-MYC. Proc Natl Acad Sci USA 2000;97:2229–34.
[48] Bouchard C, Thieke K, Maier A, Saffrich R, Hanley-Hyde J, Ansorge W, et al. Direct induction of cyclin D2 by Myc contributes to cell cycle progression and sequestration of p27. EMBO J 1999;18:5321–33.
[49] Gartel AL, Shchors K. Mechanisms of c-myc-mediated transcriptional repression of growth arrest genes. Exp Cell Res 2003;283:17–21.
[50] Daksis JI, Lu RY, Facchini LM, Marhin WW, Penn LJ. Myc induces cyclin D1 expression in the absence of de novo protein synthesis and links mitogen-stimulated signal transduction to the cell cycle. Oncogene 1994;9:3635–45.
[51] Vermeulen K, Van Bockstaele DR, Berneman ZN. The cell cycle: a review of regulation, deregulation and therapeutic targets in cancer. Cell Prolif 2003;36:131–49.
[52] Kittipatarin C, Li WQ, Bulavin DV, Durum SK, Khaled AR. Cell cycling through Cdc25A: transducer of cytokine proliferative signals. Cell Cycle 2006;5:907–12.
[53] Moriyama M, Yamochi T, Semba K, Akiyama T, Mori S. BCL-6 is phosphorylated at multiple sites in its serine- and proline-clustered region by mitogen-activated protein kinase (MAPK) in vivo. Oncogene 1997;14:2465–74.
[54] Niu H, Ye BH, Dalla-Favera R. Antigen receptor signaling induces MAP kinase-mediated phosphorylation and degradation of the BCL-6 transcription factor. Genes Dev 1998;12:1953–61.
[55] Sears R, Nuckolls F, Haura E, Taya Y, Tamai K, Nevins JR. Multiple Ras-dependent phosphorylation pathways regulate Myc protein stability. Genes Dev 2000;14:2501–14.
[56] Wang R, He G, Nelman-Gonzalez M, Ashorn CL, Gallick GE, Stukenberg PT, et al. Regulation of Cdc25C by ERK-MAP kinases during the G2/M transition. Cell 2007;128:1119–32.
[57] Youle RJ, Strasser A. The BCL-2 protein family: opposing activities that mediate cell death. Nat Rev Mol Cell Biol 2008; 9:47–59.
[58] Wen R, Wang D, McKay C, Bunting KD, Marine JC, Vanin EF, et al. Jak3 selectively regulates Bax and Bcl-2 expression to promote T-cell development. Mol Cell Biol 2001;21:678–89.
[59] Lu L, Chaudhury P, Osmond DG. Regulation of cell survival during B lymphopoiesis: apoptosis and Bcl-2/Bax content of precursor B cells in bone marrow of mice with altered expression of IL-7 and recombinase-activating gene-2. J Immunol 1999;162:1931–40.
[60] Malin S, McManus S, Cobaleda C, Novatchkova M, Delogu A, Bouillet P, et al. Role of STAT5 in controlling cell survival and immunoglobulin gene recombination during pro-B cell development. Nat Immunol 2010;11:171–9.
[61] Li WQ, Jiang Q, Khaled AR, Keller JR, Durum SK. Interleukin-7 inactivates the pro-apoptotic protein Bad promoting T cell survival. J Biol Chem 2004;279:29160–6.
[62] Li WQ, Guszczynski T, Hixon JA, Durum SK. Interleukin-7 regulates Bim proapoptotic activity in peripheral T-cell survival. Mol Cell Biol 2010;30:590–600.
[63] Huntington ND, Labi V, Cumano A, Vieira P, Strasser A, Villunger A, Di Santo JP, Alves NL. Loss of the pro-apoptotic BH3-only Bcl-2 family member Bim sustains B lymphopoiesis in the absence of IL-7. Int Immunol 2009;21:715–25.
[64] Oliver PM, Wang M, Zhu Y, White J, Kappler J, Marrack P. Loss of Bim allows precursor B cell survival but not precursor B cell differentiation in the absence of interleukin 7. J Exp Med 2004;200:1179–87.
[65] Corcoran AE, Smart FM, Cowling RJ, Crompton T, Owen MJ, Venkitaraman AR. The interleukin-7 receptor alpha chain transmits distinct signals for proliferation and differentiation during B lymphopoiesis. EMBO J 1996;15:1924–32.
[66] Dias S, Silva Jr H, Cumano A, Vieira P. Interleukin-7 is necessary to maintain the B cell potential in common lymphoid progenitors. J Exp Med 2005;201:971–9.
[67] Kikuchi K, Lai AY, Hsu CL, Kondo M. IL-7 receptor signaling is necessary for stage transition in adult B cell development through up-regulation of EBF. J Exp Med 2005;201:1197–203.
[68] Lukin K, Fields S, Guerrettaz L, Straign D, Rodriguez V, Zandi S, et al. A dose-dependent role for EBF1 in repressing non-B-cell-specific genes. Eur J Immunol 2011;41:1787–93.

[69] Zandi S, Mansson R, Tsapogas P, Zetterblad J, Bryder D, Sigvardsson M. EBF1 is essential for B-lineage priming and establishment of a transcription factor network in common lymphoid progenitors. J Immunol 2008;181:3364–72.

[70] Pongubala JM, Northrup DL, Lancki DW, Medina KL, Treiber T, Bertolino E, et al. Transcription factor EBF restricts alternative lineage options and promotes B cell fate commitment independently of Pax5. Nat Immunol 2008;9:203–15.

[71] Gyory I, Boller S, Nechanitzky R, Mandel E, Pott S, Liu E, et al. Transcription factor Ebf1 regulates differentiation stage-specific signaling, proliferation, and survival of B cells. Genes Dev 2012;26:668–82.

[72] Medina KL, Pongubala JM, Reddy KL, Lancki DW, Dekoter R, Kieslinger M, et al. Assembling a gene regulatory network for specification of the B cell fate. Dev Cell 2004;7:607–17.

[73] O'Riordan M, Grosschedl R. Coordinate regulation of B cell differentiation by the transcription factors EBF and E2A. Immunity 1999;11:21–31.

[74] Hirokawa S, Sato H, Kato I, Kudo A. EBF-regulating Pax5 transcription is enhanced by STAT5 in the early stage of B cells. Eur J Immunol 2003;33:1824–9.

[75] Goetz CA, Harmon IR, O'Neil JJ, Burchill MA, Johanns TM, Farrar MA. Restricted STAT5 activation dictates appropriate thymic B versus T cell lineage commitment. J Immunol 2005;174:7753–63.

[76] Cobaleda C, Schebesta A, Delogu A, Busslinger M. Pax5: the guardian of B cell identity and function. Nat Immunol 2007;8:463–70.

[77] Nutt SL, Thevenin C, Busslinger M. Essential functions of Pax-5 (BSAP) in pro-B cell development. Immunobiology 1997;198: 227–35.

[78] Rolink AG, Nutt SL, Melchers F, Busslinger M. Long-term in vivo reconstitution of T-cell development by Pax5-deficient B-cell progenitors. Nature 1999;401:603–6.

[79] Nutt SL, Heavey B, Rolink AG, Busslinger M. Commitment to the B-lymphoid lineage depends on the transcription factor Pax5. Nature 1999;401:556–62.

[80] Simmons S, Knoll M, Drewell C, Wolf I, Mollenkopf HJ, Bouquet C, et al. Biphenotypic B-lymphoid/myeloid cells expressing low levels of Pax5: potential targets of BAL development. Blood 2012;120:3688–98.

[81] Corcoran AE, Riddell A, Krooshoop D, Venkitaraman AR. Impaired immunoglobulin gene rearrangement in mice lacking the IL-7 receptor. Nature 1998;391:904–7.

[82] Chowdhury D, Sen R. Stepwise activation of the immunoglobulin mu heavy chain gene locus. Embo J 2001;20:6394–403.

[83] Chowdhury D, Sen R. Transient IL-7/IL-7R signaling provides a mechanism for feedback inhibition of immunoglobulin heavy chain gene rearrangements. Immunity 2003;18:229–41.

[84] Bertolino E, Reddy K, Medina KL, Parganas E, Ihle J, Singh H. Regulation of interleukin 7-dependent immunoglobulin heavy-chain variable gene rearrangements by transcription factor STAT5. Nat Immunol 2005;6:836–43.

[85] Fuxa M, Skok J, Souabni A, Salvagiotto G, Roldan E, Busslinger M. Pax5 induces V-to-DJ rearrangements and locus contraction of the immunoglobulin heavy-chain gene. Genes Dev 2004;18:411–22.

[86] Zhang Z, Espinoza CR, Yu Z, Stephan R, He T, Williams GS, et al. Transcription factor Pax5 (BSAP) transactivates the RAG-mediated V(H)-to-DJ(H) rearrangement of immunoglobulin genes. Nat Immunol 2006;7:616–24.

[87] Liu H, Schmidt-Supprian M, Shi Y, Hobeika E, Barteneva N, Jumaa H, et al. Yin Yang 1 is a critical regulator of B-cell development. Genes Dev 2007;21:1179–89.

[88] Alkhatib A, Werner M, Hug E, Herzog S, Eschbach C, Faraidun H, et al. FoxO1 induces Ikaros splicing to promote immunoglobulin gene recombination. J Exp Med 2012;209:395–406.

[89] Martensson IL, Keenan RA, Licence S. The pre-B-cell receptor. Curr Opin Immunol 2007;19:137–42.

[90] Karasuyama H, Kudo A, Melchers F. The proteins encoded by the VpreB and lambda 5 pre-B cell-specific genes can associate with each other and with mu heavy chain. J Exp Med 1990;172:969–72.

[91] Tsubata T, Reth M. The products of pre-B cell-specific genes (lambda 5 and VpreB) and the immunoglobulin mu chain form a complex that is transported onto the cell surface. J Exp Med 1990;172:973–6.

[92] Bradl H, Wittmann J, Milius D, Vettermann C, Jack HM. Interaction of murine precursor B cell receptor with stroma cells is controlled by the unique tail of lambda 5 and stroma cell-associated heparan sulfate. J Immunol 2003;171:2338–48.

[93] Gauthier L, Rossi B, Roux F, Termine E, Schiff C. Galectin-1 is a stromal cell ligand of the pre-B cell receptor (BCR) implicated in synapse formation between pre-B and stromal cells and in pre-BCR triggering. Proc Natl Acad Sci USA 2002;99:13014–9.

[94] Milne CD, Corfe SA, Paige CJ. Heparan sulfate and heparin enhance ERK phosphorylation and mediate preBCR-dependent events during B lymphopoiesis. J Immunol 2008;180:2839–47.

[95] Kohler F, Hug E, Eschbach C, Meixlsperger S, Hobeika E, Kofer J, et al. Autoreactive B cell receptors mimic autonomous pre-B cell receptor signaling and induce proliferation of early B cells. Immunity 2008;29:912–21.

[96] Guloglu FB, Roman CA. Precursor B cell receptor signaling activity can be uncoupled from surface expression. J Immunol 2006;176:6862–72.

[97] Bankovich AJ, Raunser S, Juo ZS, Walz T, Davis MM, Garcia KC. Structural insight into pre-B cell receptor function. Science 2007;316:291–4.

[98] Meixlsperger S, Kohler F, Wossning T, Reppel M, Muschen M, Jumaa H. Conventional light chains inhibit the autonomous signaling capacity of the B cell receptor. Immunity 2007;26:323–33.

[99] Ohnishi K, Melchers F. The nonimmunoglobulin portion of lambda5 mediates cell-autonomous pre-B cell receptor signaling. Nat Immunol 2003;4:849–56.

[100] Ubelhart R, Bach MP, Eschbach C, Wossning T, Reth M, Jumaa H. N-linked glycosylation selectively regulates autonomous precursor BCR function. Nat Immunol 2010;11:759–65.

[101] Kitamura D, Roes J, Kuhn R, Rajewsky K. A B cell-deficient mouse by targeted disruption of the membrane exon of the immunoglobulin mu chain gene. Nature 1991;350:423–6.

[102] Gong S, Nussenzweig MC. Regulation of an early developmental checkpoint in the B cell pathway by Ig beta. Science 1996;272: 411–4.

[103] Pelanda R, Braun U, Hobeika E, Nussenzweig MC, Reth M. B cell progenitors are arrested in maturation but have intact VDJ recombination in the absence of Ig-alpha and Ig-beta. J Immunol 2002;169:865–72.

[104] Kitamura D, Kudo A, Schaal S, Muller W, Melchers F, Rajewsky K. A critical role of lambda 5 protein in B cell development. Cell 1992;69:823–31.

[105] Mundt C, Licence S, Shimizu T, Melchers F, Martensson IL. Loss of precursor B cell expansion but not allelic exclusion in VpreB1/VpreB2 double-deficient mice. J Exp Med 2001;193:435–45.

[106] Galler GR, Mundt C, Parker M, Pelanda R, Martensson IL, Winkler TH. Surface mu heavy chain signals down-regulation of the V(D)J-recombinase machinery in the absence of surrogate light chain components. J Exp Med 2004;199:1523–32.

[107] Pelanda R, Schaal S, Torres RM, Rajewsky K. A prematurely expressed Ig(kappa) transgene, but not V(kappa)J(kappa) gene segment targeted into the Ig(kappa) locus, can rescue B cell development in lambda5-deficient mice. Immunity 1996;5:229–39.

[108] Schuh W, Meister S, Roth E, Jack HM. Cutting edge: signaling and cell surface expression of a mu H chain in the absence of lambda 5: a paradigm revisited. J Immunol 2003;171:3343–7.

[109] Su YW, Flemming A, Wossning T, Hobeika E, Reth M, Jumaa H. Identification of a pre-BCR lacking surrogate light chain. J Exp Med 2003;198:1699–706.

[110] Minegishi Y, Coustan-Smith E, Wang YH, Cooper MD, Campana D, Conley ME. Mutations in the human lambda5/14.1 gene result in B cell deficiency and agammaglobulinemia. J Exp Med 1998;187:71–7.

[111] Kurosaki T. Functional dissection of BCR signaling pathways. Curr Opin Immunol 2000;12:276–81.

[112] Flaswinkel H, Reth M. Dual role of the tyrosine activation motif of the Ig-alpha protein during signal transduction via the B cell antigen receptor. Embo J 1994;13:83–9.

[113] Rowley RB, Burkhardt AL, Chao HG, Matsueda GR, Bolen JB. Syk protein-tyrosine kinase is regulated by tyrosine-phosphorylated Ig alpha/Ig beta immunoreceptor tyrosine activation motif binding and autophosphorylation. J Biol Chem 1995;270:11590–4.

[114] Rolli V, Gallwitz M, Wossning T, Flemming A, Schamel WW, Zurn C, et al. Amplification of B cell antigen receptor signaling by a Syk/ITAM positive feedback loop. Mol Cell 2002;10:1057–69.

[115] Herzog S, Hug E, Meixlsperger S, Paik JH, DePinho RA, Reth M, et al. SLP-65 regulates immunoglobulin light chain gene recombination through the PI(3)K-PKB-Foxo pathway. Nat Immunol 2008;9:623–31.

[116] Turner M, Mee PJ, Costello PS, Williams O, Price AA, Duddy LP, et al. Perinatal lethality and blocked B-cell development in mice lacking the tyrosine kinase Syk. Nature 1995;378:298–302.

[117] Fleming HE, Paige CJ. Cooperation between IL-7 and the pre-B cell receptor: a key to B cell selection. Semin Immunol 2002;14:423–30.

[118] Storch B, Meixlsperger S, Jumaa H. The Ig-alpha ITAM is required for efficient differentiation but not proliferation of pre-B cells. Eur J Immunol 2007;37:252–60.

[119] Marshall AJ, Fleming HE, Wu GE, Paige CJ. Modulation of the IL-7 dose-response threshold during pro-B cell differentiation is dependent on pre-B cell receptor expression. J Immunol 1998;161:6038–45.

[120] Fleming HE, Paige CJ. Pre-B cell receptor signaling mediates selective response to IL-7 at the pro-B to pre-B cell transition via an ERK/MAP kinase-dependent pathway. Immunity 2001;15:521–31.

[121] Ramadani F, Bolland DJ, Garcon F, Emery JL, Vanhaesebroeck B, Corcoran AE, et al. The PI3K isoforms p110alpha and p110delta are essential for pre-B cell receptor signaling and B cell development. Sci Signal 2010;3. ra60.

[122] Osborne LC, Duthie KA, Seo JH, Gascoyne RD, Abraham N. Selective ablation of the YxxM motif of IL-7Ralpha suppresses lymphomagenesis but maintains lymphocyte development. Oncogene 2010;29:3854–64.

[123] Okkenhaug K, Vanhaesebroeck B. PI3K in lymphocyte development, differentiation and activation. Nat Rev Immunol 2003;3:317–30.

[124] Okkenhaug K. Signaling by the phosphoinositide 3-kinase family in immune cells. Annu Rev Immunol 2013;31:675–704.

[125] Aiba Y, Kameyama M, Yamazaki T, Tedder TF, Kurosaki T. Regulation of B-cell development by BCAP and CD19 through their binding to phosphoinositide 3-kinase. Blood 2008;111:1497–503.

[126] Castellano E, Downward J. RAS interaction with PI3K: more than just another effector pathway. Genes Cancer 2011;2:261–74.

[127] Mandal M, Powers SE, Ochiai K, Georgopoulos K, Kee BL, Singh H, et al. Ras orchestrates exit from the cell cycle and light-chain recombination during early B cell development. Nat Immunol 2009;10:1110–7.

[128] Werner M, Hobeika E, Jumaa H. Role of PI3K in the generation and survival of B cells. Immunol Rev 2010;237:55–71.

[129] Satterthwaite AB, Li Z, Witte ON. Btk function in B cell development and response. Semin Immunol 1998;10:309–16.

[130] Scharenberg AM, Kinet JP. PtdIns-3,4,5-P3: a regulatory nexus between tyrosine kinases and sustained calcium signals. Cell 1998;94:5–8.

[131] Marshall AJ, Niiro H, Yun TJ, Clark EA. Regulation of B-cell activation and differentiation by the phosphatidylinositol 3-kinase and phospholipase Cgamma pathway. Immunol Rev 2000;176:30–46.

[132] Tsukada S, Saffran DC, Rawlings DJ, Parolini O, Allen RC, Klisak I, et al. Deficient expression of a B cell cytoplasmic tyrosine kinase in human X-linked agammaglobulinemia. Cell 1993;72:279–90.

[133] Vetrie D, Vorechovsky I, Sideras P, Holland J, Davies A, Flinter F, et al. The gene involved in X-linked agammaglobulinaemia is a member of the src family of protein-tyrosine kinases. Nature 1993;361:226–33.

[134] Manning BD, Cantley LC. AKT/PKB signaling: navigating downstream. Cell 2007;129:1261–74.

[135] Alessi DR, Caudwell FB, Andjelkovic M, Hemmings BA, Cohen P. Molecular basis for the substrate specificity of protein kinase B; comparison with MAPKAP kinase-1 and p70 S6 kinase. FEBS Lett 1996;399:333–8.

[136] Datta SR, Dudek H, Tao X, Masters S, Fu H, Gotoh Y, et al. Akt phosphorylation of BAD couples survival signals to the cell-intrinsic death machinery. Cell 1997;91:231–41.

[137] del Peso L, Gonzalez-Garcia M, Page C, Herrera R, Nunez G. Interleukin-3-induced phosphorylation of BAD through the protein kinase Akt. Science 1997;278:687–9.

[138] Ogawara Y, Kishishita S, Obata T, Isazawa Y, Suzuki T, Tanaka K, et al. Akt enhances Mdm2-mediated ubiquitination and degradation of p53. J Biol Chem 2002;277:21843–50.

[139] Zhou BP, Liao Y, Xia W, Zou Y, Spohn B, Hung MC. HER-2/neu induces p53 ubiquitination via Akt-mediated MDM2 phosphorylation. Nat Cell Biol 2001;3:973–82.

[140] Mayo LD, Donner DB. A phosphatidylinositol 3-kinase/Akt pathway promotes translocation of Mdm2 from the cytoplasm to the nucleus. Proc Natl Acad Sci USA 2001;98:11598–603.

[141] Viglietto G, Motti ML, Bruni P, Melillo RM, D'Alessio A, Califano D, et al. Cytoplasmic relocalization and inhibition of the cyclin-dependent kinase inhibitor p27(Kip1) by PKB/Akt-mediated phosphorylation in breast cancer. Nat Med 2002;8:1136–44.

[142] Shin I, Yakes FM, Rojo F, Shin NY, Bakin AV, Baselga J, et al. PKB/Akt mediates cell-cycle progression by phosphorylation of p27(Kip1) at threonine 157 and modulation of its cellular localization. Nat Med 2002;8:1145–52.

[143] Liang J, Zubovitz J, Petrocelli T, Kotchetkov R, Connor MK, Han K, et al. PKB/Akt phosphorylates p27, impairs nuclear import of p27 and opposes p27-mediated G1 arrest. Nat Med 2002;8:1153–60.
[144] Zhou BP, Liao Y, Xia W, Spohn B, Lee MH, Hung MC. Cytoplasmic localization of p21Cip1/WAF1 by Akt-induced phosphorylation in HER-2/neu-overexpressing cells. Nat Cell Biol 2001;3:245–52.
[145] Motti ML, Califano D, Troncone G, De Marco C, Migliaccio I, Palmieri E, et al. Complex regulation of the cyclin-dependent kinase inhibitor p27kip1 in thyroid cancer cells by the PI3K/AKT pathway: regulation of p27kip1 expression and localization. Am J Pathol 2005;166:737–49.
[146] Diehl JA, Cheng M, Roussel MF, Sherr CJ. Glycogen synthase kinase-3beta regulates cyclin D1 proteolysis and subcellular localization. Genes Dev 1998;12:3499–511.
[147] Welcker M, Singer J, Loeb KR, Grim J, Bloecher A, Gurien-West M, et al. Multisite phosphorylation by Cdk2 and GSK3 controls cyclin E degradation. Mol Cell 2003;12:381–92.
[148] Yeh E, Cunningham M, Arnold H, Chasse D, Monteith T, Ivaldi G, et al. A signalling pathway controlling c-Myc degradation that impacts oncogenic transformation of human cells. Nat Cell Biol 2004;6:308–18.
[149] Huang J, Manning BD. A complex interplay between Akt, TSC2 and the two mTOR complexes. Biochem Soc Trans 2009;37:217–22.
[150] Wiza C, Nascimento EB, Ouwens DM. Role of PRAS40 in Akt and mTOR signaling in health and disease. Am J Physiol Endocrinol Metab 2012;302:E1453–60.
[151] Kovacina KS, Park GY, Bae SS, Guzzetta AW, Schaefer E, Birnbaum MJ, et al. Identification of a proline-rich Akt substrate as a 14-3-3 binding partner. J Biol Chem 2003;278:10189–94.
[152] Decker T, Hipp S, Ringshausen I, Bogner C, Oelsner M, Schneller F, et al. Rapamycin-induced G1 arrest in cycling B-CLL cells is associated with reduced expression of cyclin D3, cyclin E, cyclin A, and survivin. Blood 2003;101:278–85.
[153] Ozes ON, Mayo LD, Gustin JA, Pfeffer SR, Pfeffer LM, Donner DB. NF-kappaB activation by tumour necrosis factor requires the Akt serine-threonine kinase. Nature 1999;401:82–5.
[154] Kim AH, Khursigara G, Sun X, Franke TF, Chao MV. Akt phosphorylates and negatively regulates apoptosis signal-regulating kinase 1. Mol Cell Biol 2001;21:893–901.
[155] Coffer PJ, Burgering BM. Forkhead-box transcription factors and their role in the immune system. Nat Rev Immunol 2004;4:889–99.
[156] Vogt PK, Jiang H, Aoki M. Triple layer control: phosphorylation, acetylation and ubiquitination of FOXO proteins. Cell Cycle 2005;4:908–13.
[157] Medema RH, Kops GJ, Bos JL, Burgering BM. AFX-like Forkhead transcription factors mediate cell-cycle regulation by Ras and PKB through p27kip1. Nature 2000;404:782–7.
[158] Dijkers PF, Birkenkamp KU, Lam EW, Thomas NS, Lammers JW, Koenderman L, et al. FKHR-L1 can act as a critical effector of cell death induced by cytokine withdrawal: protein kinase B-enhanced cell survival through maintenance of mitochondrial integrity. J Cell Biol 2002;156:531–42.
[159] Brunet A, Bonni A, Zigmond MJ, Lin MZ, Juo P, Hu LS, et al. Akt promotes cell survival by phosphorylating and inhibiting a Forkhead transcription factor. Cell 1999;96:857–68.
[160] van der Vos KE, Coffer PJ. The extending network of FOXO transcriptional target genes. Antioxid Redox Signal 2011;14:579–92.
[161] Papavasiliou F, Jankovic M, Gong S, Nussenzweig MC. Control of immunoglobulin gene rearrangements in developing B cells. Curr Opin Immunol 1997;9:233–8.
[162] Jung D, Giallourakis C, Mostoslavsky R, Alt FW. Mechanism and control of V(D)J recombination at the immunoglobulin heavy chain locus. Annu Rev Immunol 2006;24:541–70.
[163] Schweighoffer E, Vanes L, Mathiot A, Nakamura T, Tybulewicz VL. Unexpected requirement for ZAP-70 in pre-B cell development and allelic exclusion. Immunity 2003;18:523–33.
[164] Grawunder U, Leu TM, Schatz DG, Werner A, Rolink AG, Melchers F, et al. Down-regulation of RAG1 and RAG2 gene expression in preB cells after functional immunoglobulin heavy chain rearrangement. Immunity 1995;3:601–8.
[165] Tze LE, Schram BR, Lam KP, Hogquist KA, Hippen KL, Liu J, et al. Basal immunoglobulin signaling actively maintains developmental stage in immature B cells. PLoS Biol 2005;3:e82.
[166] Verkoczy L, Duong B, Skog P, Ait-Azzouzene D, Puri K, Vela JL, et al. Basal B cell receptor-directed phosphatidylinositol 3-kinase signaling turns off RAGs and promotes B cell-positive selection. J Immunol 2007;178:6332–41.
[167] Koretzky GA, Abtahian F, Silverman MA. SLP76 and SLP65: complex regulation of signalling in lymphocytes and beyond. Nat Rev Immunol 2006;6:67–78.
[168] Chiu CW, Dalton M, Ishiai M, Kurosaki T, Chan AC. BLNK: molecular scaffolding through 'cis'-mediated organization of signaling proteins. Embo J 2002;21:6461–72.
[169] Watanabe D, Hashimoto S, Ishiai M, Matsushita M, Baba Y, Kishimoto T, et al. Four tyrosine residues in phospholipase C-gamma 2, identified as Btk-dependent phosphorylation sites, are required for B cell antigen receptor-coupled calcium signaling. J Biol Chem 2001;276:38595–601.
[170] Kohler F, Storch B, Kulathu Y, Herzog S, Kuppig S, Reth M, et al. A leucine zipper in the N terminus confers membrane association to SLP-65. Nat Immunol 2005;6:204–10.
[171] Feske S. Calcium signalling in lymphocyte activation and disease. Nat Rev Immunol 2007;7:690–702.
[172] Dolmetsch RE, Lewis RS, Goodnow CC, Healy JI. Differential activation of transcription factors induced by $Ca^{2+}$ response amplitude and duration. Nature 1997;386:855–8.
[173] Gao T, Furnari F, Newton AC. PHLPP: a phosphatase that directly dephosphorylates Akt, promotes apoptosis, and suppresses tumor growth. Mol Cell 2005;18:13–24.
[174] Brognard J, Sierecki E, Gao T, Newton AC. PHLPP and a second isoform, PHLPP2, differentially attenuate the amplitude of Akt signaling by regulating distinct Akt isoforms. Mol Cell 2007;25:917–31.
[175] Thompson EC, Cobb BS, Sabbattini P, Meixlsperger S, Parelho V, Liberg D, et al. Ikaros DNA-binding proteins as integral components of B cell developmental-stage-specific regulatory circuits. Immunity 2007;26:335–44.
[176] Sabbattini P, Lundgren M, Georgiou A, Chow C, Warnes G, Dillon N. Binding of Ikaros to the lambda5 promoter silences transcription through a mechanism that does not require heterochromatin formation. Embo J 2001;20:2812–22.
[177] Ma S, Pathak S, Mandal M, Trinh L, Clark MR, Lu R. Ikaros and Aiolos inhibit pre-B-cell proliferation by directly suppressing c-Myc expression. Mol Cell Biol 2010;30:4149–58.
[178] Jumaa H, Wollscheid B, Mitterer M, Wienands J, Reth M, Nielsen PJ. Abnormal development and function of B lymphocytes in mice deficient for the signaling adaptor protein SLP-65. Immunity 1999;11:547–54.

[179] Pappu R, Cheng AM, Li B, Gong Q, Chiu C, Griffin N, et al. Requirement for B cell linker protein (BLNK) in B cell development. Science 1999;286:1949–54.

[180] Trageser D, Iacobucci I, Nahar R, Duy C, von Levetzow G, Klemm L, et al. Pre-B cell receptor-mediated cell cycle arrest in Philadelphia chromosome-positive acute lymphoblastic leukemia requires IKAROS function. J Exp Med 2009;206:1739–53.

[181] Nakayama J, Yamamoto M, Hayashi K, Satoh H, Bundo K, Kubo M, et al. BLNK suppresses pre-B-cell leukemogenesis through inhibition of JAK3. Blood 2009;113:1483–92.

[182] Amin RH, Schlissel MS. Foxo1 directly regulates the transcription of recombination-activating genes during B cell development. Nat Immunol 2008;9:613–22.

[183] Dengler HS, Baracho GV, Omori SA, Bruckner S, Arden KC, Castrillon DH, et al. Distinct functions for the transcription factor Foxo1 at various stages of B cell differentiation. Nat Immunol 2008;9:1388–98.

[184] Lee J, Desiderio S. Cyclin A/CDK2 regulates V(D)J recombination by coordinating RAG-2 accumulation and DNA repair. Immunity 1999;11:771–81.

[185] Busslinger M. Transcriptional control of early B cell development. Annu Rev Immunol 2004;22:55–79.

[186] Marecki S, Fenton MJ. The role of IRF-4 in transcriptional regulation. J Interferon Cytokine Res 2002;22:121–33.

[187] Eguchi J, Wang X, Yu S, Kershaw EE, Chiu PC, Dushay J, et al. Transcriptional control of adipose lipid handling by IRF4. Cell Metab 2011;13:249–59.

[188] Ochiai K, Maienschein-Cline M, Mandal M, Triggs JR, Bertolino E, Sciammas R, et al. A self-reinforcing regulatory network triggered by limiting IL-7 activates pre-BCR signaling and differentiation. Nat Immunol 2012;13:300–7.

[189] Ma S, Turetsky A, Trinh L, Lu R. IFN regulatory factor 4 and 8 promote Ig light chain kappa locus activation in pre-B cell development. J Immunol 2006;177:7898–904.

[190] Muljo SA, Schlissel MS. A small molecule Abl kinase inhibitor induces differentiation of Abelson virus-transformed pre-B cell lines. Nat Immunol 2003;4:31–7.

[191] Johnson K, Hashimshony T, Sawai CM, Pongubala JM, Skok JA, Aifantis I, et al. Regulation of immunoglobulin light-chain recombination by the transcription factor IRF-4 and the attenuation of interleukin-7 signaling. Immunity 2008;28:335–45.

[192] Lu R, Medina KL, Lancki DW, Singh H. IRF-4,8 orchestrate the pre-B-to-B transition in lymphocyte development. Genes Dev 2003;17:1703–8.

[193] Ma S, Pathak S, Trinh L, Lu R. Interferon regulatory factors 4 and 8 induce the expression of Ikaros and Aiolos to down-regulate pre-B-cell receptor and promote cell-cycle withdrawal in pre-B-cell development. Blood 2008;111:1396–403.

[194] Calamito M, Juntilla MM, Thomas M, Northrup DL, Rathmell J, Birnbaum MJ, et al. Akt1 and Akt2 promote peripheral B-cell maturation and survival. Blood 2010;115:4043–50.

[195] Venigalla RK, McGuire VA, Clarke R, Patterson-Kane JC, Najafov A, Toth R, et al. PDK1 regulates VDJ recombination, cell-cycle exit and survival during B-cell development. Embo J 2013;32:1008–22.

[196] Conley ME, Dobbs AK, Quintana AM, Bosompem A, Wang YD, Coustan-Smith E, et al. Agammaglobulinemia and absent B lineage cells in a patient lacking the p85alpha subunit of PI3K. J Exp Med 2012;209:463–70.

[197] Minegishi Y, Rohrer J, Coustan-Smith E, Lederman HM, Pappu R, Campana D, et al. An essential role for BLNK in human B cell development. Science 1999;286:1954–7.

[198] Lam KP, Kuhn R, Rajewsky K. In vivo ablation of surface immunoglobulin on mature B cells by inducible gene targeting results in rapid cell death. Cell 1997;90:1073–83.

[199] Srinivasan L, Sasaki Y, Calado DP, Zhang B, Paik JH, DePinho RA, et al. PI3 kinase signals BCR-dependent mature B cell survival. Cell 2009;139:573–86.

[200] Delgado P, Cubelos B, Calleja E, Martinez-Martin N, Cipres A, Merida I, et al. Essential function for the GTPase TC21 in homeostatic antigen receptor signaling. Nat Immunol 2009;10:880–8.

[201] Mackay F, Schneider P. Cracking the BAFF code. Nat Rev Immunol 2009;9:491–502.

[202] Smith SH, Cancro MP. Cutting edge: B cell receptor signals regulate BLyS receptor levels in mature B cells and their immediate progenitors. J Immunol 2003;170:5820–3.

[203] Henley T, Kovesdi D, Turner M. B-cell responses to B-cell activation factor of the TNF family (BAFF) are impaired in the absence of PI3K delta. Eur J Immunol 2008;38:3543–8.

[204] Khan WN. B cell receptor and BAFF receptor signaling regulation of B cell homeostasis. J Immunol 2009;183:3561–7.

[205] Schweighoffer E, Vanes L, Nys J, Cantrell D, McCleary S, Smithers N, et al. The BAFF receptor transduces survival signals by co-opting the B cell receptor signaling pathway. Immunity 2013;38:475–88.

[206] Wardemann H, Yurasov S, Schaefer A, Young JW, Meffre E, Nussenzweig MC. Predominant autoantibody production by early human B cell precursors. Science 2003;301:1374–7.

[207] Nemazee DA, Burki K. Clonal deletion of B lymphocytes in a transgenic mouse bearing anti-MHC class I antibody genes. Nature 1989;337:562–6.

[208] Gay D, Saunders T, Camper S, Weigert M. Receptor editing: an approach by autoreactive B cells to escape tolerance. J Exp Med 1993;177:999–1008.

[209] Koralov SB, Novobrantseva TI, Konigsmann J, Ehlich A, Rajewsky K. Antibody repertoires generated by VH replacement and direct VH to JH joining. Immunity 2006;25:43–53.

[210] Goodnow CC, Crosbie J, Adelstein S, Lavoie TB, Smith-Gill SJ, Brink RA, et al. Altered immunoglobulin expression and functional silencing of self-reactive B lymphocytes in transgenic mice. Nature 1988;334:676–82.

[211] Choi J, Kim ST, Craft J. The pathogenesis of systemic lupus erythematosus-an update. Curr Opin Immunol 2012;24:651–7.

[212] Marks SD, Tullus K. Autoantibodies in systemic lupus erythematosus. Pediatr Nephrol 2012;27:1855–68.

[213] Yurasov S, Wardemann H, Hammersen J, Tsuiji M, Meffre E, Pascual V, et al. Defective B cell tolerance checkpoints in systemic lupus erythematosus. J Exp Med 2005;201:703–11.

[214] Starr TK, Jameson SC, Hogquist KA. Positive and negative selection of T cells. Annu Rev Immunol 2003;21:139–76.

[215] Eschbach C, Bach MP, Fidler I, Pelanda R, Kohler F, Rajewsky K, et al. Efficient generation of B lymphocytes by recognition of self-antigens. Eur J Immunol 2011;41:2397–403.

[216] van Loo PF, Dingjan GM, Maas A, Hendriks RW. Surrogate-light-chain silencing is not critical for the limitation of pre-B cell expansion but is for the termination of constitutive signaling. Immunity 2007;27:468–80.

[217] Martin DA, Lu L, Cascalho M, Wu GE. Maintenance of surrogate light chain expression induces developmental delay in early B cell compartment. J Immunol 2007;179:4996–5005.

[218] Allman D, Pillai S. Peripheral B cell subsets. Curr Opin Immunol 2008;20:149–57.

[219] Hardy RR. B-1 B cells: development, selection, natural autoantibody and leukemia. Curr Opin Immunol 2006;18:547–55.

[220] Pillai S, Cariappa A, Moran ST. Marginal zone B cells. Annu Rev Immunol 2005;23:161–96.

[221] McHeyzer-Williams LJ, McHeyzer-Williams MG. Antigen-specific memory B cell development. Annu Rev Immunol 2005;23:487–513.

[222] Casola S, Otipoby KL, Alimzhanov M, Humme S, Uyttersprot N, Kutok JL, et al. B cell receptor signal strength determines B cell fate. Nat Immunol 2004;5:317–27.

[223] Hayakawa K, Asano M, Shinton SA, Gui M, Allman D, Stewart CL, et al. Positive selection of natural autoreactive B cells. Science 1999;285:113–6.

[224] Hayakawa K, Asano M, Shinton SA, Gui M, Wen LJ, Dashoff J, et al. Positive selection of anti-thy-1 autoreactive B-1 cells and natural serum autoantibody production independent from bone marrow B cell development. J Exp Med 2003;197:87–99.

[225] Wen L, Brill-Dashoff J, Shinton SA, Asano M, Hardy RR, Hayakawa K. Evidence of marginal-zone B cell-positive selection in spleen. Immunity 2005;23:297–308.

[226] Zikherman J, Parameswaran R, Weiss A. Endogenous antigen tunes the responsiveness of naive B cells but not T cells. Nature 2012;489:160–4.

[227] Martin F, Kearney JF. Marginal-zone B cells. Nat Rev Immunol 2002;2:323–35.

[228] Clayton E, Bardi G, Bell SE, Chantry D, Downes CP, Gray A, et al. A crucial role for the p110delta subunit of phosphatidylinositol 3-kinase in B cell development and activation. J Exp Med 2002;196:753–63.

[229] Jou ST, Carpino N, Takahashi Y, Piekorz R, Chao JR, Carpino N, et al. Essential, nonredundant role for the phosphoinositide 3-kinase p110delta in signaling by the B-cell receptor complex. Mol Cell Biol 2002;22:8580–91.

[230] Okkenhaug K, Bilancio A, Farjot G, Priddle H, Sancho S, Peskett E, et al. Impaired B and T cell antigen receptor signaling in p110delta PI 3-kinase mutant mice. Science 2002;297:1031–4.

[231] Martin F, Kearney JF. Positive selection from newly formed to marginal zone B cells depends on the rate of clonal production, CD19, and btk. Immunity 2000;12:39–49.

[232] Rickert RC, Rajewsky K, Roes J. Impairment of T-cell-dependent B-cell responses and B-1 cell development in CD19-deficient mice. Nature 1995;376:352–5.

[233] Anzelon AN, Wu H, Rickert RC. Pten inactivation alters peripheral B lymphocyte fate and reconstitutes CD19 function. Nat Immunol 2003;4:287–94.

[234] Kuppers R. Mechanisms of B-cell lymphoma pathogenesis. Nat Rev Cancer 2005;5:251–62.

[235] Nambiar M, Kari V, Raghavan SC. Chromosomal translocations in cancer. Biochim Biophys Acta 2008;1786:139–52.

[236] Schmitz R, Young RM, Ceribelli M, Jhavar S, Xiao W, Zhang M, et al. Burkitt lymphoma pathogenesis and therapeutic targets from structural and functional genomics. Nature 2012;490:116–20.

[237] Sander S, Calado DP, Srinivasan L, Kochert K, Zhang B, Rosolowski M, et al. Synergy between PI3K signaling and MYC in Burkitt lymphomagenesis. Cancer Cell 2012;22:167–79.

[238] Kloo B, Nagel D, Pfeifer M, Grau M, Duwel M, Vincendeau M, et al. Critical role of PI3K signaling for NF-kappaB-dependent survival in a subset of activated B-cell-like diffuse large B-cell lymphoma cells. Proc Natl Acad Sci USA 2011;108:272–7.

[239] Rickert RC. New insights into pre-BCR and BCR signalling with relevance to B cell malignancies. Nat Rev Immunol 2013;13:578–91.

[240] Bichi R, Shinton SA, Martin ES, Koval A, Calin GA, Cesari R, et al. Human chronic lymphocytic leukemia modeled in mouse by targeted TCL1 expression. Proc Natl Acad Sci USA 2002;99:6955–60.

[241] Puri KD, Gold MR. Selective inhibitors of phosphoinositide 3-kinase delta: modulators of B-cell function with potential for treating autoimmune inflammatory diseases and B-cell malignancies. Front Immunol 2012;3:256.

[242] Fruman DA, Rommel C. PI3Kdelta inhibitors in cancer: rationale and serendipity merge in the clinic. Cancer Discov 2011;1:562–72.

[243] Lannutti BJ, Meadows SA, Herman SE, Kashishian A, Steiner B, Johnson AJ, et al. CAL-101, a p110delta selective phosphatidylinositol-3-kinase inhibitor for the treatment of B-cell malignancies, inhibits PI3K signaling and cellular viability. Blood 2011;117:591–4.

[244] Baudot AD, Jeandel PY, Mouska X, Maurer U, Tartare-Deckert S, Raynaud SD, et al. The tyrosine kinase Syk regulates the survival of chronic lymphocytic leukemia B cells through PKCdelta and proteasome-dependent regulation of Mcl-1 expression. Oncogene 2009;28:3261–73.

[245] Buchner M, Baer C, Prinz G, Dierks C, Burger M, Zenz T, et al. Spleen tyrosine kinase inhibition prevents chemokine- and integrin-mediated stromal protective effects in chronic lymphocytic leukemia. Blood 2010;115:4497–506.

[246] Buchner M, Fuchs S, Prinz G, Pfeifer D, Bartholome K, Burger M, et al. Spleen tyrosine kinase is overexpressed and represents a potential therapeutic target in chronic lymphocytic leukemia. Cancer Res 2009;69:5424–32.

Chapter 7

# Development and Function of B Cell Subsets

Andre M. Vale[1], John F. Kearney[2], Alberto Nobrega[5] and Harry W. Schroeder[2,3,4]

[1]Program in Immunobiology, Carlos Chagas Filho Institute of Biophysics, Federal University of Rio de Janeiro, Rio de Janeiro, Brazil, [2]Department of Microbiology, University of Alabama at Birmingham, Birmingham, AL, USA, [3]Department of Medicine, University of Alabama at Birmingham, Birmingham, AL, USA, [4]Department of Genetics, University of Alabama at Birmingham, Birmingham, AL, USA, [5]Department of Immunology, Paulo de Goes Institute of Microbiology, Federal University of Rio de Janeiro, Rio de Janeiro, Brazil

## 1. INTRODUCTION

In the bone marrow and fetal liver, precursor B lineage cells differentiate into immature B cells as they pass through a series of developmental steps that are marked by the sequential rearrangement of the immunoglobulin (Ig) heavy (H) and light (L) chain gene loci, the expression of H chain and L chain proteins, their association to create a complete Ig molecule, and tests of the function of membrane-bound surface IgM (mIgM), which together with Igα and Igβ form the B-cell receptor (BCR) for antigen [1,2] (Figure 1). As these newly formed immature B cells leave the tissues of their birth, they are subject to further selection steps as they enter into the pool of mature B lymphocytes [3]. These steps involve a series of developmental programs that include checkpoints to evaluate the composition, specificity, and reactivity of the BCR carried by each individual lymphocyte [4,5]. These programs and checkpoints eventually result in the creation of an array of mature B cells that display a diverse BCR repertoire that can react to a broad range of both ancient and novel antigens [6–8]. Cells within this B-cell pool can be segregated into subsets of mature IgM-bearing B lymphocytes on the basis of characteristic phenotypic differences in surface molecule expression, differences in anatomic location, and differences in responses to immunologic stimuli. Cells belonging to these subsets have been shown to be the product of distinctive ontogenic progenitors, developmental programs, and maintenance requirements [9–12].

Most studies of B-cell subset development and function have been performed using mouse models. While comparative studies of humans have demonstrated similar patterns of development and subset composition between mice and humans, these patterns are not identical (reviewed by Schroeder [13]). Thus, while this chapter focuses on B-cell subsets in mice, attention is given to areas where mouse and human B-cell development diverges.

## 2. B-1, MARGINAL ZONE AND FOLLICULAR B CELLS

### 2.1 Introduction to the Three Subsets

In the adult mouse, three major subsets of mature IgM-bearing B lymphocytes are recognized: B-1 cells, marginal zone (MZ) B cells, and follicular (FO) B cells. Non-B-1 cells are typically referred to as conventional B cells, or B-2 cells; thus by definition both MZ and FO B cells fall within this category. However, in practice, references to B-2 cells generally involve the study of cells of the FO lineage since these cells comprise the vast majority of B lymphocytes in the spleen and lymph nodes, as well of the recirculating B-cell pool in the bone marrow and in the blood. All of these B-cell subsets are heterogeneous in nature and can be further divided into finer subgroups based on phenotype, anatomic location, and/or function.

The phenotypic, microanatomic localization, and functional differences that characterize B-1, MZ, and FO B-cell subsets in mice suggest specialized functions linked to the niches in which they reside. A specific subset of B-1 cells, the B-1a cells, produce natural (preexisting) antibodies (NAbs) that are polyreactive, of relatively low affinity, and primarily of the IgM isotype [14]. In mice, NAbs are present at equivalent levels in the absence of foreign stimuli, suggesting a role for endogenous ligands in the selection of the NAb repertoire [15–17]. NAbs play a critical role in the innate immune response against pathogens [18–20]. A failure to produce proper NAbs can lead to increased

Molecular Biology of B Cells. http://dx.doi.org/10.1016/B978-0-12-397933-9.00007-2

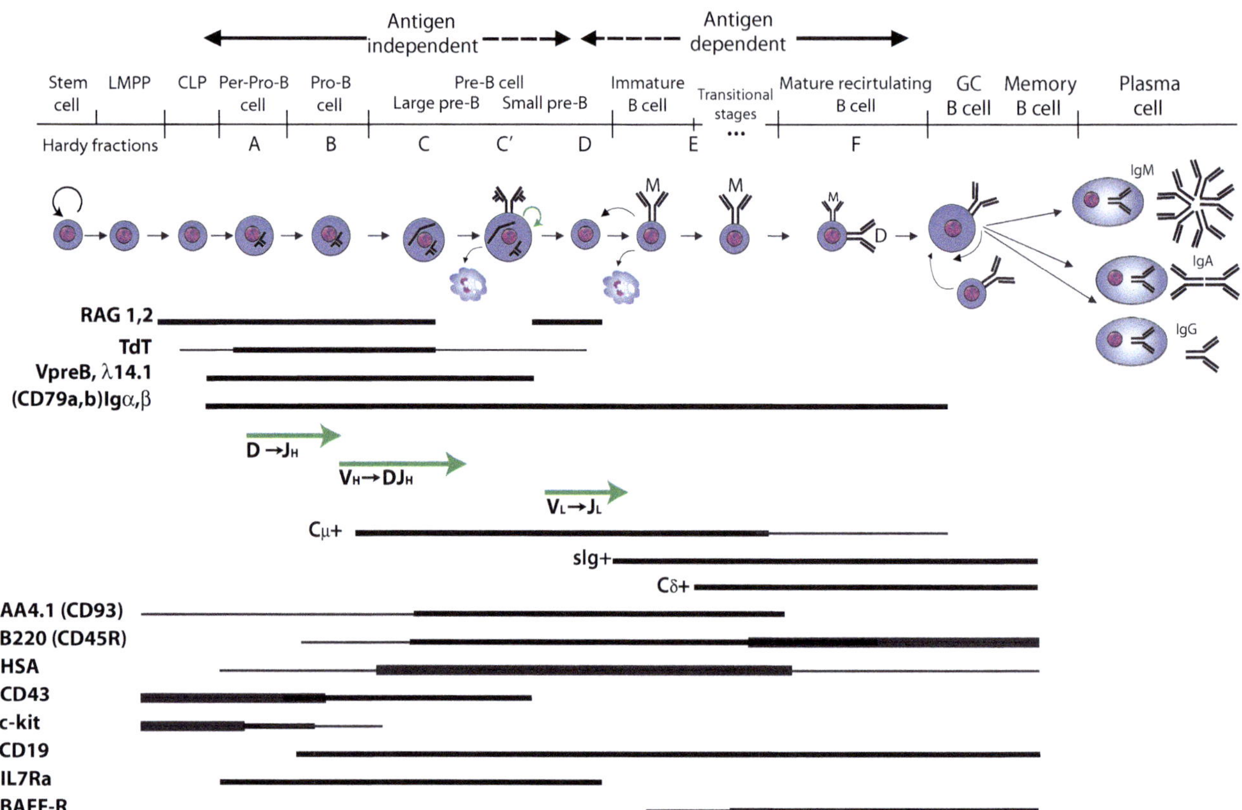


FIGURE 1 **Schematic model of B-cell differentiation.** B-cell development is typically viewed as a linear progression through different stages of differentiation. The various processes associated with the assembly of the B-cell antigen receptor complex and the expression pattern of surface molecules are illustrated (their presence or absence is noted by the black bars). The various steps in immunoglobulin (Ig) rearrangement and the pattern of expression of these surface molecules can be used to characterize stages in B-cell development and B-cell subsets. CLP, common lymphoid progenitor; GC, germinal center; LMPP, lymphoid primed multipotent progenitor.

susceptibility to microbial infections. NAbs often bind with low affinity to a variety of self-antigens, which allows them to play a role in systemic homeostasis [21,22]. B-1 B cells, mainly the B-1b subset, as well as MZ B cells, are characterized by their ability to effect early and rapid immune responses. Both B-1 and MZ B cells produce low-affinity pathogen-specific IgM or IgG3 antibodies and can be easily and quickly activated by T-cell-independent means. Their activation depends on either costimulation through innate receptors on the B cell (e.g., Toll-like receptors [TLRs]) or polymeric antigen stimulation of IgM on the cell surface (e.g., bacterial polysaccharides). These properties contribute to the apparent lower threshold of B-1 and MZ cells for activation, proliferation, and differentiation into antibody-secreting cells than recirculating FO B cells. These shared functional characteristics suggest that B-1 and MZ cells may be selected similarly. FO B cells are programmed to create high-affinity pathogen-specific antibodies that require help from T cells to produce. With the aid of T cells, this population has the capability to undergo extensive proliferation, somatic hypermutation, and class-switching, yielding IgG, IgA, and IgE antibodies of high affinity. FO B cells require days to weeks to produce class-switched antibodies, which contain Fc effector domains that often are tailored to promote optimum protection. When pathogens enter the body, all B-cell subsets are mobilized. In minutes to hours, microbes encounter circulating NAbs. In hours to days, B-1 and MZ B cells are activated in a thymus-independent (T-independent) fashion by the microbes or their products. In days to weeks, microbial antigens activate T cells as well as B cells, enabling high-affinity FO immune response (Figure 2).

B-1 cell subsets are the major subpopulation in pleural and peritoneal cavities, comprising 30–60% of total lymphocytes at these sites [23]. They can also be found, albeit at a low frequency, in the spleen and other lymphoid organs (about 0.2–1%). Their differential distribution in characteristic microenvironments is driven, at least in part, by antigen receptors, given the canonical BCRs used by some of these cells [24]. B-1 cells are self-renewing, with cell cycle and activation properties different from the bulk of recirculating B-2 cells, which are continuously being renewed from bone marrow output of newly formed B cells [25]. This pattern of strict anatomic localization with a restricted antigen receptor is similar to that seen in γδ T cells and αβ natural killer T cells, which are characterized by their

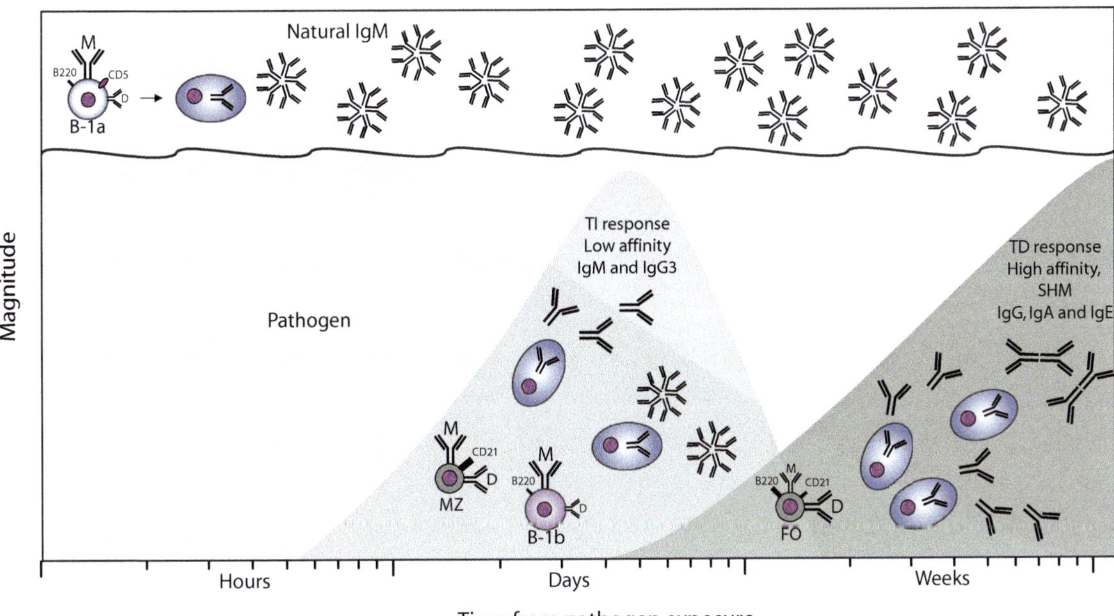


FIGURE 2 **The first line of defense against pathogens entering the body is provided by circulating natural antibodies that are continually being produced by B-1 cells.** The second line of defense initiates within hours of exposure to an antigen and reflects T-independent (TI) activation of B-1 and MZ B cells by the microbes or their products. The third line of defense takes days to weeks to materialize because it takes time for microbial antigens to activate both antigen-specific T cells and B cells, which both are required to create a high-affinity follicular (FO) immune response. Ig, immunoglobulin; MZ, marginal zone; SHM, somatic hypermutation; TD, T-dependent.

own particular geographical preferences [26]. It has been proposed that B-1 B cells derive from a specific fetal B-cell progenitor, distinct from those found in adult bone marrow. There is a strong debate about this proposal, known as the dual-lineage model of B-cell development, as discussed in more detail in Section 3.

In humans, the existence of B-1 cells has been intensively debated. CD5-expressing B cells are found in a number of human lymphoid tissues, including coelomic cavities. However, unlike mice, CD5 is not restricted to human B-1 homologs [27]. Griffin et al. [28,29] recently phenotypically identified a B-cell population as $CD20^+$ $CD27^+$ $CD43^+$ $CD70^-$ cells, with some of them expressing CD11b. Further analyses indicated that these cells display many of the features that characterize murine B-1 cells. This includes spontaneous IgM secretion, limited somatic mutations, and a skewed Ig repertoire targeted to typical B-1 cell specificities, including phosphorylcholine (PC) [28,29]. However, definitive evidence that these are the true human homologs of murine B-1 cells has yet to be obtained.

MZ B cells contribute approximately 5–10% of the B cells in the spleen. The "marginal zone" designation refers to their primary localization within the MZ of the spleen at the interface between the red and the white pulp adjacent to the marginal sinus, where—in both mice and humans—these cells can be constantly bathed in blood and its associated contents [30–32] (Figure 3). This anatomic location facilitates their role as a rapid first line of defense against blood-borne particulate antigens [33,34]. In this location, MZ B cells are intimately associated with metallophilic and marginal sinus-associated macrophages that are richly endowed with innate receptors involved in scavenging foreign and self-antigens. New evidence shows that recruitment and localization in this site is also dependent on characteristic extracellular matrix proteins that are prevalent in this part of the spleen [35].

Mouse MZ B cells are commonly confined to the spleen [33,36]. Within the spleen, however, these cells can shuttle between the MZ and the follicles. Moreover, during immune responses, MZ B cells can carry antigen to the junction between the T-cell zone and the B-cell follicle [37–39]. This enables them to transport antigen in a native state to the white pulp [40].

By contrast, in humans, B cells that share surface markers with mouse MZ B cells are not confined to the spleen: they recirculate. In addition to coexpression of IgM and IgD, which is a mark of a naive mature B cells in mice, the MZ-like cells that are found in human blood also express CD27, a marker of B-cell memory and thus a marker of antigen experience. The presence of CD27 in this human MZ-like population suggests that these cells have already encountered and responded to antigen, even though both IgM and IgD are expressed [41]. These $CD27^+IgM^+IgD^{lo}$ B cells are a major source of antipolysaccharide IgM antibodies in human serum and are thought to be responsible for controlling infection from encapsulated bacteria [42–44].

FO B cells are the most prevalent of the three subsets in the adult mouse (Figure 3). They form the major subset

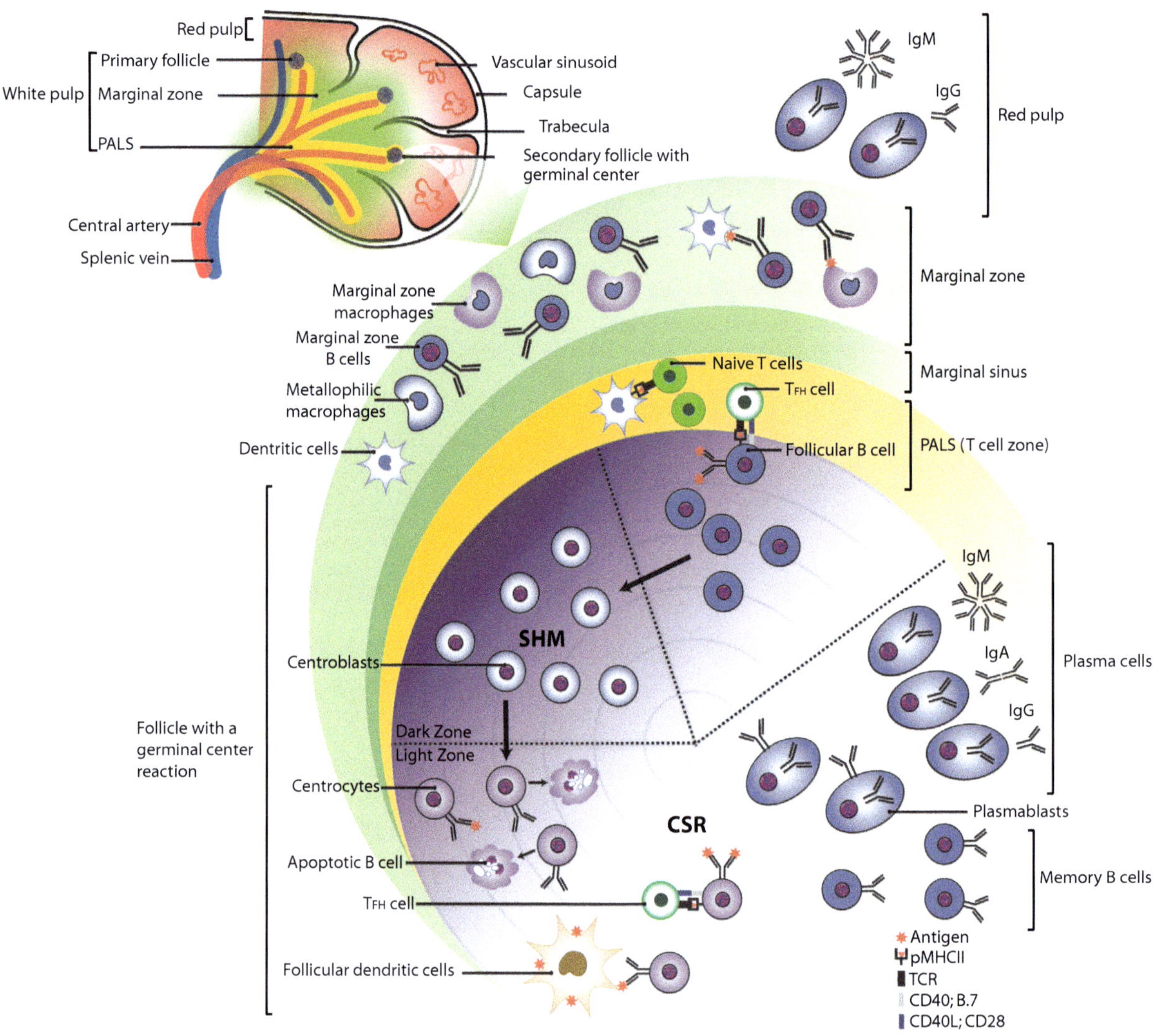


FIGURE 3 **B cells enter the spleen through the central artery and congregate in the follicles or the marginal zone (MZ), which is separated from the follicle by the marginal sinus.** Dendritic cells and macrophages help maintain MZ B cells in an activated state that enables them to respond quickly to blood-borne microbial challenge. With the aid of T follicular helper cells ($T_{FH}$), antigen-activated follicular B cells can form germinal centers, where they undergo class switching and somatic hypermutation to produce highly specific, highly reactive antibodies. CSR, class switch recombination; Ig, immunoglobulin; PALS, periarteriolar lymphoid sheath; SHM, somatic hypermutation.

of B lymphocytes in spleen and lymph nodes, comprising up to 80–90% of B cells in these organs. Their anatomic enrichment in primary follicles gives them their name. As with B-1 and MZ B cells, FO B cells are not confined to the follicles. Cells sharing the same surface markers and behaviors also predominate among the mature populations of B cells in the bone marrow, blood, and other lymphoid organs. FO B cells are continuously generated from bone marrow B-cell precursors throughout life. They are normally resting cells and are relatively long-lived, with a half-life of approximately 4 months in mice [45]. These recirculating cells are typically referred to as B-2 cells, or "conventional" B cells, even though, as noted earlier, MZ B cells can also considered B-2 cells because of their derivation from conventional, postnatal B-cell progenitors. FO B cells are mainly involved in interactions with T cells, and their response to T-cell-dependent antigens eventually cause the germinal center (GC) reactions (reviewed by Victora and Nussenzweig [46]).

All B-cell subsets are derived from newly formed B cells traveling from the fetal liver or the adult bone marrow to appropriate sites in the periphery under the influence of some known molecules, including molecules related to BCR signaling [30]. The relative contribution of known mechanisms producing the enrichment of B-cell subsets and the relative roles of self- and environmental antigen signals and survival signals that drive or maintain B cells in their proper anatomic and functional niches remains a focus of investigation. Multiple hypotheses propose BCR signaling to be crucial in the enrichment of B-1, MZ, and FO B cells in their independent niches, since this process is impaired in *xid*, CD19$^{-/-}$, CD45$^{-/-}$, and other genetically manipulated mice [47–49]. Conversely, global alteration of the antigen-binding site repertoire to enrich for specific physicochemical properties leads to very different distributions of B cells by anatomic site and B-cell subset [50,51]. The relative contribution of different mechanisms producing the enrichment of B-cell subsets and the relative roles of self- and environmental antigen signals and survival signals that drive or maintain B cells in their proper anatomic and functional niches thus remain active foci of investigation. Irrespective of which mechanism is dominant in this positioning effect, it does not alter the proposition that the immune system causes clones to be sequestered in strategically located sites where their BCR-induced functional capabilities are suited for a particular set of environmental antigens associated with a given "geographical location."

## 3. MECHANISMS FOR THE COMPARTMENTALIZATION OF B-CELL SUBSETS

### 3.1 Ontogeny

To understand better the compartmentalization of B-cell subsets, it is necessary to characterize the mechanisms involved in their development, although these are still incompletely resolved. Two divergent models have been proposed to explain the existence of the main subsets of B lymphocytes: the B-1 and B-2 (FO and MZ) cell populations. The first model proposes the existence of two distinct B-lineage-specific progenitors that give rise to B-1 and B-2 cells, respectively. This is referred to as the dual-lineage model. The selection or activation model—the alternative—holds that the specificity coupled with the intensity of BCR signaling drives the fate of the B cell to either the B-2 or the B-1 cell subsets (Figure 4). Detailed reviews describing the experimental details behind each model have been published [11,52]. If one takes into account the unique Ig repertoire expressed by B-1 cells and the polyreactive feature of the corresponding antigen receptors, it is clear that these models are not mutually exclusive.

Early in ontogeny, B lymphocytes arise from multipotent hematopoietic stem cells that first appear in the embryo within the para-aortic splanchnopleura. These cells, which include B-cell progenitors, take up residence in the liver at 11 days' fetal gestation in mice and 7 weeks' fetal gestation in humans. Shortly before birth in mice, and by the middle of the second trimester in humans, these B-cell progenitors can be found in bone marrow, which becomes their exclusive home after birth. In these primary lymphoid organs, daughter cells give rise to lymphoid-primed multipotent progenitors that produce common lymphoid precursors, which can generate T cells, B cells, natural killer cells, and dendritic cells. Definitive commitment to B-cell differentiation requires the exposure of common lymphoid progenitor daughter cells to specialized microenvironments, such as those found in the fetal liver and bone marrow. In adult bone marrow, B cells continue to be produced throughout the life of the individual, although the rate of production declines with age.

In both fetal liver and postnatal bone marrow, B-cell development is typically viewed as a stepwise process that involves passage through a series of checkpoints that are used to test the stability of newly generated Ig H and L chains, and then the initial binding characteristics of the IgM they produce [53]. Passage through these checkpoints includes testing for physicochemical properties of the antigen-binding site, including hydrophobicity and structure [54]. The initial requirement for the survival of a developing B cell is that it creates an H chain protein that can form a functional pre-B CR. Cells that survive this process repeatedly divide and then initiate rearrangement of the L chain. Production of a functional L chain permits creation of a complete IgM molecule by the immature B cell.

Advances in flow cytometry techniques and reagents made it possible to phenotypically identify and isolate cells at various stages of hematopoietic development. Thus, the common linear view of B-cell differentiation expanded beyond Ig gene products to include the expression of non-Ig cell-surface molecules and the regulated expression of specific sets of transcription factors. In this common view, the typical pathway of B-cell development begins with the differentiation of hematopoietic stem cells into early lymphoid progenitors and then common lymphoid progenitors, from which pro-B, pre-B, and finally immature B cells are produced (Figure 1). The model of bone marrow B-cell development described by Hardy et al. [55] indicates that the transition from common lymphoid progenitors into pre-pro-B cells (or fraction A) is phenotypically marked by the upregulation of CD45R (B220) and the absence of CD19, whose expression does not occur until these cells have matured into pro-B cells (or fraction B). According to the developmental model based on B-lineage-specific progenitors for B-1 and B-2 cells, the former cells appear first in ontogeny and are derived from distinct precursors enriched in the splanchnopleura of developing embryos, the fetal liver, and, more rarely, in adult bone marrow. The recent description of a fetal liver B-1-specific progenitor cell, with a phenotype not predicted by the Hardy scheme (i.e., lineage negative [lin$^{-}$ AA4.1$^{+}$ CD19$^{+}$ B220$^{lo/neg}$]), provides further supportive evidence for the lineage-specific

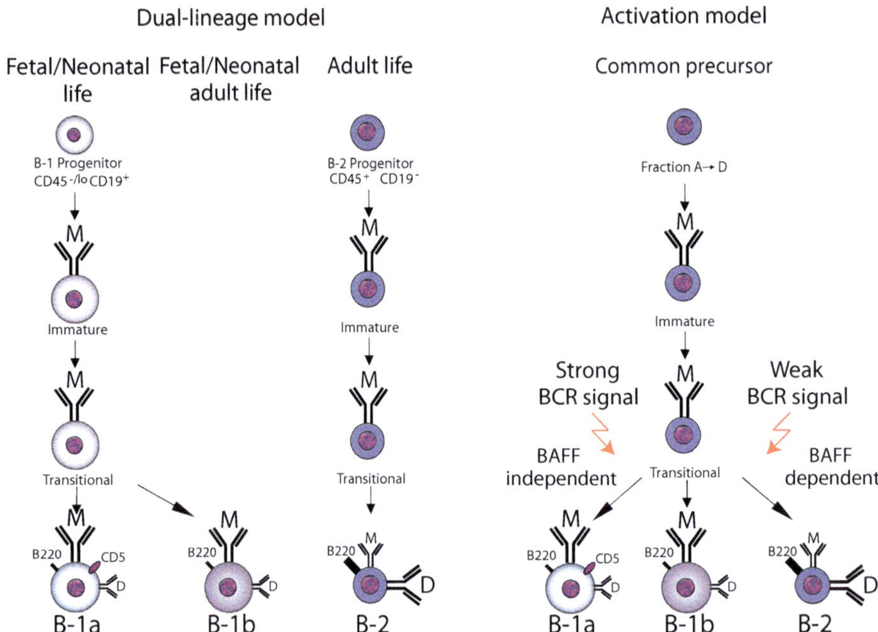


FIGURE 4 **Competing models for the origin of B-cell subsets.** The "dual-lineage model" proposes the existence of two distinct B-lineage-specific progenitors that give rise to B-1 and B-2 cells. The "activation model" proposes that the strength of BCR signaling and specificity drives the fate of transitional B cells to either the B-2 or the B-1 cell subsets. In the unified model it seems that the transitional B-cell pool consists of two types of cells, derived from separate lineages, which respond in a different manner to a BCR signal of defined strength. This signal promotes cells of the one lineage to further differentiate, whereas those of the other lineage die. BAFF, B-cell activating factor.

hypotheses [56]. Therefore, B-1 cells seem to develop from distinct progenitors that represent a majority of B cells in fetal life [57]. The identification of a B-1-specific progenitor cell, abundant in the fetal liver and rare in adult bone marrow, is in accordance with classic experiments that used irradiation chimeras. These classic studies demonstrated that the B-1 cell compartment was poorly reconstituted following bone marrow transfers but could be regenerated with fetal liver transfers [58]. Thus, under the dual-lineage model, B-1 B cells in normal mice are predicted to be primarily the product of fetal B-1-specific progenitor cells, whereas B-2 cells are viewed as derived from the classic progenitors that predominate in the adult bone marrow (Figure 4). However attractive, this model is still debatable and does not accommodate data obtained in several experimental systems that suggest a more complex process driving the formation of B-cell subsets and likely involving both different progenitors and active signaling from the BCR, as discussed in Section 3.2.

## 3.2 B-1 Cells

B-1 cells can be divided into two subsets: B-1a, which expresses CD5 on the cell surface, and B-1b, which does not. Although these two subsets may share developmental precursors, they seem to play distinct roles in immunity. For instance, in mouse models of *Streptococcus pneumoniae* infection, B-1a cells spontaneously produce NAbs important for early protection, whereas B-1b-derived antibodies are induced after exposure to an antigen, providing long-term protection (Figure 2) [59]. Both B-1a and B-1b cells can be easily generated from fetal liver precursors but are generally poorly reconstituted from the bone marrow in adult mice. Yet B-1b cells can be more efficiently reconstituted from B-1 cell–restricted precursors in adult bone marrow than the B-1a cells [60]. Irrespective of their ontogenetic origin, most B-1 cells are stimulated by self- or environmental antigens and express nonconventional membrane molecules in addition to CD5. Thus, the differentiation of B-1a and B-1b based on this marker may not be truly meaningful. During adulthood, the output of B-1 cells from the bone marrow seems to be harshly limited, as opposed to the continuous influx of B-2 cells. However, B-1 cells seem to proliferate in a homeostatic manner to stabilize B-1 cell numbers over time. This mechanism of homeostasis is termed self-replenishment [11,61,62].

To understand B-1 development and function, it is important to determine the differences between the process of BCR formation in fetal liver B-cell precursors versus adult bone marrow B-cell precursors. BCRs are the product of a complex process of V(D)J gene rearrangement events, which typically proceed in a hierarchic pattern. Pro-B cells are marked by the initiation of recombination-activating gene–mediated rearrangement of the D-$J_H$ locus. D-$J_H$ joining is followed by $V_H$-$DJ_H$ rearrangement. In both mice and humans, there are clear divergences in Ig V, D, or J gene segment use between fetal and adult B-cell precursors [63–67]. An even more critical distinction between fetal versus adult B-cell development is the expression of the enzyme terminal deoxynucleotidyl transferase (TdT), which catalyzes the addition of non-germline-encoded N nucleotides in the junctions of these rearranging gene segments and is thus critical for the development of a diverse antigen receptor repertoire [68]. When compared to postnatal bone marrow B-cell development, the activity of this enzyme is practically nonexistent during fetal B-cell development in mice and significantly reduced in humans. In the absence of N regions, the preference for VDJ rearrangement at sites of microhomology between the ends of the $V_H$ and $J_H$ gene segments and the ends of the D gene segments dominates, enriching for canonical rearranged VDJ sequences. In mice especially, developmental control of TdT activity restricts antibody diversity and enriches for junctional sequences that are essentially germline encoded early in ontogeny [69–71].

The fewer N nucleotide regions in the B-1-derived Ig sequences are consistent with the concept of a predominantly fetal/neonatal origin of the B-1 subsets. However, given the presence of significant numbers of B-1 cells bearing BCRs encoded by VDJ rearrangements containing N nucleotide insertions, the data also indicate that replenishment from adult-derived precursor cells occurs over time [71–74]. The unexpected finding by Lam and Rajewsky [75] that mice transgenic for a B-2-derived BCR, or B-1-derived BCR, only replenished the B-2 or B-1 subset, respectively, indicates that the BCR composition influences differentiation into these major subsets. These observations provide a direct challenge to the dual-lineage model while conversely providing support to the selection model, where BCR signaling plays a role on the fate of B-cell clone differentiated into the B-cell subsets. This suggests a new level of complexity to the model that explained the origin of B-1 and B-2 cells exclusively based on their derivation from different precursors. It became clear that both the B-1 and B-2 subsets do not permit the entrance of certain clonotypes. Whether adult bone marrow precursors, transgenic for a B-1 "permitted" BCR, generate B-1 cells remains an open issue; the same applies for the converse situation of a fetal B-cell precursor transgenic for a B-2 "permitted" BCR.

Irrespective of the existence of a B-1-specific fetal progenitor, B-1 cells derive primarily from early fetal and neonatal progenitors and express a repertoire with fewer N nucleotides, which reflects, in part, the aforementioned limited expression of TdT during fetal B-cell development. As a result, the B-1 BCR repertoire is enriched for germline VDJ rearrangements. After leaving the fetal liver or adult bone marrow, B-1 cells enter the spleen. Most of these cells just "pass through" and then are shunted into the peritoneal cavity. Although, B-1 cells residing in the peritoneum do not spontaneously generate IgM, they can be readily activated to migrate to the spleen, and likely to the bone marrow, to differentiate into IgM-secreting B-1 cells [22,23]. The IgM produced by these B-1 cells is heavily enriched for NAbs and does not require foreign stimuli. NAbs are polyreactive and bind with low affinity to many self-antigens, including altered self-antigens created by catabolism or environmental influences. These molecules aid in cellular homeostasis. Because of a greater reliance on germline sequence, many of these B cells create antibodies with "public" antigen binding sites that can be found in many individuals of the same species. NAbs with public antigen binding sites play a critical role in both tissue homeostasis and innate immunity, acting as a first line of defense against pathogens [76–78].

## 3.3 The Spleen

B-2 cells, or "conventional" B cells, are continuously generated from bone marrow B-cell precursors throughout life. They give rise to two distinct mature B-cell subsets in the spleen: MZ and FO B cells. After the discontinuous gut-associated lymphoid tissue, the spleen is the largest secondary lymphoid organ of the body. Histologically, the spleen is divided into the red pulp, which is rich in erythrocytes, and the white pulp, which is enriched for lymphocytes. The red pulp plays a crucial role in filtering the blood and recycling red blood cell components and iron. The blood circulates slowly through the red pulp, which aids the filtration process. The white pulp is organized as a lymphoid organ with B-cell follicles and T-cell zones [79]. The MZ is a highly specialized microanatomic site that surrounds the white pulp. It enables cells leaving the blood stream to transit from the red to the white pulp. Its close association with the red pulp enables it to sample and screen the blood for viruses and bacteria. There is considerable variation in the structure of the MZ in disparate species. It is well developed in mice and rats, intermediately developed in humans, and poorly developed in dogs and cats [80]. The central portions of the white pulp are comprised of B-cell follicles separated by T-cell-rich zones. This anatomic organization correlates with the functions of the two main splenic B-cell subsets, MZ and FO. The former, which encounters antigens directly in the blood stream, primarily engages in T-cell-independent responses; whereas the latter, which is placed in close proximity to T-cell zones, primarily engages in T-cell-dependent responses. FO B cells participate mostly in T-cell-dependent humoral response, giving rise to GCs (Figure 3). FO B cells are recirculating lymphocytes; they capture antigens brought by the lymphatic system and act as antigen-presenting cells to present antigen-derived peptides to T cells. The interaction between FO B cells and T cells occurs in the boundary of T and B zones, and this encounter is greatly facilitated by the migration dynamic of recirculating FO B cells in the spleen [81].

## 3.4 FO B Cells

Newly formed B cells exiting the bone marrow reach the spleen at a relatively immature stage; these are termed transitional B cells [82,83]. Transitional cells respond poorly to BCR triggering and are susceptible to BCR-driven apoptosis. At this stage of maturation they do not engage in B-cell immune responses. They need to complete their maturation in the spleen before entering the follicles or the MZ [84]. The generation of FO or MZ B cells in the spleen follows a maturation process in discrete steps, passing sequentially through transitional stages T1 and T2 before reaching maturity [85]. In transitional stages, B cells are submitted to negative and positive selection checkpoints [86,87]. These eliminate most of the newly formed B cells coming from bone marrow. The T1 and T2 subsets share several surface markers: $IgM^{high}$, $B220^{int}$, $CD24^{high}$, and $AA4.1^{+}$ (CD93). AA4.1 is not only a marker of transitional cells; it also is present on all recently produced bone marrow B cells. According to the original characterization by Allman and colleagues [85], the T2 subset can be discriminated from T1 by the surface expression of CD23 and by higher expression of CD21 and IgD. It is important to observe that the nomenclature describing transitional B cells has not yet reached consensus, and some research groups use different categories of transitional cell types [87]. Transitional B cells have a short life span, lasting only 1 to 5 days. It has been estimated that only 5–10% of the daily bone marrow output of $2\times10^{7}$ B cells in the adult mouse are retained in the spleen as transitional B cells [45,88,89]. B cells in the T1 stage seem to be mostly affected by negative selection, whereas cells in the T2 stage experience both positive selection and homeostatic proliferation [87,90–92]. At both transitional checkpoints, BCR signaling plays a critical role in determining death, survival, or proliferation. High-affinity recognition of self-antigens eliminates T1 B cells; those that survive negative selection progress into the T2 stage [93]. In the T2 stage, intermediate/low-affinity interactions with self-antigens are thought to promote the survival and proliferation of B cells; however, the identity of the antigens driving positive selection of T2 B cells remains largely unknown [94,95].

A third transitional subset, termed the T3 stage, has been characterized [85]. The T3 pool is thought to collect anergic B cells that have exited from either the T2 stage or the mature FO subsets [96]. However, others have suggested that T3 B cells may later integrate into the FO subset.

FO cells in the follicles of the spleen, lymph nodes, and ectopic lymphoid aggregates associated with inflammatory responses typically require help from T cells to develop into plasma cells or memory B cells [97]. FO B cells that engage in T-cell–B-cell cooperation can proliferate and form small clusters of homogeneous clonotypes. These cell clusters may generate either short-lived extrafollicular plasmablasts or else a long-lived collection of GC B cells [46,97]. High affinity for antigen or a high concentration of antigen permits FO B cells to develop into short-lived plasma cells. A lower affinity or lower concentration of antigen promotes creation of a GC. The critical element governing this choice seems to be the expression of the transcriptional repressor bcl-6. Bcl-6 suppresses the expression of Blimp-1, inhibiting the differentiation of B cells into plasmablasts and keeping the proliferative cluster united [98]. GCs are specialized microenvironments formed by a dense mesh of FO dendritic cells supporting the intense proliferation of FO B-cell clones, which undergo successive extensive rounds of activation-induced cytidine deaminase–dependent somatic hypermutation and antigen affinity-driven selection, a process termed affinity maturation. Somatic hypermutation permits massive changes to the framework scaffolding that supports the antigen binding site as well as major changes to the structure and sequence of the antigen binding site itself. Proliferating B cells in GCs interact with a specialized subset of helper T cells, named $T_{FH}$ (FO helper T cells), that produce interleukin-21, a key cytokine for the formation of GCs. $T_{FH}$ also express bcl-6, which seems to play a role in maintaining the $T_{FH}$ phenotype [99,100]. Upon antigen recognition, $T_{FH}$ deliver signals to B cells through cytokines and membrane ligands, triggering their survival, proliferation, and somatic hypermutation. The CD40-CD40L interaction plays a key role in this process. In GCs, B-cell clones compete for antigen capture and presentation. Clones that cannot find help from T cells die, whereas winning clones either undergo new rounds of proliferation and affinity maturation or leave the GC, giving rise to long-lived plasmacytes or memory B cells [46,97].

Memory B cells are not only the product of GCs; they can be generated either by T-independent or -dependent B-cell responses. In the latter case, they may derive from GCs or extrafollicular clusters. Importantly, memory B cells arising from these different processes have different life expectancies [101]. Memory B cells deriving from T-independent or extrafollicular T-dependent activation are generally short lived, although this notion has been challenged [102,103]. On the other hand, GCs derived from memory B cells can live for several months in mice and for years or decades in humans [104]. This seems to reflect selective upregulation of antiapoptotic genes, although the mechanisms by which this is accomplished have not been elucidated. It is important to note that only GC-derived memory B cells demonstrate somatic hypermutation, emphasizing the unique biological role of GCs in the immune response.

## 3.5 MZ B Cells

B cells in the T2 stage face a decision to mature either into FO or MZ B cells. That decision is relevant because these subsets differ in their biological responses. Maturation into

MZ is accompanied by functional alterations of the B cells: MZ B cells respond rapidly to TLR stimuli, participate in T-independent humoral responses, and can present lipid molecules in association with CD1d [31,105]. In contrast to FO B cells, the survival of MZ B cells depends on B-cell activating factor (BAFF) as well as on Notch 2/δ1 signaling [106]. There is now evidence suggesting that additional intermediate maturation steps exists beyond the T2 B-cell stage and the mature subsets; these include MZ precursors and FO-II, which can be characterized phenotypically based on the variation of the expression of surface markers CD93, IgM, IgD, and CD21 [107,108]. MZ B cells tend to bind self-antigens with low affinity [109]. In mice, these cells are the progeny of adult stem cells. Many MZ B cells produce Igs that bind common microbial constituents. These B cells are maintained in an activated state that enables them to respond quickly to blood-borne microbial challenge. Among these MZ B cells, Igs bearing antigen binding sites with less common physicochemical properties, including a high frequency of charged amino acids, are found more frequently [110]. Both B-1 and MZ B cells tend primarily to produce IgM without the aid of T cells.

Also present in the MZ are dendritic cells, reticular fibroblasts, and two types of specialized macrophages: metallophilic and MZ macrophages (MZMs). Marginal metallophilic macrophages (MMMs) are characterized by the expression of sialic acid–binding Ig-like lectin-1 (sialoadhesin, CD169), whereas MZMs express the C-type lectin SIGN-R1 (specific intracellular adhesion molecule-3 grabbing nonintegrin homolog-related 1), which is the mouse homolog of human DCSIGN (dendritic cell–specific ICAM-grabbing nonintegrin/CD209). MZMs and a subset of the MMMs both express the type I scavenger receptor MARCO (macrophage receptor with collagenous structure) [111]. MMMs are active in the crosspresentation of blood-borne antigens by splenic $CD8^+$ DC [112], whereas MZMs preferentially act as phagocytic cells responsible for clearing blood-borne pathogens expressing T-cell-independent antigens and apoptotic material entering the spleen [113,114].

After binding pathogens, MZMs establish direct cell–cell interactions with MZ B cells, enhancing the efficiency of the antibody response [115,116]. MZ B cells are required to maintain MZMs in the MZ. In the absence of MZ B cells, MZMs lose SIGN-R1 expression and phagocytic activity [117]. Conversely, the binding of pathogens to MZ B cells and the IgM response to polysaccharides occur only when SIGN-R1-expressing MZMs are present to first capture pathogens [118]. Similarly, in the absence of MZ B cells, the normal concentration of $CD11c^+$ dendritic cells in the MZ bridging channels is lost, and these DC and are redistributed around the MZ [119]. Thus, the MZ B cells, MZMs, MMMs, and dendritic cells together engage in a complex set of interrelated interactions to facilitate anatomic localization, maintain MZ architecture, and promote efficient responses to blood-borne pathogens.

The human spleen lacks a marginal sinus. There is a MZ around the B cell follicles, but it does not surround the periarticular lymphatic sheath, which is the T-cell-rich area. Between the MZ and the red pulp lies the so-called perifollicular zone, which has many terminal sinuses, blood-filled spaces, sheathed capillaries without endothelial lining, and scattered B cells and T cells. In this zone, antigens and leukocytes can exit the circulation and proceed to either the MZ or the red pulp [41]. The human MZ can be divided into a large inner and a small outer compartment separated by a meshwork of fibroblast-like cells expressing α-smooth muscle actin and mucosal-addressing cell adhesion molecule 1. A small ring of B cells separates the T-cell zone from the red pulp, a ring of T cells is frequently present between the inner and outer MZ, and the outer compartment is in close contact with the perifollicular zone [120].

The human MZ lacks both MZMs and MMMs. In their place, human macrophages expressing CD68 (lysosome/macrosialin) and CD169 preferentially form sheaths around capillaries in the perifollicular area. These macrophages can also be present as scattered cells expressing DC-SIGN [121]. Whether these macrophages play the same role as MZMs in trapping pathogens and interacting with MZ B cells is uncertain. Dendritic cells expressing CD11c and CD205 are found in close association with $MadCAM1^+$ cells at the inner border of the perifollicular zone. Plasmacytoid dendritic cells expressing BDCA-2 also are present in the MZ [122–124].

## 3.6 Chemokines

The molecular basis for the characteristic localization of B-cell subsets is well on the way to being defined. Studies in the field of chemokines have revealed complex mechanisms of retention, migration, and function. In the spleen, the chemokine B lymphocyte chemoattractant is responsible for the development of B-cell follicles [125,126]. Gene-targeting of *pyk-2* [127], *DOCK2* [128], and *lsc* [129], potential signaling pathways downstream of chemokine receptors, results in a drastic reduction or absence of the MZ B-cell compartment. *Pyk-2*, a tyrosine kinase, may mediate signals from G protein-coupled receptor for chemokine, lipids, integrins, and antigen receptors and clearly plays a major role in the generation of MZ B cells and the ability to respond to T-independent antigens.

In mice lacking the chemokine CXCL13 B-1 cells are deficient in peritoneal and pleural cavities but not in the spleen [130,131]. Cells in the omentum and peritoneal macrophages produce CXCL13. In adoptive transfers, B-1 cells home in on the omentum and the peritoneal cavity in a CXCL13-dependent manner. $CXCL13^{-/-}$ mice are deficient in the characteristic natural PC-specific antibodies and in their ability to mount an anti-PC response to peritoneally administered pneumococci. These clonally restricted antibody responses are produced by B-1 cells [130]. The findings of Baumgarth [22] provided the first insight into the mechanism of B-1-cell homing and compartmentalization in the body cavities and reemphasized the critical role of the B-1 cell in the production of NAbs.

B cells enter the spleen with the help of the LFA-1 and α4β1 integrins, which bind to intercellular adhesion molecule-1 and vascular cell adhesion molecule-1, respectively. Antibodies to these integrins that are administered together prevent the entry of B cells into both the MZ and follicles [132]. MZ B cells express higher levels of LFA-1 and α4β1, which may account for their increased binding to the intercellular adhesion molecule-1 and vascular cell adhesion molecule -1 ligands on a variety of endothelial and hematopoietic cells in the MZ, thus preventing their ability to pass through the endothelial lining of the MZ in the resting state. S1P1 and S1P, receptors 1 and 3 of sphingosine-1-phosphate, also contribute to the retention of MZ B cells in the MZ. Decreased expression of either chemokine receptors or integrins alters the positioning of MZ B cells and, under the influence of antigen, permits their entry into the FO area and access to the T-B border (region between the T cell and B cell areas) [38]. Antigen-induced migration of FO B cells into the T-cell areas does not depend on these integrins; however, lipopolysaccharide (LPS)-induced relocalization of MZ B cells is accompanied by integrin downregulation [133]. Thus, environmental signaling to MZ B cells or to lipid receptors from unique ligands existing in the MZ microenvironment may lead to intrinsic upregulation of these integrins and/or chemokine receptors preferentially on these MZ B cells via downstream signaling.

Before reaching the follicles, FO B cells entering the spleen or lymph nodes travel though regions enriched for antigen-presenting cells and T cells. The chemokine–chemokine receptor pairs CXCL13-CXCR5 and CCR7-CCL19/21 play an essential role in attracting B cells to follicles. Other chemoattractant receptors, such as EBI2, also play a role. CXCR5 and CCR7 are expressed by mature FO B cells; CXCL13 is made by FO stromal cells and FO dendritic cells, and CCL19/21 is present in the T-cell zone. This combination of signals guides the B cells throughout these areas and facilitates the capture of antigens and their presentation.

## 3.7 B-Cell Growth and Survival Factors

Other factors are involved in establishing and maintaining B-cell subsets; the most important is BAFF (or Blys). BAFF is a member of the tumor necrosis factor family and plays a critical role in the survival of peripheral B-cell subsets [134,135]. BAFF is a homotrimer that can be expressed either as a type II membrane protein or in a secreted form. It is mainly produced by macrophages, dendritic cells, and stromal cells in lymphoid organs. BAFF is a critical factor for the survival of mature splenic B cells, as shown by different experimental approaches. In the absence of BAFF, or the BAFF receptor (BAFF-R), both FO and MZ B-cell numbers are severely reduced [136,137]. Blockade of BAFF-R results in the death of FO and MZ B cells [138]. However, transitional B cells can be spared in the absence of BAFF, indicating either an interruption of maturation, poor survival beyond that stage, or perhaps both [139]. Interestingly, B-1 cells are not affected by the absence of BAFF, suggesting that alternative factors, not yet characterized, act to maintain these cells [137]. Signaling through BAFF-R determines the longevity of FO B cells, and the amount of available BAFF regulates the numbers of FO and MZ B cells in the spleen [140]. Notably, excess of BAFF results in significantly augmented B cell numbers, proliferation, and autoimmunity, suggesting that BAFF is involved not only in survival but also in BCR-dependent B-cell activation and that it likely influences tolerance checkpoints in transitional stages [141,142]. The exact outcome of transgenic BAFF expression depends on the promoter–enhancer combinations used [143]. In one, autoimmunity was associated with increased splenic B-1 cells. Another using a liver-specific surface and generalized soluble expression [20] favored the transitional and MZ B-cell compartments [144]. With ubiquitous expression (β-actin promoter), autoimmune manifestations were preceded by a generalized B-cell expansion [145]. The functional sites of interaction between BAFF, expressed mostly by macrophages and dendritic cells, and its receptors (BCMA and TACI) are not known, but these and other like molecules play a key role in the development, maintenance, and activation of B lineage cells [134,146].

B cells leave the bone marrow while still undergoing initial maturation, demonstrating progressively higher levels of IgD expression with a commensurate decrease of IgM. As previously mentioned, completion of this maturation process occurs in the spleen, where the relatively immature B cells pass through transitional stages before reaching maturity. This final differentiation step is a crucial checkpoint for controlling self-reactivity. In addition to BCR signaling, passage of conventional B-cell subsets through the transitional stages requires the interaction of BAFF with BAFF-R, which is expressed primarily on B cells, as well as activation of alternative nuclear factor (NF)-κB signaling [147]. Although it has been proposed that B-1 cells also develop through transitional cell intermediates and depend on the spleen to complete their maturation, B-1 transitional cells are not dependent on signaling through the BAFF-R, indicating that B-1 and B-2 transitional cells are not regulated in the same manner [148] (Figure 4). It is likely that the neonatal B-1 transitional cell wave decreases after birth, since the B-2 transitional cell wave establishes and predominates in the adult [27]. Thus, the B-1 cell compartment does not seem to be altered by the ablation of both BAFF and alternative NF-κB signaling pathways [134,149].

B-1 and B-2 cells respond differently to growth factors during their early development. The ablation of the gene encoding interleukin-7 leads to B-2 cell deficiency but does not affect B-1 cells [150]. Instead, the generation of B-1 cells seems to rely on thymic stromal lymphopoietin and the Flk-2 ligand [151,152]. These data suggest that B-1 and B-2 progenitors depend on different growth factors, reinforcing the dual-lineage model paradigm and supporting the notion of layered B-cell development and the formation of B-cell subsets and repertoires [153].

## 4. SELECTION AND DIFFERENTIAL SURVIVAL MECHANISMS: BCR SIGNALING, COMPOSITION, AND SPECIFICITY

### 4.1 BCR Signaling

The conditional introduction of loss-of-function mutations in the BCR have shown that all B cells seem to be constantly in need of some kind of BCR-mediated signal from their microenvironment for both clonal selection and their continued survival [154–156]. The origins of BCR signaling that maintain the survival of mature B cells are not well understood, although the same self-antigens involved in positive selection at the T2 stage are natural candidates to provide these stimuli. Alternatively, it has been suggested that constitutive tonic signaling from BCR would maintain the survival of mature B cells in the absence of antigen recognition [157]. Tonic signaling is thought be autonomously generated by the presence of BCR complexes in the membrane surface, possibly through self-aggregation and oligomerization, resulting in the activation of intracellular signaling cascades in sufficient levels to counteract apoptosis. Importantly, BCR tonic signaling seems to cooperate with BAFF-R to sustain the survival of FO B cells. Cooperation between BCR and BAFF-R signaling cascades is currently being elucidated and seems to involve the activation and crosstalk of both canonical (BCR) and noncanonical (BAFF-R) NF-κB pathways, depending on the activation of Syk kinase, immunoreceptor tyrosine-based *activation* motifs, and phosphoinositide 3-kinase [158,159]. It is assumed that those pathways could cooperate both in classical triggering of BCR by "bona-fide" antigens, leading to activation and proliferation, as well as in tonic survival signaling.

Growing evidence suggests that BCR signaling is not only necessary for the survival of mature B cells but also for determining the fate of the maturing B lymphocytes into the different B-cell subsets. Evidence obtained by several different experimental approaches suggests that the specificity and sequence of the BCR is critical for clonal development into B-l, FO, or MZ subsets [12]. It is known that FO, MZ and B-1 subsets differ in their repertoire of clonotypes. The causes for the physiological segregation of different repertoires in different B-cell subsets seems to be the selection process that gives access to these compartments. Key experiments show that maturation into FO, MZ, and B-1 subsets depends on the clonotypic BCR obtained from mice transgenic for the expression of rearranged H and L Ig genes. Depending on the composition of the BCR, it was observed that the majority of B-cell clones bearing a given transgenic BCR home in on either the splenic follicles, the MZ, or the B-1 compartment in the celomic cavities [75,160]. These experiments support the notion that signaling through the BCR could be a critical factor determining the fate of the maturing B-cell clonotype.

Studies of several independent Ig transgenic mice show that the density of surface BCR may also be involved in this decision by specifically modulating the intensity of clonal signaling. In anti-DNA H chain transgenic mice, normally deleted B cells enrich in the MZ, but this rescue is affected by the expression of two L chains [161]. Likewise, when surface expression of a B-1-type receptor is reduced through the expression of a second H chain, B-cell development proceeds toward the B-2 compartment [162]. Similarly, the size of the B-1 compartment is larger in homozygous anti–red blood cell transgenic mice than in heterozygous mice [163]. Mechanisms regulating B-cell density through surface BCR density not only play a role in the B-1 versus B-2 decision but also are in effect at checkpoints that act to prevent self-reactivity by editing and deletion [20,164]. Modulation of BCR activity in concert with several coreceptors and downstream molecules such as CD5, CD19, CD22, CD21, CD45, Brutons tyrosine kinase (Btk), lymphocyte-specific protein tyrosine *kinase*, and SHP-1 clearly affects the outcome of microenvironmental signals that affect B-cell development and the maintenance of B cells within the immune system [30,47,165].

The formation of B-1 cells seems to be dependent on strong BCR signaling [166] and activation of the classical NF-κB pathway. Mice with defects in the expression of NF-κB signaling mediators, such as Btk, Bcl-10, CARMA1, or NF-κB1, exhibit a severe reduction in the number of peritoneal B-1 but not B-2 cells [167]. However, normal survival and maturation of B-1 transitional cells occurs in both Btk- and NF-κB1-deficient mice [27]. Taken together, these observations suggest that strong BCR signaling and activation of canonical NF-κB pathway are critical for the survival of B-1 cells beyond the transitional stages. As discussed earlier, these observations also support the notion that BCR signaling is a determinant of the fate of B-1 cells, which argue in favor of the integration of the dual-lineage and selection models, thus reconciling both views.

The data produced by the different studies mentioned above have been brought together under a general principle acknowledging the role of the BCR in B-cell maturation. It has been proposed that the strength of BCR signaling can determine the fate of the clonotype: intermediate/low-affinity interactions with self-antigens promote maturation into the MZ subset, higher affinity interactions into the FO subset, and even higher-affinity interactions into the B-1 subset. This proposal, reviewed by Pillai et al. [86], is based on the analysis of a series of genetically engineered mice bearing different deletions of gene-encoding proteins that are involved on BCR signaling cascades or that positively or negatively modulate BCR signaling. It is important to note that the BCR-dependent segregation of maturing clonotypes into different subsets occurs in specific pathogen-free animals that are housed in clean environments, suggesting that self-antigens play a key role in repertoire selection. However, the identity of these self-antigens remains elusive [95].

## 4.2 BCR Composition

Ig, the BCR for antigen, is created by a complex process of VDJ rearrangement, N region addition, and somatic hypermutation [168]. The power of the several mechanisms of Ig diversification is immense, with a theoretical diversity in excess of $10^{15}$, even in the absence of somatic hypermutation. All of these diversification mechanisms are focused on generating the third complementarity-determining region of the H chain (CDR-H3), which lies at the center of the antigen binding site and thus often plays a critical role in defining the antigen binding characteristics of the BCR [169–171]. CDR-H3 is created de novo in the development of B-cell progenitors. Imprecise VDJ gene segment joining and variation in the extent of N nucleotide insertion create a CDR-H3 repertoire that ranges from unmodified and intact germline-encoded sequences to sequences where extensive nibbling and N addition have obscured the identity of the $D_H$ progenitor. The broad range of diversity available to CDR-H3 has functional consequences; its location at the center of the antigen binding site, as classically defined, permits this interval to play an often significant role in antigen recognition and binding [169–171].

Although at first glance the sequence composition of CDR-H3 would seem random, closer inspection reveals significant restrictions in amino acid content that translate to preferred physicochemical categories of length and hydrophobicity [54]. The essential characteristics of CDR-H3 are established early in B-cell development—well before the expression of surface IgM [54]. Transition through bone marrow checkpoints focuses the repertoire by reducing or eliminating outliers to what seems to be a preferred distribution of VDJ gene segment usage; $D_H$ reading frame preference; amino acid composition, length, predicted base, and loop structure; and hydrophobicity. This process continues in the periphery, with disparate B-cell subsets demonstrating individual preferences for global CDR-H3 content (Figure 5) [74,110].

In BALB/c mice, developmental progression from the immature B-cell stage of development (fraction E) to the

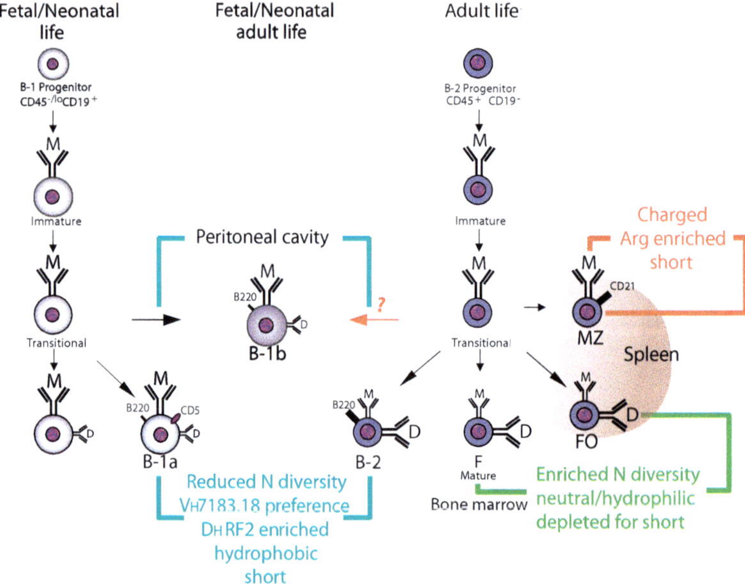


FIGURE 5 **Relationship between the three major B-cell subsets.** The fetal liver is enriched for B-1a progenitors, and the imprint of fetal development can be seen in the high prevalence of heavy chain V sequences lacking N addition. B-1b cells are produced by both fetal liver and adult bone marrow. The prevalence of B-1a and B-1b cells is highest in the peritoneal cavity (PEC). Adult bone marrow preferentially produces B-2 cells that can differentiate into either follicular (FO) or marginal zone (MZ) B cells, which are found in the spleen. B-2 cells also are found in the PEC. Selective pressure seems to operate on the composition of the BCRs expressed by these subsets in the various tissues. The third complementarity-determining region of the H chain component of the heavy chain V domain seems to regress toward a preferred category of amino acid content that best fits the "requirements" for entry into the various compartments.

recirculating, mature B-cell population (fraction F) is associated with the loss of highly nonpolar or charged CDR-H3s, as well as with reduced representation of CDR-H3s of reduced length [54]. The CDR-H3 content of the splenic transitional T1 population is similar to that of the immature bone marrow B-cell fraction, indicating that the use of nonpolar, charged, or short CDR-H3s does not prevent B cells from leaving the bone marrow and entering a primary lymphoid organ. Once in the spleen, however, the fate of B cells using these categories of CDR-H3 seems to diverge. MZ B cells seem to be tolerant of highly charged CDR-H3s; B cells belonging to either the B-1a, B-1b, or B-2 lineages in the peritoneum seem to be tolerant of highly hydrophobic CDR-H3s. FO B cells, however, seem to be just as intolerant of both highly charged and highly hydrophobic CDR-H3s as the recirculating $IgM^+IgD^+$ bone marrow B-cell pool [74,110] (Figure 5). These findings were confirmed by genetic alteration of the $D_H$ gene segment, which forces increased use of highly charged or highly hydrophobic amino acids in CDR-H3. In a set of these $D_H$-altered mice, forced use of charged amino acids in CDR-H3 led to a reduction in peritoneal B-1 cells and splenic FO cells and an increase in MZ B cells, whereas forced use of hydrophobic amino acids in CDR-H3 led to a reduction in splenic FO cells but had no effect on the numbers of B cells in the peritoneal cavity [172]. These differences in CDR-H3 content are not unique; they extend to $V_H$ utilization, as well [74,110].

## 4.3 BCR Specificity

Knowledge of the retention and migration signals for B-cell subsets to and from these sites is a key step in understanding why an anatomic separation of B-cell subsets occurs and how these cells home in to these distinctive sites. The B-1 as well as MZ B-cell populations seem to be enriched in clones that are self-reactive but that also react with bacterial antigens—typical features of the NAbs [95,173]. The recruitment and enrichment of specific clones may depend on their selective activation and survival in the specialized niche in which they reside. Canonical MZ B-cell clones survive preferentially over other clones in vivo and in vitro [47], similar to the receptor-driven selection of B-1 cells (such as the VH11Vk9 clone, which survives in culture better than B-2 cells). Thus, both MZ and B-1 cells may owe their enrichment to preferential survival mechanisms [127]. It has been previously shown that another clone with anti-phosphatidylcholine activity (VH12-Vk4) has a selective advantage in vivo over competitors at multiple checkpoints [174,175].

The existence of B-cell subsets bearing a specific BCR against self-antigens contradicts the classic paradigm of self-tolerance in which autoreactive clones are eliminated to prevent autoimmunity. Interesting enough, is the cross-reactivity against self- and pathogen-associated antigens displayed by the B-1a BCR repertoire that yields NAbs with dual function: first, in normal cellular homeostasis, helping rid the body of cellular and molecular debris [21,176,177], and second, providing one of the first lines of defense against invading pathogens [18,19]. One classic example in mice of the dual role of NAbs involves the antibody response to endogenous oxidation-specific epitopes (oxidation-specific low-density lipoproteins) and protection against *S. pneumoniae* [76,77,177–179]. NAbs bearing the germline-encoded T15 idiotype (T15-Id) seem to be selected into the peripheral B-cell repertoire and expanded in response to oxidation-specific low-density lipoproteins [178,180]. $T15\text{-}Id^+$ NAbs also bind to PC present on the cell wall of *S. pneumoniae* [18,181]. T15-Id antibodies constitute 60–80% of the natural anti-PC response [182–185]. The T15-Id is conserved across multiple mouse strains and is tightly associated with the use of a specific $V_H$ ($V_HS107.1$) and a specific $V_L$ (*Vk22*) [186]. Although many antibodies can bind PC, those containing the canonical T15-Id, lacking N-regions, confer optimal protection against lethal *S. pneumoniae* infection [187]. It seems that the BCR specificities yielding NAbs crossreactive to self- and pathogen-derived antigens have been fixed across evolution through natural selection for both homeostasis and host defense against common pathogens.

## 4.4 Microbial Exposure

A factor increasingly recognized as critical to the development of the immune system is exposure to enteric microbiota [188]. Individual microbial strains are passed to infants by their mother. The composition and diversity of the microbiota develops as a unique trait for each individual and remains stable for life. It is interesting to observe that, although germ-free mice have a marked atrophy of peripheral lymph nodes, splenic architecture and cellularity are maintained, with essentially normal numbers of FO and MZ B cells. Germ-free mice also demonstrate normal numbers of B-1a cells. However, mice colonized with only distinct communities of resident bacteria demonstrate striking defects in MZ and B-1a cell populations in the spleen and body cavity [189]. In these mice, T-independent responses, such as exposure to TNP-Ficoll or activation with LPS, yielded diminished antibody production. Treatment of these mice with antibiotics, followed by exposure to specific pathogen-free conditions, led to normal numbers of MZ and B-1a cells in the offspring. Thus, the composition and diversity of microbiota as well as exposure to antibiotics can alter patterns of MZ B cell and B-1a cell acquisition and function.

## 4.5 T-Independent Signals

The anatomic location of MZ B cells provides them direct access to antigens and particulate elements from blood, including microbial organisms. Microbial antigens derived from the microbiota in intestinal and pulmonary epithelium have access to the celomic cavities, where they come into contact with B-1 cells and stimulate them [190]. Many of these antigens of microbial origin are highly conserved molecules recognized by innate receptors, such as TLRs and NOD-like receptors. It is interesting to note that B-1a and MZ B cells respond robustly to T-cell-independent challenge and activation of innate receptors such as the TLRs [191]. For example, both B-1a and MZ B cells rapidly proliferate after LPS activation and produce high amounts of IgM, whereas FO B-2 cells are much slower to mature under these circumstances [34]. The ability to respond to TLRs also affects the T-dependent humoral response [192]. In this scenario, the ability of MZ B cells to act as antigen-presenting cells is acutely potentiated by LPS exposure, whereas FO B-2 cells are not [193]. Importantly, TLR signaling in B cells is critical for protection against intestinal inflammation [194]. These studies firmly establish the B-1a and MZ subsets as a privileged interface between microbiota and the adaptive immune system, operating through T-independent signaling.

The effect of microbiota on an organism is heavily influenced by the immune system, where MZ and B-1a cells are expected to play a significant role. This is an area of intense research. Specific mechanisms have been proposed to regulate B-1 and MZ B T-independent antibody responses. In this model, a balance arises between negative signals derived from P1/PDL-1 interactions and positive signals mediated through BLyS/TACI interactions by BCR-activated B-1 and MZ B cells. Such a regulatory network may explain how the upregulation of survival signals on B1 and MZ B cells with BCR with low avidity to self-antigens may prevent their maturation into active antibody-secreting cells and promote their maintenance and/or expansion and self-renewal.

## 5. OTHER FACTORS INVOLVED IN THE FORMATION OF B-CELL SUBSETS

### 5.1 Role of CD5

If differential responsiveness and tonic signaling through the Ig receptor are necessary for the development of B-cell subsets, what are the unique mechanisms that permit B-1 cells with higher-affinity self-interactions to survive? B-1 cells are less susceptible than both FO and MZ splenic B cells to anti-IgM-induced apoptosis in vitro. In parallel with T cells, where CD5 is involved in downregulatory functions, CD5 may also be involved in decreasing BCR-induced cell death in B-la cells [195,196]. CD5 expression on chronic lymphocytic leukemia and MZ lymphomas may reflect a relationship between self-renewing activated B-1 cells and these neoplastic B cells [197–199]. CD5 expression by B-1a cells may be associated with BCR–self-antigen interactions. However, the developmental stage or microenvironmental sites at which B-1 cells receive these proposed signals are not known.

The accumulation of self-reactive B-1 cells then occurs in the peritoneal and pleural cavities, with lower prevalence of this subset in other tissues, including the spleen. According to flow cytometry, $CD5^{+/-}$ mice have a lower apparent intensity of CD5 staining of B cells compared to $CD5^{+/+}$ littermates, suggesting that all B cells may constitutively express low levels of CD5 [196]. Indeed, in some other species, all B cells may express CD5 under appropriate conditions [200–202]. The functional deletion of CD5 does not result in dramatic abnormalities in the immune system as a whole nor in B-1 cell function. However, just as CD5 may downregulate T-cell activities, there is evidence that a similar function for CD5 may be at work in B cells [195]. The "activated" phenotype of the B-1a subset, similar to that of the MZ B-cell subset, may result from the BCR self-reactive specificities of these cells. In addition, the microenvironment in which B-1 cells are located maintains them in a state ready to react rapidly to potentially infectious organisms or gut-associated antigens [203].

### 5.2 Apoptosis

The maintenance of peripheral tolerance also involves the elimination of activated T and B cells by Fas-mediated apoptosis [204]. Although multiple pathways are involved in the apoptosis of B cells, Fas-triggered apoptosis eliminates activated B cells, including bystander B cells [205,206]. B2-cell susceptibility to Fas-mediated apoptosis is enhanced by CD40-mediated upregulation of Fas, whereas Fas susceptibility is decreased without a concomitant reduction of surface Fas expression (by signaling through the BCR) [141,207–210]. The impaired induction of Fas in B1 cells after CD40 ligation is likely responsible for the maintenance of self-reactive B cells in this subset and their tendency to give rise to chronic lymphocytic leukemia-like B-cell tumors, a proportion of which make autoimmune antibodies [211].

## 6. HOMEOSTASIS OF B-CELL SUBSETS AND REPERTOIRES

In normal adult mice, around $2–5 \times 10^7$ newly formed B cells are produced every day. Only 5–10% will survive to thrive in the secondary lymphoid organs [88]. The majority of the newly formed cells die right after leaving the primary lymphoid tissues, and only 1–3% of the emerging cells enter the peripheral B-cell pool, which consists of $\sim 10^8$ clones of, apparently, resting B cells. Only a fraction (around 10%) of all peripheral B cells contributes to the natural Ig-secreting B-cell subset. Thus, while the potential diversity of the Ig repertoire is vast, only a small subset of this repertoire is established in the body, where it is distributed in a clonal manner among the B-lymphocyte subsets. These cells, which express Igs on the membrane surface as the BCR, have the ability to differentiate into plasma cells that can secrete these Igs in large quantities as soluble antibodies. The Ig repertoire can thus be viewed as a nested set of repertoires: emerging, available, and actual [212]. The emerging repertoire consists of the Igs present on the membrane surface of the B-cell clones newly formed in the primary lymphoid organs (bone marrow and fetal liver). These cells undergo further negative and positive selections during the transitional stages. The mature B-cell clones bearing Igs populating the peripheral lymphoid organs and tissues constitute the available repertoire and are divided into different subsets: FO, MZ, and B-1. The Ig-secreting plasma cells and the antibodies present in serum and other body fluids constitute the actual repertoire [213,214].

Homeostatic control of B-cell numbers limits the cellularity of each B-cell subset in the available repertoire, as well in emerging and actual repertoires. It is a general principle controlling B-cell development and maturation and the size of each B-cell subset in the organism [94,215]. Homeostatic control can be exerted by competition for growth factors, or survival factors, which is likely the case for plasmacytes in the bone marrow [216,217]. On the other hand, as discussed earlier, homeostasis of peripheral B-cell subsets seems to depend both on survival/growth factors, such as BAFF, as well as on signaling through the BCR. We observe that tonic signaling cannot, in principle, discriminate repertoires. Thus, considering that different mature B-cell subsets show distinct repertoires, antigen-dependent selection must operate in peripheral B-cell homeostasis. If autoreactivity drives BCR signaling to promote cell survival, competition for self-antigens would automatically limits the size of clonotypes in each subset. In this scenario, which demands continuous signaling though the BCR, autoreactivity would preserve a diversified repertoire instead of being a possible source of autoimmunity.

Finally, taking into account the profound effects of microbiota on the immune system, and of the immune system on microbiota—which is mediated by cytokines produced by cells triggered by microbe-derived products and by the antibodies produced by B cells—it seems important to stress the importance of subdividing the repertoire of the different subsets, FO, MZ, and B-1, into active and resting B cells, especially regarding cytokine secretion, extending the classic subdivision of repertoires, which is centered on the secretion of Igs only. The recent abundance of data showing the effect of cytokines produced by B cells on the immune response and the general homeostasis of the organism leave no doubt about the importance of this perspective, which is currently unraveling new potential subsets of B cells, with regulatory activities, including B regulatory cells [218–220].

## 7. CONCLUSION

This chapter is not meant to be an exhaustive review of the development and function of B-cell subsets. More comprehensive reviews have been published in this area [22,23,192,221–224]. Future research will be directed at the elucidation of clonal signals and cosignals and the microenvironments within which B-cell subsets receive these developmental guides for the development of the adult repertoire. Knowledge of the chemokines and adhesion molecules that are involved in the direction of and retention of B cells within these microenvironments is forthcoming. A closely associated field involves the identification of resident cell types within the characteristic environment of each B-cell subset and the functional interactions that occur between these cells during normal development and in immunological functions and disease. Finally, the role of B-cell subsets and effector functions, mainly through cytokines, needs to be addressed in the growing field of the interface of microbiota with the immune system and the organism. The long-term effects on the adult repertoire that result from the developmental timing of antigen (e.g., microbial exposure) need to be understood.

## ACKNOWLEDGMENT

This work was supported, in part, by National Institutes of Health grants AI01478 (JFK), AI090742 (HWS), and AI090902 (HWS). AF was partially supported by CNPq, FAPERJ and FINEP. AMV was supported in part by CNPq and a fellowship from CAPES (CsF-BJT; 024/2012).

## REFERENCES

[1] Hardy RR, Hayakawa K. B cell development pathways. Annu Rev Immunol 2001;19:595–621.

[2] Hardy RR, William. Chapter 7: B lymphocyte development and biology. In: William Paul, editor. Fundamental immunology (book). 6th ed. Philadelphia: Lippincott Williams & Wilkins; 2008. pp. 237–69.

[3] Rolink AG, Andersson J, Melchers F. Molecular mechanisms guiding late stages of B-cell development. Immunol Rev 2004;197:41–50.

[4] Nemazee D, Weigert M. Revising B cell receptors. J Exp Med 2000;191:1813–7.

[5] Nemazee D. Receptor editing in lymphocyte development and central tolerance. Nat Rev Immunol 2006;6:728–40.

[6] Goodnow CC. Balancing immunity, autoimmunity, and self-tolerance. Ann NY Acad Sci 1997;815:55–66.

[7] Rolink A, Melchers F. B-cell development in the mouse. Immunol Lett 1996;54:157–61.

[8] Osmond DG, Rolink AG, Melchers F. Murine B lymphopoiesis: towards a unified model. Immunol Today 1998;19:65–8.

[9] Kantor AB, Stall AM, Adams S, Herzenberg LA, Herzenberg LA. Differential development of progenitor activity for three B-cell lineages. Proc Natl Acad Sci USA 1992;89:3320–4.

[10] Stall AM, Adams S, Herzenberg LA, Kantor AB. Characteristics and development of the murine B-1b (Ly-1 B sister) cell population. Ann NY Acad Sci 1992;651:33–43.

[11] Dorshkind K, Montecino-Rodriguez E. Fetal B-cell lymphopoiesis and the emergence of B-1-cell potential. Nat Rev Immunol 2007;7:213–9.

[12] Pillai S, Cariappa A. The follicular versus marginal zone B lymphocyte cell fate decision. Nat Rev Immunol 2009;9:767–77.

[13] Schroeder Jr HW. Similarity and divergence in the development and expression of the mouse and human antibody repertoires. Dev Comp Immunol 2005;30:119–35.

[14] Avrameas S. Natural autoantibodies: from 'horror autotoxicus' to 'gnothi seauton'. Immunol Today 1991;12:154–9.

[15] Hooijkaas H, Benner R, Pleasants JR, Wostmann BS. Isotypes and specificities of immunoglobulins produced by germ-free mice fed chemically defined ultrafiltered "antigen-free" diet. Eur J Immunol 1984;14:1127–30.

[16] Haury M, Sundblad A, Grandien A, Barreau C, Coutinho A, Nobrega A. The repertoire of serum IgM in normal mice is largely independent of external antigenic contact. Eur J Immunol 1997;27:1557–63.

[17] Bos NA, Kimura H, Meeuwsen CG, De VH, Hazenberg MP, Wostmann BS, et al. Serum immunoglobulin levels and naturally occurring antibodies against carbohydrate antigens in germ-free BALB/c mice fed chemically defined ultrafiltered diet. Eur J Immunol 1989;19:2335–9.

[18] Briles DE, Nahm M, Schroer K, Davie J, Baker P, Kearney J, et al. Antiphosphocholine antibodies found in normal mouse serum are protective against intravenous infection with type 3 *Streptococcus pneumoniae*. J Exp Med 1981;153:694–705.

[19] Benedict CL, Kearney JF. Increased junctional diversity in fetal B cells results in a loss of protective anti-phosphorylcholine antibodies in adult mice. Immunity 1999;10:607–17.

[20] Ochsenbein AF, Fehr T, Lutz C, Suter M, Brombacher F, Hengartner H, et al. Control of early viral and bacterial distribution and disease by natural antibodies. Science 1999;286:2156–9.

[21] Kaveri SV, Silverman GJ, Bayry J. Natural IgM in immune equilibrium and harnessing their therapeutic potential. J Immunol 2012;188:939–45.

[22] Baumgarth N. The double life of a B-1 cell: self-reactivity selects for protective effector functions. Nat Rev Immunol 2011;11:34–46.

[23] Baumgarth N. Innate-like B cells and their rules of engagement. Adv Exp Med Biol 2013;785:57–66.

[24] Hu L, Rezanka LJ, Mi QS, Lustig A, Taub DD, Longo DL, et al. T15-idiotype-negative B cells dominate the phosphocholine binding cells in the preimmune repertoire of T15i knockin mice. J Immunol 2002;168:1273–80.

[25] Stall AM, Wells SM, Lam KP. B-1 cells: unique origins and functions. Semin Immunol 1996;8:45–59.

[26] Bendelac A, Bonneville M, Kearney JF. Autoreactivity by design: innate B and T lymphocytes. Nat Rev Immunol 2001;1:177–86.

[27] Montecino-Rodriguez E, Dorshkind K. B-1 B cell development in the fetus and adult. Immunity 2012;36:13–21.

[28] Griffin DO, Holodick NE, Rothstein TL. Human B1 cells in umbilical cord and adult peripheral blood express the novel phenotype $CD20^+$ $CD27^+$ $CD43^+$ CD70. J Exp Med 2011;208:67–80.

[29] Griffin DO, Rothstein TL. A small $CD11b^+$ human B1 cell subpopulation stimulates T cells and is expanded in lupus. J Exp Med 2011;208(13):2591–8.

[30] Martin F, Kearney JF. Marginal-zone B cells. Nat Rev Immunol 2002;2:323–35.

[31] Pillai S, Cariappa A, Moran ST. Marginal zone B cells. Annu Rev Immunol 2005;23:161–96.

[32] Weill JC, Weller S, Reynaud CA. Human marginal zone B cells. Annu Rev Immunol 2009;27:267–85.

[33] Gray D, MacLennan IC, Bazin H, Khan M. Migrant mu+ delta+ and static $\mu^+\delta^-$ B lymphocyte subsets. Eur J Immunol 1982;12:564–9.

[34] Oliver AM, Martin F, Gartland GL, Carter RH, Kearney JF. Marginal zone B cells exhibit unique activation, proliferative and immunoglobulin secretory responses. Eur J Immunol 1997; 27:2366–74.

[35] Song J, Lokmic Z, Lammermann T, Rolf J, Wu C, Zhang X, et al. Extracellular matrix of secondary lymphoid organs impacts on B-cell fate and survival. Proc Natl Acad Sci USA 2013;110: E2915–24.

[36] Kumararatne DS, Bazin H, MacLennan IC. Marginal zones: the major B cell compartment of rat spleens. Eur J Immunol 1981;11:858–64.

[37] Attanavanich K, Kearney JF. Marginal zone, but not follicular B cells, are potent activators of naive CD4 T cells. J Immunol 2004;172:803–11.

[38] Balazs M, Martin F, Zhou T, Kearney J. Blood dendritic cells interact with splenic marginal zone B cells to initiate T-independent immune responses. [see comment] Immunity 2002;17:341–52.

[39] Martin F, Oliver AM, Kearney JF. Marginal zone and B1 B cells unite in the early response against T-independent blood-borne particulate antigens. Immunity 2001;14:617–29.

[40] Cinamon G, Zachariah MA, Lam OM, Foss Jr FW, Cyster JG. Follicular shuttling of marginal zone B cells facilitates antigen transport. Nat Immunol 2008;9:54–62.

[41] Timens W. The human spleen and the immune system: not just another lymphoid organ. Res Immunol 1991;142:316–20.

[42] Kruetzmann S, Rosado MM, Weber H, Germing U, Tournilhac O, Peter HH, et al. Human immunoglobulin M memory B cells controlling *Streptococcus pneumoniae* infections are generated in the spleen. J Exp Med 2003;197:939–45.

[43] Weller S, Braun MC, Tan BK, Rosenwald A, Cordier C, Conley ME, et al. Human blood IgM "memory" B cells are circulating splenic marginal zone B cells harboring a prediversified immunoglobulin repertoire. Blood 2004;104:3647–54.

[44] Zandvoort A, Timens W. The dual function of the splenic marginal zone: essential for initiation of anti-TI-2 responses but also vital in the general first-line defense against blood-borne antigens. Clin Exp Immunol 2002;130:4–11.

[45] Forster I, Rajewsky K. The bulk of the peripheral B-cell pool in mice is stable and not rapidly renewed from the bone marrow. Proc Natl Acad Sci USA 1990;87:4781–4.

[46] Victora GD, Nussenzweig MC. Germinal centers. Annu Rev Immunol 2012;30:429–57.

[47] Okamoto M, Murakami M, Shimizu A, Ozaki S, Tsubata T, Kumagai S, et al. A transgenic model of autoimmune hemolytic anemia. J Exp Med 1992;175:71–9.

[48] Cariappa A, Pillai S. Antigen-dependent B-cell development. Curr Opin Immunol 2002;14:241–9.

[49] Amano H, Amano E, Moll T, Marinkovic D, Ibnou-Zekri N, Martinez-Soria E, et al. The Yaa mutation promoting murine lupus causes defective development of marginal zone B cells. J Immunol 2003;170:2293–301.

[50] Ippolito GC, Schelonka RL, Zemlin M, Ivanov II, Kobayashi R, Zemlin C, et al. Forced usage of positively charged amino acids in immunoglobulin CDR-H3 impairs B cell development and antibody production. J Exp Med 2006;203:1567–78.

[51] Schelonka RL, Zemlin M, Kobayashi R, Szalai A, Ippolito GC, Zhuang Y, et al. Preferential use of D H reading frame 2 alters B cell development and antigen-specific antibody production. J Immunol 2008;181:8409–15.
[52] Berland R, Wortis HH. Origins and functions of B-1 cells with notes on the role of CD5. Annu Rev Immunol 2002;20:253–300.
[53] Vale AM, Schroeder Jr HW. Clinical consequences of defects in B-cell development. J Allergy Clin Immunol 2010;125:778–87.
[54] Ivanov II, Schelonka RL, Zhuang Y, Gartland GL, Zemlin M, Schroeder Jr HW. Development of the expressed immunoglobulin CDR-H3 repertoire is marked by focusing of constraints in length, amino acid utilization, and charge that are first established in early B cell progenitors. J Immunol 2005;174:7773–80.
[55] Hardy RR, Carmack CE, Shinton SA, Kemp JD, Hayakawa K. Resolution and characterization of pro-B and pre-pro-B cell stages in normal mouse bone marrow. J Exp Med 1991;173:1213–25.
[56] Montecino-Rodriguez E, Leathers H, Dorshkind K. Identification of a B-1 B cell-specified progenitor. Nat Immunol 2006;7:293–301.
[57] Herzenberg LA, Tung JW. B cell lineages: documented at last!. Nat Immunol 2006;7:225–6.
[58] Hayakawa K, Hardy RR, Herzenberg LA. Progenitors for Ly-1 B cells are distinct from progenitors for other B cells. J Exp Med 1985;161:1554–68.
[59] Haas KM, Poe JC, Steeber DA, Tedder TF. B-1a and B-1b cells exhibit distinct developmental requirements and have unique functional roles in innate and adaptive immunity to *S. pneumoniae*. Immunity 2005;23:7–18.
[60] Tung JW, Herzenberg LA. Unraveling B-1 progenitors. Curr Opin Immunol 2007;19:150–5.
[61] Lalor PA, Stall AM, Adams S, Herzenberg LA. Permanent alteration of the murine LY-1 B repertoire due to selective depletion of LY-1 B cells in neonatal animals. Eur J Immunol 1989;19:501–6.
[62] Herzenberg LA. B-1 cells: the lineage question revisited. Immunol Rev 2000;175:9–22.
[63] Schroeder Jr HW, Hillson JL, Perlmutter RM. Early restriction of the human antibody repertoire. Science 1987;238:791–3.
[64] Schroeder Jr HW, Mortari F, Shiokawa S, Kirkham PM, Elgavish RA, Bertrand III FE. Developmental regulation of the human antibody repertoire. Ann NY Acad Sci 1995;764:242–60.
[65] Bertrand 3rd FE, Billips LG, Schroeder Jr HW. Analysis of DHQ52 gene segment transcription and rearrangement during B-cell development in human fetal bone marrow. Ann NY Acad Sci 1995;764:228–30.
[66] Yancopoulos GD, Alt FW. Regulation of the assembly and expression of variable region genes. Annu Rev Immunol 1986;4: 339–68.
[67] Schelonka RL, Szymanska E, Vale AM, Zhuang Y, Gartland GL, Schroeder Jr HW. DH and JH usage in murine fetal liver mirrors that of human fetal liver. Immunogenetics 2010;62:653–66.
[68] Desiderio SV, Yancopoulos GD, Paskind M, Thomas E, Boss MA, Landau NR, et al. Insertion of N regions into heavy-chain gene is correlated with expression of terminal deoxytransferase in B cells. Nature 1984;311:752–5.
[69] Feeney AJ. Lack of N regions in fetal and neonatal mouse immunoglobulin V-D-J junctional sequences. J Exp Med 1990;172: 1377–90.
[70] Feeney AJ. Predominance of VH-D-JH junctions occurring at sites of short sequence homology results in limited junctional diversity in neonatal antibodies. J Immunol 1992;149:222–9.
[71] Gu H, Forster I, Rajewsky K. Sequence homologies, N sequence insertion and JH gene utilization in VH-D-JH joining: implications for the joining mechanism and the ontogenetic timing of Ly1 B cell and B-CLL progenitor generation. EMBO J 1990;9:2133–40.
[72] Tornberg UC, Holmberg D. B-1a, B-1b and B-2 B cells display unique VHDJH repertoires formed at different stages of ontogeny and under different selection pressures. EMBO J 1995;14: 1680–9.
[73] Kantor AB, Merrill CE, Herzenberg LA, Hillson JL. An unbiased analysis of V(H)-D-J(H) sequences from B-1a, B-1b, and conventional B cells. J Immunol 1997;158:1175–86.
[74] Vale AM, Tanner JM, Schelonka RL, Zhuang Y, Zemlin M, Gartland GL, et al. The peritoneal cavity B-2 antibody repertoire appears to reflect many of the same selective pressures that shape the B-1a and B-1b repertoires. J Immunol 2010;185:6085–95.
[75] Lam KP, Rajewsky K. B cell antigen receptor specificity and surface density together determine B-1 versus B-2 cell development. J Exp Med 1999;190:471–7.
[76] Binder CJ, Horkko S, Dewan A, Chang MK, Kieu EP, Goodyear CS, et al. Pneumococcal vaccination decreases atherosclerotic lesion formation: molecular mimicry between *Streptococcus pneumoniae* and oxidized LDL. Nat Med 2003;9:736–43.
[77] Chou MY, Fogelstrand L, Hartvigsen K, Hansen LF, Woelkers D, Shaw PX, et al. Oxidation-specific epitopes are dominant targets of innate natural antibodies in mice and humans. J Clin Invest 2009;119:1335–49.
[78] Vale AM, Kapoor P, Skibinski GA, Elgavish A, Mahmoud TI, Zemlin C, et al. The link between antibodies to OxLDL and natural protection against pneumococci depends on DH gene conservation. J Exp Med 2013;210:875–90.
[79] Kraal G. Cells in the marginal zone of the spleen. Int Rev Cytol 1992;132:31–74.
[80] Koike R, Nishimura T, Yasumizu R, Tanaka H, Hataba Y, Hataba Y, et al. The splenic marginal zone is absent in alymphoplastic aly mutant mice. Eur J Immunol 1996;26:669–75.
[81] Cyster JG. B cell follicles and antigen encounters of the third kind. Nat Immunol 2010;11:989–96.
[82] Carsetti R, Kohler G, Lamers MC. Transitional B cells are the target of negative selection in the B cell compartment. J Exp Med 1995;181:2129–40.
[83] Allman DM, Ferguson SE, Lentz VM, Cancro MP. Peripheral B cell maturation. II. Heat-stable antigen(hi) splenic B cells are an immature developmental intermediate in the production of long-lived marrow-derived B cells. J Immunol 1993;151:4431–44.
[84] Loder F, Mutschler B, Ray RJ, Paige CJ, Sideras P, Torres R, et al. B cell development in the spleen takes place in discrete steps and is determined by the quality of B cell receptor-derived signals. J Exp Med 1999;190:75–89.
[85] Allman D, Lindsley RC, DeMuth W, Rudd K, Shinton SA, Hardy RR. Resolution of three nonproliferative immature splenic B cell subsets reveals multiple selection points during peripheral B cell maturation. J Immunol 2001;167:6834–40.
[86] Pillai S, Cariappa A, Moran ST. Positive selection and lineage commitment during peripheral B-lymphocyte development. Immunol Rev 2004;197:206–18.
[87] Meyer-Bahlburg A, Andrews SF, Yu KO, Porcelli SA, Rawlings DJ. Characterization of a late transitional B cell population highly sensitive to BAFF-mediated homeostatic proliferation. J Exp Med 2008;205:155–68.

[88] Osmond DG. The turnover of B-cell populations. Immunol Today 1993;14:34–7.

[89] Allman D, Pillai S. Peripheral B cell subsets. Curr Opin Immunol 2008;20:149–57.

[90] Cancro MP, Kearney JF. B cell positive selection: road map to the primary repertoire? J Immunol 2004;173:15–9.

[91] Wen L, Brill-Dashoff J, Shinton SA, Asano M, Hardy RR, Hayakawa K. Evidence of marginal-zone B cell-positive selection in spleen. Immunity 2005;23:297–308.

[92] Norvell A, Mandik L, Monroe JG. Engagement of the antigen-receptor on immature murine B lymphocytes results in death by apoptosis. J Immunol 1995;154:4404–13.

[93] Su TT, Rawlings DJ. Transitional B lymphocyte subsets operate as distinct checkpoints in murine splenic B cell development. J Immunol 2002;168:2101–10.

[94] Gaudin E, Rosado M, Agenes F, McLean A, Freitas AA. B-cell homeostasis, competition, resources, and positive selection by self-antigens. Immunol Rev 2004;197:102–15.

[95] Hardy RR, Hayakawa K. Positive and negative selection of natural autoreactive B cells. Adv Exp Med Biol 2012;750:227–38.

[96] Cambier JC, Gauld SB, Merrell KT, Vilen BJ. B-cell anergy: from transgenic models to naturally occurring anergic B cells? Nat Rev Immunol 2007;7:633–43.

[97] Shlomchik MJ, Weisel F. Germinal center selection and the development of memory B and plasma cells. Immunol Rev 2012;247:52–63.

[98] Shapiro-Shelef M, Calame K. Regulation of plasma-cell development. Nat Rev Immunol 2005:230–42.

[99] King C. New insights into the differentiation and function of T follicular helper cells. Nat Rev Immunol 2009;9:757–66.

[100] Ma CS, Deenick EK, Batten M, Tangye SG. The origins, function, and regulation of T follicular helper cells. J Exp Med 2012;209:1241–53.

[101] McHeyzer-Williams M, Okitsu S, Wang N, McHeyzer-Williams L. Molecular programming of B cell memory. Nat Rev Immunol 2012;12:24–34.

[102] Bortnick A, Chernova I, Quinn 3rd WJ, Mugnier M, Cancro MP, Allman D. Long-lived bone marrow plasma cells are induced early in response to T cell-independent or T cell-dependent antigens. J Immunol 2012;188:5389–96.

[103] Foote JB, Mahmoud TI, Vale AM, Kearney JF. Long-term maintenance of polysaccharide-specific antibodies by IgM-secreting cells. J Immunol 2012;188:57–67.

[104] Tarlinton D, Good-Jacobson K. Diversity among memory B cells: origin, consequences, and utility. Science 2013;341:1205–11.

[105] Lopes-Carvalho T, Kearney JF. Development and selection of marginal zone B cells. Immunol Rev 2004;197:192–205.

[106] Cariappa A, Tang M, Parng C, Nebelitskiy E, Carroll M, Georgopoulos K, et al. The follicular versus marginal zone B lymphocyte cell fate decision is regulated by Aiolos, Btk, and CD21. Immunity 2001;14:603–15.

[107] Srivastava B, Quinn III WJ, Hazard K, Erikson J, Allman D. Characterization of marginal zone B cell precursors. J Exp Med 2005;202:1225–34.

[108] Cariappa A, Boboila C, Moran ST, Liu H, Shi HN, Pillai S. The recirculating B cell pool contains two functionally distinct, long-lived, posttransitional, follicular B cell populations. J Immunol 2007;179:2270–81.

[109] Lopes-Carvalho T, Foote J, Kearney JF. Marginal zone B cells in lymphocyte activation and regulation. Curr Opin Immunol 2005;17:244–50.

[110] Schelonka RL, Tanner J, Zhuang Y, Gartland GL, Zemlin M, Schroeder Jr HW. Categorical selection of the antibody repertoire in splenic B cells. Eur J Immunol 2007;37:1010–21.

[111] Nolte MA, Arens R, Kraus M, van Oers MH, Kraal G, van Lier RA, et al. B cells are crucial for both development and maintenance of the splenic marginal zone. J Immunol 2004;172:3620–7.

[112] Backer R, Schwandt T, Greuter M, Oosting M, Jungerkes F, Tuting T, et al. Effective collaboration between marginal metallophilic macrophages and $CD8_+$ dendritic cells in the generation of cytotoxic T cells. Proc Natl Acad Sci USA 2010;107:216–21.

[113] Aichele P, Zinke J, Grode L, Schwendener RA, Kaufmann SH, Seiler P. Macrophages of the splenic marginal zone are essential for trapping of blood-borne particulate antigen but dispensable for induction of specific T cell responses. J Immunol 2003;171:1148–55.

[114] Odermatt B, Eppler M, Leist TP, Hengartner H, Zinkernagel RM. Virus-triggered acquired immunodeficiency by cytotoxic T-cell-dependent destruction of antigen-presenting cells and lymph follicle structure. Proc Natl Acad Sci USA 1991;88:8252–6.

[115] Chen Y, Pikkarainen T, Elomaa O, Soininen R, Kodama T, Kraal G, et al. Defective microarchitecture of the spleen marginal zone and impaired response to a thymus-independent type 2 antigen in mice lacking scavenger receptors MARCO and SR-A. J Immunol 2005;175:8173–80.

[116] Yokota T, Ehlin-Henriksson B, Hansson GK. Scavenger receptors mediate adhesion of activated B lymphocytes. Exp Cell Res 1998;239:16–22.

[117] You Y, Myers RC, Freeberg L, Foote J, Kearney JF, Justement LB, et al. Marginal zone B cells regulate antigen capture by marginal zone macrophages. J Immunol 2011;186:2172–81.

[118] Koppel EA, Litjens M, van den Berg VC, van Kooyk Y, Geijtenbeek TB. Interaction of SIGNR1 expressed by marginal zone macrophages with marginal zone B cells is essential to early IgM responses against *Streptococcus pneumoniae*. Mol Immunol 2008;45:2881–7.

[119] You Y, Zhao H, Wang Y, Carter RH. Cutting edge: primary and secondary effects of CD19 deficiency on cells of the marginal zone. J Immunol 2009;182:7343–7.

[120] Steiniger B, Timphus EM, Barth PJ. The splenic marginal zone in humans and rodents: an enigmatic compartment and its inhabitants. Histochem Cell Biol 2006;126:641–8.

[121] Steiniger B, Timphus EM, Jacob R, Barth PJ. CD27+ B cells in human lymphatic organs: re-evaluating the splenic marginal zone. Immunology 2005;116:429–42.

[122] Nascimbeni M, Perie L, Chorro L, Diocou S, Kreitmann L, Louis S, et al. Plasmacytoid dendritic cells accumulate in spleens from chronically HIV-infected patients but barely participate in interferon-alpha expression. Blood 2009;113:6112–9.

[123] Pack M, Trumpfheller C, Thomas D, Park CG, Granelli-Piperno A, Munz C, et al. DEC-205/$CD205_+$ dendritic cells are abundant in the white pulp of the human spleen, including the border region between the red and white pulp. Immunology 2008;123:438–46.

[124] Steiniger B, Stachniss V, Schwarzbach H, Barth PJ. Phenotypic differences between red pulp capillary and sinusoidal endothelia help localizing the open splenic circulation in humans. Histochem Cell Biol 2007;128:391–8.

[125] Gunn MD, Ngo VN, Ansel KM, Ekland EH, Cyster JG, Williams LT. A B-cell-homing chemokine made in lymphoid follicles activates Burkitt's lymphoma receptor-1. Nature 1998;391:799–803.

[126] Cyster JG. Chemokines and cell migration in secondary lymphoid organs. Science 1999;286:2098–102.

[127] Guinamard R, Okigaki M, Schlessinger J, Ravetch JV. Absence of marginal zone B cells in Pyk-2-deficient mice defines their role in the humoral response. Nat Immunol 2000;1:31–6.
[128] Fukui Y, Hashimoto O, Sanui T, Oono T, Koga H, Abe M, et al. Haematopoietic cell-specific CDM family protein DOCK2 is essential for lymphocyte migration. Nature 2001;412:826–31.
[129] Girkontaite I, Missy K, Sakk V, Harenberg A, Tedford K, Potzel T, et al. Lsc is required for marginal zone B cells, regulation of lymphocyte motility and immune responses. Nat Immunol 2001;2: 855–62.
[130] Ansel KM, Harris RB, Cyster JG. CXCL13 is required for B1 cell homing, natural antibody production, and body cavity immunity. Immunity 2002;16:67–76.
[131] Ha SA, Tsuji M, Suzuki K, Meek B, Yasuda N, Kaisho T, et al. Regulation of B1 cell migration by signals through Toll-like receptors. J Exp Med 2006;203:2541–50.
[132] Lo CG, Lu TT, Cyster JG. Integrin-dependence of lymphocyte entry into the splenic white pulp. J Exp Med 2003;197:353–61.
[133] Lu TT, Cyster JG. Integrin-mediated long-term B cell retention in the splenic marginal zone. Science 2002;297:409–12.
[134] Mackay F, Schneider P. Cracking the BAFF code. Nat Rev Immunol 2009;9:491–502.
[135] Mackay F, Browning JL. BAFF: a fundamental survival factor for B cells. Nat Rev Immunol 2002;2:465–75.
[136] Rolink AG, Tschopp J, Schneider P, Melchers F. BAFF is a survival and maturation factor for mouse B cells. Eur J Immunol 2002;32:2004–10.
[137] Schneider P, Takatsuka H, Wilson A, Mackay F, Tardivel A, Lens S, et al. Maturation of marginal zone and follicular B cells requires B cell activating factor of the tumor necrosis factor family and is independent of B cell maturation antigen. J Exp Med 2001;194: 1691–7.
[138] Rauch M, Tussiwand R, Bosco N, Rolink AG. Crucial role for BAFF-BAFF-R signaling in the survival and maintenance of mature B cells. PloS One 2009;4:e5456.
[139] Lindsley RC, Thomas M, Srivastava B, Allman D. Generation of peripheral B cells occurs via two spatially and temporally distinct pathways. Blood 2007;109:2521–8.
[140] Kreuzaler M, Rauch M, Salzer U, Birmelin J, Rizzi M, Grimbacher B, et al. Soluble BAFF levels inversely correlate with peripheral B cell numbers and the expression of BAFF receptors. J Immunol 2012;188:497–503.
[141] Mackay F, Woodcock SA, Lawton P, Ambrose C, Baetscher M, Schneider P, et al. Mice transgenic for BAFF develop lymphocytic disorders along with autoimmune manifestations. J Exp Med 1999;190:1697–710.
[142] Thien M, Phan TG, Gardam S, Amesbury M, Basten A, Mackay F, et al. Excess BAFF rescues self-reactive B cells from peripheral deletion and allows them to enter forbidden follicular and marginal zone niches. Immunity 2004;20:785–98.
[143] Macpherson AJ, Gatto D, Sainsbury E, Harriman GR, Hengartner H, Zinkernagel RM. A primitive T cell-independent mechanism of intestinal mucosal IgA responses to commensal bacteria. Science 2000;288:2222–6.
[144] Zeng D, Lee MK, Tung J, Brendolan A, Strober S. Cutting edge: a role for CD1 in the pathogenesis of lupus in NZB/NZW mice. J Immunol 2000;164:5000–4.
[145] Ochsenbein AF, Zinkernagel RM. Natural antibodies and complement link innate and acquired immunity. Immunol Today 2000;21:624–30.
[146] Sindhava VJ, Scholz JL, Cancro MP. Roles for BLyS family members in meeting the distinct homeostatic demands of innate and adaptive B cells. Front Immunol 2013;4:37.
[147] Bossen C, Schneider P. BAFF, APRIL and their receptors: structure, function and signaling. Semin Immunol 2006;18:263–75.
[148] Casola S. Control of peripheral B-cell development. Curr Opin Immunol 2007;19:143–9.
[149] Gerondakis S, Siebenlist U. Roles of the NF-kappaB pathway in lymphocyte development and function. Cold Spring Harb Perspect Biol 2010;2:a000182.
[150] Carvalho TL, Mota-Santos T, Cumano A, Demengeot J, Vieira P. Arrested B lymphopoiesis and persistence of activated B cells in adult interleukin $7^{-/-}$ mice. J Exp Med 2001;194: 1141–50.
[151] Vosshenrich CA, Cumano A, Muller W, Di Santo JP, Vieira P. Thymic stromal-derived lymphopoietin distinguishes fetal from adult B cell development. Nat Immunol 2003;4:773–9.
[152] Jensen CT, Kharazi S, Boiers C, Liuba K, Jacobsen SE. TSLP-mediated fetal B lymphopoiesis? Nat Immunol 2007;8:897. author reply 898.
[153] Herzenberg LA. Toward a layered immune system. Cell 1989;59:953–4.
[154] Lam KP, Kuhn R, Rajewsky K. In vivo ablation of surface immunoglobulin on mature B cells by inducible gene targeting results in rapid cell death. Cell 1997;90:1073–83.
[155] Kraus M, Alimzhanov MB, Rajewsky N, Rajewsky K. Survival of resting mature B lymphocytes depends on BCR signaling via the Igalpha/beta heterodimer. Cell 2004;117:787–800.
[156] Kitamura D, Roes J, Kuhn R, Rajewsky K. A B cell-deficient mouse by targeted disruption of the membrane exon of the immunoglobulin mu chain gene. Nature 1991;350:423–6.
[157] Monroe JG. ITAM-mediated tonic signalling through pre-BCR and BCR complexes. Nat Rev Immunol 2006;6:283–94.
[158] Schweighoffer E, Vanes L, Nys J, Cantrell D, McCleary S, Smithers N, et al. The BAFF receptor transduces survival signals by co-opting the B cell receptor signaling pathway. Immunity 2013;38: 475–88.
[159] Stadanlick JE, Kaileh M, Karnell FG, Scholz JL, Miller JP, Quinn 3rd WJ, et al. Tonic B cell antigen receptor signals supply an NF-kappaB substrate for prosurvival BLyS signaling. Nat Immunol 2008;9:1379–87.
[160] Martin F, Kearney JF. Positive selection from newly formed to marginal zone B cells depends on the rate of clonal production, CD19, and btk. Immunity 2000;12:39–49.
[161] Li Y, Li H, Weigert M. Autoreactive B cells in the marginal zone that express dual receptors. J Exp Med 2002;195:181–8.
[162] Watanabe N, Ikuta K, Fagarasan S, Yazumi S, Chiba T, Honjo T. Migration and differentiation of autoreactive B-1 cells induced by activated gamma/delta T cells in antierythrocyte immunoglobulin transgenic mice. J Exp Med 2000;192:1577–86.
[163] Ohdan H, Swenson KG, Kruger Gray HS, Yang YG, Xu Y, Thall AD, et al. Mac-1-negative B-1b phenotype of natural antibody-producing cells, including those responding to Gal alpha 1,3Gal epitopes in alpha 1,3-galactosyltransferase-deficient mice. J Immunol 2000;165:5518–29.
[164] Boes M, Prodeus AP, Schmidt T, Carroll MC, Chen J. A critical role of natural immunoglobulin M in immediate defense against systemic bacterial infection. J Exp Med 1998;188:2381–6.
[165] Su I, Tarakhovsky A. B-1 cells: orthodox or conformist? Curr Opin Immunol 2000;12:191–4.

[166] Casola S, Otipoby KL, Alimzhanov M, Humme S, Uyttersprot N, Kutok JL, et al. B cell receptor signal strength determines B cell fate. Nat immunol 2004;5:317–27.

[167] Thome M. CARMA1, BCL-10 and MALT1 in lymphocyte development and activation. Nat Rev Immunol 2004;4:348–59.

[168] Tonegawa S. Somatic generation of antibody diversity. Nature 1983;302:575–81.

[169] Kabat EA, Wu TT, Perry HM, Gottesman KS, Foeller C. Sequences of proteins of immunological interest. 5th ed. Bethesda, Maryland: U.S.Department of Health and Human Services; 1991.

[170] Padlan EA. Anatomy of the antibody molecule. Mol Immunol 1994;31:169–217.

[171] Xu JL, Davis MM. Diversity in the CDR3 region of V(H) is sufficient for most antibody specificities. Immunity 2000;13:37–45.

[172] Schroeder Jr HW, Zemlin M, Khass M, Nguyen HH, Schelonka RL. Genetic control of DH reading frame and its effect on B-cell development and antigen-specifc antibody production. Crit Rev Immunol 2010;30:327–44.

[173] Kearney JF. Innate-like B cells. Springer Semin Immunopathol 2005;26:377–83.

[174] Baumgarth N, Herman OC, Jager GC, Brown LE, Herzenberg LA, Chen J. B-1 and B-2 cell-derived immunoglobulin M antibodies are nonredundant components of the protective response to influenza virus infection. J Exp Med 2000;192:271–80.

[175] Baumgarth N, Herman OC, Jager GC, Brown L, Herzenberg LA. Innate and acquired humoral immunities to influenza virus are mediated by distinct arms of the immune system. Proc Natl Acad Sci USA 1999;96:2250–5.

[176] Aprahamian T, Rifkin I, Bonegio R, Hugel B, Freyssinet JM, Sato K, et al. Impaired clearance of apoptotic cells promotes synergy between atherogenesis and autoimmune disease. J Exp Med 2004;199:1121–31.

[177] Binder CJ, Silverman GJ. Natural antibodies and the autoimmunity of atherosclerosis. Springer Semin Immunopathol 2005;26:385–404.

[178] Shaw PX, Horkko S, Chang MK, Curtiss LK, Palinski W, Silverman GJ, et al. Natural antibodies with the T15 idiotype may act in atherosclerosis, apoptotic clearance, and protective immunity. J Clin Invest 2000;105:1731–40.

[179] Chen Y, Park YB, Patel E, Silverman GJ. IgM antibodies to apoptosis-associated determinants recruit C1q and enhance dendritic cell phagocytosis of apoptotic cells. J Immunol 2009;182:6031–43.

[180] Kearney JF. Innate-like B cells. Springer Semin Immunopathol 2005;26:377–83.

[181] Potter M, Leon MA. Three IgA myeloma immunoglobulins from the BALB/mouse: precipitation with pneumococcal C polysaccharide. Science 1968;162:369–71.

[182] Gearhart PJ, Sigal NH, Klinman NR. Heterogeneity of the BALB/c antiphosphorylcholine antibody response at the precursor cell level. J Exp Med 1975;141:56–71.

[183] Sigal NH, Gearhart PJ, Klinman NR. The frequency of phosphorylcholine-specific B cells in conventional and germfree BALB/C mice. J Immunol 1975;114:1354–8.

[184] Levy M. Frequencies of phosphorylcholine-specific and T15-associated 10/13 idiotope-positive B cells within lipopolysaccharide-reactive B cells of adult BALB/c mice. Eur J Immunol 1984;14:864–8.

[185] Feeney AJ, Clarke SH, Mosier DE. Specific H chain junctional diversity may be required for non-T15 antibodies to bind phosphorylcholine. J Immunol 1988;141:1267–72.

[186] Crews S, Griffin J, Huang H, Calame K, Hood L. A single VH gene segment encodes the immune response to phosphorylcholine: somatic mutation is correlated with the class of the antibody. Cell 1981;25:59–66.

[187] Briles DE, Forman C, Hudak S, Claflin JL. Anti-phosphorylcholine antibodies of the T15 idiotype are optimally protective against *Streptococcus pneumoniae*. J Exp Med 1982;156:1177–85.

[188] Maynard CL, Elson CO, Hatton RD, Weaver CT. Reciprocal interactions of the intestinal microbiota and immune system. Nature 2012;489:231–41.

[189] Wei B, Su TT, Dalwadi H, Stephan RP, Fujiwara D, Huang TT, et al. Resident enteric microbiota and $CD8_+$ T cells shape the abundance of marginal zone B cells. Eur J Immunol 2008;38: 3411–25.

[190] Murakami M, Nakajima K, Yamazaki K, Muraguchi T, Serikawa T, Honjo T. Effects of breeding environments on generation and activation of autoreactive B-1 cells in anti-red blood cell autoantibody transgenic mice. J Exp Med 1997;185:791–4.

[191] Genestier L, Taillardet M, Mondiere P, Gheit H, Bella C, Defrance T. TLR agonists selectively promote terminal plasma cell differentiation of B cell subsets specialized in thymus-independent responses. J Immunol 2007;178:7779–86.

[192] Rawlings DJ, Schwartz MA, Jackson SW, Meyer-Bahlburg A. Integration of B cell responses through Toll-like receptors and antigen receptors. Nat Rev Immunol 2012;12:282–94.

[193] Oliver AM1, Martin F, Kearney JF. IgMhighCD21high lymphocytes enriched in the splenic marginal zone generate effector cells more rapidly than the bulk of follicular B cells. J Immunol Jun 15, 1999;162(12):7198–207.

[194] Kirkland D, Benson A, Mirpuri J, Pifer R, Hou B, DeFranco AL, et al. B cell-intrinsic MyD88 signaling prevents the lethal dissemination of commensal bacteria during colonic damage. Immunity 2012;36:228–38.

[195] Wang J, Taniuchi I, Maekawa Y, Howard M, Cooper MD, Watanabe T. Expression and function of Fas antigen on activated murine B cells. Eur J Immunol 1996;26:92–6.

[196] Fredrickson TN, Lennert K, Chattopadhyay SK, Morse 3rd HC, Hartley JW. Splenic marginal zone lymphomas of mice. Am J Pathol 1999;154:805–12.

[197] Rothstein TL, Wang JK, Panka DJ, Foote LC, Wang Z, Stanger B, et al. Protection against Fas-dependent Th1-mediated apoptosis by antigen receptor engagement in B cells. Nature 1995;374: 163–5.

[198] Lagresle C, Mondiere P, Bella C, Krammer PH, Defrance T. Concurrent engagement of CD40 and the antigen receptor protects naive and memory human B cells from APO-1/Fas-mediated apoptosis. J Exp Med 1996;183:1377–88.

[199] Hirose S, Yan K, Abe M, Jiang Y, Hamano Y, Tsurui H, et al. Precursor B cells for autoantibody production in genomically Fas-intact autoimmune disease are not subject to Fas-mediated immune elimination. Proc Natl Acad Sci USA 1997;94:9291–5.

[200] Raman C, Knight KL. $CD5_+$ B cells predominate in peripheral tissues of rabbit. J Immunol 1992;149:3858–64.

[201] Knabel M, Cihak J, Losch U. Characterization of new monoclonal antibodies identifying avian T lymphocyte antigens. Immunobiology 1993;188:415–29.

[202] Jurgens JB, Gartland LA, Du Pasquier L, Horton JD, Gobel TW, Cooper MD. Identification of a candidate CD5 homologue in the amphibian *Xenopus laevis*. J Immunol 1995;155:4218–23.

[203] Gross JA, Johnston J, Mudri S, Enselman R, Dillon SR, Madden K, et al. TACI and BCMA are receptors for a TNF homologue implicated in B-cell autoimmune disease. Nature 2000;404:995–9.

[204] Batten M, Groom J, Cachero TG, Qian F, Schneider P, Tschopp J, et al. BAFF mediates survival of peripheral immature B lymphocytes. J Exp Med 2000;192:1453–66.

[205] Tachibana I, Hemler ME. Role of transmembrane 4 superfamily (TM4SF) proteins CD9 and CD81 in muscle cell fusion and myotube maintenance. J Cell Biol 1999;146:893–904.

[206] Cook GA, Wilkinson DA, Crossno Jr JT, Raghow R, Jennings LK. The tetraspanin CD9 influences the adhesion, spreading, and pericellular fibronectin matrix assembly of Chinese hamster ovary cells on human plasma fibronectin. Exp Cell Res 1999;251:356–71.

[207] Rathmell JC, Townsend SE, Xu JC, Flavell RA, Goodnow CC. Expansion or elimination of B cells in vivo: dual roles for CD40- and Fas (CD95)-ligands modulated by the B cell antigen receptor. Cell 1996;87:319–29.

[208] Fagarasan S, Honjo T. T-Independent immune response: new aspects of B cell biology. Science 2000;290:89–92.

[209] Dighiero G, Borche L. Evidence that the B lymphocyte proliferating in B-CLL and in other B-cell malignancies is frequently committed to production of natural autoantibodies. Nouv Rev Fr Hematol 1990;32:323–6.

[210] Smith KG, Nossal GJ, Tarlinton DM. FAS is highly expressed in the germinal center but is not required for regulation of the B-cell response to antigen. Proc Natl Acad Sci USA 1995;92:11628–32.

[211] Le Naour F, Rubinstein E, Jasmin C, Prenant M, Boucheix C. Severely reduced female fertility in CD9-deficient mice. Science 2000;287:319–21.

[212] Coutinho AA. Beyond clonal selection and network. Immunol Rev 1989;110:63–87.

[213] Coutinho A, Forni L, Holmberg D, Ivars F, Vaz N. From an antigen-centered, clonal perspective of immune responses to an organism-centered, network perspective of autonomous activity in a self-referential immune system. Immunol Rev 1984;79:151–68.

[214] Grandien A, Fucs R, Nobrega A, Andersson J, Coutinho A. Negative selection of multireactive B cell clones in normal adult mice. Eur J Immunol 1994;24:1345–52.

[215] Crowley JE, Treml LS, Stadanlick JE, Carpenter E, Cancro MP. Homeostatic niche specification among naive and activated B cells: a growing role for the BLyS family of receptors and ligands. Semin Immunol 2005;17:193–9.

[216] Chu VT, Berek C. The establishment of the plasma cell survival niche in the bone marrow. Immunol Rev 2013;251:177–88.

[217] Manz RA, Hauser AE, Hiepe F, Radbruch A. Maintenance of serum antibody levels. Annu Rev Immunol 2005;23:367–86.

[218] Mauri C, Bosma A. Immune regulatory function of B cells. Annu Rev Immunol 2012;30:221–41.

[219] Bouaziz JD, Yanaba K, Tedder TF. Regulatory B cells as inhibitors of immune responses and inflammation. Immunol Rev 2008;224:201–14.

[220] Lund FE, Randall TD. Effector and regulatory B cells: modulators of CD4+ T cell immunity. Nat Rev Immunol 2010;10:236–47.

[221] Cerutti A, Cols M, Puga I. Marginal zone B cells: virtues of innate-like antibody-producing lymphocytes. Nat Rev Immunol 2013;13:118–32.

[222] Garraud O, Borhis G, Badr G, Degrelle S, Pozzetto B, Cognasse F, et al. Revisiting the B-cell compartment in mouse and humans: more than one B-cell subset exists in the marginal zone and beyond. BMC Immunol 2012;13:63.

[223] Swanson CL, Pelanda R, Torres RM. Division of labor during primary humoral immunity. Immunol Res 2013;55:277–86.

[224] Vinuesa CG, Chang PP. Innate B cell helpers reveal novel types of antibody responses. Nat Immunol 2013;14:119–26.

Chapter 8

# B Cells and Antibodies in Jawless Vertebrates

Brantley R. Herrin, Masayuki Hirano, Jianxu Li, Sabyasachi Das, Yoichi Sutoh and Max D. Cooper
*Department of Pathology and Laboratory Medicine, Emory Vaccine Center, Emory University School of Medicine, Atlanta, GA, USA*

## 1. INTRODUCTION

Before the discovery of *VLR* genes, it was anticipated that lampreys and hagfish, the nearest living phylogenetic relatives of gnathostomes, would have ancestral *BCR*, *TCR*, *MHC*, and *RAG* genes. Surprisingly, transcriptome analysis provided no evidence for the presence of these cardinal elements of the immunoglobulin (Ig)-based adaptive immune system (AIS) shared by all jawed vertebrates (Figure 1), even though orthologs for a variety of genes that lymphocytes in jawed vertebrate use for migration, proliferation, differentiation, and intracellular signaling were found [1–8]. In a revised approach, lamprey larvae were immunostimulated with a cocktail of particulate antigens and phytomitogens before an extensive transcriptome survey of the activated lymphocytes. Although Ig-based antigen receptors were not found, this search revealed a large number of sequences encoding highly diverse, leucine-rich repeat (LRR) proteins [9]. These were named variable lymphocyte receptors (VLRs) due to their lymphocyte-restricted expression and remarkable sequence diversity. Each VLR protein contains a conserved signal peptide (SP), a 27–38-residue N-terminal LRR (LRRNT), an 18-residue LRR1 followed by a variable number of diverse 24-residue LRRs (LRRV), a 13-residual connecting peptide (CP), a 48–65-residue C-terminal LRR (LRRCT), and a conserved C-terminal threonine/proline-rich stalk region with a glycosylphosphatidylinositol (GPI) anchor site and a hydrophobic tail (Figure 2(A)). Treatment of VLR-expressing cells with bacterial GPI-specific phospholipase C released the protein into the supernatant, indicative of membrane anchorage by GPI linkage [9]. Each lymphocyte was found to express a VLR of unique LRR sequence connected to the cell surface by the stalk region of invariant sequence.

## 2. LAMPREYS AND HAGFISH HAVE THREE TYPES OF *VLR* GENES

The names of the *VLR* genes have changed as new *VLR* genes were discovered, and a brief historical account may help to clarify the current nomenclature. The discovery of a *VLR* gene in lampreys led to the identification of two *VLR* genes in hagfish; these genes were provisionally named *VLRA* and *VLRB* [10]. Comparative sequence analysis of the hagfish and lamprey *VLR* genes indicated that the originally identified lamprey *VLR* gene is an ortholog of hagfish *VLRB*, hence its name was changed to *VLRB* to reflect this relationship. In a subsequent search of the draft genome database of sea lamprey, lamprey *VLRA* was identified [11]. Later, a third lamprey *VLR* gene, designated *VLRC*, was discovered through analyzing the sea lamprey expressed sequence tag database [12]. The discovery of lamprey *VLRC* raised the question of whether hagfish also have a third *VLR* gene. Analysis of a nonproductive hagfish "*VLRA*" sequence then led to the identification of a third *VLR* gene that was cloned from pacific hagfish (*Eptatretus stoutii*) [13]. Comprehensive comparative analysis of the three *VLR* genes in lampreys and hagfish indicated that the third hagfish *VLR* is the lamprey *VLRA* ortholog and that the previously identified hagfish "*VLRA*" is the lamprey *VLRC* counterpart. In summary, three orthologous *VLR* genes (*VLRA*, *VLRB*, and *VLRC*) have now been characterized in both lamprey and hagfish (Figure 2(B)), suggesting that this anticipatory receptor system evolved in a common ancestor of the two cyclostome lineages approximately 480 million years ago [13].

## 3. *VLR* GENE ASSEMBLY MECHANISM AND SEQUENCE DIVERSITY

In their germline status, the *VLR* genes are incomplete in that they have coding sequences only for the leader sequence, incomplete N- and C-terminal LRR subunits, and for the stalk region [9–13]. Each *VLR* gene has two exons, with the first exon encoding only a portion of the 5′ untranslated region. The second exon contains the rest of the 5′ untranslated region, an SP, typically a 5′ portion

Molecular Biology of B Cells. http://dx.doi.org/10.1016/B978-0-12-397933-9.00008-4

of the LRRNT, a 3′ portion of the LRRCT, and the stalk region (Figure 2(B)). The invariant stalk region of each VLR isotype has its own unique sequence; therefore, discriminating antibodies can be made with specificity for each isotype. For all *VLR* genes except lamprey *VLRB*, the 5′ LRRNT sequence is separated from the 3′ LRRCT sequence by a short noncoding intervening sequence that lacks splice donor and acceptor sites. The lamprey *VLRB* gene is more complex in that the 5′ LRRNT coding sequence is separated from the 3′ LRRCT by two large noncoding intervening sequences with a short 5′ LRRCT segment separating the two intervening sequences [9]. Each germline *VLR* gene is flanked by hundreds of different LRR-encoding sequences that are randomly selected for use as templates to add the missing LRR cassettes needed for completion of a mature *VLR* gene [9,11,14–16]. A gene conversion-like mechanism has been postulated for the complex *VLR* assembly process [11,15,16] in which the intervening sequence is replaced in a stepwise, piecewise manner by random selection of flanking *LRR* cassettes to serve as templates for adding the necessary sequences to complete a *VLR* gene (Figure 3). The assembly process can be initiated at either the 5′ LRRNT or the 3′ LRRCT end [15,16] and requires short nucleotide homology [10–30 bp] between donor and acceptor sequences that guide the copying of flanking LRR donor segments into the recipient germline gene [15]. Notably, the location of the donor LRR sequences is not changed by rearrangement during the *VLR* assembly process, in keeping with the absence of recombination signal sequences and lamprey *RAG1* and *RAG2* genes.

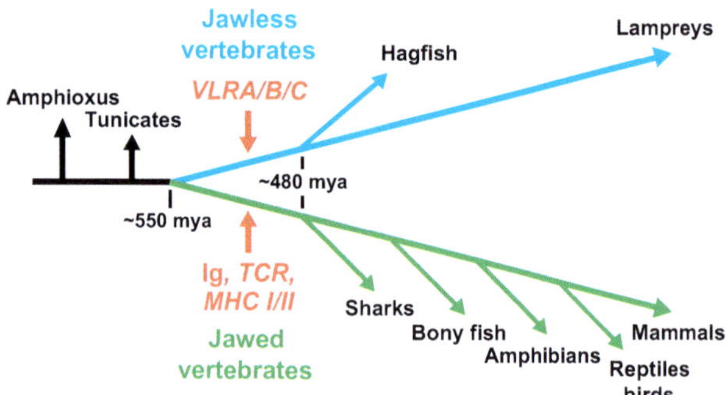


FIGURE 1 **Schematic view of the evolution of adaptive immunity.** The LRR-based *VLR* genes are found only in the extant jawless vertebrates, whereas Ig-based *BCR*, *TCR*, and *MHCI/II* genes are found only in the extant jawed vertebrates. Neither type of antigen recognition receptors has been found in amphioxus, the representative head of the chordates, or in tunicates representing a sister lineage of the vertebrate lineage. These observations suggest that the LRR-based VLRs and the Ig-based BCRs and TCRs represent convergent solutions for specific antigen recognition.

Analysis of the diversity of lamprey *VLRB* gene sequences indicates a potential repertoire of $>10^{14}$ distinct receptors, a magnitude that is comparable to the theoretical diversity of the mammalian antibody repertoire [16]. *VLR* genes are assembled on one allele at a time and allelic exclusion of *VLR* gene assembly is the general rule [9,15,17]. Hagfish *VLRC* (formerly called *VLRA*) and *VLRB* are located on the same chromosome, but they are not in proximity to each other [18], thereby allowing these *VLR* gene loci to function as separate units. Analysis of *VLR* gene assembly at the single-cell level in the inshore hagfish (*Eptatretus burgeri*) indicates that only one type of *VLR* gene, either *VLRB* or *VLRC*, is assembled and transcribed in individual hagfish lymphocytes [17].

Although the molecular basis for *VLR* gene assembly has not yet been elucidated, there are some parallels with

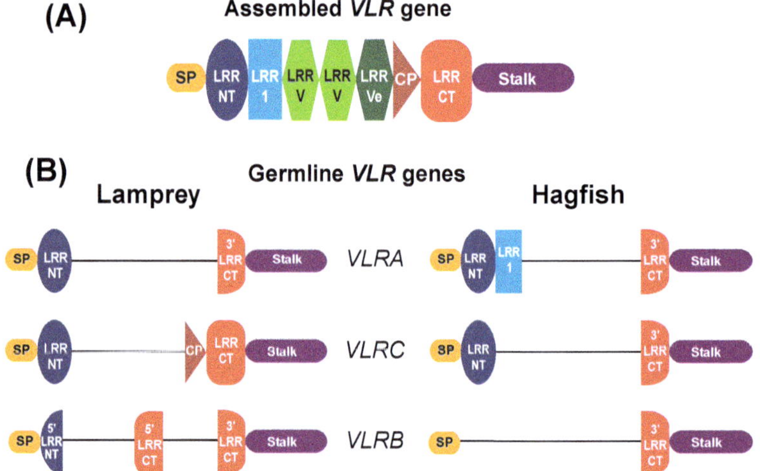


FIGURE 2 **Schematic depiction of the lamprey and hagfish *VLR* genes.** (A) Assembled "mature" *VLR*: signal peptide (SP), LRRNT, first LRR1, two variable VLR (LRRV), end LRRV (LRRVe), connecting peptide (CP), LRRCT, stalk region. (B) Lamprey and hagfish germline *VLRA*, *VLRC*, and *VLRB* genes. Lamprey *VLRB* has two large noncoding intervening sequences separating the 5′ LRRNT and 3′ LRRCT, whereas all other VLR genes have a single, short intervening sequence (not drawn to scale).

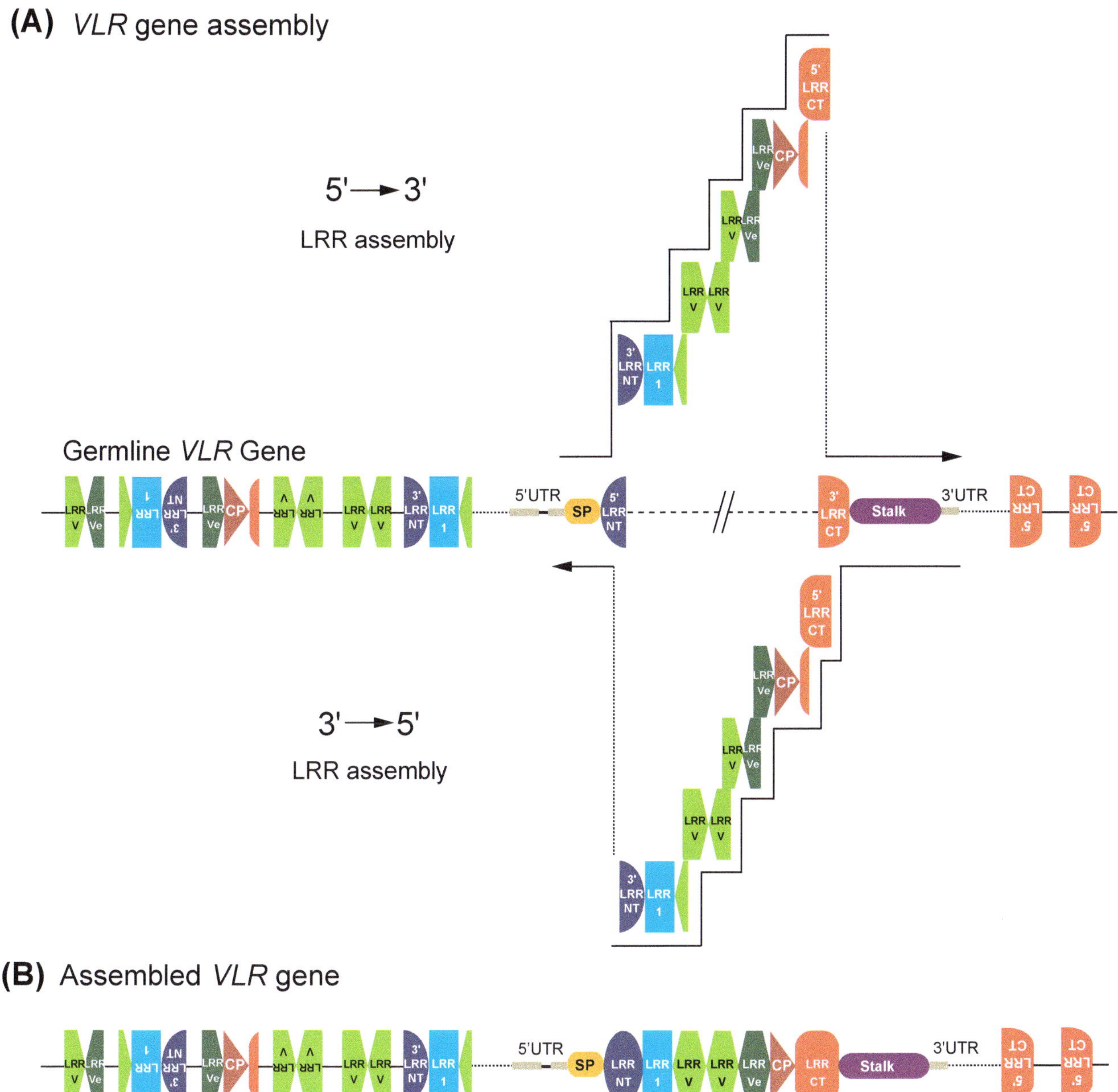


FIGURE 3 ***VLR* gene assembly.** (A) The incomplete germline *VLR* genes are flanked by hundreds of LRR cassettes. The noncoding intervening sequence between portions of the LRRNT and LRRCT is replaced by LRR fragments that are sequentially copied from randomly selected donor LRR cassettes. Mature *VLR* genes are assembled from either the LRRNT or LRRCT end in a stepwise manner that is directed by short sequence homology between the donor and recipient LRR sequences. (B) Genomic configuration after *VLR* gene assembly.

the gene conversion mechanism catalyzed by activation-induced cytidine deaminase (AID) that birds and certain mammals use to diversify their antibody repertoire [19]. In this process, sequences from flanking $V_L$ and $V_H$ pseudogenes are copied into an *IgL* or *IgH* gene initially assembled by RAG-mediated V(D)J recombination. Due to similarities with the gene conversion mechanism for Ig diversification, the involvement of an AID-like enzyme in *VLR* assembly was envisioned. In support of this hypothesis, two AID-apolipoprotein B mRNA-editing catalytic component (APOBEC) family orthologs, cytosine deaminase 1 (CDA1) and cytosine deaminase 2 (CDA2), have been identified in lampreys [11] (Figure 4). Moreover, *CDA1* expression is limited to the VLRA and VLRC lymphocyte lineages, whereas *CDA2* expression is restricted to the VLRB lymphocyte lineage [20,21]. These findings support the idea that CDA1 may catalyze *VLRA* and *VLRC* gene assemblies, whereas CDA2 plays a similar role in *VLRB* gene assembly.

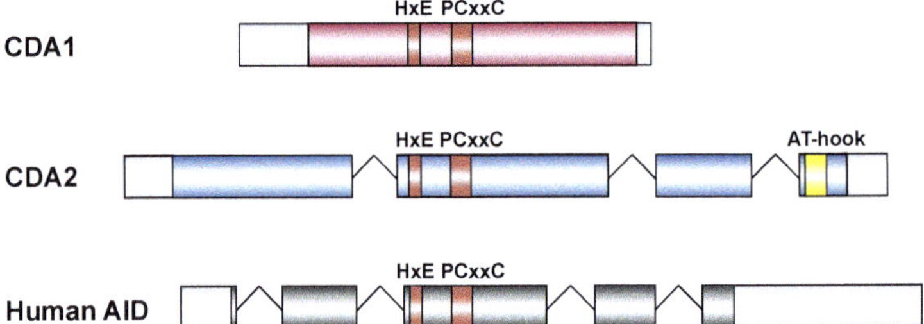


FIGURE 4 **Schematic diagram of the lamprey CDA1, CDA2, and the human AID loci.** CDA1 and CDA2 are encoded by a single exon and four exons, respectively. They share a deaminase "HxE-PCxxC" (where "x" is any amino acid) zinc-coordination motif in the core fold. CDA2 has at its C terminus an AT-hook that could be used for binding in the DNA minor groove.

## 4. LAMPREY CDA1 AND CDA2

The two AID/APOBEC orthologs in lampreys, CDA1 and CDA2, are posited to catalyze the conversion of cytosine to uracil on the DNA strand of lamprey lymphocytes, leading to *VLR* assembly through a gene conversion mechanism [11]. *CDA1* and *CDA2* are encoded by a single exon and four exons, respectively, and they share a characteristic deaminase "HxE-PCxxC" (where "x" is any amino acid) zinc-coordination motif (Figure 4). CDA2 has at its C terminus an AT-hook, a small protein motif that could be used for binding in the DNA minor groove [22]. In an in vivo mutagenesis assay, CDA1 overexpression enhanced the transition mutation rate for host yeast and *Escherichia coli* DNA. CDA1 overexpression increased the frequency of rifampicin-resistant mutants in *ung*-deficient *E. coli* by 75-fold versus a 10-fold increase that was induced by human AID. Both enzymes are predicted to have a characteristic secondary structure, six α-helices and five β-strands in the core fold, that is conserved among vertebrate AID/APOBEC family members. CDA1 and CDA2 form a node within the phylogenetic tree of the vertebrate AID/APOBEC family that is supported by significant bootstrap value [11]. So far, AID/APOBEC orthologs have not been identified in invertebrates. These results suggest that the agnathan CDA1 and CDA2 are the earliest orthologs to be derived from a common ancestor of the AID/APOBEC family [23]. Alternatively, *RAG* genes have not been identified in jawless vertebrates, implying that the RAG insertion took place after divergence of the jawed vertebrate lineage [24]. This leads to the conclusion that, before *RAG* insertion, a primordial *AID/APOBEC* gene evolved in a common ancestor of vertebrates and was selected to function in the diversification of antigen receptors in the alternative AISs of jawless and jawed vertebrates [25,26].

## 5. VLRA, VLRB, AND VLRC ARE EXPRESSED BY DIFFERENT LYMPHOCYTE POPULATIONS

Mouse monoclonal antibodies specific for the invariant portions of the lamprey VLRA, VLRB, and VLRC proteins have been useful for characterization of the lymphocytes that express the three VLR isotypes. Cell surface staining with the VLR isotype-specific antibodies indicates that VLRA, VLRB, and VLRC are expressed by three separate lymphocyte populations with distinct tissue localization patterns [20,21]. VLRC$^+$ cells are more numerous than VLRA$^+$ cells in the principal lymphoid tissues of lamprey larvae; they constitute the major lymphocyte population in typhlosole and gills, whereas VLRB$^+$ lymphocytes predominate in blood and the kidneys. Immunohistochemical staining of the tissue sections has shown that VLRB$^+$ lymphocytes are abundant in the extravascular regions of the kidneys and typhlosole as well as within blood vessels [27]. The VLRA$^+$ and VLRC$^+$ lymphocytes have a similar distribution pattern in the typhlosole, kidneys, gills, and hypopharyngeal fold [21,28]. Both lymphocyte types are round or oval shaped in blood vessels and in interstitial spaces of the kidneys and typhlosole, whereas they assume a dendritic shape in the epithelium of the gills and intestine. The VLRC$^+$ cells outnumber VLRA$^+$ cells in the epithelium, wherein VLRB$^+$ cells are rare. Of note, VLRC$^+$ cells with interdigitating morphology are the predominant lymphocyte type in the skin epidermis [21]. This epidermal prevalence of VLRC$^+$ cells in lampreys is reminiscent of the dendritic epidermal T cells in mice that express a canonical γδ T cell receptor.

## 6. CHARACTERIZATION OF B-LIKE AND TWO T-LIKE LYMPHOCYTE POPULATIONS IN CYCLOSTOMES

Lamprey VLRB cells are functionally equivalent to the B cells in jawed vertebrates [27,29]. Although members of all three lamprey lymphocyte lineages undergo proliferation and lymphoblast transformation in response to immunization with antigens, such as bacteria, viruses, fungi, and mammalian cells, only the VLRB$^+$ lymphocytes undergo plasma cell differentiation and secrete their antigen receptors as disulfide-linked, multivalent antibodies much like the IgM antibodies of jawed vertebrates [27,30] (Figure 5). Antigen-specific VLRB antibodies are detectable in the serum about 7 days after immunization with mammalian erythrocytes or bacteria. The antibody response to a primary immunization peaks at approximately three weeks after immunization and persists for at least two months [27]. A second immunization

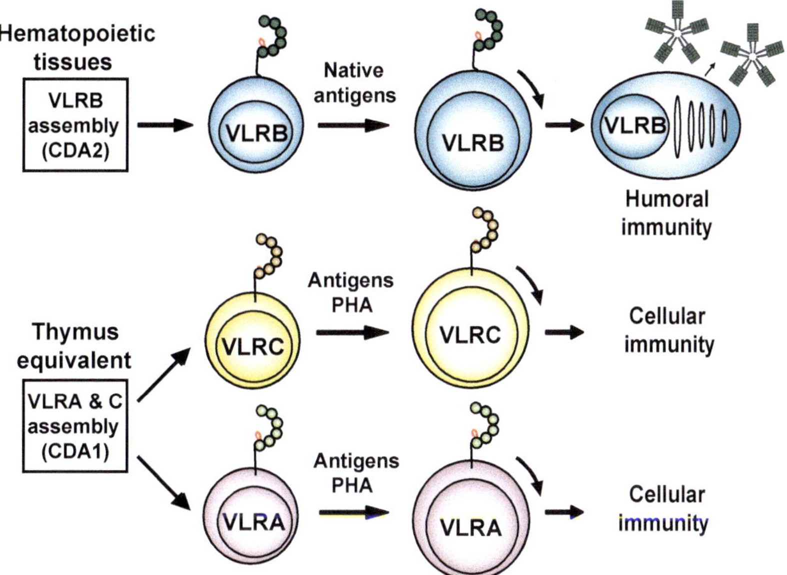


FIGURE 5 **Development of B cell-like and T cell-like lineages in lampreys.** *VLR* gene assembly may be catalyzed by AID-APOBEC family orthologs cytidine deaminase 1 (CDA1) and CDA2. CDA1 is expressed during the assembly of *VLRA* and *VLRC* genes in the thymus-equivalent (thymoid) region of the gills, whereas CDA2 is expressed during *VLRB* gene assembly in the typhlosole, the major hematopoietic tissue in lamprey larvae. Cells belonging to all three lineages may proliferate in response to immunization with foreign antigen, but VLRA and VLRC cells respond preferentially to stimulation with the plant lectin phytohemagglutinin (PHA).

two weeks after the primary immunization boosts VLRB antibody titers 10- to 100-fold higher than the primary response. VLRB antibody-secreting plasmacytes with expanded cytoplasm and secretory organelles characteristic of jawed vertebrate plasma cells are found in the blood, kidneys, and typhlosole. The VLRB antibodies may use effector mechanisms that are analogous to those used by Ig-based antibodies of jawed vertebrates, but these are not yet well defined. Specific VLRB antibodies from immunized lampreys can initiate complement-dependent lysis of bacteria, heterologous erythrocytes, and tumor cell lines via interaction with a C1q-like complex and C3 [31]. These findings indicate that orthologs of the classical complement components, which allow Ig-based antibodies to interface with innate defense mechanisms in jawed vertebrates, may have comparable functions in jawless vertebrates. However, whether or not the VLRB antibodies can bind to functional equivalents of the Ig Fc receptors on effector cells has not yet been determined.

VLRA$^+$ and VLRC$^+$ lymphocytes share many similarities with the $\alpha\beta$ and $\gamma\delta$ T cells in jawed vertebrates [20,21,28]. They respond to immunization by undergoing lymphoblastoid transformation and proliferation, but they do not secrete their antigen receptors (Figure 5). The VLRA$^+$ and VLRC$^+$ cells in lampreys are also preferentially activated by the classical T cell mitogen phytohemagglutinin, and they undergo extensive proliferation to become the predominant lymphocyte populations [20,21]. The lymphoblastoid VLRA$^+$ and VLRC$^+$ lymphocytes are relatively large, but they have much less rough endoplasmic reticulum than the VLRB-secreting plasmacytes. After immunization with either *Bacillus anthracis* exosporium or *E. coli*, antigen-binding VLRA$^+$ and VLRC$^+$ cells have not been detected either before or after immunization [20,21], whereas VLRB$^+$ bacteria-binding cells are easily detected and their frequency increases with time after immunization [20,27]. VLRBs thus can bind their cognate antigens both as cell surface receptors and as secreted VLRB antibodies, whereas it is still unclear whether VLRA$^+$ and VLRC$^+$ cells recognize native or processed antigens. However, there is one notable report in which yeast surface display of a VLRA receptor cDNA library from an immunized adult lamprey was used to identify 13 VLRA clones that bound hen egg lysozyme (HEL) with high affinity [32], and the crystal structure was characterized for one of these VLRA clones in complex with the HEL antigen [33]. This result infers that certain VLRA subsets may bind native protein antigens without the need for antigen processing and presentation. In this regard, major histocompatibility complex (MHC) receptors have not been found either by transcriptome analysis of lamprey cells [1,6,9] or in the lamprey genome [24,34]. Whether VLRA$^+$ and VLRC$^+$ cells can recognize processed antigens presented by a convergent analog of the MHC molecules thus remains an unresolved issue.

## 7. VLRA$^+$, VLRB$^+$, AND VLRC$^+$ CELLS HAVE DISTINCT GENE EXPRESSION PROFILES

VLRB$^+$ lymphocytes in jawless vertebrates preferentially express orthologs of many genes that are expressed by B cells in jawed vertebrates [20,21] (Figure 6); the

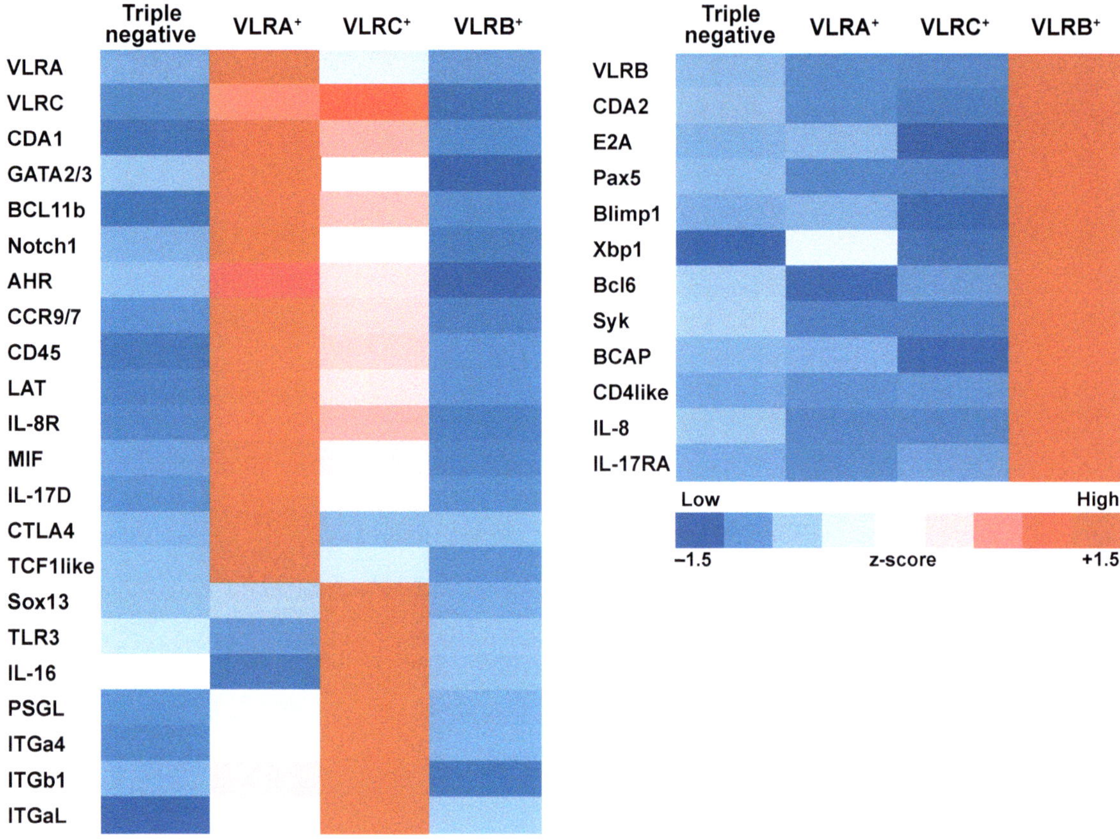


FIGURE 6 **Differential gene expression profiles of VLRA⁺, VLRB⁺, and VLRC⁺ lymphocytes.** Relative transcript levels of the indicated genes were measured by quantitative PCR for purified populations of VLRA⁺, VLRB⁺, VLRC⁺, and triple-negative lymphocytes, and the β-actin normalized value for each gene was compiled into a heatmap (z-scores) with three-color scale (blue-white-red). *The figure was modified from Ref. [21].*

orthologous genes include ones for the hematopoietic progenitor homing receptor *CXCR4*; the transcription factors *E2A*, *Pax5*, *BLIMP-1*, and *Bcl6*; the herpes virus entry mediator/tumor necrosis factor receptor superfamily member 14 (*TNFRSF14*) that binds to LIGHT on T cells; two components of the BCR-mediated signaling cascades, *Syk* and the B cell adaptor protein (*BCAP*); the chemotactic inflammatory cytokine interleukin (*IL*)-*8*; the IL-17 receptor (*IL-17R*); and the Toll-like receptor (TLR) orthologs of *TLR2*, *TLR7*, and *TLR10*. In contrast, VLRA⁺ lymphocytes preferentially express genes that are orthologous to those typically expressed by T cells in jawed vertebrates: the *GATA1/2*, *c-Rel*, aryl hydrocarbon receptor (*AHR*) and *BCL11b* transcriptional factors used for T cell differentiation, the *CCR9* chemokine receptor involved in homing of lymphocyte progenitors to the thymus, the Notch1 T cell fate-determining molecule, the *CD45* tyrosine phosphatase receptor protein that is essential for T cell development, the *CXCR2* IL-8 receptor, and the *IL-17* and *MIF* proinflammatory cytokines [20] (Figure 6). Activated VLRA⁺ cells upregulate their expression of *IL-17* and *MIF*, whereas activated VLRB⁺ cells upregulate expression of *IL-8*. These findings, coupled with the demonstration of reciprocal expression of *IL-17R* by VLRB⁺ cells and *IL-8R* by VLRA⁺ lymphocytes, attest the likelihood of functional interactions between these two lymphocyte populations.

Although the overall transcriptional profile of VLRC⁺ cells is similar to that of VLRA⁺ cells, the VLRC⁺ cells differ in their preferential expression of the SRY-box containing gene 13 (*SOX13*) encoding a fate-determining transcription factor used for γδ T-cell lineage commitment (refs); an integrin αL (*ITGAL*) ortholog of one component of the heterodimeric lymphocyte function associated antigen 1 (*LFA1*); and integrins α4 and β1 (*ITGA4* and *ITGB1*) orthologs of the two components of very late antigen 4 (*VLA4*), the expression of which correlates with the adherence of human γδ T cells to epithelial cells, *TLR3* and *IL-16*, a modulator of T-cell activation (Figure 6). Thus, in several ways, the transcriptional programs for lamprey VLRA and VLRC cells resemble those of mammalian αβ and γδ T cells, respectively [21].

## 8. GENERATION OF THE T-LIKE AND B-LIKE CELLS IN LAMPREYS

Evidence suggests that T-like and B-like lymphocytes in lampreys are generated in separate locations, much like the thymus and hematopoietic tissue origins of T and B cells in the jawed vertebrates [28]. Lymphoepithelial regions located at the tips of the gill folds and the neighboring secondary lamellae apparently serve as a thymus-equivalent microenvironment in which lymphoid progenitors migrate to undergo assembly of either their *VLRA* or *VLRC* genes. Epithelial cells in those "thymoid" sites express *FOXN1*, a marker of the thymopoietic microenvironment in jawed vertebrates, and *Delta-like Ligand B* (*DLL-B*), a Notch ligand gene. Conversely, the lymphoid cells in this location may express *Notch1*, *CDA1*, and the VLRA or VLRB proteins. Notably, the "thymoid" region is the only place in which cells with nonfunctional *VLRA* and *VLRC* assemblies are relatively abundant, due to point mutations or deletions or incomplete *VLR* gene assembly [14,21,28]. An extensive analysis of the *VLR* assembly status of "thymoid" cells suggests a differentiation model in which bipotential thymocyte precursor cells begin to undergo *VLRC* assembly before the assembly of *VLRA* [21]. If the *VLRC* assembly is nonproductive on the first allele, the assembly process may then proceed to the second allele of *VLRC* or even proceed to the *VLRA* locus to achieve a productive *VLRC* or *VLRA* assembly that allows the two types of immature T-like lymphocytes to survive as receptor positive cells.

The major generation site for the B-like cells appears to be the typhlosole, a hematopoietic tissue that extends along the entire length of the intestine in lamprey larvae. The prime evidence for their hematopoietic tissue origin is that CDA2-expressing cells are most abundant in this location and, importantly, the typhlosole is the only tissue in which cells with incomplete *VLRB* assembly can easily be found [28]. Nevertheless, *CDA2*-expressing cells are also present in the blood and kidneys, albeit in lower abundance than in the typhlosole. The *CDA2* expression by circulating VLRB$^+$ cells raises the intriguing possibility that the lamprey B-like cells may undergo somatic hypermutation to change the affinity of their VLRB receptors and antibodies.

## 9. THE UNIQUE STRUCTURE OF VLRB ANTIBODIES

The LRR-based VLRB antibodies of jawless vertebrates, although functionally analogous to Ig-based antibodies of jawed vertebrates, have a very different structure. Lamprey VLRB antibodies are secreted as multimers of identical polypeptide chains, either as tetrameric or pentameric pairs of dimers that are linked at their base by disulfide bonds between cysteine residues in the C terminus of the stalk region [27,30]. When these cysteines are artificially deleted, the VLRB antibodies are secreted as monomers. The VLRB C-terminal cysteines are removed for cell surface expression, presumably by proteolytic cleavage at the GPI-addition site. Conversely, the cysteine-rich C-terminal peptide is retained in the secreted multimeric VLRB antibodies [30]. In jawed vertebrates, C termini for either the membrane version or the secreted forms of antibodies are encoded on separate exons, and switching between the two forms is achieved by alternative RNA processing. However, there is no evidence for an additional VLRB exon or alternative processing of VLRB transcripts. Therefore, switching between cell surface and secreted forms of VLRB is likely regulated by whether or not the GPI-addition mechanism is activated.

Transmission electron microscope images show that the recombinant VLRB multimers resemble multimeric IgM antibodies in the arrangement of their antigen-binding LRR region [30]. However, the VLRB stalk region is very flexible due to the extended threonine/proline-rich sequence that connects the VLRB antigen-binding domain to the cysteine-rich C terminus. In this way the stalk region resembles the hinge region of Ig antibodies, particularly the human IgA1 hinge, that is also threonine/proline-rich, extended, and flexible. The VLRB stalk region is likely to be heavily *O*-glycosylated at the multiple serine and threonine residues, a feature that could serve to promote the extended structure of the stalk region and to protect it from protease cleavage. Although the VLRB antigen-binding domain and stalk region are functional counterparts of the Fab' and hinge region of Ig-based antibodies, a functionally Fc-equivalent region has not yet been defined in VLRB antibodies. The interaction between VLRB antibodies and the complement C1q-like complex could theoretically involve binding to the invariant stalk region or to other proteins associated with VLRB [31]. A means by which the cell surface VLRs could trigger cellular responses is also presently unknown. Neither VLRB nor any of the other VLR isotypes have a cytoplasmic tail with signaling motifs [9–13]. VLR coreceptor molecules are likely to be used for signal transduction after antigen ligation, in the same way that B cell receptors use Igα/β and T cell receptors use the CD3 coreceptor proteins to initiate their signaling pathways. The present lack of knowledge about the composition of the VLR receptors highlights the need for the development of lamprey lymphocyte cell lines.

## 10. VLRB MONOCLONAL ANTIBODIES

Because hybridoma fusion partners for lamprey lymphocytes are presently unavailable and methods to immortalize lamprey lymphocytes have not yet been developed, recombinant expression library screening approaches have been developed to isolate antigen-specific VLRB monoclonal antibodies [30,32,35]. The products of VLRB cDNAs expressed in mammalian cell lines are secreted as disulfide-linked, multivalent antibodies that can then be screened for antigen-binding specificity. Using this approach, VLRB monoclonal antibodies specific for the anthrax BclA spore-coat protein [30], H-trisaccharide of human O$^+$ erythrocytes [36], the CDR3 region of a B CLL BCR [37], and human

CD5 [38] have been isolated by screening cDNA libraries from lampreys immunized with *B. anthracis* exosporium, human O-type erythrocytes, B and T cells, respectively.

BclA-specific VLRB monoclonal antibodies can discriminate *B. anthracis* spores from those of the closely related *Bacillus thuringiensis* and *Bacillus cereus* species, and they retain their ability to bind spores after exposure to strong acids, bases, and temperatures up to 70 °C [30]. The monomeric subunits of most monoclonal VLRB antibodies isolated using this method are of relatively low affinity (micromolar $K_d$), but the multivalent antibodies bind with high avidity to antigens with repetitive epitopes. For instance, the affinity of a monomeric form of a BclA-specific clone (VLR4), is only 2.6 μM, yet the multivalent form of this lamprey antibody agglutinates anthrax spores at concentrations as low as 5 pg/ml [30,39]. Similarly, the multivalent anti-CD5 monoclonal VLRB antibody (VLR32) brightly stains human T cells by immunofluorescence, whereas staining of T cells is not detectable with the monomeric version of the antibody [38]. These monoclonal VLRB antibodies may reflect the properties of most antibodies found in the lamprey repertoire in that they have low antigen affinity but nevertheless bind with high avidity due to multivalency. In many ways, the lamprey VLRB antibodies thus resemble jawed vertebrate IgM antibodies, which are also multivalent and typically have low affinity. This is an effective immune strategy, however, because most natural pathogens, such as bacteria, viruses, and fungi, are particulate and have repetitive epitopes that are ideal targets for multivalent antibodies.

An efficient way to obtain antigen-specific monoclonal VLRB antibodies is through isolation by yeast display of VLRB libraries fused to a yeast surface protein [32,35]. The advantage of this method is that it allows for high-throughput screening of millions of VLRB clones and the selection of binders from both immune and nonimmune libraries. For example, HEL-specific VLRB monoclonal antibodies were isolated using a yeast display library derived from a lamprey that had been repeatedly immunized over a four month period with soluble HEL in complete Freund's adjuvant [32]. Although the monoclonal VLRB antibodies isolated using yeast display were also of relatively low affinity, a variant (CTMut.5) with 1300-fold higher affinity than its parent clone (HEL.2D) was generated by in vitro mutagenesis and an additional round of selection. Monoclonal VLRB antibodies specific for multiple protein and carbohydrate antigens have also been isolated by screening large nonimmune VLRB yeast display libraries, including binders to β-galactosidase, cholera toxin subunit B, R-phycoerythrin, blood group A and B trisaccharides, and the tumor-associated Thomsen–Friedenreich disaccharide antigen (TFα) [32,40,41].

VLR subcomponents have been used for protein engineering applications to create artificial LRR scaffolds and to construct chimeric proteins with enhanced stability and recombinant protein yields for use in structural studies. For instance, modular VLR scaffolds, termed repebodies, with improved soluble expression in *E. coli* have been produced by linking consensus LRR modules to an LRRCT and exchanging the LRRNT for an Internalin-B capping domain that contains an α-helical cap, rather than the native disulfide-linked antiparallel β-strands [42]. Repebody binders specific for myeloid differentiation protein 2 (MD2) and HEL were generated by rational design, and an IL-6 binder was obtained by screening a repebody phage display library. In another application, the LRRCT region of hagfish VLRB was used to cap the C terminus of truncated TLRs to produce chimeric proteins with increased stability, allowing the production of sufficient quantities of protein for crystallographic studies [43,44]. This hybrid LRR technique enabled the first crystal structures for TLR1, TLR2, TLR4, TLR5, and TLR6 to be obtained [43,45–49].

## 11. STRUCTURE OF VLR ANTIGEN-BINDING DOMAINS

The structures for the LRR portions of the lamprey and hagfish VLRA, VLRB, and VLRC proteins have been solved, as have the structures of four lamprey VLRB antibodies and one lamprey VLRA protein co-crystalized with antigen [33,36,39,40,50–53]. All three VLR isotypes form crescent-shaped solenoids, but there are differences in the capping LRR modules of the isotypes (Figure 7). The N terminus is capped by an LRRNT module encoding two β-strands that face antiparallel to one another, and it contains two disulfide bonds that facilitate structural stability. The concave surface is composed mainly of parallel β-strands, beginning with LRR1 and followed by variable numbers of LRRV modules, an LRRVe, and the LRRCP. The C terminus is capped by an LRRCT module that has a variable insert loop, which faces the concave surface, and an α-helix that forms the base of the solenoid. Like the LRRNT, the LRRCT contains two disulfide bonds that add to its stability. The LRRCT insert loop in VLRA and VLRB is highly variable in amino acid composition and in length, which can range from 9 to 13 in VLRA and from 0 to 13 in VLRB. In contrast, the VLRC LRRCT sequence is almost invariant and does not include an extended insert loop [50,53] (Figure 7(C)). In all three VLR isotypes, the LRR1, LRRV, LRRVe, and LRRCP modules each encode a highly variable β-strand; together these form a continuous β-sheet with the highest amino acid variability being located toward the center of the concave surface. For VLRB, the LRRNT β-strand adjacent to LRR1 is variable, whereas the LRRNTs of VLRA and VLRC have limited sequence variability. Although the VLRs may vary in the number of β-strands they contain, the average number of β-strands varies for each VLR isotype. In general, VLRB tends to be smaller, with an average of 2.5 LRRV-encoded β-strands, whereas VLRA and VLRC are slightly larger,

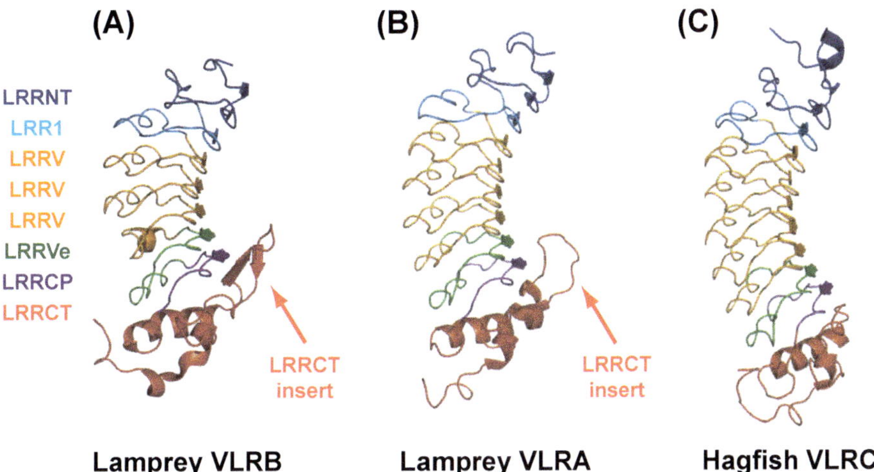


FIGURE 7 **Structural comparison of VLR isotypes.** (A) VLRB proteins are crescent shaped with sequence diversity concentrated on the concave surface. The concave surface is composed of β-strands contributed by the LRRNT (dark blue), LRR1 (light blue), variable numbers of LRRV modules (orange), LRRVe (green), LRRCP (purple), and an LRRCT (red) with a variable insert loop (arrow) (PDBID 3E6J). VLRA (PDBID 3M18) (B) and VLRC (PDBID 2O6Q) (C) are structurally similar to VLRB, except VLRC does not have an LRRCT insert.

with an average of four LRRV-encoded β-strands. Isotypic differences in LRRNT and LRRCT amino acid variability and the length of the LRRCT insert loop suggest that the different VLR isotypes may bind to antigens in different ways or to different types of antigens. Multiple examples of VLRB binding to native antigens have been described, and the VLRB LRRCT loop has consistently been shown to make essential contacts with antigens [36,39,40,52]. The relatively invariant LRRNTs of VLRA and VLRC and the invariant LRRCT of VLRC suggest that these regions either do not bind to antigens or bind only to a very restricted set of highly conserved epitopes. Alternatively, VLRA and VLRC could use these capping modules to interact with invariant regions of cognate MHC-like receptors that present processed antigens.

## 12. STRUCTURES OF VLRB ANTIBODY–ANTIGEN COMPLEXES

In four presently solved VLRB-antigen structures, the antigens make contact with multiple residues in the β-strands of the LRRCP, LRRVe, LRRV(s), and LRR1 modules, and the highly variable LRRCT insert is also critical for the interaction (Figure 8) [36,39,40,52]. The LRRNT region of VLRB antibodies has high sequence variability, but LRRNT contact with the antigen was observed in only one of the four crystal structures, VLR4-BclA [39]. The currently available structures suggest that the more C-terminal LRR β-strands (LRRCP, LRRVe, and LRRVs) typically have more antigen contact, although exceptions to this rule are likely to be found as more VLRB-antigen structures are determined. In the two structures of VLRB antibodies with their cognate protein antigens, the LRRCT insert of VLR4 binds in a shallow groove on the side of BclA (Figure 8(A)) [39], whereas the LRRCT loop of VLR.2D is inserted into the catalytic cleft of HEL (Figure 8(B)) [52]. Remarkably, two other naturally occurring, single-chain Ig antibodies, camelid $V_{HH}$ (cAb-Lys3) [54] and shark IgNAR (PBLA8) [55], also bind to a similar epitope in the catalytic cleft of HEL, whereas most conventional $V_HV_L$ mouse antibodies bind to flatter HEL epitopes away from the enzymatic site. The total surface area buried in the VLRB.2D-HEL and VLR4-BclA complexes is ~1500 Å$^2$, which is comparable to the surface area buried by most Ig antibody–protein antigen complexes [56]. Less total surface areas are buried in the glycan-specific VLRB complexes, RBC36-H-trisaccharide (Fucα1-2Galβ1-4GlcNacα) [36], and aGPA.23-TFα disaccharide (Galβ1-3GalNacα) [40], due to the smaller size of these carbohydrate antigens. Although the two anti-glycan VLRB monoclonal antibodies have the same number of LRR modules, aGPA.23 differs from RBC36 at eight of the 10 residues that contact TFα. The H-trisaccharide-specific RBC36 antibody has a 10 amino acid LRRCT insert that encodes for a β-hairpin, and the TFα disaccharide-specific aGPA.23 antibody has a nine amino acid loop LRRCT insert, but both LRRCTs contain a tryptophan residue that contacts the antigen. In the RBC36 antibody, the tryptophan residue (Trp204) is located at the end of the first β-strand, and it makes contact with the galactose of the H-trisaccharide (Figure 8(C)), whereas the tryptophan (Trp187) in the aGPA.23 antibody is located at the tip of the LRRCT insert and it makes contacts with both the galactose and *N*-acetylgalactosamine of TFα (Figure 8(D)). An important difference in antigen binding by VLRB antibodies from that by Ig-based antibodies is that the only flexible portion of the antigen-binding site of VLRB antibodies is the extended loop at the base of the antigen-binding region, which is formed

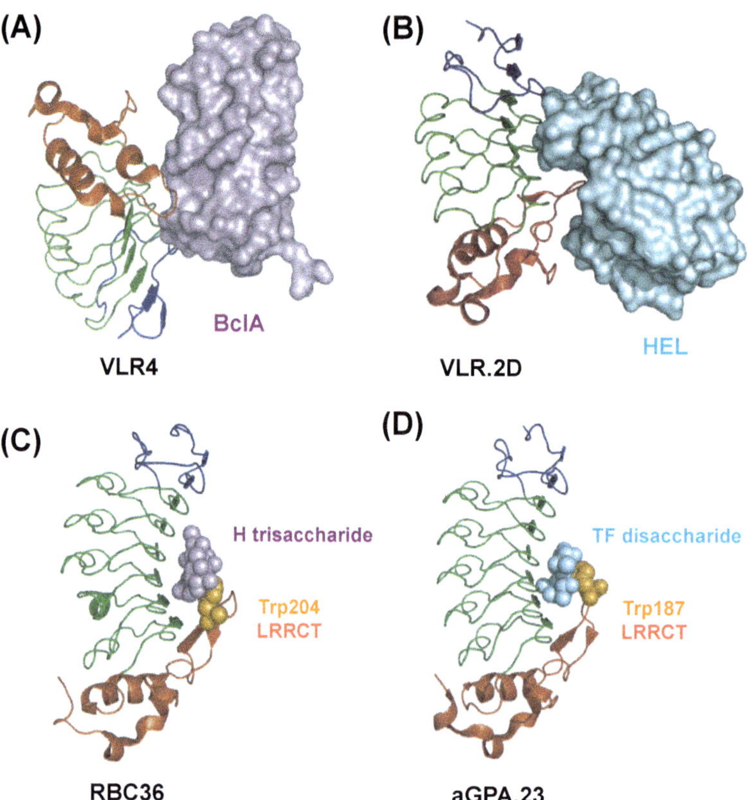


FIGURE 8 **Structures of VLRB antibody–antigen complexes.** Ribbon diagrams of VLRB–antigen complexes: LRRNT (blue); LRR1, LRRV(s), LRRVe, LRRCP (green); LRRCT (red). (A) VLR4 binds to BclA using its LRRCT loop and multiple β-strands on its concave surface (PDBID 3TWI). (B) The LRRCT loop of VLR.2D is inserted into the catalytic cleft of HEL (PDBID 3G3A). (C) RBC36 uses multiple β-strands and a tryptophan residue (Trp204) in its LRRCT loop to bind to the H trisaccharide (PDBID 3E6J). (D) aGPA.23 also uses a tryptophan residue (Trp187) in its LRRCT loop to contact the TF disaccharide (PDBID 4K79).

by the variable LRRCT insert; the larger antigen-binding region, provided by residues in the β-strands lining the VLRB concave surface, is inflexible. The precise antigen specificity of the monoclonal VLRB antibodies that have been analyzed so far may relate to these structural characteristics.

## 13. CONCLUSION

Recognition of the parallel evolution of an alternative AIS in the extant jawless vertebrates, hagfish and lampreys, has altered our view of how lymphocyte-based adaptive immunity emerged. The basic genetic program for compartmentalized lymphocyte development along functionally specialized T- and B-cell lineages is shared by both jawless and jawed vertebrates and thus constitutes a fundamental organizing principle for AISs in all vertebrates. More surprising is the discovery of the convergent evolution of very different solutions for the generation of clonally diverse repertoires of antigen receptors in the two vertebrate lineages. Whereas jawed vertebrates generate diverse antigen receptors by recombinatorial assembly of *Ig V(D)J* gene segments, the jawless vertebrates generate their VLRs through recombinatorial use of donor LRR cassettes for the assembly of their incomplete germline *VLR* genes.

Much remains to be learned about the alternative AISs in jawless vertebrates. In particular, very little is yet known about the mechanisms of antigen recognition by the T-like cells bearing either the VLRA or VLRC surface proteins. However, the functional and structural features of cognate antigen recognition by the VLRB products of the lamprey B-like cells are much better defined. It is already clear that VLRB antibodies, which are the only naturally selected alternatives to classical Ig-based antibodies, have several advantageous features that make them useful alternatives for the diagnosis, monitoring, and, possibly, therapy of human diseases. The VLRB antigen-binding domain is relatively small, modular, and easily fused to other proteins to extend their functional capabilities. The highly variable LRRCT loop in the C-terminal LRR of VLRB antibodies allows binding to epitopes that would be inaccessible to most Ig-based antibodies. VLRB antibodies retain their ability to bind antigen after exposure to acidic and basic conditions (pH 2–11) and temperatures up to 70 °C. They are encoded by a single gene, a feature that greatly facilitates VLRB characterization and genetic engineering. The VLR genes are as diverse as their Ig gene counterparts, but their unique sequences and different self-antigens for tolerance constraints predict that the VLRB repertoires will include novel specificities.

## REFERENCES


[1] Mayer WE, Uinuk-Ool T, Tichy H, Gartland LA, Klein J, Cooper MD. Isolation and characterization of lymphocyte-like cells from a lamprey. Proc Natl Acad Sci USA 2002;99:14350–5.

[2] Nagata T, Suzuki T, Ohta Y, Flajnik MF, Kasahara M. The leukocyte common antigen (CD45) of the pacific hagfish, *Eptatretus stoutii*: implications for the primordial function of CD45. Immunogenetics 2002;54:286–91.

[3] Pancer Z, Mayer WE, Klein J, Cooper MD. Prototypic T cell receptor and CD4-like coreceptor are expressed by lymphocytes in the agnathan sea lamprey. Proc Natl Acad Sci USA 2004;101:13273–8.

[4] Rothenberg EV, Pant R. Origins of lymphocyte developmental programs: transcription factor evidence. Semin Immunol 2004;16:227–38.

[5] Suzuki T, Shin IT, Kohara Y, Kasahara M. Transcriptome analysis of hagfish leukocytes: a framework for understanding the immune system of jawless fishes. Dev Comp Immunol 2004;28:993–1003.

[6] Uinuk-Ool T, Mayer WE, Sato A, Dongak R, Cooper MD, Klein J. Lamprey lymphocyte-like cells express homologs of genes involved in immunologically relevant activities of mammalian lymphocytes. Proc Natl Acad Sci USA 2002;99:14356–61.

[7] Cannon JP, Haire RN, Pancer Z, Mueller MG, Skapura D, Cooper MD, et al. Variable domains and a VpreB-like molecule are present in a jawless vertebrate. Immunogenetics 2005;56:924–9.

[8] Suzuki T, Shin IT, Fujiyama A, Kohara Y, Kasahara M. Hagfish leukocytes express a paired receptor family with a variable domain resembling those of antigen receptors. J Immunol 2005;174:2885–91.

[9] Pancer Z, Amemiya CT, Ehrhardt GR, Ceitlin J, Gartland GL, Cooper MD. Somatic diversification of variable lymphocyte receptors in the agnathan sea lamprey. Nature 2004;430:174–80.

[10] Pancer Z, Saha NR, Kasamatsu J, Suzuki T, Amemiya CT, Kasahara M, et al. Variable lymphocyte receptors in hagfish. Proc Natl Acad Sci USA 2005;102:9224–9.

[11] Rogozin IB, Iyer LM, Liang L, Glazko GV, Liston VG, Pavlov YI, et al. Evolution and diversification of lamprey antigen receptors: evidence for involvement of an AID-APOBEC family cytosine deaminase. Nat Immunol 2007;8:647–56.

[12] Kasamatsu J, Sutoh Y, Fugo K, Otsuka N, Iwabuchi K, Kasahara M. Identification of a third variable lymphocyte receptor in the lamprey. Proc Natl Acad Sci USA 2010;107:14304–8.

[13] Li J, Das S, Herrin BR, Hirano M, Cooper MD. Definition of a third VLR gene in hagfish. Proc Natl Acad Sci USA 2013;110:15013–8.

[14] Das S, Hirano M, Aghaallaei N, Bajoghli B, Boehm T, Cooper MD. Organization of lamprey variable lymphocyte receptor C locus and repertoire development. Proc Natl Acad Sci USA 2013;110:6043–8.

[15] Nagawa F, Kishishita N, Shimizu K, Hirose S, Miyoshi M, Nezu J, et al. Antigen-receptor genes of the agnathan lamprey are assembled by a process involving copy choice. Nat Immunol 2007;8:206–13.

[16] Alder MN, Rogozin IB, Iyer LM, Glazko GV, Cooper MD, Pancer Z. Diversity and function of adaptive immune receptors in a jawless vertebrate. Science 2005;310:1970–3.

[17] Kishishita N, Matsuno T, Takahashi Y, Takaba H, Nishizumi H, Nagawa F. Regulation of antigen-receptor gene assembly in hagfish. EMBO Rep 2010;11:126–32.

[18] Kasamatsu J, Suzuki T, Ishijima J, Matsuda Y, Kasahara M. Two variable lymphocyte receptor genes of the inshore hagfish are located far apart on the same chromosome. Immunogenetics 2007;59:329–31.

[19] Petersen-Mahrt S. DNA deamination in immunity. Immunol Rev 2005;203:80–97.

[20] Guo P, Hirano M, Herrin BR, Li J, Yu C, Sadlonova A, et al. Dual nature of the adaptive immune system in lampreys. Nature 2009;459:796–801.

[21] Hirano M, Guo P, McCurley N, Schorpp M, Das S, Boehm T, et al. Evolutionary implications of a third lymphocyte lineage in lampreys. Nature 2013;501:435–8.

[22] Aravind L, Landsman D. AT-hook motifs identified in a wide variety of DNA-binding proteins. Nucleic Acids Res 1998;26:4413–21.

[23] Kato L, Stanlie A, Begum NA, Kobayashi M, Aida M, Honjo T. An evolutionary view of the mechanism for immune and genome diversity. J Immunol 2012;188:3559–66.

[24] Smith JJ, Kuraku S, Holt C, Sauka-Spengler T, Jiang N, Campbell MS, et al. Sequencing of the sea lamprey (Petromyzon marinus) genome provides insights into vertebrate evolution. Nat Genet 2013;45:415–21, 421e1-2.

[25] Kasahara M, Sutoh Y. Two forms of adaptive immunity in vertebrates: similarities and differences. Adv Immunol 2014;122:59–90.

[26] Hirano M, Das S, Guo P, Cooper MD. The evolution of adaptive immunity in vertebrates. Adv Immunol 2011;109:125–57.

[27] Alder MN, Herrin BR, Sadlonova A, Stockard CR, Grizzle WE, Gartland LA, et al. Antibody responses of variable lymphocyte receptors in the lamprey. Nat Immunol 2008;9:319–27.

[28] Bajoghli B, Guo P, Aghaallaei N, Hirano M, Strohmeier C, McCurley N, et al. A thymus candidate in lampreys. Nature 2011;470:90–4.

[29] Herrin BR, Cooper MD. Alternative adaptive immunity in jawless vertebrates. J Immunol 2010;185:1367–74.

[30] Herrin BR, Alder MN, Roux KH, Sina C, Ehrhardt GR, Boydston JA, et al. Structure and specificity of lamprey monoclonal antibodies. Proc Natl Acad Sci USA 2008;105:2040–5.

[31] Wu F, Chen L, Liu X, Wang H, Su P, Han Y, et al. Lamprey variable lymphocyte receptors mediate complement-dependent cytotoxicity. J Immunol 2013;190:922–30.

[32] Tasumi S, Velikovsky CA, Xu G, Gai SA, Wittrup KD, Flajnik MF, et al. High-affinity lamprey VLRA and VLRB monoclonal antibodies. Proc Natl Acad Sci USA 2009;106:12891–6.

[33] Deng L, Velikovsky CA, Xu G, Iyer LM, Tasumi S, Kerzic MC, et al. A structural basis for antigen recognition by the T cell-like lymphocytes of sea lamprey. Proc Natl Acad Sci USA 2010;107:13408–13.

[34] Mehta TK, Ravi V, Yamasaki S, Lee AP, Lian MM, Tay BH, et al. Evidence for at least six Hox clusters in the Japanese lamprey (*Lethenteron japonicum*). Proc Natl Acad Sci USA 2013;110:16044–9.

[35] Xu G, Tasumi S, Pancer Z. Yeast surface display of lamprey variable lymphocyte receptors. Methods Mol Biol 2011;748:21–33.

[36] Han BW, Herrin BR, Cooper MD, Wilson IA. Antigen recognition by variable lymphocyte receptors. Science 2008;321:1834–7.

[37] Nakahara H, Herrin BR, Alder MN, Catera R, Yan XJ, Chiorazzi N, et al. Chronic lymphocytic leukemia monitoring with a lamprey idiotope-specific antibody. Cancer Immunol Res 2013;1:223–8.

[38] Yu C, Ali S, St-Germain J, Liu Y, Yu X, Jaye DL, et al. Purification and identification of cell surface antigens using lamprey monoclonal antibodies. J Immunol Methods 2012;386:43–9.

[39] Kirchdoerfer RN, Herrin BR, Han BW, Turnbough Jr CL, Cooper MD, Wilson IA. Variable lymphocyte receptor recognition of the immunodominant glycoprotein of Bacillus anthracis spores. Structure 2012;20:479–86.

[40] Luo M, Velikovsky CA, Yang X, Siddiqui MA, Hong X, Barchi Jr JJ, et al. Recognition of the Thomsen-Friedenreich pancarcinoma carbohydrate antigen by a lamprey variable lymphocyte receptor. J Biol Chem 2013;288:23597–606.

[41] Hong X, Ma MZ, Gildersleeve JC, Chowdhury S, Barchi Jr JJ, Mariuzza RA, et al. Sugar-binding proteins from fish: selection of high affinity "lambodies" that recognize biomedically relevant glycans. ACS Chem Biol 2013;8:152–60.

[42] Lee SC, Park K, Han J, Lee JJ, Kim HJ, Hong S, et al. Design of a binding scaffold based on variable lymphocyte receptors of jawless vertebrates by module engineering. Proc Natl Acad Sci USA 2012;109:3299–304.

[43] Hong M, Yoon SI, Wilson IA. Recombinant expression of TLR5 proteins by ligand supplementation and a leucine-rich repeat hybrid technique. Biochem Biophys Res Commun 2012;427:119–24.

[44] Jin MS, Lee JO. Application of hybrid LRR technique to protein crystallization. BMB Rep 2008;41:353–7.

[45] Kang JY, Lee JO. Structural biology of the Toll-like receptor family. Annu Rev Biochem 2011;80:917–41.

[46] Park BS, Song DH, Kim HM, Choi BS, Lee H, Lee JO. The structural basis of lipopolysaccharide recognition by the TLR4-MD-2 complex. Nature 2009;458:1191–5.

[47] Kang JY, Nan X, Jin MS, Youn SJ, Ryu YH, Mah S, et al. Recognition of lipopeptide patterns by Toll-like receptor 2-Toll-like receptor 6 heterodimer. Immunity 2009;31:873–84.

[48] Kim HM, Park BS, Kim JI, Kim SE, Lee J, Oh SC, et al. Crystal structure of the TLR4-MD-2 complex with bound endotoxin antagonist Eritoran. Cell 2007;130:906–17.

[49] Jin MS, Kim SE, Heo JY, Lee ME, Kim HM, Paik SG, et al. Crystal structure of the TLR1-TLR2 heterodimer induced by binding of a tri-acylated lipopeptide. Cell 2007;130:1071–82.

[50] Kanda R, Sutoh Y, Kasamatsu J, Maenaka K, Kasahara M, Ose T. Crystal structure of the lamprey variable lymphocyte receptor C reveals an unusual feature in its N-terminal capping module. PLoS One 2014;9:e85875.

[51] Lee JY, Kim HS, Baek IW, Back JM, Han MR, Kong SY, et al. Overexpression, crystallization and preliminary X-ray crystallographic analysis of the variable lymphocyte receptor 2913 ectodomain fused with internalin B. Acta Crystallogr Sect F Struct Biol Cryst Commun 2013;69:39–41.

[52] Velikovsky CA, Deng L, Tasumi S, Iyer LM, Kerzic MC, Aravind L, et al. Structure of a lamprey variable lymphocyte receptor in complex with a protein antigen. Nat Struct Mol Biol 2009;16:725–30.

[53] Kim HM, Oh SC, Lim KJ, Kasamatsu J, Heo JY, Park BS, et al. Structural diversity of the hagfish variable lymphocyte receptors. J Biol Chem 2007;282:6726–32.

[54] De Genst E, Silence K, Decanniere K, Conrath K, Loris R, Kinne J, et al. Molecular basis for the preferential cleft recognition by dromedary heavy-chain antibodies. Proc Natl Acad Sci USA 2006;103:4586–91.

[55] Stanfield RL, Dooley H, Flajnik MF, Wilson IA. Crystal structure of a shark single-domain antibody V region in complex with lysozyme. Science 2004;305:1770–3.

[56] Deng L, Luo M, Velikovsky A, Mariuzza RA. Structural insights into the evolution of the adaptive immune system. Annu Rev Biophys 2013;42:191–215.

Chapter 9

# The Origin of V(D)J Diversification

Ellen Hsu[1] and Susanna M. Lewis[2]

[1]*Department of Physiology and Pharmacology, The State University of New York Health Science Center at Brooklyn, Brooklyn, NY, USA,* [2]*Department of Pathology and Laboratory Medicine, Perelman School of Medicine, University of Pennsylvania, Philadelphia, PA, USA*

The vertebrate immune system can respond to a broad and ever-changing spectrum of pathogens, and V(D)J recombination plays a central role in achieving this stunning versatility. V(D)J recombination has ancient origins; its first appearance in vertebrates is estimated at about 450 million years ago. The selective forces that set the stage and directed the evolutionary process have intrigued investigators ever since the discovery that the DNA of immunoglobulin (Ig) and T cell receptor (TCR) genes rearranged. V(D)J recombination appears to have emerged apparently without antecedent in evolution, an observation suggesting that perhaps the essentials of the recombination machinery were acquired through horizontal transfer. The recent availability of fully sequenced genomes from a variety of organisms has enabled closer examination of this idea. In the first part of this chapter we examine the hypothesis that the V(D)J recombination machinery is derived from a DNA transposon that was "exapted" [1]. The second section discusses the possible effect of germline V(D)J recombination events on the creation of different Ig and TCR loci. The last part of the chapter considers the nature of the ur-V gene and reviews current thinking on how V(D)J recombination fits into the evolution of adaptive immunity as a whole.

## 1. THE ALIEN SEED

The discovery of V(D)J recombination was a breakthrough. In rapid succession, papers outlined the basic features of the process and defined sequence motifs marking recombination sites [2,3]. Noting similarities between V(D)J recombination targets and those of prokaryotic insertion sequences, Susumu Tonegawa first articulated the evolutionary implications:

*We propose that such a mechanism was initiated when an IS-like DNA element was accidentally inserted into one of the multiple V DNA copies of an ancestral polymeric gene, splitting it into two portions, one corresponding to the present day embryonic V DNA segment and the other to the J DNA segment. Thus the V J joining may well be a reversal of an ancient accidental insertion of an IS-like DNA element which was subsequently exploited by vertebrates to increase the V-region diversity within the general framework of the intron-exon structure of higher cell genes [2].*

The idea was gradually fleshed out as more was learned about V(D)J joining, but the essential features have held up (reviewed in Refs [4–6]). After entry into the genome and perhaps no longer under selection for transposon function, it is thought that the DNA element decayed. Possibly multiple transposition events were involved, but ultimately terminal recognition sites of the transposon became the recombination signal sequences (RSS) seen as present-day V, D, and J gene segments. The transposase enzyme responsible for element mobilization was recruited as the V(D)J recombinase, encoded by two recombination-activating genes, RAG1 and RAG2. The proposed correlation is shown in Figure 1. In a sense, the transposon theory is especially fitting in that an integral aspect of the adaptive immune system—chance and necessity—provided its own diversification machinery. Three mechanisms underlying the model: transposition, horizontal transfer and exaptation are presented in the sections to follow.

### 1.1 Transposons and Transposition

Transposable elements are mobile genetic elements that include (1) retrotransposons, where an RNA transcript of the element is copied by reverse transcriptase and the DNA copy is inserted at random locations into new sites of the genome, and (2) DNA transposons, which are segments of DNA able to move to new sites in a genome, generally in a "cut-and-paste" mode. Some transposons have a transient existence as infectious phage or virus-like particles when they cycle in and out of genomes, examples being bacteriophage Mu or the HERV-K human endogenous retrotransposon. In DNA transposons that contain a minimal amount of coding sequence, their sole gene encodes the

Molecular Biology of B Cells. http://dx.doi.org/10.1016/B978-0-12-397933-9.00009-6

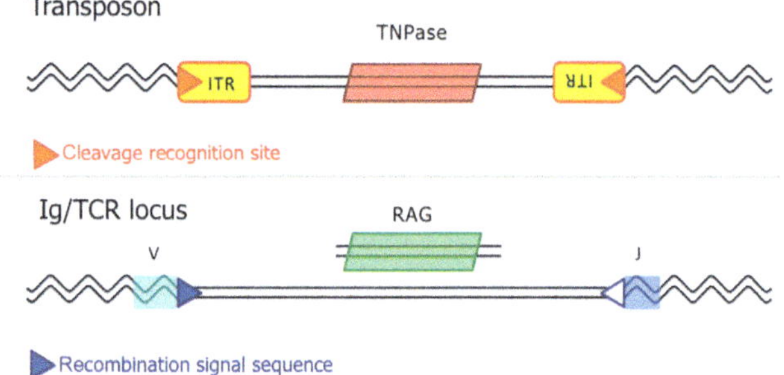


FIGURE 1 **Similarities between transposition and VDJ recombination. Top** panel shows a minimal autonomous transposon. The termini contain inverted terminal repeats (ITR, also called TIR) that contain transposase-binding sites. The outer edges of the ITRs also contain sequence motifs that define the two cleavage sites at the elements borders. In some transposons, the ITRs and the cleavage recognition motifs are the same. **Bottom** panel illustrates the basic features of a rearranging Ig or TCR locus. It has the features of a non-autonomous transposon because the transposase (RAG) is not part of the structure. The RSSs correspond to cleavage recognition sites that can still be site-specifically cleaved by RAG, encoded "in trans" as shown.

catalytic transposase, the protein(s) essential for mobility. Certain transposons have ancient origins, as implied by a ubiquitous distribution among living organisms. In a recent bioinformatic survey involving over 10 million protein-encoding genes from numerous species, transposase genes topped the list with respect to both abundance and representation in all sampled biomes [7]. Most were of the cut-and-paste class and have roots that extend far back in the history of life. The representation of certain structural folds in their transposases and the ability to bind $Mn^{2+}$ suggest they originated in an Archaean world over 2.5 billion years ago [8].

In addition to a transposase gene, autonomous or fully functional DNA transposons must also carry recognition sequences for transposase binding and cleavage, and these are situated within inverted terminal repeats (ITR) at either end of the element (Figure 1, upper panel). Although a transposon may contain other genes, their main function requires simply a transposase gene flanked by ITRs. Typically the number of transposon copies will wax and wane after invading a genome, while decayed "non-autonomous" versions of the elements steadily accumulate. The corrupted non-autonomous elements often have internal defects that destroy the transposase gene, and they can move only if transposase is available "in trans," that is, from other autonomous copies. A copy that lost one or both ITR is no longer able to move at all and subject to further mutational degeneration. Eventually, production of transposase by the remaining few autonomous copies becomes insufficient for self-mobilization in the presence of far larger numbers of non-autonomous elements. When the invasion subsides, the genome is left littered with non-autonomous element "fossils" [9].

The current understanding of the mechanism of transposition for the major class of DNA cut-and-paste transposons, including all known DNA transposons in eukaryotes [10], is based on close analysis of a few key examples [11–13]. The relevant transposases cleave and join DNA strands via a sequence of metal-dependent phosphoryl transfers with water or the 3′ OH group of a broken DNA strand serving as the nucleophile. The catalytic site of the transposases contains a "DDE" or "DDD" amino acid motif comprising three negatively charged residues (Asp, Asp, Glu/Asp) incorporated into a characteristic RNase H-like fold. Though details vary for different DDE/D transposons, a successful round of transposition invariably initiates with a pair of hydrolytic nicks at the outside edge of each cleavage motif, that is, at the left and right boundary of the ITRs. According to which set of strands is broken, 3′ hydroxyls are associated either with the flanking DNA or with the element ends. This process is illustrated in Figure 2(A) and explained in the legend. Depending on the position of the first nick, excision by transesterification will create hairpins either at the flanking DNA ends or at transposon ends. Once a transposon with exposed 3′ hydroxyls is created (Figure 2(A), step 2), DNA cut-and-paste transposases carry out the integration step the same way. A random site in the host genome is captured, and the transposon is inserted into the recipient target DNA (Figure 2(A), step 3).

Intermediates in transposition occur in the context of a protein-DNA complex that undergoes structural rearrangements during the course of the integration process. The initial paired end complex comprises a multimer of transposase molecules in contact with the transposon DNA ends. Protein–protein contacts may shift, and different DNAs may move in and out of the catalytic site(s). At the integration step, a "strand transfer complex" has formed with the genomic DNA that in some cases is so stable the transposase must be proteolytically removed [12].

Host repair factors ligate broken DNA strands left by the transposase. One leftover break remains at the exit site, and two single strand gaps are found at each 5′ end of the transposon. Target site duplications (TSDs) arise from repair of the two single strand gaps that are left at either end of the transposon (Figure 2(A), step 4). The length of the TSDs (~9 bp or less) reflects the staggered nature of the strand interruptions in the new target site and is a characteristic feature of a given transposon family. In general, the presence of TSDs helps identify novel transposons where they appear as non-annotated repeats in genome assemblies of plants, animals, and microbes.

## 1.2 Horizontal Transfer

Horizontal or "lateral" transfer is a term used when genetic material is passed between different species or even phyla instead of being vertically transmitted from parent to

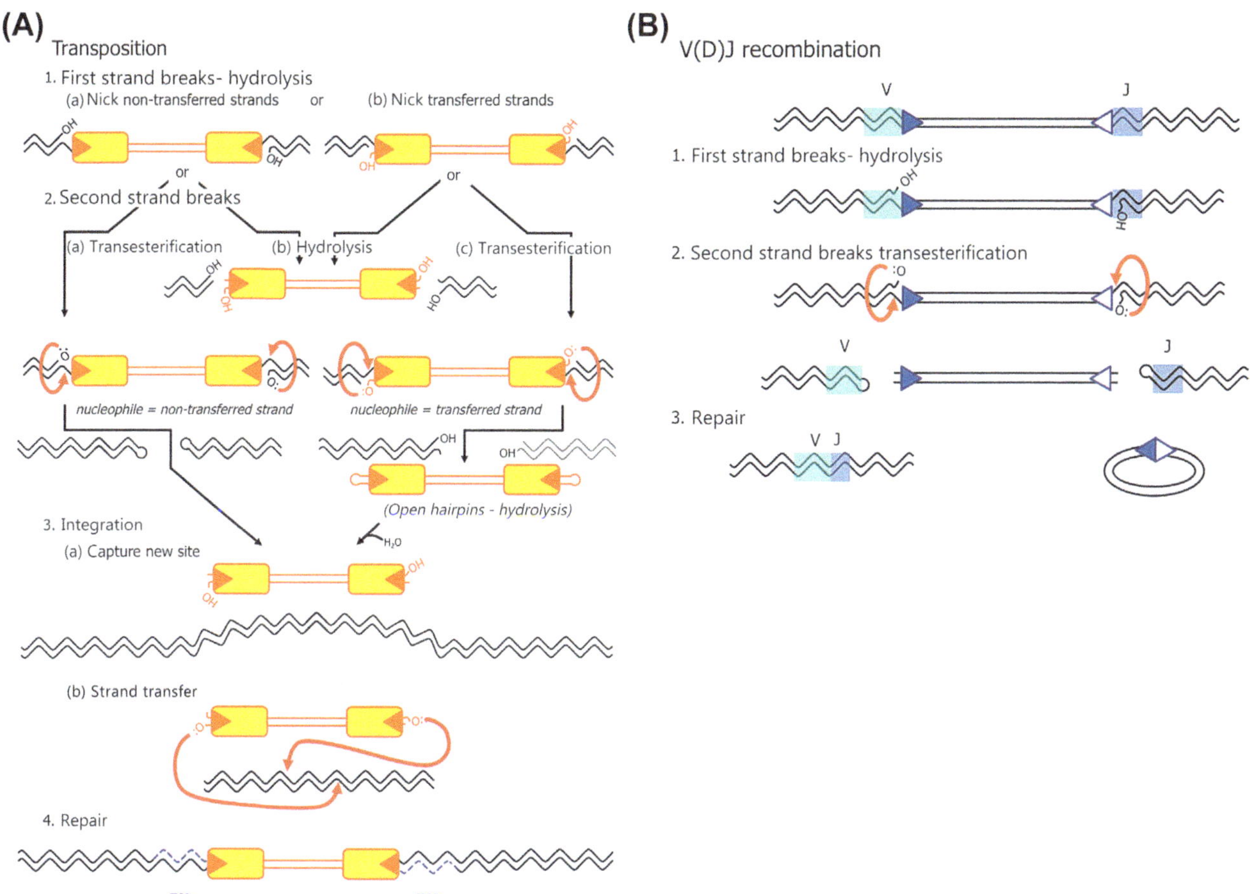


FIGURE 2 **Transposition compared to V(D)J recombination.** (A) **Transposition**. Transposons mobilized by DDE family transposases exit and enter DNA through phosphoryl transfers. Excision and integration take place in the context of a protein-DNA complex made up of the transposase (not depicted) and the DNA ends of the element. Step 1. Hydrolytic strand breaks. The two strands at either end of a transposon are differentiated by the designation "non-transferred strand" or "transferred strand," referring to which strand becomes chemically linked to the new integration site in the final strand transfer step (see below). Step 2. The second strands are cut by hydrolysis or transesterification [2(a) or (b)]. (a) If the first nick occurred on the non-transferred strand, transesterification will release a linear transposon DNA with conventional termini but leave hairpin structures at the donor break site. (b) Alternatively, the second strand break is achieved by hydrolysis. Both liberated transposon and donor break site will have 3′ hydroxyls and 5′ phosphates. (c) Where the first nick was made on the transferred strand, the transposon's 3′ hydroxyls can serve as the nucleophiles in transesterification. If so, hairpins are formed at the transposon's ends, while conventional DNA ends occur at the donor break site. An additional step (hydrolysis) is required to nick open the hairpin termini prior to integration. Step 3. Integration by transesterification. (a) Once free 3′ hydroxyls are created at each transposon end, the complex can capture the new integration target site. (b) Transesterification uses the available transposon hydroxyls to interrupt the target site. The breaks in the integration target are usually staggered, so that the integrated transposon has single strand gaps at each side, typically of 2–9 bp. Step 4. Repair is carried out by host polymerase and ligase functions, with the result that 2–9 bp TSDs, characteristic of a given transposon, are created and reflect the placement of the strand breaks at the integration step. **(B) V(D)J recombination**. Step 1. The catalytic RAG protein (not depicted) makes a first strand break by hydrolysis. The first cleavage is on the non-transferred strand; the 3′ hydroxyls are associated with the flanking V and J gene segments rather that the RSS cleavage targets. Step 2. The second break is introduced by transesterification, with the coding ends bearing hairpin termini. Step 3. Processing and joining of ends are carried out by components of the non-homologous end-joining pathway of DNA double strand break repair.

offspring [14]. Foreign DNA such as transposons can be introduced when brought in by a virus, parasite, or symbiont, or even perhaps in the diet [15]. Horizontal transfer is generally suspected when phylogenetic relationships are in conflict, such as when sequences in two different species seem more similar than anticipated or when a class of repeats has a patchy distribution, being present or wholly absent from related species in a way that is incongruent with established kinships [16]. The possibility of horizontal transfer can be tested in various ways such as by comparing mutation rates in the suspect sequences with nuclear genes of the organism.

Initially, there were two reasons to suspect that horizontal transfer played a part in creating the V(D)J recombination apparatus, though neither fits the above criteria. One reason, as mentioned, was that site-directed recombination is unusual in vertebrates, and features of the V(D)J recombination system seemed reminiscent of prokaryotic transposons. Another was that V(D)J recombination and adaptive immunity appeared to have arisen somewhat abruptly in vertebrate evolution, constituting what has been termed a "big bang" [17]. There is an estimated 50-70 million year interval between the divergence of jawless fishes and the jawed fishes, today represented respectively by lampreys and hagfish and by cartilaginous fishes; intervening lineages are extinct. In this period an immune system arose with all of the basic elements, including V(D)J recombination, and is found from sharks to human beings. Wholesale import of a pre-formed recombination capacity seemed to greatly simplify the possible schemes for how this could have taken form.

Unfortunately the theory that V(D)J recombination was acquired from elsewhere does not lend itself to standard tests for horizontal transfer. Tens of millions of years can obscure evidence of transmission so early in vertebrate evolution. There are no detectable differences between nucleotide substitution rates in RAG1 or RAG2 and the genomes in which they reside, and in fact evolutionists have used the RAG1 and RAG2 genes as representative markers in phylogeny studies [18]. However, in the next sections other observations are presented, supporting the idea that V(D)J recombination originated outside the vertebrate lineage in which it became a fixture.

## 1.3 Exaptation

In contrast to adaptation, where incremental improvements are selected to build a structure over time, exaptation describes the process of using pre-formed components to create a new function [1]. As such, transposons and/or their proteins have been exapted for new purposes. Transposons have been transformed into structural components of centromeres or telomeres, have provided transcription factors, and became recombinases in mating type switching and macronuclear differentiation as well as contributing regulatory elements [13,19]. How the putative ancestral RAG transposon was domesticated is perhaps the biggest mystery surrounding the origin of V(D)J recombination. Although we lack a detailed idea of what sort of organism was the involuntary host of the ancient RAG transposon, new information has begun to suggest profitable directions to pursue.

## 1.4 Modus Operandi

Parallels to transposition were strengthened once RAG proteins were identified and site-specific cleavage activity was reconstructed in vitro [20–24]. The first strand breaks in V(D)J recombination are hydrolytic nicks that occur at the 5′ side of the RSS edge. These nicks are positionally equivalent to an initial hydrolysis of a transposon's non-transferred strand (step 1 in Figure 2(A and B)). The second strand break is accomplished by transesterification [22,25], with the 3′hydroxyl generated at the first break acting as the acceptor. This is again consistent with a transposase-catalyzed cleavage pattern. The second break establishes hairpin connections between the top and bottom strands at the V, D, or J ends. The coding end hairpins correspond to hairpin on flanking broken ends in transposition, and as such, VDJ recombination is overall analogous to transposon excision as it occurs in the eukaryotic DNA transposons, Hermes and Hztransib [26,27].

Testing of experimentally generated RAG mutants has established that three amino acids, two aspartic and one glutamic acid, are critical for RAG1 nuclease function, and they occur with spacing consistent with a legitimate DDE motif [28–30]. DDE motifs are characteristic of enzymes that carry out metal-dependent phosphoryl transferase reactions, which accords with the $Mg^{2+}$ dependence of RAG1 and RAG2 cleavage activity in vitro.

Regardless of the parallels, a fundamental difference between V(D)J recombination and transposition is that the former involves manipulation of two DNA segments, but the latter involves three. V(D)J recombination corresponds to transposition events that stop before any integration-related activities takes place. In this view of V(D)J recombination as aborted transposition, the RSS-containing "signal ends" correspond to the liberated transposon that would normally be held in a post-cleavage complex. However, the cleaved signal ends as well as that of the V or J coding ends are processed by the cellular DNA repair operations.

Would the hypothetical RAG transposase have preserved transposition-specific functions after 450 million years of disuse? The answer was provided by two groups that independently demonstrated RAG1 and RAG2 could indeed carry out complete transposition in vitro [31,32]. Given appropriate DNA substrates and conditions, RAG proteins cleaved a pair of RSSs, excise the intervening DNA segment, captured a DNA target site, and transferred the RSS 3′ ends. The transposition products exhibited short but somewhat variable TSDs of about 5 bp. Such duplications are not associated with any aspect of normal V(D)J recombination and provided an unambiguous transposition signature. These observations were subsequently extended to the in vivo context. The occurrence of reproducible low-frequency transposition events was observed when expressing RAG1 and RAG2 in yeast cells [33], and in vivo transposition in mouse cells was confirmed with a specifically designed detection system [34]. In the latter study, the ratio of transposition to V(D)J recombination was measured at about 1: 50,000. By extrapolation to humans, an adult will have hosted 10,000 transposition events during each day of lymphocyte production; this estimate raises interesting questions about how transposition by RAG1/2 is normally curtailed [35]. In retrospect, the retention of vestigial transposase functions by the RAG proteins is probably not accidental and indicates instead an inseparability of cleavage and integration functions [12,36].

## 1.5 Structure and Linkage of RAG1 and RAG2 Genes

The RAG1 and RAG2 genes are closely linked and in convergent transcriptional orientation in animals representing a variety of taxa: chickens, rabbits, humans, mice, *Xenopus*, pufferfish, zebrafish, and sharks [37]. The coding sequences of the RAG1 and RAG2 genes are each fully contained in a single exon in most of these animals. The exception is that an intron is present in the middle of the RAG1 coding sequences of ray-finned fish [38].

Both of the above features, the tight and conserved linkage between the structurally unrelated RAG1 and RAG2 genes and a paucity of introns, are consistent with a prokaryotic origin [21,39]. In bacteria, genes for proteins belonging to the same metabolic pathway are often grouped together in coregulated operons. Their introns, especially those that are not self-splicing, are rare whereas the coding regions in human genes on average are interrupted by 7.2 introns each [40]. In short, the structural features of the RAG1/RAG2 locus are exceptional and contribute to the evidence that favors their horizontal transfer from a prokaryotic source.

## 1.6 Fossil Transposons

Evidence presented thus has led to a general acceptance of the hypothesis that key elements of the V(D)J recombination system were indeed derived from a transposon. What was the structure of the "RAG transposon" and at what point in animal evolution did the presumed horizontal transmission event(s) occur?

When genome analyses began on species representing key radiations, there arose an opportunity to pursue the origins of V(D)J recombination from a paleogenomic perspective. At present somewhere between one to two dozen eukaryotic transposon superfamilies have been discovered, one of which is called Transib [9]. Fossil Transib elements are present in insects, hydra, echinoderms, and fungi, though detected in neither plants nor vertebrates. A transposase sequence reconstructed from multiple fossil copies revealed an identifiable DDE motif and exhibited some similarity to RAG1 [41]. Notably, some Transib fossils are flanked by 5 bp duplications, as seen in the majority of in vitro and in vivo RAG1/2 transposition products. The sequences of some Transib ITRs were similar to the RSSs in that abutting the cleavage site is a heptamer-like sequence followed by nonamer-like sequence after an interval of 23 bp. Some Transib transposons carry a 23 bp RSS-like sequence in one ITR and a 12 bp spacer RSS in the other, suggestive of the 12/23 rule. RAG1 similarity has also been reported in transposons belonging to other eukaryotic transposon superfamilies, ([42] and citations therein), but additional RAG1 signature features distinguish Transib [10,41].

An intact Transib element was identified in the cotton bollworm, *Helicoverpa zea*, with terminal motifs (<u>CACGGTG</u>G) including underlined heptamer-like sequence and an open reading frame encoding 507 amino acids [43]. The Hztransib transposase was expressed, purified from *Escherichia coli*, and investigated for biochemical activities in vitro [27]. Just as for RAG1/2-mediated transposition, the Hztransib transposase was able to nick an appropriate substrate DNA between the ITR end and the flanking sequences and to transfer those ends to a new DNA site. The formation of hairpin ends at the flanking host break was also demonstrated, as were the expected 5 bp TSDs, visualized upon bacterial replication of the in vitro generated transposition products. Mutations at the DDE triad affected nicking and hairpinning activities to varying extents. The features of the Hztransib transposon suggest that some type of RAG1 transposon could have existed at one time. How RAG2 might fit into this picture is not clear.

The Transib sequences had little sequence identity (14–17%) to the RAG1 core, which is the 625 amino acid portion of RAG1 containing its catalytic activity [44]. A separate set of sequences with greater similarity to RAG1 core (26–35%) was detected in lancelet, hydra, and sea anemone; on the other hand, they lacked any evidence of ITRs or other indicators of transposition [41]. The relationship between these two sets, which are non-overlapping in their RAG1 homology, is not clear.

More than two dozen RAG1 core-like sequences were detected in sea urchin genomic sequence. They were nearly all fragmented or disrupted; however, one encoded an apparent RAG1 homolog with a large open reading frame. The gene, named SpRAG1L, is transcribed and encodes a 983 amino acid sequence that is 31% identical to the mouse RAG1 core [45]. SpRAG1L moreover possessed sequence similarities necessary for the function of the RAG1/2 complex. Not only did these include the DDE motif and a domain implicated in binding the RSS, but also conserved was a RAG1 zinc finger that is thought to be important for interaction with RAG2. These similarities and others involve a considerably larger portion of RAG1 than the Transib transposases.

Previously there had been fruitless searches for RAG2 genes in diverse organisms. Taking a different tack, Fugmann and coworkers [45] focused attention specifically on the region downstream from SpRAG1L and discovered a sequence with low but discernable identity to RAG2. This open reading frame was 3181 bp away from SpRAG1L, and as in the vertebrate RAG1/2 locus, was oriented in a convergent transcriptional orientation.

## 1.7 RAG2-like Gene in Sea Urchin

The mammalian RAG2 protein has two distinct features. In silico analysis revealed a series of "kelch" repeats in RAG2

[46], and structural studies in other systems have shown that kelch repeats can fold into a propeller-like structure that supports intermolecular protein–protein and protein–DNA interactions. Consistent with this model, spontaneous and directed mutations of RAG2 in the β propeller can reduce or fully abrogate both cleavage and RAG1/RAG2 association in vitro whereas mutations outside the β propellers tend to have a much milder effect [47,48]. A plant homeodomain (PHD) motif occurs near the C terminus of RAG2 [46]. The PHD motif occurs in a region of the protein that is not essential for in vitro V(D)J recombination [44], but mutations in this motif are pathogenic in humans. Several groups have determined that the PHD domain in RAG2 binds hypermethylated histone H3 modified on lysine 4 (H3K4), which is a general mark for active chromatin (reviewed in Ref. [49]).

The sea urchin SpRAG2L gene encodes a predicted protein that, as with vertebrate RAG2, has a six-bladed propeller structure comprising kelch repeats. C-terminal to the kelch repeats is an identifiable PHD domain. There are many proteins in the sea urchin genome that contain either a predicted propeller *or* a PHD motif, but SpRAG2L is unique in possessing both. It has been demonstrated that the PHD domain of SpRAG2L, like RAG2, binds hypermethylated lysine 4 in H3 tails, although the former recognizes the dimethyl modification, whereas the latter favors the trimethylated form [50]. The identities of the sea urchin proteins as homologs of RAG1 and RAG2 are supported by additional studies showing regulated co-expression of their genes, as well as evidence of protein–protein associations in vitro [45]. At this time the function of SpRAG1L and SpRAG2L is an unsolved question.

SpRAG1L/2L shows no evidence of having been recently brought into the sea urchin genome by a transposition event. Unlike most vertebrate RAG genes, the protein-coding region of both SpRAG1L and SpRAG2L apparently have been in place long enough to have acquired multiple introns [45]. Recent work suggests that such genes are likely to be present throughout eleutherozoan echinoderms (sea urchins to starfish) (J. Rast, pers. commun.). If the SpRAG1L/2L genes are a sea urchin equivalent of the RAG1/2 locus in jawed vertebrates, are these loci related by descent?

At the moment, no animal evolving from a lineage that diverged between echinoderms and cartilaginous fishes harbors both RAG1 and RAG2-like genes, and certainly none has evidence of a locus where they are linked in a tail-to-tail arrangement (Figure 3). The genome of amphioxus (cephalochordate) contains RAG1 core-like genes [60,61] and those of two divergent tunicate (urochordate) species, *Ciona intestinalis* and *Oikopleura dioica*, do not have any at all (Ref. [62] and L. Du Pasquier, pers. commun.). About 85% of the sea lamprey genome has been analyzed to date without uncovering RAG homologs [53], but since this is a crucial position—jawless fish are the only surviving vertebrate precedents before the emergence of cartilaginous fishes—no conclusion can be drawn until the lamprey genome assembly has been completely examined.

One recent hypothesis is that during the time RAG1 core-like sequences were inserted into various lineages by a RAG1 transposon, one integrated next to an existing RAG2 gene in an ancestral deuterostome [5,6]. The primordial RAG1 and RAG2 coupling took place with this insertion event, and SpRAG1L/2L is a descendant; in this scenario no RAG1/2 transposon ever existed as such. However, if this unique

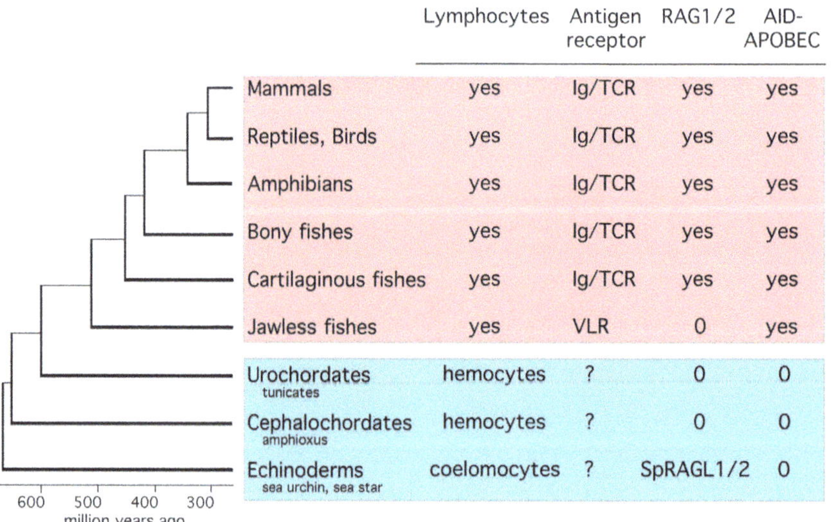


FIGURE 3 **Adaptive immune system components in deuterostome animals.** Vertebrates (top box) include gnathostomes and jawless fishes; invertebrates (bottom box) protochordates and echinoderms. The scale shows when taxa emerged according to molecular clock estimations [51,52]. Jawless fishes include both hagfish and lamprey, which express variable lymphocyte receptors (VLRs). Information on jawless fishes, protochordates and sea urchin is from Refs [45,53–59]. *After Ref. [101], p. 2, copyright (2011), with permission from Elsevier.*

RAG1/2 confection had been established at that time and was thus incapable of subsequent movement, the expectation would be to detect a RAG1/2 locus in some animals whose divergence preceded jawed vertebrates. If it turns out that there is no RAG1/2 locus in the lamprey—just as there are no Ig, TCR, or major histocompatibility (MHC) genes—then it is necessary to explain how RAG1/2 apparently was discarded or mutationally degraded in so many organisms and yet survived to evolve into a critical role in the stem vertebrate.

The alternate idea is that the RAG1/2 locus was contained in a RAG1/2 transposon that independently inserted into echinoderms and into an ancestral vertebrate [5]. Then the absence of a RAG1/2 locus in the lamprey genome, if established, would constitute the type of "patchy" distribution that is suggestive of horizontal transfer [16]. This scheme places no constraint on whether or not the RAG1/2 locus should be present in jawless fish, but asks for some explanation for its relationship to the RAG1 core-like genes in a wide variety of non-vertebrate animals [41].

## 1.8 Current Questions: Transposon Origins, Horizontal Transfer, and Exaptation

There are many aspects of RAG structure–function to solve before we can begin to understand exaptation. Biochemical experiments have shown that RAG1 and RAG2 comprise a functional transposase. How a transposase can be converted to a V(D)J recombinase without also shedding the (unused) ability to capture and integrate into DNA is a question worthy of attention, as it may provide information on protein evolution, defense mechanisms, and RAG1/2 catalysis. In tracing the steps of how these proteins came to have their current role in antigen receptor diversification, discovering the function of the sea urchin SpRAG1L and SpRAG2L gene products would be especially informative. The presence or absence of RAG homologs in the lamprey and other species may close a gap (or establish one) in the evolutionary history of V(D)J recombination and should help clarify if and when horizontal transfer took place. All of these provide a way to test notions of horizontal transfer and exaptation. Where once the field had to wait on chance observations and what might be considered circumstantial evidence, it has been transformed, and definite experimental lines of inquiry can be pursued. Ultimately, it may be possible to define the transposon contribution to V(D)J recombination at a level of detail that includes a description of the transposons involved, the recipient organisms, and the status of the immune system at the time.

# 2. THE EVOLUTION OF BCR AND TCR LOCI

At the outset, as commonly depicted, a V-like gene belonging to the Ig superfamily was interrupted by a transposon insertion (Figure 4(A)). The transposing element possessed RSS-containing termini, but may or may not have contained RAG genes as well. It is not necessary that the actual RAG1/2 transposon itself inserted into the V gene; as discussed previously, a fossil transposon with intact termini can be mobilized "in trans" by the RAG transposase. Once the ability of RAG1/2 to carry out transposition was suppressed and redirected to V region assembly, the recombinase appears to have gone to work on the germ line and likely was instrumental in forming the different Ig and TCR loci. Figure 4 sketches out the RAG-mediated manipulations that are either known or postulated to have modified the antigen receptor loci during vertebrate evolution.

A preliminary step (Figure 4(B)) would have been duplications of the split gene. This must be assumed in order to create substrates for generating different configurations. There is plenty of support for ancient and ongoing segmental duplications throughout evolution [64], and duplications are unrestricted in regard to gene segment or locus boundaries. With respect to antigen receptor loci, there exist examples that involved only a sub-region of an Ig locus [65] or alternatively an entire functional V(D)J/C unit, giving rise to Ig clusters in cartilaginous fish [66]. Once the split V-like gene was established in multiple copies, the basic elements for the creation of different loci and alternative configurations were in place.

## 2.1 Germline Coding Joint Formation: V(D)J Gene Segment Fusions

The strongest evidence that V(D)J recombination takes place in the germ line comes from examining Ig locus structures in sharks. Shark IgH genes occur as duplications of a simple [V-D-J/CH] recombination modules; likewise IgL genes are multiply reiterated [V-J/CL] clusters as depicted in Figure 4(B) [66,67]. In the genome of every elasmobranch species examined so far a number of Ig genes seem to have recombined elements [68]; for instance, some IgL genes have fused VJ instead of the "split" V-J or non-joined configuration (Figure 4(C)-(1)). It appeared as if these had recombined in the germ line. Since the leader intron was retained in the germline-rearranged IgH or IgL, they were not the result of retrotransposed mRNA. One suggestion was that these phenomena may have arisen from recombinase activity in germ cells, although V(D)J rearrangement was established as an activity crucial to lymphocyte development [63].

The germ line joining hypothesis is supported by studies that examined the nature of the VJ junctions, demonstrating that they were generated from non-joined gene segments [69]. By determining the phylogenetic relationship among joined and split IgL gene clusters among two shark species, it was possible to estimate roughly how often and when germ line joining events occurred and with respect to species divergence. One particular V to J recombination

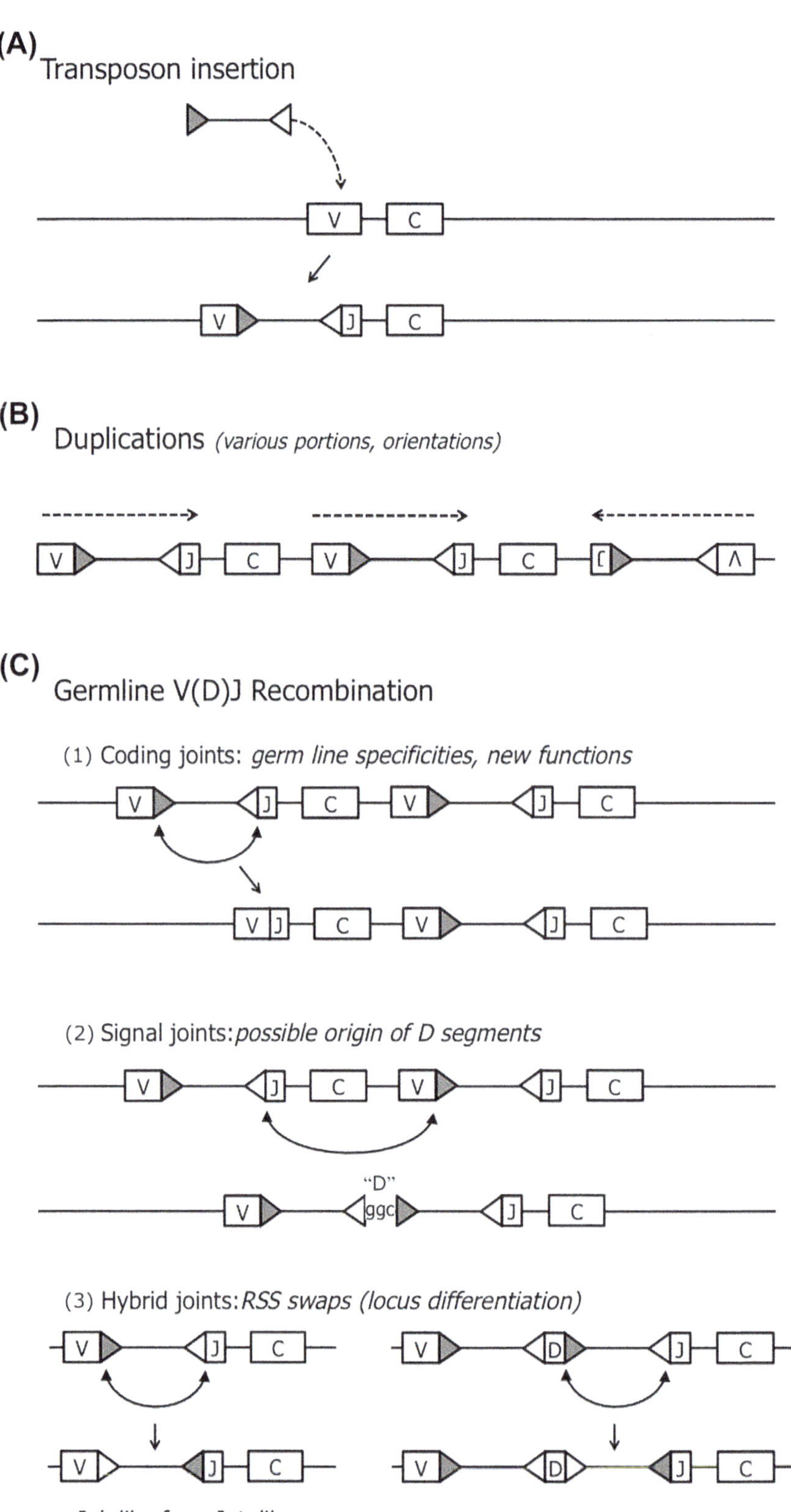


FIGURE 4 **Germ line rearrangement gives rise to different locus configurations.** (A) A transposon splits the ur-V gene, placing RSSs adjacent to the interrupted coding sequences. (B) Duplications of the interrupted unit take place. (C) (1) Standard germ line V-to-J recombination creates fused germ line genes. (2) Some types of recombination between the repeated clusters, such as J-to-V recombination as shown, can create a signal joint; signal joints are also created by V-to-J recombination between clusters that are inverted with respect to one another. Such events, with the insertion of non-templated "N-region" nucleotides into the signal joint, generate a proto-D segment. (3) A non-standard V(D)J recombination outcome, termed a "hybrid joint" is known to occur in vivo, and evidence exists that they arise in the germ line as well [63]. As depicted, hybrid joints effectively swap the 12 and 23 RSSs between two gene segments. The change in RSS affiliation brought about by aberrant hybrid joint formation may have helped drive the differentiation of Ig/TCR loci in evolution.

event was proposed to have occurred within the last 7 million years. Evidence also suggests that such germ line joining occurred not once but several times in the nurse shark genome. Thus in some species germ line V(D)J recombination, creating fused VJ and VDJ genes, is perhaps still ongoing.

What are consequences of germ line joining events? Joined genes in shark, catfish, chicken, and opossum are available for consideration. Although fully joined V(D)J segments were described in cartilaginous fishes a while ago [63], the role of their products as antibodies remains unclear. However, in the nurse shark there is one IgH cluster carrying a joined, in-frame VDJ that is unique in that its three C exons are derived but highly divergent from the four Cμ exons. The IgM1gj polypeptide associates with shark L chains, is prominent in neonatal sera, and is probably expressed exclusively as a secreted molecule [67,70]. Because it is not a surface receptor, IgM1gj is not expected to participate directly in activating the cell that expresses it. Without diversity and without the ability to induce clonal expansion, the IgM1gj has lost the main features of adaptive immunity and appears to have actively evolved to another function.

In the channel catfish there are three duplicated IgH with large numbers of V, D, and J segments at two of them [71]; at one there is an assembled VDJ [72]. This VDJ exon prevents further rearrangement at its IgH because all other D gene segments were eliminated at its formation. The VDJ has apparently fallen into disuse, as it does not seem to be expressed [73]. Instead, a leader-like sequence arose downstream of the VDJ unit, enabling a V-less C region transcript to be processed for that locus. The secreted C-alone protein binds catfish granulocyte blood cells, similar to the binding of IgD on mammalian basophils [74], raising the possibility of a specialized immune function. Like IgM1gj, the C-alone polypeptide may have passed into the realm of innate immunity. In both instances germ line rearrangement, by pre-empting somatic recombination, is the first step in initiating changes in the fate of an Ig locus. The joined gene may become defunct or alternatively the recombination event can initiate an evolutionary exploration of new functional possibilities.

Germ line joining could well have occurred in species like humans and mice that do not have redundant or duplicated recombination modules as in fish, but since tetrapods have Ig loci configured in an extended array with linked groupings of Vs, Ds and Js, germline V(D)J recombination would be expected to delete much of the locus and negatively impact the diversity of T or B cell receptors. Yet, there are two animals in which evidence of germ line joining can be seen—opossum and chickens [75,76]. In the former, a VD fusion was observed amidst a VH gene segment array in *Monodelphis domestica*. The VD can undergo rearrangement, generating unusually long CDR3s due to its extended flank, but as there is little compositional variation it is not clear what role such H chains play in the opossum antibody repertoire. In chickens this kind of VD fusion was incorporated into its unique Ig diversification process. There is only one functional, rearranging VH gene at the chicken IgH, and all others are present as VD fusions. The series of pseudo V genes are contiguous with D-like sequences and sometimes also with recognizable J segments at their 3′ ends. In chickens, the germ line-joined copies augment antibody diversity by providing lengthier donor sequence for somatic gene conversion at the IgH locus [77].

In conclusion, where germ line-joining events have not resulted in inactivation, the outcomes are remarkable for the transformative changes that are induced. Some examples described above suggest that sometimes germ line rearrangement of V gene segments does not affect the antibody repertoire so much as *catalyze other events* shaping the fate of the Ig locus or cluster. In shark and catfish, novel immune molecules were generated. In chicken, the locus configuration was altered for improved templates used in gene conversion diversification.

## 2.2 Germ Line Signal Joint Formation: the Creation of D Segments

If RAG can act on RSSs in germ cells, it also has the potential to change the organization of antigen receptor loci. It is possible that this may indeed have happened at multiple points in the ancestral vertebrate lineage. In Figure 4(C)-(2) we consider how RAG action may have driven significant alterations like creating three-segment loci (V, D, and J) from two segments (V and J).

V(D)J recombination creates a reciprocal product to coding joints, termed signal joints. These comprise RSS-to-RSS junctions, and although normally excised from the genome when formed, signal joints are chromosomally retained when rearrangement involves inverted gene segments. Inversional joining is common and occurs for the many V genes in the mouse and human Ig kappa locus that are in the opposite transcriptional orientation to the J genes. In Figure 4(C)-(2) it is suggested that V(D)J recombination in germ cells could likewise produce a chromosomally retained signal joint between inverted gene segments, and if such a signal joint includes nucleotides as in N region, it becomes structurally analogous to a D segment [78]. Thereafter, positive selection for diversity (as discussed in the next section) would work to ensure such changes became fixed in a population. There is no reason to suppose that D segments were invented only once or only rarely by germ line V(D)J recombination; N region insertion into signal joints is not uncommon [79] and inversional joining would provide a ready source of signal joint formation. Thus a three-segment locus could be derived by germ line joining from simpler kappa or lambda-like loci.

## 2.3 Germ Line Hybrid Joint Formation: RSS Swaps and Locus "Speciation"

Conflicting family relationships between Ig and TCR genes [4] are even harder to understand when the RSS affiliations of the gene segments are taken into consideration. For example, the V genes of kappa and lambda are more similar to each other (and to IgH) than they are to TCR V genes, yet Ig VH and Vλ and TCR Vβ all have 23 spacer RSSs, whereas Ig Vκ has a 12 spacer RSS. Likewise, JH and Jκ have 23 RSSs whereas 12 RSSs occur at both Jλ and Jβ. Some of this complexity can be unraveled by schemes that allow for germ line joining; then it is possible to interchange one type of RSS for another through V(D)J recombination.

Figure 4(C)-(3) illustrates how any RSS configuration can derive from another by germ line hybrid joint formation [63,78]. RSS swapping is seen in vivo in lymphoid cells where hybrid joint formation takes place instead of the usual (and productive) coding joint/signal joint outcome of V(D)J recombination [78–82]. Germ line hybrid joint formation, similarly resulting in an RSS swap, would set aside the affected gene segments for separate evolution ("speciation" at the molecular level) because swapped units constrained by the 12/23 rule, can profitably rearrange only with one another. Recombination events with any of the original gene segments will result in aberrant V–V and J–J joining.

The above schemes provide a simple basis for much of the observed variation in locus configuration; in some animals, evidence of these rearrangements has been preserved in the genome. Distinctive RSS configurations may have been selected for at some point, differentiating the various loci as regulatory mechanisms became more complex. Although the odd germ line joining event allows exploration of new functional possibilities, inducing large rearrangements in the germ line is probably hazardous, and its likelihood of being retained must be on the same order as any spontaneous mutation in the population [69].

# 3. CONSIDERATIONS ON THE UR-V GENE

The previous sections have described the possible transposon origin for V(D)J recombination and the subsequent sculpting of the Ig and TCR gene organization by germ line rearrangements during evolution. If we postulate the first rearrangeable antigen receptor gene indeed arose from an incident of splitting through transposon insertion, several questions come to mind—how and why was this one particular insertion selected for, and what might have been the nature of the archaic "ur-V gene"?

The ur-V gene would have encoded a domain—an autonomous folding unit—of the Ig superfamily, whose members include cell surface molecules (Ig, TCR, MHC, β2-microglobulin, Thy-1, CD4, CD8) as well as cell adhesion and muscle components (for reviews, see Refs [83,84]). Four sets of domain structures have been distinguished: V (variable), I (intermediary), C1, and C2 (constant). These structures are differentiated from each other by sequence patterns and lengths but they all share the immunoglobulin fold, which is formed by two β-pleated sheets that are packed face-to-face and consist of anti-parallel β-strands. The disulfide bridge characteristic of Ig-related molecules is not necessary to the formation of this sandwich structure. Figure 5 shows ribbon diagrams of the V, I, and C structures. V domains are distinguished by the presence of two strands, C′ and C″. The antigen-combining portions are mainly situated in loops, those connecting B to C strand (CDR1), F to G strand (CDR3). In the prototype Ig gene, the hypothetical transposition event would have disrupted the region encoding the loop between F and G strands in the V domain.

The relationships between the four Ig superfamily structural sets are not entirely clear as yet, but the I set is probably the earliest and the C1 the most recent in evolution. Du Pasquier [86] analyzed a receptor tyrosine kinase in the sponge *Geodia cydonium*, concluding that it possessed the V frame that is common to V and I set structures [83]. With completion of the *Caenorhabditis elegans* genome, an analysis was made to identify the repertoire of the Ig superfamily domains in the nematode. Out of 64 proteins, all 488 domains were of the I set, establishing that this was the ancestral domain structure [87]. Among molecules reported in later invertebrates, such as mollusks and arthropods, domains with V or C2 features, together or alone, could be distinguished, such as fibrinogen-related proteins (FREPs) from the snail or amalgam in Drosophila [86,88].

TCR and Ig domains consist of V and C1 type domains, and C1 domains are also part of other immune response-associated molecules like class I and class II MHC, β2-microglobulin, and tapasin, all of which also apparently have no homolog earlier than in cartilaginous fishes (for a review see Ref. [89]). Current analyses of the available sea lamprey genome revealed a few C1-like sequences (J. Rast and K. Buckley, pers. commun.) and suggest that such structures primarily evolved after divergence of jawless fishes.

## 3.1 Rearrangement Generates Sequence Length Diversity

We reason that the ur-V gene probably was one member of a multigene family. Since the immediate effect of the transposition would have been disruption of the target gene, redundancy of the gene function would have prevented immediate elimination of the to-be founder gene. Amongst multiple duplicated members that compensated for its loss of function, the split gene could have hitchhiked along, eventually to be eliminated by recombination or possibly altered by gene conversion, as has been observed in the "birth-and-death" evolution of multigene families [90]. Fixation of this novel but null (because silenced) mutant could have

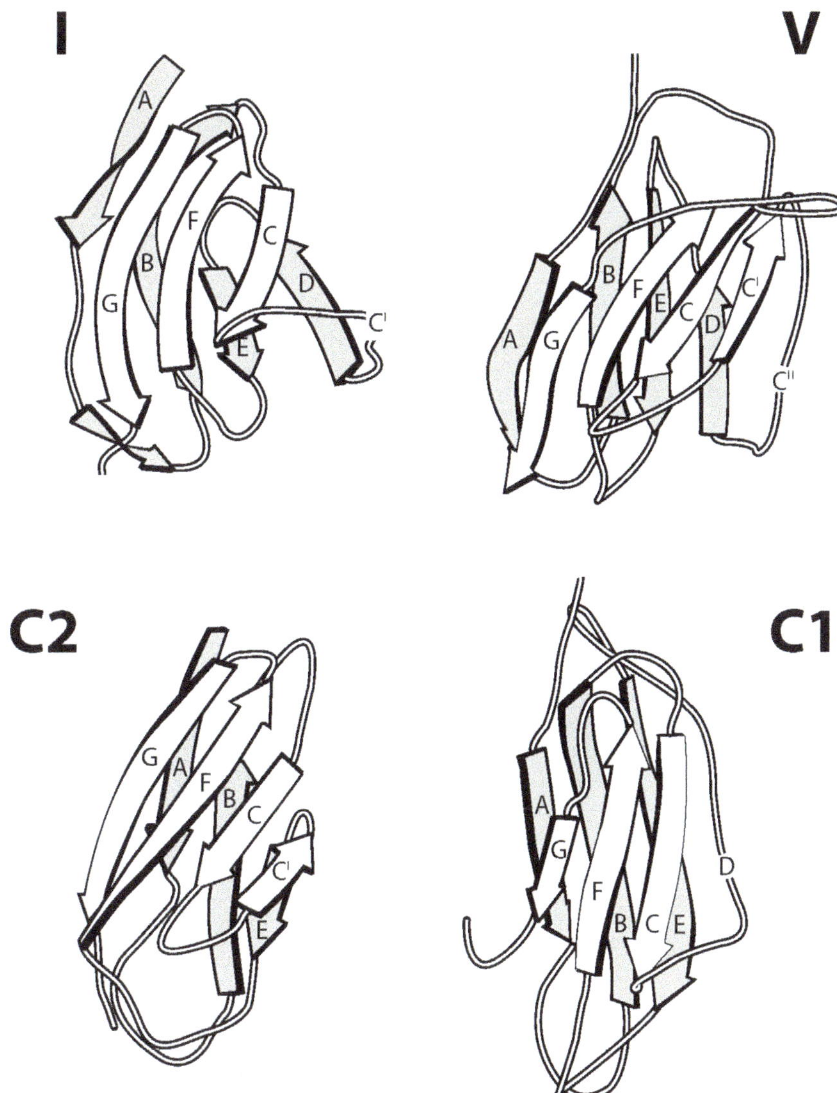


FIGURE 5 **Immunoglobulin superfamily domain structures.** The region between C and E strands varies among the sets. C1 set structures appear only late in evolution, in gnathostomes, and differ from I set by the absence of C′. V set differs the addition of C″ strand and C2 set by the deletion of D strand. Ribbon diagram of I set is based on telokin structure [83]. *(Adapted from [83] with permission from Elsevier.)* The V set is based on Fab NEW VH, C2 set on human CD4 domain 2, and C1 set on β-2 microglobulin [85]. *Adapted from [85], p. 54, copyright (1997) with permission from Elsevier.*

been by chance, but the recruitment for function must have arisen soon after the integration event. Had there not been selection for the split gene, it would have acquired unfavorable changes in the coding region, or the RSS would have mutated out of recognition for the RAG recombinase. It has been observed that TCR V pseudogenes with frameshift mutations tend to accumulate additional changes, possibly because reversion to a functional gene is less apt to occur than for V pseudogenes with point mutations that generate premature termination; and the transposition in this case is effectively a frameshift mutation.

The selecting feature would have included RAG activity, and this activity would have to be expressed in somatic cells. In the somatic cell, the RAG proteins would excise the RSSs, along with the intervening DNA, that had been incorporated into the ur-V gene in the germ line. With this removal the cleaved DNA ends at the exit site would be rejoined, most likely by RAG-independent endogenous repair mechanisms. Even discounting genetic and enzymatic elaborations, like the D gene segments and N region addition by terminal deoxynucleotidyl transferase (TdT), heterogeneity can be generated at the new junction. The opening of hairpin DNA as formed by RAG recombinase action, followed by limited exonuclease trimming and ligation, all are carried out by functions that are part of the endogenous non-homologous end-joining pathway of repair (reviewed in Ref. [91]).

The net effect of the removal of the intervening DNA from the split gene is that the re-formed gene sequences, if ligated in-frame, encode proteins with both sequence and length heterogeneity. Although there are somatic mechanisms that change DNA sequence—mutation and gene conversion among them—there are none that generate sequence length diversity at high frequency. Gene conversion will incorporate insertions or deletions, but according to the set of templates available in the germ line. Somatic hypermutation can create insertions and deletions (in/del) in V regions but does so with low efficiency in functioning genes [92,93].

Whereas in/dels occur at different locations throughout non-productive V(D)J, they have survived mainly in CDR in functional rearrangements [94]. Sequence length variability alters not only the chemical nature but, more drastically, the physical shape of the region involved. The loops best support this type variability, without interfering with the main-chain folding that creates the Ig fold (see Figure 5), and illustrate it by being the least conserved feature in sequence alignments of evolutionarily related proteins [84]. The RAG recombinase activity is site-specific, and this ensured localization of sequence length variability at both a tolerated place and on a regular basis. The subsequent creation of D gene segments (possibly via germ line joining, as above) to further increase heterogeneity at the CDR3 loop supports the idea that junctional diversity is a very strong selecting feature for rearranging genes [95].

TdT is a unique DNA polymerase whose function in developing lymphocytes is to elaborate antigen receptor junctional diversity by addition of non-templated bases [96]. TdT and polymerase μ are members of the pol X family of DNA polymerases; their similarity (61%) and identical exon-intron organization suggest a common ancestry [97] that is supported by the presence of only a single copy gene in the sea lamprey, tunicate, amphioxus, and sea urchin genomes [53,54,61,62]. Sharks represent the earliest vertebrates with both pol μ and TdT and whose TdT expression is restricted to lymphoid tissues ([98], E. Hsu unpublished data). The appearance of TdT with V(D)J recombination supports the notion that the generation of the unusual molecular heterogeneity provided by RAG action may have been the selecting factor for the split gene.

## 3.2 Considering Jawless Fishes

If the RAG transposon arrived after the divergence of jawless fishes, what sort of immune system did the ancestral jawed vertebrate possess? The existing representatives of jawless fishes (cyclostomes), hagfish and lamprey, have adaptive immune systems based on lymphocytes that express diverse molecules called variable lymphocyte receptors (VLR) that contain leucine-rich-repeat (LLR) elements (see Chapter 8). Could the jawed (gnathostome) ancestor have once used the VLR system?

Some part of the answer rests on the phylogenetic position of lampreys, that is, if the lamprey diverged from the hagfish or the stem vertebrate lineage (Figure 6(A and B), resp.) ([99] and references therein). The short interval between these events has made their relationships a difficult and highly debated issue. Since hagfish and lamprey have similar immune systems, the VLR gene system evolved prior to their diversification, and the arrows in Figure 6 indicate the latest points in the phylogeny trees it could have arisen. In Figure 6(B) the inference is that the ancestor of jawed vertebrates possessed a VLR system because it and lampreys shared lineage after divergence of hagfishes. The story is less clear in Figure 6(A); the ancestor would not have carried VLR only if VLR did not arise at an earlier point than indicated.

The question is even more interesting for what answers may reveal about the circumstances under which lymphocyte receptors evolved. There are at least two kinds of cyclostome lymphocyte subsets with reciprocally expressed VLR receptors, one subset that secretes VLR molecules like B cells whereas the other resembles T cells and derives from a thymus-like structure [55,100]. This discovery was most unexpected, because it presents two interesting possibilities; either lymphocyte subsets arose in jawed vertebrates independently from cyclostomes or else subsets were already present much earlier, in the stem vertebrate [101].

If the relationships are as depicted in Figure 6(B), the VLR-based immune system was present in gnathostomes but superseded by Ig/TCR. As inferred from Figure 6(A), the ancestors of jawed and jawless vertebrates already possessed lymphocyte subsets; the respective Ig/TCR and VLR evolved independently in the different lineages. In such case, perhaps what now appears to be convergent evolution was the result of selection for molecules most effectively transacting the existing lymphocytes' immune functions [101].

The temporal relationship of Ig/TCR to VLR, if any, may be resolved when the hagfish genome is available and can

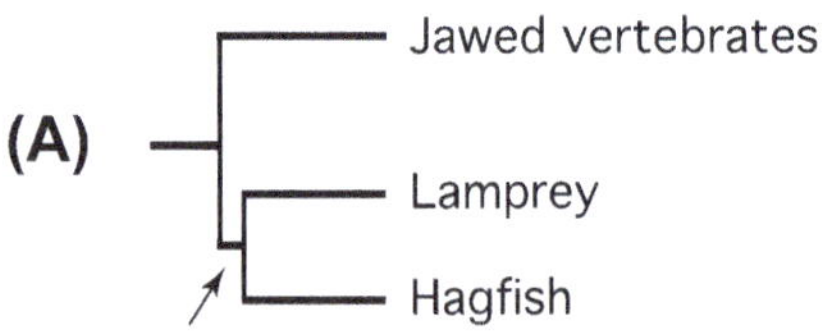


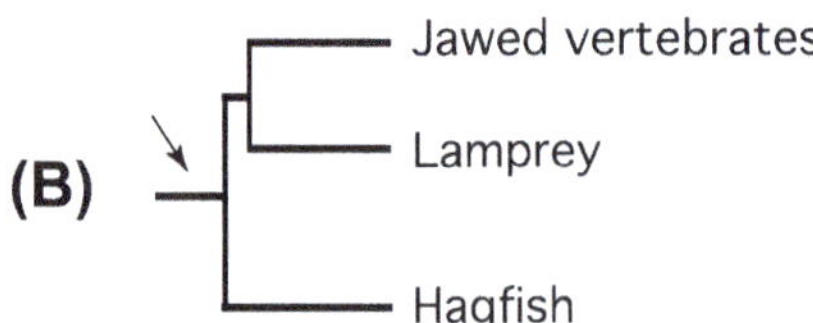


FIGURE 6 **Hypothetical trees for relationships among hagfishes, lampreys, and jawed vertebrates.** Tree A shows hagfishes and lampreys as sister groups. Tree B shows lampreys and jawed vertebrates as sister groups. Arrows indicate the latest point at which the VLR system could have arisen in order for it to exist in both hagfishes and lampreys.

be assessed relative to those of lamprey and jawed vertebrates. Janvier [102] commented that whereas physiological and morphological studies supported monophyly of lampreys and gnathostomes (Figure 6(B)), molecular analyses tended to favor the hagfish and lamprey being sister groups (Figure 6(A)). If examination of the hagfish genome eventually supports the latter hypothesis, then VLR and Ig/TCR could have evolved in parallel. The ancestral vertebrates may have had an immune system with lymphocytes subsets, perhaps using another gene system (discussed below). They may have possessed both LRR and Ig superfamily elements, the former predominating after hagfish divergence, and the other after acquisition of RAG1/2 genes.

## 3.3 Hypermutation in Evolution

If VLR was not the kind of lymphocyte antigen receptor utilized by the ancestral gnathostome, we speculate that there may have already existed a mechanism whereby the archaic non-rearranging V genes possessed the ability to somatically diversify. If the novel advantage introduced by RAG was extensive sequence diversification, the original gene function of the unsplit ur-genes would have also required this feature.

Rearrangement and somatic mutation both exist in cartilaginous fishes (sharks, skates, rays), representatives of the earliest jawed vertebrates [103]. Somatic hypermutation (SHM) at Ig genes is instigated by activation-induced cytidine deaminase (AID), which converts cytidine in DNA to uracil [104,105]. Repair of the site leads to gene conversion in some species and in others insertion of mismatches, resulting in mutated sequence. AID is a recently evolved member of the APOBEC family of cytidine deaminases and is highly conserved among jawed vertebrates [106]. Two members of the AID-APOBEC family have been identified and detected in lamprey lymphocytes. Their function is not known, but PmCDA1 has been shown to induce transition mutations when expressed in *E. coli*, which suggests a similar ability to cause deamination at C:G sites [56]. It is hypothesized that assembly of the VLR antigen receptor cassette through a homology-pairing recombination mechanism is initiated by the AID-APOBEC-induced DNA break.

No AID-APOBEC members were found in tunicate or amphioxus [56]. The human AID protein shares 66% identity and 79% similarity with shark AID and 25% identity and 50% similarity with lamprey pmCDA1. If the jawless fish and shark genes had a common AID precursor that affected DNA, it evolved in the stem vertebrate.

What could have been the role of AID in this ancestral vertebrate? SHM produces primarily point mutations, in mouse and human Ig as well as Ig from the amphibian *Xenopus* [105,107]. However, in cartilaginous fish Ig half of the substitutions are in tandem, as stretches of 2–5 bp; the difference is most likely due to species-specific processing of the AID-induced lesion [95,108]. Recent experiments have revealed that the translesion synthesis DNA polymerase zeta is capable of extending a mismatch with a mismatch during SHM [109,110]; possibly pol zeta is more prominent or of considerably higher activity during shark SHM. It is not clear why sharks have a mutational pathway that operates more efficiently per AID hit, but that it is so in cartilaginous fishes suggests there could exist a system where, in the absence of a VDJ recombination mechanism, very extensive antigen receptor sequence diversity was possible through somatic mutation [95].

We suggest that the function of the ur-V genes in the ancestral jawed vertebrate was presumably immune in nature and effected by circulating cells. In addition to expression of AID-APOBEC, both jawless and jawed vertebrate lymphocytes share the ability to undergo cell division upon activation. In the latter, antigen receptor engagement is coupled to pathways taking the lymphocyte out of $G_0$ into the $G_1$ state and upregulating genes that enable mitosis; usually a second receptor sensing pathogen presence contributes the crucial signal. In B cells there are Toll receptors or receptors for factors produced by accessory cells such as helper T cells and dendritic cells that have already been activated by pathogen. Proliferation is specific to the activated cells. The greater the antigen receptor diversity, the lower the frequency for any specificity (estimated at one in $10^5$–$10^6$) and the fewer the initial cell numbers to combat a particular pathogen. Hence, clonal proliferation is not only fundamental to adaptive immunity, but also necessary from the very beginning for a diversification mechanism to have selective value. During the evolution of both the VLR and the transposon-split V gene, certain hemocytes must already have possessed some ability to proliferate.

## 3.4 An Innate Defense V Gene

It has been proposed that a transposition event involving a RAG transposon and an archaic V gene initiated the creation of rearranging antigen receptors. We suggest that the prototype V genes was part of a multigene family that may have diversified by hypermutation, and introduction of a recombining mechanism that generated loop length variability greatly augmented diversification in a way more revolutionary than base substitutions. This was the selecting factor for the split gene created by RAG.

There have been attempts to identify prototype V lineage genes in jawless fish or other animals; some examples are referenced here ([111–114], discussed in Ref. [115]) but will not be examined. One difficulty in identifying the "actual" lineage that became disrupted in jawed vertebrates is that, with the acquisition of junctional diversification, the rest of the sequence will have evolved to accommodate or enhance the useful new structures generated by rearrangement. Hence, the modern vertebrate V gene is sufficiently diverged so that its sequence no longer serves much more than as a general guideline to searches in jawless fish or protochordates. For example, large sets of nonrearranging V-like genes were isolated from amphioxus [116], but so far their function is not clear.

We have speculated that the ancient V gene mutated and may have had some sort of defense function. Although in itself not proof of any immune function, somatic mutation should be a fairly restricted phenomenon. One candidate for a cousin of the Ig prototype may be the FREPs isolated from the hemolymph of freshwater snails, *Biomphalaria glabrata*. FREPs are a family of molecules consisting of one or two Ig superfamily V-set-like domains connected with a carbohydrate-binding lectin domain [86,117]. There is extensive sequence heterogeneity that may be produced by somatic diversification. The FREPs do not rearrange in the manner of Ig/TCR. They are a multigene family consisting of sets of tandemly arranged V domain-like exons with fibrinogen exons, and the fibrinogen domains can be construed as serving a C region-type function. Some genes are upregulated in response to infection by flatworm parasites as *Schistosoma mansoni* and *Echinostoma paraensi* [118]. These characteristics suggest that some similar kind of ancestral innate defense gene in a primitive vertebrate could have been the target of the postulated RAG transposition.

## 4. CONCLUDING REMARKS

Although we have made large claims for RAG1/2 manipulating the genome, there is little actual experimental evidence for such activity. There are few animal species where both RAG1 and RAG2 mRNA are present in ovary or testes tissue, and it is not clear that there are active proteins in the germ cells themselves. Those species possessing the possibility for ongoing germ line rearrangement, such as cartilaginous fishes, are not typical laboratory models, and their genomic content is not (yet) characterized. Defined reporter sequences are needed to monitor for changes, and a sensitive or high throughput assay system must be devised in expectation that germ line rearrangement events are of very low frequency.

If the genomic DNA in germ cells of any animal expressing RAG proteins in gonads can be closely tracked, one can try to ascertain the frequency of RAG-induced recombination per germ cell. Does RAG act at all germ cell stages? If so, could RAG-induced changes accumulate? Is all chromatin equally susceptible to RAG? In studying RAG action on germ line DNA, its role in antigen receptor evolution can be investigated: as provider of parts and mechanism of a recombining gene system, sculptor of gene organization, and mutator of genetic material to inactivation or transformation.

## ACKNOWLEDGMENTS

We would like to thank the following individuals for answering many and varied questions: Jonathan Rast, Kate Buckley, Rob Miller, Maja Nowakowski, Zahra Zakeri, Chris Amemiya. The authors also thank Jürgen Brosius for reviewing the manuscript and Louis Du Pasquier for his comments and advice throughout. This work was supported in part by the National Institutes of Health (R01-GM068095).

## REFERENCES

[1] Gould SJ, Vrba ES. Exaptation – a missing term in the science of form. Paleobiology 1982;8:4–15.

[2] Sakano H, Hüppi K, Heinrich G, Tonegawa S. Sequences at the somatic recombination sites of immunoglobulin light-chain genes. Nature 1979;280:288–94.

[3] Siu G, Clark SP, Yoshikai Y, Malissen M, Yanagi Y, Strauss E, et al. The human T cell antigen receptor is encoded by variable, diversity, and joining gene segments that rearrange to generate a complete V gene. Cell 1984;37:393–401.

[4] Flajnik MF, Du Pasquier L. Evolution of the immune system. In: Paul WE, editor. Fundamental immunology. 7th ed. Philadelphia (PA): Wollers Kluwer Health/Lippincott Williams and Wilkins; 2013. pp. 57–124.

[5] Litman GW, Rast JP, Fugmann SD. The origins of vertebrate adaptive immunity. Nat Rev Immunol 2010;10:543–53.

[6] Fugmann SD. The origins of the RAG genes-From transposition to V(D)J recombination. Semin Immunol 2010;22:10–6.

[7] Aziz RK, Breitbart M, Edwards RA. Transposases are the most abundant, most ubiquitous genes in nature. Nucleic Acids Res 2010;38:4207–17.

[8] Abrusan G, Zhang Y, Szilagyi A. Structure prediction and analysis of DNA transposon and LINE retrotransposon proteins. J Biol Chem 2013;288:16127–38.

[9] Ladeveze V, Chaminade N, Lemeunier F, Periquet G, Aulard S. General survey of hAT transposon superfamily with highlight on hobo element in Drosophila. Genetica 2012;140:375–92.

[10] Yuan YW, Wessler SR. The catalytic domain of all eukaryotic cut-and-paste transposase superfamilies. Proc Natl Acad Sci USA 2011;108:7884–9.

[11] Hickman AB, Chandler M, Dyda F. Integrating prokaryotes and eukaryotes: DNA transposases in light of structure. Crit Rev Biochem Mol Biol 2010;45:50–69.

[12] Montano SP, Rice PA. Moving DNA around: DNA transposition and retroviral integration. Curr Opin Struct Biol 2011;21:370–8.

[13] Dyda F, Chandler M, Hickman AB. The emerging diversity of transpososome architectures. Q Rev Biophys 2012;45:493–521.

[14] Bushman F. Lateral DNA transfer. Cold Spring Harbor: Cold Spring Harbor Laboratory Press; 2002.

[15] Wijayawardena BK, Minchella DJ, DeWoody JA. Hosts, parasites, and horizontal gene transfer. Trends Parasitol 2013;29:329–38.

[16] Wallau GL, Ortiz MF, Loreto ELS. Horizontal transposon transfer in Eukarya: detection, bias, and perspectives. Genome Biol Evol 2012;4:801–11.

[17] Bernstein RM, Schluter SF, Bernstein H, Marchalonis JJ. Primordial emergence of the recombination activating gene 1 (RAG1): sequence of the complete shark gene indicates homology to microbial integrases. Proc Natl Acad Sci USA 1996;93:9454–9.

[18] Shen XX, Liang D, Feng YJ, Chen MY, Zhang P. A versatile and highly efficient toolkit including 102 nuclear markers for vertebrate phylogenomics, tested by resolving the higher level relationships of the caudata. Mol Biol Evol 2013;30:2235–48.
[19] Oliver KR, Greene WK. Transposable elements: powerful facilitators of evolution. BioEssays 2009;31:703–14.
[20] Schatz DG, Oettinger MA, Baltimore D. The V(D)J recombination activating gene, RAG-1. Cell 1989;59:1035–48.
[21] Oettinger MA, Schatz DG, Gorka C, Baltimore D. Rag-1 and Rag-2, adjacent genes that synergistically activate V(D)J recombination. Science 1990;248:1517–23.
[22] McBlane JF, van Gent DC, Ramsden DA, Romeo C, Cuomo CA, Gellert M, et al. Cleavage at a V(D)J recombination signal requires only RAG1 and RAG2 proteins and occurs in two steps. Cell 1995;83:387–95.
[23] Eastman QM, Leu TMJ, Schatz DG. Initiation of V(D)J recombination in vitro obeying the 12/23 rule. Nature 1996;380:85–8.
[24] Cortes P, Weis-Garcia R, Misulovin Z, Nussenzweig A, Lai J-S, Li G, et al. In vitro V(D)J recombination: signal joint formation. Proc Natl Acad Sci USA 1996;93:14008–13.
[25] van Gent DC, Mizuuchi D, Gellert M. Similarities between initiation of V(D)J recombination and retroviral integration. Science 1996;271:1592–4.
[26] Zhou L, Mitra R, Atkinson PW, Hickman AB, Dyda F, Craig NL. Transposition of hAT elements links transposable elements and V(D)J recombination. Nature 2004;432:995–1001.
[27] Hencken CG, Li XH, Craig NL. Functional characterization of an active Rag-like transposase. Nat Struct Mol Biol 2012;19:834–6.
[28] Landree MA, Wibbenmeyer JA, Roth DB. Mutational analysis of RAG1 and RAG2 identifies three catalytic amino acids in RAG1 critical for both cleavage steps of V(D)J recombination. Genes Dev 1999;13:3059–69.
[29] Kim DR, Dai Y, Mundy CL, Yang W, Oettinger MA. Mutations of acidic residues in RAG1 define the active site of the V(D)J recombinase. Genes Dev 1999;13:3070–80.
[30] Fugmann SD, Villey IJ, Ptaszek LM, Schatz DG. Identification of two catalytic residues in RAG1 that define a single active site within the RAG1/RAG2 protein complex. Mol Cell 2000;5:97–107.
[31] Agrawal A, Eastman QM, Schatz DG. Transposition mediated by RAG1 and RAG2 and its implications for the evolution of the immune system. Nature 1998;394:744–51.
[32] Hiom K, Melek M, Gellert M. DNA transposition by the RAG1 and RAG2 proteins: a possible source of oncogenic translocations. Cell 1998;94:463–70.
[33] Clatworthy AE, Valencia MA, Haber JE, Oettinger MA. V(D)J recombination and RAG-mediated transposition in yeast. Mol Cell 2003;12:489–99.
[34] Ramsden DA, Weed BD, Reddy YV. V(D)J recombination: born to be wild. Semin Cancer Biol 2010;20:254–60.
[35] Matthews AGW, Oettinger MA. Regulation of RAG transposition. In: Ferrier P, editor. V(D)J recombination, ;vol 650; 2009. pp. 16–31.
[36] Montano SP, Pigli YZ, Rice PA. The Mu transpososome structure sheds light on DDE recombinase evolution. Nature 2012;491:413–7.
[37] Schluter SF, Marchalonis JJ. Cloning of shark RAG2 and characterization of the RAG1/RAG2 gene locus. FASEB J 2003;17:470–2.
[38] Venkatesh B, Ning Y, Brenner S. Late changes in spliceosomal introns define clades in vertebrate evolution. Proc Natl Acad Sci USA 1999;96:10267–71.
[39] Oettinger MA. Cutting apart V(D)J recombination. Curr Opin Genet Dev 1996;6:141–5.
[40] Hong X, Scofield DG, Lynch M. Intron size, abundance, and distribution within untranslated regions of genes. Mol Biol Evol 2006;23:2392–404.
[41] Kapitonov VV, Jurka J. RAG1 core and V(D)J recombination signal sequences were derived from Transib transposons. PLoS Biol 2005;3:998–1011.
[42] Panchin Y, Moroz LL. Molluscan mobile elements similar to the vertebrate – recombination-activating genes. Biochem Biophys Res Commun 2008;369:818–23.
[43] Chen S, Li X. Molecular characterization of the first intact Transib transposon from *Helicoverpa zea*. Gene 2008;408:51–63.
[44] Sadofsky MJ, Hesse JE, McBlane JF, Gellert M. Expression and V(D)J recombination activity of mutated RAG-1 proteins. Nucleic Acids Res 1993;21:5644–50.
[45] Fugmann SD, Messier C, Novack LA, Cameron RA, Rast JP. An ancient evolutionary origin of the Rag1/2 gene locus. Proc Natl Acad Sci USA 2006;103:3728–33.
[46] Callebaut I, Mornon JP. The V(D)J recombination activating protein RAG2 consists of a six-bladed propeller and a PHD finger-like domain, as revealed by sequence analysis. Cell Mol Life Sci 1998;54:880–91.
[47] Corneo B, Moshous D, Callebaut I, de Chasseval R, Fischer A, de Villartay JP. Three-dimensional clustering of human RAG2 gene mutations in severe combined immune deficiency. J Biol Chem 2000;275:12672–5.
[48] Gomez CA, Ptaszek LM, Villa A, Bozzi F, Sobacchi C, Brooks EG, et al. Mutations in conserved regions of the predicted RAG2 kelch repeats block initiation of V(D)J recombination and result in primary immunodeficiencies. Mol Cell Biol 2000;20:5653–64.
[49] Jones JM, Simkus C. The roles of the RAG1 and RAG2 "non-core" regions in V(D)J recombination and lymphocyte development. Arch Immunol Ther Exp 2009;57:105–16.
[50] Wilson DR, Norton DD, Fugmann SD. The PHD domain of the sea urchin RAG2 homolog, SpRAG2L, recognizes dimethylated lysine 4 in histone H3 tails. Dev Comp Immunol 2008;32:1221–30.
[51] Benton MJ, Donoghue PC. Paleontological evidence to date the tree of life. Mol Biol Evol 2007;24:26–53.
[52] Erwin DH, Laflamme M, Tweedt SM, Sperling EA, Pisani D, Peterson KJ. The Cambrian conundrum: early divergence and later ecological success in the early history of animals. Science 2011;334:1091–7.
[53] Smith JJ, Kuraku S, Holt C, Sauka-Spengler T, Jiang N, Campbell MS, et al. Sequencing of the sea lamprey (*Petromyzon marinus*) genome provides insights into vertebrate evolution. Nat Genet 2013;45:415–21.
[54] Hibino T, Loza-Coll M, Messier C, Majeske AJ, Cohen AH, Terwilliger DP, et al. The immune gene repertoire encoded in the purple sea urchin genome. Dev Biol 2006;300:349–65.
[55] Guo P, Hirano M, Herrin BR, Li J, Yu C, Sadlonova A, et al. Dual nature of the adaptive immune system in lampreys. Nature 2009;459:796–801.
[56] Rogozin IB, Iyer LM, Liang L, Glazko GV, Liston VG, Pavlov YI, et al. Evolution and diversification of lamprey antigen receptors: evidence for involvement of an AID-APOBEC family cytosine deaminase. Nat Immunol 2007;8:647–56.
[57] Huang S, Yuan S, Guo L, Yu Y, Li J, Wu T, et al. Genomic analysis of the immune gene repertoire of amphioxus reveals extraordinary innate complexity and diversity. Genome Res 2008;18:1112–26.

[58] Azumi K, Sabau SV, Fujie M, Usami T, Koyanagi R, Kawashima T, et al. Gene expression profile during the life cycle of the urochordate Ciona intestinalis. Dev Biol 2006;308:572–82.

[59] Denoeud F, Henriet S, Mungpakdee S, Aury JM, Da Silva C, Brinkmann H, et al. Plasticity of animal genome architecture unmasked by rapid evolution of a pelagic tunicate. Science 2010;330:1381–5.

[60] Dong ML, Fu YG, Yu CL, Su J, Huang SF, Wu XK, et al. Identification and characterisation of a homolog of an activation gene for the recombination activating gene 1 (RAG 1) in amphioxus. Fish Shellfish Immun 2005;19:165–74.

[61] Holland LZ, Albalat R, Azumi K, Benito-Gutierrez E, Blow MJ, Bronner-Fraser M, et al. The amphioxus genome illuminates vertebrate origins and cephalochordate biology. Genome Res 2008;18:1100–11.

[62] Azumi K, De Santis R, De Tomaso A, Rigoutsos I, Yoshizaki F, Pinto MR, et al. Genomic analysis of immunity in a Urochordate and the emergence of the vertebrate immune system: "waiting for Godot". Immunogenetics 2003;55:570–81.

[63] Kokubu F, Litman R, Shamblott MJ, Hinds K, Litman GW. Diverse organization of immunoglobulin VH gene loci in a primitive vertebrate. EMBO J 1988;11:3413–22.

[64] Bailey JA, Gu Z, Clark RA, Reinert K, Samonte RV, Schwartz S, et al. Recent segmental duplications in the human genome. Science 2002;297:1003–7.

[65] Matsuda F, Shin EK, Hirabayashi Y, Nagaoka H, Yochida MC, Zong SQ, et al. Organization of variable region segments of the human immunoglobulin heavy chain: duplication of the D5 cluster within the locus and interchromosomal translocation of variable region segments. EMBO J 1990;9:2501–6.

[66] Litman GW, Anderson MK, Rast JP. Evolution of antigen binding receptors. Annu Rev Immunol 1999;17:109–47.

[67] Hsu E, Pulham N, Rumfelt LL, Flajnik MF. The plasticity of immunoglobulin gene systems in evolution. Immunol Rev 2006;210: 8–26.

[68] Yoder JA, Litman GW. Immune-type diversity in the absence of somatic rearrangement. Curr Top Microbiol 2000;248:271–82.

[69] Lee SS, Fitch D, Flajnik MF, Hsu E. Rearrangement of immunoglobulin genes in shark germ cells. J Exp Med 2000;191:1637–48.

[70] Rumfelt LL, Avila D, Diaz M, Bartl S, McKinney EC, Flajnik MF. A shark antibody heavy chain encoded by a nonsomatically rearranged VDJ is preferentially expressed in early development and is convergent with mammalian IgG. Proc Natl Acad Sci USA 2001;98:1775–80.

[71] Edholm ES, Bengtén E, Wilson M. Insights into the function of IgD. Dev Comp Immunol 2011;35:1309–16.

[72] Ghaffari SH, Lobb CJ. Structure and genomic organization of a second cluster of immunoglobulin heavy chain gene segments in the channel catfish. J Immunol 1999;162:1519–29.

[73] Edholm E-S, Bengtén E, Stafford JL, Sahoo M, Taylor EB, Miller NW, et al. Identification of two IgD(+) B cell populations in channel catfish, *Ictalurus punctatus*. J Immunol 2010;185:4082–94.

[74] Chen K, Xu W, Wilson M, He B, Miller NW, Bengtén E, et al. Immunoglobulin D enhances immune surveillance by activating antimicrobial, proinflammatory and B cell-stimulating programs in basophils. Nat Immunol 2009;10:889–98.

[75] Wang X, Miller RD. Recombination, transcription, and diversity of a partially germline-joined VH in a mammal. Immunogenetics 2012;64:713–7.

[76] Reynaud CA, Dahan A, Anquez V, Weill JC. Somatic hyperconversion diversifies the single Vh gene of the chicken with a high incidence in the D region. Cell 1989;59:171–83.

[77] Reynaud C-A, Bertocci B, Dahan A, Weill J-C. Formation of the chicken B-cell repertoire: ontogenesis, regulation of Ig gene rearrangement, and diversification by gene conversion. Adv Immunol 1994;57:353–78.

[78] Lewis S, Hesse JE, Mizuuchi K, Gellert M. Novel strand exchanges in V(D)J recombination. Cell 1988;55:1099–107.

[79] Lieber MR, Hesse JE, Mizuuchi K, Gellert M. Lymphoid V(D)J recombination: nucleotide insertion at signal joints as well as coding joints. Proc Natl Acad Sci USA 1988;85:8588–92.

[80] Morzycka-Wroblewska E, Lee FEH, Desiderio SV. Unusual immunoglobulin gene rearrangement leads to replacement of recombinational signal sequences. Science 1988;242:261–3.

[81] Sollbach AE, Wu GE. Inversions produced during V(D)J rearrangement at IgH, the immunoglobulin heavy-chain locus. Mol Cell Biol 1995;15:671–81.

[82] VanDyk L, Meek K. Assembly of IgH CDR3: mechanism, regulation, and influence on antibody diversity. Int Rev Immunol 1992;8:123–33.

[83] Harpaz Y, Chothia C. Many of the immunoglobulin superfamily domains in cell adhesion molecules and surface receptors belong to a new structural set which is close to that containing variable domains. J Mol Biol 1994;238:528–39.

[84] Williams AF, Barclay AN. The immunoglobulin superfamily–domains for cell surface recognition. Annu Rev Immunol 1988;6:381–405.

[85] Barclay AN, Birkeland ML, Brown MH, Beyers AD, Davis SJ, Somoza C, et al. The leucocyte antigen facts book. London: Academic Press, Harcourt Brace Jovanovich; 1993.

[86] Du Pasquier L, Zucchetti I, De Santis R. Immunoglobulin superfamily receptors in protochordates: before RAG time. Immunol Rev 2004;198:233–48.

[87] Teichmann SA, Chothia C. Immunoglobulin superfamily proteins in *Caenorhabditis elegans*. J Mol Biol 2000;296:1367–83.

[88] Seeger MA, Haffley L, Kaufman TC. Characterization of amalgam: a member of the immunoglobulin superfamily from Drosophila. Cell 1988;55:589–600.

[89] Flajnik MF, Kasahara M. Comparative genomics of the MHC: glimpses into the evolution of the adaptive immune system. Immunity 2001;15:351–62.

[90] Ota T, Nei M. Divergent evolution and evolution by the birth-and-death process in the immunoglobulin VH gene family. Mol Biol Evol 1994;11:469–82.

[91] Lieber MR. The mechanism of double-strand DNA break repair by the nonhomologous DNA end-joining pathway. Annu Rev Biochem. 2010;79:181–211.

[92] Goossens T, Klein U, Kuppers R. Frequent occurrence of deletions and duplications during somatic hypermutation: implications for oncogene translocations and heavy chain disease. Proc Natl Acad Sci USA 1998;95:2463–8.

[93] Wilson PC, de Bouteiller O, Liu YJ, Potter K, Banchereau J, Capra JD, et al. Somatic hypermutation introduces insertions and deletions into immunoglobulin V genes. J Exp Med 1998;187:59–70.

[94] de Wildt RM, van Venrooij WJ, Winter G, Hoet RM, Tomlinson IM. Somatic insertions and deletions shape the human antibody repertoire. J Mol Biol 1999;294:701–10.

[95] Lee SS, Tranchina D, Ohta Y, Flajnik MF, Hsu E. Hypermutation in shark immunoglobulin light chain genes results in contiguous substitutions. Immunity 2002;16:571–82.

[96] Gilfillan S, Benoist C, Mathis D. Mice lacking terminal deoxynucleotidyl transferase: adult mice with a fetal antigen receptor repertoire. Immunol Rev 1995;148:201–19.

[97] Aoufouchi S, Flatter E, Dahan A, Faili A, Bertocci B, Storck S, et al. Two novel human and mouse DNA polymerases of the polX family. Nucleic Acids Res 2000;28:3684–93.

[98] Bartl S, Miracle AL, Rumfelt LL, Kepler TB, Mochon E, Litman GW, et al. Terminal deoxynucleotidyl transferases from elasmobranchs reveal structural conservation within vertebrates. Immunogenetics 2003;55:594–604.

[99] Near TJ. Conflict and resolution between phylogenies inferred from molecular and phenotypic data sets for hagfish, lampreys, and gnathostomes. J Exp Zool B Mol Dev Evol 2009;312B:749–61.

[100] Hirano M, Guo P, McCurley N, Schorpp M, Das S, Boehm T, et al. Evolutionary implications of a third lymphocyte lineage in lampreys. Nature 2013;501:435–9.

[101] Hsu E. The invention of lymphocytes. Curr Opin Immunol 2011;23:1–7.

[102] Janvier P. Modern look for ancient lamprey. Nature 2006;443:921–4.

[103] Flajnik MF. Comparative analyses of immunoglobulin genes: surprises and portents. Nat Rev Immunol 2002;2:688–98.

[104] Muramatsu M, Kinoshita K, Fagarasan S, Yamada S, Shinkai Y, Honjo T. Class switch recombination and hypermutation require activation-induced cytidine deaminase (AID), a potential RNA editing enzyme. Cell 2000;102:553–63.

[105] Di Noia JM, Neuberger MS. Molecular mechanisms of antibody somatic hypermutation. Annu Rev Biochem 2007;76:1–22.

[106] Conticello SG, Thomas CJ, Petersen-Mahrt SK, Neuberger MS. Evolution of the AID/APOBEC family of polynucleotide (deoxy) cytidine deaminases. Mol Biol Evol 2005;22:367–77.

[107] Wilson M, Hsu E, Marcuz A, Courtet M, Du Pasquier L, Steinberg C. What limits affinity maturation of antibodies in *Xenopus*–the rate of somatic mutation or the ability to select mutants? EMBO J 1992;11:4337–47.

[108] Zhu C, Hsu E. Error-prone DNA repair activity during somatic hypermutation in shark B lymphocytes. J Immunol 2010;185: 5336–59.

[109] Saribasak H, Maul RW, Cao Z, Yang WW, Schenten D, Kracker S, et al. DNA polymerase ζ generates tandem mutations in immunoglobulin variable regions. J Exp Med 2012;209:1075–81.

[110] Daly J, Bebenek K, Watt DL, Richter K, Jiang C, Zhao ML, et al. Altered Ig hypermutation pattern and frequency in complementary mouse models of DNA polymerase ζ activity. J Immunol 2012;18:5528–37.

[111] Pancer Z, Mayer WE, Klein J, Cooper MD. Prototypic T cell receptor and CD4-like coreceptor are expressed by lymphocytes in the agnathan sea lamprey. Proc Natl Acad Sci USA 2004;101:13273–8.

[112] Cannon JP, Haire RN, Pancer Z, Mueller MG, Skapura D, Cooper MD, et al. Variable domains and a VpreB-like molecule are present in a jawless vertebrate. Immunogenetics 2005;56:924–9.

[113] Suzuki T, Shin-I T, Fujiyama A, Kohara Y, Kasahara M. Hagfish leukocytes express a paired receptor family with a variable domain resembling those of antigen receptors. J Immunol 2005;174:2885–91.

[114] Haruta C, Suzuki T, Kasahara M. Variable domains in hagfish: NICIR is a polymorphic multigene family expressed preferentially in leukocytes and is related to lamprey TCR-like. Immunogenetics 2006;58:216–25.

[115] Zucchetti I, De Santis R, Grusea S, Pontarotti P, Du Pasquier L. Origin and evolution of the vertebrate leukocyte receptors: the lesson from tunicates. Immunogenetics 2009;61:463–81.

[116] Cannon JP, Haire RN, Litman GW. Identification of diversified genes that contain immunoglobulin-like variable regions in a protochordate. Nat Immunol 2002;3:1200–7.

[117] Zhang SM, Adema CM, Kepler TB, Loker ES. Diversification of Ig superfamily genes in an invertebrate. Science 2004;305:251–4.

[118] Hanington PC, Forys MA, Dragoo JW, Zhang SM, Adema CM, Loker ES. Role for a somatically diversified lectin in resistance of an invertebrate to parasite infection. Proc Natl Acad Sci USA 2010;107:21087–92.

Chapter 10

# Structure and Signaling Function of the B-Cell Antigen Receptor and Its Coreceptors

Michael Reth[1,2,3], Lars Nitschke[4], Masaki Hikida[5] and Tomohiro Kurosaki[6,7]

[1]*BIOSS Centre of Biological Signaling Studies, University of Freiburg, Freiburg im Breisgau, Germany,* [2]*Faculty of Biology, Institute of Biology III, University of Freiburg, Freiburg im Breisgau, Germany,* [3]*MPI for Immunobiology and Epigenetics, Freiburg im Breisgau, Germany,* [4]*Division of Genetics, Department of Biology, University of Erlangen-Nürnberg, Erlangen, Germany,* [5]*Center for Innovation in Immunoregulative Technology and Therapeutics, Graduate School of Medicine, Kyoto University, Kyoto, Japan,* [6]*Laboratory for Lymphocyte Differentiation, Center for Integrative Medical Sciences (IMS), RIKEN, Yokohama, Japan,* [7]*Laboratory for Lymphocyte Differentiation, WPI Immunology Frontier Research Center, Osaka University, Osaka, Japan*

## 1. INTRODUCTION

The B-cell antigen receptor (BCR) plays a central role in the clonal selection of B cells and the generation of humoral immunity. In addition from inducing a specific immune response, the expression of the BCR also is required for the proper development and maintenance of the B-cell pool in the periphery. Each mature B cell carries roughly 120,000 BCR complexes on the cell surface with identical antigen binding sites. Because of the highly variable immunoglobulin (Ig) gene rearrangement program, B cells differ from each other in the antigen-binding specificity of their BCR. It has been estimated that most of the $10^{11}$ B cells of the human immune system differ from each other in their exact binding specificity. The enormous binding variability of their antigen receptors are characteristic of both B and T cells. However, unlike T cells, whose antigen receptors are restricted in binding to a defined antigen structure, namely the major histocompatibility complex (MHC)/peptide complex, B cells can be activated by a large group of structurally highly diverse molecules. In this respect, the BCR is quite different from most other receptors on the cell surface that have only one or a few specific ligands. Any model of B-cell activation should take into account these unique properties of the BCR. In this chapter we describe the basic structure of the BCR and several proposed models for its activation. We also discuss the interaction of the BCR with signal-transducing kinases, with a focus on the Src-family kinase Lyn and the spleen tyrosine kinase (Syk). We then address the structure and function of the B-cell coreceptors that either amplify (CD19) or inhibit (CD22) BCR signaling. Once B cells are activated, signals from BCR are processed through several signaling pathways. In our discussion of these processes, we focus on the phospholipase (PL) C-γ2 calcium release machinery and the phosphoinositide 3-kinase (PI3K) signaling pathway controlling proliferation and survival of B cells. Together, these pathways drive the clonal expansion of B cells and their differentiation into antibody-producing plasma cells.

## 2. BASIC STRUCTURE OF THE BCR COMPLEX

In his famous Croonian lecture in 1900, Paul Ehrlich postulated the existence of an antigen receptor as part of his side chain theory. It was only in 1960 that membrane-bound antibodies and in 1989 that the BCR signaling subunits were discovered (Figure 1(A)). The BCRs of all major Ig classes (IgM, IgD, IgG, IgE, and IgA) have a similar structural design [1,2]. They consist of the membrane-bound Ig (mIg) molecule and the Igα/Igβ (CD79a/CD79b) heterodimer, which mediate antigen binding and signaling, respectively [3,4]. The mIg molecule is a disulfide-linked tetramer consisting of two identical heavy chains (HCs) and two identical light chains (LCs). The two components of the Igα/Igβ heterodimer share many structural features. Both proteins carry a glycosylated extracellular Ig domain, a linker region with a heterodimer-forming cysteine, an evolutionarily highly conserved transmembrane (TM), and a cytoplasmic tail sequence of either 61 (Igα) or 48 (Igβ) amino acids. While Igα is covalently bound to Igβ via a disulfide bridge,

Molecular Biology of B Cells. http://dx.doi.org/10.1016/B978-0-12-397933-9.00010-2

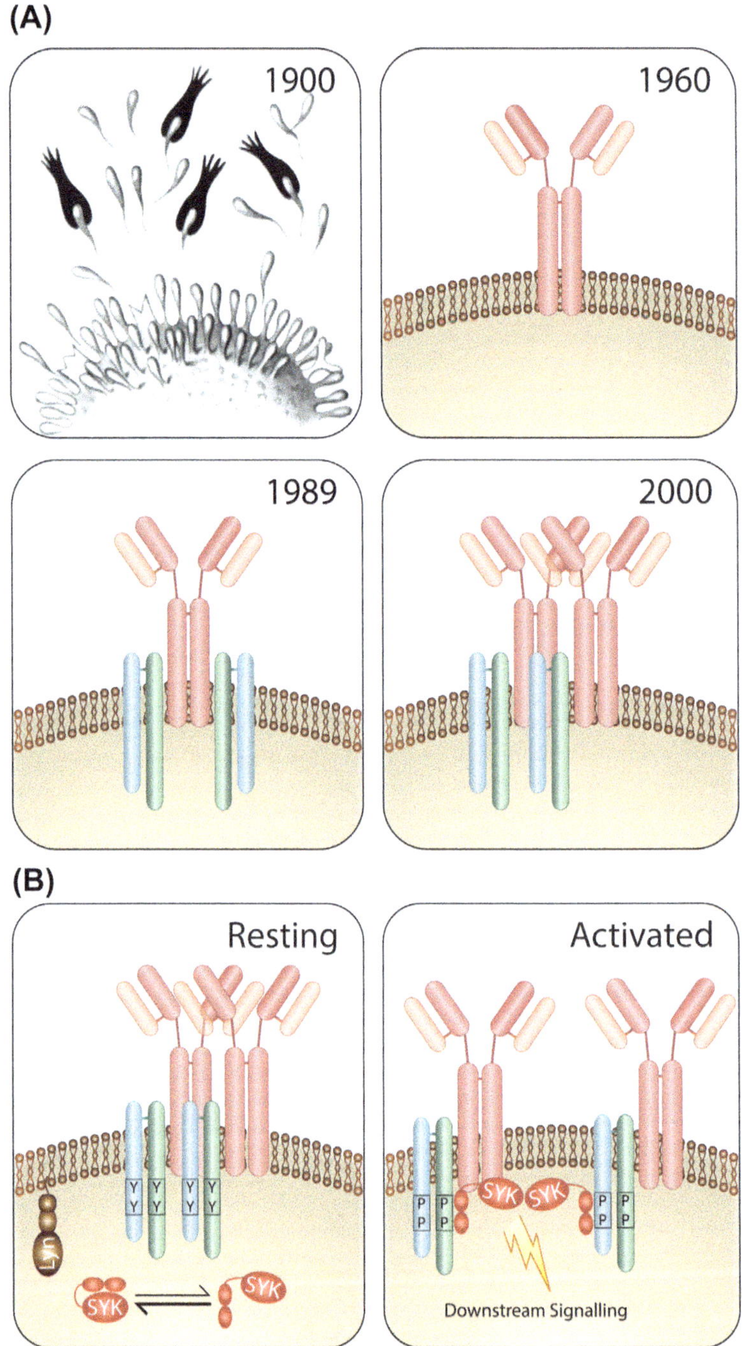


FIGURE 1 **Basic B-cell receptor (BCR) structure and activation process.** (A) History of the BCR models. In 1900, Paul Ehrlich proposed in his side chain theory the first receptor-ligand model explaining the specific activation of cells for antibody production. That B cells carry membrane-bound antibody molecules on their cell surface was discovered in 1960, and immunoglobulin (Ig) α and Igβ, the signaling subunits of the BCR, were found in 1989. The original model of the BCR suggested a symmetric 1:2 interaction between the two heavy chains of membrane-bound Ig molecule and the Igα/Igβ heterodimer. Only in 2000 was the the 1:1 interaction between mIg and Igα/Igβ shown, and it was discovered that this basic BCR complex forms higher oligomers. (B) Alterations of the BCR conformation according to the dissociation activation model hypothesis. In the resting state, the BCR forms a tight oligomer so that immunoreceptor tyrosine-based activation motif (ITAM) sequences of Igα/Igβ can not readily interact with the kinase Syk, which exists in an equilibrium between a closed autoinhibitory and an open active conformation. The Src-family kinase Lyn does not primarily act as an ITAM kinase; rather, upon BCR ligation it modifies the cytoskeleton, thus making the receptor more accessible to Syk. The BCR can be opened and activated in two different ways: either from the outside via binding by an antigen or from the inside by the phosphorylation of the ITAM tyrosines by Syk and by binding of Syk to the receptor.

the Igα/Igβ heterodimer is noncovalently associated with the mIg molecule in a 1:1 complex [5]. Thus the purification of the whole BCR complex (IgM-BCR = 440 kDa) requires special detergents such as digitonin. Other detergents such as Triton-X 100 can, depending on their concentration, disrupt the BCR complex [6]. The specific interaction of the mIg molecule with the Igα/Igβ heterodimer is mediated by the extracellular membrane–proximal CH domain and the TM region of MHC. Specifically, a tyrosine and a serine residue situated on one side of the TM region of the MHC is implicated in Igα/Igβ binding [7]. However, the whole TM sequence of mIgM is evolutionarily highly conserved [8], suggesting that it is involved in other interactions. The cytoplasmic sequences of Igα and Igβ contain a highly conserved consensus sequence D/E$x_7$D/ExxYxxL$x_7$YxxL/I, called the immunoreceptor tyrosine-based activation motif (ITAM) [9]. This motif is found in most signaling subunits of the multicomponent immune receptor family [10,11] and in proteins of viruses causing B-cell tumors [9]. The ITAM motif is required for the interaction of the BCR with downstream signaling elements [12]. The two tyrosines in the ITAM sequence have a dual function in BCR activation: they serve as substrates and as binding partners for protein tyrosine kinases [13,14]. Each BCR complex has two ITAM sequences—one in Igα and the other in the Igβ tail—with partially redundant functions. Thus a BCR with a truncation of the tail or an ITAM mutation in only one of the two signaling subunits is still able to signal and can promote B-cell activation [15,16]. By contrast, a BCR carrying ITAM mutations at both the Igα and Igβ component is completely signaling inert [17]. In addition to the ITAM sequence, most amino acids of the intracellular tails of Igα and Igβ are evolutionarily conserved, suggesting the presence of additional functions. While we lack structural information about these tails, a recent study suggests that the confirmation of the tails is altered during BCR activation [18].

## 3. BCR ACTIVATION MODELS

Currently, there are two conflicting models for BCR activation. These models differ only in the organization of the BCR monomers on resting B cells (Figure 2). Here, we call the basic 440-kDa BCR complex, comprising the Ig HCs and LCs and the Igα/Igβ heterodimer, the BCR monomer. The crosslinking hypothesis suggests that the numerous BCR complexes on the resting B-cell surface are monomers and that it is the crosslinking of two monomers that initiates the BCR activation process [19]. According to this model, the crosslinking of the BCR place BCR-associated kinases in close proximity to each other, allowing their crosswise phosphorylation and activation. This model was proposed after it was found that only dimeric F$(ab)_2$ but not monomeric Fab fragments of anti-BCR antibodies could stimulate the BCR [20]. Furthermore, Förster resonance energy transfer and live-cell imaging studies were interpreted to support BCR crosslinking [21,22]. Upon closer inspection, the crosslinking hypothesis has several problems: First, many model antigens such as ovalbumin (Ova) or hen egg lysozyme are monomeric and thus cannot crosslink the BCR, but they are able to efficiently activate the receptor [23]. This is also true for the binding of a small Ova peptide to B cells expressing an Ova-specific BCR, whereas free haptens, which also bind, cannot stimulate the BCR [24,25]. Second, given the fact that B-cell antigens are a structurally highly diverse group of molecules, it is hard to envision that these antigens can always bring two BCR monomers together in a way that supports the signaling process [26]. Third, the notion that the many BCR complexes are monomeric on the surface of a living B cell has never been rigorously proven.

An alternative model of BCR activation is the dissociation activation model (DAM) [27]. This hypothesis is based on the finding that the BCR, when solubilized under conditions of lysis with a mild detergent, is detected on native gels as an oligomer [5]. The model of an oligomeric structure of the resting BCR was further supported by a quantitative bifluorescence complementation assay showing that the BCR spontaneously forms dimers on the membrane of living cells [28] (Figure 1(A)). Furthermore, a BCR mutant that was deficient in dimer formation was identified with this assay. Interestingly, when expressed on B cells, this monomeric BCR mutant is hyperactive and rapidly internalized, suggesting that the formation of closed BCR dimers or oligomers is required for the stable expression of the BCR on the cell surface. The resting state of the BCR may be further regulated by its association with cytoskeleton elements, shielding the ITAM-containing signaling tail of the receptor from BCR-interacting kinases [1]. BCR oligomerization and confinement may thus be an important mechanism to keep the many antigen receptors on the surface of resting B cells silent and to regulate the threshold of BCR activation. The binding of an antigen, be it monomeric or polymeric, with high enough affinity may disturb and alter this preorganized BCR structure. Opening the closed BCR oligomer could increase the accessibility of the ITAM sequence, thus allowing interaction with BCR signal-transducing kinases (Figure 1(B)). The advantage of this model is that it explains why the receptor can be activated by so many structurally diverse ligands [27]. The dissociation of a preordered BCR oligomer only requires the binding of an antigen to the receptor with high enough affinity and not to a particular structure of the antigen. However, polymeric antigens can disturb and open an oligomeric BCR more efficiently than monomeric antigens; the same is true for antireceptor antibodies. This can explain the original observation that only F$(ab)_2$—and not monomeric Fab fragments—of anti-BCR antibodies were able to activate B cells [20].

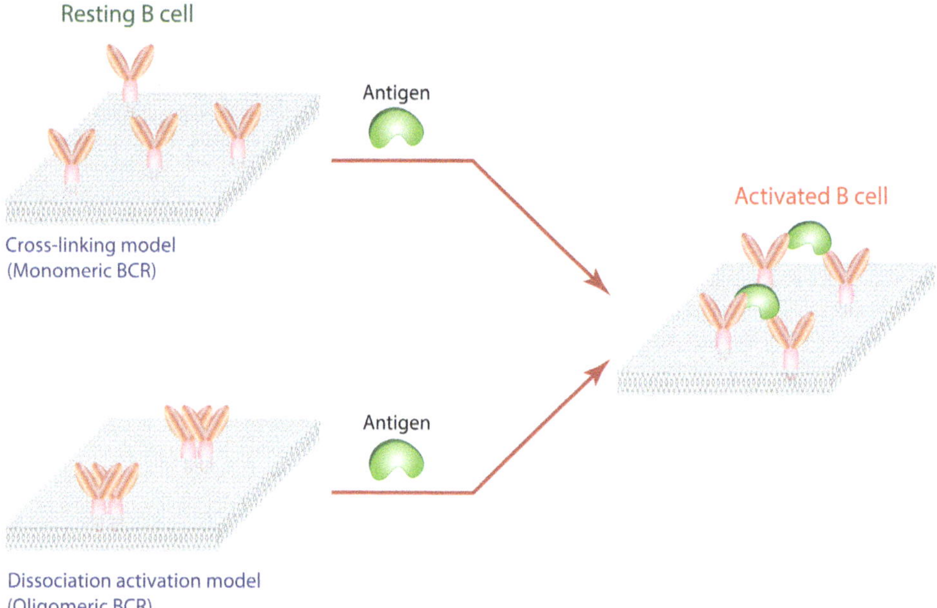


FIGURE 2 **Two models of B-cell activation.** These models differ only in the conformation of the B-cell receptor (BCR) on the surface of resting B cells. According to the crosslinking model, the BCRs on resting cells are randomly distributed monomers that are activated upon antigen binding by crosslinking. According to the dissociation activation model, the BCRs are preorganized in oligomers and reside, presumably with other receptors, in highly organized nanoscale membrane structures. The binding of antigen to the BCR results in the disruption of this organization as well as the opening and activation of the receptor.

An important difference between the two alternative models of BCR activation is the spacing of the BCR monomers before and after B-cell activation. According to the crosslinking model, the BCR monomers move closer together, whereas the DAM hypothesis predicts that they move apart from each other (Figure 2). Unfortunately, these relative BCR movements involve 10–30 nm distances and thus cannot directly be monitored by light microscopy because of the diffraction barrier of visible light around 250 nm [29]. A proximity ligation assay (PLA) has recently been developed that can monitor the nanoscale proximity between two proteins [30]. This assay uses DNA oligonucleotide-coupled anti-Ig Fab fragments (Fab-PLA) [31]. An IgM:IgM or IgD:IgD Fab-PLA study showed that BCR monomers move apart from each other upon B-cell activation, thus supporting the DAM hypothesis. These results are not in line with the predictions of the crosslinking model nor its recently proposed variation called the conformation-induced oligomerization model [32]. Furthermore, this PLA assay revealed that BCR-associated kinases play a more direct role in BCR activation than has been assumed so far.

## 4. THE RESTING STATE OF THE BCR

A resting B cells carries up to 120,000 identical copies of the BCR on its surface. How so many receptors are regulated so that they remain silent unless exposed to a cognate antigen is not clear. At present, we know more about the function and interactions of the activated BCR than of the resting BCR. Without information on the organization of the BCR on the surface of resting B cells, our knowledge of the B-cell activation process will remain incomplete. This problem applies not only to the BCR but also to most receptors on the cell surface. The fluid mosaic model described by Singer and Nicholson [33] suggested free diffusion of membrane proteins in the plasma membrane lipid bilayer. Indeed, cell fusion experiments showed free mixing of surface markers [34]. This view was challenged by single-molecule tracking experiments showing that the cytoskeleton can restrict the free diffusion of membrane proteins [35,36]. Furthermore, it was found that the plasma membrane is not uniform but rather compartmentalized into membrane areas with different lipid compositions [37,38]. An electron microscopy study of the inner leaflet of the plasma membrane by Wilson et al. [39] showed that many membrane proteins are not uniformly distributed over the whole membrane but rather are clustered in nanoscale membrane areas. Lillemeier et al. [40,41] extended this study by demonstrating that most proteins on the plasma membrane of T lymphocytes are confined in so-called protein islands with a size of 80–150 nm. These authors also showed that the actin cytoskeleton is associated with these islands and that the inhibition of actin polymerization by latrunculin alters the abundance and size of these structures. A recent study of all membrane proteins in the yeast genome showed their confinement in nanoscale membrane areas [42,43]. A nanoscale compartmentalization thus seems to be an evolutionarily conserved feature of most membrane proteins.

In spite of these findings, there are still conflicting views of the organization of the BCR on resting B cells. Live-cell total internal reflection fluorescence microscopy (TIRFM) in conjunction with single-particle tracking analyses suggested that most BCRs on resting B cells are freely diffusing and highly mobile monomers [22]. According to this study, a reduction in the movement of the BCR is associated with increased signaling [44]. Another TIRFM study found that the cytoskeleton restricts the free diffusion of the BCR on resting B cells and that B-cell activation results in increased BCR mobility [45]. This conclusion is supported by the finding that inhibition of the actin cytoskeleton by latrunculin resulted in increased BCR mobility. One problem with these live-cell imaging studies is that this technique does not resolve structures below the diffraction limit of visible light (250nm) [29]. It is thus not clear whether these studies monitor the movement of single BCR monomers or rather that of 80- to 120-nm BCR-containing protein islands. Similarly, the BCR capping and BCR micro-cluster formation detected with the TIRFM technique after B-cell activation may be due to the concatenation of protein islands rather than to the aggregation of single BCR monomers. Indeed, a recent study using the direct stochastic optical reconstruction microscopy superresolution technique found that the BCR is organized in preformed nanoclusters on the surface of resting B cells, as previously suggested [27,46]. Interestingly, the direct stochastic optical reconstruction microscopy analysis showed a preformed nanocluster organization for both the IgM-BCR and IgD-BCR, but this study did not reveal whether these two receptor classes on the B-cell surface are colocalized. A recent PLA study found that the IgM-BCR and the IgD-BCR reside in different membrane patches on the surface of resting B cells [31]. Little is known about the function of the IgD-BCR, even though it is the most abundant antigen receptor on mature B cells. This BCR class is evolutionarily highly conserved and expressed on B cells in nearly all vertebrates [47,48]. The finding of a different nanoscale organization of the IgD-BCR may shed new light on the function of this mysterious receptor.

Signaling from the BCR can be activated not only from the outside by exposure of B cells to cognate antigens or antireceptor antibodies but also from the inside via the inhibition of actin polymerization by latrunculin [49]. This finding suggests that the actin cytoskeleton is involved in silencing the BCR on resting B cells. How cytoskeletal elements are attached to the resting BCR is so far not clear. A BCR diffusion study suggests that ezrin-radixin-moesin family members establish barriers for BCR diffusion in the membrane of resting B cells [50]. Whether the BCR is directly or only indirectly associated with these cytoskeletal proteins and what the exact composition of these barriers is require further study. One protein that is localized close to the resting BCR is the protein-arginine-methyl-transferase1 (PRMT1). Interestingly, the cytosolic tail of Igα carries a highly conserved arginine (R198), which is methylated by PRMT1 in resting B cells and becomes demethylated during B-cell activation [51]. R198 methylation is the first protein modification found to be associated with the resting BCR. The finding that R198 methylation inhibits signaling from the BCR is in line with the description of PRMT1 as a negative regulator at different cellular locations it the nucleus and cytosol [52]. PRMT1 may be part of a BCR silencing complex that controls the threshold of BCR activation in resting B cells. A detailed description of this postulated silencing complex may result in the better treatment of hyperactive B cells associated with autoimmune or tumor diseases.

## 5. INTERACTION OF THE BCR WITH SIGNAL-TRANSDUCING KINASES AND ADAPTORS

Two different protein tyrosine kinases (PTKs), namely the Src family PTK Lyn as well as Syk, are involved in the activation of the BCR. These two kinases interact with the BCR in different ways. In contrast to Lyn, which predominantly phosphorylates only the first ITAM tyrosine (Y182), Syk can efficiently phosphorylate both ITAM tyrosines (Y182 and Y193) of Igα [14,53,54]. Because only the double-phosphorylated ITAM is an efficient binding partner for the tandem SH2 domains of Syk, the formation of the BCR/Syk complex requires the kinase activity of Syk [55]. The current model of BCR activation entails a sequential interaction of the BCR, first with Src family kinases such as Lyn and then with Syk. This model is not fully supported by genetic data, namely the analysis of mice deficient in these kinases. While Syk-deficient mice display a complete block of B-cell development, mice lacking Lyn, the dominant Src family kinase in B cells, have only a mild B-cell developmental defect and develop autoimmunity [56,57]. The biochemical analysis of kinase-deficient B cells shows that Syk-negative B cells are BCR-signaling inert, while BCR signaling is only delayed and not prevented in Lyn-negative B cells [58,59]. The latter finding is currently explained by the redundancy of different Src family kinases. However, it has not been shown that other members of the Src family can efficiently replace Lyn as kinases interacting with the BCR. Using a BCR/Syk rebuilding approach as well as in vitro kinase experiments, it was shown that the activity of Syk is drastically increased in the presence of an ITAM-containing peptide, as long as Syk carries its tandem SH2 domains [60]. This result, together with an analysis of ITAM mutants, shows that the two ITAM tyrosines are not only a substrate of Syk but, once phosphorylated, are also a binding partner and an allosteric activator of this kinase. This dual interaction of Syk with the ITAM tyrosines suggests that signaling at the BCR is amplified by a positive BCR/Syk feedback. A recent biochemical study questioned the notion that Syk is an ITAM kinase [61]. However, this study used only Syk inhibitors and did not study Syk-deficient B cells. To determine the role of Syk in BCR activation, it is important to directly measure the recruitment of Syk to the BCR, in combination with an analysis of the nanoscale conformation of the BCR. This was achieved in a recent Fab-PLA study that showed that only kinase-active Syk, not kinase-dead Syk, is recruited to the BCR and that Syk can activate the BCR from the inside by phosphorylating and binding the ITAM sequences [31]. This inside-out signaling by Syk is most clearly seen in B cells treated with latrunculin. The exposure of wild-type B cells to this drug results in the immediate dissociation of the BCR oligomer, whereas the BCR conformation is not altered in Syk-deficient B cells. If Syk is the dominant ITAM kinase, what then is the role of Lyn in the BCR activation process? The treatment of Lyn-deficient B cells with latrunculin results in only a delay—not the complete inhibition—of BCR signaling, as was the case in Syk-deficient B cells. Thus, in the presence of Lyn, the BCR seems to be more responsive to the action of Syk. One way to explain this is that the major action of Lyn is to remove cytoskeletal elements or other negative regulators preventing the interaction of Syk with the BCR. Taken together, these data suggest the following scheme of B-cell activation. Binding of an antigen, be it monomeric or polymeric, results in a disturbance of the closed BCR oligomer. These physical movements are sensed by Lyn, which then phosphorylates and alters as yet ill-defined cytoskeletal elements attached to the BCR, thus making the ITAM available for phosphorylation and binding of Syk. As part of a BCR/Syk complex, Syk not only becomes fully active but also is placed at the inner leaflet of the plasma membrane so that it can rapidly phosphorylate the ITAMs of neighboring BCRs, resulting in further Syk recruitment and the amplification of the BCR signal (Figure 1(B)). In this way, only a few antigen molecules can activate a large number of BCR complexes.

A time course experiment shows that the number of BCR/Syk complexes declines 5 min after the start of B-cell activation. There are presumably several mechanisms that counteract Syk activity and limit the positive BCR/Syk signaling feedback. One is the phosphorylation of negative regulators such as CD22 and CD74, which carry an immunoreceptor tyrosine-based inhibition motif (ITIM). The phosphorylation of the tyrosine in this motif results in the recruitment and activation of the protein tyrosine phosphatase SHP-1 that can efficiently dephosphorylate ITAM tyrosine and counteract Syk activation. Another way is by the rapid internalization of the activated BCR [62]. In addition, Syk can switch its activity from tyrosine-phosphorylation to serine/threonine phosphorylation during the time course of B cell activation [63]. The latter activity results in serine/threonine phosphorylation of the Igα tail and inhibition of BCR signaling [64]. Any alteration in these negative regulatory processes can promote the development of autoimmune diseases or tumor growth.

## 6. BCR CORECEPTORS CD19 AND CD22

Signaling from the BCR is modulated by accessory TM molecules or coreceptors on the B-cell surface. This modulation of signaling is achieved by recruiting additional intracellular signaling molecules. Coreceptors can not only increase or decrease the BCR signaling strength but also change the quality of the induced signaling pathways. The effect of coreceptors on the BCR signaling pathways often is regulated by ligand binding. Important examples of B cell coreceptors are CD19 and CD22, which enhance and inhibit BCR signaling, respectively.

CD19 is associated with the complement receptor CD21 and connections via complement-bound antigens to the BCR to enhance BCR signaling. CD19 also is linked to the cytoskeleton via the associated tetraspanin CD81; this association is involved in BCR regulation [65]. CD22 carries inhibitory ITIM signaling motifs that recruit the tyrosine phosphatase SHP-1, leading to the inhibition of BCR signaling. Inhibition by CD22 is regulated by the binding of CD22 ligands, which are sialic acids attached to cellular glycoproteins. The presence or absence of these ligands regulates the association of CD22 to the BCR and thus influences the strength of inhibition [66]. Since the regulation of B-cell signaling strength is important for immune responses, it is not surprising that mutations in the CD19/CD21 or CD22 pathways have been linked to either human immune deficiencies or autoimmune diseases.

## 7. CD19 FUNCTIONS IN A COMPLEX WITH CD21 AND CD81

CD19 has been identified as part of a membrane protein complex on B cells with the complement receptor 2 or CD21 (or its alternatively spliced form CD35) and the tetraspanin CD81. The association of CD19 with CD21 is crucial for transducing signals triggered by combined binding of complement C3d-coupled antigens to the BCR and CD21 [67]. The association of CD19 with CD81 is important for CD19 surface expression and links the membrane protein complex to the cytoskeleton, as described later. The CD19/CD21/CD81 complex was discovered when it was observed that coligation of CD21 and the BCR can greatly enhance BCR signaling [68]. CD19, CD81, and the protein CD225 (Leu-13), which has an unknown function, could be coprecipitated with CD21. While CD21 has a short intracellular tail, the 230-amino acid-long intracellular tail of the associated CD19 was soon discovered to be the element responsible for signal transduction. This was strongly supported by the finding that crosslinking of CD19 with the BCR enhanced both B-cell $Ca^{2+}$ signaling as well as proliferation. Thus, C3d-coupled antigens can enhance signaling by coligating CD21 with the BCR and transmitting activation signals via CD19 [67].

CD19-deficiency in mice mainly affects peripheral B-cell populations. While conventional B2 cells developed normally, there was a defect in B1 cell and marginal zone B-cell populations. In contrast, transgenic mice overexpressing human CD19 had an enlarged B1 cell population [69,70]. The B1 cells depend on BCR signaling for maintenance, and therefore these results of CD19-deficient or CD19-overexpressing mice can be explained by changes in BCR signaling. Conventional B2 cells developed normally in CD19-deficient mice but had a survival disadvantage in competitive situations such as after adoptive transfer of mixed CD19–/– and wild-type B cells [71]. Upon immunization with thymus-dependent antigens, CD19–/– mice showed a defect in the generation of switched Ig responses associated with germinal center (GC) formation [69,70]. B-cell memory also was impaired. Vesicular stomatitis virus–infected CD19-deficient animals were able to form GCs but had a defect in memory responses and high titer antibody responses [72]. CD21-deficient mice show a similar but weaker defect in GC responses to thymus-dependent antigens [73]. This may indicate that corecognition of complement Cd3-opsonized antigen by CD21 and BCR plays a role in GC formation. However, not all soluble protein antigens contain attached Cd3, and therefore the GC defect of CD19–/– mice may also be explained by impaired BCR signaling during B-cell activation.

CD19 is expressed from the pro-B-cell developmental stage onward in the bone marrow, whereas CD21 is expressed at a later stage, when transitional B cells leave the bone marrow and migrate to the spleen [65]. Therefore, CD21-independent functions of CD19 exist, particularly in early B-cell development and selection. Impaired proliferation of cells at the transition from the pro-B-cell to small resting pre-B-cell stage has been found in CD19-deficient mice [74]. These mice also show impaired positive selection processes during the transition from immature to mature B cells. Whether CD21-independent functions of CD19 require the presence of an alternative CD19 ligand to regulate BCR signal augmentation is controversial. So far, there is no convincing evidence for the existence of a special CD19 ligand. However, there could be some constitutive association of CD19 with the BCR, as indicated, for instance, by experiments showing impaired proliferation of CD19–/– B cells after stimulation of the BCR with anti-IgM [69,75]. It is not known whether this association of CD19 with the BCR is direct or indirect, for example, by recruiting both receptors to a common membrane microdomain such as lipid rafts. A recent report showed that CD19 is found on the plasma membrane in nanoclusters that are regulated by the tetraspanin CD81 and the actin cytoskeleton [46]. B-cell activation caused by the disruption of the actin cytoskeleton was shown to require the expression of the BCR as well as CD19 and CD81. This study also suggests that IgM, IgD and CD19 are organized on resting B cells in different nanoclusters that are not altered during B-cell activation [46]. In contrast, a recent Fab-PLA study showed that CD19 is specifically colocalized with the IgD-BCR in resting B cells, whereas after B-cell activation, CD19 is found in association with the IgM-BCR [31]. This, together with earlier data, suggests a relocalization of CD19 during B-cell activation [76,77].

## 8. SIGNALING BY THE CD19 TAIL

The cytoplasmic tail of CD19 carries nine tyrosine residues that are well conserved between human and mouse, some of which are phosphorylated upon BCR crosslinking [65]. The phosphorylation of the CD19 tail is augmented when CD19 is coligated to the BCR. The importance of these tyrosines for signal transduction has been studied individually with mutant cell lines, and its physiological role has been studied using transgenic mice expressing CD19 with mutated tyrosines. Of these nine tyrosines, Y482 and Y513 seem to be particularly important because Y→F mutations at these positions prevent phosphorylation of the other seven tyrosines of the CD19 cytoplasmic tail [78–80] (see Figure 3). One study showed that Lyn is activated by binding to Y513 and subsequently phosphoryates Y482 and Y513 [81]. In contrast, another study reported normal CD19 tyrosine phosphorylation in Lyn–/– mice [82]. This suggests that another tyrosine kinase, presumably Syk, can compensate when Lyn is absent. However, Lyn-deficient mice show impaired phosphorylation of inhibitory receptors and develop autoimmunity [83], suggesting a predominantly negative regulating role of this Src kinase (Figures 4 and 5).

Transgenic mice with tandem Y→F mutations in the CD19 tail were created to examine the in vivo contribution of all the CD19 tyrosines. These transgenic mice were then crossed onto the CD19-deficient background. Mice with Y482F/Y513F mutations developed the same phenotype as CD19–/– mice, whereas mutations of the other tyrosines suggested they were dispensable for the in vivo function of CD19 [84,85]. It is not very likely that this phenotype can be explained by grossly impaired Lyn activation, since Lyn–/– and CD19–/– mice have quite opposite phenotypes [83]. Y482F and Y513 are both part of YxxM PI3K binding domains and CD19–/– mice show impaired PI3K signaling [75,86]. Since the deletion of the PI3K components p85a or p110d leads to B-cell phenotypes similar to those in CD19–/– mice, the main in vivo function of CD19 is likely the augmentation of PI3K signaling [87–90]. This is supported by data from PTEN–/– mice, in which a genetic deletion in the inositol phosphatase that directly reverses the

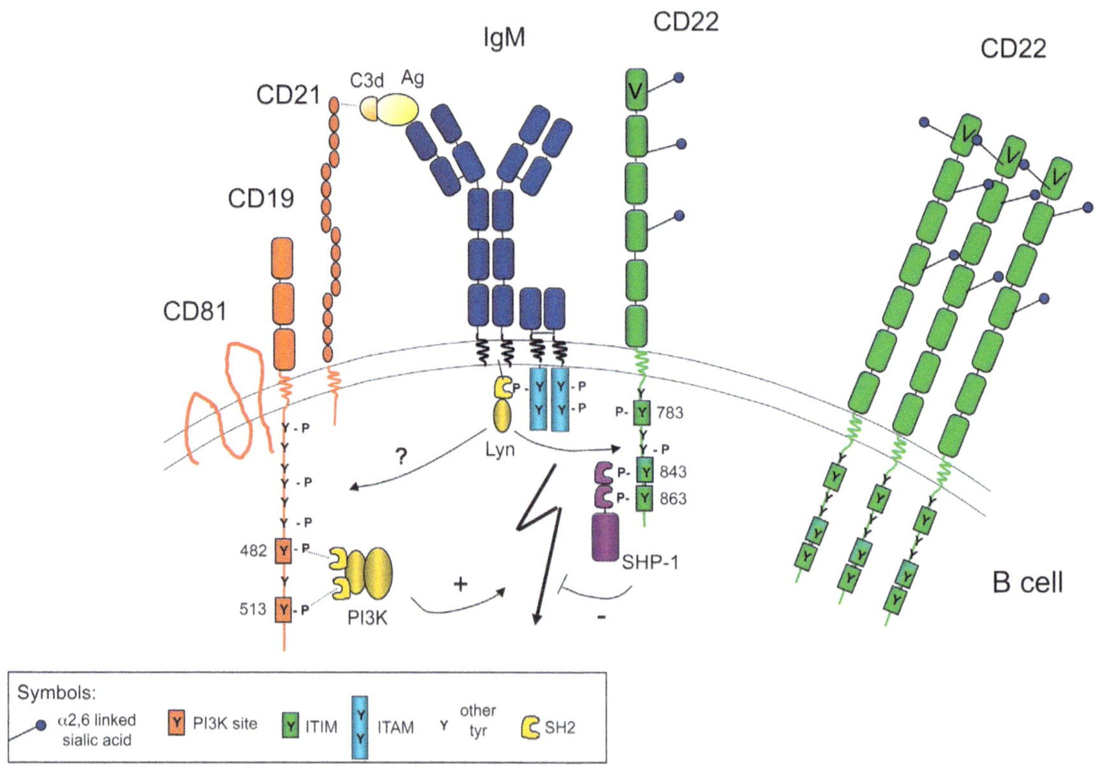


FIGURE 3 **The CD19/CD81/CD21 complex can bind via CD21 to the complement C3d component that is often found attached to an antigen (Ag).** In this way, a C3d-Ag complex can not only bind to both receptor complexes (CD19/CD81/CD21 and B-cell receptor (BCR)) but also increase coligation of the two receptors. CD21-independent interactions of CD19 with the BCR also have been shown. The CD19 tail contains nine tyrosines, which are phosphorylated by Lyn or other Src kinases. Phosphorylated Y482 and Y513 are bound by phosphoinositide 3 kinase (PI3K), the most important downstream signaling molecule bound to the CD19 tail, and this enhances signaling. CD22 engages in *cis*-interactions by binding to its ligands α2,6-linked sialic acids, forming homo-oligomers. Upon BCR stimulation, CD22 is recruited to the BCR through protein–protein interactions. The immunoreceptor tyrosine-based inhibition motifs (ITIMs) of the CD22 tail are phosphorylated by Lyn. The tyrosine phosphatase SHP-1 is subsequently recruited to CD22 and inhibits signaling. IgM, immunoglobulin M; ITAM, immunoreceptor tyrosine-based activation motif.

PI3K reaction can rescue all phenotypes of CD19-deficient mice [91]. However, CD19-independent BCR-induced activation of PI3K also exists.

## 9. HUMAN MUTATIONS IN THE CD19/CD21/CD81 COMPLEX

Since crucial functions for CD19 in regulating PI3K signaling and controlling antibody responses were found in mouse models, it is not surprising that patients with homozygous CD19 mutations develop an antibody-deficiency syndrome [92]. The same is true for patients with CD81 gene defects, which have impaired CD19 surface expression and an antibody deficiency [93]. These findings show a conservation of CD19/CD81 functions between mouse and human. CD19 gene polymorphisms also have been linked to autoimmunity. In patients with systemic sclerosis, single nucleotide polymorphisms in the CD19 gene leading to 20% higher CD19 expression levels were linked to the disease [94]. CD19 gene polymorphisms also were linked with systemic lupus erythematosus; in this case, however, they lead to lower CD19 surface expression without affecting the protein sequence [95]. Further studies are needed to understand the role of CD19 in autoimmune diseases. Reduced expression of complement receptor 2/CD21 was detected on B cells of patients with systemic lupus erythematosus [96]. CD21 also is expressed on follicular dendritic cells, where it is thought to be important for binding immune complexes. These CD19-independent functions of CD21 may also contribute to autoimmune diseases.

## 10. CD22: AN INHIBITORY RECEPTOR

CD22 is an inhibitory receptor of the Siglec family that is predominantly expressed on B cells. It binds to its ligands α2,6-linked sialic acids (2,6Sia), which are abundant terminal sugars on soluble and cell surface–expressed glycoproteins [66]. CD22 is to some extent associated with the BCR; however, this association is regulated by 2,6Sia-carrying *cis*-ligands. CD22 contains six tyrosines in its cytoplasmic tail. Three tyrosines (Y783, Y843, Y863) are part of classical inhibitory ITIM motifs, one of which (Y817) is in an ITIM-like motif, and another is needed for recruitment of Grb-2 (Y828) (Figure 3). CD22 is rapidly phosphorylated

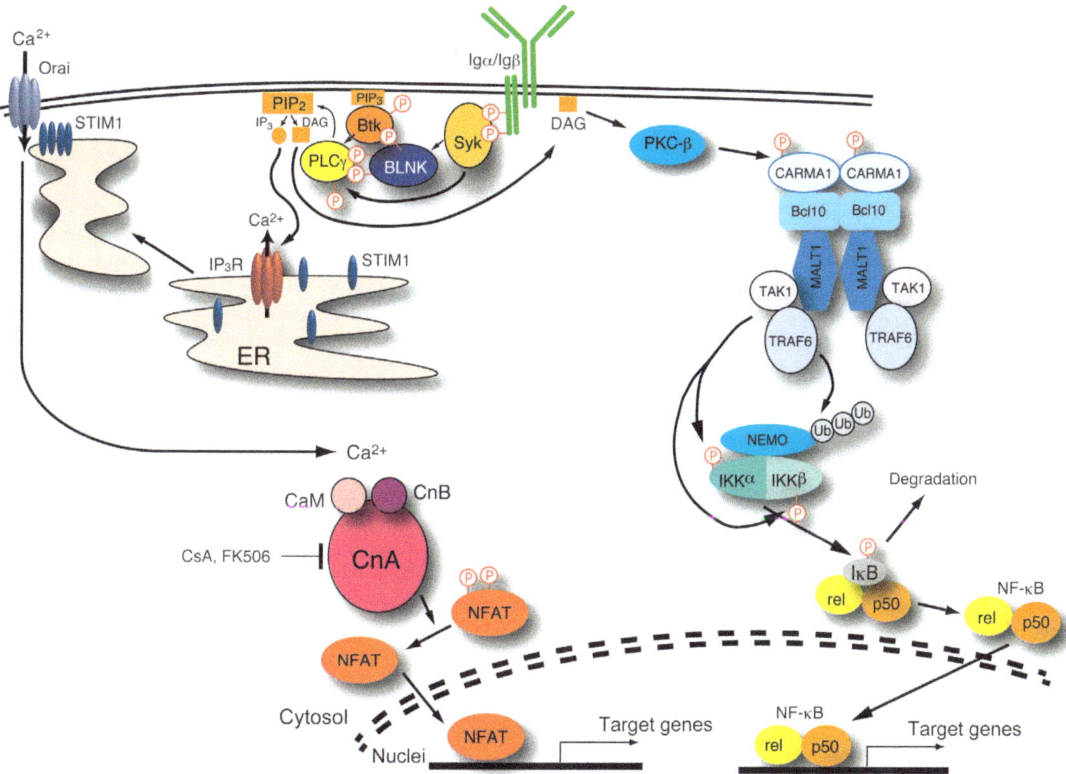


FIGURE 4 **B-cell receptor (BCR) engagement induces translocation of signaling molecules to the antigen receptor and subsequent activation.** Upon BCR activation, Syk phosphorylates and binds to the immunoreceptor tyrosine-based activation motifs (ITAMs) of Igα and Igβ. Syk then phosphorylates the adaptor SLP65/BLNK, providing a scaffold for the binding and interaction of phospholipase (PL) Cγ2 and other signaling molecules such as Btk. Btk is also recruited to the plasma membrane by association of its Pleckstrin homology domain with phosphatidylinositol-4,5-trisphosphate ($PIP_3$) in the membrane. Activated PLCγ2 hydrolyzes phosphatidylinositol-4,5-bisphosphate ($PIP_2$) to form diacylglycerol (DAG) and inositol-1,4,5-triphosphate ($IP_3$). The thus generated DAG leads to the activation of Ras/mitogen-activated protein kinase and nuclear factor (NF)-κB pathways. $IP_3$ evokes the mobilization of $Ca^{2+}$ and triggers the initial $Ca^{2+}$ release from the endoplasmic reticulum (ER) lumen. Following depletion of ER $Ca^{2+}$, STIM1 proteins are activated, translocate, and bind to ER–plasma membrane junctions, where they gate Orai1 $Ca^{2+}$ entry channels. $Ca^{2+}$ entry subsequently activates NF of activated T cells (NFAT) and the other signaling molecules, which results in the expression of appropriate target genes. DAG, generated by $PIP_2$ hydrolysis, activates protein kinase (PK) C-β, which then activates the IκB kinase (IKK) complex, presumably by promoting membrane recruitment and aggregation of the CARMA1/bcl10/MALT1 complex. TAK1 and TRAF6 are then recruited to this complex and activate the IKK complex by phosphorylation of IKKα and IKKβ, in combination with ubiquitination, followed by degradation of NEMO (IKKγ). Activation of the IKK complex results in the phosphorylation and ubiquitin-mediated degradation of IκB. The released rel/p50 complex translocates to the nucleus and activates the expression of target genes.

after BCR crosslinking [66]. The kinase Lyn is crucial for this phosphorylation; Lyn-deficient mice show strongly decreased CD22 tyrosine phosphorylation [97,98]. Upon tyrosine phosphorylation of CD22, the tyrosine phosphate SHP-1 is recruited to phosphorylated CD22 ITIM motifs [99]. Several in vitro studies showed that SHP-1 is the crucial negative regulator of inhibitory signaling by CD22 [100,101]. CD22-deficient mice show increased BCR-induced $Ca^{2+}$ signaling, which is due to the loss of this negative regulation [102–105]. Also, CD22 knock-in mice with Y→F mutations of all three ITIMs show similar increased $Ca^{2+}$ signaling [106]. As a consequence of this lack of inhibition, CD22-deficient B cells show increased tyrosine phosphorylation of important signaling molecules such as SLP65/BLNK or CD19 [107,108]. This can explain the negative regulation of the initiation of $Ca^{2+}$ signaling. CD22 also associates with a $Ca^{2+}$ pump called PMCA-4, which is involved in pumping $Ca^{2+}$ out of the cytosol after signaling is terminated. This pump is activated by CD22 via SHP-1 [100,109–111]. Thus, loss of CD22 in CD22-deficient mice affects both the initiation and the termination of $Ca^{2+}$ signaling in B cells.

In addition to the inhibitory SHP-1, CD22 also is able to bind several other intracellular signaling molecules, such as Grb2, Shc, and SHIP, or the activating signaling proteins Syk and PLCγ2, on its phosphorylated tail. All these interactions have been demonstrated in cell lines, but their physiological consequences are unknown. Membrane-bound (m) IgG has a stronger signaling capacity than mIgM, which is relevant for memory B-cell responses. One report showed that CD22 phosphorylation in mIgG+ B cell lines is impaired, suggesting that CD22 in unable to regulate signals transmitted by mIgG [112]. However, this was not confirmed in mouse models where transgenic mice expressing mIgG showed

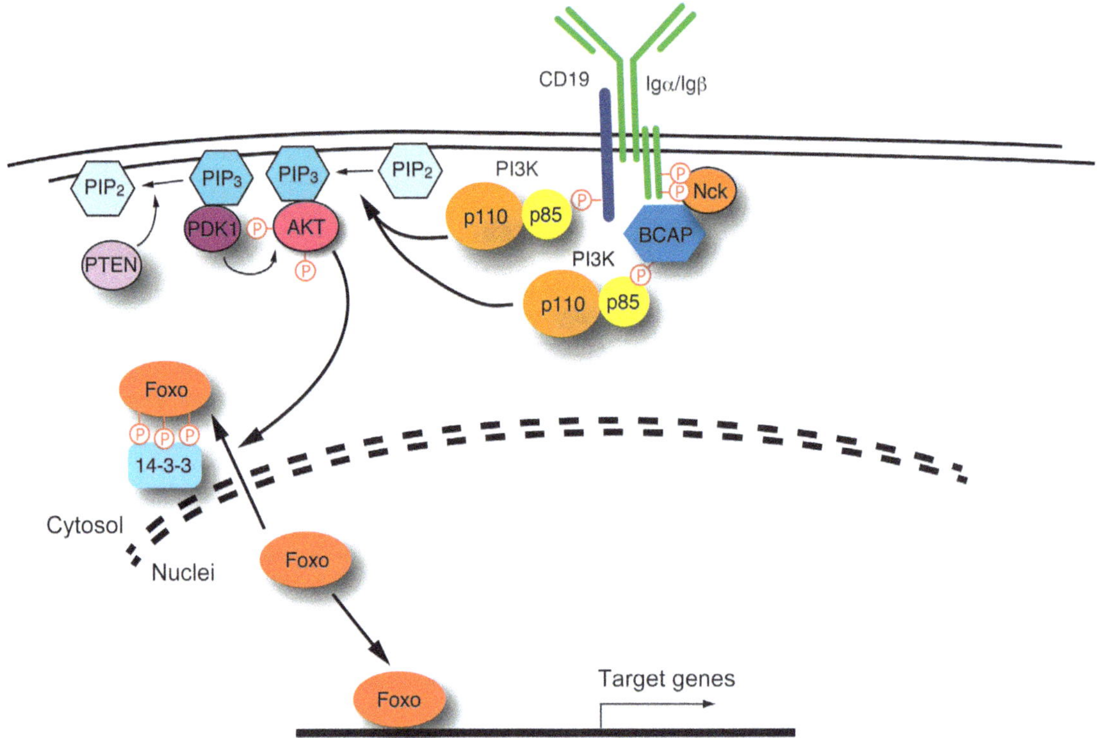


FIGURE 5 **The phosphoinositide 3 kinase (PI3K) signaling cascade induced by B-cell receptor (BCR) ligation.** Upon BCR engagement, Nck recruits BCAP to the BCR. BCAP and CD19 are phosphorylated upon BCR ligation. Phosphorylation of these molecules leads to the recruitment and activation of PI3K heterodimers, consisting of the catalytic p110 subunit and the regulatory p85α subunit. Activated PI3K subsequently phosphorylates phosphatidylinositol-4,5-bisphosphate ($PIP_2$) to form phosphatidylinositol-4,5-trisphosphate ($PIP_3$). Akt is then recruited to the generated $PIP_3$ through its Pleckstrin homology domain and is activated by PDK1. This leads to downstream signal transduction. Upon activation, Akt phosphorylates Foxo transcription factors and creates docking sites for 14-3-3. When associated with 14-3-3, Foxo is excluded from the nucleus. Phosphatase and tensin homolog dephosphorylates $PIP_3$ back to $PIP_2$, thus terminating the PI3K signal.

normal CD22 tyrosine phosphorylation and CD22 inhibition [113,114]. A recently discovered enhancing signaling motif in the cytoplasmic tail of mIgG can explain the higher signaling capacity of mIgG [115], which is independent of CD22.

## 11. REGULATION OF CD22 SIGNALING BY LIGAND INTERACTIONS

It is well established that CD22 is an inhibitory coreceptor for BCR signaling, but the question remains whether this inhibition is a constitutive or a regulated process and which physiological conditions influence the inhibition. Recent studies have shown that ligand binding to CD22 plays an important role in this regulation. The first extracellular V-set Ig domain of CD22 binds to 2,6Sia, which are sialic acids that are present on various soluble proteins in the blood plasma [116,117] but also are abundantly expressed on the surface of many cells, including B and T cells [118,119]. CD22 can therefore bind ligands in *trans* on adjacent cell surfaces or in *cis* on the same cell surface. The high expression of 2,6Sia on B cells seems to "mask" most CD22 molecules by *cis*-binding and limits its ability to *trans*-bind [120,121]. However, ligand *cis*- and *trans*-binding may be dynamic processes depending on ligand availability, accessibility, and cell–cell contact and may therefore not be mutually exclusive. CD22 has been shown to interact with mIgM and CD45 on the B-cell surface, and the proximity to the BCR is important for the inhibitory function of CD22 [122]. The association of CD22 with the BCR does not seem to be mediated via 2,6Sia binding; rather it is a protein–protein interaction. CD22 binds to itself via sialic acid interactions, forming oligomers, as demonstrated by a photoaffinity crosslinking approach [123] (Figure 3). In resting B cells, these CD22 homo-oligomers are on membrane domains distinct from the BCR [124].

Two genetic approaches showed that disruption of these CD22-ligand interactions on the cell surface results in a stronger CD22–BCR association and a stronger inhibition of $Ca^{2+}$ signaling. The first approach deleted the gene coding for ST6Gal1, the enzyme that creates 2,6Sia on the cell surface [125,126]. The second approach was a CD22 knock-in mouse line with a mutated ligand-binding domain [106]. A weaker IgM-induced $Ca^{2+}$ signal and enhanced association of CD22 to IgM was detected in both cases. This resulted in stronger CD22 phosphorylation, stronger SHP-1 recruitment, and stronger signal inhibition. Another study of CD22-knockin mice with a mutated ligand-binding domain did not report this effect on $Ca^{2+}$ signaling, possibly because the CD22 surface expression in this line was impaired by the introduced mutation [127]. In addition to *cis*-binding, CD22 was shown to be able to *trans*-bind to ligands on other cells. B-cell interactions with other cells expressing a cell-bound antigen together with 2,6Sia can lead to CD22-dependent suppression of B-cell signaling and proliferation [128]. Also, CD22 on ST6Gal1-deficient B cells can be recruited to the site of cell contact if a neighboring wild-type B cell expresses 2,6Sia [125]. These experiments show that, although CD22 is mainly engaged in *cis* interactions, it can also be engaged in *trans* interactions. This has recently been exploited by the use of liposomes with attached high-affinity CD22 ligands and protein antigens that could suppress antigen-specific B-cell responses [129].

## 12. THE ROLE OF CD22 IN PREVENTING AUTOIMMUNITY

Hyperresponsive B cells resulting from a lack of inhibition of BCR signaling can lead to the development of autoimmunity. This has been shown in mice lacking the inhibitory FcγRIIb receptor [130] and in mice deficient for important signaling molecules of inhibitory pathways such as Lyn [83] or SHP-1 [131]. CD22-deficient mice on a C57BL/6 background show normal immune responses and do not spontaneously develop autoimmunity. However, aging CD22-deficient mice on a mixed C57BL/6×129 or on a Y-linked autoimmune accelerator background do develop autoimmunity [132,133]. Also, several natural mouse strains prone to autoimmunity, such as New Zealand white, Murphy Roths large, and BXSB mice, carry the Cd22a allele with a truncated first CD22 Ig domain [134]. CD22 protein expressed from the Cd22a allele shows impaired ligand binding. These data indicate that the mouse *Cd22* gene could be a susceptibility gene for autoimmunity. In addition to CD22, mouse B cells express the CD22 homolog Siglec-G. In contrast to CD22-deficient mice, CD22×Siglec-G double-deficient mice developed autoimmunity with the formation of high-affinity autoantibodies and glomerulonephritis [135]. This indicates a redundancy of CD22 and Siglec-G, with one being able to replace the function of the other; only when both Siglecs are lost is autoimmunity strongly enhanced.

Sialic acid acetyl esterase (SIAE) modifies CD22 ligands and has been linked to autoimmunity. Mutations in the SIAE gene impairing the function of the protein have been associated to several human autoimmune diseases, and SIAE-deficient mice also develop autoimmunity [136,137]. Genome-wide association studies have not identified the human CD22 locus so far, but they have identified downstream signaling molecules of the CD22 inhibitory pathway, such as Lyn or Blk [138]. Overall, there is good evidence that loss of CD22 or its inhibitory pathway can contribute to the development of autoimmune diseases.

## 13. BCR-CONTROLLED SIGNALING PROCESSES

The ability to signal through the BCR signaling complex is mandatory for B-cell developmental progression as well as for immune responses [139,140]. The phosphorylation of the two ITAM tyrosine residues of Igα and Igβ in activated B cells by Lyn and Syk is followed by the recruitment and activation of adaptor proteins and further cytoplasmic PTKs such as the Bruton's tyrosine kinase (Btk). Together, these elements channel the BCR activation signal through different intracellular signaling pathways. In recent years work has focused on how BCR signaling pathways are quantitatively and qualitatively regulated and how these signaling pathways are integrated into cellular outcomes. Here we focus our discussion of these processes on the PLCγ2 calcium response and the PI3K pathway (Figures 3 and 4).

## 14. BCR-MEDIATED ADAPTOR AND PLCγ2 ACTIVATION

PLCγ2 is an essential component of the calcium response pathway and is required for the generation of second messengers following BCR engagement (Figure 3). PLCγ2 degrades phosphatidylinositol-4,5-bisphosphate ($PIP_2$) located in the plasma membrane to diacylglycerol (DAG) and soluble inositol-1,4,5-triphosphate ($IP_3$). Upon BCR ligation, PLCγ2 translocates to the plasma membrane, thereby gaining access to its substrate $PIP_2$. The translocation and activation of PLCγ2 is controlled by the adaptor protein SLP65/BLNK, a major substrate of Syk once this kinase is activated and bound to the BCR [12,141,142]. The importance of the adaptor–PLCγ2 interaction has been highlighted by several mutant mouse models showing that SLP65/BLNK-deficient B cells fail to induce the translocation of PLCγ2 to the plasma membrane and its subsequent activation; conversely, membrane-tethered PLCγ2 rescues the defective PLCγ2 activation in the SLP65/BLNK-deficient B cells [143,144]. The biological importance of PLCγ2 recruitment was further demonstrated by the existence of a gain-of-function mutant of PLCγ2 [145]. This mutant PLCγ2 manifested prolonged interactions with the plasma membrane, thereby causing severe spontaneous inflammation and autoimmunity, although which cell types are most responsible for the inflammatory phenotypes has not been examined in detail.

In regard to the SLP65/BLNK recruitment to the BCR and to the membrane, several—but not necessarily mutually exclusive—mechanisms have been proposed. First, it has been found that, after its phosphorylation, the fourth tyrosine (Y204) in the cytoplasmic tail of Igα is a binding target of the C-terminal SH2 domain of SLP65/BLNK [146–148]. Second, the N-terminal region of SLP65/BLNK carries a leucine zipper that is essential for its constitutive association of the adaptor with the plasma membrane [149]. Third, the C-terminal SH2 domain of SLP65/BLNK can also bind to a C-terminal tyrosine of Syk that becomes phosphorylated once the kinase is active. Which of these three mechanisms dominantly operates during B-cell activation and in what order is still a matter of debate [150,151].

Once SLP65/BLNK is recruited to the BCR complex and phosphorylated by Syk, it provides binding sites for Btk and PLCγ2. Indeed, both Btk and PLCγ2 have SH2 domains. Because PLCγ2 and Btk are placed in close proximity to each other, PLCγ2 undergoes tyrosine phosphorylation by Btk [152–156]. A detailed analysis revealed that among the three putative phosphorylation sites in PLCγ2, Y753 and Y759 are mainly phosphorylated by Btk. It also has been shown that the extent of phosphorylation of Y753 and Y759 correlates with the lipase activity of PLCγ2 [152]. In addition to Y753 and Y759, PLCγ2 undergoes phosphorylation at Y1217 in a Btk-independent manner [152]. Although the importance of this phosphorylation in the context of BCR signaling is clear, this is probably not caused by direct activation of the lipase activity. Rather, phosphorylation at Y1217 seems to induce association with as yet unidentified molecules containing SH2, thereby stabilizing the PLCγ2 residency in the plasma membrane and/or enhancing the accessibility of PLCγ2 to its substrate.

The physiological role of PLCγ2 also has been extensively examined [157–160]. In PLCγ2 knockout mice, the number of mature B cells in the spleen dramatically decreases and B1a cells in the peritoneal cavity disappear [157,160]. These results clearly indicate that PLCγ2 is essential for B-cell maturation and maintenance. That the maintenance of memory B cells also requires PLCγ2 also has been reported [158].

## 15. $IP_3$ PROMOTES CALCIUM RELEASE AND ACTIVATION OF NUCLEAR FACTOR OF ACTIVATED T CELLS

BCR-mediated PLCγ2 activation leads to the hydrolysis of phosphoinositide $PIP_2$, which results in the formation of $IP_3$ and DAG. The resulting $IP_3$ binds $IP_3$ receptors on the surface of the endoplasmic reticulum (ER) to release calcium from internal stores [161]. Three types of $IP_3$ receptors are known, namely types 1, 2, and 3. Triple knockouts of these $IP_3$ receptor isoforms completely abolish the calcium responses induced by BCR crosslinking [162]. Thus, these three isoforms of $IP_3$ receptors are essential for BCR-mediated calcium mobilization, although they are functionally redundant. This transient response to BCR ligation represents the first increase of cytosolic calcium. However, the capacity of ER stores is limited, and sensing BCR-mediated emptying of ER stores leads to the opening of calcium-permeable ion channels in the plasma membrane, termed store-operated calcium (SOC) channels. This evokes the second step of calcium mobilization, giving rise to sustained elevated calcium concentrations in the cytosol (Figure 4).

The first step in the activation of SOC channels is sensing calcium depletion in the ER. A sensor molecule for detecting ER intraluminal calcium content remained unknown for a long time until stromal interaction molecule (STIM) 1 and STIM2 were identified as calcium sensors in the ER [163,164]. This observation was soon followed by the finding that Orai1, Orai2, and Orai3 (also known as CRACM1/2/3), which are localized in the plasma membrane and contain four TM segments, function as pore-forming subunits of the SOC channel [165,166].

The importance of STIM1 for BCR-induced $Ca^{2+}$ entry was first shown by gene-targeting experiments in DT40 B cells [167]. In the absence of STIM1 expression, SOC influx induced by BCR ligation was almost completely abrogated. This study also revealed that STIM1 moves to ER–plasma membrane junctions upon $Ca^{2+}$ depletion, forming puncta and closely associating with the activation of Orai channels. These data are in agreement with other studies using non-B cells [168–172].

Recent detailed studies of STIM1 have provided more mechanistic insights into how STIM1 conformation changes upon $Ca^{2+}$ release from the ER and how it binds the Orai1 channel [167,173,174]. In cells with replete ER $Ca^{2+}$ stores, the EF-hand–sterile α motif domain in the N terminus of STIM1 (localized on the luminal side of the ER) assumes a closed configuration. Upon $Ca^{2+}$ depletion, hydrophobic regions in the EF-hand–sterile α motif domain are exposed, thus allowing neighboring N-termini of STIM1 to dimerize. Once STIM1 are dimerized at the N-terminus, conformational changes in the C-terminus of STIM1 are induced (localized on the cytoplasmic side), resulting in an elongated state and thereby enabling the C-terminus of STIM1 to bind Orai1 channels at ER–plasma membrane junctions.

It was shown that STIM1 and STIM2 also play essential roles in SOC $Ca^{2+}$ influx induced by BCR ligation in primary B cells [175]. When B cells with defective expression of both STIM1 and STIM2 were stimulated by BCR ligation in vitro, proliferation and survival were dramatically impaired. However, overall in vivo antigen-specific antibody production was not affected by STIM1/2 deficiency in either T-dependent and T-independent responses. Although the immune responses induced by immunization are not affected by the loss of SOC influx, the unique finding that loss of STIM1 and STIM2 impairs IL-10 production from B cells and leads to the onset of severe experimental autoimmune encephalomyelitis clearly shows the physiological importance of SOC influx in B cells.

It also has been shown that nuclear factor of activated T cells (NFAT) c2 is responsible for calcium-dependent IL-10 production [175,176]. The NFAT family is a well-studied calcium-dependent transcription factor downstream of calcineurin [177]. NFAT is cytosolic in resting cells and requires continuous elevated $Ca^{2+}$ to remain in the nucleus, where it activates transcription of target genes. B cells express three NFAT family members, NFATc1 (alternatively named NFAT2 or NFATc), NFATc2 (NFAT1 or NFATp), and NFATc3 (NFAT4 or NFATx), which are components of the BCR signaling cascade [178]. Intrinsic roles for NFATc1 in B cells have been analyzed recently [179]. NFATc1 ablation in B cells resulted in impaired BCR-mediated proliferation and facilitated activation-induced cell death. Also, the number of peritoneal B1a cells decreased by 5- to 10-fold. Interestingly, ablation of NFATc1 in B cells led to an increase in IL-10 production and ameliorated the clinical course of experimental autoimmune encephalomyelitis. The apparent disparity between NFATc1 and STIM1/2 ablation in B cells could be explained by the idea that NFATc2 and/or NFATc3 play a compensatory role in IL-10 production.

## 16. DAG AND NUCLEAR FACTOR-κB ACTIVATION

Nuclear factor (NF)-κB transcription factors are retained in the cytoplasm by binding to the inhibitor IκB. The activated IκB kinase (IKK) complex, consisting of IKKα, IKKβ, and IKKγ (alternatively named NEMO), phosphorylates IκB, thereby inducing its degradation [180]. In B cells, PKCβ is activated by DAG and $Ca^{2+}$, both of which are generated by PLCγ2 activation. Ablation of PKCβ leads to the defective activation of the canonical NF-κB pathway, which in turn results in defects in B-cell activation and maturation [181]. In the signaling pathway between PKCβ and IKK activation, the formation of a macromolecular NF-κB signaling complex is required. This complex is composed of adaptor molecules: CARMA1, Bcl10, and MALT1 [182]. Fine-tuned regulation of the NF-κB signaling pathway imposed by CARMA1/Bcl10/MALT1 is crucial to maintaining normal B-cell proliferation/survival and differentiation (Figure 3).

PKCβ phosphorylates CARMA1 on Ser668, which is essential for subsequent CARMA1/Bcl10/MALT1 association and for IKK activation [183,184]. IKKα and IKKβ undergo Ser phosphorylation upon BCR ligation, but PKCβ seems not to be a direct kinase on such phosphorylation because the Ser residues in the activation loop of IKKα and IKKβ do not match the PKCβ consensus phosphorylation site. It turns out that TAK1, a member of the mitogen-activated protein 3 kinase family, is responsible for phosphorylating IKK [185]. Indeed, TAK1-deficient DT40 B cells completely lack BCR-mediated IKK activation [184]. Moreover, immunoprecipitated TAK1 can phosphorylate activation-loop Ser residues of IKKβ in vitro. In addition to phosphorylation, K63-linked ubiquitination also plays a crucial role in BCR-mediated IKK activation. The CARMA1/Bcl10/MALT1 complex is associated with TRAF6 (Ub E3 ligase), which induces K63-linked ubiquitination of IKKγ [185–188]. This IKKγ ubiquitination is critical for BCR-mediated IKK activation.

## 17. ACTIVATION OF THE PI3K PATHWAY

PI3K is a lipid kinase that produces $PIP_3$ from $PIP_2$. Newly generated $PIP_3$ is involved in the activation of downstream signaling molecules such as Akt (see section 18, "Akt and Foxo Regulation") [86,189–191] (Figure 5). The class IA PI3K family is a heterodimeric molecule consisting of a 110-kDa catalytic subunit (p110α, p110β, and p110δ) and a regulatory subunit (p50α, p55α, and p85α), whereas class IB PI3K consists of a 110-kDa catalytic domain (p110γ) and a regulatory subunit (p55γ, p101, and p84). Among the three isoforms in the class IA family, p110δ is expressed primarily in hematopoietic cells, including B cells, in contrast to p110α and p110β, which are expressed ubiquitously. Before activation, the regulatory subunit inhibits the activity of the catalytic subunit. Once the tandem SH2 domains in the regulatory subunit bind to tyrosine-phosphorylated YxxM motifs in the binding partner, inhibition of the associated catalytic subunit is released.

In B cells, activation of PI3K has been thought to be mainly mediated by CD19, which possesses a binding site for the SH2 domain in the p85α regulatory subunit. Indeed, two YxxM motifs in the cytoplasmic region of CD19 have been shown to play an essential role in the recruitment of the p85α subunit [84]. However, the defect in B-cell development in p85α-deficient mice is more severe than in CD19-deficient mice [69,70,88,192], suggesting the involvement of other binding molecules in the activation of PI3K in B cells.

The B-cell adaptor for PI3K (BCAP), which is highly expressed in B cells, can bind p85α through four YxxM motifs [193]. Upon BCR ligation, BCAP undergoes tyrosine phosphorylation, thereby recruiting p85α. When tyrosine residues in these four YxxM motifs were substituted by phenylalanine, BCAP no longer bound the p85α subunit [193]. Moreover, when CD19 and BCAP genes were doubly disrupted, B-cell maturation was almost completely blocked, and this perturbation in B-cell development was partly recovered by the introduction of a constitutively active form of PI3K or PDK1 [194]. This suggests the complementary roles of BCAP and CD19 in BCR-mediated PI3K activation.

In contrast to the well-elucidated physical association of BCR with CD19, the interaction between BCR and BCAP has not been examined until recently. However, new data have clarified this connection; the adaptor protein Nck binds directly to phosphorylated Igα and recruits BCAP to the BCR signalosome by association between its SH3 domain and the second proline-rich region of BCAP [195]. Nck1 and Nck2 double mutant mice showed a greater defect in Akt phosphorylation than BCAP-deficient mice. This defect was comparable to that observed in BCAP and CD19 double-deficient mice. This could mean that Nck might also be important for CD19-mediated BCR signaling in addition to BCAP recruitment.

Vav was identified as another important regulatory molecule in the PI3K signaling pathway [196]. The Vav family proteins (Vav1, Vav2, and Vav3) are cytoplasmic guanine nucleotide exchange factors for Rho-family GTPases. These guanine nucleotide exchange factor proteins have a conserved functional unit consisting of a coupled Pleckstrin homology (PH) domain and a Dbl homology domain, which are closely associated in the native conformation [196,197]. Like other molecules with PH domains, Vav has been shown to bind membrane $PIP_3$ produced by PI3K and to exert its biological functions [198,199]. Thus, it is evident that activation of Vav is controlled by PI3K. On the other hand, it also has been shown that Vav regulates PI3K activity, presumably through Rac1, a member of Rho-family GTPases, which is known to activate the lipid kinase [200,201]. These results suggest the existence of a positive feedback loop involving PI3K-Vav-Rac in the PI3K signaling pathway. Vav1 has been shown to be critical for T-cell development and involved in mast cell activation [202–205]. B-cell developmental and functional defects are relatively mild in mice lacking either Vav1 or Vav2, whereas double knockout mice exhibit severe defects [198,202,203,206]. Vav3, highly expressed in peripheral lymphocytes and spleen cells [207,208], has been shown to modulate BCR-mediated PI3K activation [200]. These results strongly suggest that PI3K signaling events affected by the Vav family are indispensable for both T cells and B-cell development and activation.

The physiological roles of PI3K in B cells have been well examined. In p110δ-deficient mice, the development of B cells in the bone marrow is relatively normal compared to defective maturation in the periphery [87,89]. The number of follicular B2 cells in the spleen is dramatically reduced, and the development of peritoneal B1 cells and marginal zone B cells is almost completely blocked [87,209,210]. BCR-mediated proliferation in vitro depends on p110δ activity. Deletion or inactivation of p110δ or p85α dramatically reduces BCR-mediated phosphorylation of Akt, FoxO, and protein kinase D, which leads to reduced $Ca^{2+}$ flux, impaired cell cycle progression, and reduced glucose metabolism. Also, attenuated PI3K signaling results in impaired B-cell homeostasis [211]. Conversely, provision of a constitutively active PI3K molecule is sufficient to rescue B cells from apoptosis upon inducible deletion of the BCR [212]. These findings strongly suggest that PI3K is a primary component of the so-called tonic signal, which is required for the maintenance of peripheral B cells.

## 18. AKT AND FOXO REGULATION

As described earlier, one of the major signaling pathways by which PI3K exerts its physiological function is through activation of Akt, followed by modulation of downstream transcription factors [213]. Foxo transcription factors are well-known transcription factors regulated by the PI3K/Akt pathway. In B cells, activated Akt phosphorylates Foxo1 and Foxo3 at three conserved phosphorylation sites following BCR ligation, resulting in diminished DNA binding ability and exclusion from the nucleus through the formation of a complex with the 14-3-3 chaperone protein. Ectopic expression of Foxo1 or Foxo3 resulted in cell cycle arrest and increased cell death [214].

It has been shown that Foxo1 is required for both early B-cell development and peripheral B-cell function using *Foxo1* gene deletions at various stages of B-cell differentiation [215]. When *Foxo1* is deleted at a very early stage, a strong block of differentiation at the pro-B cell stage due to the failure to express the IL-7 receptor was observed. It also has been reported that Foxo1 directly regulates expression of the *Rag-1* and *Rag-2* genes throughout early B-cell development, and transcription of these genes is repressed by Akt activation [216]. In agreement with this report, when a mutant form of Foxo3 lacking the Akt phosphorylation sites was overexpressed, Igκ light chain gene rearrangement in the pre-B stage was promoted [217]. This study also showed the possibility that pre-BCR induces and/or activates SLP65/BLNK, which in turn inhibits Akt activation, thereby promoting Igκ rearrangement. In the case of peripheral B cells, deletion of Foxo1 resulted in fewer lymph node B cells because of the lower expression of L-selectin. Also, expression of activation-induced cytidine deaminase following antigen immunization was dramatically reduced, resulting in lower levels of IgG production. Together, these data strongly suggest that Foxo1 and Foxo3, controlled by the PI3K/Akt signaling pathway, are essential transcription factors involved in B-cell development, maintenance, and activation.

## ACKNOWLEDGMENT

The authors thank Peter Nielsen and Lise Leclercq for their critical reading of this manuscript and Jianying Yang for some of the drawings. This study was supported by the Excellence Initiative of the German Federal and State Governments (EXC 294), by European Research Council grant 322972, and by the Deutsche Forschungsgemeinschaft through SFB746, SFB643 and TRR130.

## REFERENCES

[1] Neuberger MS, Patel KJ, Dariavach P, Nelms K, Peaker CJ, Williams GT. The mouse B-cell antigen receptor: definition and assembly of the core receptor of the five immunoglobulin isotypes. Immunol Rev 1993;132:147–61.

[2] Reth M. Antigen receptors on B lymphocytes. Annu Rev Immunol 1992;10:97–121.

[3] Kurosaki T. Molecular dissection of B cell antigen receptor signaling (review). Int J Mol Med 1998;1:515–27.
[4] Wienands J, Engels N. Multitasking of Ig-alpha and Ig-beta to regulate B cell antigen receptor function. Int Rev Immunol 2001;20:679–96.
[5] Schamel WW, Reth M. Monomeric and oligomeric complexes of the B cell antigen receptor. Immunity 2000a;13:5–14.
[6] Schamel WW, Reth M. Stability of the B cell antigen receptor complex. Mol Immunol 2000b;37:253–9.
[7] Sanchez M, Misulovin Z, Burkhardt AL, Mahajan S, Costa T, Franke R, et al. Signal transduction by immunoglobulin is mediated through Ig alpha and Ig beta. J Exp Med 1993;178:1049–55.
[8] Reth M. Oligomeric antigen receptors: a new view on signaling for the selection of lymphocytes. Trends Immunol 2001;22:356–60.
[9] Reth M. Antigen receptor tail clue. Nature 1989;338:383–4.
[10] Cambier JC. Antigen and Fc receptor signaling. The awesome power of the immunoreceptor tyrosine-based activation motif (ITAM). J Immunol 1995a;155:3281–5.
[11] Cambier JC. New nomenclature for the Reth motif (or ARH1/TAM/ARAM/YXXL). Immunol Today 1995b;16:110.
[12] Johnson SA, Pleiman CM, Pao L, Schneringer J, Hippen K, Cambier JC. Phosphorylated immunoreceptor signaling motifs (ITAMs) exhibit unique abilities to bind and activate Lyn and Syk tyrosine kinases. J Immunol 1995;155:4596–603.
[13] Flaswinkel H, Barner M, Reth M. The tyrosine activation motif as a target of protein tyrosine kinases and SH2 domains. Semin Immunol 1995;7:21–7.
[14] Flaswinkel H, Reth M. Dual role of the tyrosine activation motif of the Ig-α protein during signal transduction via the B cell antigen receptor. EMBO J 1994;13:83–9.
[15] Reichlin A, Gazumyan A, Nagaoka H, Kirsch KH, Kraus M, Rajewsky K, et al. A B cell receptor with two Igα cytoplasmic domains supports development of mature but anergic B cells. J Exp Med 2004;199:855–65.
[16] Storch B, Meixlsperger S, Jumaa H. The Ig-α ITAM is required for efficient differentiation but not proliferation of pre-B cells. Eur J Immunol 2007;37:252–60.
[17] Kraus M, Pao LI, Reichlin A, Hu Y, Canono B, Cambier JC, et al. Interference with immunoglobulin (Ig)α immunoreceptor tyrosine-based activation motif (ITAM) phosphorylation modulates or blocks B cell development, depending on the availability of an Igbeta cytoplasmic tail. J Exp Med 2001;194:455–69.
[18] Lee WY, Tolar P. Activation of the B cell receptor leads to increased membrane proximity of the Igα cytoplasmic domain. PloS One 2013;8:e79148.
[19] Metzger H. Transmembrane signaling: the joy of aggregation. J Immunol 1992;149:1477–87.
[20] Woodruff MF, Reid B, James K. Effect of antilymphocytic antibody and antibody fragments on human lymphocytes in vitro. Nature 1967;215:591–4.
[21] Harwood NE, Batista FD. Early events in B cell activation. Annu Rev Immunol 2010;28:185–210.
[22] Tolar P, Sohn HW, Pierce SK. The initiation of antigen-induced B cell antigen receptor signaling viewed in living cells by fluorescence resonance energy transfer. Nat Immunol 2005;6:1168–76.
[23] Mukherjee S, Zhu J, Zikherman J, Parameswaran R, Kadlecek TA, Wang Q, et al. Monovalent and multivalent ligation of the B cell receptor exhibit differential dependence upon Syk and Src family kinases. Sci Signal 2013;6:ra1.
[24] Avalos AM, Bilate AM, Witte MD, Tai AK, He J, Frushicheva MP, et al. Monovalent engagement of the BCR activates ovalbumin-specific transnuclear B cells. J Exp Med 2014;211:365–79.
[25] Minguet S, Dopfer EP, Schamel WW. Low-valency, but not monovalent, antigens trigger the B-cell antigen receptor (BCR). Int Immunol 2010;22:205–12.
[26] Reth M, Wienands J, Schamel WW. An unsolved problem of the clonal selection theory and the model of an oligomeric B-cell antigen receptor. Immunol Rev 2000;176:10–8.
[27] Yang J, Reth M. The dissociation activation model of B cell antigen receptor triggering. FEBS Lett 2010a;584:4872–7.
[28] Yang J, Reth M. Oligomeric organization of the B-cell antigen receptor on resting cells. Nature 2010b;467:465–9.
[29] Reth M. Matching cellular dimensions with molecular sizes. Nat Immunol 2013;14:765–7.
[30] Soderberg O, Leuchowius KJ, Gullberg M, Jarvius M, Weibrecht I, Larsson LG, et al. Characterizing proteins and their interactions in cells and tissues using the in situ proximity ligation assay. Methods 2008;45:227–32.
[31] Klaesener K, Maity PC, Hobeika E, Yang J, Reth M, et al. B cell activation involve nanoscale receptor reorganizations and inside-out signaling by Syk, eLife 2014;3:e02069.
[32] Tolar P, Pierce SK. A conformation-induced oligomerization model for B cell receptor microclustering and signaling. Curr Top Microbiol Immunol 2010;340:155–69.
[33] Singer SJ, Nicolson GL. The fluid mosaic model of the structure of cell membranes. Science 1972;175:720–31.
[34] Frye LD, Edidin M. The rapid intermixing of cell surface antigens after formation of mouse-human heterokaryons. J Cell Sci 1970;7:319–35.
[35] Kusumi A, Nakada C, Ritchie K, Murase K, Suzuki K, Murakoshi H, et al. Paradigm shift of the plasma membrane concept from the two-dimensional continuum fluid to the partitioned fluid: high-speed single-molecule tracking of membrane molecules. Annu Rev Biophys Biomol Struct 2005;34:351–78.
[36] Sako Y, Kusumi A. Compartmentalized structure of the plasma membrane for receptor movements as revealed by a nanometer-level motion analysis. J Cell Biol 1994;125:1251–64.
[37] Simons K, Ikonen E. Functional rafts in cell membranes. Nature 1997;387:569–72.
[38] Simons K, Sampaio JL. Membrane organization and lipid rafts. Cold Spring Harb Perspect Biol 2011;3:a004697.
[39] Wilson BS, Pfeiffer JR, Oliver JM. Observing FcepsilonRI signaling from the inside of the mast cell membrane. J Cell Biol 2000;149:1131–42.
[40] Lillemeier BF, Mortelmaier MA, Forstner MB, Huppa JB, Groves JT, Davis MM. TCR and Lat are expressed on separate protein islands on T cell membranes and concatenate during activation. Nat Immunol 2010;11:90–6.
[41] Lillemeier BF, Pfeiffer JR, Surviladze Z, Wilson BS, Davis MM. Plasma membrane-associated proteins are clustered into islands attached to the cytoskeleton. Proc Natl Acad Sci USA 2006;103:18992–7.
[42] Mueller NS, Wedlich-Soldner R, Spira F. From mosaic to patchwork: matching lipids and proteins in membrane organization. Mol Membr Biol 2012;29:186–96.
[43] Spira F, Mueller NS, Beck G, von Olshausen P, Beig J, Wedlich-Soldner R. Patchwork organization of the yeast plasma membrane into numerous coexisting domains. Nat Cell Biol 2012;14:640–8.

[44] Tolar P, Sohn HW, Pierce SK. Viewing the antigen-induced initiation of B-cell activation in living cells. Immunol Rev 2008;221:64–76.
[45] Harwood NE, Batista FD. Visualizing the molecular and cellular events underlying the initiation of B-cell activation. Curr Top Microbiol Immunol 2009;334:153–77.
[46] Mattila PK, Feest C, Depoil D, Treanor B, Montaner B, Otipoby KL, et al. The actin and tetraspanin networks organize receptor nanoclusters to regulate B cell receptor-mediated signaling. Immunity 2013;38:461–74.
[47] Edholm ES, Bengten E, Wilson M. Insights into the function of IgD. Dev Comp Immunol 2011;35:1309–16.
[48] Geisberger R, Lamers M, Achatz G. The riddle of the dual expression of IgM and IgD. Immunology 2006;118:429–37.
[49] Treanor B, Depoil D, Gonzalez-Granja A, Barral P, Weber M, Dushek O, et al. The membrane skeleton controls diffusion dynamics and signaling through the B cell receptor. Immunity 2010;32:187–99.
[50] Treanor B, Depoil D, Bruckbauer A, Batista FD. Dynamic cortical actin remodeling by ERM proteins controls BCR microcluster organization and integrity. J Exp Med 2011;208:1055–68.
[51] Infantino S, Benz B, Waldmann T, Jung M, Schneider R, Reth M. Arginine methylation of the B cell antigen receptor promotes differentiation. J Exp Med 2010;207:711–9.
[52] Nicholson TB, Chen T, Richard S. The physiological and pathophysiological role of PRMT1-mediated protein arginine methylation. Pharmacol Res 2009;60:466–74.
[53] Pao LI, Cambier JC. Syk, but not Lyn, recruitment to B cell antigen receptor and activation following stimulation of CD45- B cells. J Immunol 1997;158:2663–9.
[54] Schmitz R, Baumann G, Gram H. Catalytic specificity of phosphotyrosine kinases Blk, Lyn, c-Src and Syk as assessed by phage display. J Mol Biol 1996;260:664–77.
[55] Futterer K, Wong J, Grucza RA, Chan AC, Waksman G. Structural basis for Syk tyrosine kinase ubiquity in signal transduction pathways revealed by the crystal structure of its regulatory SH2 domains bound to a dually phosphorylated ITAM peptide. J Mol Biol 1998;281:523–37.
[56] Lamagna C, Hu Y, DeFranco AL, Lowell CA. B cell-specific loss of Lyn kinase leads to autoimmunity. J Immunol 2014;192:919–28.
[57] Turner M, Mee PJ, Costello PS, Williams O, Price AA, Duddy LP, et al. Perinatal lethality and blocked B-cell development in mice lacking the tyrosine kinase Syk. Nature 1995;378:298–302.
[58] Jiang A, Craxton A, Kurosaki T, Clark EA. Different protein tyrosine kinases are required for B cell antigen receptor-mediated activation of extracellular signal-regulated kinase, c-Jun NH2-terminal kinase 1, and p38 mitogen-activated protein kinase. J Exp Med 1998;188:1297–306.
[59] Yang C, Maruyama S, Yanagi S, Wang X, Takata M, Kurosaki T, et al. Syk and Lyn are involved in radiation-induced signaling, but inactivation of Syk or Lyn alone is not sufficient to prevent radiation-induced apoptosis. J Biochem 1995;118:33–8.
[60] Rolli V, Gallwitz M, Wossning T, Flemming A, Schamel WW, Zurn C, et al. Amplification of B cell antigen receptor signaling by a Syk/ITAM positive feedback loop. Mol Cell 2002;10:1057–69.
[61] Stepanek O, Draber P, Drobek A, Horejsi V, Brdicka T. Nonredundant roles of Src-family kinases and Syk in the initiation of B-cell antigen receptor signaling. J Immunol 2013;190:1807–18.
[62] Jang C, Machtaler S, Matsuuchi L. The role of Ig-alpha/beta in B cell antigen receptor internalization. Immunol Lett 2010;134:75–82.
[63] Heizmann B, Reth M, Infantino S. Syk is a dual-specificity kinase that self-regulates the signal output from the B-cell antigen receptor. Proc Natl Acad Sci USA 2010;107:18563–8.
[64] Muller R, Wienands J, Reth M. The serine and threonine residues in the Ig-alpha cytoplasmic tail negatively regulate immunoreceptor tyrosine-based activation motif-mediated signal transduction. Proc Natl Acad Sci USA 2000;97:8451–4.
[65] Del Nagro CJ, Otero DC, Anzelon AN, Omori SA, Kolla RV, Rickert RC. CD19 function in central and peripheral B-cell development. Immunol Res 2005;31:119–31.
[66] Jellusova J, Nitschke L. Regulation of B cell functions by the sialic acid-binding receptors Siglec-G and CD22. Front Immunol 2012;2:96.
[67] Fearon DT, Carroll MC. Regulation of B lymphocyte responses to foreign and self-antigens by the CD19/CD21 complex. Annu Rev Immunol 2000;18:393–422.
[68] Carter RH, Fearon DT. CD19: lowering the threshold for antigen receptor stimulation of B lymphocytes. Science 1992;256:105–7.
[69] Engel P, Zhou LJ, Ord DC, Sato S, Koller B, Tedder TF. Abnormal B lymphocyte development, activation, and differentiation in mice that lack or overexpress the CD19 signal transduction molecule. Immunity 1995;3:39–50.
[70] Rickert RC, Rajewsky K, Roes J. Impairment of T-cell-dependent B-cell responses and B-1 cell development in CD19-deficient mice. Nature 1995;376:352–5.
[71] Otero DC, Anzelon AN, Rickert RC. CD19 function in early and late B cell development: I. Maintenance of follicular and marginal zone B cells requires CD19-dependent survival signals. J Immunol 2003;170:73–83.
[72] Fehr T, Rickert RC, Odermatt B, Roes J, Rajewsky K, Hengartner H, et al. Antiviral protection and germinal center formation, but impaired B cell memory in the absence of CD19. J Exp Med 1998;188:145–55.
[73] Ahearn JM, Fischer MB, Croix D, Goerg S, Ma M, Xia J, et al. Disruption of the Cr2 locus results in a reduction in B-1a cells and in an impaired B cell response to T-dependent antigen. Immunity 1996;4:251–62.
[74] Otero DC, Rickert RC. CD19 function in early and late B cell development. II. CD19 facilitates the pro-B/pre-B transition. J Immunol 2003;171:5921–30.
[75] Buhl AM, Pleiman CM, Rickert RC, Cambier JC. Qualitative regulation of B cell antigen receptor signaling by CD19: selective requirement for PI3-kinase activation, inositol-1,4,5-trisphosphate production and $Ca^{2+}$ mobilization. J Exp Med 1997;186:1897–910.
[76] Pesando JM, Bouchard LS, McMaster BE. CD19 is functionally and physically associated with surface immunoglobulin. J Exp Med 1989;170:2159–64.
[77] Phee H, Rodgers W, Coggeshall KM. Visualization of negative signaling in B cells by quantitative confocal microscopy. Mol Cell Biol 2001;21:8615–25.
[78] Brooks SR, Li X, Volanakis EJ, Carter RH. Systematic analysis of the role of CD19 cytoplasmic tyrosines in enhancement of activation in Daudi human B cells: clustering of phospholipase C and Vav and of Grb2 and Sos with different CD19 tyrosines. J Immunol 2000;164:3123–31.

[79] Buhl AM, Cambier JC. Phosphorylation of CD19 Y484 and Y515, and linked activation of phosphatidylinositol 3-kinase, are required for B cell antigen receptor-mediated activation of Bruton's tyrosine kinase. J Immunol 1999;162:4438–46.
[80] Fujimoto M, Poe JC, Hasegawa M, Tedder TF. CD19 amplification of B lymphocyte $Ca^{2+}$ responses: a role for Lyn sequestration in extinguishing negative regulation. J Biol Chem 2001;276:44820–7.
[81] Fujimoto M, Fujimoto Y, Poe JC, Jansen PJ, Lowell CA, DeFranco AL, et al. CD19 regulates Src family protein tyrosine kinase activation in B lymphocytes through processive amplification. Immunity 2000;13:47–57.
[82] Xu Y, Beavitt S-J, Harder KW, Hibbs ML, Tarlinton DM. The activation and subsequent regulatory roles of Lyn and CD19 after B cell receptor ligation are independent. J Immunol 2002;169:6910–8.
[83] Hibbs ML, Tarlinton DM, Armes J, Grail D, Hodgson G, Maglitto R, et al. Multiple defects in the immune system of Lyn-deficient mice, culminating in autoimmune disease. Cell 1995;83:301–11.
[84] Wang Y, Brooks SR, Li X, Anzelon AN, Rickert RC, Carter RH. The physiologic role of CD19 cytoplasmic tyrosines. Immunity 2002;17:501–14.
[85] Wang Y, Carter RH. CD19 regulates B cell maturation, proliferation, and positive selection in the FDC zone of murine splenic germinal centers. Immunity 2005;22:749–61.
[86] Otero DC, Omori SA, Rickert RC. Cd19-dependent activation of Akt kinase in B-lymphocytes. J Biol Chem 2001;276:1474–8.
[87] Clayton E, Bardi G, Bell SE, Chantry D, Downes CP, Gray A, et al. A crucial role for the p110delta subunit of phosphatidylinositol 3-kinase in B cell development and activation. J Exp Med 2002;196:753–63.
[88] Fruman DA, Snapper SB, Yballe CM, Davidson L, Yu JY, Alt FW, et al. Impaired B cell development and proliferation in absence of phosphoinositide 3-kinase p85α. Science 1999;283:393–7.
[89] Jou ST, Carpino N, Takahashi Y, Piekorz R, Chao JR, Wang D, et al. Essential, nonredundant role for the phosphoinositide 3-kinase p110delta in signaling by the B-cell receptor complex. Mol Cell Biol 2002;22:8580–91.
[90] Okkenhaug K, Bilancio A, Farjot G, Priddle H, Sancho S, Peskett E, et al. Impaired B and T cell antigen receptor signaling in p110delta PI 3-kinase mutant mice. Science 2002;297:1031–4.
[91] Anzelon AN, Wu H, Rickert RC. Pten inactivation alters peripheral B lymphocyte fate and reconstitutes CD19 function. Nat Immunol 2003;4:287–94.
[92] van Zelm MC, Reisli I, van der Burg M, Castano D, van Noesel CJ, van Tol MJ, et al. An antibody-deficiency syndrome due to mutations in the CD19 gene. N Engl J Med 2006;354:1901–12.
[93] van Zelm MC, Smet J, Adams B, Mascart F, Schandene L, Janssen F, et al. CD81 gene defect in humans disrupts CD19 complex formation and leads to antibody deficiency. J Clin Invest 2010;120:1265–74.
[94] Tsuchiya N, Kuroki K, Fujimoto M, Murakami Y, Tedder TF, Tokunaga K, et al. Association of a functional CD19 polymorphism with susceptibility to systemic sclerosis. Arthritis Rheum 2004;50:4002–7.
[95] Sato S, Hasegawa M, Fujimoto M, Tedder TF, Takehara K. Quantitative genetic variation in CD19 expression correlates with autoimmunity. J Immunol 2000;165:6635–43.
[96] Erdei A, Isaák A, Török K, Sándor N, Kremlitzka M, Prechl J, Bajtay Z. Expression and role of CR1 and CR2 on B and T lymphocytes under physiological and autoimmune conditions. Mol Immunol 2009;46:2767–73.
[97] Chan VW, Lowell CA, DeFranco AL. Defective negative regulation of antigen receptor signaling in Lyn- deficient B lymphocytes. Curr Biol 1998;8:545–53.
[98] Smith KGC, Tarlinton DM, Doody GM, Hibbs ML, Fearon DT. Inhibition of the B cell by CD22: a requirement for Lyn. J Exp Med 1998;187:807–11.
[99] Doody GM, Justement LB, Delibrias CC, Matthews RJ, Lin J, Thomas ML, et al. A role in B cell activation for CD22 and the protein tyrosine phosphatase SHP. Science 1995;269:242–4.
[100] Chen J, McLean PA, Neel BG, Okunade G, Shull GE, Wortis HH. CD22 attenuates calcium signaling by potentiating plasma membrane calcium-ATPase activity. Nat Immunol 2004;5:651–7.
[101] Nadler MJ, McLean PA, Neel BG, Wortis HH. B cell antigen receptor-evoked calcium influx is enhanced in CD22-deficient B cell lines. J Immunol 1997;159:4233–43.
[102] Nitschke L, Carsetti R, Ocker B, Kohler G, Lamers MC. CD22 is a negative regulator of B-cell receptor signalling. Curr Biol 1997;7:133–43.
[103] O'Keefe TL, Williams GT, Davies SL, Neuberger MS. Hyperresponsive B cells in CD22-deficient mice. Science 1996;274:798–801.
[104] Otipoby KL, Andersson KB, Draves KE, Klaus SJ, Farr AG, Kerner JD, et al. CD22 regulates thymus-independent responses and the lifespan of B cells. Nature 1996;384:634–7.
[105] Sato S, Miller AS, Inaoki M, Bock CB, Jansen PJ, Tang ML, et al. CD22 is both a positive and negative regulator of B lymphocyte antigen receptor signal transduction: altered signaling in CD22-deficient mice. Immunity 1996;5:551–62.
[106] Muller J, Obermeier I, Wohner M, Brandl C, Mrotzek S, Angermuller S, et al. CD22 ligand-binding and signaling domains reciprocally regulate B-cell $Ca^{2+}$ signaling. Proc Natl Acad Sci USA 2013;110:12402–7.
[107] Gerlach J, Ghosh S, Jumaa H, Reth M, Wienands J, Chan AC, et al. B cell defects in SLP65/BLNK-deficient mice can be partially corrected by the absence of CD22, an inhibitory coreceptor for BCR signaling. Eur J Immunol 2003;33:3418–26.
[108] Sato S, Jansen PJ, Tedder TF. CD19 and CD22 expression reciprocally regulates tyrosine phosphorylation of Vav protein during B lymphocyte signaling. Proc Natl Acad Sci USA 1997;94:13158–62.
[109] Law CL, Sidorenko SP, Chandran KA, Zhao Z, Shen SH, Fischer EH, et al. CD22 associates with protein tyrosine phosphatase 1C, Syk, and phospholipase C-gamma(1) upon B cell activation. J Exp Med 1996;183:547–60.
[110] Poe JC, Fujimoto M, Jansen PJ, Miller AS, Tedder TF. CD22 forms a quaternary complex with SHIP, Grb2, and Shc. A pathway for regulation of B lymphocyte antigen receptor-induced calcium flux. J Biol Chem 2000;275:17420–7.
[111] Yohannan J, Wienands J, Coggeshall KM, Justement LB. Analysis of tyrosine phosphorylation-dependent interactions between stimulatory effector proteins and the B cell co-receptor CD22. J Biol Chem 1999;274:18769–76.
[112] Wakabayashi C, Adachi T, Wienands J, Tsubata T. A distinct signaling pathway used by the IgG-containing B cell antigen receptor. Science 2002;298:2392–5.
[113] Horikawa K, Martin SW, Pogue SL, Silver K, Peng K, Takatsu K, et al. Enhancement and suppression of signaling by the conserved tail of IgG memory-type B cell antigen receptors. J Exp Med 2007;204:759–69.

[114] Waisman A, Kraus M, Seagal J, Ghosh S, Melamed D, Song J, et al. IgG1 B cell receptor signaling is inhibited by CD22 and promotes the development of B cells whose survival is less dependent on Ig alpha/beta. J Exp Med 2007;204:747–58.
[115] Engels N, Konig LM, Heemann C, Lutz J, Tsubata T, Griep S, et al. Recruitment of the cytoplasmic adaptor Grb2 to surface IgG and IgE provides antigen receptor-intrinsic costimulation to class-switched B cells. Nat Immunol 2009;10:1018–25.
[116] Adachi T, Harumiya S, Takematsu H, Kozutsumi Y, Wabl M, Fujimoto M, et al. CD22 serves as a receptor for soluble IgM. Eur J Immunol 2012;42:241–7.
[117] Hanasaki K, Powell LD, Varki A. Binding of human plasma sialoglycoproteins by the B cell-specific lectin CD22. Selective recognition of immunoglobulin M and haptoglobin. J Biol Chem 1995;270:7543–50.
[118] Bi S, Baum LG. Sialic acids in T cell development and function. Biochim Biophys Acta 2009;1790:1599–610.
[119] Engel P, Nojima Y, Rothstein D, Zhou LJ, Wilson GL, Kehrl JH, et al. The same epitope on CD22 of B lymphocytes mediates the adhesion of erythrocytes, T and B lymphocytes, neutrophils, and monocytes. J Immunol 1993;150:4719–32.
[120] Danzer CP, Collins BE, Blixt O, Paulson JC, Nitschke L. Transitional and marginal zone B cells have a high proportion of unmasked CD22: implications for BCR signaling. Int Immunol 2003;15:1137–47.
[121] Razi N, Varki A. Masking and unmasking of the sialic acid-binding lectin activity of CD22 (Siglec-2) on B lymphocytes. Proc Natl Acad Sci USA 1998;95:7469–74.
[122] Zhang M, Varki A. Cell surface sialic acids do not affect primary CD22 interactions with CD45 and surface IgM nor the rate of constitutive CD22 endocytosis. Glycobiology 2004;14:939–49.
[123] Han S, Collins BE, Bengtson P, Paulson JC. Homomultimeric complexes of CD22 in B cells revealed by protein-glycan cross-linking. Nat Chem Biol 2005;1:93–7.
[124] Collins BE, Smith BA, Bengtson P, Paulson JC. Ablation of CD22 in ligand-deficient mice restores B cell receptor signaling. Nat Immunol 2006;7:199–206.
[125] Collins BE, Blixt O, DeSieno AR, Bovin N, Marth JD, Paulson JC. Masking of CD22 by cis ligands does not prevent redistribution of CD22 to sites of cell contact. Proc Natl Acad Sci USA 2004;101:6104–9.
[126] Grewal PK, Boton M, Ramirez K, Collins BE, Saito A, Green RS, et al. ST6Gal-I restrains CD22-dependent antigen receptor endocytosis and Shp-1 recruitment in normal and pathogenic immune signaling. Mol Cell Biol 2006;26:4970–81.
[127] Poe JC, Fujimoto Y, Hasegawa M, Haas KM, Miller AS, Sanford IG, et al. CD22 regulates B lymphocyte function in vivo through both ligand-dependent and ligand-independent mechanisms. Nat Immunol 2004;5:1078–87.
[128] Lanoue A, Batista FD, Stewart M, Neuberger MS. Interaction of CD22 with α2,6-linked sialoglycoconjugates: innate recognition of self to dampen B cell autoreactivity? Eur J Immunol 2002;32:348–55.
[129] Macauley MS, Pfrengle F, Rademacher C, Nycholat CM, Gale AJ, von Drygalski A, et al. Antigenic liposomes displaying CD22 ligands induce antigen-specific B cell apoptosis. J Clin Invest 2013;123:3074–83.
[130] Bolland S, Ravetch JV. Spontaneous autoimmune disease in Fc(gamma)RIIB-deficient mice results from strain-specific epistasis. Immunity 2000;13:277–85.
[131] Pao LI, Lam KP, Henderson JM, Kutok JL, Alimzhanov M, Nitschke L, et al. B cell-specific deletion of protein-tyrosine phosphatase Shp1 promotes B-1a cell development and causes systemic autoimmunity. Immunity 2007;27:35–48.
[132] Mary C, Laporte C, Parzy D, Santiago ML, Stefani F, Lajaunias F, et al. Dysregulated expression of the Cd22 gene as a result of a short interspersed nucleotide element insertion in Cd22a lupus-prone mice. J Immunol 2000;165:2987–96.
[133] O'Keefe TL, Williams GT, Batista FD, Neuberger MS. Deficiency in CD22, a B cell-specific inhibitory receptor, is sufficient to predispose to development of high affinity autoantibodies. J Exp Med 1999;189:1307–13.
[134] Lajaunias F, Nitschke L, Moll T, Martinez-Soria E, Semac I, Chicheportiche Y, et al. Differentially regulated expression and function of CD22 in activated B- 1 and B-2 lymphocytes. J Immunol 2002;168:6078–83.
[135] Jellusova J, Wellmann U, Amann K, Winkler TH, Nitschke L. CD22 x Siglec-G double-deficient mice have massively increased B1 cell numbers and develop systemic autoimmunity. J Immunol 2010;184:3618–27.
[136] Cariappa A, Takematsu H, Liu H, Diaz S, Haider K, Boboila C, et al. B cell antigen receptor signal strength and peripheral B cell development are regulated by a 9-O-acetyl sialic acid esterase. J Exp Med 2009;206:125–38.
[137] Surolia I, Pirnie SP, Chellappa V, Taylor KN, Cariappa A, Moya J, et al. Functionally defective germline variants of sialic acid acetylesterase in autoimmunity. Nature 2010;466:243–7.
[138] Harley IT, Kaufman KM, Langefeld CD, Harley JB, Kelly JA. Genetic susceptibility to SLE: new insights from fine mapping and genome-wide association studies. Nat Rev Genet 2009;10:285–90.
[139] Kurosaki T. Functional dissection of BCR signaling pathways. Curr Opin Immunol 2000;12:276–81.
[140] Niiro H, Clark EA. Regulation of B-cell fate by antigen-receptor signals. Nat Rev Immunol 2002;2:945–56.
[141] Kurosaki T, Johnson SA, Pao L, Sada K, Yamamura H, Cambier JC. Role of the Syk autophosphorylation site and SH2 domains in B cell antigen receptor signaling. J Exp Med 1995;182: 1815–23.
[142] Rowley RB, Burkhardt AL, Chao HG, Matsueda GR, Bolen JB. Syk protein-tyrosine kinase is regulated by tyrosine-phosphorylated Igα/Igβ immunoreceptor tyrosine activation motif binding and autophosphorylation. J Biol Chem 1995;270:11590–4.
[143] Hashimoto S, Iwamatsu A, Ishiai M, Okawa K, Yamadori T, Matsushita M, et al. Identification of the SH2 domain binding protein of Bruton's tyrosine kinase as BLNK–functional significance of Btk-SH2 domain in B-cell antigen receptor-coupled calcium signaling. Blood 1999;94:2357–64.
[144] Ishiai M, Kurosaki M, Pappu R, Okawa K, Ronko I, Fu C, et al. BLNK required for coupling Syk to PLCγ2 and Rac1-JNK in B cells. Immunity 1999;10:117–25.
[145] Yu P, Constien R, Dear N, Katan M, Hanke P, Bunney TD, et al. Autoimmunity and inflammation due to a gain-of-function mutation in phospholipase Cγ2 that specifically increases external $Ca^{2+}$ entry. Immunity 2005;22:451–65.
[146] Engels N, Wollscheid B, Wienands J. Association of SLP-65/BLNK with the B cell antigen receptor through a non-ITAM tyrosine of Ig-α. Eur J Immunol 2001;31:2126–34.

[147] Kabak S, Skaggs BJ, Gold MR, Affolter M, West KL, Foster MS, et al. The direct recruitment of BLNK to immunoglobulin alpha couples the B-cell antigen receptor to distal signaling pathways. Mol Cell Biol 2002;22:2524–35.

[148] Patterson HC, Kraus M, Kim YM, Ploegh H, Rajewsky K. The B cell receptor promotes B cell activation and proliferation through a non-ITAM tyrosine in the Igα cytoplasmic domain. Immunity 2006;25:55–65.

[149] Kohler F, Storch B, Kulathu Y, Herzog S, Kuppig S, Reth M, et al. A leucine zipper in the N terminus confers membrane association to SLP65. Nat Immunol 2005;6:204–10.

[150] Miyazaki A, Yogosawa S, Murakami A, Kitamura D. Identification of CMTM7 as a transmembrane linker of BLNK and the B-cell receptor. PLoS One 2012;7:e31829.

[151] Oellerich T, Bremes V, Neumann K, Bohnenberger H, Dittmann K, Hsiao HH, et al. The B-cell antigen receptor signals through a preformed transducer module of SLP65 and CIN85. EMBO J 2011;30:3620–34.

[152] Kim YJ, Sekiya F, Poulin B, Bae YS, Rhee SG. Mechanism of B-cell receptor-induced phosphorylation and activation of phospholipase C-γ2. Mol Cell Biol 2004;24:9986–99.

[153] Rodriguez R, Matsuda M, Perisic O, Bravo J, Paul A, Jones NP, et al. Tyrosine residues in phospholipase Cγ2 essential for the enzyme function in B-cell signaling. J Biol Chem 2001;276:47982–92.

[154] Su YW, Zhang Y, Schweikert J, Koretzky GA, Reth M, Wienands J. Interaction of SLP adaptors with the SH2 domain of Tec family kinases. Eur J Immunol 1999;29:3702–11.

[155] Takata M, Kurosaki T. A role for Bruton's tyrosine kinase in B cell antigen receptor-mediated activation of phospholipase C-γ2. J Exp Med 1996;184:31–40.

[156] Tan JE, Wong SC, Gan SK, Xu S, Lam KP. The adaptor protein BLNK is required for B cell antigen receptor-induced activation of nuclear factor-κB and cell cycle entry and survival of B lymphocytes. J Biol Chem 2001;276:20055–63.

[157] Hashimoto A, Takeda K, Inaba M, Sekimata M, Kaisho T, Ikehara S, et al. Cutting edge: essential role of phospholipase C-γ2 in B cell development and function. J Immunol 2000;165:1738–42.

[158] Hikida M, Casola S, Takahashi N, Kaji T, Takemori T, Rajewsky K, et al. PLC-γ2 is essential for formation and maintenance of memory B cells. J Exp Med 2009;206:681–9.

[159] Hikida M, Johmura S, Hashimoto A, Takezaki M, Kurosaki T. Coupling between B cell receptor and phospholipase C-γ2 is essential for mature B cell development. J Exp Med 2003;198:581–9.

[160] Wang D, Feng J, Wen R, Marine JC, Sangster MY, Parganas E, et al. Phospholipase Cγ2 is essential in the functions of B cell and several Fc receptors. Immunity 2000;13:25–35.

[161] Patterson RL, Boehning D, Snyder SH. Inositol 1,4,5-trisphosphate receptors as signal integrators. Annu Rev Biochem 2004;73:437–65.

[162] Sugawara H, Kurosaki M, Takata M, Kurosaki T. Genetic evidence for involvement of type 1, type 2 and type 3 inositol 1,4,5-trisphosphate receptors in signal transduction through the B-cell antigen receptor. EMBO J 1997;16:3078–88.

[163] Liou J, Kim ML, Heo WD, Jones JT, Myers JW, Ferrell Jr JE, et al. STIM is a $Ca^{2+}$ sensor essential for $Ca^{2+}$-store-depletion-triggered $Ca^{2+}$ influx. Curr Biol 2005;15:1235–41.

[164] Roos J, DiGregorio PJ, Yeromin AV, Ohlsen K, Lioudyno M, Zhang S, et al. STIM1, an essential and conserved component of store-operated $Ca^{2+}$ channel function. J Cell Biol 2005;169:435–45.

[165] Feske S, Gwack Y, Prakriya M, Srikanth S, Puppel SH, Tanasa B, et al. A mutation in Orai1 causes immune deficiency by abrogating CRAC channel function. Nature 2006;441:179–85.

[166] Vig M, Peinelt C, Beck A, Koomoa DL, Rabah D, Koblan-Huberson M, et al. CRACM1 is a plasma membrane protein essential for store-operated $Ca^{2+}$ entry. Science 2006;312:1220–3.

[167] Baba Y, Hayashi K, Fujii Y, Mizushima A, Watarai H, Wakamori M, et al. Coupling of STIM1 to store-operated $Ca^{2+}$ entry through its constitutive and inducible movement in the endoplasmic reticulum. Proc Natl Acad Sci USA 2006;103:16704–9.

[168] Huang GN, Zeng W, Kim JY, Yuan JP, Han L, Muallem S, et al. STIM1 carboxyl-terminus activates native SOC, $I_{crac}$ and TRPC1 channels. Nat Cell Biol 2006;8:1003–10.

[169] Luik RM, Wu MM, Buchanan J, Lewis RS. The elementary unit of store-operated $Ca^{2+}$ entry: local activation of CRAC channels by STIM1 at ER-plasma membrane junctions. J Cell Biol 2006;174:815–25.

[170] Wu MM, Buchanan J, Luik RM, Lewis RS. $Ca^{2+}$ store depletion causes STIM1 to accumulate in ER regions closely associated with the plasma membrane. J Cell Biol 2006;174:803–13.

[171] Xu P, Lu J, Li Z, Yu X, Chen L, Xu T. Aggregation of STIM1 underneath the plasma membrane induces clustering of Orai1. Biochem Biophys Res Commun 2006;350:969–76.

[172] Zhang SL, Yu Y, Roos J, Kozak JA, Deerinck TJ, Ellisman MH, et al. STIM1 is a $Ca^{2+}$ sensor that activates CRAC channels and migrates from the $Ca^{2+}$ store to the plasma membrane. Nature 2005;437:902–5.

[173] Li Z, Lu J, Xu P, Xie X, Chen L, Xy T. Mapping the interacting domains of STIM1 and Orai1 in $Ca^{2+}$ release-activated $Ca^{2+}$ channel activation. J Biol Chem 2007;282:29448–56.

[174] Stathopulos PB, Zheng L, Li GY, Plevin MJ, Ikura M. Structural and mechanistic insights into STIM1-mediated initiation of store-operated calcium entry. Cell 2008;135:110–22.

[175] Matsumoto M, Fujii Y, Baba A, Hikida M, Kurosaki T, Baba Y. The calcium sensors STIM1 and STIM2 control B cell regulatory function through interleukin-10 production. Immunity 2011;34:703–14.

[176] Baba Y, Kurosaki T. Impact of $Ca^{2+}$ signaling on B cell function. Trends Immunol 2011;32:589–94.

[177] Crabtree GR, Olson EN. NFAT signaling: choreographing the social lives of cells. Cell 2002;(109 Suppl.):S67–79.

[178] Timmerman LA, Healy JI, Ho SN, Chen L, Goodnow CC, Crabtree GR. Redundant expression but selective utilization of nuclear factor of activated T cells family members. J Immunol 1997;159:2735–40.

[179] Bhattacharyya S, Deb J, Patra AK, Thuy Pham DA, Chen W, Vaeth M, et al. NFATc1 affects mouse splenic B cell function by controlling the calcineurin–NFAT signalıng network. J Exp Med 2011;208:823–39.

[180] Hacker H, Karin M. Regulation and function of IKK and IKK-related kinases. Sci STKE 2006;2006. re13.

[181] Saijo K, Mecklenbrauker I, Santana A, Leitger M, Schmedt C, Tarakhovsky A. Protein kinase C β controls nuclear factor κB activation in B cells through selective regulation of the IκB kinase α. J Exp Med 2002;195:1647–52.

[182] Thome M. CARMA1, BCL-10 and MALT1 in lymphocyte development and activation. Nat Rev Immunol 2004;4:348–59.

[183] Shinohara H, Maeda S, Watarai H, Kurosaki T. IκB kinase β-induced phosphorylation of CARMA1 contributes to CARMA1 Bcl10 MALT1 complex formation in B cells. J Exp Med 2007;204:3285–93.

[184] Shinohara H, Yasuda T, Aiba Y, Sanjo H, Hamadate M, Watarai H, et al. PKC β regulates BCR-mediated IKK activation by facilitating the interaction between TAK1 and CARMA1. J Exp Med 2005;202:1423–31.
[185] Adhikari A, Xu M, Chen ZJ. Ubiquitin-mediated activation of TAK1 and IKK. Oncogene 2007;26:3214–26.
[186] Jun JE, Wilson LE, Vinuesa CG, Lesage S, Blery M, Miosge LA, et al. Identifying the MAGUK protein Carma-1 as a central regulator of humoral immune responses and atopy by genome-wide mouse mutagenesis. Immunity 2003;18:751–62.
[187] Karin M, Ben-Neriah Y. Phosphorylation meets ubiquitination: the control of NF-κB activity. Annu Rev Immunol 2000;18:621–63.
[188] Thome M, Weil R. Post-translational modifications regulate distinct functions of CARMA1 and BCL10. Trends Immunol 2007;28:281–8.
[189] Alessi DR, James SR, Downes CP, Holmes AB, Gaffney PR, Reese CB, et al. Characterization of a 3-phosphoinositide-dependent protein kinase which phosphorylates and activates protein kinase Bα. Curr Biol 1997;7:261–9.
[190] Delcommenne M, Tan C, Gray V, Rue L, Woodgett J, Dedhar S. Phosphoinositide-3-OH kinase-dependent regulation of glycogen synthase kinase 3 and protein kinase B/AKT by the integrin-linked kinase. Proc Natl Acad Sci USA 1998;95:11211–6.
[191] Franke TF, Yang SI, Chan TO, Datta K, Kazlauskas A, Morrison DK, et al. The protein kinase encoded by the Akt proto-oncogene is a target of the PDGF-activated phosphatidylinositol 3-kinase. Cell 1995;81:727–36.
[192] Suzuki H, Terauchi Y, Fujiwara M, Aizawa S, Yazaki Y, Kadowaki T, et al. Xid-like immunodeficiency in mice with disruption of the p85α subunit of phosphoinositide 3-kinase. Science 1999;283:390–2.
[193] Okada T, Maeda A, Iwamatsu A, Gotoh K, Kurosaki T. BCAP: the tyrosine kinase substrate that connects B cell receptor to phosphoinositide 3-kinase activation. Immunity 2000;13:817–27.
[194] Aiba Y, Kameyama M, Yamazaki T, Tedder TF, Kurosaki T. Regulation of B-cell development by BCAP and CD19 through their binding to phosphoinositide 3-kinase. Blood 2008;111:1497–503.
[195] Castello A, Gaya M, Tucholski J, Oellerich T, Lu KH, Tafuri A, et al. Nck-mediated recruitment of BCAP to the BCR regulates the $PI_3K$-Akt pathway in B cells. Nat Immunol 2013;14:966–75.
[196] Han J, Luby-Phelps K, Das B, Shu X, Xia Y, Mosteller RD, et al. Role of substrates and products of PI 3-kinase in regulating activation of Rac-related guanosine triphosphatases by Vav. Science 1998;279:558–60.
[197] Rapley J, Tybulewicz VL, Rittinger K. Crucial structural role for the PH and C1 domains of the Vav1 exchange factor. EMBO Rep 2008;9:655–61.
[198] Tedford K, Nitschke L, Girkontaite I, Charlesworth A, Chan G, Sakk V, et al. Compensation between Vav-1 and Vav-2 in B cell development and antigen receptor signaling. Nat Immunol 2001;2:548–55.
[199] Das B, Shu X, Day GJ, Han J, Krishna UM, Falck JR, et al. Control of intramolecular interactions between the pleckstrin homology and Dbl homology domains of Vav and Sos1 regulates Rac binding. J Biol Chem 2000;275:15074–81.
[200] Inabe K, Ishiai M, Scharenberg AM, Freshney N, Downward J, Kurosaki T. Vav3 modulates B cell receptor responses by regulating phosphoinositide 3-kinase activation. J Exp Med 2002;195:189–200.
[201] Vigorito E, Bardi G, Glassford J, Lam EW, Clayton E, Turner M. Vav-dependent and vav-independent phosphatidylinositol 3-kinase activation in murine B cells determined by the nature of the stimulus. J Immunol 2004;173:3209–14.
[202] Tarakhovsky A, Turner M, Schaal S, Mee PJ, Duddy LP, Rajewsky K, et al. Defective antigen receptor-mediated proliferation of B and T cells in the absence of Vav. Nature 1995;374:467–70.
[203] Zhang R, Alt FW, Davidson L, Orkin SH, Swat W. Defective signalling through the T- and B-cell antigen receptors in lymphoid cells lacking the vav proto-oncogene. Nature 1995;374:470–3.
[204] Fischer KD, Zmuldzinas A, Gardner S, Barbacid M, Bernstein A, Guidos C. Defective T-cell receptor signalling and positive selection of Vav-deficient $CD4^+$ $CD8^+$ thymocytes. Nature 1995;374:474–7.
[205] Manetz TS, Gonzalez-Espinosa C, Arudchandran R, Xirasagar S, Tybulewicz V, Rivera J. Vav1 regulates phospholipase cgamma activation and calcium responses in mast cells. Mol Cell Biol 2001;21:3763–74.
[206] Doody GM, Bell SE, Vigorito E, Clayton E, McAdam S, Tooze R, et al. Signal transduction through Vav-2 participates in humoral immune responses and B cell maturation. Nat Immunol 2001;2:542–7.
[207] Movilla N, Bustelo XR. Biological and regulatory properties of Vav-3, a new member of the Vav family of oncoproteins. Mol Cell Biol 1999;19:7870–85.
[208] Zeng L, Sachdev P, Yan L, Chan JL, Trenkle T, McClelland M, et al. Vav3 mediates receptor protein tyrosine kinase signaling, regulates GTPase activity, modulates cell morphology, and induces cell transformation. Mol Cell Biol 2000;20:9212–24.
[209] Okkenhaug K, Vanhaesebroeck B. PI3K in lymphocyte development, differentiation and activation. Nat Rev Immunol 2003;3:317–30.
[210] Deane JA, Fruman DA. Phosphoinositide 3-kinase: diverse roles in immune cell activation. Annu Rev Immunol 2004;22:563–98.
[211] Delgado P, Cubelos B, Calleja E, Martinez-Martin N, Cipres A, Merida I, et al. Essential function for the GTPase TC21 in homeostatic antigen receptor signaling. Nat Immunol 2009;10:880–8.
[212] Srinivasan L, Sasaki Y, Calado DP, Zhang B, Paik JH, DePinho RA, et al. PI3 kinase signals BCR-dependent mature B cell survival. Cell 2009;139:573–86.
[213] Haftmann C, Stittrich AB, Sgouroudis E, Matz M, Chang HD, Radbruch A, et al. Lymphocyte signaling: regulation of FoxO transcription factors by microRNAs. Ann NY Acad Sci 2012;1247:46–55.
[214] Yusuf I, Zhu X, Kharas MG, Chen J, Fruman DA. Optimal B-cell proliferation requires phosphoinositide 3-kinase-dependent inactivation of FOXO transcription factors. Blood 2004;104:784–7.
[215] Dengler HS, Baracho GV, Omori SA, Bruckner S, Arden KC, Castrillon DH, et al. Distinct functions for the transcription factor Foxo1 at various stages of B cell differentiation. Nat Immunol 2008;9:1388–98.
[216] Amin RH, Schlissel MS. Foxo1 directly regulates the transcription of recombination activating genes during B cell development. Nat Immunol 2008;9:613–22.
[217] Herzog S, Hug E, Meixlsperger S, Paik JH, DePinho RA, Reth M, et al. SLP-65 regulates immunoglobulin light chain gene recombination through the $PI_3K$-PKB-Foxo pathway. Nat Immunol 2008;9:623–31.

Chapter 11

# Fc and Complement Receptors

Jeffrey V. Ravetch[1], Falk Nimmerjahn[2] and Michael C. Carroll[3]

[1]*Laboratory of Molecular Genetics and Immunology, The Rockefeller University, New York, USA,* [2]*Institute of Genetics, University of Erlangen-Nuernberg, Germany,* [3]*Program in Cell and Molecular Medicine, Boston Childrens Hospital, Harvard Medical School, Boston, Massachusetts, USA*

Both the Fc and complement systems have evolved distinct strategies for facilitating the interaction of environmental antigens with B cells to contribute to an effective antibody response. However, we have chosen to consider these two systems in parallel because they are interdependent in considering the fate of interaction of the immune complex with the adaptive immune response. In this chapter, we first deal with the significance of these systems by reviewing the phenotypes of complement and Fc receptor (FcR) deficiencies in both mice and humans. Our understanding of the mechanisms that account for these phenotypes is discussed for each pathway, and the interactions that occur between these systems are summarized. For the FcR system, we now recognize two general classes of receptors based on structural homology and mode of binding: type I receptors that include the canonical FcγRs that comprise three activating and one inhibitory member and bind in a 1:1 complex to the open Fc conformation, and type II receptors that include DC-SIGN, SIGN R1, and CD23, C-type lectins that bind immunoglobulin (Ig) G in a 2:1 complex to the closed Fc conformation (Figure 1). We focus on FcγRIIB, the inhibitory receptor expressed exclusively on both B cells and follicular dendritic cells (FDCs), for which substantial data are now available. In addition, we discuss more recent data indicating that members of the FcR-like (FcRl) family, including FcRl4 and FcRl5 and a receptor for IgM, are expressed on B cells and may also be involved in modulating B-cell function. Although much is known about the role of CD23 in regulating IgE synthesis, its role in IgG homeostasis, albeit suggestive, remains to be determined. The early components of complement, notably C1q, C4, and C3 and the complement receptors CR1 and CR2, are particularly relevant to the role of complement in afferent responses. In vivo studies of complement and Fcγ receptor systems are our primary emphasis, although in vitro data are discussed where appropriate. Complement and FcγR involvement in the efferent response is not discussed here; for discussion of those systems, please see several recent reviews [1,2].

## 1. CONSEQUENCES OF FcγRIIB DEFICIENCY

The targeted disruption of FcγRIIB results in animals with amplified antibody responses to both TI and TD antigens, with Ig titers increased by five- to ten-fold [3]. Baseline serum immunoglobulin levels are unaffected, and the mice display a normal Ig half-life. This amplification is apparent on all genetic backgrounds investigated. Backcrossing onto the C57BL/6 background further resulted in animals with severe autoimmune disease [4,5]. These animals displayed a spontaneous loss of tolerance to nuclear antigens, with high titers of IgG antibodies to double-stranded DNA (dsDNA) and histone H2A/2B by 6 months of age. Autoimmune disease was apparent in these animals, with prominent vasculitis and glomerulonephritis resulting from IC deposition. The defect in these animals appears to reside, in part, in the B-cell compartment, as determined by reconstitution of the autoimmune phenotype through the transfer of FcγRIIB-deficient bone marrow (BM) into B-cell– or T- and B-cell–deficient recipients who have otherwise normal FcγRIIB expression on their myeloid cells. Moreover, restoring FcγRIIB expression either ubiquitously or selectively on B cells was able to restore humoral tolerance in autoimmune-prone mouse strains with impaired FcγRIIB expression, including FcγRIIB knockout mice [6,7]. Of note, treatment of mice and human patients with chronic inflammatory demyelinating polyneuropathy (CIDP) with pooled polyclonal IgG preparations (intravenous immunoglobulin therapy or IVIG) was shown to upregulate the inhibitory FcR on myeloid cells and B cells, which may heighten the threshold for autoantibody production [8–11]. Further evidence that FcγRIIB-deficient animals have a defect in the maintenance of peripheral tolerance was demonstrated by the ability of these animals to develop autoantibodies to murine type II or type IV collagen after immunization with bovine type II or IV collagen, respectively, on a nonsusceptible H-2b background [12,13]. The consequence of this induced breakage of tolerance to these autoantigens was the

Molecular Biology of B Cells. http://dx.doi.org/10.1016/B978-0-12-397933-9.00011-4

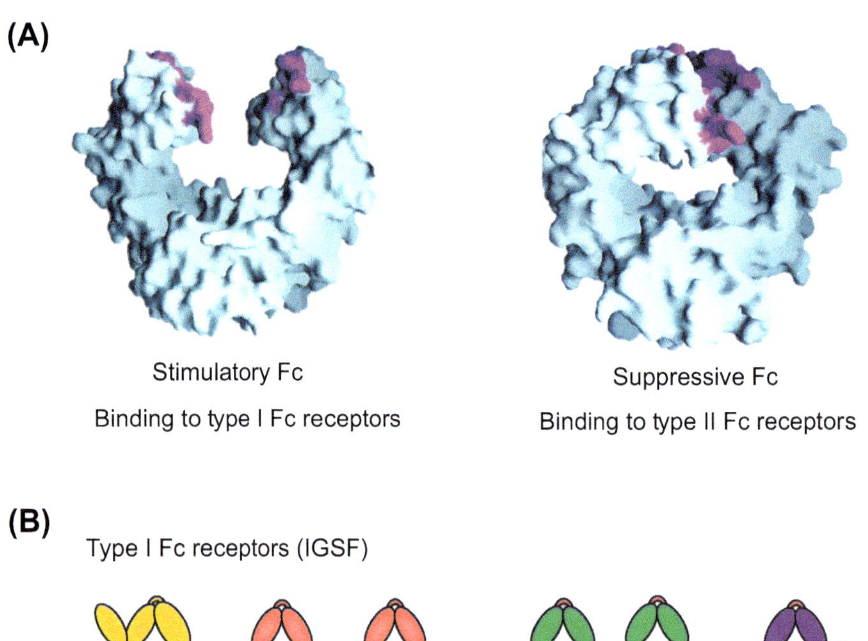


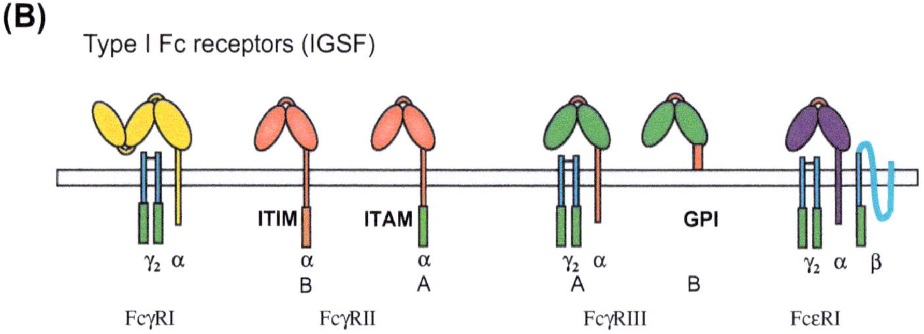


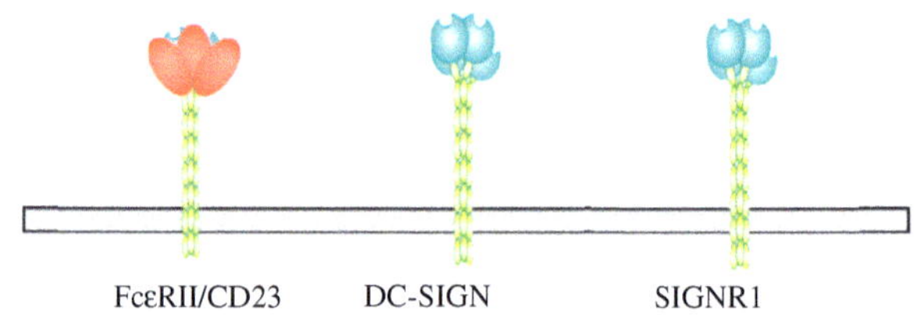


FIGURE 1 **The family of FcRs and their ligands.** Shown are the two glycosylation-dependent conformations of an IgG Fc-fragment with either immunostimulatory or inhibitory activity (A) and the two groups of FcRs recognizing these different IgG Fc conformations (B). Type I receptors belong to the immunoglobulin superfamily (IGSF), whereas type II receptors are members of the C-type lectin family.

development of collagen-induced arthritis in nonsusceptible backgrounds in response to type II collagen immunization, or Goodpasture's syndrome in the case of type IV collagen immunization. These data suggest that decreased levels of FcγRIIB correlate with a susceptibility to the development of autoimmune diseases such as systemic lupus erythematosus (SLE) and are further supported by the reports of decreased FcγRIIB levels on B cells in mouse strains predisposed to the development of antinuclear antibodies, such as NZB, BXSB, and MRL [14,15], and by its tight linkage to SLE loci on chromosome 1 in both human and mouse studies [16]. In humans, a failed upregulation of FcγRIIB from naïve to memory B cells was noted in SLE and CIDP patients [8,17], and a functionally impaired FcγRIIB variant was found to be highly associated with the development of SLE in Asian and Caucasian patient cohorts [18–21]. In this allelic variant, an isoleucine residue in the transmembrane domain of FcγRIIB is replaced by a charged threonine residue (FcγRIIBI232T), resulting in an impaired recruitment to lipid rafts and a strongly reduced function of this receptor variant. More direct evidence that this functionally impaired FcγRIIB allele indeed might be responsible for a loss of humoral tolerance in humans was afforded by studies in which immunodeficient mice were reconstituted with human hematopoietic stem cells from donors carrying either the functionally intact FcγRIIB-232I or the altered FcγRIIB-232T variant [2]. Indeed, a variety of autoantibody species including dsDNA antibodies associated with the development of autoimmune disease in humans could be detected in humanized mice, in which a human immune system developed on the nonfunctional FcγRIIB-I232T background. These data indicate that FcγRIIB expression on B cells contributes to the maintenance of peripheral tolerance to nuclear antigens and to the suppression of autoantibody production Fc and complement receptor in response to crossreactive, exogenous antigens. The mechanisms that account for these observations will be discussed in detail below.

## 2. CONSEQUENCES OF COMPLEMENT AND COMPLEMENT RECEPTOR DEFICIENCIES

As is the case for FcγRIIB, complement and complement receptors are required for the maintenance of tolerance to autoantigens, although through quite different mechanisms, as is discussed below. Deficiency in complement proteins C1q and C4 leads to an increased susceptibility to the development of lupus and the production of autoantibodies both in humans and murine models [22]. This striking observation supports a critical role for complement and its receptors in the protection from maturation of self-reactive B cells. Using a murine model (564 Igi) in which the B cells were heterozygous for insertion of rearranged Ig heavy and light chain genes specific for the ribonucleoprotein, deficiency in C4 allowed for an escape of anergy at the immature stage of differentiation (findings are discussed in more detail below) [23]. Thus, complement participates in negative selection at the transitional stage.

The development of knockout mice bearing targeted Cr2 loci provided animal models to not only directly test the importance of the two receptors but also to dissect the mechanism of B-cell regulation in vivo. Three lines of targeted mice have been reported, and in general they have a similar phenotype, although one line expresses a low level of truncated CD21. All three lines fail to develop secondary antibody responses to foreign protein antigens administered in the absence of adjuvant; a reduced number and size of germinal centers (GCs) characterize their response. Both the primary and secondary responses to certain thymus-dependent (T-D) antigens are reduced, suggesting that complement can function both in the initial activation and expansion of antigen specific B cells as well as their survival within the GC. The similar impaired response of mice deficient in either C3 or C4 with those deficient in the receptor suggests that complement mediates its effects on B-cell responses via CD21/CD35. The observation that co-crosslinking of the B-cell receptor (BCR) and coreceptor lowered the threshold for B-cell activation led to the important general concept that complement receptors link innate and adaptive immunity [24]. Thus, the innate immune system provides a novel mechanism of identifying pathogens by covalent attachment of C3 and enhancing the humoral response via CD21/CD35. This concept was tested by comparing the humoral response of mice deficient in complement proteins C3 or C4 or receptors CD21/CD35 after infection with herpes simplex virus type 1 (HSV-1). Significantly, all three groups of mice failed to develop a significant antibody response to the infectious virus despite multiple infections (Figure 2). Thus, complement and its receptors are essential in the formation of a memory B-cell response to this important human pathogen. The defect was in the B-cell compartment, because CD4+T cells were activated normally and responded to viral antigens on secondary stimulation in vitro.

Experiments involving chimeric mice support the overall conclusion that CD21/CD35 expression on B cells is essential for B-cell activation in vivo but that FDC expression is also important in the localization of antigen within the follicles and the promotion of long-term B-cell survival. Complement receptors CD21/CD35 also appear to be important in the selection or expansion and maintenance of B-1 cells, because Cr2-def mice have an altered repertoire of natural antibody to certain but not all self-antigens. Thus, Cr2-def mice are missing or have reduced levels of natural antibody involved in the induction of reperfusion injury (ischemia/reperfusion [I/R]).

In summary, complement receptors CD21 and CD35 mediate the enhancing effects of complement C3. Thus, the receptors link the innate and adaptive response that results in enhanced humoral immunity.

## 3. Fc RECEPTORS

### 3.1 Expression Pattern and Signaling Properties of FcγRIIB

The interpretation of the phenotypes displayed by FcγRIIB-deficient mice is complicated by the ubiquitous expression pattern of this inhibitory receptor. On B cells and FDCs, RIIB is the only FcR for IgG expressed. RIIB is expressed at all stages of B-cell development, including pre-, pro-, and mature populations. Of note, FcγRIIB expression is maintained on memory B cells and plasma cells, on which most of the other B lineage molecules are either absent or present at very low levels [2,25]. Expression levels of FcγRIIB are modulated on different B-cell populations, displaying higher levels on peripheral B cells and reduced expression on GC and activated B cells [26,27]. Interleukin-4 further down-regulates FcγRIIB expression on GC B cells [27]. FcγRIIB expression is induced on FDCs upon antigen stimulation. In addition, FcγRIIB is expressed on immature dendritic cells, where it accounts for >75% of surface expression of FcγRs on those cells. FcγRIIB deficiency does not appear to perturb the development of any of these lineages, suggesting that FcγRIIB functions specifically in response to

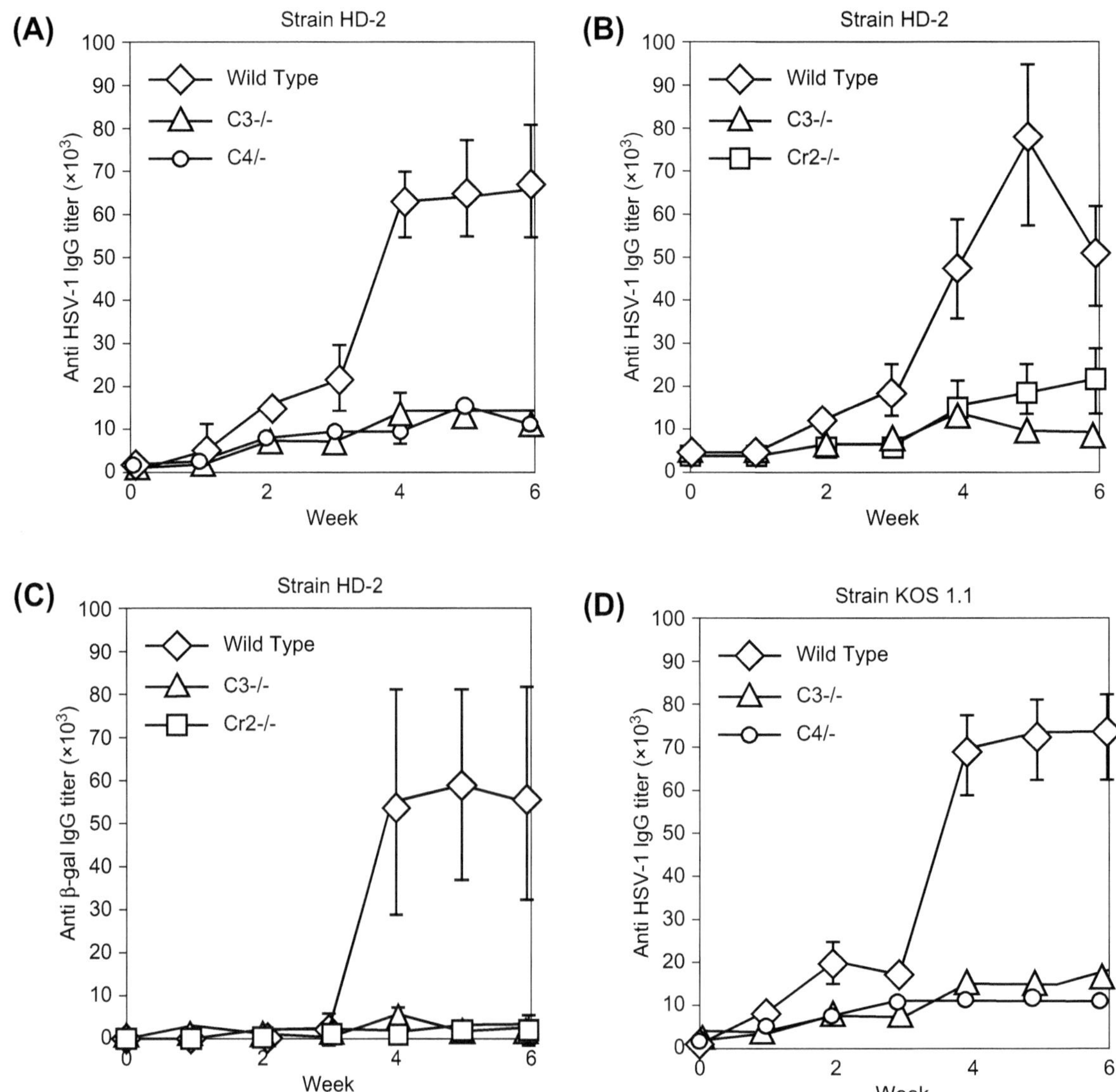


FIGURE 2 **Deficiency in complement receptors CD21 and CD35 or complement proteins C3 and C4 results in an impaired humoral response to herpes simplex virus type 1 (HSV-1).** Mice were infected intradermally with 2 × 106 plaque-forming units of either replication-deficient strain HD-2 (panels a–c) or replication-competent strain KOS 1.1 (panel d) at day 0 and challenged on week 3 with a similar dose of live virus. Results indicate a normal secondary response as expected among wild-type (WT) controls but an impaired response in mice deficient in complement receptors or proteins.

IC engagement during the active immune response. Consistent with this scenario, several studies indicate that FcγRIIB limits the expansion of autoreactive plasma blasts and plasma cells that are present at enhanced levels in autoimmune-prone mouse strains expressing reduced levels of the inhibitory FcR [2,28]. FcγRIIB is not expressed on natural killer (NK) cells, T cells, or stromal cells.

## 3.2 ITIM Pathways

The inhibitory motif, embedded in the cytoplasmic domain of the single-chain FcγRIIB molecule, was defined as a 13 amino acid sequence AENTITYSLLKHP, shown to be both necessary and sufficient to mediate the inhibition of BCR-generated calcium mobilization and cellular proliferation [29,30]. Significantly, phosphorylation of the tyrosine of this motif was shown to occur upon BCR coligation and was required for its inhibitory activity. This modification generated a Src homology 2 (SH2) recognition domain that is the binding site for the inhibitory signaling molecule SH2 domain-containing inositol polyphosphate 5-phosphatase (SHIP) [13,31]. In addition to its expression on B cells, where it is the only IgG FcR, FcγRIIB is widely expressed on macrophages, neutrophils, mast cells, dendritic cells, and FDCs, and it is absent only from T and NK cells. Studies on FcγRIIB provided the impetus to identify similar sequences in other surface molecules that mediated cellular inhibition and resulted in the description of the

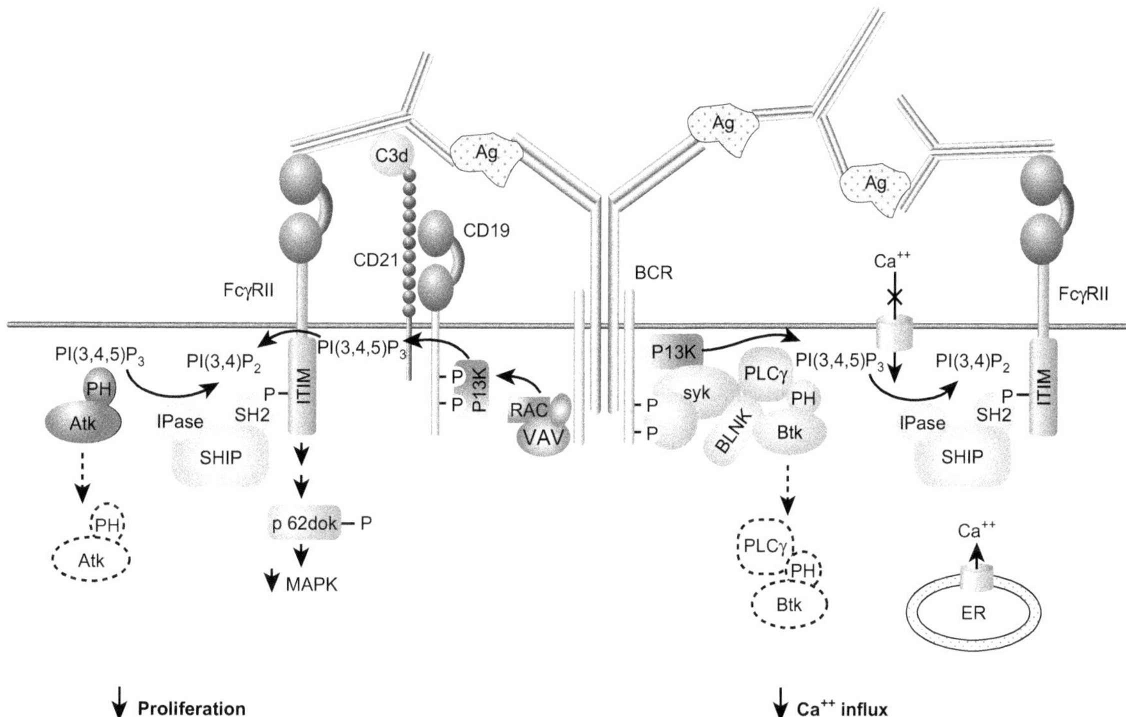


FIGURE 3 **Signaling pathways triggered by BCR–CD19/21–FcγRIIB coligation.** Cellular activation is inhibited by the recruitment of the inositol phosphatase SHIP to the FcγRIIB phosphorylated ITIM sequence.

immunoreceptor tyrosine-based activation motif (ITIM), a general feature of inhibitory receptors. FcγRIIB displays three separable inhibitory activities, two of which are dependent on the ITIM motif and one that is independent of this motif (Figures 3 and 4). Coengagement of FcγRIIB to an ITAM-containing receptor leads to tyrosine phosphorylation of the ITIM by the lyn kinase, recruitment of SHIP, and the inhibition of ITAM-triggered calcium mobilization and cellular proliferation [31,32]. These two activities result from different signaling pathways, with calcium inhibition requiring the phosphatase activity of SHIP to hydrolyze phosphatidylinositol (3,4,5)-triphosphate (PIP3) and the ensuing dissociation of pleckstrin homology (PH) domain–containing proteins such as Btk and phospholipase Cγ [33] (Figure 3). The net effect is to block calcium influx and prevent sustained calcium signaling. Calcium-dependent processes such as degranulation, phagocytosis, antibody-dependent cell-mediated cytotoxicity, cytokine release, and proinflammatory activation are all blocked. The arrest of proliferation in B cells is also dependent upon the ITIM pathway, through the activation of the adaptor protein dok and subsequent inactivation of mitogen-activated protein kinases [16,34]. The role of SHIP in this process has not been fully defined, because it can affect proliferation in several ways. SHIP, through its catalytic phosphatase domain, can prevent activation of the PH domain survival factor Akt by hydrolysis of PIP3 [35,36]. SHIP also contains phosphotyrosine-binding domains that could act to recruit dok to the membrane and provide access to the lyn kinase that is involved in its activation. Dok-deficient B cells are unable to mediate the FcγRIIB-triggered arrest of BCR-induced proliferation, while retaining their ability to inhibit a calcium influx, thus demonstrating the dissociation of these two ITIM-dependent pathways.

The third inhibitory activity displayed by FcγRIIB is independent of the ITIM sequence and is displayed upon homoaggregation of the receptor. Under these conditions of FcγRIIB clustering, a proapoptotic signal is generated through the transmembrane sequence (Figure 4). This proapoptotic signal is blocked by the recruitment of SHIP, which occurs upon coligation of FcγRIIB to the BCR, due to the Btk requirement for this apoptotic pathway. More recently it was demonstrated that in addition to Btk c-Jun $NH_2$-terminal kinase and the c-Abl kinase also were also involved in this pathway [37]. This novel activity has only been reported in B cells and has been proposed to act as a means of maintaining peripheral tolerance for B cells that have undergone somatic hypermutation. Support for this model comes from the in vivo studies of FcγRIIB-deficient mice in induced and spontaneous models of autoimmunity. Recent studies have provided evidence that FcγRIIB is involved in affinity maturation, Ig enhancement and suppression, and generation of the memory and plasma cell response [25,28,38]. These various roles contribute to the overall role of FcγRIIB in the maintenance of peripheral tolerance because perturbation in one or more of these

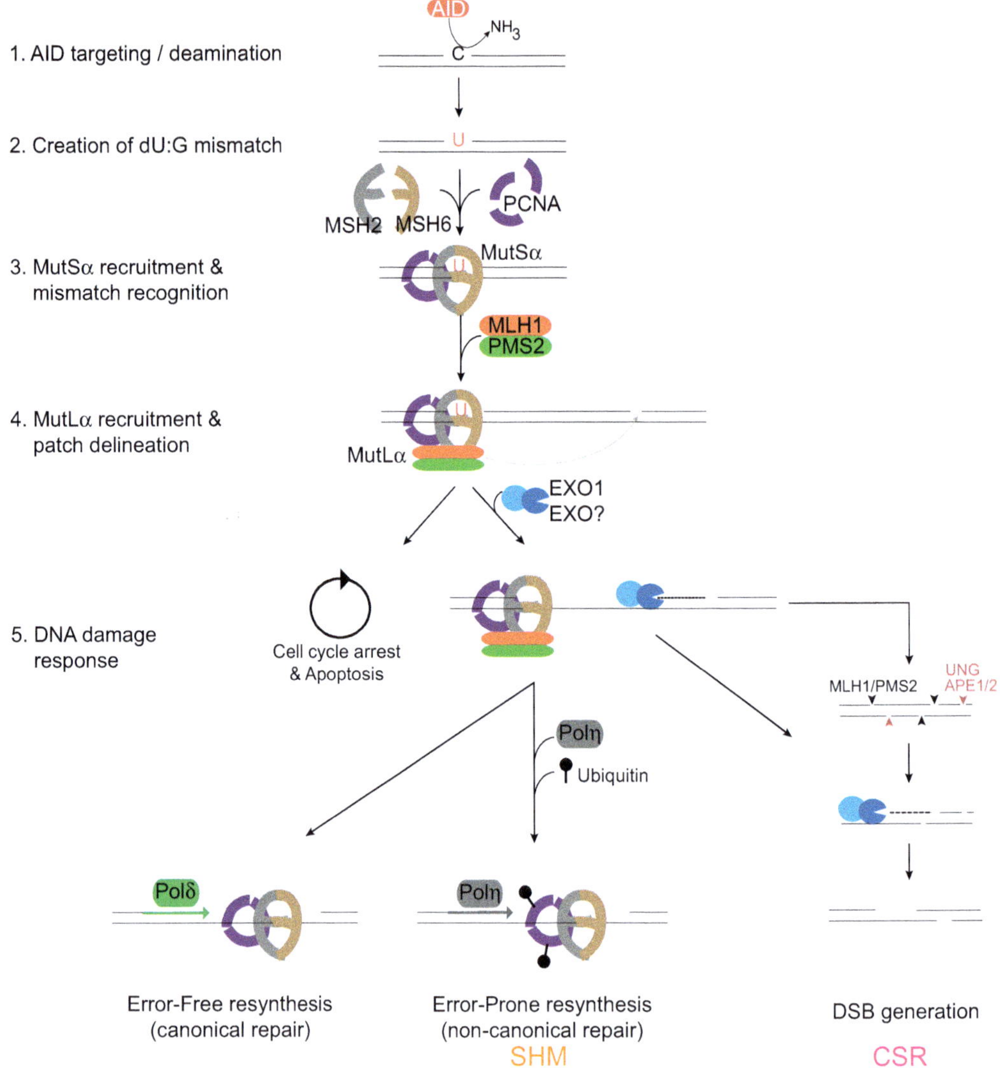


FIGURE 4 **A model for the role of FcγRIIB in affinity maturation of germinal center B cells.** Higher affinity BCRs rescue somatically hypermutated B cells from FcγRIIB-triggered apoptosis and negative selection.

pathways can contribute to the emergence and amplification of the autoimmune response.

## 3.3 Affinity Maturation

Affinity maturation occurs within the GC, where somatically mutated BCRs undergo selection on antigen retained on FDCs [39,40]. Antigen is retained in the form of ICs and involves the interaction of both complement receptors and FcγRIIB with these ICs on FDCs. B cells also express both complement and FcγRIIB. An analysis of the role of RIIB in affinity maturation thus represents the contribution of both FDCs and RIIB to this process. The role of FDC-expressed FcγRIIB has been clarified by the development of chimeric mice in which only the FDC FcγRIIB is deficient and the B-cell FcγRIIB is retained [39,41]. In these mice, potentiation of affinity maturation is seen. This potentiation likely results from the increased stringency of selection that occurs in the GC that, in turn, results from unopposed access of FDC ICs to B-cell FcγRIIB, as has been proposed previously [36]. Coligation of the BCR and FcγRIIB attenuates the BCR signal by the negative signaling role of SHIP This negative signal would impose a requirement for a higher threshold for effective BCR stimulation and thus favor higher affinity BCRs. In addition, under those conditions where coligation to BCR is ineffective, ligation of FcγRIIB alone results in an apoptotic signal, resulting in the elimination of those somatically mutated B cells with low affinity for antigen. The situation is reversed when B cells that lack FcγRIIB expression are transferred. In that case decreased apoptosis is observed, and the resulting B cells display a reduced affinity for antigen (Kalergis and Ravetch, unpublished observations). This is perhaps due to the persistence of low-affinity B cells that arise as a consequence of somatic mutation and are normally eliminated by a negative selection mechanism involving FcγRIIB crosslinking and induce apoptosis. Because FcγRIIB is regulated on both FDCs and B cells, the potential is provided for fine-tuning the survival and selection of B cells in the GC by the interaction with ICs on FDCs. This role of FcγRIIB on affinity maturation could provide part of the explanation for the loss of tolerance in RIIB-deficient animals. Because FcγRIIB would provide one mechanism for the elimination of low-affinity and potentially autoreactive BCRs that may arise through somatic mutation during the GC reaction, decreasing its expression on GC B cells could favor the persistence and eventual positive selection of these cells. This property could dominate over loss of FcγRIIB on FDCs, because its contribution to the retention of antigen is minimal compared with that of complement receptors. Enhancement of this autoimmune phenotype would be mediated dendritic cell (DC) maturation and the resulting T cell activation [40]. The analysis of conditional deficiency of FcγRIIB on B cells, FDCs, and DCs will clarify these activities.

## 3.4 Ig Enhancement and Suppression

The ability of IgG to mediate the suppression of antibody responses is perhaps the best-known clinical application of feedback suppression. The administration of IgG antierythrocyte-specific antibody completely prevents the emergence of antierythrocyte antibody response [40,42] and prevents the emergence of hemolytic disease of the newborn [43]. The responsible mechanism is still not defined, although experimental support for both epitope masking and FcγRIIB-mediated inhibition has been presented. In experimental mouse models, monoclonal IgG-anti-TNP suppressed the primary response to trinitrophenyl-conjugated sheep erythrocytes in mice that lacked FcγRIIB, suggesting an FcγRIIB-independent mechanism [44]. Other studies demonstrated decreased suppression for IgG antibodies with reduced FcR binding ability or through blocking FcR engagement by antibody blockade [45]. These differences may reflect the selective effects of decreased FcγRIIB binding on either B cells, FDCs, or antigen-presenting cells (APCs) where different effects can be elicited depending upon the pathway affected. Here, too, discrimination between these pathways is necessary to define the principal mechanism underlying IgG suppression.

The situation for enhancement by IgG is less complex. In this case, the ability of an IC, in the absence of adjuvant, to augment antigen presentation by APCs has been well documented in several systems [46]. Two activities are involved in IgG-mediated enhancement: the ability of IC to be more efficiently internalized by APCs through Fc-mediated pathways, leading to enhanced presentation [47,48]; and the ability of ICs to induce the maturation of immature DCs through activation of FcγRIII ITAM pathways [49].

This latter activity is restricted by the preferential expression of FcγRIIB on immature DCs. This pathway provides an explanation for the observation that IgG enhancement is augmented up to several hundredfold in FcγRIIB-deficient animals. In the presence of adjuvants, IgG enhancement is significantly reduced, because alternative maturation signals are provided [50], and the enhanced uptake of antigen is the primary activity being followed.

## 3.5 Memory Response

FcγRIIB has been postulated to play a role in both the recall response and in the switch from antibody-secreting cells (ASCs) to memory cells in the spleen. Experimental support for a role in these processes comes from chimeric animals in which the FDC FcγRIIB is absent. In those cases, the frequency of both ASCs and memory cells in the spleen is reduced, although the magnitude of the antibody response remained unchanged and the affinity actually increased [39]. Notably the long-term recall response was impaired. Here, too, the likely explanation points to the role of FcγRIIB on FDCs as providing a means of limiting the ability of retained ICs from interacting with FcγRIIB on B cells. The unrestricted interaction of FcγRIIB on B cells with ICs could account for the perturbation in the generation of high-affinity antibody-forming cells (AFCs) and memory cells, as discussed above. Further studies using a conditional deficiency of FcγRIIB on these two cell populations will be necessary to resolve these pathways. The relative importance of the FcγRIIB pathway on FDCs, compared with the complement pathway, is discussed below.

## 3.6 Plasma Cell Homeostasis

Upon B-cell activation and affinity maturation in the spleen and lymph nodes(LN), short-lived plasma blasts and long-lived plasma cells develop of which the latter populations homes to niches in the BM that provide factors for the long-term survival of these cells. The limited number of niches available in the BM creates the conceptual problem that once all the niches are occupied with plasma cells, no further plasma cells generated during a subsequent antigen encounter or vaccination would have the capacity to become long lived and generate protective antibody titers [51]. As this is obviously not the case, there have to be mechanisms in place that generate free niches for newly generated plasma cells. There is convincing evidence that signaling via FcγRIIB might be involved in this pathway via its pro-apoptotic function. Thus, plasma cells downregulate BCR expression while actually upregulating FcγRIIB expression levels. ICs generated during an ongoing immune response would therefore selectively trigger FcγRIIB-dependent proapoptotic signaling pathway on plasma cells that would induce cell death on at least a portion of the current plasma cell pool, allowing newly formed plasma cells to occupy these niches [25,52].

## 3.7 CD23

CD23 is the low-affinity receptor for IgE and is expressed throughout B-cell development with a very low expression on marginal zone B cells and T1 transitional B cells in the spleen. It exists in two isoforms, CD23a and CD23b, of which only the first isoform is expressed constitutively. CD23 engagement by antibody can result in either positive or negative responses. For example, injection of small soluble antigens complexed with IgE was demonstrated to enhance the antibody response in a CD23-dependent manner [42]. In contrast, IgE binding to CD23 on B cells results in reduced IgE production. As a consequence of this activity, CD23-deficient mice have elevated serum IgE levels. Apart from B cells, CD23 expression has been described on FDCs; however, the aforementioned enhancement of antibody responses was suggested to be dependent on CD23 expressed on B cells, indicating that B cells may directly act as APCs to enhance T helper (Th) cell responses [43]. Unlike FcγRIIB, CD23 does not belong to the immunoglobulin superfamily of proteins but rather to the family of C-type lectin receptors. More recently, mouse SIGNR1 and its human ortholog DC-SIGN, which also belong to this protein family, were identified to bind to IgG glycovariants rich in sialic acid. Binding of highly sialylated IgG variants to these receptors was shown to be responsible for repressing autoantibody-dependent inflammatory processes via triggering a Th2-type cytokine pattern that upregulated the inhibitory FcγRIIB on innate immune effector cells [53]. Recent studies have demonstrated that sialylated IgG can bind to CD23 in a manner analogous to IgE binding to CD23 and distinct from FcγR binding and thus may provide another mechanism for feedback regulation of IgG synthesis. This activity may explain the ability of IVIG to reduce autoantibody levels in patients treated with this preparation.

## 3.8 FCR-Like Proteins

In addition to the inhibitory FcγRIIB, several members of the family of FcRl proteins are expressed on mouse and human B cells [54,55]. As most of our current knowledge about the function of these proteins comes from human cells, we focus on the human system. The family of human FcRls consists of eight members of which FcRl1–FcRl6 are transmembrane proteins, whereas FcRlA and FcRlB may represent intracellular or secreted proteins. Similar to classical FcγRs, the FcRl proteins can be divided into activating and inhibitory family members depending on the presence of cytoplasmatic ITAM (FcRl1) or ITIM (FcRl4 and FcRl6) motifs; FcRl2, FcRl3, and FcRl5 have both, ITAM-like and ITIM motifs. On B cells expression of FcRl1–FcRl5 can be detected, although differences with respect to the various B-cell differentiation stages were noted. Whereas FcRl1, FcRl3, and FcRl5 are already expressed at the pre-B-cell stage, expression of FcRl2 and FcRl4 is limited to memory and plasma cells. In general, FcRl expression is most abundant on memory B cells, whereas on plasma cells FcRl5 expression is dominant. Consistent with the presence of ITAM motifs in FcRl1, crosslinking of this receptor on B cells was shown to enhance BCR-induced calcium flux and B-cell proliferation [56]. In contrast, triggering of FcRl4 and FcRl5 delivered potent inhibitory signals, leading to reduced BCR-induced calcium flux in an SHP-1-dependent manner, suggesting that the noncannonical ITAM motif in FcRl5 is dominated by the ITIM motif also present in this receptor [57,58]. Despite the identification of this receptor family due to their sequence similarity to classical FcγRs, so far only FcRl4 and FcRl5 have been shown to be bona fide FcRs. Thus, FcRl4 was able to bind to immunoglobulin A, whereas FcRl5 could bind IgG. Interestingly, the molecular mechanism of binding seems to differ from IgG binding to classical FcγRs [59,60]. In both cases, the binding affinity was in the micromolar range, suggesting that ICs may be the physiological ligands of these receptors. At present, one can only speculate about the role of FcRl triggering via IgA or IgG ICs, but considering the rather restricted expression of FcRl4 on a subset of memory B cells, a negative feedback regulation via IgA complexes or dimers may be anticipated. With respect to FcRl5, one may expect a broader negative feedback on B-cell developmental stages due to the ubiquitous expression pattern throughout B-cell development, but more work will be necessary to fully understand the function of these proteins.

## 3.9 Fcμ-Receptor

Although the existence of an FcR for IgM was anticipated for a long time, the cloning and molecular characterization of this molecule only started in 2009 [61]. Originally identified as a molecule modulating FAS-mediated apoptosis and hence called Fas apoptotic inhibitory molecule 3 (or TOSO), it was demonstrated to have the capacity to bind to polymeric IgM recently. The FcμR is located in proximity to the poly-Ig receptor and the Fca/μR on chromosome 1, but it only shows a distant relationship to these receptors based on amino acid sequence [61]. Its intracellular sequence does not contain any ITIM or ITAM motifs, but it contains conserved Ser and Tyr residues that become phosphorylated upon IgM binding to the receptor. The expression of the FcμR is largely restricted to T cells and B cells in humans and shows a cell-type-specific modulation upon activation. Whereas it is upregulated on B cells upon BCR activation, it becomes downmodulated on T cells after activation. In mice and humans, FcμR is expressed from the pre-B/immature B-cell to the plasma cell stage, with a notable reduction of expression on GC B cells. Besides B2-B cells, B1 cells also show a prominent expression of FcμR. In contrast to humans, FcμR expression in mice is restricted to B cells, and the first analysis of a mouse strain deficient for FcμR expression revealed no major phenotype for B-cell development [62]. However, these mice had increased levels of serum IgM and IgG3 and an increase in splenic B1 cells. In contrast, the amount of marginal zone B cells was reduced. Moreover, although responding well to nonprotein antigens such as phosphorylcholine, the antibody response to protein antigens was reduced in the absence of the FcμR. Of note, the heightened level of natural IgM antibodies was paralleled by enhanced levels of anti-chromatin and anti-DNA antibodies of IgM and IgG isotypes. Although much more work needs to be done to fully elucidate the function of FcμR on B cells, the available data suggest that it may be involved in modulating natural serum IgM and at least in part the B-cell response toward protein antigens.

## 4. COMPLEMENT RECEPTORS

The discovery of antibody as important in recognition of bacteria directly led to the identification of a heat-sensitive component of serum that was also required for the elimination of pathogens. The component was termed "complement" for its complementary role in leading to lysis of antibody-bound bacteria. Thus, the study of the complement system has paralleled that of antibody for more than a century. It is now appreciated that the complement system represents more than 20 serum and cell surface proteins that act in concert to activate in a regulated manner the central component C3 and the later membrane attack complex C5–C9 [63,64]. Whereas much of the focus of early work was on proteolytic events and assembly of the membrane attack complex, work during the past several decades has uncovered an important role of the complement system in the regulation of B cells. This portion of the chapter focuses on our current understanding of the complement receptors involved in B-cell regulation in vivo. Importantly, complement receptors act at multiple steps of B-cell activation and help to shape not only the antibody repertoire but also influence the formation of a memory response to both T-dependent and T-independent antigens [65]. Two major effects on B cells are (1) localization of antigen to lymphoid compartment; and (2) enhancement of B-cell signaling via the coreceptor. Finally, we discuss recent studies that identify a role for complement receptors in the retention of antigen on FDC.

### 4.1 Early Components of Complement Influence the Humoral Response

The finding of C3 activation products bound to the surface of B lymphocytes led to the speculation that complement receptors were important in regulating the immune response [66]. Indeed, transient depletion of C3 revealed a critical role for complement in humoral immunity to both T-D and thymus-independent (T-I) antigens [67]. These observations were confirmed and extended in humans and animals bearing natural deficiencies in C3 or the classical pathway (C1q, C2, and C4) and later in gene-targeted mice [68]. We now appreciate that the activation of complement C3 and covalent attachment to antigen "marks" the antigen as foreign and greatly enhances its immunogenicity. The classical pathway is most important for activation and binding to T-D antigens, whereas C3 binds to carbohydrate antigens, probably via the lectin or alternative pathways.

### 4.2 Regulation of Complement Receptors

Given the central importance of complement C3 in inflammation, it is not surprising that it is tightly regulated. Most cells express regulators of complement such as membrane cofactor, decay-activating factor, and complement receptor carry that function either to displace C3 convertase or act as cofactors in its proteolytic inactivation. CD35 also participates as a regulator of complement but is more limited in its distribution. In addition to its cofactor activity, CD35 binds activated C4 (C4b) and C3 (C3b) and degradation products of C3—iC3b and C3d. In humans, it is the major immune adhesion receptor expressed on red blood cells and functions in the clearance of ICs. In addition, it is found on myeloid cells (DCs and macrophages), FDCs, and B cells. CD21 is a homolog of human CD21 that binds iC3b and C3d but lacks cofactor activity. It is expressed on B cells and some T cells. In the mouse, CD35 and CD21 are expressed at a single locus (Cr2), because CD21 represents a splice product of the former [69,70]. They appear to be coexpressed primarily on B cells and FDCs. Like the regulatory receptors, CD21 and CD35 share a common structural motif termed complement control protein (CCP), and they are assembled from these repeating units. For example, murine CD21 and CD35 are assembled from 15 to 21 CCP repeats, respectively. Recent cocrystalization studies identify the binding of C3d ligand at the interface of the two N-terminal CCP of CD21 (Isenman).

In summary, most mammalian cells bear complement receptors that function to limit the activation of C3 on autologous cells. A subset of C3 receptors, CD21 and CD35, are more limited in expression and function both in binding of i.c. and adaptive immunity.

### 4.3 CD21 Forms a Co-receptor on B Cells

The B-cell coreceptor best described for human B cells represents a complex of three receptors: CD21, CD19, and CD81. CD21 provides the major recognition domain and binds C3d-coated antigens that lead to crosslinking of

CD19 and CD81. CD19 is the primary signaling receptor and becomes rapidly phosphorylated on crosslinking. Co-crosslinking of the coreceptor and BCR results in downstream events that amplify the signal so that less antigen is required for activation and provides a unique signal-enhancing B-cell activation [71]. Thus, the coreceptor functions to lower the threshold for activation of B cells. Although CD19 has intrinsic signaling properties separate from CD21, the binding of C3d-coated antigens appears to be specifically required for the enhancement of B-cell responses. For example, co-crosslinking of CD21 and BCR leads to preferential targeting of the complex into lipid rafts that enhance signaling. In mice, CD35 also forms a coreceptor with CD19 and CD81; therefore, the two receptors are treated as one (CD21/CD35) for remaining discussion in this chapter. Thus, the CD21/CD19/CD81 coreceptor provides a critical link between the innate immune system and adaptive immunity.

In summary, CD21 with CD19 and CD81 forms a coreceptor that is expressed by all mature B-cell subsets. An encounter with C3d-coated antigen coligates the coreceptor and BCR on cognate B cells and results in a reduction in the antigen signal required for a B-cell response.

### 4.4 CD21/CD35 Regulate B-Cell Responses

As predicted by Nussenzweig and colleagues three decades earlier, complement receptors are important in regulating B-cell responses. This was first demonstrated in vivo by pretreating mice with antibodies specific for CD21/CD35 or a soluble inhibitor of CD21 (sCR2) [72]. Mice administered blocking antibody or sCR2 before immunization with T-D antigens failed to respond normally, and their antibody response was impaired. The defect was in the B-cell compartment, because the T cells from treated and immunized mice were primed normally to the red blood cell antigen [73]. These studies confirmed the in vivo importance of CD21/CD35 in enhancement of B-cell immunity. However, they did not distinguish between complement receptor expression on B cells versus FDCs. Thus, uptake of C3d-coated antigen by CD21/CD35 not only crosslinks the B-cell coreceptor but also is important for localization of antigens within the lymphoid compartment.

## 5. CORECEPTOR SIGNALING VERSUS ANTIGEN LOCALIZATION TO FDCS

Over 40 years ago, FDCs were first identified as important for retention of antigen in GCs. In contrast to myeloid cells, FDCs retain antigen for long periods, up to months in mice, without degradation. Immune complexes coated with complement C3 are taken-up by FDCs through CD21/CD35 receptors and rapidly internalized into a nondegradative compartment. Periodically, ICs are cycled to the cell surface where they are accessible for acquisition by cognate B cells (Figure 5) [74]. How FcRIIB participates in antigen retention is not clear, but it could be involved in regulation of the periodic cycling. Interestingly, one important pathway for delivery of IC to FDCs within the secondary lymphoid tissues is via CD21/CD35 transport by naïve B cells (Figure 5) [75,76]. Thus, transport and delivery of C3-coated IC is mediated by the complement receptors but does not appear

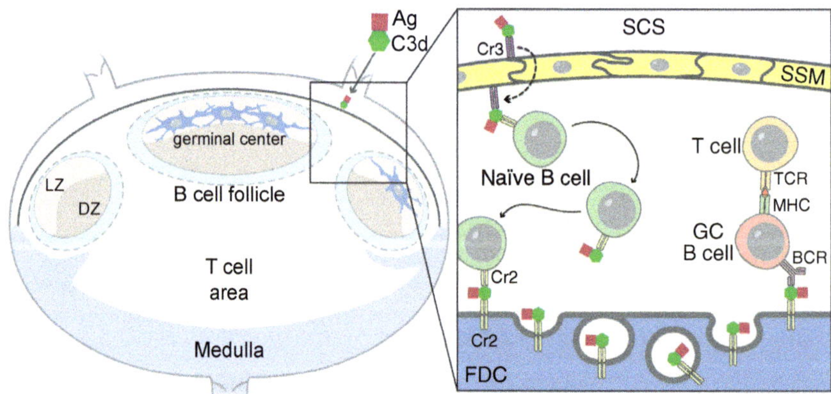


FIGURE 5 **Lymph node architecture and antigen trafficking.** (Left panel) Schematic representation of the lymph nodes (LN). Depicted are the T cell zone and the B-cell follicles, which can contain a germinal center (GC). The GC has a dark zone (DZ) and a light zone (LZ), where the follicular dendritic cells (FDC) reside. The medulla (bottom of diagram) is distal to the subcapsular sinus and includes the efferent lymphatics.(Right panel) Antigen (Ag) enters the LN through the afferent lymphatic vessels that drain into the subcapsular sinus (SCS). Lymph borne immune complexes (IC) activate complement leading to covalent attachment of activated C3 (C3d). C3d-IC can be captured by complement receptor 3 (Cr3) and FcR on subcapsular sinus macrophages (SSM). The C3d-IC is then shuttled to the underlying B-cell compartment where it is taken up by naïve B cells via complement receptor 2 (Cr2). The naïve B-cell then travels towards the FDC where the cargo is transferred to FDC in a Cr2-dependent pathway. FDC internalize the C3d-IC into a nondegrading endosomal vesicle that periodically cycles to the surface. It is important to note that the IC is protected from degradation for extensive periods in these vesicles and that the IC remains intact and bound to Cr2. On the surface of the FDC, C3d-IC can be acquired by GC B cells and internalized for presentation to helper T cells.

to involve CD19 and CD81 or the BCR as the IC remain on the surface of the B-cell until off-loaded onto the FDC.

These observations raised the general question of the relative importance of CD21/CD35 expression on B cells and FDCs in vivo. Thus, does the B-cell response depend equally on both coreceptor signaling and antigen localization? This question was addressed by the development of chimeric mice in which the B cells expressed CD21/CD35 and FDCs were deficient or vice versa. Such chimeric mice were made possible by the finding that FDCs are radioresistant and not restored by engraftment of adult BM into irradiated recipients [76]. For example, chimeric mice bearing CD21/CD35 on B cells, but deficient on FDCs, could be prepared by the reconstitution of lethally irradiated CD21-def mice with wild-type (WT) (Cr2+) BM. Alternatively, reconstitution of irradiated WT mice with BM prepared from Cr2-def mice yielded mice bearing Cr2+FDCs but Cr2-def B cells. Based on characterization of the reciprocal chimeric mice, it is clear that a full memory response requires both coreceptor and retention of antigen on FDCs [77].

## 5.1 B Cells Require Complement Receptors at Five Different Stages

### 5.1.1 *Repertoire of Natural Antibody is Altered in Cr2-def Mice*

Natural antibody is produced primarily by a subset of B cells termed B-1. They are distinguished from conventional B cells (B-2) by anatomical localization, repertoire, cell surface phenotype, activation, and lifespan (this subset of B cells is discussed more thoroughly in another chapter). B-1 cells are IgM-int, IgD-lo, and CD21-int, CD23–, and localize primarily in the peritoneal tissues, although they are also found at low levels in the spleen and lymph nodes (LNs). Two subsets have been identified based on expression of CD5: CD5+, termed B-1a, and CD5–, termed B-1b. Unlike conventional B cells, both subsets express CD43 and CD11b. Although B-1 cells can undergo isotype switch, they are not thought to enter GCs or acquire extensive somatic mutation; however, they are long lived and appear to undergo self-renewal. Their repertoire appears to be biased toward highly conserved antigens such as phosphoryl choline and phosphatidyl choline and nuclear antigens such as DNA and nuclear proteins. This bias might reflect their development, because they are selected during the early neonatal period in which terminal deoxynucleotidyl transferase is not expressed. Their repertoire is also influenced by the positive selection by self- and enteric bacterial antigens [78,79]. The development of B-1 cells requires an intact BCR, as defects in proteins involved in BCR signaling such as vav, PI-3 kinase, and CD19 result in a more profound loss of B-1 relative to conventional B-2 cells [80]. Alternatively, mutations leading to hyperresponsive BCR signaling can result in an increased frequency of B-1 cells. Whether this reflects a general requirement for intrinsic signaling of BCR or encounter with cognate antigen is not clear. However, studies identifying the requirement of cognate antigen for B-1 cell development suggests that interaction with self-antigens or enteric bacteria is important for the initial positive selection or expansion and maintenance. This is illustrated in the I/R model. I/R represents an acute inflammatory response against self after reperfusion of ischemic tissues (Zhang et al. JEM 2006). It is mediated by natural IgM (and IgG) and classical pathway complement. Not only are Ig-deficient animals protected from injury in the I/R model, but also Cr2-def mice are protected from full injury [81,82]. Thus, despite apparent normal levels of serum IgM Cr2-def mice are protected in an intestinal model of I/R. Injury can be restored by reconstitution with pooled IgM prepared from WT mice. Alternatively, reconstitution of Cr2-def mice with WT peritoneal B cells also restores their susceptibility to I/R injury. Because engrafted Cr2+ B-1 cells are maintained in Cr2-def mice in the absence of stromal expression of CD21/CD35 it seems most likely that coreceptor signaling rather than FDC binding is important in the expansion and maintenance of the B-1 subset of cells [83].

In summary, B-1 cells are a major source of natural antibody and are positively selected during early development by cognate antigen. The interaction requires complement receptors with at least some antigens to enhance antigen receptor signaling.

### 5.1.2 *Activation of Naïve B Cells*

B cells, like T cells, must be tightly regulated to circumvent the nonspecific activation of bystander cells during an ongoing infection. Antigen specificity is ensured in large part by the requirement for two signals—BCR and CD40—to promote activation and expansion against specific pathogens. B-cell encounters with antigen in the absence of T cell help (CD40L costimulation) or vice versa can result in the induction of anergy or cell death. A well-characterized pathway for regulation of peripheral lymphocytes is CD95 or Fas. Trimerization of the Fas receptor on peripheral B cells stimulated by CD40 alone leads to the assembly of the caspase death pathway and B-cell apoptosis. Crosslinking of BCR induces the expression of cFLIP (Fas ligand inhibitor protein) that blocks the caspase pathway and results in cell survival and expansion. Antigen affinity is important in this regulatory step, because coligation of CD21 and BCR can protect, whereas crosslinking of BCR alone by moderate-affinity antigen can result in cell death in vivo [84]. Thus, as predicted by Carter and Fearon, the coreceptor is important in lowering the threshold for B-cell activation (Figure 5).

In summary, the Fas pathway regulates naïve peripheral B cells to limit bystander activation during an ongoing infection. Regulation is dependent on antigen affinity and complement receptors, because low-to-moderate affinity antigens require coreceptor crosslinking to prevent Fas-dependent apoptosis. Thus, engagement of the BCR and coreceptor is important for the survival and expansion of naïve B cells after an encounter with many T-dependent antigens.

### 5.1.3 Germinal Center Survival

The GCs represent a specialized microenvironment within the B-cell follicles of lymphoid tissue [39]. They are transient in duration and disappear within 15–21 days after immunization (see chapter by Kelsoe for a more detailed discussion). Within the GC, activated B cells (termed centrocytes) undergo rapid expansion, isotype switch, and somatic hypermutation followed by antigen selection. As mutated centrocytes emerge from the dark zone, they encounter C3-coated antigen retained on FDC primarily via CD21/CD35 but also FcR. Survival of GC B cells within the GC is dependent on T cell help (CD40 ligand), antigen, and interaction with FDC (Figure 5). Coreceptor expression is also required for the survival of B cells within the GC, based on several lines of evidence. Treatment of immune mice with soluble CD21 receptor (sCR2) results in rapid loss of the GC [85]. Thus, contact between C3d-coated antigens localized on FDCs is essential for the survival of GC B cells. Whether the B cells are eliminated in a Fas- or FcγRIIB-dependent mechanism is not known. Further support comes from studies comparing GC survival of Cr2+ and Cr2-def Ig transgenic (Tg) B cells (specific for hen lysozyme). In this study, the adoptive transfer of high affinity Ig-Tg B cells into mice immunized with specific antigen identified the participation of the Cr2+ but not Cr2-def Tg B cells within the GC. Thus, expression of a high-affinity BCR is not sufficient to mediate the survival of B cells in the absence of coreceptor expression.

In summary, GC represents a specialized environment within the lymphoid compartment for expansion, isotype switch, somatic hypermutation, and antigen-dependent selection of high-affinity B cells. Survival requires the presence of antigen, T cell help, and contact with FDCs. Coreceptor signaling independent of antigen affinity is also required and explains at least part of the need for FDCs.

### 5.1.4 Memory B Cells and Persistence of Antibody Secretion

A hallmark of the adaptive immune response is the formation of memory B cells. These post-GC cells are antigen selected and of relatively high affinity. A subpopulation of memory cells (Bmem) differentiate into AFCs or plasma cells that persist over the long term, primarily in the BM but also in the secondary lymphoid compartment. The role of antigen in the maintenance of AFCs and Bmem is controversial. Earlier studies by Gray and colleagues demonstrated a critical role for antigen in long-term antibody secretion and recall [86]. These results, combined with the observations that antibody affinity continues to increase long after GC wanes, suggest that at least some post-GC AFC precursors are continually selected by antigen such that the higher affinity clones are preferentially maintained. Adoptive transfer studies using chimeric mice in which FDCs are deficient in CD21/CD35 or FcγRIIB provide further support that antigen is important in the maintenance of Bmem cells (Figure 5). Moreover, the recent finding that antigens are retained by FDCs for extensive periods in a nondegradative intracellular compartment provides additional support for the antigen-selection model for maintaining B-cell memory. An alternative view is that Bmem, like memory T cells, are long lived in the absence of antigen. This view is supported by results that suggest that Bmem cells are nondividing over long periods, and recent elegant genetic experiments demonstrating that switching of BCR specificity after formation of Bmem did not appear to alter their survival as functional memory cells. Taken together, a current view is that antigen is important for the affinity maturation and maintenance of memory B cells. However, there are long-lived plasma cells and Bmem cells that exist in the absence of antigen.

In summary, antigen is retained over long periods on FDCs primarily via CD21/CD35, but FcγRIIB also appears to be essential for effective recall responses. The presence of antigen is important for both the affinity maturation and efficient maintenance of memory B cells.

### 5.1.5 Negative Selection of Self-Reactive B Cells

The autoimmune disease SLE is characterized by antibodies specific for nuclear antigens, such as dsDNA and RNA Although full development of disease is due to multiple gene defects, dysregulation of B cells is a major factor. Defects resulting in excess BCR signaling, such as loss of negative regulators such as FcγRIIB, often lead to production of anti-dsDNA or anti-nuclear antibodies. For example, deficiency in the regulatory FcR FcγRIIB predisposes to lupus autoantibodies, as discussed above. One interpretation of these results is that the breakdown in normal regulation of self-reactive B cells within the BM and periphery results in survival of self-reactive cells that can become activated and secrete autoantibodies in the presence of T cell help. Multiple checkpoints have evolved to limit the development and activation of self-reactive B cells. Within the BM (central tolerance), B cells that rearrange and express self-reactive receptors at the immature stage are either edited (rearrangement of additional upstream VL genes) or eliminated by apoptosis (this topic is covered in greater detail in Chapter 16). How immature B cells encounter self-antigen is not clear. However, an emerging paradigm is that apoptotic debris includes ligands for Toll-like receptors such as dsDNA and RNA and internalization by autoreactive B cells can result in their activation and differentiation into autoantibody-secreting cells [87]. As discussed above, a major role for the complement system is clearance of ICs and apoptotic debris. Therefore, defects in early complement such as C1q and C4 and complement receptors could lead to excess apoptotic debris and triggering of autoreactive B cells. Thus, one explanation for the increased lupus susceptibility in C1q- or C4-deficient humans is impaired clearance of apoptotic cells. If CD21/CD35 participates in the uptake of C4-bound lupus antigens, Cr2-def mice might be expected also to display increased production of lupus autoantibodies. Support for a role for CD21/CD35 receptors comes from the genetic mapping of susceptibility loci in a lupus-prone strain of mice (NZM 2410/NZW) [88]. One of the susceptibility loci, Sle-1c, includes the *Cr2* locus, suggesting that CD21 or CD35 might be involved in the lupus phenotype. Structural analysis of the CD21 allele (NZM2410/NZW) identified several nucleotide differences, one of which results in the addition of a carbohydrate attachment site within the region of C3d binding. Functional studies of B cells that express the mutant allele confirmed that binding of C3d is reduced and that coreceptor activity is diminished. The findings from this study suggest that the *Cr2*-locus encodes the Sle-1c susceptibility gene. However, a complicating factor in use of the Cr2-def strains available is that they were constructed in strain 129 embryonic stem cells, and this region of the 129 chromosome carries additional alleles of genetic susceptibility to SLE when crossed onto the C57BL/6 background. In future studies, it will be important to develop Cr2-def mice on the pure C57BL/6 background.

## 6. FRONTIERS: COMPLEMENT VERSUS Fc RECEPTORS

Complement (CR) and FcRs are both expressed on mature B cells but play dramatically different roles in activation: CRs enhance, whereas FcRs tune down, the BCR signal. Both receptors recognize ICs, but the isotype of the Ig greatly affects the outcome. B-cell regulation via ICs formed with IgM or IgG3 requires complement and CD21/CD35, whereas IgG1-, 2a-, and 2b-containing ICs influence B cells via FcγRIIB. For example, the administration of T-D antigen plus specific IgM or IgG3 enhances B-cell response and is CD21/CD35 dependent. By contrast, the presence of IgG1 ICs can suppress the response of cognate B cells. Although these two receptor systems are often activated during different stages of the humoral response, they have intrinsic competing roles in regulating the GC reaction. An imbalance between these positive and negative regulators of BCR can have a dramatic effect on the formation of B-cell memory and affinity. For example, B-cell deficiency in CD21/CD35 results in limited survival but higher affinity, probably due to increased selection, whereas deficiency in FcγRIIB results in reduced selection and an expansion of lower affinity B cells.

Like B cells, FDCs express both CD21/CD35 and FcγRIIB. However, in contrast to B cells, both receptor systems have a positive role in formation of Bmem cells (Figure 5). Commensurate with a role in antigen retention, CD21/CD35 are constitutively expressed at relatively high levels on FDCs, whereas FcγRIIB appears to be regulated and expressed during an ongoing GC reaction. How FcγRIIB is regulated is not known, but one possibility is that expression is triggered by uptake of ICs through CD21/CD35 independently of CD19. Absence of FcγRIIB on FDCs, but its presence on lymphoid or myeloid cells, can have a dramatic effect on the humoral response to T-D antigens. For example, chimeric mice (deficient in FDC FcγRIIB) immunized with T-D antigen develop increased affinity of antibody but have an impaired long-term recall response [89]. One explanation is that lack of FcγRIIB on FDCs favors Fc ligand interaction with FcγRIIB on the B-cell and increased negative selection. Thus, in the absence of FDC expression of FcγRIIB, antigen selection would become more intense and limit the frequency of surviving B cells such that only those having higher affinity survive. The presence of CD21/CD35 on FDCs contributes to B-cell survival by providing ligand (C3d-coated antigen) for coligation of BCR and coreceptor signaling. This would be important in countering the negative signal via B-cell FcγRIIB. Saturation of the lymphoid compartment during an ongoing immune response with either specific antigen or sCR2 induces a rapid elimination of the GC, as discussed earlier. One explanation for the elimination of B cells is FcγRIIB signaling. Thus, a disruption of positive signaling via coreceptor and/or BCR would favor negative signaling by FcγRIIB and elimination of the cognate B cells. It is notable that both CR and FcR are protective against SLE despite their opposite roles in B-cell regulation. As discussed earlier, mice deficient in FcγRIIB or CD21/CD35 have an increased propensity to develop lupus autoantibodies. Although both receptor systems bind IC (or IC-C3d) and participate in clearance, an alternative role for protection against lupus is via B-cell regulation. As discussed above, deficiency in either receptor can have a dramatic effect on the fate of the B-cell in response to environmental antigens. Therefore, it is most likely that both sets of receptors participate in the regulation of self-reactive B cells. However, given the positive and negative effects of the receptors on mature B cells, it is probable that they act at different stages of B-cell development such that they both have a protective role. For example, CD21/CD35 could limit the activation of self-reactive B cells at the immature stage, where antigen encounters result in editing, cell death, or anergy. A possible role discussed earlier is coreceptor signaling in transitional B cells in response to complement-tagged self-antigen in the periphery. Further study in both these areas is needed. By contrast, FcγRIIB is more likely involved in the regulation of mature B cells in the periphery. Given the nature of its ligand and the phenotype of FcγRIIB-deficient mice, its protective role would be most prominent in the GC and at the plasma blast/plasma cell stage where FcγRIIB is expressed without concomitant expression of the BCR. An important checkpoint occurs within the GC to prevent the activation of B cells that become self-reactive as a result of somatic hypermutation. Deficiency in FcγRIIB lowers antigen selection pressure and could allow self-reactive B cells to expand and escape the GC.

In summary, in vivo studies for mice deficient in CRs and FcγRs has greatly extended our understanding of B-cell regulation and how ligands for these receptor systems can modulate responses to environmental and self-antigens. However, these studies also highlight how much is left to learn. In this chapter, we have attempted to outline some of the important questions that have emerged over the past 5 years. It is clear that a better understanding of how CRs and FcRs participate in the humoral response to environmental antigens will be useful not only in future vaccine research but also in helping to dissect the events leading to dysregulation of self-reactive B cells and the development of autoimmune disorders such as SLE.

## REFERENCES

[1] Carroll MC. The complement system in regulation of adaptive immunity. Nat Immunol 2004;5(10):981–6.
[2] Baerenwaldt A, et al. Fcγ receptor IIB (FcγRIIB) maintains humoral tolerance in the human immune system in vivo. Proc Natl Acad Sci USA 2011;108(46):18772–7.
[3] Takai T, et al. Augmented humoral and anaphylactic responses in Fc γ RII-deficient mice. Nature 1996;379:346–9.
[4] Bolland S, Ravetch JV. Spontaneous autoimmune disease in FcγRIIB-deficient mice results from strain-specific epistasis. Immunity 2000;13(2):277–85.
[5] Bolland S, et al. Genetic modifiers of systemic lupus erythematosus in FcγRIIB(-/-) mice. J Experimental Medicine 2002;195(9):1167–74.
[6] Brownlie RJ, et al. Distinct cell-specific control of autoimmunity and infection by FcγRIIb. J Experimental Medicine 2008;205(4):883–95.
[7] McGaha TL, Sorrentino B, Ravetch JV. Restoration of tolerance in lupus by targeted inhibitory receptor expression. Science 2005;307(5709):590–3.
[8] Tackenberg B, et al. Impaired inhibitory Fcγ receptor IIB expression on B cells in chronic inflammatory demyelinating polyneuropathy. Proc Natl Acad Sci USA 2009;106(12):4788–92.
[9] Bruhns P, et al. Colony-stimulating factor-1-dependent macrophages are responsible for IVIG protection in antibody-induced autoimmune disease. Immunity 2003;18(4):573–81.
[10] Kaneko Y, et al. Pathology and protection in nephrotoxic nephritis is determined by selective engagement of specific Fc receptors. J Exp Med 2006;203(3):789–97.
[11] Samuelsson A, Towers TL, Ravetch JV. Anti-inflammatory activity of IVIG mediated through the inhibitory Fc receptor. Science 2001;291(5503):484–6.
[12] Nakamura A, et al. Fcγ receptor IIB-deficient mice develop Goodpasture's syndrome upon immunization with type IV collagen: a novel murine model for autoimmune glomerular basement membrane disease. J Exp Med 2000;191(5):899–906.
[13] Tridandapani S, et al. Protein interactions of Src homology 2 (SH2) domain-containing inositol phosphatase (SHIP): association with Shc displaces SHIP from FcgammaRIIb in B cells. J Immunology 1999;162(3):1408–14.
[14] Jiang Y, et al. Polymorphisms in IgG Fc receptor IIB regulatory regions associated with autoimmune susceptibility. Immunogenetics 2000;51(6):429–35.
[15] Pritchard NR, et al. Autoimmune-prone mice share a promoter haplotype associated with reduced expression and function of the Fc receptor FcgammaRII. Curr Biol 2000;10(4):227–30.
[16] Tamir I, et al. The RasGAP-binding protein p62dok is a mediator of inhibitory FcγRIIB signals in B cells. Immunity 2000;12(3):347–58.
[17] Mackay M, et al. Selective dysregulation of the FcγIIB receptor on memory B cells in SLE. J Exp Med 2006;203(9):2157–64.
[18] Chu ZT, et al. Association of Fcγ receptor IIb polymorphism with susceptibility to systemic lupus erythematosus in Chinese: a common susceptibility gene in the Asian populations. Tissue Antigens 2004;63(1):21–7.
[19] Kyogoku C, et al. Fcγ receptor gene polymorphisms in Japanese patients with systemic lupus erythematosus: contribution of FCGR2B to genetic susceptibility. Arthritis Rheum 2002;46(5):1242–54.
[20] Siriboonrit U, et al. Association of Fcgamma receptor IIb and IIIb polymorphisms with susceptibility to systemic lupus erythematosus in Thais. Tissue Antigens 2003;61(5):374–83.
[21] Willcocks LC, et al. A defunctioning polymorphism in FCGR2B is associated with protection against malaria but susceptibility to systemic lupus erythematosus. Proc Natl Acad Sci USA 2010;107(17):7881–5.
[22] Walport MJ. Complement. First of two parts. N Engl J Med 2001;344:1140–4.
[23] Chatterjee P, et al. Complement C4 maintains peripheral B-cell tolerance in a myeloid cell dependent manner. Eur J Immunol 2013;43(9):2441–50.
[24] Fearon DT, Locksley RM. The instructive role of innate immunity in the acquired immune response. Science 1996;272:50–4.
[25] Xiang Z, et al. FcγRIIb controls bone marrow plasma cell persistence and apoptosis. Nat Immunol 2007;8(4):419–29.
[26] Rao SP, Vora KA, Manser T. Differential expression of the inhibitory IgG Fc receptor FcγRIIB on germinal center cells: implications for selection of high-affinity B cells. J Immunol 2002;169(4):1859–68.
[27] Rudge EU, et al. Interleukin 4 reduces expression of inhibitory receptors on B cells and abolishes CD22 and Fc gamma RII-mediated B-cell suppression. J Exp Med 2002;195(8):1079–85.
[28] Fukuyama H, Nimmerjahn F, Ravetch JV. The inhibitory Fcγ receptor modulates autoimmunity by limiting the accumulation of immunoglobulin G+ anti-DNA plasma cells. Nat Immunol 2005;6(1):99–106.
[29] Amigorena S, et al. Cytoplasmic domain heterogeneity and functions of IgG Fc receptors in B lymphocytes. Science 1992;256(5065):1808–12.
[30] Muta T, et al. A 13-amino-acid motif in the cytoplasmic domain of Fc gamma RIIB modulates B-cell receptor signalling. Nature 1994;369(6478):340.
[31] Ono M, et al. Role of the inositol phosphatase SHIP in negative regulation of the immune system by the receptor FcγRIIB. Nature 1996;383(6597):263–6.

[32] Daeron M, et al. The same tyrosine-based inhibition motif, in the intracytoplasmic domain of Fc gamma RIIB, regulates negatively BCR-, TCR-, and FcR-dependent cell activation. Immunity 1995;3(5):635–46.

[33] Bolland S, et al. SHIP modulates immune receptor responses by regulating membrane association of Btk. Immunity 1998;8(4):509–16.

[34] Yamanashi Y, et al. Role of the rasGAP-associated docking protein p62(dok) in negative regulation of B-cell receptor-mediated signaling. Gene Dev 2000;14(1):11–6.

[35] Aman MJ, et al. The inositol phosphatase SHIP inhibits Akt/PKB activation in B cells. J Biol Chem 1998;273(51):33922–8.

[36] Liu Q, S. T, Kozieradzki I, Wakeham A, Itie A, Dumont DJ, et al. SHIP is a negative regulator of growth factor receptor-mediated PKB/Akt activation and myeloid cell survival. Gene Dev 1999;13(7): 786–91.

[37] Tzeng SJ, et al. The B-cell inhibitory Fc receptor triggers apoptosis by a novel c-Abl family kinase-dependent pathway. J Biol Chem 2005;280(42):35247–54.

[38] Tiller T, et al. Development of self-reactive germinal center B cells and plasma cells in autoimmune FcγRIIB-deficient mice. J Exp Med 2010;207(12):2767–78.

[39] MacLennan I. Germinal centers. Ann Rev Immunol 1994;12:117–39.

[40] Tew JG, et al. Follicular dendritic cells and presentation of antigen and costimulatory signals to B cells. Immunol Rev 1997;156: 39–52.

[41] Burton GF, et al. Follicular dendritic cells (FDC) in retroviral infection: host/pathogen perspectives. Immunol Rev 1997;156:185–97.

[42] Hjelm F, et al. Antibody-mediated regulation of the immune response. Scand J Immunol 2006;64(3):177–84.

[43] Gustavsson S, Wernersson S, Heyman B. Restoration of the antibody response to IgE/antigen complexes in CD23-deficient mice by CD23+ spleen or bone marrow cells. J Immunol 2000;164(8):3990–5.

[44] Karlsson MC, Getahun A, Heyman B. FcgammaRIIB in IgG-mediated suppression of antibody responses: different impact in vivo and in vitro. J Immunol 2001;167(10):5558–64.

[45] Kumpel BM. In vivo studies of monoclonal anti-D and the mechanism of immune suppression. Transfusion clinique et biologique. J De la Soc francaise De Transfus Sang 2002;9(1):9–14.

[46] Heyman B. Regulation of antibody responses via antibodies, complement, and Fc receptors. Annu Rev Immunol 2000;18:709–37.

[47] Regnault A, et al. Fcγ receptor-mediated induction of dendritic cell maturation and major histocompatibility complex class I-restricted antigen presentation after immune complex internalization. J Exp Med 1999;189(2):371–80.

[48] Wernersson S, K. M, Dahlström J, Mattsson R, Verbeek JS, Heyman B. IgG-mediated enhancement of antibody responses is low in Fc receptor gamma chain-deficient mice and increased in Fc gamma RII-deficient mice. J Immunol 1999;163(2):618–22.

[49] Kalergis AM, R. J. Inducing tumor immunity through the selective engagement of activating Fcγ receptors on dendritic cells. J Exp Med 2002;195(12):1653–9.

[50] Banchereau J, Steinman RM. Dendritic cells and the control of immunity. Nature 1998;392(6673):245–52.

[51] Radbruch A, et al. Competence and competition: the challenge of becoming a long-lived plasma cell. Nat Rev Immunol 2006;6(10):741–50.

[52] Ravetch JV, N. M. Killing some to make way for others. Nat Immunol 2007;8(4):337–9.

[53] Anthony RM, et al. Intravenous gammaglobulin suppresses inflammation through a novel T(H)2 pathway. Nature 2011;475(7354): 110–3.

[54] Davis RS. Fc receptor-like molecules. Annu Rev Immunol 2007;25:525–60.

[55] Ehrhardt GR, et al. Fc receptor-like proteins (FCRL): immunomodulators of B-cell function. Adv Exp Med Biol 2007;596:155–62.

[56] Leu CM, et al. FcRH1: an activation coreceptor on human B cells. Blood 2005;105(3):1121–6.

[57] Haga CL, et al. Fc receptor-like 5 inhibits B-cell activation via SHP-1 tyrosine phosphatase recruitment. Proc Natl Acad Sci USA 2007;104(23):9770–5.

[58] Ehrhardt GR, et al. The inhibitory potential of Fc receptor homolog 4 on memory B cells. Proc Natl Acad Sci USA 2003;100(23): 13489–94.

[59] Wilson TJ, Fuchs A, Colonna M. Cutting edge: human FcRL4 and FcRL5 are receptors for IgA and IgG. J Immunol 2012;188(10): 4741–5.

[60] Franco A, et al. Human Fc receptor-like 5 binds intact IgG via mechanisms distinct from those of Fc receptors. J Immunol 2013;190(11):5739–46.

[61] Kubagawa H, et al. Identity of the elusive IgM Fc receptor (FcmuR) in humans. J Exp Med 2009;206(12):2779–93.

[62] Li FJ, K. Y, McCollum MK, Wilson L, Motohashi T, Bertoli LF, et al. Enhanced levels of both the membrane-bound and soluble forms of IgM Fc receptor (FcμR) in patients with chronic lymphocytic leukemia. Blood 2011;118(18):4902–9.

[63] Reid KBM, Porter RR. The proteolytic activation systems of complement. Annu Rev Biochem 1981;50:433–64.

[64] Muller-Eberhard HJ. Molecular organization and function of the complement system. Ann Rev Biochem 1988;57:321–47.

[65] Fearon D, Carroll M. Regulation of B lymphocyte responses to foreign and self-antigens by the CD19/CD21 complex. Annu Rev Immunol 2000;18:393–422.

[66] Nussenzweig V. Complement-receptor lymphocytes. Am J Pathol 1971;65(2):479–80.

[67] Pepys MB. Role of complement in induction of the allergic response. Nature: New Biol 1972;237:157–9.

[68] Carroll MC. The role of complement in B-cell activation and tolerance. Adv Immunol 2000;74:61–88.

[69] Kurtz CB, et al. The murine complement receptor gene family. IV. Alternative splicing of Cr2 gene transcripts predicts two distinct gene products that share homologous domains with both human CR2 and CR1. J Immunol 1990;144:3581–91.

[70] Molina H, et al. A molecular and immunochemical characterization of mouse CR2. Evidence for a single gene model of mouse complement receptors 1 and 2. J Immuno 1990;145(9):2974–83.

[71] Carter RH, Fearon DT. CD19: lowering the threshold for antigen receptor stimulation of B lymphocytes. Science 1992;256(5053): 105–7.

[72] Hebell T, Ahearn JM, Fearon DT. Suppression of the immune response by a soluble complement receptor of B lymphocytes. Science 1991;254(5028):102–5.

[73] Gustavsson S, Kinoshita T, Heyman B. Antibodies to murine complement receptor 1 and 2 can inhibit the antibody response in vivo without inhibiting T helper cell induction. J Immunol 1995;154:6524–8.

[74] Heesters BA, et al. Endocytosis and recycling of immune complexes by follicular dendritic cells enhances B-cell binding and activation. Immunity 2013;38(6):1164–75.

[75] Phan TG, et al. Immune complex relay by subcapsular sinus macrophages and noncognate B cells drives antibody affinity maturation. Nat Immunol 2009;38(6):786–93.

[76] Krautler NJ, et al. Follicular dendritic cells emerge from ubiquitous perivascular precursors. Cell 2012;150(1):194–206.

[77] Carroll MC, Isenman DE. Regulation of humoral immunity by complement. Immunity 2012;37(2):199–207.

[78] Hardy RR, Hayakawa K. B-cell development pathways. Annu Rev Immunol 2001;19:595–621.

[79] Herzenberg LA. B-1 cells: the lineage question revisited. Immunol Rev 2000;175:9–22.

[80] Cariappa A, Pillai S. Antigen-dependent B-cell development. Curr Opin Immunol 2002;14(2):241–9.

[81] Fleming SD, et al. Mice deficient in complement receptors 1 and 2 lack a tissue injury-inducing subset of the natural antibody repertoire. J Immunol 2002;169(4):2126–33.

[82] Reid RR, et al. Functional activity of natural antibody is altered in Cr2-deficient mice. J Immunol 2002;169(10):5433–40.

[83] Carroll M, Holers V. Innate autoimmunity. Adv Immunol 2005;86:137–57.

[84] Fischer MB, et al. Dependence of germinal center B cells on expression of CD21/CD35 for survival. Science 1998;280(5363):582–5.

[85] Fischer MB, et al. Local synthesis of C3 within the splenic lymphoid compartment can reconstitute the impaired immune response in C3-deficient mice. J Immunol 1998;160:2619–25.

[86] Gray D. A role for antigen in the maintenance of immunological memory. Nat Rev Immunol 2002;2(1):60–5.

[87] Leadbetter EA, et al. Chromatin-IgG complexes activate B cells by dual engagement of IgM and Toll-like receptors. Nature 2002;416(6881):603–7.

[88] Boackle SA, et al. Cr2, a candidate gene in the murine Sle1c lupus susceptibility locus, encodes a dysfunctional protein. Immunity 2001;15(5):775–85.

[89] Barrington RA, et al. B lymphocyte memory: role of stromal cell complement and FcγRIIB receptors. J Exp Med 2002;196(9):1189–99.

Chapter 12

# B Cell Localization and Migration in Health and Disease

Anja E. Hauser[1,2] and Uta E. Höpken[3]

[1]*Immune Dynamics, Deutsches Rheumaforschungszentrum, Berlin, Germany;* [2]*Charité Universitätsmedizin, Berlin, Germany;* [3]*Department of Tumor- and, Immunogenetics, Max-Delbrück Center for Molecular Medicine, Berlin, Germany*

## 1. INTRODUCTION

Motility is an essential feature of B cells. Already during their maturation process, B cells of various developmental stages are traveling between specialized microenvironmental tissue niches within the bone marrow. In these niches, they interact with different stromal cell types that promote their maturation at defined stages of development, a process which will be described in Section 2 of this book chapter. Once the B cells have reached maturity, they leave the bone marrow and start to recirculate via the blood flow throughout the whole body in constant surveillance for antigens. To become activated, mature B cells enter secondary lymphoid organs (SLOs) such as the spleen, lymph nodes (LNs), and Peyer's patches (PPs). These tissues have a common basic architecture: they consist of separate zones for B and T lymphocytes. The organization of these zones crucially depends on the homeostatic chemokine and lymphotoxin/tumor necrosis factor (TNF) system. Chemokines are produced by stromal and endothelial cells typical for each region, and they chemoattract the respective cell type to ensure their correct localization within the tissue. Antibody-secreting plasma cells localize separately from the B cells in areas of the SLOs that are also populated by macrophages.

Guided by chemokines, B cells are positioned in strategic locations within SLOs, which guarantees that they encounter their specific antigens and that the rare antigen-specific cells encounter cognate T helper cells. Antigen-dependent activation of the B cells results in their repositioning within the SLOs and, subsequently, in their differentiation. In T-dependent immune responses, specialized microanatomical structures called germinal centers (GCs) are formed, where B cells undergo a dynamic process of somatic hypermutation and subsequent Darwinian selection of high-affinity B cell receptor (BCR) variants, which ultimately results in the affinity maturation of the humoral immune response. The past decade has brought significant advances in understanding B cell dynamics within the intact tissue of SLOs. Many of these insights are derived from intravital microscopy experiments, especially two-photon microscopy. This technology, which uses pulsed femtosecond lasers in the near-infrared range to excite fluorophores such as fluorescent dyes and proteins, has helped significantly to understand the migration and cellular interactions of B cells in their genuine environment as well as the dynamics of antigen delivery to the B cells in the SLOs. Because of the relatively low energy that the tissues are exposed to, this method is tissue-friendly and allows the analysis of cellular dynamics deep within tissues over the course of several hours; thus, it has contributed significantly to the understanding of the processes of B cell migration, activation, and their interactions with other cells. B cell entry into SLOs as well as dynamics within the SLOs is discussed in Section 3. In addition to follicular B cells, several other B cell subsets with different functions and therefore different localization and migration properties have been described. B-1 cells are a self-renewing population abundant in the peritoneal cavity. Nonrecirculating marginal zone B cells, specialized to detect blood-borne antigens, are positioned at the interface of the blood-filled red pulp and the lymphocyte-containing white pulp of the spleen. Another subset that is highly abundant in the spleen is formed by memory B cells. Together with antibody-secreting long-lived plasma cells they constitute B cell-mediated immunological memory. Long-lived plasma cells preferentially localize in the bone marrow, where they can persist in specialized microanatomical compartments termed survival niches over extended time periods (i.e., several decades in humans). Plasmablasts, the migratory precursors of plasma cells, have to translocate from SLOs to the bone marrow to reach these niches. The details of their migration are described in Section 4 of

Molecular Biology of B Cells. http://dx.doi.org/10.1016/B978-0-12-397933-9.00012-6

this chapter. Section 5 of this chapter reviews the current knowledge on body cavity trafficking of B-1 cells, Section 6 focuses on the migration of mucosal B cells.

In addition to the physiological aspects of B cell localization, they are also found in chronically inflamed tissues, where they crucially contribute to the formation of tertiary lymphoid tissue (Section 7). The chapter ends with a section on the migration of neoplastic B cells (Section 8).

## 2. MIGRATION OF B CELLS IN THE BONE MARROW

### 2.1 Migration and Interactions of Developing B Cells

During lymphopoiesis, B cells in the bone marrow pass through a well-characterized sequence of various developmental stages. The developing B cells depend critically on stage-specific signals for their survival and differentiation, which are delivered to them in various microenvironmental niches defined by stromal cells. B cells are believed to migrate between these tissue areas during their development. The concept of fixed microenvironmental niches that are required to host cells and control their fate was first proposed in the 1970s by Schofield et al. [1] in the context of stem cells. Hematopoietic stem cells (HSCs) are believed to be surrounded by a unique microenvironment that controls their fate with respect to division, survival, and differentiation. Since then, similar concepts have been proven to apply to several hematopoietic cell types and stages of differentiation, including developing B cells and even memory plasma cells in the bone marrow.

B cell precursors in the bone marrow are generated from multipotent HSCs, which persist throughout the lifetime of an individual and are characterized by their self-renewing potential, a feature that depends on intrinsic and extrinsic factors. Until recently, it was not clear if HSCs and restricted hematopoietic progenitors (HPCs) reside within specialized microenvironmental niches that are spatially distinct from each other, and a dichotomy between osteoblastic niches on the one hand and vascular stem cell niches on the other hand evolved in the literature. HSCs have been described to localize in the vicinity of osteoblasts, which are bone-forming cells lining the border between solid bone and the marrow. Osteoblasts have a crucial function in regulating the maintenance of the stem cell pool—an increase in the number of trabecular N-cadherin + osteoblasts is concomitant with an elevated number of HSCs [2]. This effect is mediated by activation of the Notch1 pathway, triggered by osteoblast-derived Jagged1, leading to HSC proliferation [3]. Osteoblasts also secrete other factors regulating HSC homeostasis, such as thrombopoietin, angiopoietin, and the chemokine C-X-C-motif ligand-12 (CXCL12) [4]. However, endothelial cells, which often localize near the endosteum, especially in flat bones [5], have been shown to constitute another crucial component of the HSC niche [6]. Together with Leptin-receptor$^+$ perivascular cells, endothelial cells provide stem cell factor (SCF, also known as kit ligand) within the vascular niche, and its conditional deletion from these cells results in a depletion of HSCs from the bone marrow [7].

CXCL12, which is also termed "stromal cell derived factor one alpha" because of its high expression by stromal components, is a crucial chemokine for regulating HSC localization and migration in the bone marrow via its receptor, C-X-C-motif receptor-4 (CXCR4) [8–10]. Multipotent hematopoietic progenitors (MPPs) have been shown to directly contact the processes of CXCL12-expressing reticular stromal cells [11]. Experiments in which CXCL12 was selectively depleted from various cell types known to contribute to the maintenance of HSCs and HPCs (i.e., osteoblasts, perivascular and reticular stromal cells) have demonstrated that HSCs and lymphoid progenitors occupy distinct niches in the bone marrow: CXCL12 produced by endothelial and perivascular as well as mesenchymal stromal cells supports the survival of HSCs. In contrast, osteoblast-derived CXCL12 retains HPCs in the bone marrow and supports B-lineage-lymphoid progenitors [12,13].

CXCL12 is also clearly required for the development of the earliest B lineage progenitors, which are identified by the expression of c-kit, interleukin (IL)7R$\alpha$, and CD93 (AA4.1) [14]. These early pre–pro B cells migrate toward CXCL12 in vitro. In addition to acting as a chemoattractant, CXCL12 promotes their survival and proliferation, acting synergistically with SCF and IL-7. Most B220$^+$ Fms-related tyrosine kinase three/Fetal liver kinase 2$^+$ (Flt3/Flk2$^+$) pre–pro B cells contact the bodies of CXCL12-expressing stromal cells; this adhesion is mediated via $\alpha 4\beta 1$ integrin (also known as very late antigen 4, VLA4) on the B cells binding to vascular cell adhesion molecule-1 (VCAM-1) on the surface of the CXCL12$^+$ cells [11]. These CXCL12-abundant reticular cells share some morphological features with the perivascular CXCL12 producing cells supporting HSCs, but it is currently unclear whether they actually represent one population or different, specialized subsets.

As B cell development progresses with the rearrangement of the immunoglobulin (Ig) heavy chain gene variable regions, the pro-B cells (B220$^+$c-kit$^+$) are not found in direct contact with CXCL12$^+$ cells any longer. Instead, at this stage they localize in the vicinity of IL-7 producing stromal cells, which are thought to be a separate population based on histological stainings [11]. Reaching the pre-B cell stage is accompanied by another change of their microenvironment: At this point, when they produce a functional Ig$\mu$ chain that associates with the surrogate light chain to form the pre-BCR, they leave the IL-7$^+$ stromal cells and move on toward another stromal subset that specifically expresses galectin-1 (Gal1), an S-type lectin, on their surface [15]. Stromal Gal1 acts as a ligand for the pre-BCR. It is able to induce the formation of pre-BCR clusters that also contain VLA4 and lymphocyte function-associated antigen-1 (LFA1) at the contact site to the stromal cell. This synaptic contact is able to trigger intracellular tyrosine kinase activity and signal transduction of the pre-BCR, which is essential for proliferation and differentiation at the pre-BII cell stage [16]. IL-7$^+$ and Gal-1$^+$ cells represent distinct populations of mesenchymal cells in the bone marrow [15].

Taken together, this well-organized order of events within the spatial context of the bone marrow has led to the concept that there are different microenvironmental niches, defined by resident immobile stromal subsets, which control the homeostasis of the developing B cells. It implies that the B cells at least transiently become mobile to migrate between these niches during their development. It is not known how exactly the transition of the cells between those various niches is regulated, particularly which other chemotactic factors are required to fine-tune their localization, because CXCL12 is involved at multiple stages of the process.

In addition to the parenchyma, the sinusoids in the bone marrow also constitute important sites for immature B cells [17–20]. Sinusoids are venous blood vessels that branch through the marrow parenchyma, converging into a big central vein that connects to the nutrient vein, which leaves the bone via the cortex [21]. Bone marrow sinusoids are lined with a thin-walled endothelium and they represent the interface where B cells enter the circulation after finishing their maturation in the bone marrow. The chemokine receptor CXCR4 is partially responsible for the retention of B cell in the bone marrow [22]. Antagonizing CXCR4 signaling results in an increased frequency of mature and immature (IgD$^{lo}$) B cells in the marrow sinusoids and at the same time reduces their numbers in the parenchyma, indicating that this chemokine receptor mediates their translocation into the blood [23]. In addition to CXCR4, the sphingosin-1-phosphate (S1P) receptor 1 (S1P1) has also been shown to play a role. Hematopoietic cells, especially erythrocytes, have been indicated to produce S1P in the blood [24], leading to higher concentrations of this sphingolipid in the bone marrow sinusoids than in the parenchyma, thus mediating the transmigration of B cell into the sinusoids [25]. S1P1-deficient B cells are retained in the parenchyma and show a reduced egress into the sinusoids; a similar effect can be observed after blocking S1P signaling by the administration of the antagonist Fingolimod (FTY720).

Once the B cells have crossed the endothelium and are located inside of the sinusoids, they do not immediately leave the bone marrow with the bloodstream. Instead, they stay attached to and crawl along the luminal side of the sinusoidal endothelium [23]. The adhesion to the endothelium is mediated by signaling via cannabinoid receptor-2 (CB2) on the B cells as well as via VLA4- and VCAM-1-mediated adhesion. The prolonged retention of B cells in the sinusoids suggests that the sinusoids constitute a special vascular niche of their own for the B cells instead of just serving as entry sites into the circulation. This notion is supported by CB2-deficient mice, which exhibit a change in BCR repertoire [23]. A schematic overview of the localization of developing B cells in the bone marrow is depicted in Figure 1.

## 2.2 Mature B Cells and Immune Responses in the Bone Marrow

In addition to its task as a primary lymphoid organ, the bone marrow is the site of activation of mature B cells. These B cells localize close to the venous blood vessels in the bone marrow in so-called perisinusoidal niches. This position at the interface between the tissue and the bloodstream is ideally suited for surveying of pathogens in the bloodstream [26]. It is reminiscent of the position of marginal zone B cells in the spleen, a subset of cells with which they also share the ability to be activated by blood-borne T-independent (TI) antigens such as bacteria. Neither the spleen nor the bone marrow has access to lymph vessels. In contrast to marginal zone B cells, the perisinusoidal B cell subset in the bone marrow immune niches circulates in the blood. The survival of this subset in the bone marrow niches is critically dependent on CD11c$^+$, chemokine C-X3-C motif receptor 1$^+$ (CX3CR1$^+$) dendritic cells (DCs), which are abundant in the bone marrow. These DCs secrete the cytokine macrophage migration inhibitory factor that signals in B cells via the CD74-CD44 receptor complex, resulting in the activation of an antiapoptotic phosphatidylinositol-3-hydroxy kinase-Akt kinase pathway [27]. Similar to B cells in the bone marrow immune niches, marginal zone B cells also depend on DCs for their survival, but in their case the survival signals are delivered by the TNF family members B cell activating factor of the TNF family and A proliferation inducing ligand (APRIL). Hence, in addition to being the site of hematopoiesis in adult mammals, bone marrow also functions as a SLO. A third function of the bone marrow for the immune system is the maintenance of immunological memory: After their activation in SLOs, some plasma cells re-localize to the bone marrow, where they can become long-lived in survival niches that share some similarities to the developmental B cell niches (e.g., by the presence of CXCL12-producing stromal cells; for details, see the paragraph on plasma cell migration in Section 4 of this chapter).

# 3. MIGRATION OF B CELLS INTO AND WITHIN SLOS

## 3.1 Microanatomical Structure of SLOs and Mechanisms of B Cell Entry

### 3.1.1 Lymph Nodes

SLOs feature a unique compartmentalized architecture with predefined areas for different immune cell populations that

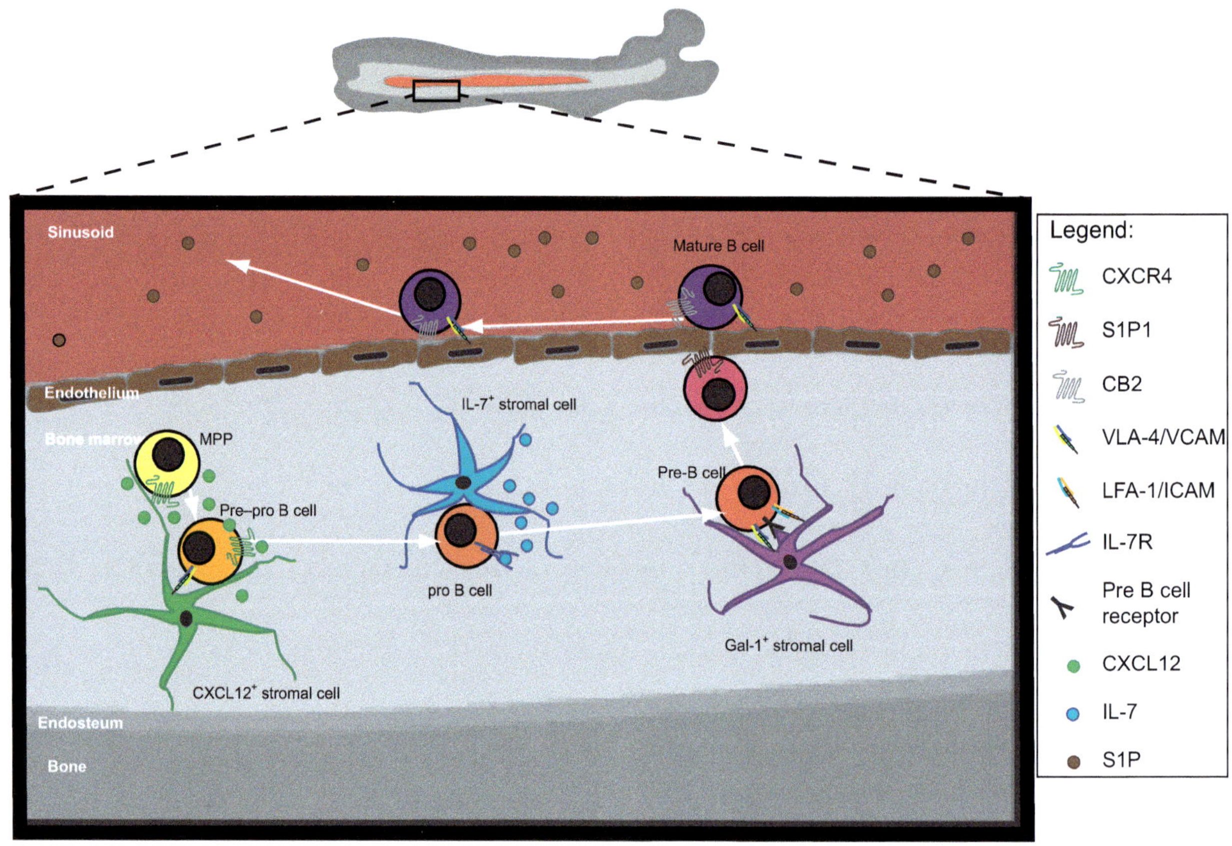


FIGURE 1 **Localization of B cells in the bone marrow during B cell development.** Multipotent hematopoietic progenitors (MPP) are contacting the processes of CXCL12+ stromal cells. Pre–pro B cells, the earliest B cell precursors, are found to localize near the cell bodies of CXCL12-producing cells. More mature pro-B cells are found in contact with IL-7+ stromal cells. Pre-B cells are attached to Galectin-1+ stromal cells via a synapse formed by the pre-B cell receptor binding to Galectin-1 on the surface of the stromal cells as well as by LFA-1/ICAM and VLA4/VCAM-mediated interactions. B cell egress into the bone marrow sinusoids is mediated by S1P1 on the B cells binding S1P, which is highly abundant in the blood. Once they have entered the sinusoids, B cells adhere to and crawl along the luminal side of the endothelium, mediated via CB2 and VLA4 on the B cells, before they are released into the bloodstream.

are organized by the presence of homeostatic chemokines produced by stromal cells, as shown in Figure 2. The structure of SLOs is optimized for promoting the adequate activation of immune cells during an immune response (e.g., by facilitating the interactions of rare, cognate T–B interactions). B cell zones within SLOs (also termed follicles) are characterized by the presence of the chemokine CXCL13 [28,29] and the corresponding chemokine receptor CXCR5, which is present on most B cells. Mice lacking CXCR5 have a severely disturbed splenic architecture with a lack of primary follicles [30,31]. CXCL13 is expressed in high amounts by follicular dendritic cells (FDCs), a specialized radio-resistant cell population in the B cell follicles [32], as well as by some other stromal subsets in this area. Recently, a population of ubiquitous perivascular platelet derived growth factor β-positive (PDGFR-β+) cells has been identified as precursors for FDCs, explaining the de novo generation of FDC-containing B cell follicles in tertiary lymphoid follicles (TLOs) during the course of chronic inflammation [33] (see also Section 7 on B cells in extralymphoid tissues).

B cell follicles are located in the cortical areas of the LN (i.e., the outer areas close to the capsule, directly underneath the subcapsular sinus). It collects the afferent lymph and is lined by lymphatic endothelium and a layer of subcapsular sinus macrophages. Between the FDC-rich area in the center of the follicles and the subcapsular sinus resides a specialized subset of stromal cells involved in the transport of antigens from the sinus into the follicles. Similar to FDCs, they express various adhesion molecules such as vascular adhesion molecule-1 (VCAM-1), intercellular adhesion molecule-1 (ICAM-1), mucosal vascular addressin cell adhesion molecule (MAdCAM-1), stroma markers such as gp38 and BP3 (CD157), and the B cell chemokine CXCL13. They differ from FDCs in their high expression of receptor activator of nuclear factor kappa B ligand and low expression of complement receptor-2 (CR2/CD21), and they seem to have role in the organogenesis of secondary lymphoid tissues [34]. Because of their location at the interface between lymphoid tissue and lymphatics, they are called marginal reticular cells (MRCs).

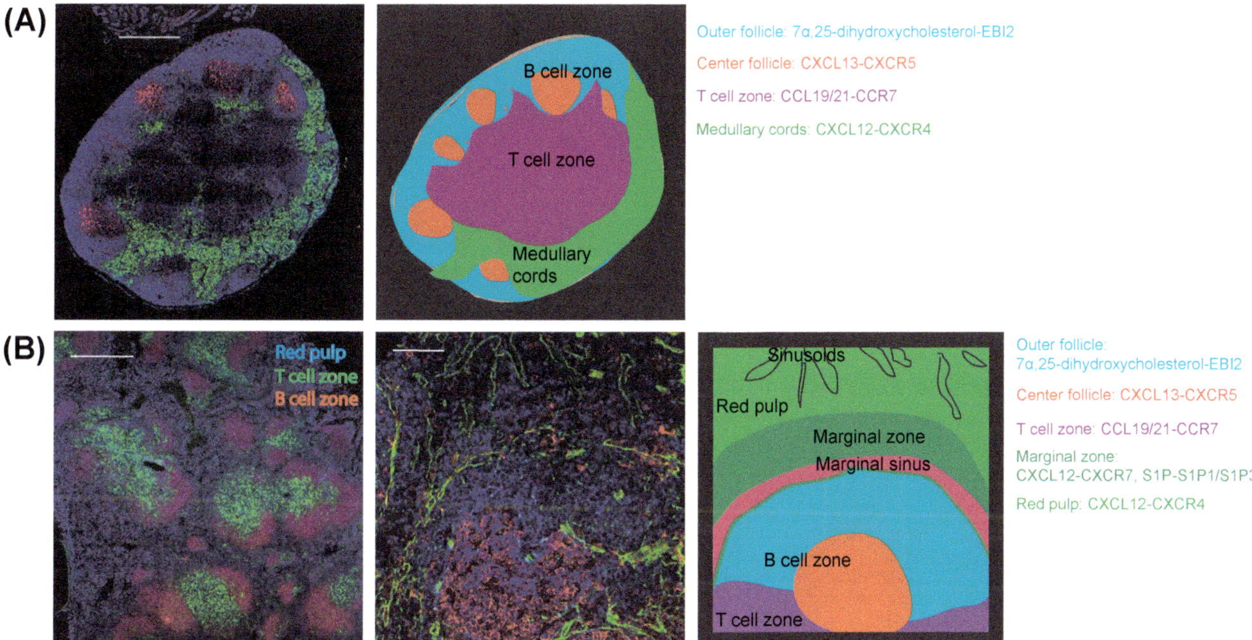


FIGURE 2 **Structure of secondary lymphoid organs.** (A) Left panel: immunofluorescence histology and schematic overview (right) of a lymph node; different colors depict the main zones as defined by chemokines. Scale bar = 100 μm. (B) Left panel: immunofluorescence histology of an overview of the spleen. Red pulp areas are shown in blue, B and T cell zones of the white pulp are depicted in red and green, respectively. Scale bar = 500 μm. Middle and right panel: histology and schematic overview of a splenic B cell follicle with neighboring T cell zone and adjacent marginal sinus, marginal zone, and red pulp. Scale bar = 50 μm.

At the basal side of the follicle, oriented toward the center of the node, the B cell zone adjoins the T cell area. This T–B border region constitutes an important site for the interaction of antigen-specific T and B cell pairs [35]. The B cells do not uniformly fill the whole area underneath the capsule but are rather organized in follicles that are surrounded by T cells, the so-called interfollicular areas. These regions are critical for the initial encounter of antigen-specific T and B cell pairs [36]. The CC-chemokine receptor-7 (CCR7) on the T cells is responsible for their localization in the T cell zone, its ligands CC-chemokine ligand (CCL) 19 and CCL21 are produced by fibroblastic reticular cells in that area.

Toward the hilus of the LN, the T cell zone borders the medullary cords, which are characterized by the presence of macrophages and antibody-secreting cells. Stromal cells in that area produce high amounts of CXCL12. LNs are penetrated by an extensive system of lymphoid vessels, the sinusoids, which are most abundant in the medullary cord area. These medullary sinuses collect lymph from cortical sinuses, which extend throughout the outer cortex of the LN, adjacent to the B cell follicles and are directly connected to the subcapsular sinus upstream. Downstream, medullary sinuses drain into the efferent lymph vessel, which leaves the organ in the hilus area, the region where the LN artery and vein also connect to the LN. Thus, the efferent lymph is highly enriched in lymphocytes in comparison to the afferent lymph [37].

B cells enter LNs and PPs via microvessels, which are unique to these organs. Unlike most blood vessels, which feature a squamous endothelium, these specialized microvessels are characterized by a cuboid endothelial cell layer and are therefore called high endothelial venules (HEVs). Erythrocytes are a prominent source of S1P [24], making the blood an S1P-rich compartment, where lymphocytes downregulate the respective receptor S1P1 because of its desensitization. This process depends on the G-protein coupled receptor kinase 2 and is crucial for the lymphocytes to override their retention in the blood to enter the lymphoid tissue [38].

HEVs are mainly found in the T cell zone in the LNs. B cells enter SLOs in a fixed sequence of checkpoints; this also holds true for T lymphocytes and has been termed "multistep model" [39]. A schematic overview of the model is depicted in Figure 3. B cells in the bloodstream first start to form loose, dynamic contacts with the vessel walls by moving along the luminal side of the endothelium, a process termed "tethering" and "rolling." Rolling adhesion requires rapid association and dissociation of molecular bonds between the cells and the endothelium. These contacts are mediated by L-selectin (CD62L), a C-type lectin on the B cell surface. CD62L binds to peripheral node addressins (PNAd; i.e., a group of mucin-type glycoproteins) with modified (i.e., sialylated, fucosylated, and sulfated) carbohydrate groups, including CD34, Glycosylation-dependent cell adhesion molecule 1, and MadCAM1, which is specifically

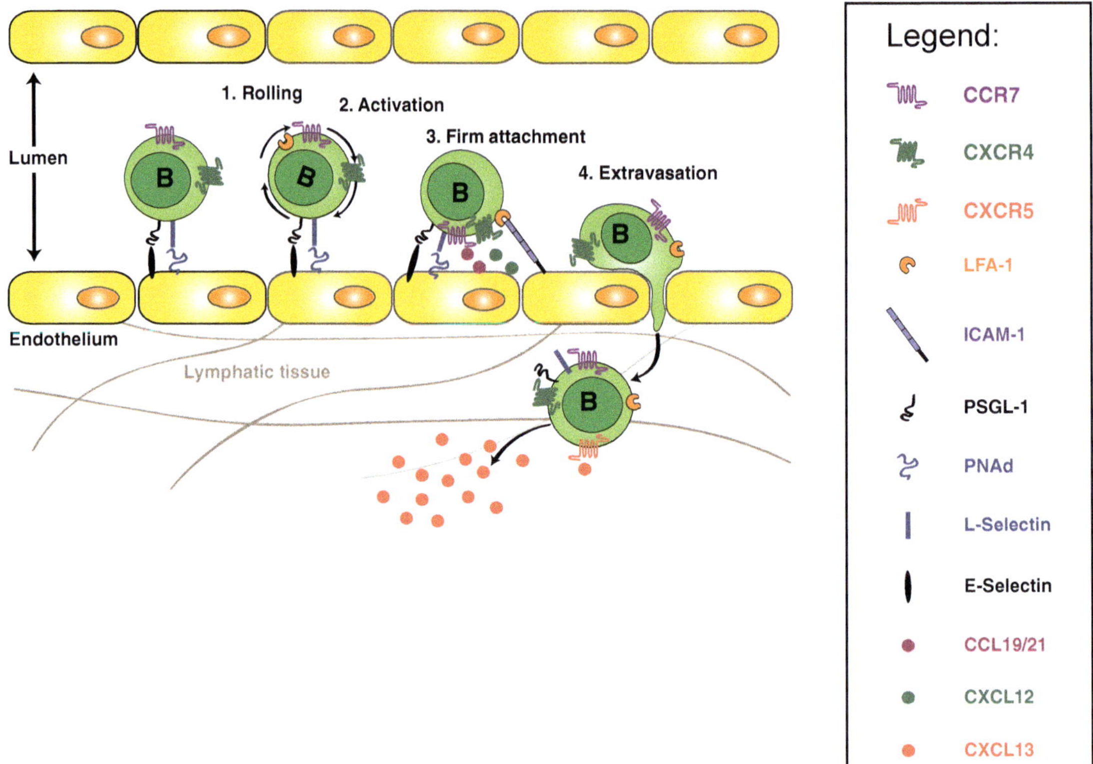


FIGURE 3 **Mechanisms of transendothelial migration of lymphocytes.** Naïve B cell migration across high endothelial venules (HEVs) into lymph nodes using L-selectin, CCR7, CXCR4, and LFA1. Tethering and rolling to the HEVs is mediated by selectins, followed by engagement of chemokine receptors and then firm attachment via integrin-mediated arrest, eventually followed by extravasation into underlying tissue.

expressed in lymphoid tissues in the mucosa [40]. Blocking CD62L binding to these oligosaccharide determinants by using the monoclonal antibody MEL-14 in vivo blocks lymphocyte extravasation into peripheral LNs [41].

In addition to L-selectin-mediated interactions, the α4 integrins α4β1 and α4β7 also contribute to the rolling process in mucosal LNs and PPs by interacting, respectively, with VCAM-1 and MadCAM-1 on the endothelial surface [42–44].

The next step of the cascade is the triggering of integrin activation on the B cell surface, which allows them to adhere firmly to the endothelium. This activation is mediated by chemokines on the luminal endothelial surface, resulting in a conformational change of the integrins, thus enabling them to undergo stable adhesion with the endothelium. Chemokine receptors are G-protein coupled receptors consisting of seven transmembrane domains. When $G_{\alpha i}$-signaling downstream of chemokine receptors is disrupted in lymphocytes by pertussis toxin treatment, their entry into LNs is completely inhibited [45], underlining the crucial role of the chemokines in this process. Various chemokines are able to provide the signals required for triggering the integrin activation on B cells in peripheral LNs: CCR7, which constitutes the main chemokine receptor involved in integrin activation in T cells, also supports most triggering events in B cells. CXCR4 can also induce integrin activation in peripheral LNs [46]. Of the two CCR7 ligands, only CCL21 is actually expressed by HEVs. CCL19 is not expressed in HEVs; instead, it is produced by stromal cells in the vicinity of the vasculature, and the endothelial cells are able to take it up and present it at their luminal side [47]. CCL21 is expressed by the HEVs of LNs and PPs [48]. Similar to CCL19, the CXCR4 ligand CXCL12 is produced by perivascular stromal cells and is present at the luminal side of the endothelium [46].

CXCR5, the receptor for the B cell chemokine CXCL13, causes integrin activation on B cells for their entry into PPs via HEVs [46]. CXCL13 has been detected on HEVs within PPs.

In peripheral LNs, firm adhesion to the endothelium is mediated mainly by the LFA1, a heterodimer consisting of CD11a and CD18 [49,50]. It binds to the Ig-superfamily member ICAM-1, which is highly abundant on the HEV endothelium [51] and to ICAM-2 [52], which is ubiquitously expressed on vascular endothelium. The α4 integrins α4β1 and α4β7 also play a role in the adhesion of lymphocytes to HEVs in peripheral LNs by binding to VCAM-1 [50].

In mesenteric lymph nodes (MLNs), the situation is different: LFA1 and α4β7 contribute equally in mediating lymphocyte adhesion, with α4β7 binding to MadCAM-1.

In PPs, α4β7-MadCAM-1 interactions account for most of the firm adhesion events [53].

To perform the final, irreversible step of crossing the endothelium, a process called diapedesis [54], B cells move in an amoeboid fashion from the apical to the basal HEV surface. Diapedesis is a process that requires polarization of the cells into a leading edge with lamellipodia and a uropod, involving rearrangement of their cytoskeleton. Transduction from chemokine receptor and integrin-mediated signals from the cell surface to the cytoskeleton is mediated by small G proteins of the Rat sarcoma (Ras) superfamily. The F-actin-binding protein Suppressor of white apricot 70 (SWAP70) is expressed in B but not in T lymphocytes and interacts with the small Rho-GTPase Ras-related C3 botulinum toxin substrate (Rac), which is crucial for cytoskeleton remodeling. B cells in SWAP70$^{-/-}$ mice accumulate in HEVs and fail to enter LNs because of impaired transmigration [55].

For crossing the HEVs, lymphocytes can either take a paracellular or a transcellular route, with the former occurring more commonly. This final step is also the least well characterized part of the multistep model, and data in the literature are mainly derived from other leukocytes than B cells. Junctional adhesion molecules (JAMs), members of the Ig superfamily, are expressed on HEVs and are concentrated at the border between endothelial cells. Their role is to stabilize endothelial cell junctions via homophilic interactions [54]. In addition to undergoing homophilic interactions, JAM-A can also bind LFA1, which results in destabilization of the junction between two endothelial cells, thereby enabling transendothelial migration [56]. Antibodies to JAM-A can inhibit T lymphocyte transendothelial migration in vitro [57]. Platelet endothelial cell adhesion molecule-1 (PECAM1, CD31) is another molecule enriched at the border of endothelial cells. Similar to JAMs, it tightens junctions between endothelial cells through homophilic interactions but can also bind to CD31 on leukocytes. PECAM1 is present in the lateral border recycling compartment, a membrane complex that is connected to the cell borders at the surface of endothelial cells [58]. When a leukocyte crosses the endothelial junction, the lateral border recycling compartment is recruited to this site and surrounds the cell during paracellular diapedesis in a kinesin-mediated, microtubule-dependent process [58,59]. In addition to CD31, JAM-A and CD99 are recruited to the lateral border recycling compartment. Similar trafficking of the lateral border recycling compartment happens during transcellular migration processes [60]. Matrix metalloproteinases have also been shown to be involved in lymphocyte transmigration across HEVs in vivo [61].

### *3.1.2 Spleen*

Because of its function in filtering the blood and its open blood system, the anatomical situation in the spleen is different from the one in SLOs. The afferent splenic artery branches into arterioles (called central arterioles). They are for the most part surrounded by lymphoid aggregates, which are termed "white pulp," because of their appearance in light microscopy and which consist of T cell zones and B cell follicles. They become terminal arterioles, which can subsequently open out into the marginal zone, a region that marks the interface between white and red pulp characterized by the presence of MRCs. It is filled with a specialized B cell subpopulation called marginal zone B cells, as well as with macrophages, which together play an important part in surveying the blood for antigens [62]. In contrast to the circulating follicular B cells, which continuously patrol the body in search for antigens spending on average 1 day in a SLO, marginal B cells form a resident population. Most blood cells including lymphocytes enter the spleen passively via this route, traveling through the marginal zone that surrounds the white pulp, and from there arriving in the red pulp, which consists of large, blood-filled venous sinusoids that are responsible for the red color of this area and that eventually drain into splenic veins [63,64]. A dense network of reticular fibers runs through the red pulp, similar to the medullary cord region in LNs. The stromal cells in this area produce high levels of CXCL12, and there are macrophages and plasma cells present. Some of the arterioles also span over the marginal zone, leading directly into the red pulp.

A fraction of B cells enters the spleen actively via the white pulp [65]; this process is mainly dependent on CXCR5 [31], although some B cells might also respond to the T zone chemokines CCL19 and CCL21. Similar to the entry in LNs and PPs, this process also requires the presence of LFA1, which binds to ICAM1. In addition, α4β1 plays a synergistic role as demonstrated by experiments using blocking antibodies against LFA1 and α4 [66].

## 3.2 Migration of B Cells within SLOs

### *3.2.1 Activation of Follicular B Cells*

After entering secondary lymphoid tissues, B cells migrate into the follicular region. This process is mediated by the B cell zone chemokine CXCL13 and its receptor CXCR5 on the follicular B cells. CXCL13 is produced by stromal cells in the B cell zone such as FDCs and MRCs [32,34]. In addition to its function in attracting B cells within mature LNs, CXCL13 is a crucial factor for the development of most peripheral LNs and PPs [31]. Spleen and MLNs are formed in CXCR5-deficient mice, but they lack B cell follicles. CXCR5-deficient B cells do not home normally into B cell zones in adoptive transfer experiments. In addition to CXCR5, follicular B cells express CCR7, the receptor for the T zone chemokines CCL19 and CCL21, on their surface [67], which probably directs their migration from HEVs, where they extravasate through the T cell zone toward the B cell follicles.

The first observations of B cells by intravital imaging were made approximately a decade ago, showing that naïve B cells in LNs migrate through the tissue in a random walk to scan for antigen [68]. This interstitial motility of B and T lymphocytes depends on signaling molecules downstream of the G-protein-coupled chemokine receptors, which control the polymerization of F-actin and the formation of lamellipodia, such as the small GTPase Rac. Accordingly, Dedicator of cytokinesis-2, a guanine exchange factor for Rac, has been shown to be crucial for the basal motility of B cells in lymphoid tissue [69].

B cells in LN follicles acquire antigen within minutes after subcutaneous injection [70]. Antigen transport into the follicles occurs through various ways. Small soluble antigens (<70 kDa) are guided into these areas via an extensive network of microanatomical channels in the SLOs [71]; these conduits consist of a core of collagen-I and microfibrils that are ensheathed by fibroblastic reticular cells. Conduit networks have first been described in the T cell zone [72,73]. Stromal cells of the B cell zone also form conduits, although the system is less pronounced than in the T cell zone [34,71]. In addition to their ability to transport antigen, conduits can channel chemokines, which are produced by fibroblastic reticular cells. Similar to DCs that are in close contact with fibroblastic reticular cells in the T cell zone [74], B cells have been shown to directly take up antigen from the conduits [71].

Larger antigens and immune complexes do not get channeled through conduits. They are taken up by subcapsular sinus macrophages and transferred to B cells that shuttle them from the sinus into the follicles [75]. This shuttling process is independent of BCR specificity; instead, it is mediated by a CR2 (CD21)-dependent mechanism [76]. The immune complexes subsequently become deposited on FDCs localized at the center of the B cell follicles. These stromal cells are characterized by long dendritic extensions that reach throughout the B cell follicles and by the surface expression of complement receptor-1 (CR1) and CR2 (CD35 and CD21, respectively) and Fc gamma receptor (FcγR) [32], which serve to retain native antigen in the form of immune complexes on their surface over extended time periods. Moreover, FDCs have been shown to internalize these immune complexes via an actin-dependent mechanism and keep them in a nondegrading intracellular compartment from where they are recycled to the FDC surface and displayed to B cells [77]. Immature FDCs have been shown to develop from PDGFR-$\beta^+$ cells, which are ubiquitously present in the organism [33], which explains how they can appear in tertiary lymphoid structures of chronically inflamed tissues such as the synovium of arthritic joints or salivary glands in patients with Sjögren's syndrome (SS) [78], where they develop into functional FDCs.

In the absence of inflammation, mature FDCs are located exclusively within primary B cell follicles of SLOs. During an immune response, lymphotoxin-α (LTα) and tumor necrosis factor alpha (TNFα) derived from activated B cells induce the maturation of the FDCs, resulting in upregulation of CD21, CD35, and FcγR, thereby enhancing the capacity of the FDCs to bind immune complexes and promote GC formation. Within GCs, cognate B cells are able to take up antigens directly from FDCs [79].

An alternative site where B cells encounter antigen in LNs is outside of the follicles, in the regions around HEVs. DCs in the T cell zone have been shown to directly transfer native antigen to B cells surveying the areas around HEVs [80]. This extrafollicular activation may play a particular role at early stages of an immune response against blood-borne antigens.

Within the first hour after a follicular B cell has encountered antigen, it upregulates the G-protein-coupled receptor Epstein-Barr-virus induced molecule-2 (EBI2, also known as GPR183 [81]). Oxysterols, oxidized cholesterol derivatives that were initially characterized as intermediates during bile synthesis, have been identified as the ligands for EBI2 [82,83]. Their synthesis from cholesterol requires two enzymes: cholesterol-25-hydroxylase (C25H, responsible for hydroxylation at position 25) and CYP7B1 (25-hydroxycholesterol 7 α-hydroxylase, causing hydroxylation at position 7α). Both enzymes are expressed in stromal cells, including fibroblastic reticular cells and MRCs present in peripheral regions of the B cell follicles [84]. In contrast, FDCs express lower amounts of C25H and are able to produce the enzyme hydroxy-delta-5-steroid dehydrogenase, which modifies oxysterols, resulting in the inactivation of their ability to trigger EBI2 and thereby creating an area with low oxysterol bioactivity in the center of the follicles. This regional distribution of oxysterols regulates the localization of naïve B cells into the periphery of the follicles [85,86]. The rapid transcriptional upregulation of EBI2 after BCR stimulation causes B cells to first move to the outer follicle areas. A subsequent increase in surface expression of CCR7, the receptor for the T zone chemokines CCL19 and CCL21 that form a gradient ranging from the T zone into the B cell follicles, results in directed B cell movement from the follicle toward the border of the T cell zone. Thereby, the likelihood of encountering a cognate T cell within the first 6 h after antigen exposure [67] is greatly increased. The T–B border region is the area where the first cognate interactions happen [87], resulting in stable, highly motile T–B conjugates that are led by the B cell and can last for at least an hour, as opposed to noncognate interactions that are stable for less than 10 min [35]. This initial CCR7-mediated encounter at the T–B border is followed by a movement of the B cells toward the interfollicular areas and the perimeter of the follicles, distant from the T cell zone [88,89], mediated by a combination of CCR7 downregulation [90] and by EBI2.

Approximately 12–24 h after their initial T cell encounter and after they have sequentially interacted with several T cells, the B cells start to proliferate, which is a process that depends on CD40 being triggered by CD154 on the T cells [87]. B and T cells start to upregulate the transcriptional repressor B cell lymphoma-6 (Bcl6) while they are still in the area at the edge of the follicles [36]. The T cells then start to express T follicular helper cell markers such as programed death-1, GL7 antigen, and the receptor for B cell zone chemokine CXCR5, which mediates their subsequent migration into the central areas of the follicles to initiate the formation of GCs. B cell migration from the periphery of the follicles into these areas is preceded by downregulation of EBI2 and takes place at approximately day 3 after antigen encounter [86].

### 3.2.2 Extrafollicular Plasma Cells

A fraction of B cells in these early phases of the immune response takes a different route: They rapidly differentiate into antibody secreting cells, located at sites of the SLOs that are distant from the B cell zones (therefore they are called extrafollicular; i.e., the red pulp of the spleen and the medullary cords in the LNs). Their translocation is mediated by a downregulation of CXCR5 that occurs upon their differentiation into plasmablasts [91], resulting in a loss of the ability to migrate toward the B cell zone chemokine CXCL13 [92]. CXCR4 expression is retained on plasmablasts, causing them to migrate toward CXCL12 (stromal cell-derived factor-1 [SDF-1]), which is highly expressed by stromal cells of LN medullary cords and the splenic red pulp [93]. CXCL12 is also abundant in the dark zone (DZ) of GCs [94,95], and it is not completely clear how the migration from the GCs to the CXCL12-rich environment in extrafollicular sites of SLOs is regulated. EBI2 also contributes to the extrafollicular positioning of plasmablasts by facilitating their migration away from the follicles to the extrafollicular foci [86].

Compared with naïve B cells and GC B cells, plasma cells display a more directed pattern of migration [96], providing an efficient way to translocate from follicles to extrafollicular sites. In contrast, at least a fraction of plasma cells in medullary cords are thought to be rather sessile [97,98]. Upregulation of the plasma cell-specific transcription factor B lymphocyte-induced maturation protein 1 (Blimp1), which goes along with terminal differentiation (i.e., transition from the plasmablast to the plasma cell stage [99]) correlates negatively with their motility [96]. In line with this, the term "plasmablast" is now widely used for a migratory antibody secreting cell, although it historically refers to an antibody secreting cell that still has the capacity to divide.

Extrafollicular plasma cells are thought to be mainly short-lived because a high degree of apoptosis has been shown to occur among them [100]. They are responsible for the first wave of specific antibodies in a particular immune response, they are mainly IgM$^+$, and they generally do not carry SHMs, although hypermutation has been shown among extrafollicular plasma cells under certain autoimmune conditions [101]. The number of extrafollicular plasma cells peaks within the first week of the immune response and then wanes rapidly; however, there are also reports about the persistence of long-lived plasma cells in the splenic red pulp [102]. Some long-lived plasma cells have been shown to occur in SLOs of New Zealand Black/White (NZB/W) mice, a model for systemic lupus erythematosus [103], as well as in chronically inflamed organs, indicating that the micromilieu in ongoing inflammatory responses fosters the survival of plasma cells [104–106]. However, most long-lived plasma cells are maintained in the bone marrow, and those are thought to be mainly generated in GC reactions [107].

### 3.2.3 Dynamics of the GC Reaction

GCs represent specialized microanatomical structures within the B cell follicles where B cells show a high degree of proliferative activity, along with SHM and class switch recombination [107]. The process of affinity maturation of the humoral immune response involves the activation-induced cytidine deaminase (AID)-dependent introduction of mutations in variable regions of the Ig genes. This results in the generation of new BCR variants, which are subsequently being selected based on their affinity to bind antigen, which is present on the surface of FDCs. GCs are highly dynamic structures that change their appearance during the maturation of the humoral immune response [108]. Immature GCs represent growing clusters of proliferating Bcl6$^+$ B cells located close to FDCs at the border to the T cell zone. As the humoral immune response matures, two histologically distinct zones can be observed within GCs at approximately 8 days after immunization, termed "dark zone" and "light zone" according to their appearance in brightfield microscopy. The DZ predominantly comprises densely packed blasting B cells (referred to as centroblasts), and it is during their extensive division that SHM occurs. The light zone (LZ) contains B cells (centrocytes), interspersed T cells, and FDCs with an extensive network of dendrites. GC B cell recognition of antigen occurs when centrocytes enter the LZ and contact FDCs [79]. It is envisioned that only those B cells that can effectively compete for antigen binding due to a high affinity of their BCR receive signals to survive [109] and proliferate [110] whereas those cells that fail to acquire antigen die by apoptosis. The presence of apoptotic B cells within the LZ and tingible body macrophages, which engulf apoptotic bodies, is also consistent with the selection by antigen occurring within that zone. In this way, GC B cells with higher affinity for a given antigen are thought to expand relative to lower affinity B cells, resulting in the affinity maturation that is the hallmark of immune responses that elicit GCs. Given that the FDC network and its evident antigen deposition are more extensive in the LZ, the selection of better binding variants is taking place in the LZ. The Ig sequences of GC B cells and computer modeling studies have indicated that the distribution of replacement and silent mutations among clonally related sequences obtained from single GCs first requires the selection of variants. This selection is followed by the proliferation of successors that further mutate and are then selected again. However, mutation within the DZ seemed incompatible with the need to acquire antigen within the LZ. Therefore, the recirculation of B cells between the DZ and LZ was proposed to explain the need for sequential events at disparate locations within the GC. In this widely accepted model termed "cyclic re-entry," GC B cells literally migrate between zones for each of these phases, repeating this interzonal migration several times [107]. In line with this, the zonal organization of GCs is mediated by chemokines produced by stromal cell populations present in the two zones: FDCs produce high amounts of CXCL13 whereas CXCR5 directs the migration of GC B cells into the LZ. In contrast, the DZ is characterized by a high abundance of the CXCR4 ligand CXCL12 [94]. Toll-like receptor-4 (TLR4) signaling also seems to contribute to B cell accumulation in the DZ [111]. Two-photon studies have revealed B cells in the GCs to be highly motile [97,98,112]. They are heterogeneous in their morphology, probing for antigen and picking up antigen from FDCs [113], and they have been shown to travel between the two zones. The migration from the LZ to DZ is regulated by signals received from T follicular helper cells [95]. This suggests a two-step selection process for B cells in the LZ, involving signals from FDCs and T cells.

The average residence time for B cells in LNs is 24 h [114,115]. B cells emigrate via the cortical sinusoids adjacent to the follicles. The process of crossing the endothelium to enter the sinusoids is dependent on S1P, and treatment with FTY720 results in retention of the B cells in the LNs [114]. During inflammatory processes, lymphocyte egress is transiently blocked by an interferon-α/β-dependent upregulation of the transmembrane C-type lectin CD69 [116], thereby increasing the retention time of the lymphocytes in lymphoid organs. CD69 has been shown to suppress S1P function by directly interacting with S1P1 [117], thereby preventing access of B and T lymphocytes to cortical sinusoids [118]. In the spleen, follicular B cells exit into the red pulp via the marginal zone in an S1P1 receptor-dependent mechanism [119].

### 3.2.4 Marginal Zone B Cells

Located at the interface between white and red pulp, the marginal zone of the spleen serves as a special microenvironment where immune cells get immediately in contact with the circulating blood. It contains specialized populations of macrophages and a subpopulation of resident B cells with the ability to rapidly respond after an antigenic stimulation and to renew themselves [120]. The microanatomical structure of the marginal zone differs between rodents and primates. In mice, some small arteries terminate in the marginal sinus, which separates the white pulp and marginal zone, whereas other arteries cross the marginal zone and connect to the venous sinuses in the red pulp. The human marginal zone can be subdivided into an inner and an outer zone separated by a layer of fibroblastic cells. Humans lack a marginal sinus. Instead, they have an additional microanatomical compartment, the perifollicular zone, surrounding the marginal zone. The perifollicular zone is characterized by the presence of a population of macrophages expressing sialic-acid-binding immunoglobulin-like lectin-1 surrounding the small arteries that terminate in this region [62].

Similar to B-1 B cells, the marginal zone B cell repertoire is enriched for B cell clones specific for TI antigens, such as phosphorylcholine, a bacterial surface molecule, enabling their rapid response to blood-borne particulate antigens [121]. Marginal zone B cells can be phenotypically distinguished from follicular B cells by a lower expression of CD23 as well as by a higher expression of complement receptors CD21/35, a high expression of surface IgM, and the expression of the nonclassical major histocompatibility complex molecule CD1d, allowing presentation of lipid antigens. They also express higher levels of LFA1 and α4β1 than follicular B cells. The ligands for these integrins, ICAM1 and VCAM-1, are abundant in the marginal zone and function to retain the marginal zone B cells in this area [122], enabling them to resist the shear forces from the circulating blood in this region [119]. A combination of $G_{\alpha i}$ protein-coupled receptors mediates the positioning of marginal zone B cells, and inhibition of $G_{\alpha i}$ signaling by treatment with pertussis toxin causes a selective loss of these cells from the spleen [123]. Among the molecules involved in marginal zone B cell localization are the receptors for S1P, S1P1 and—albeit to a lesser extent—S1P3 [124,125]. Another $G_{\alpha i}$ protein-coupled receptor crucial for marginal zone B cell positioning is CB2. CB2-deficient mice have reduced numbers of marginal zone B cells in the spleen [126] and exhibit a loss of these cells in the blood as well as reduced CD1d-restricted humoral responses. The producers of 2-arachidonylglycerol, the ligand for CB2 within the spleen, remain to be identified; however, stromal cells in the marginal zone are likely candidates.

In addition, the chemokine receptor CXCR7 has been shown to be expressed on marginal zone B cells. CXCR7 functions as a scavenger receptor for the chemokines CXCL11 and CXCL12 by mediating their internalization and subsequent degradation. Blocking CXCR7 using small molecular antagonists resulted in a reduction of marginal zone B cells. The remaining marginal zone B cells showed increased surface expression of CXCR7, indicating interrupted internalization and recycling of the receptor as well as elevated levels of CXCL12 in the serum, which is in line with a role for CXCR7 to serve as a sink for CXCL12 [127].

Despite being noncirculating, marginal zone B cells exhibit a high degree of motility within the spleen and have been shown to be involved in transporting blood-borne antigens from the marginal zone into the splenic B cell follicles by shuttling continuously between marginal zone and follicles [125]. Memory B cells have also been shown to localize in the marginal zone (see Section 3.2.5).

### 3.2.5 Memory B Cells

Memory B cells are able to persist in the human body over decades [128–131]. Thereby, they are maintaining memory for a given antigen without the need of constant antigenic stimulation or proliferation [132,133]. Memory B cells are highly abundant in the human spleen, and they make up 45% of the total B cell population in this organ [134,135]. Class switch has been considered as a characteristic feature of these antigen-experienced B cells; however, somatic mutations, indicative of a memory B cells, can also be detected in the Ig variable (IgV) genes in a fraction of IgM$^+$ B cells. In humans, B cells expressing the memory marker CD27$^+$ can be found in the marginal zone [136] whereas a population of B cells in the blood also stains positive for CD27 [137]. Reduced frequencies of memory B cells in the blood are detected in splenectomized patients [135,138]. Taken together, several lines of evidence indicate that human memory B cells might mostly reside in the spleen, and some memory B cells recirculate in the blood.

For murine memory B cells, the localization as well as their function after rechallenge is heterogeneous and partly depends on the isotype that these cells express: IgM$^+$ memory B cells are scattered throughout the follicles whereas IgG1$^+$ memory B cells have been shown to localize in the vicinity of senescent GCs [139]. Moreover, fate mapping of GC emigrants in mice expressing cre recombinase under control of the promoter for the AID gene has revealed differences in the localization and differentiation behavior of IgM$^+$ and IgG$^+$ memory B cells. After adoptive transfer of and subsequent rechallenge with a given antigen, IgM$^+$ memory B cells localize in the B cell follicles and have the potential to give rise to new GCs. In contrast, most IgG$^+$ memory B cells differentiate into antibody-secreting cells after antigen recall. These plasma cells are localized in the splenic red pulp [140,141]. The chemokine cues acting on the two memory B cell populations and their differential contributions to the long-lived bone marrow plasma cell pool have not yet been investigated.

Similar to effector/memory T cells, memory B cells can home back to the tissues where they have been generated or localize to the SLOs that drain these tissues. This is especially evident in mucosal immune reactions. For example, latent infection of humans with Epstein Barr virus (EBV) leads to the presence of the virus in memory B cells, the dispersal of which can thus be tracked from the site of infection in Waldeyer's ring (i.e., lymphoid tissue in the nasopharynx, including tonsils and adenoids). Recirculating memory B cells carrying EBV can at first be found evenly distributed in the blood, spleen, and LNs, but later after infection they preferentially accumulate in Waldeyer's ring again [142]. Likewise, memory B cells generated in a mucosal immune response against rotavirus in mice express the mucosal homing receptor $\alpha4\beta7$ and accumulate in PPs [143].

## 4. LOCATION AND MIGRATION OF ANTIBODY-SECRETING CELLS

A fraction of plasmablasts generated in systemic immune responses leave the SLOs and travel via the blood to the bone marrow. These plasmablasts are generally thought to originate from GC responses. However, long-lived plasma cells have also been shown to be induced in TI responses as well as after inhibiting the formation of GCs by blocking CD40 ligand (CD40L, CD154) in vivo [144], indicating that extrafollicular responses can contribute to the pool of long-lived plasma cells in the bone marrow. Inhibition of Bcl6 by Blimp1 (and vice versa [145]) explains why only very few Blimp1$^+$ plasma cells can be detected inside of GCs of mice and humans [96,146,147]. However, CD138$^+$ antibody-secreting cells have been detected to be located directly adjacent to GCs, lining the border between the GC DZ and T cell zone [148]. It is this region where, in the course of an immune response, the first cognate interactions between B and T cells happen [35]. The presence of plasmablasts at his location further underlines the importance of this region, not only during the initiation but also during the maintenance of the GC response.

The exact routes that these cells take to exit from spleen and LNs have not yet been uncovered, but it is known that egress of antibody-secreting cells from the spleen and from PPs requires signaling through S1P1 [149,150]. S1P, the ligand for S1P1, is produced by erythrocytes; therefore, it is highly abundant in the bloodstream [24]. Plasmablasts in the blood express higher amounts of S1P1 compared with their counterparts in SLOs. Treatment of mice with the immunosuppressant S1P antagonist FTY720 inhibits plasmablast egress into the blood in vivo [149]. In addition, $\beta2$-integrin has been shown to play a role in plasmablast exit from peripheral LNs [151].

Plasmablast entry pathways into the bone marrow are not yet fully elucidated. Presumably, plasmablasts enter via sinusoidal veins in the bone marrow, but the mechanisms that mediate their transmigration have not been investigated in detail. One possible candidate involved into this process may be CD62L because plasmablasts have been reported to express this homing receptor [99]. Another candidate is the adhesion molecule CD22, which is expressed on plasmablasts and becomes downregulated in terminally differentiated plasma cells. Sialylated ligands for CD22 have been reported to be present on sinusoidal endothelial cells of the bone marrow [152].

Homing of plasmablasts to the bone marrow is the prerequisite for them to become long-lived plasma cells. This migration process crucially depends on the chemokine receptor CXCR4. Its ligand, the chemokine CXCL12, is produced in high amounts by stromal cells of the bone marrow; therefore, it is also known as SDF-1 [153]. CXCR4-deficient plasmablasts fail to enter the bone marrow and instead accumulate in the blood [93]. The capacity of murine splenic plasmablasts to migrate to CXCL12 ex vivo is restricted to the first week of a systemic immune response [92]; after that time period, they lose their migratory capacity. Likewise, the number of antigen-specific plasmablasts is found to peak in the blood of immunized humans at day 7 after boost [154].

Inside of the marrow, long-lived plasma cells are thought to reside in specialized microenvironmental compartments called survival niches [155]. In addition to its function as a chemoattractant, CXCL12 has been shown to mediate the survival of bone marrow plasma cells [156], acting synergistically with other soluble factors and adhesion molecules. Therefore, CXCL12-producing bone marrow stromal cells are crucial for the maintenance of long-lived plasma cells, the source of long-term serum antibody titers. In addition to stromal cells, which are considered as the main organizers of the niche, other accessory cells are contributing to the niche by providing survival factors such as APRIL and IL-6. In addition to eosinophils [157] and megakaryocytes [158], other hematopoietic cell types such as DCs [159], basophils [160] and monocytes/macrophages [161] have been reported to contribute to plasma cell maintenance in the survival niches. Whether CXCL12 attracts these accessory cells to the niche remains to be investigated. An overview of plasma cell migration into the bone marrow and the factors known to contribute to the survival niche is shown in Figure 4.

Although plasmablasts did not show migratory activity against the ligands for CCR2 ex vivo, mice with CCR2-deficient plasma cells have significantly reduced long-term

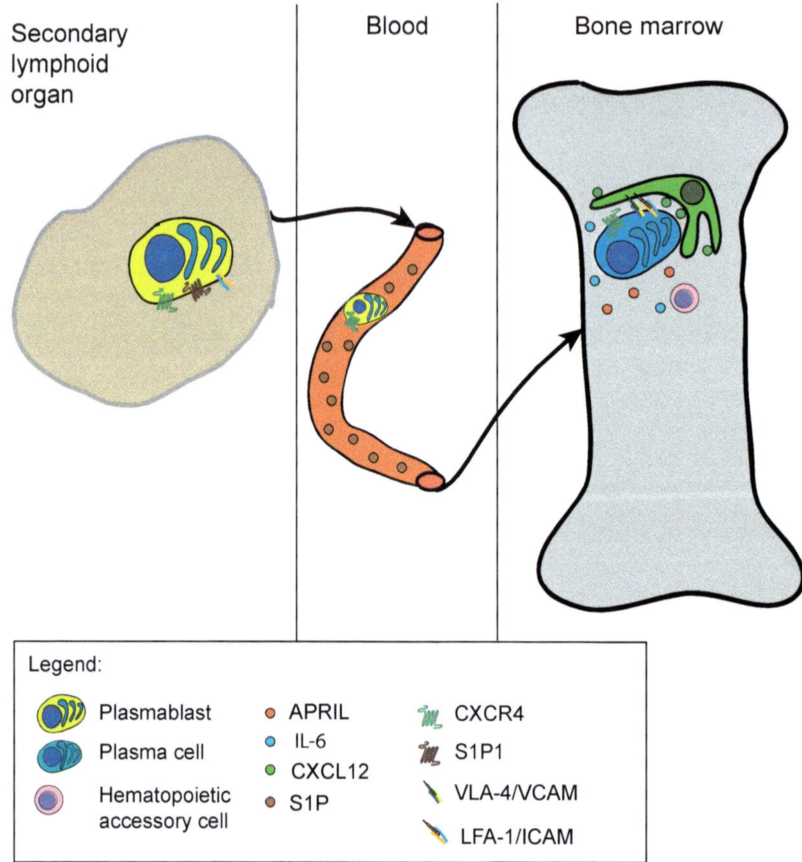


FIGURE 4 **Migration of antibody secreting cells.** Plasmablasts are generated in secondary lymphoid organs. They become migratory, leave the spleen/lymph nodes guided via S1P1, and travel via the blood to the bone marrow parenchyma. Plasmablasts reaching microenvironmental survival niches within the bone marrow, which are defined by the presence of CXCL12⁺ stromal cells along with certain hematopoietic accessory cells, can become long-lived.

antigen-specific antibody titers, which may indicate an additional role for this chemokine receptor in the retention of long-lived plasma cells in the survival niches [162].

Plasma cell adhesion in the bone marrow is mediated by VLA4 and LFA1. Administering a combination of blocking antibodies against these two molecules in vivo resulted in a 75% reduction of bone marrow plasma cell numbers [163]. Ligands for VLA4 that are present on bone marrow stroma include VCAM-1, fibronectin, and osteopontin; the LFA1 ligands ICAM-1,-2, and -3 are also present on stromal cells. Most likely, VLA4 and LFA1 act by fixing the plasma cells into their niches. Another molecule that has been suggested to promote survival of plasma cells in their niches by functioning as an adhesion molecule is CD93 [164].

A fraction of plasmablasts has also been shown to migrate to ligands of CXCR3 [92], suggesting a role of this chemokine receptor in the accumulation of antibody-secreting cells in inflamed organs. In fact, chronically inflamed kidneys of NZB/W lupus mice have been shown to contain CXCR3$^+$ antibody-secreting cells [106], and a role of CXCR3 for the entry of plasmablasts into the inflamed central nervous system has been demonstrated in a mouse model of viral encephalomyelitis [165]. Although the differentiation of plasma cells is generally accompanied by an upregulation of CXCR4, the induction of CXCR3 depends on the inflammatory cytokine interferon-$\gamma$ [166].

## 5. BODY CAVITY B-1 B CELL TRAFFICKING

In the first line of defense against various pathogens, TI primary immune responses are initiated by antigen-specific activation of B cell clones of the B-1 B cell origin [121,167]. These innate-like B cells [168] bridge innate and adaptive immunity and generate a vigorous primary plasmablast response to blood-borne antigens within the first 3 days after bacterial and viral infection and nematode parasite invasion [167]. B-1 cells are the major source of IgM natural antibodies in serum and contribute to IgA-producing plasma cells in the lamina propria (LP) in the gut [169]. B-1 B cells can be subdivided based on their CD5 expression in CD5$^+$ B-1a and CD5$^-$ B-1b cells with complementary, distinct functions. B-1a cells are primarily responsible for spontaneous production of natural IgM antibodies, whereas B-1b cells appear to be the primary source of dynamic TI antibody production and long-term protection after bacterial infections [170,171]. In addition to their functional characteristics, B-1 cells differ from conventional B-2 cells by their anatomical localization, which is preferentially in the peritoneal, pleural, and thoracic cavities [169,172]. Because B-1 cells are most prevalent in body cavities, it was previously suggested that peritoneal B-1 cells show significant spontaneous IgM secretion [173]. However, numerous other studies indicate that peritoneal B-1 B cells do not spontaneously secrete natural IgM antibodies; instead, activation signals such as cytokines or mitogenic stimuli are needed to induce antibody production [174]. Furthermore, it was shown that natural IgM antibody secretion was largely confined to B-1 B cells in the spleen and bone marrow. Consistent with this, peritoneal B-1 cells rapidly relocate to SLOs after infection, where they differentiate to cytokine and IgM-producing cells [175].

For many years it was debated if an equivalent of murine B-1 B cells in humans exists at all. However, a human B-1 B cell type that spontaneously secretes IgM and is capable to activate T cells was recently identified in the umbilical cord and in adult peripheral blood [176,177]. It has been implied that human B-1 B cells may contribute to autoimmune diseases through production of autoantibodies and T cell stimulation [178,179]. On the other hand, human B-1 B cells could be useful in treating diseases in elderly humans where protective natural antibody production is severely impaired. Characterization of the full physiological and pathophysiological function of human B-1 B cells may lead to new avenues for treating autoimmune diseases [180].

The dynamic process of B-1 B cell movement is guided by a mesothelial sheet of tissue, the omentum, lining the peritoneal cavity and connecting various organs, such as the stomach, pancreas, spleen, and transverse colon [181]. Milky spots represent cellular aggregates containing lymphocytes and macrophages in the mesothelial sheets and the fat pads of the omentum. Their associated capillary structures are thought to serve as the major route through which leukocytes including B-1 cells migrate into the peritoneal cavity. Milky spots can also be found in the mediastinal pleura, and experimental evidence suggests that pleural mesothelial surfaces can additionally serve as an entry portal for B-1 B cells [182,183].

Principal molecules mediating lymphocyte entry into SLOs (i.e., the homeostatic chemokine receptors CCR7, CXCR4, and CXCR5) are also involved in lymphocytic migration and recirculation through peripheral tissues [184–186]. Their respective ligands, the chemokines CCL19, CCL21, CXCL12, and CXCL13, are constitutively expressed in the peritoneal, pleural, and thoracic body cavities [182,187]. Genetically modified mice lacking either the chemokine receptor CXCR5 or its ligand CXCL13 exhibit a profound decrease in peritoneal B-2 and an even stronger reduction in peritoneal B-1 B cells accompanied by impaired body cavity immunity. CXCL13 was found to be expressed by peritoneal macrophages and by radiation-resistant cells within the omentum. Peritoneal mesothelial cells also express CXCL12, and inhibiting the CXCL12/CXCR4 signaling axis in mice causes the loss of B-1 and B-2 B cells in the peritoneal cavity [188]. Consequently, homeostatic chemokines affect the peritoneal micromilieu by regulating peritoneal lymphocyte recruitment and recirculation.

However, homeostasis of the peritoneal cavity is not only maintained by lymphocytic immigration but also by regulated exit from this anatomical site. Furthermore, the regulated egress of peritoneal B-1 cells is crucial for rapid antigenic clearance outside of their residential site [189].

In systemic immune compartments, S1P has been identified as a key exit molecule for lymphocyte emigration from organized SLOs and the thymus. S1P is a biologically active lysophospholipid that signals via a family of five G-protein-coupled receptors, the S1P receptors S1P1-5, of which S1P1 is preferentially expressed on lymphocytes [190,191]. S1P was also reported to regulate peritoneal B cell trafficking, and comparable expression levels of the type 1 S1P receptor were found on B-1 and B-2 B cells. Treatment of mice with FTY720, 2-amino-2-[2-(4-octylphenyl)-ethyl] propane-1,3-diol hydrochloride, which acts as an agonist for S1P receptors, increased the B cell emigration from the peritoneum and inhibited peritoneal immigration from the blood. Subsequently, this led to an impaired peritoneal B cell migration into the intestine and to decreased intestinal IgA production [192].

The effect of homeostatic chemokines on lymphocytic emigration from body cavities has been described for conventional B-2 but not for B-1 B lymphocytes. CCR7 mediates a crucial and dominant exit signal for conventional B and T lymphocytes from body cavities. Subsequently, CCR7-deficient mice exhibited profound accumulation of conventional B-2 and T but not of B-1 cells in pleural and peritoneal body cavities [193].

Direct signaling through TLRs, a family of pattern-recognition receptors, has also been described to regulate egress of B-1 B cells from body cavities [189,194].

TLRs recognize conserved molecules and patterns of microorganisms and play an essential role in innate immune responses. They are expressed by many different innate immune cells, and TLR engagement links the innate to the adaptive immune system [195,196]. Receptors expressed by immune cells of the adaptive immune system are usually highly antigen-specific and clonally rearranged. In this regard, B cells are unique because they express TLRs and an antigen-specific clonally rearranged BCR. TLR signaling can influence a broad range of functional B cell responses, among them the up- or downregulation of surface molecules that regulate the positioning and migration of B cells into SLOs [197].

B-1 B cells are particularly activated and differentiated by stimuli such as lipopolysaccharide and cytokines. Recently, it has been demonstrated that molecular signals required for B-1 B cell motility and egress from body cavities are facilitated through TLRs (i.e., TLR4). Mechanistically, TLR signaling in vitro and in vivo led to the downregulation of integrins, including α1, α6, and β1, as well as CD9 on B-1 cells, which supports their detachment from the local matrix and enhanced the migratory capacity toward chemokines [189].

In addition, it was shown that a transit of B-1 and B-2 B cells through the peritoneal cavity alters their migratory capability. Expression levels of homing molecules and integrins (i.e., CXCR4, CXCR5, and β7 integrin) adjust in response to local environmental cues. This leads to the egress of B-1 cells from the peritoneal cavity and enables the cells to enter the intestine, facilitating the transition from innate to adoptive immune responses [189,194].

Overall, peritoneal B-1 B cell inbound and outbound trafficking is not exclusively controlled by chemokines but also by a regulatory network combining chemokine-, integrin-, and S1P-mediated migration [189,192,193,198].

## 6. MUCOSAL B CELL MIGRATION

The mucosa-associated lymphoid tissue (MALT) is scattered along mucosal linings and consists of the gut-associated lymphoid tissues (GALT), the bronchus-associated lymphoid tissue (BALT), the nose-associated lymphoid tissue (NALT), the vulvovaginal-associated lymphoid tissue (VALT), and some additional lymphoid sites within accessory organs of the gut such as the parotid gland [199]. MALTs are located directly beneath the mucosal epithelia cell layer, only separated from the outside world by the follicle-associated epithelium. They serve as immune inductive sites and harbor well-organized lymphoid structures. The migration of immune cells from these mucosal inductive sites to mucosal effector tissues occurs via the lymphatic system and is a prerequisite for cellular immune responses in the gastrointestinal tract [200,201].

Secondary lymphoid structures of the small intestine include PPs and isolated lymphoid follicles (ILFs), whereas the large intestine contains ILFs and the appendix. MALT structures consist mostly of B cells that form large follicles harboring high numbers of surface IgA-positive (sIgA$^+$) B cells interspersed by interfollicular T cell regions (IFRs). The largest fraction of antibody-secreting cells in the body is located in the MALT, and virtually all plasma cells induced in mucosal immune reactions secrete antibodies of the IgA isotype. A polymeric immunoglobulin receptor-dependent mechanism allows dimeric IgA to be transported via the gut epithelium into the lumen, where IgA is present mainly as a part of the mucus layer covering the surface [202].

Class switch of B cells to IgA is induced in ILFs, PPs, and MLNs. These tissues provide an environment rich in factors such as transforming growth factor-β and APRIL, which have been shown to promote IgA responses [203]. Soluble lymphotoxin-α derived from innate lymphoid cells (ILCs) regulates T-dependent IgA induction by controlling the migration of T cells into the gut, whereas membrane-bound lymphotoxin β (LTα1β2) on RORγt+ILCs is critical for TI IgA responses [204]. In the absence of eosinophils, the numbers of IgA$^+$ LP plasma cells are significantly reduced. This is at least partly due to a defect in their generation because eosinophil-deficient mice have a defective class switch to IgA in PPs. In addition, eosinophils promote the survival of LP plasma cells in vitro; therefore, they might also be involved in their maintenance in vivo, similar to the situation described in the bone marrow [205].

T cells reside either within the IFRs (i.e., PPs) or in the center of the B follicular structures (i.e., ILFs) [201]. Intestinal antigen-presenting cells, including DCs, are found within the subepithelial area of B cell follicles but also in the IRFs [206]. PPs, which develop in the fetal small intestine independently of intestinal microbiota, structurally resemble LNs. In contrast, the formation of ILFs is induced by the intestinal flora and their numbers increase during chronic inflammation. Thus, ILFs share some features with tertiary lymphoid structures. Under steady-state conditions, intestinal bacteria trigger a gut-restricted antibody response [207–209].

Primed lymphocytes traffic as effector cells to secretory effector tissue sites through specific upregulation of gut homing receptors. Recently, cells and molecular mediators have been identified that are responsible for imprinting specialized tissue-specific trafficking programs. Environmental cues from food (vitamin A) and GALT-located DCs, which process vitamin A to its active metabolite retinoic acid, imprint lymphocyte homing to the small intestine through enhanced expression of mucosal homing receptors, foremost the adhesion molecule $\alpha4\beta7$ and the chemokine receptor CCR9 [210].

Importantly, cues for lymphocytic extravasation differ between naïve and memory/effector lymphocyte subsets, and they often appear to be tissue-specific. In this regard, MadCAM-1 is highly expressed by HEVs of human GALT and MLNs. In mice, naïve B cells bind to glycosylated MadCAM-1 via their abundant expression of CD62L (L-selectin). This initial tethering enables chemokine signaling-dependent firm adhesion of MadCAM-1 to the integrin $\alpha4\beta7$. The latter step is crucial for intestinal extravasation of naïve lymphocytes into GALT, whereas emigration via HEVs into MLNs is mediated by the peripheral LN addressin, PNAd [211].

Because activated B cells lack prominent CD62L expression, their extravasation via HEVs in GALT may be selectively dependent on the interaction between activated $\alpha4\beta7$ and MadCAM-1. In addition, emigration of naïve and effector/memory B cells can also be mediated by adhesion molecules, which are less tissue-restricted, such as LFA1, $\alpha L\beta2$, or CD11a/CD18, which bind to the endothelial ICAM-1 and ICAM-2 [212].

Homing of B cells to SLOs is regulated by the CCR7 ligands, CCL19 and CCL21, together with the CXCR4 ligand CXCL12 [213,214]. CCR7 and CXCR4 mediate endothelial B cell adhesion in LNs. B cell migration into PPs additionally depends on CXCR5. In mice, the CXCL13-CXCR5 signaling axis supports direct extravasation of B cells into the follicle of PPs via HEV-like vessels, which express CXCL13 and no CCL21 [46]. CXCL13-expressing HEV-like vessels have also been found in human GALT [215].

The dissemination of naïve or activated B cells from inductive sites occurs via draining microlymphatics, which start blindly with a fenestrated endothelium. This lymph endothelium, similar to HEVs, expresses CCL21 and specific adhesion molecules. Memory B cells (sIgD$^{-}$) together with some CD19$^{+}$CD38$^{high}$ B cell blasts both expressing high levels of $\alpha4\beta7$ are often positioned close to and sometimes within microlymphatics, which probably initiate the first exit step from the GALT toward the intestinal LP. These $\alpha4\beta7^{high}$ subsets normally express very low levels of L-selectin, which enables them to bind to unmodified MadCAM-1 expressed apically on the LP microvasculature, leading to extravasation into the small intestinal LP. Notably, a few memory B cells do express high levels of L-selectin; however, they are most likely re-entering GALT or extravasate in MNLs and peripheral LNs [216,217]. Mucosal IgA$^{+}$ plasmablasts first use lymph vessels to migrate to the MLNs and then they enter the blood circulation through the thoracic duct to finally home back to the LP.

In addition to $\alpha4\beta7$, the CCL25-CCR9 interaction is thought to play a crucial role in small intestinal tropism of effector/memory B cells. The chemokine CCL25 is selectively expressed by the crypt epithelium and endothelium of the small intestine in humans as well as in mice, but it shows low expression at other mucosal sites [218,219]. As stated before, the imprinting of small intestinal tropism is at least partially dependent on the supply of retinoic acid by regional DCs [220].

The chemokine ligand CCL28 is universally produced by various mucosal epithelial tissues (i.e., the large intestine, stomach, the respiratory tract, and mammary and salivary glands). CCL25 and CCL28 enhance the integrin $\alpha4$-dependent adhesion of IgA$^{+}$ plasmablasts to MadCAM-1 [221]. The CCL28-sensitive chemokine receptor, CCR10, is expressed on virtually all IgA$^{+}$ plasmablasts, including intestinal, nonintestinal, and mucosal effector tissues. Even IgA$^{+}$ plasmablasts in the small intestine express CCR10. However, because epithelial expression of CCL28 is higher in the colon than in the small intestine and appendix, the CCL28-CCR10 axis most likely plays an important role in compartmentalized intestinal B cell homing [211,219]. The differential expression patterns of CCR9 and CCR10 are probably responsible for differences in trafficking of antigen-specific IgA-producing cells: Restricted distribution of these cells in the upper aerodigestive tract and urogenital tract is mediated by upregulation of CCR10 expression, but not CCR9, and the expression of $\alpha4\beta1$, but not $\alpha4\beta7$, which allows them to bind to tissue sites expressing VCAM-1 and CCL28. On the other hand, upregulation of CCR9 as well as CCR10, and the additional expression of $\alpha4\beta7$ enables IgA$^{+}$ plasmablasts to migrate essentially to all mucosal tissues that express CCL28 or CCL25 together with MadCAM-1 or VCAM-1 [211,219].

The two major chemokines CCL25 and CCL28, which attract $IgA^+$ plasmablasts and are selectively expressed by epithelial cells, belong to one specific subfamily of chemokines. Other chemokines that belong to this subclass are the chemokine CCL27, a second ligand for CCR10, also known as cutaneous T-cell-attracting chemokine [219], and CCL20, the CCR6 ligand. CCL20 is expressed by lymphoid tissue-associated intestinal epithelium and by inflamed epithelial cells in the lung or skin, and its receptor CCR6 is expressed by all naïve and memory B cells. More specifically, CCR6 is downregulated after B cell antigen receptor triggering, it remains absent during differentiation into immunoglobulin-secreting plasma cells, but it is re-expressed by post-GC memory B cells. Hence, CCL20 responsiveness is apparently crucial for B-lineage maturation and antigen-driven B-cell differentiation as well as the recruitment of circulating memory B cells to sites of inflammation [219,222].

Topical antigen exposure has been described to have a deep effect on site-specific accumulation of $IgA^+$ plasma cells [223], most likely mediated via activated T cells. Hence, local antigen drives T cell activation, which then provides essential second signals for B cell retention and survival. The amount of $IgA^+$ plasma cells at a specific effector site is determined by the level of topical antigen exposure, which is enormously high in human colonic mucosa [200,224].

Antibody-secreting cells in the LP have been thought to be short-lived [225], but there is indirect evidence in the recent literature that there might also be a long-lived population present in this tissue [226,227]. Some LP plasma cells are able to produce TNFα and inducible nitric oxide synthase, revealing an unexpected heterogeneity in the antibody-secreting cell compartment in the gut [228].

Overall, activated immune cells seem to accumulate preferentially at effector sites corresponding to the inductive sites where the cells were initially primed. Hence, intestinal PCs are largely derived from B cells that have been originally activated within the GALT [211].

## 7. HOMING OF B CELLS DURING CHRONIC INFLAMMATION AND TERTIARY LYMPHOID ORGAN FORMATION

The contribution of local B cell infiltrates to chronic inflammation has been suggested in various autoimmune diseases such as rheumatoid arthritis (RA), multiple sclerosis (MS), autoimmune diabetes, SS, myasthenia gravis (MG), and Hashimoto's thyroiditis [229] as well as during chronic inflammation associated with infectious diseases such as Lyme arthritis, *Mycobacterium tuberculosis*, chronic hepatitis C (HCV), and *Helicobacter pylori* infection [229,230]. It has been described that persistent viral infections trigger the formation of tertiary lymphoid organs (TLOs), where ectopic GCs serve as inductive sites and by that support a protective, antiviral immune response. On the other hand, increasing evidence in animal models and patients suggests that TLOs triggered by viral infections can break self-tolerance and initiate the development of autoimmune diseases (i.e., prototypical mucosal autoimmune diseases, SS, RA, MS, and MG) [231].

The inflammatory lesions that occur during chronic inflammation, infections, or autoimmunity can resemble TLOs and are thought to arise by a process termed lymphoid organ neogenesis [78]. The development of these so-called ectopic lymphoid follicles or TLOs is a hallmark of chronic inflammatory autoimmune and infectious diseases, such as RA, SS, and *H. pylori*-induced gastritis [232]. In addition, de novo formation of lymphoid tissue during chronic inflammation has been associated with a higher risk for neoplastic transformation and lymphoma development (i.e., the infection-dependent formation of primary gastric MALT B cell lymphomas) [233].

Because these ectopic lymphoid follicle-like structures resemble conventional SLOs, it does not come as a surprise that the developmental program for the formation of LNs and TLOs is remarkably similar. Hence, the principal molecular mediators of lymphoid organogenesis, including inflammatory cytokines of the TNF/lymphotoxin family and homeostatic chemokines, are also involved in lymphoid tissue neogenesis in chronic inflammatory processes and autoimmune diseases [229,232,234].

In several transgenic mouse models, TLOs have been induced by tissue-specific overexpression of cytokines and homeostatic chemokines. Ectopic expression of lymphoid chemokines, such as CXCL13 and CCL21, causes the infiltration of naïve lymphocytes, which consolidate into organized lymphoid structures composed primarily of B and T cells and a few DCs [235–237]. In addition, stroma and endothelial cells from the local environment are necessary for the development of a LN-like architecture because they provide inflammatory cytokines, lymphotoxin, and TNFα [238], which support formation of a FDC network with high expression of CXCL13 and CCL21 [239,240].

In humans, inflammation has been associated with the expression of CXCL13 in the salivary glands of patients with SS, in synovial follicles of patients with RA, in MS ectopic B cell follicles, in inflammatory lesions of celiac disease, and in Helicobacter-induced MALT and MALT lymphomas. The homeostatic chemokine receptors CCL21 and CXCL12 were also found to be ectopically expressed by endothelial cells within the synovium of RA patients, within MS B cell follicles, in the salivary glands of SS, and the thyroid of Hashimoto thyroiditis patients [78].

In chronic inflammatory mouse models of RA and Helicobacter-induced gastritis, experimental evidence was provided that the CXCR5/CXCL13 signaling axis is a key molecular regulator of TLO formation. CXCR5 deficiency led to a lack of lymphoid neogenesis and an overall reduction of chronic inflammation in both disease models [241,242].

The ectopic formation of lymphoid tissue has been mainly associated with destructive inflammation and tissue damage. However, the decisive functional role of TLOs in immunity is not yet fully understood. As discussed in the beginning, local antigen presentation could be beneficial to prevent bacteremia and viremia during persistent infections [243]. On the other hand, TLOs have the propensity to develop into B cell lymphomas and may perpetuate chronic inflammation in autoimmune diseases. Within these functional lymphoid niches, autoreactive B cells can undergo in situ affinity maturation and differentiate into autoantibody-producing cells [232]. Earlier studies in patients with RA provided the first evidence regarding competent GC formation that supports antigen-driven B cell differentiation and clonal expansion [244,245]. Apart from these earlier observations, oligoclonal B cell proliferation and somatic hypermutation in the Ig variable genes have additionally been detected in TLOs of patients with primary SS [246] and MG [247]. In a murine model of autoimmune diabetes, experimental evidence was presented regarding the selection of autoreactive B lymphocytes and their ability to present antigen to T cells within the TLOs of pancreatic islets [248].

The presence of plasma cells was also observed in GCs of TLOs from RA and SS patients, although it remains unclear whether their terminal differentiation takes place within the TLOs or whether plasma cells have been recruited to these tertiary lymphoid structures [78]. In this regard, persistent viral infections, such as EBV reactivation, have been linked to the development of B cell autoreactivity within ectopic B cell follicles in the target organs of RA, MS, and MG. Another striking example of infection-induced TLO formation is the occurrence of intraportal lymphoid follicles during chronic HCV infection [249,250]. These intrahepatic TLOs play a crucial role in regulating complex viral–host interactions; however, lack of viral clearance causes chronic B cell activation, which can contribute to the development of autoimmunity and extrahepatic manifestations [231].

In conclusion, the recruitment of B cells to ectopic sites of chronic inflammation and persistent infections results in the formation of TLOs that play an important role in antimicrobial immunity, but they also support the breach of self-immunological tolerance and the development of autoimmunity.

## 8. MIGRATION OF NEOPLASTIC B CELLS

Tumor cell malignancy is characterized by its intrinsic ability to invade various tissues distant from the primary site of tumor origin. The most common type of human lymphoid malignancies consists of B cell lymphomas. During all stages of B cell development (i.e., from the first developmental stages in the bone marrow to the later stages in SLOs and in MALTs), malignant transformation of B cells can occur [251,252]. Subsequently, lymphomas are classified according to the morphologic features characteristic of the various developmental stages of normal B cell differentiation [252,253]. Furthermore, transformed B-lymphoid cells have somewhat conserved the physiological highly effective migratory capacity of their normal lymphocyte counterparts. Over nearly 3 decades the knowledge about the regulatory mechanisms of lymphocyte trafficking in health and disease has greatly advanced. Today, clinical and experimental studies strongly suggest that the sophisticated motility program of lymphocytes, governed by adhesion molecules together with the chemokine/chemokine receptor system, also controls lymphoma/leukemia cell migration and dissemination [254–256]. The high motility of lymphocytes and their malignant counterparts accounts for rapid lymphoma dissemination. Because malignant B cells migrate preferentially into tissues to which lymphocytes home physiologically, tumor cells invade a specialized microenvironment that is already well equipped to support lymphocytic differentiation, survival, and activation [257].

One of the first identified homing receptors for lymphoma cell entry into LNs was the 90-kD hyaluronate-binding glycoprotein (CD44), originally characterized by its Hermes-1 positive staining pattern. CD44 is physiologically involved in binding of lymphocytes to HEVs. Lymphomas and leukemias expressing the Hermes-1 antigen were found to involve many or all LN groups [258]. Studies with mouse lymphoma cells also supported the notion that lymphomas that effectively bind to HEVs spread early via the blood and invade all peripheral lymphoid organs [259]. These earlier findings led to the conclusion that lymphoid neoplasms spread by entering sites of normal lymphocyte traffic and that an identification of functionally important homing receptors will be crucial to understand the in vivo migratory and dissemination behavior of neoplasms [258].

To date, the homing signature of malignant lymphocytes has been extensively studied and treatments targeting chemokine receptors and adhesion molecules with antibodies or drugs have been shown to be valuable tools for therapeutic intervention in lymphoma and leukemia patients [254,255,260]. An overview of the most important adhesion molecules, chemokines and chemokine receptors, involved in tissue-specific lymphocyte homing and lymphoma dissemination is given in Table 1.

B cell-derived lymphoproliferative disorders with distinctive nodal lodging, such as acute lymphoblastic leukemia, B cell chronic lymphatic leukemia (CLL), follicular lymphoma (FL), and hairy cell leukemia, frequently express the homeostatic chemokine receptors CXCR4 and CXCR5 [255,261]. Their respective ligands, CXCL12 and CXCL13,

**TABLE 1** Putative Adhesion Molecules, Chemokines, and Chemokine Receptors Mediating Lymphocyte Homing and Lymphoma Dissemination

| Receptor | Expression on Lymphocytes | Expression on B Cell Lymphoma/Leukemia | Ligands | Predominant Site of Homing |
|---|---|---|---|---|
| | | **Adhesion Molecules** | | |
| L-selectin | Naïve B and T cells, $T_{CM}$ cells | B-CLL, MCL, MZBCL, nodal FL, DLBCL | PNAd/MadCAM-1 | Peripheral lymph node |
| αLβ2/LFA-1 | Broad expression on B and T cells | Broad expression on B cell lymphomas and MM | ICAM-1, ICAM-2, ICAM-3 | Multiple sites/ epithelium |
| α4β1/VLA-4 | α4β7 | Broad expression on B cell lymphomas and MM | VCAM-1 | Inflammatory sites and BM |
| α4β7 | IgA plasmablasts, gut-homing T cells, naïve B and T cells (low) | GI-tract MCL (MLP), GI-tract MZBCL (maltoma), GI-tract FL and PTCL (EA-TCL) | MadCAM-1/VCAM-1 | Gut |
| CD44s | Broad expression on B and T cells, splice variants on activated lymphocytes | B-CLL, nodal MCL, GI-tract (MLP and low-grade MALT, GI-tract DLBCL, nodal ALCL) | Hyaluronate, collagen IV, fibronectin, growth factors | Inflammatory sites, modulation of cell growth and motility |
| **Chemokine Receptors** | | | | |
| CCR1 | Polymorphonuclear leukocytes, activated T and B cells | B cell CLL (after BCR triggering) | CCL3, CCL5-7, CCL9/10 | Inflammatory sites |
| CCR3 | Polymorphonuclear leukocytes, activated T and B cells | MM, nonneoplastic leukocytic infiltrates of cHD | CCL5-8, CCL11, 13, 15, 24, 26-27 | Inflammatory sites |
| CCR5 | Polymorphonuclear leukocytes, activated T and B cells | B-CLL (after BCR triggering) | CCL3, CCL4, CCL5, CCL8 | Inflammatory sites |
| CCR6 | B cells, IgA plasmablasts, memory T cells, iDCs | DLBCL, MCL, MZBL, MALT lymphoma | CCL20 | Mucosal tissues, skin |
| CCR7 | Naïve B and T cells, $T_{CM}$ cells, mDCs | B cell CLL, MCL, cHD | CCL19, CCL21 | Peripheral lymph node |
| CCR9 | Gut-homing B and T cells, intraepithelial T cells, IgA plasmablasts | DLBCL (low), MLBCL | CCL25 | Intestinal mucosa |
| CCR10 | Gut- and skin-homing B and T cells, IgA plasmablasts | MM, cHD | CCL27, CCL28 | Large intestine, skin |
| CXCR3 | Activated Th1 cells, plasmablasts | B-CLL, splenic MZL, extranodal MZL | CXCL9, CXCL10, CXCL11 | Inflammatory sites |
| CXCR4 | Pre-B and mature B cells, plasma cells, T cells | Broad expression on B cell lymphomas and MM | CXCL12 | Secondary lymphoid tissues, GCs, and BM |
| CXCR5 | Mature B cells, $T_{FH}$ cells | Broad expression on B cell lymphomas | CXCL13 | B cell follicle/GCs in SLOs |
| CXCR6 | Activated B and T cells, plasma cells | MM | CXCL16 | Inflammatory sites, BM |

Adhesion molecules, chemokines, and chemokine receptors, which play a major role in B cell lymphoma/leukemia homing, are listed. ALCL, anaplastic large-cell lymphoma; B-CLL, B-cell chronic lymphocytic leukemia; BM, bone marrow; DLBCL, diffuse large B-cell lymphoma; FL, follicular lymphoma; GC, germinal center; GI, gastrointestinal tract; ICAM, intercellular adhesion molecule; LFA, lymphocyte function-associated; MadCAM, mucosal addressin cell adhesion molecule; MALT, mucosa-associated lymphoid tissue; MCL, mantle cell lymphoma; MM, multiple myeloma; MZBL, marginal zone B-cell lymphoma; PNAd, peripheral lymph node addressin; SLO, secondary lymphoid organs; VCAM, vascular cell adhesion molecule; VLA, very late antigens.

are expressed on HEVs of LNs and PPs and on HEV-like vessels at sites of lymphoid neogenesis, and they are fundamentally involved in the organization of GCs [32,214]. Within GCs, CXCL13 is produced by FDCs and guides B cells into the LZ, whereas CXCL12 attracts B cells to the DZ of GCs [94]. In line with their ectopic expression at sites of chronic inflammation and lymphoid neogenesis, expression of the receptors CXCR4 and CXCR5 is also found on

MALT lymphomas and on diffuse large B cell lymphomas (DLBCL), which is characterized by a primary intestinal localization [254]. Furthermore, autocrine expression of CXCL13 and the formation of B cell aggregates have been associated with the development of primary central nervous system lymphomas [262].

CCR6, a chemokine receptor being physiologically expressed predominantly on mantle and marginal zone B cells of SLOs, is also found on most mantle cell (MCL), marginal zone (MZL), nonmediastinal DLBCL, and MALT lymphomas [263].

It is interesting to note that extranodal lymphomas arising within the anterior mediastinum, such as mediastinal large B cell lymphoma (MLBCL), exhibit low expression of characteristic lymph node-homing receptors (i.e., CXCR5 and CCR7) [264], which may hinder their nodal but predisposes their extranodal dissemination.

Another striking example for chemokine receptor-dependent lymphoma cell lodging within a specific compartment of LNs is Hodgkin lymphoma (HL) [265]. HL is a B cell neoplasia that arises in two major forms: the frequent classical HL (cHL) and the rarer nodular lymphocyte predominant HL (NLPHD) [266,267]. The lymphocytic and histiocytic cells (L&H) in NLPHD lodge within follicular structures (B cell zone) and express CXCR4 but not CCR7, whereas the Hodgkin-Reed-Sternberg cells in cHL reside in the interfollicular or follicular mantle zone (T cell zone) and strongly express CXCR4 and CCR7, the latter chemokine receptor being crucial for entering T cell zones of LNs [265]. Functionally, a differential chemokine receptor expression pattern contributes to a specific compartmental confinement of lymphoma cells to tumor growth-promoting niches within the target organ [257,264,265,268].

When B cells differentiate into plasma cells, a coordinated change in chemokine receptor expression occurs. Expression of the homeostatic chemokine receptors CCR7 and CXCR5 is downregulated, which leads to an unresponsiveness toward T- and B- zone chemokines CCL19, CCL21, and CXCL13 [93,269]. In contrast, CXCR4 is upregulated, and constitutive expression of its ligand CXCL12 by bone marrow stromal cells causes plasma cell homing to the bone marrow [92]. Within the bone marrow microenvironment, $\alpha4\beta1$-mediated interactions with fibronectin and VCAM-1 facilitate plasma cell retention and survival. Likewise, recruitment and retention of malignant plasma cells in multiple myeloma (MM) also critically depends on the CXCL12/CXCR4 signaling axis. CXCL12 not only promotes homing to but also induces $\alpha4\beta1$-mediated interactions with fibronectin and VCAM-1 within the bone marrow niche, which mediates tumor cell survival and cell adhesion-mediated drug resistance [270].

Additional functional chemokine receptors being expressed by normal plasma cells as well as by their malignant counterparts in MM are CXCR6, CCR6, CCR3, and CCR10 [271].

Overall, CXCR4 has emerged as a key receptor in the recruitment and crosstalk between malignant B cells and their microenvironment. The CXCL12/CXCR4 signaling axis mediates metastatic spread to growth-promoting tissue niches where CXCL12 is expressed and CXCL12 itself stimulates tumor cell growth in a paracrine manner and supports tumor angiogenesis [272]. Functional CXCR4 expression has also been described for several nonhematopoietic malignancies, including brain neoplasm [261,273]. Hence, the CXCR4/CXCL12 signaling axis apparently plays a crucial role in the progression and metastasis of many hematopoietic and nonhematopoietic malignancies. Several small molecular CXCR4 antagonists have been generated initially for the treatment of HIV because CXCR4 functions as a coreceptor for X4 HIV-1 viruses. Two of these antagonists, AMD3100 and ALX40-4C, have also been successfully tested in animal tumor models and in first clinical trials in cancer patients [272].

Therapeutic targeting of the CXCR4/CXCL12 axis in B cell CLL patients revealed promising data in an ongoing trial, in which the CXCR4 antagonist plerixafor (AMD3100) is combined with the established CLL drug rituximab (anti-CD20 monoclonal antibody, depletes $CD20^+$ B cells). This approach targets leukemia cell mobilization from its protective bone marrow microenvironment and thereby renders tumor cells more accessible to conventional drugs [260].

Numerous clinical studies correlate a specific chemokine receptor expression pattern on neoplastic B cells with their dissemination to anatomic niches not only within the bone marrow but also in SLOs [264,265,274,275]. Hence, homing receptors and cytokines attribute to lymphoid tissue formation (i.e., homeostatic chemokine receptors CCR7 and CXCR5 and the lymphotoxin/TNF family), facilitate tumor stroma interactions within SLOs, and thereby emerge as novel therapeutic avenues [254,276]. In an experimental therapeutical approach it has already been shown that reciprocal crosstalk between stromal cells and lymphoma cells could be inhibited by treatment with LT$\beta$R-Ig, which substantially impaired lymphoma growth [254,276]. Likewise, targeting adhesion molecules by monoclonal antibodies, a strategy already used in patients with autoimmune diseases, may also prove to be a valid therapy for lymphoid malignancies [254,255]. Examples with proven efficiency in autoimmunity include natalizumab, a monoclonal antibody directed against integrin $\alpha4$, and MLN02, an antibody targeting $\alpha4\beta7$ [277,278]. Understanding the molecular and cellular mechanisms of lymphoma/leukemia cell dissemination and tumor-stroma crosstalk will lead to novel therapeutic targeting. An overview of the molecules that are involved in the regulation of B cell migration and lymphoma dissemination is provided in Table 1.

## 9. CONCLUSION

Taken together, B lymphocyte trafficking propensity, continuous recirculation, and homing is a prerequisite for the orchestration of systemic adaptive immune responses. The last decade has dramatically increased the knowledge on dynamics of B cells within lymphoid tissues. Technological advances in intravital microscopy have allowed for the first time to monitor cellular dynamics deep in the tissue with cellular and even subcellular resolution over time. This has literally led to a new view of B cell behavior in lymphoid tissues. In the past, based on classical immunohistological stainings, these tissues have been regarded as a static accumulation of cells. Intravital imaging has revealed lymphoid tissues to be a very dynamic entity of cells, with most B cell populations being highly motile. Chemokines and adhesion molecules of static stromal cells, which form the scaffold of lymphoid tissues, are responsible for organizing these tissues into different zones and guiding B cell motility. They enable the B cells to undergo interactions with multiple other cell types in various tissues, which are necessary for their maturation, their activation, and their differentiation. Although the main chemotactic cues for B cells have been identified, there is still a lot to find out about how B cells integrate and prioritize the various chemotactic signals that they are exposed to in the tissue.

This outstanding lymphocytic mobility reveals a dark face in the dissemination of lymphoma. It is widely accepted today that the rapid and strikingly tissue-specific lymphoma dissemination is not a consequence of tumor progression but of a very conserved physiological behavior of their lymphocyte precursors [255]. Hence, the sophisticated motility program of lymphocytes, governed by adhesion molecules together with the chemokine/chemokine receptor system, also regulate B cell lymphoma/leukemia cell trafficking and dissemination [254–256]. In addition, some chemokines and their receptors support proliferation of neoplastic cells and prevent B leukemia cells from undergoing spontaneous apoptosis [260,279]. Overall, targeting chemokines, chemokine receptors, or adhesion molecules that are part of the homing signature of neoplastic cells has emerged as a promising tool for treatment of malignant lymphoma and leukemia. In addition, expression profiles of chemokine receptors and adhesion molecules can also be implemented as novel diagnostic and prognostic markers.

Experimental therapeutical approaches in mice have already shown that preventing the homing to and reciprocal crosstalk between tumor cells and their microenvironment can substantially impair lymphoma/leukemia growth [257,268]. In patients, therapeutic approaches targeting adhesion molecules, as demonstrated for efficient antibody treatments against integrin $\alpha4$ (natalizumab) and $\alpha4\beta7$ (MLN02) in patients with autoimmune diseases, have also proven to be valid therapeutic targets for lymphoid malignancies [254,255]. Inhibitors for the chemokine receptor CXCR4 (i.e., plerixafor (AMD3100) and T140 analogs) are the only chemokine receptor inhibitors that are therapeutically used in clinical trials of B cell CLL [261]. Inhibitors for other chemokine receptors are still being developed, and along with the promising data from ongoing CXCR4 antagonist trials in B cell CLL, chemokine receptor inhibitors may lead to novel therapeutic avenues in the pathogenesis of malignant lymphoma and leukemia.

## REFERENCES

[1] Schofield R. The relationship between the spleen colony-forming cell and the haemopoietic stem cell. Blood Cells 1978;4:7–25.

[2] Zhang J, Niu C, Ye L, Huang H, He X, Tong WG, et al. Identification of the haematopoietic stem cell niche and control of the niche size. Nature 2003;425:836–41.

[3] Calvi LM, Adams GB, Weibrecht KW, Weber JM, Olson DP, Knight MC, et al. Osteoblastic cells regulate the haematopoietic stem cell niche. Nature 2003;425:841–6.

[4] Kiel MJ, Morrison SJ. Uncertainty in the niches that maintain haematopoietic stem cells. Nat Rev Immunol 2008;8:290–301.

[5] Lo Celso C, Fleming HE, Wu JW, Zhao CX, Miake-Lye S, Fujisaki J, et al. Live-animal tracking of individual haematopoietic stem/progenitor cells in their niche. Nature 2009;457:92–6.

[6] Kiel MJ, Yilmaz OH, Iwashita T, Yilmaz OH, Terhorst C, Morrison SJ. SLAM family receptors distinguish hematopoietic stem and progenitor cells and reveal endothelial niches for stem cells. Cell 2005;121:1109–21.

[7] Ding L, Saunders TL, Enikolopov G, Morrison SJ. Endothelial and perivascular cells maintain haematopoietic stem cells. Nature 2012;481:457–62.

[8] Peled A, Petit I, Kollet O, Magid M, Ponomaryov T, Byk T, et al. Dependence of human stem cell engraftment and repopulation of NOD/SCID mice on CXCR4. Science 1999;283:845–8.

[9] Petit I, Szyper-Kravitz M, Nagler A, Lahav M, Peled A, Habler L, et al. G-CSF induces stem cell mobilization by decreasing bone marrow SDF-1 and up-regulating CXCR4. Nat Immunol 2002;3:687–94.

[10] Sugiyama T, Kohara H, Noda M, Nagasawa T. Maintenance of the hematopoietic stem cell pool by CXCL12-CXCR4 chemokine signaling in bone marrow stromal cell niches. Immunity 2006;25:977–88.

[11] Tokoyoda K, Egawa T, Sugiyama T, Choi BI, Nagasawa T. Cellular niches controlling B lymphocyte behavior within bone marrow during development. Immunity 2004;20:707–18.

[12] Ding L, Morrison SJ. Haematopoietic stem cells and early lymphoid progenitors occupy distinct bone marrow niches. Nature 2013;495:231–5.

[13] Greenbaum A, Hsu YM, Day RB, Schuettpelz LG, Christopher MJ, Borgerding JN, et al. CXCL12 in early mesenchymal progenitors is required for haematopoietic stem-cell maintenance. Nature 2013;495:227–30.

[14] Egawa T, Kawabata K, Kawamoto H, Amada K, Okamoto R, Fujii N, et al. The earliest stages of B cell development require a chemokine stromal cell-derived factor/pre-B cell growth-stimulating factor. Immunity 2001;15:323–34.

[15] Mourcin F, Breton C, Tellier J, Narang P, Chasson L, Jorquera A, et al. Galectin-1-expressing stromal cells constitute a specific niche for pre-BII cell development in mouse bone marrow. Blood 2011;117:6552–61.
[16] Espeli M, Mancini SJ, Breton C, Poirier F, Schiff C. Impaired B-cell development at the pre-BII-cell stage in galectin-1-deficient mice due to inefficient pre-BII/stromal cell interactions. Blood 2009;113:5878–86.
[17] Batten SJ, Osmond DG. The localization of B lymphocytes in mouse bone marrow: radioautographic studies after in vivo perfusion of radiolabelled anti-IgM antibody. J Immunol Methods 1984;72:381–99.
[18] Osmond DG, Batten SJ. Genesis of B lymphocytes in the bone marrow: extravascular and intravascular localization of surface IgM-bearing cells in mouse bone marrow detected by electron-microscope radioautography after in vivo perfusion of $^{125}$I anti-IgM antibody. Am J Anat 1984;170:349–65.
[19] Hermans MH, Hartsuiker H, Opstelten D. An in situ study of B-lymphocytopoiesis in rat bone marrow. Topographical arrangement of terminal deoxynucleotidyl transferase-positive cells and pre-B cells. J Immunol 1989;142:67–73.
[20] Jacobsen K, Osmond DG. Microenvironmental organization and stromal cell associations of B lymphocyte precursor cells in mouse bone marrow. Eur J Immunol 1990;20:2395–404.
[21] Travlos GS. Normal structure, function, and histology of the bone marrow. Toxicol Pathol 2006;34:548–65.
[22] Ma Q, Jones D, Springer TA. The chemokine receptor CXCR4 is required for the retention of B lineage and granulocytic precursors within the bone marrow microenvironment. Immunity 1999;10: 463–71.
[23] Pereira JP, An J, Xu Y, Huang Y, Cyster JG. Cannabinoid receptor 2 mediates the retention of immature B cells in bone marrow sinusoids. Nat Immunol 2009;10:403–11.
[24] Pappu R, Schwab SR, Cornelissen I, Pereira JP, Regard JB, Xu Y, et al. Promotion of lymphocyte egress into blood and lymph by distinct sources of sphingosine-1-phosphate. Science 2007;316:295–8.
[25] Pereira JP, Xu Y, Cyster JG. A role for S1P and S1P1 in immature-B cell egress from mouse bone marrow. PLoS One 2010;5:e9277.
[26] Cariappa A, Mazo IB, Chase C, Shi HN, Liu H, Li Q, et al. Perisinusoidal B cells in the bone marrow participate in T-independent responses to blood-borne microbes. Immunity 2005;23:397–407.
[27] Sapoznikov A, Pewzner-Jung Y, Kalchenko V, Krauthgamer R, Shachar I, Jung S. Perivascular clusters of dendritic cells provide critical survival signals to B cells in bone marrow niches. Nat Immunol 2008;9:388–95.
[28] Legler DF, Loetscher M, Roos RS, Clark-Lewis I, Baggiolini M, Moser B. B cell-attracting chemokine 1, a human CXC chemokine expressed in lymphoid tissues, selectively attracts B lymphocytes via BLR1/CXCR5. J Exp Med 1998;187:655–60.
[29] Gunn MD, Ngo VN, Ansel KM, Ekland EH, Cyster JG, Williams LT. A B-cell-homing chemokine made in lymphoid follicles activates Burkitt's lymphoma receptor-1. Nature 1998;391:799–803.
[30] Voigt I, Camacho SA, de Boer BA, Lipp M, Forster R, Berek C. CXCR5-deficient mice develop functional germinal centers in the splenic T cell zone. Eur J Immunol 2000;30:560–7.
[31] Forster R, Mattis AE, Kremmer E, Wolf E, Brem G, Lipp M. A putative chemokine receptor, BLR1, directs B cell migration to defined lymphoid organs and specific anatomic compartments of the spleen. Cell 1996;87:1037–47.
[32] Cyster JG, Ansel KM, Reif K, Ekland EH, Hyman PL, Tang HL, et al. Follicular stromal cells and lymphocyte homing to follicles. Immunol Rev 2000;176:181–93.
[33] Krautler NJ, Kana V, Kranich J, Tian Y, Perera D, Lemm D, et al. Follicular dendritic cells emerge from ubiquitous perivascular precursors. Cell 2012;150:194–206.
[34] Katakai T, Suto H, Sugai M, Gonda H, Togawa A, Suematsu S, et al. Organizer-like reticular stromal cell layer common to adult secondary lymphoid organs. J Immunol 2008;181:6189–200.
[35] Okada T, Miller MJ, Parker I, Krummel MF, Neighbors M, Hartley SB, et al. Antigen-engaged B cells undergo chemotaxis toward the T zone and form motile conjugates with helper T cells. PLoS Biol 2005;3:e150.
[36] Kerfoot SM, Yaari G, Patel JR, Johnson KL, Gonzalez DG, Kleinstein SH, et al. Germinal center B cell and T follicular helper cell development initiates in the interfollicular zone. Immunity 2011;34:947–60.
[37] Gowans JL. The effect of the continuous re-infusion of lymph and lymphocytes on the output of lymphocytes from the thoracic duct of unanaesthetized rats. Br J Exp Pathol 1957;38:67–78.
[38] Arnon TI, Xu Y, Lo C, Pham T, An J, Coughlin S, et al. GRK2-dependent S1PR1 desensitization is required for lymphocytes to overcome their attraction to blood. Science 2011;333:1898–903.
[39] Butcher EC. Leukocyte-endothelial cell adhesion as an active, multi-step process: a combinatorial mechanism for specificity and diversity in leukocyte targeting. Adv Exp Med Biol 1992;323: 181–94.
[40] Rosen SD. Endothelial ligands for L-selectin: from lymphocyte recirculation to allograft rejection. Am J Pathol 1999;155:1013–20.
[41] Lepault F, Gagnerault MC, Faveeuw C, Boitard C. Recirculation, phenotype and functions of lymphocytes in mice treated with monoclonal antibody MEL-14. Eur J Immunol 1994;24:3106–12.
[42] Alon R, Kassner PD, Carr MW, Finger EB, Hemler ME, Springer TA. The integrin VLA-4 supports tethering and rolling in flow on VCAM-1. J Cell Biol 1995;128:1243–53.
[43] Berlin C, Bargatze RF, Campbell JJ, von Andrian UH, Szabo MC, Hasslen SR, et al. Alpha 4 integrins mediate lymphocyte attachment and rolling under physiologic flow. Cell 1995;80:413–22.
[44] Mazo IB, Gutierrez-Ramos JC, Frenette PS, Hynes RO, Wagner DD, von Andrian UH. Hematopoietic progenitor cell rolling in bone marrow microvessels: parallel contributions by endothelial selectins and vascular cell adhesion molecule 1. J Exp Med 1998;188: 465–74.
[45] Huang K, Im SY, Samlowski WE, Daynes RA. Molecular mechanisms of lymphocyte extravasation. III. The loss of lymphocyte extravasation potential induced by pertussis toxin is not mediated via the activation of protein kinase C. J Immunol 1989;143:229–38.
[46] Okada T, Ngo VN, Ekland EH, Forster R, Lipp M, Littman DR, et al. Chemokine requirements for B cell entry to lymph nodes and Peyer's patches. J Exp Med 2002;196:65–75.
[47] Baekkevold ES, Yamanaka T, Palframan RT, Carlsen HS, Reinholt FP, von Andrian UH, et al. The CCR7 ligand elc (CCL19) is transcytosed in high endothelial venules and mediates T cell recruitment. J Exp Med 2001;193:1105–12.
[48] Gunn MD, Tangemann K, Tam C, Cyster JG, Rosen SD, Williams LT. A chemokine expressed in lymphoid high endothelial venules promotes the adhesion and chemotaxis of naive T lymphocytes. Proc Natl Acad Sci USA 1998;95:258–63.

[49] Hamann A, Jablonski-Westrich D, Harder R, Thiele HG. Homing receptor expression and migration of activated lymphocytes. Adv Exp Med Biol 1988;237:511–8.

[50] Berlin-Rufenach C, Otto F, Mathies M, Westermann J, Owen MJ, Hamann A, et al. Lymphocyte migration in lymphocyte function-associated antigen (LFA)-1-deficient mice. J Exp Med 1999;189:1467–78.

[51] Lawrence MB, Berg EL, Butcher EC, Springer TA. Rolling of lymphocytes and neutrophils on peripheral node addressin and subsequent arrest on ICAM-1 in shear flow. Eur J Immunol 1995;25:1025–31.

[52] Lehmann JC, Jablonski-Westrich D, Haubold U, Gutierrez-Ramos JC, Springer T, Hamann A. Overlapping and selective roles of endothelial intercellular adhesion molecule-1 (ICAM-1) and ICAM-2 in lymphocyte trafficking. J Immunol 2003;171:2588–93.

[53] Bargatze RF, Jutila MA, Butcher EC. Distinct roles of L-selectin and integrins alpha 4 beta 7 and LFA-1 in lymphocyte homing to Peyer's patch-HEV in situ: the multistep model confirmed and refined. Immunity 1995;3:99–108.

[54] Muller WA. Mechanisms of leukocyte transendothelial migration. Annu Rev Pathol 2011;6:323–44.

[55] Pearce G, Angeli V, Randolph GJ, Junt T, von Andrian U, Schnittler HJ, et al. Signaling protein SWAP-70 is required for efficient B cell homing to lymphoid organs. Nat Immunol 2006;7:827–34.

[56] Wojcikiewicz EP, Koenen RR, Fraemohs L, Minkiewicz J, Azad H, Weber C, et al. LFA-1 binding destabilizes the JAM-A homophilic interaction during leukocyte transmigration. Biophys J 2009;96:285–93.

[57] Ostermann G, Weber KS, Zernecke A, Schroder A, Weber C. JAM-1 is a ligand of the beta(2) integrin LFA-1 involved in transendothelial migration of leukocytes. Nat Immunol 2002;3:151–8.

[58] Mamdouh Z, Chen X, Pierini LM, Maxfield FR, Muller WA. Targeted recycling of PECAM from endothelial surface-connected compartments during diapedesis. Nature 2003;421:748–53.

[59] Mamdouh Z, Kreitzer GE, Muller WA. Leukocyte transmigration requires kinesin-mediated microtubule-dependent membrane trafficking from the lateral border recycling compartment. J Exp Med 2008;205:951–66.

[60] Mamdouh Z, Mikhailov A, Muller WA. Transcellular migration of leukocytes is mediated by the endothelial lateral border recycling compartment. J Exp Med 2009;206:2795–808.

[61] Faveeuw C, Preece G, Ager A. Transendothelial migration of lymphocytes across high endothelial venules into lymph nodes is affected by metalloproteinases. Blood 2001;98:688–95.

[62] Mebius RE, Kraal G. Structure and function of the spleen. Nat Rev Immunol 2005;5:606–16.

[63] Brelinska R, Pilgrim C. The significance of the subcompartments of the marginal zone for directing lymphocyte traffic within the splenic pulp of the rat. Cell Tissue Res 1982;226:155–65.

[64] Ford WL. The immunological and migratory properties of the lymphocytes recirculating through the rat spleen. Br J Exp Pathol 1969;50:257–69.

[65] Nieuwenhuis P, Ford WL. Comparative migration of B- and T-Lymphocytes in the rat spleen and lymph nodes. Cell Immunol 1976;23:254–67.

[66] Lo CG, Lu TT, Cyster JG. Integrin-dependence of lymphocyte entry into the splenic white pulp. J Exp Med 2003;197:353–61.

[67] Reif K, Ekland EH, Ohl L, Nakano H, Lipp M, Forster R, et al. Balanced responsiveness to chemoattractants from adjacent zones determines B-cell position. Nature 2002;416:94–9.

[68] Miller MJ, Wei SH, Parker I, Cahalan MD. Two-photon imaging of lymphocyte motility and antigen response in intact lymph node. Science 2002;296:1869–73.

[69] Nombela-Arrieta C, Mempel TR, Soriano SF, Mazo I, Wymann MP, Hirsch E, et al. A central role for DOCK2 during interstitial lymphocyte motility and sphingosine-1-phosphate-mediated egress. J Exp Med 2007;204:497–510.

[70] Pape KA, Catron DM, Itano AA, Jenkins MK. The humoral immune response is initiated in lymph nodes by B cells that acquire soluble antigen directly in the follicles. Immunity 2007;26:491–502.

[71] Roozendaal R, Mempel TR, Pitcher LA, Gonzalez SF, Verschoor A, Mebius RE, et al. Conduits mediate transport of low-molecular-weight antigen to lymph node follicles. Immunity 2009;30:264–76.

[72] Katakai T, Hara T, Lee JH, Gonda H, Sugai M, Shimizu A. A novel reticular stromal structure in lymph node cortex: an immuno-platform for interactions among dendritic cells, T cells and B cells. Int Immunol 2004;16:1133–42.

[73] Katakai T, Hara T, Sugai M, Gonda H, Shimizu A. Lymph node fibroblastic reticular cells construct the stromal reticulum via contact with lymphocytes. J Exp Med 2004;200:783–95.

[74] Sixt M, Kanazawa N, Selg M, Samson T, Roos G, Reinhardt DP, et al. The conduit system transports soluble antigens from the afferent lymph to resident dendritic cells in the T cell area of the lymph node. Immunity 2005;22:19–29.

[75] Junt T, Moseman EA, Iannacone M, Massberg S, Lang PA, Boes M, et al. Subcapsular sinus macrophages in lymph nodes clear lymph-borne viruses and present them to antiviral B cells. Nature 2007;450:110–4.

[76] Phan TG, Grigorova I, Okada T, Cyster JG. Subcapsular encounter and complement-dependent transport of immune complexes by lymph node B cells. Nat Immunol 2007;8:992–1000.

[77] Heesters BA, Chatterjee P, Kim YA, Gonzalez SF, Kuligowski MP, Kirchhausen T, et al. Endocytosis and recycling of immune complexes by follicular dendritic cells enhances B cell antigen binding and activation. Immunity 2013;38:1164–75.

[78] Drayton DL, Liao S, Mounzer RH, Ruddle NH. Lymphoid organ development: from ontogeny to neogenesis. Nat Immunol 2006;7:344–53.

[79] Suzuki K, Grigorova I, Phan TG, Kelly LM, Cyster JG. Visualizing B cell capture of cognate antigen from follicular dendritic cells. J Exp Med 2009;206:1485–93.

[80] Qi H, Egen JG, Huang AY, Germain RN. Extrafollicular activation of lymph node B cells by antigen-bearing dendritic cells. Science 2006;312:1672–6.

[81] Kelly LM, Pereira JP, Yi T, Xu Y, Cyster JG. EBI2 guides serial movements of activated B cells and ligand activity is detectable in lymphoid and nonlymphoid tissues. J Immunol 2011;187:3026–32.

[82] Liu C, Yang XV, Wu J, Kuei C, Mani NS, Zhang L, et al. Oxysterols direct B-cell migration through EBI2. Nature 2011;475:519–23.

[83] Hannedouche S, Zhang J, Yi T, Shen W, Nguyen D, Pereira JP, et al. Oxysterols direct immune cell migration via EBI2. Nature 2011;475:524–7.

[84] Yi T, Wang X, Kelly LM, An J, Xu Y, Sailer AW, et al. Oxysterol gradient generation by lymphoid stromal cells guides activated B cell movement during humoral responses. Immunity 2012;37 :535–48.

[85] Pereira JP, Kelly LM, Xu Y, Cyster JG. EBI2 mediates B cell segregation between the outer and centre follicle. Nature 2009;460: 1122–6.

[86] Gatto D, Paus D, Basten A, Mackay CR, Brink R. Guidance of B cells by the orphan G protein-coupled receptor EBI2 shapes humoral immune responses. Immunity 2009;31:259–69.
[87] Garside P, Ingulli E, Merica RR, Johnson JG, Noelle RJ, Jenkins MK. Visualization of specific B and T lymphocyte interactions in the lymph node. Science 1998;281:96–9.
[88] Coffey F, Alabyev B, Manser T. Initial clonal expansion of germinal center B cells takes place at the perimeter of follicles. Immunity 2009;30:599–609.
[89] Pape KA, Kouskoff V, Nemazee D, Tang HL, Cyster JG, Tze LE, et al. Visualization of the genesis and fate of isotype-switched B cells during a primary immune response. J Exp Med 2003;197:1677–87.
[90] Chan TD, Gatto D, Wood K, Camidge T, Basten A, Brink R. Antigen affinity controls rapid T-dependent antibody production by driving the expansion rather than the differentiation or extrafollicular migration of early plasmablasts. J Immunol 2009;183:3139–49.
[91] Wehrli N, Legler DF, Finke D, Toellner KM, Loetscher P, Baggiolini M, et al. Changing responsiveness to chemokines allows medullary plasmablasts to leave lymph nodes. Eur J Immunol 2001;31:609–16.
[92] Hauser AE, Debes GF, Arce S, Cassese G, Hamann A, Radbruch A, et al. Chemotactic responsiveness toward ligands for CXCR3 and CXCR4 is regulated on plasma blasts during the time course of a memory immune response. J Immunol 2002;169:1277–82.
[93] Hargreaves DC, Hyman PL, Lu TT, Ngo VN, Bidgol A, Suzuki G, et al. A coordinated change in chemokine responsiveness guides plasma cell movements. J Exp Med 2001;194:45–56.
[94] Allen CD, Ansel KM, Low C, Lesley R, Tamamura H, Fujii N, et al. Germinal center dark and light zone organization is mediated by CXCR4 and CXCR5. Nat Immunol 2004;5:943–52.
[95] Victora GD, Schwickert TA, Fooksman DR, Kamphorst AO, Meyer-Hermann M, Dustin ML, et al. Germinal center dynamics revealed by multiphoton microscopy with a photoactivatable fluorescent reporter. Cell 2010;143:592–605.
[96] Fooksman DR, Schwickert TA, Victora GD, Dustin ML, Nussenzweig MC, Skokos D. Development and migration of plasma cells in the mouse lymph node. Immunity 2010;33:118–27.
[97] Allen CD, Okada T, Tang HL, Cyster JG. Imaging of germinal center selection events during affinity maturation. Science 2007;315:528–31.
[98] Schwickert TA, Lindquist RL, Shakhar G, Livshits G, Skokos D, Kosco-Vilbois MH, et al. In vivo imaging of germinal centres reveals a dynamic open structure. Nature 2007;446:83–7.
[99] Kallies A, Hasbold J, Tarlinton DM, Dietrich W, Corcoran LM, Hodgkin PD, et al. Plasma cell ontogeny defined by quantitative changes in blimp-1 expression. J Exp Med 2004;200:967–77.
[100] Smith KG, Hewitson TD, Nossal GJ, Tarlinton DM. The phenotype and fate of the antibody-forming cells of the splenic foci. Eur J Immunol 1996;26:444–8.
[101] William J, Euler C, Christensen S, Shlomchik MJ. Evolution of autoantibody responses via somatic hypermutation outside of germinal centers. Science 2002;297:2066–70.
[102] Sze DM, Toellner KM, Garcia de Vinuesa C, Taylor DR, MacLennan IC. Intrinsic constraint on plasmablast growth and extrinsic limits of plasma cell survival. J Exp Med 2000;192:813–21.
[103] Hoyer BF, Moser K, Hauser AE, Peddinghaus A, Voigt C, Eilat D, et al. Short-lived plasmablasts and long-lived plasma cells contribute to chronic humoral autoimmunity in NZB/W mice. J Exp Med 2004;199:1577–84.
[104] Cassese G, Lindenau S, de Boer B, Arce S, Hauser A, Riemekasten G, et al. Inflamed kidneys of NZB/W mice are a major site for the homeostasis of plasma cells. Eur J Immunol 2001;31:2726–32.
[105] Starke C, Frey S, Wellmann U, Urbonaviciute V, Herrmann M, Amann K, et al. High frequency of autoantibody-secreting cells and long-lived plasma cells within inflamed kidneys of NZB/W F1 lupus mice. Eur J Immunol 2011;41:2107–12.
[106] Lacotte S, Decossas M, Le Coz C, Brun S, Muller S, Dumortier H. Early differentiated CD138(high) MHCII$^+$ IgG$^+$ plasma cells express CXCR3 and localize into inflamed kidneys of lupus mice. PLoS One 2013;8:e58140.
[107] MacLennan IC. Germinal centers. Annu Rev Immunol 1994;12:117–39.
[108] Camacho SA, Kosco-Vilbois MH, Berek C. The dynamic structure of the germinal center. Immunol Today 1998;19:511–4.
[109] Liu YJ, Joshua DE, Williams GT, Smith CA, Gordon J, MacLennan IC. Mechanism of antigen-driven selection in germinal centres. Nature 1989;342:929–31.
[110] Kosco MH, Pflugfelder E, Gray D. Follicular dendritic cell-dependent adhesion and proliferation of B cells in vitro. J Immunol 1992;148:2331–9.
[111] Hwang IY, Park C, Harrison K, Kehrl JH. TLR4 signaling augments B lymphocyte migration and overcomes the restriction that limits access to germinal center dark zones. J Exp Med 2009;206:2641–57.
[112] Hauser AE, Junt T, Mempel TR, Sneddon MW, Kleinstein SH, Henrickson SE, et al. Definition of germinal-center B cell migration in vivo reveals predominant intrazonal circulation patterns. Immunity 2007;26:655–67.
[113] Andresen V, Pollok K, Rinnenthal JL, Oehme L, Gunther R, Spiecker H, et al. High-resolution intravital microscopy. PLoS One 2012;7:e50915.
[114] Sinha RK, Park C, Hwang IY, Davis MD, Kehrl JH. B lymphocytes exit lymph nodes through cortical lymphatic sinusoids by a mechanism independent of sphingosine-1-phosphate-mediated chemotaxis. Immunity 2009;30:434–46.
[115] Tomura M, Yoshida N, Tanaka J, Karasawa S, Miwa Y, Miyawaki A, et al. Monitoring cellular movement in vivo with photoconvertible fluorescence protein "Kaede" transgenic mice. Proc Natl Acad Sci USA 2008;105:10871–6.
[116] Shiow LR, Rosen DB, Brdickova N, Xu Y, An J, Lanier LL, et al. CD69 acts downstream of interferon-alpha/beta to inhibit S1P1 and lymphocyte egress from lymphoid organs. Nature 2006;440:540–4.
[117] Bankovich AJ, Shiow LR, Cyster JG. CD69 suppresses sphingosine 1-phosophate receptor-1 (S1P1) function through interaction with membrane helix 4. J Biol Chem 2010;285:22328–37.
[118] Grigorova IL, Panteleev M, Cyster JG. Lymph node cortical sinus organization and relationship to lymphocyte egress dynamics and antigen exposure. Proc Natl Acad Sci USA 2010;107:20447–52.
[119] Arnon TI, Horton RM, Grigorova IL, Cyster JG. Visualization of splenic marginal zone B-cell shuttling and follicular B-cell egress. Nature 2013;493:684–8.
[120] Martin F, Kearney JF. Marginal-zone B cells. Nat Rev Immunol 2002;2:323–35.
[121] Martin F, Oliver AM, Kearney JF. Marginal zone and B1 B cells unite in the early response against T-independent blood-borne particulate antigens. Immunity 2001;14:617–29.
[122] Lu TT, Cyster JG. Integrin-mediated long-term B cell retention in the splenic marginal zone. Science 2002;297:409–12.

[123] Guinamard R, Okigaki M, Schlessinger J, Ravetch JV. Absence of marginal zone B cells in Pyk-2-deficient mice defines their role in the humoral response. Nat Immunol 2000;1:31–6.

[124] Cinamon G, Matloubian M, Lesneski MJ, Xu Y, Low C, Lu T, et al. Sphingosine 1-phosphate receptor 1 promotes B cell localization in the splenic marginal zone. Nat Immunol 2004;5:713–20.

[125] Cinamon G, Zachariah MA, Lam OM, Foss Jr FW, Cyster JG. Follicular shuttling of marginal zone B cells facilitates antigen transport. Nat Immunol 2008;9:54–62.

[126] Ziring D, Wei B, Velazquez P, Schrage M, Buckley NE, Braun J. Formation of B and T cell subsets require the cannabinoid receptor CB2. Immunogenetics 2006;58:714–25.

[127] Wang H, Beaty N, Chen S, Qi CF, Masiuk M, Shin DM, et al. The CXCR7 chemokine receptor promotes B-cell retention in the splenic marginal zone and serves as a sink for CXCL12. Blood 2012;119:465–8.

[128] Crotty S, Felgner P, Davies H, Glidewell J, Villarreal L, Ahmed R. Cutting edge: long-term B cell memory in humans after smallpox vaccination. J Immunol 2003;171:4969–73.

[129] Hammarlund E, Lewis MW, Hansen SG, Strelow LI, Nelson JA, Sexton GJ, et al. Duration of antiviral immunity after smallpox vaccination. Nat Med 2003;9:1131–7.

[130] Amanna IJ, Carlson NE, Slifka MK. Duration of humoral immunity to common viral and vaccine antigens. N Engl J Med 2007;357:1903–15.

[131] Yu X, Tsibane T, McGraw PA, House FS, Keefer CJ, Hicar MD, et al. Neutralizing antibodies derived from the B cells of 1918 influenza pandemic survivors. Nature 2008;455:532–6.

[132] Maruyama M, Lam KP, Rajewsky K. Memory B-cell persistence is independent of persisting immunizing antigen. Nature 2000;407:636–42.

[133] Schittek B, Rajewsky K. Maintenance of B-cell memory by long-lived cells generated from proliferating precursors. Nature 1990;346:749–51.

[134] Giesecke C, Frolich D, Reiter K, Mei HE, Wirries I, Kuhly R, et al. Tissue distribution and dependence of responsiveness of human antigen-specific memory B cells. J Immunol 2014;192:3091–100.

[135] Mamani-Matsuda M, Cosma A, Weller S, Faili A, Staib C, Garcon L, et al. The human spleen is a major reservoir for long-lived vaccinia virus-specific memory B cells. Blood 2008;111:4653–9.

[136] Steiniger B, Timphus EM, Jacob R, Barth PJ. CD27$^{+}$ B cells in human lymphatic organs: re-evaluating the splenic marginal zone. Immunology 2005;116:429–42.

[137] Leyendeckers H, Odendahl M, Lohndorf A, Irsch J, Spangfort M, Miltenyi S, et al. Correlation analysis between frequencies of circulating antigen-specific IgG-bearing memory B cells and serum titers of antigen-specific IgG. Eur J Immunol 1999;29:1406–17.

[138] Martinez-Gamboa L, Mei H, Loddenkemper C, Ballmer B, Hansen A, Lipsky PE, et al. Role of the spleen in peripheral memory B-cell homeostasis in patients with autoimmune thrombocytopenia purpura. Clin Immunol 2009;130:199–212.

[139] Aiba Y, Kometani K, Hamadate M, Moriyama S, Sakaue-Sawano A, Tomura M, et al. Preferential localization of IgG memory B cells adjacent to contracted germinal centers. Proc Natl Acad Sci USA 2010;107:12192–7.

[140] Dogan I, Bertocci B, Vilmont V, Delbos F, Mégret J, Storck S, et al. Multiple layers of B cell memory with different effector functions. Nat Immunol 2009;10:1292–9.

[141] Benson MJ, Elgueta R, Schpero W, Molloy M, Zhang W, Usherwood E, et al. Distinction of the memory B cell response to cognate antigen versus bystander inflammatory signals. J Exp Med 2009;206:2013–25.

[142] Laichalk LL, Hochberg D, Babcock GJ, Freeman RB, Thorley-Lawson DA. The dispersal of mucosal memory B cells: evidence from persistent EBV infection. Immunity 2002;16:745–54.

[143] Youngman KR, Franco MA, Kuklin NA, Rott LS, Butcher EC, Greenberg HB. Correlation of tissue distribution, developmental phenotype, and intestinal homing receptor expression of antigen-specific B cells during the murine anti-rotavirus immune response. J Immunol 2002;168:2173–81.

[144] Bortnick A, Chernova I, Quinn 3rd WJ, Mugnier M, Cancro MP, Allman D. Long-lived bone marrow plasma cells are induced early in response to T cell-independent or T cell-dependent antigens. J Immunol 2012;188:5389–96.

[145] Tunyaplin C, Shaffer AL, Angelin-Duclos CD, Yu X, Staudt LM, Calame KL. Direct repression of prdm1 by Bcl-6 inhibits plasmacytic differentiation. J Immunol 2004;173:1158–65.

[146] Angelin-Duclos C, Cattoretti G, Lin KI, Calame K. Commitment of B lymphocytes to a plasma cell fate is associated with Blimp-1 expression in vivo. J Immunol 2000;165:5462–71.

[147] Cattoretti G, Buttner M, Shaknovich R, Kremmer E, Alobeid B, Niedobitek G. Nuclear and cytoplasmic AID in extrafollicular and germinal center B cells. Blood 2006;107:3967–75.

[148] Meyer-Hermann M, Mohr E, Pelletier N, Zhang Y, Victora GD, Toellner KM. A theory of germinal center B cell selection, division, and exit. Cell Rep 2012;2:162–74.

[149] Kabashima K, Haynes NM, Xu Y, Nutt SL, Allende ML, Proia RL, et al. Plasma cell S1P1 expression determines secondary lymphoid organ retention versus bone marrow tropism. J Exp Med 2006;203:2683–90.

[150] Gohda M, Kunisawa J, Miura F, Kagiyama Y, Kurashima Y, Higuchi M, et al. Sphingosine 1-phosphate regulates the egress of IgA plasmablasts from Peyer's patches for intestinal IgA responses. J Immunol 2008;180:5335–43.

[151] Pabst O, Peters T, Czeloth N, Bernhardt G, Scharffetter-Kochanek K, Forster R. Cutting edge: egress of newly generated plasma cells from peripheral lymph nodes depends on beta 2 integrin. J Immunol 2005;174:7492–5.

[152] Nitschke L, Floyd H, Ferguson DJ, Crocker PR. Identification of CD22 ligands on bone marrow sinusoidal endothelium implicated in CD22-dependent homing of recirculating B cells. J Exp Med 1999;189:1513–8.

[153] Nagasawa T, Hirota S, Tachibana K, Takakura N, Nishikawa S, Kitamura Y, et al. Defects of B-cell lymphopoiesis and bone-marrow myelopoiesis in mice lacking the CXC chemokine PBSF/SDF-1. Nature 1996;382:635–8.

[154] Odendahl M, Mei H, Hoyer BF, Jacobi AM, Hansen A, Muehlinghaus G, et al. Generation of migratory antigen-specific plasma blasts and mobilization of resident plasma cells in a secondary immune response. Blood 2005;105:1614–21.

[155] Radbruch A, Muehlinghaus G, Luger EO, Inamine A, Smith KG, Dorner T, et al. Competence and competition: the challenge of becoming a long-lived plasma cell. Nat Rev Immunol 2006;6:741–50.

[156] Cassese G, Arce S, Hauser AE, Lehnert K, Moewes B, Mostarac M, et al. Plasma cell survival is mediated by synergistic effects of cytokines and adhesion-dependent signals. J Immunol 2003;171:1684–90.

[157] Chu VT, Frohlich A, Steinhauser G, Scheel T, Roch T, Fillatreau S, et al. Eosinophils are required for the maintenance of plasma cells in the bone marrow. Nat Immunol 2011;12:151–9.
[158] Winter O, Moser K, Mohr E, Zotos D, Kaminski H, Szyska M, et al. Megakaryocytes constitute a functional component of a plasma cell niche in the bone marrow. Blood 2010;116:1867–75.
[159] Rozanski CH, Arens R, Carlson LM, Nair J, Boise LH, Chanan-Khan AA, et al. Sustained antibody responses depend on CD28 function in bone marrow-resident plasma cells. J Exp Med 2011;208: 1435–46.
[160] Rodriguez Gomez M, Talke Y, Goebel N, Hermann F, Reich B, Mack M. Basophils support the survival of plasma cells in mice. J Immunol 2010;185:7180–5.
[161] Belnoue E, Tougne C, Rochat AF, Lambert PH, Pinschewer DD, Siegrist CA. Homing and adhesion patterns determine the cellular composition of the bone marrow plasma cell niche. J Immunol 2012;188:1283–91.
[162] Delogu A, Schebesta A, Sun Q, Aschenbrenner K, Perlot T, Busslinger M. Gene repression by Pax5 in B cells is essential for blood cell homeostasis and is reversed in plasma cells. Immunity 2006;24:269–81.
[163] DiLillo DJ, Hamaguchi Y, Ueda Y, Yang K, Uchida J, Haas KM, et al. Maintenance of long-lived plasma cells and serological memory despite mature and memory B cell depletion during CD20 immunotherapy in mice. J Immunol 2008;180:361–71.
[164] Chevrier S, Genton C, Kallies A, Karnowski A, Otten LA, Malissen B, et al. CD93 is required for maintenance of antibody secretion and persistence of plasma cells in the bone marrow niche. Proc Natl Acad Sci USA 2009;106:3895–900.
[165] Marques CP, Kapil P, Hinton DR, Hindinger C, Nutt SL, Ransohoff RM, et al. CXCR3-dependent plasma blast migration to the central nervous system during viral encephalomyelitis. J Virol 2011;85:6136–47.
[166] Muehlinghaus G, Cigliano L, Huehn S, Peddinghaus A, Leyendeckers H, Hauser AE, et al. Regulation of CXCR3 and CXCR4 expression during terminal differentiation of memory B cells into plasma cells. Blood 2005;105:3965–71.
[167] Martin F, Kearney JF. B1 cells: similarities and differences with other B cell subsets. Curr Opin Immunol 2001;13:195–201.
[168] Bendelac A, Bonneville M, Kearney JF. Autoreactivity by design: innate B and T lymphocytes. Nat Rev Immunol 2001;1:177–86.
[169] Hardy RR, Hayakawa K. B cell development pathways. Annu Rev Immunol 2001;19:595–621.
[170] Alugupalli KR, Gerstein RM. Divide and conquer: division of labor by B-1 B cells. Immunity 2005;23:1–2.
[171] Haas KM, Poe JC, Steeber DA, Tedder TF. B-1a and B-1b cells exhibit distinct developmental requirements and have unique functional roles in innate and adaptive immunity to *S. pneumoniae*. Immunity 2005;23:7–18.
[172] Kantor AB, Herzenberg LA. Origin of murine B cell lineages. Annu Rev Immunol 1993;11:501–38.
[173] Tumang JR, Frances R, Yeo SG, Rothstein TL. Spontaneously Ig-secreting B-1 cells violate the accepted paradigm for expression of differentiation-associated transcription factors. J Immunol 2005;174:3173–7.
[174] Choi YS, Dieter JA, Rothaeusler K, Luo Z, Baumgarth N. B-1 cells in the bone marrow are a significant source of natural IgM. Eur J Immunol 2012;42:120–9.
[175] Baumgarth N. Innate-like B cells and their rules of engagement. Adv Exp Med Biol 2013;785:57–66.
[176] Griffin DO, Holodick NE, Rothstein TL. Human B1 cells in umbilical cord and adult peripheral blood express the novel phenotype $CD20^+$ $CD27^+$ $CD43^+$ CD70. J Exp Med 2011;208:67–80.
[177] Griffin DO, Rothstein TL. A small $CD11b^+$ human B1 cell subpopulation stimulates T cells and is expanded in lupus. J Exp Med 2011;208:2591–8.
[178] Youinou P, Renaudineau Y. The paradox of CD5-expressing B cells in systemic lupus erythematosus. Autoimmun Rev 2007;7:149–54.
[179] Baumgarth N. The double life of a B-1 cell: self-reactivity selects for protective effector functions. Nat Rev Immunol 2011;11:34–46.
[180] Rothstein TL, Griffin DO, Holodick NE, Quach TD, Kaku H. Human B-1 cells take the stage. Ann NY Acad Sci 2013;1285:97–114.
[181] Williams R, White H. The greater omentum: its applicability to cancer surgery and cancer therapy. Curr Probl Surg 1986;23:789–865.
[182] Ansel KM, Harris RB, Cyster JG. CXCL13 is required for B1 cell homing, natural antibody production, and body cavity immunity. Immunity 2002;16:67–76.
[183] Doherty NS, Griffiths RJ, Hakkinen JP, Scampoli DN, Milici AJ. Post-capillary venules in the "milky spots" of the greater omentum are the major site of plasma protein and leukocyte extravasation in rodent models of peritonitis. Inflamm Res 1995;44:169–77.
[184] Debes GF, Arnold CN, Young AJ, Krautwald S, Lipp M, Hay JB, et al. Chemokine receptor CCR7 required for T lymphocyte exit from peripheral tissues. Nat Immunol 2005;6:889–94.
[185] Bromley SK, Thomas SY, Luster AD. Chemokine receptor CCR7 guides T cell exit from peripheral tissues and entry into afferent lymphatics. Nat Immunol 2005;6:895–901.
[186] Hopken UE, Wengner AM, Loddenkemper C, Stein H, Heimesaat MM, Rehm A, et al. CCR7 deficiency causes ectopic lymphoid neogenesis and disturbed mucosal tissue integrity. Blood 2007;109:886–95.
[187] Muller G, Hopken UE, Stein H, Lipp M. Systemic immunoregulatory and pathogenic functions of homeostatic chemokine receptors. J Leukoc Biol 2002;72:1–8.
[188] Foussat A, Balabanian K, Amara A, Bouchet-Delbos L, Durand-Gasselin I, Baleux F, et al. Production of stromal cell-derived factor 1 by mesothelial cells and effects of this chemokine on peritoneal B lymphocytes. Eur J Immunol 2001;31:350–9.
[189] Ha SA, Tsuji M, Suzuki K, Meek B, Yasuda N, Kaisho T, et al. Regulation of B1 cell migration by signals through Toll-like receptors. J Exp Med 2006;203:2541–50.
[190] Cyster JG. Chemokines, sphingosine-1-phosphate, and cell migration in secondary lymphoid organs. Annu Rev Immunol 2005;23:127–59.
[191] Rosen H, Goetzl EJ. Sphingosine 1-phosphate and its receptors: an autocrine and paracrine network. Nat Rev Immunol 2005;5:560–70.
[192] Kunisawa J, Kurashima Y, Gohda M, Higuchi M, Ishikawa I, Miura F, et al. Sphingosine 1-phosphate regulates peritoneal B-cell trafficking for subsequent intestinal IgA production. Blood 2007;109:3749–56.
[193] Hopken UE, Winter S, Achtman AH, Kruger K, Lipp M. CCR7 regulates lymphocyte egress and recirculation through body cavities. J Leukoc Biol 2010;87:671–82.
[194] Berberich S, Forster R, Pabst O. The peritoneal micromilieu commits B cells to home to body cavities and the small intestine. Blood 2007;109:4627–34.

[195] Iwasaki A, Medzhitov R. Toll-like receptor control of the adaptive immune responses. Nat Immunol 2004;5:987–95.

[196] Pasare C, Medzhitov R. Toll-like receptors: linking innate and adaptive immunity. Adv Exp Med Biol 2005;560:11–8.

[197] Meyer-Bahlburg A, Rawlings DJ. Differential impact of Toll-like receptor signaling on distinct B cell subpopulations. Front Biosci (Landmark Ed) 2012;17:1499–516.

[198] Fagarasan S, Shinkura R, Kamata T, Nogaki F, Ikuta K, Honjo T. Mechanism of B1 cell differentiation and migration in GALT. Curr Top Microbiol Immunol 2000;252:221–9.

[199] McDermott MR, Bienenstock J. Evidence for a common mucosal immunologic system. I. Migration of B immunoblasts into intestinal, respiratory, and genital tissues. J Immunol 1979;122:1892–8.

[200] Mayer L. Mucosal immunity. Immunol Rev 2005;206:5.

[201] McGhee JR, Fujihashi K. Inside the mucosal immune system. PLoS Biol 2012;10:e1001397.

[202] Brandtzaeg P. Mucosal immunity: induction, dissemination, and effector functions. Scand J Immunol 2009;70:505–15.

[203] Cerutti A. The regulation of IgA class switching. Nat Rev Immunol 2008;8:421–34.

[204] Kruglov AA, Grivennikov SI, Kuprash DV, Winsauer C, Prepens S, Seleznik GM, et al. Nonredundant function of soluble LTalpha3 produced by innate lymphoid cells in intestinal homeostasis. Science 2013;342:1243–46.

[205] Chu VT, Beller A, Rausch S, Strandmark J, Zanker M, Arbach O, et al. Eosinophils promote generation and maintenance of immunoglobulin-a-expressing plasma cells and contribute to gut immune homeostasis. Immunity 2014;40:582–93.

[206] Iwasaki A. Mucosal dendritic cells. Annu Rev Immunol 2007;25:381–418.

[207] Hooper LV, Macpherson AJ. Immune adaptations that maintain homeostasis with the intestinal microbiota. Nat Rev Immunol 2010;10:159–69.

[208] Hooper LV, Littman DR, Macpherson AJ. Interactions between the microbiota and the immune system. Science 2012;336:1268–73.

[209] Chinen T, Rudensky AY. The effects of commensal microbiota on immune cell subsets and inflammatory responses. Immunol Rev 2012;245:45–55.

[210] Sigmundsdottir H, Butcher EC. Environmental cues, dendritic cells and the programming of tissue-selective lymphocyte trafficking. Nat Immunol 2008;9:981–7.

[211] Brandtzaeg P, Johansen FE. Mucosal B cells: phenotypic characteristics, transcriptional regulation, and homing properties. Immunol Rev 2005;206:32–63.

[212] Butcher EC, Picker LJ. Lymphocyte homing and homeostasis. Science 1996;272:60–6.

[213] Forster R, Schubel A, Breitfeld D, Kremmer E, Renner-Muller I, Wolf E, et al. CCR7 coordinates the primary immune response by establishing functional microenvironments in secondary lymphoid organs. Cell 1999;99:23–33.

[214] Muller G, Hopken UE, Lipp M. The impact of CCR7 and CXCR5 on lymphoid organ development and systemic immunity. Immunol Rev 2003;195:117–35.

[215] Carlsen HS, Baekkevold ES, Johansen FE, Haraldsen G, Brandtzaeg P. B cell attracting chemokine 1 (CXCL13) and its receptor CXCR5 are expressed in normal and aberrant gut associated lymphoid tissue. Gut 2002;51:364–71.

[216] Brandtzaeg P, Farstad IN, Haraldsen G. Regional specialization in the mucosal immune system: primed cells do not always home along the same track. Immunol Today 1999;20:267–77.

[217] Kunkel EJ, Butcher EC. Chemokines and the tissue-specific migration of lymphocytes. Immunity 2002;16:1–4.

[218] Bowman EP, Kuklin NA, Youngman KR, Lazarus NH, Kunkel EJ, Pan J, et al. The intestinal chemokine thymus-expressed chemokine (CCL25) attracts IgA antibody-secreting cells. J Exp Med 2002;195:269–75.

[219] Kunkel EJ, Butcher EC. Plasma-cell homing. Nat Rev Immunol 2003;3:822–9.

[220] Iwata M, Hirakiyama A, Eshima Y, Kagechika H, Kato C, Song SY. Retinoic acid imprints gut-homing specificity on T cells. Immunity 2004;21:527–38.

[221] Hieshima K, Kawasaki Y, Hanamoto H, Nakayama T, Nagakubo D, Kanamaru A, et al. CC chemokine ligands 25 and 28 play essential roles in intestinal extravasation of IgA antibody-secreting cells. J Immunol 2004;173:3668–75.

[222] Krzysiek R, Lefevre EA, Bernard J, Foussat A, Galanaud P, Louache F, et al. Regulation of CCR6 chemokine receptor expression and responsiveness to macrophage inflammatory protein-3alpha/CCL20 in human B cells. Blood 2000;96:2338–45.

[223] Gerdts V, Uwiera RR, Mutwiri GK, Wilson DJ, Bowersock T, Kidane A, et al. Multiple intestinal 'loops' provide an in vivo model to analyse multiple mucosal immune responses. J Immunol Methods 2001;256:19–33.

[224] Mayer KD, Mohrs K, Crowe SR, Johnson LL, Rhyne P, Woodland DL, et al. The functional heterogeneity of type 1 effector T cells in response to infection is related to the potential for IFN-gamma production. J Immunol 2005;174:7732–9.

[225] Kamata T, Nogaki F, Fagarasan S, Sakiyama T, Kobayashi I, Miyawaki S, et al. Increased frequency of surface IgA-positive plasma cells in the intestinal lamina propria and decreased IgA excretion in hyper IgA (HIGA) mice, a murine model of IgA nephropathy with hyperserum IgA. J Immunol 2000;165:1387–94.

[226] Hapfelmeier S, Lawson MA, Slack E, Kirundi JK, Stoel M, Heikenwalder M, et al. Reversible microbial colonization of germ-free mice reveals the dynamics of IgA immune responses. Science 2010;328:1705–9.

[227] Mesin L, Di Niro R, Thompson KM, Lundin KE, Sollid LM. Long-lived plasma cells from human small intestine biopsies secrete immunoglobulins for many weeks in vitro. J Immunol 2011;187:2867–74.

[228] Fritz JH, Rojas OL, Simard N, McCarthy DD, Hapfelmeier S, Rubino S, et al. Acquisition of a multifunctional IgA$^+$ plasma cell phenotype in the gut. Nature 2012;481:199–203.

[229] Hjelmstrom P. Lymphoid neogenesis: de novo formation of lymphoid tissue in chronic inflammation through expression of homing chemokines. J Leukoc Biol 2001;69:331–9.

[230] Shomer NH, Fox JG, Juedes AE, Ruddle NH. Helicobacter-induced chronic active lymphoid aggregates have characteristics of tertiary lymphoid tissue. Infect Immun 2003;71:3572–7.

[231] Lucchesi D, Bombardieri M. The role of viruses in autoreactive B cell activation within tertiary lymphoid structures in autoimmune diseases. J Leukoc Biol 2013;94:1191–9.

[232] Aloisi F, Pujol-Borrell R. Lymphoid neogenesis in chronic inflammatory diseases. Nat Rev Immunol 2006;6:205–17.

[233] Mazzucchelli L, Blaser A, Kappeler A, Scharli P, Laissue JA, Baggiolini M, et al. BCA-1 is highly expressed in *Helicobacter pylori*-induced mucosa-associated lymphoid tissue and gastric lymphoma. J Clin Invest 1999;104:R49–54.

[234] Luther SA, Lopez T, Bai W, Hanahan D, Cyster JG. BLC expression in pancreatic islets causes B cell recruitment and lymphotoxin-dependent lymphoid neogenesis. Immunity 2000;12:471–81.

[235] Luther SA, Ansel KM, Cyster JG. Overlapping roles of CXCL13, interleukin 7 receptor alpha, and CCR7 ligands in lymph node development. J Exp Med 2003;197:1191–8.
[236] Fan L, Reilly CR, Luo Y, Dorf ME, Lo D. Cutting edge: ectopic expression of the chemokine TCA4/SLC is sufficient to trigger lymphoid neogenesis. J Immunol 2000;164:3955–9.
[237] Chen SC, Vassileva G, Kinsley D, Holzmann S, Manfra D, Wiekowski MT, et al. Ectopic expression of the murine chemokines CCL21a and CCL21b induces the formation of lymph node-like structures in pancreas, but not skin, of transgenic mice. J Immunol 2002;168:1001–8.
[238] Cupedo T, Jansen W, Kraal G, Mebius RE. Induction of secondary and tertiary lymphoid structures in the skin. Immunity 2004;21:655–67.
[239] Mebius RE. Organogenesis of lymphoid tissues. Nat Rev Immunol 2003;3:292–303.
[240] Fu YX, Chaplin DD. Development and maturation of secondary lymphoid tissues. Annu Rev Immunol 1999;17:399–433.
[241] Wengner AM, Hopken UE, Petrow PK, Hartmann S, Schurigt U, Brauer R, et al. CXCR5- and CCR7-dependent lymphoid neogenesis in a murine model of chronic antigen-induced arthritis. Arthritis Rheum 2007;56:3271–83.
[242] Winter S, Loddenkemper C, Aebischer A, Rabel K, Hoffmann K, Meyer TF, et al. The chemokine receptor CXCR5 is pivotal for ectopic mucosa-associated lymphoid tissue neogenesis in chronic *Helicobacter pylori*-induced inflammation. J Mol Med (Berl) 2010;88:1169–80.
[243] von Herrath MG, Harrison LC. Antigen-induced regulatory T cells in autoimmunity. Nat Rev Immunol 2003;3:223–32.
[244] Berek C, Kim HJ. B-cell activation and development within chronically inflamed synovium in rheumatoid and reactive arthritis. Semin Immunol 1997;9:261–8.
[245] Schroder AE, Sieper J, Berek C. Antigen-dependent B cell differentiation in the synovial tissue of a patient with reactive arthritis. Mol Med 1997;3:260–72.
[246] Salomonsson S, Jonsson MV, Skarstein K, Brokstad KA, Hjelmstrom P, Wahren-Herlenius M, et al. Cellular basis of ectopic germinal center formation and autoantibody production in the target organ of patients with Sjogren's syndrome. Arthritis Rheum 2003;48:3187–201.
[247] Sims GP, Shiono H, Willcox N, Stott DI. Somatic hypermutation and selection of B cells in thymic germinal centers responding to acetylcholine receptor in myasthenia gravis. J Immunol 2001;167:1935–44.
[248] Kendall PL, Yu G, Woodward EJ, Thomas JW. Tertiary lymphoid structures in the pancreas promote selection of B lymphocytes in autoimmune diabetes. J Immunol 2007;178:5643–51.
[249] Mosnier JF, Degott C, Marcellin P, Henin D, Erlinger S, Benhamou JP. The intraportal lymphoid nodule and its environment in chronic active hepatitis C: an immunohistochemical study. Hepatology 1993;17:366–71.
[250] Murakami J, Shimizu Y, Kashii Y, Kato T, Minemura M, Okada K, et al. Functional B-cell response in intrahepatic lymphoid follicles in chronic hepatitis C. Hepatology 1999;30:143–50.
[251] Kuppers R, Klein U, Hansmann ML, Rajewsky K. Cellular origin of human B-cell lymphomas. N Engl J Med 1999;341:1520–9.
[252] Kuppers R. Mechanisms of B-cell lymphoma pathogenesis. Nat Rev Cancer 2005;5:251–62.
[253] Seifert M, Scholtysik R, Kuppers R. Origin and pathogenesis of B cell lymphomas. Methods Mol Biol 2013;971:1–25.
[254] Hopken UE, Rehm A. Homeostatic chemokines guide lymphoma cells to tumor growth-promoting niches within secondary lymphoid organs. J Mol Med (Berl) 2012;90:1237–45.
[255] Pals ST, de Gorter DJ, Spaargaren M. Lymphoma dissemination: the other face of lymphocyte homing. Blood 2007;110:3102–11.
[256] Drillenburg P, Pals ST. Cell adhesion receptors in lymphoma dissemination. Blood 2000;95:1900–10.
[257] Hopken UE, Rehm A. Homeostatic chemokines guide lymphoma cells to tumor growth-promoting niches within secondary lymphoid organs. J Mol Med Berl 2012;90:1237–45.
[258] Jalkanen S, Wu N, Bargatze RF, Butcher EC. Human lymphocyte and lymphoma homing receptors. Annu Rev Med 1987;38:467–76.
[259] Bargatze RF, Wu NW, Weissman IL, Butcher EC. High endothelial venule binding as a predictor of the dissemination of passaged murine lymphomas. J Exp Med 1987;166:1125–31.
[260] Burger JA. Chemokines and chemokine receptors in chronic lymphocytic leukemia (CLL): from understanding the basics towards therapeutic targeting. Semin Cancer Biol 2010;20:424–30.
[261] Burger JA. Chemokines and chemokine receptors in chronic lymphocytic leukemia (CLL): from understanding the basics towards therapeutic targeting. Semin Cancer Biol 2010;20:424–30.
[262] Smith JR, Braziel RM, Paoletti S, Lipp M, Uguccioni M, Rosenbaum JT. Expression of B-cell-attracting chemokine 1 (CXCL13) by malignant lymphocytes and vascular endothelium in primary central nervous system lymphoma. Blood 2003;101:815–21.
[263] Rodig SJ, Jones D, Shahsafaei A, Dorfman DM. CCR6 is a functional chemokine receptor that serves to identify select B-cell non-Hodgkin's lymphomas. Hum Pathol 2002;33:1227–33.
[264] Rehm A, Anagnostopoulos I, Gerlach K, Broemer M, Scheidereit C, Johrens K, et al. Identification of a chemokine receptor profile characteristic for mediastinal large B-cell lymphoma. Int J Cancer 2009;125:2367–74.
[265] Hopken UE, Foss HD, Meyer D, Hinz M, Leder K, Stein H, et al. Up-regulation of the chemokine receptor CCR7 in classical but not in lymphocyte-predominant Hodgkin disease correlates with distinct dissemination of neoplastic cells in lymphoid organs. Blood 2002;99:1109–16.
[266] Sextro M, Diehl V, Franklin J, Hansmann ML, Anagnostopoulos I, Marafioti T, et al. Lymphocyte predominant Hodgkin's disease – a workshop report. European Task Force on Lymphoma. Ann Oncol 1996;7(Suppl. 4):61–5.
[267] Kaufman D, Longo DL. Hodgkin's disease. Crit Rev Oncol Hematol 1992;13:135–87.
[268] Rehm A, Mensen A, Schradi K, Gerlach K, Wittstock S Winter S, et al. Cooperative function of CCR7 and lymphotoxin in the formation of a lymphoma-permissive niche within murine secondary lymphoid organs. Blood 2011;118:1020–33.
[269] Cyster JG. Homing of antibody secreting cells. Immunol Rev 2003;194:48–60.
[270] Damiano JS, Cress AE, Hazlehurst LA, Shtil AA, Dalton WS. Cell adhesion mediated drug resistance (CAM-DR): role of integrins and resistance to apoptosis in human myeloma cell lines. Blood 1999;93:1658–67.
[271] Nakayama T, Hieshima K, Izawa D, Tatsumi Y, Kanamaru A, Yoshie O. Cutting edge: profile of chemokine receptor expression on human plasma cells accounts for their efficient recruitment to target tissues. J Immunol 2003;170:1136–40.
[272] Burger JA, Kipps TJ. CXCR4: a key receptor in the crosstalk between tumor cells and their microenvironment. Blood 2006;107:1761–7.

[273] Hattermann K, Mentlein R, . An infernal trio: the chemokine CXCL12 and its receptors CXCR4 and CXCR7 in tumor biology. Ann Anat 2013;195:103–10.

[274] Lopez-Giral S, Quintana NE, Cabrerizo M, Alfonso-Perez M, Sala-Valdes M, De Soria VG, et al. Chemokine receptors that mediate B cell homing to secondary lymphoid tissues are highly expressed in B cell chronic lymphocytic leukemia and non-Hodgkin lymphomas with widespread nodular dissemination. J Leukoc Biol 2004;76:462–71.

[275] Trentin L, Cabrelle A, Facco M, Carollo D, Miorin M, Tosoni A, et al. Homeostatic chemokines drive migration of malignant B cells in patients with non-Hodgkin lymphomas. Blood 2004;104:502–8.

[276] Rehm A, Mensen A, Schradi K, Gerlach K, Wittstock S, Winter S, et al. Cooperative function of CCR7 and lymphotoxin in the formation of a lymphoma-permissive niche within murine secondary lymphoid organs. Blood 2011;118:1020–33.

[277] Thomas S, Baumgart DC, Targeting leukocyte migration and adhesion in Crohn's disease and ulcerative colitis. Inflammopharmacology 2012;20:1–18.

[278] Ghosh N, Chaki R, Mandal SC., Inhibition of selective adhesion molecules in treatment of inflammatory bowel disease. Int Rev Immunol 2012;31:410–27.

[279] Schrottner P, Leick M, Burger M. The role of chemokines in B cell chronic lymphocytic leukaemia: pathophysiological aspects and clinical impact. Ann Hematol 2010;89:437–46.

Chapter 13

# B Cells as Regulators

Ping Shen, Ellen Hilgenberg, Andreia C. Lino, Van Duc Dang, Stefanie Ries, Imme Sakwa and Simon Fillatreau
*Deutsches Rheuma-Forschungszentrum, Leibniz Institute, Berlin, Germany*

## 1. INTRODUCTION

B cells can contribute to immunity through production of antibodies, presentation of antigen to T cells, and secretion of cytokines. Through production of interleukin (IL)-10, B cells can limit immunopathology in mouse models of ulcerative colitis (UC), experimental autoimmune encephalomyelitis (EAE), and collagen-induced arthritis [1–3]. Although of potential benefit in immune-driven disorders, the suppressive functions of B cells can be deleterious during infectious diseases [4,5]. These B cell-mediated antiinflammatory functions are provided by activated rather than resting B cells. Indeed, B cells require signals provided by the B cell receptor (BCR), toll-like receptors (TLR), CD40, and the IL-21 receptor to achieve an efficient control of inflammatory diseases [2,6,7]. There is accumulating evidence that IL-10-producing B cells are in vivo a subset of $CD19^{+}CD138^{hi}$ plasmablasts or plasma cells expressing BLIMP1, in line with the fact that suppression is mediated by activated B cells. Human B cells might also regulate immunity through IL-10 production. Therefore, a better understanding of the suppressive functions of B cells might result in the development of novel strategies for the manipulation of the immune system in diseases. Here, we review current knowledge on the regulatory functions of B cells in mice, including the role of IL-35 as a mediator of B cell-mediated suppression, and on antibody-secreting cells (ASCs) as major sources of IL-10 and IL-35 in vivo. Finally, we discuss data available on the phenotype and function of IL-10-producing B cells in humans.

## 2. REGULATORY ROLE OF B CELLS IN UC

UC is a chronic inflammatory disease of the large intestine and a major subtype of inflammatory bowel disease (IBD) [8,9]. Indeed, predisposition to UC is strongly influenced by polymorphisms in immune-related genes [10]. For example, IL-10 protects against inflammatory diseases of the intestine, and severe IBD rapidly develops in patients with complete loss-of-function mutations in the *IL-10* receptor gene [11]. There is evidence that the development of inflammatory diseases of the intestine is associated with B cell hyperactivity; however, the roles of B cells during human IBD remain poorly understood [8].

Several mouse models of IBD have been developed based on disruption of selected genes, which facilitates investigating how immune cells and cytokines contribute to intestinal inflammation [12]. Mice deficient in the gene coding for the TCRα chain, which lack conventional $CD4^{+}$ T cells but harbor atypical $CD4^{+}$ T cells expressing TCR composed of TCRβ homodimers, spontaneously develop a chronic inflammatory disease of the intestine that shares some histological and immunological features with UC [13,14]. Signs of disease include hyperplasia of the lamina propria with infiltration of inflammatory cells, decrease of the number of goblet cells, elongation of crypts, and development of crypt abscesses [13,14]. The disease is associated with an increased differentiation of the TCRββ $CD4^{+}$ T cells present in these mice toward a $T_H2$-like phenotype, and IL-4 is a key disease-driving cytokine [15,16]. These T cells are essential for disease development because UC can be prevented by T cell depletion therapy using anti-TCRβ antibodies [14]. This disease also involves a B cell hyperactivation leading to markedly increased B and plasma cell numbers in the intestine [14] as well as heightened production of antibodies toward food antigens and autoantigens such as tropomyosin, DNA, and histones [14,17]. These B cell responses are largely T cell-dependent [14].

Before discussing the role of B cells in this model of UC, it is important to emphasize that this disease can be markedly influenced by changes at the interface between the immune system and the gut environment. In particular, it is modulated by environmental factors including microbiota and diet [18]. TCRα-deficient mice fed an elementary diet containing chemically defined amino acids, dextrin, a small amount of soybean oil, and vitamins but no protein did not develop IBD whereas 85% of the mice fed with

Molecular Biology of B Cells. http://dx.doi.org/10.1016/B978-0-12-397933-9.00013-8

a normal diet containing proteins from soybean, wheat, and fish developed colitis [18]. The feeding of mice with an elementary diet resulted in an almost complete disappearance of IL-4 production by CD4+ T cells in the colon lamina propria in comparison to mice that received a regular diet [18]. These differently fed mice also displayed distinct intestinal microbial contents. For example, 88% of sick TCRα-deficient mice fed a normal diet harbored *Bacteroides vulgatus*, but this bacteria was absent from the TCRα-deficient mice that received an elementary diet [18]. This bacteria might directly facilitate UC because its rectal administration in TCRα-deficient mice fed an elementary diet was sufficient to provoke the appearance of clinical signs of colitis such as hyperplasia of the lamina propria region, elongation of the crypts and microvilli, infiltration of inflammatory cells in the lamina propria, and a decrease in the number of goblet cells [18]. *Bacteroides*-reactive antibodies were detected in the sick mice, suggesting that B cells might control this microbial driver of the disease. Any mutation that may influence the control of such a microbe might consequently influence the development of UC.

The role of B cells in UC was investigated using TCRα-deficient mice on a B cell-deficient background [19]. Remarkably, in the absence of B cells, TCRα-deficient mice developed an earlier and more severe UC than TCRα-deficient mice with B cells [19]. The incidence of UC was 100% in the B cell-deficient mice at 8 weeks of age, when only a few of the mice with B cells had any disease [19]. Several mechanisms could account for the protective effect of B cells against UC. First, B cells might provide an important mechanism of host defense against intestinal microbes having pathogenic roles in UC such as *Bacteroides*. Second, a lack of B cells could lead to an aberrant accumulation and activation of atypical TCRββ CD4+ T cells, the homeostatic control mechanisms of which are unknown in these mice. Third, B cells could have an antiinflammatory function limiting the pathogenic activities of the atypical T cells driving UC. The establishment of this notion would require that the underlying molecular mechanism is uncovered.

A unique function of B cells is the production of antibodies. Antibodies play a critical role in host defense against microbes, and in TCRα-deficient mice lacking B cells the administration of antibodies purified from the serum of TCRα-deficient mice reduced UC incidence [19]. The beneficial effect of antibodies correlated with a reduction of the number of apoptotic cells in the colon and spleen of treated mice. One possible interpretation is that antibodies controlled a component of the gut microbiota that promoted the development of UC, and that the lower number of apoptotic cells in the less sick mice was a consequence of the milder disease. However, as suggested by the authors of that study, it was also possible that these antibodies promoted the clearance of apoptotic cells, which could promote an aggravation of inflammation if not appropriately and promptly removed from the host. In support of this latter possibility, the injection of five monoclonal antibodies reacting against colonic tissue also led to an amelioration of UC in the treated mice [19]. However, it was not excluded that their protective effect involved a direct effect on the microbiota.

B cells also mediate protective functions independent of antibodies in this disease [20]. Adoptive transfer of B cells limited UC development in healthy TCRα-deficient mice lacking B cells, and such cell therapy was also effective in mice with ongoing disease [20]. Indeed, administration of B cells into mice suffering from severe chronic colitis for approximately 10 weeks led to a rapid amelioration of disease within 10 days after treatment compared with untreated mice [20]. In contrast, transfer of serum antibodies could not improve UC course in mice with established disease, suggesting the existence of antibody-independent protective functions of B cells with therapeutic potential. Activated B cells mediated these activities because CD86 and CD40 (but not CD80) contributed positively to their protective effect [20].

This protective function of B cells was further characterized with an analysis of the B cell response in mice with UC. The B cell compartment undergoes a marked expansion (~20-fold increase in absolute cell number) in mesenteric lymph nodes (MLNs) in TCRα-deficient mice after colitis development [1]. The B cells found in MLNs from sick mice are $IgM^+CD23^{hi}CD21^{int}CD62L^{low}$ and homogenously express CD1d at a higher level than B cells from the MLNs of wild-type mice [1]. This increased expression of CD1d is also found on B cells from the colonic lamina propria but not the spleen, indicating that it corresponds to a local B cell response at the inflamed site [1]. CD1d upregulation was selectively observed on MLN B cells in sick but not in healthy TCRα-deficient mice, indicating a direct implication of colitis in this process. However, it is not a general consequence of intestinal inflammation. For instance, CD1d was not upregulated on B cells from MLNs of IL-10-deficient mice that also developed IBD [1]. In TCRα-deficient mice, CD1d directly contributed to the suppressive function of B cells, because unlike their CD1d-sufficient counterparts, B cells isolated from mice with a genetic defect in CD1d did not ameliorate the course of UC in recipient mice upon adoptive transfer [1]. A key function of CD1d was to promote an increased IL-10 expression in MLN B cells after the onset of UC [1]. This was essential for the B cell-mediated protective effect because IL-10-deficient B cells were unable to ameliorate UC upon administration in recipient mice [1]. The B cell-mediated protective pathway implicating CD1d and IL-10 operated independently of antibodies because a lack of CD1d did not have any major effect on antibody production in TCRα-deficient mice.

Collectively, these elegant analyses establish that in TCRα-deficient mice, during the development of UC, a population of CD1d$^{hi}$ B cells accumulates in MLNs, which limits the progression of the disease in an IL-10-dependent manner. Although this attenuated the progression of established disease, it had no effect on the disease incidence [1]. Disease incidence was similar in TCRα-deficient mice and in mice lacking TCRα and CD1d [1]. In contrast, disease incidence was markedly higher in TCRα-deficient mice lacking B cells [1]. Thus, B cells also provide a protection from UC onset via a mechanism that is CD1d-independent and could be related to antibodies, as discussed previously.

## 3. PROTECTIVE FUNCTION OF B CELLS IN EAE

Multiple sclerosis (MS) is a chronic demyelinating disease of the central nervous system (CNS) associated with an infiltration of leukocytes at lesion sites. In most cases, MS progression initially follows a relapsing-remitting course, indicating the existence of endogenous mechanisms that can transiently halt disease progression. Although its etiology is unknown, the immune system certainly plays an important role in this disease because immune-related genes influence susceptibility to MS [21]. EAE is the primary animal model to study the mechanisms regulating immune-mediated attacks of the CNS in MS. EAE can be induced in susceptible mouse strains by immunization with peptides from CNS proteins such as myelin oligodendrocyte glycoprotein (MOG). After immunization, activated myelin-reactive T cells accumulate in the CNS, where they recruit and activate myeloid cells, which ultimately leads to clinical disease.

The regulatory function of B cells in EAE was identified with the observation that mice having a deficiency in IL-10 expression only in B cells developed a chronic form of EAE after immunization with mouse MOG, whereas mice with wild-type B cells rapidly recovered from disease after a short episode of paralysis [2]. This difference in disease course was associated with a stronger autoreactive CD4$^+$ T cell response in mice with IL-10-deficient B cells compared with control mice, suggesting that the establishment of a chronic disease resulted from a stronger T cell-driven attack in the target organ [2]. In line with this notion, depletion of CD4$^+$ T cells in B cell-deficient mice when they reached the peak of EAE severity resulted in a rapid recovery whereas untreated mice developed a chronic disease [22]. In contrast, depletion of CD4$^+$ T cells at later stages failed to drive resolution of symptoms, suggesting that the chronic paralysis was then maintained independently of CD4$^+$ T cells. These results indicate that B cells limit via IL-10 the aggressiveness of the autoreactive T cell reaction, leading to a self-limiting episode of paralysis instead of an irreversible disease.

Understanding how such IL-10-mediated regulatory activities are induced in B cells might open novel therapeutic strategies to limit pathogenic inflammation in MS. This is especially pertinent because B cells from MS patients secreted less IL-10 compared with B cells from healthy individuals (see Section 6). Several signaling pathways have been identified that are crucial for the protective functions of B cells in EAE. The BCR plays an essential role because mice in which all B cells expressed a single BCR of irrelevant antigen specificity developed a chronic disease similar to mice with IL-10-deficient B cells [2]. The antigen specificity of the B cells involved in the suppression of disease was not defined, but it seems likely that autoreactive B cells were involved because B cells isolated from wild-type mice after recovery from EAE produced IL-10 upon restimulation with MOG in the presence of an agonistic antibody to CD40 in vitro [2]. Of note, mice with a CD40 deficiency restricted to B cells also developed chronic EAE, similar to mice with IL-10-deficient B cells [2]. BCR and CD40 signaling might contribute to the suppressive function of B cells by promoting their IL-10 secretion. However, coengagement of BCR and CD40 was insufficient to induce IL-10 secretion by naïve B cells in vitro, suggesting that a distinct signaling pathway initiated the IL-10-mediated suppressive activity of B cells. Agonists of TLR such as lipopolysaccharides (LPS) from Gram-negative bacteria were found to have a unique capacity to trigger IL-10 secretion by naïve B cells in vitro [6]. Moreover, mice with a B cell-restricted deficiency in the signaling adaptor MyD88, which plays a major role in signaling via all TLR except TLR3, developed a chronic disease [6]. TLR4 was subsequently identified as the main TLR implicated in the protective functions of B cells during this disease [23]. In contrast, mice with a MyD88 deficiency in all cells were completely resistant to EAE, indicating that this adaptor controlled the initiation and the resolution of EAE by signaling in distinct cell types [6].

Collectively, the data discussed above on the signals controlling the regulatory functions of B cells suggest a two-step model of B cell-mediated suppression: B cells are first primed in an innate manner through TLR, which initiates their expression of IL-10, and then in a second step, those primed B cells expressing a MOG-reactive BCR capture the autoantigen and engage in cognate interaction with autoreactive CD4$^+$ T cells [6,24]. During this interaction, these T cells stimulate through triggering of CD40 the clonal expansion of the MOG-reactive B cells and amplify their production of IL-10. According to this model, the first suppression step is not antigen-restricted but instructed by the amount of TLR agonists present, and the second phase of antigen-specific suppression is directly dependent on the number of autoreactive CD4$^+$ T cells available. Therefore, B cells would provide a mechanism of immune counter-regulation directly proportional to the strength of the inflammatory pathways. In line with this model, a recent study identified a molecular mechanism that is selectively involved in the second phase of IL-10 expression and is essential for the protective function of

B cells during EAE. The calcium sensor stromal interaction molecules (STIM) 1 and STIM2 selectively contribute to the amplification of IL-10 expression in B cells after BCR stimulation, but they do not regulate their initial IL-10 secretion after TLR engagement [25]. As expected, mice with a B cell-restricted deficiency in STIM1 and STIM2 developed more severe EAE than control mice [25]. Further support for a two-step model was provided by experiments in which B cells were first activated in vitro before adoptive transfer in recipient mice to assess their protective value [7]. The transferred B cells had to express CD40 and major histocompatibility complex-II (MHC-II) to suppress EAE in recipient mice [7]. A possible explanation is that the few B cells of relevant antigenic specificity in these adoptive transfer experiments (polyclonal B cell preparations were injected) had to expand in the recipient mice to reach a sufficient number to ameliorate the disease course. Supporting this explanation, it was also found that MHC-II was important on B cells for their protective function against UC, but it was shown to be required for the expansion of the transferred B cells in recipient mice rather than for their suppressive activity per se because B cells could regulate disease independently of MHC-II provided that a higher number of cells were administered [1]. In addition to TLR, BCR, CD40, and MHC-II, the IL-21 receptor also contributes to the protective activity of B cells in EAE [7]. Remarkably, IL-21 is implicated in B cell differentiation into ASCs [26,27], and plasmablasts/plasma cells are the main subset of IL-10 expressing B cells in EAE [23]. The fact that B cells engaged in terminal ASC differentiation provide IL-10 is consistent with the fact that all of the major pathways involved B cell activation (BCR, CD40, TLR) contribute positively to their regulatory activities in vivo.

The cell type targeted by B cell-derived IL-10, and responsible for their beneficial effect in EAE, has not yet been identified. In addition to IL-10-producing B cells, $CD4^+Foxp3^+$ regulatory T cells (Tregs) are also involved in remission from EAE, and Treg-depleted mice develop chronic EAE similar to that observed in mice in which only B cells cannot express IL-10 [28]. Tregs are the predominant cell type producing IL-10 in the CNS during EAE, suggesting that they promote remission by limiting inflammation in the target tissue [28]. This led to the notion that B cells might facilitate recovery from EAE by stimulating the suppressive function of Tregs. However, B cell-deficient mice displayed normal activation of Tregs in secondary lymphoid organs, normal accumulation of Tregs in CNS, and a normal increase in the suppressive function of Tregs in the CNS during EAE [29]. Likewise, the depletion of B cells before EAE induction, which resulted in development of chronic disease, did not negatively affect the Tregs compartment during EAE [30]. The notion that Tregs can undergo efficient activation in the absence of B cells is further supported by clinical studies showing that patients treated by B cell-depletion therapy did not display any reduction in Tregs cell number during the period of B cell depletion. In fact, some patients even showed increased Tregs cell frequencies and numbers in peripheral blood [31,32]. Taken together, these studies suggest that Tregs cells can function normally in the absence of B cells, although they might be insufficient to induce disease remission in this case. An alternative candidate cell type that could be an important target of IL-10 produced by B cells is dendritic cells (DCs). IL-10 has potent suppressive activities on DCs [33], and IL-10 produced by TLR-activated B cells inhibits the expression of the proinflammatory cytokines IL-6, IL-12, IL-23, and tumor necrosis factor-α (TNF-α) by DCs in vitro [6]. Moreover, DCs from B cell-deficient mice expressed higher amounts of IL-12 and promoted stronger $T_H1$ responses than DCs from wild-type mice [34]. Interestingly, plasmablasts/plasma cells accumulate together with some DCs in extrafollicular foci in spleen, suggesting that these areas could represent the locale where B cell-mediated suppression takes place [35]. The determination of the localization of IL-10-producing plasmablasts/plasma cells, and of their proximal cellular microenvironment in vivo, should facilitate identifying how they regulate disease.

The protective function of B cells is more complex during EAE. B cells do not only limit disease severity via IL-10, but also through provision of IL-35 [23]. IL-35 is a heterodimeric cytokine of the IL-12 cytokine family made of p35 and Epstein Barr-induced 3 (EBi3) [23]. It seems associated with immune suppression because it was first discovered in trophoblast extracts from human placenta [36] and was later found to be expressed by Tregs cells and to be essential for their suppressive functions [37,38]. B cells constitutively expressed *Il-12p35* mRNA and upregulated *Ebi3* expression, and they secreted IL-35 upon stimulation via CD40 plus TLR4 in vitro [23]. It is worth noting that B cells did not upregulate EBi3 expression after stimulation via TLR4 alone, underlying the unique role of CD40 in induction of IL-35 expression. Mice in which only B cells lacked either p35 or EBi3 developed a markedly exacerbated EAE after immunization with mouse MOG compared with mice with wild-type B cells. Because p35 can also associate with p40 to form IL-12, and EBi3 can pair with p28 to make IL-27, mice with deficiencies in p40 or p28 restricted to B cells were additionally used as controls. These mice displayed disease courses comparable to those of control mice with wild-type B cells, indicating that B cells did not regulate EAE through provision of IL-12 or IL-27 [23]. These results emphasized the unique protective role of IL-35 from B cells in limitation of disease severity [23]. To address whether single B cells had to express IL-10 and IL-35 to achieve a protective effect or whether cell-extrinsic cytokine is sufficient, mice were developed in which half of the B cells could make IL-10 but not IL-35 whereas the other half could produce IL-35 but not IL-10 [23]. These mice displayed a normal EAE course, indicating that B cells could achieve a suppressive effect when independently providing IL-10 and IL-35. Remarkably, plasmablasts/plasma cells were the main B cell subtype expressing the two IL-35 subunits EBi3 and p35. These data demonstrate that some plasmablasts/plasma cells have key regulatory roles through provision of the antiinflammatory cytokines IL-10 and IL-35 during EAE.

## 4. REGULATORY ROLES OF B CELLS IN SYSTEMIC LUPUS ERYTHEMATOSUS

Systemic lupus erythematosus (SLE) is a systemic autoimmune disease that can affect multiple vital organs such as brain, blood, and kidney [39]. SLE involves several immunologic disturbances, among which B cell hyperactivity is prominent [40]. Indeed, active SLE is associated with the expansion of a unique population of $CD19^{+}CD20^{-}CD27^{hi}$ plasma cells in blood [41], and the levels of circulating $CD27^{hi}$ plasma cells correlate with disease activity [40]. IL-10 has important roles in SLE, and variations in the *IL-10* gene have repeatedly been associated with susceptibility to SLE [42,43]. The frequency of IL-10 expressing B cells was found to be higher in SLE patients than in healthy individuals [44]. Although the pathogenic roles of IL-10 might be more prominent than its protective functions in some SLE patients, as suggested by the positive outcome in one IL-10 blocking therapy trial [45], it is likely that IL-10 differentially affects the disease course depending on where, when, and by which cell type it is produced. Several mouse models of SLE have been used to explore the physiological relevance of IL-10 production by B cells in this disease.

Mice with a null mutation in the *Lyn* gene spontaneously develop a chronic SLE-like autoimmune disease characterized by autoantibody production, lymphocyte activation, immune complex deposition, and nephritis [46]. IL-10 production is increased during development of the disease and has key protective roles because mice deficient in both Lyn and IL-10 develop enhanced splenomegaly and lymphadenopathy (primarily due to accumulation of myeloid and T cells); increased leukocyte infiltration in kidney, liver, and lungs; and a markedly reduced life span [47]. Remarkably, an analysis of the cell types expressing IL-10 during the course of the disease using IL-10 reporter mice revealed that B cells were the major source of this cytokine, in absolute numbers and in frequencies, compared with T cells and myeloid cells [47]. Most IL-10.eGFP-expressing B cells had a plasmablast/plasma cell phenotype with high expression of CD138 [47]. The induction of such IL-10-producing plasmablasts/plasma cells was under the control of the environment because wild-type B cells developed into IL-10 expressing cells, most of them (>60%) with a plasmablast/plasma cell phenotype, upon adoptive transfer in recipient Lyn-deficient mice [47]. Importantly, adoptive transfer of wild-type B cells resulted in an amelioration of disease signs including splenomegaly, myeloid cell accumulation, and T cell activation in recipient Lyn/IL-10 double-deficient mice, demonstrating that IL-10-secreting B cells can ameliorate the disease course in the absence of any other IL-10-expressing cell type. In contrast, adoptive transfer of Tregs cells had no effect on the disease course. Collectively, these data show that B cells can suppress a spontaneous systemic autoimmune disease through provision of IL-10.

The mode of action of B cell-derived IL-10 was not investigated in detail in this study, but it correlated with an alteration of the Tregs phenotype. Mice lacking Lyn and IL-10 had more Tregs than Lyn-deficient mice, but these Tregs expressed lower Foxp3 levels than their counterparts in Lyn-deficient mice [47]. Adoptive transfer of B cells reverted this downregulation of Foxp3. This could reflect a direct action of B cells on Tregs cells, or an indirect effect via modulation of inflammation. Myeloid cells from mice with deficiency in Lyn and IL-10 displayed increased IL-6 production compared with Lyn-deficient mice [47], which could result in a downregulation of Foxp3 expression and a loss of regulatory functions in Tregs cells [48]. IL-6 secretion by DC can also inhibit the regulatory function of Tregs cells [49]. The importance of IL-6 in this disease is emphasized by the fact that IL-6 deficiency abrogates tissue damage in Lyn-deficient mice [50].

B cells are not always a source of protective IL-10 in SLE. IL-10 also has protective functions in the $MRL.Fas^{lpr}$ mouse model of SLE [51]. Analyses done using an IL-10 reporter mouse strain indicated that B cells were a minor source of IL-10 in this disease in comparison to T cells and macrophages [52]. Indeed, approximately 80% of IL-10 producing cells were T cells, 4–10% were macrophages, and only 1–4% were B cells [52]. Within B cells, IL-10 was most commonly found in plasmablasts/plasma cells [52]. The effect of IL-10 expression by B cells in that model was addressed using mice with a Cre-mediated deletion of the *IL-10* gene based on a floxed allele [52]. Deletion of IL-10 expression in B cells did not affect any of the disease signs examined or the survival of the mice [52]. Nonetheless, there is evidence that IL-10-producing B cells can ameliorate the disease course in this SLE model because adoptive transfer of B cells activated to express IL-10 with an agonistic antibody to CD40 markedly improved proteinuria and survival of recipient $MRL.Fas^{lpr}$ mice [53]. This B cell-mediated effect was associated with a reduced production of interferon-γ (IFN-γ) and an increased secretion of IL-10 by T cells as well as a reduced level of antibody deposition in the kidney [53]. B cells isolated from $MRL.Fas^{lpr}$ mice did not have any beneficial effect if not activated via CD40 before transfer, suggesting that B cells do not receive the signals necessary for their differentiation into protective IL-10-expressing cells in these mice in vivo [53]. This interpretation is consistent with the observation that only a minority of B cells naturally expresses IL-10 in this model.

Taken together, these data demonstrate that B cells can limit the progression of a spontaneous systemic autoimmune disease depending on the proportion of B cells induced to express IL-10. The comparison of these distinct SLE models, which are respectively associated with high and low frequencies of IL-10-producing B cells, might provide an opportunity to identify the parameters controlling the induction of antiinflammatory activities in B cells during natural autoimmune diseases.

## 5. REGULATORY ROLE OF B CELLS IN BACTERIAL INFECTIONS

The implication of intrinsic TLR signaling in the suppressive functions of B cells, and the observation that B cells secreted IL-10 in a TLR-dependent manner upon exposure to bacteria or microbes [4,54,55], suggested that B cell-mediated regulation might play a role in infectious diseases. This has been examined in infections with *Salmonella typhimurium* and *Listeria monocytogenes*. These infections remain the leading causes of foodborne illnesses in humans. They usually result in a self-limiting gastroenteritis, but they can have fatal consequences in some cases. For example, the bacterium *Salmonella typhi* can cause typhoid fever, which is associated with approximately 600,000 deaths each year [56]. *Listeria* infection can lead to meningitis in immunocompromised individuals and to stillbirth in pregnant women [57].

The Gram-negative bacteria *S. typhimurium* triggered B cells to produce IL-10 via a TLR2/TLR4 and MyD88 dependent-mechanism in vitro, and IL-10 expressing B cells appeared in the spleen of infected mice at 24 h after infection [4]. Remarkably, B cells were the main source of IL-10 at this stage, and all IL-10-expressing cells also expressed the cell surface molecule CD138, which is a marker of plasmablasts and plasma cells [4]. This B cell-derived IL-10 had a significant effect on the host response because mice with an IL-10-deficiency restricted to B cells displayed enhanced activation of NK cells and neutrophils, as well as prolonged survival, after *Salmonella* infection compared with mice with wild-type B cells [4]. Similar observations were made in mice with a deficiency in MyD88 restricted to B cells, indicating that this plasma cell response was directly controlled by intrinsic TLR signaling. In line with this, the frequency of CD19$^+$CD138$^+$ B cells expressing IL-10 was proportional to the amount of bacteria administered to the mice. IL-10 expression by B cells also regulated the *Salmonella*-reactive CD4$^+$ T cell response because mice with an IL-10-deficiency restricted to B cells displayed a two-fold increase in their number of *Salmonella*-reactive IFN-γ- or TNF-α-producing CD4$^+$ T cells compared with controls after challenge. Similar observations were made in mice carrying a MyD88-deficiency restricted to B cells.

IL-35 also contributed to the suppressive function of B cells in mice infected with *Salmonella* [23]. Thus, mice with a B cell-restricted deficiency in either p35 or EBi3 displayed an improved survival compared with mice with control B cells, which was associated with a stronger accumulation of macrophages in spleen [23]. Thus, B cells use IL-10 and IL-35 to regulate anti-*Salmonella* immunity [23]. The B cell subsets involved in this process were further analyzed by tracking the phenotype of IL-35-expressing cells in mice infected with *Salmonella* [23]. Remarkably, *Ebi3* mRNA was selectively upregulated in CD19$^+$CD138$^{hi}$ plasma cells, but not in CD19$^+$CD138$^-$ B cells, in the spleen of infected mice. Subdivision of plasma cells into distinct subsets according to their expression levels of CD138 and CD22 indicated that only the most mature CD138$^{hi}$CD22$^-$ and CD138$^{hi}$CD22$^+$ cells, which had the highest capacity to produce antibodies and contained the highest level of *Blimp-1* and *Irf4* mRNA, coexpressed *Ebi3* and *p35* mRNA at the single cell level [23]. As expected, all of the cells coexpressing *Ebi3* and *p35* also cotranscribed Blimp-1 in single-cell polymerase chain reaction analyses [23]. Only very few plasmablasts/plasma cells cotranscribed *Il-10*, *p35*, and *Ebi3*, suggesting that distinct sets of cells provided IL-10 and IL-35 during *Salmonella* infection.

The antiinflammatory role of B cells was also observed in mice infected by the Gram-positive bacteria *L. monocytogenes* [5]. After infection with virulent *Listeria*, B cell-deficient mice survived better than control wild-type mice, which correlated with better control of the bacteria on day 3 after infection in spleen and liver [5]. In wild-type mice, the number of bacteria inversely correlated with the amount of IL-10 in spleen on day 1.5 after challenge, suggesting that IL-10 plays a major role in this disease. Splenocytes from B cell-deficient mice produced less IL-10 and more IFN-γ than control mice [5]. The B cell subset implicated in this phenomenon was further analyzed using adoptive transfer of selected B cell subsets in recombination activating gene-deficient mice [5]. Adoptive transfer of marginal zone B cells resulted in increased IL-10 levels and a 10-fold increase in bacteria number, which did not happen when follicular B cells (inefficient IL-10 producers) were instead administered. Marginal zone B cells similarly reduced via an IL-10-dependent mechanism the control of the bacteria in IL-10-deficient mice [5]. These data show that marginal zone B cells (or a differentiated product of these cells) suppressed via IL-10 the innate response to *Listeria* via a mechanism independent of Tregs cells or other IL-10-producing cell types. Further supporting this concept, mice with a selective deficiency in marginal zone B cells due to an ablation of the gene coding for Notch2 selectively in B cells were also more resistant to *Listeria* infection than wild-type mice [5]. Remarkably, mice selectively deficient in marginal zone B cells had a significantly reduced number of bacteria in the spleen already on day 1.5 after challenge compared with controls, highlighting that the B cell-mediated regulatory process occurs early during the course of the disease. On day 3 after infection, a deficiency in marginal zone B cells resulted in a 70-fold reduction in the number of bacteria in the spleen compared with control mice [5]. These data are consistent with those obtained in *Salmonella* infection, indicating that B cell-mediated suppression starts early after challenge, can affect innate immune cells without involving any other IL-10-producing cell type or Tregs cells, and is a function of plasmablasts/plasma cells possibly differentiated from marginal zone B cells.

## 6. CHARACTERIZATION AND FUNCTION OF IL-10-PRODUCING B CELLS IN HUMANS

Human B cell lines and normal B cells can produce IL-10 [58,59] and subsequently inhibit the activation of T and myeloid cells in vitro [60,61]. The notion that human B cells may have regulatory activities in vivo could explain the observation that in a cohort of patients with rheumatoid arthritis, individuals with a higher number of circulating B cells displayed a milder course of disease [62].

One of the first studies highlighting a reduced production of IL-10 by B cells from patients with an autoimmune disease compared with healthy donors was done in the context of MS [63]. Reduced IL-10 production was observed upon B cell activation via CD40, or BCR plus CD40 [63], or TLR9 [64], indicating a general modification of the function of B cells in this disease rather than a defect in a specific stimulation pathway. B cells from patients with type 1 diabetes were also found to produce less IL-10 than B cells from healthy subjects upon activation [65]. The notion that defects in this B cell regulatory process facilitate the progression of these diseases is in agreement with the fact that conditions associated with increased IL-10 production by B cells were associated with a less severe MS course. For example, patients with relapsing-remitting MS who were therapeutically infected by helminth parasites displayed a higher production of IL-10 by B cells and a lower frequency of MS flares compared with uninfected patients [66,67]. Moreover, some medications currently used to treat MS were found to stimulate IL-10 expression in human B cells. The MS drugs interferon β-1b and mitoxantrone promoted an increased expression of IL-10 by B cells from healthy individuals or MS patients in vitro [63,68]. It is believed that defects in vitamin D can promote MS, and 1,25-dihydroxyvitamin $D_3$ (calcitriol) increased the secretion of IL-10 by human B cells in vitro [69]. Although these data should be interpreted prudently because they indirectly address the role of IL-10 production by B cells in autoimmunity, it is conceivable that novel drugs could be developed to specifically promote the IL-10-mediated suppressive functions of B cells in patients with autoimmune diseases such as MS.

The notion that human B cells might have regulatory functions providing protection from autoimmune diseases might explain why some exacerbations of inflammatory disease were observed in certain rare patients treated with the B cell-depleting agent rituximab. For example, B cell-depletion therapy was associated with a severe exacerbation of disease, which temporally correlated with depletion of B cells and IL-10 from the intestinal mucosa, in a patient with UC [70]. It also led to unwanted effects in some patients suffering from neuropathy [71,72], to the development of UC in a patient with Grave's disease, and to the development of psoriasis in some rheumatoid arthritis or SLE patients [73,74]. Likewise, administration of rituximab to renal transplant patients on days 0 and 7 led to acute rejection within the first weeks after transplantation in five of the six patients who received rituximab, leading to a termination of this study, which was supposed to enroll 120 patients [75]. In contrast, only one of seven patients who received the anti-CD25 daclizumab as "control antibody" developed such a rejection reaction [75]. It is important to note that the acute rejection rate was higher in the rituximab-treated group (83%) than in previous studies for patients who did not receive any induction therapy [76]. A possible explanation is that such treatment depleted B cells with antiinflammatory activities, subsequently facilitating the induction of a deleterious T cell response.

The characterization of the phenotype of the human IL-10-producing B cells should facilitate assessing its function and relevance in immune-mediated diseases. This has been approached using two types of approaches. First, B cell subsets were isolated according to expression of cell surface receptors classically used to separate previously known B cell subpopulations and were subsequently stimulated to compare their capacity to produce IL-10. Second, B cells were separated into IL-10 secretors and nonsecretors using a cytokine secretion capture assay, and then were compared using global transcriptome approaches without any *a priori* assumptions concerning their phenotype. Using the first approach, it was found that $CD19^+CD27^-$ naïve-like B cells secreted more IL-10 than $CD19^+CD27^+$ B cells after 48 h of stimulation via BCR and CD40 [63]. Within $CD19^+CD27^-$ B cells, those carrying a $CD24^{hi}CD38^{hi}$ phenotype produced approximately two-fold more IL-10 after stimulation via CD40 for 72 h than their $CD24^{int}CD38^{int}$ counterparts [77]. In contrast, $CD24^{hi}CD27^+$ B cells were the dominant IL-10-secreting cell type when peripheral blood B cells were stimulated for 5 h with LPS in the presence of phorbol 12-myristate 13-acetate plus ionomycin and subsequently analyzed by flow cytometry [60]. The phenotype of the most effective IL-10-producing B cells might also differ according to the organ sampled. In the spleen, B cells competent for IL-10 production were predominantly $CD27^+$, but they did not differentially express CD24 compared with IL-10 negative cells [60]. Distinct results were also obtained using an IL-10 cytokine capture assay. In this case after TLR9 stimulation, B cells secreting IL-10 did not differ from non-secreting cells by their expression of CD24, CD27, or CD38 but rather by their higher expression of CD25 and the transferrin receptor CD71 as well as lower levels of ecto-5-prime nucleotidase CD73 [78]. To identify the molecular properties associated with the acquisition of IL-10 expression, a distinct study separated IL-10-positive from negative cells after activation with anti-CD40, CpG, and IL-4 for 48 h to compare their global transcriptomes [79]. Remarkably, 30% of the transcripts upregulated in IL-10-positive B cells were associated with ASC differentiation (e.g., prdm1) whereas genes coding for B cell identity (e.g., Pax5) were down-regulated [79]. Thus, the acquisition of IL-10 expression by human B cells might be associated with their differentiation toward a plasmablast/plasma cell phenotype, as previously observed for mouse cells [4]. The possibility that human and mouse B cell differentiation into IL-10-producing cells follows shared processes is further supported by the fact that IL-21 can promote IL-10 expression and ASC differentiation in human and mouse B cells [80].

## 7. CONCLUDING REMARKS

It is now accepted that B cells can regulate immunity through the provision of IL-10 in mice, and evidence is accumulating that a similar pathway may be present in humans. Therefore, it is important to elucidate the relevance of these processes in human diseases. B cell-mediated suppression might function similarly in humans and mice. In mouse strains that spontaneously develop SLE, it was found that the frequency of IL-10-producing B cells varied markedly depending on the autoimmune disease model. Expectedly, B cell-mediated suppression had a significant effect on the disease course when B cells were a dominant IL-10-producing cell type whereas it had little effect when IL-10-producing B cells were a minor population. These observations indicate that B cells can acquire effector functions without adopting a regulatory profile. Moreover, a reduced production of IL-10 by B cells might facilitate the progression of the disease because B cells could achieve a therapeutic effect via an IL-10-dependent mechanism upon adoptive transfer in recipient mice with few endogenous IL-10-producing cells.

The factors controlling whether B cells will produce IL-10 or not upon activation are incompletely understood. The signals currently identified as important for eliciting suppressive functions in B cells include BCR, TLR, CD40, and IL-21 receptor. These pathways are also known for their key roles in humoral immunity. The notion that the regulatory activities of B cells might be intimately linked to some aspects of B-cell-mediated immunity is further supported by the recent characterization of IL-10-producing B cells as plasmablasts/plasma cells in several disease models. Likewise, plasmablasts/plasma cells were the main source of B cell-derived IL-35 in EAE and *Salmonella* infection. Why do some plasmablasts/plasma cells produce IL-10 or IL-35 and other do not? It will be important to elucidate the molecular and cellular mechanisms underlying these different outcomes.

The observation that plasmablasts/plasma cells were the main source of B cell-derived IL-10 is intriguing considering that IL-10 can be a potent stimulatory factor for ASC differentiation. IL-10-production by ASC could have deleterious effect in some autoimmune diseases depending on how IL-10-expressing ASC contribute to the pathogenic humoral response. The same reasoning applies to IL-35, which is relatively poorly known. The characterization of the antigen specificity of IL-10- and IL-35-producing B cells and the properties of the antibodies they produce will be critical to assess their various possible effects on diseases.

## CONFLICT OF INTEREST

The authors declare no financial or commercial conflict of interest.

## ACKNOWLEDGMENTS

The laboratory of S.F. is supported by grants from the Deutsche Forschungsgemeinschaft (SFB-650, TRR-36, TRR-130, FI-1238/02), Hertie Stiftung, and an advanced grant from the Merieux Institute. The authors thank Randall Lindquist for critical reading of the manuscript.

## REFERENCES

[1] Mizoguchi A, Mizoguchi E, Takedatsu H, Blumberg RS, Bhan AK. Chronic intestinal inflammatory condition generates IL-10-producing regulatory B cell subset characterized by CD1d upregulation. Immunity 2002;16:219–30.

[2] Fillatreau S, Sweenie CH, McGeachy MJ, Gray D, Anderton SM. B cells regulate autoimmunity by provision of IL-10. Nat Immunol 2002;3:944–50.

[3] Mauri C, Gray D, Mushtaq N, Londei M. Prevention of arthritis by interleukin 10-producing B cells. J Exp Med 2003;197:489–501.

[4] Neves P, Lampropoulou V, Calderon-Gomez E, Roch T, Stervbo U, Shen P, et al. Signaling via the MyD88 adaptor protein in B cells suppresses protective immunity during *Salmonella typhimurium* infection. Immunity 2010;33:777–90.

[5] Lee CC, Kung JT. Marginal zone B cell is a major source of Il-10 in *Listeria monocytogenes* susceptibility. J Immunol 2012;189:3319–27.

[6] Lampropoulou V, Hoehlig K, Roch T, Neves P, Calderon Gomez E, Sweenie CH, et al. TLR-activated B cells suppress T cell-mediated autoimmunity. J Immunol 2008;180:4763–73.

[7] Yoshizaki A, Miyagaki T, DiLillo DJ, Matsushita T, Horikawa M, Kountikov EI, et al. Regulatory B cells control T-cell autoimmunity through IL-21-dependent cognate interactions. Nature 2012;491:264–8.

[8] Abraham C, Cho JH. Inflammatory bowel disease. N Engl J Med 2009;361:2066–78.

[9] Bamias G, Kaltsa G, Ladas SD. Cytokines in the pathogenesis of ulcerative colitis. Discov Med 2011;11:459–67.

[10] Barrett JC, Hansoul S, Nicolae DL, Cho JH, Duerr RH, Rioux JD, et al. Genome-wide association defines more than 30 distinct susceptibility loci for Crohn's disease. Nat Genet 2008;40:955–62.

[11] Glocker EO, Kotlarz D, Boztug K, Gertz EM, Schaffer AA, Noyan F, et al. Inflammatory bowel disease and mutations affecting the interleukin-10 receptor. N Engl J Med 2009;361:2033–45.

[12] Strober W, Ehrhardt RO. Chronic intestinal inflammation: an unexpected outcome in cytokine or T cell receptor mutant mice. Cell 1993;75:203–5.

[13] Mombaerts P, Mizoguchi E, Grusby MJ, Glimcher LH, Bhan AK, Tonegawa S. Spontaneous development of inflammatory bowel disease in T cell receptor mutant mice. Cell 1993;75:274–82.

[14] Takahashi I, Kiyono H, Hamada S. $CD4^+$ T-cell population mediates development of inflammatory bowel disease in T-cell receptor alpha chain-deficient mice. Gastroenterology 1997;112:1876–86.

[15] Mizoguchi A, Mizoguchi E, Bhan AK. The critical role of interleukin 4 but not interferon gamma in the pathogenesis of colitis in T-cell receptor alpha mutant mice. Gastroenterology 1999;116:320–6.

[16] Mizoguchi A, Mizoguchi E, Chiba C, Spiekermann GM, Tonegawa S, Nagler-Anderson C, et al. Cytokine imbalance and autoantibody production in T cell receptor-alpha mutant mice with inflammatory bowel disease. J Exp Med 1996;183:847–56.

[17] Takahashi I, Iijima H, Katashima R, Itakura M, Kiyono H. Clonal expansion of $CD4^+$ $TCR\beta\beta^+$ T cells in TCR alpha-chain- deficient mice by gut-derived antigens. J Immunol 1999;162:1843–50.

[18] Kishi D, Takahashi I, Kai Y, Tamagawa H, Iijima H, Obunai S, et al. Alteration of V beta usage and cytokine production of $CD4^+$ $TCR\beta\beta$ homodimer T cells by elimination of *Bacteroides vulgatus* prevents colitis in TCR alpha-chain-deficient mice. J Immunol 2000;165:5891–9.

[19] Mizoguchi A, Mizoguchi E, Smith RN, Preffer FI, Bhan AK. Suppressive role of B cells in chronic colitis of T cell receptor alpha mutant mice. J Exp Med 1997;186:1749–56.

[20] Mizoguchi E, Mizoguchi A, Preffer FI, Bhan AK. Regulatory role of mature B cells in a murine model of inflammatory bowel disease. Int Immunol 2000;12:597–605.

[21] Oksenberg JR, Baranzini SE, Sawcer S, Hauser SL. The genetics of multiple sclerosis: SNPs to pathways to pathogenesis. Nat Rev Genet 2008;9:516–26.

[22] Hoehlig K, Shen P, Lampropoulou V, Roch T, Malissen B, O'Connor R, et al. Activation of $CD4^+$ $Foxp3^+$ regulatory T cells proceeds normally in the absence of B cells during EAE. Eur J Immunol 2012;42:1164–73.

[23] Shen P, Roch T, Lampropoulou V, O'Connor RA, Stervbo U, Hilgenberg E, et al. IL-35-producing B cells are critical regulators of immunity during autoimmune and infectious diseases. Nature 2014;507:366–70.

[24] Lampropoulou V, Calderon-Gomez E, Roch T, Neves P, Shen P, Stervbo U, et al. Suppressive functions of activated B cells in autoimmune diseases reveal the dual roles of toll-like receptors in immunity. Immunol Rev 2010;233:146–61.

[25] Matsumoto M, Fujii Y, Baba A, Hikida M, Kurosaki T, Baba Y. The calcium sensors STIM1 and STIM2 control B cell regulatory function through interleukin-10 production. Immunity 2011;34:703–14.

[26] Ozaki K, Spolski R, Ettinger R, Kim HP, Wang G, Qi CF, et al. Regulation of B cell differentiation and plasma cell generation by IL-21, a novel inducer of Blimp-1 and Bcl-6. J Immunol 2004;173:5361–71.

[27] Ettinger R, Sims GP, Fairhurst AM, Robbins R, da Silva YS, Spolski R, et al. IL-21 induces differentiation of human naive and memory B cells into antibody-secreting plasma cells. J Immunol 2005;175:7867–79.

[28] McGeachy MJ, Stephens LA, Anderton SM. Natural recovery and protection from autoimmune encephalomyelitis: contribution of $CD4^+CD25^+$ regulatory cells within the central nervous system. J Immunol 2005;175:3025–32.

[29] Calderon-Gomez E, Lampropoulou V, Shen P, Neves P, Roch T, Stervbo U, et al. Reprogrammed quiescent B cells provide an effective cellular therapy against chronic experimental autoimmune encephalomyelitis. Eur J Immunol 2011;41:1696–708.

[30] Matsushita T, Yanaba K, Bouaziz JD, Fujimoto M, Tedder TF. Regulatory B cells inhibit EAE initiation in mice while other B cells promote disease progression. J Clin Invest 2008;118:3420–30.

[31] Feuchtenberger M, Muller S, Roll P, Waschbisch A, Schafer A, Kneitz C, et al. Frequency of regulatory T cells is not affected by transient B cell depletion using anti-CD20 antibodies in rheumatoid arthritis. Open Rheumatol J 2008;2:81–8.

[32] Sfikakis PP, Souliotis VL, Fragiadaki KG, Moutsopoulos HM, Boletis JN, Theofilopoulos AN. Increased expression of the FoxP3 functional marker of regulatory T cells following B cell depletion with rituximab in patients with lupus nephritis. Clin Immunol 2007;123:66–73.

[33] Moore KW, de Waal Malefyt R, Coffman RL, O'Garra A. Interleukin-10 and the interleukin-10 receptor. Annu Rev Immunol 2001;19: 683–765.

[34] Moulin V, Andris F, Thielemans K, Maliszewski C, Urbain J, Moser M. B lymphocytes regulate dendritic cell (DC) function in vivo: increased interleukin 12 production by DCs from B cell-deficient mice results in T helper cell type 1 deviation. J Exp Med 2000;192:475–82.

[35] Garcia De Vinuesa C, Gulbranson-Judge A, Khan M, O'Leary P, Cascalho M, Wabl M, et al. Dendritic cells associated with plasmablast survival. Eur J Immunol 1999;29:3712–21.

[36] Devergne O, Birkenbach M, Kieff E. Epstein-Barr virus-induced gene 3 and the p35 subunit of interleukin 12 form a novel heterodimeric hematopoietin. Proc Natl Acad Sci USA 1997;94:12041–6.

[37] Collison LW, Workman CJ, Kuo TT, Boyd K, Wang Y, Vignali KM, et al. The inhibitory cytokine IL-35 contributes to regulatory T-cell function. Nature 2007;450:566–9.

[38] Niedbala W, Wei XQ, Cai B, Hueber AJ, Leung BP, McInnes IB, et al. IL-35 is a novel cytokine with therapeutic effects against collagen-induced arthritis through the expansion of regulatory T cells and suppression of Th17 cells. Eur J Immunol 2007;37:3021–9.

[39] Meister RH, Berger IH, Schwille PO. Cholecystokinin and glucose-induced insulinaemia in dogs with and without pancreatic acinar atrophy. J Clin Chem Clin Biochem Z Klin Chem Klin Biochem 1990;28:379–90.

[40] Jacobi AM, Odendahl M, Reiter K, Bruns A, Burmester GR, Radbruch A, et al. Correlation between circulating $CD27^{high}$ plasma cells and disease activity in patients with systemic lupus erythematosus. Arthritis Rheum 2003;48:1332–42.

[41] Odendahl M, Jacobi A, Hansen A, Feist E, Hiepe F, Burmester GR, et al. Disturbed peripheral B lymphocyte homeostasis in systemic lupus erythematosus. J Immunol 2000;165:5970–9.

[42] Nath SK, Harley JB, Lee YH. Polymorphisms of complement receptor 1 and interleukin-10 genes and systemic lupus erythematosus: a meta-analysis. Hum Genet 2005;118:225–34.

[43] Gateva V, Sandling JK, Hom G, Taylor KE, Chung SA, Sun X, et al. A large-scale replication study identifies TNIP1, PRDM1, JAZF1, UHRF1BP1 and IL10 as risk loci for systemic lupus erythematosus. Nat Genet 2009;41:1228–33.

[44] Amel Kashipaz MR, Huggins ML, Lanyon P, Robins A, Powell RJ, Todd I. Assessment of Be1 and Be2 cells in systemic lupus erythematosus indicates elevated interleukin-10 producing CD5+ B cells. Lupus 2003;12:356–63.

[45] Llorente L, Richaud-Patin Y, Garcia-Padilla C, Claret E, Jakez-Ocampo J, Cardiel MH, et al. Clinical and biologic effects of anti-interleukin-10 monoclonal antibody administration in systemic lupus erythematosus. Arthritis Rheum 2000;43:1790–800.

[46] Xu Y, Harder KW, Huntington ND, Hibbs ML, Tarlinton DM. Lyn tyrosine kinase: accentuating the positive and the negative. Immunity 2005;22:9–18.

[47] Scapini P, Lamagna C, Hu Y, Lee K, Tang Q, DeFranco AL, et al. B cell-derived IL-10 suppresses inflammatory disease in Lyn-deficient mice. Proc Natl Acad Sci USA 2011;108:E823–32.

[48] Wan YY, Flavell RA. Regulatory T-cell functions are subverted and converted owing to attenuated Foxp3 expression. Nature 2007;445:766–70.

[49] Wan S, Xia C, Morel L. IL-6 produced by dendritic cells from lupus-prone mice inhibits CD4+CD25+ T cell regulatory functions. J Immunol 2007;178:271–9.

[50] Gutierrez T, Halcomb KE, Coughran AJ, Li QZ, Satterthwaite AB. Separate checkpoints regulate splenic plasma cell accumulation and IgG autoantibody production in Lyn-deficient mice. Eur J Immunol 2010;40:1897–905.

[51] Yin Z, Bahtiyar G, Zhang N, Liu L, Zhu P, Robert ME, et al. IL-10 regulates murine lupus. J Immunol 2002;169:2148–55.

[52] Teichmann LL, Kashgarian M, Weaver CT, Roers A, Muller W, Shlomchik MJ. B cell-derived IL-10 does not regulate spontaneous systemic autoimmunity in MRL.Fas(lpr) mice. J Immunol 2012;188:678–85.

[53] Blair PA, Chavez-Rueda KA, Evans JG, Shlomchik MJ, Eddaoudi A, Isenberg DA, et al. Selective targeting of B cells with agonistic anti-CD40 is an efficacious strategy for the generation of induced regulatory T2-like B cells and for the suppression of lupus in MRL/lpr mice. J Immunol 2009;182:3492–502.

[54] Belperron AA, Dailey CM, Bockenstedt LK. Infection-induced marginal zone B cell production of Borrelia hermsii-specific antibody is impaired in the absence of CD1d. J Immunol 2005;174:5681–6.

[55] Bankoti R, Gupta K, Levchenko A, Stager S. Marginal zone B cells regulate antigen-specific T cell responses during infection. J Immunol 2012;188:3961–71.

[56] Mittrucker HW, Kaufmann SH. Immune response to infection with *Salmonella typhimurium* in mice. J Leukoc Biol 2000;67:457–63.

[57] Pamer EG. Immune responses to *Listeria monocytogenes*. Nat Rev Immunol 2004;4:812–23.

[58] Benjamin D, Knobloch TJ, Dayton MA. Human B-cell interleukin-10: B-cell lines derived from patients with acquired immunodeficiency syndrome and Burkitt's lymphoma constitutively secrete large quantities of interleukin-10. Blood 1992;80:1289–98.

[59] Duddy ME, Alter A, Bar-Or A. Distinct profiles of human B cell effector cytokines: a role in immune regulation? J Immunol 2004;172:3422–7.

[60] Iwata Y, Matsushita T, Horikawa M, Dilillo DJ, Yanaba K, Venturi GM, et al. Characterization of a rare IL-10-competent B-cell subset in humans that parallels mouse regulatory B10 cells. Blood 2011;117:530–41.

[61] Bouaziz JD, Calbo S, Maho-Vaillant M, Saussine A, Bagot M, Bensussan A, et al. IL-10 produced by activated human B cells regulates CD4+ T-cell activation in vitro. Eur J Immunol 2010;40:2686–91.

[62] Wagner U, Kaltenhauser S, Pierer M, Wilke B, Arnold S, Hantzschel H. B lymphocytopenia in rheumatoid arthritis is associated with the DRB1 shared epitope and increased acute phase response. Arthritis Res 2002;4:R1.

[63] Duddy M, Niino M, Adatia F, Hebert S, Freedman M, Atkins H, et al. Distinct effector cytokine profiles of memory and naive human B cell subsets and implication in multiple sclerosis. J Immunol 2007;178:6092–9.

[64] Hirotani M, Niino M, Fukazawa T, Kikuchi S, Yabe I, Hamada S, et al. Decreased IL-10 production mediated by toll-like receptor 9 in B cells in multiple sclerosis. J Neuroimmunol 2010;221:95–100.

[65] Jagannathan M, McDonnell M, Liang Y, Hasturk H, Hetzel J, Rubin D, et al. Toll-like receptors regulate B cell cytokine production in patients with diabetes. Diabetologia 2010;53:1461–71.

[66] Correale J, Farez M, Razzitte G. Helminth infections associated with multiple sclerosis induce regulatory B cells. Ann Neurol 2008;64:187–99.

[67] Correale J, Farez M. Association between parasite infection and immune responses in multiple sclerosis. Ann Neurol 2007;61:97–108.

[68] Ramgolam VS, Sha Y, Marcus KL, Choudhary N, Troiani L, Chopra M, et al. B cells as a therapeutic target for IFN-β in relapsing-remitting multiple sclerosis. J Immunol 2011;186:4518–26.

[69] Heine G, Niesner U, Chang HD, Steinmeyer A, Zugel U, Zuberbier T, et al. 1,25-dihydroxyvitamin $D_3$ promotes IL-10 production in human B cells. Eur J Immunol 2008;38:2210–8.

[70] Goetz M, Atreya R, Ghalibafian M, Galle PR, Neurath MF. Exacerbation of ulcerative colitis after rituximab salvage therapy. Inflamm Bowel Dis 2007;13:1365–8.

[71] Broglio L, Lauria G. Worsening after rituximab treatment in anti-mag neuropathy. Muscle Nerve 2005;32:378–9.

[72] Renaud S, Gregor M, Fuhr P, Lorenz D, Deuschl G, Gratwohl A, et al. Rituximab in the treatment of polyneuropathy associated with anti-MAG antibodies. Muscle Nerve 2003;27:611–5.

[73] El Fassi D, Nielsen CH, Kjeldsen J, Clemmensen O, Hegedus L. Ulcerative colitis following B lymphocyte depletion with rituximab in a patient with Graves' disease. Gut 2008;57:714–5.

[74] Dass S, Vital EM, Emery P. Development of psoriasis after B cell depletion with rituximab. Arthritis Rheum 2007;56:2715–8.

[75] Clatworthy MR, Watson CJ, Plotnek G, Bardsley V, Chaudhry AN, Bradley JA, et al. B-cell-depleting induction therapy and acute cellular rejection. N Engl J Med 2009;360:2683–5.

[76] Vincenti F, Kirkman R, Light S, Bumgardner G, Pescovitz M, Halloran P, et al. Interleukin-2-receptor blockade with daclizumab to prevent acute rejection in renal transplantation. Daclizumab Triple Therapy Study Group. N Engl J Med 1998;338:161–5.

[77] Blair PA, Norena LY, Flores-Borja F, Rawlings DJ, Isenberg DA, Ehrenstein MR, et al. $CD19^{+}CD24^{hi}CD38^{hi}$ B cells exhibit regulatory capacity in healthy individuals but are functionally impaired in systemic Lupus erythematosus patients. Immunity 2010;32:129–40.

[78] van de Veen W, Stanic B, Yaman G, Wawrzyniak M, Sollner S, Akdis DG, et al. IgG4 production is confined to human IL-10-producing regulatory B cells that suppress antigen-specific immune responses. J Allergy Clin Immunol 2013;131:1204–12.

[79] Heine G, Drozdenko G, Grun JR, Chang HD, Radbruch A, Worm M. Autocrine IL-10 promotes human B-cell differentiation into IgM or IgG-secreting plasmablasts. Eur J Immunol 2014;44:1615–1621.

[80] Good KL, Bryant VL, Tangye SG. Kinetics of human B cell behavior and amplification of proliferative responses following stimulation with IL-21. J Immunol 2006;177:5236–47.

Chapter 14

# B Cell Memory and Plasma Cell Development

Chapter 14a

## GENERATION OF MEMORY B CELLS

Toshitada Takemori
*RIKEN Center for Integrative Medical Sciences, Yokohama, Japan*

## 1. MEMORY B CELL MARKERS

There are no known specific markers for memory B cells in mice, although studies suggest that the CD38$^{low}$ and CD38$^{high}$ phenotypes are indicative of isotype-switched germinal center (GC) and memory B cells in the mouse, respectively [1]. Further analysis suggested that IgG1$^{+}$CD38$^{high}$ B cells, but not IgG1$^{+}$CD38$^{low}$ B cells, are capable of inducing a significant IgG1 secondary response in the adoptive hosts [2], demonstrating that the CD38$^{low}$ and CD38$^{high}$ phenotype distinction can be used to monitor the development of the antigen-specific memory B cells in the T cell-dependent (TD) response. In humans, approximately 30–50% of the peripheral blood B cells are CD27$^{+}$, and CD27 has been identified as a good surface marker for human memory B cells [3–5]. IgD$^{-}$CD27$^{+}$ cells in the peripheral blood have already undergone class switch recombination (CSR) and have accumulated somatic hypermutations (SHMs) in their $V_H$ genes in comparison to CD27$^{-}$ B cells, which have not. Polyclonal stimulation of B cells from anthrax vaccine adsorbed (AVA) vaccinated individuals generated AVA-specific IgG$^{+}$ antibody-secreting cells (ASCs) in vitro, but the deletion of CD27$^{+}$ B cells abrogated the response [6], indicating that human memory B cells are present in the CD27$^{+}$ B cell compartment.

## 2. PROPERTIES OF MEMORY B CELLS

Naive B cells activated after primary antigen encounter initially produce antigen-specific immunoglobulin (Ig) M, and later IgG, followed by GC formation. The response reaches its maximum at approximately 2 weeks after stimulation, accompanied by increased antibody affinity as mutations accumulate in the IgV regions. Memory B cells respond promptly and much more efficiently upon subsequent exposure to minute amounts of antigens, and they rapidly differentiate into ASCs that produce substantially higher levels of protective Ig than naive B cells [7–10]. How can memory B cells rapidly expand and accumulate mutations in the secondary response? It is conceivable that, after challenge, memory B cells enter into preexisting GCs that were formed in the primary response, where they can expand and accumulate mutations [11]. The capacity of B cells to help to establish protective immunity is a central trait of CD4 T cells and CD4$^{+}$ memory T cells that may play a pivotal role in humoral immunity by controlling the terminal differentiation of memory B cells [12,13]. Class-switched memory B cells express high levels of the co-receptors required for T cell interaction compared with naïve and GC B cells [10,14,15], improving memory cell response potency. Furthermore, memory B cells have a potent antigen-presenting cell (APC) activity as compared with naïve B cells, which provides an effective activation of cognate helper T cells [16], resulting in increased efficacy of memory B cell activation.

In the context of class-switched memory cells, it has been proposed that the unique cytoplasmic domain of IgG causes the prompt activation of antigen-experienced IgG memory B cells [17]. However, one recent study suggested that, upon challenge, IgG1 B cells that have never encountered antigen did not rapidly differentiate into plasma cells (PCs) upon secondary challenge, signifying the importance of the stimulation history [18].

## 3. DIVERSITY WITHIN THE MEMORY COMPARTMENT

The immune system has evolved distinct differentiation pathways to select two classes of antibody binding sites into the memory compartment: those that have been subjected to evolutionary selection without SHM and others selected for high-affinity antigen binding in the GC reaction through a process of SHM (see Section 5) [19]. Memory B cells are also heterogeneous according to their surface phenotype. It has

Molecular Biology of B Cells. http://dx.doi.org/10.1016/B978-0-12-397933-9.00014-X

been suggested that not only class-switched but also IgM+ memory B cells are generated in humans [20] and in mice in the TD and T cell-independent response [4,14,21–25]. There are no specific markers to distinguish IgM memory cells from naïve B cells, whereas an activation-induced cytidine deaminase reporter mouse makes it possible to allow tracking of B cells involved in GC reactions in response to TD antigens. This analysis showed that GC reactions can produce long-lived, somatically mutated IgM and isotype-switched memory B cells [23]. Alternatively, analysis of the cell division history of IgM+ B cells transferred into recipients after immunization supports the long-lived persistence of antigen-experienced IgM+ B cells [24]. A recall response defines a memory cell, but a mutated population of IgM+ long-term B cells responds poorly to a recall antigen [10,23,24]. Somatically mutated IgM+ B cells expand in GCs during secondary responses whereas isotype-switched memory B cells differentiate into effector cells, promoting the idea that each isotype may have a distinct role during a pathogen-elicited memory response [23,24].

## 4. DYNAMICS IN B CELL RESPONSE TOWARD MEMORY FORMATION

In TD B cell responses, antigen-specific T cells are activated through communication with dendritic cells (DCs) in the T cell zone. Subsequently, T cells interact with antigen-specific B cells at the border of the B and T cell areas in lymphoid organs. Activated B cells then move to the periphery of the follicle and proliferate, with a fraction of these cells undergoing Ig CSR [25,26]. Activated B cells upregulate expression of the orphan G protein-coupled receptor Epstein-Barr virus-induced gene-2 (EBI2) [27,28], which recognizes a ligand, 7α, 25-dihydroxycholesterol synthesized in lymphoid stromal cells and at the follicle perimeter, leading EBI2+ activated B cells to localize there [29]. Memory B cells upregulate EBI2, but this expression is downregulated in GC B cells.

Some of the activated B cells mediate the primary antibody response through differentiation into plasma and memory B cells [10,19,28,30]. Subsequently, other activated B cells form GC; this is accompanied by upregulation of the transcriptional repressor B cell lymphoma-6 protein (Bcl6), on which GC B cell differentiation depends [31,32]. GCs develop from day 5 to day 6 and expand within the B cell follicles, resulting in the establishment of histologically discrete structures at 2 weeks of response [33]. GC initiation is required for cognate interaction with T cells and APC; thereby, it is disrupted in the absence of signaling downstream of CD40 or inducible co-stimulator (ICOS) [34–36]. GCs consist predominantly of B cells, together with minor populations of follicular dendritic cells (FDCs) and CD4 follicular helper (Tfh) cells [37]. FDCs, nonhematopoietic cells derived from perivascular precursor cells [38], retain and present antigens on their surfaces.

GC development is required for Tfh cells [35,39]. Tfh cells arise by a distinct developmental pathway dependent on the transcription factors Ascl2 and Bcl6 as well as cognate B cells [39–42]. Bcl6-deficient CD4+ T cells fail to develop into Tfh cells, resulting in impaired GC B cell development and high-affinity memory response [10,39–41].

Tfh cells are distinctive in their migratory activity into B cell follicles because of their upregulation of chemokine (C-X-C motif) receptor-5 (CXCR5) and ICOS as well as their downregulation of chemokine (C–C motif) receptor-7 expression to provide essential help to B cells for the GC formation [37]. Tfh cells interact with GC B cells and help in the selection of high-affinity B cells [43]. Loss of Tfh cells by Bcl6 deletion in CD4 T cells reduced the number and size of GCs in the follicle at the early immune response and shortened the duration of GC reactions, resulting in a gross reduction in the generation of mutated GCs and memory B cells [10]. GC B cells help in the full development of Tfh cells, demonstrating that GC B cells and Tfh cells are mutually required for their development. The cytokine most closely associated with Tfh cells is interleukin (IL)-21, directing GC B cells to maintain proliferation [44,45].

## 5. SELECTION OF HIGH-AFFINITY MEMORY B CELLS IN GCs

Within the GC itself, B cells undergo massive proliferation accompanied by CSR and SHM of their rearranged IgV region genes, a process in which cells acquire mutations that likely increase antibody affinity for the immunizing antigen [8,46,47]. The V gene mutation and subsequent positive selection increasing the average affinity of the B cell population over time is referred to as *affinity maturation* [8].

The V gene selection process critically depends on antigen being presented to the B cells by GC FDCs and presented by the B cells themselves in the form of antigenic peptides to antigen-specific Tfh cells [43]. Higher-affinity B cell receptor is translated into improved antigen acquisition by GC B cell variants and thus a greater possibility of extensive cognate interactions with Tfh cells in the GCs. This process may provide the proliferative and differentiation signals that enable the expansion of clones with improved affinity.

The selection of high-affinity B cells within the GCs is impaired in Fas-deficient *lpr* mice or in mutant mice with conditional deletion of Fas in B cells in the primary response as a consequence of inefficient negative selection [48,49]. Fas deficiency expands the memory compartment through the increased recruitment of newly generated precursors from the late GCs, resulting in the accumulation of heavily mutated memory B cells at high frequency. These results suggest that Fas is required for clonal selection within the GCs and the establishment of the memory B cell repertoire.

The selected, high-affinity GC B cells are then believed to differentiate into memory B and PCs, coinciding with the downregulation of Bcl6 expression. A large fraction of GC-derived memory and antibody-forming cells (AFCs) express somatically mutated IgV region genes, and these cells persist for long periods of time after termination of the GC response [8,50]. Affinity-matured memory B cells and AFCs then proceed to emigrate from the GCs.

## 6. HOW MEMORY B CELLS DEVELOP IN THE GC REACTION

Within the GC, high-affinity B cells are selected and differentiate into memory cells; however, the mechanism behind this process remains obscure. We know that Bcl6 expression blocks GC B cells from undergoing PC and memory B cell differentiation. We also know that expression of the Blimp-1 transcription factor encoded by PRDM1 is critical for promoting B cell differentiation into PCs. BCL6 has been suggested to repress the PRDM1 locus [51], probably through indirect binding to the PRDM1 promoter by association with AP-1 and BACH2 [52,53]. Accordingly, downregulation of Bcl6 is important for memory and PC differentiation. Although the precise mechanism underlying Bcl6 downregulation in the GC reaction remains obscure, it has been suggested that GC B cell division results in the asymmetric distribution of Bcl6 in offspring [54], which may generate daughter cells with preferential Bcl6 downregulation. Such cells may differentiate into memory cells and PCs by Pax5 downregulation and upregulation of Blimp1 [55]. However, how Bcl6-less cells transition toward memory cells remains unknown.

## 7. GC-INDEPENDENT MEMORY B CELLS

Although involvement of the GC reaction in memory B cell development is broadly accepted, the fact that not all memory B cells carry IgV regions with somatic mutations [14,21,30,43,47,56] has led to the proposal that some memory cells emerge via GC-independent pathways. In fact, unmutated memory cells can be generated in irradiated mice reconstituted with Bcl6-deficient bone marrow (BM) [57].

However, because Bcl6 germline deletion causes multiple immunological dysfunctions, the role of Bcl6 in GC B cell formation has been further assessed by using mutant mice with Bcl6 deletion in their B cell compartment [10]. Conditional *Bcl6* deletion in the B cell lineage affected neither B cell development and subset distribution in terms of numbers nor the initial expansion of antigen-activated B cells after immunization; however, it did not allow the development of GC B cells from day 5 after immunization, the earliest time point at which such cells became apparent in wild-type (WT) mice. Under these conditions, IgG1$^+$ memory B cells developed from day 5 after immunization like in WT mice, evidently independently of the GC reaction. These findings were corroborated using a different model that did not rely on genetic manipulation [10,58]. In addition, IgM memory B cells were also generated in a GC-independent pathway, whereas GC-independent IgM memory B cells did not contribute much to the overall antibody production in the secondary response [10], as observed in GC-dependent IgM memory B cells [21,24,58].

In the mouse, the memory compartment primarily contains a very large proportion of nonmutated cells. However, the frequency of nonmutated memory B cells gradually decreases as the GC reaction progresses whereas blocking the GC reaction prevents the recruitment of mutated cells into the memory compartment [10,48]. These results suggest that nonmutated, GC-independent memory B cells develop early in the response in WT and conditional *Bcl6*-deficient mice and are maintained for long periods of time. Despite the recruitment of substantial numbers of GC B cell progeny into the memory compartment, in due course, the absolute numbers of memory cells were similar between WT mice and GC response-deficient mice [10]. This raises the possibility that the lymphoid environment has some limited capacity to sustain memory B cells, irrespective of whether they are GC-derived or not.

Although memory B cells promptly respond to a minute amount of antigen, memory B cells are sustained in the periphery in the presence of antigen at the early immune response. This may reflect the localization of memory B cells in B cell follicles (Figure 1) [10,14], where effector T cells are rarely detected [59]. It has been predicted that memory B cells should rapidly expand and differentiate into PCs under the cognate control of memory T cells, which move to the B cell follicle upon antigen reexposure [60].

## 8. GC-INDEPENDENT AND -DEPENDENT MEMORY B CELLS DEVELOP WITH THE HELP OF DIFFERENT T-CELL SUBSETS

The conditional deletion of *Bcl6* through CD4-cre did not affect T and B cell numbers or their phenotype in naïve animals, but it impaired Tfh development in mice immunized with TD antigen [10]. Mutant mice had significantly reduced numbers of IgG1$^+$ GC B cells and mutated memory B cells. The absence of Bcl-6 in B and T cells did not affect the formation of GC-independent memory B cells after immunization, demonstrating that a non-Tfh subset of CD4 T cells can support the generation of these cells [10]. Thus, Tfh depletion impaired the development of mutated memory cells because of a failure in expansion and/or maintenance of GC B cells [10,40,41], but it did not affect the generation of unmutated memory cells. These

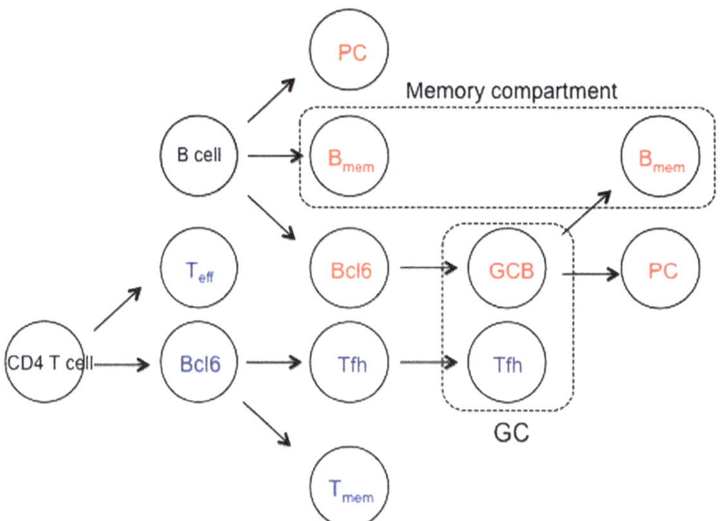


FIGURE 1 **The developmental pathways of germinal center (GC)-independent and GC-dependent memory B cells, with the help by T follicular helper (Tfh) and non-Tfh CD4 T cells.** The capacity to help B cells and establish long-term protective immunity is a central trait of CD4 T cells. Simultaneously, cognate B cells help CD4 T cells to develop into $CXCR5^{high}PD1^{high}$ follicular T helper (Tfh) and memory T cells ($T_{mem}$). Early in the immune response, at the T-B cell border, antigen-engaged B cells are activated and migrate into the follicle, where the B cells proliferate and can undergo class switch recombination (CSR) prior to the GC reaction. T-cell dependent B cell activation involves serial contacts of varying duration among the interacting cells in the follicles, providing ample opportunity for the branching of differentiation into memory B cells ($B_{mem}$) or extrafollicular plasma cells (PC) prior to the GC reaction. Subsequently, a number of B cells upregulate Bcl6 and expand to form GCs in the follicle center, in parallel with CD4 T-cell differentiation into $CXCR5^{high}PD1^{high}$ follicular T helper (Tfh) cells within follicles through increased Bcl6 protein expression. Tfh cells form a complex with GC B cells and this cognate interaction is essential for maintenance of Tfh cells. GC B cells proliferate and accumulate mutations in their V genes and differentiate into high-affinity mutated memory cells and PC with the help of Tfh cells. In the GC-independent pathway of TD memory B cell development, Tfh cells are not required, at least early on in the response. Thus antigen-engaged B cells are driven into the GC-independent memory pathway with the help of T cells other than Tfh. In order for CD4 T cells to develop into memory cells, both Bcl6 expression and cognate B cell interaction are required. Thus mutual cognate interactions between B and T cells establish immunological memory, in association with or without Bcl6 expression.

studies provided definitive evidence for the existence of a novel TD memory B cell generation pathway, which is distinct from the classical generation of memory B cells in GCs, which has been predicted from previous studies [21,30,45,56–58,61]. The memory cells generated in this pathway appear earlier after immunization than their mutated counterparts, express germline-encoded antibodies, and persist in the peripheral B cell compartment over long periods of time and in numbers comparable to those of memory cells of GC origin.

## 9. GC-INDEPENDENT AND -DEPENDENT MEMORY B CELLS ARE INVOLVED IN THE SECONDARY RESPONSE

High-affinity memory B cells are preferentially selected during secondary responses and rapidly differentiate into antibody-producing cells. A recent study suggests that mutated and unmutated $IgG1^+$ memory B cells respond to a secondary challenge and expand while accumulating somatic mutations in their $IgV_H$ genes [11,62]. In this study, both types of memory cells subsequently established a $V_H$ gene repertoire, which was distinct from the original repertoire before antigen reexposure. Mutated and unmutated $IgG1^+$ memory cells equally contribute to set up a new antibody repertoire through a dynamic process of mutation and selection, becoming optimally adapted to the antigen upon secondary challenge [11].

The high-affinity somatic mutants drive the immune system to strongly focus on the initiating antigen epitope. In contrast, it seems that the germline-encoded repertoire sustains cross-reacting specificities and provides flexibility to create a new diverse antibody repertoire upon antigen reexposure, probably via the GC reaction. This property may guarantee rapid generation of antibodies optimally adapted to the invading pathogen and the variants that can arise during the course of an infection. Thus, the activity of the memory compartment consists of two types of cells and is vital for eliminating pathogens and toxins that have not efficiently been previously eliminated by preexisting circulating antibodies.

## REFERENCES

[1] Ridderstad A, Tarlinton DM. Kinetics of establishing the memory B cell population as revealed by CD38 expression. J Immunol 1998;160:4688–95.

[2] Takahashi Y, Dutta PR, Cerasoli DM, Kelsoe G. In situ studies of the primary immune response to (4-hydroxy-3-nitrophenyl)acetyl. V. Affinity maturation develops in two stages of clonal selection. J Exp Med 1998;187:885–95.

[3] Agematsu K, Nagumo H, Shinozaki K, Hokibara S, Yasui K, Terada K, et al. Absence of IgD$^-$CD27$^+$ memory B cell population in X-linked hyper-IgM syndrome. J Clin Invest 1998;102:853–60.
[4] Klein U, Rajewsky K, Küppers R. Human immunoglobulin (Ig)M$^+$IgD$^+$ peripheral blood B cells expressing the CD27 cell surface antigen carry somatically mutated variable region genes: CD27 as a general marker for somatically mutated (memory) B cells. J Exp Med 1998;188:1679–89.
[5] Tangye SG, Liu YJ, Aversa G, Phillips JH, de Vries JE. Identification of functional human splenic memory B cells by expression of CD148 and CD27. J Exp Med 1988;188:1691–703.
[6] Crotty S, Aubert RD, Glidewell J, Ahmed R. Tracking human antigen-specific memory B cells: a sensitive and generalized ELISPOT system. J Immunol Methods 2004;286:111–22.
[7] Ahmed R, Gray D. Immunological memory and protective immunity: understanding their relation. Science 1996;272:54–60.
[8] Rajewsky K. Clonal selection and learning in the antibody system. Nature 1996;381:751–8.
[9] Tangye SG, Tarlinton DM. Memory B cells: effectors of long-lived immune responses. Eur J Immunol 2009;39:2065–75.
[10] Kaji T, Ishige A, Hikida M, Taka J, Hijikata A, Kubo M, et al. Distinct cellular pathways select germline-encoded and somatically mutated antibodies into immunological memory. J Exp Med 2012;209: 2079–97.
[11] Kaji T, Furukawa K, Ishige A, Toyokura I, Nomura M, Okada M, et al. Both mutated and unmutated memory B cells accumulate mutations in the course of the secondary response and develop a new antibody repertoire optimally adapted to the secondary stimulus. Int Immunol 2013;12:683–95.
[12] Fazilleau N, McHeyzer-Williams LJ, Rosen H, McHeyzer-Williams MG. The function of follicular helper T cells is regulated by the strength of T cell antigen receptor binding. Nat Immunol 2009;10:375–84.
[13] MacLeod MK, David A, McKee AS, Crawford F, Kappler JW, Marrack P. Memory CD4 T cells that express CXCR5 provide accelerated help to B cells. J Immunol 2011;186:2889–96.
[14] Anderson SM, Tomayko MM, Ahuja A, Haberman AM, Shlomchik MJ. New markers for murine memory B cells that define mutated and unmutated subsets. J Exp Med 2007;204:2103–14.
[15] Good-Jacobson KL, Szumilas CG, Chen L, Sharpe AH, Tomayko MM, Shlomchik MJ. PD-1 regulates germinal center B cell survival and the formation and affinity of long-lived plasma cells. Nat Immunol 2010;11:535–42.
[16] Yuseff MI, Pierobon P, Reversat A, Lennon-Dumenil AM. How B cells capture, process and present antigens: a crucial role for cell polarity. Nat Rev Immunol 2013;13:475–86.
[17] Tangye SG, Avery DT, Deenick EK, Hodgkin PD. Intrinsic differences in the proliferation of naive and memory human B cells as a mechanism for enhanced secondary immune responses. J Immunol 2003;170:686–94.
[18] Kometani K, Nakagawa R, Shinnakasu R, Kaji T, Rybouchkin A, Moriyama S, et al. Repression of the transcription factor Bach2 contributes to predisposition of IgG1 memory B cells toward plasma cell differentiation. Immunity 2013;39:136–47.
[19] Takemori T, Kaji T, Takahashi Y, Shimoda M, Rajewsky K. Generation of memory B cells inside and outside germinal centers. Eur J Immunol 2014;44(5):1258–64.
[20] Ma CS, Pittaluga S, Avery DT, Hare NJ, Maric I, Klion AD, et al. Selective generation of functional somatically mutated IgM$^+$CD27$^+$, but not Ig isotype-switched, memory B cells in X-linked lymphoproliferative disease. J Clin Invest 2006;116:322–33.
[21] Weller S, Faili A, Garcia C, Braun MC, Le Deist FF, de Saint Basile GG, et al. CD40-CD40L independent Ig gene hypermutation suggests a second B cell diversification pathway in humans. Proc Natl Acad Sci USA 2001;98:1166–70.
[22] Obukhanych TV, Nussenzweig MC. T-independent type II immune responses generate memory B cells. J Exp Med 2006;203:305–10.
[23] Dogan I, Bertocci B, Vilmont V, Delbos F, Mégret J, Storck S, et al. Multiple layers of B cell memory with different effector functions. Nat Immunol 2009;10:1292–9.
[24] Pape KA, Taylor JJ, Maul RW, Gearhart PJ, Jenkins MK. Different B cell populations mediate early and late memory during an endogenous immune response. Science 2011;331:1203–7.
[25] Garside P, Ingulli E, Merica RR, Johnson JG, Noelle RJ, Jenkins MK. Visualization of specific B and T lymphocyte interactions in the lymph node. Science 1998;281:96–9.
[26] Okada T, Miller MJ, Parker I, Krummel MF, Neighbors M, Hartley SB, et al. Antigen-engaged B cells undergo chemotaxis toward the T zone and form motile conjugates with helper T cells. PLoS Biol 2005;3:e150.
[27] Gatto D, Paus D, Basten A, Mackay CR, Brink R. Guidance of B cells by the orphan G protein-coupled receptor EBI2 shapes humoral immune responses. Immunity 2009;21:259–69.
[28] Pereira JP, Kelly LM, Cyster JG. Finding the right niche: B-cell migration in the early phases of T-dependent antibody responses. Int Immunol 2010;22:413–9.
[29] Yi T, Wang X, Kelly LM, An J, Xu Y, Sailer AW, et al. Oxysterol gradient generation by lymphoid stromal cells guides activated B cell movement during humoral responses. Immunity 2013;21: 535–48.
[30] Chan TD, Gatto D, Wood K, Camidge T, Basten A, Brink R. Antigen affinity controls rapid T-dependent antibody production by driving the expansion rather than the differentiation or extrafollicular migration of earlyplasmablasts. J Immunol 2009;183: 3139–49.
[31] Ye BH, Cattoretti G, Shen Q, Zhang J, Hawe N, de Waard R, et al. The BCL-6 proto-oncogene controls germinal-centre formation and Th2-type inflammation. Nat Genet 1997;16:161–70.
[32] Dent AL, Shaffer AL, Yu X, Allman D, Staudt LM. Control of inflammation, cytokine expression, and germinal center formation by BCL-6. Science 1997;276:589–92.
[33] Jacob J, Kelsoe G. In situ studies of the primary immune response to (4-hydroxy-3-nitrophenyl)acetyl. II. A common clonal origin for periarteriolar lymphoid sheath-associated foci and germinal centers. J Exp Med 1992;176:679–87.
[34] Kawabe T, Naka T, Yoshida K, Tanaka T, Fujiwara H, Suematsu S, et al. The immune responses in CD40-deficient mice: impaired immunoglobulin class switching and germinal center formation. Immunity 1994;1:167–78.
[35] McAdam AJ, Greenwald RA, Levin MA, Chernova T, Malenkovich N, Ling V, et al. ICOS is critical for CD40-mediated antibody class switching. Nature 2001;409:102–5.
[36] Tafuri A, Shahinian A, Bladt F, Yoshinaga SK, Jordana M, Wakeham A, et al. ICOS is essential for effective T-helper-cell responses. Nature 2001;409:105–9.
[37] Zotos D, Tarlinton DM. Determining germinal centre B cell fate. Trends Immunol 2012;33:281–8.
[38] Krautler NJ, Kana V, Kranich J, Tian Y, Perera D, Lemm D, et al. Follicular dendritic cells emerge from ubiquitous perivascular precursors. Cell 2012;150:194–206.

[39] Crotty S. Follicular helper CD4 T cells (TFH). Ann Rev Immunol 2011;29:621–63.
[40] Johnston RJ, Poholek AC, DiToro D, Yusuf I, Eto D, Barnett B, et al. Bcl6 and Blimp-1 are reciprocal and antagonistic regulators of T follicular helper cell differentiation. Science 2009;325: 1006–10.
[41] Yu D, Rao S, Tsai LM, Lee SK, He Y, Sutcliffe EL, et al. The transcriptional repressor Bcl-6 directs T follicular helper cell lineage commitment. Immunity 2009;31:457–68.
[42] Liu X, Chen X, Zhong B, Wang A, Wang X, Chu F, et al. Transcription factor achaete-scute homologue 2 initiates follicular T-helper-cell development. Nature 2014;507:513–8.
[43] Victora GD, Schwickert TA, Fooksman DR, Kamphorst AO, Meyer-Hermann M, Dustin ML, et al. Germinal center dynamics revealed by multiphoton microscopy with a photoactivatable fluorescent reporter. Cell 2010;143:592–605.
[44] Linterman MA, Beaton L, Yu D, Ramiscal RR, Srivastava M, Hogan JJ, et al. IL-21 acts directly on B cells to regulate Bcl-6 expression and germinal center responses. J Exp Med 2010;207: 353–63.
[45] Zotos D, Coquet JM, Zhang Y, Light A, D'Costa K, Kallies A, et al. IL-21 regulates germinal center B cell differentiation and proliferation through a B cell–intrinsic mechanism. J Exp Med 2010;207:365–78.
[46] Berek C, Berger A, Apel M. Maturation of the immune response in germinal centers. Cell 1991;67:1121–9.
[47] Jacob J, Kelsoe G, Rajewsky K, Weiss U. Intraclonal generation of antibody mutants in germinal centres. Nature 1991;354: 389–92.
[48] Takahashi Y, Ohta H, Takemori T. Fas is required for clonal selection in germinal centers and the subsequent establishment of the memory B cell repertoire. Immunity 2001;14:181–92.
[49] Hao Z, Duncan GS, Seagal J, Su YW, Hong C, Haight J, et al. Fas receptor expression in germinal-center B cells is essential for T and B lymphocyte homeostasis. Immunity 2008;29:615–27.
[50] Tarlinton DM. B-cell memory: are subsets necessary? Nat Rev Immunol 2006;6:785–90.
[51] Shapiro-Shelef M, Lin KI, McHeyzer-Williams LJ, Liao J, McHeyzer-Williams MG, Calame K. Blimp-1 is required for the formation of immunoglobulin secreting plasma cells and pre-plasma memory B cells. Immunity 2003;19:607–20.
[52] Vasanwala FH, Kusam S, Toney LM, Dent AL. Repression of AP-1 function: a mechanism for the regulation of Blimp-1 expression and B lymphocyte differentiation by the B cell lymphoma-6 protooncogene. J Immunol 2002;169:1922–9.
[53] Huang C, Geng H, Boss I, Wang L, Melnick A. Cooperative transcriptional repression by BCL6 and BACH2 in germinal center B-cell differentiation. Blood 2014;123:1012–20.
[54] Barnett BE, Ciocca ML, Goenka R, Barnett LG, Wu J, Laufer TM, et al. Asymmetric B cell division in the germinal center reaction. Science 2012;335:342–4.
[55] Kallies A, Hasbold J, Fairfax K, Pridans C, Emslie D, McKenzie BS, et al. Initiation of plasma-cell differentiation is independent of the transcription factor Blimp-1. Immunity 2007;26:555–66.
[56] Blink EJ, Light A, Kallies A, Nutt SL, Hodgkin PD, Tarlinton DM. Early appearance of germinal center-derived memory B cells and plasma cells in blood after primary immunization. J Exp Med 2005;201:545–54.
[57] Toyama H, Okada S, Hatano M, Takahashi Y, Takeda N, Ichii H, et al. Memory B cells without somatic hypermutation are generated from Bcl6-deficient B cells. Immunity 2002;17:329–39.
[58] Taylor JJ, Pape KA, Jenkins MK. A germinal center-independent pathway generates unswitched memory B cells early in the primary response. J Exp Med 2012;209:597–606.
[59] Pepper M, Pagan AJ, Igyarto BZ, Taylor JJ, Jenkins MK. Opposing signals from the Bcl6 transcription factor and the interleukin-2 receptor generate T helper 1 central and effector memory cells. Immunity 2011;35:583–95.
[60] McHeyzer-Williams LJ, McHeyzer-Williams MG. Antigen-specific memory B cell development. Annu Rev Immunol 2005;23: 487–513.
[61] Inamine A, Takahashi Y, Baba N, Miyake K, Tokuhisa T, Takemori T, et al. Two waves of memory B-cell generation in the primary immune response. Int Immunol 2005;17:581–9.
[62] Vora KA, Tumas-Brundage K, Manser T. Contrasting the in situ behavior of a memory B cell clone during primary and secondary immune responses. J Immunol 1999;163:4315–27.

Chapter 14b

# Plasma Cell Biology

**David Tarlinton**

*The Walter and Eliza Hall Institute of Medical Research, Parkville, VIC, Australia*

## 1. INTRODUCTION

Although all surface immunoglobulin (Ig)-positive B cells have the potential to generate antibody-secreting plasma cells, it is apparent that not all plasma cells are identical. There are differences in plasma cell life span, expression of cell surface markers, and amount and isotype of antibody secreted. These differences between plasma cells reflect potentially the nature of the antigen that triggered their formation, the organ and environment in which they developed, and where they reside. Thus, in understanding plasma cell biology, one needs to consider whether the response was one that was dependent on T cell help or not; whether it was a self or foreign antigen; and whether the response occurred in the gut, spleen, tonsil, lymph node, or a tissue subjected to autoimmune attack.

Related to the manner of plasma cell formation are issues of survival. Although it is now clear that all plasma cells use ultimately the same anti-apoptotic protein to sustain themselves, it is unclear what uniformity there is in the mechanisms by which these survival proteins are induced and then maintained. That is, the intrinsic short lifespan of plasma cells requires them to continually receive signals from their environment that trigger biochemical signaling pathways that ultimately result in the expression of particular prosurvival proteins of the Bcl2 family of proteins. The requirement for external factors in plasma cell maintenance, which can last for decades, means that the ability of plasma cells to receive such signals is crucial in determining the persistence of any individual cell. It is now generally accepted that such prosurvival signals are secreted by various cell types that collectively constitute survival niches for plasma cells. The distribution, composition, and occupancy of these niches are fundamental in determining plasma cell persistence, frequency, and turnover.

Considerable information is available on the changes in gene expression that accompany the differentiation of B cells into plasma cells. Analysis of the gene expression program of plasma cells reveals it to be fundamentally different from that of B cells, representing in many respects a change of fate for the cell. Such a dramatic change is appropriate given the cell biological changes that occur during this differentiation. The B cell loses expression of many cell surface markers while gaining others associated with the localization and survival strategy of plasma cells. In addition, there are changes in genes encoding components of the signaling pathways associated with these cell surface receptors. Plasma cells also extensively gain protein production capacity and expand their secretory apparatus. The transition from a B cell to a plasma cell may not always be due to the uniform up- and downregulation of key transcription factors, although a core group is absolutely crucial; the process of regulating the change may vary according to the nature of the immune response (e.g., whether it involves T cell-mediated signals or not).

## 2. PLASMA CELL SUBSETS

There are currently two recognized subsets of antibody-secreting cells (ASCs): plasmablasts (PBs) and plasma cells [1,2]. Plasma cells can be further divided into short- and long-lived and, in a related scheme, immature and mature. As discussed below, the developmental relationships between these stages remain somewhat uncertain, as do the cues that prompt progression along this purported developmental pathway.

### 2.1 Plasmablasts

B cells activated by antigen and receiving appropriate additional stimuli through antigen, T cell contact, and/or cytokines will initiate proliferation, in which the cells enlarge and appear as blast cells. Among ASCs, there is a clear stage of proliferating B cells secreting antibody [3]. Cells in this stage of development are called PBs. PBs express many markers of the B cell lineage, including the B cell receptor (BCR), co-stimulatory molecules such as B220 (a B-cell restricted isoform of CD45), and CD80/86, and the mechanism for antigen presentation through expression of major histocompatibility complex II PB secreting each of the Ig isotypes are detectable [1].

The development of PBs has been well characterized in in vitro culture experiments with mouse and human B cells [4,5]. In both cases, B cells stimulated with mitogen and cytokines initiate proliferation and then generate PBs with a division-linked frequency that follows a probabilistic function. That is, there appears to be an intrinsic program for proliferating B cells to differentiate into ASCs such that per division, PBs appear at a certain frequency. This frequency depends on the mitogen—be that a T cell-derived signal, such as stimulation through CD40, or one that stimulates B cells through toll-like receptors (TLRs), such as lipopolysaccharide (LPS), through TLR4—but it can be additionally modulated by cytokines. For example, interleukin (IL)-5 is a potent enhancer of PB formation in mouse B cell cultures, as is IL-21 in human B cell cultures [6,7]. Observing PBs in vivo is a more complex process; typically, one observes inferential characteristics of proliferation such as expression of the cell division marker, Ki-67, on a population known to contain ASCs. Another attribute of PBs apparent in in vitro cultures with important implications for in vivo biology is that once B cell blasts begin antibody secretion, they no longer undergo isotype switching [4]. However, the reciprocal does not occur; differentiation into PBs is independent of isotype, meaning that the division-based probability of B cell differentiation into ASCs is independent of their isotype [4].

PBs are inherently short lived irrespective of the nature of the antigen. In the case of an immunological boost in a secondary response, there is a relatively synchronous appearance of PBs in the peripheral lymphoid organs and blood. For example, humans receiving an annual flu vaccine or a tetanus toxoid boost produce a "wave" of PBs in the blood approximately 7 days after injection [8,9]. The subsequent rapid disappearance of these cells is an indicator of their intrinsically short lifespan. A similar wave of ASCs appears after boost immunization in mice, showing similarly short lifespans. In mice, it is possible to examine the appearance of ASCs in organs during primary immune responses [10]. In this circumstance, proliferating, secreting B cells develop in the first days after immunization. In the case of T cell-dependent (TD) antigens (those requiring the participation of CD4 T cells to mount an immune response), these initial ASCs accumulate at the edges of the B cell areas of the lymphoid organs in which the response is occurring. These aggregates of ASCs, including PBs and plasma cells, disappear within several days of their appearance, again indicative of a short lifespan [11].

### 2.2 Plasma Cells

Plasma cells are postmitotic ASCs [3]. They exist in all lymphoid organs, although the frequency is quite varied, ranging from extremely low in peripheral lymph nodes up to a high of approximately 0.5% of nucleated cells in the spleen of a mouse from a standard animal facility. After immunization, the local frequency can increase quite sharply, although such an increase is due to PBs and plasma cells [12]. Plasma cells are also present in significant numbers, albeit low frequency, in bone marrow. Indeed, bone marrow is considered the site for the long-term maintenance of plasma cells arising from immunization [12].

Stages of plasma cell maturation have been defined by numerous observations in different locations and at different stages of immune responses [1]. This ranges from alterations in cell surface markers, sensitivity to chemotactic signals, and the capacity to secrete antibody. However, the relationship between these potentially different stages remains uncertain in that it is not possible to say with certainty that any two such stages are necessarily sequential. However, immature plasma cells generally remain responsive to chemotactic signals and express B cell markers at lower levels than PBs and plasma cell markers at lower levels than the most mature plasma cells. The defining feature of plasma cells is their expanded endoplasmic reticulum (ER), and although this is best measured using electron microscopy, its changing distribution is probably the most accurate reflection of plasma cell maturation in that PBs have significantly less well-developed ER compared with plasma cells whereas changes in ER content within plasma cells are also likely.

Until recently, the ability to distinguish types or compartments of plasma cells was poorly developed. Early studies using incorporation of nucleotide analogs to measure population turnover determined that there were short-lived and long-lived plasma cells that were unequally distributed amongst lymphoid tissues in that there was a preponderance of short-lived plasma cells in the periphery and long-lived plasma cells in bone marrow [13]. These studies concluded also that most IgM plasma cells were short-lived irrespective of location, suggesting a relationship between isotype and lifespan. More recently, with the identification of multiple cell surface proteins that identify plasma cells and B cells at different stages of development, it is possible to identify and recover immature and mature plasma cells from various locations [12,14,15]. This allows for functional studies on these cells after isolation, including migration and gene expression analysis, although such analyses have yet to define a pattern that uniquely resolves mature and immature plasma cells [1] (Figure 2).

Additional clarity to the question of plasma cell maturation has been provided by the use of genetically engineered reporter mice [12,15]. One such strain, in which a cDNA encoding the green fluorescent protein (GFP) was inserted into the *Prdm1* locus encoding the transcriptional repressor Blimp1, permitted the reliable identification and isolation of all ASCs by fluorescence activated cell sorting [12]. Furthermore, it is apparent from analyzing these mice that among the steady-state plasma cell populations of mice, two amounts of Blimp1 are expressed: intermediate ($Blimp1^{int}$) and high ($Blimp1^{hi}$). Furthermore, cells within the $Blimp1^{int}$ population had many of the characteristics previously attributed to PBs and immature plasma cells whereas the $Blimp1^{hi}$ population had characteristics of mature plasma cells [12]. Chief among these properties were lifespan, with the $Blimp1^{int}$ being predominantly short lived and $Blimp1^{hi}$ being long lived; proliferation; and phenotype, with markers such as B220 (CD45) expressed on $Blimp1^{int}$ but at much reduced amounts on $Blimp1^{hi}$ plasma cells. Finally, using Blimp1 as a marker of plasma cells revealed a preponderance of $Blimp1^{hi}$ plasma cells in the bone marrow, an equal distribution of $Blimp1^{int}$ and $Blimp1^{hi}$ in the spleen, and an excess of $Blimp1^{int}$ in the blood, all consistent with previous information from multiple studies that the bone marrow is the site of the most mature plasma cells and that immature plasma cells are the migratory plasma cells.

## 3. CELLULAR ASPECTS OF PLASMA CELL DIFFERENTIATION

There are three routes to plasma cell formation. First is what is referred to as "natural antibody production," which is Ig production in the absence of immunization. The second and third routes to plasma cell formation are in response to the TD and T cell-independent (TI) antigenic stimulation of B cells. Each of these pathways elicits a distinct type of B cell response, distinct in the course of the response and the nature of the B cell responding. The plasma cells derived from these situations are themselves also potentially distinct.

### 3.1 B1 Cells, Natural Antibody, and Plasma Cell Differentiation

Natural antibody (nAb) is defined as that produced in the absence of deliberate immunization or exposure to pathogens [16]. It comprises mainly IgM and shows enrichment for reactivity to antigens such as phospholipids and polysaccharides. Many such antibodies show polyreactivity and will bind self-antigens. The role of this antibody in the immune system remains a matter of research, but it is apparent that pre-existing IgM can be crucial to the efficient and rapid initiation of responses to exogenous antigens. There are proposals that nAb also has a regulatory role in suppressing autoimmune disease and maintaining tissue homeostasis through clearance of apoptotic cells. Although the location and biology of the plasma cells producing these antibodies is uncertain, a significant body of opinion is that they derive from the B1 cell lineage in mouse and its equivalent in humans.

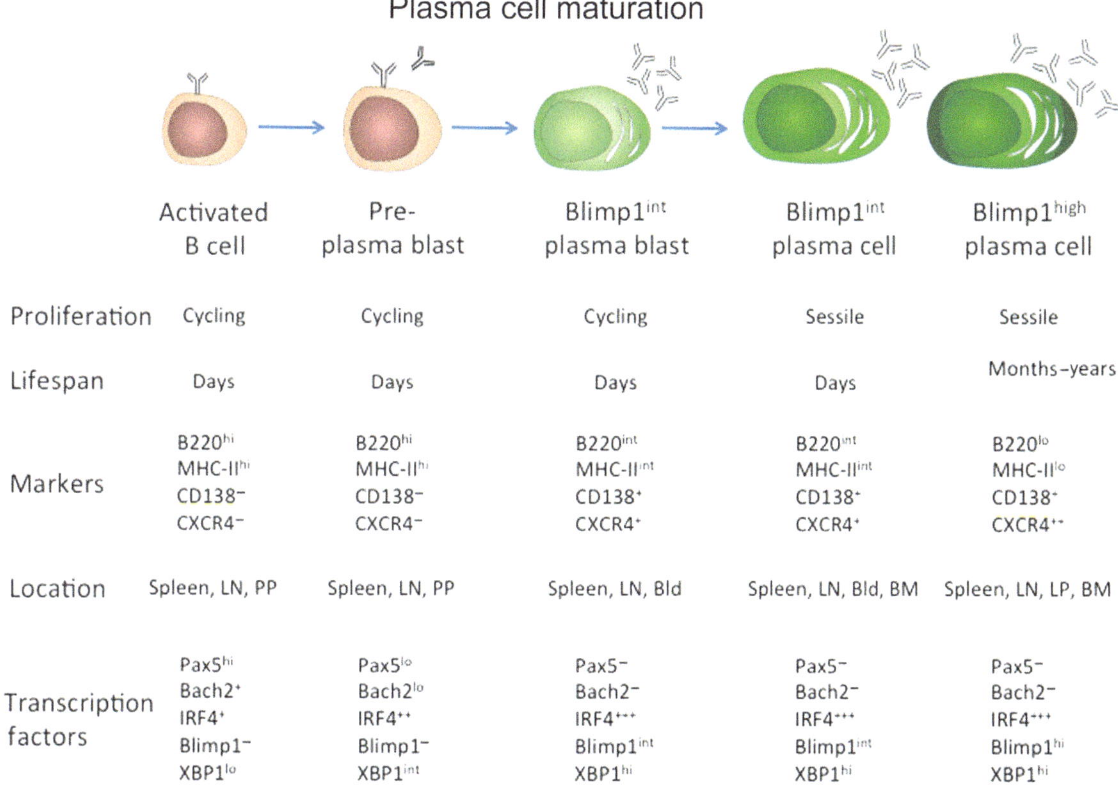


FIGURE 2 **Schematic representation of plasma cell maturation defined by Blimp1 expression and proliferation.** Markers differentially expressed refer primarily to work in mice. Abbreviations: LN, lymph node; PP, Peyer's patches; Bld, blood; BM, bone marrow; LP, lamina propria. "Lo," "int," and "hi" are used as relative terms to describe the amount of expression within plasma cell differentiation. See text for details on stages of plasma cell development and exceptions to some of the generalities depicted in the figure.

The B1 cells producing nAb reside primarily in the spleen and bone marrow and most probably result from the relocation of B1 cells stimulated to differentiate through a TLR-mediated process in other locations [16]. It has been proposed that B1 cells spontaneously and constitutively secrete antibody, and that this can occur in the absence of Blimp1 expression [17]. However, these observations were not verified by the Blimp1$^{gfp}$ reporter line, which, although revealing that the B1 cells in the peritoneum expressed Blimp1 albeit at low amounts, failed to detect spontaneous or constitutive secretion. In addition, stimulation of B1 cells into ASCs resulted in Blimp1-expressing cells more or less indistinguishable from those derived from B2 cells [18]. This does not mean that those B1 cells that differentiate into plasma cells in vivo in the absence of deliberate antigenic stimulation are not distinct or unique in ways yet to be described.

## 3.2 TI Immune Responses

TI antigens, capable of inducing antibody production without the presence or involvement of T cells, are divided into one class that is itself a polyclonal B cell mitogen and a second class that stimulates B cells only through the BCR.

TI type 1 antigens (TI-1) are typified by LPS, a cell wall component of Gram-negative bacteria that stimulates B cell proliferation and activates an inherent differentiation program through stimulation of TLR4. This process acts independently of the BCR; thus, such antigens typically induce a significant polyclonal B cell response upon introduction into an animal. PBs that develop in vivo after LPS stimulation are inherently short lived and secrete predominantly IgM with some switching to IgG3 in mouse. Marginal zone (MZ) B cells appear to be particularly rapid in differentiating in response to LPS [1].

Type 2 TI (TI-2) antigens are typified by long-chain polysaccharides from the cell walls of encapsulated bacteria such as *Streptococcus pneumoniae*. Model TI-2 antigens are the polysaccharides ficoll and dextran and the synthetic polymer acrylamide. Immunization of mice with these antigens elicits a PB response, again from predominantly MZ B cells with some contribution from B1 cells, particularly to dextran. These PBs initially appear as foci at the bridging channels of the white pulp in the spleen, sustained for a relatively brief period by the provision of the B cell survival protein, BAFF (B-cell activating factor belonging to the tumor necrosis factor family), itself produced by dendritic cells that probably migrate to this location as a consequence of the same immunization [19]. Therefore, this reaction is considered to be self-limiting. One reported exception to a transient TI-2 PB reaction was the continuous production of such PBs in the spleens in lymphopenic mice that were reconstituted by transfer of peritoneal B1 cells followed by immunization with hapten coupled to a TI-2 carrier [20]. Although this system resulted in the continuous production of PBs for months and clearly revealed this potential in B1 cells, it is important to remember that the environment surrounding the response was artificial. In general, responses occur in B cell-replete mice with circulating Ig. However, in such replete animals, persistent antibody production after immunization with haptenated TI-1 antigen was recently reported to be due to the presence of bone marrow resident antigen-specific plasma cells [21]. This response was elicited in T cell-deficient mice and reveals the potential for plasma cells of any origin to gain access to locations in which they can persist for considerable periods.

## 3.3 αGalCer

At the interface of TI and TD antigens are those involving the glycolipid αGalCer (αGC). Although this has long been known as a potent agonist of natural killer T (NKT) cells, the immunization of mice with conventional B cell antigens coupled to αGC revealed a remarkable potency for NKT cells to stimulate B cells capable of presenting αGC antigens through CD1 in a Bcl6- and IL-21 dependent manner [22,23]. Such immunizations resulted in copious amounts of IgM and switched isotype antibodies derived from PBs located in foci at the bridging channels of the spleen. Although this response also initiated germinal centers (GCs), these were short lived and nonproductive in terms of memory and affinity maturation. Given that only MZ B cells express the CD1 restriction element, it appears most likely that this B-cell population is the one from which these responses originate. Again, the ASCs derived from this reaction are short lived, with little evidence for migration to and occupation of survival niches in the bone marrow.

## 3.4 TD Immune Responses and Extrafollicular Foci

Protein-containing antigens require the involvement of cognate CD4 T cells to initiate and propagate the immune response [1,19]. Antigen-activated B and T cells migrate toward the border of their respective areas of the lymphoid organ in which the response is occurring, driven by the altered chemokine receptor expression that follows antigen priming [24,25]. For B cells, this means increased expression of CCR7, the ligands for which, CCL19 and CCL21, are expressed in the T cell areas of the organs. This migration is balanced by continued expression of CXC chemokine receptor (CXCR)-5, the receptor for the follicular homing chemokine, CXC chemokine ligand (CXCL)-13, thereby ensuring that the B cells remain at the follicle boundary rather than moving onto the T cell area. T cells increase expression of CXCR5, thereby promoting their migration toward the B cell follicles. The co-localization of antigen-activated B and T cells increases significantly the probability of their interacting, which, when it occurs, is long lasting. These interactions upregulate the G-protein-coupled receptor Epstein–Barr virus-induced protein-2 (EBI2) on B cells while also diminishing CCR7, driving the B cell to the outer follicle or to the intrafollicular region. The T cells co-migrate with the B cells, continuing to interact and indeed inducing the first rounds of B cell proliferation. Some of these proliferating B cells upregulate the key GC transcription factor, Bcl6, which represses EBI2, causing in those B cells migration to the center of the follicle, positioning on the follicular dendritic cell network, and what is the beginning of a GC. Some of the activated T cells, now identifiable as specialized follicular helper T cells (Tfh), also migrate into the follicle [26]. Of the B cells remaining in the outer follicle, some migrate toward the bridging channels in the spleen or the perivascular areas in lymph nodes, giving rise to foci of ASCs.

Foci of ASCs develop within days after TD immunization [19]. They initially secrete IgM, with other isotypes appearing subsequently, but there is no evidence for affinity maturation among these PBs, and they are of limited survival capacity, dying in situ by apoptosis within several days of their formation [11]. The nature of the T cells that provide help in this process remains incomplete, but there are insights from studies of normal and diseased mice. Bcl6-deficient T cells are effectively incapable of generating foci of PBs, and this is mediated in part by defective production of IL-21 [27]. A role for IL-21 was already suggested from studies of IL-21 receptor-deficient mice and by the identification in the autoimmune MRL(lpr) strain of an IL-21-producing subset of T helper cells that was required for extrafollicular foci formation [28,29]. Although development of these cells in MRL mice was dependent on inducible co-stimulator, they were distinguishable from Tfh by expressing low amounts of CD162 while being CXCR4$^{+}$, a feature that presumably explained their co-localization with PBs [29].

PBs in extrafollicular foci progress to be short-lived plasma cells. They downregulate markers associated with proliferation such as Ki 67 and assume a phenotype consistent with plasma cells, such as increased expression of the survival factor receptor B cell maturation antigen (BCMA). The progression of PB to plasma cell documents an immune response to exogenous antigen in a lymph node, where the differentiation accompanied their migration from the outer T cell zone to the medulla of the node [30]. This migration, along a CXCL12 gradient, localized the plasma cells in a region of the node enriched for monocytes expressing the ligand for BCMA, APRIL (a proliferation-inducing ligand), a well-defined plasma cell survival factor. Despite this co-localization, such plasma cells rarely persist for more than 7–14 days, dying in situ by apoptosis that can be blocked by enforced expression of *Bcl2* [31]. Short-lived plasma cells of the foci can, in the right circumstances, contribute to the long-lived plasma cell population. In experiments where other forms of plasma cell production were blocked, it was possible to detect foci-derived plasma cells in the bone marrow [21]. Again, although showing what is possible, this phenomenon is in all probability quite rare in normal circumstances where most long-lived plasma cells derive from the GC. However, one should note the influence of antigen and adjuvant on this process. Vaccination with the yellow fever vaccine establishes an exceedingly long-lived plasma cell compartment in the draining lymph nodes, probably because of the TLRs stimulated by that vaccine [32].

It has been proposed that existing high-affinity B cells are preferentially recruited into foci of ASCs at the initiation of the TD response, leaving cells of more modest affinity to enter the GC, where they are subjected to affinity maturation [33]. Although making a degree of teleological sense—one can imagine the benefit of taking the "best" B cells activated by antigen and using them for immediate antibody production—it appears the situation is not quite like this. In reality, it appears that PB proliferation during extrafollicular differentiation is very sensitive to the avidity of the BCR-antigen interaction, much more so than for B cells entering the GC [34]. That is, under conditions of high avidity, antigen-specific B cells differentiating along the PB pathway undergo significantly more proliferation than their peers undergoing GC differentiation, whereas in cases of low avidity, there is more equal expansion along each pathway, giving an appearance of affinity-driven preferential PB differentiation.

## 3.5 TD Immune Responses and GCs

The alternative differentiation pathway for B cells activated by protein antigens and in receipt of T cell help is the formation of GCs [35]. These structures are the sites of intensive B cell proliferation, activation-induced cytidine deaminase-mediated Ig V gene somatic hypermutation, and the selective survival and expansion of B cells with BCR of improved affinity for the immunizing antigen. The expansion of GC B cells with improved antigen binding underlies the phenomenon of affinity maturation in which the average affinity for antigen increases during the course of an immune response. This increase can be of orders of magnitude, reflecting the power of this process that selects and expands the best antigen-binding B cells. Of course, having some of these cells differentiate into plasma cells is fundamental to the success of the immune response because GC-derived plasma cells producing high-affinity antibody are the basis of long-term immunity after infection and vaccination. The selective differentiation of high-affinity B cells from among the population of GC B cells and their subsequent retention as long-lived plasma cells—for decades in humans—in survival niches in the bone marrow are among the most intensively researched areas in plasma cell biology. Although survival is addressed below, the process of differentiation will be considered here.

Within the GC, B cells with V gene combinations that confer high-affinity binding are preferentially induced to differentiate into plasma cells [1]. This was first observed with certainty in the analysis of the response to the hapten, nitrophenyl [36]. In this response, which has limited capacity for affinity maturation (~10 fold), it was apparent in comparing the frequency and distribution of somatic mutations in the V genes of antigen-specific plasma cells from the bone marrow and those of GC B cells from the spleen at the same early time that those plasma cells that were mutated (1) were mutated at a very low frequency per V gene and (2) were greatly enriched for a single mutation that improved affinity by the maximal 10-fold. That is, there was very strong selection for those GC B cells with mutations that improved affinity to differentiate into plasma cells in the early stages of the response when mutations were rare. This result, confirmed in other systems [37], now underpins much of the thinking about how GCs operate to generate plasma cells.

In contemplating that the differentiation of a GC B cell into a plasma cell could be a selective process based on affinity, one may consider several possible mechanisms [38]. The BCR on the high-affinity GC B cell could transduce a distinct signal that stimulates differentiation over continued proliferation. A problem with this mechanism is that BCR signaling is apparently not crucial to GC B cell selection and differentiation. For example, mutations in BCR signaling components such as CD45 do not affect affinity maturation of GC B cells or serum Ig amounts. In addition, it also appears that signaling from the BCR is suppressed in GC B cells [39], making it difficult, but not impossible, to see how such a signal could selectively trigger plasma cell differentiation. When one considers in vitro differentiation of B cells into PBs, the factors that best promote this outcome are T cell-derived signals such as CD40L, IL5, and IL21 [4,40]. In this setting, the probability of a B cell differentiating into a plasma cell is related to the amount of T cell help it receives. When one transfers that observation back to the in vivo GC, it may be that the crucial link between affinity for antigen and plasma cell differentiation is the capacity of the B cell to recruit T cell help. High-affinity B cells presumably capture, process, and/or present more antigen than their low-affinity peers, allowing for more frequent and longer interactions with GC T cells. It is further possible that the duration or intensity of these B–T interactions influences the nature and quantity of cytokines produced by the GC T cells, which in turn affects differentiation.

It remains uncertain which of the cytokines produced by Tfh actually influence GC B cell differentiation. In vitro, IL5 acts as a potent amplifier of plasma cell differentiation of mouse B cells, but mice deficient in IL5 show no obvious defect in plasma cell formation or function. Others have reported on the crucial role for Tfh-produced IL4 in the activity of the GC [41], although in vitro work suggests it is a potent switch and survival factor rather than a differentiation factor. IL21 is the obvious candidate, being the unique cytokine produced by Tfh, and mice deficient in IL21 or the IL21 receptor show severe deficits in antibody production and it clearly has this role for human B cells. However, IL21 deficiency is rather more complex than a simple failure to promote differentiation. There is a 10- to 100-fold deficiency in extrafollicular PB formation in the absence of IL21 in a model immune response, and it is also clear that bone marrow resident plasma cells are significantly reduced in IL21-deficient mice [28,42]. What is not clear is whether these deficiencies result from a reduction in B cell proliferation or a premature termination of the GC reaction itself. That is, there is a clear relationship between B cell proliferation and differentiation; therefore, a reduced frequency of plasma cells in the extrafollicular foci and the bone marrow in the absence of IL21 signaling could reflect diminished proliferation rather than a missing differentiation signal or it could be a combination of both. Unfortunately, the contribution of IL21 to plasma cell formation in mouse has been difficult to quantify in vitro because it induces apoptosis in stimulated B cells. Despite this, it remains an excellent candidate. In humans, IL21 acting in combination with IL2 is particularly potent in inducing B cell differentiation, and patients with IL21 receptor deficiency show the expected failure to produce antigen-specific antibodies, confirming an in vivo role [43].

The timing of plasma cell formation from the GC is another important variable in immunity to infection and immunization. Although many models show that GCs produce memory B cells and plasma cells continuously during the course of the response, an alternative was recently proposed suggesting that production might be biphasic [44]. Early studies examining the accumulation of V-gene somatic mutations in bone marrow plasma cells and memory B cells in conventional mice showed both to increase during the course of the response, although the frequency of mutations in bone marrow plasma cells did not reach that in memory B cells, being approximately 50% and suggesting that memory B cell production by the GC might continue beyond that of bone marrow plasma cells [36]. However, the more recent proposal posits that the GC first produces memory B cells and that bone marrow plasma cells are a later product. This proposal is based on kinetic data of when long-term, persistent memory B cells and plasma cells arise in an immune response using BCR transgenic B cells. The phenomena of early and late production need to be shown in additional systems before being accepted as the norm, and if correct, the question of what triggers differentiation and switching between plasma cell and memory B cell production remains unanswered. One potential advantage of the early plasma cell model of GC operation is that it provides a means for the rapid appearance of high-affinity antibody that can in effect feed back into the GC reaction to sequester antigen and thereby drive affinity maturation and later curtail the response. Although often speculated, there is recent evidence to support the idea of antibody feedback into the GC [45].

The final consideration of plasma cell recruitment out of the GC concerns not the affinity but specificity. A mouse study using West Nile virus as an infection model revealed that the antibody secreted from the plasma cells in bone marrow was capable of neutralizing the original infecting strain but not a viral escape mutant [46]. However, it was remarkable that the memory B cell compartment contained cells capable of producing plasma cells that could neutralize the escape variant. That is, the plasma cell response had acquired high affinity but with narrow specificity, suggesting that selection in the GC of plasma cells and memory B cells operates by distinct mechanisms. The reduced emphasis on affinity in the memory B cell compartment fits with an earlier comparison of the affinity of the antibody produced by individual memory B cells with that from plasma cells that revealed more stringent, affinity-based selection into the bone marrow plasma cell compartment compared with the memory B cell compartment [36].

The possible segregation of function into discrete memory B cell compartments has received a significant fillip with the recent examination of IgM-expressing memory B cells in mice. Although long identified in humans using CD27 as a pan-memory B cell marker, the existence of such cells in mice was difficult to prove. Using elegant genetic and cell isolation strategies, IgM memory B cells were recently characterized in mice [47,48]. It is interesting to note that the IgM compartment was identified as being predisposed to form GC upon re-exposure to cognate antigen. On the other hand, IgG memory B-cells preferentially differentiated into plasma cells. Using very elegant experimental techniques, a mouse was created with naïve B cells expressing an IgG1 rather than IgM BCR. However, such naïve IgG1 B cells did not show any propensity to differentiate into plasma cells, revealing that such a phenomenon was a property of being a memory B cell rather than expressing IgG per se [49]. These experiments also generated data suggesting that the possible molecular basis for this predisposition was the low amounts of the transcription factor Bach2, a key inhibitor of plasma cell differentiation, expressed in IgG memory B cells.

## 4. MOLECULAR BIOLOGY OF PLASMA CELL DIFFERENTIATION

The transformation from B cell to plasma cell requires a dramatic change in the transcription program of the cell [50]. The B cell has to silence genes that define B cell identity and function and express the corresponding genes for plasma cell identity and function. Each state is relatively stable and maintained by what are referred to as master regulators such as Pax5 and Bcl6 for the B cell state and interferon regulatory factor-4 (IRF4) and Blimp1 for the plasma cell state. These states are mutually antagonistic such that the transcription factors promoting the plasma cell pattern inhibit expression of the B cell pattern whereas those that maintain the B cell pattern repress those that promote plasma cell differentiation. Although such a system makes it relatively easy to understand the stability of each state, it does not easily explain the initiation of differentiation because a B cell has to effectively lose its identity before assuming that of a plasma cell.

Blimp1 is often referred to as the master regulator of plasma cell differentiation based in part on its ubiquitous expression in plasma cells and on the severe defect in plasma cell development in Blimp1-deficient mice [50,51]. Blimp1 acts primarily as a transcriptional repressor, and among its targets are *Pax5* and *Bcl6*, genes that sustain B cell identity in general and in GCs, respectively. This activity of Blimp1 has led to a relatively simple model for explaining plasma cell differentiation. Induction of Blimp1 by whatever means turns off Pax5, a repressor of Blimp1, thereby silencing the B cell program and allowing the plasma cell program to develop [51]. This scheme, although still frequently invoked, fails to account for several key features of the molecular regulation of plasma cell differentiation.

Blimp1-deficient B cells can be induced to initiate differentiation into what are termed pre-PBs [52]. These cells, attempting to express Blimp1 but unable to do so, actually gain some characteristics of plasma cells and lose some of those of B cells. The pre-PBs increase expression of *Xbp1* and *Igj*, both ASC-associated genes, and show reduced activity of Pax5. Despite initiating plasma cell differentiation without Blimp1 expression, they do not complete the process. Given the crucial observation in this system of the reduced activity of Pax5 in the pre-PBs, an alternative model of differentiation has been developed based on the loss of Pax5 activity as the initiating event [50]. This by itself acts to de-repress a multitude of genes, including *Xbp1* and *Igj*, while reducing expression of others including *Bach2*, another transcription repressor, the expression of which is dependent on Pax5. One of the crucial targets of Bach2 repression is Blimp1 itself. Thus, reduced Bach2 promotes Blimp1 expression, which then amplifies the differentiation process by further repressing Pax5 and further reducing Bach2 until the change from B cell to plasma cell is complete. In this model, a key early event is the reduction in Pax5, although what induces this remains uncertain, although IRF4 appears as an attractive candidate. Finally, why IgG memory B-cells preferentially differentiate into plasma cells in vivo, mentioned earlier, becomes clear with their inability to induce Bach2 expression [49].

Bcl6 is often labeled as a key negative regulator of plasma cell differentiation, functioning particularly in GCs to suppress differentiation while affinity maturation occurs [51]. It is often drawn in a mutually antagonistic relationship with Blimp1, and there is evidence to support this proposal. However, it is not clear how universal this relationship is or whether silencing Bcl6 is obligatory for plasma cell differentiation. Although Blimp1 has the capacity to repress Bcl6 expression, there are examples of co-expression of Bcl6 and Blimp1 in cell lines, and the in vitro differentiation of Bcl6-deficient B cells into ASCs is relatively unaffected, just as a Bcl6 transgene in B cells does not unduly repress plasma cell formation [50]. That is, the reciprocal relationship between Blimp1 and Bcl6 may reflect an association rather than a dependency.

A transcription factor that is absolutely crucial in plasma cell formation is IRF4, which is also required for class switching and GC B cell development [50]. The involvement of IRF4 in these processes appears to be dependent on concentration, with naïve B cells expressing relatively low amounts of IRF4, intermediate amounts triggering class switch recombination and GC formation, and high amounts appearing to activate Blimp1, thereby triggering plasma cell differentiation [53,54]. High amounts of IRF4 can also repress Bcl6, although, as mentioned above, the relevance of this to differentiation is uncertain. In naïve B-cells, IRF4 expression is restrained by another transcription factor, Mitf, such that Mitf-deficient B-cells differentiate spontaneously into plasma cells. What regulates Mitf remains undetermined [50].

The pivotal role of IRF4 in plasma cell differentiation and the fact that it is upstream of Blimp1 suggests that increasing the concentration of this factor may be the crucial event initiating plasma cell differentiation [50]. However, how this change in IRF4 is mediated is currently unclear, as is any connection between IRF4 and Pax5. Once IRF4 induces Blimp1, the balance in gene expression is probably irreversibly tipped in favor of differentiation. This proposal fits well with the putative function of IRF4 in the regulation of GC B cells. In this model, IRF4 is induced by T cell stimulation through CD40; it represses Bcl6 [55]; and, potentially if induced sufficiently, it induces plasma cell differentiation. Such a role would link the strength or duration of T cell signaling with plasma cell formation in the GC. It is interesting to note that IL21, the cytokine produced by Tfh in GC, induces phosphorylated signal transducer and activator of transcription-3, which was found to bind to a regulatory region of the Blimp1 gene as part of a complex that included and depended on IRF4. This again suggests a key role for IL21 in GC differentiation.

It remains to be seen what roles these factors have in maintaining plasma cells. It appears based on studies with multiple myeloma lines that IRF4 may be crucial for plasma cell survival [56] whereas preliminary studies on the deletion of Blimp1 in existing bone marrow plasma cells suggests that it is also continually required [57]. However, this latter experiment may not have distinguished between persistence and formation after transfer, leaving the issue unresolved.

## 5. PLASMA CELL SURVIVAL

Despite persisting for decades in humans and many months in mice, plasma cells are intrinsically short lived. That is, if removed from their in vivo location and placed in culture in vitro, most die within a day. Clearly exogenous signals are continually provided to plasma cells to sustain them in vivo. Equally clearly, given the discussion on short- and long-lived plasma cells, the provision of these signals may vary in quality, quantity, or duration depending on location and possibly on the plasma cell origin. One also has to consider whether all of the stimuli that sustain plasma cells are equal in their ability to do so and the nature of the cells providing these signals. Finally, one should consider the processing of these survival signals by plasma cells into a molecular outcome that sustains the cells as viable, antibody-secreting entities.

Factors that can sustain plasma cells have been defined by a combination of in vitro and in vivo experimentation. The in vitro work has taken the approach of culturing plasma cells in media with defined survival factors and measuring the viability [1,58–60]. The in vivo work relies on ablation of a factor by antibody treatment or genetic engineering to mutate the gene encoding either the factor or its receptor. These strategies have identified potential roles for soluble and membrane-bound factors. IL6, tumor necrosis factor-α, CXCL12, APRIL, and BAFF all have the ability to support plasma cell viability, as do vascular cell adhesion molecule-1 and intercellular adhesion molecule-1, binding respectively to very late antigen-4 and lymphocyte function-associated antigen-1, and hyaluronic acid and fibronectin binding to CD44. In addition, there are survival factors that are considered to operate in inflammatory situations, specifically CXCL9, 10, and 11 binding to CXCR3 [58]. Recently, the T cell co-stimulatory molecule CD28 has been identified as providing survival signals to plasma cells by binding its ligands CD80/86 [61]. It is interesting to note that CD28 is repressed in B cells by Pax5, so its early re-expression in plasma cells is fully consistent with their loss of Pax5 activity [62].

Given the range of factors that able to support plasma cell survival, it is not surprising that there is redundancy. Thus, deletion of IL6 or treatment with blocking antibody has only a minimal effect on plasma cell frequency or persistence in mice. Likewise, loss of CD44, CD28, or blocking VLA4 have a relatively modest effect on plasma cell frequency, indicating that whatever their in vivo roles might be, they are able to be replaced by any or several of the remaining survival molecules. Equally, loss of CXCR4, the receptor for CXCL12, leads to a significant reduction in plasma cell frequency in the periphery and the bone marrow, but to what extent this latter defect reflects chemotaxis rather than survival has not been determined. The pathway that appears unique and for which differences in production, migration, and survival have been resolved is APRIL binding to its receptor BCMA. This pathway appears to be a crucial mediator of long-term survival of plasma cells, especially in bone marrow [64].

Animals deficient in either APRIL or BCMA fail to establish a normal bone marrow compartment, and treatment of mice with blocking reagents (e.g., TACI receptor fused to human Ig) produced a similar outcome. The origin of APRIL varies according to the tissue. In peripheral lymphoid organs it is produced by a subset of macrophages whereas in bone marrow it is produced by a range of cells including monocytes, eosinophils, and megakaryocytes, all of which have been identified as being important for plasma cell survival [60]. In humans, use of anti-APRIL reagents to diminish antibody production in autoimmune patients has some clear effects, but it does not lead to the loss of all antibody.

The long-term survival of plasma cells requires their co-localization with cells producing survival factors. One current proposal is that many of the cells producing survival factors are responsive to the chemokine CXCL12, as are plasma cells themselves in their immature state [65]. In this way, plasma cells and the cells that support them will naturally congregate around CXCL12-expressing stromal cells, thereby forming a niche in which the survival factors will be available. Indeed, examination of the localization of plasma cells in bone marrow revealed 95% of IgG plasma cells to be adjacent to CXCL12-expressing cells, clearly suggesting the importance of this factor in organizing the niche, especially when coupled with the corresponding defect in the CXCR4 knockouts. Whether multiple different APRIL-producing cell types need to be in all niches is unclear, but flexibility in this may explain the various observations that have been made on the composition of survival niches [58].

The ability of immature plasma cells to migrate along a CXCL12 gradient may well be crucial not only to their survival but also to maintaining plasma cell homeostasis. It has been observed that mature plasma cells lose the ability for chemotaxis, meaning that if they are displaced from a survival niche, they will be unable to re-establish themselves [8,58]. Thus, newly formed plasma cells will have the ability to displace and replace existing plasma cells at some frequency. Evidence for the phenomenon of new plasma cells pushing out old plasma cells can be seen in the blood of boosted humans in which after the boost, a population of phenotypically mature plasma cells circulating in the blood can be seen, which are presumed to be plasma cells recently displaced from survival niches. This turnover presumably allows for new specificities to enter into the long-lived pool but at a rate that prevents the sudden loss of all previous plasma cell memory.

## 6. PLASMA CELL SURVIVAL MOLECULES

To persist, plasma cells have to convert the binding of extrinsic factors to their receptors into a molecular state that sustains the cells. However, the various factors capable of sustaining plasma cells and the various receptors and associated signaling pathways involved provides a daunting challenge. Although the cytokine IL6 clearly induces activation of STAT molecules that are in turn associated with increased expression of prosurvival molecules of the Bcl2 family, whether this is the actual basis of the IL6 survival signal remains unclear. Likewise, APRIL signaling results in the activation of NFκB transcription factors, which are known to promote transcription of Bcl2 family member prosurvival genes in a multitude of cell types, but again whether this is the crucial component of APRIL-mediated survival of plasma cells remains unclear.

It is clear from many studies that prosurvival members of the Bcl2 family of transcription factors influence plasma cell survival. Early studies of Bcl2 transgenic mice revealed significantly enhanced plasma cell survival in vitro and in vivo, suggesting that one of the limits on plasma cell persistence could be expression of Bcl2 family members [66]. The Bcl2 family comprises several prosurvival members such as Bcl2, $Bclx_L$, and Mcl1 that function to impede the activation of death-promoting members of the family such as Bim [67]. This is done by the binding of the prosurvival members to the death-promoting members through a specific interaction involving the BH3 domain on the death-promoting partner binding to a hydrophobic groove on its prosurvival partner. The specificity of this interaction has allowed for the development of mimetics of the BH3 domain that can act as small molecular triggers of apoptosis by uncoupling prosurvival prodeath interactions. One of these molecular triggers of apoptosis is the compound ABT-737, which has specificity for interactions involving the prosurvival members Bcl2, $Bclx_L$, and Bclw [68]. Thus, treatment with ABT-737 will induce the death of cells dependent on Bcl2, $Bclx_L$, or Bclw for their survival. This feature has been used to probe the survival mechanism of plasma cells, and it was found that plasma cells generated in a previous immune response and located in spleen and bone marrow were resistant to ABT-737 treatment because their frequency was unaffected by treatment [69]. However, the migration or establishment of plasma cells in the bone marrow was sensitive to treatment, indicating that at some stage in development or movement to the bones plasma cells did depend on one or several of Bcl2, $Bclx_L$, or Bclw for survival.

The identification of the prosurvival protein plasma cells required for their persistence once in a niche was revealed by in situ gene deletion experiments in which Mcl1, another prosurvival member of the Bcl2 family, was deleted from existing plasma cells. This resulted in the almost immediate death of greater than 90% of plasma cells in spleen and bone marrow [70]. Similar deletion of $Bclx_L$ had no obvious effect, revealing the specificity in this role for Mcl1. Mcl1 is highly expressed in plasma cells of mouse and human, suggesting strongly that its expression is the ultimate endpoint of the signaling pathways stimulated by the binding of the various exogenous prosurvival factors that constitute the plasma cell survival niche to their receptors. This may well provide a means of identifying the biochemical intermediates in survival signaling and afford a means of interfering with these processes so as to modulate plasma cell frequency, particularly in situations of plasma cell disease.

## REFERENCES


[1] Oracki SA, Walker JA, Hibbs ML, Corcoran LM, Tarlinton DM. Plasma cell development and survival. Immunol Rev 2010;237:140–59.

[2] Tarlinton D, Radbruch A, Hiepe F, Dorner T. Plasma cell differentiation and survival. Curr Opin Immunol 2008;20(2):162–9.

[3] Fagraeus A. The plasma cellular reaction and its relation to the formation of antibodies in vitro. J Immunol 1948;58:1–13.

[4] Hasbold J, Corcoran LM, Tarlinton DM, Tangye SG, Hodgkin PD. Evidence from the generation of immunoglobulin G-secreting cells that stochastic mechanisms regulate lymphocyte differentiation. Nat Immunol 2004;5:55–63.

[5] Tangye SG, Hodgkin PD. Divide and conquer: the importance of cell division in regulating B-cell responses. Immunology 2004;112:509–20.

[6] Ettinger R, Sims GP, Fairhurst AM, Robbins R, da Silva YS, Spolski R, et al. IL-21 induces differentiation of human naive and memory B cells into antibody-secreting plasma cells. J Immunol 2005;175:7867–79.

[7] Bryant VL, Ma CS, Avery DT, Li Y, Good KL, Corcoran LM, et al. Cytokine-mediated regulation of human B cell differentiation into Ig-secreting cells: predominant role of IL-21 produced by $CXCR5^+$ T follicular helper cells. J Immunol 2007;179:8180–90.

[8] Odendahl M, Mei H, Hoyer BF, Jacobi AM, Hansen A, Muehlinghaus G, et al. Generation of migratory antigen-specific plasma blasts and mobilization of resident plasma cells in a secondary immune response. Blood 2005;105:1614–21.

[9] Wrammert J, Smith K, Miller J, Langley WA, Kokko K, Larsen C, et al. Rapid cloning of high-affinity human monoclonal antibodies against influenza virus. Nature 2008;453:667–71.
[10] Jacob J, Kassir R, Kelsoe G. In situ studies of the primary immune response to (4-hydroxy-3- nitrophenyl)acetyl. I. The architecture and dynamics of responding cell populations. J Exp Med 1991;173: 1165–75.
[11] Smith KG, Hewitson TD, Nossal GJ, Tarlinton DM. The phenotype and fate of the antibody-forming cells of the splenic foci. Eur J Immunol 1996;26:444–8.
[12] Kallies A, Hasbold J, Tarlinton DM, Dietrich W, Corcoran LM, Hodgkin PD, et al. Plasma cell ontogeny defined by quantitative changes in blimp-1 expression. J Exp Med 2004;200:967–77.
[13] Ho F, Lortan JE, MacLennan IC, Khan M. Distinct short-lived and long-lived antibody-producing cell populations. Eur J Immunol 1986;16:1297–301.
[14] Manz RA, Thiel A, Radbruch A. Lifetime of plasma cells in the bone marrow. Nature 1997;388:133–4.
[15] Fooksman DR, Schwickert TA, Victora GD, Dustin ML, Nussenzweig MC, Skokos D. Development and migration of plasma cells in the mouse lymph node. Immunity 2010;33:118–27.
[16] Baumgarth N. Innate-like B cells and their rules of engagement. Adv Exp Med Biol 2013;785:57–66.
[17] Tumang JR, Frances R, Yeo SG, Rothstein TL. Spontaneously Ig-secreting B-1 cells violate the accepted paradigm for expression of differentiation-associated transcription factors. J Immunol 2005;174:3173–7.
[18] Fairfax KA, Corcoran LM, Pridans C, Huntington ND, Kallies A, Nutt SL, et al. Different kinetics of blimp-1 induction in B cell subsets revealed by reporter gene. J Immunol 2007;178:4104–11.
[19] MacLennan IC, Toellner KM, Cunningham AF, Serre K, Sze DM, Zuniga E, et al. Extrafollicular antibody responses. Immunol Rev 2003;194:8–18.
[20] Hsu MC, Toellner KM, Vinuesa CG, Maclennan IC. B cell clones that sustain long-term plasmablast growth in T-independent extrafollicular antibody responses. Proc Natl Acad Sci USA 2006;103:5905–10.
[21] Bortnick A, Allman D. What is and what should always have been: long-lived plasma cells induced by T cell-independent antigens. J Immunol 2013;190:5913–8.
[22] Chang PP, Barral P, Fitch J, Pratama A, Ma CS, Kallies A, et al. Identification of Bcl-6-dependent follicular helper NKT cells that provide cognate help for B cell responses. Nat Immunol 2012;13:35–43.
[23] King IL, Fortier A, Tighe M, Dibble J, Watts GF, Veerapen N, et al. Invariant natural killer T cells direct B cell responses to cognate lipid antigen in an IL-21-dependent manner. Nat Immunol 2012;13:44–50.
[24] Pereira JP, Kelly LM, Cyster JG. Finding the right niche: B-cell migration in the early phases of T-dependent antibody responses. Int Immunol 2010;22:413–9.
[25] Gatto D, Brink R. B cell localization: regulation by EBI2 and its oxysterol ligand. Trends Immunol 2013;34:336–41.
[26] Okada T, Moriyama S, Kitano M. Differentiation of germinal center B cells and follicular helper T cells as viewed by tracking Bcl6 expression dynamics. Immunol Rev 2012;247:120–32.
[27] Lee SK, Rigby RJ, Zotos D, Tsai LM, Kawamoto S, Marshall JL, et al. B cell priming for extrafollicular antibody responses requires Bcl-6 expression by T cells. J Exp Med 2011;208:1377–88.
[28] Zotos D, Coquet JM, Zhang Y, Light A, D'Costa K, Kallies A, et al. IL-21 regulates germinal center B cell differentiation and proliferation through a B cell-intrinsic mechanism. J Exp Med 2010;207:365–78.
[29] Odegard JM, Marks BR, DiPlacido LD, Poholek AC, Kono DH, Dong C, et al. ICOS-dependent extrafollicular helper T cells elicit IgG production via IL-21 in systemic autoimmunity. J Exp Med 2008;205:2873–86.
[30] Mohr E, Serre K, Manz RA, Cunningham AF, Khan M, Hardie DL, et al. Dendritic cells and monocyte/macrophages that create the IL-6/APRIL-rich lymph node microenvironments where plasmablasts mature. J Immunol 2009;182:2113–23.
[31] Smith KG, Light A, O'Reilly LA, Ang SM, Strasser A, Tarlinton D. bcl-2 transgene expression inhibits apoptosis in the germinal center and reveals differences in the selection of memory B cells and bone marrow antibody-forming cells. J Exp Med 2000;191:475–84.
[32] Kasturi SP, Skountzou I, Albrecht RA, Koutsonanos D, Hua T, Nakaya HI, et al. Programming the magnitude and persistence of antibody responses with innate immunity. Nature 2011;470:543–7.
[33] Paus D, Phan TG, Chan TD, Gardam S, Basten A, Brink R. Antigen recognition strength regulates the choice between extrafollicular plasma cell and germinal center B cell differentiation. J Exp Med 2006;203:1081–91.
[34] Chan TD, Gatto D, Wood K, Camidge T, Basten A, Brink R. Antigen affinity controls rapid t-dependent antibody production by driving the expansion rather than the differentiation or extrafollicular migration of early plasmablasts. J Immunol 2009;183(5):3139–49.
[35] Victora GD, Nussenzweig MC. Germinal centers. Annu Rev Immunol 2012;30:429–57.
[36] Smith KG, Light A, Nossal GJ, Tarlinton DM. The extent of affinity maturation differs between the memory and antibody-forming cell compartments in the primary immune response. EMBO J 1997;16:2996–3006.
[37] Phan TG, Paus D, Chan TD, Turner ML, Nutt SL, Basten A, et al. High affinity germinal center B cells are actively selected into the plasma cell compartment. J Exp Med 2006;203:2419–24.
[38] Tarlinton DM, Smith KG. Dissecting affinity maturation: a model explaining selection of antibody-forming cells and memory B cells in the germinal centre. Immunol Today 2000;21:436–41.
[39] Khalil AM, Cambier JC, Shlomchik MJ. B cell receptor signal transduction in the GC is short-circuited by high phosphatase activity. Science 2012;336:1178–81.
[40] Avery DT, Bryant VL, Ma CS, de Waal Malefyt R, Tangye SG. IL-21-induced isotype switching to IgG and IgA by human naive B cells is differentially regulated by IL-4. J Immunol 2008;181:1767–79.
[41] Reinhardt RL, Liang HE, Locksley RM. Cytokine-secreting follicular T cells shape the antibody repertoire. Nat Immunol 2009;10:385–93.
[42] Linterman MA, Beaton L, Yu D, Ramiscal RR, Srivastava M, Hogan JJ, et al. IL-21 acts directly on B cells to regulate Bcl-6 expression and germinal center responses. J Exp Med 2010;207:353–63.
[43] Moens L, Tangye SG. Cytokine-mediated regulation of plasma cell generation: IL-21 takes center stage. Front Immunol 2014;5:65.
[44] Shlomchik MJ, Weisel F. Germinal center selection and the development of memory B and plasma cells. Immunol Rev 2012;247:52–63.
[45] Zhang Y, Meyer-Hermann M, George LA, Figge MT, Khan M, Goodall M, et al. Germinal center B cells govern their own fate via antibody feedback. J Exp Med 2013;210:457–64.
[46] Purtha WE, Tedder TF, Johnson S, Bhattacharya D, Diamond MS. Memory B cells, but not long-lived plasma cells, possess antigen specificities for viral escape mutants. J Exp Med 2011;208:2599–606.
[47] Dogan I, Bertocci B, Vilmont V, Delbos F, Megret J, Storck S, et al. Multiple layers of B cell memory with different effector functions. Nat Immunol 2009;10:1292–9.

[48] Pape KA, Taylor JJ, Maul RW, Gearhart PJ, Jenkins MK. Different B cell populations mediate early and late memory during an endogenous immune response. Science 2011;331:1203–7.
[49] Kometani K, Nakagawa R, Shinnakasu R, Kaji T, Rybouchkin A, Moriyama S, et al. Repression of the transcription factor Bach2 contributes to predisposition of IgG1 memory B cells toward plasma cell differentiation. Immunity 2013;39:136–47.
[50] Nutt SL, Taubenheim N, Hasbold J, Corcoran LM, Hodgkin PD. The genetic network controlling plasma cell differentiation. Semin Immunol 2011;23:341–9.
[51] Shapiro-Shelef M, Calame K. Regulation of plasma-cell development. Nature reviews. Immunology 2005;5:230–42.
[52] Kallies A, Hasbold J, Fairfax K, Pridans C, Emslie D, McKenzie BS, et al. Initiation of plasma-cell differentiation is independent of the transcription factor Blimp-1. Immunity 2007;26:555–66.
[53] Sciammas R, Shaffer AL, Schatz JH, Zhao H, Staudt LM, Singh H. Graded expression of interferon regulatory factor-4 coordinates isotype switching with plasma cell differentiation. Immunity 2006;25:225–36.
[54] Ochiai K, Maienschein-Cline M, Simonetti G, Chen J, Rosenthal R, Brink R, et al. Transcriptional regulation of germinal center B and plasma cell fates by dynamical control of IRF4. Immunity 2013;38:918–29.
[55] Saito M, Gao J, Basso K, Kitagawa Y, Smith PM, Bhagat G, et al. A signaling pathway mediating downregulation of BCL6 in germinal center B cells is blocked by BCL6 gene alterations in B cell lymphoma. Cancer Cell 2007;12:280–92.
[56] Shaffer AL, Emre NC, Lamy L, Ngo VN, Wright G, Xiao W, et al. IRF4 addiction in multiple myeloma. Nature 2008;454:226–31.
[57] Shapiro-Shelef M, Lin KI, Savitsky D, Liao J, Calame K. Blimp-1 is required for maintenance of long-lived plasma cells in the bone marrow. J Exp Med 2005;202:1471–6.
[58] Winter O, Dame C, Jundt F, Hiepe F. Pathogenic long-lived plasma cells and their survival niches in autoimmunity, malignancy, and allergy. J Immunol 2012;189:5105–11.
[59] Tangye SG. Staying alive: regulation of plasma cell survival. Trends Immunol 2011;32:595–602.
[60] Chu VT, Berek C. The establishment of the plasma cell survival niche in the bone marrow. Immunol Rev 2013;251:177–88.
[61] Rozanski CH, Arens R, Carlson LM, Nair J, Boise LH, Chanan-Khan AA, et al. Sustained antibody responses depend on CD28 function in bone marrow-resident plasma cells. J Exp Med 2011;208:1435–46.
[62] Delogu A, Schebesta A, Sun Q, Aschenbrenner K, Perlot T, Busslinger M. Gene repression by Pax5 in B cells is essential for blood cell homeostasis and is reversed in plasma cells. Immunity 2006;24:269–81.
[63] Hargreaves DC, Hyman PL, Lu TT, Ngo VN, Bidgol A, Suzuki G, et al. A coordinated change in chemokine responsiveness guides plasma cell movements. J Exp Med 2001;194:45–56.
[64] O'Connor BP, Raman VS, Erickson LD, Cook WJ, Weaver LK, Ahonen C, et al. BCMA is essential for the survival of long-lived bone marrow plasma cells. J Exp Med 2004;199:91–8.
[65] Belnoue E, Tougne C, Rochat AF, Lambert PH, Pinschewer DD, Siegrist CA. Homing and adhesion patterns determine the cellular composition of the bone marrow plasma cell niche. J Immunol 2012;188:1283–91.
[66] Strasser A, Whittingham S, Vaux DL, Bath ML, Adams JM, Cory S, et al. Enforced BCL2 expression in B-lymphoid cells prolongs antibody responses and elicits autoimmune disease. Proc Natl Acad Sci USA 1991;88:8661–5.
[67] Peperzak V, Vikstrom IB, Tarlinton DM. Through a glass less darkly: apoptosis and the germinal center response to antigen. Immunol Rev 2012;247:93–106.
[68] Oltersdorf T, Elmore SW, Shoemaker AR, Armstrong RC, Augeri DJ, Belli BA, et al. An inhibitor of Bcl-2 family proteins induces regression of solid tumours. Nature 2005;435:677–81.
[69] Carrington EM, Vikstrom IB, Light A, Sutherland RM, Londrigan SL, Mason KD, et al. BH3 mimetics antagonizing restricted prosurvival Bcl-2 proteins represent another class of selective immune modulatory drugs. Proc Natl Acad Sci USA 2010;107:10967–71.
[70] Peperzak V, Vikstrom I, Walker J, Glaser SP, Lepage M, Coquery CM, et al. Mcl-1 is essential for the survival of plasma cells. Nat Immunol 2013;14:290–7.

Chapter 14c

# Memory Plasma Cells

**Falk Hiepe[1] and Radbruch Andreas[2]**

*[1]Department of Rheumatology and Clinical Immunology, Charité University Hospital, Berlin, Germany; [2]German Rheumatism Research Center Berlin, a Leibniz Institute, Berlin, Germany*

Memory plasma cells are mature long-lived plasma cells surviving in niches in bone marrow or inflamed tissues, where they secrete antibodies independently of antigen stimulation, T cell help, and memory B cells [1,2]. Two independent groups first proved their existence in 1997 and 1998 [3,4]. They fulfill a crucial function in the homeostasis of humoral immunity by maintaining antibody levels, which also means that they are extremely well defended from endogenous and exogenous influences [5]. However, as a result of a pathological immune reaction, memory plasma cells can significantly contribute to pathogenic antibody responses, such as in autoimmune diseases and allergies. Using New Zealand black/white (NZB/W) mice as a model of systemic lupus erythematosus (SLE), the key role of memory plasma cells in autoimmunity was demonstrated for the first time. In the spleen and, in particular, in the bone marrow, a substantial proportion of plasma cells did not incorporate the nucleotide analog bromodeoxyuridine (BrdU) into DNA despite supplementation with BrdU for 3 months, as determined by flow cytometry. This lack of incorporation during BrdU pulse chase shows that the plasma cells do not undergo DNA synthesis; therefore, they are nondividing, long-lived memory cells [6]. Accordingly, these long-lived plasma cells are resistant to immunosuppression with cyclophosphamide and glucocorticoids [6,7] as well as irradiation [4]. It could be also shown in an ovalbumin-induced murine model of allergic asthma that inhalation of aerosolized allergen generates memory allergen-specific immunoglobulin (Ig)G1-, IgA- and IgE-secreting plasma cells that are resistant to an antiproliferative treatment with cyclophosphamide [8].

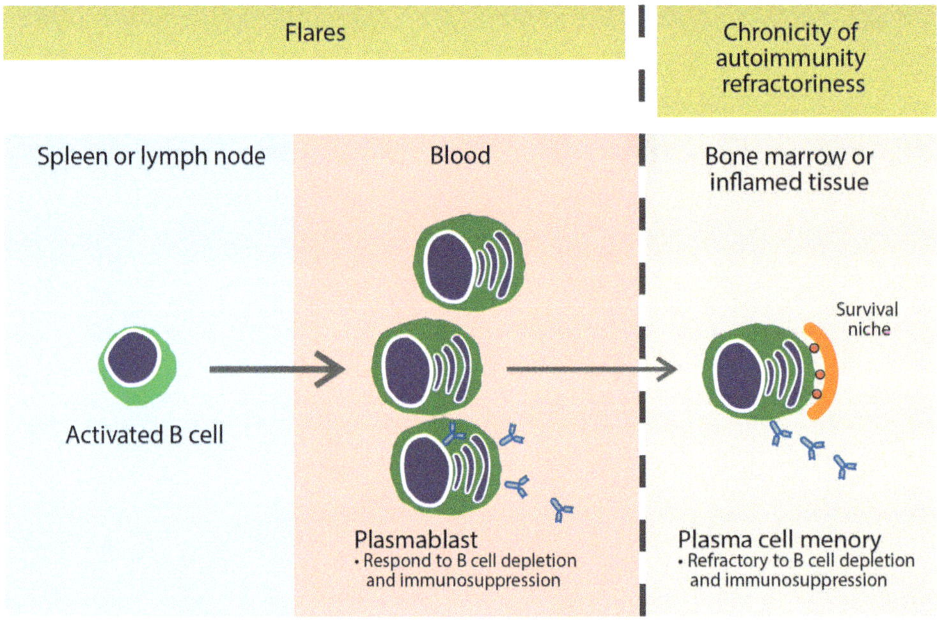


FIGURE 3 **Contribution of short-lived plasmablasts and long-lived memory plasma cells to autoimmunity.**

It was also recently demonstrated that pathology can be driven exclusively by autoantibodies secreted by memory plasma cells. For this purpose, plasma cells from lupus-prone NZB/W mice were adoptively transferred to immunodeficient Rag1$^{-/-}$ mice. All plasma cells that survived this procedure became long-lived memory cells in bone marrow and spleen, which resulted in the development of stable titers of anti-dsDNA autoantibodies, proteinuria, and immune complex nephritis detected by renal immunohistology in recipient mice [9].

Because CD20 is not expressed on plasma cells, B-cell depletion with the anti-CD20 antibody rituximab is suited to distinguish between antibodies secreted by short-lived and long-lived memory plasma cells. Studies in mice confirm that mature and memory B cells are not required for maintaining splenic and bone marrow plasma cell populations [2,10]. Protective antibody titers have remained extremely stable for many years [11] and do not change after rituximab therapy [12,13]. By contrast, the responses of autoantibody levels are much more heterogeneous. Some remain unchanged by rituximab, others are ablated, and some decrease but do not disappear completely. This means that autoantibodies that decrease or disappear after B cell depletion therapy are secreted by short-lived plasmablasts and plasma cells, whereas persisting autoantibodies are secreted by memory plasma cells.

The fact that pathogenic antibodies secreted by memory plasma cells are resistant to conventional immunosuppression or therapies targeting B cells explains at least partly the chronicity of antibody-mediated diseases, why some patients are extremely refractory to conventional immunosuppression and B cell depletion therapies, and why these treatments are unable to cure autoimmune diseases (Figure 3) [14]. Therefore, memory plasma cells are an important constituent of the pathogenic immunological memory, which then renders them a challenging therapeutic target.

The role of IgG4$^+$ plasma cells in the pathogenesis of IgG4-related disease, a fibroinflammatory disorder characterized by tumefactive lesions, a dense lymphoplasmacytic infiltrate rich in IgG4$^+$ plasma cells, storiform fibrosis, and, often but not always, elevated serum IgG4 concentrations [15,16], is unclear so far. The disease was first described in Japan [17,18] and can affect almost any organ in the body. Commonly, the disease responds well to treatments with glucocorticoids or B cell depletion therapy with rituximab. The rapid and selective rituximab-induced decline in IgG4 levels suggests that the bulk of IgG4 in this disease originates from short-lived plasma cells [15]. However, it is not excluded that the IgG4$^+$ plasma cells in the affected organs are long-lived memory plasma cells. It seems to be possible that the main effect of glucocorticoids and rituximab is due to resolution of inflammation, which in turn interrupts the survival of these IgG4$^+$ memory plasma cells [19].

## 1. HOW DOES THE PATHOGENIC PLASMA CELL MEMORY ARISE?

Memory plasma cells arise as a result of a secondary immune response. So far, no qualitative differences between protective and pathological memory plasma cells have been identified.

B cells differentiate first into plasmablasts, which are proliferating cells and are able to migrate toward a chemokine gradient in the bone marrow or inflamed tissues where they can mature to nonproliferating plasma cells and become long-lived memory cells in case they can occupy a niche. That means autoreactive plasmablasts have to displace existing memory plasma cells by competition from their niches to become themselves memory plasma cells [19]. In active autoimmune diseases, such as during SLE flares, excessive numbers of plasmablasts are generated because of B cell hyperactivity [20,21]. In contrast to secondary immunization (e.g., with tetanus toxoid), which leads to a flood of plasmablasts that lasts for only 1–2 days [22], autoreactive plasmablasts are detectable during the whole period of active disease. However, little is known about the recruitment efficiency of pathogenic plasmablasts into the memory plasma cell compartment in human antibody-mediated diseases. It is conceivable that the permanent supply of pathogenic plasmablasts could contribute to the gradual extrusion of old protective memory plasma cells from their niches, resulting in their replacement by pathogenic plasma cells (Figure 4).

In addition to potentially replacing old cells from established niches, newly generated pathogenic plasmablasts can also encounter new niches. Active diseases may increase inflammation; in turn, inflamed tissues form niches for plasma cells. Because pathogenic plasmablasts usually outnumber their protective counterparts during disease flares, most plasma cells that home in the inflamed tissue niches are likely to be pathogenic. The pool of pathogenic memory plasma cells is extended by these mechanisms, thus contributing to a vicious circle [14].

## 2. PLASMA CELL NICHES

Memory plasma cells need an environment that secures them and provides optimal conditions for their survival. This seems to be much more important for their survival than intrinsic factors [23]. Memory plasma cells regularly reside in the bone marrow. The niche is organized by stroma cells expressing chemokine (C-X-C motif) ligand-12 (CXCL12), the ligand of chemokine (C-X-C motif) receptor-4 (CXCR4), which is on plasma cells [24,25]. In addition, several other cells such as megakaryocytes and eosinophils are components of the niche, probably by the production of the survival factors a proliferation-inducing ligand (APRIL) and interleukin (IL)-6 [26–28]. Plasma cells display a panel of

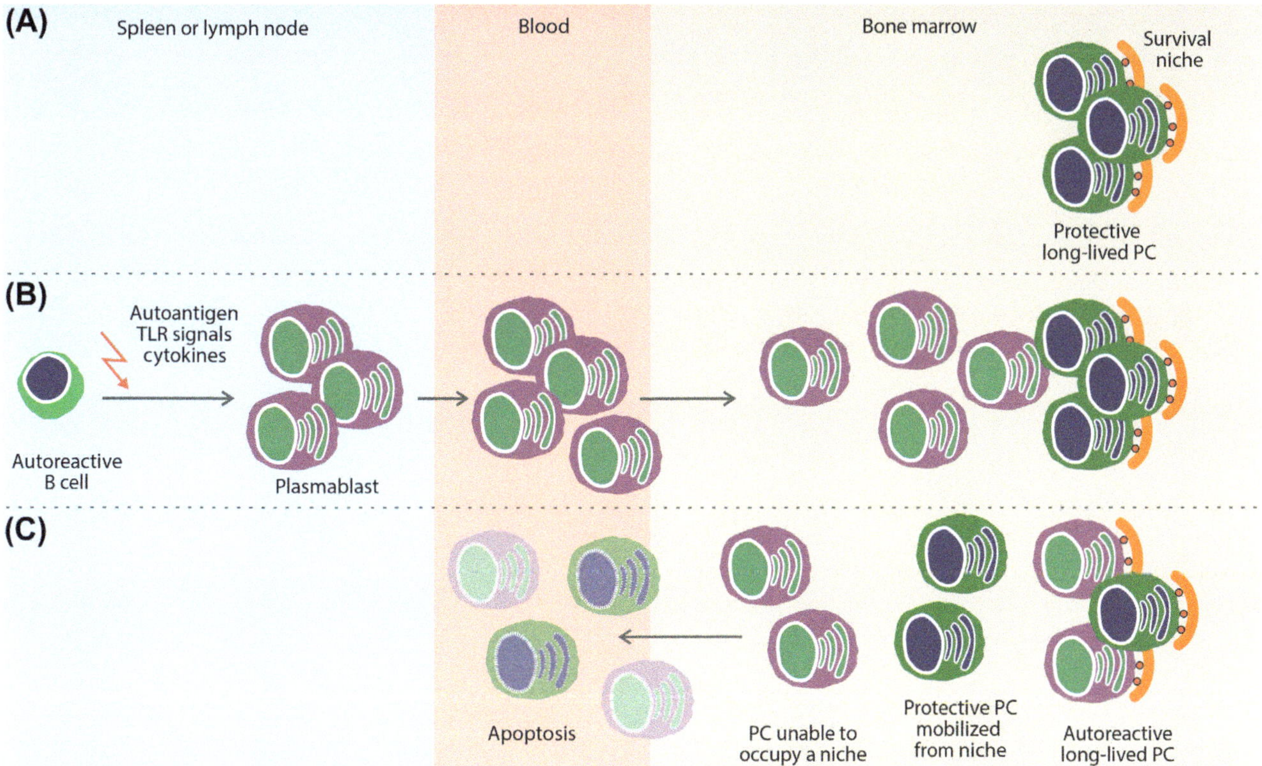


FIGURE 4 **Development of autoreactive long-lived plasma cells.** Protective long-lived PCs (in blue) are usurped from immunological niches by newly generated autoreactive long-lived PCs (in red). (A) PC memory: long-lived plasma cells secrete antibodies to foreign antigens and survive in bone marrow niches. (B) After breaking tolerance, autoreactive B cells differentiate into plasmablasts that emerge from B cell compartments and migrate toward the bone marrow, where they compete for niches occupied by protective long-lived PCs. Because of the accumulation of waves of autoreactive plasmablasts, some of the long-lived PCs are displaced from their niches by competition for survival and adherence signals, and the niches are taken over by autoreactive plasmablasts. These immature cells become mature long-lived PCs that secrete autoantibodies. (C) PCs mobilized from their niches, and newly generated autoreactive plasmablasts or PCs not occupying a niche, soon undergo apoptosis and die. Abbreviations: PC, plasma cell; TLR, toll-like receptor.

adhesion molecules, such as very late antigen-4 and lymphocyte function-associated antigen-1, CD44, P-selectin ligand 1, and CD93, which are involved in their survival [29–31].

B cell maturation antigen of memory plasma cells delivers a decisive survival signal, enhancing the expression of mcl1 [32]. Both of its ligands, B cell activating factor of the tumor necrosis factor (BAFF) and APRIL, are apparently expressed in the bone marrow, and neutralization of both is required to eliminate plasma cells from the bone marrow [33]. For most bone marrow plasma cells, APRIL is delivered by eosinophilic granulocytes, which thus form an essential "accessory cell" of the bone marrow niche for memory plasma cells [28]. In other tissues and for some memory plasma cells in the bone marrow, other accessory cells can provide either BAFF or APRIL to let the plasma cells survive (e.g., megakaryocytes) [27].

The number of plasma cell niches is limited in bone marrow so that normally the memory plasma cells do not exceed 0.5% of the total mononuclear cell count [34,35]. However, in pathological conditions, niches that are provided by inflamed tissues can extend the number of memory plasma cells. This can be beneficial in case of an infection if memory plasma cells locally secrete antibodies against the causative microbe on the site of infection. In such a situation, the memory plasma cells disappear after successful resolution of inflammation, which was in turn supported by local antibody production. In contrast, in an autoimmune-driven chronic inflammation, the inflamed tissue can home memory plasma cells, including autoreactive plasma cells, which may contribute to the persistence or amplifying of inflammation by local autoantibody production.

Less is known about the components of the plasma cell niche in inflamed tissues. It is noticeable that plasma cells are—as in bone marrow—in contact with cells expressing CXCL12. In patients with Sjögren's syndrome, plasma cells are located in close proximity to ductal and acinar epithelia as well as mononuclear cells within the focal infiltrate expressing CXCL12 and IL-6 [28]. CXCL12 is also overexpressed in inflamed kidneys [36] whereas the CXCL10-CXCR3 interaction seems not to be crucial for plasma cell homing because infiltration of plasma cells lacking CXCR3 in inflamed kidneys was not blocked [37]. In autoimmune thrombocytopenia, the spleen provides niches for autoreactive memory plasma cells [38]. In autoimmune disease, plasma cells themselves can contribute to their survival. It was shown that plasma cells in the nephritic kidneys derived from SLE patients aberrantly express APRIL and BAFF [39].

## 3. HOW TO TARGET MEMORY PLASMA CELLS

Figure 5 summarizes the options to affect the appearance and survival of memory plasma cells. It includes the targeting of precursors of memory plasma cells, the inhibition of migration of newly generated plasmablasts into the bone marrow or inflamed tissues, the targeting of the survival niche, and the direct targeting of plasma cells.

## 4. DIRECT PLASMA CELL TARGETING

Because of their refractoriness to conventional treatments, memory plasma cells might be one of the keys to find out ways to cure antibody-mediated diseases. This is supported by findings after immunoablation followed by autologous stem cell transplantation that performed in refractory autoimmune diseases. Long-term remissions without continuation of immunosuppressive drugs were observed in up to 50% of patients suffering from these severe diseases. The immunoablation mainly done with antithymocyte globulin may lead to a complete deletion of the pathological immunological memory including the autoreactive plasma cell memory. This is the basis for the development of a tolerant immune system. There are at least two possible reasons why some patients will relapse: (1) there are indications that the depletion of the autoreactive plasma cell memory is not sufficient enough in some patients, and (2) the degree of the genetic background varies in complex autoimmune diseases. Although monogenetic causes of autoimmune diseases, which spontaneously develop the disease, are rare, it seems to be clear that the strength of the genetic influence differs from patient to patient. Patients with a greater susceptibility for autoimmunity need only fewer and weaker triggers for inducing relapse whereas the disease will probably never recur in patients with a weak genetic constellation for autoimmunity and the absence of the disease-inducing trigger. However, immunoablation attacks the whole adaptive immune system, which includes the protective and the pathological immunological memory that consists of B, T, and plasma cells.

Plasma cells including memory plasma cells can be selectively depleted by proteasome inhibitors whereas other cells of the adaptive immune system are not or are hardly affected [40]. Bortezomib is a reversible, highly selective, and potent inhibitor of the 26S proteasome and is approved for treatment of multiple myeloma. This malignant plasma cell disease is associated with excessive production of monoclonal Igs. The proapoptotic effect of bortezomib in multiple myeloma is mainly due to the accumulation of unfolded proteins, which must be degraded by the ubiquitin-proteasome system that bortezomib inhibits in cells with high protein biosynthesis [41]. Bortezomib was successfully used to deplete splenic and bone marrow plasma cells in murine models of lupus. It drastically prolonged the survival of lupus mice by preventing nephritis in mice when the treatment was started in a preclinical phase before high titers of anti-dsDNA antibodies appeared and before onset of nephritis. However, it also was effective in mice with

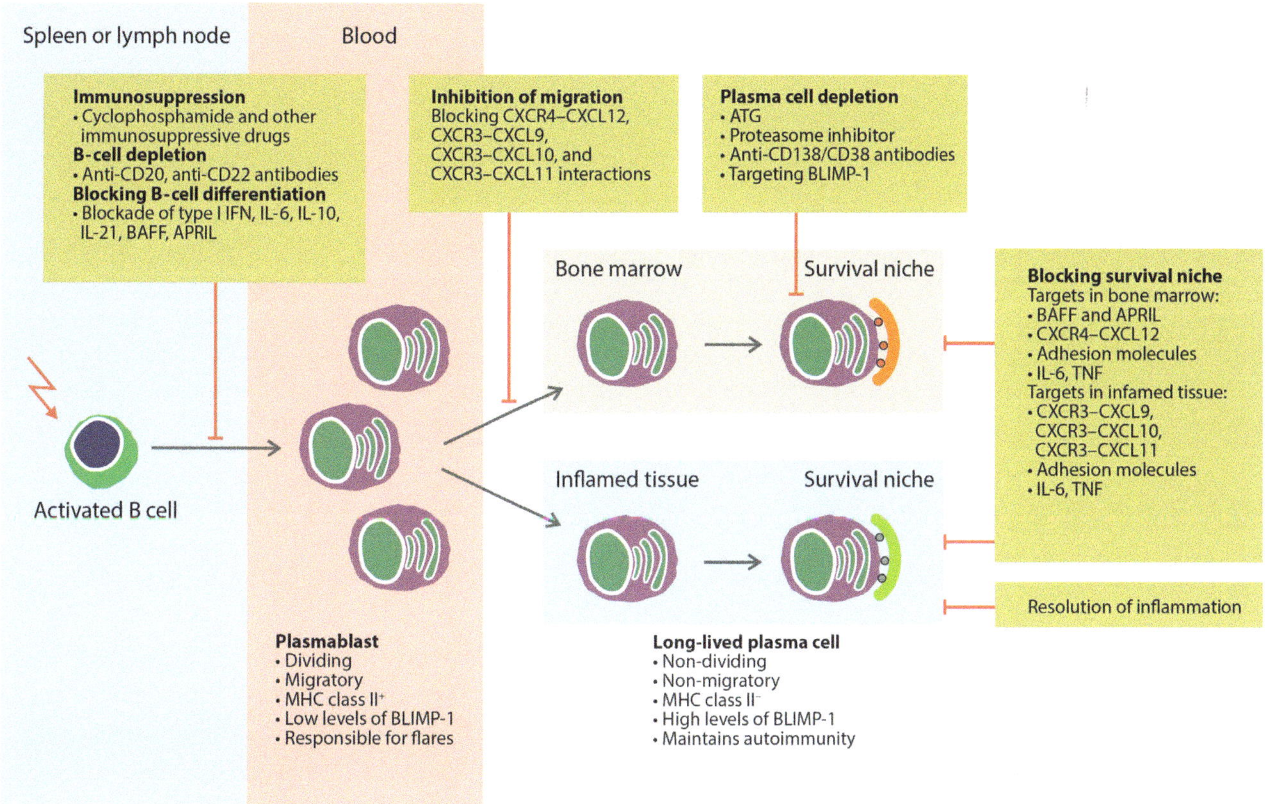


FIGURE 5 **Strategies for regulating and targeting long-lived plasma cells.** Conventional immunosuppression blocks proliferation of B cells and plasmablasts. B cell depletion therapy inhibits the supply of newly generated plasma cells. Differentiation of B cells into plasma cells is promoted by several cytokines, including type I IFN, BAFF, APRIL, IL-6, IL-10, and IL-21, which might serve as therapeutic targets. It might be possible to impede the migration of plasmablasts into plasma cell niches in the bone marrow and inflamed areas by inhibiting the interactions of chemokines and their receptors on plasmablasts (CXCL12 and CXCR4 in bone marrow; CXCL9, CXCL10, CXCL11, and CXCR3 in inflamed tissues). These approaches restrict the supply of newly generated plasma cells but do not affect the established long-lived plasma cell pools in bone marrow and inflamed tissues, which can only be depleted by direct targeting of plasma cells (e.g., by immunoablation using ATG, proteasome inhibition, or targeting of survival niche factors). Some approaches might target the fresh supply of plasma cells and the plasma cell survival mechanisms. Abbreviations: APRIL, a proliferation-inducing ligand; ATG, antithymocyte globulin; BAFF, B-cell-activating factor; CXCR, CXC-chemokine receptor; CXCL, CXC-chemokine ligand; IFN, interferon; IL, interleukin; TNF, tumor necrosis factor.

manifest disease leading to a significant reduction of proteinuria and anti-dsDNA antibody levels [42]. In animals with an experimental autoimmune myasthenia gravis, bortezomib efficiently reduced the increase of anti-acetylcholinesterase receptor autoantibody titers, prevented ultrastructural damage of the postsynaptic membrane, improved neuromuscular transmission, and decreased myasthenic symptoms [43]. In SLE patients who did not respond to glucocorticoids and immunosuppressive drugs such as cyclophosphamide or mycophenolate mofetil, bortezomib resulted in a significant improvement of disease activity. This was related to a drop of anti-dsDNA autoantibody, which did not decrease upon the immunosuppressive treatment done before, and of vaccine titers against mumps, measles, and tetanus toxoid, indicating that bortezomib can deplete memory plasma cells in men [44]. However, in patients with persistent B cell hyperactivity, a rapid recurrence of autoreactive memory plasma cells seems to be possible [44]. This is also observed in NZB/W mice in which B cell hyperactivity is genetically determined [42,45]. The targeting of the precursors of memory plasma cells followed after plasma cell depletion with bortezomib can prevent the regeneration of the autoreactive memory plasma cell pool in NZB/W mice (Taddeo et al., submitted for publication).

There are few case reports describing the induction by bortezomib of clinical remission in patients with refractory autoantibody-mediated diseases such as thrombotic thrombocytopenic purpura [46,47] and autoimmune hemolytic anemia [48,49]. This remission was accompanied by a striking reduction of autoantibody levels.

Plasma cell depletion with bortezomib has also emerged as a therapeutic approach in transplantation medicine where alloantibodies induce acute rejection [50–52]. There is one case report of a patient with an IgG4-related disease refractory to glucocorticoids who responded to plasma cell depletion with bortezomib [49].

## REFERENCES

[1] Manz RA, Löhning M, Cassese G, Thiel A, Radbruch A. Survival of long-lived plasma cells is independent of antigen. Int Immunol 1998;10(11):1703–11.
[2] Ahuja A, Anderson SM, Khalil A, Shlomchik MJ. Maintenance of the plasma cell pool is independent of memory B cells. Proc Natl Acad Sci USA 2008;105(12):4802–7.
[3] Manz RA, Thiel A, Radbruch A. Lifetime of plasma cells in the bone marrow. Nature 1997;388(6638):133–4.
[4] Slifka MK, Antia R, Whitmire JK, Ahmed R. Humoral immunity due to long-lived plasma cells. Immunity 1998;8(3):363–72.
[5] Manz R, Hauser A, Hiepe F, Radbruch A. Maintenance of serum antibody levels. Annu Rev Immunol 2005;23:367–86.
[6] Hoyer B, Moser K, Hauser A, Peddinghaus A, Voigt C, Eilat D, et al. Short-lived plasmablasts and long-lived plasma cells contribute to chronic humoral autoimmunity in NZB/W mice. J Exp Med 2004;199(11):1577–84.
[7] Mumtaz IM, Hoyer BF, Panne D, Moser K, Winter O, Cheng QY, et al. Bone marrow of NZB/W mice is the major site for plasma cells resistant to dexamethasone and cyclophosphamide: implications for the treatment of autoimmunity. J Autoimmun 2012;39(3):180–8.
[8] Luger EO, Fokuhl V, Wegmann M, Abram M, Tillack K, Achatz G, et al. Induction of long-lived allergen-specific plasma cells by mucosal allergen challenge. J Allergy Clin Immunol 2009;124(4):819–26.
[9] Cheng Q, Mumtaz IM, Khodadadi L, Radbruch A, Hoyer BF, Hiepe F. Autoantibodies from long-lived 'memory' plasma cells of NZB/W mice drive immune complex nephritis. Ann Rheum Dis 2013;72(12):2011–7.
[10] DiLillo DJ, Hamaguchi Y, Ueda Y, Yang K, Uchida J, Haas KM, et al. Maintenance of long-lived plasma cells and serological memory despite mature and memory B cell depletion during CD20 immunotherapy in mice. J Immunol 2008;180(1):361–71.
[11] Amanna IJ, Carlson NE, Slifka MK. Duration of humoral immunity to common viral and vaccine antigens. N Engl J Med 2007;357(19):1903–15.
[12] Cambridge G, Leandro MJ, Edwards JC, Ehrenstein MR, Salden M, Bodman-Smith M, et al. Serologic changes following B lymphocyte depletion therapy for rheumatoid arthritis. Arthritis Rheum 2003;48(8):2146–54.
[13] Ferraro AJ, Drayson MT, Savage CO, MacLennan IC. Levels of autoantibodies, unlike antibodies to all extrinsic antigen groups, fall following B cell depletion with Rituximab. Eur J Immunol 2008;38(1):292–8.
[14] Hiepe F, Dörner T, Hauser AE, Hoyer BF, Mei H, Radbruch A. Long-lived autoreactive plasma cells drive persistent autoimmune inflammation. Nat Rev Rheumatol 2011;7(3):170–8.
[15] Mahajan VS, Mattoo H, Deshpande V, Pillai SS, Stone JH. IgG4-related disease. Annu Rev Pathol 2014;9:315–47.
[16] Stone JH, Zen Y, Deshpande V. IgG4-related disease. N Engl J Med 2012;366(6):539–51.
[17] Hamano H, Kawa S, Horiuchi A, Unno H, Furuya N, Akamatsu T, et al. High serum IgG4 concentrations in patients with sclerosing pancreatitis. N Engl J Med 2001;344(10):732–8.
[18] Kamisawa T, Funata N, Hayashi Y, Eishi Y, Koike M, Tsuruta K, et al. A new clinicopathological entity of IgG4-related autoimmune disease. J Gastroenterol 2003;38(10):982–4.
[19] Radbruch A, Muehlinghaus G, Luger EO, Inamine A, Smith KG, Dörner T, et al. Competence and competition: the challenge of becoming a long-lived plasma cell. Nat Rev Immunol 2006;6(10):741–50.
[20] Odendahl M, Jacobi A, Hansen A, Feist E, Hiepe F, Burmester GR, et al. Disturbed peripheral B lymphocyte homeostasis in systemic lupus erythematosus. J Immunol 2000;165(10):5970–9.
[21] Jacobi AM, Mei H, Hoyer BF, Mumtaz IM, Thiele K, Radbruch A, et al. HLA-DRhigh/CD27high plasmablasts indicate active disease in patients with systemic lupus erythematosus. Ann Rheum Dis 2010;69(1):305–8.
[22] Odendahl M, Mei H, Hoyer BF, Jacobi AM, Hansen A, Muehlinghaus G, et al. Generation of migratory antigen-specific plasma blasts and mobilization of resident plasma cells in a secondary immune response. Blood 2005;105(4):1614–21.
[23] Manz RA, Arce S, Cassese G, Hauser AE, Hiepe F, Radbruch A. Humoral immunity and long-lived plasma cells. Curr Opin Immunol 2002;14(4):517–21.
[24] Tokoyoda K, Egawa T, Sugiyama T, Choi BI, Nagasawa T. Cellular niches controlling B lymphocyte behavior within bone marrow during development. Immunity 2004;20(6):707–18.
[25] Tokoyoda K, Hauser AE, Nakayama T, Radbruch A. Organization of immunological memory by bone marrow stroma. Nat Rev Immunol 2010;10(3):193–200.
[26] Winter O, Dame C, Jundt F, Hiepe F. Pathogenic long-lived plasma cells and their survival niches in autoimmunity, malignancy, and allergy. J Immunol 2012;189(11):5105–11.
[27] Winter O, Moser K, Mohr E, Zotos D, Kaminski H, Szyska M, et al. Megakaryocytes constitute a functional component of a plasma cell niche in the bone marrow. Blood 2010;116(11):1867–75.
[28] Szyszko EA, Brokstad KA, Oijordsbakken G, Jonsson MV, Jonsson R, Skarstein K. Salivary glands of primary Sjogren's syndrome patients express factors vital for plasma cell survival. Arthritis Res Ther 2011;13(1):R2.
[29] Cassese G, Arce S, Hauser AE, Lehnert K, Moewes B, Mostarac M, et al. Plasma cell survival is mediated by synergistic effects of cytokines and adhesion-dependent signals. J Immunol 2003;171(4):1684–90.
[30] Underhill GH, Minges Wols HA, Fornek JL, Witte PL, Kansas GS. IgG plasma cells display a unique spectrum of leukocyte adhesion and homing molecules. Blood 2002;99(8):2905–12.
[31] Chevrier S, Genton C, Kallies A, Karnowski A, Otten LA, Malissen B, et al. CD93 is required for maintenance of antibody secretion and persistence of plasma cells in the bone marrow niche. Proc Natl Acad Sci USA 2009;106(10):3895–900.
[32] Peperzak V, Vikström I, Walker J, Glaser SP, LePage M, Coquery CM, et al. Mcl-1 is essential for the survival of plasma cells. Nat Immunol 2013;14(3):290–7.
[33] Benson MJ, Dillon SR, Castigli E, Geha RS, Xu S, Lam KP, et al. Cutting edge: the dependence of plasma cells and independence of memory B cells on BAFF and APRIL. J Immunol 2008;180(6):3655–9.
[34] Haaijman JJ, Schuit HR, Hijmans W. Immunoglobulin-containing cells in different lymphoid organs of the CBA mouse during its lifespan. Immunology 1977;32(4):427–34.
[35] Brieva JA, Roldan E, De la Sen ML, Rodriguez C. Human in vivo-induced spontaneous IgG-secreting cells from tonsil, blood and bone marrow exhibit different phenotype and functional level of maturation. Immunology 1991;72(4):580–3.
[36] Balabanian K, Couderc J, Bouchet-Delbos L, Amara A, Berrebi D, Foussat A, et al. Role of the chemokine stromal cell-derived factor 1 in autoantibody production and nephritis in murine lupus. J Immunol 2003;170(6):3392–400.

[37] Moser K, Kalies K, Szyska M, Humrich JY, Amann K, Manz RA. CXCR3 promotes the production of IgG1 autoantibodies but is not essential for the development of lupus nephritis in NZB/NZW mice. Arthritis Rheum 2012;64(4):1237–46.

[38] Mahevas M, Patin P, Huetz F, Descatoire M, Cagnard N, Bole-Feysot C, et al. B cell depletion in immune thrombocytopenia reveals splenic long-lived plasma cells. J Clin Invest 2013;123(1):432–42.

[39] Chu VT, Enghard P, Schurer S, Steinhauser G, Rudolph B, Riemekasten G, et al. Systemic activation of the immune system induces aberrant BAFF and APRIL expression in B cells in patients with systemic lupus erythematosus. Arthritis Rheum 2009;60(7):2083–93.

[40] Lang VR, Mielenz D, Neubert K, Bohm C, Schett G, Jack HM, et al. The early marginal zone B cell-initiated T-independent type 2 response resists the proteasome inhibitor bortezomib. J Immunol 2010;185(9):5637–47.

[41] Meister S, Schubert U, Neubert K, Herrmann K, Burger R, Gramatzki M, et al. Extensive immunoglobulin production sensitizes myeloma cells for proteasome inhibition. Cancer Res 2007;67(4):1783–92.

[42] Neubert K, Meister S, Moser K, Weisel F, Maseda D, Amann K, et al. The proteasome inhibitor bortezomib depletes plasma cells and protects mice with lupus-like disease from nephritis. Nat Med 2008;14(7):748–55.

[43] Gomez AM, Vrolix K, Martinez-Martinez P, Molenaar PC, Phernambucq M, van der Esch E, et al. Proteasome inhibition with bortezomib depletes plasma cells and autoantibodies in experimental autoimmune myasthenia gravis. J Immunol 2011;186(4):2503–13.

[44] Alexander T, Sarfert R, Klotsche J, Kühl AA, Rubbert-Roth A, Lorenz HM, Rech J, Hoyer BF, Cheng Q, Waka A, Taddeo A, Wiesener M, Schett G, Burmester GR, Radbruch A, Hiepe EF, Voll RE. The proteasome inhibitior bortezomib depletes plasma cells and ameliorates clinical manifestations of refractory systemic lupus erythematosus. Ann Rheum Dis 2014, in press.

[45] Taddeo A, Gerl V, Hoyer BF, Chang HD, Kohler S, Schaffert H, Thiel A, Radbruch A, Hiepe F. Selection and depletion of plasma cells based on the specificity of the secreted antibody. Eur J Immunol 2014, in press.

[46] Mazepa MA, Raval JS, Moll S, Ma A, Park YA. Bortezomib induces clinical remission and reduction of ADAMTS13 inhibitory antibodies in relapsed refractory idiopathic thrombotic thrombocytopenic purpura. Br J Haematol 2014;164(6):900–2.

[47] van Balen T, Schreuder MF, de Jong H, van de Kar NC. Refractory thrombotic thrombocytopenic purpura in a 16-year-old girl: successful treatment with bortezomib. Eur J Haematol 2014;92(1):80–2.

[48] Carson KR, Beckwith LG, Mehta J. Successful treatment of IgM-mediated autoimmune hemolytic anemia with bortezomib. Blood 2010;115(4):915.

[49] Khan ML, Colby TV, Viggiano RW, Fonseca R. Treatment with bortezomib of a patient having hyper IgG4 disease. Clin Lymphoma Myeloma Leuk 2010;10(3):217–9.

[50] Ejaz NS, Alloway RR, Halleck F, Durr M, Budde K, Woodle ES. Review of bortezomib treatment of antibody-mediated rejection in renal transplantation. Antioxid Redox Signal 2014 Apr 28 [Epub ahead of print]. http://dx.doi.org/0.1089/ars.2014.5892.

[51] Walsh RC, Alloway RR, Girnita AL, Woodle ES. Proteasome inhibitor-based therapy for antibody-mediated rejection. Kidney Int 2012;81(11):1067–74.

[52] Woodle ES, Alloway RR, Girnita A. Proteasome inhibitor treatment of antibody-mediated allograft rejection. Curr Opin Organ Transpl 2011;16(4):434–8.

Chapter 15

# The Role of the BAFF and Lymphotoxin Pathways in B Cell Biology

Fabienne Mackay[1] and Jennifer L. Gommerman[2]

[1]Department of Immunology, Monash University, Clayton, VIC, Australia; [2]Department of Immunology, University of Toronto, Toronto, ON, Canada

## 1. BAFF/APRIL: IMPORTANT REGULATORS OF B CELL SURVIVAL, HOMEOSTASIS, AND FUNCTION

### 1.1 Introduction

For many years, immunologists have believed that expression of a B cell receptor (BCR) was sufficient and essential to allow B cells to develop and survive in the periphery. This notion was later reconsidered after the discovery of the B cell activating factor belonging to the tumor necrosis factor (TNF) family (BAFF; also known as BLyS, TALL-1, zTNF-4, THANK, and TNFSF13b; reviewed in [1,2]), which emerged as an equally essential factor for B cell maturation and survival in the periphery. Not only was the discovery of BAFF a major paradigm shift in our understanding of B cell development and survival, a decade of studies on this factor led to the development of an inhibitor approved for use in the clinic as a novel B cell therapy treating a subset of patients with systemic lupus erythematosus (SLE) [3,4]. Apart from TNF, BAFF is the second ligand from the TNF superfamily that has led to bench-to-bedside outcomes.

### 1.2 The BAFF–APRIL System

#### 1.2.1 The Ligands

BAFF and a related ligand, a proliferation-inducing ligand (APRIL; also known as TALL-2, TRDL-1, or TNFSF13a), are heterotrimeric ligand members of the TNF superfamily (reviewed in [1,2,5]); as such, they are typically produced as type II transmembrane proteins that are proteolytically cleaved at a furin protease site and released in a soluble form [6]. BAFF has also been described as membrane-bound ligand. In contrast, APRIL is cleaved in the Golgi before release and exists only in a soluble form. However, recently, APRIL-δ, a malignant isoform of APRIL that lacks a furin cleavage site, appears to remain membrane-bound on the surface of leukemia cell precursors [7]. The 285-amino-acid BAFF is cleaved by furin convertase between amino acids R133 and A134, which are situated at the end of a consensus furin cleavage motif (R-X-(R/K)-R) in the stalk region, and BAFF is released as a soluble 152-amino-acid, 17-kDa protein [8,9]. Likewise, 250-amino-acid APRIL is cleaved between R104 and A105 and is released as a 146-amino-acid, 17-kDa protein [10].

The human *baff* gene, encoded at 13q32-q34, contains a transmembrane domain and flanking regions in exon 1, a furin-processing site in exon 2, and a TNF homology domain (THD) in exons 3–6 [1]. The THD facilitates binding to the BAFF receptors (BAFF-R). The murine *baff* gene contains an additional exon encoding 30 amino acid residues between the furin processing site and the THD. Mouse and human APRIL contain a furin-processing site in exon 2 [1]. BAFF is an active ligand as a homotrimer [11]. The trimer contains two magnesium ions interacting with the side chains of Q234, N235, and N243 [12], and it is the main form of BAFF found in the circulation [13]. However, a 60-mer form of BAFF has been obtained at physiological pH conditions, and this form is also able to bind to its receptors [14]. BAFF 60-mers arise by the assembly of 20 BAFF trimers into a virus-like cluster. BAFF contains a "flap" region, which is absent in APRIL and is essential for the 60-mer assembly [1,14]. The "flap" region is also referred to as the DE loop because it is formed by five amino acids inserted between the β-strands D and E [15]. Interactions between BAFF trimers are extensive and complex because these involve hydrogen bonds, salt bridges, and hydrophobic contacts [14]. For this reason, the formation of 60-mer BAFF is not reversible at physiological pH, whereas dissociation of the 60-mer into trimers can occur at acidic pH. The strong dependency on a physiological pH for the formation of the 60-mer suggests that BAFF 60-mer assembly is not due to random aggregation. Despite this argument, other studies have claimed that assembly of the BAFF 60-mer was merely an artifact involving fusion of recombinant N-terminal tags inserted to the BAFF

Molecular Biology of B Cells. http://dx.doi.org/10.1016/B978-0-12-397933-9.00015-1

proteins [11]. However, to address this criticism, another study was conducted using a construct of BAFF with no amino-terminal tag that was also able to form 60-mers in solution [16], hence ruling out the possibility that BAFF 60-mer assembly is an artifact caused by the fusion of protein tags. The function of BAFF 60-mer may involve the recruitment of numerous receptors; hence, clustering these receptors on the cell surface to increase the local concentration of receptor/ligand interaction thus enhancing signaling [14]. The BAFF 60-mer is the more active form of BAFF; therefore, it is likely to be physiologically relevant. In biological situations in which the local concentration of BAFF greatly increases, a corresponding increase in the formation of BAFF 60-mers might be expected, along with increased propensity for signaling through the BAFF/APRIL receptor transmembrane activator and cyclophilin ligand interactor (TACI; also known as TNFRSF13B), which requires the 60-mer form of BAFF to signal in some cases [13].

BAFF and APRIL are capable of forming heterotrimers when co-expressed [17]. These are biologically active and were detected in the serum of healthy humans and patients with systemic autoimmune rheumatic diseases [17]. Heterotrimers have been reported to be upregulated in a subset of patients with rheumatic diseases; however, it is unclear whether this may reflect a limitation in the anti-BAFF and anti-APRIL antibodies used for the capture and detection of heterotrimers [17]. However, purely based on structure prediction, it is also possible that other members of the TNF family are similarly capable of forming heterotrimers with BAFF and/or APRIL [17]. Indeed, a more recent study demonstrated that BAFF/APRIL complexes have biological activity and can modulate B1 B cell function and are frequently elevated in SLE but not rheumatoid arthritis (RA) patient sera [18].

ΔBAFF is a splice variant of BAFF that was identified in humans and mice [19,20]. The DNA sequence of ΔBAFF lacks the exon encoding the first β-sheet of the THD, which is exon 3 in humans and exon 4 in mice [19]. ΔBAFF is not released from the cell membrane, but it can form multimers with BAFF such that the release of BAFF from the cell is inhibited. As such, ΔBAFF appears to act as a negative regulator of BAFF [20]. Expression of ΔBAFF is less abundant than that of the full-length isoform of BAFF in lymphatic and central nervous system (CNS) tissues [21]. More recently, two new transcript variants of BAFF have been identified in peripheral blood mononuclear cells [22], but functional studies are required to determine the structural and biological significance of these new variants. Another study discovered yet another splice variant form of BAFF in mice and humans. This last variant form of BAFF was discovered using a new set of primers that generated a polymerase chain reaction band missing a 114-base-pair segment encoded by exon 4. It is interesting to note that early studies suggested that this new variant of BAFF, named Δ4BAFF, associated with DNA and possibly acted as a transcription factor for the full-length form of BAFF [23].

The APRIL locus is situated adjacent to the locus of TWEAK, another member of the TNF superfamily. A splice variation event at the APRIL/TWEAK locus leads to the production of an APRIL/TWEAK hybrid ligand, named TWE-PRIL. This ligand is a fusion of the APRIL extracellular domain with the TWEAK transmembrane portion; therefore, it is suspected to be biologically active similarly to APRIL [24]. However, the biological activity and regulation of this particular ligand remains unknown.

### 1.2.2 Sources of BAFF and APRIL Production

BAFF and APRIL are produced in various hematopoietic and nonhematopoietic cell lineages (Table 1) [2]. Myeloid cells in particular produce high amounts of BAFF, and cell types producing BAFF and APRIL include monocytes [25], macrophages, dendritic cells (DCs) [6,26], neutrophils [27–29], mast cells and eosinophils [27], and astrocytes [21]. These immune cells participate in innate immune responses, which act to defend the host from infection in a non-specific manner. They do not confer long-lasting protective immunity, but they have a more immediate action than cells of the adaptive immune system. However, many of these cells can also act as antigen-presenting cells, and production of BAFF by these cells can stimulate T cell activation and survival [27,30]. BAFF also influences the activation of innate cells; in particular, it induces monocyte survival, proinflammatory cytokine secretion, and co-stimulatory molecule expression [31]. In addition, BAFF production by innate cells is triggered upon stimulation by proinflammatory factors and co-stimulatory molecules. For example, BAFF production is upregulated by interferon-γ (IFN-γ), interleukin (IL)-10, and CD40L [32]. Pathogen-associated molecular patterns, such as lipopolysaccharide (LPS) or peptidoglycans, upregulate the expression of BAFF and APRIL in macrophages and DCs via toll-like receptors (TLRs) [33–35]. It is interesting to note that BAFF and APRIL expression is upregulated within B cells upon TLR activation [36]. TLRs are a means by which cells of the innate immune system recognize pathogenic material or pathogen-associated molecular patterns and initiate a T cell-independent immune response. TLR are especially important in marginal zone (MZ) B cells, which are situated adjacent to the splenic MZ sinus where the blood carrying pathogens enters the spleen. As such, MZ B cell activation via TLR is a strategic host response to blood-borne pathogens, and this may explain why BAFF plays an important role in upregulating TLR expression on the surface of these cells [33].

In the synovium of patients with RA, BAFF is produced by B and T cells, but also by synovial sublining cells [37].

**TABLE 1 Cellular Sources of B Cell Activating Factor of Tumor Necrosis Factor (BAFF) and a Proliferation-Inducing Ligand (APRIL) Production**

| | Hematopoietic Producers | Nonhematopoietic Producers |
|---|---|---|
| BAFF | Dendritic cells (DCs)<br>Monocytes<br>Macrophages<br>Neutrophils<br>T cells<br>Natural killer cells<br>B cells<br>Leukemic B cells<br>Promyelocytic leukemia cells<br>Bone marrow (BM) CD14+ cells and osteoclasts | Salivary gland epithelial cells<br>Fibroblast-like synoviocytes<br>Cytotrophoblast cells and stromal cells in human placenta<br>Astrocytes<br>Renal cell carcinoma |
| APRIL | DCs<br>Monocytes<br>Macrophages<br>Myeloid precursor cells<br>Neutrophils<br>B cells<br>Leukemic B cells<br>T cells<br>BM CD14+ cells and osteoclasts | Intestinal epithelial cells<br>Cytotrophoblast cells and stromal cells in human placenta<br>Renal cell carcinoma<br>Most tumor cells |

Healthy fibroblast-like synoviocytes produce large amounts of BAFF in response to IFN-γ or TNF stimulation [38,39]. Synovial fibroblasts from RA patients constitutively express BAFF and APRIL, and they strongly upregulate the expression of these two factors upon stimulation of TLR3 and to a lesser extent TLR4 but not TLR2 [40].

APRIL expression in the bone marrow (BM) is critical for plasmablast survival because APRIL-deficient mice were unable to sustain a transferred population of plasmablasts [41]. BM-resident macrophages are an important source of this survival signal. In this context, APRIL binds preferentially to heparin sulfate proteoglycan (HSPG) on the surface of CD138+ plasma cells (PCs), upregulating expression of the prosurvival factor Bcl-$X_L$ [41]. However, more recently, it was shown that eosinophils in the BM are also a critical source of PC survival factors including APRIL [42]. Furthermore, it is now understood that BAFF and APRIL are mutually redundant sources of survival factors for BM PCs [43]. Tumor-associated macrophages are also a major source of APRIL in gastric mucosa-associated lymphoid tissue (MALT) lymphoma [44]. Activin A, a member of the transforming growth factor-β (TGF-β) family, strongly activates APRIL expression by mouse macrophages [45]. A mechanism for the role B cell maturation antigen (BCMA) in PC maintenance has recently been described. Indeed, BCMA signaling promotes the expression of Mcl1, a factor essential for PC survival [46].

BAFF and APRIL are produced by astrocytes as part of their active role in cerebral innate immunity [21,47]. Astrocytes in vitro produce high amounts of BAFF when stimulated with inflammatory cytokines [47], suggesting that they may be required to supply cerebral immune cells with survival signals. BAFF, together with other soluble glial-derived cytokines, may be responsible for triggering adaptive immunity in an inflamed CNS [47]. It is interesting to note that B cells may play an important pathogenic role in multiple sclerosis (MS) because patients responded to depletion of B cells using anti-CD20 (rituximab) with reduced inflammatory brain lesions and relapses [48]. BAFF production in the brain may contribute to the pathogenic effect of B cells in MS. Surprisingly and contrary to expectations, one attempt to block BAFF but also APRIL with atacicept (a soluble recombinant TACI protein fused with the Fc portion of a human antibody used as decoy) led to adverse effects and intensification of the inflammatory activity [49], suggesting a more complicated function of the BAFF/APRIL system beyond that of B cell survival.

Activated T cells produce APRIL [50,51] and TWEPRIL [24], but BAFF expression from T cells has been somewhat controversial. BAFF mRNA has been detected within resting and stimulated T cells [26,27], stromal cells [52], and follicular dendritic cells (FDCs) [53]. A recent study has demonstrated the follicular helper T cells ($T_{fh}$) in the germinal center (GC) are an important source of BAFF [54]. Confirmation of BAFF expression at the protein level in these cell types in humans has been difficult to obtain [55]. Given that T cells co-express APRIL, it is possible that T cells produce BAFF/APRIL heterotrimers, which might complicate BAFF protein detection by anti-BAFF monoclonal antibodies [56].

BAFF and APRIL directly modulate T cell function by enhancing human T cell response to anti-CD3 activation [26,27] and T cell survival [51]. The addition of exogenous BAFF to activated splenic T cells increased anti-apoptotic Bcl-2 expression, suggesting that BAFF may have a positive influence on the survival of T cells [27]. Modification of the B cell compartment in BAFF Tg mice also indirectly affects the quality of T cell responses [33].

### 1.2.3 Receptors for BAFF and APRIL

BAFF and APRIL share two receptors: TACI and BCMA (also known as TNFRSF17; reviewed in [1,2,57]). APRIL and BAFF have different affinities for these receptors, with APRIL binding strongly to BCMA but moderately to TACI, and BAFF binding weakly to BCMA but strongly to TACI [1,58]. In addition, BAFF has a high affinity for the BAFF-R (also known as TNFRSF13C) [1,2]. BAFF binds preferentially to BAFF-R and TACI, and with relatively weaker affinity for BCMA (100-fold or 1000-fold weaker, respectively); APRIL binds comparably to TACI and BCMA [1,59]. Because TACI is the only receptor to bind relatively well to APRIL and BAFF, heterotrimers of these ligands are thought to be specific for TACI (Figure 1) [17].

The DxL motif on BAFF-R [60], TACI, and BCMA [61] is critical for binding to BAFF and APRIL. In the case of TACI, binding is favored by an additional two amino acids for BAFF and APRIL. Four residues on BCMA (Y13, I22, Q25, and R27) have been implicated in this ligand specificity [62]. Interaction between BAFF/APRIL and their receptors occurs via conserved cysteine-rich domains (CRDs) in the extracellular regions of the receptors [61]. TNF receptor (TNFR) superfamily proteins typically contain three to six CRDs, TACI contains two CRDs, BCMA contains a single CRD, and BAFF-R contains a partial CRD [61]. The complete CRDs are each approximately 40 amino acid residues in length. A form of TACI exists in which the N-terminal CRD is removed by alternate splicing, and this form is fully capable of ligand-induced cell signaling [63]. This capability suggests that the second CRD of TACI alone mediates complete affinity for BAFF and APRIL [61]. This notion is supported by the extent of evolutionary conservation, which

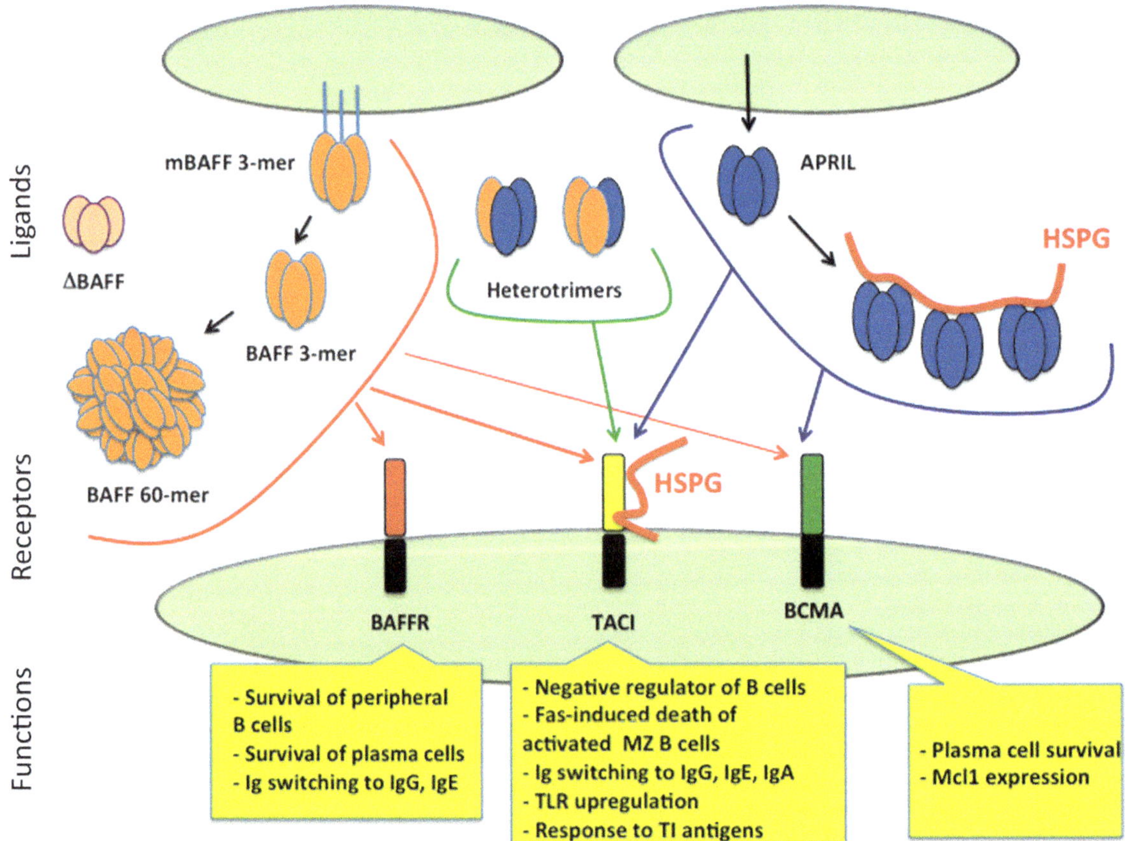


FIGURE 1 **Ligands and receptors of the BAFF/APRIL system and functions.** The upper part of the figure represents the various forms of the ligands. Arrows indicate the binding specificity of these different ligands to corresponding receptors (middle section of the figure). The red line binding APRIL ligands or TACI represents heparin sulfate proteoglycan (HSPG). Call boxes indicate the function of each of the receptors. Abbreviations: BAFF, B cell activating factor of tumor necrosis factor; mBAFF, membrane-bound BAFF; TI, thymus-independent; TLR, toll like receptor; APRIL, a proliferation-inducing ligand; TACI, transmembrane activator and cyclophilin ligand interactor; BCMA, B cell maturation antigen.

is higher in the second CRD than the first CRD, because highly conserved amino acid residues tend to be critical for protein function [61].

The expression pattern of these receptors on B cells differs depending on the type of B cells and their stage of maturation and activation. BAFF-R expression is not detectable on B cell precursors in the BM, but it appears on immature B cells as they progress toward their differentiation into splenic transitional type 1 peripheral B cells and upon acquiring a functional BCR [27,64,65]. BAFF-R is the dominant receptor expressed on mature B cells [27,65]. BAFF-R is essential for the survival and maturation of immature B cells [66,67]. BAFF-R expression is also upregulated on activated T cells [27] and constitutively expressed on regulatory T cells [68]. The role of BAFF-R on activated T cells appears to be to promote Bcl2 expression and T helper-1 (Th1) cytokine production in effector CD4 T cells [27]. The role of BAFF-R on regulatory T cells remains unclear.

Mouse TACI is expressed most highly on mature innate-like B cells such as MZ B cells and B-1 B cells, which are very important B cell subsets responding efficiently to type I and type II T-independent antigens [33,69]. It is interesting to note that T-independent type II and type I antibody responses are impaired in mice lacking TACI [70,71]. TACI is an important promoter of class switch recombination (CSR) in B cells, and this function involves direct interaction between TACI and the signaling adaptor myeloid differentiation primary response gene 88 [72]. Another feature of mice lacking TACI is a greater number of B cells, and some TACI-deficient mouse lines develop autoimmunity and B cell lymphoma [73]. Together, these observations suggest a negative regulation of B cell numbers by TACI, and some studies have suggested a role for TACI in driving B cell apoptosis [73]. However, TACI, unlike other receptors from the TNFR superfamily such as TNFR-1 or Fas [74], does not have a death domain in its intracellular portion, and currently described signaling elements downstream of TACI are not obviously linked to a death pathway [72,75]. A very recent study has shed some light on this aspect, demonstrating that the role of TACI in regulating B cells is indirect and involves indirect upregulation of the Fas and Fas ligand (FasL; a TNF–TNFR pair triggering apoptosis in lymphocytes) on the surface of LPS-activated MZ B cells [76]. TACI$^{-/-}$ MZ B cells fail to die after LPS activation and remain activated for long periods of time (Figure 2) [76]. TACI expression on B cells varies throughout life and is decreased in newborn humans and mice [77] and on GC B cells [27]. It is interesting to note that in TACI-deficient mice, numbers of $T_{fh}$ and GC B cells are increased, suggesting a role for TACI in regulating these cells [78], although TACI seems to be more important in antibody-secreting cells (ASCs) in which it normally downregulates the proapoptotic molecule Bim, allowing ASC formation [78]. TACI signaling also supports Blimp-1 expression, a transcriptional regulator important for ASC differentiation [79].

TACI is expressed on human memory B cells, which in healthy individuals also express higher levels of Fas, consistent with a general correlation seen between TACI and Fas expression [36,76,80]. TACI is also expressed by a subpopulation of activated CD27$^{neg}$ B cells [36,80] and macrophages, and it influences their cell survival [31]. TACI mRNA is expressed by a subset of T cells [81]; however, polyclonal and monoclonal antibodies against TACI and gene profiling experiments failed to confirm TACI expression in T cells [27]. Many new T cell subsets have been identified in recent years, and it remains to be seen whether TACI can be expressed on these cells in some circumstances.

TACI expression is strongly upregulated on the surface of B cells in response to the activation of many TLRs, an aspect that is consistent with a key role for TACI in T-independent B cell responses [33,35,82,83]. TACI and APRIL interact with HSPG, forming a complex presumably designed to facilitate crosslinking of TACI by APRIL via stabilizing ligand/receptor interaction and receptor crosslinking, thus allowing robust signaling (Figure 1) [84–86].

BCMA expression is limited to GC B cells, plasmablasts, and PCs [15,53,87]. BCMA is not expressed on splenic B cell subsets, but BCMA expression is high on BM PCs [88]. In fact, BCMA is important for the survival of long-lived PCs in the BM [89]. Recent work showed that BCMA signaling is critical for the expression of the prosurvival molecule Mcl-1 in PCs [46]. BCMA may play a role in antigen presentation by B cells [90] because, in contrast to TACI and BAFF-R, only BCMA can activate the JNK pathway, which leads to the induction of the antigen (Ag) presentation response [90]. However, further studies revealed that BAFF-induced MHCII upregulation (an indicator of the Ag presentation response) was clearly dependent on BAFF-R and TACI [13].

A new receptor for BAFF, Nogo 66 receptor has been described, but it appears to be expressed in the CNS on neurites but not on B cells [91]. However, independent further confirmation of BAFF interaction with Nogo 66 receptor is needed.

## 1.3 The Role of the BAFF–APRIL System in B Cells in Health

### 1.3.1 *B Cell Survival and Maturation*

The primary function of BAFF is to ensure B cell survival during B cell maturation in the spleen [66,92,93]. B cells acquire BAFF-R expression at the immature stage and must undergo intermediate differentiation before reaching the mature B cells stage [15,67]. The intermediate differentiation stage that takes place in the spleen is called the

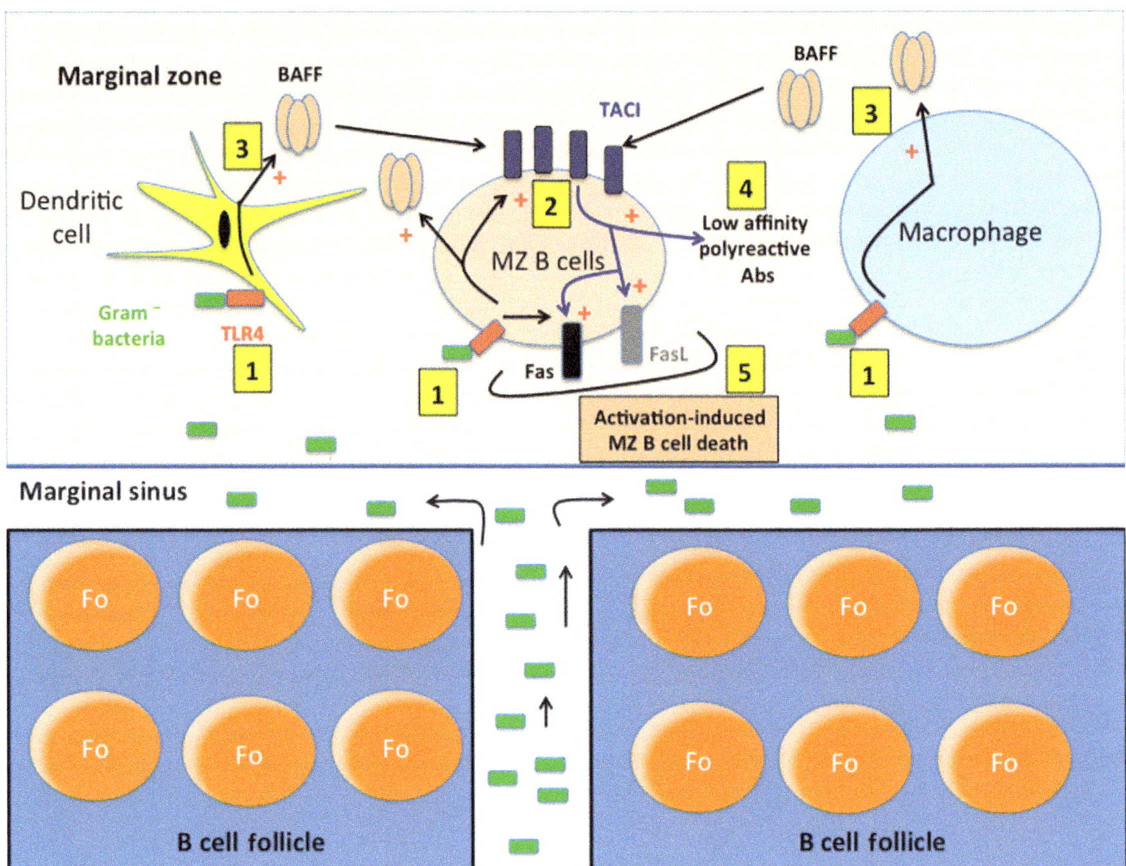


FIGURE 2 **Activation-induced B cell death controlled by TACI on marginal zone (MZ) B cells.** This figure depicts the anatomy of the mouse spleen with the outer part of the white pulp represented by the mouse splenic MZ (top in pink), the MZ sinus (middle T shape structure in white), and the B cell follicles (bottom with follicular B cell (Fo) in orange) in the inner part of the white pulp. In green are represented gram-negative bacteria entering the spleen via the MZ sinus. (1) LPS on the surface of gram-negative bacteria entering the MZ activate TLR4 on the surface of dendritic cells (DCs), MZ B cells, and MZ macrophage. (2) Activation of TLR4 in MZ B cells leads to the strong upregulation of TACI expression on the surface of these cells. (3) In response to TLR4 activation, DCs, MZ B cells, and macrophages produce BAFF, which binds to TACI on MZ B cells. (4) TACI activation is very important for the production of low-affinity polyreactive antibody in response to LPS activation of MZ B cells. (5) At the same time, LPS-activated MZ B cells upregulate Fas and Fas ligand (FasL) on their surface in a TLR4/TACI-dependent fashion, leading to activation-induced MZ B cell death. BAFF, B cell activating factor of tumor necrosis factor; TLR, toll like receptor; TACI, transmembrane activator and cyclophilin ligand interactor.

immature transitional B cell stage, with several transitional stages described over the years from type 1, 2, and 3 [94,95]. Type 2 and 3 B cells are particularly dependent on BAFF signaling via BAFF-R for survival [2,66]. Mice lacking either BAFF or BAFF-R have no mature B cells apart from B-1 B cells and a small percentage of $IgD^{hi}/IgM^{lo}$ B cells in the spleen [92]. Expression of Bcl2, a prosurvival oncogene, can rescue B cell maturation of all B cells in BAFF-deficient animals except MZ B cells, suggesting that the role of BAFF is not limited to the support of B cell survival but also plays a role in the differentiation of MZ B cells [96].

Activation of the survival receptor BAFF-R leads to the recruitment of TRAF2 and TRAF3, and ultimately NF-κB2 activation, and transcriptional activation of anti-apoptotic genes of the Bcl2 family [97,98]. BAFF also mediates B cell survival by blocking the BCR-induced upregulation of Bim (a proapoptotic molecule) and preventing the translocation of the proapoptotic protein kinase Cδ (PKCδ) to the nucleus [99,100]. Treatment of B cells with BAFF results in increases in cell size and metabolic activity of the B cell as well as upregulation of Akt [101]. Upstream of Akt, BAFF-mediated B cell survival requires the catalytic subunit of PI3K, p110δ [102]. More recent results suggest that BCR and BAFF signaling converge high on the signaling pathway, at the level of the Src-family tyrosine kinase [103,104]. Together, this work strongly suggests that signaling via the BCR is critical for optimal BAFF survival signals, although the exact regulation of BAFF-R signals by BCR signals still remains sketchy [105].

### *1.3.2 B Cell Activation*

Initial work showed that BAFF also promotes the activation of B cells in culture [9]. However, how important this aspect of BAFF function is relative to other known B cell co-stimulators such as CD40L and IL-4 is not fully understood in vivo. B-1 B cells, as mentioned above, do not require BAFF for development or survival [92]; however, B-1 B cell numbers are increased in BAFF Tg mice, suggesting that the co-stimulatory effect of BAFF may have contributed to their expansion and proliferation postactivation [106]. The receptor TACI triggers IgG and IgA CSR [107,108]. It is interesting to note that the cytoplasmic portion of TACI interacts with the signaling adaptor myeloid differentiation primary response gene 88 [72]. Upon activation of TACI, the DNA-editing enzyme activation-induced cytidine deaminase (AID) is expressed, allowing CSR [72]. It is interesting to note that BAFF Tg mice lacking T cell help produce IgG isotypes typical of T-dependent B cell responses, suggesting that BAFF overexpression can act as a surrogate T cell-like co-stimulator of CSR via TACI [33]. BAFF Tg mice also produce high levels of IgA [33,109,110]; however, increased IgA levels in BAFF Tg mice are dependent on a T cell mechanism [33]. TACI also plays a role in controlling Blimp1 expression and as such contributes to generation of PCs [79]. BCMA plays a role at the end stage of B cell activation [111] and promotes the survival of a subset of PCs [89]. BCMA strongly promotes the expression of the prosurvival factor Mcl1 in the BM [46], which contributes to long-term survival of PCs in this compartment but not the spleen. APRIL expression in the BM is also critical for PC survival [5,41], especially in the murine setting, where BCMA exhibits poor affinity for BAFF [58].

## 1.4 The Role of the BAFF–APRIL System on B Cells in Disease

### 1.4.1 BAFF and B Cell Tolerance

The immune system generates lymphocytes daily [112,113]. Many newly generated lymphocytes are self-reactive and from the BM to the periphery, many immune checkpoints exist to eliminate or neutralize self-reactive and potentially harmful immune cells [114,115]. This process is called acquired immune tolerance and is driven by the recognition of self-antigen by self-reactive lymphocytes [114,116]. B cell immune tolerance takes place at various sites in the body, including the BM and the spleen [114–116]. Many different immune tolerance mechanisms have been described for B cells, from negative selection to receptor editing, which will allow the expression of an alternative BCR that is no longer self-reactive [114,115]. Another mechanism observed in the periphery is B cell anergy, in which self-reactive B cells become nonresponsive and fail to be activated upon BCR contact with self-antigens [114,115]. Exclusion of self-reactive B cells from important niches in the spleen, where survival factors are present, in particular when competing with non-self-reactive B cells, is another mechanism known to eliminate these potentially harmful B cells [117].

As mentioned above, BAFF is a B cell survival factor, and fine regulation of B cell survival is critical through the process of B cell maturation. Indeed, too much B cell survival can allow self-reactive B cells to survive and prevent negative selection [2]. Illustrating this point, BAFF Tg mice develop B cell-dependent autoimmune disorders [75,109]. How excessive BAFF breaks immune tolerance has been tested in vivo using an animal model in which all B cells recognize an experimental antigen, hen egg lysozyme (HEL), which has been intercrossed with mice expressing HEL as a transgene, hence generating progeny born with HEL as a self-antigen [118,119]. Experiments using this model have shown that self-reactive anti-HEL B cells have a greater dependence on BAFF for their survival [119]. It is interesting to note that HEL-specific B cells developing in an animal expressing HEL are generally more immature than HEL-specific B cells developing in the absence of the self-antigen [118]. Immature transitional B cells are known to be more dependent on BAFF for survival irrespective of their specificity [66,92]. The missing experiment would be the following: a transfer of transitional HEL-specific B cells (generated in wild-type (WT) or HEL-expressing animals) into HEL-expressing animals to check their BAFF requirement for survival. It is likely that in this case little difference will be seen.

Another study further clarified the role of BAFF in this model [118]. When HEL-specific B cells develop in a BAFF Tg mouse expressing HEL as a self-antigen, they expand in the periphery rather than being eliminated, when HEL is expressed as a soluble form (sHEL) [118]. However, when HEL is expressed as a membrane-bound form with strong affinity to the HEL-specific BCR, B cell negative selection in the BM occurs [120], and this is not affected by BAFF overexpression [118]. This result was expected because developing B cells in the BM do not express BAFF-R [27,64,65]. In contrast, the observed expansion of HEL-specific B cells in BAFF Tg mice expressing sHEL is an indication that excessive BAFF production corrupts peripheral B cell tolerance driven by sHEL [118]. However, the exceptionally high affinity of the HEL-specific BCR and the absence of competition with polyclonal non-self-reactive B cells are complicating factors in these experiments. To address these issues, mice were generated to express the HEL-specific heavy chain but not the light chain to generate high- and low-affinity HEL-specific B cells, and BM from these mice was used combined with WT BM to make chimeric mice in which only 10% of the B cell repertoire was HEL-specific and the recipients expressed sHEL and BAFF [118]. In these chimeric mice, high-affinity but not low-affinity HEL-specific B cells were eliminated despite the mice producing large amounts of BAFF; however, the number of low-affinity HEL-specific B cells, many of which exhibited a MZ phenotype, was greatly enhanced [118]. In the case of low-affinity HEL-specific B cells, unlike their high-affinity counterparts, these cells are allowed to mature to a stage when they express sufficient BAFF-R on the cell surface to be rescued by excessive BAFF production, survive, and expand [118]. In conclusion, a strong interaction between the HEL-specific BCR and HEL remains an effective way of negatively selecting self-reactive B cells that is not altered by excessive BAFF production. In contrast, excessive BAFF will alter the number of low-affinity self-reactive B cells in the periphery and increase their numbers.

As mentioned above, many of the low self-reactive B cells that are not eliminated because their affinity for the self-antigen is below the threshold that normally negatively selects self-reactive B cells populate the MZ B cell compartment [118]. MZ B cells are known to be innate-like B cells because they respond very efficiently to TLR activation, in particular TLR4 activation in response to LPS [121–123]. Moreover, MZ B cells are very good antigen-presenting cells to naïve B cells and as such form a bridge between the innate and the adaptive immune system [124]. However, considering that LPS can activate TLR4 on all B cells regardless of whether these cells are self-reactive or not, and MZ B cells are potentially self-reactive, there is a danger of generating activated autoreactive MZ B cells that can potentially access T cell help, increase their affinity in a GC, and eventually differentiate into antibody-forming cells producing high-affinity autoantibodies. However, most infections with gram-negative bacteria do not lead to an autoimmune disease, thereby raising the question as to how the activation of MZ B cells is regulated. Autoantibody production, such as anti-nuclear autoantibodies (ANAs), has been commonly observed in the clinic in the serum of individuals suffering from infections, but this production is usually transient and disappears when the infection is cleared [125]. In general, B cell response to type I antigens such as LPS is short-lived [126]. All of these observations point to a regulatory mechanism terminating the response of innate B cells such as MZ B cells to nonspecific activation with TLR ligands, but details of this mechanism are not fully understood.

Recent work has shed new light on one of such regulatory mechanism [76]. It is interesting to note that MZ B cells express much higher levels of TACI on the cell surface when compared with other B cell subsets in the spleen [33,69]. Moreover, exposure to LPS in vivo very strongly upregulates TACI on MZ B cells [76]. The reason for such a specific LPS-induced higher TACI expression on MZ B cells has remained unclear. It is interesting to note that LPS and BAFF cooperate to induce Fas expression on the surface of MZ B cells, a process dependent on TACI expression [76,82]. TACI expression on MZ B cells is essential for optimal induction of Fas and FasL expression on these cells [76]. Further work demonstrated that LPS leads to rapid activation of MZ B cells in vivo [123] but also induces MZ B cell apoptosis [76]. MZ B cell apoptosis was deficient in mice lacking TACI or the membrane form of FasL, which is the active form driving apoptosis via Fas [76].

TACI signaling has another key function contributing to the sensitivity of LPS-activated MZ B cells to apoptosis because it represses the expression of many inhibitors of the Fas signaling pathway, and in particular FLIP (p43) [76]. FLIP (p43) forms heterodimers with caspase-8, which interact directly or indirectly with other factors, such as TRAF-family adapters, rat insulin promoter (RIP) kinases, and possibly RAF-1 [127]. This results in the activation of NF-κB and certain other signaling pathways that promote cell survival and proliferation as well as in the suppression of RIPK3-mediated necroptosis [128,129]. Consistent with these findings, TACI-deficient MZ B cells but not WT MZ B cells are positive for the proliferation marker Ki67 [76]. Together, recent findings suggest that TACI signaling in MZ B cells serves two major purposes: (1) it upregulates Fas and FasL on the surface of these cells, and (2) it represses the expression of intracellular anti-apoptotic factors. TACI signaling sensitizes MZ B cells to apoptosis. BAFF and APRIL can trigger this effect [76]. Moreover, TLR4 and the downstream signaling element Mal are very important for Fas and FasL upregulation on LPS-activated MZ B cells [76]. It is interesting to note that the MZ B cell compartment is enlarged in TLR4- and Mal-deficient animals [76], suggesting a role for these two molecules in regulating this particular B cell compartment. This effect is very specific for TLR4, TACI, and MZ B cells [76]. Other TLR can induce Fas on the surface of B cells, but this effect is TACI-independent [76]. B-1 B cells, which also express high levels of TACI, do not respond like MZ B cells to LPS in vivo. These differences suggest that possible BAFF-independent mechanisms exist to sensitize other B cells to apoptosis in response to innate activation.

TACI is the first receptor on B cells known to drive innate activation-induced B cell death, and this effect explains the short-lived nature of MZ B cell responses to type I independent antigens [126]. Upon contact with gram-negative bacteria entering the spleen via the marginal sinus, cells in the MZ such as macrophages, LPS-activated DC, and MZ B cells respond by producing BAFF and APRIL [2]. LPS also triggers TLR4 on the surface of MZ B cells, which in response strongly upregulates TACI on the cell surface [76]. LPS activation of MZ B cells leads to rapid and effective production of low-affinity nonspecific polyreactive antibodies that recognize conserved epitopes on bacteria [123,130]. These antibodies can opsonize bacteria and act as a first line of defense while the adaptive immune system prepares itself for an antigen-specific response. This first wave of low-affinity and nonspecific antibody production by LPS-activated MZ B cells is TACI dependent because these responses are deficient in TACI-deficient animals [71].

However, as soon as MZ B cells have been activated by LPS, these cells upregulate Fas and FasL on the cell surface and downregulate the expression of anti-Fas signaling factors such as XIAP and cFLIP, allowing these cells to die before they can migrate to the T cell zone, gain T cell help, and increase their affinity in a GC, a process that would not require TACI and is normal in TACI-deficient animals [71,131]. In conclusion, TACI plays a role in regulating and terminating activation of MZ B cells to innate activation.

The question is whether or not TACI is important for B cell tolerance as a whole. A recent study with common variable immune deficiency (CVID) patients carrying a mutation in the *TACI* gene confirmed that indeed B cell tolerance in these patients is defective [132], confirming in vivo the important role of TACI in not just regulating B cell responses to innate activation but for the maintenance of a healthy B cell repertoire in humans.

### 1.4.2 BAFF and B Cell Autoimmunity

In BAFF Tg mice, excessive BAFF production leads to the expansion of the B cell compartment and in particular MZ B cells [75,109,133–135]. In addition, BAFF Tg mice produce a wide range of autoantibodies, from rheumatoid factors (RFs), anti-double stranded and anti-single stranded DNA autoantibodies to ANAs, and anti-histone autoantibodies [75,109,133–135]. As BAFF Tg mice age, they develop autoimmune symptoms such as nephritis and sialadenitis, reminiscent of human SLE and Sjögren's syndrome (SS), respectively [75,109,133–135]. Mechanisms leading to tissue inflammation and destruction in BAFF Tg mice have been extensively studied [33,75,109,133–135]. Dysregulation of T and B lymphocyte functions contributes to development of autoimmune disorders such as SLE [136]. BAFF Tg mice have a greater proportion of activated T cells and spontaneously develop large GCs, which are T-dependent immune structures designed to select B cells with an improved affinity for their specific antigen or in this case potential autoantigens [33,109]. Moreover, BAFF is produced by activated T cells, which also upregulate BAFF-R after activation [27,56]. BAFF binds to BAFF-R on T cells, which induces the expression of prosurvival factors and the production of Th1 cytokines such as IFN-$\gamma$ [27,56]. Th1 cells, a subset of IFN-$\gamma$-producing effector T cells, play a pathogenic role in autoimmunity, and, in BAFF Tg mice, it appeared that both of the activated T and B cell compartments are dysregulated, suggesting a strong role for T and B cell cooperation, leading to strong GC-dependent autoantibody production as the main underlying cause for autoimmune disorders. This model fitted the views that T cells provide help for B cells and B cells can serve as antigen-presenting cells to T cells, perpetuating a vicious cycle leading to inflammation and tissue destruction [136]. However, this model was challenged by the observation that BAFF Tg mice lacking GC develop autoimmune disorders undistinguishable to that of BAFF Tg mice developing GC spontaneously [106,135], ruling out a role for B cell affinity maturation as an important pathogenic event. Moreover, although the number of activated T cells is elevated in BAFF Tg mice [109], so is the number of Foxp3$^+$ regulatory T cells [33], questioning the exact role of T cells in driving disease in the BAFF Tg mouse model of SLE. It is surprising to note that T cell-deficient BAFF Tg mice also develop an autoimmune disease undistinguishable from that of T cell-sufficient BAFF Tg mice [33], suggesting that T cells are dispensable for the autoimmune disorders developing in BAFF Tg mice. Disease in T cell-deficient BAFF Tg mice was characterized by the deposition of IgG1, 2b, and 2c autoantibodies in the kidneys of these animals and trapping of C3 complement, which triggers tissue inflammation and destruction [33]. IgG1, 2b, and 2c isotype production in the absence of T cells is unusual because normally isotype switching to these isotypes requires T cell help. As mentioned above, BAFF and APRIL are produced by activated T cells [56], and signaling via TACI is important for isotype class switching [72]. It is possible that excessive BAFF production in BAFF Tg mice has mimicked an overproduction of BAFF by T cells and led to switching to and production of IgG1, 2b, and 2c autoantibodies.

It is interesting to note that IgA levels are elevated in BAFF Tg mice [109,110,137], and this aspect is T cell-dependent because IgA levels become normalized in T cell-deficient BAFF Tg mice [33]. However, in T cell-sufficient BAFF Tg mice, elevated IgA levels play a significant role in the kidney pathology developing in BAFF Tg mice [110]. Indeed, in IgA-deficient BAFF Tg mice, the renal pathology was reduced [110]. In addition, commensal flora-reactive IgA were abnormally detected in the blood of BAFF Tg mice, suggesting an imbalance with the microflora and a defect of the normal mucosal-peripheral compartmentalization in these animals [110]. IgA nephropathy in humans is a male-dominant disease and correlates with higher APRIL serum levels [110]. It is interesting to note that IgA-associated nephropathy in BAFF Tg mice is also more severe in male mice when compared with females [110], suggesting a possible gender-biased development of fatal nephropathy in BAFF Tg mice likely to be associated with different BAFF transgene expression in males and female mice. In conclusion, disease in BAFF Tg mice occurs in the absence of T cells; however, the nature of the renal disease is different than that of T cell-sufficient BAFF Tg mice. These findings shed some light on the contribution of T cells in kidney pathologies.

### 1.4.3 The Role of the BAFF/APRIL System in B Cell Immunodeficiencies

CVID is one of the most common immunodeficiency syndromes and is characterized by a reduction in antibody production, in particular IgG, IgA, and IgM, as well as poor responses to vaccinations and vulnerability to infections, in particular those of the respiratory track (reviewed in [138]). CVID is also characterized by autoimmune manifestations in approximately 20% of affected patients [138]. The etiology of CVID remains unclear, but it appears to be linked to several genetic defects affecting immune functions, although the inheritance pattern of these genetic modifications does not always fit predictions [139]. Mutations or deletions of several immune genes affecting B cell function have been associated with CVID, such as the inducible co-stimulator (ICOS), CD19, CD81, and CD20 [139].

A recent advance in our understanding of CVID emerged with the discovery of a series of mutations in the *TACI* gene that were found in approximately 8–10% of CVID patients [140–144]. Homozygosity for variants C104R, A181E, and S144X of the *TACI* gene was observed in several CVID patients [140,141]. Heterozygosity for variants C104R, A181E, S194X, and R202H has also been reported in CVID patients [140,141]. The most common variants are C104R affecting the extracellular domain and A181E located in the transmembrane domain, and for these, heterozygosity is more common than homozygosity [140,141]. One report described experiments establishing how the C104R mutation impairs TACI function [145]. It appears that this mutated form of TACI interferes with TACI signaling by associating with WT TACI in the absence of any ligand, and as such, forming signaling-defective oligomeric receptor complexes [145]. Later work looking at the equivalent TACI mutation in mice (C76R) demonstrated that B cell function was impaired in mice heterozygous for the mutation, suggesting that this mutation impairs TACI function in heterozygotes via haploinsufficiency [146].

Mice lacking TACI develop defects resembling that of CVID patients such as reduced IgA levels and impaired responses to T-independent antigens [71,131], a problem which appears to be linked in part to the role of TACI in maintaining Blimp-1 expression, essential for PC generation and persistence [79]. However, relatives of CVID patients can often carry the same TACI mutations without any immunological abnormalities [147,148]. It is interesting to note that both B cells from CVID patients and healthy relatives are defective in vitro, in particular in response to TLR9 activation and TACI stimulation [147]. In contrast, in vitro IgG and IgA production from healthy relative-derived B cells was normal when compared with B cells from CVID relatives [147]. Therefore, mutation in the *TACI* gene is not the sole contributing factor, but it may facilitate emergence of CVID in combination with other contributing factors.

BAFF-R is essential for B cell survival [67]; therefore, a dominant mutation in the *BAFF-R* gene is likely to lead to immunodeficiency. Indeed, BAFF-R variants, present at the heterozygous state, have been identified in CVID patients with low B cell numbers [149]. These variants appeared to be polymorphic and do not affect BAFF-R expression [149]. It is unclear whether these variants can signal [149]. Another report described siblings carrying a homozygous *BAFF-R* gene deletion preventing BAFF-R protein expression [150]. Although only one sibling had recurrent infections, the other remained relatively asymptomatic. Both siblings had reduced B cell numbers and reduced IgM and IgG levels, but IgA levels remained within a normal range [150]. TACI (which was not mutated in these patients) is critical for IgA production in mice [71,131], which may explain this difference. Therefore, similar to individuals treated with the B cell-depleting agent belimumab (anti-BAFF monoclonal antibody, also known as Benlystat®), the incidence of infection in BAFF-R-null patients may vary from one individual to another depending on additional factors and/or exposure to pathogens [151]. For this reason, although defects in BAFF-R or TACI function will certainly increase an individual's vulnerability to immunodeficiency, additional factors are required to precipitate clear CVID clinical manifestations.

### 1.4.4 The Role of the BAFF/APRIL System in B Cell Malignancies

The classical picture of the BAFF system involved in B cell survival is slowly evolving into a more complex picture of interwoven regulatory function and downstream signaling designed to alter cellular survival, growth, and migration. Although the central role of BAFF in autoimmunity has now been well established, work is still ongoing trying to understand the contribution of the BAFF system to malignancies. The current conceptual picture suggests that BAFF and APRIL may be pathogenic players in microenvironments of hematological tumors of B cell origin such as chronic lymphocytic leukemia (CLL).

Indeed, there is strong evidence that the BAFF/APRIL system plays an important role in the progression of CLL, particularly promoting paracrine CLL survival because these cells produce BAFF and APRIL and express BAFF and APRIL receptors [152–154]. Other cells described in CLL, such as nurse-like cells, also produce BAFF and can contribute to CLL progression [155]. It is interesting to note that, unlike normal transitional B cells, BAFF and APRIL support the survival of CLL cells via activation of the classical but not the alternate NF-κB pathway, suggesting a different signaling pathway allowing persistence of CLL cells.

Overexpression of BAFF has been shown to enhance c-myc expression and promote CLL-like disease via activation of the canonical IKK in Myc/BAFF Tg mice [156]. BAFF production by stromal support cells, such as microvascular endothelial cells, has been shown to be triggered by CD40L aberrantly expressed on the surface of CLL B cells, further supporting the importance of support stroma for cancer cell survival and explaining observed ex vivo apoptosis [157]. BAFF Tg mice lacking TNF also develop B cell lymphoma and MALT lymphomas [106]. B lymphoma development has also been observed in TACI-deficient mice [73]. Moreover, APRIL transgenic mice develop a disease similar to CLL [158], and anti-APRIL blocking antibodies were effective at stopping the CLL-like pathology in these mice [159]. A picture is emerging in which too much BAFF or APRIL can allow inadequate survival of neoplastic B cells. This may occur via two mechanisms: one involving excessive survival via BAFF-R upon BAFF binding or defective TACI-mediated B cell regulation via B cell death.

Given these observations, measurement of serum BAFF levels were compared in relation to existing scoring systems of CLL [160]. Prognostic use of BAFF levels has already shown promise in CLL when combined with CD38, ZAP70 expression, and the mutational status [161] and in follicular lymphoma, in which expression of BAFF and BAFF-R are elevated and coincide with inferior progression-free survival, with some suggestion that overexpression of BAFF is secondary to elevated expression of BAFF-R and therefore increased sensitivity to BAFF [162].

It is interesting to note that soluble BCMA isoforms have been identified, and although their function is not yet elucidated, it suggests the exciting prospect of natural decoys produced to regulate levels of free BAFF or APRIL [7]. Most interestingly, overexpression of APRIL was found to coincide with overexpression of HSPGs [163], which are known co-receptors for APRIL when binding TACI. This has renewed speculation that HSPGs may act as primary receptors for APRIL in addition to co-receptor function. BAFF-R has been implicated in nuclear interactions with IKKβ and NF-κB/c-Rel in healthy and neoplastic B cells, including non-Hodgkin's lymphoma (NHL) B cells, which enhance histone H3 phosphorylation and binding of NF-κB on promoters for inducible genes, including Bcl-xl, CD40L, and most strikingly BAFF [164]. This complicates further the mechanism regulating the BAFF system, in which the receptor itself is a transcriptional regulator and exerts its effects via chromatin remodeling in addition to classical NF-κB activation at the membrane surface [164].

Mutations in BAFF-Rs have also emerged as critical factors in disease etiopathology and have emerging prognostic power. Substitution mutations in the *BAFF-R* gene at His159Tyr in patients with NHL increased recruitment of TRAF3, TRAF2, and, surprisingly, TRAF6, demonstrating for the first time that TRAF6 is an essential component of downstream BAFF-R signaling in NHL and in health [165].

Inhibition of BAFF or its receptors has been a strong focal point for therapeutic development. Merck's Phase Ib trial with the atacicept inhibitor of BAFF and APRIL has shown that the agent is tolerated at moderate dosage, but currently no data on the clinical activity in CLL trials are available [166]. A strategy using mutant BAFF fused with Pin2/TRF1-interacting protein X1, a telomerase reverse transcriptase inhibitor, has also been effective at selectively killing BAFF-R expressing Burkitt lymphoma lines [167]. The latter example further reinforces growing evidence that BAFF function is not confined to NF-κB activation and prosurvival transcriptional activity, but that BAFF also directs alteration at immunosenescence checkpoints of target cells at the chromatin level to alter proliferative capacity and ultimately survival. These developments will preface development of a new generation of cancer therapeutics.

Over a decade of research on the BAFF and APRIL system has clearly demonstrated how central this system is in B cell biology in health and disease states. This work led to the development of the second most important biologic treatment targeting B cells after rituximab. However, several questions remain, in particular differences in the function of these ligands and their receptors in mouse versus humans and how this may affect the development of targeting strategies in this system.

## 2. THE LYMPHOTOXIN PATHWAY: SHAPING B CELL ENVIRONMENTS

### 2.1 Introduction

Although BAFF and APRIL predominantly exert their functions by acting directly on B cells, other TNF superfamily members, in particular the lymphotoxin (LT) pathway, play a crucial role in shaping environments in which B cells are activated. Furthermore, expression of the LT-αβ ligand by B cells plays an important role in the homeostasis of several immune accessory cells in the steady state and during infection/inflammation. This section focuses on the contribution of the LT pathway to B cell biology.

### 2.2 Receptors and Ligands of the LT Pathway

LT-α was discovered as a soluble cytotoxic factor in 1968, approximately 7 years before the subsequent discovery of TNFα [168,169]. Homotrimers of LTα were found to induce cytotoxic activity, and these soluble homotrimers were later shown to bind to TNFRs I and II. Because $LT\alpha_3$ and TNFα shared common receptors, for quite some time it was assumed that the biologies of these two cytokine pathways were mutually redundant. Indeed, after the discovery of TNFα, LTα was (temporarily) renamed TNFβ [170].

However, two major discoveries revealed that the LT pathway was a unique axis that governs biological processes distinct from TNFα/TNFR signaling. First, the discovery that LTα could form heteromers with another TNF superfamily molecule called LTβ to form a heterodimer ($LT\alpha_1\beta_2$) revealed greater complexity in this system [171]. Secondly, the discovery of the LTβ-receptor (LTβR) that could propagate signals, in particular the activation of the NF-κB pathway, revealed a TNFR-independent counter-receptor for $LT\alpha_1\beta_2$ [172]. In fact, $LT\alpha_1\beta_2$ does not bind to TNFR1/TNFRII.

Further cementing the concept that $LT\alpha_1\beta_2$/LTβR signaling provides unique biological function, genetic ablation of LTα, LTβ, and LTβR, but not TNFα, results in the absence of most lymph nodes (LNs). In addition to the $LT\alpha_1\beta_2$ heterotrimer, a second ligand for the LTβR was discovered called LIGHT (which stands for homologous to LTs, exhibits Inducible expression, and competes with HSV gD for Herpesvirus entry mediator, a receptor expressed by T lymphocytes) [173]. This ligand can bind to another receptor called HVEM, which itself can bind to other molecules. As can be imagined, one of the challenges in studying the LT pathway in the context of B cells has been to isolate the role of particular ligands in the context of specific cell types. This challenge can be addressed using chimeric mice, tissue-specific deletion of the LTβR and its ligands, and adoptive transfer approaches.

### 2.2.1 Expression Pattern of LT Molecules

The LTβ moiety of the $LT\alpha_1\beta_2$ heterotrimer effectively anchors the $LT\alpha_1\beta_2$ complex to the cell surface. This provides a means for direct cell:cell communication between $LT\alpha_1\beta_2$-expressing cells and LTβR expressing cells [174]. In general, $LT\alpha_1\beta_2$ is not secreted, although it can be cleaved by matrix metalloproteinases in pathological environments such as in the RA synovium [175]. The expression of $LT\alpha_1\beta_2$ is largely limited to lymphocytes and NK cells. It is interesting to note that although expression of $LT\alpha_1\beta_2$ on T cells requires activation of the T cell (i.e., through the T cell receptor) [176], expression of $LT\alpha_1\beta_2$ on B cells is constitutive, albeit expressed at very low levels [177]. $LT\alpha_1\beta_2$ is further upregulated on B cells that have been activated by antigen and express GC B cell markers such as GL7 and FAS [178]. Some reports have also demonstrated the expression of $LT\alpha_1\beta_2$ on DCs [179,180]. LIGHT is also expressed on activated T cells as well as B cells, NK cells, platelets, and DC [181].

With respect to the LTβR, the expression pattern is quite different, being expressed primarily on "accessory" cells of the immune system including stromal cells within lymphoid tissues, macrophages, DCs, and FDCs. Given that the LTβR ligands are membrane-anchored and are upregulated on activated lymphocytes, this suggests dynamic crosstalk between LTβR-expressing accessory cells and lymphocytes recruited into immune responses. Indeed, the induced expression of $LT\alpha_1\beta_2$ on antigen-specific T cells is required for DC activation and cross-priming of $CD8^+$ T cell responses [182]. Furthermore, expression $LT\alpha_1\beta_2$ on innate lymphoid cells (ILCs) in the gut performs a similar crosstalk function by activating mucosal DCs to mediate host defense against intestinal pathogens [183,184]. Although expression of $LT\alpha_1\beta_2$ on T cells is important for optimizing DC activation and cytokine secretion, as will be explained later, the role of $LT\alpha_1\beta_2$ on B cells in the immune response is less direct.

### 2.2.2 Genetic Targeting of the TNF/LT Locus

The *tnf*, *LTa*, and *LTb* genes are closely linked on mouse chromosome 17 (chromosome 6 in humans). This has caused difficulties in interpreting data derived from some of the original strains of $TNF^{-/-}$ mice, which in some cases had a concomitant disruption of *LTa*. Indeed, TNF null mice constructed using Cre-loxP technology have a very different Peyer's patch (PP) phenotype compared with conventionally generated $TNF^{-/-}$ mice [185]. Moreover, the *tnf*, *LTa*, and *LTb* genes are located within the H-2 locus. Thus, knockout mice generated using 129 ES cells will have a 129 MHC, potentially resulting in altered immune responses and incompatibilities due to minor mismatches. Such issues must be kept in mind when interpreting experiments that have used adoptive transfer or BM chimera models because the knockout cells could potentially be rejected by host immune responses.

Despite these caveats, consistent phenotypes have emerged in LT-deficient mice with respect to lymphoid tissue homeostasis. As previously mentioned, deletion of *LTa* [186], *LTb* [187], or *LTbR* [188] genes results in the abnormal development of PPs and most LNs. Although LTβR and LTα-deficient mice lack all LNs, $LT\beta^{-/-}$ mice retain mesenteric LNs, which vary in size depending on the colony. Residual cervical LNs can also be observed in $LT\beta^{-/-}$ mice. The residual LNs in $LT\beta^{-/-}$ mice may be due to the compensatory triggering of LTβR signaling by LIGHT because mice that lack LTβ and LIGHT do not have mesenteric LNs [189]. However, LIGHT is not necessary for LN development because $LIGHT^{-/-}$ mice have a normal complement of LNs and PPs.

In addition to abnormal LN and PP development, several abnormalities in the finer structure of the spleen and LN are apparent. In the spleen, the MZ is located at the interface of the red pulp and the lymphocyte-rich white pulp. The MZ contains specialized MZ B cells as well as $CD169^+$ MZ metallophilic macrophages. These populations, as well as the MZ sinus itself, require LTβR signaling, and the MZ effectively collapses in LT-deficient mice or mice treated with lymphotoxin inhibitors [188,190]. In addition to effects on the MZ, stromal cells that support lymphocytes in the spleen are dysregulated in LT-deficient mice. In particular, a subset of stromal cells called FDCs, which are located in the B cell follicle, are not detectable based on their expression of markers such as CD35 and FDC-M1/MFG-E8. Corresponding to the reduction in FDC, the normally crisp separation between B cells in the follicle and T cells in the periarteriol lymphoid sheath in the spleen is blurred in LT-deficient mice [188]. This is likely due to a strong reduction in chemokine expression within the spleen, in particular CXCL13, CCL19, and CCL21, which are produced by LT-sensitive stromal cells [191]. Reduction in these chemokines results in suboptimal segregation of T cells and B cells into their physiological splenic niches.

In LT-deficient mice, the architecture and cellular composition of the LNs is also compromised. For example, the subcapsular sinus, which envelopes the LN, also hosts a population of CD169+ macrophages that are absent in LT-deficient mice [192]. However, the segregation of B cells and T cells within the LN appears to be relatively preserved [193]. Within the T cell zone, the high-endothelial venules (HEVs), which are the portals of entry for naïve lymphocytes, do not appropriately display the adhesion molecule PNAd on the HEV lumen; thus, LNs in LT-deficient mice are hypocellular [194]. As in the spleen, FDCs are also absent. It is interesting to note that conditional ablation of the LTβR in another stromal cell subset called fibroblastic reticular cells (FRCs) results in some functional abnormalities of these cells, such as poor chemokine/cytokine production, although many physical attributes of FRCs, such as the ability to produce elaborate extracellular matrix networks, is intact [195].

Lastly, in LT-deficient mice, the gut-associated lymphoid tissue (GALT) organization is also affected. As mentioned, PPs are absent and mesenteric LNs are absent in LTα- and LTβR-deficient mice. Discrete follicles of lymphocytes, called isolated lymphoid follicles (ILFs), which depend on the host microbiota, are located in the lamina propria of the gut. ILFs host dense collections of B cells and ILCs. ILFs are absent in LT-deficient settings [196]. Because the ILF are more of an "inducible" nature, this finding has implications in considering ectopic "tertiary" follicles that can appear in the context of inflammation, which will be discussed later.

## 2.3 The Role of B Cell-Associated $LT\alpha_1\beta_2$ in Immune Cell Homeostasis

### 2.3.1 *Expression of $LT\alpha_1\beta_2$ on B Cells and Maintenance of Lymphoid Tissue Accessory Cells*

LTβR signaling is well recognized for its important role in the maintenance of various stromal cell types. In the developing embryo, LTβR signaling on lymphoid tissue organizing (LTo) cells is critical for the development of an LN anlage. In the case of embryonic development, the source of $LT\alpha_1\beta_2$ is derived from lymphoid tissue inducer (LTi) cells, and iterative interactions between LTo and LTi cells are required to induce LT-dependent chemokines that allow for the population of developing LN [197]. This same LT-dependent paradigm that plays out during development is recapitulated in the adult, with B cells being an important reservoir of $LT\alpha_1\beta_2$ for the constitutive LTβR signals that must take place in stromal cells for their maintenance. The most notable stromal cells that require constitutive LTβR signaling to be maintained in a mature state are FDCs [198], and the relevance of expression of $LT\alpha_1\beta_2$ on B cells for FDC differentiation will be discussed in the next section. Other stromal cells play a role in the peripheral immune response with variable dependency on the expression of $LT\alpha_1\beta_2$ on B cells. Notably, in responses against mouse cytomegalovirus, B cell-derived $LT\alpha_1\beta_2$ is critical for an early wave of type I IFN that is produced by stromal cells [199]. This early type I IFN response, which is not dependent on TLR signaling, is crucial for the early resolution of the virus. The nature of the stroma that produces this early wave of type I IFN has not been resolved. In addition, FRCs express chemokines and other cytokines that augment immunity. FRCs do not appear to be dependent on $LT\alpha_1\beta_2$ derived from B cells in the spleen, and in the LN, FRCs do not require LTβR signaling for the production of extracellular matrix. However, their ability to enhance an immune response through the production of IL-7 is indeed LT-dependent [195]. The cellular source of $LT\alpha_1\beta_2$ for FRC function has not been determined.

As mentioned, the integrity of the MZ and the subcapsular sinus of the spleen depends on LTβR signaling. It is unclear how LTβR signaling maintains the microarchitecture of the splenic MZ, which contains marginal reticular cells that are less well-characterized than the other stromal cell types [200]. Looking more closely at the MZ, the MAdCAM+ Sialodhesin+ sinus is ablated in the absence of LTβR signaling, and deletion of $LT\alpha_1\beta_2$ only on B cells results in a partial effect on the sinus [201], suggesting that $LT\alpha_1\beta_2$ on B cells plays a role in maintaining the MZ sinus, but other cell types expressing $LT\alpha_1\beta_2$ may partially compensate.

Particular macrophages within the splenic MZ and the subcapsular sinus are important for antigen capture because these zones are the first point of entry for antigen into the spleen and LN, respectively. CD169+ macrophages require LTβR signaling for their persistence. These macrophages are largely, but not completely, absent in mice that lack $LT\alpha_1\beta_2$ only on B cells [201], suggesting that expression of $LT\alpha_1\beta_2$ on B cells plays an important role in maintaining these accessory cells. Given that the MZ is host to a large reservoir of MZ B cells, and the MZ and subcapsular sinus are adjacent to the B cell-rich follicle, it makes sense that B cells would provide an important source of $LT\alpha_1\beta_2$ for the maintenance of macrophages in these niches. Macrophage intrinsic LTβR signaling and LTβR signaling in radio-resistant cells appears to be required for macrophage maintenance because BM chimeras in which LTβR is deleted only in hematopoietic cells exhibit a partial reduction in $CD169^+$ macrophages whereas complete LTβR signaling blockade results in a more dramatic effect [202].

In addition to macrophages, DCs also rely on LTβR signaling; however, unlike macrophages, in which LTβR signaling must occur in macrophages as well as within the radio-resistant compartment, DC-intrinsic LTβR signaling is sufficient to maintain their homeostasis by promoting DC proliferation [203,204]. The source of $LT\alpha_1\beta_2$ required to maintain splenic $CD11b^+$ DC is derived from B cells, suggesting an important cross-talk between these cell types [204]. Subsequent work has shown that LT-sensitive DCs have a particular phenotype in the spleen ($CD11b^+ESAM^{hi}$) and in the gut ($CD11b^+CD103^+$) [205]. The interaction between cells expressing $LT\alpha_1\beta_2$ and DCs in the mucosa will be further discussed in Section 2.5.

Collectively, these examples reveal that the expression of $LT\alpha_1\beta_2$ on B cells is important for the homeostatic maintenance of various "accessory" cell types such as macrophages, DCs, and FDCs. The signaling mechanism by which LTβR maintains these cell types remains unclear.

## 2.4 The Role of B Cell-Associated $LT\alpha_1\beta_2$ and the GC Reaction

GCs are dynamic structures that facilitate the production of high-affinity antibodies and, as described elsewhere in this chapter, can be the source of autoreactive B cells in the context of excessive levels of BAFF. Please refer to the Chapter "B cell localization and migration in health and disease" by Anja Hauser and Uta Höpken for a complete review of the biology of the GC reaction. As mentioned, the expression of $LT\alpha_1\beta_2$ on B cells is necessary for the presence of FDCs in the LNs and the spleen. In the case of GCs, a very high level of $LT\alpha_1\beta_2$ is expressed on GC B cells, presumably to further aggregate FDCs into a tight GC structure that facilitates Ag trapping and presentation to GC B cells during the GC reaction [177,178]. This extra high level of $LT\alpha_1\beta_2$ on GC B cells is mediated via CD40 signaling in B cells, another TNF family member that is essential for B cell class switching to IgG and for T-dependent GC reactions. The source of CD40-ligand likely comes from $T_{FH}$ cells [206]. ICOS is also involved in the GC response, and indeed LT-deficient and ICOS-deficient mice phenocopy in terms of their GC defects [178]. Thus, a multistep pathway involving ICOS and CD40L expression on T cells and subsequent T·B communication between $T_{FH}$ and GC B cells inducing the expression of very high levels of $LT\alpha_1\beta_2$ is required to optimize the establishment of a GC niche.

Nevertheless, despite the induced expression of high levels of $LT\alpha_1\beta_2$ on GC B cells, the question of whether this is actually necessary for a GC response has been debated. Disruption of LTβR signaling certainly has profound effects on GC responses in the spleen. Significant evidence demonstrates the elimination of LTβR signaling and, as a consequence, mature $CD35^+FDC\text{-}M1^+$ FDCs result in a very poor GC response provoked by systemic antigens such as sheep red blood cells administered via the intraperitoneal route [190,207]. Furthermore, defective GC reactions in response to protein antigens such as the hapten nitrophenyl have been noted in LT-deficient mice, in which clusters of activated B cells expressing markers such as GL7, FAS, and peanut agglutinin in the spleen are not observed. Defects in GC responses in these mice also correspond with poor humoral immunity, with lower amounts of antibody (Ab) produced, and lower quality of Ab due to poor affinity maturation, particularly at low doses of Ag [178,186].

On the other hand, LTβ deficient mice exhibit reasonable GC in the residual mesenteric LNs [208]. This begs the question of what is the role of LTβR signaling in the GC response in superficial peripheral LNs. To address this question, BM chimeric mice have been examined in the context of a local subcutaneous administration of a model antigen phycoerytherin (PE). PE responses in the local draining LNs of LTβ chimeric mice are surprisingly normal, with GC B cells upregulating AID and clustering in what appear to be GC niches within B cell follicles. Furthermore, sorting on PE-specific, AID-expressing GC B cells from either WT or LT-deficient LNs revealed that somatic hypermutation occurs normally, at least in the first 9 days of the immune response. However, by day 12 of the immune response, clusters of AID-expressing PE-specific GC B cells are no longer apparent in the LNs of LTβ chimeric mice compared with WT chimeric mice. The abrupt collapse of the GC is followed by a diminution in the affinity of the antibody response at late time points. Thus, it appears that the LT pathway, and by extension mature FDCs, are not required for the initiation of a GC response. Presumably other scaffold cells such as FRCs may substitute for the lack of FDCs and, indeed, FRCs are readily observed within FDC-less GCs in LTβ chimeric mice [202]. Although FRCs do not express the receptors that would present Ag–Ab complexes (complement receptor CD35, Fc receptors), noncognate B cells could presumably step in and provide this function through their own Fc and complement receptors (Figure 3). However, as the GC response continues, presumably these non-FDC "replacement" elements become insufficient to continue to support an ongoing GC response. This is consistent with a study that showed that total ablation of FDCs using diptheria toxin at the peak of the immune response results in an abrupt collapse of the GC in a draining LN [209].

In FDC-less mice, it is unclear from where the source of CXCL13 required to draw B cells in the GC is derived. As mentioned, FDCs express the LTβR, and signaling through this receptor is required for stromal cells that secrete optimal levels of CXCL13, at least in the spleen. In that absence of LTβR signaling, FDCs that express CD35 and FDC-M1 are notably absent, and CXCL13 levels in the spleen are strongly reduced. However, despite the loss of LTβR signaling, B cell localization to splenic follicles is not completely disorganized, and in the LN it is actually quite normal. Therefore,

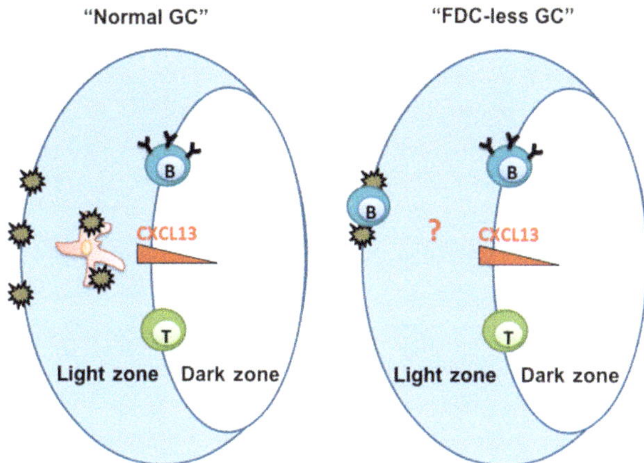


FIGURE 3 **The key ingredients for a GC: B cells, T cells, and antigen (Ag).** Following exposure to pathogen, Ag is transported from the SCS by SCS macrophages and noncognate B cells into the GC. Under normal circumstances, FDCs produce CXCL13 in the light zone of the GC and can display Ag on the surface of FDC extensions, thus providing a reservoir of Ag for the purposes of an affinity maturation template. T cells provide necessary help in the form of cytokines and CD40L. However, in an LN where LTβR signaling has been ablated, mature FDCs are not observed, but GC clusters appear. We hypothesize that an alternative reticular cell type, such as an FRC, may permit the display of CXCL13, but cannot retain Ag. On the other hand, noncognate B cells may provide a means for displaying Ag, albeit for limited periods of time. These minimal elements are sufficient for promoting a short-term GC response as well as somatic hypermutation of the BCR. FDC, follicular dendritic cells; BCR, B cell receptor; GC, germinal center; FRC, fibroblastic reticular cells; LN, lymph nodes.

there are likely other compensatory migratory cues and/or there are LT-independent alternative mechanisms for the induction of CXCL13, particularly in the LNs.

## 2.5 LT and Mucosal B Cell Biology

### 2.5.1 *The Expression of LTαβ on Mucosal B Cells and PP Architecture*

PPs are an important part of the mucosal immune system in which most priming of B cells occurs. In the absence of LTβR signaling, PPs do not develop, and if the LT pathway is blocked in the adult animal with LTβR-Ig, PPs tend to shrink in size [210]. It was later shown that $LT\alpha_1\beta_2$ expressed by B cells is required for the development of lymphoid follicles and optimal production of CXCL13 within follicle regions in PPs [211]. Because PPs are a rich source of B cells, this is a possible explanation for their smaller size in adult WT mice in which the LT pathway is blocked pharmacologically. It is interesting to note that the follicle-associated epithelium is intact in LT-deficient mice [211], suggesting that other pathways are important for maintaining this barrier (which includes antigen-transporting M cells). RANK, another TNF family member, is a likely candidate for this function [212]. Not unlike the LNs and spleen, the detection of mature FDCs in PPs is dependent on the expression of $LT\alpha_1\beta_2$ on B cells. FDCs in the PPs may play a particular role in shaping IgA responses because, in response to bacterial antigens, they can produce CXCL13 and BAFF and can secrete matrix metalloproteinases that activate TGFβ, a major IgA switch factor [213].

### 2.5.2 *The Generation of ILFs—A Reservoir for Mucosal B Cells*

Similar to PPs, ILFs are also found in the gastrointestinal mucosa of the small intestine. These ILFs resemble PPs and contain a dense cluster of $B220^+$ B cells and $CD11c^+$ cells. However, in contrast to PPs, ILFs can develop after birth, and in the absence of microbial Ags, the ILFs remain quite small. In the presence of luminal bacteria, ILFs become more organized and larger, taking on a more "mature" morphology. Expression of $LT\alpha_1\beta_2$ on B cells is sufficient to induce ILFs in LT-deficient mice [196]. However, in an LT-competent animal, ILCs are an important source of $LT\alpha_1\beta_2$ for inducing and maintaining ILFs. Thus, it is likely that B cells and ILCs may provide overlapping sources of $LT\alpha_1\beta_2$ for the generation of ILFs. Functions for ILFs and the colonic version of ILFs (colonic patches) in coordinating immune responses are emerging. For example, colonic patches orchestrate interactions between DCs and ILCs in an LT-dependent manner to provoke an innate response to *Citrobacter rodentium* [183,184]. In some cases, ILFs may also be an important site for IgA CSR [214], particularly for T-independent IgA CSR. The expression of AID in this location further argues that ILFs can support IgA CSR.

### 2.5.3 *LT-Dependent Induction of IgA CSR*

Significant amounts of IgA are produced in the GALT. B cells in PPs receive T cell help in GCs, and the cytokine milieu in this location promotes CSR to the IgA isotype [213]. IgA-producing B cells then leave the PPs as IgA-producing PCs, circulating through lymphatics to the thoracic duct and into the blood, ultimately entering the gut LP. Within the LP, $IgA^+$ PCs secrete copious amounts of IgA and can produce other molecules that may be involved in local defense mechanisms and/or homeostasis of the $IgA^+$ PC pool [215].

LT-deficient mice have a profound reduction in IgA in their serum and fecal pellets, although the reason for this impairment remains unclear [216]. It is important to point out that although LT-deficient mice lack LNs and PPs, these structures are dispensable for the generation of IgA that can occur in the ILFs of the LPs [214]. Thus, unraveling the role of LTβR signaling requires an assessment of how IgA is made in the presence and absence of these organized lymphoid structures. In the absence of LNs and PPs, LTβR signaling is required for the expression of CXCL13 in radio-resistant cells to recruit B cells into the gut lamina propria and transplantation of LTβR-sufficient gut sections into LTβR-replete intestine is sufficient to induce a local IgA response [216]. On the other hand, B cells can still migrate, albeit in lower numbers, to the LTβR-deficient gut. Thus, there are likely other LT-dependent events that are required to optimize IgA CSR.

It is important to keep in mind that in the presence of PPs, LTβR signaling may actually lie downstream of other signals. Indeed, a recent study showed that the TNFR/LTα axis is required for IgA production. Specifically, the authors deleted LTα expression on ILCs and observed a profound reduction in B cell accumulation and IgA production in the gut. A role for LTβR/LTαβ signaling was also defined in T-independent IgA CSR [217]. Consistent with this observation, LTα-deficient mice have a strongly impaired immune response to *Rotavirus*, although curiously, the IgA response to *Rotavirus* eventually emerges with time [218]. It is possible that without $LT\alpha_1\beta_2$/LTβR and LTα/TNFR signaling, the LIGHT/LTβR axis may eventually compensate. Indeed, overexpression of LIGHT results in a hyper-IgA phenotype [219]. Thus, although a role for the LTα/TNFR signaling axis in IgA induction has been clearly demonstrated, unraveling the role of LTβR signaling in T-dependent (PP sufficient) and T-independent (PP deficient) settings remains a challenging question.

### 2.5.4 Expression of $LT\alpha_1\beta_2$ on Mucosal B Cells is Required for Orchestrating Th2 Responses

In addition to its role in IgA responses, B cell-associated LT can also play a role in mucosal T cell responses. As mentioned, the interaction between $LT\alpha_1\beta_2$-expressing T cells and LTβR-expressing DCs is important for cross-priming a $CD8^+$ T cell response to protein antigen [182]. However, in the context of infection with a parasite, *Heligmosomoides polygyrus*, Lund and colleagues showed that expression of $LT\alpha_1\beta_2$ by B cells is important for B cell:DC interactions in the interfollicular regions of mesenteric LNs. These regions are typically not rich in B cells; however, upon infection with *H. polygyrus*, local production of CXCL13 in this region draws $CXCR5^+$ B cells and a subset of $CXCR5^+$ DCs into the interfollicular region. These interactions are critical for *H. polygyrus*-induced Th2 development. The Th2 program induced by *H. polygyrus* infection is prevented by predepletion of follicular B cells with anti-CD20 antibodies, and it is the expression of $LT\alpha_1\beta_2$ on B cells that is required to orchestrate the encounters between $CXCR5^+$ DCs and T cells to drive the Th2 response. Therefore, in the context of an intestinal infection with *H. polygyrus*, expression of $LT\alpha_1\beta_2$ on B cells is critical for coordinating a Th2 response [220].

However, it is likely that other factors can also guide the movement of B cells to interfollicular regions. In particular, the G protein-coupled receptor Epstein–Barr virus-induced-2 (EBI2) is important for guiding B cells to the T/B interface during immune responses. Specialized stroma in these regions produce the enzymes required to convert cholesterol to 7α,25-dihydroxycholesterol, the ligand for EBI2 [221]. Such stromal cells are also present in the so-called bridging channels of the spleen, where DCs are localized. The localization of DCs in this region may be important for two reasons: (1) so that DCs may be positioned close to the MZ where Ag is first encountered upon systemic infection and (2) to position DCs close to B cells that express $LT\alpha_1\beta_2$ so that they may receive constitutive LTβR-mediated homeostatic signals [222].

### 2.5.5 The Role of $LT\alpha_1\beta_2$ on B Cells during Pathological Gut Inflammation

Given its role in the mucosal immune system, it is not surprising that the LT pathway can shape the inflammatory environment of the gut during disease. Inhibition of LTβR signaling has a beneficial effect in several models of mucosal inflammation. For example, the transfer of $CD45RB^{high}CD4^+$ T cells into lymphocyte-replete (SCID) mice results in colitis that is likewise inhibited by LTβR-Ig treatment, and similar results were obtained with a BM transplantation colitis model [223]. Th2-driven colitis initiated by the hapten TNBS can also be attenuated using LTβR pathway blockade [224,225]. Injection of dextran sodium sulfate (DSS) induces a chronic wasting disease and colitic inflammation that is often used as a mouse model for human colitis. It is interesting to note that although in one study LTβR-Ig treatment was shown to ameliorate DSS colitis and reduce inflammation concomitant with the reduction of the expression of homing marker MAdCAM-1 in the GALT [226], the opposite results were obtained in a second study in which LTβR-Ig treatment seemed to exacerbate colitis [227]. As of yet, no explanation has been provided for the discrepancy of the two studies. However, what was interesting about the second study was that the expression of $LT\alpha_1\beta_2$ on B cells was disease promoting whereas the expression of $LT\alpha_1\beta_2$ on T cells prevented disease. The authors speculate that interactions between of $LT\alpha_1\beta_2$ on T cells and LTβR on myeloid/DC may drive anti-inflammatory mediators whereas the expression of $LT\alpha_1\beta_2$ on B cells may promote inflammation by supporting the generation of tertiary structures in the colon (i.e., colonic patches).

Perhaps one way to resolve these contradictory data is to refer to another system of colonic inflammation: Infection with *C. rodentium* results in an inflammatory condition that recapitulates some aspects of colitis. Treatment of *C. rodentium* infected mice with LTβR-Ig increases disease-related mortality [228]. Specifically, weight loss is exacerbated, there are more intestinal bacterial abscesses, and these were associated with a higher burden of *C. rodentium* in the periphery (spleen, liver). Similar results are observed with $LT\beta R^{-/-}$ and $LT\beta^{-/-}$ mice. In addition, deletion of LTβR in DCs [183,184] can attenuate inflammation associated with *C. rodentium* infection, and this is also the case if LTβR is deleted in intestinal epithelial cells [229]. However, it is important to point out that in the case of *C. rodentium*, the generation of an antipathogen immune response is required to resolve the infection, and it would make sense that an unresolved infection would generate colonic inflammation. Indeed, $RAG^{-/-}$ mice are highly susceptible to *C. rodentium* infection [230]. Given that LTβR signaling is required for IgA responses and the T cell:DC collaboration, it is not surprising that LT-deficient mice have problems clearing a *C. rodentium* infection. Thus, the inflammation observed in the *C. rodentium* setting in LT-deficient mice may be due to poor pathogen handling. Likewise, in the DSS colitis context in which host microbiota can penetrate the injured epithelial layer and cause inflammation, defective IgA responses or impaired T cell:DC communication could lead to local inflammation, as was observed in the aforementioned DSS studies performed in LT-deficient settings.

## 2.6 LT and Tertiary Lymphoid Structures

### 2.6.1 The LT Pathway and TLS Formation

The role of B cells in initiating tertiary lymphoid structures (TLS) has been well described using several different experimental systems, and the mechanisms of TLS formation and their role in disease are discussed in another chapter. The RIP has been used to drive ectopic expression of $LT\alpha_1\beta_2$ in the context of pancreatic tissue. Expression of LTα and LTβ is required to generate fully mature TLS in the pancreas with delineated T and B cell areas and fully mature HEVs [231]. Therefore, LTβR signaling on some kind of pancreas-resident cell type can be initiated by the RIP-induced $LT\alpha_1\beta_2$. However, to uncover whether B cells could be capable of providing an important source of $LT\alpha_1\beta_2$ to induce TLS in the pancreas, transgenic expression of CXCL13 by the RIP was also tested. RIP-CXCL13 mice also exhibited TLS in the pancreas, suggesting that B cells are an important initiating cell for TLS in nonlymphoid tissues [231]. Because the injection of LT inhibitors blocked follicle formation in RIP-CXCL13 mice, the main function for B cells in this context was via provision of $LT\alpha_1\beta_2$.

### 2.6.2 The LT Pathway and FDC Precursors

The implication of these studies is that during inflammation, immune cells, and particularly B cells that express $LT\alpha_1\beta_2$, can function to induce local LTβR signaling to promote the formation of TLS. However, this begs the question of what cell types express LTβR to initiate TLS formation. Recent studies by Aguzzi and colleagues have shown that TLS can be initiated by so-called "FDC precursor" cells (Figure 4). FDC precursors were revealed to originate from radioresistant cell types that express plasma-derived growth factor receptor-β (PDGFRβ). Although mature FDCs are absent in $TNFR^{-/-}$ mice, such pre-FDC-expressing Mfge8 (also known as FDC-M1) and CXCL13 can be observed in the marginal sinus and white pulp areas. However, these

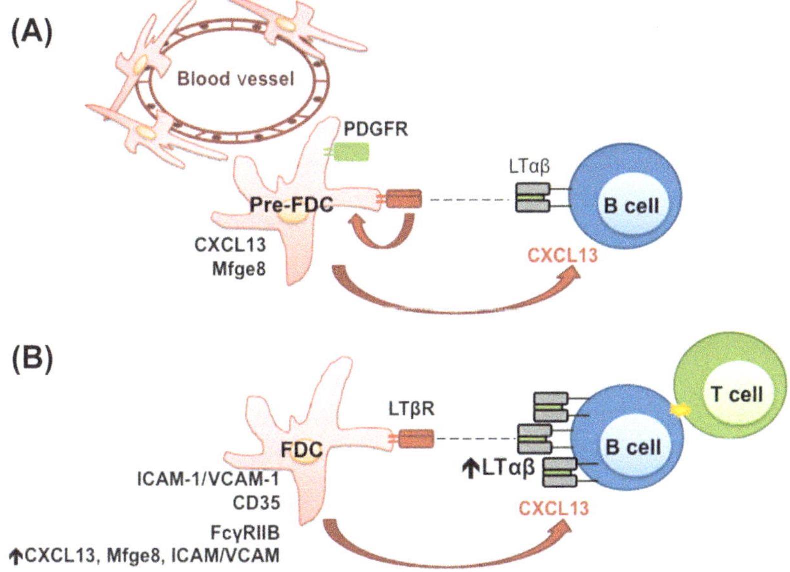


FIGURE 4 **Dynamic lymphotoxin (LT)-dependent maturation of follicular dendritic cells (FDC).** (A) FDCs are derived from sessile precursor mural cells associated with blood vessels. When these pre-FDCs make contact with $LT\alpha_1\beta_2$-expressing cells, particularly B cells, upon triggering of the LTβR the pre-FDC express markers such as Mfge8 and produce CXCL13. (B) Further differentiation of FDCs is required for them to reach a mature status. This can be achieved through TNFR1, or during a GC reaction exceptionally high levels of $LT\alpha_1\beta_2$ expression on GC B cells can also induce FDC maturation to express Mfge8 and high levels of CD35. PDGFR, plasma-derived growth factor receptor.

pre-FDCs are greatly decreased and dispersed in mice that lack LTβR signaling or lack lymphoid cells (RAG2$^{-/-}$ γc$^{-/-}$ mice). Introduction of B cells into RAG2$^{-/-}$ γc$^{-/-}$ mice induced the formation of mature FDCs at perivascular locations in the spleen, suggesting that a precursor cell in this location could give rise to mature CD35$^{+}$FcgR$^{+}$ FDC. Such Mfge8$^{+}$ precursors were also found to be positive for PDGFRβ and were derived from cells (perhaps pericytes) that contact the vascular endothelium. It is important to note that by growing pericytes in a collagen-based matrix and transplanting into the kidney capsule of LTβR$^{-/-}$ mice, the authors were able to show that mature FDCs can be generated in the kidneys of recipient mice that supported the recruitment of lymphocytes. Therefore, FDC precursors are presumably sessile mural cell types that reside in perivascular locations throughout the body, and this provides a mechanism for why TLS may arise in a wide variety of tissues [232]. During inflammation, presumably the expression of $LT\alpha_1\beta_2$ on infiltrating activated immune cells (particularly B cells) is sufficient to trigger a maturation program in these pericytes to establish TLS within the inflamed environment. Stimulation of LTβR on such perivascular cells induces the upregulation of CXCL13 and Mfge8, and further maturation of these cells induces the expression of Ag capture machinery.

However, TLS have also been observed to form in the absence of LTβR signaling. In the case of TLS in the lungs (so-called inducible bronchial lymphoid tissues (iBALT) structures), expression of CXCL13 can occur independent of the LT pathway, although in the absence of LTβR signaling, the iBALT were disorganized [233]. This implies that other pathways can compensate for the absence of LT in inducing CXCL13 for the recruitment of B cells.

Although LTβR signaling is dispensable for the formation of iBALT, Th17 cells (i.e., T cells that produce the inflammatory cytokines IL-17 and IL-22) are required to induce iBALT in the lungs [233]. Furthermore, it has been shown that TLS in the spinal cord that emerge during EAE can be induced by Th17 cells [233]. Th17 cells are a particularly abundant source of $LT\alpha_1\beta_2$ expression, and they can also produce TNFα. Thus, it makes sense that they may stimulate local stromal cells through TNFR/LTβR to produce the necessary chemokines required to recruit additional lymphocytes to the site of inflammation, particularly B cells.

## 2.7 LT—Therapeutic Perspectives that Pertain to B Cells

### 2.7.1 Therapeutic Rationale for Targeting B Cells and the LT Pathway

The targeting of B cells for the treatment of disease has been a relatively new frontier for immunotherapeutics, with B cell depleting agents (anti-CD20) having efficacy in MS, RA, SLE, and lymphoma [234]. Anti-CD20 (rituximab/ocrelizumab) effectively couples with antibody-dependent cellular cytotoxicity mechanisms to deplete mature follicular B cells, sparing PCs and some subsets of B cells. However, ablation of mature B cells using anti-CD20 treatment can potentially leave patients vulnerable to infection, and not all B cells have pathogenic potential. In fact, anti-CD20 treatment depletes IL-10-producing B cells in mice, and this can worsen clinical outcomes in mouse models of MS [235].

Of course, an alternative strategy is to inhibit specific pathogenic functions attributable to B cells. As such, the LT inhibitor LTβR-Ig (baminercept) may have promise in disease settings in which B cells are playing a pathogenic role via their expression of $LT\alpha_1\beta_2$. As mentioned, significant evidence shows that $LT\alpha_1\beta_2$ expressed by B cells is not only important in FDC biology in the lymphoid tissues, and in some cases can cause a decay in the GC response, but expression of $LT\alpha_1\beta_2$ on naïve and Ag-activated B cells serves to stimulate LTβR on FDC precursor cells that exist in perivascular locations [232].

### 2.7.2 Targeting the LT Pathway in Autoimmunity

Inhibitors of the LT pathway have been tested in several preclinical animal models of disease [236]. Because of (1) the wide range of cells that express LTβR; (2) the complex role for LTβR in LN development, maintenance of lymphoid tissue architecture, DC/macrophage homeostasis and function, and TLS formation; (3) the different ligands ($LT\alpha_1\beta_2$, LIGHT) that activate the LTβR; and (4) the interdigitating pathways (LTβR/$LT\alpha_1\beta_2$, TNFR/LTα, HVEM/LIGHT etc.), it has been very difficult to discern the mechanism of action of LTβR-Ig in disease, and indeed, different disease models likely invoke different mechanisms.

Some good examples of the effect of LTβR-Ig in disease settings where the B cell compartment has been affected are animal models of RA and SS. With respect to RA, prophylactic and therapeutic treatment of DBA-1 mice injected with collagen in complete Freund's adjuvant reduces disease burden. Efficacy in this collagen-induced arthritis (CIA) model correlates with a reduction in collagen titers in the periphery, rather than a decreased T cell response, suggesting that LTβR-Ig may affect the anti-collagen B cell response. Indeed, the FDC networks in the draining LNs were obliterated in this mouse model. Furthermore, the induction of RA by the adoptive transfer of complement containing immune complexes cannot be inhibited by LTβR-Ig treatment, suggesting that this therapeutic will not work in a setting in which the GC stage has been bypassed [237]. However, in a second study, LTβR-Ig treatment was not found to ameliorate CIA, but rather depletion of Th1/Th17 cells using an anti-LTα monoclonal antibody approach worked to reduce disease. Given its multiple functions in orchestrating immune responses, it is likely that the LT pathway can influence the CIA model by modulating T and B cell responses [238].

In the case of SS, the NOD mice develop a Sjögren's-like disease as they age. Treatment with LTβR-Ig at 9 weeks of age partially restores salivary flow in these mice, and this was associated with reduced FDC networks in the submandibulary glands [239]. More recent work has shown that the B cell content in the lacrimal glands of the Sjögren's model is greatly reduced with LTβR-Ig treatment, and this corresponds with reductions in HEV structures as well as chemokines such as CXCL13. A clinical trial testing the efficacy of LTβR-Ig (baminercept) in Sjögren's patients is ongoing (please see http://clinicaltrials.gov/show/NCT01552681).

Although these autoimmune disease examples suggest an involvement for B cells, the precise stage of B cell activation at which the LT pathway may interfere has not been determined. As mentioned, disruption of FDC by inhibition of the LT pathway can have variable effects on the GC response depending on the tissue in which the GC is taking place, with the spleen GC being especially LT-sensitive and the LN GC less so. Although FDC precursors are associated with perivascular areas throughout the body, the simulation of LTβR on these precursors is required for their differentiation into mature FDCs. Even in the presence of LTβR signaling, FDCs differ in terms of their capacity to express an antigen capture machine and to support local GC responses. So-called "primary" FDCs express less CD35 whereas "secondary" FDCs express high levels of CD35 and Fcγ receptors that aid in the capture of antibody/complement-coated antigen as well as in the opsonization of apoptotic bodies [240]. Such secondary FDCs are typically associated with GC B cells that express AID, and indeed, the presence of secondary FDCs in ectopic lymphoid structures during autoimmunity is strongly associated with AID expression in B cells in the salivary glands of SS patients [241] and in the synovium of RA patients [242]. Perhaps the regulation of AID, although relatively intact in an LT-deficient LN, is much more sensitive to LTβR-dependent cell types in nonlymphoid tissues.

### 2.7.3 Targeting the LT Pathway in Cancer

The role of LT biology in cancer progression is complex because of divergent signaling outcomes of triggering LTβR in the tumor itself, versus the tumor microenvironment—and a full dissection of the LT pathway in tumorigenesis is beyond the scope of this chapter. Focusing on the tumor microenvironment, it is increasingly appreciated that chronic inflammation can have a negative outcome on tumor growth and metastasis, and this negative outcome is associated with local NF-κB activation within the tumor. Indeed, prostate cancer progression was shown to be associated with local NF-κB activation triggered by the LTβR, and the source of $LT\alpha_1\beta_2$in this scenario was shown to be derived from B cells [243]. However, in other cases, localization of B cells within tumor environments may also have beneficial outcomes, and the role of B cells in different tumors will likely be dictated by their effector or regulatory functions [244].

Follicular lymphomas, which can be supported by stromal cell networks within lymphoid tissues, may also depend on LTβR signaling. In vitro evidence using mesenchymal stem cells (MSCs) shows that the stimulation of LTβR on MSCs induces the adoption of an FRC phenotype [245], and, in vivo, expression of $LT\alpha_1\beta_2$on B cells allows for a coordinate interaction between lymphoma cells and the FRC network [246]. Lymphomas that are found in association with an FRC network are more aggressive with a significant survival advantage. Thus, one strategy to treat lymphoma may be to ablate LT-dependent networks within the lymphoid tissue to liberate lymphomas from microenvironments that are permissive to their growth.

## 3. CONCLUSIONS AND PERSPECTIVES

The coordinate interactions among TNF superfamily members shape immune responses. In this chapter, we have focused on two TNF superfamily pathways: the BAFF/APRIL pathway and the LT pathway. Given the role of the BAFF/APRIL and LT pathways in regulating B cell homeostasis, the GC response, the fine balance between B cell tolerance and immunity, and innate B cell function and survival, it is reasonable to consider these pathways as important players in B cell biology. However, other TNF superfamily members, notably CD40, TNFα, and OX40, play important roles in the B cell response that are beyond the scope of this chapter to discuss. Nevertheless, targeting the TNF superfamily has been used to treat rheumatic B cell-mediated diseases as well as B cell malignancies, and it is expected that additional TNF superfamily therapies will emerge.

## REFERENCES


[1] Bossen C, Schneider P. BAFF, APRIL and their receptors: structure, function and signaling. Semin Immunol 2006;18:263–75.

[2] Mackay F, Schneider P. Cracking the BAFF code. Nat Rev Immunol 2009;9:491–502.

[3] Wallace DJ, et al. A phase II, randomized, double-blind, placebo-controlled, dose-ranging study of belimumab in patients with active systemic lupus erythematosus. Arthritis Rheum 2009;61:1168–78.

[4] Sanz I, Yasothan U, Kirkpatrick P. Belimumab. Nat Rev Drug Discov 2011;10:335–6.

[5] Dillon SR, Gross JA, Ansell SM, Novak AJ. An APRIL to remember: novel TNF ligands as therapeutic targets. Nat Rev Drug Discov 2006;5:235–46.

[6] Nardelli B, et al. Synthesis and release of B-lymphocyte stimulator from myeloid cells. Blood 2001;97:198–204.

[7] Maia S, et al. Aberrant expression of functional baff-system receptors by malignant B-cell precursors impacts leukemia cell survival. PLoS One 2011;6:e20787.

[8] Moore PA, et al. BLyS: member of the tumor necrosis factor family and B lymphocyte stimulator. Science 1999;285:260–3.

[9] Schneider P, et al. BAFF, a novel ligand of the tumor necrosis factor family, stimulates B cell growth. J Exp Med 1999;189:1747–56.

[10] Lopez-Fraga M, Fernandez R, Albar JP, Hahne M. Biologically active APRIL is secreted following intracellular processing in the Golgi apparatus by furin convertase. EMBO Rep 2001;2: 945–51.

[11] Zhukovsky EA, Lee JO, Villegas M, Chan C, Chu S, Mroske C. TNF ligands: is TALL-1 a trimer or a virus-like cluster? Nature 2004;427:413–4. discussion 414.

[12] Oren DA, et al. Structural basis of BLyS receptor recognition. Nat Struct Biol 2002;9:288–92.

[13] Bossen C, et al. TACI, unlike BAFF-R, is solely activated by oligomeric BAFF and APRIL to support survival of activated B cells and plasmablasts. Blood 2008;111:1004–12.

[14] Liu Y, Xu L, Opalka N, Kappler J, Shu HB, Zhang G. Crystal structure of sTALL-1 reveals a virus-like assembly of TNF family ligands. Cell 2002;108:383–94.

[15] Mackay F, Schneider P, Rennert P, Browning J. BAFF AND APRIL: a tutorial on B cell survival. Annu Rev Immunol 2003;21:231–64.

[16] Cachero TG, et al. Formation of virus-like clusters is an intrinsic property of the tumor necrosis factor family member BAFF (B cell activating factor). Biochemistry 2006;45:2006–13.

[17] Roschke V, et al. BLyS and APRIL form biologically active heterotrimers that are expressed in patients with systemic immune-based rheumatic diseases. J Immunol 2002;169:4314–21.

[18] Stohl W. Systemic lupus erythematosus and its ABCs (APRIL/BLyS complexes). Arthritis Res Ther 2010;12:111.

[19] Gavin AL, Ait-Azzouzene D, Ware CF, Nemazee D. DeltaBAFF, an alternate splice isoform that regulates receptor binding and biopresentation of the B cell survival cytokine, BAFF. J Biol Chem 2003;278:38220–8.

[20] Gavin AL, et al. deltaBAFF, a splice isoform of BAFF, opposes full-length BAFF activity in vivo in transgenic mouse models. J Immunol 2005;175:319–28.

[21] Krumbholz M, et al. BAFF is produced by astrocytes and up-regulated in multiple sclerosis lesions and primary central nervous system lymphoma. J Exp Med 2005;201:195–200.

[22] Smirnova AS, Andrade-Oliveira V, Gerbase-DeLima M. Identification of new splice variants of the genes BAFF and BCMA. Mol Immunol 2008;45:1179–83.

[23] Youinou P, Pers JO. The late news on baff in autoimmune diseases. Autoimmun Rev 2010;9:804–6.

[24] Pradet-Balade B, et al. An endogenous hybrid mRNA encodes TWE-PRIL, a functional cell surface TWEAK-APRIL fusion protein. EMBO J 2002;21:5711–20.

[25] Lavie F, Miceli-Richard C, Ittah M, Sellam J, Gottenberg JE, Mariette X. B-cell activating factor of the tumour necrosis factor family expression in blood monocytes and T cells from patients with primary Sjogren's syndrome. Scand J Immunol 2008;67:185–92.

[26] Huard B, et al. BAFF production by antigen-presenting cells provides T cell co-stimulation. Int Immunol 2004;16:467–75.

[27] Ng LG, et al. B cell-activating factor belonging to the TNF family (BAFF)-R is the principal BAFF receptor facilitating BAFF costimulation of circulating T and B cells. J Immunol 2004;173:807–17.

[28] Scapini P, et al. G-CSF-stimulated neutrophils are a prominent source of functional BLyS. J Exp Med 2003;197:297–302.

[29] Scapini P, et al. Proinflammatory mediators elicit secretion of the intracellular B-lymphocyte stimulator pool (BLyS) that is stored in activated neutrophils: implications for inflammatory diseases. Blood 2005;105:830–7.

[30] Sutherland AP, et al. BAFF augments certain Th1-associated inflammatory responses. J Immunol 2005;174:5537–44.

[31] Chang SK, Arendt BK, Darce JR, Wu X, Jelinek DF. A role for BLyS in the activation of innate immune cells. Blood 2006;108:2687–94.

[32] Morimoto S, et al. Expression of B-cell activating factor of the tumour necrosis factor family (BAFF) in T cells in active systemic lupus erythematosus: the role of BAFF in T cell-dependent B cell pathogenic autoantibody production. Rheumatology (Oxford) 2007;46:1083–6.

[33] Groom JR, et al. BAFF and MyD88 signals promote a lupuslike disease independent of T cells. J Exp Med 2007;204:1959–71.

[34] Hardenberg G, et al. Specific TLR ligands regulate APRIL secretion by dendritic cells in a PKR-dependent manner. Eur J Immunol 2007;37:2900–11.

[35] Katsenelson N, Kanswal S, Puig M, Mostowski H, Verthelyi D, Akkoyunlu M. Synthetic CpG oligodeoxynucleotides augment BAFF- and APRIL-mediated immunoglobulin secretion. Eur J Immunol 2007;37:1785–95.

[36] Chu VT, Enghard P, Riemekasten G, Berek C. In vitro and in vivo activation induces BAFF and APRIL expression in B cells. J Immunol 2007;179:5947–57.

[37] Nakajima K, et al. Expression of BAFF and BAFF-R in the synovial tissue of patients with rheumatoid arthritis. Scand J Rheumatol 2007;36:365–72.

[38] Nagatani K, et al. Rheumatoid arthritis fibroblast-like synoviocytes express BCMA and are stimulated by APRIL. Arthritis Rheum 2007;56:3554–63.

[39] Ohata J, et al. Fibroblast-like synoviocytes of mesenchymal origin express functional B cell-activating factor of the TNF family in response to proinflammatory cytokines. J Immunol 2005;174: 864–70.

[40] Bombardieri M, et al. A BAFF/APRIL-dependent TLR3-stimulated pathway enhances the capacity of rheumatoid synovial fibroblasts to induce AID expression and Ig class-switching in B cells. Ann Rheum Dis 2011;70:1857–65.

[41] Belnoue E, et al. APRIL is critical for plasmablast survival in the bone marrow and poorly expressed by early-life bone marrow stromal cells. Blood 2008;111:2755–64.

[42] Chu VT, et al. Eosinophils are required for the maintenance of plasma cells in the bone marrow. Nat Immunol 2011;12:151–9.

[43] Benson MJ, et al. Cutting edge: the dependence of plasma cells and independence of memory B cells on BAFF and APRIL. J Immunol 2008;180:3655–9.

[44] Munari F, et al. Tumor-associated macrophages as major source of APRIL in gastric MALT lymphoma. Blood 2011;117:6612–6.

[45] Lee H-J, Seo G-Y, Kim J-H, Lee M-R, Kim P-H. Activin A stimulates mouse macrophages to express APRIL via the Smad3 and ERK/CREB pathways. Immunol Lett 2011;140:92–6.

[46] Peperzak V, et al. Mcl-1 is essential for the survival of plasma cells. Nat Immunol 2013;14:290–7.
[47] Farina C, Aloisi F, Meinl E. Astrocytes are active players in cerebral innate immunity. Trends Immunol 2007;28:138–45.
[48] Hauser SL, et al. B-cell depletion with rituximab in relapsing-remitting multiple sclerosis. N Engl J Med 2008;358:676–88.
[49] Hartung H-P, Kieseier BC. Atacicept: targeting B cells in multiple sclerosis. Ther Adv Neurol Disord 2010;3:205–16.
[50] Hahne M, et al. APRIL, a new ligand of the tumor necrosis factor family, stimulates tumor cell growth. J Exp Med 1998;188:1185–90.
[51] Stein JV, et al. APRIL modulates B and T cell immunity. J Clin Invest 2002;109:1587–98.
[52] Gorelik L, Gilbride K, Dobles M, Kalled SL, Zandman D, Scott ML. Normal B cell homeostasis requires B cell activation factor production by radiation-resistant cells. J Exp Med 2003;198:937–45.
[53] Zhang M, et al. Expression and function of TNF family member B cell-activating factor in the development of autoimmune arthritis. Int Immunol 2005;17:1081–92.
[54] Goenka R, et al. Local BLyS production by T follicular cells mediates retention of high affinity B cells during affinity maturation. J Exp Med 2014;211:45–56.
[55] Tangye SG, Bryant VL, Cuss AK, Good KL. BAFF, APRIL and human B cell disorders. Semin Immunol 2006;18:305–17.
[56] Mackay F, Leung H. The role of the BAFF/APRIL system on T cell function. Semin Immunol 2006;18:284–9.
[57] Rennert P, et al. A soluble form of B cell maturation antigen, a receptor for the tumor necrosis factor family member APRIL, inhibits tumor cell growth. J Exp Med 2000;192:1677–84.
[58] Schneider P, et al. Maturation of marginal zone and follicular B cells requires B cell activating factor of the tumor necrosis factor family and is independent of B cell maturation antigen. J Exp Med 2001;194:1691–7.
[59] Day ES, et al. Selectivity of BAFF/BLyS and APRIL for binding to the TNF family receptors BAFFR/BR3 and BCMA. Biochemistry 2005;44:1919–31.
[60] Liu Y, et al. Ligand-receptor binding revealed by the TNF family member TALL-1. Nature 2003;423:49–56.
[61] Hymowitz SG, et al. Structures of APRIL-receptor complexes: like BCMA, TACI employs only a single cysteine-rich domain for high affinity ligand binding. J Biol Chem 2005;280:7218–27.
[62] Patel DR, et al. Engineering an APRIL-specific B cell maturation antigen. J Biol Chem 2004;279:16727–35.
[63] Wang R, et al. Expression and characterization of a variant of TACI (CRD2-shortTACIFc) in *Pichia pastoris*. Protein Pept Lett 2012;19:315–25.
[64] Cancro MP. The BLyS family of ligands and receptors: an archetype for niche-specific homeostatic regulation. Immunol Rev 2004;202:237–49.
[65] Zhang X, et al. BAFF supports human B cell differentiation in the lymphoid follicles through distinct receptors. Int Immunol 2005;17:779–88.
[66] Batten M, et al. BAFF mediates survival of peripheral immature B lymphocytes. J Exp Med 2000;192:1453–66.
[67] Thompson JS, et al. BAFF-R, a newly identified TNF receptor that specifically interacts with BAFF. Science 2001;293:2108–11.
[68] Ye Q, et al. BAFF binding to T cell-expressed BAFF-R costimulates T cell proliferation and alloresponses. Eur J Immunol 2004;34:2750–9.
[69] Ng LG, et al. BAFF costimulation of Toll-like receptor-activated B-1 cells. Eur J Immunol 2006;36:1837–46.
[70] Yan M, Marsters SA, Grewal IS, Wang H, Ashkenazi A, Dixit VM. Identification of a receptor for BLyS demonstrates a crucial role in humoral immunity. Nat Immunol 2000;1:37–41.
[71] von Bulow GU, van Deursen JM, Bram RJ. Regulation of the T-independent humoral response by TACI. Immunity 2001;14:573–82.
[72] He B, et al. The transmembrane activator TACI triggers immunoglobulin class switching by activating B cells through the adaptor MyD88. Nat Immunol 2010;11:836–45.
[73] Seshasayee D, Valdez P, Yan M, Dixit VM, Tumas D, Grewal IS. Loss of TACI causes fatal lymphoproliferation and autoimmunity, establishing TACI as an inhibitory BLyS receptor. Immunity 2003;18:279–88.
[74] Bodmer J-L, Schneider P, Tschopp J. The molecular architecture of the TNF superfamily. Trends Biochem Sci 2002;27:19–26.
[75] Gross JA, et al. TACI and BCMA are receptors for a TNF homologue implicated in B-cell autoimmune disease. Nature 2000;404:995–9.
[76] Figgett WA, et al. The TACI receptor regulates T-cell-independent marginal zone B cell responses through innate activation-induced cell death. Immunity 2013;39:573–83.
[77] Kanswal S, Katsenelson N, Selvapandiyan A, Bram RJ, Akkoyunlu M. Deficient TACI expression on B lymphocytes of newborn mice leads to defective Ig secretion in response to BAFF or APRIL. J Immunol 2008;181:976–90.
[78] Ou X, Xu S, Lam KP. Deficiency in TNFRSF13B (TACI) expands T-follicular helper and germinal center B cells via increased ICOS-ligand expression but impairs plasma cell survival. Proc Natl Acad Sci USA 2012;109:15401–6.
[79] Tsuji S, Cortesao C, Bram RJ, Platt JL, Cascalho M. TACI deficiency impairs sustained Blimp-1 expression in B cells decreasing long-lived plasma cells in the bone marrow. Blood 2011;118:5832–9.
[80] Darce JR, Arendt BK, Wu X, Jelinek DF. Regulated expression of BAFF-binding receptors during human B cell differentiation. J Immunol 2007;179:7276–86.
[81] von Bulow GU, Russell H, Copeland NG, Gilbert DJ, Jenkins NA, Bram RJ. Molecular cloning and functional characterization of murine transmembrane activator and CAML interactor (TACI) with chromosomal localization in human and mouse. Mamm Genome 2000;11:628–32.
[82] Acosta-Rodriguez EV, et al. BAFF and LPS cooperate to induce B cells to become susceptible to CD95/Fas-mediated cell death. Eur J Immunol 2007;37:990–1000.
[83] Treml LS, et al. TLR stimulation modifies BLyS receptor expression in follicular and marginal zone B cells. J Immunol 2007;178:7531–9.
[84] Hendriks J, et al. Heparan sulfate proteoglycan binding promotes APRIL-induced tumor cell proliferation. Cell Death Differ 2005;12:637–48.
[85] Ingold K, et al. Identification of proteoglycans as the APRIL-specific binding partners. J Exp Med 2005;201:1375–83.
[86] Sakurai D, Hase H, Kanno Y, Kojima H, Okumura K, Kobata T. TACI regulates IgA production by APRIL in collaboration with HSPG. Blood 2007;109:2961–7.
[87] Avery DT, et al. BAFF selectively enhances the survival of plasmablasts generated from human memory B cells. J Clin Invest 2003;112:286–97.

[88] Stadanlick JE, et al. Tonic B cell antigen receptor signals supply an NF-kappaB substrate for prosurvival BLyS signaling. Nat Immunol 2008;9:1379–87.

[89] O'Connor BP, et al. BCMA is essential for the survival of long-lived bone marrow plasma cells. J Exp Med 2004;199:91–8.

[90] Yang M, Hase H, Legarda-Addison D, Varughese L, Seed B, Ting AT. B cell maturation antigen, the receptor for a proliferation-inducing ligand and B cell-activating factor of the TNF family, induces antigen presentation in B cells. J Immunol 2005;175:2814–24.

[91] Zhang L, et al. Identification of BLyS (B lymphocyte stimulator), a non-myelin-associated protein, as a functional ligand for Nogo-66 receptor. J Neurosci 2009;29:6348–52.

[92] Gross JA, et al. TACI-Ig neutralizes molecules critical for B cell development and autoimmune disease. impaired B cell maturation in mice lacking BLyS. Immunity 2001;15:289–302.

[93] Schiemann B, et al. An essential role for BAFF in the normal development of B cells through a BCMA-independent pathway. Science 2001;293:2111–4.

[94] Allman D, Pillai S. Peripheral B cell subsets. Curr Opin Immunol 2008;20:149–57.

[95] Hayakawa K, Li YS, Wasserman R, Sauder S, Shinton S, Hardy RR. B lymphocyte developmental lineages. Ann N Y Acad Sci 1997;815:15–29.

[96] Tardivel A, et al. The anti-apoptotic factor Bcl-2 can functionally substitute for the B cell survival but not for the marginal zone B cell differentiation activity of BAFF. Eur J Immunol 2004;34:509–18.

[97] Kayagaki N, et al. BAFF/BLyS receptor 3 binds the B cell survival factor BAFF ligand through a discrete surface loop and promotes processing of NF-kappaB2. Immunity 2002;17:515–24.

[98] Gardam S, Sierro F, Basten A, Mackay F, Brink R. TRAF2 and TRAF3 signal adapters act cooperatively to control the maturation and survival signals delivered to B cells by the BAFF receptor. Immunity 2008;28:391–401.

[99] Craxton A, Draves KE, Gruppi A, Clark EA. BAFF regulates B cell survival by downregulating the BH3-only family member Bim via the ERK pathway. J Exp Med 2005;202:1363–74.

[100] Mecklenbrauker I, Kalled SL, Leitges M, Mackay F, Tarakhovsky A. Regulation of B-cell survival by BAFF-dependent PKCdelta-mediated nuclear signalling. Nature 2004;431:456–61.

[101] Patke A, Mecklenbrauker I, Erdjument-Bromage H, Tempst P, Tarakhovsky A. BAFF controls B cell metabolic fitness through a PKCβ- and Akt-dependent mechanism. J Exp Med 2006;203: 2551–62.

[102] Henley T, Kovesdi D, Turner M. B-cell responses to B-cell activation factor of the TNF family (BAFF) are impaired in the absence of PI3K delta. Eur J Immunol 2008;38:3543–8.

[103] Schweighoffer E, et al. The BAFF receptor transduces survival signals by co-opting the B cell receptor signaling pathway. Immunity 2013;38:475–88.

[104] Infantino S, Tarlinton DM. Feeling a little SYK after mixing BAFF with BCR. Immunity 2013;38:406–8.

[105] Mackay F, Figgett WA, Saulep D, Lepage M, Hibbs ML. B-cell stage and context-dependent requirements for survival signals from BAFF and the B-cell receptor. Immunol Rev 2010;237:205–25.

[106] Batten M, et al. TNF deficiency fails to protect BAFF transgenic mice against autoimmunity and reveals a predisposition to B cell lymphoma. J Immunol 2004;172:812–22.

[107] Castigli E, et al. TACI is mutant in common variable immunodeficiency and IgA deficiency. Nat Genet 2005;37:829–34.

[108] Castigli E, et al. TACI and BAFF-R mediate isotype switching in B cells. J Exp Med 2005;201:35–9.

[109] Mackay F, et al. Mice transgenic for BAFF develop lymphocytic disorders along with autoimmune manifestations. J Exp Med 1999;190:1697–710.

[110] McCarthy DD, et al. Mice overexpressing BAFF develop a commensal flora-dependent, IgA-associated nephropathy. J Clin Invest 2011;121:3991–4002.

[111] Thompson JS, et al. BAFF binds to the tumor necrosis factor receptor-like molecule B cell maturation antigen and is important for maintaining the peripheral B cell population. J Exp Med 2000;192:129–35.

[112] Griesemer AD, Sorenson EC, Hardy MA. The role of the thymus in tolerance. Transplantation 2010;90:465–74.

[113] Hardy RR, Hayakawa K. B cell development pathways. Ann Rev Immunol 2001;19:595–621.

[114] Goodnow CC, et al. Self-tolerance checkpoints in B lymphocyte development. Adv Immunol 1995;59:279–368.

[115] Cornall RJ, Goodnow CC, Cyster JG. The regulation of self-reactive B cells. Curr Opin Immunol 1995;7:804–11.

[116] Glynne R, Ghandour G, Rayner J, Mack DH, Goodnow CC. B-lymphocyte quiescence, tolerance and activation as viewed by global gene expression profiling on microarrays. Immunol Rev 2000;176:216–46.

[117] Cyster JG, Hartley SB, Goodnow CC. Competition for follicular niches excludes self-reactive cells from the recirculating B-cell repertoire. Nature 1994;371:389–95.

[118] Thien M, et al. Excess BAFF rescues self-reactive B cells from peripheral deletion and allows them to enter forbidden follicular and marginal zone niches. Immunity 2004;20:785–98.

[119] Lesley R, et al. Reduced competitiveness of autoantigen-engaged B cells due to increased dependence on BAFF. Immunity 2004;20: 441–53.

[120] Brink R, et al. Immunoglobulin M and D antigen receptors are both capable of mediating B lymphocyte activation, deletion, or anergy after interaction with specific antigen. J Exp Med 1992;176: 991–1005.

[121] Kearney JF. Innate-like B cells. Springer Semin Immunopathol 2005;26:377–83.

[122] Martin F, Kearney JF. Marginal-zone B cells. Nat Rev Immunol 2002;2:323–35.

[123] Oliver AM, Martin F, Gartland GL, Carter RH, Kearney JF. Marginal zone B cells exhibit unique activation, proliferative and immunoglobulin secretory responses. Eur J Immunol 1997;27:2366–74.

[124] Attanavanich K, Kearney JF. Marginal zone, but not follicular B cells, are potent activators of naive CD4 T cells. J Immunol 2004;172:803–11 (Baltimore, MD: 1950).

[125] Rose NR. Autoimmunity, infection and adjuvants. Lupus 2010;19:354–8.

[126] Liu YJ, Zhang J, Lane PJ, Chan EY, MacLennan IC. Sites of specific B cell activation in primary and secondary responses to T cell-dependent and T cell-independent antigens. Eur J Immunol 1991;21:2951–62.

[127] Hyer ML, Samuel T, Reed JC. The FLIP-side of Fas signaling. Clin Cancer Res 2006;12:5929–31.

[128] Kaiser WJ, et al. RIP3 mediates the embryonic lethality of caspase-8-deficient mice. Nature 2011;471:368–72.
[129] Oberst A, et al. Catalytic activity of the caspase-8-FLIP(L) complex inhibits RIPK3-dependent necrosis. Nature 2011;471:363–7.
[130] Chen X, Martin F, Forbush KA, Perlmutter RM, Kearney JF. Evidence for selection of a population of multi-reactive B cells into the splenic marginal zone. Int Immunol 1997;9:27–41.
[131] Yan M, et al. Activation and accumulation of B cells in TACI-deficient mice. Nat Immunol 2001;2:638–43.
[132] Romberg N, et al. CVID-associated TACI mutations affect autoreactive B cell selection and activation. J Clin Invest 2013;123(10): 4283–93.
[133] Groom J, et al. Association of BAFF/BLyS overexpression and altered B cell differentiation with Sjogren's syndrome. J Clin Invest 2002;109:59–68.
[134] Fletcher CA, Groom JR, Woehl B, Leung H, Mackay C, Mackay F. Development of autoimmune nephritis in genetically asplenic and splenectomized BAFF transgenic mice. J Autoimmun 2011;36:125–34.
[135] Fletcher CA, et al. Development of nephritis but not sialadenitis in autoimmune-prone BAFF transgenic mice lacking marginal zone B cells. Eur J Immunol 2006;36:2504–14.
[136] Nagy G, Koncz A, Perl A. T- and B-cell abnormalities in systemic lupus erythematosus. Crit Rev Immunol 2005;25:123–40.
[137] McCarthy DD, Chiu S, Gao Y, Summers-deLuca LE, Gommerman JL. BAFF induces a hyper-IgA syndrome in the intestinal lamina propria concomitant with IgA deposition in the kidney independent of LIGHT. Cell Immunol 2006;241:85–94.
[138] Scharenberg AM, Hannibal MC, Torgerson T, Ochs HD, Rawlings DJ. Common variable immune deficiency overview. In: Pagon RA, Bird TD, Dolan CR, Stephens K, Adam MP, editors. GeneReviews. Seattle (WA): University of Washington; 2006.
[139] Woo YJ, et al. Regulation of B cell activating factor (BAFF) receptor expression by NF-κB signaling in rheumatoid arthritis B cells. Exp Mol Med 2011;43:350–7.
[140] Salzer U, et al. Mutations in TNFRSF13B encoding TACI are associated with common variable immunodeficiency in humans. Nat Genet 2005;37:820–8.
[141] Zhang L, et al. Transmembrane activator and calcium-modulating cyclophilin ligand interactor mutations in common variable immunodeficiency: clinical and immunologic outcomes in heterozygotes. J Allergy Clin Immunol 2007;120:1178–85.
[142] Martinez-Pomar N, et al. Role of TNFRSF13B variants in patients with common variable immunodeficiency. Blood 2009;114:2846–8.
[143] Pan-Hammarstrom Q, et al. Reexamining the role of TACI coding variants in common variable immunodeficiency and selective IgA deficiency. Nat Genet 2007;39:429–30.
[144] Garibyan L, Lobito AA, Siegel RM, Call ME, Wucherpfennig KW, Geha RS. Dominant-negative effect of the heterozygous C104R TACI mutation in common variable immunodeficiency (CVID). J Clin Invest 2007;117:1550–7.
[145] Lee JJ, et al. The C104R mutant impairs the function of transmembrane activator and calcium modulator and cyclophilin ligand interactor (TACI) through haploinsufficiency. J Allergy Clin Immunol 2010;126:1234–41. e1232.
[146] Bacchelli C, et al. The C76R transmembrane activator and calcium modulator cyclophilin ligand interactor mutation disrupts antibody production and B-cell homeostasis in heterozygous and homozygous mice. J Allergy Clin Immunol 2011;127:1253–63.
[147] Martinez-Gallo M, Radigan L, Almejun MB, Martinez-Pomar N, Matamoros N, Cunningham-Rundles C. TACI mutations and impaired B-cell function in subjects with CVID and healthy heterozygotes. J Allergy Clin Immunol 2013;131:468–76.
[148] Park JH, Resnick ES, Cunningham-Rundles C. Perspectives on common variable immune deficiency. Ann N Y Acad Sci 2011;1246:41–9.
[149] Losi CG, et al. Mutational analysis of human BAFF receptor TNFRSF13C (BAFF-R) in patients with common variable immunodeficiency. J Clin Immunol 2005;25:496–502.
[150] Warnatz K, et al. B-cell activating factor receptor deficiency is associated with an adult-onset antibody deficiency syndrome in humans. Proc Natl Acad Sci USA 2009;106:13945–50.
[151] Jacobi AM, et al. Effect of long-term belimumab treatment on B cells in systemic lupus erythematosus: extension of a phase II, double-blind, placebo-controlled, dose-ranging study. Arthritis Rheum 2010;62:201–10.
[152] Kern C, et al. Involvement of BAFF and APRIL in the resistance to apoptosis of B-CLL through an autocrine pathway. Blood 2004;103: 679–88.
[153] Haiat S, Billard C, Quiney C, Ajchenbaum-Cymbalista F, Kolb JP. Role of BAFF and APRIL in human B-cell chronic lymphocytic leukaemia. Immunology 2006;118:281–92.
[154] Sawicka-Powierza J, Jablonska E, Kloczko J, Piszcz J, Garley M, Ratajczk-Wrona W. Evaluation of TNF superfamily molecules release by neutrophils and B leukemic cells of patients with chronic B-cell lymphocytic leukemia. Neoplasma 2011;58:45–50.
[155] Nishio M, et al. Nurselike cells express BAFF and APRIL, which can promote survival of chronic lymphocytic leukemia cells via a paracrine pathway distinct from that of SDF-1α. Blood 2005;106: 1012–20.
[156] Zhang W, et al. B-cell activating factor and v-Myc myelocytomatosis viral oncogene homolog (c-Myc) influence progression of chronic lymphocytic leukemia. Proc Natl Acad Sci USA 2010;107:18956–60.
[157] Cols M, et al. Stromal endothelial cells establish a bidirectional crosstalk with chronic lymphocytic leukemia cells through the TNF-related factors BAFF, APRIL, and CD40L. J Immunol 2012;188:6071–83.
[158] Planelles L, et al. APRIL promotes B-1 cell-associated neoplasm. Cancer Cell 2004;6:399–408.
[159] Guadagnoli M, et al. Development and characterization of APRIL antagonistic monoclonal antibodies for treatment of B-cell lymphomas. Blood 2011;117:6856–65.
[160] Ferrer G, et al. Combined analysis of levels of serum B-cell activating factor and a proliferation-inducing ligand as predictor of disease progression in patients with chronic lymphocytic leukemia. Leuk Lymphoma 2011;52:2064–8.
[161] Molica S, et al. Baff serum level predicts time to first treatment in early chronic lymphocytic leukemia. Eur J Haematol 2010;85:314–20.
[162] Li YJ, et al. Expression of BAFF and BAFF-R in follicular lymphoma: correlation with clinicopathologic characteristics and survival outcomes. PLoS One 2012;7:e50936.
[163] Moreaux J, et al. APRIL and TACI interact with syndecan-1 on the surface of multiple myeloma cells to form an essential survival loop. Eur J Haematol 2009;83:119–29.
[164] Fu L, Lin-Lee YC, Pham LV, Tamayo AT, Yoshimura LC, Ford RJ. BAFF-R promotes cell proliferation and survival through interaction with IKKβ and NF-κB/c-Rel in the nucleus of normal and neoplastic B-lymphoid cells. Blood 2009;113:4627–36.

[165] Hildebrand JM, et al. A BAFF-R mutation associated with non-Hodgkin lymphoma alters TRAF recruitment and reveals new insights into BAFF-R signaling. J Exp Med 2010;207:2569–79.
[166] Kofler DM, Gawlik BB, Elter T, Gianella-Borradori A, Wendtner CM, Hallek M. Phase 1b trial of atacicept, a recombinant protein binding BLyS and APRIL, in patients with chronic lymphocytic leukemia. Leukemia 2012;26:841–4.
[167] Zhang L, et al. Selective killing of Burkitt's lymphoma cells by mBAFF-targeted delivery of PinX1. Leukemia 2011;25:331–40.
[168] Granger GA, Williams TW. Lymphocyte cytotoxicity in vitro: activation and release of a cytotoxic factor. Nature 1968;218:1253–4.
[169] Ruddle NH, Waksman BH. Cytotoxicity mediated by soluble antigen and lymphocytes in delayed hypersensitivity. 3. Analysis of mechanism. J Exp Med 1968;128:1267–79.
[170] Ware CF, Van Arsdale TL, Crowe PD, Browning JL. The ligands and receptors of the lymphotoxin system. Curr Top Microbiol Immunol 1995;198:175–218.
[171] Browning JL, et al. Lymphotoxin β, a novel member of the TNF family that forms a heteromeric complex with lymphotoxin on the cell surface. Cell 1993;72:847–56.
[172] Crowe PD, et al. A lymphotoxin-beta-specific receptor. Science 1994;264:707–10.
[173] Mauri DN, et al. LIGHT, a new member of the TNF superfamily, and lymphotoxin α are ligands for herpesvirus entry mediator. Immunity 1998;8:21–30.
[174] Ware CF. Network communications: lymphotoxins, LIGHT, and TNF. Annu Rev Immunol 2005;23:787–819.
[175] Young J, et al. Lymphotoxin-αβ heterotrimers are cleaved by metalloproteinases and contribute to synovitis in rheumatoid arthritis. Cytokine 2010;51:78–86.
[176] Ware CF, Crowe PD, Grayson MH, Androlewicz MJ, Browning JL. Expression of surface lymphotoxin and tumor necrosis factor on activated T, B, and natural killer cells. J Immunol 1992;149:3881–8.
[177] Ansel KM, et al. A chemokine-driven positive feedback loop organizes lymphoid follicles. Nature 2000;406:309–14.
[178] Vu F, Dianzani U, Ware CF, Mak T, Gommerman JL. ICOS, CD40, and lymphotoxin β receptors signal sequentially and interdependently to initiate a germinal center reaction. J Immunol 2008;180:2284–93.
[179] GeurtsvanKessel CH, et al. Dendritic cells are crucial for maintenance of tertiary lymphoid structures in the lung of influenza virus-infected mice. J Exp Med 2009;206:2339–49.
[180] Moussion C, Girard JP. Dendritic cells control lymphocyte entry to lymph nodes through high endothelial venules. Nature 2011;479:542–6.
[181] del Rio ML, Lucas CL, Buhler L, Rayat G, Rodriguez-Barbosa JI. HVEM/LIGHT/BTLA/CD160 cosignaling pathways as targets for immune regulation. J Leukoc Biol 2010;87:223–35.
[182] Summers deLuca L, Ng D, Gao Y, Wortzman ME, Watts TH, Gommerman JL. LTβR signaling in dendritic cells induces a type I IFN response that is required for optimal clonal expansion of $CD8^+$ T cells. Proc Natl Acad Sci USA 2011;108:2046–51.
[183] Ota N, et al. IL-22 bridges the lymphotoxin pathway with the maintenance of colonic lymphoid structures during infection with *Citrobacter rodentium*. Nat Immunol 2011;12:941–8.
[184] Tumanov AV, et al. Lymphotoxin controls the IL-22 protection pathway in gut innate lymphoid cells during mucosal pathogen challenge. Cell Host Microbe 2011;10:44–53.
[185] Kuprash DV, et al. Novel tumor necrosis factor-knockout mice that lack Peyer's patches. Eur J Immunol 2005;35:1592–600.
[186] Matsumoto M, et al. Affinity maturation without germinal centres in lymphotoxin-α-deficient mice. Nature 1996;382:462–6.
[187] Koni PA, Sacca R, Lawton P, Browning JL, Ruddle NH, Flavell RA. Distinct roles in lymphoid organogenesis for lymphotoxins α and β revealed in lymphotoxin β-deficient mice. Immunity 1997;6:491–500.
[188] Futterer A, Mink K, Luz A, Kosco-Vilbois MH, Pfeffer K. The lymphotoxin β receptor controls organogenesis and affinity maturation in peripheral lymphoid tissues. Immunity 1998;9:59–70.
[189] Scheu S, Alferink J, Potzel T, Barchet W, Kalinke U, Pfeffer K. Targeted disruption of LIGHT causes defects in costimulatory T cell activation and reveals cooperation with lymphotoxin β in mesenteric lymph node genesis. J Exp Med 2002;195:1613–24.
[190] Mackay F, Majeau GR, Lawton P, Hochman PS, Browning JL. Lymphotoxin but not tumor necrosis factor functions to maintain splenic architecture and humoral responsiveness in adult mice. Eur J Immunol 1997;27:2033–42.
[191] Ngo VN, et al. Lymphotoxin α/β and tumor necrosis factor are required for stromal cell expression of homing chemokines in B and T cell areas of the spleen. J Exp Med 1999;189:403–12.
[192] Phan TG, Green JA, Gray EE, Xu Y, Cyster JG. Immune complex relay by subcapsular sinus macrophages and noncognate B cells drives antibody affinity maturation. Nat Immunol 2009;10:786–93.
[193] Fu YX, Huang G, Matsumoto M, Molina H, Chaplin DD. Independent signals regulate development of primary and secondary follicle structure in spleen and mesenteric lymph node. Proc Natl Acad Sci USA 1997;94:5739–43.
[194] Browning JL, et al. Lymphotoxin-β receptor signaling is required for the homeostatic control of HEV differentiation and function. Immunity 2005;23:539–50.
[195] Chai Q, et al. Maturation of lymph node fibroblastic reticular cells from myofibroblastic precursors is critical for antiviral immunity. Immunity 2013;38:1013–24.
[196] Lorenz RG, Chaplin DD, McDonald KG, McDonough JS, Newberry RD. Isolated lymphoid follicle formation is inducible and dependent upon lymphotoxin-sufficient B lymphocytes, lymphotoxin β receptor, and TNF receptor I function. J Immunol 2003;170:5475–82.
[197] van de Pavert SA, Mebius RE. New insights into the development of lymphoid tissues. Nat Rev Immunol 2010;10:664–74.
[198] Mackay F, Browning JL. Turning off follicular dendritic cells. Nature 1998;395:26–7.
[199] Schneider K, et al. Lymphotoxin-mediated crosstalk between B cells and splenic stroma promotes the initial type I interferon response to cytomegalovirus. Cell Host Microbe 2008;3:67–76.
[200] Katakai T. Marginal reticular cells: a stromal subset directly descended from the lymphoid tissue organizer. Front Immunology 2012;3:200.
[201] Tumanov A, et al. Distinct role of surface lymphotoxin expressed by B cells in the organization of secondary lymphoid tissues. Immunity 2002;17:239–50.
[202] Boulianne B, et al. AID-expressing germinal center B cells cluster normally within lymph node follicles in the absence of FDC-$M1^+$ $CD35^+$ follicular dendritic cells but dissipate prematurely. J Immunol 2013;191:4521–30.
[203] De Trez C, et al. The inhibitory HVEM-BTLA pathway counter regulates lymphotoxin receptor signaling to achieve homeostasis of dendritic cells. J Immunol 2008;180:238–48.

[204] Kabashima K, Banks TA, Ansel KM, Lu TT, Ware CF, Cyster JG. Intrinsic lymphotoxin-β receptor requirement for homeostasis of lymphoid tissue dendritic cells. Immunity 2005;22: 439–50.

[205] Lewis KL, et al. Notch2 receptor signaling controls functional differentiation of dendritic cells in the spleen and intestine. Immunity 2011.

[206] Vinuesa CG, Linterman MA, Goodnow CC, Randall KL. T cells and follicular dendritic cells in germinal center B-cell formation and selection. Immunol Rev 2010;237:72–89.

[207] Fu YX, Molina H, Matsumoto M, Huang G, Min J, Chaplin DD. Lymphotoxin-α (LTα) supports development of splenic follicular structure that is required for IgG responses. J Exp Med 1997;185:2111–20.

[208] Koni PA, Flavell RA. Lymph node germinal centers form in the absence of follicular dendritic cell networks. J Exp Med 1999;189:855–64.

[209] Wang X, et al. Follicular dendritic cells help establish follicle identity and promote B cell retention in germinal centers. J Exp Med 2011;208:2497–510.

[210] Debard N, Sierro F, Browning J, Kraehenbuhl JP. Effect of mature lymphocytes and lymphotoxin on the development of the follicle-associated epithelium and M cells in mouse Peyer's patches. Gastroenterology 2001;120:1173–82.

[211] Tumanov AV, Kuprash DV, Mach JA, Nedospasov SA, Chervonsky AV. Lymphotoxin and TNF produced by B cells are dispensable for maintenance of the follicle-associated epithelium but are required for development of lymphoid follicles in the Peyer's patches. J Immunol 2004;173:86–91.

[212] Knoop KA, et al. RANKL is necessary and sufficient to initiate development of antigen-sampling M cells in the intestinal epithelium. J Immunol 2009;183:5738–47.

[213] Suzuki K, et al. The sensing of environmental stimuli by follicular dendritic cells promotes immunoglobulin A generation in the gut. Immunity 2010;33:71–83.

[214] Tsuji M, et al. Requirement for lymphoid tissue-inducer cells in isolated follicle formation and T cell-independent immunoglobulin A generation in the gut. Immunity 2008;29:261–71.

[215] Fritz JH, et al. Acquisition of a multifunctional IgA$^+$ plasma cell phenotype in the gut. Nature 2012;481:199–203.

[216] Kang HS, et al. Signaling via LTβR on the lamina propria stromal cells of the gut is required for IgA production. Nat Immunol 2002;3:576–82.

[217] Kruglov AA, et al. Nonredundant function of soluble LTα3 produced by innate lymphoid cells in intestinal homeostasis. Science 2013;342:1243–6.

[218] Lopatin U, Blutt SE, Conner ME, Kelsall BL. Lymphotoxin alpha-deficient mice clear persistent rotavirus infection after local generation of mucosal IgA. J Virology 2013;87:524–30.

[219] Wang J, et al. Dysregulated LIGHT expression on T cells mediates intestinal inflammation and contributes to IgA nephropathy. J Clin Invest 2004;113:826–35.

[220] Leon B, Ballesteros-Tato A, Browning JL, Dunn R, Randall TD, Lund FE. Regulation of T(H)2 development by CXCR5$^+$ dendritic cells and lymphotoxin-expressing B cells. Nat Immunol 2012;13:681–90.

[221] Yi T, et al. Oxysterol gradient generation by lymphoid stromal cells guides activated B cell movement during humoral responses. Immunity 2012;37:535–48.

[222] Yi T, Cyster JG. EBI2-mediated bridging channel positioning supports splenic dendritic cell homeostasis and particulate antigen capture. Elife 2013;2:e00757.

[223] Mackay F, et al. Both the lymphotoxin and tumor necrosis factor pathways are involved in experimental murine models of colitis. Gastroenterology 1998;115:1464–75.

[224] An MM, et al. Lymphtoxin β receptor-Ig ameliorates TNBS-induced colitis via blocking LIGHT/HVEM signaling. Pharmacol Res 2005;52:234–44.

[225] Dohi T, et al. Elimination of colonic patches with lymphotoxin β receptor-Ig prevents Th2 cell-type colitis. J Immunol 2001;167: 2781–90.

[226] Stopfer P, et al. Blocking lymphotoxin-β receptor activation diminishes inflammation via reduced mucosal addressin cell adhesion molecule-1 (MAdCAM-1) expression and leucocyte margination in chronic DSS-induced colitis. Clin Exp Immunol 2004;136:21–9.

[227] Jungbeck M, Stopfer P, Bataille F, Nedospasov SA, Mannel DN, Hehlgans T. Blocking lymphotoxin β receptor signalling exacerbates acute DSS-induced intestinal inflammation—opposite functions for surface lymphotoxin expressed by T and B lymphocytes. Mol Immunol 2008;45:34–41.

[228] Spahn TW, et al. The lymphotoxin-β receptor is critical for control of murine *Citrobacter rodentium*-induced colitis. Gastroenterology 2004;127:1463–73.

[229] Wang Y, et al. Lymphotoxin beta receptor signaling in intestinal epithelial cells orchestrates innate immune responses against mucosal bacterial infection. Immunity 2010;32:403–13.

[230] Bergstrom KS, et al. Modulation of intestinal goblet cell function during infection by an attaching and effacing bacterial pathogen. Infect Immun 2008;76:796–811.

[231] Luther SA, Lopez T, Bai W, Hanahan D, Cyster JG. BLC expression in pancreatic islets causes B cell recruitment and lymphotoxin-dependent lymphoid neogenesis. Immunity 2000;12:471–81.

[232] Krautler NJ, et al. Follicular dendritic cells emerge from ubiquitous perivascular precursors. Cell 2012;150:194–206.

[233] Rangel-Moreno J, et al. The development of inducible bronchus-associated lymphoid tissue depends on IL-17. Nat Immunol 2011;12:639–46.

[234] Browning JL. B cells move to centre stage: novel opportunities for autoimmune disease treatment. Nat Rev Drug Discov 2006;5:564–76.

[235] Matsushita T, Yanaba K, Bouaziz JD, Fujimoto M, Tedder TF. Regulatory B cells inhibit EAE initiation in mice while other B cells promote disease progression. J Clin Invest 2008;118:3420–30.

[236] Gommerman JL, Browning JL. Lymphotoxin/light, lymphoid microenvironments and autoimmune disease. Nat Rev Immunol 2003;3:642–55.

[237] Fava RA, et al. A role for the lymphotoxin/LIGHT axis in the pathogenesis of murine collagen-induced arthritis. J Immunol 2003;171: 115–26.

[238] Chiang EY, et al. Targeted depletion of lymphotoxin-α-expressing TH1 and TH17 cells inhibits autoimmune disease. Nat Med 2009;15:766–73.

[239] Gatumu MK, Skarstein K, Papandile A, Browning JL, Fava RA, Bolstad AI. Blockade of lymphotoxin-beta receptor signaling reduces aspects of Sjogren's syndrome in salivary glands of non-obese diabetic mice. Arthritis Res Ther 2009;11:R24.

[240] El Shikh ME, Pitzalis C. Follicular dendritic cells in health and disease. Front Immunol 2012;3:292.

[241] Bombardieri M, et al. Activation-induced cytidine deaminase expression in follicular dendritic cell networks and interfollicular large B cells supports functionality of ectopic lymphoid neogenesis in autoimmune sialoadenitis and MALT lymphoma in Sjogren's syndrome. J Immunol 2007;179:4929–38.

[242] Humby F, et al. Ectopic lymphoid structures support ongoing production of class-switched autoantibodies in rheumatoid synovium. PLoS Med 2009;6:e1.

[243] Ammirante M, Luo JL, Grivennikov S, Nedospasov S, Karin M. B-cell-derived lymphotoxin promotes castration-resistant prostate cancer. Nature 2010;464:302–5.

[244] Nelson BH. $CD20^+$ B cells: the other tumor-infiltrating lymphocytes. J Immunol 2010;185:4977–82.

[245] Ame-Thomas P, et al. Human mesenchymal stem cells isolated from bone marrow and lymphoid organs support tumor B-cell growth: role of stromal cells in follicular lymphoma pathogenesis. Blood 2007;109:693–702.

[246] Rehm A, et al. Cooperative function of CCR7 and lymphotoxin in the formation of a lymphoma-permissive niche within murine secondary lymphoid organs. Blood 2011;118:1020–33.

Chapter 16

# The Mucosal Immune System: Host–Bacteria Interaction and Regulation of Immunoglobulin A Synthesis

Sidonia Fagarasan[1], Giuliana Magri[2] and Andrea Cerutti[2,3]

[1]*Laboratory for Mucosal Immunity, RIKEN Center for Integrative Medical Sciences, RIKEN Yokohama, Tsurumi, Yokohama, Japan,* [2]*Institut Hospital del Mar d'Investigacions Mèdiques, Barcelona, Spain,* [3]*Immunology Institute, Department of Medicine, Icahn School of Medicine at Mount Sinai, New York, NY, USA*

## 1. INTRODUCTION

Mucosal membranes regulate our relationship with the external environment by providing a dynamic interface that facilitates the absorption of beneficial substances, while blocking the penetration of noxious agents and microbes. Given their persistent exposure to external attacks, mucosal organs have evolved multiple layers of defensive mechanisms characterized by increased specificity. Besides physical, mechanical, and chemical defensive strategies, mucosal surfaces use sophisticated immunological mechanisms to repel toxins and microbes. These complex protective activities have at their center epithelial cells. In addition to forming chemical and mechanical barriers, delivering antimicrobial compounds, and transporting secretory antibodies, epithelial cells regulate the signaling networks that connect the mucosal immune system with the external environment, including bacteria.

Mammals have coevolved with microbes for over 150 million years and thus have developed symbiotic and mutualistic relationships with many prokaryotic species, particularly in the intestine. Despite exceeding the number of eukaryotic cells forming our body by at least an order of magnitude, bacteria peacefully live as commensals in the intestinal mucosa [1]. These commensals process indigestible polysaccharides, synthesize essential vitamins, stimulate the maturation of the gut immune system, and form an ecological niche that prevents the growth of pathogens. Conversely, the lumen of the gut provides commensals with a stable habitat rich in energy derived from the ingested food. In this mutualistic relationship, microbial signals stimulate the intestinal mucosa to generate a noninflammatory homeostatic balance that is characterized by hyporesponsiveness against commensals but active readiness against pathogens [2].

The confinement of commensals in the intestinal lumen involves the transepithelial release of immunoglobulin A (IgA) into mucosal secretions [3,4]. IgA is mostly produced by B cells located in gut-associated lymphoid follicles and constitutes the most abundant antibody class in the intestine [5]. Due to the vast area covered by the intestinal mucosa, IgA is also the most abundant antibody in our body. Of note, IgA establishes a state of armed peace in the homeostatic interaction between the host and commensal bacteria [6]. When commensals or pathogens trespass the epithelial barrier, a state of open war breaks out and IgA receives help from the antibody isotype IgG to repel microbial intruders. In this life-threatening situation, IgG provides a second line of defense that controls microbial dissemination by eliciting a robust inflammatory reaction. This chapter discusses the fundamental principles underlying the regulation and function of IgA responses in the intestinal mucosa and the role of IgA in host–bacteria interaction.

## 2. GEOGRAPHY, REGULATION, AND PROPERTIES OF GUT IMMUNOGLOBULIN A

Mature B cells acquire IgA antibodies endowed with powerful mucosal effector functions by replacing the constant region μ (Cμ) gene with the Cα gene in the human Ig heavy chain locus through class switch recombination (CSR) [7]. This DNA recombinatorial event targets intronic DNA sequences termed switch (S) regions, located upstream of Cμ and Cα genes [8]. Germline transcription of Sμ and Sα

Molecular Biology of B Cells. http://dx.doi.org/10.1016/B978-0-12-397933-9.00016-3

regions opens their chromatin to make them substrates of activation-induced cytidine deaminase (AID), an enzyme essential for the initiation of CSR [7–9]. In both mice and humans, CSR usually involves the activation of B cells by antigen-activated CD4$^+$ T cells expressing the TNF family member CD40 ligand (CD40L) and cytokines, which are critical to transcriptionally activate a specific $C_H$ gene [7]. Germline Cα gene transcription is highly dependent on the cytokine transforming growth factor-β1 (TGF-β1) [10]. Indeed, mice with B cell-specific TGF-β1 receptor deficiency have compromised mucosal IgA responses [11].

Germline transcription generates an Iα-Sα-Cα transcript that associates with the template strand of the DNA to form a stable DNA–RNA hybrid specifically recognized by AID [8,9]. This DNA-editing enzyme is essential for CSR, as mice or humans lacking AID do not generate class-switched antibodies and develop a severe antibody deficiency called hyper-IgM (HIGM) syndrome [12]. The enzyme AID deaminates cytosine residues on both strands of the transcribed Sα region [13], thus generating multiple DNA lesions that are processed into double-stranded DNA breaks [14]. Fusion of double-stranded DNA breaks at Sα and Sμ through the nonhomologous end-joining pathway induces excision of the noncoding intervening (I) DNA, thereby juxtaposing $V_HDJ_H$ to Cα [7]. This event generates a chromosomal $V_HDJ_H$-Cα sequence that encodes the IgA protein as well as an extrachromosomal Sα-Sμ switch circle that encodes an Iα-Cμ switch circle transcript [7,15].

In humans, IgA includes two IgA1 and IgA2 subclasses that are named after their relative concentration in the serum [7]. IgA2 has two allotypic variants, termed IgA2m(1) and IgA2m(2) [16]. Circulating IgA is mostly composed of IgA1, whereas mucosal IgA includes both IgA1 and IgA2 [7]. In mucosal secretions, the relative production of IgA1 and IgA2 varies among different mucosal tissues, IgA2 being particularly abundant at sites heavily colonized by bacteria, such as the large intestine and the female urogenital tract [5]. It has been proposed that the different distribution of IgA1 and IgA2 may reflect their different sensitivity to bacterial proteases. Indeed, the IgA2 molecule may be more resistant to bacterial degradation due to its shorter hinge region [17].

As expected, the differential distribution of IgA subclasses in various body fluids reflects a similar distribution of IgA1-secreting and IgA2-secreting plasma cells in various tissues. IgA1-secreting plasma cells predominate in the bone marrow and the upper part of the digestive and respiratory tracts, whereas IgA2-secreting plasma cells are as abundant as or more abundant than IgA1-secreting plasma cells in the distal digestive tract and female genital tract [18].

Unlike other immunoglobulins, IgA can be found in both monomeric and polymeric forms that have a characteristic distribution. In healthy individuals, 80–90% of serum IgA consists of monomeric IgA, whereas mucosal IgA predominantly consists of dimers and oligomers held together by a plasma cell-derived polypeptide, termed joining (J) chain [19]. In the intestinal lamina propria, polymeric IgA is transported across intestinal epithelial cells (IECs) through a process referred to as transcytosis. This process requires binding of the J chain to the polymeric Ig receptor (pIgR) expressed on the basolateral surface of IECs. During transcytosis, pIgR is intracellularly cleaved into the secretory component, which remains bound to the J chain of polymeric IgA to form a secretory IgA (SIgA) complex [20].

Secretory component increases the stability of SIgA and its localization in the mucosal environment [21]. Indeed, three-dimensional studies of SIgA show that secretory component wraps the Cα2 and Cα3 domains of IgA monomers to make a compact SIgA complex that is resistant to the degradative activity of intestinal IgA proteases [22]. The stability of SIgA further increases after its interaction with antigens, including bacteria [23]. Secretory component also confers hydrophilic properties to the Fcα portions of the SIgA complex via seven surface exposed N-linked oligosaccharides equipped with sialic acid residues. This property is important for the interaction of SIgA with the mucus layer that protects the apical surface of IECs [24].

## 3. SYNTHESIS OF GUT IMMUNOGLOBULIN A

In the gut, IgA can be generated through multiple pathways, within or outside organized follicular structures belonging to the gut-associated lymphoid tissues (GALT) and with or without cognate help from CD4$^+$ T cells [25]. Two functional compartments known as inductive and effector sites characterize the geography of IgA production and function in the intestine [26]. Primary inductive sites include the organized follicular structures of Peyer's patches (PPs), which specifically develop in the small intestine during fetal life. Additional inductive sites termed isolated lymphoid follicles (ILFs) develop after birth in both small and large intestinal segments as a result of gut colonization by commensal bacteria [27]. Additional IgA is produced in the follicles of mesenteric lymph nodes (MLNs). While PPs, ILFs, and MLNs function as IgA inductive sites, the intestinal lamina propria (LP) serves as an IgA effector site, in that it mostly supports the maturation and survival of IgA-producing plasmablasts and plasma cells emerging from PPs and possibly MLNs [25].

## 4. T CELL-DEPENDENT IMMUNOGLOBULIN A INDUCTION

PPs are large lymphoid structures consisting of multiple B cell follicles built on a network of follicular dendritic cells (FDCs) and separated by interfollicular areas containing

T cells and dendritic cells (DCs). PPs develop independently of the gut microbiota through a series of coordinated interactions between stromal cells (SCs) and bone marrow-derived lymphoid tissue-inducing (LTi) cells expressing the transcription factor Id2 and retinoic acid-related orphan receptor γt (RORγt) [28–30], also known as group 3 innate lymphoid cells (ILCs) [31]. Such interactions involve key cytokines and chemokines, including lymphotoxin (LT), IL-7, and CXCL13, which bind to LTβ receptor (LTβR), IL-7R, and CXCR5, respectively [32,33]. After colonization with bacteria, PPs become larger and form germinal centers (GCs), which are unique microenvironments where antigen-activated B cells interact with cognate CD4$^+$ T cells to generate a large and highly diverse repertoire of IgA antibodies. The number of PPs varies from 7 to 10 in mice and from 100 to 200 in humans, depending on the age of the individual [34]. Immunosenescence and chronic inflammation lead to atrophy of PPs, which causes an impairment of IgA responses and a decreased diversity of gut bacterial communities [34].

## 4.1 Germinal Center Reaction

GCs arise following antigenic stimulation and represent the hallmark of T cell-dependent (TD) antibody responses [35–38]. GCs are special structures that facilitate the interaction of B cells with antigens displayed on the surface of FDCs and CD4$^+$ T cells. A defining characteristic of GC B cells is their expression of AID, an enzyme that mediates antibody diversification through CSR and somatic hypermutation (SHM) [9,39]. These processes are required for the generation of long-lived plasma cells secreting affinity-matured IgA (Figure 1).

### *4.1.1 Germinal Center B Cells*

The GC reaction is highly dependent on the initial activation of B cells by antigen through the B cell receptor (BCR, or surface Ig). Antigen-activated B cells upregulate the chemokine receptor CCR7 and thereafter follow chemotactic gradients established by CCL19 and CCL21 to migrate to the T–B border, which is located between the B cell follicle and the T cell extrafollicular area. At the T–B border, B cells interact with antigen-activated CD4$^+$ T cells expressing CD40 ligand (CD40L or CD154) [40,41]. Engagement of CD40 on B cells by CD40L on T cells triggers two distinct differentiation pathways. In the extrafollicular pathway, B cells move to the extrafollicular area, where they proliferate and differentiate in short-lived IgM-secreting plasmablasts. In the follicular pathway, B cells migrate to the center of the follicle, where they proliferate within a meshwork of FDCs to form the GC [42]. The factors controlling these pathways are not completely understood, but the affinity of the BCR and various environmental cues may drive distinct transcriptional programs that direct the differentiation of antigen-specific T cell-activated B cells toward either short-lived plasma cells or GC B cells [43–45].

Accordingly, the B cells that maintain the expression of the orphan G protein-couple receptor Epstein–Barr virus-induced gene (EBI2 or GPR183) and upregulate the B lymphocyte-induced maturation protein 1 (BLIMP-1) transcription factor remain in the outer follicle and differentiate into short-lived plasmablasts [46,47]. In contrast, the B cells that downregulate EBI2 and express the B cell lymphoma-6 (Bcl-6) transcription factor move to the center of the follicle and differentiate into GC B cells [47,48]. FDCs and the migration inhibitory sphingosine-1-phosphate (S1P) receptor S1PR2 are required for the confinement of B cells within the GC [49]. PP B cells preferentially use this GC pathway, because only a few IgA plasmablasts and plasma cells are usually observed in and around PPs under steady-state conditions. However, when the organization of follicular structures is disrupted by inflammation, including *Salmonella* infection, PP B cells seem to adopt an "emergency" pathway and rapidly differentiate into IgA-secreting plasmablasts [50].

### *4.1.2 Germinal Center T Cells: $T_{FH}$ and $T_{FR}$ Cells*

The GC contains two major subsets of CD4$^+$ T cells, known as CXCR5$^+$PD-1$^{hi}$Foxp3$^-$ T follicular helper ($T_{FH}$) cells and CXCR5$^+$PD-1$^{hi}$Foxp3$^+$ T follicular regulatory cells ($T_{FR}$). Collectively referred to as GC T cells, $T_{FH}$ and $T_{FR}$ cells require Bcl-6 for their generation [51–53] and migrate to the B cell follicle in response to CXCL13, a chemokine released by FDCs that binds to CXCR5 on $T_{FH}$ and $T_{FR}$ cells [54,55].

In addition to CXCR5, $T_{FH}$ cells highly express the co-stimulatory molecules CD40L, inducible co-stimulator (ICOS), programmed cell death-1 (PD-1), and OX40 and abundantly produce powerful B cell-helper cytokines such as IL-21 and IL-4. In general, $T_{FH}$ cells are essential for the maintenance and function of GCs and for the generation of both memory B cells and plasma cells [36].

The generation of $T_{FH}$ cells requires their stimulation through the T cell receptor, stable interactions with B cells through signaling lymphocyte-associated molecules (SLAMs), induction of Bcl-6 expression, and downregulation of BLIMP-1 [56,57]. Bcl-6 appears to be critical for $T_{FH}$ cell development, because it leads to the upregulation of CXCR5 through suppression of miRNAs such as miR 17-92, which antagonizes CXCR5 expression [53]. Analogous to B cells, the upregulation of CXCR5 and the concomitant downregulation of CCR7 facilitate the homing of $T_{FH}$ cells into the FDC network [51–53,58–60]. Other transcription factors such as IRF4, c-Maf, Batf, Ascl2, and STAT3/5 further control the initial activation events leading to the development of $T_{FH}$ cells [61,62].

$T_{FR}$ cells represent a less abundant but perhaps equally important GC CD4$^+$ T cell subset. $T_{FR}$ cells share

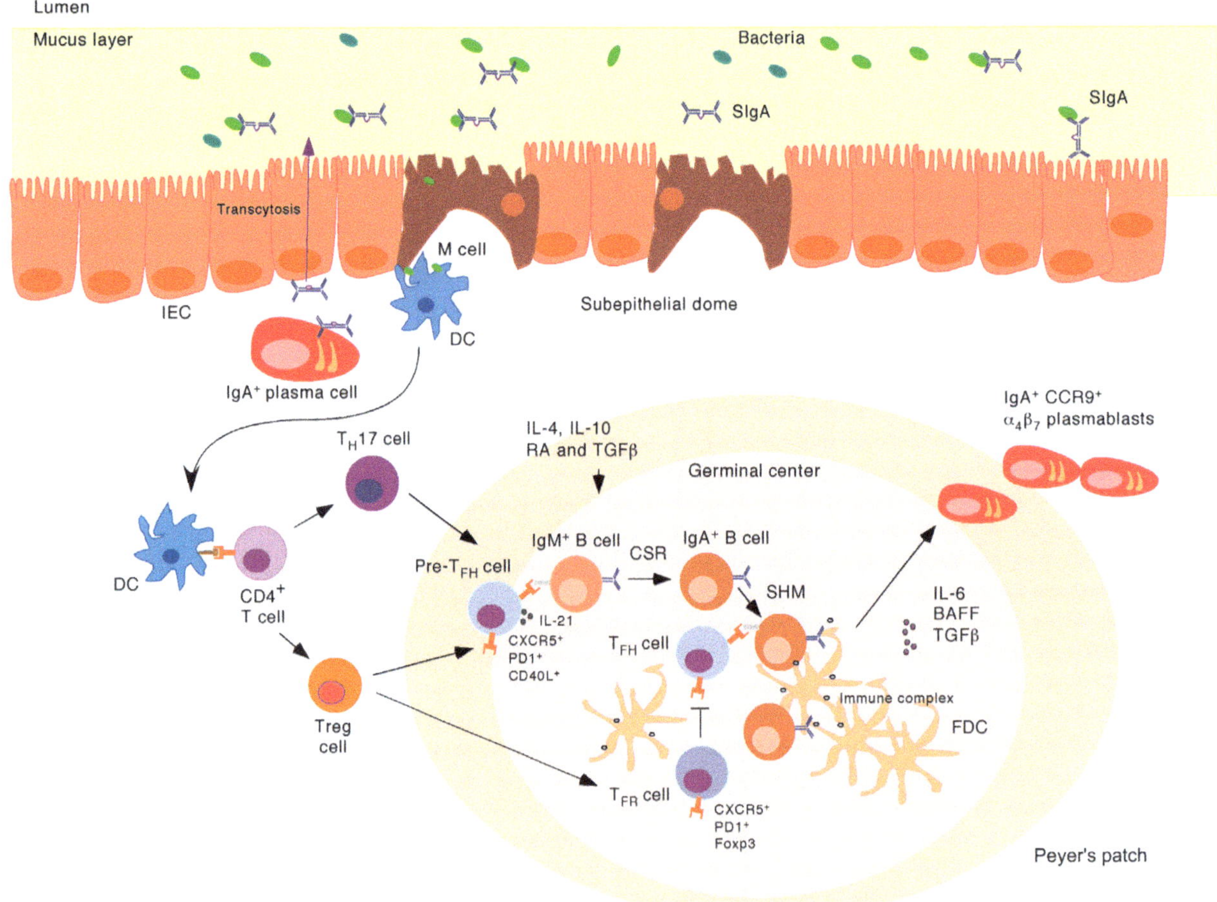


FIGURE 1 **IgA production through the TD pathway.** Intraluminal antigens are captured by microfold (M) cells and taken up by gut DCs located in the subepithelial dome of PPs. DCs may also directly capture free or SIgA-bound microbial antigens. After processing antigen, DCs enter interfollicular areas to activate naïve T cells and initiate a transcriptional program that generates $T_{FH}$ cells. In PPs, these $T_{FH}$ cells can originate from either Foxp3⁺Treg cells or $T_H17$ precursors. Pre-$T_{FH}$ cells interact with antigen-specific follicular B cells at the T–B border and induce B cell clonal expansion and CSR from IgM to IgA by expressing CD40L and secreting TGF-β and IL-21. After upregulating CXCR5, both mature $T_{FH}$ cells and IgA class-switched B cells migrate to the GC in response to the CXCR5 ligand CXCL13, which is produced by FDCs. In the GC, follicular B cells complete IgA CSR, SHM, and affinity maturation by interacting with $T_{FH}$ cells and FDCs. Besides exposing immunocomplexes to select follicular B cells expressing high-affinity BCRs, FDCs enhance IgA CSR by releasing BAFF, APRIL, and TGF-β in response to dietary RA and microbiota-derived TLR ligands. Of note, GC B cell responses are regulated by $T_{FR}$ cells, which inhibit $T_{FH}$ cell expansion and cytokine secretion and directly suppress B cells. Class-switched IgA⁺ plasmablasts that emerge from the GC reaction express BLIMP-1 and upregulate the gut homing receptors α4β7 and CCR9 in response to RA from TLR-stimulated DCs. This process permits the migration of plasmablasts to the intestinal LP, where they become long-lived plasma cells. Most of these plasma cells are mutated and selected and release a very diverse repertoire of IgA antibodies that contribute to the establishment of symbiotic relationships with the microbiota.

several features with $T_{FH}$ cells, including high expression of CXCR5 and PD-1 and preferential location within the GC. Similar to $T_{FH}$ cells, $T_{FR}$ cell development depends on Bcl-6 expression, presence of B cells, and intact signaling through the adaptor SLAM-associated protein (SAP), which points to an important role of T–B interaction for their entry into the GC [63–65]. Yet, unlike $T_{FH}$ cells, $T_{FR}$ cells express the transcription factor Foxp3, GITR, and CTLA-4 but lack CD40L, and they have far less potential to secrete B-helper cytokines other than IL-10. In the case of systemic GCs, $T_{FR}$ cells seem to derive from thymic Foxp3⁺ T regulatory (Treg) cells. In the case of intestinal GCs, the origin of $T_{FR}$ cells remains unclear, but they may derive from Foxp3⁺ Treg cells generated either in the thymus or directly in the gut [66].

### 4.1.3 Follicular Dendritic Cells

FDCs are a subset of radio-resistant SCs specialized in the capture of large amounts of antigen in the form of immune complexes [67,68]. Upon antigenic stimulation, both lymphocyte and SC compartments undergo extensive maturation and remodeling. By increasing the expression of VCAM-1, ICAM-1, and MadCAM-1 adhesion molecules as well as low-affinity CD23 (also known as FcεRII) and CD32 (also known as FcγRIIb) Fc receptors, FDCs facilitate the deposition of immune complexes and the development of the GC [69]. Of note, in peripheral lymph nodes (pLNs) or spleen, FDCs archive antigens on their surface over extended periods of time and therefore serve as an antigen reservoir during the GC reaction. Accordingly, lack of FDCs causes a rapid collapse of the GC [70].

In addition to trapping immunocomplexes and exposing antigen to B cells, FDCs express adhesion molecules, cytokines, and chemokines that help the localization and retention of both B and T cells in the GC. Indeed, FDCs lacking VCAM-1 and ICAM-1 adhesion molecules are unable to support the GC reaction, including affinity maturation [71]. Importantly, FDCs produce IL-6 and BAFF, which may play a role in the GC reaction, including gut IgA production [67,68,72–75]. Furthermore, FDCs produce abundant CXCL13, which helps the compartmentalization of the GC besides serving as CXCR5 ligand for antigen-activated B, $T_{FH}$, and $T_{FR}$ cells.

## 4.2 Regulation of Germinal Centers

The biology of the GC is controlled at multiple levels [36,38,76]. The proliferation, survival, and differentiation of GC B cells into memory B cells or plasma cells involves signals from BCR receptor molecules such as Ig, CD19, CD21, CD79, and CD81; $T_{FH}$ cell co-stimulatory molecules such as CD40L, PD-1, IL-21, CD95L, and ICOS; B-T and B-FDC interaction molecules such as SAP, CD84, and DOCK8; and B cell transcription factors such as Bcl-6, PAX-5, E2A, IRF-4, IRF-8, SPIB, MEF2c, BLIMP-1, and XBP-1 [54,57]. In this section we will briefly discuss how an abnormally high availability of $T_{FH}$ cell help leads to an inappropriate selection of GC B cells that deteriorates the quality of IgA-secreting plasma cells.

Recent studies point to a critical role of $T_{FR}$ cells in controlling GC responses. In systemic GCs, the lack of $T_{FR}$ cells results in an increased number of $T_{FH}$ cells and GC B cells that produce polyspecific/polyreactive rather than specific/monoreactive antibodies [63,64]. The migration of Foxp3$^+$ Treg cells and their differentiation into $T_{FR}$ cells also appears critical for the selection of IgA in PPs and the generation of intestinal IgA-secreting plasma cells (Fagarasan S., unpublished data). Indeed, the absence of $T_{FR}$ cells in GCs from PPs causes excessive expansion of $T_{FH}$ cells, which in turn decreases the frequency and quality of IgA plasma cells in the LP. Exactly how $T_{FR}$ cells exert their regulatory function is unknown, but $T_{FR}$ cells may act directly on $T_{FH}$ cells by controlling their expansion and cytokine secretion.

A "slowdown" of proliferation might be required for appropriate cytokine production by $T_{FH}$ cells, as different cytokines are produced at distinct stages of the cell cycle in activated T cells [77]. $T_{FR}$ cells may also act directly on B cells [78,79]. Indeed, antigen-activated CD4$^+$CD25$^+$CD69$^-$ cells (mainly Foxp3$^+$ cells) upregulate CXCR5 expression and suppress B cell antibody production induced by $T_{FH}$ cells. This suppression requires the interaction between B and T cells. In systemic GCs, the number of $T_{FR}$ cells increases at a later stage of the GC response and their presence coincides with a decline of the number of $T_{FH}$ cells that may constrain the longevity of the GC reaction [65]. Therefore, the $T_{FH}/T_{FR}$ cell ratio may change as the GC reaction develops and likely reflects the maturation and longevity of a GC [55]. In PPs, the dynamics of GCs are quite different from those of GCs from systemic lymphoid organs and the $T_{FH}/T_{FR}$ ratio may predominantly impact the IgA selection process.

Another layer of GC regulation depends on the inhibitory receptor programmed cell death 1 (PD-1). Beyond its use as a marker for GC T cells, PD-1 is known to play critical roles in shutting down ineffective immune responses and maintaining immune tolerance [80]. PD-1 interacts with two ligands, known as PD-L1 and PD-L2 [81]. In general, PD-L1 is ubiquitously expressed, whereas PD-L2 is more selectively expressed by certain DCs and macrophages and by GC and memory B cells. In PPs, PD-L1 expression is poor on nonswitched IgM$^+$ GC B cells and switched IgA$^+$ GC B cells, but increases on IgA$^+$ plasma cells. In contrast, PD-L2 expression is high on nonswitched IgM$^+$ B cells and switched IgA$^+$ GC B cells, which are predominantly located in the FDC area of the GC [82]. Thus, PD-1 on $T_{FH}$ cells may be engaged by either PD-L1 or PD-L2, depending on the nature of the interacting B cell. The ensuing signal inhibits $T_{FH}$ cells, as the absence of PD-1 increases both the frequency and number of $T_{FH}$ cells in PPs.

Also, the properties of PP $T_{FH}$ cells change in the absence of PD-1. Indeed, PD-1-deficient $T_{FH}$ cells not only show increased expression of Bcl-6 and decreased expression of IRF-4, but also produce reduced amounts of B cell-activating IL-21 and excessive amounts of proinflammatory IFN-γ and TNF-α. The resulting increased number of $T_{FH}$ cells with a proinflammatory phenotype impairs the selection and clonal expansion of IgA$^+$ GC B cells in PPs, which leads to a deterioration of the quality of IgA-secreting plasma cells in the LP. Similar observations were made in systemic GCs, where the expansion of $T_{FH}$ cells in the absence of restraining signals from PD-1 leads to a reduced formation of IgG plasma cells in the bone morrow [83].

## 4.3 Specific Characteristics of Gut Germinal Centers

Several characteristics distinguish GCs in PPs from GCs in pLNs [84]. For instance, GCs are always detected in PPs due to their constant stimulation by commensal bacteria. Indeed, animals kept in germ-free conditions or treated with antibiotics that eliminate the indigenous flora have extremely few GCs in PPs [85]. By contrast, GCs in pLNs develop only after deliberate immunization or upon systemic infection. PP GCs also differ from pLN GCs in regard to the isotype and amount of their antibody response. Thus, PP GCs preferentially induce IgA, whereas pLN GCs predominantly generate IgG. In the gut, B cells may be skewed toward IgA production due to the presence of specific immunological and metabolic signals generated by local IECs, DCs, B cells, and T cells [6,86].

As discussed earlier, TGF-β1 is essential to direct IgA class switching, because mice with B cells lacking the TGF-βRII have extremely few $IgA^+$ B cells in PPs [87]. In these structures, TGF-β1 likely induces IgA class switching in cooperation with CD40L and IL-21. Indeed, IL-21 enhances the generation and expansion of IgA-producing B cells and its deficiency leads to a loss of IgA-secreting plasma cells [88–90]. Ex vivo, the addition of IL-21 restores antibody production by B cells from patients with common variable immunodeficiency (CVID) or selective IgA deficiency (SIgAD) [91]. The production of TGF-β1 and IL-21 likely depends on the nature of the $T_{FH}$ cell [92]. In PPs, many $T_{FH}$ cells originate from regulatory or effector $CD4^+$ T cells induced in the context of gut antigens, including $Foxp3^+$ Treg cells and Th17 cells [93]. While $Foxp3^+$ T cells release TGF-β1, Th17 cells produce abundant IL-21. In the presence of TGF-β1, IL-21 enhances the CD40-dependent differentiation of IgA-expressing but not IgG-expressing B cells, which likely contributes to the preferential production of IgA in PPs [88,94]. In these structures, a proper balance between TGF-β1 and IL-21 may also be important to regulate the clonal expansion of $IgA^+$ B cells [82].

$T_{FH}$ cells are not the only source of intestinal TGF-β1, as this cytokine is also produced by DCs and IECs in response to bacterial products such as Toll-like receptor (TLR) ligands. DCs and IECs also release retinoic acid (RA), a metabolite of dietary vitamin A that synergizes with TGF-β1 and IL-6 to enhance IgA CSR and production [86,95–98]. RA is also abundantly produced by FDCs from PPs, which are conditioned by the gut microenvironment to support the preferential generation of IgA [74]. Accordingly, FDCs express TLRs and RA receptors (RARs) that cooperatively induce FDC expression of molecules involved in the activation and release of TGF-β1 [74]. These molecules include latent TGF-β-binding proteins (LTBP1, LTBP2, LTBP3), matrix metalloproteinases (MMP2, MMP9), bone morphogenetic protein 2 (BMP2), and integrin αv [74].

In addition to TGF-β1, PP FDCs produce abundant CXCL13 and BAFF that further facilitate IgA production. The IgA-inducing program of PP FDCs largely depends on their stimulation by microbial signals and RA, as the lack of either TLR signaling or vitamin A causes loss of GC B cells and impaired generation of $IgA^+$ B cells within PPs [74]. In summary, PPs are characterized by highly organized and dynamic microcompartments that provide a unique environment for a strongly biased generation of large amounts of affinity matured IgA antibodies specific for gut antigens.

## 5. T CELL-INDEPENDENT IMMUNOGLOBULIN A INDUCTION

Mice lacking the GALT retain some intestinal IgA production [6]. This response is less efficient, slower, and less effective than in GALT-sufficient mice, because it occurs in the absence of the selection steps that normally occur in PPs. However, extrafollicular IgA responses may be important soon after an infection to provide protection until more effective and specific follicular IgA responses develop [99]. Of note, some intestinal IgA production can also occur in mice lacking T cells [100,101]. Such IgA responses predominantly take place within ILFs, but can also occur in MLNs and PPs and in the diffuse lymphoid tissue of the LP (Figure 2).

### 5.1 Immunoglobulin A Generation in Isolated Lymphoid Follicles

An ILF is a cluster of B cells surrounded by a thick rim of DCs-macrophages containing very few T cells interspersed between B cells [27,85]. The large majority of ILFs in normal mice lack GCs. Unlike PPs, which develop in the sterile fetal environment, ILFs develop only after bacteria colonization of the intestine from smaller "anlagen" structures called cryptopatches (CPs). The size and cellular composition of ILFs depend on the bacterial load of the intestine [27,85,102,103]. Bacterial activation not only increases the size of ILFs, but also stimulates the B cells located in ILFs to express AID, undergo CSR from IgM to IgA, and differentiate to IgA-producing plasma cells. Exactly how ILF B cells become activated to generate IgA is not clear, but it is likely that the bias of ILF B cells for IgA class switching involves factors derived from bacteria-activated DCs and SCs [100].

Similar to PPs, the formation and maturation of ILFs require the presence of $ROR\gamma t^+$ ILCs, which are also known as LTi cells [4,100,104,105]. ILCs interact with SCs through TNF and LT and, in the presence of TLR signals from the microbiota, promote the recruitment of B cells and macrophage-DCs through CXCL13, CCL19, and CCL20 [100]. Furthermore, ILCs stimulate macrophage-DCs to release the active form of TGF-β1. Of note, gut macrophage-DCs uniquely express abundant TNF following activation by bacteria and this expression causes induction of MMPs that trigger the release of active TGF-β1 from an inactive precursor.

In addition to TNF and TGF-β1, gut macrophages-DCs and SCs secrete BAFF and its homologue, a proliferation-inducing ligand (APRIL), which cooperate with TGF-β and bacterial TLR ligands to induce IgA CSR and production in B cells independent of cognate help from T cells [100,101,106–108]. This T cell-independent (TI) pathway likely explains why ILFs can support IgA production in the absence of a GC reaction [100].

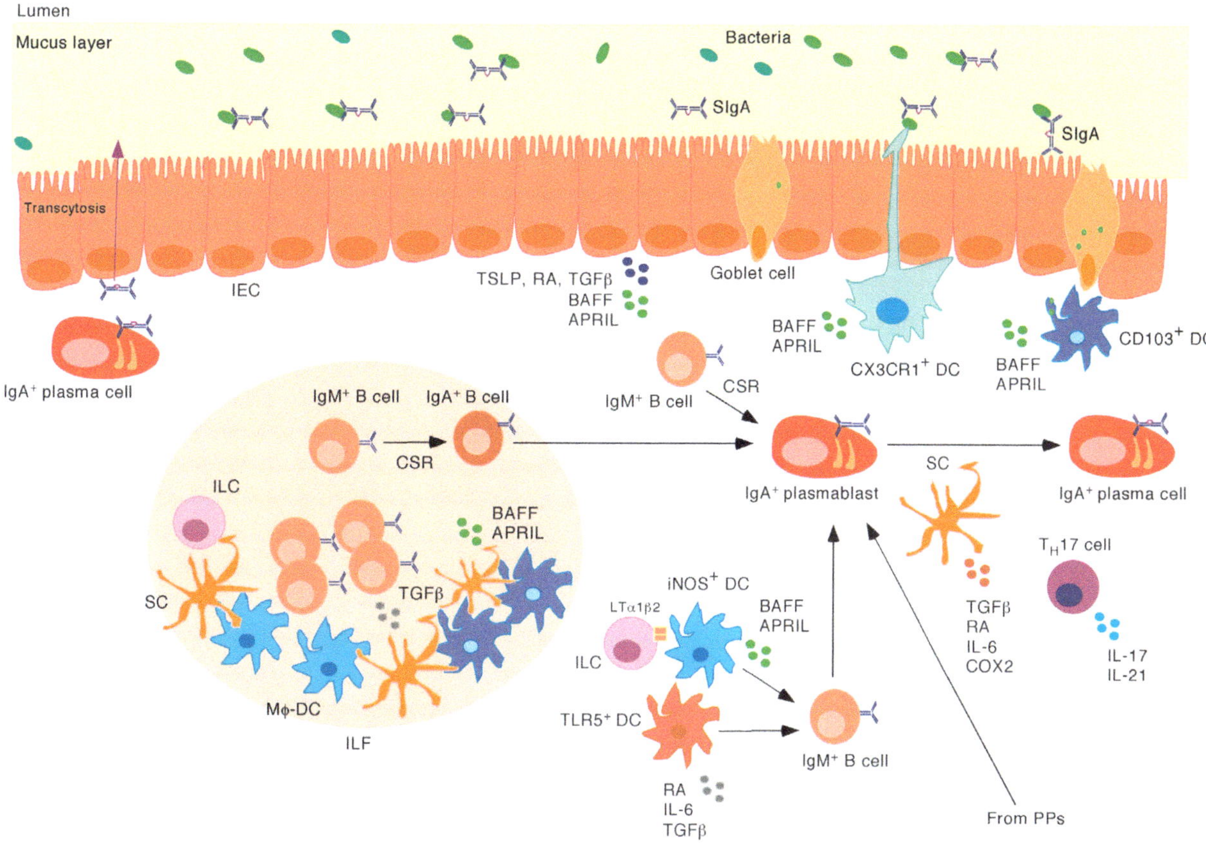


FIGURE 2 **IgA production through the TI pathway.** B cells located in the ILFs and LP can produce IgA even in the absence of help from T cells. In ILFs, soluble factors such as BAFF, APRIL, and TGF-β produced by bacteria-activated DCs and RORγt+ ILC-activated SCs stimulate B cells to undergo preferential CSR from IgM to IgA. RORγt+ ILC further enhance IgA CSR by inducing nitric oxide secretion by extrafollicular iNOS-expressing DCs (also known as TipDCs) through a mechanism involving membrane-bound LTα1β2. Nitric oxide is thought to stimulate DC expression of BAFF and APRIL. In the intestinal LP, multiple subsets of conventional myeloid DCs contribute to IgA synthesis upon receiving TLR signals from bacteria. Intraluminal bacteria are sampled by CX3CR1+ DCs through transepithelial projections and thereafter can transfer antigen to CX3CR1− DCs with IgA-inducing function. These DCs include TipDCs, which release BAFF and APRIL, as well as TLR5+ DCs, which release IL-6 and RA. IgA-inducing cytokines such as BAFF, APRIL, RA, and TGF-β are also produced by IECs and help plasma cell survival and possibly differentiation in addition to IgA CSR and production. IECs may further enhance IgA responses by stimulating DC production of APRIL via TSLP. The differentiation and survival of plasma cells emerging from TI (or TD) pathways are also increased by TLR-activated SCs via RA, TGF-β, IL-6, and cyclooxygenase 2 and by $T_H17$ cells via IL-21 and IL-17. This last cytokine may also increase IgA transcytosis across IECs. In general, the IgA antibodies generated in ILFs and LP lack or have reduced SHM and are thought to function as a first-line defense against mucosal pathogens.

## 5.2 Immunoglobulin A Generation in Mesenteric Lymph Nodes and Peyer's Patches

IgA production can also occur in MLNs and PPs through a TI pathway involving plasmacytoid DCs (pDCs) [101]. By producing type I interferon in response to TLR signals from the microbiota, SCs stimulate pDC release of BAFF and APRIL, which thereafter trigger IgA CSR and production in B cells. FDCs may further enhance this TI IgA response by stimulating B cells through BAFF, APRIL, and TGF-β1. Indeed, imprinting signals from RA and the microbiota stimulate FDC release of BAFF and CXCL13 as well as FDC processing of TGF-β1 via MMP9 and MMP2 [74]. The expression of these TI IgA-inducing signals by FDCs might explain why mice lacking CD40 retain some IgA production in PPs [109,110].

## 5.3 Immunoglobulin A Generation in the Intestinal Lamina Propria

Mice that completely lack the GALT as a result of a deficiency of ILCs due to the lack of RORγt or Id2 transcription factors retain some intestinal IgA production [98,100]. This evidence indicates that some IgA can be generated also from the few B cells present in the LP. Along the same lines, mice lacking GCs due to a deletion of the OCA-B transcription factor retain some B cells expressing IgA and AID in the LP [111]. Similarly, AID can be detected in IgA+ B cells from the intestinal LP of either healthy individuals or immunodeficient patients with an insufficiency of CD40L–CD40 interaction [112]. Some of these AID-expressing LP B cells can be detected in subepithelial areas in close proximity to DCs-macrophages expressing CD11c [112], further supporting the existence of a primitive follicle-independent pathway for local IgA production [6]. This pathway is less efficient and effective compared to the one occurring in PPs and ILFs. However, it may help the protection of the mucosal barrier at an early phase of an infection [99], until more effective and specific IgA responses develop at canonical follicular inductive sites.

Despite these findings, the notion that some B cells undergo IgA class switching in the LP remains debated [113]. Indeed, some studies show that the LP contains neither AID nor germline Iα-Cα nor switch Iα-Cμ circle transcripts [110,114,115]. Other studies conclude that the LP lacks IgA class switching because, compared with PPs, the LP contains little or no AID, Iα-Cα, and/or Iα-Cμ transcripts [116,117]. One possible explanation for these discordant results is that LP B cells are more scattered and express fewer AID transcripts than PP B cells and thus IgA CSR can be easily underestimated or pass completely unrecognized in the LP, unless appropriate comparisons are established and sensitive methodologies [113].

In general, the possibility that the LP supports TI pathways for IgA production is in agreement with works showing that lower vertebrates such as trout have no GALT but mount antibody responses to gut microbes through an IgA-like mucosal antibody termed IgT [118]. This pathway involves antibody-secreting cells similar to B-1 cells, a mouse subset of innate-like CD5$^+$ B cells that express polyreactive BCRs and mainly reside in the peritoneal cavity [119]. Similar to splenic marginal zone B cells, B-1 cells rapidly generate low-affinity and poorly specific IgM and IgA antibodies in the absence of help from CD4$^+$ T cells [120]. B-1 cells may mount intestinal TI antibody responses after migrating from the peritoneal cavity to the intestinal LP through the omentum in response to microbial signals [121]. Disagreement over the involvement of B-1 cells in gut antibody production might result from phenotypic changes occurring in B-1 cells following their homing to the LP [122–124]. Alternatively, TI antibody responses may mostly involve conventional B-2 cells [100], perhaps including a unique subset of GALT-independent B cells recently identified in the LP [125].

### 5.3.1 Immunoglobulin A-Inducing Signals from Dendritic Cells in the Lamina Propria

Macrophage-DC subsets present in the LP and in proximity to ILFs would initiate B cell activation and IgA class switching after sampling antigen through transepithelial projections or transient antigen-challenging passages generated by mucus-secreting goblet cells [126,127]. After acquiring antigen, macrophage-DCs may interact with B cells positioned beneath the intestinal epithelium. Alternatively, LP B cells may acquire antigen from M cells located in proximity to ILFs [128]. DCs and M cells might convey antigen-containing immune complexes into nondegradative intracellular compartments by using FcγRIIB, DC-SIGN, and/or dectin-1 receptors [129,130]. Subsequent recycling of antigen-containing vesicles to the cell surface would allow DCs and M cells to make antigen available to B cells [131–133].

In addition to exposing antigen to B cells, DCs-macrophages release CD40L-related B cell-helper factors such as BAFF and APRIL, which engage transmembrane activator and calcium modulator and cyclophylin ligand interactor (TACI) on B cells [106,107,132,134,135]. TACI cooperates with TLR ligands, TGF-β1, and IL-10 to induce IgA class switching and plasma cell differentiation [74,100,136]. This pathway involves MyD88, an adaptor protein that functions downstream of both TLRs and TACI and stimulates intestinal B cell production of both IgA and IgM [136,137].

In mice, an additional CD103$^+$TLR5$^+$ DC subset highly sensitive to the bacterial TLR5 ligand flagellin may trigger IgA responses via RA and IL-6 [138]. In addition to inducing gut-homing receptors in B cells emerging from PP GCs, RA has been shown to enhance IgA CSR and production in B cells by cooperating with TLR ligands, BAFF, APRIL, and TGF-β [6,97,139].

Recently, an additional follicle-independent pathway for IgA production has been identified in the LP and would involve RORγt$^+$ ILCs expressing membrane-bound LTαβ2 [140]. This TNF family member may induce IgA production in B cells by triggering nitric oxide release by iNOS-expressing DCs [140]. In the presence of TLR signals, nitric oxide would enhance IgA responses by stimulating DC expression of BAFF and APRIL [108].

### 5.3.2 Immunoglobulin A-Inducing Signals from T Cells, Intestinal Epithelial Cells, and Stromal Cells in the Lamina Propria

Besides DCs, the gut LP harbors Foxp3$^+$ T cells that produce TGF-β1 and IL-10 as well as RORγt$^+$ Th17 cells that produce IL-17 [141,142]. When combined with BAFF, IL-17 enhances B cell survival, proliferation, and differentiation, at least in systemic B cells and under inflammatory conditions [143]. Whether this pathway is also operative in gut B cells remains unknown. Of note, Foxp3$^+$Treg cells and Th17 cells also express CD40L and their migration to the gut LP is under the control of LTα3 from RORγt$^+$ ILCs [140,144].

LP B cells would receive additional IgA-inducing signals from BAFF, APRIL, TGF-β1, RA, and IL-6 produced by IECs and/or SCs [112,145,146]. In particular, IECs produce very abundant amounts of BAFF and APRIL, which may cooperate with BAFF and APRIL from DCs to support IgA induction and plasma cell survival in the gut LP. In addition, IECs release thymic stromal lymphopoietin (TSLP), an IL-7-like cytokine that stimulates DC production of BAFF, APRIL, and IL-10 [112,147,148]. Together with DCs, SCs express cyclooxygenase-2 (COX-2), an enzyme that may further facilitate the generation of IgA-producing plasma cells in the LP either directly or indirectly through induction of Foxp3+ Treg cells [149,150](Fagarasan et al., unpublished data). Thus, the generation of IgA in the gut LP likely involves both TI and TD pathways governed by both immunological and stromal signals.

## 6. FUNCTION OF IMMUNOGLOBULIN A

SIgA impedes the attachment of intestinal antigens to intestinal IECs through a process known as "immune exclusion." In humans, orally delivered specific IgA protects against pathogens such as *Salmonella typhimurium* [151,152], *Vibrio cholera* [153], *Shigella flexneri* [154], and *Helicobacter pylori* [155]. Conversely, mice lacking SIgA have increased mucosal infections [156–160]. However, the precise mechanisms underlying immune exclusion are not completely understood. In vitro studies show that SIgA binds to bacterial antigens and thereafter prevents the interaction of bacteria with bacterial receptors on IECs [153]. In addition to impeding the adhesion of pathogenic bacteria to IECs, SIgA interferes with their virulence by reducing both motility and invasiveness [161]. For instance, subagglutinating concentrations of IgA specific for the O-antigen of lipopolysaccharide can reduce the motility of *S. typhimurium* and consequently affect its capacity to invade IECs [162].

IgA can provide additional protection after microbial invasion of IECs. By intercepting microbes that have invaded the epithelium, IgA can facilitate their release as nonvirulent immune complexes [163–165]. IgA can also capture microbes intruding in the LP and thereafter promote their expulsion through crypt IECs [166]. Moreover, IgA binding to bacteria can induce changes that affect the viability and/or pathogenicity of bacteria [21].

In addition to protecting against mucosal pathogens, SIgA also plays important roles in the control of the indigenous microbiota. AID-deficient mice, which lack IgA, have deregulated bacterial communities in the gut [85]. Indeed, an abnormal expansion of uncultured anaerobes such as segmented filamentous bacteria (spore-forming gram-positive bacteria belonging to the Firmicutes phylum) develops in all gut segments, but particularly in the small intestine of AID-deficient mice [85]. The aberrant expansion of a few Firmicutes species causes hyperactivation of both intestinal and systemic B cells [85]. Of note, reconstitution of AID-deficient mice with normal levels of IgA decreases the number of segmented filamentous bacteria and restores the normal composition of the gut microbiota [85,167].

Not only the quantity but also the quality of SIgA is important to control gut homeostasis. $AID^{G23S}$ is a mutant form of AID that has normal CSR but impaired SHM activity [168]. Thus, knock-in mice expressing $AID^{G23S}$ show limited diversity of the IgA repertoire due to a severe defect of SHM, but the concentration of serum and fecal IgG and IgA antibodies is equivalent to that of WT mice [168]. Nevertheless, $AID^{G23S}$ mice show signs of systemic immune activation, including GC B cell hyperplasia, as well as hyperresponsiveness to oral cholera toxin [168]. In addition, $AID^{G23S}$ mice have an abnormal expansion of Proteobacteria and skewed bacterial composition toward Firmicutes over Bacteroidetes [168], suggesting that the mutation and antigen-driven selection of SIgA is critical for the control of the gut microbiota.

As discussed earlier, PD-1 deficiency increases the number of PP $T_{FH}$ cells and impairs the selection of IgA-producing B cells in the GCs of PPs, leading to a reduction in fecal bacteria coated with IgA [82]. The microbiota composition is also altered, because PD-1-deficient mice have a marked reduction in the number of "healthy" short-chain fatty acid producing-bacteria, such as *Bifidobacterium* [82,169]. In contrast, *Enterobacteriaceae*, which are only marginally represented in the small intestine of healthy mice, are significantly increased in PD-1-deficient mice [82,169]. Similar to AID-deficient mice, PD-1-deficient mice develop hyperactivation of systemic lymphoid organs, including T and B cell hyperplasia and presence of serum IgG antibodies against commensal bacteria [82,169]. These alterations can be mitigated by treatment with antibiotics, which indicates that inhibitory signals from PD-1 are critical to achieve an adequate mucosal-systemic compartmentalization of bacteria via IgA [170]. Despite these advances, how IgA maintains a normal compartmentalization and composition of the microbiota remains unclear.

## 7. CLINICAL RELEVANCE

In addition to neutralizing mucosal pathogens and toxins, IgA modulates the interaction of commensal bacteria with the mucosal immune system to mitigate the overall inflammatory tone of the intestine [171]. Indeed, the majority of intestinal IgA recognizes commensal microbes and enteric pathogens in addition to self-antigens [172]. Therefore, it is not surprising that some patients with primary antibody disorders such as CVID and HIGM syndrome develop not only gastrointestinal infections but also inflammatory bowel disease and autoimmune disorders [173–176].

In humans, SIgAD causes an isolated defect of IgA, whereas CVID impairs IgA along with IgM and IgG production [177,178]. In the majority of patients, SIgAD often goes unnoticed, probably because overproduction of IgM and IgG compensates for the lack of IgA [179]. Yet, some SIgAD and CVID patients suffer from respiratory and gastrointestinal infections, respond poorly to vaccines, and develop lymphoproliferative disorders, autoimmunity, and allergy [173,175,179,180]. In addition, some CVID patients develop malabsorption, inflammatory bowel disease, celiac disease, and nodular lymphoid hyperplasia, a benign lymphoproliferative disorder that consists of multiple nodular lesions made up of lymphoid aggregates usually confined to the LP of the small intestine (SI) [176,181].

Nodular lymphoid hyperplasia probably originates from polyclonal activation of LP B cells by commensal bacteria undergoing aberrant expansion in the SI [182]. Accordingly, patients with SIgAD and CVID develop small bowel bacteria overgrowth syndrome, which leads to heterogeneous clinical manifestations associated with malabsorption [173–176]. The molecular basis of impaired mucosal IgA responses and gastrointestinal disorders in SIgAD and CVID remain largely unknown, but some patients have defective TACI signaling [183–186].

The intestinal nodules observed in patients with primary antibody deficiency resemble ILFs and may arise from the hyperactivation of mucosal follicular B cells by an altered microbiota. Indeed, similar nodules can be detected in the SI of mice with AID deficiency [85]. In these mice, the lack of AID impairs CSR and SHM and thus prevents the formation of high-affinity mucosal IgA. The ensuing inappropriate expansion of commensal anaerobic bacteria such as segmented filamentous bacteria causes hyperactivation of B cells in the SI [85]. Over time, this process may lead to the aberrant expansion of allergen-reactive, autoreactive, and clonal B cells, which could contribute to the increased frequency of allergy, autoimmunity (including celiac disease, hemolytic anemia, and immune thrombocytopenic purpura), and B cell tumors (mostly non-Hodgkin lymphoma) in patients with SIgAD, CVID, and HIGM.

Primary IgA deficiencies characterized by specific gene defects can also help to gain new insights into the regulation of gut IgA production. In HIGM syndrome caused by CD40 deficiency, follicles cannot form GCs, and yet the LP from both intestinal and upper respiratory tracts contains B cells that express IgA and some AID [107,112]. Similar observations can be made in human immunodeficiency virus-infected patients with acquired immunodeficiency syndrome who rapidly undergo a virtually complete loss of mucosal $CD4^+$ T cells [187]. Overall, these findings suggest that human B cells can produce some IgA through GC-independent and possibly TI pathways that do not require help to B cells by $CD4^+$ T cells.

## 8. CONCLUSIONS

The intestine is a major site of pathogen entry, contains trillions of commensal bacteria, and is exposed to large amounts of food antigens. To protect the integrity of the epithelial barrier and avoid potentially catastrophic inflammatory reactions, SIgA not only restricts the access of microbes and other antigens to the mucosal surface, but also modulates the sampling of antigens and the quality of the local immune response. Although better known for its ability to neutralize toxins and some pathogens, IgA also plays an important role in the selection and maintenance of a diverse and spatially diversified community of commensal bacteria. To achieve these functions, the intestine has developed multiple follicular and extrafollicular IgA-inducing strategies that follow both TD and TI pathways. The precise cellular and signaling components of these pathways and their relative contribution to mucosal immunity and homeostasis remain to be fully elucidated. Further studies are also needed to characterize how intestinal IgA discriminates commensals from pathogens and whether specific commensals are needed to optimize homeostatic IgA responses. This information may not only facilitate the development of novel vaccine strategies but also contribute to a better understanding of intestinal disorders, including inflammatory bowel diseases, food allergies, and metabolic diseases.

## REFERENCES

[1] Macpherson AJ, Harris NL. Interactions between commensal intestinal bacteria and the immune system. Nat Rev Immunol 2004;4:478–85.

[2] Sansonetti PJ. War and peace at mucosal surfaces. Nat Rev Immunol 2004;4:953–64.

[3] Brandtzaeg P, Baekkevold ES, Morton HC. From B to A the mucosal way. Nat Immunol 2001;2:1093–4.

[4] Fagarasan S. Intestinal IgA synthesis: a primitive form of adaptive immunity that regulates microbial communities in the gut. Curr Top Microbiol Immunol 2006;308:137–53.

[5] Mestecky J, Russell MW, Elson CO. Intestinal IgA: novel views on its function in the defence of the largest mucosal surface. Gut 1999;44:2–5.

[6] Fagarasan S, Kawamoto S, Kanagawa O, Suzuki K. Adaptive immune regulation in the gut: T cell-dependent and T cell-independent IgA synthesis. Annu Rev Immunol 2010;28:243–73.

[7] Cerutti A. The regulation of IgA class switching. Nat Rev Immunol 2008;8:421–34.

[8] Chaudhuri J, Alt FW. Class-switch recombination: interplay of transcription, DNA deamination and DNA repair. Nat Rev Immunol 2004;4:541–52.

[9] Muramatsu M, Kinoshita K, Fagarasan S, Yamada S, Shinkai Y, Honjo T. Class switch recombination and hypermutation require activation-induced cytidine deaminase (AID), a potential RNA editing enzyme. Cell 2000;102:553–63.

[10] Coffman RL, Lebman DA, Shrader B. Transforming growth factor β specifically enhances IgA production by lipopolysaccharide-stimulated murine B lymphocytes. J Exp Med 1989;170:1039–44.

[11] Borsutzky S, Cazac BB, Roes J, Guzman CA. TGF-beta receptor signaling is critical for mucosal IgA responses. J Immunol 2004;173:3305–9.

[12] Revy P, Muto T, Levy Y, Geissmann F, Plebani A, Sanal O, et al. Activation-induced cytidine deaminase (AID) deficiency causes the autosomal recessive form of the Hyper-IgM syndrome (HIGM2). Cell 2000;102:565–75.

[13] Chaudhuri J, Tian M, Khuong C, Chua K, Pinaud E, Alt FW. Transcription-targeted DNA deamination by the AID antibody diversification enzyme. Nature 2003;422:726–30.

[14] Stavnezer J, Guikema JE, Schrader CE. Mechanism and regulation of class switch recombination. Annu Rev Immunol 2008;26:261–92.

[15] Kinoshita K, Harigai M, Fagarasan S, Muramatsu M, Honjo T. A hallmark of active class switch recombination: transcripts directed by I promoters on looped-out circular DNAs. Proc Natl Acad Sci USA 2001;98:12620–3.

[16] Griffin CC, Linhardt RJ, Van Gorp CL, Toida T, Hileman RE, Schubert 2nd RL, et al. Isolation and characterization of heparan sulfate from crude porcine intestinal mucosal peptidoglycan heparin. Carbohydr Res 1995;276:183–97.

[17] Flanagan JG, Lefranc MP, Rabbitts TH. Mechanisms of divergence and convergence of the human immunoglobulin α1 and α2 constant region gene sequences. Cell 1984;36:681–8.

[18] Pabst O. New concepts in the generation and functions of IgA. Nat Rev Immunol 2012;12:821–32.

[19] Brandtzaeg P. Presence of J chain in human immunocytes containing various immunoglobulin classes. Nature 1974;252:418–20.

[20] Mostov KE, Simister NE. Transcytosis. Cell 1985;43:389–90.

[21] Corthesy B. Multi-faceted functions of secretory IgA at mucosal surfaces. Front Immunol 2013;4:185.

[22] Bonner A, Almogren A, Furtado PB, Kerr MA, Perkins SJ. The nonplanar secretory IgA2 and near planar secretory IgA1 solution structures rationalize their different mucosal immune responses. J Biol Chem 2009;284:5077–87.

[23] Duc M, Johansen FE, Corthesy B. Antigen binding to secretory immunoglobulin A results in decreased sensitivity to intestinal proteases and increased binding to cellular Fc receptors. J Biol Chem 2010;285:953–60.

[24] Phalipon A, Cardona A, Kraehenbuhl JP, Edelman L, Sansonetti PJ, Corthesy B. Secretory component: a new role in secretory IgA-mediated immune exclusion in vivo. Immunity 2002;17:107–15.

[25] Suzuki K, Kawamoto S, Maruya M, Fagarasan S. GALT: organization and dynamics leading to IgA synthesis. Adv Immunol 2010;107:153–85.

[26] Macpherson AJ, McCoy KD, Johansen FE, Brandtzaeg P. The immune geography of IgA induction and function. Mucosal Immunol 2008;1:11–22.

[27] Hamada H, Hiroi T, Nishiyama Y, Takahashi H, Masunaga Y, Hachimura S, et al. Identification of multiple isolated lymphoid follicles on the antimesenteric wall of the mouse small intestine. J Immunol 2002;168:57–64.

[28] Yoshida H, Honda K, Shinkura R, Adachi S, Nishikawa S, Maki K, et al. IL-7 receptor alpha+ CD3(-) cells in the embryonic intestine induces the organizing center of Peyer's patches. Int Immunol 1999;11:643–55.

[29] Eberl G, Littman DR. The role of the nuclear hormone receptor RORgammat in the development of lymph nodes and Peyer's patches. Immunol Rev 2003;195:81–90.

[30] Eberl G, Marmon S, Sunshine MJ, Rennert PD, Choi Y, Littman DR. An essential function for the nuclear receptor RORgamma(t) in the generation of fetal lymphoid tissue inducer cells. Nat Immunol 2004;5:64–73.

[31] Spits H, Artis D, Colonna M, Diefenbach A, Di Santo JP, et al. Innate lymphoid cells–a proposal for uniform nomenclature. Nat Rev Immunol 2013;13:145–9.

[32] van de Pavert SA, Mebius RE. New insights into the development of lymphoid tissues. Nat Rev Immunol 2010;10:664–74.

[33] Finke D, Meier D. Molecular networks orchestrating GALT development. Curr Top Microbiol Immunol 2006;308:19–57.

[34] Cornes JS. Number, size, and distribution of Peyer's patches in the human small intestine: Part I the development of Peyer's patches. Gut 1965;6:225–9.

[35] Liu YJ, Arpin C. Germinal center development. Immunol Rev 1997;156:111–26.

[36] Vinuesa CG, Linterman MA, Goodnow CC, Randall KL. T cells and follicular dendritic cells in germinal center B-cell formation and selection. Immunol Rev 2010;237:72–89.

[37] Victora GD, Nussenzweig MC. Germinal centers. Annu Rev Immunol 2012;30:429–57.

[38] Klein U, Dalla-Favera R. Germinal centres: role in B-cell physiology and malignancy. Nat Rev Immunol 2008;8:22–33.

[39] Muramatsu M, Sankaranand VS, Anant S, Sugai M, Kinoshita K, Davidson NO, et al. Specific expression of activation-induced cytidine deaminase (AID), a novel member of the RNA-editing deaminase family in germinal center B cells. J Biol Chem 1999; 274:18470–6.

[40] Foy TM, Laman JD, Ledbetter JA, Aruffo A, Claassen E, Noelle RJ. gp39-CD40 interactions are essential for germinal center formation and the development of B cell memory. J Exp Med 1994;180:157–63.

[41] Xu J, Foy TM, Laman JD, Elliott EA, Dunn JJ, Waldschmidt TJ, et al. Mice deficient for the CD40 ligand. Immunity 1994;1:423–31.

[42] Allen CD, Okada T, Cyster JG. Germinal-center organization and cellular dynamics. Immunity 2007;27:190–202.

[43] Benson MJ, Dillon SR, Castigli E, Geha RS, Xu S, Lam KP, et al. Cutting edge: the dependence of plasma cells and independence of memory B cells on BAFF and APRIL. J Immunol 2008;180:3655–9.

[44] Chan TD, Gatto D, Wood K, Camidge T, Basten A, Brink R. Antigen affinity controls rapid T-dependent antibody production by driving the expansion rather than the differentiation or extrafollicular migration of early plasmablasts. J Immunol 2009;183:3139–49.

[45] Paus D, Phan TG, Chan TD, Gardam S, Basten A, Brink R. Antigen recognition strength regulates the choice between extrafollicular plasma cell and germinal center B cell differentiation. J Exp Med 2006;203:1081–91.

[46] Gatto D, Paus D, Basten A, Mackay CR, Brink R. Guidance of B cells by the orphan G protein-coupled receptor EBI2 shapes humoral immune responses. Immunity 2009;31:259–69.

[47] Pereira JP, Kelly LM, Xu Y, Cyster JG. EBI2 mediates B cell segregation between the outer and centre follicle. Nature 2009;460:1122–6.

[48] Lee SC, Bottaro A, Chen L, Insel RA. Mad1 is a transcriptional repressor of Bcl-6. Mol Immunol 2006;43:1965–71.

[49] Green JA, Suzuki K, Cho B, Willison LD, Palmer D, Allen CD, et al. The sphingosine 1-phosphate receptor S1P(2) maintains the homeostasis of germinal center B cells and promotes niche confinement. Nat Immunol 2011;12:672–80.

[50] Martinoli C, Chiavelli A, Rescigno M. Entry route of Salmonella typhimurium directs the type of induced immune response. Immunity 2007;27:975–84.

[51] Johnston RJ, Poholek AC, DiToro D, Yusuf I, Eto D, Barnett B, et al. Bcl6 and Blimp-1 are reciprocal and antagonistic regulators of T follicular helper cell differentiation. Science 2009;325:1006–10.

[52] Nurieva RI, Chung Y, Martinez GJ, Yang XO, Tanaka S, Matskevitch TD, et al. Bcl6 mediates the development of T follicular helper cells. Science 2009;325:1001–5.

[53] Yu D, Rao S, Tsai LM, Lee SK, He Y, Sutcliffe EL, et al. The transcriptional repressor Bcl-6 directs T follicular helper cell lineage commitment. Immunity 2009;31:457–68.

[54] Crotty S. Follicular helper CD4 T cells (TFH). Annu Rev Immunol 2011;29:621–63.

[55] Ramiscal RR, Vinuesa CG. T-cell subsets in the germinal center. Immunol Rev 2013;252:146–55.

[56] Choi YS, Yang JA, Crotty S. Dynamic regulation of Bcl6 in follicular helper CD4 T (Tfh) cells. Curr Opin Immunol 2013;25:366–72.

[57] Vinuesa CG, Cyster JG. How T cells earn the follicular rite of passage. Immunity 2011;35:671–80.

[58] Forster R, Mattis AE, Kremmer E, Wolf E, Brem G, Lipp M. A putative chemokine receptor, BLR1, directs B cell migration to defined lymphoid organs and specific anatomic compartments of the spleen. Cell 1996;87:1037–47.

[59] Hardtke S, Ohl L, Forster R. Balanced expression of CXCR5 and CCR7 on follicular T helper cells determines their transient positioning to lymph node follicles and is essential for efficient B-cell help. Blood 2005;106:1924–31.

[60] Haynes NM, Allen CD, Lesley R, Ansel KM, Killeen N, Cyster JG. Role of CXCR5 and CCR7 in follicular Th cell positioning and appearance of a programmed cell death gene-1high germinal center-associated subpopulation. J Immunol 2007;179:5099–108.

[61] Liu X, Chen X, Zhong B, Wang A, Wang X, Chu F, et al. Transcription factor achaete-scute homologue 2 initiates follicular T-helper-cell development. Nature 2014;507:513–8.

[62] Liu X, Nurieva RI, Dong C. Transcriptional regulation of follicular T-helper (Tfh) cells. Immunol Rev 2013;252:139–45.

[63] Chung Y, Tanaka S, Chu F, Nurieva RI, Martinez GJ, Rawal S, et al. Follicular regulatory T cells expressing Foxp3 and Bcl-6 suppress germinal center reactions. Nat Med 2011;17:983–8.

[64] Linterman MA, Pierson W, Lee SK, Kallies A, Kawamoto S, Rayner TF, et al. Foxp3+ follicular regulatory T cells control the germinal center response. Nat Med 2011;17:975–82.

[65] Wollenberg I, Agua-Doce A, Hernandez A, Almeida C, Oliveira VG, Faro J, et al. Regulation of the germinal center reaction by Foxp3+ follicular regulatory T cells. J Immunol 2011;187:4553–60.

[66] Kato LM, Kawamoto S, Maruya M, Fagarasan S. Gut TFH and IgA: key players for regulation of bacterial communities and immune homeostasis. Immunol Cell Biol 2014;92:49–56.

[67] Kosco-Vilbois MH. Are follicular dendritic cells really good for nothing? Nat Rev Immunol 2003;3:764–9.

[68] Tew JG, Kosco MH, Burton GF, Szakal AK. Follicular dendritic cells as accessory cells. Immunol Rev 1990;117:185–211.

[69] Allen CD, Cyster JG. Follicular dendritic cell networks of primary follicles and germinal centers: phenotype and function. Semin Immunol 2008;20:14–25.

[70] Wang X, Cho B, Suzuki K, Xu Y, Green JA, An J, et al. Follicular dendritic cells help establish follicle identity and promote B cell retention in germinal centers. J Exp Med 2011;208:2497–510.

[71] Victoratos P, Lagnel J, Tzima S, Alimzhanov MB, Rajewsky K, Pasparakis M, et al. FDC-specific functions of p55TNFR and IKK2 in the development of FDC networks and of antibody responses. Immunity 2006;24:65–77.

[72] Kopf M, Herren S, Wiles MV, Pepys MB, Kosco-Vilbois MH. Interleukin 6 influences germinal center development and antibody production via a contribution of C3 complement component. J Exp Med 1998;188:1895–906.

[73] Nishikawa Y, Hikida M, Magari M, Kanayama N, Mori M, Kitamura H, et al. Establishment of lymphotoxin beta receptor signaling-dependent cell lines with follicular dendritic cell phenotypes from mouse lymph nodes. J Immunol 2006;177:5204–14.

[74] Suzuki K, Maruya M, Kawamoto S, Sitnik K, Kitamura H, Agace WW, et al. The sensing of environmental stimuli by follicular dendritic cells promotes immunoglobulin A generation in the gut. Immunity 2010;33:71–83.

[75] Wu Y, El Shikh ME, El Sayed RM, Best AM, Szakal AK, Tew JG. IL-6 produced by immune complex-activated follicular dendritic cells promotes germinal center reactions, IgG responses and somatic hypermutation. Int Immunol 2009;21:745–56.

[76] Vinuesa CG, Sanz I, Cook MC. Dysregulation of germinal centres in autoimmune disease. Nat Rev Immunol 2009;9:845–57.

[77] Bird JJ, Brown DR, Mullen AC, Moskowitz NH, Mahowald MA, Sider JR, et al. Helper T cell differentiation is controlled by the cell cycle. Immunity 1998;9:229–37.

[78] Lim HW, Hillsamer P, Banham AH, Kim CH. Cutting edge: direct suppression of B cells by CD4+ CD25+ regulatory T cells. J Immunol 2005;175:4180–3.

[79] Lim HW, Hillsamer P, Kim CH. Regulatory T cells can migrate to follicles upon T cell activation and suppress GC-Th cells and GC-Th cell-driven B cell responses. J Clin Invest 2004;114:1640–9.

[80] Okazaki T, Honjo T. PD-1 and PD-1 ligands: from discovery to clinical application. Int Immunol 2007;19:813–24.

[81] Okazaki T, Chikuma S, Iwai Y, Fagarasan S, Honjo T. A rheostat for immune responses: the unique properties of PD-1 and their advantages for clinical application. Nat Immunol 2013;14:1212–8.

[82] Kawamoto S, Tran TH, Maruya M, Suzuki K, Doi Y, Tsutsui Y, et al. The inhibitory receptor PD-1 regulates IgA selection and bacterial composition in the gut. Science 2012;336:485–9.

[83] Good-Jacobson KL, Szumilas CG, Chen L, Sharpe AH, Tomayko MM, Shlomchik MJ. PD-1 regulates germinal center B cell survival and the formation and affinity of long-lived plasma cells. Nat Immunol 2010;11:535–42.

[84] Suzuki K, Fagarasan S. How host-bacterial interactions lead to IgA synthesis in the gut. Trends Immunol 2008;29:523–31.

[85] Fagarasan S, Muramatsu M, Suzuki K, Nagaoka H, Hiai H, Honjo T. Critical roles of activation-induced cytidine deaminase in the homeostasis of gut flora. Science 2002;298:1424–7.

[86] Cerutti A, Rescigno M. The biology of intestinal immunoglobulin A responses. Immunity 2008;28:740–50.

[87] Cazac BB, Roes J. TGF-β receptor controls B cell responsiveness and induction of IgA in vivo. Immunity 2000;13:443–51.

[88] Avery DT, Bryant VL, Ma CS, de Waal Malefyt R, Tangye SG. IL-21-induced isotype switching to IgG and IgA by human naive B cells is differentially regulated by IL-4. J Immunol 2008;181:1767–79.

[89] Avery DT, Deenick EK, Ma CS, Suryani S, Simpson N, Chew GY, et al. B cell-intrinsic signaling through IL-21 receptor and STAT3 is required for establishing long-lived antibody responses in humans. J Exp Med 2010;207:155–71.

[90] Bryant VL, Ma CS, Avery DT, Li Y, Good KL, Corcoran LM, et al. Cytokine-mediated regulation of human B cell differentiation into Ig-secreting cells: predominant role of IL-21 produced by CXCR5+ T follicular helper cells. J Immunol 2007;179:8180–90.

[91] Borte S, Pan-Hammarstrom Q, Liu C, Sack U, Borte M, Wagner U, et al. Interleukin-21 restores immunoglobulin production ex vivo in patients with common variable immunodeficiency and selective IgA deficiency. Blood 2009;114:4089–98.

[92] Seo GY, Youn J, Kim PH. IL-21 ensures TGF-beta 1-induced IgA isotype expression in mouse Peyer's patches. J Leukoc Biol 2009;85:744–50.

[93] Hirota K, Turner JE, Villa M, Duarte JH, Demengeot J, Steinmetz OM, et al. Plasticity of Th17 cells in Peyer's patches is responsible for the induction of T cell-dependent IgA responses. Nat Immunol 2013;14:372–9.
[94] Dullaers M, Li D, Xue Y, Ni L, Gayet I, Morita R, et al. A T cell-dependent mechanism for the induction of human mucosal homing immunoglobulin A-secreting plasmablasts. Immunity 2009;30: 120–9.
[95] Mora JR, Iwata M, Eksteen B, Song SY, Junt T, Senman B, et al. Generation of gut-homing IgA-secreting B cells by intestinal dendritic cells. Science 2006;314:1157–60.
[96] Sato A, Hashiguchi M, Toda E, Iwasaki A, Hachimura S, Kaminogawa S. CD11b+ Peyer's patch dendritic cells secrete IL-6 and induce IgA secretion from naive B cells. J Immunol 2003;171: 3684–90.
[97] Tokuyama H, Tokuyama Y. The regulatory effects of all-trans-retinoic acid on isotype switching: retinoic acid induces IgA switch rearrangement in cooperation with IL-5 and inhibits IgG1 switching. Cell Immunol 1999;192:41–7.
[98] Uematsu S, Fujimoto K, Jang MH, Yang BG, Jung YJ, Nishiyama M, et al. Regulation of humoral and cellular gut immunity by lamina propria dendritic cells expressing Toll-like receptor 5. Nat Immunol 2008;9:769–76.
[99] Lopatin U, Blutt SE, Conner ME, Kelsall BL. Lymphotoxin alpha-deficient mice clear persistent rotavirus infection after local generation of mucosal IgA. J Virol 2013;87:524–30.
[100] Tsuji M, Suzuki K, Kitamura H, Maruya M, Kinoshita K, Ivanov II , et al. Requirement for lymphoid tissue-inducer cells in isolated follicle formation and T cell-independent immunoglobulin A generation in the gut. Immunity 2008;29:261–71.
[101] Tezuka H, Abe Y, Asano J, Sato T, Liu J, Iwata M, et al. Prominent role for plasmacytoid dendritic cells in mucosal T cell-independent IgA induction. Immunity 2011;34:247–57.
[102] Bouskra D, Brezillon C, Berard M, Werts C, Varona R, Boneca IG, et al. Lymphoid tissue genesis induced by commensals through NOD1 regulates intestinal homeostasis. Nature 2008;456:507–10.
[103] Pabst O, Herbrand H, Friedrichsen M, Velaga S, Dorsch M, Berhardt G, et al. Adaptation of solitary intestinal lymphoid tissue in response to microbiota and chemokine receptor CCR7 signaling. J Immunol 2006;177:6824–32.
[104] Eberl G, Littman DR. Thymic origin of intestinal alphabeta T cells revealed by fate mapping of RORgammat+ cells. Science 2004;305:248–51.
[105] Ivanov II, Diehl GE, Littman DR. Lymphoid tissue inducer cells in intestinal immunity. Curr Top Microbiol Immunol 2006;308:59–82.
[106] Litinskiy MB, Nardelli B, Hilbert DM, He B, Schaffer A, Casali P, et al. DCs induce CD40-independent immunoglobulin class switching through BLyS and APRIL. Nat Immunol 2002;3:822–9.
[107] Xu W, Santini PA, Matthews AJ, Chiu A, Plebani A, He B, et al. Viral double-stranded RNA triggers Ig class switching by activating upper respiratory mucosa B cells through an innate TLR3 pathway involving BAFF. J Immunol 2008;181:276–87.
[108] Tezuka H, Abe Y, Iwata M, Takeuchi H, Ishikawa H, Matsushita M, et al. Regulation of IgA production by naturally occurring TNF/iNOS-producing dendritic cells. Nature 2007;448:929–33.
[109] Bergqvist P, Gardby E, Stensson A, Bemark M, Lycke NY. Gut IgA class switch recombination in the absence of CD40 does not occur in the lamina propria and is independent of germinal centers. J Immunol 2006;177:7772–83.
[110] Bergqvist P, Stensson A, Lycke NY, Bemark M. T cell-independent IgA class switch recombination is restricted to the GALT and occurs prior to manifest germinal center formation. J Immunol 2010;184:3545–53.
[111] Crouch EE, Li Z, Takizawa M, Fichtner-Feigl S, Gourzi P, Montano C, et al. Regulation of AID expression in the immune response. J Exp Med 2007;204:1145–56.
[112] He B, Xu W, Santini PA, Polydorides AD, Chiu A, Estrella J, et al. Intestinal bacteria trigger T cell-independent immunoglobulin A2 class switching by inducing epithelial-cell secretion of the cytokine APRIL. Immunity 2007;26:812–26.
[113] Cerutti A. Location, location, location: B-cell differentiation in the gut lamina propria. Mucosal Immunol 2008;1:8–10.
[114] Shikina T, Hiroi T, Iwatani K, Jang MH, Fukuyama S, Tamura M, et al. IgA class switch occurs in the organized nasopharynx- and gut-associated lymphoid tissue, but not in the diffuse lamina propria of airways and gut. J Immunol 2004;172:6259–64.
[115] Barone F, Patel P, Sanderson JD, Spencer J. Gut-associated lymphoid tissue contains the molecular machinery to support T-cell-dependent and T-cell-independent class switch recombination. Mucosal Immunol 2009;2:495–503.
[116] Boursier L, Gordon JN, Thiagamoorthy S, Edgeworth JD, Spencer J. Human intestinal IgA response is generated in the organized gut-associated lymphoid tissue but not in the lamina propria. Gastroenterology 2005;128:1879–89.
[117] Lin M, Du L, Brandtzaeg P, Pan-Hammarstrom Q. IgA subclass switch recombination in human mucosal and systemic immune compartments. Mucosal Immunol 2013;7.
[118] Zhang YA, Salinas I, Li J, Parra D, Bjork S, Xu Z, et al. IgT, a primitive immunoglobulin class specialized in mucosal immunity. Nat Immunol 2010;11:827–35.
[119] Macpherson AJ, Gatto D, Sainsbury E, Harriman GR, Hengartner H, Zinkernagel RM. A primitive T cell-independent mechanism of intestinal mucosal IgA responses to commensal bacteria. Science 2000;288:2222–6.
[120] Suzuki K, Maruya M, Kawamoto S, Fagarasan S. Roles of B-1 and B-2 cells in innate and acquired IgA-mediated immunity. Immunol Rev 2010;237:180–90.
[121] Ha SA, Tsuji M, Suzuki K, Meek B, Yasuda N, Kaisho T, et al. Regulation of B1 cell migration by signals through Toll-like receptors. J Exp Med 2006;203:2541–50.
[122] Rosado MM, Aranburu A, Capolunghi F, Giorda E, Cascioli S, Cenci F, et al. From the fetal liver to spleen and gut: the highway to natural antibody. Mucosal Immunol 2009;2:351–61.
[123] Stoel M, Jiang HQ, van Diemen CC, Bun JC, Dammers PM, et al. Restricted IgA repertoire in both B-1 and B-2 cell-derived gut plasmablasts. J Immunol 2005;174:1046–54.
[124] Thurnheer MC, Zuercher AW, Cebra JJ, Bos NA. B1 cells contribute to serum IgM, but not to intestinal IgA, production in gnotobiotic Ig allotype chimeric mice. J Immunol 2003;170:4564–71.
[125] Shimomura Y, Ogawa A, Kawada M, Sugimoto K, Mizoguchi E, Shi HN, et al. A unique B2 B cell subset in the intestine. J Exp Med 2008;205:1343–55.
[126] McDole JR, Wheeler LW, McDonald KG, Wang B, Konjufca V, Knoop KA, et al. Goblet cells deliver luminal antigen to CD103+ dendritic cells in the small intestine. Nature 2012;483:345–9.
[127] Niess JH, Brand S, Gu X, Landsman L, Jung S, McCormick BA, et al. CX3CR1-mediated dendritic cell access to the intestinal lumen and bacterial clearance. Science 2005;307:254–8.

[128] Jang MH, Kweon MN, Iwatani K, Yamamoto M, Terahara K, Sasakawa C, et al. Intestinal villous M cells: an antigen entry site in the mucosal epithelium. Proc Natl Acad Sci USA 2004;101:6110–5.

[129] Rochereau N, Drocourt D, Perouzel E, Pavot V, Redelinghuys P, Brown GD, et al. Dectin-1 is essential for reverse transcytosis of glycosylated SIgA-antigen complexes by intestinal M cells. PLoS Biol 2013;11:e1001658.

[130] Baumann J, Park CG, Mantis NJ. Recognition of secretory IgA by DC-SIGN: implications for immune surveillance in the intestine. Immunol Lett 2010;131:59–66.

[131] Bergtold A, Desai DD, Gavhane A, Clynes R. Cell surface recycling of internalized antigen permits dendritic cell priming of B cells. Immunity 2005;23:503–14.

[132] Massacand JC, Kaiser P, Ernst B, Tardivel A, Burki K, Schneider P, et al. Intestinal bacteria condition dendritic cells to promote IgA production. PLoS One 2008;3:e2588.

[133] Pone EJ, Zan H, Zhang J, Al-Qahtani A, Xu Z, Casali P. Toll-like receptors and B-cell receptors synergize to induce immunoglobulin class-switch DNA recombination: relevance to microbial antibody responses. Crit Rev Immunol 2010;30:1–29.

[134] Nardelli B, Belvedere O, Roschke V, Moore PA, Olsen HS, Migone TS, et al. Synthesis and release of B-lymphocyte stimulator from myeloid cells. Blood 2001;97:198–204.

[135] Craxton A, Magaletti D, Ryan EJ, Clark EA. Macrophage- and dendritic cell–dependent regulation of human B-cell proliferation requires the TNF family ligand BAFF. Blood 2003;101:4464–71.

[136] He B, Santamaria R, Xu W, Cols M, Chen K, Puga I, et al. The transmembrane activator TACI triggers immunoglobulin class switching by activating B cells through the adaptor MyD88. Nat Immunol 2010;11:836–45.

[137] Kirkland D, Benson A, Mirpuri J, Pifer R, Hou B, DeFranco AL, et al. B cell-intrinsic MyD88 signaling prevents the lethal dissemination of commensal bacteria during colonic damage. Immunity 2012;36:228–38.

[138] Uematsu S, Jang MH, Chevrier N, Guo Z, Kumagai Y, Yamamoto M, et al. Detection of pathogenic intestinal bacteria by Toll-like receptor 5 on intestinal CD11c+ lamina propria cells. Nat Immunol 2006;7:868–74.

[139] Mora JR, von Andrian UH. Differentiation and homing of IgA-secreting cells. Mucosal Immunol 2008;1:96–109.

[140] Kruglov AA, Grivennikov SI, Kuprash DV, Winsauer C, Prepens S, Seleznik GM, et al. Nonredundant function of soluble LTalpha3 produced by innate lymphoid cells in intestinal homeostasis. Science 2013;342:1243–6.

[141] Gaboriau-Routhiau V, Rakotobe S, Lecuyer E, Mulder I, Lan A, Bridonneau C, et al. The key role of segmented filamentous bacteria in the coordinated maturation of gut helper T cell responses. Immunity 2009;31:677–89.

[142] Ivanov II, Atarashi K, Manel N, Brodie EL, Shima T, Karaoz U, et al. Induction of intestinal Th17 cells by segmented filamentous bacteria. Cell 2009;139:485–98.

[143] Doreau A, Belot A, Bastid J, Riche B, Trescol-Biemont MC, Ranchin B, et al. Interleukin 17 acts in synergy with B cell-activating factor to influence B cell biology and the pathophysiology of systemic lupus erythematosus. Nat Immunol 2009;10:778–85.

[144] Cong Y, Feng T, Fujihashi K, Schoeb TR, Elson CO. A dominant, coordinated T regulatory cell-IgA response to the intestinal microbiota. Proc Natl Acad Sci USA 2009;106:19256–61.

[145] Iliev ID, Mileti E, Matteoli G, Chieppa M, Rescigno M. Intestinal epithelial cells promote colitis-protective regulatory T-cell differentiation through dendritic cell conditioning. Mucosal Immunol 2009;2:340–50.

[146] Iliev ID, Spadoni I, Mileti E, Matteoli G, Sonzogni A, Sampietro GM, et al. Human intestinal epithelial cells promote the differentiation of tolerogenic dendritic cells. Gut 2009;58:1481–9.

[147] Xu W, He B, Chiu A, Chadburn A, Shan M, Buldys M, et al. Epithelial cells trigger frontline immunoglobulin class switching through a pathway regulated by the inhibitor SLPI. Nat Immunol 2007;8:294–303.

[148] Rimoldi M, Chieppa M, Salucci V, Avogadri F, Sonzogni A, Sampietro GM, et al. Intestinal immune homeostasis is regulated by the crosstalk between epithelial cells and dendritic cells. Nat Immunol 2005;6:507–14.

[149] Broere F, du Pre MF, van Berkel LA, Garssen J, et al. Cyclooxygenase-2 in mucosal DC mediates induction of regulatory T cells in the intestine through suppression of IL-4. Mucosal Immunol 2009;2:254–64.

[150] Newberry RD, Stenson WF, Lorenz RG. Cyclooxygenase-2-dependent arachidonic acid metabolites are essential modulators of the intestinal immune response to dietary antigen. Nat Med 1999;5:900–6.

[151] Iankov ID, Petrov DP, Mladenov IV, Haralambieva IH, Kalev OK, Balabanova MS, et al. Protective efficacy of IgA monoclonal antibodies to O and H antigens in a mouse model of intranasal challenge with *Salmonella enterica* serotype Enteritidis. Microbes Infect 2004;6:901–10.

[152] Michetti P, Mahan MJ, Slauch JM, Mekalanos JJ, Neutra MR. Monoclonal secretory immunoglobulin A protects mice against oral challenge with the invasive pathogen Salmonella typhimurium. Infect Immun 1992;60:1786–92.

[153] Apter FM, Michetti P, Winner 3rd LS, Mack JA, Mekalanos JJ, Neutra MR. Analysis of the roles of antilipopolysaccharide and anti-cholera toxin immunoglobulin A (IgA) antibodies in protection against Vibrio cholerae and cholera toxin by use of monoclonal IgA antibodies in vivo. Infect Immun 1993;61:5279–85.

[154] Phalipon A, Kaufmann M, Michetti P, Cavaillon JM, Huerre M, Sansonetti P, et al. Monoclonal immunoglobulin A antibody directed against serotype-specific epitope of Shigella flexneri lipopolysaccharide protects against murine experimental shigellosis. J Exp Med 1995;182:769–78.

[155] Blanchard TG, Czinn SJ, Maurer R, Thomas WD, Soman G, Nedrud JG. Urease-specific monoclonal antibodies prevent Helicobacter felis infection in mice. Infect Immun 1995;63:1394–9.

[156] Blutt SE, Miller AD, Salmon SL, Metzger DW, Conner ME. IgA is important for clearance and critical for protection from rotavirus infection. Mucosal Immunol 2012;5:712–9.

[157] Davids BJ, Palm JE, Housley MP, Smith JR, Andersen YS, Martin MG, et al. Polymeric immunoglobulin receptor in intestinal immune defense against the lumen-dwelling protozoan parasite Giardia. J Immunol 2006;177:6281–90.

[158] Lycke N, Erlandsson L, Ekman L, Schon K, Leanderson T. Lack of J chain inhibits the transport of gut IgA and abrogates the development of intestinal antitoxic protection. J Immunol 1999;163:913–9.

[159] Schwartz-Cornil I, Benureau Y, Greenberg H, Hendrickson BA, Cohen J. Heterologous protection induced by the inner capsid proteins of rotavirus requires transcytosis of mucosal immunoglobulins. J Virol 2002;76:8110–7.

[160] Sun K, Johansen FE, Eckmann L, Metzger DW. An important role for polymeric Ig receptor-mediated transport of IgA in protection against Streptococcus pneumoniae nasopharyngeal carriage. J Immunol 2004;173:4576–81.

[161] Cravioto A, Tello A, Villafan H, Ruiz J, del Vedovo S, Neeser JR. Inhibition of localized adhesion of enteropathogenic *Escherichia coli* to HEp-2 cells by immunoglobulin and oligosaccharide fractions of human colostrum and breast milk. J Infect Dis 1991;163:1247–55.

[162] Forbes SJ, Eschmann M, Mantis NJ. Inhibition of Salmonella enterica serovar typhimurium motility and entry into epithelial cells by a protective antilipopolysaccharide monoclonal immunoglobulin A antibody. Infect Immun 2008;76:4137–44.

[163] Corthesy B, Benureau Y, Perrier C, Fourgeux C, Parez N, Greenberg H, et al. Rotavirus anti-VP6 secretory immunoglobulin A contributes to protection via intracellular neutralization but not via immune exclusion. J Virol 2006;80:10692–9.

[164] Fujioka H, Emancipator SN, Aikawa M, Huang DS, Blatnik F, Karban T, et al. Immunocytochemical colocalization of specific immunoglobulin A with sendai virus protein in infected polarized epithelium. J Exp Med 1998;188:1223–9.

[165] Mazanec MB, Coudret CL, Fletcher DR. Intracellular neutralization of influenza virus by immunoglobulin A anti-hemagglutinin monoclonal antibodies. J Virol 1995;69:1339–43.

[166] Robinson JK, Blanchard TG, Levine AD, Emancipator SN, Lamm ME. A mucosal IgA-mediated excretory immune system in vivo. J Immunol 2001;166:3688–92.

[167] Suzuki K, Meek B, Doi Y, Muramatsu M, Chiba T, Honjo T, et al. Aberrant expansion of segmented filamentous bacteria in IgA-deficient gut. Proc Natl Acad Sci USA 2004;101:1981–6.

[168] Wei M, Shinkura R, Doi Y, Maruya M, Fagarasan S, Honjo T. Mice carrying a knock-in mutation of Aicda resulting in a defect in somatic hypermutation have impaired gut homeostasis and compromised mucosal defense. Nat Immunol 2011;12:264–70.

[169] Fukuda S, Toh H, Hase K, Oshima K, Nakanishi Y, Yoshimura K, et al. Bifidobacteria can protect from enteropathogenic infection through production of acetate. Nature 2011;469:543–7.

[170] Maruya M, Kawamoto S, Kato LM, Fagarasan S. Impaired selection of IgA and intestinal dysbiosis associated with PD-1-deficiency. Gut Microbes 2013;4:165–71.

[171] Peterson DA, McNulty NP, Guruge JL, Gordon JI. IgA response to symbiotic bacteria as a mediator of gut homeostasis. Cell Host Microbe 2007;2:328–39.

[172] Benckert J, Schmolka N, Kreschel C, Zoller MJ, Sturm A, Wiedenmann B, et al. The majority of intestinal IgA+ and IgG+ plasmablasts in the human gut are antigen-specific. J Clin Invest 2011;121:1946–55.

[173] Washington K, Stenzel TT, Buckley RH, Gottfried MR. Gastrointestinal pathology in patients with common variable immunodeficiency and X-linked agammaglobulinemia. Am J Surg Pathol 1996;20:1240–52.

[174] Jesus AA, Duarte AJ, Oliveira JB. Autoimmunity in hyper-IgM syndrome. J Clin Immunol 2008;28(Suppl 1):S62–66.

[175] Malamut G, Verkarre V, Suarez F, Viallard JF, Lascaux AS, Cosnes J, et al. The enteropathy associated with common variable immunodeficiency: the delineated frontiers with celiac disease. Am J Gastroenterol 2010;105:2262–75.

[176] Knight AK, Cunningham-Rundles C. Inflammatory and autoimmune complications of common variable immune deficiency. Autoimmun Rev 2006;5:156–9.

[177] de Villartay JP, Fischer A, Durandy A. The mechanisms of immune diversification and their disorders. Nat Rev Immunol 2003;3:962–72.

[178] Cunningham-Rundles C, Ponda PP. Molecular defects in T- and B-cell primary immunodeficiency diseases. Nat Rev Immunol 2005;5:880–92.

[179] Cunningham-Rundles C. Physiology of IgA and IgA deficiency. J Clin Immunol 2001;21:303–9.

[180] Agarwal S, Smereka P, Harpaz N, Cunningham-Rundles C, Mayer L. Characterization of immunologic defects in patients with common variable immunodeficiency (CVID) with intestinal disease. Inflamm Bowel Dis 2010;17:251–9.

[181] Bastlein C, Burlefinger R, Holzberg E, Voeth C, Garbrecht M, Ottenjann R. Common variable immunodeficiency syndrome and nodular lymphoid hyperplasia in the small intestine. Endoscopy 1988;20:272–5.

[182] Fagarasan S, Honjo T. Intestinal IgA synthesis: regulation of front-line body defences. Nat Rev Immunol 2003;3:63–72.

[183] Castigli E, Wilson S, Garibyan L, Rachid R, Bonilla F, Schneider L, et al. Reexamining the role of TACI coding variants in common variable immunodeficiency and selective IgA deficiency. Nat Genet 2007;39:430–1.

[184] Castigli E, Wilson SA, Garibyan L, Rachid R, Bonilla F, Schneider L, et al. TACI is mutant in common variable immunodeficiency and IgA deficiency. Nat Genet 2005;37:829–34.

[185] Salzer U, Chapel HM, Webster AD, Pan-Hammarstrom Q, Schmitt-Graeff A, Schlesier M, et al. Mutations in TNFRSF13B encoding TACI are associated with common variable immunodeficiency in humans. Nat Genet 2005;37:820–8.

[186] Zhang L, Radigan L, Salzer U, Behrens TW, Grimbacher B, Diaz G, et al. Transmembrane activator and calcium-modulating cyclophilin ligand interactor mutations in common variable immunodeficiency: clinical and immunologic outcomes in heterozygotes. J Allergy Clin Immunol 2007;120:1178–85.

[187] Xu W, Santini PA, Sullivan JS, He B, Shan M, Ball SC, et al. HIV-1 evades virus-specific IgG2 and IgA responses by targeting systemic and intestinal B cells via long-range intercellular conduits. Nat Immunol 2009;10:1008–17.

Chapter 17

# Gut Microbiota and Their Regulation

Frederick W. Alt and Duane R. Wesemann

*Howard Hughes Medical Institute, Program in Cellular and Molecular Medicine, Boston Children's Hospital, Boston, MA, USA; Department of Genetics, Harvard Medical School, Boston, MA, USA*

## 1. MICROBIOTA

Evolutionary pressures over millennia have resulted in complex interrelationships between animals and the microorganisms that inhabit them. Mutualistic microorganisms are particularly abundant in the gut lumen, where an estimated $10^{14}$ bacteria reside [1]. These microbes represent over 1000 different prokaryotic species belonging to a limited set of a dozen taxa and dominated by gram-positive anaerobes [2]. The term "microbiota" is defined as the collection of microorganisms in a distinct location, such as the human gut. The term "microbiome" generally refers to the collective genes contained within the microbiota, which in the case of the human gut includes some three million unique genes [3] that are mostly from bacteria of relatively few of the known bacterial phyla—the two most prominent being the Firmicutes and the Bacteroidetes [2–4]. Microbiota in general have been shown to be stable, resilient, and specific [5,6]. The microbiota that inhabit animal hosts are frequently described as commensal (one partner benefits, whereas the other is unaffected); however, as more light has been shed on animal/microbe interdependence, the term mutualist (both partners derive benefit) has been suggested to provide a more accurate description [2,7]. Sequencing efforts have produced large data sets revealing a diverse array of bacterial communities across various human body sites, diets, and populations and delineating the range of functional and structural configurations in healthy populations [3,4,8,9]. These studies show substantial variability in microbial community structure between individuals in terms of microbial taxa, but a remarkable similarity between individuals in terms of the metagenomic makeup of imputed metabolic functions [10].

The trillions of commensal bacteria that make the human body their home are required for optimal host physiology. Microorganisms provide defense functionalities to hosts by delivering signals that lead to optimal host immune system development in addition to crowding out potential pathogens [11]. Mutualistic bacteria have also been shown to provide vertebrate hosts with metabolic capabilities that enhance energy and nutrient uptake from the diet [2]. In addition, based on the strong associations of the specific intestinal microbiota with diseases such as allergy, inflammatory bowel disease, diabetes, and obesity [12], microbial composition is beginning to be appreciated as a key regulator of human health. In this regard, animal hosts derive benefit by exerting control over the composition of their mutualistic microbial partners [2].

Microbes have evolved species-specific physical interactions with the host that mediate stable bacterial colonization [13], and optimal host immune system development may require specific groups of commensal species. For example, despite containing similar phyla and microbe abundance, a human microbiota was not as effective as a mouse microbiota in providing the signals required for optimal mouse immune maturation as measured by T cell proliferation, dendritic cell numbers, and antimicrobial peptide expression [14]. Although the mechanisms underlying this effect remain to be worked out, these observations indicate that specific host species anticipate a specific microbiota and that optimal immune development does not occur without it. In this context, a fundamental goal in modern immunology is to understand the factors that contribute to the development and maintenance of the homeostatic balance between commensal microbes and the host immune system.

The mucosal immune system carries out the complex function of containing the enormous microbial load while limiting the potential harmful effects of chronic activation of an inflammatory response to commensal microbes that inhabit the gut lumen [15]. In this regard, there is a bidirectional dialogue between commensal microbes and the host immune system—commensals play a key role in shaping the host immune system, whereas the host immune system plays a role in shaping the ecological structure of the luminal microbe content [15]. Microflora in general, as well as specific species and families of bacteria, have been shown to shape mucosal T cell subsets. For example, isolated lymphoid follicles, normally present in the intestinal mucosa, do not develop in germ-free mice [15]. Additionally, colonization of mice with a specific bacterial species—namely, segmented filamentous bacteria (SFB)—induces the accumulation of

**Molecular Biology of B Cells. http://dx.doi.org/10.1016/B978-0-12-397933-9.00017-5**

proinflammatory $T_H17$ cells [16], which have been shown to play a role in host systemic inflammatory diseases [17]. In contrast to the SFB–$T_H17$ axis, colonization of mice with closteridial strains induces the expansion of anti-inflammatory, IL-10-secreting regulatory T cells ($T_{regs}$) in the gut lamina propria (LP) [18], which promotes an environment of immune quiescence and tolerance. Additionally, polysaccharide A of *Bacteroides fragilis* leads to T cell IL-10 production, which limits the expansion of proinflammatory $T_H17$ cells [19]—a property probably enabling this species to coexist with the host as a commensal [15]. Therefore, gut microbes impart both pro- and anti-inflammatory signals and shape the host T cell profile, but microbial effects on B cell development and function are also beginning to be understood.

## 2. MICROBES, PRIMARY Ig DIVERSIFICATION, AND EARLY LIFE B CELL SELECTION

A potential link between the microbiota and primary B cell development was implicated when it was first demonstrated that the bursa of Fabricius is key for Ig production in chickens [20]. The bursa of Fabricius is an out-pouching of the avian hindgut that involutes during puberty, and its removal early in life results in the absence of Ig and B cells [20,21]. Finding related gut-associated follicular structures only in prepubertal rabbits and sheep led to suggestions that there may be a mammalian equivalent of the avian bursa [21,22]. Early B cell development and primary Ig diversification in rabbits are particularly similar to those in chickens in that they both generate a very limited recombination-activating gene (RAG)-mediated V(D)J repertoire that is subsequently diversified through activation-induced cytidine deaminase (AID)-mediated activities in hindgut structures [23,24]. The observations that chickens and rabbits undergo substantial primary Ig diversification in gut-associated structures early in life raises the notion that the gut environment in the young may be beneficial in shaping nascent Ig repertoires. Although mechanisms differ between organisms, many vertebrates seem to also share a link between the gut, microbes, and early B cell selection events associated with windows of time early in life. Known examples of these associations are reviewed here.

In rabbits, V(D)J recombination generates a very limited repertoire due to heavily biased $V_H$ gene segment usage owing to preferential rearrangement of D-proximal $V_H$ gene segments [25]. Shortly after birth, B cells with this limited repertoire migrate to gut-associated lymphoid tissues—namely, the sacculus rotundus and the appendix—where they undergo proliferation and AID-mediated preimmune Ig repertoire diversification to develop a full preimmune repertoire in the first 1–2 months of postnatal life [26]. These hindgut structures subsequently undergo changes to resemble secondary lymphoid tissue in which conventional immune responses occur [27]. The role of microbe exposure in primary Ig diversification in newborn rabbits has been addressed in germ-free animals [28–31] as well as in model systems in which exposure to microbiota was blocked by the removal of the sacculus rotundus and microsurgical closure of the appendix at birth [31–33]. All of these models demonstrated that prevention of microbial exposure reduced gut-associated follicle development, B cell numbers, and peripheral Ig diversity in young rabbits.

Whereas the extent to which direct microbial encounters with surface Ig influence primary Ig diversification in these models remains an intriguing question, the microbial effects on early Ig specificity during primary rabbit B cell diversification were implicated in studies using mutant *Alicia* rabbits. *Alicia* rabbit B cells express IgM of the $V_Hn$ allotype almost exclusively at birth. However, during the first weeks of life, the $V_Hn$ representation diminishes in favor of increased $V_Ha$ allotype representation. When intestinal microbiota interactions were prevented by surgical sequestration of the sacculus rotundus and appendix from microbial exposure, the peripheral repertoire throughout the body remained largely of the $V_Hn$ allotype [34], indicating that the presence of microbes may select for Ig specificity during primary development in young rabbits. Based on the finding of conserved amino acids in $V_Ha$ framework regions positioned in areas that could react with a putative antigen, it was proposed that a microbial superantigen may be positively selecting for the $V_Ha$-expressing B cells in the young rabbit intestine [34]. In addition, more $V_Ha$ cells were found to proliferate and fewer to die compared to $V_Hn$ cells in the gut-associated lymphoid tissue [35], in agreement with a potential commensal microbe-dependent positive selection occurring while primary diversification is taking place in these specialized gut structures early in the rabbit's life.

Further insights into the potential role of microbiota in primary Ig diversification have been obtained from studies in chickens, in which primary Ig diversification occurs in the bursa after RAG-mediated generation of a very limited Ig repertoire. Cells committed to the B lineage with severely limited Ig repertoire migrate to the bursa of Fabricius, an out-pouching of epithelium connected to the distal intestine in birds required for B cell development [36–39]. The chicken IgH locus contains a single functional $V_H$ segment and several related $V_H$ pseudogenes [38,40]. Before the bird hatches, B-lineage cells with limited V(D)J repertoire migrate to the bursa primordium in a single wave during embryonic life [41,42]. There, these B-lineage cells undergo both expansion and diversification of their Ig genes via AID-mediated gene conversion utilizing pseudogene V segments as donor sequences [37,43].

Although chicken Ig diversification takes place in a gut-associated structure, it occurs to a substantial degree during embryonic development (before hatching), in contrast to that in rabbits, arguing against a role for microbiota at least in the initial phase of chicken Ig diversification. However, microbes appear to influence early repertoires soon after hatching. The role of Ig specificity in chicken B cell development and Ig diversification was addressed in a 1999 study utilizing a retroviral gene transfer model, which employed a truncated Igμ heavy chain (Tμ) that lacks the antigen-binding VDJ and C1 segments [44] (Figure 1). Tμ does not require the presence of light chains for its surface expression of B cell precursors and can be distinguished from endogenous IgM [44]. After induction of Tμ expression in chicken embryos, B cell expansion and primary Ig diversification occurred in $T\mu^+$ cells in the absence of endogenous IgM (Figure 1). Expression of the Tμ protein was able to support rapid cell division and Ig diversification via gene conversion during embryonic life, similar to that observed with endogenous IgM expression [45]. However, bursal cells expressing Tμ demonstrated substantially reduced rates of cell division and increased levels of apoptosis after hatching [46] (Figure 1). These studies indicated that the specificity of the Ig is dispensable for the early stages of bursal B-lineage cell diversification, but suggest a role for antigen encounter in avian B cell selection events shortly after hatching. Thus, sterile Ig diversification before hatching appears to produce the substrate that influences the early avian Ig repertoire shortly after hatching via environmental agents such as the microbiota. As primary antibody diversification continues to occur in the chicken bursa for a period of at least 6 weeks after birth, luminal microbes may have a substantial impact on the shaping of the primary Ig repertoires.

Sheep intestines harbor two distinct types of Peyer's patches that differ in their ontogeny, cell composition, and physiology [47]. Ileal Peyer's patches resemble the young rabbit sacculus rotundus and chick bursa in that they are made up of ~95% B cells that are essentially all $IgM^+$ and <0.5% $CD4^+$ T cells [48–50]. In contrast, jejunal Peyer's patches consist of ~40% $IgM^+$ B cells, 10–15% $CD4^+$

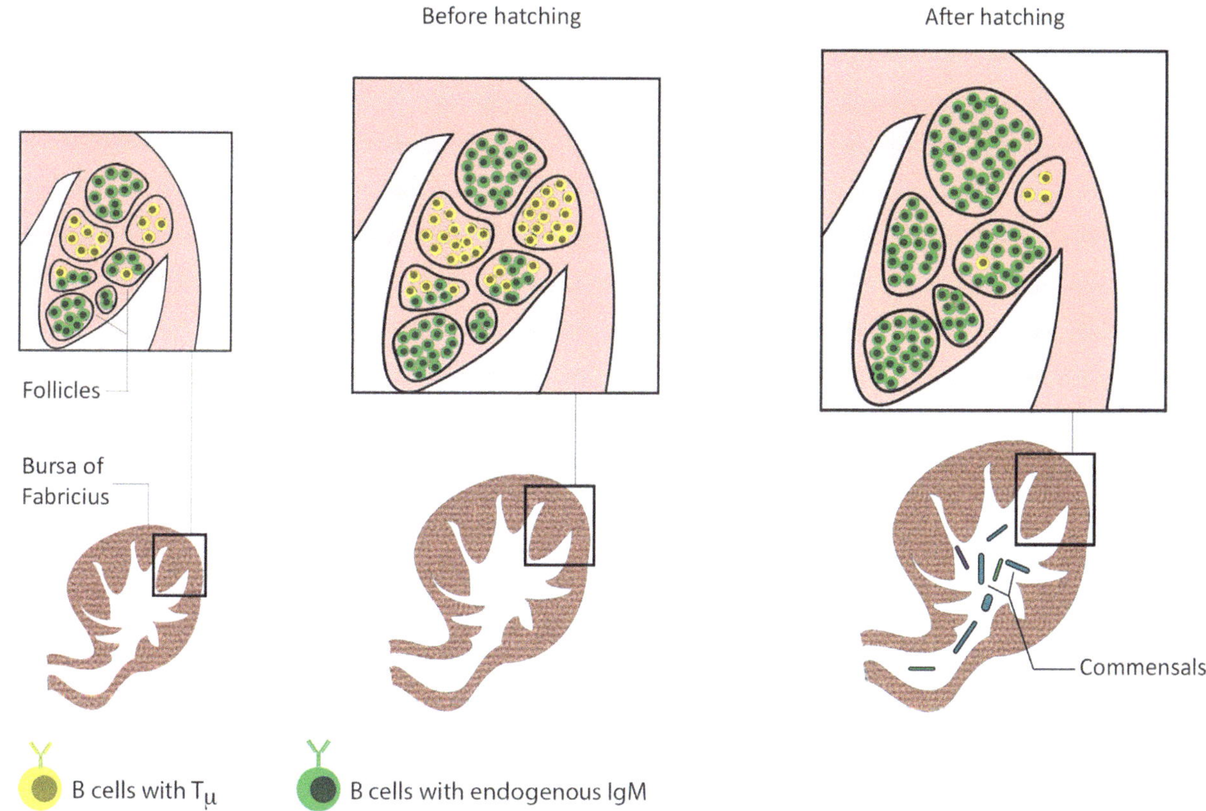


FIGURE 1 **Commensal microbes influence B cell development within bursal follicles in chicken.** In the bursal follicles of chicken, B cell precursors expand and diversify their immunoglobulins via gene conversion. Through this process, bursal follicles expand both before and after the time of hatching. Although the B cells do not encounter exogenous antigens prior to hatching, they do encounter commensal microbes present in the bursal lumen posthatching. The role of commensals in the development of bursal follicles becomes apparent in the presence of B cells that encode a truncated μ heavy chain (Tμ) instead of endogenous IgM. The Tμ molecules, generated by retroviral gene transfer, lack the $VDJ_H$ and $C\mu 1$ domains and thus lack antigen specificity. Despite the absence of endogenous IgM, B cell precursors that express Tμ are capable of exclusively colonizing follicles within the bursa prior to hatching. They also support gene conversion as evidenced by the $VJ_L$ from neonatal bursal cells expressing Tμ, which have been shown to undergo as much diversification as the bursal cells expressing surface IgM. However, after the chicken hatches, the B cells with Tμ protein cannot expand as rapidly as those that have an intact IgM and also undergo apoptosis at a higher rate, with the most marked apoptotic cells observed in the bursal follicles exclusively colonized in these cells. This fundamental difference in the need for VDJ-encoded determinants and an intact IgM molecule highlights the significance of commensal microbes in B cell development in the chicken bursa.

T cells, and 4–6% plasma cells and 35–40% isotype-switched B cells [48–50]. In addition, ileal Peyer's patches are present at birth and involute between 6 and 15 months of age, whereas jejunal Peyer's patches are present throughout the entire life [21,51]. These observations stimulated the hypothesis that jejunal Peyer's patches may be a significant site for the induction of mucosal immunity in the form of secondary lymphoid tissue, similar to Peyer's patches in rodents and primates, whereas ileal Peyer's patches in neonates and young lambs may serve a function more in line with primary Ig repertoire modulation, perhaps similar to young rabbits and chicks [52,53]. In this regard, the observation that only ~5% of B cells in ileal Peyer's patches survive, with the remainder dying in situ by apoptosis [51,54,55], indicates that a substantial selection process occurs there before involution early in life. The localization of the ileal Peyer's patches just above the ileal–cecal junction, where the highest concentrations of bacteria accumulate, suggests that luminal microbes probably influence this selection process.

The potential role of microbes in the development of the primary Ig diversification in sheep ileal Peyer's patches was addressed in experiments in which B cell somatic mutation of the V segment of the λ light chain was analyzed from sterile fragments of the ileal loops surgically sequestered from microbes during fetal life, as well as in germ-free sheep 6 and 8 weeks after birth [52,53]. The size of ileal Peyer's patches was decreased in the surgical ileal loops as well as in the germ-free sheep after birth, indicating that luminal microbes played a role in ileal follicle size. Earlier studies reported high levels of somatic mutation in the surgical loops and germ-free animals, suggesting that microbes may not be required to induce primary Ig diversification in sheep [52]; however, the discovery of more Vλ gene segments revealed that many of the unique nucleotide patterns initially thought to be a result of mutation in early studies were actually hard-coded germ-line variants [56], thus clarifying that somatic mutation occurs almost exclusively after birth [56], consistent with a potential role for microbes and/or dietary substances in this process. Ileal and jejunal Peyer's patches in piglets are structurally very similar to those of lambs and it has been shown that piglet ileal Peyer's patch B cell somatic mutation is dependent upon microbial exposure early in life as well [57].

Although the extent to which postnatal ileal Peyer's patch somatic mutation is physiologically and functionally distinct from that occurring in jejunal Peyer's patches remains to be fully resolved, the role of ileal versus jejunal Peyer's patches was examined in experiments in which antibody responses to a model antigen, namely, a vector expressing glycoprotein-D of bovine herpesvirus 1 (gD), were assessed after antigen challenge in sterile loops of jejunum versus ileum in young lambs. Anti-gD antibodies and B cell proliferation were strongly elicited in the jejunal Peyer's patches, but not those in the ileum [58]. However, gD-reactive antibodies were discovered in the spleens of ileum loop-injected sheep, and not in loops that did not contain follicles, suggesting that the gD antigen may have had a positive selection effect on emigrant lymphocytes from the ileal Peyer's patches [58], reminiscent of the putative microbe-dependent positive-selection activities that may occur in young rabbits [34]. Additionally, involution of ileal lymphoid follicles was delayed after gD antigen exposure, supporting the notion that B cells there sense and respond to the luminal environment.

Owing to distinct diversification and selection mechanisms, the hindgut B cell structures in hatchling birds, young rabbits, lambs, and piglets may not allow a simple categorization of the B cell activities occurring in these species as the same process [57]. However, there are three aspects that are common among gut B cell activities in the young of these species. First, some B cell developmental and selection events occurring in the gut are distinct from B cell activation and clonal expansion events typical of conventional inflammatory responses. Second, these selection events appear to be influenced by gut microbes. And finally, there appears to be a window of time early in life in which these selection/diversification events take place in each of these species. Together these aspects raise the notion that the gut microbiota may be of some value to shaping the burgeoning Ig repertoire early in life, but the nature of this benefit, and the degree to which this phenomenon extends to other species, is not fully understood. In this regard, mice and humans do not appear to have gut-associated B cell follicles that involute a few months after birth, as are found in the avian bursa, rabbit sacculus rotundus, or artiodactyl ileal Peyer's patches. However, the human fetal small intestinal LP was shown to harbor pre-B cells with readily detectable V pre-B expression [59]. As RAG-dependent V(D)J recombination is essentially the sole driver of the primary human Ig repertoire, the finding of pre-B cells in the human fetus raises the question whether early B cell development may serve a purpose similar to that of the gut-resident primary Ig diversification in chickens and rabbits. The extent to which early B cell development and primary Ig diversification occur in the intestines of human infants and children has not been addressed. However, recent studies have addressed this in mice.

Similar to humans, RAG-mediated V(D)J recombination during early B cell development is the major driver of preimmune diversification in mice. Consistent with the notion that the gut may provide some benefit to the process of early B cell selection and preimmune Ig diversification early in life, RAG-expressing early developing B-lineage cells, undergoing active V(D)J recombination, have been observed in the small intestinal LP of weanling mice [60]. Based on studies using a mouse model that contains a functional RAG2–green fluorescent protein fusion gene within the endogenous

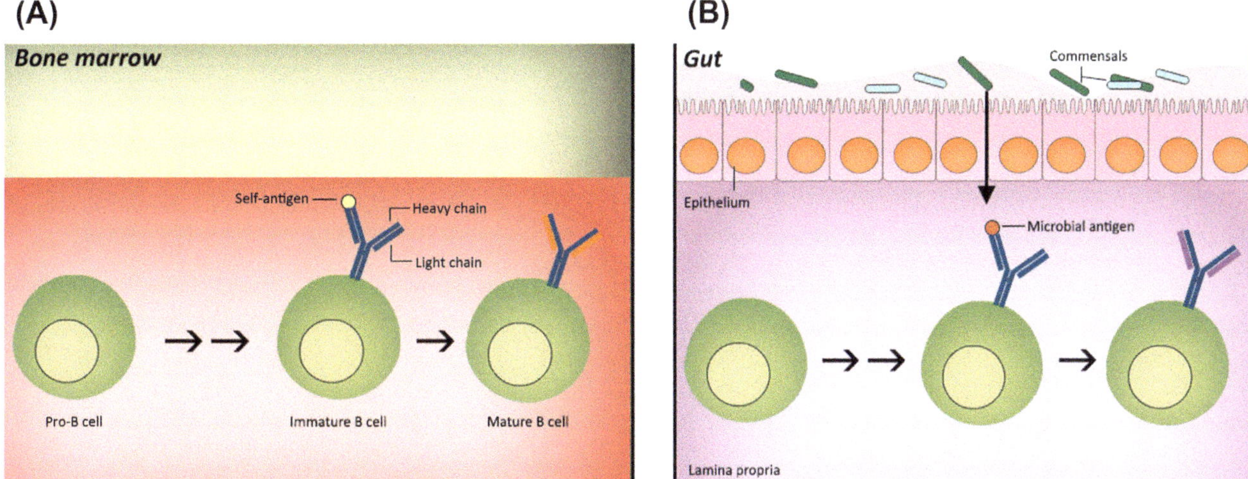


FIGURE 2 **B cell receptor editing in the small intestine.** (A) B-cell development occurs in the bone marrow and is regulated by interactions with self-antigens. As a pre-B cell develops into an immature B cell, it starts to express a B cell receptor, comprising immunoglobulin heavy chains and immunoglobulin light (IgL) chains, on its surface. Cells expressing IgM molecules can continue to express RAG proteins and undergo continued assembly of IgL chain genes to replace the previously assembled IgL through receptor editing. (B) B cell receptor editing also occurs in the lamina propria of the gut, where microbes may influence the preimmune Ig repertoire. *(Adapted from Ref. [66]).*

RAG2 locus, RAG-expressing $CD19^{+}B220^{low}$ cells were shown to make up ~4% of total $CD19^{+}$ cells at weaning age (18–24 days) despite being essentially undetectable in the first week after birth. RAG-expressing B cells then decreased to background levels by postnatal day 35. The coincident timing of early gut B cell accumulation with weaning age implied a role for the gut microbiota in this process given that gut microbes expand markedly as mice are weaned off of IgA found in maternal milk [61]. In addition, the amount of $RAG2^{+}$ cells in the small intestinal LP increased when germ-free mice were colonized with microbes by cohousing with conventionally raised mice [60].

In addition to affecting the amount of early B cell progenitors in the gut mucosa, colonization of germ-free mice led to an increase in the Igλ/Igκ ratio specifically in the LP of the small intestine, and not in any other tissues such as the spleen or bone marrow [60]. As increased Igλ usage in the B cell repertoire has been used to as a marker for increased receptor editing [62–64], these observations are consistent with the notion that commensal microflora influence the Ig repertoire by affecting BCR editing of immature gut B cells. In this regard, B-lineage cells with phenotypic characteristics of editing cells—namely, $RAG2^{+}B220^{low}IgM^{low}$—were observed in the small intestinal LP of weanling mice, and comparison of the emerging Ig repertoire in $RAG2^{+}$ gut LP cells with that in $RAG2^{+}$ cells from the bone marrow revealed significant differences in Vκ gene segment usage despite similar $V_H$ usage, which would be expected if LP B cells were undergoing a selection process like BCR editing owing to the distinct antigenic environment in the gut LP.

As B cell receptor editing is a mechanism of antigen-mediated shaping of primary Ig repertoires, these findings are consistent with a model suggesting that luminal contents may play a role in shaping preimmune repertoires during a window of opportunity early in life [65] (Figure 2). Although mechanistically distinct, early B cell development and microbe-dependent primary Ig repertoire modification in weanling mice may be functionally related to the process of gut-resident B cell receptor selection/diversification that occurs in young chicks, rabbits, piglets, and lambs. The mechanism underlying these effects as well as the physiological relevance of the early life timing and gut locale are questions that await further elucidation.

## 3. MICROBIAL INFLUENCE ON IgA PRODUCTION

IgA is the largest IgH isotype produced by the body as a whole, representing over three-quarters of the IgH produced [67,68]. The gut mucosal system is home to the largest activated B cell pool in the body, containing at least 80% of all plasmablasts and plasma cells, mostly producing dimers of IgA [69]. IgA production has been shown to be dependent upon the presence of microbes, as germ-free animals have extremely low levels of IgA [70–72]. In addition, most intestinal IgA is directed against intestinal flora [73].

Although mechanistic details underlying how microbes influence IgA production have yet to be fully uncovered, unique advances have been made in studies using a mutant *Escherichia coli* strain (HA107) engineered to allow reversible colonization of germ-free mice. The HA107 bacteria are alive, but unable to divide, and become undetectable

72 h after installation into the mouse gut by gavage, thus enabling the animals to receive known quantities of live organisms, become germ-free again, and be rechallenged [74]. Gut challenge with differing amounts of HA107 revealed that the intestinal barrier appears to sample only a tiny fraction of the live luminal bacteria, as up to 100 million ($10^8$) colony-forming units (CFU) of HA107 bacteria produced no measurable IgA responses, and greater amounts ($10^9$ and $10^{10}$ CFU) provided comparable levels of specific IgA responses [74]. Although highly specific, the IgA response against HA107 demonstrated an additive effect in the response to sequential bacterial exposures, as opposed to a synergistic effect seen with classical systemic immune memory responses [74]. In addition, although specific IgA responses can last over 16 weeks after initial challenge with HA107, exposure of HA107-treated, germ-free mice to other commensal bacterial species resulted in rapid abrogation of HA107-specific IgA and generation of an IgA response that matched the existing commensal content [74], thus indicating that IgA responses can continually evolve to react against the most prominent luminal species.

The extent to which IgA responses to *E. coli*/HA107 colonization reflect IgA responses to bacteria in general remains to be elucidated, as different bacterial species are known to have varying effects on mucosal immune responses. In particular, colonization with SFB has been shown to be exceptionally effective at producing large amounts of IgA [75]. Notably, stimulation with these bacteria is also associated with the induction of $T_H17$ cells [16]. With the use of IL-17 fate reporter mice, it was shown that $T_H17$ cells that expand and home to the small intestine in response to SFB colonization migrate into the Peyer's patches and take on a T follicular helper phenotype to assist B cells in undergoing activation toward IgA-producing cells [76]. In addition, mice that lacked $T_H17$ cells were deficient in their ability to mount an antigen-specific IgA response when immunized with cholera toxin. Although the mechanism of how SFB specifically leads to $T_H17$ accumulation in the gut is not fully understood, these studies provide a link between how SFB is connected to T-cell-dependent IgA production (Figure 3).

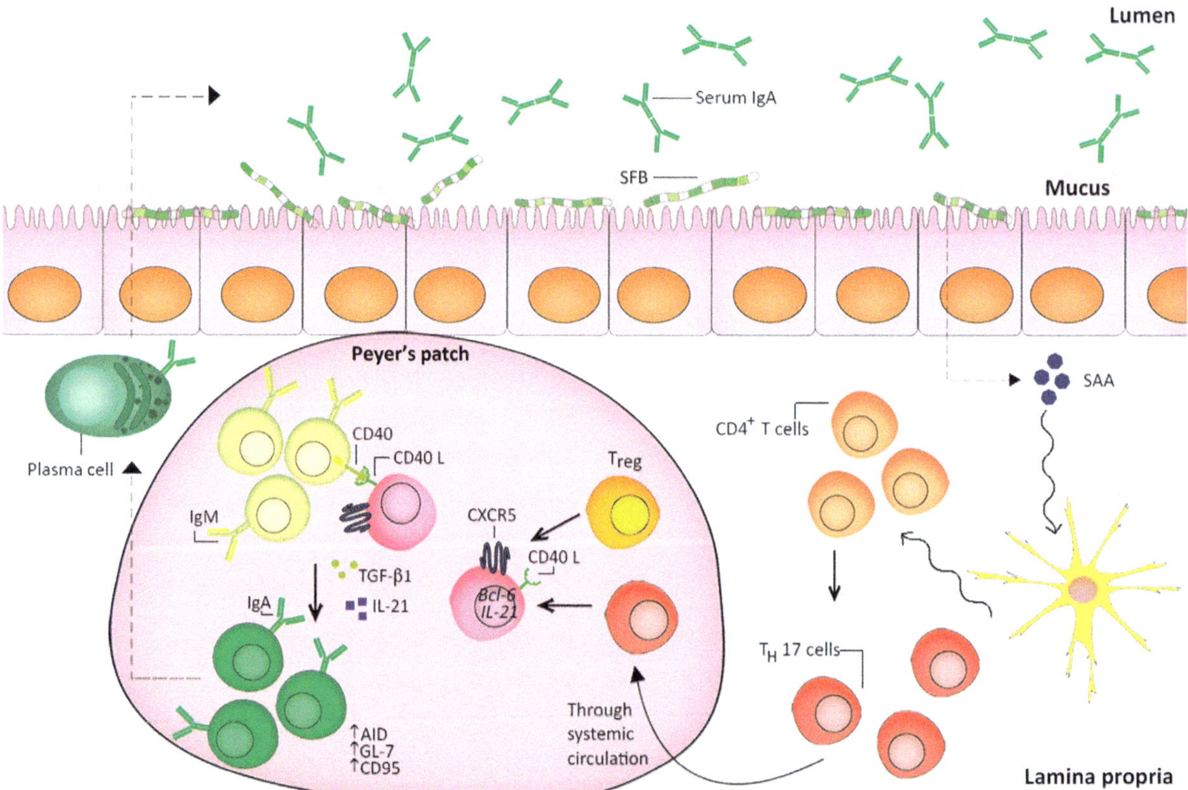


FIGURE 3 **The role of specific commensals in shaping the immune system.** Segmented filamentous bacteria (SFB) are commensals that adhere tightly to the epithelium in the ileum and are known to correlate with lower abundance of intestinal pathogens. SFB induce the formation of serum amyloid A (SAA) in the ileum, which in turn stimulates dendritic cells to induce differentiation of $T_H17$ cells in the lamina propria. When $T_H17$ cells are adoptively transferred into mice that lack them, they preferentially home to the gut lamina propria and Peyer's patches. $T_H17$ cells are important for B-cell-derived immunity because in Peyer's patches, $T_H17$ cells differentiate into T follicular helper ($T_{FH}$) cells. $T_{FH}$ cells express CXCR5 and CD40L, which allows them to position themselves into B cell niches in the germinal center and interact with B cells and induce them to express AID and germinal center markers such as GL-7 and CD95. This in turn is necessary in B cells for class-switch recombination into IgA.

## 4. MICROBIAL INFLUENCE ON IgE PRODUCTION

Although microbes are required for induction of IgA, it seems that the opposite is true for IgE, which is elevated at baseline for unknown reasons in germ-free mice and decreases upon colonization with commensal microbes [77,78]. Germ-free mice begin to produce IgE shortly after weaning age to levels 1000–10,000 times those of conventionally raised mice by 64 days of life and maintain this level unless colonized with microbes within the first week of life [77]. Notably, colonization with one or two species was not sufficient to protect from hyper-IgE despite gavage with high amounts of a single strain of bacteria (multiple rounds of $10^9$–$10^{10}$ CFU), but a more diverse set of microbes representing 7–40 different phyla was required to provide a robust protection against hyper-IgE seen at levels observed in germ-free mice [77]. Based on experiments measuring sterile Iε germ-line transcripts, it was shown that IgH class-switch recombination was potentially occurring at mucosal sites and not in the spleen or lymph nodes. Additionally, the elevated IgE in germ-free mice required $CD4^+$ T cells, IL-4, and the presence of organized lymphoid mucosal tissues, but was not influenced by or dependent upon dietary antigens [77]. Microbe-dependent depression of IgE levels was abrogated when MyD88—a key mediator of innate recognition of microbes—was conditionally deleted in B-lineage cells, thus indicating that direct sensing of microbial products by B cells is required in the process of microbial-mediated abrogation of IgE [78].

Although the potential physiologic rationale underlying elevated IgE in mice lacking signals from a diverse microbiota remains a mystery, germ-free-dependent hyper-IgE status preconditions animals for enhanced propensity for $T_H2$ inflammation [77,78] (Figure 4). In this regard, mice with reduced commensal microbial exposure either through antibiotic treatment or through germ-free conditions have an IgE-dependent increase in basal levels of circulating basophils and $T_H2$-activated T cells and an increased density of FcεRI-bound IgE on basophils [78]. Elevated IgE was also noted to increase the maturation of basophil precursors by

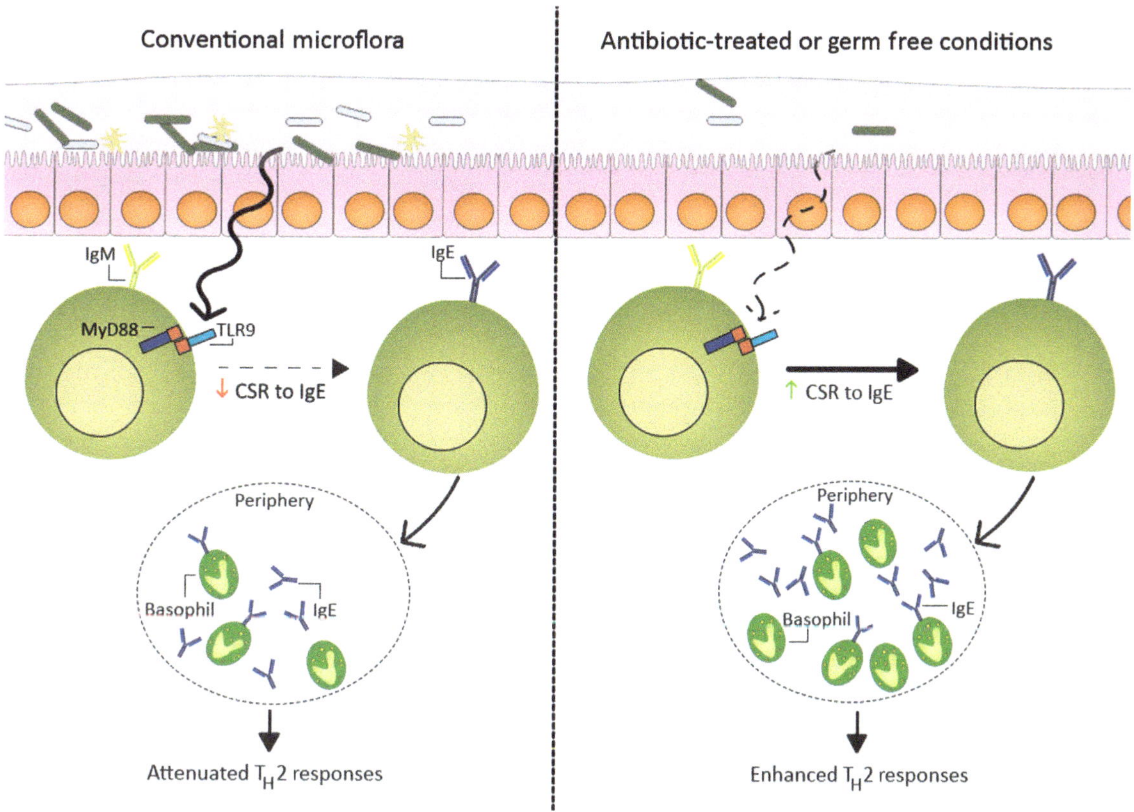


FIGURE 4 **Microbes regulate IgE.** Commensal microbes in the gut help maintain B cell homeostasis by preventing excess IgE production. Microbial products such as CpG DNA moderate class-switch recombination (CSR) to IgE through TLR–MyD88 signaling pathways, thus reducing accumulation of IgE in the lymphatic system. When mice are kept under germ-free conditions or are treated with antibiotics to deplete the microbial population in the gut, the lack of a strong MyD88 signal leads to greater frequency of CSR to IgE, which is manifested as a higher IgE concentration in the serum. The increase in IgE proportion correlates with an expansion of the circulating basophil population, a phenotype also observed in humans with hyperimmunoglobulinemia E syndrome. It was found that commensals help maintain a steady basophil number in the circulation by preventing proliferation of basophil precursors in the bone marrow through an IgE-mediated mechanism. Because an increased amount of circulating basophils is characteristic of enhanced $T_H2$ cytokine-dependent reactions, this finding provides evidence for a mechanism by which commensals influence allergic responses.

increasing their IL-3 responsiveness [78]. Consistent with previous reports that mast cell homeostasis is regulated by IgE levels [79], elevated IgE due to germ-free status is also associated with increased amounts of surface-bound IgE on mast cells and exaggerated oral-induced systemic anaphylaxis [77]. Epidemiologic studies have identified associations between alterations in the composition of communities of commensal bacteria and allergic disease [80–82]. Key questions remain regarding why IgE seems to be the default pathway in the absence of a diversity of microbes and what is the extent to which similar mechanisms are at work in affecting susceptibility to human allergic disease.

## 5. B-LINEAGE CELL INFLUENCE ON COMMENSAL MICROBES

Although a full understanding of the physiologic roles of B cells and Ig production in the gut is incomplete, it is clear that B cells exert a substantial effect in regulating host:microbe homeostasis through mucosal IgA, which has been shown to carry out varied functions. In addition to restricting bacterial access to the epithelium [73,83,84], IgA can influence the expression of microbial molecules [85] and has been shown to promote the survival of specific bacteria [86]. IgA has also been shown to regulate the metabolic and defense responses of intestinal epithelial cells in a trialogue between B-cell-produced IgA, intestinal epithelium, and gut luminal microbes [87].

By way of its unique ability to influence microbial communities in the gut lumen, IgA plays an important role in allowing host metabolic activities to be carried out by the intestinal epithelium. In this regard, intestinal epithelial cells from B-cell-deficient mice upregulate genes related to defense, inflammatory, and interferon-inducible responses, whereas genes involved in metabolic processes such as oxidation and reduction reactions associated with energy generation and steroid and cholesterol metabolism become significantly downregulated [87]. The ~100-fold elevated levels of IgM naturally found in IgA-deficient mice do not seem to protect the intestinal epithelium from these changes in transcriptional profiles [87]. Gene network analysis has revealed two interacting gene systems within intestinal epithelial cells, one governing lipid metabolism and another governing innate immunity (Figure 5). These two epithelial functions appear to be coordinately regulated and inversely connected via a small number of genes, with the main ones being Gbp6, an interferon-inducible gene suggested to have antibacterial function [88], and Gata4, which regulates metabolic functions in intestinal epithelium.

Layers of immunologic redundancy exist to keep a distance between the intestinal mucosal barrier and commensals,

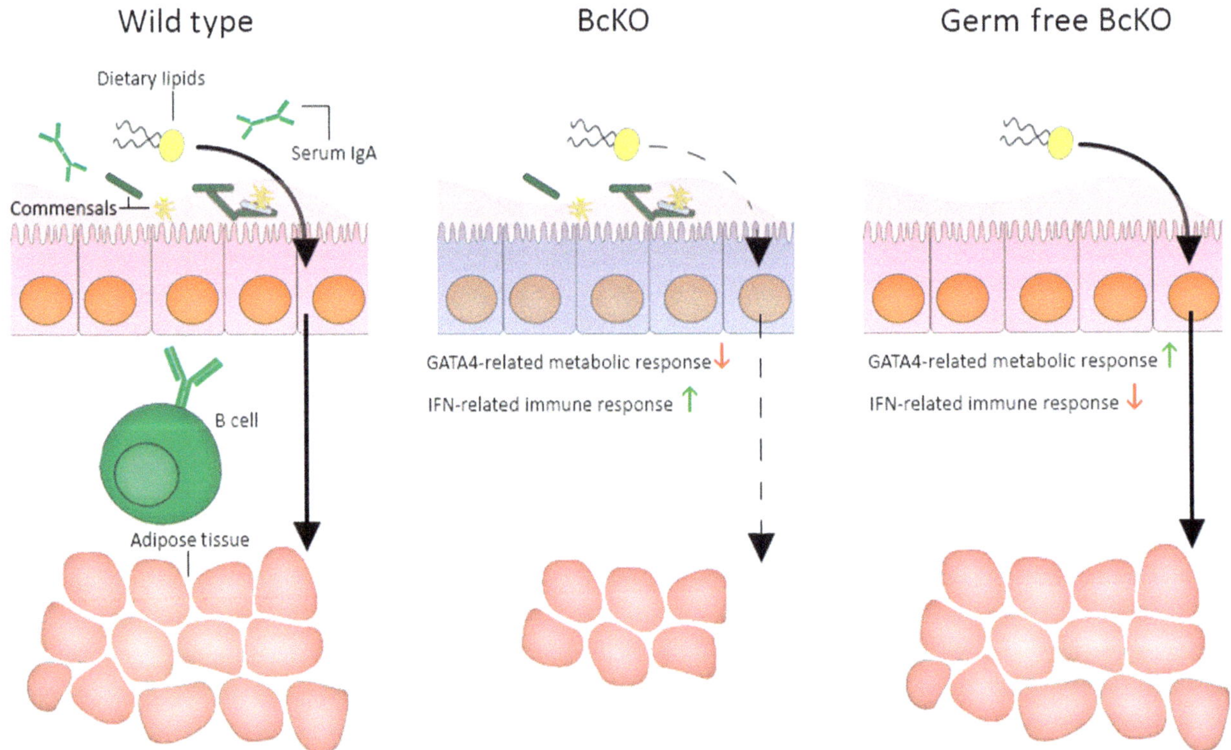


FIGURE 5 **Microbes and IgA govern the balance between the metabolic and the immune functions of the intestinal epithelium.** In the absence of IgA, the gut epithelium compensates for the lack of B-cell-derived immunity from the microbiota by turning on its own immune functions via upregulation of genes responsible for inflammatory and interferon (IFN)-inducible responses. However, it does so at the expense of its routine metabolic functions, most of which appear to be related to the transcription factor GATA4. One of the main effects of impaired GATA4 function was poor lipid and cholesterol absorption and storage. This functional switch in the intestinal epithelium is not seen in the absence of microbes in germ-free mice.

mutualists, and pathogens alike, but the recruitment of secondary layers of defense does not come without a cost. In the example above, upon failure of the B cell–IgA system, epithelial defense mechanisms sacrifice metabolic functions in favor of defense. In this regard, observations that the lack of IgA is responsible for an inability to properly process nutrients from the diet in mice are clinically relevant for diseases of immune deficiency in which Ig production is defective, such as common variable immune deficiency (CVID). As the intestinal epithelium may activate defense programs at the expense of metabolic functions in the absence of IgA, the immune system may essentially function as a shield to allow gut barrier tissue to perform metabolic functions. Biopsies from human small-intestine CVID patients have increased sets of interferon-induced genes and a decrease in groups of genes involved in lipid and carbohydrate metabolism and micronutrient transport, in agreement with the findings in mice [87]. Consistent with a role for Ig in allowing for optimal metabolic functions, patients with antibody deficiency syndromes such as CVID indeed have problems with malabsorption and low weight gain [89–91]. As current therapeutic Ig replacement protocols for CVID provide only systemic IgG, these findings suggest that there may be some rationale for developing a gut luminal IgA replacement protocol or microbial control treatment for these patients [87].

Studies using mice lacking IgA in the setting of AID deficiency have played a key role in elucidating the functions of IgA and somatic hypermutation in the maintenance of commensal bacterial homeostasis [92–94]. AID-deficient mice have an accumulation of intestinal IgM plasma cells and a microbe-dependent enlargement of intestinal follicular structures [92]. Gut luminal microbes mediate the intestinal follicular hyperplasia, as antibiotic treatments normalize the phenotype. In addition, IgA plays a role in shaping the microbial ecology in the intestinal lumen, as AID-deficient mice appear to harbor a large increase in small intestinal anaerobic bacteria [92,93]. Notably, the microbial community-shaping function of IgA seems to be restricted to the small intestine.

Because AID is required for both IgH class-switch recombination (CSR) and somatic hypermutation, the contributory roles of these two individual processes cannot be determined in AID-deficient mice. These two pathways were dissected in mice carrying a knock-in mutation of AID$^{G23S}$. Although mice with this AID mutation have defects in both switch recombination and somatic hypermutation, the defect in somatic hypermutation is much more pronounced in vivo [94]. In this context, AID$^{G23S}$ mice are able to compensate for the reduced CSR efficiency in terms of total levels of Ig. In this regard, they have normal levels of all Ig isotypes, including IgA, in the serum and feces, but somatic hypermutation is attenuated, reaching levels only ~10–20% of those found in wild-type mice [94]. Both the total number of mutations per clone and the number of mutated clones appear to be reduced in both Peyer's patch germinal center B cells and IgA$^+$ plasma cells in the intestinal LP. This mouse model thus enabled investigations into whether secretion of nonsomatically mutated IgA is sufficient to maintain bacterial homeostasis and to mount protective mucosal immunity. Despite normal levels of IgA in intestinal secretions, AID$^{G23S}$ mice manifest phenotypes similar to those of AID-deficient mice. Notably, the AID$^{G23S}$ microbiota was shown to have overrepresentation of anaerobes such as SFB [94] (Figure 6). Thus, somatic

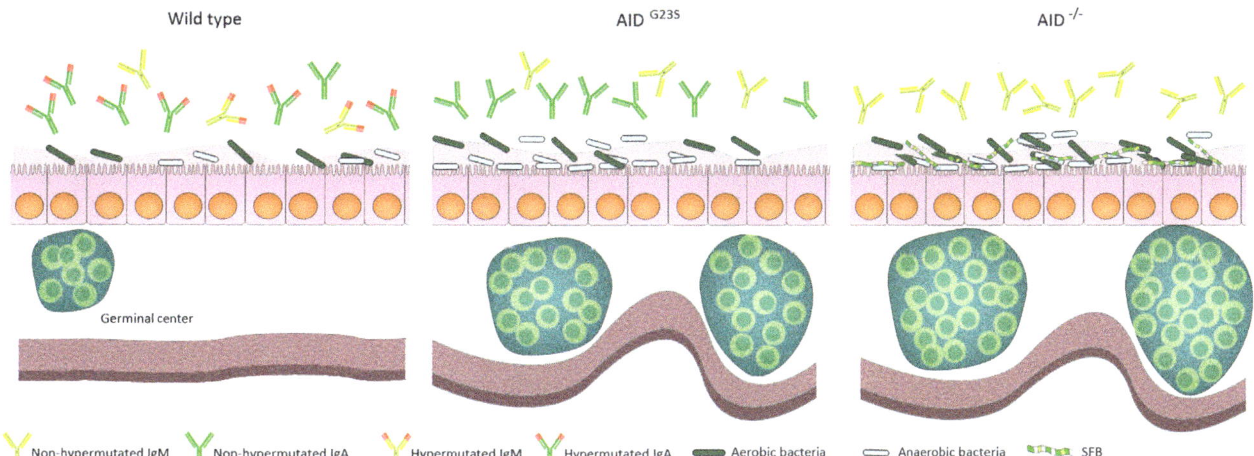


FIGURE 6 **Significance of B cell somatic hypermutation (SHM) in maintaining a steady microbial population.** AID$^{G23S}$ mice carry a knocked-in mutation of the gene encoding AID, which causes severe defects in SHM but has a very mild effect on class-switch recombination (CSR). As a result, these mice harbor proportions of IgA comparable to those in wild-type mice but lack high-affinity immunoglobulins accumulated via SHM. Absence of SHM results in expansion of gut flora, with the most marked increase in the number of anaerobic bacteria. The expansion of the microbial community in turn triggers the germinal centers (GCs) in the gut resulting in hyperplasia of the GCs in Peyer's patches and the lamina propria. Similar phenotypes, albeit more pronounced, are also observed in AID$^{-/-}$ mice, which lack the gene encoding AID and thus have only IgM that lacks SHM. The AID$^{-/-}$ mice also contain a large number of segmented filamentous bacteria (SFB) in their small intestine. Isolating the effect of SHM from that of CSR suggests that SHM is responsible for the disturbance in the gut microbial population and the resultant stimulation of gut GCs.

hypermutation in concert with IgA plays a key role in controlling the ecology of the luminal bacterial consortium.

## 6. CONCLUSION

Nearly half a century has passed since the discovery of the link between Ig development and a chicken hindgut structure—after which the B lymphocyte lineage is named [20]. With regard to the gut, data generated since that time suggest that the gut environment, in particular the microbiota, plays a special role in shaping the Ig repertoire, but potentially only during a limited window of time early in life. Future work will certainly provide more data to advance our understanding of the functional relevance of early life microbial exposures in shaping the fitness of humoral immune responses. In addition, as specific ecologies of intestinal microbiota may predispose to disease [12], years to come will assuredly uncover the role of B-lineage cells in establishing and maintaining healthy relationships with microbes as well as the optimal microbial membership key to optimal host health.

## REFERENCES

[1] Savage DC. Microbial ecology of the gastrointestinal tract. Annu Rev Microbiol 1977;31:107–33.

[2] Ley RE, Peterson DA, Gordon JI. Ecological and evolutionary forces shaping microbial diversity in the human intestine. Cell 2006;124:837–48.

[3] Qin J, Li R, Raes J, Arumugam M, Burgdorf KS, Manichanh C, et al. A human gut microbial gene catalogue established by metagenomic sequencing. Nature 2010;464:59–65.

[4] Human Microbiome Project, C. Structure, function and diversity of the healthy human microbiome. Nature 2012;486:207–14.

[5] Zaretsky I, Polonsky M, Shifrut E, Reich-Zeliger S, Antebi Y, Aidelberg G, et al. Monitoring the dynamics of primary T cell activation and differentiation using long term live cell imaging in microwell arrays. Lab Chip 2012;12:5007–15.

[6] Palmer C, Bik EM, DiGiulio DB, Relman DA, Brown PO. Development of the human infant intestinal microbiota. PLoS Biol 2007;5:e177.

[7] McFall-Ngai M, Hadfield MG, Bosch TC, Carey HV, Domazet-Loso T, Douglas AE, et al. Animals in a bacterial world, a new imperative for the life sciences. Proc Natl Acad Sci U. S. A 2013;110:3229–36.

[8] Turnbaugh PJ, Ley RE, Hamady M, Fraser-Liggett CM, Knight R, Gordon JI. The human microbiome project. Nature 2007;449:804–10.

[9] Yatsunenko T, Rey FE, Manary MJ, Trehan I, Dominguez-Bello MG, Contreras M, et al. Human gut microbiome viewed across age and geography. Nature 2012;486:222–7.

[10] Lozupone CA, Stombaugh JI, Gordon JI, Jansson JK, Knight R. Diversity, stability and resilience of the human gut microbiota. Nature 2012;489:220–30.

[11] Ferriere R, Bronstein JL, Rinaldi S, Law R, Gauduchon M. Cheating and the evolutionary stability of mutualisms. Proc Biol Sci 2002;269:773–80.

[12] de Vos WM, de Vos EA. Role of the intestinal microbiome in health and disease: from correlation to causation. Nutr Rev 2012;70 (Suppl. 1):S45–56.

[13] Lee SM, Donaldson GP, Mikulski Z, Boyajian S, Ley K, Mazmanian SK. Bacterial colonization factors control specificity and stability of the gut microbiota. Nature 2013;501:426–9.

[14] Chung H, Pamp SJ, Hill JA, Surana NK, Edelman SM, Troy EB, et al. Gut immune maturation depends on colonization with a host-specific microbiota. Cell 2012;149:1578–93.

[15] Hooper LV, Littman DR, Macpherson AJ. Interactions between the microbiota and the immune system. Science 2012;336:1268–73.

[16] Ivanov II , Atarashi K, Manel N, Brodie EL, Shima T, Karaoz U, et al. Induction of intestinal Th17 cells by segmented filamentous bacteria. Cell 2009;139:485–98.

[17] Wu HJ, Ivanov II , Darce J, Hattori K, Shima T, Umesaki Y, et al. Gut-residing segmented filamentous bacteria drive autoimmune arthritis via T helper 17 cells. Immunity 2010;32:815–27.

[18] Atarashi K, Tanoue T, Shima T, Imaoka A, Kuwahara T, Momose Y, et al. Induction of colonic regulatory T cells by indigenous Clostridium species. Science 2011;331:337–41.

[19] Round JL, Lee SM, Li J, Tran G, Jabri B, Chatila TA, et al. The toll-like receptor 2 pathway establishes colonization by a commensal of the human microbiota. Science 2011;332:974–7.

[20] Glick G, Chang TS, Jaap RG. The bursa of Fabricius and antibody production. Poult Sci 1956;35:224–34.

[21] Reynolds JD, Morris B. The evolution and involution of Peyer's patches in fetal and postnatal sheep. Eur J Immunol 1983;13:627–35.

[22] Cooper MD. A life of adventure in immunobiology. Annu Rev Immunol 2010;28:1–19.

[23] Perey DY, Frommel D, Hong R, Good RA. The mammalian homologue of the avian bursa of Fabricius. II. Extirpation, lethal x-irradiation, and reconstitution in rabbits. Effects on humoral immune responses, immunoglobulins, and lymphoid tissues. Lab Invest 1970;22:212–27.

[24] Vajdy M, Sethupathi P, Knight KL. Dependence of antibody somatic diversification on gut-associated lymphoid tissue in rabbits. J Immunol 1998;160:2725–9.

[25] Knight KL. Restricted VH gene usage and generation of antibody diversity in rabbit. Annu Rev Immunol 1992;10:593–616.

[26] Crane MA, Kingzette M, Knight KL. Evidence for limited B-lymphopoiesis in adult rabbits. J Exp Med 1996;183:2119–21.

[27] Weinstein PD, Mage RG, Anderson AO. The appendix functions as a mammalian bursal equivalent in the developing rabbit. Adv Exp Med Biol 1994;355:249–53.

[28] Stepankova R, Kovaru F. Immunoglobulin-producing cells in lymphatic tissues of germfree and conventional rabbits as detected by an immunofluorescence method. Folia Microbiol (Praha) 1985;30:291–4.

[29] Stepankova R, Kovaru F, Kruml J. Lymphatic tissue of the intestinal tract of germfree and conventional rabbits. Folia Microbiol (Praha) 1980;25:491–5.

[30] Tlaskalova-Hogenova H, Stepankova R. Development of antibody formation in germ-free and conventionally reared rabbits: the role of intestinal lymphoid tissue in antibody formation to *E. coli* antigens. Folia Biol (Praha) 1980;26:81–93.

[31] Lanning D, Sethupathi P, Rhee KJ, Zhai SK, Knight KL. Intestinal microflora and diversification of the rabbit antibody repertoire. J Immunol 2000;165:2012–9.

[32] Perey DY, Good RA. Experimental arrest and induction of lymphoid development in intestinal lymphoepithelial tissues of rabbits. Lab Invest 1968;18:15–26.

[33] Stramignoni A, Mollo F, Rua S, Palestro G. Development of the lymphoid tissue in the rabbit appendix isolated from the intestinal tract. J Pathol 1969;99:265–9.

[34] Rhee KJ, Jasper PJ, Sethupathi P, Shanmugam M, Lanning D, Knight KL. Positive selection of the peripheral B cell repertoire in gut-associated lymphoid tissues. J Exp Med 2005;201:55–62.
[35] Pospisil R, Young-Cooper GO, Mage RG. Preferential expansion and survival of B lymphocytes based on VH framework 1 and framework 3 expression: "positive" selection in appendix of normal and VH-mutant rabbits. Proc Natl Acad Sci U. S. A 1995;92:6961–5.
[36] Weill JC, Reynaud CA. The chicken B cell compartment. Science 1987;238:1094–8.
[37] Ratcliffe MJ. Generation of immunoglobulin heavy chain diversity subsequent to cell surface immunoglobulin expression in the avian bursa of Fabricius. J Exp Med 1989;170:1165–73.
[38] McCormack WT, Tjoelker LW, Thompson CB. Avian B-cell development: generation of an immunoglobulin repertoire by gene conversion. Annu Rev Immunol 1991;9:219–41.
[39] Ratcliffe MJ, Jacobsen KA. Rearrangement of immunoglobulin genes in chicken B cell development. Semin Immunol 1994;6:175–84.
[40] Reynaud CA, Anquez V, Dahan A, Weill JC. A single rearrangement event generates most of the chicken immunoglobulin light chain diversity. Cell 1985;40:283–91.
[41] Houssaint E, Belo M, Le Douarin NM. Investigations on cell lineage and tissue interactions in the developing bursa of Fabricius through interspecific chimeras. Dev Biol 1976;53:250–64.
[42] Houssaint E, Torano A, Ivanyi J. Ontogenic restriction of colonization of the bursa of Fabricius. Eur J Immunol 1983;13:590–5.
[43] Reynaud CA, Anquez V, Grimal H, Weill JC. A hyperconversion mechanism generates the chicken light chain preimmune repertoire. Cell 1987;48:379–88.
[44] Sayegh CE, Demaries SL, Iacampo S, Ratcliffe MJ. Development of B cells expressing surface immunoglobulin molecules that lack V(D) J-encoded determinants in the avian embryo bursa of fabricius. Proc Natl Acad Sci U. S. A 1999;96:10806–11.
[45] Reynolds JD. Mitotic rate maturation in the Peyer's patches of fetal sheep and in the bursa of Fabricius of the chick embryo. Eur J Immunol 1987;17:503–7.
[46] Sayegh CE, Ratcliffe MJ. Perinatal deletion of B cells expressing surface Ig molecules that lack V(D)J-encoded determinants in the bursa of Fabricius is not due to intrafollicular competition. J Immunol 2000;164:5041–8.
[47] Griebel PJ, Hein WR. Expanding the role of Peyer's patches in B-cell ontogeny. Immunol Today 1996;17:30–9.
[48] Griebel P, Ferrari G. CD40 signalling in ileal Peyer's patch B cells: implications for T cell-dependent antigen selection. Int Immunol 1995;7:369–79.
[49] Hein WR, Dudler L, Mackay CR. Surface expression of differentiation antigens on lymphocytes in the ileal and jejunal Peyer's patches of lambs. Immunology 1989;68:365–70.
[50] Larsen HJ, Landsverk T. Distribution of T and B lymphocytes in jejunal and ileocaecal Peyer's patches of lambs. Res Vet Sci 1986;40:105–11.
[51] Reynolds JD. Evidence of extensive lymphocyte death in sheep Peyer's patches. I. A comparison of lymphocyte production and export. J Immunol 1986;136:2005–10.
[52] Reynaud CA, Garcia C, Hein WR, Weill JC. Hypermutation generating the sheep immunoglobulin repertoire is an antigen-independent process. Cell 1995;80:115–25.
[53] Reynaud CA, Mackay CR, Muller RG, Weill JC. Somatic generation of diversity in a mammalian primary lymphoid organ: the sheep ileal Peyer's patches. Cell 1991;64:995–1005.
[54] Pabst R, Reynolds JD. Peyer's patches export lymphocytes throughout the lymphoid system in sheep. J Immunol 1987;139:3981–5.
[55] Motyka B, Reynolds JD. Apoptosis is associated with the extensive B cell death in the sheep ileal Peyer's patch and the chicken bursa of Fabricius: a possible role in B cell selection. Eur J Immunol 1991;21:1951–8.
[56] Jenne CN, Kennedy LJ, McCullagh P, Reynolds JD. A new model of sheep Ig diversification: shifting the emphasis toward combinatorial mechanisms and away from hypermutation. J Immunol 2003;170:3739–50.
[57] Butler JE, Sinkora M. The enigma of the lower gut-associated lymphoid tissue (GALT). J Leukoc Biol 2013;94:259–70.
[58] Mutwiri G, Watts T, Lew L, Beskorwayne T, Papp Z, Baca-Estrada ME, et al. Ileal and jejunal Peyer's patches play distinct roles in mucosal immunity of sheep. Immunology 1999;97:455–61.
[59] Golby S, Hackett M, Boursier L, Dunn-Walters D, Thiagamoorthy S, Spencer J. B cell development and proliferation of mature B cells in human fetal intestine. J Leukoc Biol 2002;72:279–84.
[60] Wesemann DR, Portuguese AJ, Meyers RM, Gallagher MP, Cluff-Jones K, Magee JM, et al. Microbial colonization influences early B-lineage development in the gut lamina propria. Nature 2013;501:112–5.
[61] Mackie RI, Sghir A, Gaskins HR. Developmental microbial ecology of the neonatal gastrointestinal tract. Am J Clin Nutr 1999;69: 1035S–45S.
[62] Tiegs SL, Russell DM, Nemazee D. Receptor editing in self-reactive bone marrow B cells. J Exp Med 1993;177:1009–20.
[63] Retter MW, Nemazee D. Receptor editing occurs frequently during normal B cell development. J Exp Med 1998;188:1231–8.
[64] Hertz M, Nemazee D. BCR ligation induces receptor editing in IgM+IgD- bone marrow B cells in vitro. Immunity 1997;6: 429–36.
[65] Bird L. B cell development: a window of opportunity. Nat Rev Immunol 2013;13:707.
[66] Schlissel M. Immunology: B-cell development in the gut. Nature 2013;501:42–3.
[67] Delacroix DL, Hodgson HJ, McPherson A, Dive C, Vaerman JP. Selective transport of polymeric immunoglobulin A in bile. Quantitative relationships of monomeric and polymeric immunoglobulin A, immunoglobulin M, and other proteins in serum, bile, and saliva. J Clin Invest 1982;70:230–41.
[68] Conley ME, Delacroix DL. Intravascular and mucosal immunoglobulin A: two separate but related systems of immune defense? Ann Intern Med 1987;106:892–9.
[69] Brandtzaeg P, Halstensen TS, Kett K, Krajci P, Kvale D, Rognum TO, et al. Immunobiology and immunopathology of human gut mucosa: humoral immunity and intraepithelial lymphocytes. Gastroenterology 1989;97:1562–84.
[70] Benveniste J, Lespinats G, Salomon J. Serum and secretory IgA in axenic and holoxenic mice. J Immunol 1971;107:1656–62.
[71] Benveniste J, Lespinats G, Adam C, Salomon JC. Immunoglobulins in intact, immunized, and contaminated axenic mice: study of serum IgA. J Immunol 1971;107:1647–55.
[72] Shroff KE, Meslin K, Cebra JJ. Commensal enteric bacteria engender a self-limiting humoral mucosal immune response while permanently colonizing the gut. Infect Immun 1995;63:3904–13.
[73] Hooper LV, Macpherson AJ. Immune adaptations that maintain homeostasis with the intestinal microbiota. Nat Rev Immunol 2010;10:159–69.

[74] Hapfelmeier S, Lawson MA, Slack E, Kirundi JK, Stoel M, Heikenwalder M, et al. Reversible microbial colonization of germ-free mice reveals the dynamics of IgA immune responses. Science 2010;328:1705–9.

[75] Talham GL, Jiang HQ, Bos NA, Cebra JJ. Segmented filamentous bacteria are potent stimuli of a physiologically normal state of the murine gut mucosal immune system. Infect Immun 1999;67:1992–2000.

[76] Hirota K, Turner JE, Villa M, Duarte JH, Demengeot J, Steinmetz OM, et al. Plasticity of Th17 cells in Peyer's patches is responsible for the induction of T cell-dependent IgA responses. Nat Immunol 2013;14:372–9.

[77] Cahenzli J, Koller Y, Wyss M, Geuking MB, McCoy KD. Intestinal microbial diversity during early-life colonization shapes long-term IgE levels. Cell Host Microbe 2013;14:559–70.

[78] Hill DA, Siracusa MC, Abt MC, Kim BS, Kobuley D, Kubo M, et al. Commensal bacteria-derived signals regulate basophil hematopoiesis and allergic inflammation. Nat Med 2012;18:538–46.

[79] Kitaura J, Song J, Tsai M, Asai K, Maeda-Yamamoto M, Mocsai A, et al. Evidence that IgE molecules mediate a spectrum of effects on mast cell survival and activation via aggregation of the FcepsilonRI. Proc Natl Acad Sci U. S. A 2003;100:12911–6.

[80] Kalliomaki M, Kirjavainen P, Eerola E, Kero P, Salminen S, Isolauri E. Distinct patterns of neonatal gut microflora in infants in whom atopy was and was not developing. J Allergy Clin Immunol 2001;107:129–34.

[81] Kummeling I, Stelma FF, Dagnelie PC, Snijders BE, Penders J, Huber M, et al. Early life exposure to antibiotics and the subsequent development of eczema, wheeze, and allergic sensitization in the first 2 years of life: the Koala birth cohort study. Pediatrics 2007;119:e225–31.

[82] Marra F, Marra CA, Richardson K, Lynd LD, Kozyrskyj A, Patrick DM, et al. Antibiotic use in children is associated with increased risk of asthma. Pediatrics 2009;123:1003–10.

[83] Cong Y, Feng T, Fujihashi K, Schoeb TR, Elson CO. A dominant, coordinated T regulatory cell-IgA response to the intestinal microbiota. Proc Natl Acad Sci U. S. A 2009;106:19256–61.

[84] Macpherson AJ, Hunziker L, McCoy K, Lamarre A. IgA responses in the intestinal mucosa against pathogenic and non-pathogenic microorganisms. Microbes Infect 2001;3:1021–35.

[85] Peterson DA, McNulty NP, Guruge JL, Gordon JI. IgA response to symbiotic bacteria as a mediator of gut homeostasis. Cell Host Microbe 2007;2:328–39.

[86] Obata T, Goto Y, Kunisawa J, Sato S, Sakamoto M, Setoyama H, et al. Indigenous opportunistic bacteria inhabit mammalian gut-associated lymphoid tissues and share a mucosal antibody-mediated symbiosis. Proc Natl Acad Sci U. S. A 2010;107:7419–24.

[87] Shulzhenko N, Morgun A, Hsiao W, Battle M, Yao M, Gavrilova O, et al. Crosstalk between B lymphocytes, microbiota and the intestinal epithelium governs immunity versus metabolism in the gut. Nat Med 2011;17:1585–93.

[88] Degrandi D, Konermann C, Beuter-Gunia C, Kresse A, Wurthner J, Kurig S, et al. Extensive characterization of IFN-induced GTPases mGBP1 to mGBP10 involved in host defense. J Immunol 2007;179:7729–40.

[89] Hughes WS, Cerda JJ, Holtzapple P, Brooks FP. Primary hypogammaglobulinemia and malabsorption. Ann Intern Med 1971;74:903–10.

[90] Agarwal S, Mayer L. Pathogenesis and treatment of gastrointestinal disease in antibody deficiency syndromes. J Allergy Clin Immunol 2009;124:658–64.

[91] Owens SR, Greenson JK. The pathology of malabsorption: current concepts. Histopathology 2007;50:64–82.

[92] Fagarasan S, Muramatsu M, Suzuki K, Nagaoka H, Hiai H, Honjo T. Critical roles of activation-induced cytidine deaminase in the homeostasis of gut flora. Science 2002;298:1424–7.

[93] Suzuki K, Meek B, Doi Y, Muramatsu M, Chiba T, Honjo T, et al. Aberrant expansion of segmented filamentous bacteria in IgA-deficient gut. Proc Natl Acad Sci U. S. A 2004;101:1981–6.

[94] Wei M, Shinkura R, Doi Y, Maruya M, Fagarasan S, Honjo T. Mice carrying a knock-in mutation of Aicda resulting in a defect in somatic hypermutation have impaired gut homeostasis and compromised mucosal defense. Nat Immunol 2011;12:264–70.

Chapter 18

# Molecular Mechanisms of AID Function

Nasim A. Begum, Hitoshi Nagaoka, Maki Kobayashi and Tasuku Honjo
*Department of Immunology and Genomic Medicine, Kyoto University Graduate School of Medicine, Kyoto, Japan*

## 1. INTRODUCTION

How antibody memory is generated is one of the core questions about the mechanisms of acquired immunity. Two phenomena, somatic hypermutation (SHM) and class switch recombination (CSR), are critical for generating antigen-induced antibody memory. Both of these mechanisms depend on genetic alterations in the immunoglobulin (Ig) locus. Because SHM is assumed to be the product of error-prone DNA repair after a single-strand break (SSB) [1,2], SHM's molecular mechanism is divided into two phases: the target-specific SSB and the subsequent error-prone DNA synthesis that is probably mediated by translesion polymerases (TLPs). Efficient SHM, including the putative direct deamination of C to T on the DNA, requires the suppression of error-free repair at the DNA damage site. CSR, on the other hand, is mediated by the recombination of double-strand breaks (DSBs) in two S regions located several kilobases or a few megabases apart. Because SSB is the initial cleavage product in the switch (S) region [3], CSR requires the processing of the SSB to a DSB, the appropriate pairing and synapsis of the two ends, and their recombination.

From a mechanistic view, many researchers originally concluded that the enzymes that mediate SHM are distinct from those required for CSR, and proposed mutasomes and recombinatosomes [4]. However, the discovery of activation-induced cytidine deaminase (AID) led to the unexpected revelation that a single enzyme is responsible for both SHM and CSR in the Ig locus [5,6]. As is usually the case with scientific progress, this discovery gave rise to more questions than answers: for instance, whether AID deaminates cytidine (C) in DNA, or in RNA. Two hypotheses on this subject, the DNA deamination model and the RNA editing model, are still extensively debated [5,7–9]; in this chapter, we will summarize these hypotheses and the evidence for or against them. Another particular mystery is how AID, which has 198 residues, can select specific cleavage targets in the whole genome. AID is not absolutely specific to the Ig genes, but neither is it wholly promiscuous [10–12]. Clearly, the number of genes outside the Ig locus that are mutated or translocated after AID activation is limited, and several hypotheses have been proposed to explain how AID limits its targets [11,13,14]. The two hypotheses of C's deamination in DNA versus RNA propose different mechanisms by which AID determines its specific targets. This chapter reviews experimental data that bear on these critical questions, from the two different views of AID's function.

CSR and SHM are accomplished by distinct molecular mechanisms, because CSR requires not only DNA cleavage, but also its recombination. A series of genetic findings indicated that the AID protein molecule itself, although relatively small, mediates these two reactions in CSR [5,15]. Natural and artificial AID mutagenesis revealed that C-terminal mutations specifically abolish CSR without affecting SHM [16]. Some C-terminally truncated AID mutants enhance SHM but almost completely lose CSR, which suggests that AID's C-terminus is dispensable for its DNA cleavage activity [17]. On the other hand, N-terminal mutations affect SHM more severely than CSR, because AID's ability to cleave DNA is compromised [18]. These findings were interpreted by the RNA editing model to indicate that AID may interact with at least two different RNA–cofactor complexes, generating two or more edited RNA products that mediate DNA cleavage and recombination separately [19,20]. In contrast, the DNA deamination hypothesis proposes that DNA cleavage alone is sufficient to induce CSR, because it assumes that recombination is mediated by the endogenous DNA repair system. Although a CSR defect caused by a C-terminal mutation has been explained by the disruption of AID's stability or nuclear–cytoplasmic transport [21], these proposed mechanisms are not sufficient to explain the different effects of AID mutations on SHM and CSR.

Finally, there is the critical question of how AID expression is restricted to activated B-cells. AID expression must be strictly regulated, because its target specificity is not absolute, and its aberrant expression can induce

Molecular Biology of B Cells. http://dx.doi.org/10.1016/B978-0-12-397933-9.00018-7

wide-ranging genome instability. In fact, dysregulated or ectopic AID expression can cause various types of tumors [20,22].

Whereas some of the questions that emerged with AID's discovery 14 years ago have been answered, many others are still a matter of extensive debate. This chapter summarizes the current evidence supporting the two differing views of basic AID function—that it deaminates C in DNA, or in RNA

## 2. AID STRUCTURE AND FUNCTION

### 2.1 General Structural Features and Functional Associations of AID

The AID gene encodes a small protein of 26 kDa. The primary structure of AID is simple. Its N-terminal and C-terminal halves are connected by a cytidine deaminase (CD) catalytic domain and a linker sequence [16,23]. How such a small molecule conducts such highly sophisticated genomic alterations in B-lymphocytes has yet to be understood. The genetic analysis of hyper-IgM (HIGM)-syndrome type II patients [24] and their families provided structure–function clues indicating that AID is responsible for Ig-isotype deficiencies [6,25]. In these patients, point mutations were detected all over the coding region of AID, including the nuclear localization signal (NLS) motif at the N terminus, the nuclear export signal (NES) motif at the C-terminus, the central catalytic domain containing the CD motif, and the linker sequence. In addition to loss-of-function mutations at the catalytic domain, the wide distribution of loss-of-function mutations indicates that AID's structural integrity is required for its physiological function. AID mutants can be clearly grouped into two categories: C-terminal mutants that correlate strongly with a loss of CSR function, and N-terminal mutants that cause the loss of both CSR and SHM (Figure 1).

#### 2.1.1 *AID's N-terminal Region Is Required for DNA Cleavage in CSR and SHM*

Although the AID point mutations found in French and Japanese HIGM-II cases provided many clues to the consequences of those mutations [6,16], detailed structural–functional correlation studies require functional analyses in a model system. Initial efforts to establish a suitable model system included the generation of a library of mouse AID mutants and their functional assays in AID-deficient spleen cells. Shinkura et al. [18] reported several AID N-terminal mutants (Y13H, V18R, V18SR19V, W20K, and G23S), in which AID's CD catalytic activity was not disturbed, that displayed intact CSR but decreased SHM. Shivarov et al. [26] conducted extensive AID mutant studies comparing in vitro DNA deamination activity with in vivo CSR and SHM function, and found a disassociation between in vitro and in vivo activities. The same study also found that most N-terminal mutants are defective in both CSR and SHM. Another study reported that an N-terminal AID mutant was defective in introducing DNA cleavage in the variable (V) and S regions [17], a prerequisite function for AID-induced SHM and CSR. Because no AID mutants with defective SHM but intact CSR have been identified, it is likely that

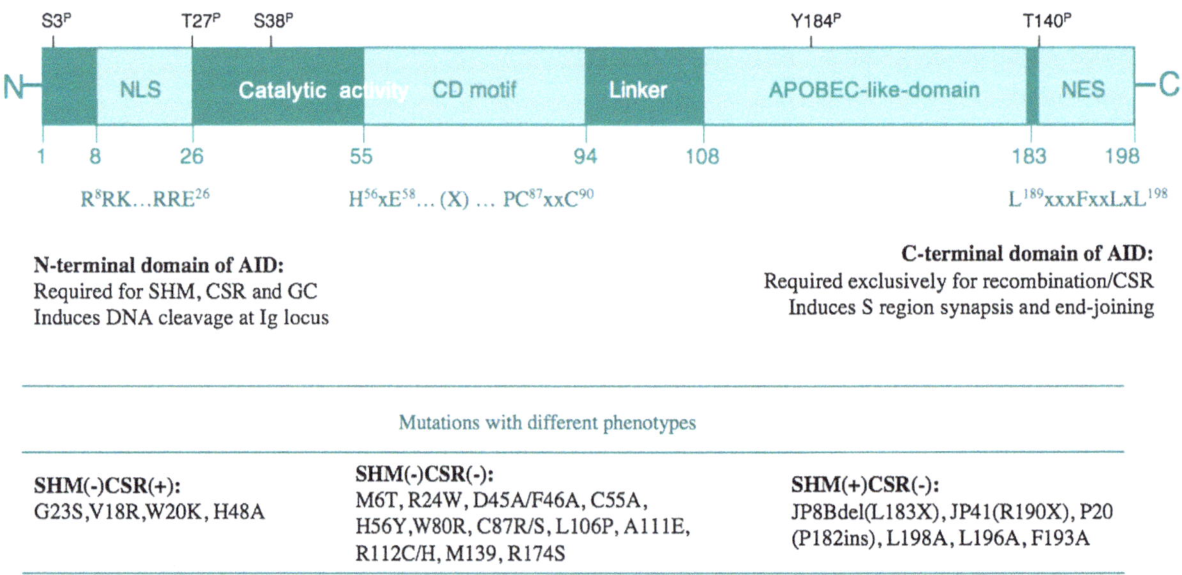


FIGURE 1 **General structural features and functionally important mutants of AID.** The schematic diagram shows the hAID structure with various functional domains and motifs. The numbers below the scheme indicate the amino acid residues, and the potential phospho-Ser and -Thr residues are indicated above the scheme. Consensus sequence motifs for NLS, CD, and NES are shown below their respective locations. Selected AID mutants that significantly contributed to the understanding of the functions of AID in CSR and SHM are grouped into three categories as shown under the scheme.

the same mechanism for AID-induced DNA cleavage is used in the V and S regions. The disassociation of AID's functions in CSR and SHM was demonstrated more clearly by in vivo G23S-mutation studies; in AID-G23S knock-in mice, CSR is fully proficient but SHM is severely defective [27]. Except for impaired DNA cleavage activity, no other mechanistic defect of AID-G23S is known at present.

### *2.1.2 AID's C-terminal Region Is Required for Recombination*

The analysis of several C-terminal AID mutants identified in Japanese patients revealed that any truncation, frame shift, or defect in AID's C-terminus causes CSR to fail, but leaves SHM intact [16]. When expressed in AID-deficient splenic B-cells, the AID mutants P20 (hAID with a 34-aa insertion at residue 182), JP41 (hAID190X), JP8B (hAID frameshift at 183), and JP8Bdel (hAID183X) showed severe CSR defects, whereas SHM was intact or enhanced (Figure 1). Because a C-terminal AID mutant (Δ189–198) retained its full ability to deaminate DNA in *ung-1 Escherichia coli* cells [28], it is intriguing that C-terminal–defective AID mutants are unable to induce CSR in vivo.

Another interesting observation is that mutants of the AID NES, which is located at residues 189–198, are defective in CSR and accumulate in the nucleus [29,30]. However, the 34-residue insertion in the P20 mutant did not destroy the NES motif, and this mutant is fully competent for nucleocytoplasmic shuttling but is defective in CSR function [16,31]. All of the C-terminal AID mutants tested in these various studies showed a normal or augmented ability to cleave DNA in both the V and S regions [17]. These results suggest that DNA cleavage alone is not sufficient for CSR, and that the C-terminal region may be responsible for an as yet unknown mechanism that is required for CSR.

In agreement with this idea, whereas AID supported Sμ–Sα synapse formation when IgM-to-IgA switching was activated in CH12F3-2A cells, the P20 mutant failed to do so [31]. Furthermore, the CSR junctions in residual switched clones were predominantly repaired by the alternative end joining (A-EJ) pathway, because classical nonhomologous end joining (C-NHEJ) factor–like Ku80 recruitment was reduced in the S region [32,33]. Conversely, accumulation of the SS-gap (SSG) repair binding protein PARP1 was enhanced, which suggests that the AID C-terminal mutants efficiently generated SSB but not DSB. Therefore, it is likely that the DNA end processing required for the efficient generation and subsequent recombination of DSB is impaired in C-terminal AID mutants.

Taken together, it is clear that AID has two distinct roles: (1) it induces DNA cleavage in the IgH locus (V and S regions), for which AID's N-terminal region is responsible, and (2) it induces recombination by processing SSB to DSB and by forming the synapse; for these, the AID C terminus is required.

### *2.1.3 AID Is a Nucleocytoplasmic Shuttling Protein*

One of AID's puzzling properties is its preferential localization in the cytoplasm, even though AID-induced DNA cleavage and recombination are restricted to the nucleus. Although it has not been experimentally demonstrated whether AID's migration to the nucleus is mandatory, detailed functional analyses of AID mutants revealed how AID is imported into the nucleus and exported out to the cytoplasm [29,30]. AID has the unique signatures of a shuttling protein: namely, bipartite NLS and NES motifs, which reside at the AID's N-terminal (residues 8–26) and C-terminal (residues 189–198) regions, respectively. Deleting the C-terminal region or modifying the NES motif causes AID to accumulate in the nucleus. Because leptomycin-B (LMB) [34] treatment causes AID–GFP fusion proteins to accumulate in the nucleus [29,30,35], it is likely that AID's import into the nucleus is active rather than occurring by passive diffusion [30,35]. Consistent with this idea, importin-α/β and other nuclear import factors can be found in association with both wild-type (WT) and constitutively nuclear forms of AID. Oxidative stress, which blocks importin-α–dependent nuclear import, inhibits the LMB-mediated nuclear accumulation of AID. Furthermore, in cellular energy depletion assays, AID mutants that accumulated in the nucleus became dispersed and diffused out of the nucleus. Not only AID's NLS, but also additional N-terminal residues, located between residues 19 and 54, have been shown to interact with the nuclear import complex [35]. Therefore, in addition to the bipartite NLS motif, other structural features of the entire N-terminal region may contribute to AID's active transport into the nucleus. A putative nuclear localization motif overlapping with the NLS has also been reported [36].

AID's NES resides at the very end of the C-terminus, and consists of more than 10 residues [17,29,37]. Most AID C-terminal mutants accumulate in the nucleus; in other words, they fail to diffuse out of the nucleus. In addition, it is clear that the NES-mediated nuclear export activity is not the sole requirement for CSR, because a heterologous NES that preserves interaction with the export factor CRM1 does not support CSR [17,37]. Similarly, a mutation in the NES motif that abolishes CSR function does not abrogate AID's export function [37]. This finding is further supported by the fact that some AID C-terminal mutants with intact NES are CSR defective. In fact, the P20 mutant, which has an intact WT NES, has severe defects in CSR, probably because the insertion perturbs the C-terminal–specific structural integrity required for CSR, by separating the NES from the rest of AID through an insertion [16].

It has also been suggested that the NES region might possess an as yet unidentified cytoplasmic retention signal that functions independently of AID's nuclear import and export [35,38], or that AID's interaction with a large RNA–protein complex or with other cytoskeletal proteins might contribute to its cytoplasmic retention [39]. How AID remains in the cytoplasm in the steady state is not fully understood, although it is certain that there is an active mechanism that continuously exports AID from the nucleus to the cytoplasm [38].

## 2.2 Proteins That Interact with AID

### *2.2.1 Homomeric AID Interaction: Homology Modeling and Biochemical Assays*

AID's 3-dimensional (3D) structure has not been solved, because purifying the AID protein in large enough quantities for crystallization and structural analysis has proven difficult. Most CD enzymes are monomeric, although dimeric and tetrameric CD enzymes have also been reported [40,41]. To understand the mechanisms of AID's functions, it is important to examine how AID or its active form exists and behaves in the cell under physiological conditions. Because it is still debated whether AID–APOBEC–family proteins [42] act on an RNA or DNA substrate, determining their 3D structures may provide critical information regarding their substrate recognition and specificity determination, whether directly or in conjunction with other cellular proteins.

Initial studies using small amounts of purified N-terminally streptavidin-tagged AID showed that the purified AID protein migrates as a tetramer [43], but whether the tetrameric protein was functional was not clear. Later analyses of the activity of GST- or His–tagged AID purified by size-exclusion chromatography showed that AID fractions containing the predicted tetrameric size had the highest CD activity [44]. Unfortunately, this study did not use monomeric GST, and because GST inherently forms a dimer, the contribution of the GST fusion partner in the AID–AID multimer formation cannot be excluded. Nevertheless, AID's tetrameric structure gains support from the crystal structure of APOBEC2 [45], which shares significant amino acid sequence similarity with AID. APOBEC2's crystal structure suggests that it forms a tetramer composed of the head-to-head interaction of two dimers. On the other hand, contrary to the proposed tetrameric structure of AID, data from an atomic force microscopy (AFM) study of AID indicated that AID could be monomeric in solution, and that the monomeric form is catalytically active [46]. However, one group interpreted the AFM dataset differently, and concluded that AID could be tetrameric, like APOBEC2 [47]. APOBEC2's solution structure also suggests that it is a monomer [48]. Despite many efforts to demonstrate APOBEC2's catalytic activity, it has not been successfully demonstrated [49,50]. Thus, predicted models of AID based on APOBEC2 or APOBEC3G may not be reliable for inferring AID's actual state.

When AID proteins with two different tags were coexpressed, immunoprecipitation (IP) by either tag pulled down AID proteins with the other tag [16]. Although this result clearly indicates an AID–AID association, it neither excludes the possibility of a large-complex formation nor distinguishes the dimer from other multimers. Recently, bimolecular fluorescence complementation (BiFC) assays [51,52], which enable a fluorescent reporter protein to be reconstituted through dimerization with fused bait proteins, have been used to explore AID's interaction properties in living cells, and revealed that truncating AID at either the N- or C-terminus causes defects in AID–AID interaction (Mondal S, NAB TH, our unpublished data). Therefore, it is possible that AID can form head-to-tail dimers or multimers. Orientation-dependent monomeric and multimeric states might unmask and mask the NES/NLS motifs, as has been observed for other nucleocytoplasmic shuttling proteins. Interestingly, the P20 mutant is also unable to show BiFC-based dimerization in live cells, which supports the idea that AID's C-terminus has functional aspects other than CRM1-mediated export by the NES motif.

Using semipurified AID from HEK293T cells, it was shown that AID proteins with CD activity are found in a high molecular weight (HMW) complex of 500 kDa [53]. Glycerol-gradient sedimentation also showed that cytoplasmic AID protein expressed as a GFP-fusion protein in CH12F3-2A cells migrated as a high molecular mass complex of 11S [54]. Another sedimentation and migration study identified a similar cytoplasmic complex of ~300 kDa of Flag-AID expressed from the endogenous locus of DT40 cells [39]. The HMW complex is highly sensitive to high salt treatment [53,54], which suggests that it may include many proteins that are loosely or indirectly associated with AID. Interestingly, an AID-chromatin complex isolated from DT40 cells also appears to exist as an HMW complex [55]. It is not known whether these large complexes contain monomeric or oligomeric AID. The BiFC assays show that high salt or RNAase treatment also disrupts AID's homomeric interactions (Mondal S, NAB, TH, unpublished data). A comparison with APOBEC2's structure suggests that AID has more positively charged surface-exposed residues, which might be the source of AID's intrinsic affinity for RNA. Nonspecific RNA binding may in turn facilitate the formation of higher-order RNA-protein (RNP) complexes with monomeric and/or oligomeric AID, giving rise to various higher-order structures in the glycerol gradient. This formation of RNP complexes is a potential cause for the difficulties encountered in purifying AID or analyzing its structure and function in a straightforward manner.

### 2.2.2 AID Interacts with Numerous Partners

Extensive functional analyses of AID mutants have revealed the importance of AID's catalytic, N-, and C-terminal domains. However, these analyses do not provide clear mechanistic insights as to how CSR and SHM are differentially regulated, or how its target-locus specificity is determined. A number of studies have attempted to clarify AID's function by identifying a large number of cofactor candidates that might be relevant or irrelevant to AID's function. Many of these factors have been proposed to recruit AID to specific DNA targets based on the DNA deamination model. The RNA editing hypothesis, however, postulates that cofactors capture AID's specific target RNAs, which would be analogous to APOBEC1's requirement of ACF for the RNA editing of ApoB100 mRNA [56]. The proteins that have been identified and investigated in detail to determine their relationship with AID can be broadly divided into the following categories: (1) transcription-associated and DNA-binding proteins; (2) factors associated with splicing and RNA processing; (3) intracellular transporting factors; (4) factors that stabilize and modify proteins, and (5) DNA repair proteins.

#### 2.2.2.1 Transcription-Associated and DNA-Binding Proteins

An early co-immunoprecipitation (co-IP) study revealed that AID interacts with RNA polymerase II (RNAPII), which suggests that AID interacts functionally with the transcriptional complex [57]. Later large-scale RNAi screening and various AID co-IP approaches, including IP from B-cell chromatin fractions, identified RNAPII-associated transcription elongation factors such as SPT5, SPT6, TRIM28/KAP1, the PAF and FACT complexes, and others [54,55,58,59]. Depleting any of these proteins dramatically reduced CSR in CH12cells, which agrees with the transcription-coupled recruitment of AID to the Ig locus during SHM and CSR [58,60]. Subsequent studies revealed that many of these factors are involved in chromatin regulation of the Ig locus [61–63]. However, whether each of these proteins interacts directly with AID or indirectly through RNAPII has not been investigated in detail.

Co-immunoprecipitation analyses suggested that AID's N terminus is critical for its interaction with SPT6, whereas its C-terminus interacts with TRIM28 [54,59]. Depleting SPT6 or TRIM28 not only reduced CSR, but also affected the AID gene expression in CH12F3-2A cells. AID expression was also severely reduced in mb1-cre-Trim28-CKO mice [54]. However, using CD19-cre-Trim28 CKO mice, it was concluded that Trim28 regulates CSR but not SHM [59].

Two DNA-binding proteins, RPA and 14-3-$\gamma$, are reported to interact with AID's N-terminus (phospho-S38) and C-terminus, respectively [64,65]. Whereas RPA is known to bind ssDNA nonspecifically, 14-3-3$\gamma$ recognizes a sequence-specific DNA motif [66]. Both RPA and 14-3-3$\gamma$ have been proposed to promote DNA deamination by AID, based on in vitro assay results (see Section 4).

#### 2.2.2.2 Factors Associated with Splicing and RNA Processing

Yeast two-hybrid screening using AID as bait identified CTNNBL1, a splicing-associated factor that interacts with AID through its NLS [67,68]. Although knockdown (KD) of CTNNBL1 decreased Ig-gene conversion in DT40 cells, no CSR defect was observed in CTNNBL1-null spleen B or CH12F3-2A cells [69]. By applying an in-cell protein–biotinylation strategy followed by AID–IP, Nowak et al. [70] later identified the splicing factor PTBP2 as another of AID's interactors. This study reported that PTBP2 depletion in CH12F3-2A cells reduces CSR and proposed that PTBP2 acts as an AID-targeting factor to the S region for CSR, but a detailed mechanistic study was not performed. Basu et al. [71] reported that 11 subunits of the RNA–exosome complex associate with AID, and proposed that these components enhance CSR by increasing DNA deamination on both DNA strands, based on in vitro data. The results from depleting the exosome subunits in CH12 cells varied [61]. Several other RNA processing and splicing-associated proteins (including RNAPII, TFIID, snRNP8, snRNP200, PRP6, PRP9, DDX3X, DDX1, DDX15, Nucleolin, and SMN) have also been found in AID–IP complexes isolated from chromatin fractions of DT40 cells or by other methods [55,72].

#### 2.2.2.3 Intracellular Transport Factors

CRM1 and proteins of the karyopherin family have central roles in nucleocytoplasmic transport [73,74]. AID co-IP and in vitro interaction assays detected importin $\alpha/\beta$ proteins, which may be involved in the mechanism that imports AID into the nucleus via NLS. AID's interaction with exportin CRM1 was verified by LMB sensitivity tests and by mutagenesis of the NES motif [17,29,30]. The EEF1 protein, a translation elongation factor, was proposed to be involved in AID's cytoplasmic retention [39]; EEF1 was shown to interact directly with AID's C-terminus with a 1:1 stoichiometry. Apart from bona fide subcellular trafficking factors, CTNNBL1 and GANP have also been reported to influence AID's intracellular localization. Overexpression of the germinal-center nuclear protein (GANP) relocalized AID from the cytoplasm to the nucleus [29]. The AID residue D143 was found to be important for GANP interaction both in vitro and in vivo [75]. Conversely, the AID-mediated transport regulation of other proteins has been reported; AID overexpression was observed to relocalize nuclear Tet-family enzymes to the cytoplasm [76].

#### 2.2.2.4 Factors That Stabilize and Modify Proteins

AID associates with molecular chaperones belonging to the heat-shock protein family (HSP40, SHP70, and HSP90) in the cytoplasm. AID's interaction with the HSP90 system protects it from polyubiquitination-mediated proteasomal degradation [77,78]. A deficiency of HSP40 DnaJa1 destabilizes and decreases the AID protein, with corresponding proportional decreases in CSR and SHM. The degradation of nuclear AID is faster than that of cytoplasmic AID, and occurs through both ubiquitin-dependent and -independent mechanisms. Although the pathway is not fully understood, an interaction with nuclear REG-γ has been shown to promote AID's nuclear degradation [21]. CSRis enhanced in REG-γ–deficient mice, possibly owing to the extended stabilization of the AID protein. AID has also been shown to interact with two E3 ubiquitin ligases, MDM2 and RNF126, through its residues 189–198 and 88–166, respectively [79,80]. However, MDM2 deficiency did not lead to significant defects in Ig-gene conversion in DT40 cells. Similarly, RNF126's functional significance has yet to be determined. AID co-IP has also revealed interactions with posttranslational protein-modification enzymes belonging to the protein kinase and phosphatase families [65,80,81].

#### 2.2.2.5 DNA Repair–Associated Factors

Screening based on AID protein solubility identified 127 proteins that interact with AID, including RNF126 [80]. Some of these proteins were previously unknown. Reciprocal affinity purification screening confirmed AID's interaction with 36 of these proteins. Interestingly, this screening detected AID's association with DNA-repair factors such as Rad51, FEN1, and RRM2 [80]. Previously, co-IP with AID identified DNA PKcs, a critical CSR-specific DNA repair factor [82,83]. Moreover, DNA damage signals in CSR induced the interaction of APE1, a base excision repair enzyme, with a phosphorylated form of AID (p-Ser38) [84].

Because AID possesses an intrinsic affinity for RNA, protein complexes related to RNA processing might easily interact with AID in co-IP experiments or during affinity purification. In addition, SPT5- and SPT6-like proteins are able to form a macromolecular complex, because they associate with exosomes, RNAPII, and other elongation-associated factors. AID may also be able to form multiprotein RNP complexes in nuclear and cytoplasmic subcompartments. Innovative approaches will be needed to resolve the complexity and functional significance of the formation of nuclear, chromatin, and cytoplasmic AID complexes. Taken together, the exact functions of many AID-binding proteins have not been clarified for either the DNA deamination model or the RNA editing model. Because it is unlikely that most such diverse proteins are relevant to AID function, it is not clear why AID interacts with so many different proteins (Table 1).

### *2.2.3 Similarities of AID with the APOBEC Family, and an Evolutionary Perspective*

AID is a member of the large apolipoprotein B mRNA-editing catalytic-component (APOBEC) protein family, which carries a Zn-dependent CD domain [42,92]. The AID CD's active site consists of two cysteines, a histidine residue for Zn coordination and a glutamic acid that serves as a proton donor during rC/dC deamination. Currently, 11 APOBEC family members (APOBEC1, APOBEC2, APOBEC3A–3H, APOBEC4, and AID) are known in humans. All of these members possess either a single or double catalytic active site for CD activity, and a yet-uncharacterized domain called the APOBEC domain. Whereas AID contains one active CD domain, APOBEC3 proteins, which evolved through duplication and fusion in primates, possess two domains, one of which is often catalytically inert [93,94].

Despite their structural similarity, the APOBEC family's physiological functions are diverse and distinct, and AID cannot be substituted by other family members. Bacterial and yeast CDs convert free cytidine nucleotide to uridine, whereas APOBEC1 converts cytosine to uracil in ApoB mRNA with the help of the RNA binding protein cofactor ACF, causing the generation of a truncated protein with a different function [95,96]. APOBEC1 was later found to convert dC to dU in DNA in vitro as well as in *E. coli*, which shows that AID-APOBEC family proteins have the potential to edit both RNA and DNA substrates [49,97]. Whereas APOBEC3G proteins act on the human immunodeficiency virus genome and APOBEC3A acts nonspecifically on the host genome [98,99], we do not know the physiological substrates for the deamination activity of APOBEC family members other than APOBEC1. Interestingly, the APOBEC3 family members are able to restrict viral and retrotransposon activity independently of their CD function, although the mechanism is poorly defined [100,101]. Because most of these studies were conducted by protein overexpression, the potential for genome-wide mutagenesis through the CD activity of the APOBEC-family proteins, including AID, in a physiological context remains to be elucidated.

AID–like genes have been identified in bony fish and cartilaginous fish, the earliest jawed vertebrates, based on sequence conservation [42]. This finding correlates well with the origin of the antibody diversification process that appeared in jawed vertebrates [19,102,103]. The lamprey, a non-jawed vertebrate, has two AID-like genes, pmcda1 and pmcda2 [104]. The diversification of immune receptors in lampreys occurs via a gene conversion–like process reminiscent of the Ig-gene conversion process in chicken B-cells, which is AID-dependent. Phylogenetic tree analysis suggests that the ancestor of the AID–APOBEC family of CDs might have appeared first in non-jawed vertebrates. The APOBEC3 family members are thought to have expanded through the duplication and amplification of an ancestral AID-like gene during primate evolution [105,106]. It has been proposed that this massive expansion took place to combat the evolving nature of viral pathogens.

**TABLE 1** Activation-Induced Cytidine Deaminase (AID) Interacting Proteins

| Interacting Proteins | Interaction Region | General Function/ Property | Associated with | | | Specific Functions/Proposed Mechanisms | Refs |
|---|---|---|---|---|---|---|---|
| HSP40 | AID internal | Protein chaperone | CSR | SHM | GC | Stabilizes AID protein in the cytoplasm | [78] |
| HSP90 | AID (19–84 aa) | Protein chaperone | CSR | SHM | GC | Stabilizes AID protein in the cytoplasm | [77] |
| REG-γ | N-terminal | Proteasome activator subunit | CSR | nd | nd | Reduces nuclear resident time of AID via its degradation | [21] |
| YY1 | N-terminal | Transcription factor | CSR | nd | nd | Regulates stability of AID in the nucleus | [403] |
| EEF1A | N- and C-termini | Translation elongation complex component | CSR | nd | nd | Involved in AID retention in the cytoplasm | [39] |
| Importin-α3 | N-terminal/NLS | Nuclear import of proteins with NLS | nd | nd | nd | May be involved in nuclear import of AID | [35] |
| CTNNBL1 | N-terminal/NLS | Pre-mRNA processing associated factor | – | nd | GC | Interact with the NLS and may have redundant role in trafficking | [67–69] |
| CRM1 | C-terminal/NES | Transport of NES-containing proteins | CSR | SHM | nd | Exports AID out from the nucleus to the cytoplasm | [30,37] |
| GANP | D143 position | Transcription/mRNP export | – | SHM | nd | Targeting AID to V(D)J; transports AID into the nucleus | [75,86] |
| TRIM28 | C-terminal | Transcriptional regulator | CSR | – | nd | Targeting AID to H3K9me3-enriched S region | [54,59] |
| HP1 | nd | Transcriptional corepressor | CSR | nd | nd | Targeting AID to H3K9me3-enriched S region | [59] |
| RNAP II | nd | Transcription | nd | nd | nd | Targeting of AID in the transcribe locus | [55,57] |
| RPA | AID p-S38 | ssDNA binding protein | CSR | SHM | nd | Targeting AID p-S38 to the transcribed DNA | [64,65] |
| SPT5 | nd | Transcription elongation factor | CSR | SHM | nd | Targeting AID to the transcriptionally paused regions; target chromatin regulation and DNA repair | [58,87] |
| SPT6 | N-terminal | Histone chaperone, histone-PTM regulator | CSR | SHM | nd | Target chromatin regulation; present in AID-associated complex | [54,55] |
| FACT complex | nd | Histone chaperone, histone-PTM regulator | CSR | SHM | nd | Target chromatin regulation; present in AID and Top1-associated chromatin complex | [55,62,87] |
| PAF complex | N-terminal | Transcription elongation | CSR | nd | nd | Targeting AID in the transcription elongation complex | [55,87] |
| 14-3-3 proteins | C-terminal | Adapter/scaffold protein; signal transduction | CSR | nd | nd | Targeting AID to S-region AGCT motif and H3K4me3.K9ac.S10P mark | [66,88,89] |
| RNA exosomes | nd | RNA processing and degradation | CSR | SHM | nd | Targeting AID to both strands of the transcribed DNA | [71] |
| PTBP2 | nd | Pre-mRNA splicing associated factor | CSR | nd | nd | Targeting AID to S region through GLT | [70] |
| PKA | N-terminal | Site-specific protein phosphorylation | CSR | SHM | GC | Phosphorylation of AID at S38 and promotes RPA association; tethers to the S region | [65,81,84,90] |

*Continued*

**TABLE 1** Activation-Induced Cytidine Deaminase (AID) Interacting Proteins—cont'd

| Interacting Proteins | Interaction Region | General Function/ Property | Associated with | | | Specific Functions/Proposed Mechanisms | Refs |
|---|---|---|---|---|---|---|---|
| DNA-PKcs | C-term, CDD | DSB repair in NHEJ/ recombination | CSR | – | GC | Promotes repair and protects from AID-induced apoptosis | [59,82] |
| APE1 | AID *p*-S38 | Repair and transcription | CSR | – | nd | Abasic site cleavage | [84] |
| MDM2 | C-terminus | E3 ubiquitin protein ligase | nd | – | – | Dispensable for AID-associated function | [79] |
| CIB1 | nd | Calcium and integrin binding protein | – | nd | – | Dispensable for AID-associated function | [404] |
| SMN; U2AF65 | nd | Splicing-associated factor | nd | nd | nd | Unknown | [72] |
| RNF126 | AID (88–166) | E3 ubiquitin protein ligase | nd | nd | nd | Unknown | [80] |

| Other Proteins Identified in AID co-immunoprecipitations (co-IPs) | Refs |
|---|---|
| TOP2B, RAD50, SMCA5, GCN1L1, TCP1-g, epsilon, Zta, TBB3, Ac-Co carboxylase, Q3U2N8, Niban, Symplekin, Dynine, Q3TML0, Mre11, ZF protein, DDX47, GALK1,RL4,Trip12, AK129105, DNL1, SMRC2, DPOD1, IRF8, NF-κB inhibitor, N-pac homolog, PCD4, ARP3, RBBP4, AMPM1, γ-Actin like, ORC5, PDK3 | [70] |
| DDX17, DDX20, hnRNPA1, IGF2BP1, PABPC1, REG-g/PSME3, AFF4. | [54] |
| SPT5, SF3A, CTR9, RNAPII subunits, PAF1, SPT6, LaminB1-2, Top1, DDX3Y, IL3IF, LEO1, MCM5, SMU1, YBOX1, PRP6, 72kD unknown, PRP4, SPT16, SSRP1, DDX1, C17orf85, SF3B, DDX15 | [55] |
| hnRNP-M, hnRNP-A0, DDX21, DDX19, LGTN/Sui1, CPSF73, XRN2, EEF1A1, RBM39, NIP30, NOL11, CTNNBL1, Nup93, Importin b1, RAD51, RRM2B, IER5, FEN1, Actin, RDH1, PGAM1, IMPDH2, UbcH7, PPP2R1A, RPA1, TCP1-eta, HSPA5/BiP, HERPUD2, C14orf94, C22orf28, CXorf9, ZNF44, DAZAP2, ZMIZ22 | [80] |
| DNAJA1-2-3, HSPA8, HSP90AA1-BB1, AHSA1, BAG2, TCP1, eta, delta epsilon, PSMD1,2,6, PSMC2 | [77,78] |
| MSH2, MSH6, DNA-PKcs, CRM1, Ikaros | [59] |

# 3. AID'S MOLECULAR MECHANISM OF DNA CLEAVAGE AND RECOMBINATION

## 3.1 Properties of AID-Induced DNA Cleavage

Several unique features characterize the mechanism for AID-initiated DNA cleavage. First, the cleavage target depends completely on transcription of the locus [107]. In addition to the ordinary Ig promoter located 5′ to the V region, the Ig heavy-chain (H) locus contains an I promoter at the 5′ region of each constant (C) gene except for C delta [108–110]. Transcripts from the I promoter are designated as germline transcripts (GLT); a series of in vivo mutation analyses did not find the GLTs themselves to be involved in any DNA cleavage mechanism [111,112]. The current consensus is that transcription itself modulates the target-locus chromatin to make it accessible to DNA-cleaving enzymes [20,113,114]. There is clear evidence that transcription efficiency is correlated with CSR efficiency [115,116]. Transcription of the variable (V) region gene is also required for SHM [60]. Elegant knock-in experiments clearly showed that the RNA polymerase I (RNAPI) promoter can substitute for the RNAPII promoter to induce SHM in the Ig V region [117,118]. These results indicate that transcription-coupled modulation of the chromatin and the DNA structure of the target locus, but not the RNA polymerase protein per se, is critical for determining AID's target specificity for DNA cleavage.

Second, the initial cleavage induced by AID appears to be an SSB. This was first indicated by an experiment to isolate CSR inversion-type products using an artificial switch substrate in CH12F3-2 cells [3]. This experiment showed that most of the inverted junctions contained long insertions and deletions, which are likely to occur in the recombination of two ends carrying overhangs generated by staggered nicks on two strands. Recombination between blunt-ended DSB ends should have few insertions or deletions. Because

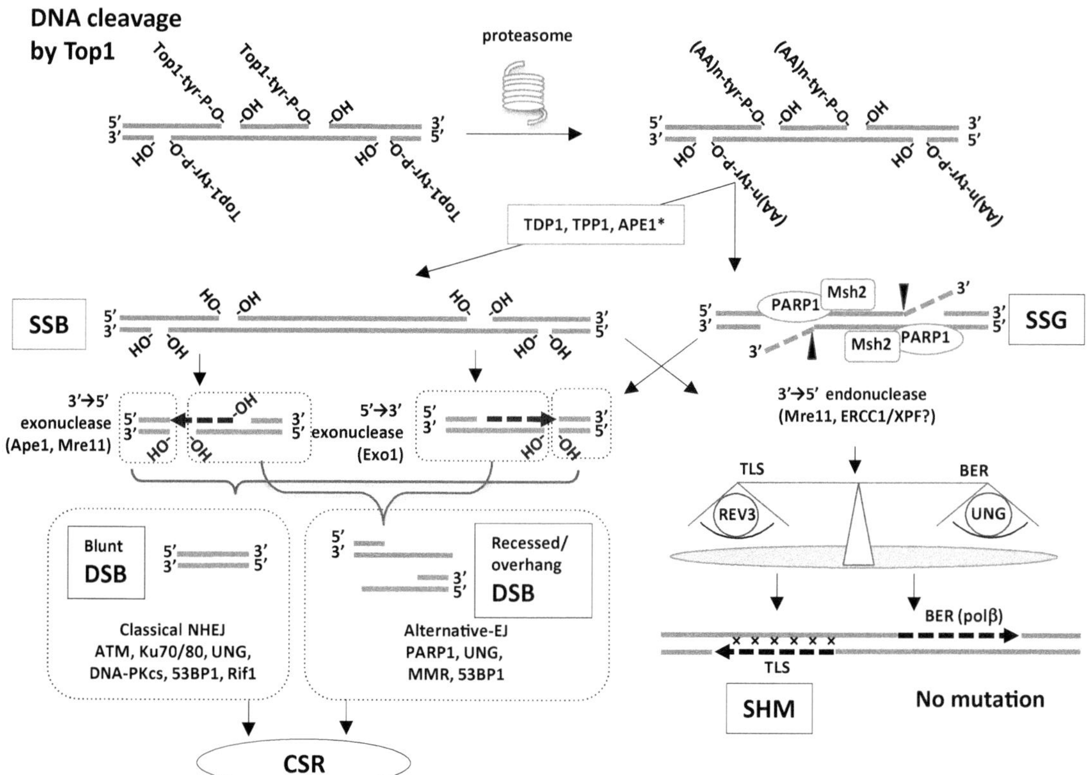


FIGURE 2 **DNA end processing in AID targets after DNA single-strand cleavage by topoisomerase-1.** SSB, single-strand break; SSG, single-strand gap; TLS, translesion synthesis; BER, base excision repair; NHEJ, non-homologous end joining; MMR, mismatch repair; ×, mutated bases. * Ape1 removes 3′-tyrosyl residue or peptides (Wilson, D.M. J Mol Biol 330, 1027–1037, 2003) *Modified from Zhang Y-W., Pommier Y. NAR 39(9), 3607–3620, 2011.*

it is unlikely that AID employs different cleavage mechanisms for CSR and SHM, SHM of the V region is probably initiated by an SSB followed by gap generation, after which the gap is filled with error-prone DNA polymerases and then sealed with DNA ligase [2]. It should be noted that SHM would have to be accompanied by as yet unknown mechanisms to suppress error-free repair, because most DNA damage in non–B-cells is repaired correctly [119]. Extensive SHM accumulations appear in the switch (S) region of the H-chain locus, which is the primary cleavage target for CSR [120]. Hypermutation accumulations have been observed even in B-cells that failed to switch [16,28]. These results indicate that not all the nicks and SSGs induced in the S region are processed to DSBs. So far, the mechanisms of SHM in the V and S regions are indistinguishable [17,114].

Because a DSB is essential for recombination, the SSB must be processed into a DSB to accomplish CSR, and this process requires several biochemical pathways. If two nicking sites on the opposite strands are reasonably close, the local exonucleolytic or endonucleolytic cleavage from the nicked site can generate a DSB. Such exonucleases are divided into two groups: 3′→5′ exonucleases such as Ape1, Ape2, and Mre11, and 5′→3′ exonucleases such as Exo1 [121,122]. Exonuclease digestion could generate one DSB with an overhang and another DSB with a blunt end (Figure 2). Endonucleases may also be involved in processing SSBs, specifically, 3′-5′ endonucleases such as the MRN complex (Mre11, RAd50, and NBS1) [123] and the ERCC1–XPF complex [124], and 5′-3′ endonucleases such as FEN1 [125]. MRN is known to be involved in both SHM and CSR [126], probably because the MRN complex has both exonuclease and endonuclease activity [127,128]. Ape1 and Ape2 appear to be important in CSR and SHM, respectively [129,130]. FEN1 is critical for gene conversion [131], but not for CSR or SHM [132]. ERCC1/XPF's roles in SHM and CSR are controversial [133,134] (Table 2). The different nuclease requirement between SHM and CSR suggests that the mechanisms for processing the SSB are distinct in the two genetic events.

An SSB can also be processed to a DSB through a mechanism mediated by helicases, which unwind a single strand on the staggered nicked region to separate DNA into two strands with overhangs. Again, the two nicks should be located within a reasonable distance to separate the two ends. Yet another mechanism of generating a DSB from an SSB depends on DNA synthesis from the 3′ OH end of the SSB. Such DNA synthesis is likely to be mediated by TLPs, and to generate blunt-end DNA when newly synthesized ss DNA meets a nick on the opposite strand with a 5′ overhang

**TABLE 2** DNA Repair Factors and Their Enrollment in Immunoglobulin Diversification

| Repair Pathway | Gene | V/JH4-SHM | Sμ-SHM | CSR | c-Myc Translocation | GC[a] |
|---|---|---|---|---|---|---|
| ATM | H2AX$^{-/-}$ mouse | No change [135] | ~60% [135] | ~30% [135–137] | No change [137] | ND |
| | 53BP1$^{-/-}$ mouse | No change [85] | No change [85] | ~10% [91,137] | No change [137] | 120% 53BP1$^{-/-}$ embryonic cells[b] [91] |
| | ATM$^{-/-}$ mouse | ~80% [138] | ~60% [138] | ~30% [137–139] | ~700% [137] | ~25%[c] [140] |
| | MDC1$^{-/-}$ mouse | ND | ND | ~50% [141] | ND | ND |
| NHEJ | DNA PKcs mutant mouse, inhibitor and/or knockout | No change [142] in scid/scid mouse | ND | 30~70% [143] in DNA-PKcs$^{-/-}$HL B cells, ~100% by inhibitor, PKi NU7026 | 180%[d] [143] ATM$^{-/-}$ PKi cells | 1000% [144] (DNA-PKcs$^{-/-/-}$ DT40) |
| | Ku70$^{-/-}$ or Ku80$^{-/-}$ mouse | ND | ~20% (in Ku80$^{-/-}$ mouse) [135] | ~10% (in Ku80$^{-/-}$ mouse) [135,145] | 250% In Ku70$^{-/-}$ mouse ES cells [146] | Increased [144] (Ku70$^{-/-}$ DT40) |
| NHEJ/AEJ | Mre11-deficient patient or $^{-/-}$ mouse | No change (ATLD patients) [147] | ND | ~20% in MRE11$^{-/-}$ mouse[e] [148] | | ND |
| | NBS1-deficient patient or $^{-/-}$ mouse | Lower than control (NBS1 patients) [147] | ~22% (in Nbs1Δ/- mouse) [126] | 40–50% (In Nbs1Δ/- mouse) [126] | ~500% (in Nbs1Δ/- mouse) [137] | ND |
| | Overexpressed.NBS1 | 290% (in Ramos cells) [149] | ND | ND | ND | 300% [149] (in DT40) |
| | Rif1 Δ/Δ mouse | ND | ND | ~20% [150] | 150% [150] | 50%[f] (in U2OS cells) [151] |
| | Cernunnos-deficient patient or $^{-/-}$ mouse | ND | 40–50% In human [152] | Decreased CSR with longer microhomology in human [152], 40% in Δ/Δ mouse [153] | Chromosome/chromatic breaks↑ [154] | ND |
| | XRCC4 Δ/Δ mouse | ND | ND | 50% [155] | | 400–700% (in CHO cells)[g] [156] |

| | | | | | | |
|---|---|---|---|---|---|---|
| AEJ/BER | UNG$^{-/-}$ | ~200% [157,158] | 200–600% [159–161] | 5–25% [137,157] | ~0% [137,161] | <5% (UNG$^{-/-}$DT40) [162] |
| | PARP1$^{-/-}$ mouse and/or inhibitor | ND | ND | No change (PARP1$^{-/-}$ mouse B-cells)[h] [376] 50%[i] (mouse B-cells with KU0058948) | No change,[j] (PARP1$^{-/-}$ mouse) [376] ~10%[k] (in mouse ES cells) [377] | ~7% (PARP1$^{-/-}$ DT40) [378] |
| | PARP2$^{-/-}$ mouse and/or inhibitor | ND | ND | No change (PARP2$^{-/-}$ mouse) [376], ~50% (PARP2$^{-/-}$ mouse cells with KU0058948) [376] | ~900%[l] (PARP2$^{-/-}$ mouse) [376] | ND |
| | XRCC1$^{+/-}$ mouse | 150–200% [379] | ND | No change [379] | ~50% [379] | ND |
| | Exo1$^{-/-}$ mouse | No change [380] | ND | ~30% [380] | | ND |
| | CtIP siRNA | ND | ND | ~50%[m] [381] (CH12 cells) | ~30%[n] [382] | ND |
| | APE1$^{-/-/-}$ in CH12F3-2A cells | ND | No change[o] | ~20% [129] | No change[o] | ND |
| | Pol beta-deficient cells or $^{-/-}$ mouse | ~700% BL2 Pol beta deficient clone [383] | ~170% in pol beta$^{-/-}$ B-cells [384] | No change in pol beta$^{-/-}$ B-cells[p] [384] | ND | ND |
| | FEN1$^{E160D/E160D}$[q] mouse | No change [132] | | No change [132] | | ~30% (FEN1$^{-/-}$DT40) [131] |
| MMR | MSH2$^{-/-}$ mouse | ~20% (AT↓) [385] | No change [386] | Decreased[r] [387], or 30% to ~60% [388] | ND | No change in CHO cells[s,] [389] |
| | MSH6$^{-/-}$ mouse | <15% (AT↓) [390] | (AT↓) [390] | ~30% [390] | ND | No change (Msh6$^{-/-}$ DT40) [391] |
| | PMS2 $^{-/-}$ mouse | No change [392] | Increased microhomology [392] | ~30% [392] | ND | No change (PMS2$^{-/-}$DT-40) [391] |
| NER | ERCC1$^{-/-}$ mouse | ND | No change [134] | No change [133] or 60–80% [134] | ND | ND |

*Continued*

**TABLE 2 DNA Repair Factors and Their Enrollment in Immunoglobulin Diversification—cont'd**

| | | | | | | |
|---|---|---|---|---|---|---|
| TLS | | | | | | |
| (Y-family) | Pol eta$^{-/-}$ mouse | ~130% (AT↓GC↑) [393] | ~200% (AT↓GC↑) [393] | No change [393] | ND | ~50% (Pol eta$^{-/-}$DT40) [394] |
| | Pol iota null cells or $^{-/-}$ mouse | 13–27% (pol iota null BL2) [178], No change (pol iota$^{-/-}$ mouse) [395,396] | ND | No change (pol iota$^{-/-}$ mouse) [395] | ND | ND |
| | Pol kappa$^{-/-}$ mouse | No change [397] | ND | No change [397] | ND | ND |
| | Rev1 mutant or Rev1$^{-/-}$ mouse | 65%[t] (Rev1 mutant mouse) [398] | 130% (transition↑) (Rev1$^{-/-}$ mouse) [160] | 40–50% (Rev1$^{-/-}$ mouse) [160] | ND | 30% (REV1$^{-/-}$DT40) [402] |
| (A-family) | Pol theta polymerase (–) helicase(+) mutant | ~80% (GC↓) [399] | ND | Slightly reduced[u] [399] | ND | No change (pol theta$^{-/-}$ DT 40) [400] |
| | Pol nu$^{-/-}$ DT 40 | ND | ND | ND | ND | No change (pol nu$^{-/-}$ DT 40) [400] |
| | Pol theta$^{-/-}$ Pol nu$^{-/-}$ DT4 | ND | ND | ND | ND | 20% [400] |
| (B-family) | REV3 (Pol zeta) Δ/Δ mouse | ~5% [177,401] | ND | 35–40% [177,401] | ND | No change (REV3$^{-/-}$DT40) [402] |
| | REV3 L2610F mutant mouse | ~170% [177] | ND | No change [177] | ND | ND |

ND, not done; Δ, conditional gene deletion by cre-loxP system.
[a]*The evaluation system of GC is described for each gene.*
[b]*I-SceI-mediated HR substrate.*
[c]*I-SceI-mediated HR subtrate in HeLa cells. siATM + ATM inhibitor KU55933.*
[d]*ATM$^{-/-}$DNA-PKcs inhibitor vs. ATM$^{-/-}$.*
[e]*20% in MRE11$^{-/-}$ and 50% in MRE-/H129N cells.*
[f]*Homology directed repair system, by siRNA.*
[g]*I-SceI-mediated gene-conversion/homologous recombination substrate.*
[h]*PARP1$^{-/-}$ spleen B-cells with PARP inhibitor KU0058948.*
[i]*Our unpublished data.*
[j]*~300% of c-myc-IgH translocation in PARP1$^{-/-}$ spleen B-cells with PARP inhibitor KU0058948.*
[k]*Olaparib in p5pF reporter system.*
[l]*300% in PARP2$^{-/-}$ spleen B-cells with PARP inhibitor.*
[m]*CtIP knockdown caused ~60% AID expression.*
[n]*pCr15 translocation reporter.*
[o]*JX, MK, and TH, our unpublished data.*
[p]*Except for IgG2a.*
[q]*FEN1$^{E160D/E160D}$ is a catalytic-deficient mutant.*
[r]*Decreased IgG3 switching in vitro cultured.*
[s]*Recombination substrate.*
[t]*In REV1D568AE569A mutant.*
[u]*High-affinity antibody is decreased to 70%.*

on the other end (Figure 2); the overhang could be removed by FEN1. Thus, all of these enzymes involved in converting SSBs into DSBs are redundant, and their defects would partially damage CSR without severely inhibiting SHM.

## 3.2 Evidence for AID's Cytidine Deamination Activity on DNA or RNA

There is clear consensus about AID's deamination activity. First, recombinant AID deaminates DNA in vitro [53,163]. Interestingly, this in vitro DNA deamination is highly specific for ssDNA [53,163] and markedly reduced for dsDNA, especially in the form of chromatin [164,165]. No evidence has been reported to indicate that AID exerts cytidine deamination activity on naked RNA as a substrate in vitro. Second, AID overexpression in *E. coli* or yeast can introduce mutations in a wide spectrum of genes [9,166]. Interestingly, such DNA deamination activity is not unique to AID, but is shared by other members of the APOBEC gene family [49]. APOBEC 1, a bona fide RNA-editing enzyme, deaminates DNA in vitro and in *E. coli* [97]. Recently, AID was shown to deaminate both RNA and DNA when encapsulated in the hepatitis B virus [167]. Collectively, AID can deaminate C on ss nucleic acids regardless of whether they are found in DNA or RNA.

## 3.3 Two Separate AID Functions

As described in Section 2, AID genes isolated from patients with HIGM type II syndrome contained mutations spread all over the molecule. Among these, mutations in the C-terminal region, such as P20 and JP8B, greatly reduced CSR but not SHM activity [16,28]. Similar CSR-specific loss-of-function mutants were generated by in vitro AID mutagenesis. These mutants, especially those with mutations near the C-terminal NES motif, had clear defects in CSR, but had enhanced SHM and DNA cleavage activity [17]. Importantly, the P20 protein and WT AID differ little in stability. These results clearly indicate that AID's DNA cleavage activity is not sufficient for efficient CSR. In addition, in a knock-in mouse with the AID N-terminal mutation G23S, CSR was relatively intact, but SHM was reduced owing to a decrease in DNA cleavage [27]. Taken together, AID appears to have two separate functions: DNA cleavage, which requires an intact N-terminus, and CSR-specific activity, located in AID's C-terminal region. The idea that AID has two separate functions is still controversial. However, the P20 mutant, which has specific defects in CSR, shows normal nucleocytoplasmic shuttling [29]. Any hypothesis for AID function must explain the fact that C-terminal mutations of AID have intact DNA cleavage activity, but no CSR activity.

## 3.4 DNA Deamination Hypothesis

### 3.4.1 The Model

AID was proposed to have DNA-deamination activity based on its in vitro cytidine deamination activity on ssDNA and G/C-specific mutations when overexpressed in *E. coli* [9,168]. The DNA deamination model postulates that AID's deaminase activity generates U from C on DNA, introducing mismatched U/G base pairs. Uracil DNA glycosylase (UNG) is proposed to remove the Us and create abasic sites, after which Ape1 cleaves the phosphodiester bond to produce nicking cleavage. The generation of a DSB requires additional nicking on the opposite strand, exonucleolytic or endonucleolytic digestion, or helicase or DNA synthesis, as discussed previously. It has also been proposed that U/G mismatches are recognized and cleaved by the mismatch repair (MMR) system of the Msh2/6 complex. Somatic hypermutation has been explained by either U/G accumulation or incorrect repair of Ape1-nicking by error-prone DNA polymerases. The DNA deamination hypothesis postulates that AID is responsible only for DNA cleavage, and that the end processing and recombination in CSR are mediated by ubiquitous DNA repair systems such as NHEJ.

### 3.4.2 Evidence For and Against the DNA Deamination Model

First, AID's DNA-deamination activity is well supported by several in vitro experiments, as described previously [43,53,163,169]. However, the in vitro deamination activities of various AID mutants do not correlate with their in vivo SHM and CSR activity [26]. In particular, the AID mutation N51A loses in vitro DNA-deamination activity but retains CSR activity. A mutation at the homologous position N57 in APOBEC1 also abolishes its ability to deaminate DNA in vitro while retaining its ability to edit RNA.

Second, a deficiency in the critical enzyme UNG dramatically reduces CSR efficiency to less than 10% of WT activity, but importantly, not to zero [157,158]. This residual activity may be explained by MMR activity, which also introduces SSBs by recognizing mismatched base pairs (U/G) [170]. However, an Msh2 deficiency reduces CSR to 20–50% of that of WT, and Msh2$^{-/-}$ UNG$^{-/-}$ mice still retain 2% of the WT CSR [170]. Curiously, UNG deficiency does not decrease SHM efficiency, but rather augments it [159,171,172]. It was recently demonstrated that UNG deficiency does not prevent high-affinity antibody generation [158]. Moreover, overexpressing UNG in B-cells suppresses SHM at the S region [173]. Thus, UNG functions differently in CSR and SHM, and this difference cannot be explained by UNG's function as proposed by the DNA deamination model.

Furthermore, in a series of UNG-mutant experiments using UNG-deficient splenic B-cells, loss-of-catalysis UNG mutations had little effect on the activity that supports CSR [174]. A plot of UNG's U-removal activity versus CSR-supporting activity is totally discordant [175]. Furthermore, UNG's suppression of SHM is independent of its catalytic activity [173]. A plot of various loss-of-catalysis UNG mutants' activities clearly showed an absence of correlation between SHM-suppressing and CSR-supporting activities, which confirms that UNG's roles are distinct in CSR and SHM (Figure 2). To suppress SHM, UNG appears to compete with error-prone polymerases to bind to cleaved sites, and serves as a scaffold to recruit the error-free repair complex of base excision repair (BER) enzymes such as Polβ, XRCC1, *Ape1*, FEN1, and others. Consistent with this scenario, UNG with a loss-of-catalysis mutation has been shown to bind to SSB [173]. These results are contradictory to UNG's function as proposed by the DNA deamination model.

Third, an Ape1 deficiency in CH12F3-2A cells reduces CSR to one-third or one-fifth of that in the original CH12F3-2 cells, and this reduction is rescued by Ape1 transfection [129]. Although these results have been interpreted to support the DNA deamination model, the residual CSR activity suggests that DNA-cleaving mechanisms other than Ape1 are involved. In fact, SHM efficiency at the S region was not affected in the Ape1-deficient CH12F3-2A cells (Xu, MK, TH, our unpublished data), which suggests that Ape1 may not be the enzyme responsible for generating the SSBs that are essential for SHM. Because Ape1 deficiency partially reduces CSR, it is likely that Ape1 is involved in processing the SSB to a DSB. In fact, Ape1 is known to remove 3′-tyrosyl residue (or peptides). Because redundant enzymes are involved in processing SSBs to DSBs, it is reasonable that Ape1 deficiency alone cannot completely block DSB formation, and thus cannot completely block CSR.

Fourth, U base accumulation in the Ig locus after the stimulation of UNG-deficient DT40 chicken B cells was reported in experiments designed to detect U bases on DNA by in vitro hydrolysis using UNG and Ape1, followed by PCR amplification to monitor the loss of Ig gene signals. The position of U, detected by linker ligation at the cleaved sites, was scattered in the $J_H$ locus. Interestingly, the frequency of U generation, as assessed by in vitro cleavage and linker ligation, did not differ in the S region DNA from $UNG^{-/-}$ and $Msh2^{-/-}$ $UNG^{-/-}$ splenic B-cells, which suggests that Msh2 does not have a significant role in removing U [176]. The frequency of U generation is as high as one per 500 bp, suggesting $2 \times 10^{-3}$ mutations even without the involvement of TLP, a major enzyme known to incorporate mutations [177,178]. In fact, recent genetic manipulation of error-prone polymerase Rev3 indicates that most mutations are introduced by error-prone polymerases [177]. Such a high frequency of U accumulation in UNG-deficient cells should affect SHM. However, a recent in vivo SHM study showed that UNG is not required for high-affinity antibody formation by SHM [158]. In addition, the lifetime mutation frequency in UNG-deficient mice is not high, which suggests that U/G mutations are efficiently repaired by UNG paralogs such as SMUG1 [179]. It remains to be shown whether there is a mechanism that suppresses error-free repair by BER and MMR only when AID is expressed. Taken together, the evidence supporting in vivo DNA deamination by AID should be reexamined.

Fifth, the DNA deamination model postulates that the recombination step for CSR is mediated by ubiquitous DNA repair systems, without AID's help. To examine this possibility, knock-in mice containing two I-SceI sites, one in Sμ and the other in Sγ1, were generated [180]. These animals clearly showed CSR when the I-Sce1 enzyme was expressed, which is consistent with the proposed DNA deamination model. However, the efficiency of CSR was only about 10–20% of that of WT, despite the much higher efficiency of DSBs induced by I-Sce1 than by AID. These results suggest that AID is likely to have an additional function to support efficient CSR, beyond generating the DSB. This possibility is supported by the specific defect of CSR in C-terminal AID mutants [16,28]. Collectively, so far, there is no direct evidence to prove the DNA deamination hypothesis in vivo. In addition, there are more data that cannot be explained by the DNA deamination model.

## 3.5 RNA Editing Hypothesis

### 3.5.1 The Model

The alternative model for AID function proposes that AID deaminates C in RNA. This model postulates that AID edits two types of RNA, and that the edited RNAs mediate two separate AID functions: namely, DNA cleavage and recombination. AID is assumed to interact with at least two separate cofactor proteins that capture separate RNA substrates to be edited. One such RNA is proposed to be an micro-RNA (miRNA) precursor that can regulate mRNA translation. More specifically, AID is proposed to modify miRNAs, and the modified miRNAs enhance the down-regulation of topoisomerase-1 (Top1) mRNA translation, which reduces the Top1 protein. The reduction of Top1 facilitates the formation of non-B DNA structure in the V and S regions of the Ig locus when they are actively transcribed. Top1 forms a transient covalent bond with the 3′ phosphate of DNA through its tyrosine residue, and generates a nick. Normally, Top1 rotates DNA around the DNA helix to normalize negative superhelix accumulated in the region downstream of the transcription machinery complex. However, Top1 is trapped in the non-B structure, which inhibits its rotation around the helix, resulting in irreversible cleavage. This Top1 covalent binding with a nick is the mechanism for AID-induced DNA cleavage.

The bound Top1 should be removed by a series of enzymes, including proteasomes, TDP1, and TPP1, before the repair and the process to accomplish SHM or CSR begins (Figure 2). AID is likely to edit not just one, but multiple miRNA precursors to cleave DNA, because Top1 KD alone enhances but does not initiate DNA cleavage. The similar translation and regulation of mRNA-encoding proteins involved in target specificity might be mediated by miRNA that has been edited by AID. The other RNA is an unknown mRNA; its editing will encode a new protein responsible for CSR recombination by processing the SSB–DSB and forming the DSB synapse. As described in Section 2, AID's C-terminal region is responsible for editing this unknown mRNA, which generates a novel protein to mediate the synapse formation of cleaved ends.

### 3.5.2 Evidence For and Against the RNA Editing Model

#### 3.5.2.1 Top1 Is the Cleaving Enzyme Regulated by AID

First, CSR is blocked by camptothecin, a highly specific Top1 inhibitor. CH12F3-2A cells were incubated with camptothecin at an extremely low concentration (30 nM) for 3 h to block CSR [181]. This treatment causes irreversible nicking by the intercalation of the camptothecin–Top1 complex. The formation of this stable complex suppresses the processing of covalently bound Top1 from DNA, which is required for CSR.

Second, AID expression reduces the amount of Top1 protein by suppressing its mRNA translation [181]. Loss-of-catalysis AID mutants cannot reduce the Top1 mRNA translation. Consistent with this finding, Top1 KD enhances DNA cleavage, as assessed by linker-mediated qPCR, and the accumulation of gH2AX focus formation, thereby enhancing CSR and SHM [181]. Furthermore, SHM frequency is enhanced in Peyer patch B-cells from Top1-heterozygous mice [182]. Taken together, these results indicate that Top1 itself is involved in DNA cleavage when the amount of Top1 protein is reduced by AID.

Third, Top1 has been shown to introduce mutations in DNA sequences prone to form non-B structure in yeast and in an *E. coli* system. In the well-known phenomenon of transcription-associated mutagenesis (TAM) mutations, the mutations, especially microdeletions, tend to accumulate at repetitive sequences. These mutations were shown to be Top1-dependent [183,184]. Furthermore, the triplet-repeat contraction and expansion that causes genetic disorders such as Huntington disease is proposed to depend on Top1 activity [185]. These findings indicate that Top1 introduces irreversible cleavage at the repetitive sequence region, where non-B DNA structures are prone to form during extensive transcription. Consistent with this possibility, the Ig S and V regions where DNA cleavage frequently occurs are enriched in inverted or tandem repeats that can form non-B structures [11,186]. This scenario is supported by several other observations. For instance, all AID-preferred targets contain repetitive sequences that are prone to form non-B structures [11]. Furthermore, Top1 reduction by antisense oligonucleotides causes genome instability, including chromosomal translocation, in cell lines [187,188].

Fourth, the RNA editing hypothesis proposes that AID edits miRNA precursors or long noncoding RNAs, and modulates their quantity or quality. Some of these edited products associate more efficiently or abundantly with Top1 mRNA and reduce its translation efficiency. Recently, Kobayashi et al. (our unpublished data) showed that Ago2-bound miRNA binds to a specific site of Top1 mRNA when AID is activated. Top1 reduction alone can induce genome instability, including c-myc-Ig translocation, and Top1 reduction enhances AID-induced DNA cleavage and CSR. The RNA editing hypothesis offers an explanation: namely, that AID edits not just miRNA for Top1 mRNA translation regulation but also several other miRNAs. Because Top1 reduction alone can induce genome instability, the other miRNAs affected by AID may be involved in the target specificity of Top1 cleavage.

#### 3.5.2.2 AID Edits mRNA to Generate a New Protein That is Specifically Required for Recombination in CSR

First, CSR depends on de novo protein synthesis [189]. When the protein-synthesis inhibitor cycloheximide was used to block new protein synthesis immediately after AID activation, even a single hour of exposure strongly suppressed CSR in CH12F3-2A cells. A similar treatment before AID activation did not affect CSR. Leptomycin B, a potent inhibitor of nuclear export that is known to bind to exportin 1 and block the transport of newly synthesized mRNA from the nucleus, also blocks CSR strongly [29].

Second, AID binds poly A–containing RNA; the AID protein was trapped by oligo dT beads, which trap poly $(A)^+$ RNA [190]. This did not occur when AID was replaced with a C-terminal–truncated form able to induce SHM but unable to support CSR. This result suggests that AID's C-terminal region is responsible for its interaction with poly $(A)^+$ RNA. This idea is supported by a recent finding that AID forms a dimer, and that the formation of this dimer depends on AID's C-terminal region. The dimer is also responsible for forming a large RNP complex identified by glycerol density gradient sedimentation (Mondal S, Begum NA, and Honjo T, unpublished data).

Third, AID's C-terminal region is likely to produce a new protein that is responsible for recombination. Analysis of CSR defects in C-terminal AID mutants showed that new proteins derived from these mutants appear to regulate the processing of SSBs–DSBs, and probably recruit several nucleases [31,191]. This defect causes the accumulation of SSBs that are not processed to DSBs. Thus, SHM is maintained even when CSR is severely defective. When an

AID C-terminal mutant is expressed, the Ig S region accumulates more SSB-binding proteins such as PARP1 and MSH2, whereas the accumulation of DSB blunt-end binding proteins such as Ku and XRCC4 is drastically reduced. The C-terminal AID mutant also fails to form synapses between two cleaved S regions. Consistent with these findings, 53BP1, DNA PKcs, and UNG, which are involved in S region synapse formation, do not accumulate in the S region in these mutants [31]. Thus, AID's C-terminal region is involved in synthesizing the new protein required to process an SSB into DSB blunt ends, prerequisite to the synapsis that is critical to NHEJ and the completion of CSR.

To summarize, the DNA deamination hypothesis, which is based on DNA cleavage through UNG and APE1 function, must now be reconsidered (Table 3). On the other hand, the evidence for the RNA editing hypothesis has been strengthened by detailed data about Top1 activity and AID's C-terminal function. Nonetheless, the RNA editing hypothesis has not been proven directly by identifying the miRNA or mRNA used as an AID target.

## 4. THE MECHANISM OF AID'S SPECIFICITY DETERMINATION FOR DNA CLEAVAGE

### 4.1 Cleavage Target Sequence: *cis*-Recognition and Transcription

CSR takes place between a universal donor S region (Sμ) and one of the acceptor switch regions (Sγ/Sα/Sε) located upstream of each constant region [113,198]. The I promoter located immediately upstream of each S region drives active transcription through the recombining S region. The S regions are 2–12 kb in length, with highly repetitive units of 25–80 bp that are arranged in tandem and contain motifs such as TGGGG, GGGGT, GGGCT, GAGCT, and AGCT [199,200].

To determine whether AID-induced DNA cleavage requires specific DNA sequences, assay systems were developed by introducing various artificial switch substrates into CH12F3-2A B-cells or by genetic manipulation of the switch regions in mice [201–204]. In the earliest effort, the Sμ and Sα regions and their derivatives were used to construct artificial switch substrates for switching in CH12F3-2A cells; results showed that S regions from different species and isotypes, as well as S regions with an inverted orientation, all showed efficient CSR [3]. Surprisingly, artificial multiple linker sequences could also serve as a CSR target. Computational secondary structure analysis showed that S regions have the potential to form non-B DNA structures including stem loop structures. Although the *Xenopus* S region is AT-rich, it harbors AGCT motifs and secondary structures with inverted repeats [205]. Replacing the mouse S region with the *Xenopus* S region does not prevent efficient CSR [165]; however, CSR efficiency is reduced if the region is replaced with a mouse S region with altered orientation [202,204]. More recently, a similar replacement study in the CH12F3-2A cell line demonstrated that transcriptional orientation and the presence of successive AGCT motifs are critical for CSR [206].

The transcription-coupled DNA structural changes in S region sequences have been suggested to be important for AID targeting, given that CSR is abolished when S regions or their promoters are deleted. Studies using bisulfite sequencing also demonstrated that the transcription of repetitive GC-rich sequences in vivo and in vitro generate non-B structures, such as R-loops and G-quartets (G4) [186,207–210]. In fact, alternating the sequence motifs, such as changing from AGCT to ACGT, can dramatically reduce GLT as well as CSR [206], which suggests that the individual S region repeats are naturally set for the optimal transcription and secondary structure formation to initiate DNA cleavage and recombination.

R-loops can be generated in an S region when the template strand hybridizes with GLT, causing the nontemplate strand to loop out. The nontemplate strand in an R-loop provides a long stretch of exposed ssDNA, which is proposed by the DNA deamination model to be an ideal AID substrate. Although AID deaminates the nontemplate strand of DNA in vitro as well as in bacteria and yeast, AID-induced SHM occurs on both strands in B-cells in vivo [211,212]. AID is also reported to deaminate dCs on both strands in supercoiled but not relaxed plasmid DNA [213,214]. However, RNaseH1 overexpression did not augment CSR in CH12F3-2A cells [116].

Like CSR, SHM is tightly coupled with transcription. In SHM, however, the transcribed V region does not form an R-loop, but generates an abundant single-stranded DNA structure [214,215]. Although the V region lacks the prototype repetitive sequence of the S region, inverted repeats and triplet-like motifs (AGC, TAC, GCT, and GTA) were found in the V region [216,217]. Although these triplets are not consecutively arranged as a triplet track, their abundant clusters are frequent targets of mutation in the V region. Moreover, SHM is enhanced by the introduction of the transcription factor E2A-binding sequence CAGCTC (E-box motif), which may be prone to forming secondary non-B DNA structures, into the V region or the GFP gene [218–220]. The transcribed GFP transgene is often used as an efficient SHM substrate owing to its high GC content and abundant AGCT motifs [107]. Intriguingly, CAGCTC motifs are found frequently in the S region.

A genome-wide approach to identifying AID-induced DNA break points in BL2, a human Burkitt lymphoma line, identified several known and novel non-Ig AID target loci, including MYC, MALAT1, and SNHG3 [11]. Similar to the IgH locus, all of the identified non-Ig loci showed frequent DNA breaks and characteristic SHM in the vicinity of the break points. All of these loci are highly transcribed, and the break regions harbor abundant repetitive sequences, including AGCT sequences [11,13].

**TABLE 3 Comparison of Two Hypotheses for AID Deamination**

| Supportive Evidence | Contradictory Data (or Other Explanation) |
|---|---|
| **DNA Deamination** | |
| DNA Cleavage Mechanism | |
| • DNA deamination in vitro, in *E. coli* and yeast [9,166] | • In vitro deamination activities of various AID mutants do not correlate with in vivo SHM and CSR activity [26]<br>• Even APOBEC1 can deaminate ssDNA in vitro [49] |
| • $UNG^{-/-}$ reduces CSR [157,171] | • UNG's catalytic activity is dispensable for DNA breaks and CSR [174,175] |
| • APE1 deficiency reduces CSR [129] | • UNG promotes end-joining and S-S synapsis in CSR [173].<br>• APE1-deficient cells have DNA breaks, c-myc-IgH translocation, and somatic hypermutation at comparable levels with APE1-proficient cells (JX, MK and TH, our unpublished data) |
| Target-Specificity Determination | |
| • AID preferentially edits the WRC motif [192]<br>• First phase of somatic hypermutation is C to U by AID; the second is from A and T by error-prone repair enzymes [193]<br>• The AID-recruiting proteins such as RPA and the others (Table 1) recruit AID to the specific target DNA region | • Most mutations are incorporated by translesion polymerases (TLP) [177,401] |
| Role of Transcription | |
| • R-loop formation at switch regions in stimulated B-cells generates single-strand DNA for AID substrates [194] | • RNaseH1 overexpression does not change CSR efficiency [116]<br>• Transcription is necessary for non-B DNA structure [181,182] |
| Domain Functions of AID | |
| • Nuclear localization signal in AID's N terminus is required for AID to enter the nucleus to deaminate DNA [36] | • AID's N terminus is responsible for DNA breaks |
| • Phosphorylation of S38 by PKA augments AID's function by facilitating interaction with RPA [65,195] | • RPA requirement is not shown in vivo |
| • I-SceI cleavage alone produces CSR (= AID is necessary for DNA breakage but dispensable for recombination) [180] | • CSR by I-SceI cleavage is much less efficient than by wild-type AID activation, although the cutting efficiency is higher than AID |
| • AID's C-terminus maintains AID's stability and regulates CSR [196] | • DNA breaks and SHM are enhanced in a C-terminal deletion mutant, but CSR is decreased [16,17,28] |
| Role of UNG in hypermutation | |
| • UNG deficiency promotes G/C transition (G to A/C to U) [157] | • UNG overexpression suppresses SHM [173] |
| **RNA Deamination** | |
| DNA Cleavage Mechanism | |
| • Topoisomerase 1 (Top1) inhibitor decreases CSR and SHM [181,182]<br>• Knockdown/heterozygote of Top1 increases CSR and SHM [182]<br>• AID decreases Top1 protein [181]<br>• H3K4me3 localization matches the highly mutated SHM region [11]<br>• Top1 interacts with FACT complex and H3K4me3 through chromatin remodeling complex (AH, NAB, MK and TH, our unpublished data) | • Top1 protein decrease alone is not sufficient for DNA breaks [181] |
| Target-Specificity Determination | |
| • AID targets are highly repetitive non-B-prone sequences [11]<br>• The FACT complex is necessary for H3K4me3 of the S region [62]<br>• Genome-wide, AID-dependent DSBs frequently occur within non-B-prone repeated sequences [197]<br>• TAM in yeast and triplet repeats' instability shares similar mechanism including non-B DNA and Top1 [183–185] | |

*Continued*

**TABLE 3 Comparison of Two Hypotheses for AID Deamination—cont'd**

| Supportive Evidence | Contradictory Data (or Other Explanation) |
|---|---|
| Role of Transcription | |
| • Germline transcription is necessary for non-B DNA structure formation [181,182]<br>• Transcriptional activity in AID-target regions regulated by epigenetic marking [62] | |
| Domain Functions of AID | |
| • AID's C-terminus is necessary for DNA end processing and DNA synapse formation [31] through polyA mRNA editing [190], but is dispensable for DNA breaks [17]<br>• AID localizes to the RNA complex [54] | |
| Hypermutation | |
| • TLP introduces SHM [177] | |
| AID can deaminate RNA | |
| • Hepatitis B virus RNA is deaminated by AID [167] | • AID alone cannot edit RNA in vitro |

These findings support the concept that transcription-induced non-B structure formation in a repetitive sequence is a critical factor for inducing breaks in both Ig and non-Ig loci during AID-induced genomic instability [13,19]. The RNA editing hypothesis proposes that the non-B DNA structure and epigenetic marks discussed subsequently constitute the target specificity determining factor by Top1.

## 4.2 The DNA Deamination Model Requires AID-Targeting Cofactors

Mechanistically, to execute dC deamination at specific targets, AID must be recruited to a transcriptionally active or accessible locus that will form the preferred ssDNA substrate [114,221]. AID's nuclear import is not sufficient to explain how it specifically deaminates dCs in the Ig V region during SHM and in the recombining S regions during CSR. Therefore, V and S region–specific AID-recruiting cofactors have been extensively pursued [39,222–224].

Based on AID's association with RNAPII, one early study proposed that AID was functionally associated with the transcription machinery [57]. Later, the ssDNA-binding protein RPA was proposed to be a cofactor that interacts specifically with PKA-phosphorylated AID (p-S38) [64,65]. Because RPA enhances dC deamination by AID in an in vitro transcription-coupled system using T7 polymerase, it was proposed that RPA stabilizes AID (p-S38) on the transcribed S region to promote AID-induced DNA deamination [221]. Supporting this view, the pharmacological inhibition of PKA impairs CSR, and both CSR and SHM are reduced in AID-S38A mice. [65,225]. However, the S38A AID mutant was proficient in dC deamination on ssDNA and in a mutagenic catalytic activity assay in *E. coli* [65,195]. Subsequent studies revealed that RPA deposition in the S region is AID-dependent [81]. Strand-specific RPA ChIP analysis of the Ig locus also demonstrated that RPA coats ssDNA generated after CSR-associated DSBs [226]. A recent study proposed that DNA deamination in the S region occurs in two phases. Initially, dC is deaminated at a low level independently of AID phosphorylation, producing a DNA damage signal that induces AID phosphorylation and RPA recruitment in the S regions; this leads to a second, elevated phase of dC deamination that enhances DSBs [84]. Basu et al. [71] proposed a mechanism by which AID can deaminate dCs in the template strand, which normally remains inaccessible because of RNA–DNA hybrid formation during transcription. The presence of nine RNA-exosome subunits, without the core catalytic unit, was shown to be sufficient for the robust targeting of AID to both strands of SHM substrates transcribed by T7 polymerase in vitro.

Normally, AID-induced mutations cluster at the promoter proximal region in the transcribed V region. Mutations usually begin to appear approximately 100–120 bp downstream of the transcription start site, and spread 1–2 kb further downstream [227–229]. Although RPA-related studies successfully linked transcription and AID-induced dC deamination or mutations on the transcribed DNA, they were unable to provide much insight into the characteristic feature (promoter proximal) of the natural AID target loci. In this regard, Pavri et al. [58,231] proposed that AID interacts with SPT5, an RNAPII-stalling factor, which in turn enhances AID's interaction with the stalled RNAPII. In eukaryotes, hypophosphorylated RNAPII initiates transcription with the help of basal transcription factors (TFs) and displaces the first nucleosome from the promoter; it then stalls at the promoter proximal region, where its C-terminal domain undergoes extensive phosphorylation (p-Ser5) [230, 232,233]. Elongating RNAPII (p-Ser2) begins transcription when the pausing factor DRB sensitivity-inducing factor (DSIF) complex (SPT5-SPT4) dissociates from the negative elongation factor (NELF) upon positive transcription elongation factor β (pTEF-β)–dependent phosphorylation [233,234]. Hence, AID's association with the poised RNAPII gives rise to the hypothesis that as AID moves along with the elongating RNAPII complex, repeated pausing may allow sufficient time for deamination in the transcribed track [114,235–237].

AID is reported to interact with other factors associated with the elongating RNAPII complexes, such as PAF, FACT, SPT5, and SPT6 [54,55,230]. All the subunits of the PAF complex have been detected in AID-IP experiments. Knockdown of LEO1, one of the PAF-complex components, reduces AID recruitment in the Ig locus [55].

Currently, the exact mechanism of AID's specific association with elongating RNAPII, pausing repeatedly in the target region, is not known. Sun and Basu [238] proposed that a macromolecular association exists among the RNA exosome, SPT5, AID, and NEDD4 in the stalling RNAPII complex, in which GLTs undergo exosome-mediated degradation. The RNA degradation process may release the complex from pausing to initiate AID-induced dC deamination on both strands of the DNA.

Although AID-induced mutations are preferentially restricted to the V- and I-promoter downstream 1–2 kb regions, core S regions also undergo extensive DNA breaks [227,239,240]. In fact, RNAPII stalls throughout the S regions; accumulations of stalling factor SPT5 as well as AID can be detected from the S regions all the way to the C regions [58,241]. Recently, the 14-3-3 series of adapter proteins was reported to bind the 5′-AGCT-3′ sequence motif as well as AID. Because S regions are enriched in AGCT motifs, it has also been suggested that 14-3-3 family proteins promote S region deamination by recruiting AID [88]. The chromatin marks H3K9acH3S10p and H3K9me3 are also implicated in recruiting AID through 14-3-3γ and TRIM28/KAP1, respectively [59,89]. Both proteins are unable to interact with a CSR-defective but SHM-proficient AID mutant with a C-terminal truncation. Interestingly, C-terminally truncated AID fused to 14-3-3γ completely rescued the defective CSR in AID-deficient B-cells [224]. These authors [66] also concluded that the HIV accessory protein Vpr, which is reported to interact with UNG and inhibit CSR [159], disrupts the S region macromolecular complex, which contains AID, RPA, and PKA, along with other molecules.

Additional modes of the locus-specific AID recruitment mechanism have been reported. Nowak et al. [70] suggested that antisense transcripts produced from the S region bind PTBP2, a cofactor that interacts with AID, and thus RNA-mediated AID targeting occurs in the S region for CSR. A similar idea was proposed with regard to the AID-cofactor GANP, which binds to the V transcript and thereby recruits AID specifically to the V region; this explains why mice lacking GANP are defective in SHM but not in CSR [86]. If this is the case, it would indicate that DNA cleavage is differentially regulated in the V and S regions.

Taken together, there are currently three types of models to explain targeted DNA deamination by AID [242]: (1) AID recruitment through ssDNA binding proteins and/or transcription elongation machinery associated components, such as SPT5, PAF1, and RNAPII; (2) target specificity determination by a locus-specific antisense RNA that mediates binding between a cofactor (GANP or PTBP2) and AID; and (3) AID recruitment by adapter proteins (TRIM28 and 14-3-3) that recognize specific histone marks, DNA sequence motifs, or both. To evaluate these models, it is important to note that SHM occurs even in mice in which the RNAPII promoter is replaced with the RNAPI promoter in the V region locus [118]. Furthermore, the S region can be replaced by multiple links, which indicates that the primary sequences are irrelevant to DNA cleavage.

## 4.3 Target Chromatin: Accessibility and Specificity

### 4.3.1 Characteristics of the Target Chromatin Nucleosome

Chromatin consists of repeating units of nucleosomes. Individual nucleosomes consist of a core H3-H4 tetramer flanked on both sides by H2A-H2B dimers [243]. Although nucleosomes can be displaced by RNAPII, in many transcribed genes they are restored by various mechanisms that are not fully understood [244–246]. Therefore, the cleaving enzyme needs to cross the nucleosomal barrier to reach the DNA substrate, after its recruitment to the target chromatin, regardless of whether the enzyme is AID or Top1.

In the DNA deamination model, it is not clear whether AID deaminates free DNA hanging between the nucleosomes, or if it is able to access the nucleosome after RNAPII migration destabilizes the nucleosome core [232,244]. AID's ability to deaminate nucleosomal DNA or chromatinized substrate was recently investigated in an in vitro transcription system [164,165]. The chromatin transcription template was constructed using plasmid DNA with a strong nucleosome positioning sequence (NPS) introduced. AID failed to induce dC deamination in nucleosome-occluded areas, but deamination sites were accessible when the plasmid DNA was transcribed. The same NPS sequence, when introduced into the V region in DT40 cells, severely reduced SHM efficiency [164]. The introduced NPS may have failed to establish the native chromatin architecture and the associated epigenetic state.

There is little information on nucleosomal organization or behavior in the context of the natural V or S region. In CH12F3-2A cells, Stanlie et al. [62] mapped the position of 2.5 nucleosomes at the 5′ end of the core Sμ, a region that contains hotspots for AID-induced mutations. The target chromatin appeared to undergo rapid nucleosome assembly, which indicates that the integrity of the locus relies on specialized histone chaperones [61,62]. Nucleosome positioning and stability are known to be greatly influenced by the nature of the target DNA sequence. Because the strong NPS did not favor AID-induced mutations, the Ig locus may naturally contain flexible nucleosomes as a result of the repetitious nature of the S region. In addition, the histone variant H3.3 was detected in the transcribing V region [87], which suggests that the locus is subjected to a high rate of histone exchange; this can make the locus labile and available for both DNA cleavage and repair. A recent study also suggested that H3.3 promotes recovery from the transient transcription shutdowns that frequently occur when a transcribed locus encounters DNA breaks [247].

Taken together, despite active transcription of the Ig locus, the rapid nucleosome reassembly mechanism presumably preserves its epigenetic integrity, which is required for recruiting various effector proteins involved in DNA cleavage and repair. On the other hand, the execution of complex DNA transactions requires adequate access to the target locus, which can be achieved through dynamic nucleosomal reorganization and/or histone-variant exchange in the target chromatin.

### 4.3.2 Histone Epigenetics for Target Determination

AID-induced DNA cleavage is always restricted to the promoter (V and I)-proximal 1–2 kb region, and to the repetitive S regions (1–10 kb) [200,201,248]. Notably, the S regions, but not the C regions, are the substrate of AID-induced DNA breaks, although the C regions are enriched with SPT5 and RNAPII (p-Ser5) and are transcribed along the same transcriptional unit [114]. Several characteristics of the local DNA and the transcribed chromatin may lead to the zonal restriction of AID-induced cleavage. For instance, transcription induces non-B DNA structures in repetitive DNA sequences that are present in the S regions, but not in the C regions. This signature is proposed to be critical for the RNA editing model [13,19,181]. Another decisive factor for the cleavage target could be the restricted distribution of the open-chromatin–associated histone epigenetic signatures, which change the transcribed chromatin into distinct boundaries, such as to distinguish the promoter proximal and promoter distal regions [249–251]. Recent studies also show a strong correlation between non-B structure–associated DNA breaks and epigenetic marks on the break point chromatin [61,62,87,181,182].

The chromatin dynamics of the transcribed Ig locus are important regardless of the model of AID function, and whether RNA or DNA is proposed as the primary target. It is important to distinguish the epigenetic marks for active transcription and the unique marks for AID-induced DNA cleavage. The DNA deamination model proposes that some epigenetic marks are required to recruit AID to the target chromatin through AID's indirect interaction with histone marks. It is not completely agreed upon as to whether AID cofactors recognize unique DNA features (R-loops or unknown primary sequences), epigenetic marks, or both. In contrast, the RNA editing model postulates that Top1 is more stably and specifically recruited to non-B DNA-enriched chromatin through combinatorial epigenetic marks.

In general, the promoter proximal sites (1 kb from the TSS) of actively transcribed loci are enriched with trimethyl histone H3 lysine 4 (H3K4me3), hyperacetylated H3 and H4, and RNAPII (p-Ser5). In contrast, the distal coding-region zones are enriched with H3K36me3 and RNAPII (p-Ser2). In the context of the Ig locus, H3K4me3, H3Ac, and H4Ac are elevated downstream of the active I promoters and throughout the transcribed S regions, whereas C regions have more H3K36me3 and H4K20me1 [241,252]. Active histone marks are generally confined to the promoter proximal region of the transcriptionally active loci. Strikingly, however, the transcribed S regions show unusually extended H3K4me3 tracks, whereas the C regions are completely devoid of H3K4me3 [61,62,241]. These results strongly suggest that H3K4me3 may be one of the epigenetic markers for targeting AID-induced cleavage.

In IgM-expressing B cells, Sμ is constitutively active and thus is enriched with active histone marks and RNA-PII, even before CSR is induced [61,62,241,253]. In contrast to Sμ, the downstream acceptor S regions accumulate H3K4me3, H3K36me3, and RNAPII only after isotype-specific CSR stimulation and transcription [253]. Extensive H3K4me3 accumulation, spreading up to 7 kb, is observed in the activated Sγ3 region in response to lipopolysaccharide (LPS) stimulation. Interestingly, when the Sμ tandem-repeat sequences were deleted in mice, H3K4me3 and H3 acetylation marks accumulated at the upstream of Sμ, and AID-induced cleavage shifted accordingly [254,255]. PAX interaction with transcription-activation domain protein-1–deficient mice did not accumulate H3K4me3, H3-acetylation, or RNAPII in the Sγ3, Sγ2b, and Sγ1, and transcription from these loci was completely inhibited [253]. Thus, there is a close connection between active transcription and specific histone epigenetic signatures, which are both essential for AID-induced DNA cleavage. A recent ChIP-seq study also shows that AID's localization significantly correlates with active histone marks, including H3K4me3 and RNA-PII (p-Ser5), in the Ig locus [231]. Although these results have been interpreted to support DNA deamination, these markers are ubiquitous to active genes.

H3K9acS10p has been proposed to serve as an additional histone code on H3K4me3-marked chromatin, which may provide additional specificity [89]. This concept is supported by the finding that H3K9acS10p can be found in both Sμ and Sγ3, but not Sγ1, in response to lipopolysaccharide (LPS) activation, which promotes switching to Sγ3. Similarly, LPS and IL4 induction, which is required for IgG1 switching, induce H3K9acS10p specifically in Sμ and Sγ1 but not in Sγ3. However, this chromatin mark is not exclusive to the actively transcribed S regions [256]. Furthermore, the presence of H3K9me3, a repressive chromatin mark, was reported specifically in the Sμ [59]. Consequently, CSR is defective in mice lacking the H3K9 methyltransferase SUV39H1 [257].

## 4.4 Chromatin Regulators for AID-Induced Cleavage and Repair

Regardless of whether DNA cleavage occurs by DNA or RNA editing, target chromatin specificity and access to the chromatinized DNA are essential, and the underlying mechanism and regulatory chromatin proteins involved remain largely unknown. Open-chromatin–specific histone post-translational modifications (PTMs) thus create an accessible chromatin environment before AID-induced DNA breakage by altering the chromatin structure, and by generating new recognition sites for effector proteins.

An analogous chromatin regulatory mechanism is necessary for efficient DNA transactions at the post-break repair phase of CSR and SHM [113,258,259]. In these events, the DNA damage signal rapidly accumulates γH2AX phosphorylation in the Ig locus [135,136], which in turn recruits MDC1 followed by 53BP1 [260,261]. The Tudor domain and the 53BP1 C terminus anchor H4K20me and H2AK15ub, respectively, on the chromatin, and the entire repair complex gradually forms around 53BP1 [262,263]. To identify critical regulators for target DNA recognition, cleavage, and repair, a large set of candidate genes were screened against siRNAs. These efforts successfully identified critical histone chaperones, elongation factors, and chromatin modulators that affect AID-induced DNA cleavage and repair.

### 4.4.1 Target-Chromatin Regulation by FACT

The histone chaperone FACT, a heterodimeric protein complex composed of SSRP1 and SPT16, is known to assist transcription elongation by RNAPII on a chromatin template [264,265]. Knockdown of either SSRP1 or SPT16 severely inhibited IgM-to-IgA class switching in CH12F3-2A cells [62]. The requirement of SSRP1 and SPT16 in CSR as part of a heterodimeric transcription elongation complex was further confirmed by introducing disruptive mutations at the interaction surface. Thus, the FACT complex was found to be an essential chromatin regulator for CSR. A mechanistic study revealed that SSRP1 depletion significantly reduced the AID-induced DNA breaks and γH2AX accumulation in both donor and acceptor S regions. Interestingly, no transcriptional defects were observed for μGLT, αGLT, or AID, and μGLT was actually enhanced when FACT components were depleted. This phenomenon was unexpected, considering that the FACT complex is known to promote RNAPII transcription through chromatin [266,267]. However, in FACT-depleted cells, the chromatin landscape of the S regions showed two distinct changes: Defects in histone stability and reduced histone PTMs are both associated with the transcribed locus. The dramatic loss of all four core histones was especially observed in Sμ, whereas active histone epigenetic marks were downregulated in both Sμ and Sα. Under FACT-depleted conditions, the H3K4me3 modification was significantly reduced in both Sμ and Sα, which correlated well with the impaired DNA cleavage observed at the respective locus [62].

Intriguingly, depleting H3K4me3 by knocking down WDR5 and ASH2, the core components of H3K4me3-histone methyltransferase (HMTase), recapitulated the results of FACT depletion—intact GLT and AID transcription, but a strong blockade of S region DNA cleavage [62]. SHMis also significantly correlated with FACT and H3K4me3 enrichment in the V region [87]. In contrast to Sμ but similar to Sα, the V regions show altered histone PTMs with no histone loss. However, FACT-enriched V and other non-Ig SHM target loci strongly accumulate the H3.3 histone variant [87,268], which is present in highly transcribed regions in a replication-independent manner [269]. Sequential ChIP analysis suggested that H3K4me3 is present in both H3 and H3.3 in V and S region chromatin (NAB, TH, unpublished data).

Although H3K4me3 is generally associated with an actively transcribing locus, it also has a critical role in marking meiotic recombination hotspots to be cleaved by Spo11 [270,271]. The same epigenetic mark is recognized by the PHD domain of RAG2, and promotes RAG-induced DNA cleavage in V(D)J recombination [272]. Therefore, the specific histone epigenetic mark H3K4me3 appears to be critical in all three recombination systems. In the context of CSR and SHM, FACT-mediated nucleosome assembly, H3K4me3 regulation, and histone-variant exchange appear to be the critical parameters in accessible DNA-cleavage complex formation [62,87]. In addition, Top1 physically interacts with FACT and H3K4me3, which serve as chromatin marks for DNA cleavage by Top1 (Husain A, TH, our unpublished data) (Figure 3).

### 4.4.2 Target-Chromatin Regulation by SPT6

Among the transcription elongation-associated histone chaperones examined, suppressor of Ty6 homolog

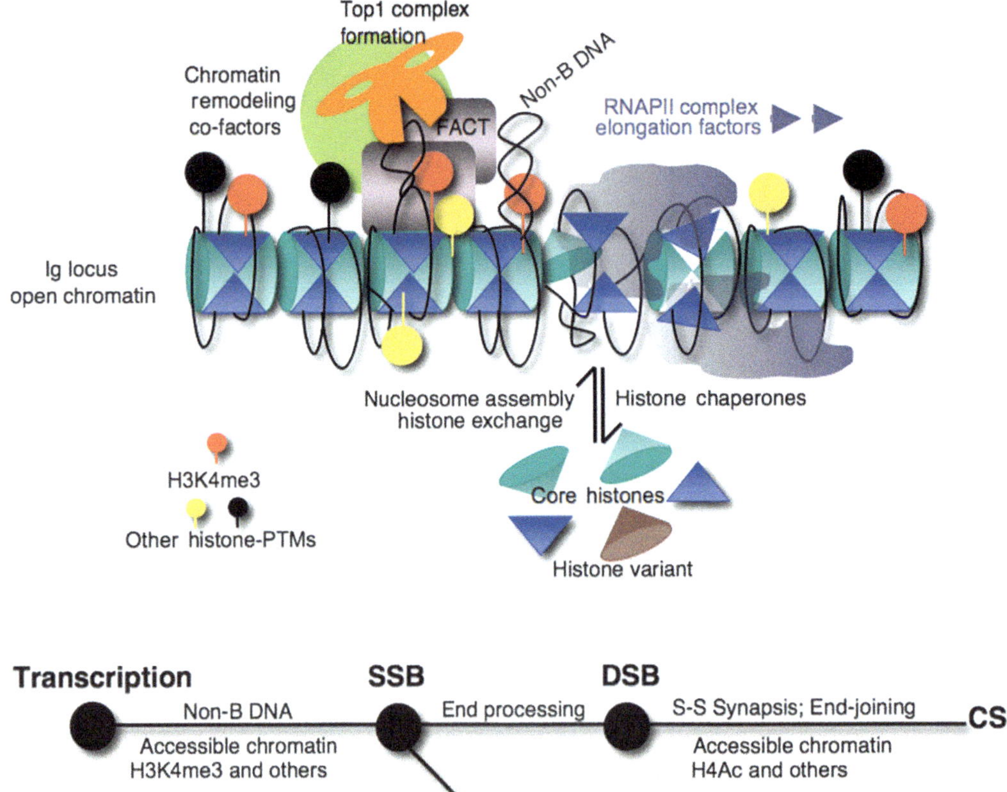


FIGURE 3 **Proposed model of Top1 and non-B DNA cleavage complex formation on the transcribed chromatin.** Actively transcribed chromatin of the Ig locus is enriched with various histone-PTMs, including H3K4me3 and others. Nucleosome assembly and histone exchange by histone chaperones help maintaining the dynamic and accessible state of the chromatin. Site-specific Top1 complex formation can be facilitated by combinatorial histone PTMs and associated chromatin regulators including FACT. Negative supercoil generated behind the RNAPII produces non-B DNA structures, the potential substrates of Top1. The scheme at the bottom highlights the key events associated with AID for the initiation and completion of SHM and CSR.

(SPT6) was found to be another essential histone chaperon involved in CSR [54,61]. SPT6, which was originally identified in *Saccharomyces cerevisiae,* is required for transcription elongation and chromatin structure maintenance. SPT6 KD in CH12F3-2A cells reduced the αGLT and AID transcription to some extent, but strongly inhibited CSR; this inhibition was not caused by altered AID transcription, because the CSR was inhibited even when AID was overexpressed [54]. Global gene expression analysis showed that only ~5% of genes were affected, none of which were genes known to be required for CSR. As with FACT KD, SPT6 depletion elevated the μGLT transcription and Sμ-cryptic transcripts. Consistent with these findings, there was no alteration of the amount of elongating RNAPII in the S regions. Deregulated histone PTMs such as H3K4me3 and H3K36me3 were evident in Sμ and Sα, and histone loss was observed in Sα. As SPT6 depletion hampered H3K4me3 formation in the S and V regions [61], both CSR and SHM were reduced, which correlated with impaired DNA cleavage at the target loci. SPT6 depletion also reduced SHM frequency in two noncoding genes, MALAT1 and SNHG3, with a concomitant decrease in H3K4me3 at these loci [61]. FACT and SPT6 are also involved in S region H3K36me3 regulation, but the KD of individual H3K36me3 HMTases (SETD2, WHSC1, and NSD1) did not dramatically affect CSR. Interestingly, SPT6 was involved in histone epigenetic regulation (H3K4me3 and H3K36me3) at the AID locus, but FACT was not [61]. AID gene expression correlated well with SPT6-mediated H3K36me3 formation, but not with H3K4me3 modulation. Thus, SPT6 appears to have dual functions in CSR: regulation of AID expression, and S region DNA cleavage, by preserving the histone epigenetic integrity of their respective genomic loci.

To understand how SPT6 regulates H3K4me3 status at the IgH locus, various SPT6 mutants were analyzed based on their ability to restore H3K4me3 upon SPT6 depletion. The C-terminal tandem SH2 domains of SPT6 were found to be necessary for recruiting SET1A H3K4me3-HMTase in both universal donor Sμ and acceptor Sα regions. Consistent with this observation, depleting the SET1-specific cofactor CXXC1 strongly inhibited CSR [61,273].

### 4.4.3 Distinct and Multiple Functions of SPT5 and SPT4

Screening with siRNA also revealed that the DSIF [274] components SPT5 and SPT4 are critical for CSR but not SHM [63]. SPT5 and SPT4 constitute the large and small subunits of the DSIF complex, respectively. Because DSIF is required for transcription elongation, KD of either subunit was expected to cause CSR defects owing to disruption of the complex. However, contrary to this expectation, SPT4 and SPT5 have distinct yet novel mechanistic roles in CSR [63]. The first indication of their dissimilarity came from global transcription analysis in CH12F3-2A cells in the absence of either SPT4 or SPT5. Although SPT4 or SPT5 depletion affected only a small number of CSR-irrelevant genes, most of the affected transcripts did not correlate well between the SPT4 and SPT5 KD samples. In addition, cryptic switch transcripts derived from the intronic region of Sμ were dramatically enhanced in SPT4 but not SPT5 KD samples, also indicating the difference in transcriptional regulation. Interestingly, SPT5 but not SPT4 KD slightly reduced the amount of AID protein without affecting its transcripts, which suggests that SPT5 is involved in AID's protein stability.

Although SPT4 depletion did not affect the H3K4me3 signal in Sμ and Sα, SPT5 depletion blocked H3K4me3 formation specifically in Sα. Consistent with this observation, SPT5 deficiency reduced Sα DNA cleavage, whereas SPT4 did not. Although it is not known why SPT5 regulates Sα chromatin more strongly than Sμ chromatin, it could be ascribed to differences in the inducible or constitutive promoter activity, nucleosome-free regions (NFR) or a non-NFR arrangement, or the specific role of the H2B ubiquitin ligase Bre1 in the histone trans-modification cascade, and thus in H3K4me3, at Sα [62,275,276]. Because αGLT transcription is more sensitive to DRB inhibition than μGLT transcription (NB, TH, unpublished), DSIF may affect the poised state in the Iα promoter more robustly than in the constitutively active Iμ promoter.

Studies on SPT4 and SPT5 also highlight their novel functions in the DNA repair phase of CSR, because both the NHEJ and HR pathways were deregulated in the absence of either of the DSIF components [63]. Likewise, earlier studies showed that SPT4 or SPT5 mutants in yeast resulted in methanesulfonate (MMS) sensitivity [277], which indicates defective DNA repair and recombination. Because proteins involved in NHEJ or homologous recombination, such as Ku70/80, DNAPKcs, and RAD51, associate with the RNAPII complex [278], DNA breaks appear to be repaired through transcription-coupled processes at some loci. Indeed, DSBs take longer to repair when located in heterochromatin, where transcription is silenced [279]. Taken together, DSIF subunits can function independently in multiple transcription-coupled steps of CSR.

Pavri et al. [58] showed that SPT5 is required for efficient CSR and is associated with more than 9000 AID-targeted loci, and concluded that SPT5 is a factor that guides AID for direct deamination of the target. However, it has not been determined whether the extremely large number of loci bound by SPT5 reflects AID's physiological relevance as a mutator, particularly because a particular protein's propensity to bind to or graze a specific DNA region does not always reflect the fundamental physiological outcome. Thus, it is likely that another as yet unknown marker is required. This concept is clearly reflected in the case of VDJ recombination; although the RAG proteins directly involved in VDJ DNA cleavage bind to both putative and cryptic recombination signal sequences (RSSs), meaningful RAG-mediated cleavage can occur only at some, and most likely not all, of these sequences [280–283].

### 4.4.4 The BRD4–Histone–Acetyl Complex in Repair and Recombination

Whereas several chromatin-associated factors involved in CSR's DNA cleavage phase have been identified, the involvement of chromatin adaptors that regulate the CSR step of DNA repair and recombination is less explored. To identify these novel players, proteins that can recognize or "read"acetylated histones were screened, because this particular histone PTM is implicated in various DNA repair systems [284,285]. RNAi screening specifically targeted to bromodomain-containing proteins, the efficient acetylated histone readers, led to the identification of BRD4 [286], a BET family protein, as a CSR regulator [287]. BRD4 depletion from the S region chromatin by either siRNA or by JQ1, a chemical inhibitor [288], significantly reduced CSR in both CH12F3-2A and cultured splenic B-cells [287]. Interestingly, the genome-wide transcripts, DSB formation in the S regions, and SHM frequency at AID target loci are largely unperturbed in BRD4-depleted compared with untreated samples, which suggests that BRD4's role in CSR is solely restricted to the postcleavage phase, that is, DNA repair and recombination. This is especially apparent because BRD4 tends to accumulate in the vicinity of the DNA breaks induced by AID.

Indeed, phenomena observed in the S regions in the absence of BRD4 include altered microhomology lengths of the switch junctions, reduced IgH/c-myc translocation efficiency, and depleted 53BP1 occupancy with enhanced MSH2/MLH1 accumulation at the S regions, which suggest that DNA repair is indeed defective. These events coincide well with those observed under UNG-ablated conditions [137,157,173,174]. Not surprisingly, then, BRD4 reduction also negatively affected UNG occupancy in the S regions. This finding indicates the importance of BRD4 in DNA repair and recombination at the S region, because UNG, through its non-canonical function, is implicated

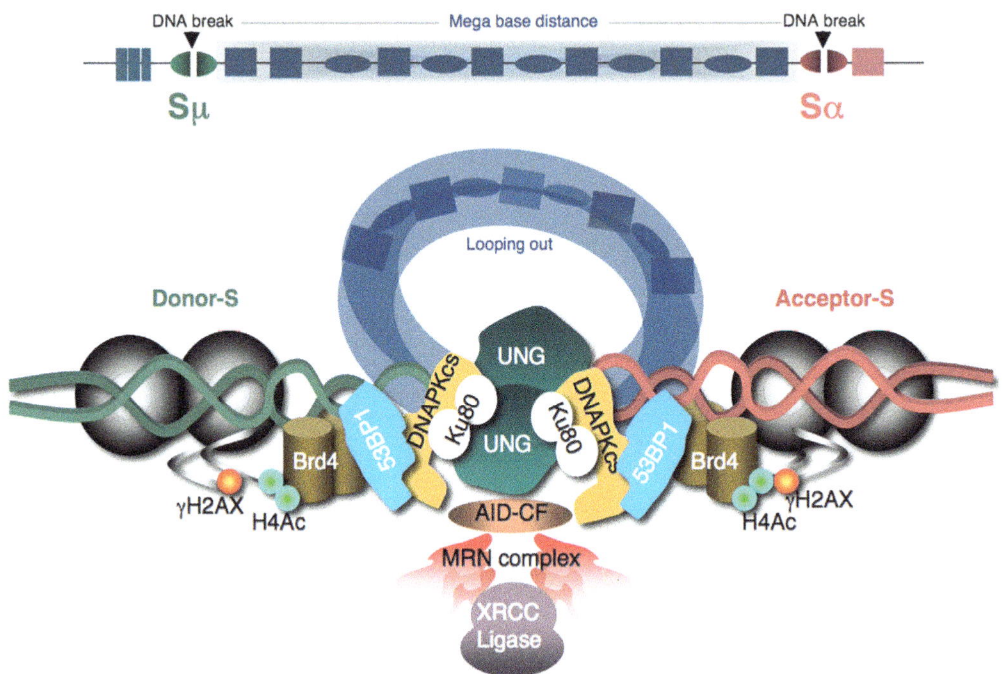


FIGURE 4 **Recombination step of CSR requires DNA end joining and switch synapse.** Distantly located donor and acceptor S regions come into close proximity to be repaired and ligated by NHEJ. This CSR-specific recombination requires various repair enzymes as well as scaffold proteins to support the complex. RNA deamination model also predicts the requirement of a novel factor (AID-CF) whose production depends on AID C-terminal domain. The chromatin-associated protein Brd4, acetylated histones and DNA damage-associated histone epigenetic mark are essential for the efficient repair–recombination complex formation.

in mediating DNA repair and CSR after DNA cleavage (Figure 4).

BRD4's involvement in DNA repair was directly demonstrated in an assay using an artificial I-SceI–dependent NHEJ, the major DNA repair pathway in CSR [33,289]. The reduction of well-known NHEJ players such as 53BP1 was also clearly reflected in the assay system. Consistent with these results, BRD4 KD not only reduced NHEJ efficiency, but also depleted the formation of 53BP1 at I-SceI cleavage sites. These results, together with the recombination defects observed in the S regions, clearly delineate BRD4's importance during the NHEJ repair and recombination step in CSR [287]. As to how BRD4 regulates key factors in CSR, sequential ChIP assays indicated that BRD4 tethers 53BP1 and UNG to the recombining S region chromatin through acetylated histones in response to γH2AX formation. Thus, BRD4 serves as a histone-reader platform that is required for recruiting CSR repair components.

It has become apparent that the chromatin landscape of a particular locus is crucial to AID-induced DNA cleavage and repair processes. A distinct set of chromatin regulators, including histone-PTM writers and readers, work in concert to provide chromatin features that promote cleavage and recombination during transcription elongation. In the presence of the appropriate chromatin environment, transcription-induced non-B structures may transform or convert target substrates that are optimal for AID-induced cleavage. Certainly, H3K4me3, H3.3, H2A.Z, and acetylated histones are not V or S region–specific epigenetic marks, and combinatorial histone codes and several associated reader proteins might be involved in conjunction with site-specific *cis*-elements such as non-B structures.

# 5. REGULATION OF AID EXPRESSION

## 5.1 AID's Expression Profile

### *5.1.1 AID Expression in Normal Tissues*

AID is expressed in activated B-cells in GCs and in the extrafollicular regions of the secondary lymphoid organ where SHM and CSR take place [5,15,290]. AID is virtually undetectable in resting mature B-cells, but is detectable in immature and pro-/pre-B-cells at low levels [5,291–294]. Although AID's expression levels in immature B-cells are orders of magnitude lower than those in GC B-cells, this low amount of AID appears to support limited levels of SHM and CSR in vivo [293,295,296]. AID's role in the immature stage of B-cells is still unclear. AID might be involved in antigen-independent Ig diversification. In addition, AID's involvement in B-cell central tolerance has been suggested in both humans and mice, although the mechanisms remain elusive [297,298].

AID expression is not absolutely restricted to B-cells; T-cells appear to express a low level of AID by environmental stimulation in the periphery [292,299]. Although AID expression has been observed in non-lymphoid tissues, such as in the ovary and testis [292,300,301], AID's physiological roles in non-B-cells are unknown. It has been proposed to be involved in epigenetic gene regulation through its ability to deaminate 5-methyl-cytosine on the DNA [302,303]. However, it should be noted that homozygous AID-deficient mice have been maintained without any obvious developmental disorder since 2000 [5].

### 5.1.2 AID Expression in Pathological Conditions

#### 5.1.2.1 Expression in Tumors

Many malignant B-cells express AID. Most, though not all, of GC-derived human B-cell lymphomas, such as diffuse large B-cell lymphoma, follicular B-cell lymphoma, and Burkitt lymphoma, express AID constitutively [15,178,304–307]. AID is also expressed in non-GC–derived B-cell malignancies [308–312]. The AID expression in B-cell tumors is required for ongoing SHM in the *IgV* gene or in oncogenes, although the AID levels in these systems do not always correlate with SHM [307,310,312,313]. The gross mutation efficiency may be influenced by other factors, such as the transcription efficiency of the target gene. Interestingly, AID expression correlates with a poor prognosis in several human B-cell lymphomas and leukemias [307,310,312–317]. Because AID affects the efficiency of hypermutation and/or translocation of oncogenes in animal models, it may have been substantially involved in tumor evolution in humans through its genotoxicity [318–326]. Interestingly, the BCR-ABL1 kinase encoded by the Philadelphia chromosome, which is a hallmark of acute lymphatic leukemia (ALL), can induce AID expression, and the prognosis is worse for those cases of ALL that express AID [327,328]. AID expression has also been reported in non-lymphoid tumors, epithelial breast cancer cell lines, and hepatoma [329,330].

#### 5.1.2.2 Expression by Pathogen Infection and Inflammation

Some pathogens can induce AID gene (*Aicda*) expression. The Epstein–Barr virus protein latent membrane protein 1, which mimics continuous CD40 signaling, induces AID expression [331,332]. Infection with the hepatitis C virus or transfection of the viral genome induced AID in B-lymphocytes and hepatocytes, respectively [333,334]. AID is also expressed in some adult T-cell leukemia cells after human T-lymphotropic virus-1 infection [335]. Infection with mouse retroviruses, such as Abelson murine leukemia virus and Moloney murine leukemia virus, induces AID expression in mouse B-cells [336,337]. Bacteria can also induce AID; for example, *H. Pylori* induces AID expression in gastric epithelium [338]. Products made by pathogens can induce AID directly or through the inflammatory cytokine tumor necrosis factor (TNF)-α [334]. AID expression is also observed in colitis-associated colorectal cancers [339]. Because many pathogens that induce AID expression are associated with tumors, it is plausible that AID is substantially involved in tumor processes. A requirement of AID for virus-induced tumor development was examined in mouse models with transgenic lines expressing the HCV core protein or HPV oncogene crossed with AID deficiency [340]; however, the results were disappointingly negative. It is possible that the mechanism of tumorigenesis is different between humans and mice. Alternatively, the association between tumors and AID expression could be coincidental, or it might be that various pathogens induce AID indirectly through inflammation, rather than directly.

## 5.2 *Aicda* Regulation

Because of its potential genotoxic activity, AID expression must be tightly regulated to maintain genomic integrity in both B-cells and non-B-cells [20]. It is especially critical to regulate AID's expression under the control of B-cell and activation-specific promoters, to avoid its misfiring by accidental signaling.

### 5.2.1 Signals and TFs Required for AID Expression

B-cells exposed to antigens and cytokines secreted from helper T-cells undergo CSR and SHM, both of which require AID expression. BCR consists of IgH, IgL, and two signaling components, Ig-α and Ig-β. Although BCR signaling is essential for specific immune responses, BCR signaling alone cannot induce appreciable AID expression. Rather, the BCR signal synergistically enhances *Aicda* expression with other stimuli, such as through CD40 and TLR [341,342].

The engagement of CD154:CD40 activates the NF-κB pathways. CD40, a member of the TNF-receptor superfamily, is a 48-kDa transmembrane glycoprotein expressed on the surface of B-cells, DCs, macrophages, epithelial cells, and activated T-cells [343]. TNF-α receptor–associated factors (Trafs) bind to the cytoplasmic tail, thereby mediating NF-κB activation. When engaged by BAFF and APRIL, other TNF-receptor families such as BAFF-R, TACI, and BCMA, activate NF- κB pathways that synergize with TLR and BCR signals to enhance *Aicda* expression and CSR [344–346]. Consistent with these findings, transcription of the *Aicda* gene depends on NF-κB [224,347]. Both canonical (p65) and non-canonical (p52) NF-κB pathways are involved in *Aicda* regulation [341,348,349].

TLRs are pattern recognition receptors that detect and bind microbial components [350]. The engagement of TLRs with their ligands elicits signal transduction mediated by adaptor proteins, including MyD88, Tirap, Trif, and Tram, thereby activating TFs such as NF-κB, p38, and IRFs. TLR and BCR signaling synergistically enhance CSR [341] and integrate to activate the canonical and non-canonical NF-κB pathways, which regulate *Aicda* expression. Because many TLR ligands are constituents of the bacterial outer cell wall and membrane, B-cell clones whose BCRs are reactive to bacteria should frequently receive signals mediated by both BCR and TLR signals.

Cytokines are another important component in regulating AID expression. Interleukin-4 (IL-4) directs B-cells to switch their Ig class to IgG4 and IgE in humans, and to IgG1 and IgE in the mouse. Interleukin-4R signaling efficiently induces *Aicda* in combination with TLR- or CD40-signaling [342,351,352]. The binding of IL-4 to its receptor activates multiple signal transduction molecules, including Stat6, IRS-2, Shc, SHP-1, and SHIP Stat6, is important in activating IL-4–induced gene expression, including *Aicda* expression [353,354]. Transforming growth factor-β critically contributes to inducing IgA in mucosae, and the transcription factor Smad is involved in TGF-β signaling [355]. Estrogen is another humoral factor that enhances *Aicda* expression. Estrogen is reported to induce *Aicda* indirectly by upregulating HoxC4, and the estrogen receptor is also reported to bind directly to the *Aicda* promoter [356,357].

### 5.2.2 *cis*-Regulatory Elements of *Aicda*

Four major evolutionarily conserved *cis*-regulatory regions influence *Aicda* expression [347]. Region 1 consists of the *Aicda* promoter and the region immediately upstream; region 2 contains the regulatory elements in the first intron; region 3 is the area around 7 and 24 kb downstream from exon 5 in the mouse and human genomes, respectively; and region 4 is the area around 8 and 16 kb upstream of the transcription start site in the mouse and human genomes, respectively [291,347,358]. Deleting region 2, region 3, or region 4 on a bacterial artificial chromosome (BAC) that contains the entire *Aicda* locus eliminates *Aicda* expression from the BAC transgenes, which demonstrates the importance of these elements [291,359].

Luciferase assays in both B- and non-B-cells showed that the *Aicda* promoter alone is basically active [348,358]. This promoter contains well-conserved binding motifs for Oct1 and/or Oct2 (5′-ATTTGAAT-3′), including the HoxC4 binding motif (5′-ATTT-3′). It also contains NF-κB (non-canonical)- and Sp1/Sp3-binding motifs [349]. HoxC4 is a homeodomain transcription factor that is expressed in B-cells and is upregulated by CD40 signaling. AID expression is impaired in HoxC4-knockout mice, which suggests that HoxC4 upregulation contributes to AID's induction in activated B cells [349]. HoxC4 appears to be necessary but not sufficient for AID expression, because other regulatory regions are also required for normal *Aicda* regulation. A CSR-competent mouse lymphoma cell line, CH12F3-2, strongly induces AID expression in response to stimuli by CD40L, IL-4, and TGF-β. A luciferase reporter assay in CH12F3-2 cells indicated that region 1 alone does not support strong AID induction, even though it carries the HoxC4 binding site [348]. These findings suggest that the coordinated regulation of HoxC4 with other *cis*-regulatory elements results in strong, finely tuned AID expression.

Region 4, the upstream regulatory element, contains binding motifs for Stat6, NF-κB, and Smad proteins, which are downstream regulators for IL-4, CD40L, and TGF-β stimulation, respectively. In addition, tandem C/EBP binding sites are required for this region's enhancer activity [348,359]. Region 2, the intronic regulatory element, contains binding sites for Myb, Pax5, E2A, and E2F. The E47 protein, an isoform of E2A, positively regulates *Aicda* expression, and Id2 or Id3 substantially represses *Aicda* expression [360,361]. Imatinab, an anticancer drug, inhibits *Aicda* expression by reducing E2A expression [362]. Because E2A and Pax5 are both essential for B-cell development, these proteins may contribute to strong *Aicda* expression specifically in B-lineage cells. In contrast, the E2f and c-Myb binding motifs show silencer activity on the promoter [348]. The deletion of these motifs increased the percentage of *Aicda*-expressing cells in vivo [359], which indicates that the silencer activity *on Aicda* regulation occurs in vivo. Region 3, the downstream regulatory element, contains a Batf binding site. Batf deficiency or the deletion of region 3 from the BAC transgene severely impairs *Aicda* expression [291,363].

In summary, the regulation of the *Aicda* gene is governed by the balance between activators and repressors that bind to sites located in four major *cis*-regulatory regions in and around the *Aicda* gene locus (Figure 5). The promoter activity can be increased by HoxC4 and non-canonical NF-κB binding. The upstream region 4 is mainly responsible for responding to stimulatory signals through cytokine receptors or costimulatory molecules. Region 2, which is in the first intron, harbors binding sites for Pax5 and E2A; these are responsible for B-lineage maintenance as well as *Aicda* activation. It also contains the E2F and c-Myb motifs that serve as a brake for *Aicda*. Batf binding to the downstream region 3 is another positive *Aicda* regulator.

Many other TFs, including Bach2, Bcl6, Blimp1, and IRF4, are reported to be important for normal AID expression. These factors seem to regulate *Aicda* via indirect mechanisms [15,364–367], although it is difficult to exclude the possibility of direct regulation completely.

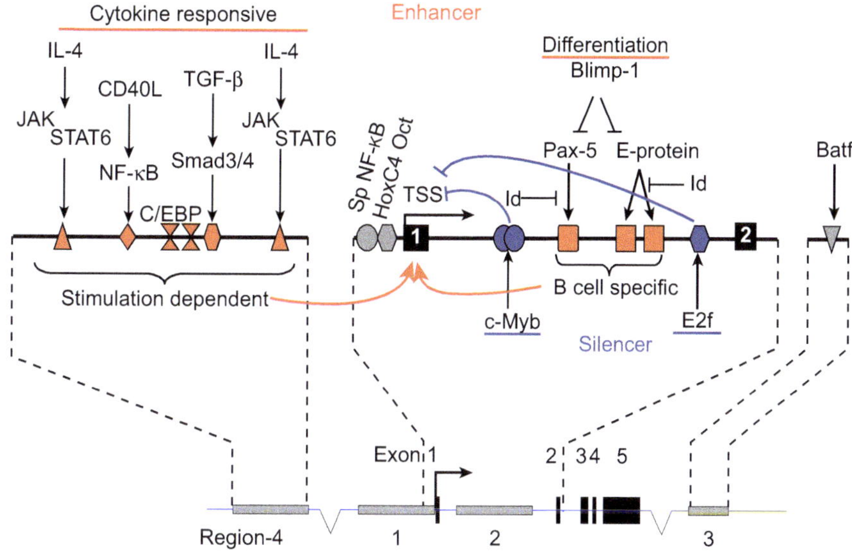


FIGURE 5 **A proposed model for the transcriptional regulation of *Aicda* in B-cells.** *cis*-Regulatory elements in the *Aicda* locus are illustrated. The binding sites for cytokine-responsive and differentiation-dependent transcription factors fulfill regulatory roles as enhancers (red) or silencers (blue) that are indicated by symbols. Binding of c-Myb and E2f independently repress AID expression. When B-cells are activated by appropriate stimuli that induce binding of STAT6, Smad3/4, C/EBP, or NF-κB, AID expression is derepressed by their activity. HoxC4 binding at the promoter region is required for the normal *Aicda* expression. The Batf binding site in the downstream region also controls *Aicda* in vivo. B cell–specific factors, namely, Pax5 and E protein, contribute to restricting the induction of AID to B-lineage cells. Gray boxes indicate the position of the first and second *Aicda* exons. TSS, transcription start site.

## 5.3 Post-Transcriptional and Post-Translational AID Regulation

### 5.3.1 MicroRNAs

Micro-RNAs are short, single-stranded noncoding RNAs that suppress the gene expression of a specific target. Several miRs, including miR-155, miR-181b, miR-361, and miR-93, regulate AID expression by binding to evolutionarily conserved sites in the 3′UTR of AID mRNA [365,368–372]. The miR-mediated translation suppression of AID has been reported in both activated B- and resting B-cells. The expression of miR155 is low in resting B-cells, but increases along with AID expression in mouse B-cells stimulated with LPS and IL-4; miR155 appears to counteract increased AID activity in these cells [368]. In contrast, miR181b is expressed in resting B-cells, but its expression is decreased by LPS and IL-4 stimulation [370]. Bcl-6, a transcription repressor essential for forming GC B-cells, represses miR-155 and miR-361 expression, and thereby contributes to inducing AID in GC B-cells [365].

### 5.3.2 Phosphorylation

AID is phosphorylated at serine 3, threonine 27, serine 38, threonine 140, and tyrosine 184 [64,90,195,373,374]. The roles of these phosphorylations are not fully understood, although many of them influence AID's activity. The phosphorylation of S38 and T140 increases AID's activity [195,225]. Phosphorylation at S38 is thought to facilitate AID's interaction with chromatin, possibly by binding to replication protein A [65,195]. In contrast, S3 phosphorylation is reported to reduce AID's activity [374]. Both protein kinase C and cyclic AMP–dependent protein kinase A can phosphorylate AID, but the major signaling pathway that functions in vivo is still unclear [65,90,195].

### 5.3.3 Regulation of Protein Stability

AID is a nucleocytoplasmic shuttling protein [29,30]. Nuclear AID undergoes rapid degradation, and two degradation mechanisms have been reported: the polyubiquitination and proteasome pathway, and an ubiquitin- and ATP-independent but proteasome-dependent degradation pathway [21,375]. The ubiquitin-independent pathway relies on the nuclear protein RDG-γ. Class switching upon stimulation with LPS and IL-4 was increased in REG-γ-deficient B-cells [21]. Because of the potential genotoxic activity of nuclear AID, these dual rapid-degradation mechanisms may represent another level of control of AID activity.

## 6. CONCLUDING REMARKS

Although the exact molecular mechanisms of AID's functions have been extensively debated, there are several points of consensus regarding experimental observations that have yet to be explained by any model of AID function. The first point, which is critical, is that AID has at least two functions, one for both SHM and CSR, and the other only for CSR. Because there are no AID mutants that abolish SHM without affecting CSR, the common AID function in SHM and CSR is probably DNA cleavage, and it is unlikely that the mechanism of that cleavage is distinct for SHM and CSR. AID's CSR-specific function is probably associated with recombination.

The second important point is that AID cannot itself specify the target to be cleaved. Several biochemical events may be involved in specifying the AID target, including transcription of the locus and unique DNA structures, such as R-loops, non-B DNA structures, and epigenetic marks. To recognize these modifications of the locus, AID obviously requires cofactors. The DNA deamination model proposes that cofactors recruit AID to the marked loci, meaning that a complex of cofactors must interact with AID, locus-specific transcripts, DNA structure, and/or epigenetic marks. In contrast, the RNA editing model proposes that cofactors capture target RNA, which is edited by AID or by part of the RNA-editosome complex. The edited RNA is assumed to generate conditions that facilitate Top1's action in SHM and CSR. The critical difference between AID and Top1 targeting is that AID must be recruited to the target locus by cofactors, whereas Top1 is already present, because it is required for general purposes such as transcription and splicing. Because global Top1 reduction by siRNA elevates SHM and CSR, minimal Top1 must ensure target-specific DNA cleavage, especially when the conditions trigger excessive substrate availability (negative supercoil-induced non-B structures). It can be speculated that only a fraction of the Top1 available is normally involved in the context of CSR or SHM. For instance, Top1 may associate with the FACT complex or an as yet unknown modulator to gain localized and controlled access to target DNA at the H3K4me3-marked chromatin. It is likely that multiple chromatin factors are involved in forming global or local Top1 cleavage complexes, for general purpose DNA transactions as well as specialized CSR or SHM functions. In addition, in cooperation with histone chaperones, Top1 has a tremendous influence on the density of the DNA supercoil along the transcribed target.

Third, SHM mostly depends on error-prone DNA synthesis. In this context, there must be a mechanism to suppress the error-free mechanisms in B-cells, because most DNA damage is correctly repaired in non-B-cells. A recent report that UNG suppresses TLP [173] provided a clue to solving this puzzle, at least in part.

Finally, it is important to recognize that biological mechanisms must evolve from related mechanisms. So far, there are no other established DNA deamination systems with important physiological functions. On the other hand, the RNA-editing mechanism is well established by cytidine deamination by APOBEC1 family members, and by adenosine deamination by ADAR I and II. An evolutionary consideration of the molecular mechanism of AID may be critical in resolving the controversies surrounding this important molecule.

## REFERENCES


[1] Faili A, Aoufouchi S, Gueranger Q, Zober C, Leon A, Bertocci B, et al. AID-dependent somatic hypermutation occurs as a DNA single-strand event in the BL2 cell line. Nat Immunol 2002;3:815–21.

[2] Kong Q, Maizels N. DNA breaks in hypermutating immunoglobulin genes: evidence for a break-and-repair pathway of somatic hypermutation. Genetics 2001;158:369–78.

[3] Kinoshita K, Tashiro J, Tomita S, Lee CG, Honjo T. Target specificity of immunoglobulin class switch recombination is not determined by nucleotide sequences of S regions. Immunity 1998;9:849–58.

[4] Weill JC, Reynaud CA. Rearrangement/hypermutation/gene conversion: when, where and why? Immunol Today 1996;17:92–7.

[5] Muramatsu M, Kinoshita K, Fagarasan S, Yamada S, Shinkai Y, Honjo T. Class switch recombination and hypermutation require activation-induced cytidine deaminase (AID), a potential RNA editing enzyme. Cell 2000;102:553–63.

[6] Revy P, Muto T, Levy Y, Geissmann F, Plebani A, Sanal O, et al. Activation-induced cytidine deaminase (AID) deficiency causes the autosomal recessive form of the Hyper-IgM syndrome (HIGM2). Cell 2000;102:565–75.

[7] Honjo T, Muramatsu M, Fagarasan S. AID: how does it aid antibody diversity? Immunity 2004;20:659–68.

[8] Di Noia J, Neuberger MS. Altering the pathway of immunoglobulin hypermutation by inhibiting uracil-DNA glycosylase. Nature 2002;419:43–8.

[9] Petersen-Mahrt SK, Harris RS, Neuberger MS. AID mutates *E. coli* suggesting a DNA deamination mechanism for antibody diversification. Nature 2002;418:99–103.

[10] Liu M, Duke JL, Richter DJ, Vinuesa CG, Goodnow CC, Kleinstein SH, et al. Two levels of protection for the B cell genome during somatic hypermutation. Nature 2008;451:841–5.

[11] Kato L, Begum NA, Burroughs AM, Doi T, Kawai J, Daub CO, et al. Nonimmunoglobulin target loci of activation-induced cytidine deaminase (AID) share unique features with immunoglobulin genes. Proc Natl Acad Sci USA 2012;109:2479–84.

[12] Hasham MG, Snow KJ, Donghia NM, Branca JA, Lessard MD, Stavnezer J, et al. Activation-induced cytidine deaminase-initiated off-target DNA breaks are detected and resolved during S phase. J Immunol 2012;189:2374–82.

[13] Begum NA, Honjo T. Evolutionary comparison of the mechanism of DNA cleavage with respect to immune diversity and genomic instability. Biochemistry 2012;51:5243–56.

[14] Alt FW, Zhang Y, Meng FL, Guo C, Schwer B. Mechanisms of programmed DNA lesions and genomic instability in the immune system. Cell 2013;152:417–29.

[15] Muto T, Muramatsu M, Taniwaki M, Kinoshita K, Honjo T. Isolation, tissue distribution, and chromosomal localization of the human activation-induced cytidine deaminase (AID) gene. Genomics 2000;68:85–8.

[16] Ta VT, Nagaoka H, Catalan N, Durandy A, Fischer A, Imai K, et al. AID mutant analyses indicate requirement for class-switch-specific cofactors. Nat Immunol 2003;4:843–8.

[17] Doi T, Kato L, Ito S, Shinkura R, Wei M, Nagaoka H, et al. The C-terminal region of activation-induced cytidine deaminase is responsible for a recombination function other than DNA cleavage in class switch recombination. Proc Natl Acad Sci USA 2009;106:2758–63.

[18] Shinkura R, Ito S, Begum NA, Nagaoka H, Muramatsu M, Kinoshita K, et al. Separate domains of AID are required for somatic hypermutation and class-switch recombination. Nat Immunol 2004;5:707–12.

[19] Kato L, Stanlie A, Begum NA, Kobayashi M, Aida M, Honjo T. An evolutionary view of the mechanism for immune and genome diversity. J Immunology 2012;188:3559–66.

[20] Honjo T, Kobayashi M, Begum N, Kotani A, Sabouri S, Nagaoka H. The AID dilemma: infection, or cancer? Adv Cancer Res 2012;113:1–44.

[21] Uchimura Y, Barton LF, Rada C, Neuberger MS. REG-gamma associates with and modulates the abundance of nuclear activation-induced deaminase. J Exp Med 2011;208:2385–91.

[22] Robbiani DF, Nussenzweig MC. Chromosome translocation, B cell lymphoma, and activation-induced cytidine deaminase. Annu Rev Pathol 2013;8:79–103.

[23] Chaudhuri J, Alt FW. Class-switch recombination: interplay of transcription, DNA deamination and DNA repair. Nat Rev Immunol 2004;4:541–52.

[24] Durandy A, Peron S, Fischer A. Hyper-igm syndromes. Curr Opin Rheumatol 2006;18:369–76.

[25] Durandy A, Peron S, Taubenheim N, Fischer A. Activation-induced cytidine deaminase: structure-function relationship as based on the study of mutants. Hum Mutat 2006;27:1185–91.

[26] Shivarov V, Shinkura R, Honjo T. Dissociation of in vitro DNA deamination activity and physiological functions of AID mutants. Proc Natl Acad Sci USA 2008;105:15866–71.

[27] Wei M, Shinkura R, Doi Y, Maruya M, Fagarasan S, Honjo T. Mice carrying a knock-in mutation of Aicda resulting in a defect in somatic hypermutation have impaired gut homeostasis and compromised mucosal defense. Nat Immunol 2011;12:264–70.

[28] Barreto V, Reina-San-Martin B, Ramiro AR, McBride KM, Nussenzweig MC. C-terminal deletion of AID uncouples class switch recombination from somatic hypermutation and gene conversion. Mol Cell 2003;12:501–8.

[29] Ito S, Nagaoka H, Shinkura R, Begum N, Muramatsu M, Nakata M, et al. Activation-induced cytidine deaminase shuttles between nucleus and cytoplasm like apolipoprotein B mRNA editing catalytic polypeptide 1. Proc Natl Acad Sci USA 2004;101:1975–80.

[30] McBride KM, Barreto V, Ramiro AR, Stavropoulos P, Nussenzweig MC. Somatic hypermutation is limited by CRM1-dependent nuclear export of activation-induced deaminase. J Exp Med 2004;199:1235–44.

[31] Sabouri S, Kobayashi M, Begum NA, Xu J, Hirota K, Honjo T. C-terminal region of activation-induced cytidine deaminase (AID) is required for efficient class switch recombination and gene conversion. Proc Natl Acad Sci USA 2014;111:2253–8.

[32] Yan CT, Boboila C, Souza EK, Franco S, Hickernell TR, Murphy M, et al. IgH class switching and translocations use a robust non-classical end-joining pathway. Nature 2007;449:478–82.

[33] Boboila C, Alt FW, Schwer B. Classical and alternative end-joining pathways for repair of lymphocyte-specific and general DNA double-strand breaks. Adv Immunol 2012;116:1–49.

[34] Kudo N, Wolff B, Sekimoto T, Schreiner EP, Yoneda Y, Yanagida M, et al. Leptomycin B inhibition of signal-mediated nuclear export by direct binding to CRM1. Exp Cell Res 1998;242:540–7.

[35] Patenaude AM, Orthwein A, Hu Y, Campo VA, Kavli B, Buschiazzo A, et al. Active nuclear import and cytoplasmic retention of activation-induced deaminase. Nat Struct Mol Biol 2009;16:517–27.

[36] Hu Y, Ericsson I, Torseth K, Methot SP, Sundheim O, Liabakk NB, et al. A combined nuclear and nucleolar localization motif in activation-induced cytidine deaminase (AID) controls immunoglobulin class switching. J Mol Biol 2013;425:424–43.

[37] Ellyard JI, Benk AS, Taylor B, Rada C, Neuberger MS. The dependence of Ig class-switching on the nuclear export sequence of AID likely reflects interaction with factors additional to Crm1 exportin. Eur J Immunol 2011;41:485–90.

[38] Patenaude AM, Di Noia JM. The mechanisms regulating the subcellular localization of AID. Nucleus 2010;1:325–31.

[39] Hasler J, Rada C, Neuberger MS. Cytoplasmic activation-induced cytidine deaminase (AID) exists in stoichiometric complex with translation elongation factor 1alpha (eEF1A). Proc Natl Acad Sci USA 2011;108:18366–71.

[40] Johansson E, Mejlhede N, Neuhard J, Larsen S. Crystal structure of the tetrameric cytidine deaminase from Bacillus subtilis at 2.0 A resolution. Biochemistry 2002;41:2563–70.

[41] Costanzi S, Vincenzetti S, Cristalli G, Vita A. Human cytidine deaminase: a three-dimensional homology model of a tetrameric metallo-enzyme inferred from the crystal structure of a distantly related dimeric homologue. J Mol Graph Modell 2006;25:10–6.

[42] Conticello SG, Thomas CJ, Petersen-Mahrt SK, Neuberger MS. Evolution of the AID/APOBEC family of polynucleotide (deoxy) cytidine deaminases. Mol Biol Evol 2005;22:367–77.

[43] Dickerson SK, Market E, Besmer E, Papavasiliou FN. AID mediates hypermutation by deaminating single stranded DNA. J Exp Med 2003;197:1291–6.

[44] Larijani M, Petrov AP, Kolenchenko O, Berru M, Krylov SN, Martin A. AID associates with single-stranded DNA with high affinity and a long complex half-life in a sequence-independent manner. Mol Cell Biol 2007;27:20–30.

[45] Prochnow C, Bransteitter R, Klein MG, Goodman MF, Chen XS. The APOBEC-2 crystal structure and functional implications for the deaminase AID. Nature 2007;445:447–51.

[46] Brar SS, Sacho EJ, Tessmer I, Croteau DL, Erie DA, Diaz M. Activation-induced deaminase, AID, is catalytically active as a monomer on single-stranded DNA. DNA Repair (Amst) 2008; 7:77–87.

[47] Bhagwat AS, Carpenter MA, Bujnicki JM. Is AID a monomer in solution? DNA Repair (Amst) 2008;7:349–50. author reply 351–2.

[48] Krzysiak TC, Grabe M, Gilbert SP. Getting in sync with dimeric Eg5. Initiation and regulation of the processive run. J Biol Chem 2008;283:2078–87.

[49] Harris RS, Petersen-Mahrt SK, Neuberger MS. RNA editing enzyme APOBEC1 and some of its homologs can act as DNA mutators. Mol Cell 2002;10:1247–53.

[50] Lada AG, Krick CF, Kozmin SG, Mayorov VI, Karpova TS, Rogozin IB, et al. Mutator effects and mutation signatures of editing deaminases produced in bacteria and yeast. Biochem (Moscow) 2011;76:131–46.

[51] Kerppola TK. Bimolecular fluorescence complementation: visualization of molecular interactions in living cells. Methods Cell Biol 2008;85:431–70.

[52] Kerppola TK. Bimolecular fluorescence complementation (BiFC) analysis as a probe of protein interactions in living cells. Annu Rev Biophys 2008;37:465–87.

[53] Chaudhuri J, Tian M, Khuong C, Chua K, Pinaud E, Alt FW. Transcription-targeted DNA deamination by the AID antibody diversification enzyme. Nature 2003;422:726–30.

[54] Okazaki IM, Okawa K, Kobayashi M, Yoshikawa K, Kawamoto S, Nagaoka H, et al. Histone chaperone Spt6 is required for class switch recombination but not somatic hypermutation. Proc Natl Acad Sci USA 2011;108:7920–5.

[55] Willmann KL, Milosevic S, Pauklin S, Schmitz KM, Rangam G, Simon MT, et al. A role for the RNA pol II-associated PAF complex in AID-induced immune diversification. J Exp Med 2012;209:2099–111.

[56] Mehta A, Kinter MT, Sherman NE, Driscoll DM. Molecular cloning of apobec-1 complementation factor, a novel RNA-binding protein involved in the editing of apolipoprotein B mRNA. Mol Cell Biol 2000;20:1846–54.

[57] Nambu Y, Sugai M, Gonda H, Lee CG, Katakai T, Agata Y, et al. Transcription-coupled events associating with immunoglobulin switch region chromatin. Science 2003;302:2137–40.

[58] Pavri R, Gazumyan A, Jankovic M, Di Virgilio M, Klein I, Ansarah-Sobrinho C, et al. Activation-induced cytidine deaminase targets DNA at sites of RNA polymerase II stalling by interaction with Spt5. Cell 2010;143:122–33.

[59] Jeevan-Raj BP, Robert I, Heyer V, Page A, Wang JH, Cammas F, et al. Epigenetic tethering of AID to the donor switch region during immunoglobulin class switch recombination. J Exp Med 2011;208:1649–60.

[60] Peters A, Storb U. Somatic hypermutation of immunoglobulin genes is linked to transcription initiation. Immunity 1996;4:57–65.

[61] Begum NA, Stanlie A, Nakata M, Akiyama H, Honjo T. The histone chaperone SPT6 is required for AID Target determination through H3K4me3 regulation. J Biol Chem 2012;287(39):32415–29.

[62] Stanlie A, Aida M, Muramatsu M, Honjo T, Begum NA. Histone3 lysine4 trimethylation regulated by the facilitates chromatin transcription complex is critical for DNA cleavage in class switch recombination. Proc Natl Acad Sci USA 2010;107:22190–5.

[63] Stanlie A, Begum NA, Akiyama H, Honjo T. The DSIF Subunits Spt4 and Spt5 have Distinct Roles at Various Phases of Immunoglobulin Class Switch Recombination. PLoS Genetics 2012;8(4): e1002675.

[64] Chaudhuri J, Khuong C, Alt FW. Replication protein A interacts with AID to promote deamination of somatic hypermutation targets. Nature 2004;430:992–8.

[65] Basu U, Chaudhuri J, Alpert C, Dutt S, Ranganath S, Li G, et al. The AID antibody diversification enzyme is regulated by protein kinase A phosphorylation. Nature 2005;438:508–11.

[66] Lam T, Thomas LM, White CA, Li G, Pone EJ, Xu Z, et al. Scaffold functions of 14-3-3 adaptors in B cell immunoglobulin class switch DNA recombination. PLoS One 2013;8:e80414.

[67] Conticello SG, Ganesh K, Xue K, Lu M, Rada C, Neuberger MS. Interaction between antibody-diversification enzyme AID and spliceosome-associated factor CTNNBL1. Mol Cell 2008;31:474–84.

[68] Ganesh K, Adam S, Taylor B, Simpson P, Rada C, Neuberger M. CTNNBL1 is a novel nuclear localization sequence-binding protein that recognizes RNA-splicing factors CDC5L and Prp31. J Biol Chem 2011;286:17091–102.

[69] Han L, Masani S, Yu K. Cutting edge: CTNNBL1 is dispensable for Ig class switch recombination. J Immunol 2010;185:1379–81.

[70] Nowak U, Matthews AJ, Zheng S, Chaudhuri J. The splicing regulator PTBP2 interacts with the cytidine deaminase AID and promotes binding of AID to switch-region DNA. Nat Immunol 2011;12:160–6.

[71] Basu U, Meng FL, Keim C, Grinstein V, Pefanis E, Eccleston J, et al. The RNA exosome targets the AID cytidine deaminase to both strands of transcribed duplex DNA substrates. Cell 2011;144:353–63.

[72] Hu Y, Ericsson I, Doseth B, Liabakk NB, Krokan HE, Kavli B. Activation-induced cytidine deaminase (AID) is localized to subnuclear domains enriched in splicing factors. Exp Cell Res 2014;322(1): 178–92.

[73] Fukuda M, Asano S, Nakamura T, Adachi M, Yoshida M, Yanagida M, et al. CRM1 is responsible for intracellular transport mediated by the nuclear export signal. Nature 1997;390:308–11.

[74] Dong X, Biswas A, Chook YM. Structural basis for assembly and disassembly of the CRM1 nuclear export complex. Nat Struct Mol Biol 2009;16:558–60.

[75] Maeda K, Singh SK, Eda K, Kitabatake M, Pham P, Goodman MF, Sakaguchi N. GANP-mediated recruitment of activation-induced cytidine deaminase to cell nuclei and to immunoglobulin variable region DNA. J Biol Chem 2010;285:23945–53.

[76] Arioka Y, Watanabe A, Saito K, Yamada Y. Activation-induced cytidine deaminase alters the subcellular localization of Tet family proteins. PLoS One 2012;7:e45031.

[77] Orthwein A, Patenaude AM, Affar el B, Lamarre A, Young JC, Di Noia JM. Regulation of activation-induced deaminase stability and antibody gene diversification by Hsp90. J Exp Med 2010;207: 2751–65.

[78] Orthwein A, Zahn A, Methot SP, Godin D, Conticello SG, Terada K, et al. Optimal functional levels of activation-induced deaminase specifically require the Hsp40 DnaJa1. EMBO J 2012;31:679–91.

[79] MacDuff DA, Neuberger MS, Harris RS. MDM2 can interact with the C-terminus of AID but it is inessential for antibody diversification in DT40 B cells. Mol Immunol 2006;43:1099–108.

[80] Delker RK, Zhou Y, Strikoudis A, Stebbins CE, Papavasiliou FN. Solubility-based genetic screen identifies RING finger protein 126 as an E3 ligase for activation-induced cytidine deaminase. Proc Natl Acad Sci USA 2013;110:1029–34.

[81] Vuong BQ, Lee M, Kabir S, Irimia C, Macchiarulo S, McKnight GS, et al. Specific recruitment of protein kinase A to the immunoglobulin locus regulates class-switch recombination. Nat Immunol 2009;10:420–6.

[82] Wu X, Geraldes P, Platt JL, Cascalho M. The double-edged sword of activation-induced cytidine deaminase. J Immunol 2005;174:934–41.

[83] Cook AJ, Oganesian L, Harumal P, Basten A, Brink R, Jolly CJ. Reduced switching in SCID B cells is associated with altered somatic mutation of recombined S regions. J Immunol 2003;171:6556–64.

[84] Vuong BQ, Herrick-Reynolds K, Vaidyanathan B, Pucella JN, Ucher AJ, Donghia NM, et al. A DNA break- and phosphorylation-dependent positive feedback loop promotes immunoglobulin class-switch recombination. Nat Immunol 2013;14:1183–9.

[85] Reina-San-Martin B, Chen J, Nussenzweig A, Nussenzweig MC. Enhanced intra-switch region recombination during immunoglobulin class switch recombination in 53BP1$^{-/-}$ B cells. Eur J Immunol 2007;37:235–9.

[86] Singh SK, Maeda K, Eid MM, Almofty SA, Ono M, Pham P, et al. GANP regulates recruitment of AID to immunoglobulin variable regions by modulating transcription and nucleosome occupancy. Nat Commun 2013;4:1830.

[87] Aida M, Hamad N, Stanlie A, Begum NA, Honjo T. Accumulation of the FACT complex, as well as histone H3.3, serves as a target marker for somatic hypermutation. Proc Natl Acad Sci USA 2013;110:7784–9.

[88] Xu Z, Fulop Z, Wu G, Pone EJ, Zhang J, Mai T, et al. 14-3-3 adaptor proteins recruit AID to 5′-AGCT-3′-rich switch regions for class switch recombination. Nat Struct Mol Biol 2010;17:1124–35.
[89] Li G, White CA, Lam T, Pone EJ, Tran DC, Hayama KL, et al. Combinatorial H3K9acS10ph histone modification in IgH locus S regions targets 14-3-3 adaptors and AID to specify antibody class-switch DNA recombination. Cell Rep 2013;5:702–14.
[90] Pasqualucci L, Kitaura Y, Gu H, Dalla-Favera R. PKA-mediated phosphorylation regulates the function of activation-induced deaminase (AID) in B cells. Proc Natl Acad Sci USA 2006;103:395–400.
[91] Ward IM, Reina-San-Martin B, Olaru A, Minn K, Tamada K, Lau JS, et al. 53BP1 is required for class switch recombination. J Cell Biol 2004;165:459–64.
[92] Conticello SG, Langlois MA, Yang Z, Neuberger MS. DNA deamination in immunity: AID in the context of its APOBEC relatives. Adv Immunol 2007;94:37–73.
[93] Navarro F, Bollman B, Chen H, Konig R, Yu Q, Chiles K, et al. Complementary function of the two catalytic domains of APOBEC3G. Virology 2005;333:374–86.
[94] Li X, Ma J, Zhang Q, Zhou J, Yin X, Zhai C, et al. Functional analysis of the two cytidine deaminase domains in APOBEC3G. Virology 2011;414:130–6.
[95] Teng BB, Ochsner S, Zhang Q, Soman KV, Lau PP, Chan L. Mutational analysis of apolipoprotein B mRNA editing enzyme (APOBEC1). structure-function relationships of RNA editing and dimerization. J Lipid Res 1999;40:623–35.
[96] Chester A, Scott J, Anant S, Navaratnam N. RNA editing: cytidine to uridine conversion in apolipoprotein B mRNA. Biochim Biophys Acta 2000;1494:1–13.
[97] Petersen-Mahrt SK, Neuberger MS. In vitro deamination of cytosine to uracil in single-stranded DNA by apolipoprotein B editing complex catalytic subunit 1 (APOBEC1). J Biol Chem 2003;278:19583–6.
[98] Harris RS, Liddament MT. Retroviral restriction by APOBEC proteins. Nat Rev Immunol 2004;4:868–77.
[99] Mussil B, Suspene R, Aynaud MM, Gauvrit A, Vartanian JP, Wain-Hobson S. Human APOBEC3A isoforms translocate to the nucleus and induce DNA double strand breaks leading to cell stress and death. PLoS One 2013;8:e73641.
[100] Ikeda T, Abd El Galil KH, Tokunaga K, Maeda K, Sata T, Sakaguchi N, et al. Intrinsic restriction activity by apolipoprotein B mRNA editing enzyme APOBEC1 against the mobility of autonomous retrotransposons. Nucleic Acids Res 2011;39:5538–54.
[101] Belanger K, Savoie M, Rosales Gerpe MC, Couture JF, Langlois MA. Binding of RNA by APOBEC3G controls deamination-independent restriction of retroviruses. Nucleic Acids Res 2013;41:7438–52.
[102] Alder MN, Rogozin IB, Iyer LM, Glazko GV, Cooper MD, Pancer Z. Diversity and function of adaptive immune receptors in a jawless vertebrate. Science 2005;310:1970–3.
[103] Rogozin IB, Iyer LM, Liang L, Glazko GV, Liston VG, Pavlov YI, et al. Evolution and diversification of lamprey antigen receptors: evidence for involvement of an AID-APOBEC family cytosine deaminase. Nat Immunol 2007;8:647–56.
[104] Pancer Z, Amemiya CT, Ehrhardt GR, Ceitlin J, Gartland GL, Cooper MD. Somatic diversification of variable lymphocyte receptors in the agnathan sea lamprey. Nature 2004;430:174–80.
[105] Harris RS, Bishop KN, Sheehy AM, Craig HM, Petersen-Mahrt SK, Watt IN, et al. DNA deamination mediates innate immunity to retroviral infection. Cell 2003;113:803–9.
[106] Conticello SG. The AID/APOBEC family of nucleic acid mutators. Genome Biol 2008;9:229.
[107] Yoshikawa K, Okazaki IM, Eto T, Kinoshita K, Muramatsu M, Nagaoka H, et al. AID enzyme-induced hypermutation in an actively transcribed gene in fibroblasts. Science 2002;296:2033–6.
[108] Stavnezer-Nordgren J, Sirlin S. Specificity of immunoglobulin heavy chain switch correlates with activity of germline heavy chain genes prior to switching. Embo J 1986;5:95–102.
[109] Lutzker S, Alt FW. Structure and expression of germ line immunoglobulin gamma 2b transcripts. Mol Cell Biol 1988;8:1849–52.
[110] Berton MT, Uhr JW, Vitetta ES. Synthesis of germ-line gamma 1 immunoglobulin heavy-chain transcripts in resting B cells: induction by interleukin 4 and inhibition by interferon gamma. Proc Natl Acad Sci USA 1989;86:2829–33.
[111] Harriman GR, Bradley A, Das S, Rogers-Fani P, Davis AC. IgA class switch in I alpha exon-deficient mice. Role of germline transcription in class switch recombination. J Clin Invest 1996; 97:477–85.
[112] Hein K, Lorenz MG, Siebenkotten G, Petry K, Christine R, Radbruch A. Processing of switch transcripts is required for targeting of antibody class switch recombination. J Exp Med 1998; 188:2369–74.
[113] Stavnezer J, Guikema JE, Schrader CE. Mechanism and regulation of class switch recombination. Annu Rev Immunol 2008; 26:261–92.
[114] Pavri R, Nussenzweig MC. AID targeting in antibody diversity. Adv Immunol 2011;110:1–26.
[115] Bachl J, Carlson C, Gray-Schopfer V, Dessing M, Olsson C. Increased transcription levels induce higher mutation rates in a hypermutating cell line. J Immunol 2001;166:5051–7.
[116] Lee CG, Kinoshita K, Arudchandran A, Cerritelli SM, Crouch RJ, Honjo T. Quantitative regulation of class switch recombination by switch region transcription. J Exp Med 2001;194:365–74.
[117] Jacobs H, Puglisi A, Rajewsky K, Fukita Y. Tuning somatic hypermutation by transcription. Curr Top Microbiol Immunol 1999;246:149–58. discussion 159.
[118] Fukita Y, Jacobs H, Rajewsky K. Somatic hypermutation in the heavy chain locus correlates with transcription. Immunity 1998;9:105–14.
[119] Kumar S, Subramanian S. Mutation rates in mammalian genomes. Proc Natl Acad Sci USA 2002;99:803–8.
[120] Nagaoka H, Muramatsu M, Yamamura N, Kinoshita K, Honjo T. Activation-induced deaminase (AID)-directed hypermutation in the immunoglobulin Smu region: implication of AID involvement in a common step of class switch recombination and somatic hypermutation. J Exp Med 2002;195:529–34.
[121] Shevelev IV, Hubscher U. The 3′–5′ exonucleases. Nat Rev Mol Cell Biol 2002;3:364–76.
[122] Burkovics P, Szukacsov V, Unk I, Haracska L. Human Ape2 protein has a 3′-5′ exonuclease activity that acts preferentially on mismatched base pairs. Nucleic Acids Res 2006;34:2508–15.
[123] Majka J, Alford B, Ausio J, Finn RM, McMurray CT. ATP hydrolysis by RAD50 protein Switches MRE11 enzyme from endonuclease to exonuclease. J Biol Chem 2012;287:2328–41.
[124] Gossage L, Madhusudan S. Current status of excision repair cross complementing-group 1 (ERCC1) in cancer. Cancer Treat Rev 2007;33:565–77.
[125] Liu Y, Kao H-I, Bambara RA. FLAP endonuclease 1: a Central component of DNA Metabolism. Annu Rev Biochem 2004;73:589–615.

[126] Reina-San-Martin B, Nussenzweig MC, Nussenzweig A, Difilippantonio S. Genomic instability, endoreduplication, and diminished Ig class-switch recombination in B cells lacking Nbs1. Proc Natl Acad Sci USA 2005;102:1590–5.

[127] Paull TT, Gellert M. The 3′ to 5′ exonuclease activity of Mre11 facilitates repair of DNA double-strand breaks. Mol Cell 1998;1: 969–79.

[128] Trujillo KM, Sung P. DNA structure-specific nuclease activities in the *Saccharomyces cerevisiae* Rad50·Mre11 complex. J Biol Chem 2001;276:35458–64.

[129] Masani S, Han L, Yu K. Apurinic/Apyrimidinic endonuclease 1 is the essential nuclease during immunoglobulin class switch recombination. Mol Cell Biol 2013;33:1468–73.

[130] Sabouri Z, Okazaki IM, Shinkura R, Begum N, Nagaoka H, Tsuchimoto D, et al. Apex2 is required for efficient somatic hypermutation but not for class switch recombination of immunoglobulin genes. Int Immunol 2009;21:947–55.

[131] Kikuchi K, Taniguchi Y, Hatanaka A, Sonoda E, Hochegger H, Adachi N, et al. Fen-1 facilitates homologous recombination by removing divergent sequences at DNA break ends. Mol Cell Biol 2005;25:6948–55.

[132] Larsen E, Kleppa L, Meza TJ, Meza-Zepeda LA, Rada C, Castellanos CG, et al. Early-onset lymphoma and extensive embryonic apoptosis in two domain-specific Fen1 mice mutants. Cancer Res 2008;68:4571–9.

[133] Winter AG, Samuel K, Hsia KT, Melton DW. The repair and recombination enzyme ERCC1 is not required for immunoglobulin class switching. DNA Repair (Amst) 2003;2:561–9.

[134] Schrader CE, Vardo J, Linehan E, Twarog MZ, Niedernhofer LJ, Hoeijmakers JH, et al. Deletion of the nucleotide excision repair gene Ercc1 reduces immunoglobulin class switching and alters mutations near switch recombination junctions. J Exp Med 2004;200:321–30.

[135] Reina-San-Martin B, Difilippantonio S, Hanitsch L, Masilamani RF, Nussenzweig A, Nussenzweig MC. H2AX is required for recombination between immunoglobulin switch regions but not for intra-switch region recombination or somatic hypermutation. J Exp Med 2003;197:1767–78.

[136] Petersen S, Casellas R, Reina-San-Martin B, Chen HT, Difilippantonio MJ, Wilson PC, et al. AID is required to initiate Nbs1/gamma-H2AX focus formation and mutations at sites of class switching. Nature 2001;414:660–5.

[137] Ramiro AR, Jankovic M, Callen E, Difilippantonio S, Chen HT, McBride KM, et al. Role of genomic instability and p53 in AID-induced c-myc-Igh translocations. Nature 2006;440:105–9.

[138] Reina-San-Martin B, Chen HT, Nussenzweig A, Nussenzweig MC. ATM is required for efficient recombination between immunoglobulin switch regions. J Exp Med 2004;200:1103–10.

[139] Lumsden JM, McCarty T, Petiniot LK, Shen R, Barlow C, Wynn TA, et al. Immunoglobulin class switch recombination is impaired in Atm-deficient mice. J Exp Med 2004;200:1111–21.

[140] Köcher S, Rieckmann T, Rohaly G, Mansour WY, Dikomey E, Dornreiter I, et al. Radiation-induced double-strand breaks require ATM but not Artemis for homologous recombination during S-phase. N Acid Res 2012;40:8336–47.

[141] Lou Z, Minter-Dykhouse K, Franco S, Gostissa M, Rivera MA, Celeste A, et al. MDC1 maintains genomic stability by participating in the amplification of ATM-dependent DNA damage signals. Molecular Cell 2006;21:187–200.

[142] Bemark M, Sale JE, Kim HJ, Berek C, Cosgrove RA, Neuberger MS. Somatic hypermutation in the absence of DNA-dependent protein kinase catalytic subunit (DNA-PK(cs)) or recombination-activating gene (RAG)1 activity. J Exp Med 2000;192:1509–14.

[143] Callen E, Jankovic M, Wong N, Zha S, Chen HT, Difilippantonio S, et al. Essential role for DNA-PKcs in DNA double-strand break repair and apoptosis in ATM-deficient lymphocytes. Mol Cell 2009;34:285–97.

[144] Tang ES, Martin A. NHEJ-deficient DT40cells have increased levels of immunoglobulin gene conversion: evidence for a double strand break intermediate. Nucleic Acids Res 2006;34:6345–51.

[145] Casellas R, Nussenzweig A, Wuerffel R, Pelanda R, Reichlin A, Suh H, et al. Ku80 is required for immunoglobulin isotype switching. Embo J 1998;17:2404–11.

[146] Weinstock DM, Brunet E, Jasin M. Formation of NHEJ-derived reciprocal chromosomal translocations does not require Ku70. Nat Cell Biol 2007;9:978–81.

[147] Du L, Dunn–Walters DK, Chrzanowska KH, Stankovic T, Kotnis A, Li X, et al. A Regulatory Role for NBS1 in Strand-Specific Mutagenesis during Somatic Hypermutation. PLoS ONE 2008;3:e2482.

[148] Dinkelmann M, Spehalski E, Stoneham T, Buis J, Wu Y, Sekiguchi JM, et al. Multiple functions of MRN in end-joining pathways during isotype class switching. Nat Struct Mol Biol 2009;16:808–13.

[149] Yabuki M, Fujii MM, Maizels N. The MRE11-RAD50-NBS1 complex accelerates somatic hypermutation and gene conversion of immunoglobulin variable regions. Nat Immunol 2005;6:730–6.

[150] Di Virgilio M, Callen E, Yamane A, Zhang W, Jankovic M, Gitlin AD, et al. Rif1 Prevents Resection of DNA Breaks and Promotes Immunoglobulin Class Switching. Science 2013;339:711–5.

[151] Buonomo SBC, Wu Y, Ferguson D, de Lange T. Mammalian Rif1 contributes to replication stress survival and homology-directed repair. J Cell Biol 2009;187:385–98.

[152] Du L, Peng R, Björkman A, Filipe de Miranda N, Rosner C, Kotnis A, et al. Cernunnos influences human immunoglobulin class switch recombination and may be associated with B cell lymphomagenesis. J Exp Med 2012;209:291–305.

[153] Li G, Alt FW, Cheng HL, Brush JW, Goff PH, Murphy MM, et al. Lymphocyte-specific compensation for XLF/cernunnos end-joining functions in V(D)J recombination. Mol cell 2008;31:631–40.

[154] Zha S, Alt FW, Cheng H-L, Brush JW, Li G. Defective DNA repair and increased genomic instability in Cernunnos-XLF-deficient murine ES cells. Proc Natl Acad Sci 2007;104:4518–23.

[155] Soulas-Sprauel P, Le Guyader G, Rivera-Munoz P, Abramowski V, Olivier-Martin C, et al. Role for DNA repair factor XRCC4 in immunoglobulin class switch recombination. J Exp Med 2007;204: 1717–27.

[156] Delacôte F, Han M, Stamato TD, Jasin M, Lopez BS. An xrcc4 defect or Wortmannin stimulates homologous recombination specifically induced by double–strand breaks in mammalian cells. Nucleic Acids Res 2002;30:3454–63.

[157] Rada C, Williams GT, Nilsen H, Barnes DE, Lindahl T, Neuberger MS. Immunoglobulin isotype switching is inhibited and somatic hypermutation perturbed in UNG-deficient mice. Curr Biol 2002;12:1748–55.

[158] Zahn A, Daugan M, Safavi S, Godin D, Cheong C, Lamarre A, et al. Separation of function between isotype switching and affinity Maturation in vivo during acute immune responses and Circulating Autoantibodies in ung-deficient mice. J Immunol 2013;190: 5949–60.

[159] Begum NA, Izumi N, Nishikori M, Nagaoka H, Shinkura R, Honjo T. Requirement of non-canonical activity of uracil DNA glycosylase for class switch recombination. J Biol Chem 2007;282:731–42.
[160] Zan H, White CA, Thomas LM, Mai T, Li G, Xu Z, et al. Rev1 recruits ung to switch regions and e,nhance,s du glycosylation for immunoglobulin class switch DNA recombination. Cell Rep 2012;2:1220–32.
[161] Kovalchuk AL, Ansarah-Sobrinho C, Hakim O, Resch W, Tolarová H, Dubois W, et al. Mouse model of endemic Burkitt translocations reveals the long-range boundaries of Ig-mediated oncogene deregulation. Proc Natl Acad Sci 2012;109:10972–7.
[162] Saribasak H, Saribasak NN, Ipek FM, Ellwart JW, Arakawa H, Buerstedde JM. Uracil DNA glycosylase disruption blocks Ig gene conversion and induces transition mutations. J Immunol 2006;176:365–71.
[163] Bransteitter R, Pham P, Scharff MD, Goodman MF. Activation-induced cytidine deaminase deaminates deoxycytidine on single-stranded DNA but requires the action of RNase. Proc Natl Acad Sci USA 2003;100:4102–7.
[164] Kodgire P, Mukkawar P, North JA, Poirier MG, Storb U. Nucleosome stability dramatically impacts the targeting of somatic hypermutation. Mol Cell Biol 2012;32:2030–40.
[165] Shen HM, Poirier MG, Allen MJ, North J, Lal R, Widom J, et al. The activation-induced cytidine deaminase (AID) efficiently targets DNA in nucleosomes but only during transcription. J Exp Med 2009;206:1057–71.
[166] Mayorov VI, Rogozin IB, Adkison LR, Frahm C, Kunkel TA, Pavlov YI. Expression of human AID in yeast induces mutations in context similar to the context of somatic hypermutation at G-C pairs in immunoglobulin genes. BMC Immunol 2005;6:10.
[167] Liang G, Kitamura K, Wang Z, Liu G, Chowdhury S, Fu W, et al. RNA editing of hepatitis B virus transcripts by activation-induced cytidine deaminase. Proc Natl Acad Sci USA 2013;110:2246–51.
[168] Neuberger MS, Harris RS, Di Noia J, Petersen-Mahrt SK. Immunity through DNA deamination. Trends Biochem Sci 2003;28:305–12.
[169] Yu K, Huang FT, Lieber MR. DNA substrate length and surrounding sequence affect the activation-induced deaminase activity at cytidine. J Biol Chem 2004;279:6496–500.
[170] Rada C, Di Noia JM, Neuberger MS. Mismatch recognition and uracil excision provide complementary paths to both Ig switching and the A/T-focused phase of somatic mutation. Mol Cell 2004;16:163–71.
[171] Imai K, Slupphaug G, Lee WI, Revy P, Nonoyama S, Catalan N, et al. Human uracil-DNA glycosylase deficiency associated with profoundly impaired immunoglobulin class-switch recombination. Nat Immunol 2003;4:1023–8.
[172] Shen HM, Tanaka A, Bozek G, Nicolae D, Storb U. Somatic hypermutation and class switch recombination in $Msh6^{-/-}Ung^{-/-}$ double-knockout mice. J Immunol 2006;177:5386–92.
[173] Yousif AS, Stanlie A, Honjo T, Begum NA. Differential regulation of somatic hypermutation and class switch recombination by non-canonical funcitons of uracil DNA glycosylase. Proc Natl Acad Sci USA 2014;111:E1016–24.
[174] Begum NA, Kinoshita K, Kakazu N, Muramatsu M, Nagaoka H, Shinkura R, et al. Uracil DNA glycosylase activity is dispensable for immunoglobulin class switch. Science 2004;305:1160–3.
[175] Begum NA, Stanlie A, Doi T, Sasaki Y, Jin HW, Kim YS, et al. Further evidence for involvement of a noncanonical function of uracil DNA glycosylase in class switch recombination. Proc Natl Acad Sci USA 2009;106:2752–7.
[176] Maul RW, Saribasak H, Martomo SA, McClure RL, Yang W, Vaisman A, et al. Uracil residues dependent on the deaminase AID in immunoglobulin gene variable and switch regions. Nat Immunol 2011;12:70–6.
[177] Daly J, Bebenek K, Watt DL, Richter K, Jiang C, Zhao ML, et al. Altered Ig hypermutation pattern and frequency in complementary mouse models of DNA polymerase zeta activity. J Immunol 2012;188:5528–37.
[178] Faili A, Aoufouchi S, Flatter E, Gueranger Q, Reynaud CA, Weill JC. Induction of somatic hypermutation in immunoglobulin genes is dependent on DNA polymerase iota. Nature 2002;419:944–7.
[179] Nilsen H, Rosewell I, Robins P, Skjelbred CF, Andersen S, Slupphaug G, et al. Uracil-DNA glycosylase (UNG)-deficient mice reveal a primary role of the enzyme during DNA replication. Mol Cell 2000;5:1059–65.
[180] Zarrin AA, Del Vecchio C, Tseng E, Gleason M, Zarin P, Tian M, et al. Antibody class switching mediated by yeast endonuclease-generated DNA breaks. Science 2007;315:377–81.
[181] Kobayashi M, Aida M, Nagaoka H, Begum NA, Kitawaki Y, Nakata M, et al. AID-induced decrease in topoisomerase 1 induces DNA structural alteration and DNA cleavage for class switch recombination. Proc Natl Acad Sci USA 2009;106:22375–80.
[182] Kobayashi M, Sabouri Z, Sabouri S, Kitawaki Y, Pommier Y, Abe T, et al. Decrease in topoisomerase I is responsible for activation-induced cytidine deaminase (AID)-dependent somatic hypermutation. Proc Natl Acad Sci USA 2011;108:19305–10.
[183] Takahashi T, Burguiere-Slezak G, Van der Kemp PA, Boiteux S. Topoisomerase 1 provokes the formation of short deletions in repeated sequences upon high transcription in Saccharomyces cerevisiae. Proc Natl Acad Sci USA 2011;108:692–7.
[184] Lippert MJ, Kim N, Cho JE, Larson RP, Schoenly NE, O'Shea SH, et al. Role for topoisomerase 1 in transcription-associated mutagenesis in yeast. Proc Natl Acad Sci USA 2011;108:698–703.
[185] Hubert Jr L, Lin Y, Dion V, Wilson JH. Topoisomerase 1 and single-strand break repair modulate transcription-induced CAG repeat contraction in human cells. Mol Cell Biol 2011;31:3105–12.
[186] Tashiro J, Kinoshita K, Honjo T. Palindromic but not G-rich sequences are targets of class switch recombination. Int Immunol 2001;13:495–505.
[187] Miao ZH, Player A, Shankavaram U, Wang YH, Zimonjic DB, Lorenzi PL, et al. Nonclassic functions of human topoisomerase I: genome-wide and pharmacologic analyses. Cancer Res 2007;67:8752–61.
[188] Tuduri S, Crabbe L, Conti C, Tourriere H, Holtgreve-Grez H, Jauch A, et al. Topoisomerase I suppresses genomic instability by preventing interference between replication and transcription. Nat Cell Biol 2009;11:1315–24.
[189] Doi T, Kinoshita K, Ikegawa M, Muramatsu M, Honjo T. De novo protein synthesis is required for the activation-induced cytidine deaminase function in class-switch recombination. Proc Natl Acad Sci USA 2003;100:2634–8.
[190] Nonaka T, Doi T, Toyoshima T, Muramatsu M, Honjo T, Kinoshita K. Carboxy-terminal domain of AID required for its mRNA complex formation in vivo. Proc Natl Acad Sci USA 2009;106:2747–51.
[191] Zahn A, Eranki AK, Patenaude A-M, Methot SP, Fifield H, Cortizas EM, et al. Activation induced deaminase C-terminal domain links DNA breaks to end protection and repair during class switch recombination. Proc Natl Acad Sci USA 2014;111:E988–97.

[192] Bransteitter R, Pham P, Calabrese P, Goodman MF. Biochemical analysis of hypermutational targeting by wild type and mutant activation-induced cytidine deaminase. J Biol Chem 2004;279: 51612–21.

[193] Spencer J, Dunn–Walters DK. Hypermutation at A-T base pairs: the A nucleotide replacement spectrum is affected by adjacent nucleotides and there is no reverse complementarity of sequences flanking mutated A and T nucleotides. J Immunol 2005;175:5170–7.

[194] Yu K, Chedin F, Hsieh CL, Wilson TE, Lieber MR. R-loops at immunoglobulin class switch regions in the chromosomes of stimulated B cells. Nat Immunol 2003;4:442–51.

[195] McBride KM, Gazumyan A, Woo EM, Schwickert TA, Chait BT, Nussenzweig MC. Regulation of class switch recombination and somatic mutation by AID phosphorylation. J Exp Med 2008;205:2585–94.

[196] Geisberger R, Rada C, Neuberger MS. The stability of AID and its function in class-switching are critically sensitive to the identity of its nuclear-export sequence. Proc Natl Acad Sci USA 2009;106:6736–41.

[197] Staszewski O, Baker RE, Ucher AJ, Martier R, Stavnezer J, Guikema JE. Activation-induced cytidine deaminase induces reproducible DNA breaks at many non-Ig Loci in activated B cells. Mol Cell 2011;41:232–42.

[198] Honjo T, Kinoshita K, Muramatsu M. Molecular mechanism of class switch recombination: linkage with somatic hypermutation. Annu Rev Immunol 2002;20:165–96.

[199] Dunnick W, Hertz GZ, Scappino L, Gritzmacher C. DNA sequences at immunoglobulin switch region recombination sites. Nucleic Acids Res 1993;21:365–72.

[200] Nikaido T, Nakai S, Honjo T. Switch region of immunoglobulin Cmu gene is composed of simple tandem repetitive sequences. Nature 1981;292:845–8.

[201] Kinoshita K, Honjo T. Linking class-switch recombination with somatic hypermutation. Nat Rev Mol Cell Biol 2001;2: 493–503.

[202] Shinkura R, Tian M, Smith M, Chua K, Fujiwara Y, Alt FW. The influence of transcriptional orientation on endogenous switch region function. Nat Immunol 2003;4:435–41.

[203] Zarrin AA, Alt FW, Chaudhuri J, Stokes N, Kaushal D, Du Pasquier L, et al. An evolutionarily conserved target motif for immunoglobulin class-switch recombination. Nat Immunol 2004;5:1275–81.

[204] Haddad D, Oruc Z, Puget N, Laviolette-Malirat N, Philippe M, Carrion C, et al. Sense transcription through the S region is essential for immunoglobulin class switch recombination. Embo J 2011;30:1608–20.

[205] Mussmann R, Courtet M, Schwager J, Du Pasquier L. Microsites for immunoglobulin switch recombination breakpoints from Xenopus to mammals. Eur J Immunol 1997;27:2610–9.

[206] Han L, Masani S, Yu K. Overlapping activation-induced cytidine deaminase hotspot motifs in Ig class-switch recombination. Proc Natl Acad Sci USA 2011;108:11584–9.

[207] Huang FT, Yu K, Hsieh CL, Lieber MR. Downstream boundary of chromosomal R-loops at murine switch regions: implications for the mechanism of class switch recombination. Proc Natl Acad Sci USA 2006;103:5030–5.

[208] Huang FT, Yu K, Balter BB, Selsing E, Oruc Z, Khamlichi AA, et al. Sequence dependence of chromosomal R-loops at the immunoglobulin heavy-chain Smu class switch region. Mol Cell Biol 2007;27:5921–32.

[209] Dempsey LA, Hanakahi LA, Maizels N. A specific isoform of hnRNP D interacts with DNA in the LR1 heterodimer: canonical RNA binding motifs in a sequence-specific duplex DNA binding protein. J Biol Chem 1998;273:29224–9.

[210] Duquette ML, Handa P, Vincent JA, Taylor AF, Maizels N. Intracellular transcription of G-rich DNAs induces formation of G-loops, novel structures containing G4 DNA. Genes Dev 2004;18: 1618–29.

[211] Milstein C, Neuberger MS, Staden R. Both DNA strands of antibody genes are hypermutation targets. Proc Natl Acad Sci USA 1998;95:8791–4.

[212] Li Z, Woo CJ, Iglesias-Ussel MD, Ronai D, Scharff MD. The generation of antibody diversity through somatic hypermutation and class switch recombination. Genes Dev 2004;18:1–11.

[213] Shen HM, Storb U. Activation-induced cytidine deaminase (AID) can target both DNA strands when the DNA is supercoiled. Proc Natl Acad Sci USA 2004;101:12997–3002.

[214] Parsa JY, Ramachandran S, Zaheen A, Nepal RM, Kapelnikov A, Belcheva A, et al. Negative supercoiling creates single-stranded patches of DNA that are substrates for aid-mediated mutagenesis. PLoS Genet 2012;8:e1002518.

[215] Ronai D, Iglesias-Ussel MD, Fan M, Li Z, Martin A, Scharff MD. Detection of chromatin-associated single-stranded DNA in regions targeted for somatic hypermutation. J Exp Med 2007;204:181–90.

[216] Smith DS, Creadon G, Jena PK, Portanova JP, Kotzin BL, Wysocki LJ. Di- and trinucleotide target preferences of somatic mutagenesis in normal and autoreactive B cells. J Immunol 1996;156: 2642–52.

[217] Shapiro GS, Aviszus K, Ikle D, Wysocki LJ. Predicting regional mutability in antibody V genes based solely on di- and trinucleotide sequence composition. J Immunol 1999;163:259–68.

[218] Michael N, Shen HM, Longerich S, Kim N, Longacre A, Storb U. The E box motif CAGGTG enhances somatic hypermutation without enhancing transcription. Immunity 2003;19:235–42.

[219] Tanaka A, Shen HM, Ratnam S, Kodgire P, Storb U. Attracting AID to targets of somatic hypermutation. J Exp Med 2010;207:405–15.

[220] McDonald JJ, Alinikula J, Buerstedde JM, Schatz DG. A critical context-dependent role for E boxes in the targeting of somatic hypermutation. J Immunol 2013;191:1556–66.

[221] Chaudhuri J, Basu U, Zarrin A, Yan C, Franco S, Perlot T, et al. Evolution of the immunoglobulin heavy chain class switch recombination mechanism. Adv Immunol 2007;94:157–214.

[222] Storck S, Aoufouchi S, Weill JC, Reynaud CA. AID and partners: for better and (not) for worse. Curr Opin Immunol 2011;23: 337–44.

[223] Vuong BQ, Chaudhuri J. Combinatorial mechanisms regulating AID-dependent DNA deamination: interacting proteins and post-translational modifications. Semin Immunol 2012;24:264–72.

[224] Xu Z, Zan H, Pone EJ, Mai T, Casali P. Immunoglobulin class-switch DNA recombination: induction, targeting and beyond. Nat Rev Immunol 2012;12:517–31.

[225] Cheng HL, Vuong BQ, Basu U, Franklin A, Schwer B, Astarita J, et al. Integrity of the AID serine-38 phosphorylation site is critical for class switch recombination and somatic hypermutation in mice. Proc Natl Acad Sci USA 2009;106:2717–22.

[226] Yamane A, Robbiani DF, Resch W, Bothmer A, Nakahashi H, Oliveira T, et al. RPA accumulation during class switch recombination represents 5′-3′ DNA-end resection during the S-G2/M phase of the cell cycle. Cell Rep 2013;3:138–47.

[227] Lebecque SG, Gearhart PJ. Boundaries of somatic mutation in rearranged immunoglobulin genes: 5′ boundary is near the promoter, and 3′ boundary is approximately 1 kb from V(D)J gene. J Exp Med 1990;172:1717–27.
[228] Rada C, Gonzalez-Fernandez A, Jarvis JM, Milstein C. The 5′ boundary of somatic hypermutation in a V kappa gene is in the leader intron. Eur J Immunol 1994;24:1453–7.
[229] Rada C, Yelamos J, Dean W, Milstein C. The 5′ hypermutation boundary of kappa chains is independent of local and neighbouring sequences and related to the distance from the initiation of transcription. Eur J Immunol 1997;27:3115–20.
[230] Pavri R, Zhu B, Li G, Trojer P, Mandal S, Shilatifard A, et al. Histone H2B monoubiquitination functions cooperatively with FACT to regulate elongation by RNA polymerase II. Cell 2006;125:703–17.
[231] Yamane A, Resch W, Kuo N, Kuchen S, Li Z, Sun HW, et al. Deep-sequencing identification of the genomic targets of the cytidine deaminase AID and its cofactor RPA in B lymphocytes. Nat Immunol 2011;12:62–9.
[232] Sims 3rd RJ, Belotserkovskaya R, Reinberg D. Elongation by RNA polymerase II: the short and long of it. Genes Dev 2004;18:2437–68.
[233] Adelman K, Lis JT. Promoter-proximal pausing of RNA polymerase II: emerging roles in metazoans. Nat Rev Genet 2012;13:720–31.
[234] Yamaguchi Y, Shibata H, Handa H. Transcription elongation factors DSIF and NELF: promoter-proximal pausing and beyond. Biochim Biophys Acta 2013;1829:98–104.
[235] Besmer E, Market E, Papavasiliou FN. The transcription elongation complex directs activation-induced cytidine deaminase-mediated DNA deamination. Mol Cell Biol 2006;26:4378–85.
[236] Kenter AL. AID targeting is dependent on RNA polymerase II pausing. Semin Immunol 2012;24:281–6.
[237] Kodgire P, Mukkawar P, Ratnam S, Martin TE, Storb U. Changes in RNA polymerase II progression influence somatic hypermutation of Ig-related genes by AID. J Exp Med 2013;210:1481–92.
[238] Sun J, Keim CD, Wang J, Kazadi D, Oliver PM, Rabadan R, et al. E3-ubiquitin ligase Nedd4 determines the fate of AID-associated RNA polymerase II in B cells. Genes Dev 2013;27:1821–33.
[239] Xue K, Rada C, Neuberger MS. The in vivo pattern of AID targeting to immunoglobulin switch regions deduced from mutation spectra in $msh^{-/-}$ $ung^{-/-}$ mice. J Exp Med 2006;203:2085–94.
[240] Rush JS, Fugmann SD, Schatz DG. Staggered AID-dependent DNA double strand breaks are the predominant DNA lesions targeted to S mu in Ig class switch recombination. Int Immunol 2004;16:549–57.
[241] Wang L, Wuerffel R, Feldman S, Khamlichi AA, Kenter AL. S region sequence, RNA polymerase II, and histone modifications create chromatin accessibility during class switch recombination. J Exp Med 2009;206:1817–30.
[242] Matthews AJ, Zheng S, Dimenna LJ, Chaudhuri J. Regulation of immunoglobulin class-switch recombination: choreography of noncoding transcription, targeted DNA deamination, and long-range DNA repair. Adv Immunol 2014;122:1–57.
[243] Kornberg RD, Lorch Y. Twenty-five years of the nucleosome, fundamental particle of the eukaryote chromosome. Cell 1999;98:285–94.
[244] Kulaeva OI, Gaykalova DA, Pestov NA, Golovastov VV, Vassylyev DG, Artsimovitch I, et al. Mechanism of chromatin remodeling and recovery during passage of RNA polymerase II. Nat Struct Mol Biol 2009;16:1272–8.
[245] Hsieh FK, Kulaeva OI, Patel SS, Dyer PN, Luger K, Reinberg D, et al. Histone chaperone FACT action during transcription through chromatin by RNA polymerase II. Proc Natl Acad Sci USA 2013;110:7654–9.
[246] Petesch SJ, Lis JT. Overcoming the nucleosome barrier during transcript elongation. Trends Genet 2012;28:285–94.
[247] Adam S, Polo SE, Almouzni G. Transcription recovery after DNA damage requires chromatin priming by the H3.3 histone chaperone HIRA. Cell 2013;155:94–106.
[248] Stavnezer J. Immunoglobulin class switching. Curr Opin Immunol 1996;8:199–205.
[249] Shilatifard A. Chromatin modifications by methylation and ubiquitination: implications in the regulation of gene expression. Annu Rev Biochem 2006;75:243–69.
[250] Li B, Carey M, Workman JL. The role of chromatin during transcription. Cell 2007;128:707–19.
[251] Berger SL. The complex language of chromatin regulation during transcription. Nature 2007;447:407–12.
[252] Wang L, Whang N, Wuerffel R, Kenter AL. AID-dependent histone acetylation is detected in immunoglobulin S regions. J Exp Med 2006;203:215–26.
[253] Daniel JA, Santos MA, Wang Z, Zang C, Schwab KR, Jankovic M, et al. PTIP promotes chromatin changes critical for immunoglobulin class switch recombination. Science 2010;329:917–23.
[254] Balter BB, Ciccone DN, Oettinger MA, Selsing E. Mice lacking Smu tandem repeats maintain RNA polymerase patterns but exhibit histone modification pattern shifts linked to class switch site locations. Mol Immunol 2012;52:1–8.
[255] Min IM, Rothlein LR, Schrader CE, Stavnezer J, Selsing E. Shifts in targeting of class switch recombination sites in mice that lack mu switch region tandem repeats or Msh2. J Exp Med 2005;201:1885–90.
[256] Zippo A, Serafini R, Rocchigiani M, Pennacchini S, Krepelova A, Oliviero S. Histone crosstalk between H3S10ph and H4K16ac generates a histone code that mediates transcription elongation. Cell 2009;138:1122–36.
[257] Bradley SP, Kaminski DA, Peters AH, Jenuwein T, Stavnezer J. The histone methyltransferase suv39h1 increases class switch recombination specifically to IgA. J Immunol 2006;177:1179–88.
[258] Downs JA, Nussenzweig MC, Nussenzweig A. Chromatin dynamics and the preservation of genetic information. Nature 2007;447:951–8.
[259] Daniel JA, Nussenzweig A. The AID-induced DNA damage response in chromatin. Mol Cell 2013;50:309–21.
[260] Franco S, Gostissa M, Zha S, Lombard DB, Murphy MM, Zarrin AA, et al. H2AX prevents DNA breaks from progressing to chromosome breaks and translocations. Mol Cell 2006;21:201–14.
[261] Manis JP, Morales JC, Xia Z, Kutok JL, Alt FW, Carpenter PB. 53BP1 links DNA damage-response pathways to immunoglobulin heavy chain class-switch recombination. Nat Immunol 2004;5:481–7.
[262] Botuyan MV, Lee J, Ward IM, Kim JE, Thompson JR, Chen J, et al. Structural basis for the methylation state-specific recognition of histone H4-K20 by 53BP1 and Crb2 in DNA repair. Cell 2006;127:1361–73.
[263] Fradet-Turcotte A, Canny MD, Escribano-Diaz C, Orthwein A, Leung CC, Huang H, et al. 53BP1 is a reader of the DNA-damage-induced H2A Lys 15 ubiquitin mark. Nature 2013;499:50–4.
[264] Belotserkovskaya R, Oh S, Bondarenko VA, Orphanides G, Studitsky VM, Reinberg D. FACT facilitates transcription-dependent nucleosome alteration. Science 2003;301:1090–3.

[265] Reinberg D, Sims 3rd RJ. de FACTo nucleosome dynamics. J Biol Chem 2006;281:23297–301.
[266] Orphanides G, LeRoy G, Chang CH, Luse DS, Reinberg D. FACT, a factor that facilitates transcript elongation through nucleosomes. Cell 1998;92:105–16.
[267] Orphanides G, Wu WH, Lane WS, Hampsey M, Reinberg D. The chromatin-specific transcription elongation factor FACT comprises human SPT16 and SSRP1 proteins. Nature 1999;400:284–8.
[268] Aida M, Honjo T. FACT and H3.3: new markers for the somatic hypermutation. Cell Cycle 2013;12:2923–4.
[269] Tagami H, Ray-Gallet D, Almouzni G, Nakatani Y. Histone H3.1 and H3.3 complexes mediate nucleosome assembly pathways dependent or independent of DNA synthesis. Cell 2004;116:51–61.
[270] Borde V, Robine N, Lin W, Bonfils S, Geli V, Nicolas A. Histone H3 lysine 4 trimethylation marks meiotic recombination initiation sites. Embo J 2009;28:99–111.
[271] Smagulova F, Gregoretti IV, Brick K, Khil P, Camerini-Otero RD, Petukhova GV. Genome-wide analysis reveals novel molecular features of mouse recombination hotspots. Nature 2011;472:375–8.
[272] Matthews AG, Kuo AJ, Ramon-Maiques S, Han S, Champagne KS, Ivanov D, et al. RAG2 PHD finger couples histone H3 lysine 4 trimethylation with V(D)J recombination. Nature 2007;450:1106–10.
[273] Lee JH, Skalnik DG. CpG-binding protein (CXXC finger protein 1) is a component of the mammalian set1 histone H3-Lys4 methyltransferase complex, the analogue of the yeast Set1/COMPASS complex. J Biol Chem 2005;280:41725–31.
[274] Wada T, Takagi T, Yamaguchi Y, Ferdous A, Imai T, Hirose S, et al. DSIF, a novel transcription elongation factor that regulates RNA polymerase II processivity, is composed of human Spt4 and Spt5 homologs. Genes Dev 1998;12:343–56.
[275] Zhou K, Kuo WH, Fillingham J, Greenblatt JF. Control of transcriptional elongation and cotranscriptional histone modification by the yeast BUR kinase substrate Spt5. Proc Natl Acad Sci USA 2009;106:6956–61.
[276] Fingerman IM, Du HN, Briggs SD. Controlling histone methylation via trans-histone pathways. Epigenetics 2008;3:237–42.
[277] Winston F, Chaleff DT, Valent B, Fink GR. Mutations affecting Ty-mediated expression of the HIS4 gene of *Saccharomyces cerevisiae*. Genetics 1984;107:179–97.
[278] Maldonado E, Shiekhattar R, Sheldon M, Cho H, Drapkin R, Rickert P, et al. A human RNA polymerase II complex associated with SRB and DNA-repair proteins. Nature 1996;381:86–9.
[279] Goodarzi AA, Noon AT, Deckbar D, Ziv Y, Shiloh Y, Lobrich M, et al. ATM signaling facilitates repair of DNA double-strand breaks associated with heterochromatin. Mol Cell 2008;31:167–77.
[280] Merelli I, Guffanti A, Fabbri M, Cocito A, Furia L, Grazini U, et al. RSSsite: a reference database and prediction tool for the identification of cryptic recombination signal sequences in human and murine genomes. Nucleic Acids Res 2010;38:W262–7.
[281] Zhang M, Swanson PC. V(D)J recombinase binding and cleavage of cryptic recombination signal sequences identified from lymphoid malignancies. J Biol Chem 2008;283:6717–27.
[282] Tsai AG, Lu H, Raghavan SC, Muschen M, Hsieh CL, Lieber MR. Human chromosomal translocations at CpG sites and a theoretical basis for their lineage and stage specificity. Cell 2008;135: 1130–42.
[283] Marculescu R, Le T, Simon P, Jaeger U, Nadel B. V(D)J-mediated translocations in lymphoid neoplasms: a functional assessment of genomic instability by cryptic sites. J Exp Med 2002;195:85–98.
[284] Polo SE, Jackson SP. Dynamics of DNA damage response proteins at DNA breaks: a focus on protein modifications. Genes Dev 2011;25:409–33.
[285] Gong F, Miller KM. Mammalian DNA repair: HATs and HDACs make their mark through histone acetylation. Mutat Res 2013;750:23–30.
[286] Dey A, Chitsaz F, Abbasi A, Misteli T, Ozato K. The double bromodomain protein Brd4 binds to acetylated chromatin during interphase and mitosis. Proc Natl Acad Sci USA 2003;100:8758–63.
[287] Stanlie A, Yousif AS, Akiyama H, Honjo T, Begum NA. Chromatin reader Brd4 functions in Ig class switching as a repair complex adaptor of non-homologous end joining. Mol Cell 2014; 55(1):97–110.
[288] Filippakopoulos P, Qi J, Picaud S, Shen Y, Smith WB, Fedorov O, et al. Selective inhibition of BET bromodomains. Nature 2010;468:1067–73.
[289] Ogiwara H, Ui A, Otsuka A, Satoh H, Yokomi I, Nakajima S, et al. Histone acetylation by CBP and p300 at double-strand break sites facilitates SWI/SNF chromatin remodeling and the recruitment of non-homologous end joining factors. Oncogene 2011;30:2135–46.
[290] Cattoretti G, Buttner M, Shaknovich R, Kremmer E, Alobeid B, Niedobitek G. Nuclear and cytoplasmic AID in extrafollicular and germinal center B cells. Blood 2006;107:3967–75.
[291] Crouch EE, Li Z, Takizawa M, Fichtner-Feigl S, Gourzi P, Montano C, et al. Regulation of AID expression in the immune response. J Exp Med 2007;204:1145–56.
[292] Qin H, Suzuki K, Nakata M, Chikuma S, Izumi N, Thi Huong L, et al. Activation-induced cytidine deaminase expression in CD4 T cells is associated with a unique IL-10-producing subset that increases with Age. PLoS One 2011;6:e29141.
[293] Kuraoka M, Liao D, Yang K, Allgood SD, Levesque MC, Kelsoe G, et al. Activation-induced cytidine deaminase expression and activity in the absence of germinal centers: insights into hyper-IgM syndrome. J Immunol 2009;183:3237–48.
[294] Ueda Y, Liao D, Yang K, Patel A, Kelsoe G. T-independent activation-induced cytidine deaminase expression, class-switch recombination, and antibody production by immature/transitional 1 B cells. J Immunol 2007;178:3593–601.
[295] Mao C, Jiang L, Melo-Jorge M, Puthenveetil M, Zhang X, Carroll MC, et al. T cell-independent somatic hypermutation in murine B cells with an immature phenotype. Immunity 2004;20:133–44.
[296] Han JH, Akira S, Calame K, Beutler B, Selsing E, Imanishi-Kari T. Class switch recombination and somatic hypermutation in early mouse B cells are mediated by B cell and Toll-like receptors. Immunity 2007;27:64–75.
[297] Kuraoka M, Holl TM, Liao D, Womble M, Cain DW, Reynolds AE, et al. Activation-induced cytidine deaminase mediates central tolerance in B cells. Proc Natl Acad Sci USA 2011;108:11560–5.
[298] Meyers G, Ng YS, Bannock JM, Lavoie A, Walter JE, Notarangelo LD, et al. Activation-induced cytidine deaminase (AID) is required for B-cell tolerance in humans. Proc Natl Acad Sci USA 2011;108:11554–9.
[299] Rommel PC, Bosque D, Gitlin AD, Croft GF, Heintz N, Casellas R, et al. Fate mapping for activation-induced cytidine deaminase (AID) marks non-lymphoid cells during mouse development. PLoS One 2013;8:e69208.
[300] Morgan HD, Dean W, Coker HA, Reik W, Petersen-Mahrt SK. Activation-induced cytidine deaminase deaminates 5-methylcytosine in DNA and is expressed in pluripotent tissues: implications for epigenetic reprogramming. J Biol Chem 2004;279:52353–60.

[301] Schreck S, Buettner M, Kremmer E, Bogdan M, Herbst H, Niedobitek G. Activation-induced cytidine deaminase (AID) is expressed in normal spermatogenesis but only infrequently in testicular germ cell tumours. J Pathol 2006;210:26–31.

[302] Pastor WA, Aravind L, Rao A. TETonic shift: biological roles of TET proteins in DNA demethylation and transcription. Nat Rev Mol Cell Biol 2013;14:341–56.

[303] Fritz EL, Papavasiliou FN. Cytidine deaminases: AIDing DNA demethylation? Genes Dev 2010;24:2107–14.

[304] Greeve J, Philipsen A, Krause K, Klapper W, Heidorn K, Castle BE, et al. Expression of activation-induced cytidine deaminase in human B-cell non-Hodgkin lymphomas. Blood 2003;101:3574–80.

[305] Hardianti MS, Tatsumi E, Syampurnawati M, Furuta K, Saigo K, Nakamachi Y, et al. Activation-induced cytidine deaminase expression in follicular lymphoma: association between AID expression and ongoing mutation in FL. Leukemia 2004;18:826–31.

[306] Pasqualucci L, Neumeister P, Goossens T, Nanjangud G, Chaganti RS, Kuppers R, et al. Hypermutation of multiple proto-oncogenes in B-cell diffuse large-cell lymphomas. Nature 2001;412:341–6.

[307] Smit LA, Bende RJ, Aten J, Guikema JE, Aarts WM, van Noesel CJ. Expression of activation-induced cytidine deaminase is confined to B-cell non-Hodgkin's lymphomas of germinal-center phenotype. Cancer Res 2003;63:3894–8.

[308] Albesiano E, Messmer BT, Damle RN, Allen SL, Rai KR, Chiorazzi N. Activation-induced cytidine deaminase in chronic lymphocytic leukemia B cells: expression as multiple forms in a dynamic, variably sized fraction of the clone. Blood 2003;102:3333–9.

[309] Cerutti A, Zan H, Kim EC, Shah S, Schattner EJ, Schaffer A, et al. Ongoing in vivo immunoglobulin class switch DNA recombination in chronic lymphocytic leukemia B cells. J Immunol 2002;169:6594–603.

[310] Heintel D, Kroemer E, Kienle D, Schwarzinger I, Gleiss A, Schwarzmeier J, et al. High expression of activation-induced cytidine deaminase (AID) mRNA is associated with unmutated IGVH gene status and unfavourable cytogenetic aberrations in patients with chronic lymphocytic leukaemia. Leukemia 2004;18:756–62.

[311] Oppezzo P, Vuillier F, Vasconcelos Y, Dumas G, Magnac C, Payelle-Brogard B, et al. Chronic lymphocytic leukemia B cells expressing AID display dissociation between class switch recombination and somatic hypermutation. Blood 2003;101:4029–32.

[312] McCarthy H, Wierda WG, Barron LL, Cromwell CC, Wang J, Coombes KR, et al. High expression of activation-induced cytidine deaminase (AID) and splice variants is a distinctive feature of poor-prognosis chronic lymphocytic leukemia. Blood 2003;101:4903–8.

[313] Pasqualucci L, Guglielmino R, Houldsworth J, Mohr J, Aoufouchi S, Polakiewicz R, et al. Expression of the AID protein in normal and neoplastic B cells. Blood 2004;104:3318–25.

[314] Palacios F, Moreno P, Morande P, Abreu C, Correa A, Porro V, et al. High expression of AID and active class switch recombination might account for a more aggressive disease in unmutated CLL patients: link with an activated microenvironment in CLL disease. Blood 2010;115:4488–96.

[315] Reiniger L, Bodor C, Bognar A, Balogh Z, Csomor J, Szepesi A, et al. Richter's and prolymphocytic transformation of chronic lymphocytic leukemia are associated with high mRNA expression of activation-induced cytidine deaminase and aberrant somatic hypermutation. Leukemia 2006;20:1089–95.

[316] Lossos IS, Levy R, Alizadeh AA. AID is expressed in germinal center B-cell-like and activated B-cell-like diffuse large-cell lymphomas and is not correlated with intraclonal heterogeneity. Leukemia 2004;18:1775–9.

[317] Dijkman R, Tensen CP, Buettner M, Niedobitek G, Willemze R, Vermeer MH. Primary cutaneous follicle center lymphoma and primary cutaneous large B-cell lymphoma, leg type, are both targeted by aberrant somatic hypermutation but demonstrate differential expression of AID. Blood 2006;107:4926–9.

[318] Okazaki IM, Hiai H, Kakazu N, Yamada S, Muramatsu M, Kinoshita K, et al. Constitutive expression of AID leads to tumorigenesis. J Exp Med 2003;197:1173–81.

[319] Robbiani DF, Bunting S, Feldhahn N, Bothmer A, Camps J, Deroubaix S, et al. AID produces DNA double-strand breaks in non-Ig genes and mature B cell lymphomas with reciprocal chromosome translocations. Mol Cell 2009;36:631–41.

[320] Rucci F, Cattaneo L, Marrella V, Sacco MG, Sobacchi C, Lucchini F, et al. Tissue-specific sensitivity to AID expression in transgenic mouse models. Gene 2006;377:150–8.

[321] Takai A, Toyoshima T, Uemura M, Kitawaki Y, Marusawa H, Hiai H, et al. A novel mouse model of hepatocarcinogenesis triggered by AID causing deleterious p53 mutations. Oncogene 2009;28:469–78.

[322] Takizawa M, Tolarova H, Li Z, Dubois W, Lim S, Callen E, et al. AID expression levels determine the extent of cMyc oncogenic translocations and the incidence of B cell tumor development. J Exp Med 2008;205:1949–57.

[323] Pasqualucci L, Bhagat G, Jankovic M, Compagno M, Smith P, Muramatsu M, et al. AID is required for germinal center-derived lymphomagenesis. Nat Genet 2008;40:108–12.

[324] Morisawa T, Marusawa H, Ueda Y, Iwai A, Okazaki IM, Honjo T, et al. Organ-specific profiles of genetic changes in cancers caused by activation-induced cytidine deaminase expression. Int J Cancer 2008;123:2735–40.

[325] Kovalchuk AL, duBois W, Mushinski E, McNeil NE, Hirt C, Qi CF, et al. AID-deficient Bcl-xL transgenic mice develop delayed atypical plasma cell tumors with unusual Ig/Myc chromosomal rearrangements. J Exp Med 2007;204:2989–3001.

[326] Kotani A, Kakazu N, Tsuruyama T, Okazaki IM, Muramatsu M, Kinoshita K, et al. Activation-induced cytidine deaminase (AID) promotes B cell lymphomagenesis in Emu-cmyc transgenic mice. Proc Natl Acad Sci USA 2007;104:1616–20.

[327] Feldhahn N, Henke N, Melchior K, Duy C, Soh BN, Klein F, et al. Activation-induced cytidine deaminase acts as a mutator in BCR-ABL1-transformed acute lymphoblastic leukemia cells. J Exp Med 2007;204:1157–66.

[328] Iacobucci I, Lonetti A, Messa F, Ferrari A, Cilloni D, Soverini S, et al. Different isoforms of the B-cell mutator activation-induced cytidine deaminase are aberrantly expressed in BCR-ABL1-positive acute lymphoblastic leukemia patients. Leukemia 2010;24:66–73.

[329] Komori J, Marusawa H, Machimoto T, Endo Y, Kinoshita K, Kou T, et al. Activation-induced cytidine deaminase links bile duct inflammation to human cholangiocarcinoma. Hepatology 2008;47:888–96.

[330] Kou T, Marusawa H, Kinoshita K, Endo Y, Okazaki IM, Ueda Y, et al. Expression of activation-induced cytidine deaminase in human hepatocytes during hepatocarcinogenesis. Int J Cancer 2007;120:469–76.

[331] He B, Raab-Traub N, Casali P, Cerutti A. EBV-encoded latent membrane protein 1 cooperates with BAFF/BLyS and APRIL to induce T cell-independent Ig heavy chain class switching. J Immunol 2003;171:5215–24.

[332] Uchida J, Yasui T, Takaoka-Shichijo Y, Muraoka M, Kulwichit W, Raab-Traub N, et al. Mimicry of CD40 signals by Epstein–Barr virus LMP1 in B lymphocyte responses. Science 1999;286:300–3.

[333] Machida K, Cheng KT, Sung VM, Shimodaira S, Lindsay KL, Levine AM, et al. Hepatitis C virus induces a mutator phenotype: enhanced mutations of immunoglobulin and protooncogenes. Proc Natl Acad Sci USA 2004;101:4262–7.

[334] Endo Y, Marusawa H, Kinoshita K, Morisawa T, Sakurai T, Okazaki IM, et al. Expression of activation-induced cytidine deaminase in human hepatocytes via NF-kappaB signaling. Oncogene 2007;26:5587–95.

[335] Ishikawa C, Nakachi S, Senba M, Sugai M, Mori N. Activation of AID by human T-cell leukemia virus tax oncoprotein and the possible role of its constitutive expression in ATL genesis. Carcinogenesis 2011;32:110–9.

[336] Gourzi P, Leonova T, Papavasiliou FN. A role for activation-induced cytidine deaminase in the host response against a transforming retrovirus. Immunity 2006;24:779–86.

[337] Gourzi P, Leonova T, Papavasiliou FN. Viral induction of AID is independent of the interferon and the toll-like receptor signaling pathways but requires NF-kappaB. J Exp Med 2007;204:259–65.

[338] Matsumoto Y, Marusawa H, Kinoshita K, Endo Y, Kou T, Morisawa T, et al. Helicobacter pylori infection triggers aberrant expression of activation-induced cytidine deaminase in gastric epithelium. Nat Med 2007;13:470–6.

[339] Endo Y, Marusawa H, Kou T, Nakase H, Fujii S, Fujimori T, et al. Activation-induced cytidine deaminase links between inflammation and the development of colitis-associated colorectal cancers. Gastroenterology 2008;135:889–98. 898 e1–3.

[340] Nguyen T, Xu J, Chikuma S, Hiai H, Kinoshita K, Moriya K, et al. Activation-induced cytidine deaminase is dispensable for virus-mediated liver and skin tumor development in mouse models. Int Immunol. 2014;26(7):397–406.

[341] Pone EJ, Zhang J, Mai T, White CA, Li G, Sakakura JK, et al. BCR-signalling synergizes with TLR-signalling for induction of AID and immunoglobulin class-switching through the non-canonical NF-kappaB pathway. Nat Commun 2012;3:767.

[342] Schrader CE, Linehan EK, Mochegova SN, Woodland RT, Stavnezer J. Inducible DNA breaks in Ig S regions are dependent on AID and UNG. J Exp Med 2005;202:561–8.

[343] Quezada SA, Jarvinen LZ, Lind EF, Noelle RJ. CD40/CD154 interactions at the interface of tolerance and immunity. Annu Rev Immunol 2004;22:307–28.

[344] Bossen C, Schneider P. BAFF, APRIL and their receptors: structure, function and signaling. Semin Immunol 2006;18:263–75.

[345] He B, Santamaria R, Xu W, Cols M, Chen K, Puga I, et al. The transmembrane activator TACI triggers immunoglobulin class switching by activating B cells through the adaptor MyD88. Nat Immunol 2010;11:836–45.

[346] Bombardieri M, Kam NW, Brentano F, Choi K, Filer A, Kyburz D, et al. A BAFF/APRIL-dependent TLR3-stimulated pathway enhances the capacity of rheumatoid synovial fibroblasts to induce AID expression and Ig class-switching in B cells. Ann Rheum Dis 2011;70:1857–65.

[347] Nagaoka H, Tran TH, Kobayashi M, Aida M, Honjo T. Preventing AID, a physiological mutator, from deleterious activation: regulation of the genomic instability that is associated with antibody diversity. Int Immunol 2010;22:227–35.

[348] Tran TH, Nakata M, Suzuki K, Begum NA, Shinkura R, Fagarasan S, et al. B cell-specific and stimulation-responsive enhancers derepress Aicda by overcoming the effects of silencers. Nat Immunol 2010;11:148–54.

[349] Park SR, Zan H, Pal Z, Zhang J, Al-Qahtani A, Pone EJ, et al. HoxC4 binds to the promoter of the cytidine deaminase AID gene to induce AID expression, class-switch DNA recombination and somatic hypermutation. Nat Immunol 2009;10:540–50.

[350] Kawai T, Akira S. The role of pattern-recognition receptors in innate immunity: update on toll-like receptors. Nat Immunol 2010;11:373–84.

[351] Muramatsu M, Sankaranand VS, Anant S, Sugai M, Kinoshita K, Davidson NO, et al. Specific expression of activation-induced cytidine deaminase (AID), a novel member of the RNA-editing deaminase family in germinal center B cells. J Biol Chem 1999;274:18470–6.

[352] Jabara H, Laouini D, Tsitsikov E, Mizoguchi E, Bhan A, Castigli E, et al. The binding site for TRAF2 and TRAF3 but not for TRAF6 is essential for CD40-mediated immunoglobulin class switching. Immunity 2002;17:265–76.

[353] Dedeoglu F, Horwitz B, Chaudhuri J, Alt FW, Geha RS. Induction of activation-induced cytidine deaminase gene expression by IL-4 and CD40 ligation is dependent on STAT6 and NFkappaB. Int Immunol 2004;16:395–404.

[354] Nelms K, Keegan AD, Zamorano J, Ryan JJ, Paul WE. The IL-4 receptor: signaling mechanisms and biologic functions. Annu Rev Immunol 1999;17:701–38.

[355] Heldin CH, Miyazono K, ten Dijke P. TGF-beta signalling from cell membrane to nucleus through SMAD proteins. Nature 1997;390:465–71.

[356] Pauklin S, Sernandez IV, Bachmann G, Ramiro AR, Petersen-Mahrt SK. Estrogen directly activates AID transcription and function. J Exp Med 2009;206:99–111.

[357] Mai T, Zan H, Zhang J, Hawkins JS, Xu Z, Casali P. Estrogen receptors bind to and activate the HOXC4/HoxC4 promoter to potentiate HoxC4-mediated activation-induced cytosine deaminase induction, immunoglobulin class switch DNA recombination, and somatic hypermutation. J Biol Chem 2010;285:37797–810.

[358] Yadav A, Olaru A, Saltis M, Setren A, Cerny J, Livak F. Identification of a ubiquitously active promoter of the murine activation-induced cytidine deaminase (AICDA) gene. Mol Immunol 2006;43:529–41.

[359] Huong le T, Kobayashi M, Nakata M, Shioi G, Miyachi H, Honjo T, et al. In vivo analysis of Aicda gene regulation: a critical balance between upstream enhancers and intronic silencers governs appropriate expression. PLoS One 2013;8:e61433.

[360] Gonda H, Sugai M, Nambu Y, Katakai T, Agata Y, Mori KJ, et al. The balance between Pax5 and Id2 activities is the key to AID gene expression. J Exp Med 2003;198:1427–37.

[361] Sayegh CE, Quong MW, Agata Y, Murre C. E-proteins directly regulate expression of activation-induced deaminase in mature B cells. Nat Immunol 2003;4:586–93.

[362] Kawamata T, Lu J, Sato T, Tanaka M, Nagaoka H, Agata Y, et al. Imatinib mesylate directly impairs class switch recombination through down-regulation of AID: its potential efficacy as an AID suppressor. Blood 2012;119:3123–7.

[363] Ise W, Kohyama M, Schraml BU, Zhang T, Schwer B, Basu U, et al. The transcription factor BATF controls the global regulators of class-switch recombination in both B cells and T cells. Nat Immunol 2011;12:536–43.

[364] De Silva NS, Simonetti G, Heise N, Klein U. The diverse roles of IRF4 in late germinal center B-cell differentiation. Immunol Rev 2012;247:73–92.

[365] Basso K, Schneider C, Shen Q, Holmes AB, Setty M, Leslie C, et al. BCL6 positively regulates AID and germinal center gene expression via repression of miR-155. J Exp Med 2012;209:2455–65.

[366] Shaffer AL, Lin KI, Kuo TC, Yu X, Hurt EM, Rosenwald A, et al. Blimp-1 orchestrates plasma cell differentiation by extinguishing the mature B cell gene expression program. Immunity 2002; 17:51–62.

[367] Xu Z, Pone EJ, Al-Qahtani A, Park SR, Zan H, Casali P. Regulation of aicda expression and AID activity: relevance to somatic hypermutation and class switch DNA recombination. Crit Rev Immunol 2007;27:367–97.

[368] Teng G, Hakimpour P, Landgraf P, Rice A, Tuschl T, Casellas R, et al. MicroRNA-155 is a negative regulator of activation-induced cytidine deaminase. Immunity 2008;28:621–9.

[369] Dorsett Y, McBride KM, Jankovic M, Gazumyan A, Thai TH, Robbiani DF, et al. MicroRNA-155 suppresses activation-induced cytidine deaminase-mediated Myc-Igh translocation. Immunity 2008;28:630–8.

[370] de Yebenes VG, Belver L, Pisano DG, Gonzalez S, Villasante A, Croce C, et al. miR-181b negatively regulates activation-induced cytidine deaminase in B cells. J Exp Med 2008;205:2199–206.

[371] Borchert GM, Holton NW, Larson ED. Repression of human activation induced cytidine deaminase by miR-93 and miR-155. BMC Cancer 2011;11:347.

[372] Landgraf P, Rusu M, Sheridan R, Sewer A, Iovino N, Aravin A, et al. A mammalian microRNA expression atlas based on small RNA library sequencing. Cell 2007;129:1401–14.

[373] McBride KM, Gazumyan A, Woo EM, Barreto VM, Robbiani DF, Chait BT, et al. Regulation of hypermutation by activation-induced cytidine deaminase phosphorylation. Proc Natl Acad Sci USA 2006;103:8798–803.

[374] Gazumyan A, Timachova K, Yuen G, Siden E, Di Virgilio M, Woo EM, et al. Amino-terminal phosphorylation of activation-induced cytidine deaminase suppresses c-myc/IgH translocation. Mol Cell Biol 2011;31:442–9.

[375] Aoufouchi S, Faili A, Zober C, D'Orlando O, Weller S, Weill JC, et al. Proteasomal degradation restricts the nuclear lifespan of AID. J Exp Med 2008;205:1357–68.

[376] Robert I, Dantzer F, Reina-San-Martin B. Parp1 facilitates alternative NHEJ, whereas Parp2 suppresses IgH/c-myc translocations during immunoglobulin class switch recombination. J Exp Med 2009;206:1047–56.

[377] Wray J, Williamson EA, Singh SB, Wu Y, Cogle CR, Weinstock DM, et al. PARP1 is required for chromosomal translocations. Blood 2013;121:4359–65.

[378] Paddock MN, Buelow BD, Takeda S, Scharenberg AM. The BRCT domain of PARP-1 Is required for immunoglobulin gene conversion. PLoS Biol 2010;8:e1000428.

[379] Saribasak H, Maul RW, Cao Z, McClure RL, Yang W, McNeill DR, et al. XRCC1 suppresses somatic hypermutation and promotes alternative nonhomologous end joining in Igh genes. J Exp Med 2011;208:2209–16.

[380] Bardwell PD, Woo CJ, Wei K, Li Z, Martin A, Sack SZ, et al. Altered somatic hypermutation and reduced class-switch recombination in exonuclease 1-mutant mice. Nat Immunol 2004;5: 224–9.

[381] Lee-Theilen M, Matthews AJ, Kelly D, Zheng S, Chaudhuri J. CtIP promotes microhomology-mediated alternative end joining during class-switch recombination. Nat Struct Mol Biol 2011;18:75–9.

[382] Zhang Y, Jasin M. An essential role for CtIP in chromosomal translocation formation through an alternative end-joining pathway. Nat Struct Mol Biol 2011;18:80–4.

[383] Poltoratsky V, Prasad R, Horton JK, Wilson SH. Down-regulation of DNA polymerase beta accompanies somatic hypermutation in human BL2 cell lines. DNA Repair (Amst) 2007;6:244–53.

[384] Wu X, Stavnezer J. DNA polymerase beta is able to repair breaks in switch regions and plays an inhibitory role during immunoglobulin class switch recombination. J Exp Med 2007;204:1677–89.

[385] Rada C, Ehrenstein MR, Neuberger MS, Milstein C. Hot spot focusing of somatic hypermutation in MSH2-deficient mice suggests two stages of mutational targeting. Immunity 1998;9:135–41.

[386] Schrader CE, Bradley SP, Vardo J, Mochegova SN, Flanagan E, Stavnezer J. Mutations occur in the Ig Smu region but rarely in Sgamma regions prior to class switch recombination. Embo J 2003;22:5893–903.

[387] Ehrenstein MR, Neuberger MS. Deficiency in Msh2 affects the efficiency and local sequence specificity of immunoglobulin class-switch recombination: parallels with somatic hypermutation. Embo J 1999;18:3484–90.

[388] Schrader CE, Edelmann W, Kucherlapati R, Stavnezer J. Reduced isotype switching in splenic B cells from mice deficient in mismatch repair enzymes. J Exp Med 1999;190:323–30.

[389] Smith JA, Bannister LA, Bhattacharjee V, Wang Y, Waldman BC, Waldman AS. Accurate homologous recombination is a prominent double-strand break repair pathway in mammalian chromosomes and is modulated by mismatch repair protein Msh2. Mol Cell Biol 2007;27:7816–27.

[390] Martomo SA, Yang WW, Gearhart PJ. A role for Msh6 but not Msh3 in somatic hypermutation and class switch recombination. J Exp Med 2004;200:61–8.

[391] Campo VA, Patenaude A-M, Kaden S, Horb L, Firka D, Jiricny J, Di Noia JM. MSH6- or PMS2-deficiency causes re-replication in DT40 B cells, but it has little effect on immunoglobulin gene conversion or on repair of AID-generated uracils. Nucleic Acids Res 2013;41:3032–46.

[392] Ehrenstein MR, Rada C, Jones AM, Milstein C, Neuberger MS. Switch junction sequences in PMS2-deficient mice reveal a microhomology-mediated mechanism of Ig class switch recombination. Proc Natl Acad Sci USA 2001;98:14553–8.

[393] Martomo SA, Yang WW, Wersto RP, Ohkumo T, Kondo Y, Yokoi M, et al. Different mutation signatures in DNA polymerase eta- and MSH6-deficient mice suggest separate roles in antibody diversification. Proc Natl Acad Sci USA 2005;102:8656–61.

[394] Hirota K, Sonoda E, Kawamoto T, Motegi A, Masutani C, Hanaoka F, et al. Simultaneous disruption of two DNA polymerases, Polη and Polζ, in Avian DT40 cells unmasks the role of Polη in cellular response to various DNA lesions. PLoS Genet 2010;6:e1001151.

[395] Martomo SA, Yang WW, Vaisman A, Maas A, Yokoi M, Hoeijmakers JH, et al. Normal hypermutation in antibody genes from congenic mice defective for DNA polymerase iota. DNA Repair (Amst) 2006;5:392–8.

[396] McDonald JP, Frank EG, Plosky BS, Rogozin IB, Masutani C, Hanaoka F, et al. 129-derived Strains of Mice Are Deficient in DNA Polymerase ι and Have Normal Immunoglobulin Hypermutation. J Exp Med 2003;198:635–43.

[397] Schenten D, Gerlach VL, Guo C, Velasco-Miguel S, Hladik CL, White CL, et al. DNA polymerase kappa deficiency does not affect somatic hypermutation in mice. Eur J Immunol 2002;32:3152–60.

[398] Masuda K, Ouchida R, Li Y, Gao X, Mori H, Wang JY. A critical role for REV1 in regulating the induction of C:G transitions and A:T mutations during Ig gene hypermutation. J Immunol 2009;183:1846–50.

[399] Masuda K, Ouchida R, Takeuchi A, Saito T, Koseki H, Kawamura K, et al. DNA polymerase theta contributes to the generation of C/G mutations during somatic hypermutation of Ig genes. Proc Natl Acad Sci USA 2005;102:13986–91.

[400] Kohzaki M, Nishihara K, Hirota K, Sonoda E, Yoshimura M, Ekino S, et al. DNA polymerases ν and θ are required for efficient immunoglobulin V gene diversification in chicken. J Cell Biol 2010;189:1117–27.

[401] Schenten D, Kracker S, Esposito G, Franco S, Klein U, et al. Pol zeta ablation in B cells impairs the germinal center reaction, class switch recombination, DNA break repair, and genome stability. J Exp Med 2009;206:477–90.

[402] Okada T, Sonoda E, Yoshimura M, Kawano Y, Saya H, Kohzaki M, Takeda S. Multiple roles of vertebrate REV genes in DNA repair and recombination. Mol Cell Biol 2005;25:6103–11.

[403] Zaprazna K, Atchison ML. YY1 controls immunoglobulin class switch recombination and nuclear activation-induced deaminase levels. Mol Cell Biol 2012;32:1542–54.

[404] Demorest ZL, MacDuff DA, Brown WL, Morham SG, Parise LV, Harris RS. The interaction between AID and CIB1 is nonessential for antibody gene diversification by gene conversion or class switch recombination. PLoS One 2010:5. e11660.

Chapter 19

# The Mechanism of IgH Class Switch Recombination

Fei-Long Meng[1,2], Frederick W. Alt[1,2] and Ming Tian[1,2]

[1]Howard Hughes Medical Institute, Program in Cellular and Molecular Medicine, Boston Children's Hospital, Boston, MA, USA, [2]Department of Genetics, Harvard Medical School, Boston, MA, USA

## 1. ANTIBODY CLASS

Antibodies are composed of immunoglobulin-heavy (IgH) and -light (IgL) chains (Figure 1). Structurally, IgH and IgL can be divided into a series of immunoglobulin (Ig) domains. Because of its enormous diversity among different antibodies, the N-terminal IgH and IgL domain is called the variable (V) region. The V regions of IgH and IgL associate to form the antigen-binding region. The remaining portions of IgH and IgL are called constant (C) regions. The IgH C regions interact with other components of the immune system to provide antibody effector functions. In the mouse, there are eight different IgH C regions (Cμ, Cδ, Cγ3, Cγ1, Cγ2b, Cγ2a, Cε, and Cα) and two different families of IgL chains (κ and λ). The IgH $C_H$ region defines the antibody class (sometimes referred to as the isotype); for example, an antibody with μ C region is referred to as IgM, whereas an antibody with a γ1 C region is referred to as IgG1. Each of the IgH C regions prescribes different effector functions. For example, IgM can unleash the power of a complement system to destroy bacteria. Immunoglobulin G helps macrophages to attack pathogens. Immunoglobulin A shields mucosal surfaces from pathogenic invasion. Specialization of IgH classes is a common strategy to strengthen the immune system across species from amphibians and birds to mammals. In this chapter, we focus on IgH class switching in mice, where most relevant studies were carried out; but the process overlaps with that in humans.

## 2. ORGANIZATION OF MOUSE IgH LOCUS

Immunoglobulin H V regions are encoded by V(D)J exons, which are assembled from Variable ($V_H$), Diversity (D) and Joining ($J_H$) gene segments via V(D)J recombination during early stages of B-cell development in the bone marrow [1]. The eight sets of $C_H$ exons lie downstream of the V(D)J exon in the following order: 5′-V(D)J-μ-δ-γ3-γ1-γ2b-γ2a-ε-α-3′ [2](Figure 2(A)). Several B cell–specific enhancers are present in the IgH locus: the intronic enhancer (Eμ) lies between $J_H$ regions and Cμ, and a 30-kb IgH regulatory region ("IgH3′RR") is downstream of Cα and contains a series of enhancers that may have different functions in both class switch recombination (CSR) and somatic hypermutation (SHM) [3–7] (Figure 2(A)). Transcription of the IgH gene initiates from a promoter ($V_H$ promoter) immediately upstream of $V_H$ and terminates downstream of Cμ or Cδ; in the former situation, the V(D)J exon is spliced to Cμ to form the mRNA for either the membrane-bound or secreted version of the μ heavy chain; whereas in the latter situation, the V(D)J exon is spliced to Cδ to form the mRNA for the membrane-bound or secreted version of the δ heavy chain. Without antigenic stimulation, resting B-cells express exclusively IgM and IgD. At this stage, IgM and IgD are retained on the cell surface, because the Cμ and Cδ exons are spliced to the downstream exons encoding a hydrophobic membrane anchor [8,9]. The surface IgM and IgD serve as B-cell receptors (BCR) that recognize antigen [10].

### 2.1 Activation-Induced Cytidine Deaminase, SHM, and IgH CSR

To monitor infection, resting B-cells circulate in peripheral lymphoid tissues including the spleen, lymph nodes, and Peyer patches. The binding of antigen to BCR triggers B-cell activation. Activated B-cells can undergo two forms of genomic alteration at the Ig loci: CSR at the C region portion of the IgH locus and SHM of both IgH and IgL V region exons [11,12]. Both CSR and SHM are initiated by activation-induced cytidine deaminase (AID), which is a single-strand (ss) DNA-specific cytidine deaminase that preferentially deaminates cytidine in the context of "AID

Molecular Biology of B Cells. http://dx.doi.org/10.1016/B978-0-12-397933-9.00019-9

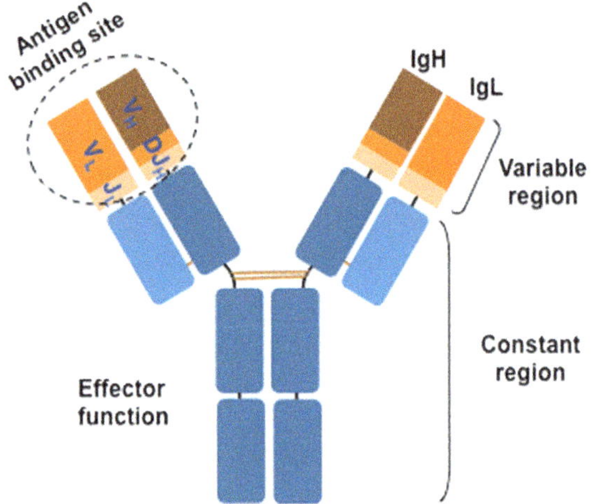


FIGURE 1 **Antibody structure.** The rectangles represent Ig domains that constitute the structural units of immunoglobulin-heavy (IgH) and -light (IgL) chains. Variable regions are shown in brown and constant regions are in blue. See text for more details.

targeting" motifs (DGYW or WRCH; D=A/G/T, Y=C/T, W=A/T, H=T/C/A, R=A/G) [13,14]. During SHM, AID specifically deaminates cytidines in the V(D)J exons of IgH and IgL; the deamination products are then further processed through pathways that involve normal base excision or mismatch repair (MMR) into mutations and/or small deletion/insertions [12]. By generating such mutations, SHM diversifies the repertoires of antigen-binding sites of the BCR; B-cells expressing BCRs with higher antigen affinity are positively selected in the light zone of germinal centers (GCs) to effect affinity maturations of the immune response [15,16].

CSR can take place both inside and outside GCs [17]. For CSR, AID is targeted to long (1- to 10-kb) repetitive S regions that lie just upstream of each set of $C_H$ exons except Cδ [18]. During CSR, Sμ, which is referred to as a donor S region, is fused to one of the other downstream acceptor S regions through a breakage and joining mechanism [19]. As

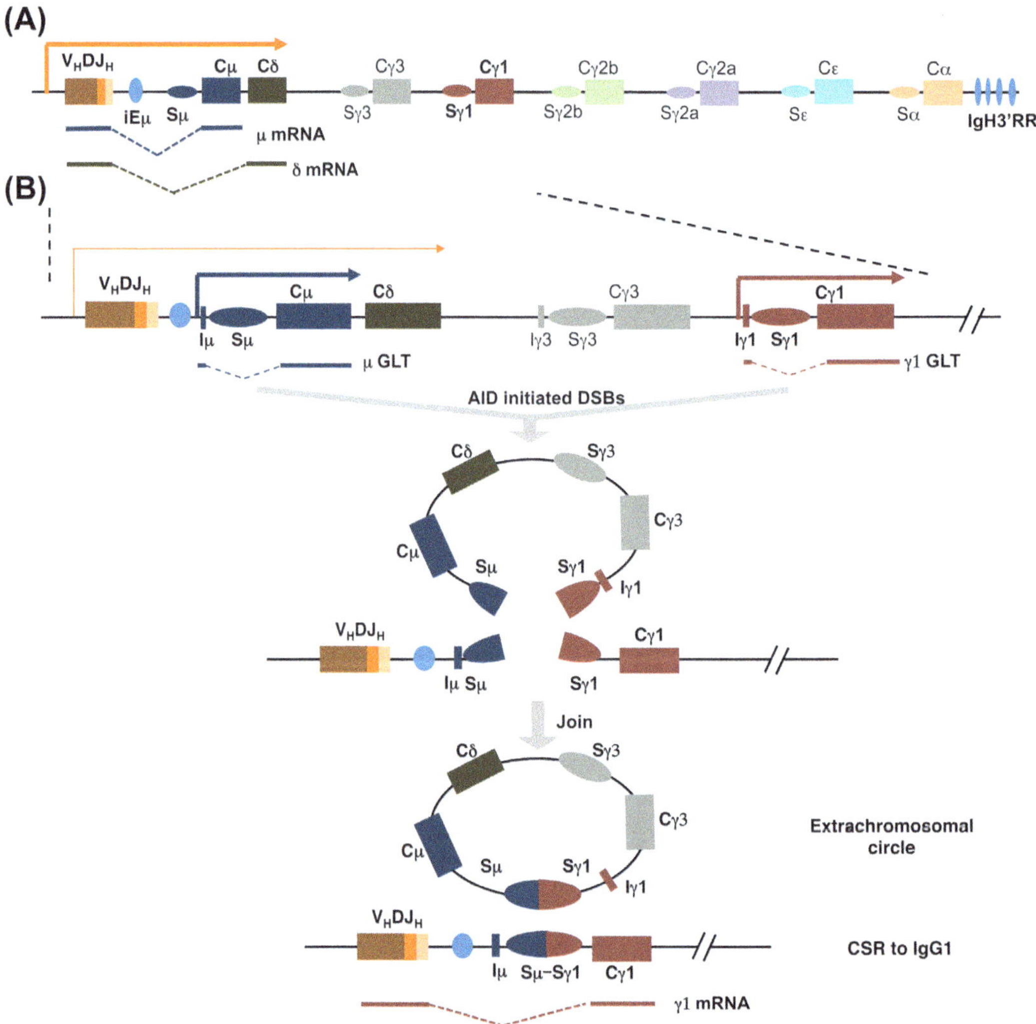


FIGURE 2 **Immunoglobulin-heavy (IgH) class switch recombination.** (A) Organization of IgH constant (C) region. Rectangles represent exons that encode the variable region ($V_H(D)J_H$) and C regions (Cμ through Cα) of IgH; the ovals in frond of the C regions represent S regions. Blue ovals represent IgH enhancers: iEμ, μ intronic enhancer; IgH3′RR, IgH 3′ regulatory region. The mRNAs for μ and δ are shown below the corresponding gene; the dashed lines depict splicing events that join the $V_HDJ_H$ exon to C exons. (B) Class switch recombination (CSR) between Sμ and Sγ1 is shown as an example. The germline transcripts and mRNAs for γ1 are shown. See text for more details. AID, activation induced cytidine deaminase.

a result, the sequences between the two S regions are thought to be deleted from the IgH locus as an extrachromosomal circle [20–22]; through breakage and joining of two S regions, a downstream $C_H$ replaces Cμ, leading to the expression of a new $C_H$, such as γ, α, or ε, and correspondingly, a new class of BCR/antibody (e.g., IgG, IgA, or IgE) (Figure 2(B)). In the case of IgE expression, CSR can occur directly between Sμ and Sε, or, alternatively, can occur subsequent to IgG switching via sequential CSR between the fused Sμ/Sγ1 to Sε [23]. Because the V(D)J exon remains the same after any type of CSR, this process alters the effector function of antibody but not the antigen-binding specificity.

## 3. A TWO-STEP MODEL OF CSR

CSR conforms to a two-step reaction scheme. First, AID-dependent DNA double-strand breaks (DSBs) can be introduced into multiple locations along the long donor Sμ and along one of the long downstream acceptor S regions. Second, the upstream end of a DSB in Sμ must be joined to a downstream end of a DSB in the acceptor S region to replace Cμ with the downstream $C_H$; in addition, the other ends of the DSBs must be joined together if the intervening region is deleted as a circle (Figure 2(B)). In this reaction, the S region DSBs are joined by general DNA repair factors, primarily by the classical non-homologous end-joining pathway (C-NHEJ), but also by alternative end-joining (A-EJ) pathways, that join general DSBs in all cells. In this regard, CSR may be distinct from V(D)J recombination, which assembles the V, D, and J gene segments into V(D)J exons. During V(D)J recombination, the lymphocyte-specific RAG1/2 endonuclease initiates that reaction by generating DSBs between recombination signal sequences (RSSs) and flanking V, D, or J gene segments [24,25a]. Subsequently, RAG1/2 holds these ends in a postcleavage synaptic complex and then passes the ends exclusively to the C-NHEJ pathway [1]. Several studies suggest that the C-terminal end of the AID protein provides activities that may influence downstream end joining [25b,26]. However, unlike RAG, such activities do not exclude pathways beyond C-NHEJ; in that regard, repair of AID DSBs is more reminiscent of the repair of general DSBs, and CSR joining potentially mechanistically overlaps with mechanisms that promote the interstitial deletion types of intrachromosomal translocation reactions (see subsequent discussion) [27,28]. Finally, neither AID nor RAG1/2 qualifies as a recombinase, in the traditional sense that they cannot mediate the whole recombination reaction by themselves.

## 4. MECHANISMS BY WHICH AID INITIATES CSR AND SHM

As the name implies, AID is expressed in response to B-cell activation [29]. Whereas neither RAG nor AID is a recombinase, AID is not even an endonuclease. Thus, AID belongs to the APOBEC family of cytidine deaminases, which, depending on the family member, catalyze the conversion of cytosine into uracil in the context of either DNA or RNA [30] (Figure 3(A)). Yet, genetic studies firmly established AID as required for the initiation of both CSR and SHM. In mice, ablation of AID completely abolishes CSR and SHM [31]. In humans, mutations in AID cause the hyper-IgM syndrome immunodeficiency disease; because of a CSR defect, these patients have high levels of IgM but little IgG and IgA in serum, and also do not show antibody affinity maturation owing to defective SHM [32].

An early hypothesis posited that AID may function in CSR and SHM via an RNA editing mechanism, based on the paradigm of its close homolog, Apobec-1. Apobec-1 catalyzes the deamination of a specific cytidine in the mRNA for apolipoprotein-B; the RNA editing changes a CAA codon for glutamine into a UAA stop codon [33–35]. The edited mRNA encodes a truncated version of apolipoprotein-B that has different functions from the full-length counterpart in lipid metabolism. By analogy, AID was hypothesized to edit the mRNAs to produce separate factors: one that leads to CSR and another that leads to SHM [36]. However, thus far there has been no concrete evidence to support a putative RNA editing role for AID in CSR at any level, and much current evidence is not consistent with RNA editing models. The alternative early hypothesis about AID function, which was based on extensive comparison of the nature of the lesions that lead to both SHM and CSR, was that AID initiates both CSR and SHM by deaminating cytosines in S region and V(D)J exon DNA, respectively [37,38].

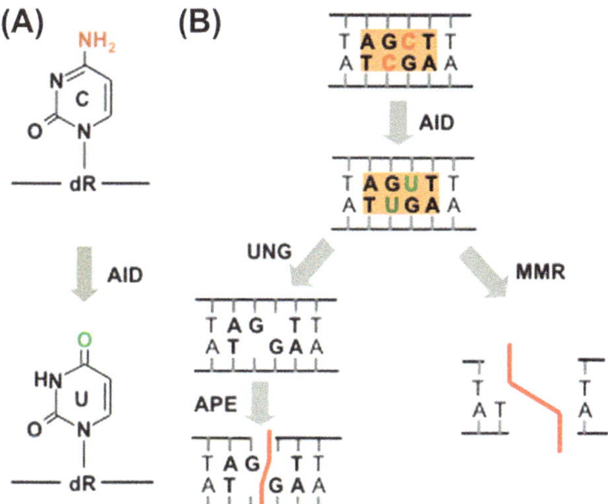


FIGURE 3 **Mechanisms of activation-induced cytidine deaminase (AID)-initiated genomic alterations.** (A) AID catalyzes deamination of cytosine at the N4 position, marked in red; "dR" denotes deoxyribose. (B) AID-mediated deamination products are further processed by base excision repair and mismatch repair pathways to double-strand breaks (DSBs). *Adapted from reference [52].* See text for more details. UNG, uracil DNA glycosylase; MMG, mismatch repair.

Numerous biochemical studies further proved that purified or recombinant AID has robust cytidine deamination activity on ssDNA substrates [39–41]. Genetic studies clearly demonstrated that AID deamination activity in vivo leads to DNA strand breaks in two ways [42–46] (Figure 3(B)). First, uracil DNA glycosylase (UNG) can excise the deamination product, uracil, to generate abasic site; then, apurinic and apyrimidic endonuclease can cleave the phosphodiester bond at the 5′ side of the abasic site [47]. Second, MMR factors can recognize the U:G mismatch and excise a patch of DNA surrounding the mismatch [48]. In both cases, strand breaks are introduced into S regions as DNA repair machinery attempts to remove the uracil mismatches (Figure 3(B)). Correspondingly, UNG- or MMR-deficient mice have substantial CSR defects [42–46] and mutations in UNG or MMR cause hyper-IgM syndrome in human patients [49–51]. Indeed, CSR is completely abolished in mice when both UNG and MMR are inactivated and high levels of C>T mutations accumulate in S regions, which serve as clear evidence of deamination products of AID [52,53]. Uracil DNA glycosylase and MMR deficiencies affect SHM in a manner that also exactly fits the predictions of the DNA deamination model [12,54]. On the strength of this compelling evidence, the DNA deamination model for AID function in CSR and SHM is now widely accepted by the field.

## 5. GERMLINE S REGION TRANSCRIPTION TARGETS AID ACTIVITY DURING CSR

Although AID is able to deaminate cytosines in naked single-strand DNA in vitro, it exhibits no activity on double-strand DNA [39–41]. Thus, for AID to act on S regions during CSR and V(D)J exons during SHM, there needs to be a mechanism to provide ssDNA substrate within AID targets in vivo. In addition, AID also needs to overcome potential barriers provided by chromatin to gain access to S region DNA in vivo. Transcription through S regions has long been correlated with CSR [55,56], where it could potentially have a major role in targeting particular acceptor S regions for CSR in a manner that provides an appropriate B-cell response for a given form of infection [57]. In this context, transcription serves two roles: directing AID to specific target sequences and providing requisite ssDNA substrates.

Each S region is part of a transcription unit that initiates from a promoter (I promoter) upstream of the S region and terminates downstream of the $C_H$ exons (Figure 2(B)). Transcription is initiated only from I region promoters upstream of acceptor S regions targeted for CSR, which is referred to as germline transcription [58–62]. During the processing of a germline $C_H$ transcript, S region sequences are spliced out as an intron, and the upstream I exon is joined to the downstream $C_H$ exons. The mature germline transcript contains no long open reading frames, and there is no evidence that it programs the synthesis of any protein. Nonetheless, germline transcription is critical for CSR: Deletion of an I promoter abolishes recombination to the downstream S region, and replacement of an I region promoter with a different promoter can drive CSR under nonphysiological conditions [63–66]. Deficiencies for epigenetic regulators or transcription factors that affect transcription from particular I region promoters in the activated B-cells also cause defects in CSR to the corresponding $C_H$ exons [67–69].

Under physiological conditions, CSR to a specific $C_H$ exons is controlled at the level of I promoter activity, which is in turn regulated by cytokines. For example, B-cells activated through the CD40/CD40L pathway undergo CSR to IgG1 and IgE in the presence of interleukin 4 (IL4), because this cytokine stimulates the germline transcription from the Iγ1 and Iε promoters [62,70,71]. Likewise, interferon-gamma (IFN-γ) promotes CSR to IgG2a [72], and transforming growth factor-β (TGF-β) stimulates CSR to IgA by similarly inducing transcription from the corresponding I region promoters [73a]. The formation of functional IgH V region exons by V(D)J recombination is subject to allelic exclusion, which leads to assembly of a productive IgH gene that encodes and IgH protein on only one allele; the second allele contains either a $DJ_H$ intermediate or an out-of-frame nonproductive V(D)J exon [73b]. However, after a particular form of B-cell activation, germline transcription from target I promoters is activated on both IgH alleles [74]. Thus, activated B-cells switch often to the same $C_H$ gene on both their productive and nonproductive IgH alleles [73c,73d], which nicely demonstrates that CSR is actively targeted to a specific S region or set of S regions by a particular form of CSR activation [62,70–72,73a,73d]. In summary, depending on the type of antigen and the context in which it is encountered, other cells of the immune system can direct B-cells to undergo CSR to generate a particular class of antibody most suited for antigen elimination.

The I promoters work in conjunction with enhancers in the IgH 3′RR downstream of Cα. A set of four IgH 3′RR enhancers are marked with DNase I hypersensitivity sites that are arranged in the order 5′-hs3a-hs1,2-hs3b-hs4-3′ [5] (Figure 2(A)). Knockout models of IgH3′RR have clearly demonstrated the important functions of this region in enhancing transcription from I region promoters, thereby, activating CSR. Deletion of hs3b and 4 impairs germline transcription and CSR to most $C_H$s except Cγ1, but has no effect on SHM of the V(D)J exon [75]; whereas elimination of the entire IgH 3′RR severely impairs CSR to all $C_H$s and also impairs SHM [3,4]. Mechanisms by which IgH3′RR regulates GLT and CSR have been proposed, including a promoter competition model suggesting that owing to proximity or strength, certain I region promoters may exclude others from interacting with the IgH3'RR to modulate particular CSR patterns [76], including sequential CSR from IgG1 to IgE. In this regard, loops and ternary complexes between IgH3′RR enhancer and I promoters have been demonstrated [5,77,78], with the physical interaction between DNA elements in the enhancers and AID target sequences potentially contributing to targeting AID activity, as suggested for the function of DIVAC and other regulatory elements in the chicken IgL enhancer [79–83].

## 6. ROLE OF TRANSCRIPTION STALLING IN AID TARGETING

To gain access to its genomic DNA substrates, AID exploits interactions with the transcription machinery. Soon after its identification, AID was found to associate with RNA polymerase II (Pol II) complex during CSR [84]. However, not all transcribed genes are AID targets in activated B-cells. Chromatin immunoprecipitation studies in B-cells activated for CSR revealed a higher Pol II density within transcribed S regions relative to adjacent regions [85]; similarly, nuclear run-on transcription assays revealed an accumulation of Pol II in S regions compared with the rest of the germline transcription unit [86]. These observations suggested that transcriptional stalling might occur at a high frequency in transcribed S region during germline transcription, leading to the proposal that such stalling may contribute to directing AID to S regions. In the context of transcriptional stalling and AID targeting, long before AID was discovered, it was proposed that targeting of SHM to V(D)J exons involved the association of a mutator (now known to be AID) with the elongating RNA polymerase [87,88]. Direct evidence linking transcriptional stalling to AID targeting came from identification of the role of Suppressor of Ty5 homolog (Spt5) in CSR by a genome-wide shRNA screen for factors required for normal CSR in the CH12F3 B-cell lymphoma line [89]. Suppressor of Ty5, a transcription elongation and stalling factor, was found to interact with AID, and chromatin association patterns implicated this factor in targeting AID to transcription templates by bridging stalled Pol II and AID [89].

### 6.1 S Region Functions in CSR

After being recruited to chromatin targets, AID faces another difficulty. Activation-induced cytidine deaminates the amino group at the N4 position of cytosine (Figure 3(A)). Because the amino group is engaged in hydrogen bonding with guanosine in duplex DNA, AID can only deaminate cytosine in the context of single-stranded DNA [39–41,90]. The double-stranded genomic DNA must be converted into a single-stranded form for AID function. During transcription, 8–9 base pairs of template DNA are unwound into a transcription bubble, in which the nontemplate strand is single-stranded [91] (Figure 4(A)). In addition, the elongating RNA polymerase II (Pol II) leaves behind a wave of negative supercoils that facilitate DNA unwinding [92] (Figure 4(B)). However, such strand unwinding is transient; AID could take advantage of such a fleeting opportunity only if it is close to the site of RNA synthesis and/or the ssDNA is generated through other mechanisms and also stabilized [93,94].

To induce CSR efficiently, germline transcription needs to work in conjunction with S regions. Deletion of S regions severely impairs CSR to the corresponding $C_H$, and unrelated sequences fail to substitute for the function of S regions in CSR [74,95,96]. One unusual feature of mouse and human S regions is that their two DNA strands have markedly different base compositions. As oriented with respect to germline transcription, the nontemplate strand is enriched for guanosine (G); for example, the nontemplate

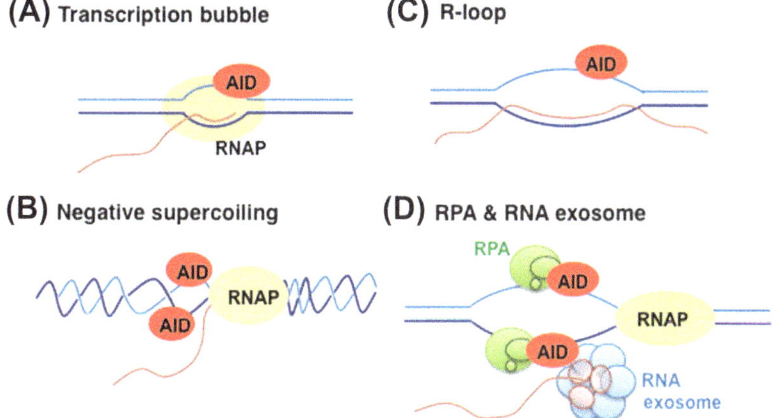


FIGURE 4 **Transcriptional targeting of activation-induced cytidine deaminase (AID) activity.** The light and dark blue lines represent the nontemplate and template strands, respectively. The red line represent RNA transcript. RNAP stands for RNA polymerase together with associated elongation factors. See text for more details. The diagram in panel (A) describes how AID gains access to the non-template strand in the context of transcription bubble. The diagram in panel (B) explains how negative supercoils associated with elongating RNA polymerase could facilitate AID access to both strands of the transcription template. The diagram in panel (C) depicts how R-loop structure provides single-stranded DNA substrate in the non-template strand for AID. The diagram in panel (D) illustrates how RPA and RNA exosome facilitate AID targeting to both strands of transcribed DNA. RPA, replication Protein A.

strand of Sμ contains 47% G. Likewise, the S region transcript is also G-rich. Because RNA/DNA hybrid containing G-rich RNA is more stable than the corresponding DNA duplex [97], transcription of S regions induces the formation of R-loops, in which the G-rich S region transcript is annealed to the template strand, and the nontemplate strand is displaced as a single-stranded loop [98–102] (Figure 4(C)). R-loops have been detected in actively transcribed S regions during CSR in vivo [103–106]. Indeed, because the nontemplate strand is single-stranded in an R-loop, it is a robust AID target compared with the template strand in biochemical assays of the transcribed S region [39]. R-loop formation at transcribed mammalian S regions could expose long stretches of nontemplate ssDNA substrates for AID deamination [103].

A second special feature of S regions from both mammals and amphibians is that they contain abundant AID hot spot motifs, which have been defined as AID targets for SHM [107]. In this context, AID prefers to deaminate cytosines within the context of WRCH or its complement DGYW [13,14]. In particular, S regions of mammals and some other species are highly enriched for AGCT, the optimal version of AID hot spot motifs, partly because it is a palindrome that can be recognized by AID on both DNA strands, and therefore, through mechanisms outlined previously, contribute to nicks or gaps on both strands that when proximal could promote the DSB intermediates required for CSR [96]. To address the sequence requirements for CSR, the mammalian Sγ1 was replaced with the *Xenopus* Sμ, which contains abundant AGCT motifs but is not GC-rich and does not form R-loops. In such CSR assays, the *Xenopus* Sμ functioned similarly as inverted Sγ1, which also does not form R-loops [74,108]. Such results are consistent with the idea that R-loops facilitate CSR, but that other yet-to-be determined S region features are sufficient to attract AID in the absence of R-loops. Indeed, V region exons that do not form R-loops can target AID during physiological SHM in GCs, which clearly demonstrates other means of generating ssDNA substrates for AID activity in vivo. In this regard, it is conceivable that R-loop–forming ability arose in mammalian S regions to further enhance AID targeting, although it remains possible that these structures function in other ways [11]. In addition, it is also possible that small R-loops, for example in the context of transcription bubbles, may contribute to AID targeting during CSR and SHM [109–111].

## 7. AID COFACTORS FACILITATE AID ACCESS TO ITS ssDNA SUBSTRATES

Biochemical studies showed that replication protein A (RPA) can facilitate AID deamination of the nontemplate strand of transcribed sequences that do not form R-loops [112]. In this regard, AID interacts with the 32-kDa subunit (RPA32) of the trimeric RPA complex, and this interaction depends on the phosphorylation of AID at S38 by protein kinase A [112,113]. In the transcription-coupled deamination assays in vitro, RPA enables AID to deaminate DNA templates that cannot form stable R-loops, in theory because RPA, which tightly binds single-stranded DNA and has an important role in DNA replication and recombination [114], traps the transiently unwound DNA during transcription in the single-stranded configuration [112,115]. Consistent with the biochemical evidence, an S38A mutation disrupted the interaction between endogenous AID and RPA and concomitantly impaired CSR and SHM substantially, although it did not affect AID biochemical catalytic activity [113,116,117].

Both R-loops and RPA can facilitate nontemplate strand DNA deamination by AID in a transcription-coupled manner in biochemical assays. However, in these assays, the template strand is largely untouched by AID [39,112,113]. Similarly, AID deaminates predominantly the nontemplate strand of transcribed DNA in *Escherichia coli* [118]. The asymmetrical deamination pattern clearly is different from that observed in vivo during CSR and SHM, where AID deamination events occur on both strands [119]. Thus, in the context of UNG and MMR double deficiencies, the original footprint of AID-induced deamination at cytidine residues is retained in S regions; in this setting, examination of C>U mutation revealed no asymmetry between the two strands [120]. How does AID gain access to the template strand in vivo? The RNA exosome may contribute to such activity [121]. The RNA exosome is a multisubunit complex that degrades RNA from the 3′ end [122]. AID interacts with the RNA exosome, and in this context, AID robustly deaminates both strands of transcribed substrates that either do or do not form R-loops equally when assayed via in vitro transcription reactions [121]. Although the RNase subunits of the RNA exosome did not appear to be required for such in vitro activities, it remains possible that such nuclease functions may still contribute in vivo by degrading primary noncoding transcripts annealed to DNA [121]. A working model on AID and RPA function proposes that once the RNA exosome displaces (or degrades) the RNA nascent transcript from the template strand, RPA might bind to and stabilize the ssDNA and facilitate S38-phosphorylated AID accumulation to enhance the deamination process [121] (Figure 4(D)).

## 8. DIFFERENTIAL AID TARGETING AND OUTCOMES DURING CSR AND SHM

CSR and SHM require AID; thus, most AID mutations affect CSR and SHM similarly. However, there are notable exceptions. Certain mutations in the N-terminus of AID have a more negative impact on SHM than on CSR [123,124]. Conversely, some mutations in the C-terminus of AID impair CSR but not SHM in both mice and humans

[26,125–128]. It has been suggested that the N- and C-terminal mutations may have disrupted the interaction of AID with cofactors that are important for SHM and CSR, respectively [123,127,128]. However, a full account of the differential function of AID in CSR and SHM awaits identification of the putative cofactors in these two processes. AID acts on distinct targets during CSR and SHM: S regions in CSR versus V(D)J exons in SHM. The ability of B-cells to differentially target AID over short distances is well illustrated during B-cell activation for CSR under in vitro conditions.. Thus, when B-cells are stimulated with appropriate cytokines in culture, they undergo robust CSR; for example, in the presence of anti-CD40 antibody plus IL-4, up to 40% or more of IgM$^+$ B-cells can switch to IgG1 after 4 days of stimulation. In striking contrast, there are few signs of SHM over the V(D)J exon in these activated B-cells, even though the V(D)J exon is transcribed [119,129]. Because the V(D)J exon and S regions are exposed to the same pool of *trans*-acting factors and closely linked on the chromosome (Figure 2), differential targeting of AID must result from differences in the target sequence, adjacent regulatory elements, and/or as yet-to-be determined aspects of transcription beyond transcription per se.

AID–generated deamination products are finally converted into DSBs (and mutations) within S regions during CSR and mutations at V(D)J exons. Base excision repair and mismatch repair pathways are involved in processing AID deamination products during both CSR and SHM [12,130]. However, how these two repair mechanisms cause DSB formation for CSR, but generate mutation for SHM, remains to be fully determined. One possible mechanism is the density of SHM target motifs, which is much higher in S regions than in V(D)J exons [107]. In this regard, S regions are also highly enriched for the AGCT motif, which can target a high density of AID lesions on both DNA strands, and thereby contribute to the generation of DSBs. However, other aspects of target DNA sequences and/or as yet uncharacterized downstream factors could also contribute to the different aberrant repair outcomes of SHMs versus DSBs.

## 9. LONG-RANGE JOINING OF S REGION BREAKS

DSBs in S regions can be joined in different ways to form distinct recombination products. This phenomenon occurs because AID can deaminate many cytosines within a single S region [120]; as a result, each S region may sustain multiple DSBs. Productive CSR requires the joining of the upstream end of a DSB in Sμ with the downstream end of a DSB in another S region (Figure 2(B)). Sμ is separated from the downstream S regions by 60–160 kb and individual S regions range from 1 to 12 kb. In competition with productive CSR, the two ends of a DSB can be joined directly back together; in addition, they can also be joined back together after resection or joined to another DSB in the same S region, with both of the latter joining events forming intra-S region deletions (ISD) (Figure 5) [27,131]. Therefore, productive CSR requires mechanisms that promote efficient long-range joining of DSBs in two S regions so that they occur frequently enough to compete with DSB rejoining, ISDs, and even translocations and allow the high levels of CSR observed in activated B-cells [132] (Figure 5).

Various mechanisms may contribute to the frequent long range joining outcome between two DSBs that is observed during CSR. Clearly, two S regions must be synapsed (physically together) and both must be simultaneously broken to allow them to be joined together. In this regard, during V(D)J recombination, the RAG endonuclease binds an appropriate pair of V, D, or J sequences, cleaves them, and via binding to their RSSs holds them to maintain synapsis until the appropriate ends are joined [24,25]. In this regard, owing to their unusual structures, AID or transcribed S regions may contribute to such a synapsis role in CSR [133,134]. However, B-cells in which either Sγ1 or both the Sμ and Sγ1 regions were replaced in lymphocytes with cassettes containing yeast I-SceI meganuclease cleavage sites were found to undergo I-SceI–dependent recombinational class switching from IgM to IgG1 [132], in the first case by joining I-SceI DSBs at the Sγ1 location to AID-initiated DSBs in Sμ, and in the second case by joining I-SceI DSBs to each other in both locations. Indeed, when a cassette harboring 28 I-SceI sites (to promote more frequent DSBs) was used in place of Sγ1, I-SceI–dependent IgH class switching occurred at levels approaching those mediated by AID-dependent DSBs at Sγ1 [28].

These types of high joining frequencies for DSBs separated by about 100 kb were also found for B lineage cells that did not express AID [132]. Thus, neither S regions nor

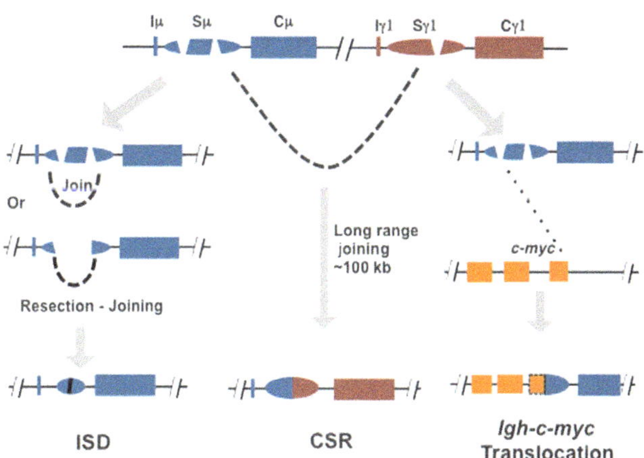


FIGURE 5 **Generation of intra-S region deletions (ISD) and chromosomal translocations versus class switch recombination (CSR).** Pathways that can lead to either ISD or chromosomal translocations CSR are illustrated and described in more detail in the text and in reference [148].

AID appear to be required for joining of DSBs in the position of S regions at levels sufficient to promote substantial CSR. In this regard, similar levels of long-range joining (e.g., over 100 kb or longer distances) were found between IgH I-SceI cassettes in T-cells and also between I-SceI and Cas-9/CRISPR–generated DSBs in the *c-myc* locus [28]. Likewise, high-throughput genome-wide translocation sequencing (HTGTS) studies showed that DSBs within megabase regions on the same chromosome are joined together more frequently than those more widely separated [27,135]. Together, these observations indicate that long-range joining of DSBs separated by up to hundreds of kilobase pairs is a general property of mammalian DSB repair rather than a special privilege of end joining during CSR.

Several nonmutually exclusive factors could contribute to this general property of joining DSBs separated by several 100 kb in mammalian chromosomes. Ataxia telangiectasia mutated (ATM)-dependent DSB response (DSBR) factors (see subsequent discussion) have been shown to form foci that extend in chromatin for more than 100 kbs surrounding DNA DSBs [136]. Because of the scale of DSBR, it is conceivable that DNA repair foci at two distant DSBs within several 100 kb may be coalesced into one DNA repair domain, thereby contributing to the more stable synapsis of two widely separated DSBs [27,137]. The widespread organization of nuclear chromatin into megabase or submegabase topological domains may also have a role in promoting the actual synapsis of S regions [138–141]. In this regard, DNA sequences in chromatin can undergo diffusion of up to 1 μ [141], a property that was proposed to underlie potential frequent joining of DSBs separated by submegabase distances found in HTGTS studies [135]. If the IgH locus were already organized into loops—for example, via interactions promoted by enhancers, I region promoters, and/or other *cis*-acting elements [142]—the distance at which S regions would need to diffuse to come into physical proximity would be even shorter than in an extended locus. In this case, frequent synapsis of S regions via diffusion together with frequent introduction of DSBs into S regions by AID and potential stabilization of the DSB synapsis via DSBRs all might contribute to DSB joining between two S regions at levels sufficient to generate physiological CSR levels. Such factors may also contribute to DSB synapsis during V(D)J recombination, where distant V genes are brought into proximity with downstream D and Js via locus contraction [143,144] and to the frequent interstitial deletions found in certain cancers, such as T-cell acute lymphocytic leukemias [145].

## 10. CLASSICAL NONHOMOLOGOUS END JOINING

In the second stage of CSR, DSBs in two broken, synapsed S regions are joined to form complete CSR. Because of sequence divergence between Sμ and acceptor S regions, most CSR junctions involve either no microhomology or just a few base pairs of microhomology [146]. Such DSB junctions are typical products of C-NHEJ [147,148]. The discovery of the C-NHEJ pathway was derived from studies of RAG-initiated V(D)J recombination [149]. After cleavage between the V, D, or J sequences and the RSS of an appropriate pair of V, D, or J segments to be joined, RAG holds the cleaved segments and RSSs in a postcleavage synaptic complex and the ends are then joined exclusively by the C-NHEJ pathway [150]. The exclusivity of joining by this pathway appears to be promoted by sequences in RAG-2 [151], but also may be ensured by V(D)J recombination occurring only in the G1 cell cycle phase when homologous recombination (which also repairs DSBs) is inactive. Because of the reliance of V(D)J recombination on C-NHEJ, defects for factors in this pathway cause severe V(D)J recombination defects that allowed them to be identified [149].

The core components of c-NHEJ include the heterodimer of Ku70 and Ku80 (Ku), X-ray Repair Cross-Complementing protein 4 (XRCC4), and DNA ligase 4 (Lig4), which are conserved through evolution [147]. During c-NHEJ, Ku binds to DSB ends, and the XRCC4/Lig4 complex ligates the ends (Figure 6). Besides these core components, Ku recruits the DNA-dependent protein kinase catalytic subunit (DNA-PKcs) to DNA DSBs [152]. DNA-PKcs phosphorylates a large number of proteins that may contribute to the reaction and also may serve an independent role in tethering

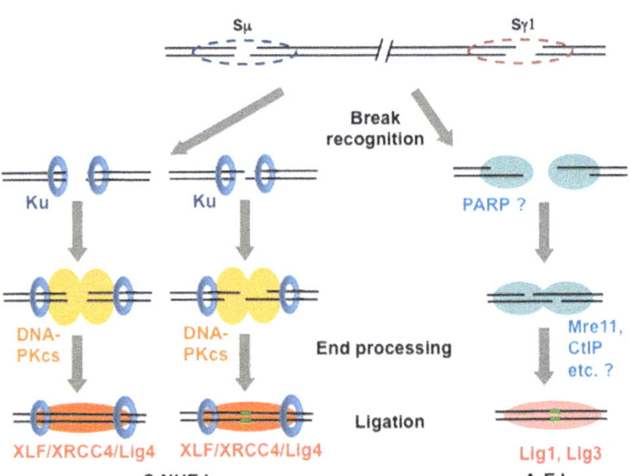


FIGURE 6 **Different end-joining pathways of double-strand breaks (DSB) repair.** Major steps of end-joining pathways, including DSB recognition, end processing, and ligation, are shown together with C-NHEJ factors known to be involved and potential alternative end joining (A-EJ) factors that have been implicated. Please see text for more detailed descriptions and note that both classical nonhomologous end-joining pathway (C-NHEJ) and A-EJ can perform direct (blunt) and MH-mediated joins, with the degree to which one type versus the other occurs, perhaps depending on substrate DNA ends or their sequence environment [148,163]. The use of MH may also be influenced by the subpathway of A-EJ employed [148,164] (see text for more details).

DSB ends [153] (Figure 6). Mutations in DNA-PKcs underlie the mouse severe combined immune deficiency (SCID) [154,155]. DNA-PKcs activates the Artemis endonuclease [156], identified based on a human SCID mutation [157], to facilitate end processing before joining. The XRCC4-like factor (XLF), also identified via a human SCID mutation [158,159], interacts with XRCC4 and has been speculated to have several potential roles in C-NHEJ [160,161], but roles for XLF in V(D)J recombination in developing lymphocytes are only clearly apparent in the absence of ATM-dependent DSBR factors [162].

Because the core C-NHEJ factors, as well as ATM and Artemis, are required for V(D)J recombination, the assessment of their roles in IgH CSR requires the use of mice (HL mice) that contain preassembled IgH V(D)J exon and IgL knock-ins in their germline to allow development of mature B-cells. Analyses of HL mice that are deficient for the various core C-NHEJ factors confirmed a role for all of these factors in CSR joining, but not an absolute role as observed in V(D)J recombination [131,163,164]. Thus, CSR to most $C_H$s is reduced to about 30–50% of WT levels in the absence of core Ku70, Ku80, XRCC4, and Lig4 [163–165], and unjoined S region breaks accumulate as IgH chromosomal breaks that can contribute to chromosomal translocations [163]. Ablation of DNA-PKcs has a variable effect on CSR, with CSR to Cγ1 being relatively spared [166–169], and ablation of Artemis has nearly no effect on CSR [170]. In both cases, however, a clear role in CSR for these proteins is from the accumulation of a modest level of IgH chromosomal breaks and translocations in CSR-activated B-cells that lack them [171]. It is possible the relatively unimpaired CSR to certain $C_H$s in DNA-PKcs–deficient cells may reflect compensation for DNA-PKcs by redundant functions of ATM (see subsequent discussion) [172]. Artemis likely is relatively dispensable for CSR because S region break ends may not require the type of processing contributed by this factor. Loss of XLF alone, despite having little effect on V(D)J recombination [173], reduces CSR with combined deficiencies for XLF and ATM further reducing the levels to those found with deficiencies for core C-NHEJ factors consistent with functional redundancy between XLF and ATM more generally in C-NHEJ beyond V(D)J recombination [162].

## 11. ALTERNATIVE END JOINING

CSR at up to 50% of normal levels in the absence of core C-NHEJ components indicates that unlike V(D)J recombination, CSR joining can occur through an A-EJ pathway or pathways [148,150]. Currently, the nature of A-EJ pathways is emerging, and the working definition of A-EJ is any type of end joining that occurs in the absence of core C-NHEJ factors [148,174]. Alternative EJ of DSBs has sometimes been confused with the label "micro-homology (MH) mediated end joining." However, C-NHEJ frequently uses very short MH (1–2 bp), and in many cases, A-EJ of certain DSBs (e.g., I-SceI–mediated DSBs) can frequently join ends without any MH [175]. Yet, A-EJ during CSR frequently employs MH [163]. In normal cells, approximately 30–50% of the CSR joints contain no overlapping bases between the two partner S regions and the MH-mediated joins mostly contain very short MHs characteristic of C-NHEJ [163,176]. In contrast, in XRCC4 and DNA ligase IV-deficient B-cells, the vast majority (greater than 95%) of CSR joints contain MH, with junctions often having MH lengths that exceed those observed for normal CSR joins [163]. However, although Ku-deficient B-cells are also heavily biased for MH, about 10% of the CSR junctions are direct and lack MH [164]. One possibility for the abundant use of MH during joining of S region breaks during CSR is that the sequences flanking these ends provide a much richer source of MH for joining and that the use of MH by different mammalian A-EJ pathways is influenced by the sequence environment of the DSBs that are joined [163]. Also, when MH are used in A-EJ, they are often much longer than those used during C-NHEJ, potentially providing a signature of A-EJ [146,163]. In this regard, some CSR junctions in normal cells display relatively longer MH than those used by C-NHEJ, which suggests that A-EJ may work in this context even in the presence of C-NHEJ [146,148,163]. Finally, a limited set of sequences of ISDs in C-NHEJ–deficient cells revealed predominantly MH-mediated joining, of which some MHs were very long; this suggests that ISDs may be promoted more frequently than CSR events by A-EJ owing to the rich source of repetitive sequences within a given type of S region [131].

Comparison of S region junctions in cells deficient for Lig4, Ku70, and both Lig4 and Ku70 led to the notion that there are at least two A-EJ pathways [148]. One is a Lig4-independent A-EJ pathway, which may use Ku for DSB recognition along with other C-NHEJ factors but with a different ligase, and as such also be considered an alternative C-NHEJ pathway [147]. The other is a Ku-independent A-EJ pathway, which mediates end joining in the absence of Ku or the combined absence of Ku and Lig4 (the recognition and joining components of C-NHEJ) and is a distinct pathway from C-NHEJ [147,164]. Alternative EJ is poorly understood at the mechanistic level, but likely is initiated with the recognition of DSB by a sensor, often continued with end processing and then completed by ligation of the DSB [148,150] (Figure 6).

DNA ligase 4–independent A-EJ may simply be an alternative C-NHEJ, as outlined previously. However, Ku (and Ku plus Lig4)-independent A-EJ requires a completely different set of components. Poly (ADP-Ribose) polymerase 1 (PARP1) has been implicated in A-EJ and is a likely candidate for the DSB sensor [177,178]. Mre11 and CtIP have also been implicated and could be involved in end processing [179–184], and either Lig1 or Lig3 (which has also been implicated) could ligate the ends [185,186] (Figure 6). Because all of these potential A-EJ factors have established functions in other DNA repair pathways, A-EJ may be intertwined with other DNA repair pathways. Evidence suggests that CSR DSBs may be initiated in G1 [187,188]. However, unanswered questions are whether A-EJ pathways, like C-NHEJ, function in the G1 phase of the cell cycle, and whether all CSR junctions are generated in G1. Finally, various studies have implicated A-EJ as an error-prone pathway that can generate chromosomal translocations [150,189,190]. Indeed, in the absence of C-NHEJ and the p53-mediated G1 checkpoint, A-EJ can generate oncogenic IgH locus translocations that contribute to B-cell lymphomas [191]. However, it remains to be proven that A-EJ is translocation prone in the presence of C-NHEJ and/or whether increased A-EJ–mediated translocations in the absence of C-NHEJ reflect increased frequencies of unjoined DSBs that can drive translocations (see subsequent discussion) [27,148].

### 11.1 ATM-Dependent DNA Repair Response

CSR DSBs activated the ATM-dependent DNA DSBR. ATM is a serine/threonine protein kinase that belongs to the phosphatidylinositol 3-kinase–related kinases (PI3K-related kinases) superfamily, which includes DNA-PKcs [192]. In response to a DSB, ATM can phosphorylate a large number of downstream proteins that form foci over megabase regions on either side of a DSB, and which can mediate cell cycle arrest and contribute to DSB repair [136,137,193,194]. Roles of these extended foci in DSB repair may include end tethering to promote more efficient C-NHEJ (and possibly A-EJ) and to promote rejoining of DSBs within a chromosome, as opposed to their translocation to different chromosomes [27,136]. Consistent with an important role of ATM in repair of CSR DSBs by end joining, ATM-deficient B-cells show significantly decreased CSR (30–50% of control) and accumulate substantial IgH chromosome breaks when activated for CSR in vitro [195,196]. In this context, ATM and DNA-PKcs have redundant roles during CSR, because inhibition of DNA-PKcs kinase activity in ATM-deficient B-cells showed further decreased CSR levels and severe genome instability [172]. During the DSBR response, ATM phosphorylates S139 near the C-terminus of Histone H2AX to form γH2AX [197], which further recruits other DDR factors including Mediator of DNA damage checkpoint protein 1 (MDC1) [198]. ATM also phosphorylates MDC1, which is recognized by the ubiquitin ligase RNF-8, and the ubiquitylation signal is further amplified by another ubiquitin ligase RNF16 [199]. ATM-dependent phosphorylation of 53BP1 recruits Rapl-interacting factor 1 (RIF1) to the DSB sites to counteract end resection [199–204]. CSR also is impaired (30–50% of control levels) in B-cells that are deficient in H2AX, MDC1, RNF-8, and RNF-168, but loss of 53BP1 or Rif1 causes a more severe defect in CSR [195,205–210].

Among the downstream ATM-dependent DSBR factors, 53BP1 appears to have unique roles in CSR, because deficiency for this DSBR factor leads to the most dramatic decrease in CSR. In 53BP1-deficient B-cells, the CSR level is dramatically decreased to 5–10% of wild-type levels [209,210]. Yet, cytogenetic examination of genome instabilities by fluorescein isothiocyanate (FISH) analysis has shown comparable increase in IgH breaks in $53BP1^{-/-}$ as well as in $ATM^{-/-}$ and $H2AX^{-/-}$ CSR-activated B-cells, indicating that the role of 53BP1 in CSR may go beyond end joining [195,211]. Notably, 53BP1 deficiency does not appear to elevate the extent of general genomic instability outside of IgH locus, as observed in ATM or H2AX-deficient CSR-activated B-cells [195]. Thus, 53BP1 has a unique function, or more likely has unique functions in CSR beyond those it serves in the context of the ATM-dependent DSBR. 53BP1 has been suggested to have several roles potentially relevant to CSR, including synapsis of long-range joins [212,213], repair pathway choice [214], and protection of ends from resection [214]. Rapl-interacting factor 1 deficiency also impairs CSR [200,201]. In this context, 53BP1 works together with RIF1 to protect broken DNA ends, including S region DSB ends from resection [200,201,204,215,216], which could contribute to decreased CSR levels by impairing C-NEHJ, as outlined previously. Currently, the overall functions of 53BP1 that contribute to its unique requirement for CSR compared with ATM or H2AX have not been fully elucidated.

## 12. CHROMOSOMAL TRANSLOCATION IN LYMPHOMAS CAUSED BY ABERRANT CSR

More than 85% of all lymphomas are of B-cell origin, and many of them are derived from mature B-cells [217]. A recurrent feature of many human mature B-cell lymphomas is reciprocal chromosomal translocations that involve the Ig loci and a variety of partners, often proto-oncogenes such as *c-Myc* in Burkitt lymphoma [218]. The strong enhancers in the IgH locus have the potential to activate translocated proto-oncogenes, giving B-cells harboring such translocations growth or survival advantages during the progression into B-cell lymphomas [218]. In this regard, the IgH3′RR can activate *c-Myc* over long distances subsequent to its translocation into IgH in a mouse lymphoma model [219], and has been strongly suggested to have similar roles in human B-cell lymphomas [220]. Indeed, drugs that target this enhancer may provide therapies for B-cell lymphomas with oncogenic translocations dependent on this enhancer [221]. Translocations that fuse IgH S regions to cellular oncogenes have been found in many human B-cell lymphomas, especially those that arise from GCs [218,222]. The nature of these tumors and the location of the translocations in S region have strongly implicated S region DSBs generated by AID as partners in the translocations [222]. That AID initiated DSBs can lead to translocations has been proven in high-throughput studies of mouse B-cells, and studies in B-cell lymphoma models have strongly supported a role of AID-mediated DSBs as the precursors to oncogenic translocations [218,222].

Chromosomal translocations require end joining of two separate DSBs on different chromosomes [27]. In B-cell lymphoma translocations, both Sμ and downstream S regions have been found to be sources of DSBs that generate one translocation partner sequence [218,222]. DSBs that generate non-Ig translocation partners of IgH S regions could come from various sources, but in many cases may involve AID off-target activities [27,222]. In this regard, although AID activity is specifically targeted and tightly regulated, it can also initiate DSBs and mutations in off-target sites including various proto-oncogenes. Genome wide studies have implicated a number of genes as being AID off-targets and have shown that the vast majority of translocation hot spots in activated B-cells have DSBs generated by AID off-targets [223,224]. Because of heterogeneity of the three-dimensional organization of the genome at the level of individual cells, the high frequency of AID-initiated DSBs in S regions and in AID off-targets in the same cells may be sufficient to drive recurrent translocations between these regions even if they are not physically proximal in most cells [27]. Moreover, deficiencies for ATM DSB response proteins or C-NHEJ factors may contribute to driving translocations at such regions by allowing the AID-initiated DSBs to persist longer in an unjoined state [225], thereby increasing the likelihood of two DSBs occurring simultaneously in a switch region and a given AID off-target sequence, which happens to be physically proximal in a given cell [27].

## 13. EVOLUTION OF THE IgH CSR MECHANISM

The process of SHM appears to precede that of IgH CSR in evolution [226]. Thus, bony and cartilaginous fish have functional AID and appear to undergo SHM of Ig V region exons [227,228]. Indeed, preferred target sequences of FISH AID appears similar to that of mammalian AID, and in that regard, Fish AID can mediate IgH CSR when expressed in place of AID in mouse B-cells [229,230]. Although more work needs to be done, CSR is currently thought to have first appeared in amphibians, such as *Xenopus laevis* [227].

As discussed earlier, *Xenopus* S regions are different from those of mammals; most notably, whereas both have a high density of AGCT motifs, which could promote DSBs, *Xenopus* S regions are A:T-rich and do not form stable R-loops upon transcription [108,231]. In this regard, evolution of CSR from SHM may have required initially the duplication of two V regions, which contain AID targeting motifs such as AGCT motifs, along with flanking duplicated $C_H$ exons. The further evolution of such primitive AID target sequences could have led to primitive S regions that function mainly based on high AGCT density and potentially target AID similarly to V region exons [11].

The CSR evolutionary model outlined previously is economical, because it requires only duplication of existing sequence regions. In this context, the mechanism could exploit an already existing propensity for joining DSBs over several 100 kb genomic distance based on mechanisms outlined earlier and to join them with the aid of the preexisting DSBR and C-NHEJ (and A-EJ) pathways. Other evolutionary solutions to sequences that serve as S regions may have arisen. In this regard, examination of S regions from additional species shows that S regions from duck and chicken are even more divergent from human and mouse counterparts than those of *Xenopus*, with substantially lower densities of AGCTs [107]. With respect to base composition, duck Sμ is GC-rich, but G:C content is equal between the two strands; on the other hand, the putative duck Sα [232] has an almost even distribution of the four bases on both strands, with only a minor enrichment of G on one strand. If duck or chicken S regions supported substantial CSR in mouse cells, new insights into the role of substrate sequences with respect to AID targeting would likely emerge.

Mammalian S regions could have further evolved to introduce R-loops, perhaps to enhance AID targeting or to perform other functions, perhaps including synapsis. Mammalian B-cells may also have evolved their *cis*-acting elements (promoters, enhancers, and other sequences) to fulfill roles in organizing the $C_H$ region of the IgH locus to promote CSR in the context of AID recruitment or in the context of three-dimensional proximity effects, as outlined earlier, to promote joining. Likewise, they could have evolved to exploit specialized functions of DSBR factors such as 53BP1. Finally, an additional aspect of interest in the evolution of mammalian S regions comes from the observation that duck Sα appears to normally mediate CSR via inversion of intervening sequences rather than deletion [232]. In mammalian cells, productive CSR must occur through a deletional mechanism. Yet, nonproductive inversional CSR events have been described in transformed B lineage tumors and lines [56,233–236]; although the actual frequency at which they occur in normal B-cells has not been determined. Thus, a lingering question regarding CSR in mammalian B-cells is to confirm whether they have evolved mechanisms to ensure that joining predominantly occurs in the productive deletional orientation, and if so to elucidate those mechanisms.

## ACKNOWLEDGMENTS

The work is supported by NIH Grant AI077595 to F.W.A. F.W.A. is an Investigator of the Howard Hughes Medical Institute. F.L.M is a Lymphoma Research Foundation fellow.


## REFERENCES


[1] Jung D, Giallourakis C, Mostoslavsky R, Alt FW. Mechanism and control of V(D)J recombination at the immunoglobulin heavy chain locus. Annu Rev Immunol 2006;24:541–70.

[2] Honjo T, Kataoka T. Organization of immunoglobulin heavy chain genes and allelic deletion model. Proc Natl Acad Sci USA 1978;75:2140–4.

[3] Vincent-Fabert C, Fiancette R, Pinaud E, Truffinet V, Cogne N, Cogne M, et al. Genomic deletion of the whole IgH 3′ regulatory region (hs3a, hs1,2, hs3b, and hs4) dramatically affects class switch recombination and Ig secretion to all isotypes. Blood 2010;116:1895–8.

[4] Rouaud P, Vincent-Fabert C, Saintamand A, Fiancette R, Marquet M, Robert I, et al. The IgH 3′ regulatory region controls somatic hypermutation in germinal center B cells. J Exp Med 2013;210:1501–7.

[5] Pinaud E, Marquet M, Fiancette R, Peron S, Vincent-Fabert C, Denizot Y, et al. The IgH locus 3′ regulatory region: pulling the strings from behind. Adv Immunol 2011;110:27–70.

[6] Cogne M, Lansford R, Bottaro A, Zhang J, Gorman J, Young F, et al. A class switch control region at the 3′ end of the immunoglobulin heavy chain locus. Cell 1994;77:737–47.

[7] Manis JP, van der Stoep N, Tian M, Ferrini R, Davidson L, Bottaro A, et al. Class switching in B cells lacking 3′ immunoglobulin heavy chain enhancers. J Exp Med 1998;188:1421–31.

[8] Alt FW, Bothwell AL, Knapp M, Siden E, Mather E, Koshland M, et al. Synthesis of secreted and membrane-bound immunoglobulin mu heavy chains is directed by mRNAs that differ at their 3′ ends. Cell 1980;20:293–301.

[9] Rogers J, Early P, Carter C, Calame K, Bond M, Hood L, et al. Two mRNAs with different 3′ ends encode membrane-bound and secreted forms of immunoglobulin mu chain. Cell 1980;20:303–12.

[10] Reth M. B cell antigen receptors. Curr Opin Immunol 1994;6:3–8.

[11] Chaudhuri J, Basu U, Zarrin A, Yan C, Franco S, Perlot T, et al. Evolution of the immunoglobulin heavy chain class switch recombination mechanism. Adv Immunol 2007;94:157–214.

[12] Di Noia JM, Neuberger MS. Molecular mechanisms of antibody somatic hypermutation. Annu Rev Biochem 2007;76:1–22.

[13] Rogozin IB, Diaz M. Cutting edge: DGYW/WRCH is a better predictor of mutability at G: C bases in Ig hypermutation than the widely accepted RGYW/WRCY motif and probably reflects a two-step activation-induced cytidine deaminase-triggered process. J Immunol 2004;172:3382–4.

[14] Peled JU, Kuang FL, Iglesias-Ussel MD, Roa S, Kalis SL, Goodman MF, et al. The biochemistry of somatic hypermutation. Annu Rev Immunol 2008;26:481–511.

[15] Victora GD, Nussenzweig MC. Germinal centers. Annu Rev Immunol 2012;30:429–57.

[16] MacLennan IC. Germinal centers. Annu Rev Immunol 1994;12: 117–39.

[17] McHeyzer-Williams LJ, McHeyzer-Williams MG. Antigen-specific memory B cell development. Annu Rev Immunol 2005;23:487–513.

[18] Honjo T. Immunoglobulin genes. Annu Rev Immunol 1983;1: 499–528.

[19] Kataoka T, Kawakami T, Takahashi N, Honjo T. Rearrangement of immunoglobulin gamma 1-chain gene and mechanism for heavy-chain class switch. Proc Natl Acad Sci USA 1980;77:919–23.

[20] von Schwedler U, Jack HM, Wabl M. Circular DNA is a product of the immunoglobulin class switch rearrangement. Nature 1990;345:452–6.

[21] Iwasato T, Shimizu A, Honjo T, Yamagishi H. Circular DNA is excised by immunoglobulin class switch recombination. Cell 1990;62:143–9.

[22] Matsuoka M, Yoshida K, Maeda T, Usuda S, Sakano H. Switch circular DNA formed in cytokine-treated mouse splenocytes: evidence for intramolecular DNA deletion in immunoglobulin class switching. Cell 1990;62:135–42.

[23] Xiong H, Dolpady J, Wabl M, Curotto de Lafaille MA, Lafaille JJ. Sequential class switching is required for the generation of high affinity IgE antibodies. J Exp Med 2012;209:353–64.

[24] Schatz DG, Spanopoulou E. Biochemistry of V(D)J recombination. Curr Top Microbiol Immunol 2005;290:49–85.

[25] [a] Jones JM, Gellert M. The taming of a transposon: V(D)J recombination and the immune system. Immunol Rev 2004;200:233–48. [b] Doi T, Kato L, Ito S, Shinkura R, Wei M, Nagaoka H, et al. The C-terminal region of activation-induced cytidine deaminase is responsible for a recombination function other than DNA cleavage in class switch recombination. Proc Natl Acad Sci USA 2009;106:2758–63.

[26] Zahn A, Eranki AK, Patenaude AM, Methot SP, Fifield H, Cortizas EM, et al. Activation induced deaminase C-terminal domain links DNA breaks to end protection and repair during class switch recombination. Proc Natl Acad Sci USA 2014;111:E988–97.

[27] Alt FW, Zhang Y, Meng FL, Guo C, Schwer B. Mechanisms of programmed DNA lesions and genomic instability in the immune system. Cell 2013;152:417–29.

[28] Gostissa M, Schwer B, Chang A, Dong J, Meyers RM, Marecki GT, et al. IgH class switching exploits a general property of two DNA breaks to be joined in cis over long chromosomal distances. Proc Natl Acad Sci USA 2014;111:2644–9.

[29] Muramatsu M, Sankaranand VS, Anant S, Sugai M, Kinoshita K, Davidson NO, et al. Specific expression of activation-induced cytidine deaminase (AID), a novel member of the RNA-editing deaminase family in germinal center B cells. J Biol Chem 1999;274:18470–6.

[30] Smith HC, Bennett RP, Kizilyer A, McDougall WM, Prohaska KM. Functions and regulation of the APOBEC family of proteins. Semin Cell Dev Biol 2012;23:258–68.

[31] Muramatsu M, Kinoshita K, Fagarasan S, Yamada S, Shinkai Y, Honjo T. Class switch recombination and hypermutation require activation-induced cytidine deaminase (AID), a potential RNA editing enzyme. Cell 2000;102:553–63.

[32] Revy P, Muto T, Levy Y, Geissmann F, Plebani A, Sanal O, et al. Activation-induced cytidine deaminase (AID) deficiency causes the autosomal recessive form of the Hyper-IgM syndrome (HIGM2). Cell 2000;102:565–75.

[33] Teng B, Burant CF, Davidson NO. Molecular cloning of an apolipoprotein B messenger RNA editing protein. Science 1993;260: 1816–9.

[34] Navaratnam N, Morrison JR, Bhattacharya S, Patel D, Funahashi T, Giannoni F, et al. The p27 catalytic subunit of the apolipoprotein B mRNA editing enzyme is a cytidine deaminase. J Biol Chem 1993;268:20709–12.

[35] Johnson DF, Poksay KS, Innerarity TL. The mechanism for apo-B mRNA editing is deamination. Biochem Biophys Res Commun 1993;195:1204–10.

[36] Honjo T, Nagaoka H, Shinkura R, Muramatsu M. AID to overcome the limitations of genomic information. Nat Immunol 2005;6:655–61.

[37] Di Noia J, Neuberger MS. Altering the pathway of immunoglobulin hypermutation by inhibiting uracil-DNA glycosylase. Nature 2002;419:43–8.

[38] Martin A, Bardwell PD, Woo CJ, Fan M, Shulman MJ, Scharff MD. Activation-induced cytidine deaminase turns on somatic hypermutation in hybridomas. Nature 2002;415:802–6.

[39] Chaudhuri J, Tian M, Khuong C, Chua K, Pinaud E, Alt FW. Transcription-targeted DNA deamination by the AID antibody diversification enzyme. Nature 2003;422:726–30.

[40] Pham P, Bransteitter R, Petruska J, Goodman MF. Processive AID-catalysed cytosine deamination on single-stranded DNA simulates somatic hypermutation. Nature 2003;424:103–7.

[41] Dickerson SK, Market E, Besmer E, Papavasiliou FN. AID mediates hypermutation by deaminating single stranded DNA. J Exp Med 2003;197:1291–6.

[42] Rada C, Williams GT, Nilsen H, Barnes DE, Lindahl T, Neuberger MS. Immunoglobulin isotype switching is inhibited and somatic hypermutation perturbed in UNG-deficient mice. Curr Biol 2002;12:1748–55.

[43] Ehrenstein MR, Neuberger MS. Deficiency in Msh2 affects the efficiency and local sequence specificity of immunoglobulin class-switch recombination: parallels with somatic hypermutation. EMBO J 1999;18:3484–90.

[44] Ehrenstein MR, Rada C, Jones AM, Milstein C, Neuberger MS. Switch junction sequences in PMS2-deficient mice reveal a microhomology-mediated mechanism of Ig class switch recombination. Proc Natl Acad Sci USA 2001;98:14553–8.

[45] Schrader CE, Edelmann W, Kucherlapati R, Stavnezer J. Reduced isotype switching in splenic B cells from mice deficient in mismatch repair enzymes. J Exp Med 1999;190:323–30.

[46] Li Z, Scherer SJ, Ronai D, Iglesias-Ussel MD, Peled JU, Bardwell PD, et al. Examination of Msh6- and Msh3-deficient mice in class switching reveals overlapping and distinct roles of MutS homologues in antibody diversification. J Exp Med 2004;200:47–59.

[47] Kim YJ, Wilson 3rd DM. Overview of base excision repair biochemistry. Curr Mol Pharmacol 2012;5:3–13.

[48] Kunkel TA, Erie DA. DNA mismatch repair. Annu Rev Biochem 2005;74:681–710.

[49] Imai K, Slupphaug G, Lee WI, Revy P, Nonoyama S, Catalan N, et al. Human uracil-DNA glycosylase deficiency associated with profoundly impaired immunoglobulin class-switch recombination. Nat Immunol 2003;4:1023–8.

[50] Peron S, Metin A, Gardes P, Alyanakian MA, Sheridan E, Kratz CP, et al. Human PMS2 deficiency is associated with impaired immunoglobulin class switch recombination. J Exp Med 2008;205:2465–72.

[51] Gardes P, Forveille M, Alyanakian MA, Aucouturier P, Ilencikova D, Leroux D, et al. Human MSH6 deficiency is associated with impaired antibody maturation. J Immunol 2012;188:2023–9.

[52] Rada C, Di Noia JM, Neuberger MS. Mismatch recognition and uracil excision provide complementary paths to both Ig switching and the A/T-focused phase of somatic mutation. Mol Cell 2004;16:163–71.

[53] Shen HM, Tanaka A, Bozek G, Nicolae D, Storb U. Somatic hypermutation and class switch recombination in Msh6(-/-)Ung(-/-) double-knockout mice. J Immunol 2006;177:5386–92.

[54] Cascalho M, Wong J, Steinberg C, Wabl M. Mismatch repair coopted by hypermutation. Science 1998;279:1207–10.

[55] Stavnezer J. Presence of a polyadenylated RNA fragment encoding the membrane domain for immunoglobulin alpha chain indicates that mRNAs for both secreted and membrane-bound alpha chains can be produced from the same RNA transcript. Nucleic Acids Res 1986;14:6129–44.

[56] Yancopoulos GD, DePinho RA, Zimmerman KA, Lutzker SG, Rosenberg N, Alt FW. Secondary genomic rearrangement events in pre-B cells: VHDJH replacement by a LINE-1 sequence and directed class switching. EMBO J 1986;5:3259–66.

[57] Manis JP, Tian M, Alt FW. Mechanism and control of class-switch recombination. Trends Immunol 2002;23:31–9.

[58] Lutzker S, Rothman P, Pollock R, Coffman R, Alt FW. Mitogen- and IL-4-regulated expression of germ-line Ig gamma 2b transcripts: evidence for directed heavy chain class switching. Cell 1988;53:177–84.

[59] Stavnezer J, Radcliffe G, Lin YC, Nietupski J, Berggren L, Sitia R, et al. Immunoglobulin heavy-chain switching may be directed by prior induction of transcripts from constant-region genes. Proc Natl Acad Sci USA 1988;85:7704–8.

[60] Xu M, Stavnezer J. Structure of germline immunoglobulin heavy-chain gamma 1 transcripts in interleukin 4 treated mouse spleen cells. Dev Immunol 1990;1:11–7.

[61] Rothman P, Lutzker S, Gorham B, Stewart V, Coffman R, Alt FW. Structure and expression of germline immunoglobulin gamma 3 heavy chain gene transcripts: implications for mitogen and lymphokine directed class-switching. Int Immunol 1990;2:621–7.

[62] Gauchat JF, Lebman DA, Coffman RL, Gascan H, de Vries JE. Structure and expression of germline epsilon transcripts in human B cells induced by interleukin 4 to switch to IgE production. J Exp Med 1990;172:463–73.

[63] Jung S, Rajewsky K, Radbruch A. Shutdown of class switch recombination by deletion of a switch region control element. Science 1993;259:984–7.

[64] Zhang J, Bottaro A, Li S, Stewart V, Alt FW. A selective defect in IgG2b switching as a result of targeted mutation of the I gamma 2b promoter and exon. EMBO J 1993;12:3529–37.

[65] Bottaro A, Lansford R, Xu L, Zhang J, Rothman P, Alt FW. S region transcription per se promotes basal IgE class switch recombination but additional factors regulate the efficiency of the process. EMBO J 1994;13:665–74.

[66] Harriman GR, Bradley A, Das S, Rogers-Fani P, Davis AC. IgA class switch in I alpha exon-deficient mice. Role of germline transcription in class switch recombination. J Clin Invest 1996;97:477–85.

[67] Daniel JA, Santos MA, Wang Z, Zang C, Schwab KR, Jankovic M, et al. PTIP promotes chromatin changes critical for immunoglobulin class switch recombination. Science 2010;329:917–23.

[68] Linehan LA, Warren WD, Thompson PA, Grusby MJ, Berton MT. STAT6 is required for IL-4-induced germline Ig gene transcription and switch recombination. J Immunol 1998;161:302–10.

[69] Kashiwada M, Levy DM, McKeag L, Murray K, Schroder AJ, Canfield SM, et al. IL-4-induced transcription factor NFIL3/E4BP4 controls IgE class switching. Proc Natl Acad Sci USA 2010;107:821–6.

[70] Rothman P, Chen YY, Lutzker S, Li SC, Stewart V, Coffman R, et al. Structure and expression of germ line immunoglobulin heavy-chain epsilon transcripts: interleukin-4 plus lipopolysaccharide-directed switching to C epsilon. Mol Cell Biol 1990;10:1672–9.

[71] Esser C, Radbruch A. Rapid induction of transcription of unrearranged S gamma 1 switch regions in activated murine B cells by interleukin 4. EMBO J 1989;8:483–8.

[72] Collins JT, Dunnick WA. Germline transcripts of the murine immunoglobulin gamma 2a gene: structure and induction by IFN-gamma. Int Immunol 1993;5:885–91.

[73] [a] Lebman DA, Lee FD, Coffman RL. Mechanism for transforming growth factor beta and IL-2 enhancement of IgA expression in lipopolysaccharide-stimulated B cell cultures. J Immunol 1990;144:952–9.
[b] Mostoslavsky R, Alt FW, Rajewsky K. The lingering enigma of the allelic exclusion mechanism. Cell 2004;118:539–44.
[c] Radbruch A, Sablitzky F. Deletion of Cmu genes in mouse B lymphocytes upon stimulation with LPS. EMBO J 1983;2:1929–35.
[d] Radbruch A, Muller W, Rajewsky K. Class switch recombination is IgG1 specific on active and inactive IgH loci of IgG1-secreting B-cell blasts. Proc Natl Acad Sci USA 1986;83:3954–7.

[74] Shinkura R, Tian M, Smith M, Chua K, Fujiwara Y, Alt FW. The influence of transcriptional orientation on endogenous switch region function. Nat Immunol 2003;4:435–41.

[75] Pinaud E, Khamlichi AA, Le Morvan C, Drouet M, Nalesso V, Le Bert M, et al. Localization of the 3′ IgH locus elements that effect long-distance regulation of class switch recombination. Immunity 2001;15:187–99.

[76] Seidl KJ, Manis JP, Bottaro A, Zhang J, Davidson L, Kisselgof A, et al. Position-dependent inhibition of class-switch recombination by PGK-neor cassettes inserted into the immunoglobulin heavy chain constant region locus. Proc Natl Acad Sci USA 1999;96:3000–5.

[77] Ju Z, Chatterjee S, Birshtein BK. Interaction between the immunoglobulin heavy chain 3′ regulatory region and the IgH transcription unit during B cell differentiation. Mol Immunol 2011;49:297–303.

[78] Ju Z, Volpi SA, Hassan R, Martinez N, Giannini SL, Gold T, et al. Evidence for physical interaction between the immunoglobulin heavy chain variable region and the 3′ regulatory region. J Biol Chem 2007;282:35169–78.

[79] Blagodatski A, Batrak V, Schmidl S, Schoetz U, Caldwell RB, Arakawa H, et al. A cis-acting diversification activator both necessary and sufficient for AID-mediated hypermutation. PLoS Genet 2009;5:e1000332.

[80] Kohler KM, McDonald JJ, Duke JL, Arakawa H, Tan S, Kleinstein SH, et al. Identification of core DNA elements that target somatic hypermutation. J Immunol 2012;189:5314–26.

[81] Kim Y, Tian M. NF-kappaB family of transcription factor facilitates gene conversion in chicken B cells. Mol Immunol 2009;46:3283–91.

[82] Kim Y, Tian M. The recruitment of activation induced cytidine deaminase to the immunoglobulin locus by a regulatory element. Mol Immunol 2010;47:1860–5.

[83] Kothapalli N, Norton DD, Fugmann SD. Cutting edge: a cis-acting DNA element targets AID-mediated sequence diversification to the chicken Ig light chain gene locus. J Immunol 2008;180:2019–23.

[84] Nambu Y, Sugai M, Gonda H, Lee CG, Katakai T, Agata Y, et al. Transcription-coupled events associating with immunoglobulin switch region chromatin. Science 2003;302:2137–40.

[85] Wang L, Wuerffel R, Feldman S, Khamlichi AA, Kenter AL. S region sequence, RNA polymerase II, and histone modifications create chromatin accessibility during class switch recombination. J Exp Med 2009;206:1817–30.

[86] Rajagopal D, Maul RW, Ghosh A, Chakraborty T, Khamlichi AA, Sen R, et al. Immunoglobulin switch mu sequence causes RNA polymerase II accumulation and reduces dA hypermutation. J Exp Med 2009;206:1237–44.

[87] Storb U. Why does somatic hypermutation by aid require transcription of its target genes? Adv Immunol 2014;122:253–77.

[88] Peters A, Storb U. Somatic hypermutation of immunoglobulin genes is linked to transcription initiation. Immunity 1996;4:57–65.

[89] Pavri R, Gazumyan A, Jankovic M, Di Virgilio M, Klein I, Ansarah-Sobrinho C, et al. Activation-induced cytidine deaminase targets DNA at sites of RNA polymerase II stalling by interaction with Spt5. Cell 2010;143:122–33.

[90] Bransteitter R, Pham P, Scharff MD, Goodman MF. Activation-induced cytidine deaminase deaminates deoxycytidine on single-stranded DNA but requires the action of RNase. Proc Natl Acad Sci USA 2003;100:4102–7.

[91] Nudler E. RNA polymerase active center: the molecular engine of transcription. Annu Rev Biochem 2009;78:335–61.

[92] Liu LF, Wang JC. Supercoiling of the DNA template during transcription. Proc Natl Acad Sci USA 1987;84:7024–7.

[93] Shen HM, Storb U. Activation-induced cytidine deaminase (AID) can target both DNA strands when the DNA is supercoiled. Proc Natl Acad Sci USA 2004;101:12997–3002.

[94] Parsa JY, Ramachandran S, Zaheen A, Nepal RM, Kapelnikov A, Belcheva A, et al. Negative supercoiling creates single-stranded patches of DNA that are substrates for AID-mediated mutagenesis. PLoS Genet 2012;8:e1002518.

[95] Khamlichi AA, Glaudet F, Oruc Z, Denis V, Le Bert M, Cogne M. Immunoglobulin class-switch recombination in mice devoid of any S mu tandem repeat. Blood 2004;103:3828–36.

[96] Han L, Masani S, Yu K. Overlapping activation-induced cytidine deaminase hotspot motifs in Ig class-switch recombination. Proc Natl Acad Sci USA 2011;108:11584–9.

[97] Ratmeyer L, Vinayak R, Zhong YY, Zon G, Wilson WD. Sequence specific thermodynamic and structural properties for DNA.RNA duplexes. Biochemistry 1994;33:5298–304.

[98] Reaban ME, Griffin JA. Induction of RNA-stabilized DNA conformers by transcription of an immunoglobulin switch region. Nature 1990;348:342–4.

[99] Daniels GA, Lieber MR. RNA: DNA complex formation upon transcription of immunoglobulin switch regions: implications for the mechanism and regulation of class switch recombination. Nucleic Acids Res 1995;23:5006–11.

[100] Tian M, Alt FW. Transcription-induced cleavage of immunoglobulin switch regions by nucleotide excision repair nucleases in vitro. J Biol Chem 2000;275:24163–72.

[101] Mizuta R, Iwai K, Shigeno M, Mizuta M, Uemura T, Ushiki T, et al. Molecular visualization of immunoglobulin switch region RNA/DNA complex by atomic force microscope. J Biol Chem 2003;278:4431–4.

[102] Roy D, Yu K, Lieber MR. Mechanism of R-loop formation at immunoglobulin class switch sequences. Mol Cell Biol 2008;28:50–60.

[103] Yu K, Chedin F, Hsieh CL, Wilson TE, Lieber MR. R-loops at immunoglobulin class switch regions in the chromosomes of stimulated B cells. Nat Immunol 2003;4:442–51.

[104] Huang FT, Yu K, Balter BB, Selsing E, Oruc Z, Khamlichi AA, et al. Sequence dependence of chromosomal R-loops at the immunoglobulin heavy-chain Smu class switch region. Mol Cell Biol 2007;27:5921–32.

[105] Huang FT, Yu K, Hsieh CL, Lieber MR. Downstream boundary of chromosomal R-loops at murine switch regions: implications for the mechanism of class switch recombination. Proc Natl Acad Sci USA 2006;103:5030–5.
[106] Kao YP, Hsieh WC, Hung ST, Huang CW, Lieber MR, Huang FT. Detection and characterization of R-loops at the murine immunoglobulin Salpha region. Mol Immunol 2013;54:208–16.
[107] Hackney JA, Misaghi S, Senger K, Garris C, Sun Y, Lorenzo MN, et al. DNA targets of AID evolutionary link between antibody somatic hypermutation and class switch recombination. Adv Immunol 2009;101:163–89.
[108] Zarrin AA, Alt FW, Chaudhuri J, Stokes N, Kaushal D, Du Pasquier L, et al. An evolutionarily conserved target motif for immunoglobulin class-switch recombination. Nat Immunol 2004;5:1275–81.
[109] Li X, Manley JL. Inactivation of the SR protein splicing factor ASF/SF2 results in genomic instability. Cell 2005;122:365–78.
[110] Gomez-Gonzalez B, Aguilera A. Activation-induced cytidine deaminase action is strongly stimulated by mutations of the THO complex. Proc Natl Acad Sci USA 2007;104:8409–14.
[111] Aguilera A, Garcia-Muse T. Causes of genome instability. Annu Rev Genet 2013;47:1–32.
[112] Chaudhuri J, Khuong C, Alt FW. Replication protein A interacts with AID to promote deamination of somatic hypermutation targets. Nature 2004;430:992–8.
[113] Basu U, Chaudhuri J, Alpert C, Dutt S, Ranganath S, Li G, et al. The AID antibody diversification enzyme is regulated by protein kinase A phosphorylation. Nature 2005;438:508–11.
[114] Wold MS. Replication protein A: a heterotrimeric, single-stranded DNA-binding protein required for eukaryotic DNA metabolism. Annu Rev Biochem 1997;66:61–92.
[115] Sikorski TW, Ficarro SB, Holik J, Kim T, Rando OJ, Marto JA, et al. Sub1 and RPA associate with RNA polymerase II at different stages of transcription. Mol Cell 2011;44:397–409.
[116] Cheng HL, Vuong BQ, Basu U, Franklin A, Schwer B, Astarita J, et al. Integrity of the AID serine-38 phosphorylation site is critical for class switch recombination and somatic hypermutation in mice. Proc Natl Acad Sci USA 2009;106:2717–22.
[117] McBride KM, Gazumyan A, Woo EM, Schwickert TA, Chait BT, Nussenzweig MC. Regulation of class switch recombination and somatic mutation by AID phosphorylation. J Exp Med 2008;205:2585–94.
[118] Ramiro AR, Stavropoulos P, Jankovic M, Nussenzweig MC. Transcription enhances AID-mediated cytidine deamination by exposing single-stranded DNA on the nontemplate strand. Nat Immunol 2003;4:452–6.
[119] Liu M, Schatz DG. Balancing AID and DNA repair during somatic hypermutation. Trends Immunol 2009;30:173–81.
[120] Xue K, Rada C, Neuberger MS. The in vivo pattern of AID targeting to immunoglobulin switch regions deduced from mutation spectra in msh2-/- ung-/- mice. J Exp Med 2006;203:2085–94.
[121] Basu U, Meng FL, Keim C, Grinstein V, Pefanis E, Eccleston J, et al. The RNA exosome targets the AID cytidine deaminase to both strands of transcribed duplex DNA substrates. Cell 2011;144:353–63.
[122] Lykke-Andersen S, Brodersen DE, Jensen TH. Origins and activities of the eukaryotic exosome. J Cell Sci 2009;122:1487–94.
[123] Shinkura R, Ito S, Begum NA, Nagaoka H, Muramatsu M, Kinoshita K, et al. Separate domains of AID are required for somatic hypermutation and class-switch recombination. Nat Immunol 2004;5:707–12.
[124] Wei M, Shinkura R, Doi Y, Maruya M, Fagarasan S, Honjo T. Mice carrying a knock-in mutation of Aicda resulting in a defect in somatic hypermutation have impaired gut homeostasis and compromised mucosal defense. Nat Immunol 2011;12:264–70.
[125] Ta VT, Nagaoka H, Catalan N, Durandy A, Fischer A, Imai K, et al. AID mutant analyses indicate requirement for class-switch-specific cofactors. Nat Immunol 2003;4:843–8.
[126] Imai K, Zhu Y, Revy P, Morio T, Mizutani S, Fischer A, et al. Analysis of class switch recombination and somatic hypermutation in patients affected with autosomal dominant hyper-IgM syndrome type 2. Clin Immunol 2005;115:277–85.
[127] Barreto V, Reina-San-Martin B, Ramiro AR, McBride KM, Nussenzweig MC. C-terminal deletion of AID uncouples class switch recombination from somatic hypermutation and gene conversion. Mol Cell 2003;12:501–8.
[128] Ellyard JI, Benk AS, Taylor B, Rada C, Neuberger MS. The dependence of Ig class-switching on the nuclear export sequence of AID likely reflects interaction with factors additional to Crm1 exportin. Eur J Immunol 2011;41:485–90.
[129] Nagaoka H, Muramatsu M, Yamamura N, Kinoshita K, Honjo T. Activation-induced deaminase (AID)-directed hypermutation in the immunoglobulin Smu region: implication of AID involvement in a common step of class switch recombination and somatic hypermutation. J Exp Med 2002;195:529–34.
[130] Stavnezer J, Guikema JE, Schrader CE. Mechanism and regulation of class switch recombination. Annu Rev Immunol 2008;26:261–92.
[131] Boboila C, Jankovic M, Yan CT, Wang JH, Wesemann DR, Zhang T, et al. Alternative end-joining catalyzes robust IgH locus deletions and translocations in the combined absence of ligase 4 and Ku70. Proc Natl Acad Sci USA 2010;107:3034–9.
[132] Zarrin AA, Del Vecchio C, Tseng E, Gleason M, Zarin P, Tian M, et al. Antibody class switching mediated by yeast endonuclease-generated DNA breaks. Science 2007;315:377–81.
[133] Wuerffel R, Wang L, Grigera F, Manis J, Selsing E, Perlot T, et al. S-S synapsis during class switch recombination is promoted by distantly located transcriptional elements and activation-induced deaminase. Immunity 2007;27:711–22.
[134] Larson ED, Duquette ML, Cummings WJ, Streiff RJ, Maizels N. MutSalpha binds to and promotes synapsis of transcriptionally activated immunoglobulin switch regions. Curr Biol 2005;15:470–4.
[135] Zhang Y, McCord RP, Ho YJ, Lajoie BR, Hildebrand DG, Simon AC, et al. Spatial organization of the mouse genome and its role in recurrent chromosomal translocations. Cell 2012;148:908–21.
[136] Rogakou EP, Boon C, Redon C, Bonner WM. Megabase chromatin domains involved in DNA double-strand breaks in vivo. J Cell Biol 1999;146:905–16.
[137] Bassing CH, Alt FW. H2AX may function as an anchor to hold broken chromosomal DNA ends in close proximity. Cell Cycle 2004;3:149–53.
[138] Dixon JR, Selvaraj S, Yue F, Kim A, Li Y, Shen Y, et al. Topological domains in mammalian genomes identified by analysis of chromatin interactions. Nature 2012;485:376–80.
[139] Nagano T, Lubling Y, Stevens TJ, Schoenfelder S, Yaffe E, Dean W, et al. Single-cell Hi-C reveals cell-to-cell variability in chromosome structure. Nature 2013;502:59–64.
[140] Naumova N, Imakaev M, Fudenberg G, Zhan Y, Lajoie BR, Mirny LA, et al. Organization of the mitotic chromosome. Science 2013;342:948–53.

[141] Nora EP, Lajoie BR, Schulz EG, Giorgetti L, Okamoto I, Servant N, et al. Spatial partitioning of the regulatory landscape of the X-inactivation centre. Nature 2012;485:381–5.

[142] Kumar S, Wuerffel R, Achour I, Lajoie B, Sen R, Dekker J, et al. Flexible ordering of antibody class switch and V(D)J joining during B-cell ontogeny. Genes Dev 2013;27:2439–44.

[143] Bossen C, Mansson R, Murre C. Chromatin topology and the regulation of antigen receptor assembly. Annu Rev Immunol 2012;30:337–56.

[144] Hewitt SL, Chaumeil J, Skok JA. Chromosome dynamics and the regulation of V(D)J recombination. Immunol Rev 2010;237:43–54.

[145] O'Neil J, Look AT. Mechanisms of transcription factor deregulation in lymphoid cell transformation. Oncogene 2007;26:6838–49.

[146] Stavnezer J, Bjorkman A, Du L, Cagigi A, Pan-Hammarstrom Q. Mapping of switch recombination junctions, a tool for studying DNA repair pathways during immunoglobulin class switching. Adv Immunol 2010;108:45–109.

[147] Lieber MR. The mechanism of double-strand DNA break repair by the nonhomologous DNA end-joining pathway. Annu Rev Biochem 2010;79:181–211.

[148] Boboila C, Alt FW, Schwer B. Classical and alternative end-joining pathways for repair of lymphocyte-specific and general DNA double-strand breaks. Adv Immunol 2012;116:1–49.

[149] Taccioli GE, Rathbun G, Oltz E, Stamato T, Jeggo PA, Alt FW. Impairment of V(D)J recombination in double-strand break repair mutants. Science 1993;260:207–10.

[150] Deriano L, Roth DB. Modernizing the nonhomologous end-joining repertoire: alternative and classical NHEJ share the stage. Annu Rev Genet 2013;47:433–55.

[151] Deriano L, Chaumeil J, Coussens M, Multani A, Chou Y, Alekseyenko AV, et al. The RAG2 C terminus suppresses genomic instability and lymphomagenesis. Nature 2011;471:119–23.

[152] Gottlieb TM, Jackson SP. The DNA-dependent protein kinase: requirement for DNA ends and association with Ku antigen. Cell 1993;72:131–42.

[153] Meek K, Gupta S, Ramsden DA, Lees-Miller SP. The DNA-dependent protein kinase: the director at the end. Immunol Rev 2004;200:132–41.

[154] Blunt T, Finnie NJ, Taccioli GE, Smith GC, Demengeot J, Gottlieb TM, et al. Defective DNA-dependent protein kinase activity is linked to V(D)J recombination and DNA repair defects associated with the murine scid mutation. Cell 1995;80:813–23.

[155] Kirchgessner CU, Patil CK, Evans JW, Cuomo CA, Fried LM, Carter T, et al. DNA-dependent kinase (p350) as a candidate gene for the murine SCID defect. Science 1995;267:1178–83.

[156] Ma Y, Pannicke U, Schwarz K, Lieber MR. Hairpin opening and overhang processing by an Artemis/DNA-dependent protein kinase complex in nonhomologous end joining and V(D)J recombination. Cell 2002;108:781–94.

[157] Moshous D, Callebaut I, de Chasseval R, Corneo B, Cavazzana-Calvo M, Le Deist F, et al. Artemis, a novel DNA double-strand break repair/V(D)J recombination protein, is mutated in human severe combined immune deficiency. Cell 2001;105:177–86.

[158] Buck D, Malivert L, de Chasseval R, Barraud A, Fondaneche MC, Sanal O, et al. Cernunnos, a novel nonhomologous end-joining factor, is mutated in human immunodeficiency with microcephaly. Cell 2006;124:287–99.

[159] Ahnesorg P, Smith P, Jackson SP. XLF interacts with the XRCC4-DNA ligase IV complex to promote DNA nonhomologous end-joining. Cell 2006;124:301–13.

[160] Tsai CJ, Chu G. Cooperative assembly of a protein-DNA filament for nonhomologous end joining. J Biol Chem 2013;288:18110–20.

[161] Hammel M, Rey M, Yu Y, Mani RS, Classen S, Liu M, et al. XRCC4 protein interactions with XRCC4-like factor (XLF) create an extended grooved scaffold for DNA ligation and double strand break repair. J Biol Chem 2011;286:32638–50.

[162] Zha S, Guo C, Boboila C, Oksenych V, Cheng HL, Zhang Y, et al. ATM damage response and XLF repair factor are functionally redundant in joining DNA breaks. Nature 2011;469:250–4.

[163] Yan CT, Boboila C, Souza EK, Franco S, Hickernell TR, Murphy M, et al. IgH class switching and translocations use a robust non-classical end-joining pathway. Nature 2007;449:478–82.

[164] Boboila C, Yan C, Wesemann DR, Jankovic M, Wang JH, Manis J, et al. Alternative end-joining catalyzes class switch recombination in the absence of both Ku70 and DNA ligase 4. J Exp Med 2010;207:417–27.

[165] Soulas-Sprauel P, Le Guyader G, Rivera-Munoz P, Abramowski V, Olivier-Martin C, Goujet-Zalc C, et al. Role for DNA repair factor XRCC4 in immunoglobulin class switch recombination. J Exp Med 2007;204:1717–27.

[166] Manis JP, Dudley D, Kaylor L, Alt FW. IgH class switch recombination to IgG1 in DNA-PKcs-deficient B cells. Immunity 2002;16:607–17.

[167] Cook AJ, Oganesian L, Harumal P, Basten A, Brink R, Jolly CJ. Reduced switching in SCID B cells is associated with altered somatic mutation of recombined S regions. J Immunol 2003;171:6556–64.

[168] Bosma GC, Kim J, Urich T, Fath DM, Cotticelli MG, Ruetsch NR, et al. DNA-dependent protein kinase activity is not required for immunoglobulin class switching. J Exp Med 2002;196:1483–95.

[169] Kiefer K, Oshinsky J, Kim J, Nakajima PB, Bosma GC, Bosma MJ. The catalytic subunit of DNA-protein kinase (DNA-PKcs) is not required for Ig class-switch recombination. Proc Natl Acad Sci USA 2007;104:2843–8.

[170] Rooney S, Alt FW, Sekiguchi J, Manis JP. Artemis-independent functions of DNA-dependent protein kinase in Ig heavy chain class switch recombination and development. Proc Natl Acad Sci USA 2005;102:2471–5.

[171] Franco S, Murphy MM, Li G, Borjeson T, Boboila C, Alt FW. DNA-PKcs and Artemis function in the end-joining phase of immunoglobulin heavy chain class switch recombination. J Exp Med 2008;205:557–64.

[172] Callen E, Jankovic M, Wong N, Zha S, Chen HT, Difilippantonio S, et al. Essential role for DNA-PKcs in DNA double-strand break repair and apoptosis in ATM-deficient lymphocytes. Mol Cell 2009;34:285–97.

[173] Li G, Alt FW, Cheng HL, Brush JW, Goff PH, Murphy MM, et al. Lymphocyte-specific compensation for XLF/cernunnos end-joining functions in V(D)J recombination. Mol Cell 2008;31:631–40.

[174] Zhang Y, Gostissa M, Hildebrand DG, Becker MS, Boboila C, Chiarle R, et al. The role of mechanistic factors in promoting chromosomal translocations found in lymphoid and other cancers. Adv Immunol 2010;106:93–133.

[175] Guirouilh-Barbat J, Huck S, Bertrand P, Pirzio L, Desmaze C, Sabatier L, et al. Impact of the KU80 pathway on NHEJ-induced genome rearrangements in mammalian cells. Mol Cell 2004;14:611–23.

[176] Dunnick W, Hertz GZ, Scappino L, Gritzmacher C. DNA sequences at immunoglobulin switch region recombination sites. Nucleic Acids Res 1993;21:365–72.

[177] Wang M, Wu W, Wu W, Rosidi B, Zhang L, Wang H, et al. PARP-1 and Ku compete for repair of DNA double strand breaks by distinct NHEJ pathways. Nucleic Acids Res 2006;34:6170–82.
[178] Sfeir A, de Lange T. Removal of shelterin reveals the telomere end-protection problem. Science 2012;336:593–7.
[179] Zhang Y, Jasin M. An essential role for CtIP in chromosomal translocation formation through an alternative end-joining pathway. Nat Struct Mol Biol 2011;18:80–4.
[180] Yun MH, Hiom K. CtIP-BRCA1 modulates the choice of DNA double-strand-break repair pathway throughout the cell cycle. Nature 2009;459:460–3.
[181] Lee-Theilen M, Matthews AJ, Kelly D, Zheng S, Chaudhuri J. CtIP promotes microhomology-mediated alternative end joining during class-switch recombination. Nat Struct Mol Biol 2011;18:75–9.
[182] Dinkelmann M, Spehalski E, Stoneham T, Buis J, Wu Y, Sekiguchi JM, et al. Multiple functions of MRN in end-joining pathways during isotype class switching. Nat Struct Mol Biol 2009;16:808–13.
[183] Xie A, Kwok A, Scully R. Role of mammalian Mre11 in classical and alternative nonhomologous end joining. Nat Struct Mol Biol 2009;16:814–8.
[184] Zha S, Boboila C, Alt FW. Mre11: roles in DNA repair beyond homologous recombination. Nat Struct Mol Biol 2009;16:798–800.
[185] Wang H, Rosidi B, Perrault R, Wang M, Zhang L, Windhofer F, et al. DNA ligase III as a candidate component of backup pathways of nonhomologous end joining. Cancer Res 2005;65:4020–30.
[186] Audebert M, Salles B, Calsou P. Involvement of poly(ADP-ribose) polymerase-1 and XRCC1/DNA ligase III in an alternative route for DNA double-strand breaks rejoining. J Biol Chem 2004;279:55117–26.
[187] Petersen S, Casellas R, Reina-San-Martin B, Chen HT, Difilippantonio MJ, Wilson PC, et al. AID is required to initiate Nbs1/gamma-H2AX focus formation and mutations at sites of class switching. Nature 2001;414:660–5.
[188] Schrader CE, Guikema JE, Linehan EK, Selsing E, Stavnezer J. Activation-induced cytidine deaminase-dependent DNA breaks in class switch recombination occur during G1 phase of the cell cycle and depend upon mismatch repair. J Immunol 2007;179:6064–71.
[189] Betermier M, Bertrand P, Lopez BS. Is non-homologous end-joining really an inherently error-prone process? PLoS Genet 2014;10:e1004086.
[190] Simsek D, Jasin M. Alternative end-joining is suppressed by the canonical NHEJ component Xrcc4-ligase IV during chromosomal translocation formation. Nat Struct Mol Biol 2010;17:410–6.
[191] Wang JH, Alt FW, Gostissa M, Datta A, Murphy M, Alimzhanov MB, et al. Oncogenic transformation in the absence of Xrcc4 targets peripheral B cells that have undergone editing and switching. J Exp Med 2008;205:3079–90.
[192] Shiloh Y. ATM and related protein kinases: safeguarding genome integrity. Nat Rev Cancer 2003;3:155–68.
[193] Nakamura AJ, Rao VA, Pommier Y, Bonner WM. The complexity of phosphorylated H2AX foci formation and DNA repair assembly at DNA double-strand breaks. Cell Cycle 2010;9:389–97.
[194] Bassing CH, Alt FW. The cellular response to general and programmed DNA double strand breaks. DNA Repair (Amst) 2004;3:781–96.
[195] Franco S, Gostissa M, Zha S, Lombard DB, Murphy MM, Zarrin AA, et al. H2AX prevents DNA breaks from progressing to chromosome breaks and translocations. Mol Cell 2006;21:201–14.
[196] Ramiro AR, Jankovic M, Callen E, Difilippantonio S, Chen HT, McBride KM, et al. Role of genomic instability and p53 in AID-induced c-myc-Igh translocations. Nature 2006;440:105–9.
[197] Burma S, Chen BP, Murphy M, Kurimasa A, Chen DJ. ATM phosphorylates histone H2AX in response to DNA double-strand breaks. J Biol Chem 2001;276:42462–7.
[198] Ciccia A, Elledge SJ. The DNA damage response: making it safe to play with knives. Mol Cell 2010;40:179–204.
[199] Panier S, Boulton SJ. Double-strand break repair: 53BP1 comes into focus. Nat Rev Mol Cell Biol 2014;15:7–18.
[200] Chapman JR, Barral P, Vannier JB, Borel V, Steger M, Tomas-Loba A, et al. RIF1 is essential for 53BP1-dependent nonhomologous end joining and suppression of DNA double-strand break resection. Mol Cell 2013;49:858–71.
[201] Di Virgilio M, Callen E, Yamane A, Zhang W, Jankovic M, Gitlin AD, et al. Rif1 prevents resection of DNA breaks and promotes immunoglobulin class switching. Science 2013;339:711–5.
[202] Escribano-Diaz C, Orthwein A, Fradet-Turcotte A, Xing M, Young JT, Tkac J, et al. A cell cycle-dependent regulatory circuit composed of 53BP1-RIF1 and BRCA1-CtIP controls DNA repair pathway choice. Mol Cell 2013;49:872–83.
[203] Feng L, Fong KW, Wang J, Wang W, Chen J. RIF1 counteracts BRCA1-mediated end resection during DNA repair. J Biol Chem 2013;288:11135–43.
[204] Zimmermann M, Lottersberger F, Buonomo SB, Sfeir A, de Lange T. 53BP1 regulates DSB repair using Rif1 to control 5′ end resection. Science 2013;339:700–4.
[205] Reina-San-Martin B, Difilippantonio S, Hanitsch L, Masilamani RF, Nussenzweig A, Nussenzweig MC. H2AX is required for recombination between immunoglobulin switch regions but not for intra-switch region recombination or somatic hypermutation. J Exp Med 2003;197:1767–78.
[206] Ramachandran S, Chahwan R, Nepal RM, Frieder D, Panier S, Roa S, et al. The RNF8/RNF168 ubiquitin ligase cascade facilitates class switch recombination. Proc Natl Acad Sci USA 2010;107:809–14.
[207] Santos MA, Huen MS, Jankovic M, Chen HT, Lopez-Contreras AJ, Klein IA, et al. Class switching and meiotic defects in mice lacking the E3 ubiquitin ligase RNF8. J Exp Med 2010;207:973–81.
[208] Li L, Halaby MJ, Hakem A, Cardoso R, El Ghamrasni S, Harding S, et al. Rnf8 deficiency impairs class switch recombination, spermatogenesis, and genomic integrity and predisposes for cancer. J Exp Med 2010;207:983–97.
[209] Manis JP, Morales JC, Xia Z, Kutok JL, Alt FW, Carpenter PB. 53BP1 links DNA damage-response pathways to immunoglobulin heavy chain class-switch recombination. Nat Immunol 2004;5:481–7.
[210] Ward IM, Reina-San-Martin B, Olaru A, Minn K, Tamada K, Lau JS, et al. 53BP1 is required for class switch recombination. J Cell Biol 2004;165:459–64.
[211] Ramiro A, Reina San-Martin B, McBride K, Jankovic M, Barreto V, Nussenzweig A, et al. The role of activation-induced deaminase in antibody diversification and chromosome translocations. Adv Immunol 2007;94:75–107.
[212] Reina-San-Martin B, Chen J, Nussenzweig A, Nussenzweig MC. Enhanced intra-switch region recombination during immunoglobulin class switch recombination in 53BP1-/- B cells. Eur J Immunol 2007;37:235–9.
[213] Difilippantonio S, Gapud E, Wong N, Huang CY, Mahowald G, Chen HT, et al. 53BP1 facilitates long-range DNA end-joining during V(D)J recombination. Nature 2008;456:529–33.

[214] Bothmer A, Robbiani DF, Feldhahn N, Gazumyan A, Nussenzweig A, Nussenzweig MC. 53BP1 regulates DNA resection and the choice between classical and alternative end joining during class switch recombination. J Exp Med 2010;207:855–65.
[215] Bunting SF, Callen E, Wong N, Chen HT, Polato F, Gunn A, et al. 53BP1 inhibits homologous recombination in Brca1-deficient cells by blocking resection of DNA breaks. Cell 2010;141: 243–54.
[216] Bothmer A, Robbiani DF, Di Virgilio M, Bunting SF, Klein IA, Feldhahn N, et al. Regulation of DNA end joining, resection, and immunoglobulin class switch recombination by 53BP1. Mol Cell 2011;42:319–29.
[217] Kuppers R, Dalla-Favera R. Mechanisms of chromosomal translocations in B cell lymphomas. Oncogene 2001;20:5580–94.
[218] Gostissa M, Alt FW, Chiarle R. Mechanisms that promote and suppress chromosomal translocations in lymphocytes. Annu Rev Immunol 2011;29:319–50.
[219] Gostissa M, Yan CT, Bianco JM, Cogne M, Pinaud E, Alt FW. Long-range oncogenic activation of Igh-c-myc translocations by the Igh 3′ regulatory region. Nature 2009;462:803–7.
[220] Janz S. Myc translocations in B cell and plasma cell neoplasms. DNA Repair (Amst) 2006;5:1213–24.
[221] Delmore JE, Issa GC, Lemieux ME, Rahl PB, Shi J, Jacobs HM, et al. BET bromodomain inhibition as a therapeutic strategy to target c-Myc. Cell 2011;146:904–17.
[222] Robbiani DF, Nussenzweig MC. Chromosome translocation, B cell lymphoma, and activation-induced cytidine deaminase. Annu Rev Pathol 2013;8:79–103.
[223] Chiarle R, Zhang Y, Frock RL, Lewis SM, Molinie B, Ho YJ, et al. Genome-wide translocation sequencing reveals mechanisms of chromosome breaks and rearrangements in B cells. Cell 2011;147:107–19.
[224] Klein IA, Resch W, Jankovic M, Oliveira T, Yamane A, Nakahashi H, et al. Translocation-capture sequencing reveals the extent and nature of chromosomal rearrangements in B lymphocytes. Cell 2011;147:95–106.
[225] Callen E, Jankovic M, Difilippantonio S, Daniel JA, Chen HT, Celeste A, et al. ATM prevents the persistence and propagation of chromosome breaks in lymphocytes. Cell 2007;130:63–75.
[226] Litman GW, Rast JP, Fugmann SD. The origins of vertebrate adaptive immunity. Nat Rev Immunol 2010;10:543–53.
[227] Stavnezer J, Amemiya CT. Evolution of isotype switching. Semin Immunol 2004;16:257–75.
[228] Cannon JP, Haire RN, Rast JP, Litman GW. The phylogenetic origins of the antigen-binding receptors and somatic diversification mechanisms. Immunol Rev 2004;200:12–22.
[229] Barreto VM, Pan-Hammarstrom Q, Zhao Y, Hammarstrom L, Misulovin Z, Nussenzweig MC. AID from bony fish catalyzes class switch recombination. J Exp Med 2005;202:733–8.
[230] Wakae K, Magor BG, Saunders H, Nagaoka H, Kawamura A, Kinoshita K, et al. Evolution of class switch recombination function in fish activation-induced cytidine deaminase, AID. Int Immunol 2006;18:41–7.
[231] Mussmann R, Courtet M, Schwager J, Du Pasquier L. Microsites for immunoglobulin switch recombination breakpoints from *Xenopus* to mammals. Eur J Immunol 1997;27:2610–9.
[232] Lundqvist ML, Middleton DL, Hazard S, Warr GW. The immunoglobulin heavy chain locus of the duck. Genomic organization and expression of D, J, and C region genes. J Biol Chem 2001;276:46729–36.
[233] Esser C, Radbruch A. Immunoglobulin class switching: molecular and cellular analysis. Annu Rev Immunol 1990;8:717–35.
[234] Jack HM, McDowell M, Steinberg CM, Wabl M. Looping out and deletion mechanism for the immunoglobulin heavy-chain class switch. Proc Natl Acad Sci USA 1988;85:1581–5.
[235] Lenz G, Nagel I, Siebert R, Roschke AV, Sanger W, Wright GW, et al. Aberrant immunoglobulin class switch recombination and switch translocations in activated B cell-like diffuse large B cell lymphoma. J Exp Med 2007;204:633–43.
[236] Vaandrager JW, Schuuring E, Kluin-Nelemans HC, Dyer MJ, Raap AK, Kluin PM. DNA fiber fluorescence in situ hybridization analysis of immunoglobulin class switching in B-cell neoplasia: aberrant CH gene rearrangements in follicle center-cell lymphoma. Blood 1998;92:2871–8.

Chapter 20

# Somatic Hypermutation: The Molecular Mechanisms Underlying the Production of Effective High-Affinity Antibodies

Alberto Martin[1], Richard Chahwan[2], Jahan Yar Parsa[1] and Matthew D. Scharff[3]

[1]*Department of Immunology, University of Toronto, Toronto, ON, Canada,* [2]*Department of Biosciences, University of Exeter, Exeter, UK,* [3]*Department of Cell Biology, Albert Einstein College of Medicine, Bronx, NY, USA*

## 1. INTRODUCTION

To protect against the extremely large number of highly pathogenic and foreign substances that are encountered by vertebrates on a regular basis, the immune system must generate an equally diverse repertoire of antibodies that can recognize, neutralize, and dispose of these potentially toxic agents. Before antigenic stimulation, it is estimated that at any given time, the antibody repertoire consists of about $10^{10}$ different antibodies in humans and other mammals [1]. Perhaps most interestingly, this large repertoire is achieved even with a limited number of antibody genes. To create this large pool of antibodies, B-cells undergo a process termed V(D)J recombination, which is initiated by the RAG1/2 protein complex, and allows B-cells to assemble the variable regions of the antibodies from shorter segments, termed variable (V), diversity (D), and joining (J) segments, into single coding exons for the heavy and light chain V regions. However, the large antibody repertoire generated by V(D)J rearrangements and the association of different heavy and light chain V regions is not sufficient to effectively protect from pathogens and toxic substances because most of the germline-encoded antibodies lack adequate affinity for the antigen. To deal with this, B-cells undergo a second round of antibody diversification that generates antibodies with increased affinity and changes in fine specificity and changes the class of antibodies they are producing, thereby altering the antibodies' effector functions. The processes involved in this secondary antibody diversification are termed somatic hypermutation (SHM), gene conversion (GCV), and class switch recombination (CSR). In this chapter, we will focus our discussion on the molecular mechanisms of the SHM process, highlighting some of the relevant research publications in this field that have contributed to a better understanding of this complex process.

### 1.1 Somatic Hypermutation

Somatic hypermutation is a process in which point mutations accumulate in the antibody V-regions of both the heavy and light chains, at rates that are about $10^6$-fold higher than the background mutation rates observed in other genes (Figure 1). This accumulation of mutations at the V-region genes occurs at the centroblast stage of B-cell differentiation in the germinal centers of secondary lymphoid organs. Whereas the overall goal of this process is to produce high-affinity antibodies, in the absence of selection, SHM does not distinguish between favorable and unfavorable mutations and can produce antibodies with (1) higher affinity for antigen, (2) lower affinity for antigen, and (3) no change in affinity for antigen. Somatic hypermutation can also lead to nonfunctional antibodies, such as antibodies that cannot fold correctly, or antibody genes that harbor premature stop codons [2,3]. Whereas SHM of the antibody V-region does not always produce a higher-affinity antibody, the selection process for antigen binding that occurs in the light zone of the germinal center selects for B-cells that produce the highest-affinity antibodies. Mutations tend to accumulate in the complementarity determining regions (CDRs) of the antibody V genes. Because the CDRs are the locations that directly contact the antigen, it is not surprising that these regions would have the most mutations after selection. Evolutionary selection over millions of years has facilitated this process by enriching the CDRs for codons with activation-induced cytidine deaminase (AID) hot spots that result in replacement mutations, whereas the codon usage in the frameworks of V genes is more likely to lead to silent or conservative mutations [4–6].

Somatic mutations of the V-region have been observed to accumulate between about 100 base pairs and about 2 kb downstream of the promoter of the rearranged V(D)J

Molecular Biology of B Cells. http://dx.doi.org/10.1016/B978-0-12-397933-9.00020-5

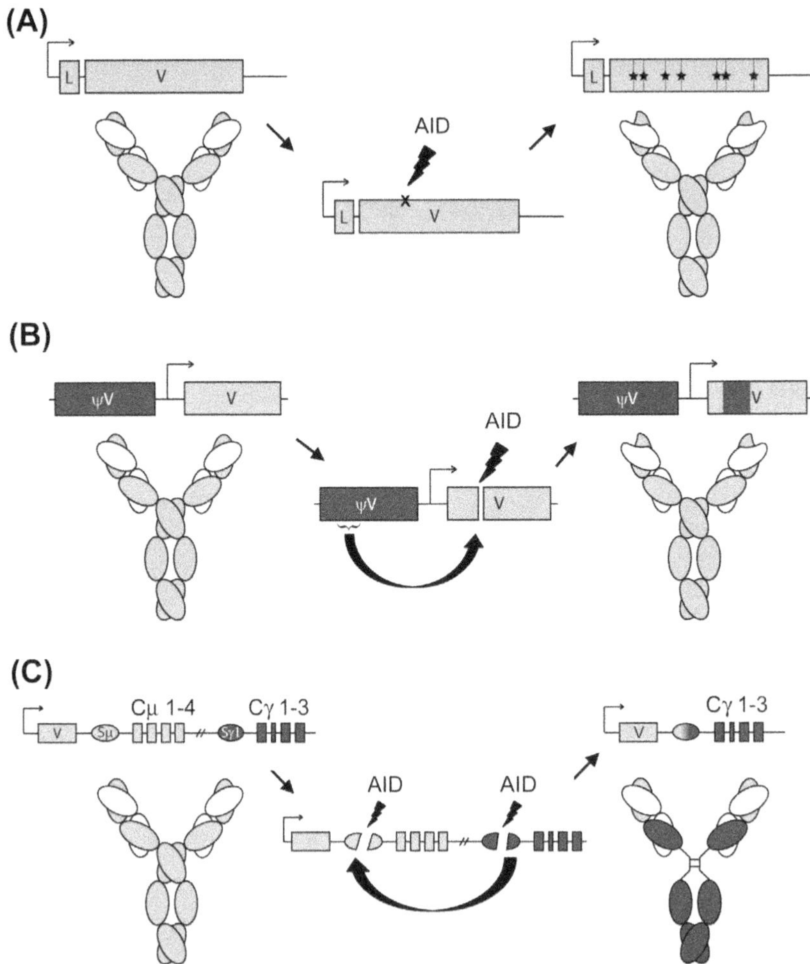


FIGURE 1 **Secondary antibody diversification processes.** (A) Somatic hypermutation: the antibody V-region accumulates point mutations at rates that are 1 million-fold higher than the background mutation rate to alter the specificity of the antibody. (B) Gene conversion: a diversification process that uses homologous recombination to introduce sequences from upstream ψV region sequences into the rearranged V region to change the antibody affinity. (C) Class switch recombination: a recombination process in which one downstream constant region (Cγ1 depicted) replaces an upstream constant region (Cμ depicted) through a double-stranded break intermediate.

gene, with the peak of mutation accumulation occurring within the V(D)J exon [7–10]. However, although the rest of the genome is normally protected from AID-induced mutations, low levels of SHM have been observed at non-antibody genes [11–17], and indeed, mutations at some of these genes have been associated with B-cell malignancies. Because the overexpression of AID leads to malignancies in T-cells and other tissues, but not in B-cells [18,19], it appears that B-cells have evolved to tolerate high mutation rates without eliciting a commensurate DNA damage response, cell cycle arrest, or apoptosis; but how this is accomplished remains an open question. We will address a number of possible explanations during this chapter.

## 1.2 Gene Conversion

Vertebrates, including mice and humans, rely on SHM to produce higher-affinity B-cells after selection in the germinal center. However, several mammalian species (e.g., rabbits) as well as birds use a separate method of altering the affinity of their antibodies, known as GCV [20,21] (Figure 1). Although V(D)J recombination is required to generate antibodies in these organisms, most diversity is obtained through the process of GCV [22]. In chickens, for example, the light chain locus contains about 25 pseudo (ψ) V-region genes that are upstream of the rearranged V-region [22]. These ψ V-regions do not undergo V(D)J recombination because they do not contain promoters or recombination signal sequences; instead, these ψ V-regions are used in the process of GCV. The initiation of GCV requires the formation of a dsDNA break within the rearranged V(D)J region [23,24]. This allows for a homologous recombination-like process to take place in which a DNA sequence from an upstream ψ V-region is copied into the rearranged V-region, thus changing the specificity and affinity of the antibody [25] (Figure 1(B)). In fact, multiple GCV events from different ψ V-regions can occur in a single B-cell clone, providing additional opportunities to generate sequence diversity and higher-affinity antibodies.

## 1.3 Class Switch Recombination

Whereas the processes of SHM and GCV allow for the selection of higher-affinity antibody-producing B-cells, the resulting antibodies will still be of the IgM isotype and have a very short half-life and limited functional capacity. The process of CSR allows for the substitution of one isotype for another, thus providing the antibody with new functions and a different distribution throughout the body [26] (Figure 1). Induction of CSR can occur in different cytokine milieus that result from cellular interactions of the B-cell with different types of T-cells that direct CSR toward a specific isotype [27] (see Chapter 12, this volume).

Class switch recombination occurs exclusively at the heavy-chain antibody locus, because this is the location of the different constant region genes that encode each isotype (Figure 1(C)). Class switch recombination requires specific sequences of DNA that lie upstream of each set of constant region genes, known as switch regions (S-regions). Each S-region is also associated with an upstream promoter and an I exon, which allows for the production of sterile (non-coding) germline transcripts through the S-region. Germline transcription is an absolute requirement for the CSR process [28,29] and for formation of dsDNA breaks at the S-region. Although S-regions are present upstream of each set of the constant region genes, the IgD constant region lacks a defined S-region. Rather than using CSR, IgD is generated through an alternative splicing mechanism with the μ constant region. However, a cryptic S-region has been discovered upstream the IgD constant regions genes that can allow for an inefficient CSR event to IgD [30,31].

# 2. ACTIVATION-INDUCED CYTIDINE DEAMINASE IN SOMATIC HYPERMUTATION

The processes of SHM, GCV, and CSR, although mechanistically distinct, all require the same B cell–specific factor, known as activation-induced cytidine deaminase (AID). The discovery of AID and the fact that this single B cell–specific factor could induce all three of these important and distinct pathways in B-cells was a paradigm-changing discovery in our understanding of the generation of antibody diversity. Initially discovered by subtractive cDNA cloning in the murine CH12F3-2 cell line [32], AID was later shown to be absolutely required for SHM and CSR in humans and mice [32,33] and GCV in chickens [34,35].

Activation-induced cytidine deaminase is a 198-amino-acid protein that has strong sequence homology to the APOBEC family of cytidine deaminases, most strongly to APOBEC-1 [32]. Because mRNA is the substrate for the APOBEC-1 enzyme, it was initially proposed that AID acted on mRNA. Along with the catalytic motif present in the active site of the enzyme, other motifs exist in the protein structure of AID, such as a linker region, an APOBEC1-like domain, an N-terminal nuclear localization signal (NLS), and a C-terminal nuclear export signal (NES) [36–38] (Figure 2). A more in-depth description of the AID protein, structure-function, regulation, and posttranslational modifications can be found in Chapter 20. Interestingly, though, the processes of SHM and CSR are in part directed by different regions within AID itself; mutation of the N-terminal region of AID decreases SHM activity, while still permitting CSR [39]. Conversely, C-terminal mutations of AID reduce CSR activity, but not SHM activity [40–44]. Activation-induced cytidine deaminase harbors a nuclear export sequence in its C-terminal tail [38,45], and that sequence is necessary for CSR, but this motif has other functions because heterologous nuclear export sequences promote AID export, but not CSR [45]. It is still unclear exactly how these domains of AID differentially influence SHM and CSR, and attempts to co-immunoprecipitate specific AID subcomplexes that function either in SHM or CSR have been elusive.

## 2.1 Substrate Specificity of AID

At the time of its discovery, the homology of AID to APOBEC-1, an RNA editing enzyme, suggested that the substrate for AID was mRNA [33]. APOBEC-1 has a key role in the RNA-editing event that produces two variants of the apolipoprotein B (apoB) protein [46,47], apoB100 and apoB48, both of which have distinct activities in lipid metabolism. APOBEC-1 in association with other factors mediates a specific RNA deamination editing step in the mRNA at nucleotide 6666, mutating a glutamine (CAA) residue to a translation stop codon (UAA), resulting in the variant form of the apoB enzyme, apoB48 [48,49]. Indeed, one of the original hypotheses suggested that AID itself was not the mutator involved in SHM and CSR; instead, it was thought that AID acted as an RNA editor [33], much like APOBEC-1, inducing a deamination event in a specific mRNA molecule.

Support for the RNA-editing hypothesis for AID comes from experiments showing that in cells expressing the estrogen receptor-AID fusion protein, which allows the translocation of AID into the nucleus to carry out its function, the protein synthesis inhibitor cycloheximide inhibited CSR; this suggested that de novo protein synthesis is required for CSR [50,51]. This result would be consistent with an RNA-editing function of AID that requires translation of a mutator de novo. In addition, AID and APOBEC family members associated with RNA in mammalian cells, and AID does so through a cofactor that interacts with the carboxy-terminus of AID [52]. Despite these findings, no specific transcript has been identified to date that is edited by AID and facilitates the SHM and CSR process.

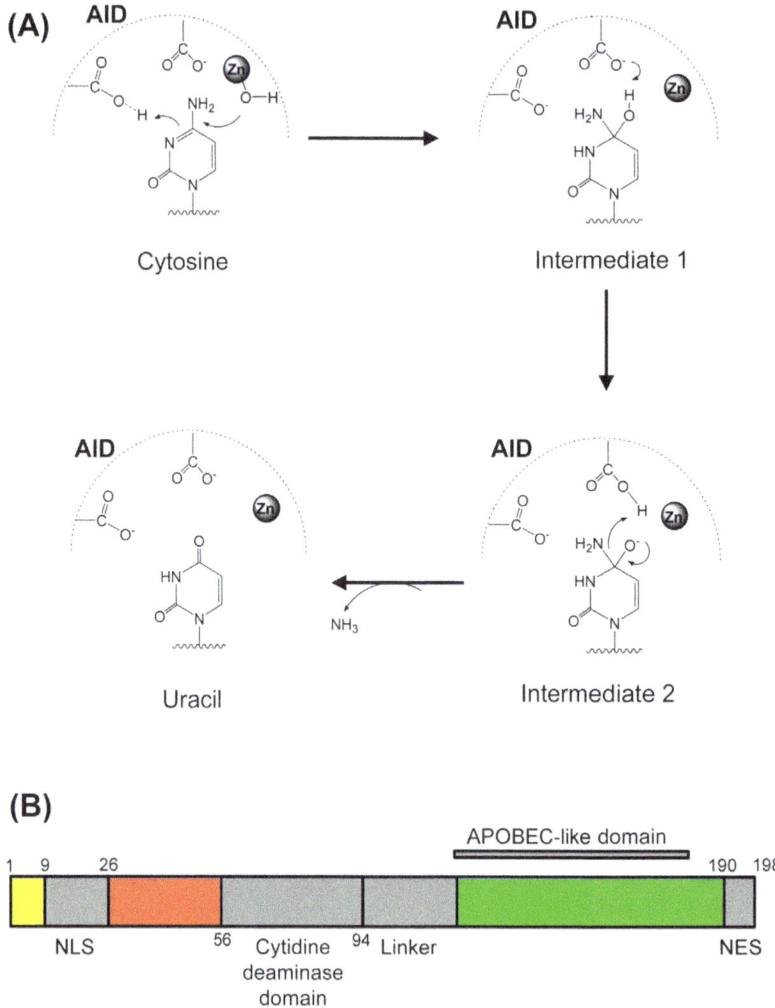


FIGURE 2 **Activation-induced deaminase structure and function.** (A) The catalytic action of AID. A histidine and 2 cysteine residues in AID coordinate $Zn^{2+}$, allowing for nucleophilic attack of C-4 of the cytidine ring. (B) Structure of AID with the location of domains involved in its activity. NLS = nuclear localization signal; NES = nuclear export signal.

On the other hand, evidence for a model in which AID acts directly on DNA was proposed [53–56] and over the years has been supported by many genetic and biochemical findings. Recombinant AID has been shown to deaminate pyrimidine (dC) in single-stranded DNA [57–63]. Furthermore, expression of AID in *Escherichia coli* induces mutations in the *E. coli* genome [64]. Because the likelihood is low that *E. coli* expresses the same mRNA molecule as vertebrate species that would be edited by AID to generate a DNA mutator, this strongly supported the DNA model of AID deamination. In fact, expression of AID in any cell type is sufficient to induce a mutator phenotype [56,64–68], indicating that AID is the only B cell–specific factor required for SHM/CSR. In addition, deficiencies in uracil-DNA glycosylase (UNG), an important factor in the base excision repair (BER) pathway (allowing for the removal of uridine [dU] in DNA) required for SHM and CSR, results in defective antibody diversification processes [53,64,69–73]. Recently, Maul et al. observed the accumulation of AID-induced dU within the antibody genes [74]. These data support the notion that the introduction of dU in DNA is a critical intermediate in the processes of SHM and CSR. Thus, there is strong evidence that AID, unlike APOBEC-1, carries out its mutator function by directly deaminating dC within DNA.

Several reports, including in vitro studies using purified AID on in vitro transcribed DNA, showed that AID has a strong preference for deaminating dC on ssDNA, perhaps because of the availability of the free primary amino group on the dC base to interact directly with AID [57–59,62,63,75]. Because transcription is positively associated with SHM rates (see subsequent discussion), it was initially hypothesized that the transcription bubble might produce the ssDNA substrate for AID activity [76]. Of course, there are potentially other sources of ssDNA in the cell, such as DNA repair and replication intermediates; however, these structures have not yet been shown to be AID substrates. Supporting the notion that the transcription bubble is the target of AID, AID was found to preferentially mutate the non-template strand on plasmid DNA in bacteria [44,77], which is the exposed ssDNA strand that is not complexed with the nascent transcript. Because RNA-DNA hybrids are poor substrates for AID [57], the findings that AID mutates the non-template strand makes sense. However, other studies in *E. coli* showed that both DNA strands can be mutated by AID [78,79]. Furthermore, in mammalian cells, AID mutates both template and non-template DNA strands at equal frequencies at the V- and S-regions [71,74,80–82], and hence the bacterial system might not represent an ideal model for AID action in mammalian cells. Biochemical studies showed that AID requires at least five nucleotides of ssDNA within a bubble for efficient deamination [57,62,63], and the dC needs to be at least one nucleotide away from the single-stranded/double-stranded DNA junction [62].

However, the fact that ssDNA largely resides within the RNA polymerase II complex [83,84], even when complexed with Spt5 (see subsequent discussion [85,86]) raises the question of whether the transcription bubble is the in vivo substrate of AID in mammalian cells. Storb and colleagues showed that purified AID can mutate both strands when the DNA is supercoiled [87]. In fact, localized negative supercoiling occurs 5′ of the RNA polymerase II complex [88,89], which is associated with formation of secondary structures in DNA [90,91]. Short patches of ssDNA that are on average seven to 11 nucleotides in length are observed to occur on both strands in the V-region, S-region, and other genes, and these structures require transcription [92,93]. Because the V-region primarily undergoes sense transcription [94] and these ssDNA patches occur on both strands at equal frequency, this suggests that these patches of ssDNA are not caused by the transcription bubble, because the nascent transcript should protect the template strand from bisulfite conversion. In bacterial systems, ssDNA patches of similar lengths are observed during transcription-induced supercoiling of plasmid DNA [93], and in this case, AID mutates negatively supercoiled plasmid DNA on both strands [87,93], just as it does in mammalian cells. These data suggest that this structure could plausibly be the in vivo substrate for AID.

## 2.2 Sequence Specificity of AID-Induced Mutations

Preceding the debate about the preferred substrate of AID, and in fact before the discovery of AID itself, much work had been conducted to unveil the preferred sequence specificity for the process of SHM. Indeed, we can now look back on these data and ascribe this sequence specificity to AID. An analysis of 324 somatic mutations in 14 different V-region genes, as well as 296 mutations in regions flanking these genes, revealed a strong preference for the DNA mutator to mutate a specific motif WRCY (where W=A/T; R=A/G; C=the dC that is deaminated; and Y=C/T) [95]. These motifs are found to be mutated in both the V- and S-regions of immunoglobulin genes [96,97]. Various studies later showed that purified AID preferentially mutates dC within the WRC motif in vitro [59,61–63,82,98], which further supports the notion that AID is the DNA mutator during SHM. In addition, the pattern of mutation of purified AID on naked single-stranded V region DNA results in a distribution and relative frequency of mutation throughout the V region that has a 50% correlation with the mutations that occur in vivo in that same V-region [99]. This result indicates that the intrinsic catalytic properties of AID accounts for a large proportion of its activity in vivo.

In the absence of crystallographic diffraction data for AID complexed with its substrate, the molecular details for the basis of the WRC specificity of AID have to be obtained using different approaches. The catalytic specificity of AID for WRC motifs suggests specific interactions with purine moieties at the R position within the WRC motif, which was found to be the case using nucleotide analogs [62]. The specificity for W within the WRC motif is more complex because it involves recognition of one purine (dA) and one pyrimidine (dT), but not the other purine (dG) or pyrimidine (dC). One interesting feature of the WRC motif is that it needs to be fully single-stranded for efficient AID deamination [62]. Hence, in contrast to sequence-specific dsDNA recognition enzymes, such as restriction endonucleases and transcription factors, AID is not capable of discerning nucleotide features in dsDNA.

The sequence specificity of AID and, for that matter, all other APOBEC family members is imparted by residues that are located at amino acid positions 113–123, carboxy terminal to the catalytic site [100,101]. This loop and its sequence specificity can be interchanged among different APOBEC family members. Further supporting the DNA deamination model of AID, mutants of AID that show altered mutation spectrum biochemically produce a similar mutation spectrum when expressed in cells [101].

## 2.3 Catalytic Mechanism for AID

Generally, cytidine deaminases share a well-conserved catalytic site motif in which a histidine and 2 carboxylate-containing residues (glutamate or aspartate) chelate $Zn^{2+}$, activating nucleophilic water to attack C-4 of the cytidine ring, and subsequently producing ammonia [102,103]. Meanwhile, two non-metal chelating carboxylate-containing residues are used to deliver protons (Figure 2) [104–106].

Activation-induced cytidine deaminase exhibits a processive behavior in vitro when catalyzing deaminations, much like that of RNA and DNA polymerases, in that it preferentially mutates dC on the same DNA strand [59], but its behavior is more like that of restriction enzymes in that it tends to slide and jump along that strand [107]. Apobec3G, a homolog of AID, also exhibits these characteristics [108]. The finding that most V-regions only undergo one deamination/cell division in vivo [109–111] suggests that processivity does not usually occur when AID interacts with ssDNA in the B-cell [112]. The processive-like property of AID in vitro might be ascribed to its inherent affinity for ssDNA [63,113,114], which might result in AID remaining on the same substrate rather than jumping onto another substrate. Although the binding-strength of AID is correlated with its deamination rate [113], AID forms a long complex half-life with ssDNA [63], which should have a negative consequence on its catalytic rate. Indeed, AID is an inefficient enzyme, carrying out a deamination every few minutes [63,107]. However, these studies do not take into account that AID's catalytic rate might be enhanced in vivo, for example, through its association with a specific cofactor(s), although this has yet to be shown.

Activation-induced cytidine deaminase suppresses translation of Top1 mRNA, leading to altered DNA structure at the V- and S-regions and increased cleavage during SHM and CSR by Top1 itself [115,116]. How AID inhibits Top1 mRNA translation and how selective it is as a translation inhibitor is unclear. Although this finding is highly intriguing, these data could be interpreted differently. For example, inhibition of Top1 could lead to increased RNA polymerase II stalling owing to increased DNA supercoiling at immunoglobulin genes. This, in turn, would lead to increased production of ssDNA, which acts as a substrate for AID. Indeed, TopA-deficient *E. coli* are increased for supercoiled DNA, ssDNA patches, and AID-mediated mutagenesis [93].

## 3. TARGETING OF THE SHM

Once inside the nucleus, AID associates with many different genes in activated B-cells [117] but mutates the antibody V- and S-regions at a much higher frequency than other genes [11]. Discovering a single perhaps B cell–specific factor that regulates the targeting of AID mutation to particular parts of the antibody genes has been of much interest in the field, but has been difficult, perhaps owing to the multiple safeguards and mechanisms that are in place to control the activity of AID. These are discussed subsequently.

### 3.1 The Role of Transcription in AID Deamination

As mentioned previously, the 5′ boundary of SHM falls ~100 base pairs downstream of the V-region promoter, which suggests a role for transcription during this process, because AID-induced mutations are rare upstream of the promoter [7,9,118]. Furthermore, the constant region genes of the Ig locus are shielded from the mutagenic activity of AID [119]. Using a modified κ light chain transgene consisting of a V-region promoter upstream of both the rearranged VJ segment and the constant region exon, Peters et al. showed that both regions underwent high levels of SHM [119,120]. Likewise, moving the promoter 750 bp upstream of its normal site led to SHM of the inserted sequence [121]. Altering the transcription rates in reporter transgenes has also been used to demonstrate that AID mutation rates are tightly linked to transcription rates [67,122,123]. In vitro studies have also shown that the process of transcription is responsible for producing ssDNA that allows for AID activity [75,77,79]. These observations provide one level of regulating AID activity once in the nucleus because genes within heterochromatin are protected from the mutagenic activity of AID. However, if transcription alone were the only means by which AID selected its target, all transcribed genes in B-cells expressing high levels of AID would be subject to its mutagenic potential. Although many transcribed genes are mutated by AID [11], the rates of mutation at these genes are more than 100-fold lower than at the V-region, which indicates that transcription alone is not sufficient to target AID to the antibody genes. Furthermore, several loci that are not transcribed can acquire AID-induced dsDNA breaks [124], which suggests that mechanisms beyond transcription can produce ssDNA, a prerequisite for AID activity.

What is the precise role for transcription in SHM? Activation-induced cytidine deaminase was shown to associate with the RNA polymerase II complex [125], but this interaction might be indirect through association with the Spt5 cofactor [126] or other factors involved in RNA processing [127,128] (see subsequent discussion). Hence, the transcriptional machinery can recruit AID to transcribed genes, and produce the ssDNA substrate for AID-mediated deamination. What is still unknown is the nature of the approximately 2-kb window of mutagenesis at the V-region and other genes mutated by AID. Although transcriptional terminators reduce downstream mutations [129,130], the region downstream of the V-region is transcribed at high rates, but is refractory to AID-induced mutagenesis. It is possible that RNA polymerase stalling, which may help to recruit AID (see subsequent discussion), occurs less frequently in this region. If this were the case, would this apply to other genes that are also mutated by AID and also exhibit a similar-sized window of mutagenesis [12,14,17]? As will be further discussed, there are other potential explanations for this phenomenon, such as the occurrence of the histone mark H3K4me3, a mark that exists at the 5′-end of transcribed genes and has been found to overlap with sites of SHM. In addition, it is unclear why mutations do not occur in the first 100 nucleotides from the transcription initiation site. This suggests that mechanisms involved in RNA processing are likely involved in AID targeting, and in fact, many factors that are associated with transcription elongation and processing of RNA have been found to associate with AID (see subsequent discussion).

## 3.2 Histone Modifications and AID Accessibility

Chromatin-associated histones undergo alterations that introduce a host of different modifications that permit new interactions to occur between them and proteins that recognize and bind these new modifications. There are many examples of proteins that are recruited to chromatin, based on their ability to bind modified histones. RAG-2, for example, has been observed to preferentially bind histone 3 once it has been trimethylated at its lysine 4 residue (H3K4me3) [131]. Thus, the idea that AID specifically targets the Ig locus based on specific histone modifications was appealing. Hyperacetylation of histones 3 and 4 was found to be present in the V-region but not in the constant region of the heavy-chain Ig locus in a cell line undergoing SHM [132]. Furthermore, acetylation of H3K9 and H3K14 correlates with CSR, and manipulation of this histone acetylation mark with the HDAC inhibitor trichostatin A increases CSR [133]. Although these data support a role for histone acetylation in promoting chromatin accessibility for AID, no differences were observed between the histone acetylation status at the λ light chain locus between germinal center and non-germinal center B-cells [134]. On the other hand, the same group showed that the phosphorylated form of histone H2B (pSer14) correlates with SHM and CSR [134], and several other histone modifications and modifying enzymes have been implicated in CSR [125,135–137].

One of the strongest candidates for a histone modification involved with SHM targeting is the H3K4me3, the same mark used by the RAG complex during V(D)J recombination [131]. Elongating RNA polymerase II recruits the histone methylase Set 1, which produces the H3K4me3 mark that persists even after transcription termination, and thereby provides molecular memory of recent transcription activity [138,139]. The Paf1 component of the RNA polymerase II complex is also important for Set1 recruitment [138], and this factor has been shown to interact with AID [128] (see subsequent discussion). The H3K4me3 level is associated with DNA cleavage at the S-region [133,140–142], and reduction of this marker by Spt6-knockdown led to a reduction in mutations in the V- and S-regions [143]. Non-immunoglobulin genes mutated by AID are also enriched for H3K4me3 [143,144], which indicates that this marker is a good predictor of AID-induced mutations. Future work will likely flesh out details regarding how this histone mark facilitates SHM and CSR at immunoglobulin genes.

In some of these cases, chromatin modifications are likely the downstream result of AID activity. Indeed, AID has been observed to indirectly stimulate the modification of histones as a result of the damage it has initiated in the DNA, such as the acetylation of histone 4 in S-regions [145]; the phosphorylation of histone H2AX, which is associated with the dsDNA breaks formed during CSR [146,147]; and the phosphorylation of histone H2B during SHM and CSR [134]. In these cases, the modified histone is indirectly a result of AID activity. To date, the role of histone modification in recruiting AID to its target loci is still unclear, and more work is required to assess the mechanism by which AID directly or indirectly interacts with these histone marks [133].

## 3.3 Nucleosomal Remodeling, AID Accessibility, and Somatic Hypermutation

Beyond the inhibitory effect of double-stranded DNA on AID accessibility, nucleosome packing provides yet another barrier for entry. Studies have suggested that nucleosomes indeed block AID activity, but this effect is reversed if the packed stretch of DNA undergoes transcription. This seems to apply to both exogenous DNA in a cell-free system as well to V-regions in activated B-cells [148,149]. An elegant experiment has taken advantage of a strong nucleosome positioning sequence (MP2) and a weaker counterpart (M5), introduced within V-region sequence, to show that both nucleosomal stability and positioning in V-regions have a marked effect on AID accessibility to generate efficient SHM [149]. It would be interesting to verify whether the Ig V-region has evolved to maintain low nucleosomal stability in addition to other active nucleosomal modifications mediated by GANP [150] and other chromatin modifying factors to maintain increased propensity for ssDNA exposure at V-regions.

## 3.4 Cis-Acting Elements in Somatic Hypermutation

Based on the relatively strong selectivity of AID for Ig loci, one attractive model put forth by many groups is that the heavy and light chain Ig genes harbor sequences that specifically attract AID. Indeed, several elements within the Ig genes have been scrutinized for their ability to specifically recruit AID. One possibility is that the primary sequence of the V-regions made them more prone to SHM, and indeed, Ig transgene constructs undergo SHM, albeit at reduced frequency compared with the endogenous Ig locus [151–155]. However, the V-region segment in these constructs, when exchanged for the human β-globin gene or the prokaryotic *neo* or *gpt* genes, resulted in no change in SHM frequencies [97]. Because transcription has a central role in the activity of AID not only at the V-region, but also during the process of CSR at the S-regions (i.e., production of sterile transcripts), it is possible that Ig promoters contain sequence elements that act to recruit AID. However, several non-Ig promoters such as promoters for the B29 [121] and chicken β-globin [156] genes can support SHM. Viral promoters such as those from cytomegalovirus (CMV) can substitute for Ig promoters during SHM, as well [66]. Class switch recombination can also be driven by non–S-region promoters [65,157]. However, not all promoters that drive high levels of transcription are able to recruit SHM to the V-regions. In the chicken DT-40 cell line, which undergoes constitutive GCV and SHM, replacing the Ig light chain V-region promoter with either the chicken β-actin promoter or the human elongation factor 1-α (EF1-α) promoter results in higher than endogenous levels of transcription, but only the chicken β-actin promoter could sustain wild-type levels of GCV/SHM [158]. Collectively, the data argue that Ig promoters do not harbor sequences that specifically recruit the SHM machinery, but nevertheless have important roles by inducing high rates of transcription that produces the ssDNA substrate for AID.

Several studies have suggested a role for the enhancers that are present within the Ig loci in targeting AID. Transgenic studies using the κ light chain, which contains a kappa intronic enhancer (iEκ) as well as a 3′-kappa enhancer (3′Eκ), suggested that the targeting of SHM to this transgene required the presence of both enhancer elements [151,154]. However, deletion of the intronic μ enhancer within the endogenous Ig heavy chain locus did not have a dramatic effect on SHM levels in vivo [159]. Replacing either set of enhancers in the κ or heavy chain loci with viral enhancers resulted in typical levels of SHM [160,161]. What additionally complicates this issue is that the viral and Ig enhancers share many motifs, and enhancers are required to maintain high levels of transcription at immunoglobulin genes; consequently, altering these enhancers might reduce transcription rates, which ultimately alter SHM rates. However, three reports show that DNA elements that act to recruit AID for SHM have no effect on transcription. A 222-bp fragment within the chicken IgL locus recruits AID but has no effect on immunoglobulin transcription [162]. Two recent studies revealed that the 3′ regulatory region, which lies downstream of the α-constant region, is necessary for SHM, but not for transcription. In one study, deletion of the 3′ regulatory region within a bacterial artificial chromosome (BAC) containing an intact immunoglobulin heavy chain gene led to a reduction of SHM and CSR in mice [163]. In another study, the 3′ regulatory region was deleted from the endogenous heavy chain locus leading to a reduction in SHM with only a minor effect on transcription [164]. Hence, both studies strongly support a role for this enhancer element in recruiting AID to the heavy chain locus. Future work will be needed to determine the precise mechanism used by the 3′ regulatory region to recruit AID, such as the identification of DNA binding factors that bind to this region, and assess whether this region interacts with the V-region through looping, as has been shown to occur between the 3′ regulatory region and S-regions during CSR [165].

Finally, perhaps the most compelling example of a cis-acting element having a role in the targeting of AID is that of the E-box motif (CAGGTG). The E-box motif, which is a motif that is bound by helix-loop-helix proteins (such as E47 and E2A), was found to contribute to high SHM rates of a κ transgene, and mutation of the motif from CAGGTG to AAGGTG resulted in a 5-fold decrease in SHM rates with no effects on the transcription rates of the transgene [109,166]. In support of this work, E-box motifs were also found to have a critical role in targeting AID to the endogenous light chain locus in DT40 cells, although sequences surrounding the E-box motifs were also shown to be necessary [167]. These data show that E-box motifs are critical for targeting of AID to immunoglobulin genes, either through a direct interaction of AID with the motif itself, which is unlikely because AID lacks any helix-loop-helix domains, or by binding of AID or one of its associated proteins to an E-box binding protein in immunoglobulin loci. These data are compelling, but once again, E-box motifs are not unique to immunoglobulin genes and are present throughout the genome. Thus, whereas E-box motifs may have an important role in the targeting or activation of AID, other layers of regulation of AID activity must exist to prevent off-target mutations.

## 3.5 Trans-Acting Factors in Targeting Somatic Hypermutation

As described earlier, the targeting of AID to immunoglobulin sequences might be facilitated by protein–protein interactions that recruit AID to specific sequences (such as the E-box motif). Indeed, several proteins that bind to AID have been suggested to have a role in targeting AID. One of the first proteins to be documented to interact with AID is RNA polymerase II [125], as discussed previously. The ssDNA binding protein replication protein A (RPA) was also shown to interact with AID [168], and this interaction is facilitated by the phosphorylation of S38 on AID, because mutant AID that cannot be phosphorylated at this residue does not interact with RPA [169,170]. This interaction is consistent with what is known about the activity of AID. That is, AID only deaminates ssDNA, and because RPA is known to specifically bind to ssDNA, RPA might maintain the single-strandedness of such sites, making them accessible to AID. In cell-free SHM assays, although AID could deaminate ssDNA, it could also deaminate dsDNA in the presence of RPA during transcription [168]. Protein replication protein A is known to bind UNG, a factor in the BER pathway that has been observed to have a critical role in the processing of AID-induced deaminations (see subsequent discussion) [171]. Thus, RPA could potentially recruit AID for the initial deamination event, but then could recruit factors for the downstream processing events that take place in SHM, GCV, and CSR. On the other hand, AID forms a relatively stable complex with ssDNA [63] and might remain at the site even after it has mutated it, which suggests that AID itself might function to recruit downstream repair factors, or RPA itself. Indeed, RPA localization to Ig regions is reduced in AID S38A and T140A mutants [117], which suggests that RPA may specifically act downstream of AID activity during the repair of AID-induced DNA damage, and might not be directly recruiting AID.

Because transcription is tightly associated with the activity of AID, factors involved in the transcription process seemed likely to be potential candidates as trans-acting factors that facilitate the targeting of AID. Activation-induced cytidine deaminase interacts with RNA polymerase II in mammalian cells [125]. However, because the transcription complex is large, and several other factors can interact with the transcription machinery, it is possible that AID is directly interacting with these other factors. One such protein is Spt5, a stalling factor that associates with a stalled RNA polymerase II complex. Spt5 interacts with AID, and is required for mutations at switch regions and CSR [126]. Indeed, AID-induced mutations at off-target genes contain a high density of stalled RNA polymerase II in association with Spt5 [126]. Activation-induced cytidine deaminase was also found to associate with PAF1 and LEO1, which are part of the RNA polymerase II elongation complex [128]. As described above, PAF1 has an important role in recruiting SET1, a histone methylase that produces the H3K4me3 mark that is associated with SHM and CSR. The RNA exosome, a multi-subunit complex that has a role in several RNA processing events, has been shown to be able to target AID to both template and non-template strands of DNA, although the mechanism by which this occurs is currently unknown [127]. Interestingly, targeting of AID with the RNA exosome was observed to occur in the absence of RPA [127], a factor (described earlier) that has been suggested to allow AID to mutate transcribed genes in cell-free systems and in vivo [168]. Collectively, these experiments suggest important roles for Spt5 and the RNA exosome in AID targeting. However, details about how Spt5 facilitates AID-induced mutations, especially in the V-region, are still lacking. In addition, it is unclear how Spt5 and the RNA exosome produce the ssDNA substrate for AID-induced mutagenesis. The ssDNA that is exposed in Spt5-RNA Pol II complexes is the non-template (top) strand, whereas the template (bottom) strand is complexed with the nascent transcript [85,86]. This is in marked contrast to the known activity of AID on both strands of DNA [71,74,80–82]. In addition, crystallographic studies show that the ssDNA that exists in Spt5-RNA polymerase II complex is minimal and mostly buried within the complex [85,86]. It is possible that the Spt5-RNA polymerase II crystal structures do not accurately reflect the structure of this complex in cells. For example, ssDNA might occur on the 3′ side of the stalled RNA polymerase II complexed with Spt5 owing to retrograde movement of the complex after stalling [172]. Alternatively, this stalled complex could stabilize melted DNA that occurs 5′ of the complex owing to the transient negative supercoiling in this region [87,93].

Several other factors have been suggested to also have a role in targeting AID to the V-regions for SHM and GCV, and the S-regions for CSR: (1) KAP1 and HP1, which were shown to cooperatively tether AID to donor S-regions via H3K9me3 [173], a histone modification previously shown to be associated with CSR despite being a mark for heterochromatin [174]; and (2) the 14.3.3 adaptor proteins that directly interact with and target AID to AGCT repeats present in the S-regions during CSR [175]. Ranjit et al. observed that the binding of AID with factors of the BER and mismatch repair (MMR) pathways at the Ig S-regions occurred through an interaction with AID's C-terminus [176]. Findings that the N-terminus and C-terminus of AID are responsible for SHM and CSR, respectively [39–43], suggest that separate factors are associated with these regions and facilitate micro-targeting of AID to the V-region or S-region. In fact, human Burkitt lymphoma cell lines carry out SHM, but not CSR [2,56,132,177] (data not shown), whereas CH12F3-2 cells and lipopolysaccharide-stimulated primary mouse B-cells efficiently undergo CSR, but not SHM [178]. This micro-targeting of AID is remarkable because the V-region and S-region are separated by only a few thousand base pairs.

In summary, although significant efforts have been undertaken to understand the means by which AID preferentially mutates immunoglobulin Ig genes, many different mechanisms have emerged as being important in AID targeting. Perhaps this reflects the complex nature of this process, and that the answer is that multiple layers of regulation for AID localization exist. Nevertheless, there are circumstances in which these safeguards have been breached, resulting in genome instability and ultimately in oncogenesis (see Chapter 25).

## 4. ACTIVATION-INDUCED CYTIDINE DEAMINASE AND DOWNSTREAM REPAIR PATHWAYS

Soon after the mutagenic cytosine deamination process catalyzed by AID takes effect on V-regions, a number of potential scenarios could unfold: As described in the previous sections, the dU:G mismatches is not detected by DNA repair proteins and gets replicated over during S-phase. Whether such mismatches actively evade repair detection or whether they are residues that have exceeded the capabilities of the DNA repair machinery is still unclear. Such a mutagenic scenario, also known as Phase 1a of SHM [72,179], accounts for dC to dT mutations that represent less than half of the mutations seen in human and murine V regions in vivo (Figure 3). The other half of the mutations are not dC to dT mutations, but result in the mutation of dC to the other bases owing to BER (see subsequent discussion) or are mutations at dA and dT and are not a result of the direct action of AID. The dU:dG mismatches created by AID could be faithfully repaired, making the dC eligible to be a target for AID once again. Although this canonical process predominates across the genome and can occur at the Ig locus, it would be counterproductive for SHM because it would nullify the effects of AID-induced mutagenesis. Rather, it appears that most of the dU:dGs caused by AID are repaired in a noncanonical error-prone manner so that the AID-induced base changes are not only preserved as mutations, but also amplify the mutational process by promoting the noncanonical error-prone pathways of BER and MMR (Figure 3) [179–188]. The recruitment of error-prone repair by AID-induced and regulated dU:dG lesions is especially interesting because they mimic the formation of the spontaneously or chemically provoked DNA mismatch such as the 70–200 dUs thought to be spontaneously generated during each S-phase of the cell cycle [189]. The distinctive difference at V-regions is that the AID-mediated dU:dG mismatches manage to recruit the noncanonical pathways of these two repair mechanisms to work in competitive synergy at the Ig locus to promote and amplify the mutation frequency of AID to reach $10^{-3}$ mutations per base pair per generation [110]. Subsequently, we will dissect these two repair pathways and provide a rationale of how they compete and collaborate to promote efficient somatic hypermutation at V-regions.

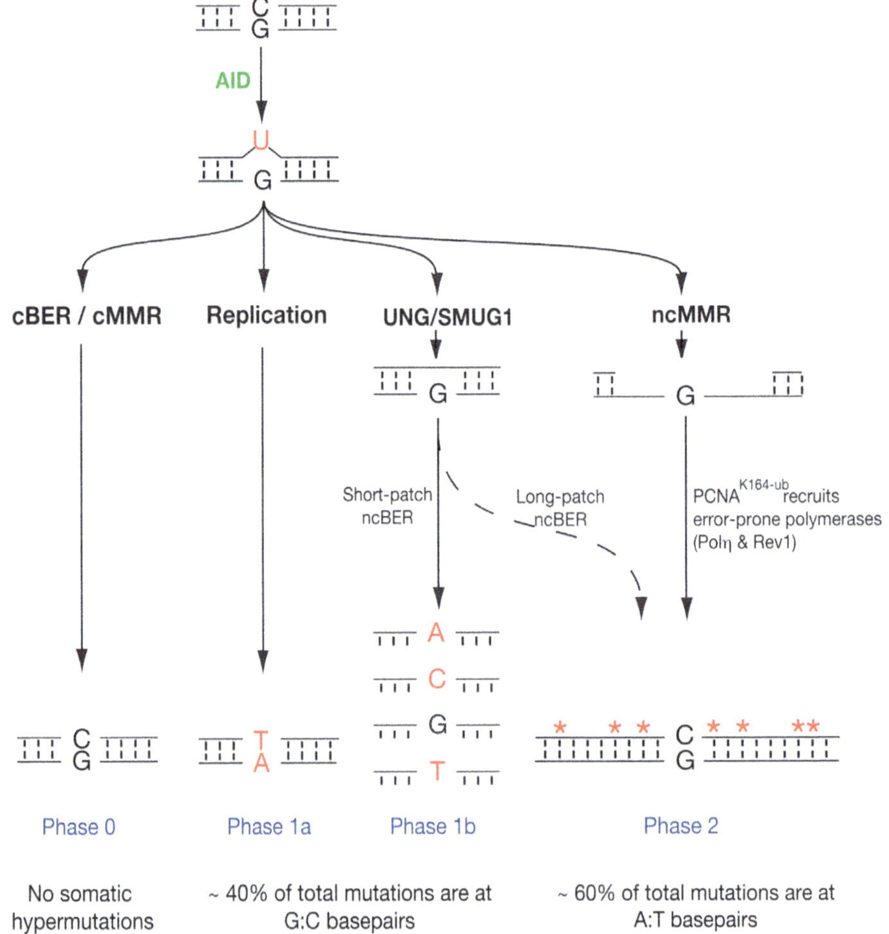


FIGURE 3 **Model of somatic hypermutation.** Activation-induced cytidine deaminase deaminates cytosines within WRC hot spots in ssDNA. Cytosine deamination by AID creates a uridine:guanine (dU:dG) mismatch that can be resolved in at least three different, yet interdependent manners. Throughout the genome, including at the Ig locus, dU:dG mismatches are detected and repaired faithfully by the cMMR and cBER machineries (Phase 0). Because such pathways leave no distinguishable byproducts, the quantification of the frequency of error-free repair during SHM is technically challenging. If the dU:dG mismatch escapes DNA repair detection, the replicative machinery will pair the intact dU with an A, and one of the daughter cells will acquires a C-to-T transition mutation (Phase 1a). The BER uracil N-glycosylases UNG, and to a lesser extent, SMUG1, detect and cleave dU bases from the DNA, eliciting the formation of an abasic site. Short-patch ncBER can then erroneously fill the abasic site, indiscriminately leading to transitions and transversions at G:C hot spots (Phase 1b). Alternatively, long-patch ncBER (right) or ncMMR (far right) can excise a tract of DNA surrounding either the abasic site or the dU:dG mismatch, respectively, to create an ssDNA gap that then engages mono-ubiquitinated PCNA to recruit error-prone polymerases leading to transition and transversion mutations at A:T sites (Phase 2).

## 5. MISMATCH REPAIR IN SOMATIC HYPERMUTATION

Canonical MMR (cMMR) is a highly conserved cellular process [190,191] that faithfully repairs erroneous insertions, deletions, and mis-paired bases that often arise during DNA replication and recombination. Canonical MMR increases replicative fidelity in mammalian cells about 100- to 1000-fold [192], and the importance of its role in maintaining genomic stability is demonstrated by the fact that defects in MMR proteins cause hereditary non-polyposis colorectal cancer (HNPCC) and contribute to many other malignancies [193,194]. Working in conjunction with other cellular proteins, such as PCNA, MMR detects and repairs mismatches by excising mis-paired bases along with a significant tract of neighboring DNA (Figure 3 and 4). Long-patch cMMR can repair all types of mismatches (although it is primarily replication associated) and can excise tracts up to a few kilobases long, whereas short-patch MMR handles only specific mismatches and removes lengths of around 10 nucleotides [190,195]. In either case, cMMR's high-fidelity polymerases fill the excised single-stranded tract and ligases seal the end. Although in bacteria successful MMR relies on the hemi-methylation of the DNA to differentiate between parent and daughter strand [196,197], it is not yet clear how the distinction is achieved in higher eukaryotes. Recent evidence suggests that ribonucleotide misincorporation into DNA may provide a strand-discrimination signal during MMR [198,199]. Noncanonical MMR (ncMMR) differs from cMMR largely by the recruitment of low-fidelity DNA polymerases, which enhance the mutagenesis of AID during the final stages of the repair process.

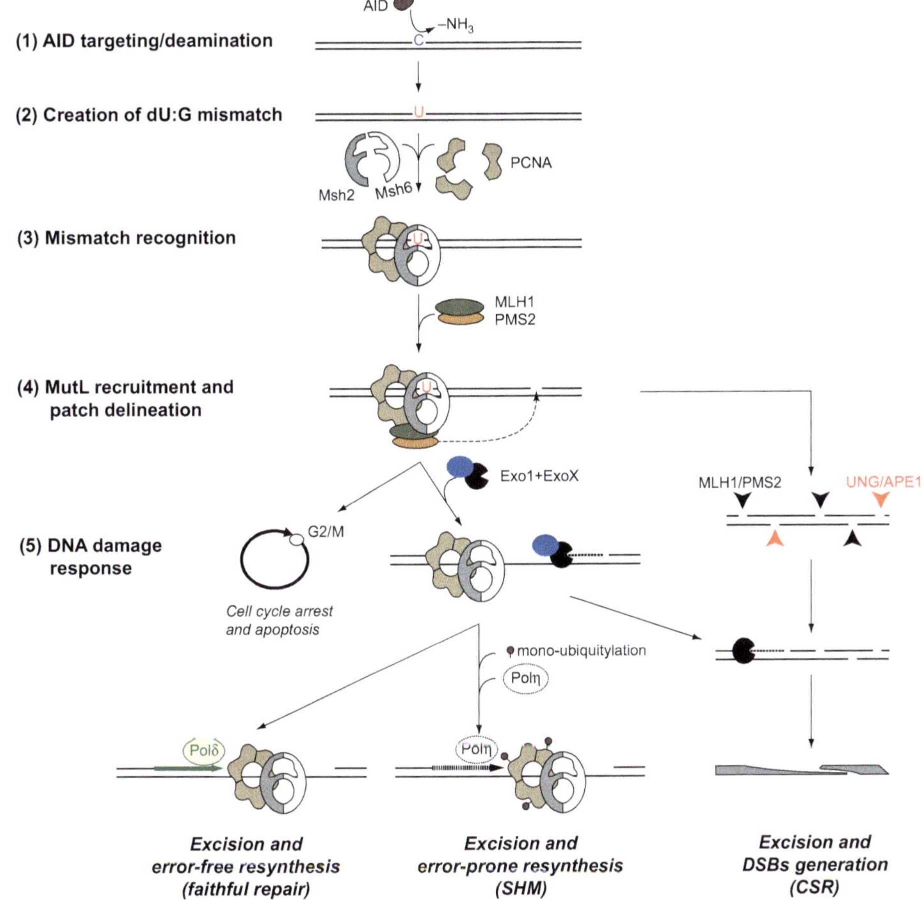


FIGURE 4 **The ncMMR pathway differentially mediates somatic hypermutation and class switch recombination.** The mutagenic process initiated by AID and cytosine deamination is followed by the recognition of the dU:dG mismatch by the MutSα (MSH2 and MSH6), which can work in synergy with PCNA during mismatch detections. After this initial sensing stage, MutSα can recruit MutLα (MLH1 and PMS2) to mediate: (1) cell cycle arrest and apoptosis if the damage is deemed irreparable (left); (2) cMMR in conjunction with Exo1 and an unknown 5′-3′ exonuclease to mediate ssDNA formation and resynthesis of DNA with error-free polymerases recruited by non-ubiquitinated PCNA (bottom left); (3) ncMMR in conjunction with, Exo1, and an unknown 5′-3′ exonuclease—assisted by the nicking activity of UNG/SMUG1 glycosylases and APE1 endonuclease—to promote the formation of DNA DSBs, which are then joined to downstream switch regions by NHEJ during CSR (bottom right). Notably, ncMMR may also (4) mediate SHM—independently of MutLα—to promote ssDNA formation and resynthesis of DNA by error-prone polymerases recruited by mono-ubiquitinated $PCNA^{K164}$ (bottom center).

## 5.1 Detecting dU:G Mismatch at V-Regions by MMR

At V-regions, much like throughout the genome, MMR recognizes single mismatched bases through the MSH2/MSH6 heterodimer complex, also called MutSα (Figure 3). Larger mismatches—conceivably caused by deamination of cytosine clusters or by insertions/deletions—are recognized by MSH2/MSH3 (MutSβ). However, the loss of MSH3 has little impact on SHM and CSR [200,201], which suggests that MutSβ does not have an important role in SHM (Table 1) although it has a functional role in MMR and DNA double-strand break repair in other instances [237]. To address this and to verify whether MSH2 and MSH6 could have independent functions in MMR, MSH2/MSH6 doubly defective and MSH2/MSH6/MSH3 triply defective mice were generated and analyzed for SHM [202]. Because the SHM and CSR phenotypes of these mice were indistinguishable from the single *Msh2*$^{-/-}$ or *Msh6*$^{-/-}$ mice [202], it was concluded that: (1) MSH6 does not have functions distinct from its role in the MutSα complex in SHM [202]; and that (2) the heterodimer formed by MSH2 and MSH6, but not MSH3, is critical for detecting AID-generated mismatches at the Ig locus and has a major role activating the downstream mechanisms that introduce many mutations at the dA:dT bases surrounding the initial dU:dG mismatch (Figure 3).

Some studies have suggested that MutSα detects mismatches by constantly scanning the genome; others propose binding DNA in situ as mismatches arise [187,195,238–240]. Because both Msh2 and Msh6 can bind PCNA, it is believed that PCNA might also be involved in the mismatch scanning process at V-regions, although this remains to be confirmed. However, studies have shown that PCNA has additional downstream effects that seem to regulate the recruitment of error-free (cMMR) versus error-prone (ncMMR) polymerases through ubiquitination (discussed subsequently). At the molecular level, MSH2 and MSH6 have an array of protein domains that cooperate to detect dU:dG mismatches and signal downstream effectors (reviewed in Refs [239,241]). The most relevant domains at this stage are the DNA-binding, PCNA-binding, and ATPase domains present in MSH2 and MSH6. Notably, MSH6 has two additional closely spaced protein regions at its flexible N-terminus [241]. The first is a region including an amino acid involved in direct DNA mismatch recognition (MSH6$^{E433}$) and that is separate from MSH6's general DNA-binding ability [242]. The other is a PWWP motif [243] that has recently been shown to bind histone H3K36-trimethyl mark (H3K36me3) [244]. Because H3K36me3 is a mark for active transcription, and because V-region mutation depends on RNA Pol II processivity, it is likely that the MSH6 mismatch recognition site and the PWWP domain have important roles in early mismatch detection. It is hoped that future studies will better address the contribution of these particular protein regions that are not shared by MSH2.

Active sensing of a dU:dG mismatch causes a conformational change in MutSα and triggers a cascade of ATP-dependent events [190,195]. Both MSH2 and MSH6 have non-redundant ATPase domains (Table 1) essential for MMR. Activated mouse B-cells harboring MSH2$^{G674A}$ or MSH6$^{T1217D}$ ATPase point mutations modeled after human colorectal cancer patients [245,246] have varying reductions in SHM compared with MSH2 or Msh6 null B-cells, respectively (Table 1) [208,209]. Neither of these mutations affects protein levels, DNA binding, or mismatch recognition, but they interfere with the processing of ATP hydrolysis [247]. The fact that mutations in the ATPase sites of MSH2 and MSH6 lead to a decrease in SHM confirms that the downstream events triggered by the conformational changes in MutSα are required for SHM. Like *Msh2*$^{-/-}$ mice, V-region mutations in *Msh2*$^{G674A/G674A}$ mice were primarily in G:C pairs within AID hot spot motifs and there was a significant decrease in dA:dT mutations. Similarly, *Msh6*$^{T1217D/T1217D}$ and *Msh6*$^{-/-}$ mice demonstrate a decrease in SHM largely attributable to a loss in dA:dT mutations. However, the actual decrease in SHM in *Msh6*$^{T1217D/T1217D}$ mice is intermediary between the wild-type and the knockout, which suggests that the MSH6 protein has an ATP-independent function that does not require the repair of the mutation, but is possibly involved in the early stages of mismatch detection, as discussed earlier. Interestingly, in *Msh6*$^{-/-}$ B-cells, mutations in V-regions are more focused than in wild-type mice or *Msh2*$^{-/-}$ mice, with fewer hot spots being targeted. In addition, the actual hot spots that are mutated in dG:dC in *MSH6*$^{T1217D/T1217D}$ mice are different from those observed in *Msh6*$^{-/-}$ or wild-type mice and more AID cold spots were targeted, some of which do not undergo mutation in either wild-type or knockout mice [209]. Because the MSH6$^{T1217D}$ protein remains associated with DNA in vitro, and there is a dominant negative phenotype for cancer [248], one of many possibilities is that this mutant protein binds to AID-induced mismatches and blocks subsequent deamination of neighboring bases by AID in the next round of AID-induced mutations. Recent studies on the possible functional interaction of AID with MSH2 and MSH6 seem to support some sort of role for MutSα in recruiting AID to hot spots [176]. These MSH2 and MSH6 ATPase point mutations are also informative because they retained the ability of MutSα to signal to the cell if too much damage had occurred, for example, from exposure to DNA damaging agents, and to trigger the G2/M check point and induce apoptosis [247,248]. These separation-of-function mutations revealed that MMR was not modifying SHM by signaling for apoptosis. Furthermore, because cells that lacked MutSα do not differ from wild-type cells in their speed of replication as assayed by CFSE in primary B-cells stimulated to isotype switch ex vivo [209,249], it is unlikely

**TABLE 1** MMR and BER Murine Mutations in SHM and CSR

| | | Factor | Mutation | Affected Domain(s) | Affected Binding Partners | SHM G:C | SHM A:T | CSR Frq | CSR S–S | Refs |
|---|---|---|---|---|---|---|---|---|---|---|
| MMR | MutSα | MSH2 | $Msh2^{-/-}$ | DNA; PIP box; Msh3/6; ATPase | All | ↑ | ⇊ | ↓ | ↓ | [202–207] |
| | | | $Msh2^{G674A}$ | ATPase | Unk | ↑ | ⇊ | ↓ | ↑ | [208] |
| | | MSH6 | $Msh6^{-/-}$ | PIP box; DNA; Msh2; ATPase | All | ↑ | ⇊ | ↓ | = | [200,201] |
| | | | $Msh6^{T1217D}$ | ATPase | Unk | ↑ | ⇊ | ↓ | = | [209] |
| | | MSH2–MSH6 | $Msh2^{-/-}$–$Msh6^{-/-}$ | All | All | ↑ | ⇊ | ↓ | ↑ | [202] |
| | MSH3 | | $Msh3^{-/-}$ | DNA; Msh2; ATPase | All | = | = | = | = | [200,201] |
| | MSH2-MSH6-MSH3 | | $Msh2^{-/-}$-$Msh6^{-/-}$–$Msh3^{-/-}$ | All | All | ↑ | ⇊ | ↓ | ↓ | [200,201] |
| | MutLα | MLH1 | $Mlh1^{-/-}$ | ATPase; Msh2/6; MLH3/Pms1/2 | All | = | = | ↓ | ↑ | [203,210,211] |
| | | | $Mlh1^{G167R}$ | ATPase | NS | = | = | ↓ | ↑ | [212] |
| | | PMS2 | $Pms2^{-/-}$ | ATPase; Mlh1; endonuclease | All | = | = | ↓ | ↑ | [213] |
| | | | $Pms2^{E702K}$ | Endonuclease | NS | = | = | ↓ | = | [214] |
| | MLH3 | | $Mlh3^{-/-}$ | ATPase; Mlh1 | All | ↓ | ↑ | = | ↓ | [215,216] |
| | PMS1 | | $Pms1^{-/-}$ | ATPase; Mlh1; HMG-box (DNA binding) | All | NA | NA | NA | NA | NA |
| | PCNA | | $Pcna^{-/-}$ | DNA binding; protein binding | All | NA | NA | NA | NA | [180] |
| | | | $Pcna^{K164R}$ | Error-prone; protein binding | $Pol_\eta$; other | ↑ | ⇊ | ↓ | ↓ | [180,217] |
| | MBD4 | | $Mbd4^{-/-}$ | Mlh1; $^{me}$CpG; uracil N-glycosylase | All | = | = | = | NA | [218] |
| | EXO1 | | $Exo1^{-/-}$ | Exonuclease; scaffold | All | ↑ | ⇊ | ↓ | NA | [219] |
| | | | $Exo1^{E109K}$ | 5′-3′ exonuclease defective | NS | = | = | = | NA | [219] |
| | | | $Exo1^{del\text{-}ex6}$ | Exonuclease; scaffold | All | ↑ | ⇊ | ↓ | ↓ | [220] |
| | LIG1 | | $Lig1^{-/-}$ | DNA ligase; DNA binding | All | NA | NA | NA | NA | [221] |

Continued

**TABLE 1 MMR and BER Murine Mutations in SHM and CSR—cont'd**

| | Factor | Mutation | Affected Domain(s) | Affected Binding Partners | SHM | | CSR | | Refs |
|---|---|---|---|---|---|---|---|---|---|
| | | | | | G:C | A:T | Frq | S–S | |
| BER | UNG | $Ung^{-/-}$ | Uracil N glycosylase; others | All | ↑ | ↓ | ↓↓↓ | NA | [72,222] |
| | SMUG1 | $Smug1^{-/-}$ | Uracil N-glycosylase; others | All | NA | NA | ~↓ | NA | [222] |
| | TDG | $Tdg^{-/-}$ | Uracil N-glycosylase; others | All | NA | NA | NA | NA | [223] |
| | APE1 | $Ape1^{-/-}$ | AP endonuclease; others | All | = | = | ↓ | NA | [224–226] |
| | APE2 | $Ape2^{-/-}$ | AP endonuclease; others | All | = | = | = | NA | [224–226] |
| | POLβ | $Pol\beta^{-/-}$ | DNA polymerase; others | All | ~ | ~↑ | NA | NA | [227] |
| | POLλ | $Pol\lambda^{-/-}$ | DNA polymerase; others | All | ~ | ~ | NA | NA | [227] |
| | LIG1 | $Lig1^{-/-}$ | DNA ligase; DNA binding | All | NA | NA | NA | NA | [221,228] |
| | XRCC1 | $XRCC1^{+/-}$ | DNA ligase scaffold; others | All | = | = | ↑ | ↓ | [229] |
| | LIG3 | $Lig3^{-/-}$ | DNA ligase; DNA binding | All | NA | NA | NA | NA | [230,231] |
| MMR + BER | POLη | $Pol\eta^{-/-}$ | DNA polymerase; others | All | ↑↑ | ↓↓ | = | = | [232] |
| | POLζ | $Pol\zeta^{+/-}$ | DNA polymerase; others | All | = | = | = | NA | [233] |
| | REV1 | *Rev1* | DNA polymerase; others | All | = | = | ~↓ | NA | [234] |
| | MSH2-UNG | $Msh2^{-/-}$–$Ung^{-/-}$ | All | All | ↑↑↑ | ↓↓↓ | ↓↓↓ | NA | [235] |
| | MSH6-UNG | $Msh6^{-/-}$–$Ung^{-/-}$ | All | All | ↑↑↑ | ↓↓↓ | ↓↓↓ | NA | [235] |
| | PCNA$^{K164R}$-UNG | $Pcna^{K164R}$–$Ung^{-/-}$ | All | All | ↑↑↑ | ↓↓↓ | NA | NA | [235,236] |

G:C = mutations at G and C bases; A:T = mutations at A and T bases; Frq = frequency of CSR; S–S = switch–switch region micro-homologies; Unk: unknown; NA: not available; NS: none suspected; arrows indicate increase or decrease compared with wild-type cohorts; ~ signifies minor or nonsignificant changes observed; = signifies no difference.

that sensing of the AID-induced point mutations in the V- or switch regions perturbs B-cell replication.

## 5.2 Mismatch Repair Signaling during SHM

At least three diverging MMR signaling responses take place downstream of the sensor MutSα targeted Ig genes. The first is mediated by the MutLα complex and is effective at switch region repeats to promote DSB formation and CSR (Figure 4). The second takes place at V-regions during SHM and is orchestrated by PCNA mono-ubiquitination, which recruits error-prone polymerases to amplify the AID-initiated missense mutations (Figure 4). The third is the classical cMMR pathway, taking place globally throughout the genome [250–252] (Figure 4). The exact mechanisms for such signaling divergence remain under intense investigation [40,157,158,174,253–257].

The key adaptor molecules or matchmakers during cMMR are the heterodimers MLH1/PMS2 (MutLα), MLH1/PMS1 (MutLβ), or MLH1/MLH3 (MutLγ). Mutations in human MLH1 account for a significant portion of patients with HNPCC, which suggests that MLH1 is critical for cMMR. The MutL heterodimers typically relay recognition of the mismatch by MutSα to the factors that carry out excision and repair by DNA nucleases, polymerases, and other downstream elements [258,259] (Figure 4). Surprisingly, none of these MutL complexes has yet been attributed a role during SHM, despite a documented minor increase in mutations in MLH3 null B cells (Table 1) [215]. Even more confounding is why mice deficient in MLH1 or PMS2 (MutLα) have a significant defect in CSR but undergo normal SHM ([203,210,211,213]; Table 1). Because such phenotypes are in stark contrast to those attributed to the loss of either MSH2 or MSH6, these results suggest that MutLα could be a major contributor to the divergence of CSR and SHM signaling [239]. One possibility is that MutLα actively mediates the formation of DSBs, an essential intermediary product during CSR, but are dispensable for SHM, which relies on SSB formation instead [179,203,210,211]. Another possibility is that MutLα complex regulates the processivity of exonuclease resection in vivo. Interactome studies using human MutLα identified a number of proteins that are not fully characterized but could affect MMR signaling at the Ig locus [260].

Whereas MutLα orchestrates the signaling cascade toward CSR, PCNA seems to mediate the choice between cMMR and ncMMR during SHM by orchestrating the recruitment of high- or low-fidelity DNA polymerases. The discovery of error-prone polymerases in mammalian cells [261] and subsequent studies in yeast showing that PCNA ubiquitylation at K164 binds and orchestrates the recruitment of low-fidelity DNA polymerases to lesions, including dG:dU mismatches [262–266], lent further credence to this idea. This has been further supported by PCNA mutation studies in birds and mammals. Chicken DT40 cells [267–269] and murine B-cells [180,217] deficient in $PCNA^{K164}$ mono-ubiquitylation have a compromised SHM response and a decrease in V-region mutations (Figure 3). However, because dA:dT mutations are not completely abolished in $PCNA^{K164R}$ B cells, either PCNA mono-ubiquitination on alternative lysine residues can compensate for Polη recruitment [251,252], or other low-fidelity polymerases can compensate for Polη [233].

The pathways mediating PCNA mono- and poly-ubiquitylation and the relay of these modifications to downstream signaling events are just beginning to be understood. RAD6/RAD18-mediated, site-specific mono-ubiquitylation of PCNA at lysine residue 164 [263,270] and de-ubiquitylation primarily by USP1 [271] are thought to control the recruitment of low-fidelity polymerases such as Polη, which is a major contributor to SHM after the recognition of dU:dG by MutSα [272]. Interestingly, K63-linked poly-ubiquitylation of PCNA activates a proteasome-independent error-free pathway of DNA repair that remains to be elucidated and that might counteract dA:dT hypermutation [270,273,274]. This process requires the E2 ubiquitin ligases Ubc13/Mms2, but the known E3 ligases HLTF or SHPRH that can poly-ubiquitylate PCNA [275,276] are not essential for SHM, which suggests the existence of an alternative E3 ligase yet to be discovered [274].

## 5.3 Resolution of the dU:dG Mismatch and Generation of dA:dT Mutations

Before the ncMMR signaling cascade takes effect at V-regions, resection of ssDNA patches is an essential prerequisite for the subsequent introduction of dA:dT mutations. Resection is believed to be initiated by a DNA nicking activity 5′ of the mismatch. Bacterial MutH endonuclease introduces nicks in the strand that contains the mismatched base depending on its DNA methylation status, but a homolog of MutH in eukaryotes remains unidentified [190,238]. The biochemical discovery of a latent endonuclease activity in PMS2 (and MLH3) and the confirmation of its biological relevance in yeast [259,277] provide compelling evidence that single-stranded nicks precede resection in mammalian cells [258,259]. Resection is then mediated by 5′-3′ exonuclease activity that clears the mismatch as well as a long tract of undamaged DNA bases. It has long been thought that resection is mediated by the catalytic activity of EXO1, the only known MMR protein possessing a 5′-3′ exonuclease domain. In vitro studies previously supported this idea because the combination of MutSα and EXO1 was sufficient to trigger resection of an artificially synthesized DNA mismatch [278–280]. However, recent in vivo data using mice with EXO1 deletion (*Exo1*$^{-/-}$) or exonuclease deficient point mutation (*Exo1*$^{E109K/E109K}$) [219] provide evidence that although EXO1 protein is indeed required for

DNA resection in cMMR and SHM, EXO1 catalytic activity may be dispensable for both processes. This necessitates a revisit of the classical MMR paradigm and suggests that EXO1 might act as a structural non-catalytic adaptor during MMR and SHM. As of yet, it remains unclear how the resection is really mediated during SHM. Although biochemical and in vivo studies of human MMR suggested that DNA synthesis–driven strand displacement followed by endonuclease function of MutLα [214,281] could contribute to resection during cMMR, this cannot be true for SHM, because, as mentioned earlier, MutLα is dispensable for SHM. Another speculative possibility is that during SHM, EXO1 might act as a structural noncatalytic adapter for another 5′ to 3′ exonuclease, such as FAN1, which was identified in an MutLα interactome study [260,282]. Such an exonuclease could in turn mediate resection during SHM.

After the mediation of ssDNA gap formation, monoubiquitinated $PCNA^{K164}$ signals the recruitment of low-fidelity DNA polymerases that amplify dG:dC mutations initiated by AID at WRC hot spots to include dA:dT mutations primarily at Polη-mediated WA hot spots. Although reporter gene experiments suggest that the cellular environment of germinal center B-cells enhances dA:dT base mutations genome-wide, the frequency of mutations observed in V-regions remains significantly higher than at any other loci [283,284]. This seems to confirm that the MMR pathway is exceedingly mutagenic at AID-targeted loci, because it is not mutagenic, and in fact corrects mutations at a *lacI* transgene that is not targeted by AID [285]. This dichotomy in MMR activity might be attributed to peculiarities in the signaling cascades that accompany AID activity, genetic, and/or epigenetic activities at V-regions. In addition, MMR factors themselves display unusual levels of plasticity in B-cells to orchestrate and regulate the ultimate outcome of the repair processes triggered by AID. Alternatively, the requirement for mono-ubiquitinated $PCNA^{K164}$ and the translesional polymerase DNA Polη to generate dA:dT mutations within the V-region suggest that cMMR might convert to ncMMR during the synthesis phase of MMR, for example, by the introduction of a lesion that requires lesion bypass. This possibility is further explored subsequently.

### 5.4 Human Mutations in MMR Factors and Their Effect on SHM

As mentioned earlier, mutations in MMR proteins contribute to a spectrum of colorectal cancers in humans. Often, the mutations occur sporadically within intestinal cells, but they can also be inherited in a homozygous manner throughout the body [194,246,286]. Surprisingly, modeling the same mutations in mice often leads to lymphomas, and mice usually die before they can develop colorectal tumors [287]. This unexpected difference in phenotype between human and murine genetics has been under intense investigation in recent years. Interestingly, a conditional deletion of a floxed Msh2 allele, using villin-Cre, specifically in the intestine of mice recapitulated many of the features of human colorectal tumors [287]. This suggests that although the broad signaling cascades affecting tumorigenesis are conserved within mammals, it is likely that different cell types in different species, or even within the same species, might have subtle differences in their thresholds to withstand the loss of MMR proteins. Conversely, whereas MMR loss in mice shows significant reductions in SHM and CSR, human patients with similar mutations did not manifest severe immunodeficiencies. Yet, a recent report suggested that MMR mutations in humans might be associated in the pathogenesis of common variable immunodeficiency diseases [288].

## 6. BASE EXCISION REPAIR IN SOMATIC HYPERMUTATION

Canonical BER (cBER) is a cellular mechanism that faithfully repairs DNA base modifications, abasic sites, and single-stranded nicks. Base excision repair involves flipping the mutated base out of the DNA helix and attempting to repair the base alone [289–291]. There are two key enzymes used in this process: DNA glycosylases and AP endonucleases. A large family of DNA N-glycosylases exists and each member has the ability to predominantly recognize specific base lesions, many of which are generated during normal cellular metabolism, including oxidative damage. For example, the TDG glycosylase detects and removes T:G mismatches and Ogg1 removes 8-oxoG from DNA caused by oxidative damage (i.e., reactive oxygen species) of dG. N-Glycosylases are used to break the β-N glycosidic bond to create an AP site which AP endonucleases recognize and subsequently nick the damaged DNA 5′ of the AP site, creating a free 3′ hydroxyl group. Subsequent reactions involving high-fidelity DNA polymerases can fill in a single nucleotide (short-patch BER) or few nucleotides (long-patch BER). This is followed by sealing of the newly synthesized bases by DNA ligases [292,293]. Non-canonical BER (ncBER) differs from cBER largely by the recruitment of low-fidelity DNA polymerases that enhance the mutagenesis of AID. The signaling in ncBER is elaborated subsequently.

### 6.1 Detection of the dU:dG Mismatch at V-Regions by BER

Parallel to the detection of AID-induced dU:dG mismatches by MutSα, BER has the ability to scan the DNA for uracil residues using four different uracil N-glycosylase proteins: uracil N-glycosylase (UNG or UDG), single-stranded-selective monofunctional UDG1 (SMUG1), thymidine DNA glycosylase (TDG or MUG), and methyl-CpG binding endonuclease (MED1 or MDB4). To varying degrees, all four of these N-glycosylases have the ability to detect and cleave uracil from DNA. However, genetic studies have established the dominant involvement of UNG in the process of antibody diversification [189,235,294,295] and excluded a significant involvement of MED1/MBD4, despite its direct interaction with MLH1 and its recognition of dU:dG mismatches [218]. On the other hand, SMUG1 has been attributed a marginal role in uracil excision as a UNG backup [222,296]. The role of TDG in SHM has not been directly tested because TDG null mice are embryonic lethal [223]. However, given the dominant effect of UNG and the complementary effect of SMUG1 in antibody diversification, it is unlikely that TDG has a significant contribution in detecting uracil residues in V-regions [189].

Like MutSα, UNG interacts with RPA [171]; however, it is still unclear whether UNG's DNA scanning ability is RPA dependent. In stark contrast to MMR, though, UNG cleaves dU from either ssDNA or dsDNA irrespective of whether it creates a mismatch. In effect, UNG can equally detect dU, dU:dG and dU:dA [297,298]. Initially, UNG binds nonspecifically to the DNA strand containing a suspected uracil residue. It then forms polar interactions with the DNA backbone through its proline-rich and glycine–serine loops that flank the uracil-binding motif (GVLLLN), which are all positioned medially in the primary protein sequence. This compression brings the suspected base into closer contact with UNG, whose C-terminal intercalation loop then pushes into the DNA's minor groove to flip the suspected base out of the helix. In doing so, the dU becomes fully accessible to the specific binding of UNG's uracil-binding motif, hence confirming and completing the detection process. Finally UNG's N-terminal water-activating loop mediates the cleavage of the uracil base, leaving the DNA with an abasic site in one strand [297,298].

## 6.2 Signaling BER at the Ig Locus

Once an abasic site is created, two BER-associated AP endonucleases (APE1 and APE2) can detect the kinked DNA gap. It has been a matter of debate whether both APE1 and APE2 are involved in SHM and CSR; this is difficult to resolve because APE1 endonuclease is embryonic lethal in mice [224,225]. However, recent studies in CH12F3 murine B-cells showed that only APE1 is involved in CSR [226]. APE1 endonucleases then trigger the cleavage of the phosphodiester DNA backbone to produce a single-strand break (SSB) with a free 5′ phosphate and a 3′ aldehyde group, which is further processed before DNA end ligation. The SSB byproduct is toxic particularly to B-cells undergoing replication and is further processed by proteins recruited by APE1 such as FEN1, PCNA, XRCC1, and DNA ligase I [299–303]. Because of its broad scaffolding activity, APE1 seems to be critical in mediating the BER signaling cascade at V-regions, although a role for its homolog APE2 has not been ruled out. Although it has not yet been shown, it is likely that, much like ncMMR, ncBER could be propagated by the recruitment of either non-modified or mono-ubiquitinated PCNA, which in turn could recruit high- or low-fidelity DNA polymerases, respectively.

## 6.3 Resolution of the dU:dG Mismatch via BER

Although not conclusively proven at V-regions, it is believed that a DNA lyase activity cleaves one base (short-patch BER) or the 5′-3′ exonuclease FEN1 cleaves two to 10 bases (long-patch BER) to create an ssDNA gap. Usually during short-patch cBER, polβ and/or polλ synthesize the missing base while DNA ligase I seals the end. During long-patch cBER, polδ and polε along with PCNA will fill the gap as ligase III and its cofactor XRCC1 seal the ends. It is conceivable that short-patch ncBER is propagated by polβ and/or polλ, whereas long-patch ncBER is mediated by mono-ubiquitinated PCNA, which instead recruits Polη to mediate dA:dT mutations. Recent studies on $Pol\beta^{-/-}$ and $Pol\lambda^{-/-}$ singly and doubly deficient mice showed that either or both polymerases result in a modest but significant decrease in V-region SHM but without specificity, which suggests that these polymerases may not directly contribute to SHM but indirectly affect GC B-cell viability or proliferation. As a result, short-patch BER is unlikely to be important during V-region SHM [227]. On the other hand, XRCC1 seems to have a negative role in signaling the BER pathway to promote SHM, because XRCC1 heterozygous mice have a higher frequency for SHM than wild-type cohorts [229]. This might further illustrate the reason behind the minimal contribution of BER in SHM.

## 6.4 Human Mutations in BER Factors and Their Effect on SHM

Unlike MMR proteins, mutations in UNG in both mice and humans show comparable immunodeficiencies in the form of a hyper-IgM syndrome type5 (HIGM5). This marked phenotype is also comparable to the loss of AID, which causes hyper-IgM syndrome type 2 (HIGM2) in humans. However, the HIGM phenotype of UNG is mainly mediated by a profound defect in CSR and marginal or minimal impact on SHM. In addition, to the best of our knowledge, no other BER protein has been attributed to HIGM or any other human immunodeficiency syndromes. We note this because UNG's profound immunodeficiency often leads to the assumption that BER is the most critical DNA repair mechanism for both SHM and CSR. In fact, UNG is critical mainly for CSR [304], but not in a redundant manner with MMR proteins, which are still the major DDR mechanism for generating dA:dT mutations in SHM and are also responsible for up to 70% of CSR, which suggests that BER and MMR might work synergistically to propagate CSR and perhaps antagonistically during SHM. That UNG$^{-/-}$ mice have increased hypermutation frequency and increased antibody affinity seems to support such an argument [304].

## 6.5 Crosstalk between MMR and BER during SHM

Initial studies suggested that dA:dT mutations observed in the V-region could be attributed to error-prone MMR and long-patch BER [53,71,72,200,204,220]. However, further analysis of UNG-deficient mice showed minimal effects on dA:dT mutations [72], which suggests that ncMMR predominates during SHM. In fact, MSH2-, MSH6-, and EXO1-deficient B-cells lack 80–90% of all A:T mutations [200,204,220]. Yet, analysis of *Msh2/Ung* [71] and *Msh6/Ung* [235] double-deficient mice revealed that the residual 10–20% of dA:dT mutations could be attributed to UNG-dependent repair. Therefore, an alternative long-patch error-prone BER may be responsible for some of the A:T mutations of SHM. What is not yet fully understood is whether both MMR and BER mechanisms collaborate or compete in resolving AID-generated dU:dG mispairs. It is possible that the relative roles of MMR and BER may vary during different phases of the cell cycle. Recently, it was suggested that not only are both pathways noncompetitive, that they might even collaborate [81,236]. This collaboration will create situations in which cMMR-mediated resynthesis of an ssDNA gap can encounter a BER-generated abasic site on the complementary strand, leading to conversion to the ncMMR and mutations at dA:dT sites. These observations suggest that ncBER/MMR cooperation might be responsible for some of the dA:dT mutations that arise after AID-induced mutations [202].

# 7. CONCLUSION

Antibody B-cells have evolved an extraordinary set of mechanisms to increase the affinity and further diversify the binding of the large repertoire of antigen-binding sites generated by V(D)J rearrangement. This process of somatic hypermutation is initiated by the highly mutagenic properties of AID that likely mutate dC to dU in single-stranded DNA and then are expanded through the recruitment of noncanonical forms of BER and MMR that are error-prone. Although much of the biochemistry of this hypermutational process is now largely understood and described in this chapter, there is still much to be learned, such as how the SHM process is mostly restricted to the immunoglobulin genes and differentially targeted to the variable and switch regions of the Ig locus. These types of questions should keep researchers engaged in this field of study for the foreseeable future.

# REFERENCES


[1] Glanville J, Zhai W, Berka J, Telman D, Huerta G, Mehta GR, et al. Precise determination of the diversity of a combinatorial antibody library gives insight into the human immunoglobulin repertoire. Proc Natl Acad Sci USA 2009;106:20216–21.

[2] Zhang W, Bardwell PD, Woo CJ, Poltoratsky V, Scharff MD, Martin A. Clonal instability of V region hypermutation in the Ramos Burkitt's lymphoma cell line. Int Immunol 2001;13:1175–84.

[3] Bemark M, Neuberger MS. By-products of immunoglobulin somatic hypermutation. Genes Chromosomes Cancer 2003;38:32–9.

[4] Jolly CJ, Wagner SD, Rada C, Klix N, Milstein C, Neuberger MS. The targeting of somatic hypermutation. Semin Immunol 1996;8:159–68.

[5] Wagner SD, Milstein C, Neuberger MS. Codon bias targets mutation [letter]. Nature 1995;376:732.

[6] Zheng NY, Wilson K, Jared M, Wilson PC. Intricate targeting of immunoglobulin somatic hypermutation maximizes the efficiency of affinity maturation. J Exp Med 2005;201:1467–78.

[7] Lebecque SG, Gearhart PJ. Boundaries of somatic mutation in rearranged immunoglobulin genes: 5' boundary is near the promoter, and 3' boundary is approximately 1 kb from V(D)J gene. J Exp Med 1990;172:1717–27.

[8] Umar A, Schweitzer PA, Levy NS, Gearhart JD, Gearhart PJ. Mutation in a reporter gene depends on proximity to and transcription of immunoglobulin variable transgenes. Proc Natl Acad Sci USA 1991;88:4902–6.

[9] Both GW, Taylor L, Pollard JW, Steele EJ. Distribution of mutations around rearranged heavy-chain antibody variable-region genes. Mol Cell Biol 1990;10:5187–96.

[10] Longerich S, Tanaka A, Bozek G, Nicolae D, Storb U. The very 5' end and the constant region of Ig genes are spared from somatic mutation because AID does not access these regions. J Exp Med 2005;202:1443–54.

[11] Liu M, Duke JL, Richter DJ, Vinuesa CG, Goodnow CC, Kleinstein SH, et al. Two levels of protection for the B cell genome during somatic hypermutation. Nature 2008;451:841–5.

[12] Pasqualucci L, Migliazza A, Fracchiolla N, William C, Neri A, Baldini L, et al. BCL-6 mutations in normal germinal center B cells: evidence of somatic hypermutation acting outside Ig loci. Proc Natl Acad Sci USA 1998;95:11816–21.

[13] Pasqualucci L, Neri A, Baldini L, Dalla-Favera R, Migliazza A. BCL-6 mutations are associated with immunoglobulin variable heavy chain mutations in B-cell chronic lymphocytic leukemia. Cancer Res 2000;60:5644–8.

[14] Shen HM, Peters A, Baron B, Zhu X, Storb U. Mutation of BCL-6 gene in normal B cells by the process of somatic hypermutation of Ig genes. Science 1998;280:1750–2.

[15] Muschen M, Re D, Jungnickel B, Diehl V, Rajewsky K, Kuppers R. Somatic mutation of the CD95 gene in human B cells as a side-effect of the germinal center reaction. J Exp Med 2000;192:1833–40.

[16] Gordon MS, Kanegai CM, Doerr JR, Wall R. Somatic hypermutation of the B cell receptor genes B29 (Igβ, CD79b) and mb1 (Igα, CD79a). Proc Natl Acad Sci USA 2003;100:4126–31.

[17] Pasqualucci L, Neumeister P, Goossens T, Nanjangud G, Chaganti RS, Kuppers R, et al. Hypermutation of multiple proto-oncogenes in B-cell diffuse large-cell lymphomas. Nature 2001;412:341–6.

[18] Komeno Y, Kitaura J, Watanabe-Okochi N, Kato N, Oki T, Nakahara F, et al. AID-induced T-lymphoma or B-leukemia/lymphoma in a mouse BMT model. Leukemia 2010;24:1018–24.

[19] Kotani A, Okazaki IM, Muramatsu M, Kinoshita K, Begum NA, Nakajima T, et al. A target selection of somatic hypermutations is regulated similarly between T and B cells upon activation-induced cytidine deaminase expression. Proc Natl Acad Sci USA 2005;102:4506–11.

[20] Knight KL, Winstead CR. Generation of antibody diversity in rabbits. Curr Opin Immunol 1997;9:228–32.

[21] Ratcliffe MJ. Antibodies, immunoglobulin genes and the bursa of Fabricius in chicken B cell development. Dev Comp Immunol 2006;30:101–18.

[22] Reynaud CA, Anquez V, Grimal H, Weill JC. A hyperconversion mechanism generates the chicken light chain preimmune repertoire. Cell 1987;48:379–88.

[23] Tang ES, Martin A. NHEJ-deficient DT40 cells have increased levels of immunoglobulin gene conversion: evidence for a double strand break intermediate. Nucleic Acids Res 2006;34:6345–51.

[24] Cook AJ, Raftery JM, Lau KK, Jessup A, Harris RS, Takeda S, et al. DNA-dependent protein kinase inhibits AID-induced antibody gene conversion. PLoS Biol 2007;5:e80.

[25] Tang ES, Martin A. Immunoglobulin gene conversion: synthesizing antibody diversification and DNA repair. DNA Repair (Amst) 2007;6:1557–71.

[26] Chaudhuri J, Alt FW. Class-switch recombination: interplay of transcription, DNA deamination and DNA repair. Nat Rev Immunol 2004;4:541–52.

[27] Stavnezer J, Guikema JE, Schrader CE. Mechanism and regulation of class switch recombination. Annu Rev Immunol 2008;26:261–92.

[28] Yancopoulos GD, DePinho RA, Zimmerman KA, Lutzker SG, Rosenberg N, Alt FW. Secondary genomic rearrangement events in pre-B cells: VHDJH replacement by a LINE-1 sequence and directed class switching. Embo J 1986;5:3259–66.

[29] Stavnezer J, Radcliffe G, Lin YC, Nietupski J, Berggren L, Sitia R, et al. Immunoglobulin heavy-chain switching may be directed by prior induction of transcripts from constant-region genes. Proc Natl Acad Sci USA 1988;85:7704–8.

[30] White MB, Word CJ, Humphries CG, Blattner FR, Tucker PW. Immunoglobulin D switching can occur through homologous recombination in human B cells. Mol Cell Biol 1990;10:3690–9.

[31] Yasui H, Akahori Y, Hirano M, Yamada K, Kurosawa Y. Class switch from mu to delta is mediated by homologous recombination between sigma mu and sigma mu sequences in human immunoglobulin gene loci. Eur J Immunol 1989;19:1399–403.

[32] Muramatsu M, Sankaranand VS, Anant S, Sugai M, Kinoshita K, Davidson NO, et al. Specific expression of activation-induced cytidine deaminase (AID), a novel member of the RNA-editing deaminase family in germinal center B cells. J Biol Chem 1999;274:18470–6.

[33] Muramatsu M, Kinoshita K, Fagarasan S, Yamada S, Shinkai Y, Honjo T. Class switch recombination and hypermutation require activation-induced cytidine deaminase (AID), a potential RNA editing enzyme. Cell 2000;102:553–63.

[34] Harris RS, Sale JE, Petersen-Mahrt SK, Neuberger MS. AID is essential for immunoglobulin V gene conversion in a cultured B cell line. Curr Biol 2002;12:435–8.

[35] Arakawa H, Hauschild J, Buerstedde JM. Requirement of the activation-induced deaminase (AID) gene for immunoglobulin gene conversion. Science 2002;295:1301–6.

[36] Brar SS, Watson M, Diaz M. Activation-induced cytosine deaminase (AID) is actively exported out of the nucleus but retained by the induction of DNA breaks. J Biol Chem 2004;279:26395–401.

[37] Ito S, Nagaoka H, Shinkura R, Begum N, Muramatsu M, Nakata M, et al. Activation-induced cytidine deaminase shuttles between nucleus and cytoplasm like apolipoprotein B mRNA editing catalytic polypeptide 1. Proc Natl Acad Sci USA 2004;101:1975–80.

[38] McBride KM, Barreto V, Ramiro AR, Stavropoulos P, Nussenzweig MC. Somatic hypermutation is limited by CRM1-dependent nuclear export of activation-induced deaminase. J Exp Med 2004;199:1235–44.

[39] Shinkura R, Ito S, Begum NA, Nagaoka H, Muramatsu M, Kinoshita K, et al. Separate domains of AID are required for somatic hypermutation and class-switch recombination. Nat Immunol 2004;5:707–12.

[40] Barreto V, Reina-San-Martin B, Ramiro AR, McBride KM, Nussenzweig MC. C-terminal deletion of AID uncouples class switch recombination from somatic hypermutation and gene conversion. Mol Cell 2003;12:501–8.

[41] Revy P, Muto T, Levy Y, Geissmann F, Plebani A, Sanal O, et al. Activation-induced cytidine deaminase (AID) deficiency causes the autosomal recessive form of the Hyper-IgM syndrome (HIGM2). Cell 2000;102:565–75.

[42] Ta VT, Nagaoka H, Catalan N, Durandy A, Fischer A, Imai K, et al. AID mutant analyses indicate requirement for class-switch-specific cofactors. Nat Immunol 2003;4:843–8.

[43] Zhu Y, Nonoyama S, Morio T, Muramatsu M, Honjo T, Mizutani S. Type two hyper-IgM syndrome caused by mutation in activation-induced cytidine deaminase. J Med Dent Sci 2003;50:41–6.

[44] Bransteitter R, Pham P, Calabrese P, Goodman MF. Biochemical analysis of hypermutational targeting by wild type and mutant activation-induced cytidine deaminase. J Biol Chem 2004;279:51612–21.

[45] Geisberger R, Rada C, Neuberger MS. The stability of AID and its function in class-switching are critically sensitive to the identity of its nuclear-export sequence. Proc Natl Acad Sci USA 2009;106:6736–41.

[46] Teng B, Burant CF, Davidson NO. Molecular cloning of an apolipoprotein B messenger RNA editing protein. Science 1993;260:1816–9.

[47] Navaratnam N, Morrison JR, Bhattacharya S, Patel D, Funahashi T, Giannoni F, et al. The p27 catalytic subunit of the apolipoprotein B mRNA editing enzyme is a cytidine deaminase. J Biol Chem 1993;268:20709–12.

[48] Chen SH, Habib G, Yang CY, Gu ZW, Lee BR, Weng SA, et al. Apolipoprotein B-48 is the product of a messenger RNA with an organ-specific in-frame stop codon. Science 1987;238:363–6.

[49] Powell LM, Wallis SC, Pease RJ, Edwards YH, Knott TJ, Scott J. A novel form of tissue-specific RNA processing produces apolipoprotein-B48 in intestine. Cell 1987;50:831–40.

[50] Doi T, Kinoshita K, Ikegawa M, Muramatsu M, Honjo T. De novo protein synthesis is required for the activation-induced cytidine deaminase function in class-switch recombination. Proc Natl Acad Sci USA 2003;100:2634–8.

[51] Nagaoka H, Ito S, Muramatsu M, Nakata M, Honjo T. DNA cleavage in immunoglobulin somatic hypermutation depends on de novo protein synthesis but not on uracil DNA glycosylase. Proc Natl Acad Sci USA 2005;102:2022–7.

[52] Nonaka T, Doi T, Toyoshima T, Muramatsu M, Honjo T, Kinoshita K. Carboxy-terminal domain of AID required for its mRNA complex formation in vivo. Proc Natl Acad Sci USA 2009.

[53] Di Noia J, Neuberger MS. Altering the pathway of immunoglobulin hypermutation by inhibiting uracil-DNA glycosylase. Nature 2002;419:43–8.

[54] Neuberger MS, Harris RS, Di Noia J, Petersen-Mahrt SK. Immunity through DNA deamination. Trends Biochem Sci 2003;28:305–12.

[55] Poltoratsky V, Goodman MF, Scharff MD. Error-prone candidates vie for somatic mutation. J Exp Med 2000;192:F27–30.

[56] Martin A, Bardwell PD, Woo CJ, Fan M, Shulman MJ, Scharff MD. Activation-induced cytidine deaminase turns on somatic hypermutation in hybridomas. Nature 2002;415:802–6.

[57] Bransteitter R, Pham P, Scharff MD, Goodman MF. Activation-induced cytidine deaminase deaminates deoxycytidine on single-stranded DNA but requires the action of RNase. Proc Natl Acad Sci USA 2003;100:4102–7.

[58] Dickerson SK, Market E, Besmer E, Papavasiliou FN. AID mediates hypermutation by deaminating single stranded DNA. J Exp Med 2003;197:1291–6.

[59] Pham P, Bransteitter R, Petruska J, Goodman MF. Processive AID-catalysed cytosine deamination on single-stranded DNA simulates somatic hypermutation. Nature 2003;424:103–7.

[60] Sohail A, Klapacz J, Samaranayake M, Ullah A, Bhagwat AS. Human activation-induced cytidine deaminase causes transcription-dependent, strand-biased C to U deaminations. Nucleic Acids Res 2003;31:2990–4.

[61] Larijani M, Frieder D, Basit W, Martin A. The mutation spectrum of purified AID is similar to the mutability index in Ramos cells and in ung(-/-)msh2(-/-) mice. Immunogenetics 2005;56:840–5.

[62] Larijani M, Martin A. Single-stranded DNA structure and positional context of the target cytidine determine the enzymatic efficiency of AID. Mol Cell Biol 2007;27:8038–48.

[63] Larijani M, Petrov AP, Kolenchenko O, Berru M, Krylov SN, Martin A. AID associates with single-stranded DNA with high affinity and a long complex half-life in a sequence-independent manner. Mol Cell Biol 2007;27:20–30.

[64] Petersen-Mahrt SK, Harris RS, Neuberger MS. AID mutates *E. coli* suggesting a DNA deamination mechanism for antibody diversification. Nature 2002;418:99–104.

[65] Okazaki IM, Kinoshita K, Muramatsu M, Yoshikawa K, Honjo T. The AID enzyme induces class switch recombination in fibroblasts. Nature 2002;416:340–5.

[66] Martin A, Scharff MD. Somatic hypermutation of the AID transgene in B and non-B cells. Proc Natl Acad Sci USA 2002;99:12304–8.

[67] Yoshikawa K, Okazaki IM, Eto T, Kinoshita K, Muramatsu M, Nagaoka H, et al. AID enzyme-induced hypermutation in an actively transcribed gene in fibroblasts. Science 2002;296:2033–6.

[68] Poltoratsky VP, Wilson SH, Kunkel TA, Pavlov YI. Recombinogenic phenotype of human activation-induced cytosine deaminase. J Immunol 2004;172:4308–13.

[69] Di Noia JM, Neuberger MS. Immunoglobulin gene conversion in chicken DT40 cells largely proceeds through an abasic site intermediate generated by excision of the uracil produced by AID-mediated deoxycytidine deamination. Eur J Immunol 2004;34:504–8.

[70] Imai K, Slupphaug G, Lee WI, Revy P, Nonoyama S, Catalan N, et al. Human uracil-DNA glycosylase deficiency associated with profoundly impaired immunoglobulin class-switch recombination. Nat Immunol 2003;4:1023–8.

[71] Rada C, Di Noia JM, Neuberger MS. Mismatch recognition and uracil excision provide complementary paths to both Ig switching and the A/T-focused phase of somatic mutation. Mol Cell 2004;16:163–71.

[72] Rada C, Williams GT, Nilsen H, Barnes DE, Lindahl T, Neuberger MS. Immunoglobulin isotype switching is inhibited and somatic hypermutation perturbed in ung-deficient mice. Curr Biol 2002;12:1748–55.

[73] Saribasak H, Saribasak NN, Ipek FM, Ellwart JW, Arakawa H, Buerstedde JM. Uracil DNA glycosylase disruption blocks Ig gene conversion and induces transition mutations. J Immunol 2006;176:365–71.

[74] Maul RW, Saribasak H, Martomo SA, McClure RL, Yang W, Vaisman A, et al. Uracil residues dependent on the deaminase AID in immunoglobulin gene variable and switch regions. Nat Immunol 2011;12:70–6.

[75] Chaudhuri J, Tian M, Khuong C, Chua K, Pinaud E, Alt FW. Transcription-targeted DNA deamination by the AID antibody diversification enzyme. Nature 2003;422:726–30.

[76] Martin A, Scharff MD. AID and mismatch repair in antibody diversification. Nat Rev Immunol 2002;2:605–14.

[77] Ramiro AR, Stavropoulos P, Jankovic M, Nussenzweig MC. Transcription enhances AID-mediated cytidine deamination by exposing single-stranded DNA on the nontemplate strand. Nat Immunol 2003;4:452–6.

[78] Shen HM, Ratnam S, Storb U. Targeting of the activation-induced cytosine deaminase is strongly influenced by the sequence and structure of the targeted DNA. Mol Cell Biol 2005;25:10815–21.

[79] Besmer E, Market E, Papavasiliou FN. The transcription elongation complex directs activation-induced cytidine deaminase-mediated DNA deamination. Mol Cell Biol 2006;26:4378–85.

[80] Milstein C, Neuberger MS, Staden R. Both DNA strands of antibody genes are hypermutation targets. Proc Natl Acad Sci USA 1998;95:8791–4.

[81] Frieder D, Larijani M, Collins C, Shulman M, Martin A. The concerted action of Msh2 and UNG stimulates somatic hypermutation at A. T base pairs. Mol Cell Biol 2009;29:5148–57.

[82] Xue K, Rada C, Neuberger MS. The in vivo pattern of AID targeting to immunoglobulin switch regions deduced from mutation spectra in msh2-/- ung-/- mice. J Exp Med 2006;203:2085–94.

[83] Kettenberger H, Armache KJ, Cramer P. Complete RNA polymerase II elongation complex structure and its interactions with NTP and TFIIS. Mol Cell 2004;16:955–65.

[84] Andrecka J, Treutlein B, Arcusa MA, Muschielok A, Lewis R, Cheung AC, et al. Nano positioning system reveals the course of upstream and nontemplate DNA within the RNA polymerase II elongation complex. Nucleic Acids Res 2009;37:5803–9.

[85] Martinez-Rucobo FW, Sainsbury S, Cheung AC, Cramer P. Architecture of the RNA polymerase-Spt4/5 complex and basis of universal transcription processivity. Embo J 2011;30:1302–10.

[86] Klein BJ, Bose D, Baker KJ, Yusoff ZM, Zhang X, Murakami KS. RNA polymerase and transcription elongation factor Spt4/5 complex structure. Proc Natl Acad Sci USA 2011;108:546–50.

[87] Shen HM, Storb U. Activation-induced cytidine deaminase (AID) can target both DNA strands when the DNA is supercoiled. Proc Natl Acad Sci USA 2004;101:12997–3002.

[88] Liu LF, Wang JC. Supercoiling of the DNA template during transcription. Proc Natl Acad Sci USA 1987;84:7024–7.

[89] Rahmouni AR, Wells RD. Direct evidence for the effect of transcription on local DNA supercoiling in vivo. J Mol Biol 1992;223:131–44.

[90] Dayn A, Malkhosyan S, Mirkin SM. Transcriptionally driven cruciform formation in vivo. Nucleic Acids Res 1992;20:5991–7.

[91] Krasilnikov AS, Podtelezhnikov A, Vologodskii A, Mirkin SM. Large-scale effects of transcriptional DNA supercoiling in vivo. J Mol Biol 1999;292:1149–60.
[92] Ronai D, Iglesias-Ussel MD, Fan M, Li Z, Martin A, Scharff MD. Detection of chromatin-associated single-stranded DNA in regions targeted for somatic hypermutation. J Exp Med 2007;204:181–90.
[93] Parsa JY, Ramachandran S, Zaheen A, Nepal RM, Kapelnikov A, Belcheva A, et al. Negative supercoiling creates single-stranded patches of DNA that are substrates for AID-mediated mutagenesis. PLoS Genet 2012;8:e1002518.
[94] Zhao Y, Dunn-Walters DK, Barone F, Spencer J. Antisense transcripts of V(D)J rearrangements; artifacts caused by false priming? Mol Immunol 2009;46:2357–62.
[95] Rogozin IB, Kolchanov NA. Somatic hypermutagenesis in immunoglobulin genes. II. Influence of neighbouring base sequences on mutagenesis. Biochim Biophys Acta 1992;1171:11–8.
[96] Betz AG, Rada C, Pannell R, Milstein C, Neuberger MS. Passenger transgenes reveal intrinsic specificity of the antibody hypermutation mechanism: clustering, polarity, and specific hot spots. Proc Natl Acad Sci USA 1993;90:2385–8.
[97] Yelamos J, Klix N, Goyenechea B, Lozano F, Chui YL, Gonzalez Fernandez A, et al. Targeting of non-Ig sequences in place of the V segment by somatic hypermutation. Nature 1995;376:225–9.
[98] Yu K, Huang FT, Lieber MR. DNA substrate length and surrounding sequence affect the activation-induced deaminase activity at cytidine. J Biol Chem 2004;279:6496–500.
[99] MacCarthy T, Kalis SL, Roa S, Pham P, Goodman MF, Scharff MD, et al. V-region mutation in vitro, in vivo, and in silico reveal the importance of the enzymatic properties of AID and the sequence environment. Proc Natl Acad Sci USA 2009;106:8629–34.
[100] Kohli RM, Abrams SR, Gajula KS, Maul RW, Gearhart PJ, Stivers JT. A portable hot spot recognition loop transfers sequence preferences from APOBEC family members to activation-induced cytidine deaminase. J Biol Chem 2009;284:22898–904.
[101] Wang M, Rada C, Neuberger MS. Altering the spectrum of immunoglobulin V gene somatic hypermutation by modifying the active site of AID. J Exp Med 2010;207:141–53.
[102] Navaratnam N, Bhattacharya S, Fujino T, Patel D, Jarmuz AL, Scott J. Evolutionary origins of apoB mRNA editing: catalysis by a cytidine deaminase that has acquired a novel RNA-binding motif at its active site. Cell 1995;81:187–95.
[103] Scott J, Navaratnam N, Carter C. Molecular modelling of the biosynthesis of the RNA-editing enzyme APOBEC-1, responsible for generating the alternative forms of apolipoprotein B. Exp Physiol 1999;84:791–800.
[104] Hall RS, Fedorov AA, Xu C, Fedorov EV, Almo SC, Raushel FM. Three-dimensional structure and catalytic mechanism of cytosine deaminase. Biochemistry 2011;50:5077–85.
[105] Betts L, Xiang S, Short SA, Wolfenden R, Carter Jr CW. Cytidine deaminase. The 2.3 A crystal structure of an enzyme: transition-state analog complex. J Mol Biol 1994;235:635–56.
[106] Papavasiliou FN, Schatz DG. The activation-induced deaminase functions in a postcleavage step of the somatic hypermutation process. J Exp Med 2002;195:1193–8.
[107] Pham P, Calabrese P, Park SJ, Goodman MF. Analysis of a single-stranded DNA-scanning process in which activation-induced deoxycytidine deaminase (AID) deaminates C to U haphazardly and inefficiently to ensure mutational diversity. J Biol Chem 2011;286:24931–42.
[108] Chelico L, Pham P, Calabrese P, Goodman MF. APOBEC3G DNA deaminase acts processively 3' --> 5' on single-stranded DNA. Nat Struct Mol Biol 2006;13:392–9.
[109] Michael N, Shen HM, Longerich S, Kim N, Longacre A, Storb U. The e box motif CAGGTG enhances somatic hypermutation without enhancing transcription. Immunity 2003;19:235–42.
[110] Jacob J, Przylepa J, Miller C, Kelsoe G. In situ studies of the primary immune response to (4-hydroxy-3-nitrophenyl)acetyl. III. The kinetics of V region mutation and selection in germinal center B cells. J Exp Med 1993;178:1293–307.
[111] Jacob J, Kelsoe G, Rajewsky K, Weiss U. Intraclonal generation of antibody mutants in germinal centres. Nature 1991;354:389–92.
[112] Storb U, Shen HM, Nicolae D. Somatic hypermutation: processivity of the cytosine deaminase AID and error-free repair of the resulting uracils. Cell Cycle 2009;8:3097–101.
[113] Dancyger AM, King JJ, Quinlan MJ, Fifield H, Tucker S, Saunders HL, et al. Differences in the enzymatic efficiency of human and bony fish AID are mediated by a single residue in the C terminus modulating single-stranded DNA binding. FASEB J 2012;26:1517–25.
[114] Mu Y, Prochnow C, Pham P, Chen XS, Goodman MF. A structural basis for the biochemical behavior of activation-induced deoxycytidine deaminase class-switch recombination-defective hyper-IgM-2 mutants. J Biol Chem 2012;287:28007–16.
[115] Kobayashi M, Aida M, Nagaoka H, Begum NA, Kitawaki Y, Nakata M, et al. AID-induced decrease in topoisomerase 1 induces DNA structural alteration and DNA cleavage for class switch recombination. Proc Natl Acad Sci USA 2009;106:22375–80.
[116] Kobayashi M, Sabouri Z, Sabouri S, Kitawaki Y, Pommier Y, Abe T, et al. Decrease in topoisomerase I is responsible for activation-induced cytidine deaminase (AID)-dependent somatic hypermutation. Proc Natl Acad Sci USA 2011;108:19305–10.
[117] Yamane A, Resch W, Kuo N, Kuchen S, Li Z, Sun HW, et al. Deep-sequencing identification of the genomic targets of the cytidine deaminase AID and its cofactor RPA in B lymphocytes. Nat Immunol 2010;12:62–9.
[118] Rada C, Gonzalez-Fernandez A, Jarvis JM, Milstein C. The 5' boundary of somatic hypermutation in a V kappa gene is in the leader intron. Eur J Immunol 1994;24:1453–7.
[119] Peters A, Storb U. Somatic hypermutation of immunoglobulin genes is linked to transcription initiation. Immunity 1996;4:57–65.
[120] Storb U, Peters A, Klotz E, Kim N, Shen HM, Kage K, et al. Somatic hypermutation of immunoglobulin genes is linked to transcription. Curr Top Microbiol Immunol 1998;229:11–9.
[121] Tumas-Brundage K, Manser T. The transcriptional promoter regulates hypermutation of the antibody heavy chain locus. J Exp Med 1997;185:239–50.
[122] Fukita Y, Jacobs H, Rajewsky K. Somatic hypermutation in the heavy chain locus correlates with transcription. Immunity 1998;9:105–14.
[123] Bachl J, Carlson C, Gray-Schopfer V, Dessing M, Olsson C. Increased transcription levels induce higher mutation rates in a hypermutating cell line. J Immunol 2001;166:5051–7.
[124] Staszewski O, Baker RE, Ucher AJ, Martier R, Stavnezer J, Guikema JE. Activation-induced cytidine deaminase induces reproducible DNA breaks at many non-Ig Loci in activated B cells. Mol Cell 2011;41:232–42.

[125] Nambu Y, Sugai M, Gonda H, Lee CG, Katakai T, Agata Y, et al. Transcription-coupled events associating with immunoglobulin switch region chromatin. Science 2003;302:2137–40.

[126] Pavri R, Gazumyan A, Jankovic M, Di Virgilio M, Klein I, Ansarah-Sobrinho C, et al. Activation-induced cytidine deaminase targets DNA at sites of RNA polymerase II stalling by interaction with Spt5. Cell 2010;143:122–33.

[127] Basu U, Meng FL, Keim C, Grinstein V, Pefanis E, Eccleston J, et al. The RNA exosome targets the AID cytidine deaminase to both strands of transcribed duplex DNA substrates. Cell 2011;144: 353–63.

[128] Willmann KL, Milosevic S, Pauklin S, Schmitz KM, Rangam G, Simon MT, et al. A role for the RNA pol II-associated PAF complex in AID-induced immune diversification. J Exp Med 2012;209:2099–111.

[129] Kodgire P, Mukkawar P, Ratnam S, Martin TE, Storb U. Changes in RNA polymerase II progression influence somatic hypermutation of Ig-related genes by AID. J Exp Med 2013;210:1481–92.

[130] Haddad D, Oruc Z, Puget N, Laviolette-Malirat N, Philippe M, Carrion C, et al. Sense transcription through the S region is essential for immunoglobulin class switch recombination. EMBO J 2011;30:1608–20.

[131] Ji Y, Resch W, Corbett E, Yamane A, Casellas R, Schatz DG. The in vivo pattern of binding of RAG1 and RAG2 to antigen receptor loci. Cell 2010;141:419–31.

[132] Woo CJ, Martin A, Scharff MD. Induction of somatic hypermutation is associated with modifications in immunoglobulin variable region chromatin. Immunity 2003;19:479–89.

[133] Wang L, Wuerffel R, Feldman S, Khamlichi AA, Kenter AL. S region sequence, RNA polymerase II, and histone modifications create chromatin accessibility during class switch recombination. J Exp Med 2009;206:1817–30.

[134] Odegard VH, Kim ST, Anderson SM, Shlomchik MJ, Schatz DG. Histone modifications associated with somatic hypermutation. Immunity 2005;23:101–10.

[135] Bradley SP, Kaminski DA, Peters AH, Jenuwein T, Stavnezer J. The histone methyltransferase Suv39h1 increases class switch recombination specifically to IgA. J Immunol 2006;177:1179–88.

[136] Kaminski DA, Stavnezer J. Stimuli that enhance IgA class switching increase histone 3 acetylation at S α, but poorly stimulate sequential switching from IgG2b. Eur J Immunol 2007;37:240–51.

[137] Li Z, Luo Z, Scharff MD. Differential regulation of histone acetylation and generation of mutations in switch regions is associated with Ig class switching. Proc Natl Acad Sci USA 2004;101:15428–33.

[138] Ng HH, Robert F, Young RA, Struhl K. Targeted recruitment of Set1 histone methylase by elongating Pol II provides a localized mark and memory of recent transcriptional activity. Mol Cell 2003;11:709–19.

[139] Ruthenburg AJ, Allis CD, Wysocka J. Methylation of lysine 4 on histone H3: intricacy of writing and reading a single epigenetic mark. Mol Cell 2007;25:15–30.

[140] Stanlie A, Begum NA, Akiyama H, Honjo T. The DSIF subunits Spt4 and Spt5 have distinct roles at various phases of immunoglobulin class switch recombination. PLoS Genet 2012;8:e1002675.

[141] Daniel JA, Santos MA, Wang Z, Zang C, Schwab KR, Jankovic M, et al. PTIP promotes chromatin changes critical for immunoglobulin class switch recombination. Science 2010;329:917–23.

[142] Daniel JA, Nussenzweig A. Roles for histone H3K4 methyltransferase activities during immunoglobulin class-switch recombination. Biochim Biophys Acta 2012;1819:733–8.

[143] Begum NA, Stanlie A, Nakata M, Akiyama H, Honjo T. The histone chaperone Spt6 is required for activation-induced cytidine deaminase target determination through H3K4me3 regulation. J Biol Chem 2012;287:32415–29.

[144] Kato L, Begum NA, Burroughs AM, Doi T, Kawai J, Daub CO, et al. Nonimmunoglobulin target loci of activation-induced cytidine deaminase (AID) share unique features with immunoglobulin genes. Proc Natl Acad Sci USA 2012;109:2479–84.

[145] Wang L, Whang N, Wuerffel R, Kenter AL. AID-dependent histone acetylation is detected in immunoglobulin S regions. J Exp Med 2006;203:215–26.

[146] Begum NA, Kinoshita K, Kakazu N, Muramatsu M, Nagaoka H, Shinkura R, et al. Uracil DNA glycosylase activity is dispensable for immunoglobulin class switch. Science 2004;305:1160–3.

[147] Petersen S, Casellas R, Reina-San-Martin B, Chen HT, Difilippantonio MJ, Wilson PC, et al. AID is required to initiate Nbs1/γ-H2AX focus formation and mutations at sites of class switching. Nature 2001;414:660–5.

[148] Shen HM, Poirier MG, Allen MJ, North J, Lal R, Widom J, et al. The activation-induced cytidine deaminase (AID) efficiently targets DNA in nucleosomes but only during transcription. J Exp Med 2009;206:1057–71.

[149] Kodgire P, Mukkawar P, North JA, Poirier MG, Storb U. Nucleosome stability dramatically impacts the targeting of somatic hypermutation. Mol Cell Biol 2012;32:2030–40.

[150] Singh SK, Maeda K, Eid MM, Almofty SA, Ono M, Pham P, et al. GANP regulates recruitment of AID to immunoglobulin variable regions by modulating transcription and nucleosome occupancy. Nat Commun 2013;4:1830.

[151] O'Brien RL, Brinster RL, Storb U. Somatic hypermutation of an immunoglobulin transgene in kappa transgenic mice. Nature 1987;326:405–9.

[152] Rogerson B, Hackett Jr J, Peters A, Haasch D, Storb U. Mutation pattern of immunoglobulin transgenes is compatible with a model of somatic hypermutation in which targeting of the mutator is linked to the direction of DNA replication. EMBO J 1991;10:4331–41.

[153] Storb U, Peters A, Klotz E, Kim N, Shen HM, Hackett J, et al. Immunoglobulin transgenes as targets for somatic hypermutation. Int J Dev Biol 1998;42:977–82.

[154] Sharpe MJ, Milstein C, Jarvis JM, Neuberger MS. Somatic hypermutation of immunoglobulin kappa may depend on sequences 3' of C kappa and occurs on passenger transgenes. EMBO J 1991;10:2139–45.

[155] Gonzalez-Fernandez A, Milstein C. Analysis of somatic hypermutation in mouse Peyer's patches using immunoglobulin kappa light-chain transgenes. Proc Natl Acad Sci USA 1993;90:9862–6.

[156] Betz AG, Milstein C, Gonzalez-Fernandez A, Pannell R, Larson T, Neuberger MS. Elements regulating somatic hypermutation of an immunoglobulin kappa gene: critical role for the intron enhancer/matrix attachment region. Cell 1994;77:239–48.

[157] Bottaro A, Lansford R, Xu L, Zhang J, Rothman P, Alt FW. S region transcription per se promotes basal IgE class switch recombination but additional factors regulate the efficiency of the process. EMBO J 1994;13:665–74.

[158] Yang SY, Fugmann SD, Schatz DG. Control of gene conversion and somatic hypermutation by immunoglobulin promoter and enhancer sequences. J Exp Med 2006;203:2919–28.

[159] Perlot T, Alt FW, Bassing CH, Suh H, Pinaud E. Elucidation of IgH intronic enhancer functions via germ-line deletion. Proc Natl Acad Sci USA 2005;102:14362–7.

[160] Bachl J, Wabl M. Enhancers of hypermutation. Immunogenetics 1996;45:59–64.
[161] Lin MM, Green NS, Zhang W, Scharff MD. The effects of E μ, 3'α (hs 1,2) and 3'kappa enhancers on mutation of an Ig-VDJ-Cγ2a Ig heavy gene in cultured B cells. Int Immunol 1998;10:1121–9.
[162] Kothapalli NR, Collura KM, Norton DD, Fugmann SD. Separation of mutational and transcriptional enhancers in Ig genes. J Immunol 2011;187:3247–55.
[163] Dunnick WA, Collins JT, Shi J, Westfield G, Fontaine C, Hakimpour P, et al. Switch recombination and somatic hypermutation are controlled by the heavy chain 3' enhancer region. J Exp Med 2009;206:2613–23.
[164] Rouaud P, Vincent-Fabert C, Saintamand A, Fiancette R, Marquet M, Robert I, et al. The IgH 3' regulatory region controls somatic hypermutation in germinal center B cells. J Exp Med 2013;210(8):1501–7.
[165] Wuerffel R, Wang L, Grigera F, Manis J, Selsing E, Perlot T, et al. S-S synapsis during class switch recombination is promoted by distantly located transcriptional elements and activation-induced deaminase. Immunity 2007;27:711–22.
[166] Tanaka A, Shen HM, Ratnam S, Kodgire P, Storb U. Attracting AID to targets of somatic hypermutation. J Exp Med 2010;207:405–15.
[167] McDonald JJ, Alinikula J, Buerstedde JM, Schatz DG. A critical context-dependent role for E boxes in the targeting of somatic hypermutation. J Immunol 2013;191(4):1556–66.
[168] Chaudhuri J, Khuong C, Alt FW. Replication protein A interacts with AID to promote deamination of somatic hypermutation targets. Nature 2004;430:992–8.
[169] Basu U, Chaudhuri J, Alpert C, Dutt S, Ranganath S, Li G, et al. The AID antibody diversification enzyme is regulated by protein kinase A phosphorylation. Nature 2005;438:508–11.
[170] Cheng HL, Vuong BQ, Basu U, Franklin A, Schwer B, Astarita J, et al. Integrity of the AID serine-38 phosphorylation site is critical for class switch recombination and somatic hypermutation in mice. Proc Natl Acad Sci USA 2009;106(8):2717–22.
[171] Nagelhus TA, Haug T, Singh KK, Keshav KF, Skorpen F, Otterlei M, et al. A sequence in the N-terminal region of human uracil-DNA glycosylase with homology to XPA interacts with the C-terminal part of the 34-kDa subunit of replication protein A. J Biol Chem 1997;272:6561–6.
[172] Wang D, Bushnell DA, Huang X, Westover KD, Levitt M, Kornberg RD. Structural basis of transcription: backtracked RNA polymerase II at 3.4 angstrom resolution. Science 2009;324:1203–6.
[173] Jeevan-Raj BP, Robert I, Heyer V, Page A, Wang JH, Cammas F, et al. Epigenetic tethering of AID to the donor switch region during immunoglobulin class switch recombination. J Exp Med 2011;208:1649–60.
[174] Kuang FL, Luo Z, Scharff MD. H3 trimethyl K9 and H3 acetyl K9 chromatin modifications are associated with class switch recombination. Proc Natl Acad Sci USA 2009;106:5288–93.
[175] Xu Z, Fulop Z, Wu G, Pone EJ, Zhang J, Mai T, et al. 14-3-3 adaptor proteins recruit AID to 5'-AGCT-3'-rich switch regions for class switch recombination. Nat Struct Mol Biol 2010;17:1124–35.
[176] Ranjit S, Khair L, Linehan EK, Ucher AJ, Chakrabarti M, Schrader CE, et al. AID binds cooperatively with UNG and Msh2-Msh6 to Ig switch regions dependent upon the AID C terminus. J Immunol 2011;187:2464–75.
[177] Sale JE, Neuberger MS. TdT-accessible breaks are scattered over the immunoglobulin V domain in a constitutively hypermutating B cell line. Immunity 1998;9:859–69.
[178] Reina-San-Martin B, Difilippantonio S, Hanitsch L, Masilamani RF, Nussenzweig A, Nussenzweig MC. H2AX is required for recombination between immunoglobulin switch regions but not for intra-switch region recombination or somatic hypermutation. J Exp Med 2003;197:1767–78.
[179] Peled JU, Kuang FL, Iglesias-Ussel MD, Roa S, Kalis SL, Goodman MF, et al. The biochemistry of somatic hypermutation. Annu Rev Immunol 2008;26:481–511.
[180] Roa S, Avdievich E, Peled JU, Maccarthy T, Werling U, Kuang FL, et al. Ubiquitylated PCNA plays a role in somatic hypermutation and class-switch recombination and is required for meiotic progression. Proc Natl Acad Sci USA 2008;105:16248–53.
[181] Pavri R, Nussenzweig MC. AID targeting in antibody diversity. Adv Immunol 2011;110:1–26.
[182] Rosenberg BR, Papavasiliou FN. Beyond SHM and CSR: AID and related cytidine deaminases in the host response to viral infection. Adv Immunol 2007;94:215–44.
[183] Chaudhuri J, Basu U, Zarrin A, Yan C, Franco S, Perlot T, et al. Evolution of the immunoglobulin heavy chain class switch recombination mechanism. Adv Immunol 2007;94:157–214.
[184] Durandy A, Peron S, Taubenheim N, Fischer A. Activation-induced cytidine deaminase: structure-function relationship as based on the study of mutants. Hum Mutat 2006;27:1185–91.
[185] Xu Z, Fulop Z, Zhong Y, Evinger III AJ, Zan H, Casali P. DNA lesions and repair in immunoglobulin class switch recombination and somatic hypermutation. Ann N Y Acad Sci 2005;1050: 146–62.
[186] Pena-Diaz J, Bregenhorn S, Ghodgaonkar M, Follonier C, Artola-Boran M, Castor D, et al. Noncanonical mismatch repair as a source of genomic instability in human cells. Mol Cell 2012;47:669–80.
[187] Pena-Diaz J, Jiricny J. Mammalian mismatch repair: error-free or error-prone? Trends Biochem Sci 2012;37:206–14.
[188] Zlatanou A, Despras E, Braz-Petta T, Boubakour-Azzouz I, Pouvelle C, Stewart GS, et al. The hMsh2-hMsh6 complex acts in concert with monoubiquitinated PCNA and Pol eta in response to oxidative DNA damage in human cells. Mol Cell 2011;43:649–62.
[189] Kavli B, Otterlei M, Slupphaug G, Krokan HE. Uracil in DNA–general mutagen, but normal intermediate in acquired immunity. DNA Repair (Amst) 2007;6:505–16.
[190] Iyer RR, Pluciennik A, Burdett V, Modrich PL. DNA mismatch repair: functions and mechanisms. Chem Rev 2006;106:302–23.
[191] Kunkel TA, Erie DA. DNA mismatch repair. Annu Rev Biochem 2005;74:681–710.
[192] Hsieh P. DNA mismatch repair: Dr. Jekyll and Mr. Hyde? Mol Cell 2012;47:665–6.
[193] Vasen HF, Blanco I, Aktan-Collan K, Gopie JP, Alonso A, Aretz S, et al. Revised guidelines for the clinical management of Lynch syndrome (HNPCC): recommendations by a group of European experts. Gut 2013;62:812–23.
[194] Steinke V, Engel C, Buttner R, Schackert HK, Schmiegel WH, Propping P. Hereditary nonpolyposis colorectal cancer (HNPCC)/ Lynch syndrome. Dtsch Arztebl Int 2013;110:32–8.
[195] Jiricny J. Postreplicative mismatch repair. Cold Spring Harb Perspect Biol 2013;5:a012633.
[196] Harfe BD, Jinks-Robertson S. DNA polymerase zeta introduces multiple mutations when bypassing spontaneous DNA damage in *Saccharomyces cerevisiae*. Mol Cell 2000;6:1491–9.
[197] Harfe BD, Jinks-Robertson S. DNA mismatch repair and genetic instability. Annu Rev Genet 2000;34:359–99.

[198] Ghodgaonkar MM, Lazzaro F, Olivera-Pimentel M, Artola-Boran M, Cejka P, Reijns MA, et al. Ribonucleotides misincorporated into DNA act as strand-discrimination signals in eukaryotic mismatch repair. Mol Cell 2013;50:323–32.

[199] Lujan SA, Williams JS, Clausen AR, Clark AB, Kunkel TA. Ribonucleotides are signals for mismatch repair of leading-strand replication errors. Mol Cell 2013;50:437–43.

[200] Wiesendanger M, Kneitz B, Edelmann W, Scharff MD. Somatic mutation in MSH3, MSH6, and MSH3/MSH6-deficient mice reveals a role for the MSH2-MSH6 heterodimer in modulating the base substitution pattern. J Exp Med 2000;191:579–84.

[201] Li Z, Scherer SJ, Ronai D, Iglesias-Ussel MD, Peled JU, Bardwell PD, et al. Examination of Msh6- and Msh3-deficient mice in class switching reveals overlapping and distinct roles of MutS homologues in antibody diversification. J Exp Med 2004;200:47–59.

[202] Roa S, Li Z, Peled JU, Zhao C, Edelmann W, Scharff MD. MSH2/MSH6 complex promotes error-free repair of AID-induced dU: G mispairs as well as error-prone hypermutation of A: T sites. PLoS One 2010;5:e11182.

[203] Schrader CE, Edelmann W, Kucherlapati R, Stavnezer J. Reduced isotype switching in splenic B cells from mice deficient in mismatch repair enzymes. J Exp Med 1999;190:323–30.

[204] Rada C, Ehrenstein MR, Neuberger MS, Milstein C. Hot spot focusing of somatic hypermutation in MSH2-deficient mice suggests two stages of mutational targeting. Immunity 1998;9:135–41.

[205] Phung QH, Winter DB, Cranston A, Tarone RE, Bohr VA, Fishel R, et al. Increased hypermutation at G and C nucleotides in immunoglobulin variable genes from mice deficient in the MSH2 mismatch repair protein. J Exp Med 1998;187:1745–51.

[206] Ehrenstein MR, Neuberger MS. Deficiency in msh2 affects the efficiency and local sequence specificity of immunoglobulin class-switch recombination: parallels with somatic hypermutation. EMBO J 1999;18:3484–90.

[207] Min IM, Schrader CE, Vardo J, Luby TM, D'Avirro N, Stavnezer J, et al. The Smu tandem repeat region is critical for Ig isotype switching in the absence of Msh2. Immunity 2003;19:515–24.

[208] Martin A, Li Z, Lin D, Bardwell PD, Iglesias-Ussel MD, Edelmann W, et al. Msh2 ATPase activity is essential for somatic hypermutation at A-T basepairs and for efficient class switch recombination. J Exp Med 2003;198:1171–8.

[209] Li Z, Zhao C, Iglesias-Ussel MD, Polonskaya Z, Zhuang M, Yang G, et al. The mismatch repair protein Msh6 influences the in vivo AID targeting to the Ig locus. Immunity 2006;24:393–403.

[210] Schrader CE, Vardo J, Stavnezer J. Mlh1 can function in antibody class switch recombination independently of Msh2. J Exp Med 2003;12:12.

[211] Schrader CE, Vardo J, Stavnezer J. Role for mismatch repair proteins Msh2, Mlh1, and Pms2 in immunoglobulin class switching shown by sequence analysis of recombination junctions. J Exp Med 2002;195:367–73.

[212] Chahwan R, van Oers JM, Avdievich E, Zhao C, Edelmann W, et al. The ATPase activity of MLH1 is required to orchestrate DNA double-strand breaks and end processing during class switch recombination. J Exp Med 2012;209:671–8.

[213] Peron S, Metin A, Gardes P, Alyanakian MA, Sheridan E, Kratz CP, et al. Human PMS2 deficiency is associated with impaired immunoglobulin class switch recombination. J Exp Med 2008;205:2465–72.

[214] van Oers JM, Roa S, Werling U, Liu Y, Genschel J, Hou Jr H, et al. PMS2 endonuclease activity has distinct biological functions and is essential for genome maintenance. Proc Natl Acad Sci USA 2010.

[215] Li Z, Peled JU, Zhao C, Svetlanov A, Ronai D, Cohen PE, et al. A role for Mlh3 in somatic hypermutation. DNA Repair (Amst) 2006;5:675–82.

[216] Wu X, Tsai CY, Patam MB, Zan H, Chen JP, Lipkin SM, et al. A role for the MutL mismatch repair Mlh3 protein in immunoglobulin class switch DNA recombination and somatic hypermutation. J Immunol 2006;176:5426–37.

[217] Langerak P, Nygren AO, Krijger PH, van den Berk PC, Jacobs H. A/T mutagenesis in hypermutated immunoglobulin genes strongly depends on PCNAK164 modification. J Exp Med 2007;204: 1989–98.

[218] Bardwell PD, Martin A, Wong E, Li Z, Edelmann W, Scharff MD. Cutting edge: the G-U mismatch glycosylase methyl-CpG binding domain 4 is dispensable for somatic hypermutation and class switch recombination. J Immunol 2003;170:1620–4.

[219] Schaetzlein S, Chahwan R, Avdievich E, Roa S, Wei K, Eoff RL, et al. Mammalian Exo1 encodes both structural and catalytic functions that play distinct roles in essential biological processes. Proc Natl Acad Sci USA 2013;110:E2470–9.

[220] Bardwell PD, Woo CJ, Wei K, Li Z, Martin A, Sack SZ, et al. Altered somatic hypermutation and reduced class-switch recombination in exonuclease 1-mutant mice. Nat Immunol 2004;5:224–9.

[221] Bentley DJ, Harrison C, Ketchen AM, Redhead NJ, Samuel K, Waterfall M, et al. DNA ligase I null mouse cells show normal DNA repair activity but altered DNA replication and reduced genome stability. J Cell Sci 2002;115:1551–61.

[222] Kemmerich K, Dingler FA, Rada C, Neuberger MS. Germline ablation of SMUG1 DNA glycosylase causes loss of 5-hydroxymethyluracil- and UNG-backup uracil-excision activities and increases cancer predisposition of Ung-/-Msh2-/- mice. Nucleic Acids Res 2012;40:6016–25.

[223] Cortazar D, Kunz C, Selfridge J, Lettieri T, Saito Y, MacDougall E, et al. Embryonic lethal phenotype reveals a function of TDG in maintaining epigenetic stability. Nature 2011;470:419–23.

[224] Schrader CE, Guikema JE, Wu X, Stavnezer J. The roles of APE1, APE2, DNA polymerase beta and mismatch repair in creating S region DNA breaks during antibody class switch. Philos Trans R Soc Lond B Biol Sci 2009;364:645–52.

[225] Guikema JE, Linehan EK, Tsuchimoto D, Nakabeppu Y, Strauss PR, Stavnezer J, et al. APE1- and APE2-dependent DNA breaks in immunoglobulin class switch recombination. J Exp Med 2007;204:3017–26.

[226] Masani S, Han L, Yu K. Apurinic/apyrimidinic endonuclease 1 is the essential nuclease during immunoglobulin class switch recombination. Mol Cell Biol 2013;33:1468–73.

[227] Schrader CE, Linehan EK, Ucher AJ, Stavnezer J. DNA polymerases beta and lambda do not directly affect Ig variable region somatic hypermutation although their absence reduces the frequency of mutations. DNA Repair (Amst) 2013;12:1087–93.

[228] Bentley D, Selfridge J, Millar JK, Samuel K, Hole N, Ansell JD, et al. DNA ligase I is required for fetal liver erythropoiesis but is not essential for mammalian cell viability. Nat Genet 1996;13:489–91.

[229] Saribasak H, Maul RW, Cao Z, McClure RL, Yang W, McNeill DR, et al. XRCC1 suppresses somatic hypermutation and promotes alternative nonhomologous end joining in Igh genes. J Exp Med 2011;208:2209–16.

[230] Simsek D, Furda A, Gao Y, Artus J, Brunet E, Hadjantonakis AK, et al. Crucial role for DNA ligase III in mitochondria but not in Xrcc1-dependent repair. Nature 2011;471:245–8.

[231] Gao Y, Katyal S, Lee Y, Zhao J, Rehg JE, Russell HR, et al. DNA ligase III is critical for mtDNA integrity but not Xrcc1-mediated nuclear DNA repair. Nature 2011;471:240–4.

[232] Faili A, Aoufouchi S, Weller S, Vuillier F, Stary A, Sarasin A, et al. DNA polymerase eta is involved in hypermutation occurring during immunoglobulin class switch recombination. J Exp Med 2004;199:265–70.

[233] Saribasak H, Maul RW, Cao Z, Yang WW, Schenten D, Kracker S, et al. DNA polymerase zeta generates tandem mutations in immunoglobulin variable regions. J Exp Med 2012;209:1075–81.

[234] Krijger PH, Tsaalbi-Shtylik A, Wit N, van den Berk PC, de Wind N, Jacobs H. Rev1 is essential in generating G to C transversions downstream of the Ung2 pathway but not the Msh2+Ung2 hybrid pathway. Eur J Immunol 2013;43(10):2765–70.

[235] Shen HM, Tanaka A, Bozek G, Nicolae D, Storb U. Somatic hypermutation and class switch recombination in Msh6(-/-)Ung(-/-) double-knockout mice. J Immunol 2006;177:5386–92.

[236] Krijger PH, Langerak P, van den Berk PC, Jacobs H. Dependence of nucleotide substitutions on Ung2, Msh2, and PCNA-Ub during somatic hypermutation. J Exp Med 2009.

[237] van Oers JM, Edwards Y, Chahwan R, Zhang W, Smith C, et al. The MutSβ complex is a modulator of p53-driven tumorigenesis through its functions in both DNA double-strand break repair and mismatch repair. Oncogene 2013.

[238] Jiricny J. The multifaceted mismatch-repair system. Nat Rev Mol Cell Biol 2006;7:335–46.

[239] Chahwan R, Edelmann W, Scharff MD, Roa S. Mismatch-mediated error prone repair at the immunoglobulin genes. Biomed Pharmacother 2011;65:529–36.

[240] Chahwan R, Edelmann W, Scharff MD, Roa S. AIDing antibody diversity by error-prone mismatch repair. Semin Immunol 2012;24:293–300.

[241] Edelbrock MA, Kaliyaperumal S, Williams KJ. Structural, molecular and cellular functions of MSH2 and MSH6 during DNA mismatch repair, damage signaling and other noncanonical activities. Mutat Res 2013;743-744:53–66.

[242] Warren JJ, Pohlhaus TJ, Changela A, Iyer RR, Modrich PL, Beese LS. Structure of the human MutSα DNA lesion recognition complex. Mol Cell 2007;26:579–92.

[243] Laguri C, Duband-Goulet I, Friedrich N, Axt M, Belin P, Callebaut I, et al. Human mismatch repair protein MSH6 contains a PWWP domain that targets double stranded DNA. Biochemistry 2008;47:6199–207.

[244] Li F, Mao G, Tong D, Huang J, Gu L, Yang W, et al. The histone mark H3K36me3 regulates human DNA mismatch repair through its interaction with MutSα. Cell 2013;153:590–600.

[245] Peltomaki P. Lynch syndrome genes. Fam cancer 2005;4:227–32.

[246] Peltomaki P, Vasen H. Mutations associated with HNPCC predisposition – update of ICG-HNPCC/INSiGHT mutation database. Dis Markers 2004;20:269–76.

[247] Lin DP, Wang Y, Scherer SJ, Clark AB, Yang K, Avdievich E, et al. An Msh2 point mutation uncouples DNA mismatch repair and apoptosis. Cancer Res 2004;64:517–22.

[248] Yang G, Scherer SJ, Shell SS, Yang K, Kim M, Lipkin M, et al. Dominant effects of an Msh6 missense mutation on DNA repair and cancer susceptibility. Cancer Cell 2004;6:139–50.

[249] Martin A, Li Z, Lin DP, Bardwell PD, Iglesias-Ussel MD, Edelmann W, et al. Msh2 ATPase activity is essential for somatic hypermutation at A-T basepairs and for efficient class switch recombination. J Exp Med 2003;198:1171–8.

[250] Kleczkowska HE, Marra G, Lettieri T, Jiricny J. hMSH3 and hMSH6 interact with PCNA and colocalize with it to replication foci. Genes Dev 2001;15:724–36.

[251] Das-Bradoo S, Nguyen HD, Bielinsky AK. Damage-specific modification of PCNA. Cell Cycle 2010;9:3674–9.

[252] Das-Bradoo S, Nguyen HD, Wood JL, Ricke RM, Haworth JC, Bielinsky AK. Defects in DNA ligase I trigger PCNA ubiquitylation at Lys 107. Nat Cell Biol 2010;12:74–9. sup pp. 1–20.

[253] Ju Z, Volpi SA, Hassan R, Martinez N, Giannini SL, Gold T, et al. Evidence for physical interaction between the immunoglobulin heavy chain variable region and the 3' regulatory region. J Biol Chem 2007;282:35169–78.

[254] Garrett FE, Emelyanov AV, Sepulveda MA, Flanagan P, Volpi S, Li F, et al. Chromatin architecture near a potential 3' end of the igh locus involves modular regulation of histone modifications during B-Cell development and in vivo occupancy at CTCF sites. Mol Cell Biol 2005;25:1511–25.

[255] Manis JP, Gu Y, Lansford R, Sonoda E, Ferrini R, Davidson L, et al. Ku70 is required for late B cell development and immunoglobulin heavy chain class switching. J Exp Med 1998;187:2081–9.

[256] Sleckman BP, Gorman JR, Alt FW. Accessibility control of antigen-receptor variable-region gene assembly: role of cis-acting elements. Annu Rev Immunol 1996;14:459–81.

[257] McBride KM, Gazumyan A, Woo EM, Barreto VM, Robbiani DF, Chait BT, et al. Regulation of hypermutation by activation-induced cytidine deaminase phosphorylation. Proc Natl Acad Sci USA 2006;103:8798–803.

[258] Kadyrov FA, Dzantiev L, Constantin N, Modrich P. Endonucleolytic function of MutLα in human mismatch repair. Cell 2006;126: 297–308.

[259] Kadyrov FA, Holmes SF, Arana ME, Lukianova OA, O'Donnell M, Kunkel TA, et al. *Saccharomyces cerevisiae* MutLα is a mismatch repair endonuclease. J Biol Chem 2007;282:37181–90.

[260] Cannavo E, Gerrits B, Marra G, Schlapbach R, Jiricny J. Characterization of the interactome of the human MutL homologues MLH1, PMS1, and PMS2. J Biol Chem 2007;282:2976–86.

[261] Tippin B, Goodman MF. A new class of errant DNA polymerases provides candidates for somatic hypermutation. Philos Trans R Soc Lond B Biol Sci 2001;356:47–51.

[262] Friedberg EC. Reversible monoubiquitination of PCNA: a novel slant on regulating translesion DNA synthesis. Mol Cell 2006;22:150–2.

[263] Garg P, Burgers PM. Ubiquitinated proliferating cell nuclear antigen activates translesion DNA polymerases eta and REV1. Proc Natl Acad Sci USA 2005;102:18361–6.

[264] Lehmann AR, Niimi A, Ogi T, Brown S, Sabbioneda S, Wing JF, et al. Translesion synthesis: Y-family polymerases and the polymerase switch. DNA Repair (Amst) 2007;6:891–9.

[265] Acharya N, Brahma A, Haracska L, Prakash L, Prakash S. Mutations in the ubiquitin binding UBZ motif of DNA polymerase eta do not impair its function in translesion synthesis during replication. Mol Cell Biol 2007;27:7266–72.

[266] Haracska L, Unk I, Prakash L, Prakash S. Ubiquitylation of yeast proliferating cell nuclear antigen and its implications for translesion DNA synthesis. Proc Natl Acad Sci USA 2006;103:6477–82.

[267] Simpson LJ, Ross AL, Szuts D, Alviani CA, Oestergaard VH, Patel KJ, et al. RAD18-independent ubiquitination of proliferating-cell nuclear antigen in the avian cell line DT40. EMBO Rep 2006;7:927–32.

[268] Bachl J, Ertongur I, Jungnickel B. Involvement of Rad18 in somatic hypermutation. Proc Natl Acad Sci USA 2006;103:12081–6.

[269] Arakawa H, Moldovan GL, Saribasak H, Saribasak NN, Jentsch S, Buerstedde JM. A role for PCNA ubiquitination in immunoglobulin hypermutation. PLoS Biol 2006;4:e366.

[270] Hoege C, Pfander B, Moldovan GL, Pyrowolakis G, Jentsch S. RAD6-dependent DNA repair is linked to modification of PCNA by ubiquitin and SUMO. Nature 2002;419:135–41.

[271] Ulrich HD. Deubiquitinating PCNA: a downside to DNA damage tolerance. Nat Cell Biol 2006;8:303–5.

[272] Delbos F, Aoufouchi S, Faili A, Weill JC, Reynaud CA. DNA polymerase eta is the sole contributor of A/T modifications during immunoglobulin gene hypermutation in the mouse. J Exp Med 2007;204:17–23.

[273] Kannouche PL, Wing J, Lehmann AR. Interaction of human DNA polymerase eta with monoubiquitinated PCNA: a possible mechanism for the polymerase switch in response to DNA damage. Mol Cell 2004;14:491–500.

[274] Krijger PH, Lee KY, Wit N, van den Berk PC, Wu X, Roest HP, et al. HLTF and SHPRH are not essential for PCNA polyubiquitination, survival and somatic hypermutation: existence of an alternative E3 ligase. DNA Repair (Amst) 2011.

[275] Motegi A, Liaw HJ, Lee KY, Roest HP, Maas A, Wu X, et al. Polyubiquitination of proliferating cell nuclear antigen by HLTF and SHPRH prevents genomic instability from stalled replication forks. Proc Natl Acad Sci USA 2008;105:12411–6.

[276] Motegi A, Sood R, Moinova H, Markowitz SD, Liu PP, Myung K. Human SHPRH suppresses genomic instability through proliferating cell nuclear antigen polyubiquitination. J Cell Biol 2006;175:703–8.

[277] Erdeniz N, Nguyen M, Deschenes SM, Liskay RM. Mutations affecting a putative MutLα endonuclease motif impact multiple mismatch repair functions. DNA Repair (Amst) 2007;6:1463–70.

[278] Dzantiev L, Constantin N, Genschel J, Iyer RR, Burgers PM, Modrich P. A defined human system that supports bidirectional mismatch-provoked excision. Mol Cell 2004;15:31–41.

[279] Genschel J, Modrich P. Mechanism of 5'-directed excision in human mismatch repair. Mol Cell 2003;12:1077–86.

[280] Genschel J, Bazemore LR, Modrich P. Human exonuclease I is required for 5' and 3' mismatch repair. J Biol Chem 2002;277:13302–11.

[281] Kadyrov FA, Genschel J, Fang Y, Penland E, Edelmann W, Modrich P. A possible mechanism for exonuclease 1-independent eukaryotic mismatch repair. Proc Natl Acad Sci USA 2009;106:8495–500.

[282] Ciccia A, McDonald N, West SC. Structural and functional relationships of the XPF/MUS81 family of proteins. Annu Rev Biochem 2008;77:259–87.

[283] Ukai A, Ishimaru K, Ouchida R, Mori H, Kano C, Moritan T, et al. Induction of a:T mutations is dependent on cellular environment but independent of mutation frequency and target gene location. J Immunol 2008;181:7835–42.

[284] Ouchida R, Ukai A, Mori H, Kawamura K, Dolle ME, Tagawa M, et al. Genetic analysis reveals an intrinsic property of the germinal center B cells to generate A: T mutations. DNA Repair (Amst) 2008;7:1392–8.

[285] Green B, Belcheva A, Nepal RM, Boulianne B, Martin A. The mismatch repair pathway functions normally at a non-AID target in germinal center B cells. Blood 2011;118:3013–8.

[286] Edelmann L, Edelmann W. Loss of DNA mismatch repair function and cancer predisposition in the mouse: animal models for human hereditary nonpolyposis colorectal cancer. Am J Med Genet C Semin Med Genet 2004;129C:91–9.

[287] Kucherlapati MH, Lee K, Nguyen AA, Clark AB, Hou Jr H, Rosulek A, et al. An Msh2 conditional knockout mouse for studying intestinal cancer and testing anticancer agents. Gastroenterology 2010;138:993–1002 e1.

[288] Duvvuri B, Duvvuri VR, Grigull J, Martin A, Pan-Hammarstrom Q, Wu GE, et al. Altered spectrum of somatic hypermutation in common variable immunodeficiency disease characteristic of defective repair of mutations. Immunogenetics 2011;63:1–11.

[289] Zharkov DO. Base excision DNA repair. Cell Mol Life Sci 2008;65:1544–65.

[290] Krokan HE, Bjoras M. Base excision repair. Cold Spring Harb Perspect Biol 2013;5:a012583.

[291] David SS, O'Shea VL, Kundu S. Base-excision repair of oxidative DNA damage. Nature 2007;447:941–50.

[292] Hoeijmakers JH. Genome maintenance mechanisms for preventing cancer. Nature 2001;411:366–74.

[293] Wood RD. DNA repair in eukaryotes. Annu Rev Biochem 1996;65:135–67.

[294] Kavli B, Sundheim O, Akbari M, Otterlei M, Nilsen H, Skorpen F, et al. hUNG2 is the major repair enzyme for removal of uracil from U: a matches, U: G mismatches, and U in single-stranded DNA, with hSMUG1 as a broad specificity backup. J Biol Chem 2002;277:39926–36.

[295] Kavli B, Andersen S, Otterlei M, Liabakk NB, Imai K, Fischer A, et al. B cells from hyper-IgM patients carrying UNG mutations lack ability to remove uracil from ssDNA and have elevated genomic uracil. J Exp Med 2005;201:2011–21.

[296] Di Noia JM, Rada C, Neuberger MS. SMUG1 is able to excise uracil from immunoglobulin genes: insight into mutation versus repair. EMBO J 2006;25:585–95.

[297] Parikh SS, Putnam CD, Tainer JA. Lessons learned from structural results on uracil-DNA glycosylase. Mutat Res 2000;460:183–99.

[298] Zharkov DO, Mechetin GV, Nevinsky GA. Uracil-DNA glycosylase: structural, thermodynamic and kinetic aspects of lesion search and recognition. Mutat Res 2010;685:11–20.

[299] Bennett RA, Wilson III DM, Wong D, Demple B. Interaction of human apurinic endonuclease and DNA polymerase beta in the base excision repair pathway. Proc Natl Acad Sci USA 1997;94:7166–9.

[300] Dianova II , Bohr VA, Dianov GL. Interaction of human AP endonuclease 1 with flap endonuclease 1 and proliferating cell nuclear antigen involved in long-patch base excision repair. Biochemistry 2001;40:12639–44.

[301] Vidal AE, Boiteux S, Hickson ID, Radicella JP. XRCC1 coordinates the initial and late stages of DNA abasic site repair through protein-protein interactions. EMBO J 2001;20:6530–9.

[302] Ranalli TA, DeMott MS, Bambara RA. Mechanism underlying replication protein a stimulation of DNA ligase I. J Biol Chem 2002;277:1719–27.

[303] Ranalli TA, Tom S, Bambara RA. AP endonuclease 1 coordinates flap endonuclease 1 and DNA ligase I activity in long patch base excision repair. J Biol Chem 2002;277:41715–24.

[304] Zahn A, Daugan M, Safavi S, Godin D, Cheong C, Lamarre A, et al. Separation of function between isotype switching and affinity maturation in vivo during acute immune responses and circulating autoantibodies in UNG-deficient mice. J Immunol 2013;190:5949–60.

Chapter 21

# Aberrant AID Expression by Pathogen Infection

Atsushi Takai, Hiroyuki Marusawa and Tsutomu Chiba
*Department of Gastroenterology and Hepatology, Graduate School of Medicine, Kyoto University, Kyoto, Japan*

## 1. INTRODUCTION

The hallmark of acquired immunity is the specificity for various pathogenic antigens. One mechanism that contributes to this specificity is the generation of diverse antigen receptors on both T and B cells. Germline antigen receptor genes comprise a collection of variable (V), diverse (D), and joining (J) regions. These regions are randomly assembled and form unique V(D)J recombinations for each lymphocyte, thereby contributing to the production of unlimited diverse antigen receptors [1,2]. In addition, mature B cells have other mechanisms for producing a variety of antibodies after antigens are recognized by the receptors, called somatic hypermutation (SHM) and class-switch recombination (CSR), that contribute to diversification of the immunoglobulin (Ig) gene. Thus, DNA mutations induced at specific sequences play a central role in producing a variety of antibodies. Although this radical mechanism should be regulated to work only at Ig genes, "off-target" mutations also exist at least in B cells.

In contrast, it is now widely accepted that cancer is a genetic disease resulting from the accumulation of genetic alterations in several critical genes that are responsible for cell growth or cell death [3]. Sjoblom et al. [4] showed that a single cancer cell generally contains 70–90 nucleotide alterations, 10–15 of which affect genes that are closely associated with cancer development. Several exogenous altering mechanisms, such as irradiation and DNA-damaging chemicals, are proposed causes of carcinogenesis, but they play a main role in only a few types of cancer. Alternative mechanisms for mutagenesis that cause cancer development should account for a large number of genetic alterations observed in cancer cells. Accumulating evidence suggests that a common molecule is involved in both the generation of genetic diversity in Ig genes for producing a variety of antibodies in B cells and the induction of genetic aberrations in tumor-related genes required for tumorigenesis associated with pathogen infection.

## 2. PHYSIOLOGIC ROLE OF ACTIVATION-INDUCED CYTIDINE DEAMINASE

SHM and CSR are unique genetic events specifically observed in B cells. SHM is defined as the introduction of nucleotide alterations at the V regions of heavy and light chain genes with a high frequency, thereby enabling the selection of B cells producing high-affinity antibodies [5–7]. In contrast, CSR exchanges the heavy chain constant domain ($C_H$) for one of a set of downstream $C_H$ exons, allowing B cells to produce different antibody isotypes (IgG, IgA, or IgE) with distinct physiologic functions, but without changing antigen specificity [8,9]. Although these two events occur independently [10,11], they are similar in that they basically take place in the germinal center of antigen-stimulated B cells [12–14]. These B cell–specific genetic modification events are initiated by the same molecule, activation-induced cytidine deaminase (AID) [15,16]. AID is a member of the apolipoprotein B mRNA-editing enzyme catalytic polypeptide (APOBEC) family comprising nucleotide-editing enzymes that insert nucleotide alterations into target DNA or RNA sequences through cytidine deamination [17]. AID is expressed only at the germinal center of activated B cells under physiologic conditions [15,18]. No hypermutations or gene-switching events at the Ig loci are found in patients with congenital AID deficiency [16,19]. Furthermore, the fact that AID-deficient B cells fail to undergo SHM and CSR confirms that AID is essential for both of these events [16,20,21]. AID was first considered to be an RNA-editing enzyme that edits target mRNA sequences to initiate either SHM or CSR [15,16,22]. Recent studies, however, support a role of AID in the deamination of DNA sequences [23–34].

Molecular Biology of B Cells. http://dx.doi.org/10.1016/B978-0-12-397933-9.00021-7

## 3. AID INDUCTION IN B CELLS

In B cells, AID expression is efficiently induced by a T cell–dependent pathway. CD40 is a member of the tumor necrosis factor (TNF) receptor family that is constitutively expressed on the surface of B cells. CD154, which is a ligand of CD40 and expressed on the surface of activated CD4$^+$T cells, binds to CD40 and activates downstream signaling cascade [35–37]. Stimulation of CD40 leads to the activation of IκB kinases that, in turn, activate the nuclear factor κB (NF-κB) canonical pathway [38]. Activated NF-κB p65/p50 heterodimers translocate to the nucleus and contribute to AID upregulation. Conversely, IgG and IgA antibodies could be produced in the early stage of viral and bacterial infection before T cells are activated, suggesting that CSR also occurs in a T cell–independent manner [39]. Indeed, mice deficient in T cells or CD154/CD40 can produce specific antibodies that are class switched and effective for some pathogens [39,40]. Several groups have revealed that Toll-like receptors (TLRs), which are highly expressed in B cells, have a putative role in CSR. Lipopolysaccharides, TLR4 ligands, and CpG, a TLR9 ligand, can significantly boost the production of specific antibodies against proteinic antigens [41–43]. The TLR pathway can also activate the canonical NF-κB pathway, thereby inducing AID expression. Pone et al. [39] showed that TLR signaling synergizes with B cell receptor signaling to induce CSR. B cell receptor signaling activates the noncanonical NF-κB pathway and enhances the TLR-induced canonical NF-κB pathway.

## 4. REGULATION OF AID EXPRESSION IN B CELLS

AID is a critical molecule for SHM and CSR, two indispensable events for producing various types of antibodies. Alternatively, AID expression must be tightly regulated to restrict its role to the appropriate cell type, loci, and time because of its intrinsic mutagenic potential. Several studies show that AID overexpression causes a high rate of mutations in exogenous green fluorescent protein genes [30,44,45]. In addition, AID has a critical role for *c-myc*/IgH translocations in B cells, which are often observed in human Burkitt's lymphoma [46–48]. To date, the sophisticated AID regulation mechanisms can be classified into two types: transcriptional regulation and posttranscriptional regulation [49].

### 4.1 Transcriptional Regulation of AID

Four well-conserved regions are related to transcriptional regulation around the *AICDA* loci and regions 1, 2, and 4 contain functional transfactor-binding motifs based on a luciferase reporter assay [50]. Region 1 is located just upstream of the transcription start site. This region contains a TATA-less promoter, but it does not regulate AID expression specifically because it promotes AID production, even in non-B cells [50,51]. The HoxC4-Oct motif located in region 1 is necessary for higher expression of AID [52]. Indeed, HoxC4 is highly expressed in the B cells of germinal center and HoxC4 deficiency causes impaired CSR because of the reduced AID expression [52]. The HoxC4-Oct element itself, however, does not respond to stimulation by CD40L, interleukin (IL)-4, and/or transforming growth factor (TGF)-β, suggesting that it may not play a major role in the induction of AID expression by stimuli that lead to CSR [50].

Region 2 is located in the first intron and contains 14 transcription factor-binding motifs, such as those for E proteins (E-boxes) and Pax5, motifs have an important role in B cell development [53,54]. The transcription reporter construct containing these binding sites enhances luciferase activity in B cells compared with non-B cells, suggesting that E-boxes and Pax-5 binding sites contribute to the positive regulation of AID expression specifically in B cells [50]. In contrast, c-Myb- and E2f-binding sites are important for negative regulation [50].

Region 4 is located approximately 8 kb upstream of the transcription start site and contains the binding sites for NF-κB, STAT6, C/EBP, and SMAD3/4. Only a reporter construct containing both regions 1 and 4 shows enhanced luciferase activity when stimulated by CD40L, IL-4, and/or TGF-β, suggesting that region 4 is mainly responsible for the upregulation of AID by environmental stimuli [50].

### 4.2 Posttranscriptional Regulation of AID

AID expression is regulated not only transcriptionally but also posttranscriptionally. MicroRNAs are noncoding, single-strand RNAs of ~22 nucleotides that function as negative gene regulators in a sequence-specific manner [55]. Several microRNAs, such as miR-155, miR-181b, miR-361, and miR-93, suppress AID expression by directly binding to the target sites in the 3′-untranslated region of AID mRNA [56–60]. For example, Dorsett et al. [59] revealed that mice with a mutation in the putative miR-155 target site in the 3′-untranslated region of AID showed an increase in steady-state AID expression and caused a high degree of *c-myc*/IgH translocation.

The enzymatic activity of AID is also regulated by phosphorylation. AID expressed at the B cell nucleus is phosphorylated on serine 38 (Ser38) [61,62]. The c-AMP–dependent protein kinase A (PKA) carries out Ser38 phosphorylation, which is important for modulating AID activity. Indeed, the Ser38 residue is located in a consensus c-AMP–dependent PKA phosphorylation site that is conserved in the AID sequence across all species that undergo CSR [61]. Mutant AID in which Ser38 is converted to alanine shows a lower frequency of SHM and CSR [63]. AID interacts with replication protein A, which is a single-strand DNA-binding protein involved in replication, recombination, and repair [62]. The recruitment of RPA to the Ig loci is facilitated by the phosphorylation of Ser38 of AID [64,65]. Phosphorylation on threonine 140 also plays an important role in SHM but not in CSR [66].

AID is expressed in both the nucleus and cytoplasm of B cells, but the amount of AID in the nucleus is tightly regulated. A bipartite nuclear localization signal and a nuclear export signal, located in the N and C terminals of AID, respectively, regulate shuttling between the nucleus and cytoplasm [67–69]. The C-terminal portion of AID is required for stable AID expression and CSR, but not for SHM [68–71], suggesting that the nuclear export signal is not only required for nuclear export but also is critical for stabilizing AID and CSR activity. In contrast, the role of the nuclear localization signal in the cytoplasm has not yet been independently verified [67,69].

In the nucleus, the half-life of AID is rather short because AID expression is unstable. Aoufouchi et al. [72] showed that the destabilization of nuclear AID expression is accompanied by polyubiquitination. REG-γ, a nuclear protein also known as PSME3, PA28γ, or 11S regulator, interacts with nuclear AID and accelerates its proteasomal degradation in a ubiquitin- and ATP-independent manner [73]. Furthermore, the transcription factor YY1 physically interacts with AID and regulates its accumulation in the nucleus [74]. In the cytoplasm, AID is not free but exists as part of a high-molecular-mass complex [75]. Orthwein et al. [76] showed that cytoplasmic AID expression is stabilized by heat shock protein 90. The interaction between heat shock protein 90 and AID prevents AID from polyubiquitination and proteasomal degradation. Most of the mechanisms of this regulation remain unclear, but these molecules might have an important role in regulating the subcellular localization of AID.

## 5. ABERRANT AID EXPRESSION BY PATHOGEN INFECTION AND TUMORIGENESIS

### 5.1 Aberrant AID Expression in Hematopoietic Cells

AID is physiologically expressed at the germinal center in B cells, but its expression in B cells is also induced outside of the germinal center by stimulation with various pathogens [77]. Especially, AID is induced by several oncogenic viruses. Epstein–Barr virus (EBV) is a virus of the herpes family associated with the development of Burkitt's lymphoma. EBV infection in B cells promotes AID expression [78,79]. Latent membrane protein 1, which is an EBV product, functionally mimics CD40 and strongly activates the NF-κB canonical pathway [80,81]. He et al. [79] showed that EBV-infected lymphoblastoid B cells express AID and have enhanced CSR activity. Furthermore, AID and pol-η, one type of error-prone translesion synthesis polymerase, is highly expressed in primary B cells infected with EBV. These B cells accumulate mutations in proto-oncogenes such as *BCL6* and *p53*, which are involved in the development of B cell lymphoma [78]. These findings indicate that EBV induces genetic alterations in B cells through AID dysregulation. Indeed, as described previously, *cmyc*/IgH translocation and several mutations that are thought to be induced by AID are observed in Burkitt's lymphoma B cells [46–48,82]. Moreover, Klemm et al. [83] revealed that AID expression contributes to the fatal B lymphoid blast crisis in chronic myeloid leukemia. They showed that lymphoid blast crisis cells but not chronic myeloid leukemia cells express AID and that AID induces *BCR-ABL1* mutations that cause imatinib resistance and progression to lymphoid blast crisis. Thus, aberrant AID upregulation causes genetic changes in various tumor-related genes in B cells, leading to their transformation.

The adverse mutagenic activity of AID is observed not only in B cells but also in T cells [84,85]. Ishikawa et al. [86] revealed that human T-cell leukemia virus type 1 (HTLV-1)–infected T cells, which cause the development of adult T-cell leukemia (ATL), express higher levels of AID than uninfected T cells and normal peripheral blood mononuclear cells. One HTLV-1 product called Tax, a key factor for the leukemogenesis of HTLV-1–infected cells [87], transcriptionally activates the AID gene through the NF-κB canonical pathway. The fact that ATL cells also constitutively express AID implies that mutations induced by Tax-activated AID contribute to tumorigenesis. Indeed, all transgenic mice with constitutive AID expression develop T-cell lymphomas with a number of point mutations in non-Ig genes such as *c-myc* and T cell receptor genes [44,88].

### 5.2 Aberrant AID Expression in Epithelial Cells

As described in Section 3, AID expression in B cells is mainly enhanced by activation of the NF-κB canonical pathway. Alternatively, the NF-κB signaling pathway is deeply associated with various types of inflammatory responses and links inflammation to tumorigenesis in epithelial organs [89–91]. Recent evidence suggests that AID is aberrantly expressed not only in B cells but also in various types of cells and contributes to infection and/or inflammation-associated tumorigenesis.

#### 5.2.1 Hepatocytes

Hepatocellular carcinoma (HCC) is one of the most common types of cancer, with a dismal outcome. Most HCC develops in patients with chronic liver disease, such as chronic hepatitis and liver cirrhosis caused by hepatitis

C virus (HCV) or hepatitis B virus (HBV) infection. The cumulative incidence rate of HCC is 18.1% for 5 years and 45.6% for 10 years in chronic hepatitis patients with HCV infection [92]. Furthermore, most HCC tumors show multicentricity and develop from damaged liver tissue caused by chronic inflammation. These clinical features of HCC indicate that persistent inflammation is closely associated with the malignant transformation of hepatocytes and plays a critical role in hepatocarcinogenesis [93]. Using immunohistochemistry, Kou et al. [94] revealed that AID immunostaining is detected in not only infiltrating lymphocytes but also in the hepatocytes of liver tissue underlying chronic hepatitis or cirrhosis caused by HCV infection. In contrast, AID expression is not observed in normal hepatocytes. These findings suggest that AID expression is induced in hepatocytes by HCV infection and/or under inflammatory conditions. Consistent with the immunohistochemistry on human clinical tissue specimens, AID expression is upregulated in response to stimulation by proinflammatory cytokines, such as TNF-α or IL-1β, or expression of HCV core protein in cultured human hepatocytes [95,96]. This AID upregulation caused by TNF-α or viral protein is dependent on the NF-κB canonical pathway, consistent with previous findings that NF-κB activation is critical for the enhancement of AID expression in B cells. Indeed, the super-repressor form of IκBα, which is a specific NF-κB inhibitor, decreases TNF-α–mediated AID expression [95]. Furthermore, constitutive AID upregulation in hepatocytes induces nucleotide alterations in various non-Ig genes, including tumor-related genes such as *p53* and *c-myc*. Interestingly, transgenic mice with AID overexpression frequently develop HCC accompanied by the accumulation of *p53* mutations [88,97]. These findings suggest that ectopic AID expression in hepatocytes contributes to enhance the susceptibility of mutagenesis, resulting in HCC development with *p53* mutations.

The physiologic implications of aberrant AID expression in inflamed hepatocytes are unclear at present. Interestingly, Liang et al. [98] recently reported that C/G to T/A mutations are frequently observed in HBV nucleocapsid DNA when AID is overexpressed in hepatocytes with HBV replication. This accumulation of mutations is also detected in viral nucleocapsid RNA but not in RNA outside the nucleocapsid. Similarly, overexpression of chicken AID causes hypermutation in association with reduction of duck HBV covalently closed circular DNA levels [99]. Interestingly, IL-1β and TNF-α remarkably reduce host cell susceptibility to HBV infection through AID activity in hepatocytes [96]. These findings suggest that AID may have a role in antiviral defense in the HBV-infected liver and could regulate the activity of HBV replication, just as APOBEC3 family members are capable of inducing hypermutations in the genome of human immunodeficiency virus-1 [100,101] and HBV [102,103].

### 5.2.2 Gastric Epithelial Cells

Most human gastric cancers develop in the stomach with chronic gastritis. The most important cause of chronic gastric inflammation is *Helicobacter pylori* (*H. pylori*) infection, defined as a class 1 carcinogen for human gastric cancer [104,105]. *H. pylori* is subclassified to "*cag*" pathogenicity island (*cag*PAI)–positive and *cag*PAI-negative strains based on the presence of *cag*PAI, a 40-kb genome fragment containing 31 genes [106,107]. The *cag*PAI-positive isolates are highly pathogenic to gastric epithelial cells compared with *cag*PAI-negative strains [108] and are a high risk factor for gastric cancer development [108–111]. In human stomach tissue, AID is aberrantly expressed in the gastric epithelial cells of stomach tissues with chronic gastritis induced by *H. pylori* infection, as well as in gastric adenocarcinoma cells. In contrast, normal gastric mucosa without inflammation does not express AID [112]. A recent clinical study showed that AID expression is strongly induced in *H. pylori*–infected gastric mucosa, especially in the antrum, and it is suppressed by *H. pylori* eradication therapy [113]. This ectopic AID expression level correlates with the severity of the gastritis and intestinal metaplasia. An in vitro study revealed that aberrant AID expression in gastric epithelial cells is strongly induced by TNF-α stimulation via activation of the NF-κB canonical pathway. AID expression is also directly triggered by *H. pylori* infection. The *cag*PAI-positive strain is capable of introducing bacterial virulence factors into host gastric epithelial cells through a type IV secretion system, leading to NF-κB activation [104,114]. These findings indicate that both bacterial factors introduced into epithelial cells and inflammatory mediators such as TNF-α induced by *H. pylori* infection cooperatively enhance AID expression in an NF-κB–dependent manner. In addition, continuous AID expression in gastric epithelial cells results in the accumulation of somatic mutations in *p53* genes, which is thought to be important for gastric cancer development. The cyclin-dependent kinase inhibitor (CDKN)2B-CDKN2A locus, which encodes the potent suppressor proteins $p16^{INK4a}$, $p15^{INK4b}$, and $p14^{ARF}$, was also identified as a target for AID [115]. Indeed, aberrant AID expression induces nucleotide alterations at the CDKN2B-CDKN2A locus in gastric epithelial cells. Moreover, AID causes preferential deletion of the CDKN2B-CDKN2A locus in vitro, and oral *H. pylori* infection also triggers a reduction in the copy number of *Cdkn2b-Cdkn2a* in the gastric mucosa of wild-type mice, whereas it does not in AID-deficient mice [115]. Consistent with these findings, AID expression is necessary for the introduction of single-strand DNA breaks in the CDKN2B gene in leukemia cells [116], and lymphoid blast crisis leukemia cells with strong AID expression frequently show deletion of the CDKN2B-CDKN2A locus [83]. These findings indicate that AID expression induced by *H. pylori* infection with the resultant chronic inflammation contributes to not only the accumulation of nucleotide alterations but also to submicroscopic deletions in tumor-related genes (Figure 1).

### 5.2.3 Colonic Epithelial Cells

Colitis-associated cancer (CAC) was first reported as a complication of ulcerative colitis (UC) in 1925 and is now well known to be associated with inflammatory bowel disease. It is estimated that CAC is the cause of approximately 15% of all deaths of patients with inflammatory bowel disease [117]. Several clinical studies revealed that the cumulative risk of CAC development increases according to the disease duration and severity of inflammation [118–120]. In contrast to the "adenoma-carcinoma sequence," which is a series of multistep genetic events that occurs in sporadic colorectal cancer development [121], *p53* mutations are frequently detected in the early stage of CAC and are already present in the colonic mucosa of UC patients [122–124]. Endo et al. [125] showed that AID is aberrantly expressed in the inflamed colonic mucosa and CAC tissues of UC patients but not in normal colonic mucosa without inflammation. As is the case in HCC and gastric cancer, this AID expression is induced by TNF-α via NF-κB activation in colonic cells. It is well recognized that T helper cell (Th) 2 cytokines contribute to chronic inflammation in UC patients [126,127]. The excessive Th2 response leads to an increase in the secretion of specific cytokines, such as IL-4 and IL-13 that share a common receptor unit and activate the transcription factor STAT6 by phosphorylation to enhance the expression of various downstream molecules [128–130]. Interestingly, AID expression is also enhanced by treatment with IL-4 or IL-13 in a STAT6-dependent manner in human colonic cells [125]. Conversely, Th1 cytokines such as IL-12 are activated dominantly in the intestinal tissues of patients with Crohn's disease [131]. AID expression is also induced by IL-12 treatment in human colonic cells. Furthermore, the expression levels of both AID and IL-12, which are strongly upregulated in the colonic mucosa of IL-10–deficient mice, are suppressed by neutralizing IL-12p40 monoclonal antibody [132]. These findings suggest that AID expression is induced not only by NF-κB activation but also by Th1 and Th2 cytokines in inflamed colonic epithelial cells. Because colonic epithelial cells are constitutively exposed to bacterial flora in the intestine, it is possible to assume that direct action of gut bacterium also may contribute to the aberrant AID expression in colonic epithelial cells.

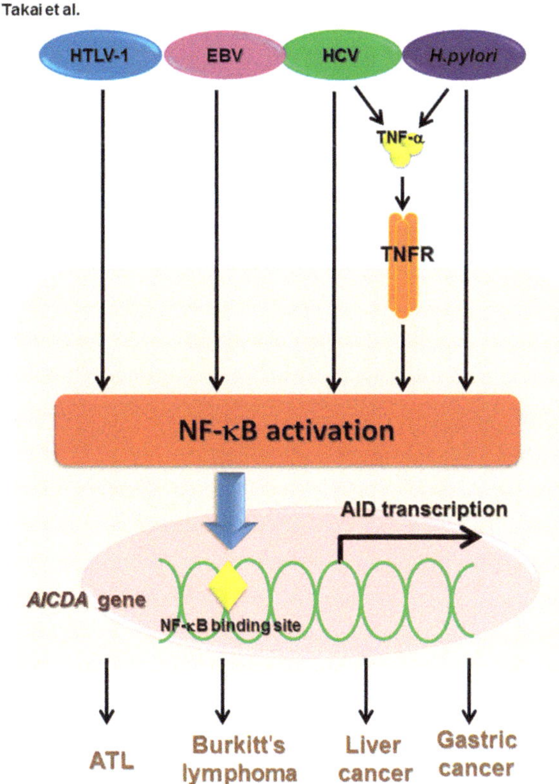


FIGURE 1 **Various types of pathogens are capable of inducing aberrant AID expression in their targeted cells.** Infection of EBV, HTLV-1, HCV, or *H. pylori* triggers upregulation of AID transcription through the activation of NF-κB canonical pathway. The infection with these pathogens itself directly induces NF-κB activation. Moreover, the resultant inflammatory response caused by HCV or *H. pylori* infection induces the production of proinflammatory cytokines such as TNF-α, which also contribute to NF-κB activation. Activated NF-κB complex is translocated into nucleus and leads to the enhancement of AID transcription. Constitutive AID expression could induce the genetic alterations in various genes and contribute to the development of tumor cells.

## 6. CONCLUSION

Aberrant expression of AID in response to pathogen infection with the resultant chronic inflammation provides a novel connection between inflammation and tumorigenesis. Other than the tumors mentioned above, several studies also have demonstrated aberrant AID expression in various epithelial tissues. For example, AID is closely associated with the development of cholangiocarcinoma and Barrett's esophageal cancer [133,134]. TNF-α induces aberrant AID expression in mammary epithelial cells and is necessary for the epithelial–mesenchymal transition of breast cancer cells [135]. Similarly, AID expression is also induced in response to TNF-α in oral epithelium and contributes to the initiation of oral squamous cell carcinoma [136,137]. AID expression has been detected in a subset of human lung cancers [137]. Furthermore, recent studies revealed that not only AID but also other APOBEC family molecules act as DNA mutators and contribute to tumorigenesis in various organs. Burns et al. [138] showed that APOBEC 3B is upregulated in a variety of cancers, such as bladder, cervical, breast, head and neck, and lung, and is responsible for the mutagenesis in these tumors. Roberts et al. [139] analyzed multiple whole-genome and exome mutations datasets and found that many mutation clusters have the same alteration patterns as those induced by APOBEC. These findings suggest that the nucleotide-editing enzymes have critical roles in inducing mutations in various cells, leading to tumorigenesis. Further studies of nucleotide-editing enzymes including AID will likely elucidate the mechanism of infection and/or inflammation-associated carcinogenesis and contribute to the development of novel cancer therapies.

## REFERENCES


[1] Bassing CH, Swat W, Alt FW. The mechanism and regulation of chromosomal V(D)J recombination. Cell 2002;109(Suppl):S45–55.
[2] Jung D, Alt FW. Unraveling V(D)J recombination; insights into gene regulation. Cell 2004;116:299–311.
[3] Lengauer C, Kinzler KW, Vogelstein B. Genetic instabilities in human cancers. Nature 1998;396:643–9.
[4] Sjoblom T, Jones S, Wood LD, Parsons DW, Lin J, Barber TD, et al. The consensus coding sequences of human breast and colorectal cancers. Science 2006;314:268–74.
[5] Martin A, Scharff MD. AID and mismatch repair in antibody diversification. Nat Rev Immunol 2002;2:605–14.
[6] Neuberger MS. Antibody diversification by somatic mutation: from Burnet onwards. Immunol Cell Biol 2008;86:124–32.
[7] Papavasiliou FN, Schatz DG. Somatic hypermutation of immunoglobulin genes: merging mechanisms for genetic diversity. Cell 2002;109(Suppl):S35–44.
[8] Honjo T, Kinoshita K, Muramatsu M. Molecular mechanism of class switch recombination: linkage with somatic hypermutation. Annu Rev Immunol 2002;20:165–96.
[9] Manis JP, Tian M, Alt FW. Mechanism and control of class-switch recombination. Trends Immunol 2002;23:31–9.
[10] Liu YJ, Malisan F, de Bouteiller O, Guret C, Lebecque S, Banchereau J, et al. Within germinal centers, isotype switching of immunoglobulin genes occurs after the onset of somatic mutation. Immunity 1996;4:241–50.
[11] Siekevitz M, Kocks C, Rajewsky K, Dildrop R. Analysis of somatic mutation and class switching in naive and memory B cells generating adoptive primary and secondary responses. Cell 1987;48: 757–70.
[12] Berek C, Berger A, Apel M. Maturation of the immune response in germinal centers. Cell 1991;67:1121–9.
[13] Jacob J, Kelsoe G, Rajewsky K, Weiss U. Intraclonal generation of antibody mutants in germinal centres. Nature 1991;354:389–92.
[14] Pascual V, Liu YJ, Magalski A, de Bouteiller O, Banchereau J, Capra JD. Analysis of somatic mutation in five B cell subsets of human tonsil. J Exp Med 1994;180:329–39.
[15] Muramatsu M, Sankaranand VS, Anant S, Sugai M, Kinoshita K, Davidson NO, et al. Specific expression of activation-induced cytidine deaminase (AID), a novel member of the RNA-editing deaminase family in germinal center B cells. J Biol Chem 1999;274:18470–6.
[16] Muramatsu M, Kinoshita K, Fagarasan S, Yamada S, Shinkai Y, Honjo T. Class switch recombination and hypermutation require activation-induced cytidine deaminase (AID), a potential RNA editing enzyme. Cell 2000;102:553–63.
[17] Conticello SG. The AID/APOBEC family of nucleic acid mutators. Genome Biol 2008;9:229.
[18] Schrader CE, Linehan EK, Mochegova SN, Woodland RT, Stavnezer J. Inducible DNA breaks in Ig S regions are dependent on AID and UNG. J Exp Med 2005;202:561–8.
[19] Revy P, Muto T, Levy Y, Geissmann F, Plebani A, Sanal O, et al. Activation-induced cytidine deaminase (AID) deficiency causes the autosomal recessive form of the Hyper-IgM syndrome (HIGM2). Cell 2000;102:565–75.
[20] Arakawa H, Hauschild J, Buerstedde JM. Requirement of the activation-induced deaminase (AID) gene for immunoglobulin gene conversion. Science 2002;295:1301–6.
[21] Harris RS, Sale JE, Petersen-Mahrt SK, Neuberger MS. AID is essential for immunoglobulin V gene conversion in a cultured B cell line. Curr Biol 2002;12:435–8.
[22] Muramatsu M, Nagaoka H, Shinkura R, Begum NA, Honjo T. Discovery of activation-induced cytidine deaminase, the engraver of antibody memory. Adv Immunol 2007;94:1–36.
[23] Barreto VM, Ramiro AR, Nussenzweig MC. Activation-induced deaminase: controversies and open questions. Trends Immunol 2005;26:90–6.
[24] Chaudhuri J, Tian M, Khuong C, Chua K, Pinaud E, Alt FW. Transcription-targeted DNA deamination by the AID antibody diversification enzyme. Nature 2003;422:726–30.
[25] Dickerson SK, Market E, Besmer E, Papavasiliou FN. AID mediates hypermutation by deaminating single stranded DNA. J Exp Med 2003;197:1291–6.
[26] Faili A, Aoufouchi S, Flatter E, Gueranger Q, Reynaud CA, Weill JC. Induction of somatic hypermutation in immunoglobulin genes is dependent on DNA polymerase iota. Nature 2002;419:944–7.
[27] Fritz EL, Rosenberg BR, Lay K, Mihailovic A, Tuschl T, Papavasiliou FN. A comprehensive analysis of the effects of the deaminase AID on the transcriptome and methylome of activated B cells. Nat Immunol 2013;14:749–55.
[28] Longerich S, Basu U, Alt F, Storb U. AID in somatic hypermutation and class switch recombination. Curr Opin Immunol 2006;18: 164–74.
[29] Peled JU, Kuang FL, Iglesias-Ussel MD, Roa S, Kalis SL, Goodman MF, et al. The biochemistry of somatic hypermutation. Annu Rev Immunol 2008;26:481–511.
[30] Petersen-Mahrt SK, Harris RS, Neuberger MS. AID mutates *E. coli* suggesting a DNA deamination mechanism for antibody diversification. Nature 2002;418:99–103.
[31] Pham P, Bransteitter R, Petruska J, Goodman MF. Processive AID-catalysed cytosine deamination on single-stranded DNA simulates somatic hypermutation. Nature 2003;424:103–7.
[32] Ramiro AR, Stavropoulos P, Jankovic M, Nussenzweig MC. Transcription enhances AID-mediated cytidine deamination by exposing single-stranded DNA on the nontemplate strand. Nat Immunol 2003;4:452–6.
[33] Shen HM, Storb U. Activation-induced cytidine deaminase (AID) can target both DNA strands when the DNA is supercoiled. Proc Natl Acad Sci USA 2004;101:12997–3002.
[34] Stavnezer J, Guikema JE, Schrader CE. Mechanism and regulation of class switch recombination. Annu Rev Immunol 2008;26: 261–92.
[35] Bishop GA, Hostager BS. The CD40-CD154 interaction in B cell-T cell liaisons. Cytokine Growth Factor Rev 2003;14:297–309.
[36] Graham JP, Arcipowski KM, Bishop GA. Differential B-lymphocyte regulation by CD40 and its viral mimic, latent membrane protein 1. Immunol Rev 2010;237:226–48.

[37] Sun SC, Liu ZG. A special issue on NF-kappaB signaling and function. Cell Res 2011;21:1–2.
[38] Oeckinghaus A, Hayden MS, Ghosh S. Crosstalk in NF-kappaB signaling pathways. Nat Immunol 2011;12:695–708.
[39] Pone EJ, Zan H, Zhang J, Al-Qahtani A, Xu Z, Casali P. Toll-like receptors and B-cell receptors synergize to induce immunoglobulin class-switch DNA recombination: relevance to microbial antibody responses. Crit Rev Immunol 2010;30:1–29.
[40] Zhan Y, Brown LE, Deliyannis G, Seah S, Wijburg OL, Price J, et al. Responses against complex antigens in various models of CD4 T-cell deficiency: surprises from an anti-CD4 antibody transgenic mouse. Immunol Res 2004;30:1–14.
[41] Eckl-Dorna J, Batista FD. BCR-mediated uptake of antigen linked to TLR9 ligand stimulates B-cell proliferation and antigen-specific plasma cell formation. Blood 2009;113:3969–77.
[42] Kasturi SP, Skountzou I, Albrecht RA, Koutsonanos D, Hua T, Nakaya HI, et al. Programming the magnitude and persistence of antibody responses with innate immunity. Nature 2011;470:543–7.
[43] Pasare C, Medzhitov R. Control of B-cell responses by Toll-like receptors. Nature 2005;438:364–8.
[44] Okazaki IM, Hiai H, Kakazu N, Yamada S, Muramatsu M, Kinoshita K, et al. Constitutive expression of AID leads to tumorigenesis. J Exp Med 2003;197:1173–81.
[45] Yoshikawa K, Okazaki IM, Eto T, Kinoshita K, Muramatsu M, Nagaoka H, et al. AID enzyme-induced hypermutation in an actively transcribed gene in fibroblasts. Science 2002;296:2033–6.
[46] Ramiro AR, Jankovic M, Callen E, Difilippantonio S, Chen HT, McBride KM, et al. Role of genomic instability and p53 in AID-induced c-myc-Igh translocations. Nature 2006;440:105–9.
[47] Ramiro AR, Jankovic M, Eisenreich T, Difilippantonio S, Chen-Kiang S, Muramatsu M, et al. AID is required for c-myc/IgH chromosome translocations in vivo. Cell 2004;118:431–8.
[48] Robbiani DF, Bothmer A, Callen E, Reina-San-Martin B, Dorsett Y, Difilippantonio S, et al. AID is required for the chromosomal breaks in c-myc that lead to c-myc/IgH translocations. Cell 2008;135:1028–38.
[49] Nagaoka H, Tran TH, Kobayashi M, Aida M, Honjo T. Preventing AID, a physiological mutator, from deleterious activation: regulation of the genomic instability that is associated with antibody diversity. Int Immunol 2010;22:227–35.
[50] Tran TH, Nakata M, Suzuki K, Begum NA, Shinkura R, Fagarasan S, et al. B cell-specific and stimulation-responsive enhancers derepress Aicda by overcoming the effects of silencers. Nat Immunol 2010;11:148–54.
[51] Yadav A, Olaru A, Saltis M, Setren A, Cerny J, Livak F. Identification of a ubiquitously active promoter of the murine activation-induced cytidine deaminase (AICDA) gene. Mol Immunol 2006;43:529–41.
[52] Park SR, Zan H, Pal Z, Zhang J, Al-Qahtani A, Pone EJ, et al. HoxC4 binds to the promoter of the cytidine deaminase AID gene to induce AID expression, class-switch DNA recombination and somatic hypermutation. Nat Immunol 2009;10:540–50.
[53] Gonda H, Sugai M, Nambu Y, Katakai T, Agata Y, Mori KJ, et al. The balance between Pax5 and Id2 activities is the key to AID gene expression. J Exp Med 2003;198:1427–37.
[54] Sayegh CE, Quong MW, Agata Y, Murre C. E-proteins directly regulate expression of activation-induced deaminase in mature B cells. Nat Immunol 2003;4:586–93.
[55] Esquela-Kerscher A, Slack FJ. Oncomirs - microRNAs with a role in cancer. Nat Rev Cancer 2006;6:259–69.
[56] Basso K, Schneider C, Shen Q, Holmes AB, Setty M, Leslie C, et al. BCL6 positively regulates AID and germinal center gene expression via repression of miR-155. J Exp Med 2012;209:2455–65.
[57] Borchert GM, Holton NW, Larson ED. Repression of human activation induced cytidine deaminase by miR-93 and miR-155. BMC Cancer 2011;11:347.
[58] de Yebenes VG, Belver L, Pisano DG, Gonzalez S, Villasante A, et al. miR-181b negatively regulates activation-induced cytidine deaminase in B cells. J Exp Med 2008;205:2199–206.
[59] Dorsett Y, McBride KM, Jankovic M, Gazumyan A, Thai TH, Robbiani DF, et al. MicroRNA-155 suppresses activation-induced cytidine deaminase-mediated Myc-Igh translocation. Immunity 2008;28:630–8.
[60] Teng G, Hakimpour P, Landgraf P, Rice A, Tuschl T, Casellas R, et al. MicroRNA-155 is a negative regulator of activation-induced cytidine deaminase. Immunity 2008;28:621–9.
[61] Basu U, Chaudhuri J, Alpert C, Dutt S, Ranganath S, Li G, et al. The AID antibody diversification enzyme is regulated by protein kinase A phosphorylation. Nature 2005;438:508–11.
[62] Chaudhuri J, Khuong C, Alt FW. Replication protein A interacts with AID to promote deamination of somatic hypermutation targets. Nature 2004;430:992–8.
[63] McBride KM, Gazumyan A, Woo EM, Barreto VM, Robbiani DF, Chait BT, et al. Regulation of hypermutation by activation-induced cytidine deaminase phosphorylation. Proc Natl Acad Sci USA 2006;103:8798–803.
[64] Cheng HL, Vuong BQ, Basu U, Franklin A, Schwer B, Astarita J, et al. Integrity of the AID serine-38 phosphorylation site is critical for class switch recombination and somatic hypermutation in mice. Proc Natl Acad Sci USA 2009;106:2717–22.
[65] Yamane A, Resch W, Kuo N, Kuchen S, Li Z, Sun HW, et al. Deep-sequencing identification of the genomic targets of the cytidine deaminase AID and its cofactor RPA in B lymphocytes. Nat Immunol 2011;12:62–9.
[66] McBride KM, Gazumyan A, Woo EM, Schwickert TA, Chait BT, Nussenzweig MC. Regulation of class switch recombination and somatic mutation by AID phosphorylation. J Exp Med 2008;205:2585–94.
[67] Brar SS, Watson M, Diaz M. Activation-induced cytosine deaminase (AID) is actively exported out of the nucleus but retained by the induction of DNA breaks. J Biol Chem 2004;279:26395–401.
[68] Ito S, Nagaoka H, Shinkura R, Begum N, Muramatsu M, Nakata M, et al. Activation-induced cytidine deaminase shuttles between nucleus and cytoplasm like apolipoprotein B mRNA editing catalytic polypeptide 1. Proc Natl Acad Sci USA 2004;101:1975–80.
[69] McBride KM, Barreto V, Ramiro AR, Stavropoulos P, Nussenzweig MC. Somatic hypermutation is limited by CRM1-dependent nuclear export of activation-induced deaminase. J Exp Med 2004;199:1235–44.
[70] Durandy A, Revy P, Imai K, Fischer A. Hyper-immunoglobulin M syndromes caused by intrinsic B-lymphocyte defects. Immunol Rev 2005;203:67–79.
[71] Geisberger R, Rada C, Neuberger MS. The stability of AID and its function in class-switching are critically sensitive to the identity of its nuclear-export sequence. Proc Natl Acad Sci USA 2009;106:6736–41.
[72] Aoufouchi S, Faili A, Zober C, D'Orlando O, Weller S, Weill JC, et al. Proteasomal degradation restricts the nuclear lifespan of AID. J Exp Med 2008;205:1357–68.

[73] Uchimura Y, Barton LF, Rada C, Neuberger MS. REG-gamma associates with and modulates the abundance of nuclear activation-induced deaminase. J Exp Med 2011;208:2385–91.

[74] Zaprazna K, Atchison ML. YY1 controls immunoglobulin class switch recombination and nuclear activation-induced deaminase levels. Mol Cell Biol 2012;32:1542–54.

[75] Hasler J, Rada C, Neuberger MS. The cytoplasmic AID complex. Semin Immunol 2012;24:273–80.

[76] Orthwein A, Patenaude AM, Affar el B, Lamarre A, Young JC, Di Noia JM. Regulation of activation-induced deaminase stability and antibody gene diversification by Hsp90. J Exp Med 2010;207: 2751–65.

[77] Perez-Duran P, de Yebenes VG, Ramiro AR. Oncogenic events triggered by AID, the adverse effect of antibody diversification. Carcinogenesis 2007;28:2427–33.

[78] Epeldegui M, Hung YP, McQuay A, Ambinder RF, Martinez-Maza O. Infection of human B cells with Epstein-Barr virus results in the expression of somatic hypermutation-inducing molecules and in the accrual of oncogene mutations. Mol Immunol 2007;44:934–42.

[79] He B, Raab-Traub N, Casali P, Cerutti A. EBV-encoded latent membrane protein 1 cooperates with BAFF/BLyS and APRIL to induce T cell-independent Ig heavy chain class switching. J Immunol 2003;171:5215–24.

[80] Uchida J, Yasui T, Takaoka-Shichijo Y, Muraoka M, Kulwichit W, Raab-Traub N, et al. Mimicry of CD40 signals by Epstein-Barr virus LMP1 in B lymphocyte responses. Science 1999;286:300–3.

[81] Young LS, Rickinson AB. Epstein-Barr virus: 40 years on. Nat Rev Cancer 2004;4:757–68.

[82] Robbiani DF, Bunting S, Feldhahn N, Bothmer A, Camps J, Deroubaix S, et al. AID produces DNA double-strand breaks in non-Ig genes and mature B cell lymphomas with reciprocal chromosome translocations. Mol Cell 2009;36:631–41.

[83] Klemm L, Duy C, Iacobucci I, Kuchen S, von Levetzow G, Feldhahn N, et al. The B cell mutator AID promotes B lymphoid blast crisis and drug resistance in chronic myeloid leukemia. Cancer Cell 2009;16:232–45.

[84] Nakamura M, Iwata S, Kimura H, Tokura Y. Elevated expression of activation-induced cytidine deaminase in T and NK cells from patients with chronic active Epstein-Barr virus infection. Eur J Dermatol 2011;21:780–2.

[85] Nakamura M, Sugita K, Sawada Y, Yoshiki R, Hino R, Tokura Y. High levels of activation-induced cytidine deaminase expression in adult T-cell leukaemia/lymphoma. Br J Dermatol 2011;165:437–9.

[86] Ishikawa C, Nakachi S, Senba M, Sugai M, Mori N. Activation of AID by human T-cell leukemia virus Tax oncoprotein and the possible role of its constitutive expression in ATL genesis. Carcinogenesis 2011;32:110–9.

[87] Grassmann R, Aboud M, Jeang KT. Molecular mechanisms of cellular transformation by HTLV-1 Tax. Oncogene 2005;24:5976–85.

[88] Morisawa T, Marusawa H, Ueda Y, Iwai A, Okazaki IM, Honjo T, et al. Organ-specific profiles of genetic changes in cancers caused by activation-induced cytidine deaminase expression. Int J Cancer 2008;123:2735–40.

[89] Greten FR, Eckmann L, Greten TF, Park JM, Li ZW, Egan LJ, et al. IKKbeta links inflammation and tumorigenesis in a mouse model of colitis-associated cancer. Cell 2004;118:285–96.

[90] Karin M, Greten FR. NF-kappaB: linking inflammation and immunity to cancer development and progression. Nat Rev Immunol 2005;5:749–59.

[91] Pikarsky E, Porat RM, Stein I, Abramovitch R, Amit S, Kasem S, et al. NF-kappaB functions as a tumour promoter in inflammation-associated cancer. Nature 2004;431:461–6.

[92] Ikeda K, Marusawa H, Osaki Y, Nakamura T, Kitajima N, Yamashita Y, et al. Antibody to hepatitis B core antigen and risk for hepatitis C-related hepatocellular carcinoma: a prospective study. Ann Intern Med 2007;146:649–56.

[93] Ikeda A, Shimizu T, Matsumoto Y, Fujii Y, Eso Y, Inuzuka T, et al. Leptin receptor somatic mutations are frequent in HCV-infected cirrhotic liver and associate with hepatocellular carcinoma. Gastroenterology 2014;146:222–332.

[94] Kou T, Marusawa H, Kinoshita K, Endo Y, Okazaki IM, Ueda Y, et al. Expression of activation-induced cytidine deaminase in human hepatocytes during hepatocarcinogenesis. Int J Cancer 2007;120:469–76.

[95] Endo Y, Marusawa H, Kinoshita K, Morisawa T, Sakurai T, Okazaki IM, et al. Expression of activation-induced cytidine deaminase in human hepatocytes via NF-kappaB signaling. Oncogene 2007;26:5587–95.

[96] Watashi K, Liang G, Iwamoto M, Marusawa H, Uchida N, Daito T, et al. Interleukin-1 and tumor necrosis factor-alpha trigger restriction of hepatitis B virus infection via a cytidine deaminase (AID). J Biol Chem 2013;288:31715–27.

[97] Takai A, Toyoshima T, Uemura M, Kitawaki Y, Marusawa H, Hiai H, et al. A novel mouse model of hepatocarcinogenesis triggered by AID causing deleterious p53 mutations. Oncogene 2009;28:469–78.

[98] Liang G, Kitamura K, Wang Z, Liu G, Chowdhury S, Fu W, et al. RNA editing of hepatitis B virus transcripts by activation-induced cytidine deaminase. Proc Natl Acad Sci USA 2013;110:2246–51.

[99] Chowdhury S, Kitamura K, Simadu M, Koura M, Muramatsu M. Concerted action of activation-induced cytidine deaminase and uracil-DNA glycosylase reduces covalently closed circular DNA of duck hepatitis B virus. FEBS Lett 2013;587:3148–52.

[100] Zhang H, Yang B, Pomerantz RJ, Zhang C, Arunachalam SC, Gao L. The cytidine deaminase CEM15 induces hypermutation in newly synthesized HIV-1 DNA. Nature 2003;424:94–8.

[101] Mangeat B, Turelli P, Caron G, Friedli M, Perrin L, Trono D. Broad antiretroviral defence by human APOBEC3G through lethal editing of nascent reverse transcripts. Nature 2003;424:99–103.

[102] Noguchi C, Hiraga N, Mori N, Tsuge M, Imamura M, Takahashi S, et al. Dual effect of APOBEC3G on Hepatitis B virus. J Gen Virol 2007;88:432–40.

[103] Suspene R, Guetard D, Henry M, Sommer P, Wain-Hobson S, Vartanian JP. Extensive editing of both hepatitis B virus DNA strands by APOBEC3 cytidine deaminases in vitro and in vivo. Proc Natl Acad Sci USA 2005;102:8321–6.

[104] Peek Jr RM, Blaser MJ. *Helicobacter pylori* and gastrointestinal tract adenocarcinomas. Nat Rev Cancer 2002;2:28–37.

[105] Chiba T, Seno H, Marusawa H, Wakatsuki Y, Okazaki K. Host factors are important in determining clinical outcomes of *Helicobacter pylori* infection. J Gastroenterol 2006;41:1–9.

[106] Blaser MJ. *Helicobacter pylori* and gastric diseases. BMJ 1998;316:1507–10.

[107] Tomb JF, White O, Kerlavage AR, Clayton RA, Sutton GG, Fleischmann RD, et al. The complete genome sequence of the gastric pathogen *Helicobacter pylori*. Nature 1997;388:539–47.

[108] Blaser MJ, Perez-Perez GI, Kleanthous H, Cover TL, Peek RM, Chyou PH, et al. Infection with *Helicobacter pylori* strains possessing cagA is associated with an increased risk of developing adenocarcinoma of the stomach. Cancer Res 1995;55:2111–5.

[109] Crabtree JE, Wyatt JI, Sobala GM, Miller G, Tompkins DS, Primrose JN, et al. Systemic and mucosal humoral responses to *Helicobacter pylori* in gastric cancer. Gut 1993;34:1339–43.

[110] Parsonnet J, Friedman GD, Orentreich N, Vogelman H. Risk for gastric cancer in people with CagA positive or CagA negative *Helicobacter pylori* infection. Gut 1997;40:297–301.

[111] Peek Jr RM, Miller GG, Tham KT, Perez-Perez GI, Zhao X, Atherton JC, et al. Heightened inflammatory response and cytokine expression in vivo to cagA+ *Helicobacter pylori* strains. Lab Invest 1995;73:760–70.

[112] Matsumoto Y, Marusawa H, Kinoshita K, Endo Y, Kou T, Morisawa T, et al. *Helicobacter pylori* infection triggers aberrant expression of activation-induced cytidine deaminase in gastric epithelium. Nat Med 2007;13:470–6.

[113] Nagata N, Akiyama J, Marusawa H, Shimbo T, Liu Y, Igari T, et al. Enhanced expression of activation-induced cytidine deaminase in human gastric mucosa infected by *Helicobacter pylori* and its decrease following eradication. J Gastroenterol 2014;49:427–35.

[114] Viala J, Chaput C, Boneca IG, Cardona A, Girardin SE, Moran AP, et al. Nod1 responds to peptidoglycan delivered by the *Helicobacter pylori* cag pathogenicity island. Nat Immunol 2004;5:1166–74.

[115] Matsumoto Y, Marusawa H, Kinoshita K, Niwa Y, Sakai Y, Chiba T. Up-regulation of activation-induced cytidine deaminase causes genetic aberrations at the CDKN2b-CDKN2a in gastric cancer. Gastroenterology 2010;139:1984–94.

[116] Feldhahn N, Henke N, Melchior K, Duy C, Soh BN, Klein F, et al. Activation-induced cytidine deaminase acts as a mutator in BCR-ABL1-transformed acute lymphoblastic leukemia cells. J Exp Med 2007;204:1157–66.

[117] Munkholm P. Review article: the incidence and prevalence of colorectal cancer in inflammatory bowel disease. Aliment Pharmacol Ther 2003;18(Suppl. 2):1–5.

[118] Rutter M, Saunders B, Wilkinson K, Rumbles S, Schofield G, Kamm M, et al. Severity of inflammation is a risk factor for colorectal neoplasia in ulcerative colitis. Gastroenterology 2004;126:451–9.

[119] Gupta RB, Harpaz N, Itzkowitz S, Hossain S, Matula S, Kornbluth A, et al. Histologic inflammation is a risk factor for progression to colorectal neoplasia in ulcerative colitis: a cohort study. Gastroenterology 2007;133:1099–105. quiz 1340–1091.

[120] Eaden JA, Abrams KR, Mayberry JF. The risk of colorectal cancer in ulcerative colitis: a meta-analysis. Gut 2001;48:526–35.

[121] Vogelstein B, Fearon ER, Hamilton SR, Kern SE, Preisinger AC, Leppert M, et al. Genetic alterations during colorectal-tumor development. N Engl J Med 1988;319:525–32.

[122] Yin J, Harpaz N, Tong Y, Huang Y, Laurin J, Greenwald BD, et al. p53 point mutations in dysplastic and cancerous ulcerative colitis lesions. Gastroenterology 1993;104:1633–9.

[123] Kern SE, Redston M, Seymour AB, Caldas C, Powell SM, Kornacki S, et al. Molecular genetic profiles of colitis-associated neoplasms. Gastroenterology 1994;107:420–8.

[124] Brentnall TA, Crispin DA, Rabinovitch PS, Haggitt RC, Rubin CE, Stevens AC, et al. Mutations in the p53 gene: an early marker of neoplastic progression in ulcerative colitis. Gastroenterology 1994;107:369–78.

[125] Endo Y, Marusawa H, Kou T, Nakase H, Fujii S, Fujimori T, et al. Activation-induced cytidine deaminase links between inflammation and the development of colitis-associated colorectal cancers. Gastroenterology. 2008;135:889–98. 898 e881–883.

[126] Fuss IJ, Neurath M, Boirivant M, Klein JS, de la Motte C, Strong SA, et al. Disparate CD4+ lamina propria (LP) lymphokine secretion profiles in inflammatory bowel disease. Crohn's disease LP cells manifest increased secretion of IFN-gamma, whereas ulcerative colitis LP cells manifest increased secretion of IL-5. J Immunol 1996;157:1261–70.

[127] Inoue S, Matsumoto T, Iida M, Mizuno M, Kuroki F, Hoshika K, et al. Characterization of cytokine expression in the rectal mucosa of ulcerative colitis: correlation with disease activity. Am J Gastroenterol 1999;94:2441–6.

[128] Hebenstreit D, Wirnsberger G, Horejs-Hoeck J, Duschl A. Signaling mechanisms, interaction partners, and target genes of STAT6. Cytokine Growth Factor Rev 2006;17:173–88.

[129] Bouma G, Strober W. The immunological and genetic basis of inflammatory bowel disease. Nat Rev Immunol 2003;3:521–33.

[130] Takeda K, Tanaka T, Shi W, Matsumoto M, Minami M, Kashiwamura S, et al. Essential role of Stat6 in IL-4 signalling. Nature 1996;380:627–30.

[131] Erdman SE, Poutahidis T. Roles for inflammation and regulatory T cells in colon cancer. Toxicol Pathol 2010;38:76–87.

[132] Takai A, Marusawa H, Minaki Y, Watanabe T, Nakase H, Kinoshita K, et al. Targeting activation-induced cytidine deaminase prevents colon cancer development despite persistent colonic inflammation. Oncogene 2012;31:1733–42.

[133] Morita S, Matsumoto Y, Okuyama S, Ono K, Kitamura Y, Tomori A, et al. Bile acid-induced expression of activation-induced cytidine deaminase during the development of Barrett's oesophageal adenocarcinoma. Carcinogenesis 2011;32:1706–12.

[134] Komori J, Marusawa H, Machimoto T, Endo Y, Kinoshita K, Kou T, et al. Activation-induced cytidine deaminase links bile duct inflammation to human cholangiocarcinoma. Hepatology 2008;47:888–96.

[135] Munoz DP, Lee EL, Takayama S, Coppe JP, Heo SJ, Boffelli D, et al. Activation-induced cytidine deaminase (AID) is necessary for the epithelial-mesenchymal transition in mammary epithelial cells. Proc Natl Acad Sci USA 2013;110:E2977–2986.

[136] Nakanishi Y, Kondo S, Wakisaka N, Tsuji A, Endo K, Murono S, et al. Role of activation-induced cytidine deaminase in the development of oral squamous cell carcinoma. PLoS One 2013;8:e62066.

[137] Shinmura K, Igarashi H, Goto M, Tao H, Yamada H, Matsuura S, et al. Aberrant expression and mutation-inducing activity of AID in human lung cancer. Ann Surg Oncol 2011;18:2084–92.

[138] Burns MB, Temiz NA, Harris RS. Evidence for APOBEC3B mutagenesis in multiple human cancers. Nat Genet 2013;45:977–83.

[139] Roberts SA, Lawrence MS, Klimczak LJ, Grimm SA, Fargo D, Stojanov P, et al. An APOBEC cytidine deaminase mutagenesis pattern is widespread in human cancers. Nat Genet 2013;45:970–6.

Chapter 22

# Molecular Pathogenesis of B Cell Lymphomas

Riccardo Dalla-Favera[1,2,3,4] and Laura Pasqualucci[1,2]

[1]*Institute for Cancer Genetics, Herbert Irving Comprehensive Cancer Center, Columbia University, New York, NY, USA;* [2]*Department of Pathology and Cell Biology, Columbia University, New York, NY, USA;* [3]*Department of Genetics and Development, Columbia University, New York, NY, USA;* [4]*Department of Microbiology and Immunology, Columbia University, New York, NY, USA*

## 1. INTRODUCTION

B cell lymphomas (BCLs) represent a heterogeneous group of biologically and clinically distinct neoplasms, including over 40 subtypes derived from the malignant transformation of mature B cells. This chapter focuses on the molecular pathogenesis of the most common and well-characterized types of BCL, including mantle cell lymphoma (MCL), follicular lymphoma (FL), diffuse large B cell lymphoma (DLBCL), Burkitt lymphoma (BL), and chronic lymphocytic leukemia (CLL), which also derives from mature B cells. Emphasis is given to the mechanisms of genetic lesion and the nature of the involved genes relative to the normal biology of B lymphocytes.

## 2. THE CELL OF ORIGIN OF LYMPHOMAS

### 2.1 BCLs and the Germinal Center Reaction

With the exception of MCL, most BCLs derive from the malignant expansion of B cells arrested at various stages of germinal center (GC) transit. Although we refer to other chapters in this book for a detailed description of the GC, we summarize here aspects that are relevant to BCL pathogenesis.

GCs are specialized structures that form in peripheral lymphoid organs when mature, naive B cells encounter a foreign antigen for the first time, in the context of signals delivered by CD4+ T cells and antigen-presenting cells (Figure 1) [1–3]. GCs are highly dynamic structures characterized by two histologically well-defined zones: the dark zone (DZ), which consists of rapidly proliferating centroblasts (CBs) (doubling time 6–12 h), and the light zone (LZ), which consists of more quiescent cells termed centrocytes (CCs), amid a network of resident accessory cells (follicular dendritic cells, FDCs, and follicular T helper cells) [4,5]. The DZ is the site where GC B cells modify the variable region of their immunoglobulin genes (*IGV*) by the process of somatic hypermutation (SHM), which introduces mostly single-nucleotide substitutions to change their affinity for the antigen [2,3,6–8]. Conversely, the LZ is the site of selection based on affinity to the antigen. A critical regulator of the GC reaction is BCL6 [9,10], a transcriptional repressor [11] whose expression in the mature B cell compartment is restricted to the GC and which is required for GC formation. BCL6 negatively regulates the expression of >400 genes, including those involved in B cell receptor (BCR) and CD40 signaling [12,13], T-cell-mediated B cell activation [12], induction of apoptosis [12,14], response to DNA damage (by modulation of genes involved in the sensing and execution of DNA damage responses) [15–18], various cytokine and chemokine signaling pathways (e.g., those triggered by interferon and TGFβ) [12,14], and plasma cell differentiation, via suppression of PRDM1/BLIMP1 [19–23]. This transcriptional program suggests that BCL6 is critical to establishing the proliferative status of CBs and allowing the execution of antigen-specific DNA modification processes (SHM and class-switch recombination (CSR)) without eliciting responses to DNA damage. In addition, BCL6 prevents the reception of a variety of signaling pathways that could lead to premature activation and differentiation prior to selection for the survival of cells producing high-affinity antibodies.

After SHM in the DZ, CBs move to the LZ, where they encounter the antigen again through the interaction with CD4+ T cells and FDCs [1,3]. CCs expressing a BCR with high affinity for the antigen are selected for survival and differentiation into memory cells and plasma cells [1,24] or reenter the DZ after stimulation by a variety of signals [4,5]. In the GC, CCs also undergo CSR, a DNA remodeling event that confers distinct effector functions to antibodies with identical specificities [24]. SHM and CSR represent B-cell-specific

Molecular Biology of B Cells. http://dx.doi.org/10.1016/B978-0-12-397933-9.00022-9

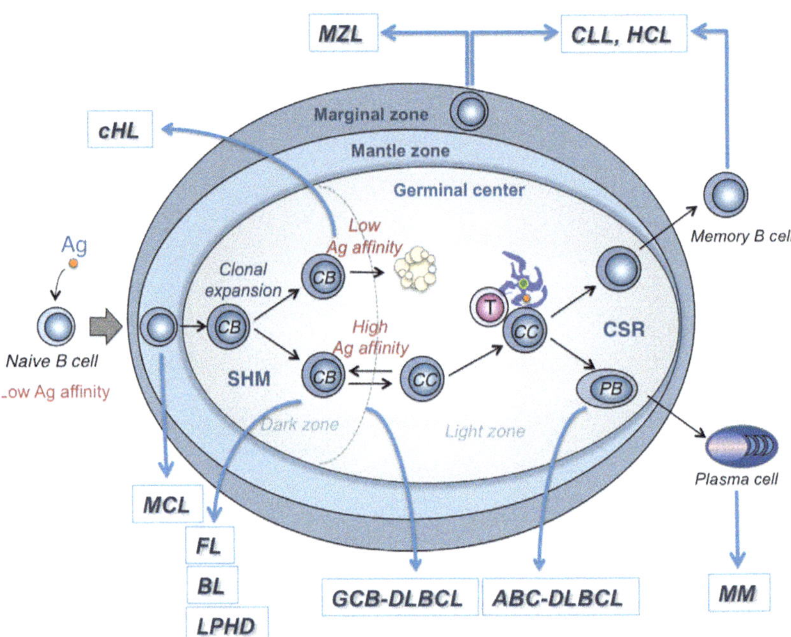


FIGURE 1 **The germinal center (GC) reaction and lymphomagenesis.** Schematic representation of a lymphoid follicle illustrating the GC, the mantle zone, and the surrounding marginal zone. B cells that have successfully rearranged their *IG* genes in the bone marrow move to peripheral lymphoid organs as naive B cells. Upon encountering a T-cell-dependent antigen, B cells become proliferating centroblasts (CB) in the dark zone of the GC and subsequently transition into centrocytes (CC) in the light zone (LZ); these cells shuttle continuously between the dark and the LZ while undergoing iterative rounds of somatic hypermutation (SHM) and selection. Only GC B cells with high affinity for the antigen are positively selected to leave the GC and terminally differentiate into plasma cells or memory B cells; low-affinity clones are eliminated by apoptosis. Thick, blue arrows link various lymphoma subtypes to their postulated normal counterpart, as identified based on the expression of phenotypic markers and the presence of somatically mutated *IGV* genes. MCL, mantle cell lymphoma; FL, follicular lymphoma; BL, Burkitt lymphoma; DLBCL, diffuse large B cell lymphoma (GCB, germinal center B-cell-like; ABC, activated B-cell-like); MZL, marginal zone lymphoma of mucosa-associated lymphoid tissue; CLL, chronic lymphocytic leukemia; HCL, hairy cell leukemia; cHL, classical Hodgkin lymphoma; LPHD, lymphocyte predominant Hodgkin disease; MM, multiple myeloma.

functions that modify the genome of B cells via mechanisms involving single- or double-strand breaks and depend on the activity of the activation-induced cytidine deaminase (AID) enzyme [25–27], a notion that will become important in the understanding of the mechanisms generating genetic alterations in BCL. Once these processes are completed, two critical signals for licensing GC exit are represented by engagement of the BCR by the antigen and activation of the CD40 receptor by the CD40 ligand present on $CD4^+$ T cells. These signals induce the activation of multiple pathways and also lead to downregulation of BCL6 at both the translational and the transcriptional level, thus restoring DNA damage responses, activation, and differentiation capabilities.

This oversimplified description of the GC reaction is useful to introduce two basic concepts for the understanding of BCL pathogenesis. First, the presence of SHM-mediated mutations in their rearranged *IGV* genes led to the conclusion that most BCL types derive from the clonal expansion of GC-experienced B cells [28]. Second, most of the genetic lesions associated with BCL—namely chromosomal translocations and aberrant somatic hypermutation (ASHM)—are due to mistakes in the machinery that normally diversifies the *IG* genes during B lymphocyte differentiation, further supporting the GC origin of most BCL (Figure 2) [29].

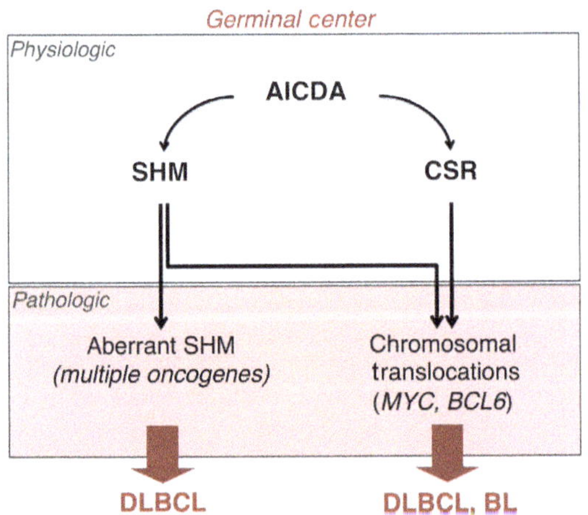


FIGURE 2 **Model for the initiation of chromosomal translocations and aberrant somatic hypermutation (ASHM) during lymphomagenesis.** B cell non-Hodgkin lymphoma (B-NHL)-associated genetic lesions are favored by mistakes occurring during the physiologic processes of SHM and class-switch recombination (CSR) in the highly proliferative environment of the germinal center (top). These events lead to chromosomal translocations, which in most cases juxtapose the *IG* genes to one of several proto-oncogenes (e.g., *BCL2* or *MYC*), and ASHM of multiple target genes, contributing to the pathogenesis of lymphoma. AICDA, activation-induced cytidine deaminase; DLBCL, diffuse large B cell lymphoma; BL, Burkitt lymphoma.

# 3. MECHANISMS OF GENETIC LESION IN LYMPHOMA

## 3.1 Chromosomal Translocations

Chromosomal translocations represent the genetic hallmark of malignancies derived from the hematopoietic system. These events are generated through the reciprocal and balanced recombination of two chromosomes and are often recurrently associated with a given tumor type, in which they are clonally represented in each tumor case. In B cell malignancies, chromosomal translocations occur at least in part as a consequence of mistakes in *IG* gene remodeling mechanisms and, based on the characteristics of the chromosomal breakpoint, can be broadly divided into three groups: (1) translocations derived from mistakes of the RAG-mediated V(D)J recombination process, as it is the case for translocations involving *IGH* and *CCND1* in MCL or *IGH* and *BCL2* in FL [29–31]; (2) translocations mediated by errors in the AID-dependent CSR process, such as those involving the *IG* genes and *MYC* in sporadic BL [29]; and (3) translocations occurring as by-products of the AID-mediated SHM mechanism, which also generates DNA breaks, such as those joining the *IG* and *MYC* loci in endemic BL [29]. In lymphoma-prone mouse models, removal of AID was sufficient to abrogate the generation of *MYC–IGH* translocations in normal B cells undergoing CSR [32,33] and to prevent the development of GC-derived B cell non-Hodgkin lymphoma (B-NHL) [34,35], thus providing conclusive experimental evidence for the involvement and requirement of *IG* gene remodeling mechanisms in the pathogenesis of BCL.

In all BCL-associated chromosomal translocations, the chromosomal recombination site occurs in the proximity of a proto-oncogene. In contrast to acute leukemias, the coding domain of the oncogene is not affected by the translocation and no gene fusion is generated. Conversely, heterologous regulatory sequences derived from the partner chromosome are juxtaposed in front of the oncogene, deregulating its pattern of expression (proto-oncogene deregulation) (Figure 3). This process is defined as homotopic when a proto-oncogene whose expression is tightly regulated in the normal tumor counterpart becomes constitutively expressed in the lymphoma cell, and heterotopic when the proto-oncogene is not expressed in the putative normal counterpart of the tumor cell and undergoes ectopic expression in the lymphoma. The heterologous regulatory sequences responsible for proto-oncogene deregulation are often derived from antigen receptor loci, which are expressed at high levels in the target tissue [29]. However, in certain translocations, such as the ones involving *BCL6* in DLBCL, different promoter regions from distinct chromosomal sites can be found juxtaposed to the proto-oncogene in individual tumor cases, leading to so-called "promiscuous" translocations [36–43]. The molecular cloning of the genetic loci involved in most recurrent translocations has led to the identification of a number of proto-oncogenes that play important roles lymphomagenesis.

## 3.2 Copy Number Gains and Amplifications

In addition to chromosomal translocations, the structure of proto-oncogenes and their pattern of expression can be

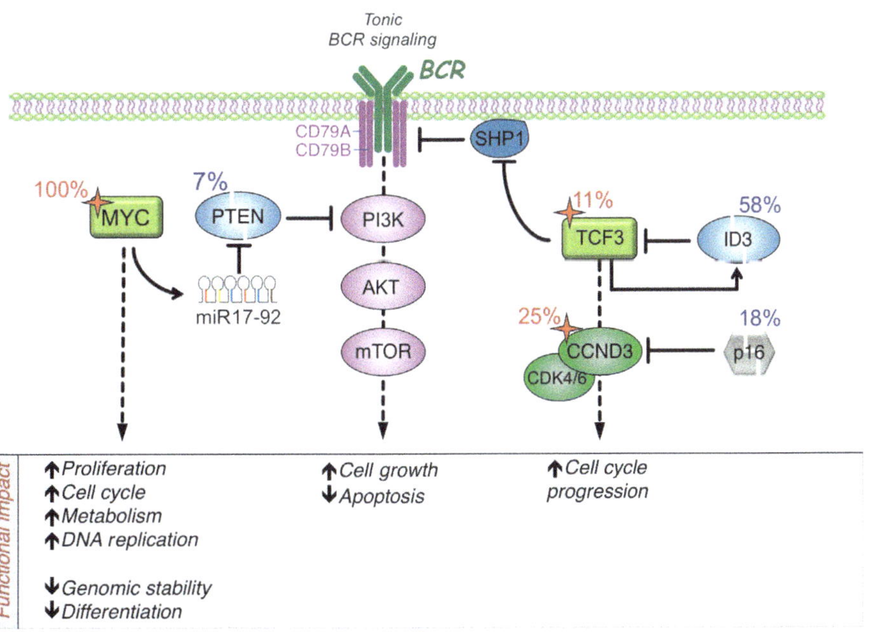


FIGURE 3 **Molecular basis of Burkitt lymphoma (BL) pathogenesis.** The most common genetic lesions associated with BL are shown. Stars indicate genes targeted by activating mutations and breaks denote genes targeted by inactivating events.

altered by copy number (CN) gains and amplifications, which lead to overexpression of an intact protein. Compared to epithelial cancers, only a few genes have been identified so far as specific targets of amplification in BCL, including *REL* and *BCL2* in DLBCL [44–47] and the genes encoding PD-1 ligands in primary mediastinal B cell lymphoma (PMBCL) [48].

## 3.3 Gain-of-Function Mutations

The biological properties of a proto-oncogene can also be altered by somatic point mutations affecting its coding sequences, which may lead to stabilization or constitutive activation of its protein product by several mechanisms. The use of massively parallel sequencing technologies has revolutionized the field by allowing the identification of numerous previously unappreciated targets of somatic mutations in cancer, including lymphomas. These genes are discussed here in individual disease sections. Of note, mutations in the *RAS* genes, a very frequent proto-oncogene alteration in human neoplasia, are rare in lymphomas [49].

## 3.4 Inactivating Mutations and Deletions

Inactivation of the *TP53* tumor suppressor gene, possibly the most common target of genetic alteration in human cancer [50], is found at relatively low frequencies and restricted to specific disease subtypes, such as BL and DLBCL derived from the transformation of FL or CLL [51,52]. The mechanism of *TP53* inactivation in BCL is analogous to that observed in human neoplasia in general, entailing point mutation of one allele and chromosomal deletion or mutation of the second allele. Recent efforts taking advantage of genome-wide technologies revealed several additional candidate tumor suppressor genes that are lost in BCL through specific chromosomal deletions and/or deleterious mutations. These include the *PRDM1/BLIMP1* gene on 6q21, which is biallelically inactivated in ~25% of activated B-cell-like (ABC) DLBCL cases [53–55], and the gene encoding the negative NF-κB regulator A20 on chromosome 6q23, which is inactivated in ABC-DLBCL, PMBCL, and subtypes of marginal-zone and Hodgkin lymphoma (HL) [56–59]. Monoallelic inactivating mutations and deletions were found to affect the acetyltransferase genes *CREBBP* and *EP300* in a significant proportion of DLBCL and FL, suggesting a role as haploinsufficient tumor suppressors [60]. These two lymphoma types also harbor truncating mutations of *MLL2*, which encodes a methyltransferase and is mutated in multiple cancer types [61]. In over 50% of CLL cases, CN losses affect the *DLEU2/miR15-a/16.1* cluster on chromosome 13q14.3 [62–66], whereas the *CDKN2A/B* (*p16/INK4a*) locus is inactivated by focal homozygous deletions in a substantial fraction of transformed FL (tFL), Richter syndrome (RS), and ABC-DLBCL cases and more rarely by epigenetic transcriptional silencing in various BCLs [67–71].

## 3.5 Aberrant SHM

The term "aberrant somatic hypermutation" refers to a malfunction in the physiologic SHM process, leading to the mutation of multiple non-*IG* genes [72] in specific subtypes of BCL, and most commonly in DLBCL, in which over 10% of actively transcribed genes have been found mutated as a consequence of ASHM.

In normal GC B cells, the process of SHM is tightly regulated and introduces mutations only in the rearranged *IgV* genes [6] as well as in the 5′ region of a few other genes, including *BCL6* and the *CD79* components of the BCR [73–76], although the functional role of mutations found in non-*IG* genes remains obscure. Conversely, multiple mutational events can be found in numerous loci in over half of DLBCL cases and, at lower frequencies. in a few other lymphoma types [72,77–81]. The identified target loci include several well-known proto-oncogenes such as *PIM1, PAX5*, and *MYC* [72]. These mutations are typically distributed within ~2 kb from the transcription initiation site (the hypermutable domain in the *IG* locus) and, depending on the genomic configuration of the target gene, may affect nontranslated as well as coding regions, with the potential of altering their function [72]. Indeed, a significant number of amino acid substitutions in the MYC protein were documented to have functional consequences in activating its oncogenic potential [82]. Nonetheless, the mechanism involved in ASHM and a comprehensive genome-wide characterization of its consequences are still lacking.

## 3.6 Infectious Agents

Genetic changes, by virtue of the introduction of viral genomes in B cells, are also associated with the pathogenesis of lymphoma. At least two viruses are associated with specific BCLs by direct infection of B cells: Epstein–Barr virus (EBV) and human herpesvirus-8 (HHV-8/KSHV). Other infectious agents, including human immunodeficiency virus (HIV), hepatitis C virus, *Helicobacter pylori*, and *Chlamydophila psittaci*, have an indirect role in NHL pathogenesis by impairing the immune system and/or providing chronic antigenic stimulation.

EBV was initially identified in cases of endemic African BL [83,84] and subsequently detected also in a fraction of sporadic BL, HIV-related lymphomas, and primary effusion lymphomas (PEL) [85–90]. Much evidence supports a pathogenetic role for this virus in NHL. First, EBV is able to significantly alter the growth of B cells [91]; second, EBV-infected lymphomas usually display a single form of the otherwise highly heterogeneous fused EBV genome termini, suggesting that the lymphoma cell population derived from the clonal expansion of a single infected cell [85,86]. Nonetheless, the role of EBV in lymphomagenesis remains unclear, because the virus infects virtually all humans at some point in their life and its transforming genes are not commonly expressed in the tumor cells of BL.

The HHV-8 gamma-herpesvirus was initially identified in tissues of HIV-related Kaposi sarcoma [92] and subsequently found to infect PEL cells as well as a large proportion of multicentric Castleman disease [93–96]. Phylogenetic analysis has shown that the closest relative of HHV-8 is Herpesvirus saimiri, a gamma-2 herpesvirus of primates associated with T cell lymphoproliferative disorders [97]. Like other gamma-herpesviruses, HHV-8 is also lymphotropic and can infect lymphocytes both in vitro and in vivo [92,95,96]. Lymphoma cells naturally infected by HHV-8 harbor the viral genome in its episomal configuration and display a marked restriction of viral gene expression, suggesting a pattern of latent infection [97].

## 4. MOLECULAR PATHOGENESIS OF MOST COMMON LYMPHOMA TYPES

### 4.1 Mantle Cell Lymphoma

MCL comprises ~5% of all BCL diagnoses and is generally regarded as incurable [67,98]. Based on its mature B cell phenotype, and the absence of SHM-associated mutations in the majority of cases, MCL has been historically considered as derived from naive, pre-GC peripheral B cells located in the inner mantle zone of secondary follicles (Figure 1). More recently, the identification of cases that harbor mutated *IGHV* genes (15–40% of all diagnoses) has revealed the existence of distinct molecular subtypes, including one influenced by the GC environment.

MCL is typically associated with the t(11;14)(q13;q32) translocation, which juxtaposes the *IGH* gene to a region containing the *CCND1* gene on chromosome 11q13 [99–101]. The translocation causes heterotopic deregulation and overexpression of cyclin D1, a member of the D-type $G_1$ cyclins that regulates the early phases of the cell cycle and is normally not expressed in resting B cells [102–104]. In addition to t(11;14), ~10% of MCLs overexpress aberrant or shorter cyclin D1 transcripts, as a consequence of secondary rearrangements, microdeletions, or point mutations in the gene 3′ untranslated region [105–107]. These alterations lead to cyclin D1 overexpression through the removal of destabilizing sequences and the consequent increase in the mRNA half-life, and are more commonly observed in cases characterized by high proliferative activity and a more aggressive clinical course [108]. The pathogenic role of cyclin D1 deregulation in human neoplasia is suggested by the ability of the overexpressed protein to transform cells in vitro and to promote B cell lymphomagenesis in transgenic mice, although only when combined to other oncogenic events [109,110]; however, an animal model that faithfully recapitulates the features of the human MCL is still lacking. Importantly, the frequency and specificity of lesions causing the ectopic expression of cyclin D1 in the tumor cells provide an excellent immunohistochemical marker for MCL diagnosis [98].

Other genetic alterations involved in MCL include genomic deletions and mutations inactivating the *ATM* gene (50% of SOX11-positive cases) [111], loss of *TP53* (20% of patients, typically associated with adverse prognosis) [112], and inactivation of the *CDKN2A* gene by deletions, point mutations, or promoter hypermethylation [113]. Also associated with aggressive tumors are mutations activating the Notch signaling pathway, including *NOTCH1* (12% of clinical samples) and *NOTCH2* (5% of samples). These alterations are mutually exclusive and mostly consist of truncating events that remove the PEST sequences required for protein degradation, leading to protein stabilization [114,115]. In a small number of cases, *BMI1* is amplified and/or overexpressed, possibly as an alternative mechanism to the loss of the *CDKN2A* cell cycle regulator gene [116]. Finally, less common, yet recurrent and thus presumably functionally significant mutations involve genes encoding the antiapoptotic protein BIRC3, the Toll-like receptor 2 (TLR2), the chromatin modifiers WHSC1 and MLL2, and the MEF2B transcription factor [114].

### 4.2 Burkitt Lymphoma

BL is an aggressive lymphoma deriving from GC B cells and resembling in particular transformed DZ CBs, as documented by the presence of highly mutated *IGHV* sequences [117–120] and the expression of a related transcriptional signature [121,122]. BL includes three clinical variants: sporadic BL (sBL), endemic BL (eBL), and HIV-associated BL, which is often diagnosed as the initial manifestation of acquired immunodeficiency syndrome (AIDS) [98].

All BL cases are associated with chromosomal translocations involving the *MYC* gene on region 8q24 and one of the *IG* loci on the partner chromosome [123,124]. In ~80% of cases, this is represented by the *IGH* locus, leading to the characteristic t(8;14)(q24;q32), whereas in the remaining 20% of cases either *IGκ* (2p12) or *IGλ* (22q11) is involved [123–126]. Although fairly homogeneous at the microscopic level, these translocations show a high degree of molecular heterogeneity, because the breakpoints are located 5′ and centromeric to *MYC* in t(8;14) but map 3′ to *MYC* in t(2;8) and t(8;22) [127]. Further molecular heterogeneity derives from the exact breakpoint sites observed on chromosomes 8 and 14 in t(8;14): translocations of eBL tend to involve sequences at an undefined distance (>1000 kb) 5′ to *MYC* on chromosome 8 and sequences within or in proximity to the *IG* $J_H$ region on chromosome 14 [128,129]. In sBL, t(8;14) preferentially involves sequences within or immediately

5′ to *MYC* (<3 kb) on chromosome 8 and within the *IG*-switch regions on chromosome 14 [128,129]. The different molecular architecture of these translocations is thought to reflect the distinct mechanism of *IG* gene remodeling involved in their generation, i.e., CSR in sBL and AIDS-BL and SHM in eBL [29].

The common consequence of t(8;14), t(2;8), and t(8;22) is the ectopic and constitutive overexpression of the MYC proto-oncogene [130–132], which is normally absent in the majority of proliferating GC B cells [9], in part because of BCL6-mediated transcriptional repression [133]. Oncogenic activation of MYC in BL is mediated by at least three distinct mechanisms: (1) juxtaposition of the *MYC* coding sequences to heterologous enhancers derived from the *IG* loci [130–132]; (2) structural alterations of the gene 5′ regulatory sequences, which alter the responsiveness to cellular factors controlling its expression [134]; in particular, the *MYC* exon 1/intron 1 junction encompasses critical regulatory elements that are either decapitated by the translocation or mutated in the translocated alleles; (3) amino acid substitutions within the gene exon 2 encoding the protein transactivation domain [135,136]; these mutations can abolish the ability of p107, a nuclear protein related to RB1, to suppress MYC activity [137] or can increase protein stability [82,138].

MYC is a nuclear phosphoprotein that functions as a sequence-specific DNA-binding transcriptional regulator controlling the transcription of target genes with diverse roles in regulating cell growth by affecting DNA replication, energy metabolism, protein synthesis, and telomere elongation [139–141]. The deregulated expression of this function is typically involved in malignant transformation. Furthermore, deregulated MYC expression is thought to cause genomic instability and thus contribute to tumor progression by facilitating the occurrence of additional genetic lesions [142]. Dysregulation of MYC expression in a number of transgenic mouse models leads to the development of aggressive BCLs with high penetrance and short latency [82,143,144]. In particular the deregulated expression of MYC in mature B cells leads to lymphomagenesis with features highly reminiscent of human BL [144].

Exome sequencing has revealed additional oncogenic mechanisms that cooperate with MYC in the development of this aggressive lymphoma. Mutations of the transcription factor TCF-3 (10–25%) and its negative regulator ID3 (35–58%) are highly recurrent in all three subtypes of BL, in which they promote tonic (antigen-independent) BCR signaling and sustain survival of the tumor cell by engaging the phosphoinositide 3-kinase pathway (Figure 4) [145]. In addition, TCF-3 can promote cell-cycle progression by transactivating CCND3. Notably, *CCND3* is itself a target of gain-of-function mutations in 38% of sBL; these events affect conserved residues in the carboxyl terminus of this D-type cyclin, which are implicated in the control of protein stability, leading to higher expression levels. Other common genetic lesions include loss of *TP53* by mutation and/or deletion (35% of both sBL and eBL cases) [51], inactivation of *CDKN2B* by deletion or hypermethylation (17% of samples) [71], and deletions of 6q, detected in ~30% of cases, independent of the clinical variant [146]. Finally, one contributing factor to the development of BL is monoclonal EBV infection, present in virtually all cases of eBL and in ~30% of sBL and AIDS-BL [83,85,147,148]. However, BL cells lack the expression of both EBV transforming antigens LMP1 and EBNA2, thus casting doubts on the pathogenetic role of this virus in BL [149].

## 4.3 Follicular Lymphoma

FL, the second most common type of B-NHL (~20% of diagnoses), is an indolent but largely incurable disease,

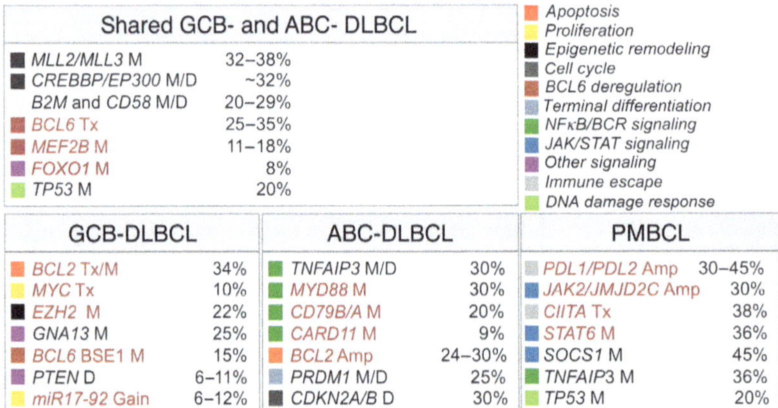


FIGURE 4 **Genetic lesions associated with DLBCL.** The most common genetic alterations identified in GCB-DLBCL, ABC-DLBCL, and PMBCL are shown; some of these lesions can be variably found in both phenotypic subtypes of the disease, whereas others are preferentially segregated with individual molecular subtypes. Loss-of-function alterations are in black and gain-of-function events are in red. The biological functions/signaling pathways affected by the lesions are indicated by color-coded squares as explained in the key. DLBCL, diffuse large B cell lymphoma; GCB, germinal center B cell-like; ABC, activated B-cell-like; PMBCL, primary mediastinal B cell lymphoma; M, mutation; D, deletion; Tx, translocation; Amp, amplification.

characterized by a continuous pattern of progression and relapse that often culminates in its histologic transformation to an aggressive lymphoma with a DLBCL phenotype and a dismal prognosis (20–30% of cases) [150,151]. The ontogeny of FL from a GC B cell is supported by the expression of specific GC B cell markers such as BCL6 and CD10, together with the presence of somatically mutated *IG* genes showing evidence of ongoing SHM activity [1].

Chromosomal translocations affecting the *IG* and *BCL2* loci on chromosome band 18q21 are detected in 80–90% of cases [100,152–155]. These rearrangements join the 3′ untranslated region of *BCL2* to an *IG* $J_H$ segment, resulting in the ectopic expression of BCL2 in GC B cells [152,153, 156–160], in which its transcription is normally repressed by BCL6 [14,23]. Approximately 70% of the breakpoints on chromosome 18 cluster within the major breakpoint region, and the remaining 5–25% map to the more distant minor cluster region, located ~20 kb downstream of the *BCL2* gene [152,153,156,157]. Rearrangements involving the 5′ flanking region of *BCL2* have been described in a minority of cases [161]. The *BCL2* gene encodes a 26-kDa integral membrane protein that controls the cell apoptotic threshold by preventing programed cell death and may thus contribute to lymphomagenesis by inducing apoptosis resistance in tumor cells independent of antigen selection. Additional genetic aberrations include mutations in multiple epigenetic modifiers, including the methyltransferase *MLL2* (80–90% of cases), the Polycomb-group oncogene *EZH2* (7% of patients), the acetyltransferases *CREBBP* and *EP300* (40% of cases), and multiple linker histone genes, which may all contribute to transformation by remodeling the epigenetic landscape of the precursor tumor cell [60,61,162,163]. A major role is also played by chronic antigen stimulation [164,165].

Exome sequencing and CN analysis of sequential, clonally related FL and tFL biopsies has allowed the characterization of the molecular events that are specifically acquired during histologic progression to DLBCL. Transformed-FL-specific lesions include inactivation of *CDKN2A/B* through deletion, mutation, and hypermethylation (one-third of patients) [68,166]; rearrangements and amplifications of *MYC* [167]; *TP53* mutations/deletions (25–30% of cases) [52,168–170]; loss of chromosome 6 (20%) [146]; ASHM [68]; and, although larger cohorts of patients will need to be studied, biallelic inactivation of the gene encoding *B2M* [68]. The last leads to loss of HLA class I expression on the cell surface of the tumor cells, suggesting that escape from immune surveillance may be important for FL transformation to DLBCL.

## 4.4 Diffuse Large B Cell Lymphoma

DLBCL accounts for ~40% of all new diagnoses in adulthood, including cases that arise de novo and cases that derive from the clinical evolution of various, less aggressive B-NHL types such as FL and CLL [98,171]. Based on gene expression profile analysis, at least three well-characterized molecular subtypes have been recognized within this diagnostic entity, and they reflect the derivation from various stages of B cell development [172]. Germinal center B cell-like (GCB) DLBCL appears to derive from GC B cells with a phenotype intermediate between CB and CC [172,173]; ABC-DLBCL shows a transcriptional signature related to BCR-activated B cells or to B cells committed to plasmablastic differentiation; and PMBCL is postulated to arise from post-GC thymic B cells; the remaining 15–30% of cases remain unclassified [172,174–176]. Stratification according to gene expression profiles has prognostic value, because patients diagnosed with GCB-DLBCL display better overall survival compared to those with ABC-DLBCL [47,177,178].

The heterogeneity of DLBCL is reflected by the multitude of genetic lesions that are associated with its pathogenesis. Compared to other B cell malignancies, DLBCL shows a significantly higher degree of genomic complexity, harboring on average ~50–100 lesions/case, with significant diversity across patients [61,179,180]. Although many of the lesions identified can be variably found in both molecular subtypes of the disease, consistent with a general role during transformation, others seem to be preferentially or exclusively associated with individual DLBCL subtypes, indicating that GCB- and ABC-DLBCL utilize distinct oncogenic pathways (Figure 5).

The most prominent program disrupted in DLBCL, independent of subtype, is represented by epigenetic modifications due to mutations in the *CREBBP/EP300* acetyltransferase genes (35% of cases) and the *MLL2* H3K4 trimethyltransferase (~30% of all DLBCLs) [60,61]. These lesions may favor tumor development by reprogramming the cancer epigenome and, in the case of *CREBBP/EP300*, by altering the balance between the activity of the BCL6 oncogene, which is typically inactivated by acetylation, and that of the tumor suppressor p53, which requires acetylation at specific residues for its function [60].

Deregulated activity of the BCL6 oncoprotein by a multitude of genetic lesions is also a major contributor to DLBCL pathogenesis, in both GCB- and ABC-DLBCL. Chromosomal rearrangements of the *BCL6* gene at band 3q27 are observed in up to 35% of cases [181–183], although with a twofold higher frequency in the ABC-DLBCL subtype [184]. These rearrangements juxtapose the intact coding domain of *BCL6* downstream and in the same transcriptional orientation to heterologous sequences derived from the partner chromosome, including *IGH* (14q23), *IGκ* (2p12), *IGλ* (22q11), and at least 20 other chromosomal sites unrelated to the *IG* loci [36–43]. The majority of these translocations result in a fusion transcript in which the promoter region and the first noncoding exon of *BCL6* are replaced by sequences derived from the partner gene [37,185]. Because the common

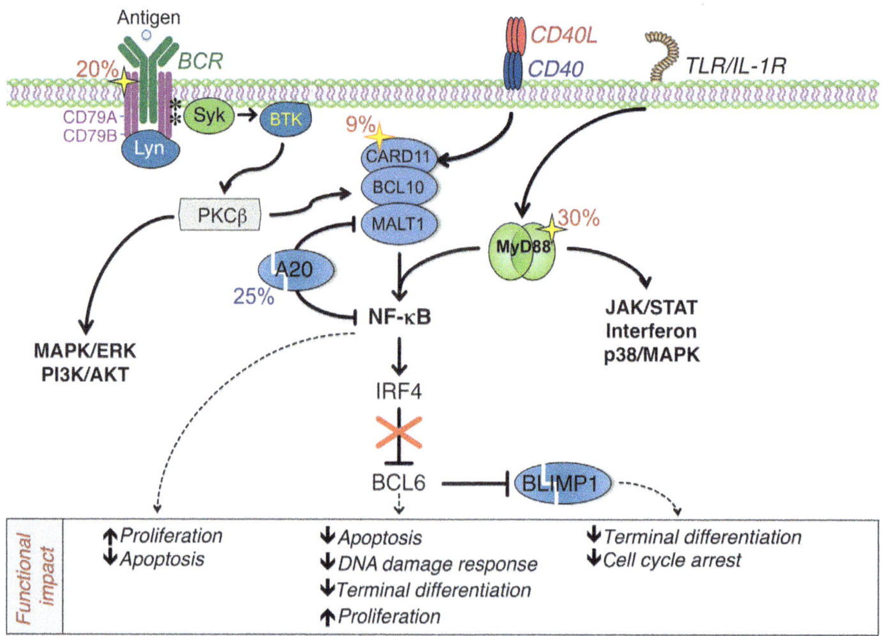


FIGURE 5 **Pathway lesions in ABC-DLBCL.** Schematic representation of the signaling pathways induced after engagement of the BCR by the antigen (sphere), CD40–CD40L interaction, and activation of the TLR. All these signals converge on activation of the NF-κB pathway and lead to upregulation of multiple genes including IRF4 and A20. IRF4, in turn, downregulates BCL6 expression, allowing the release of BLIMP1 expression. This pathway is disrupted at multiple levels in ABC-DLBCL owing to genetic lesions that favor the antiapoptotic and pro-proliferative function of NF-κB, induce chronic active BCR and JAK/STAT3 signaling, and block terminal B cell differentiation through mutually exclusive deregulation of BCL6 and inactivation of BLIMP1. DLBCL, diffuse large B cell lymphoma; ABC, activated B-cell-like; BCR, B cell receptor; TLR, Toll-like receptor.

denominator of these promoters is a broader spectrum of activity throughout B cell development, including expression in the post-GC differentiation stage, the translocation is thought to prevent the downregulation of BCL6 expression that is normally associated with differentiation into post-GC cells. Deregulated expression of an intact *BCL6* gene product is also sustained by a variety of indirect mechanisms, including gain-of-function mutations in its positive regulator MEF2B (~11% of cases) [186], inactivating mutations/deletions of *CREBBP/EP300* [60], and mutations/deletions of *FBXO11* (~5%) [187], which encodes a ubiquitin ligase involved in the control of BCL6 protein degradation. These lesions play a critical role in lymphomagenesis by enforcing the proliferative phenotype typical of GC cells, by suppressing proper DNA damage responses, and by blocking terminal differentiation. Accordingly, a mouse model of enforced deregulated BCL6 expression in GC B cells develops DLBCL [188].

DLBCL cells have also acquired the ability to escape both arms of immune surveillance, including cytotoxic T lymphocyte-mediated cytotoxicity (through genetic loss of the *B2M/HLA-I* genes) and NK cell-mediated death (through genetic loss of the CD58 molecule) [189]. Analogous effects may be achieved in PMBCL by disruption of the MHC-II transactivator CIITA and amplification of the genes encoding the immunomodulatory proteins *PDL1/PDL2* [190]. Finally, approximately 50% of all DLBCLs are associated with ASHM [72]. The number and identity of the genes that accumulate mutations in their coding and noncoding regions due to this mechanism varies in different cases and is still largely undefined. ASHM may therefore contribute to the heterogeneity of DLBCL via the alteration of different cellular pathways in different cases.

### 4.4.1 GCB-DLBCL

Genetic lesions specific to GCB-DLBCL include the t(14;18) and t(8;14) translocations, which deregulate the BCL2 and MYC oncogenes in 34% and up to 10% of cases, respectively [23,47,191–193]. Also exquisitely restricted to this subtype are mutations of the *EZH2* gene [162], which encodes a histone methyltransferase responsible for trimethylating Lys27 of histone H3; mutations of the S1PR2 adaptor protein GNA13 [61,179], which is involved in the ability of DLBCL to spread; deletions of the tumor suppressor *PTEN* [69,194]; and mutations affecting an autoregulatory domain within the BCL6 untranslated exon 1 [184,195,196]. The 5′ sequence of *BCL6* is mutated in up to 75% of DLBCL cases [75,197,198] reflecting the activity of the physiologic SHM mechanism that operates in normal GC B cells [75,199]. However, functional analysis of numerous mutated *BCL6* alleles uncovered a subset of mutations that are specifically associated with GCB-DLBCL [195] and deregulate BCL6 transcription by disrupting an autoregulatory circuit through which the BCL6 protein controls its own expression levels by binding to the promoter region of the gene [195,196] or by preventing CD40-induced BCL6 downregulation in post-GC B cells [200]. However, the full extent of mutations deregulating BCL6 expression has not been characterized, and therefore the fraction of DLBCL cases carrying abnormalities in BCL6 remains unclear.

### 4.4.2 ABC-DLBCL

A predominant feature of ABC-DLBCL is the constitutive activation of the NF-κB signaling pathway, initially evidenced by the selective expression of a signature enriched in NF-κB target genes and by the requirement of NF-κB for proliferation and survival in ABC-DLBCL cell lines [201]. Constitutive NF-κB activation is caused by a variety of alterations affecting positive and negative regulators of NF-κB. In up to 30% of cases, the *TNFAIP3* gene encoding the negative regulator A20 is biallelically inactivated by mutations and/or deletions, thus preventing termination of NF-κB responses [56,57]. The tumor suppressor role of A20 was documented by the observation that reconstitution of A20-knockout cell lines with a wild-type protein induces apoptosis and blocks proliferation, in part due to suppression of NF-κB activity [56,57]. In an additional ~10% of ABC-DLBCL cases, the *CARD11* gene is targeted by oncogenic mutations clustering in the protein coiled-coil domain and enhancing its ability to transactivate NF-κB target genes [202]. Finally, nearly 30% of ABC-DLBCL cases recurrently show a hot-spot mutation (L265P) in the intracellular Toll/interleukin-1 receptor domain of the MyD88 adaptor molecule, which has the potential to activate NF-κB as well as JAK/STAT3 transcriptional responses [203]. At lower frequencies, mutations were found in a number of genes that encode NF-κB components, overall accounting for over half of all ABC-DLBCLs [56] and suggesting that yet unidentified lesions may be responsible for the NF-κB activity in the remaining fraction of cases.

ABC-DLBCLs also display evidence of chronic active BCR signaling (leading to NF-κB activation), which is associated with somatic mutations affecting the immunoreceptor tyrosine-based activation motif signaling modules of *CD79B* and *CD79A* in 20% of ABC-DLBCL samples, but rarely in other DLBCLs [204]. Moreover, silencing of several BCR proximal and distal subunits is toxic to ABC-DLBCL. These findings provided genetic evidence supporting the development of therapies targeting BCR signaling [204]. Indeed, kinase inhibitors that interfere with this signaling pathway, and in particular the Bruton tyrosine kinase inhibitor Ibrutinib, are emerging as a new treatment paradigm for ABC-DLBCL.

Terminal B cell differentiation is disrupted in ~25% of ABC-DLBCL by biallelic truncating or missense mutations and/or genomic deletions involving the *PRDM1/BLIMP1* gene, and in another 25% of cases by transcriptional repression of BLIMP1 through constitutively active BCL6 alleles involved in chromosomal translocations [53–55]. The *PRDM1* gene encodes a zinc finger transcriptional repressor that is required for terminal B cell differentiation [205]. Thus, BLIMP1 inactivation contributes to lymphomagenesis by blocking post-GC B cell differentiation. Consistently, translocations deregulating the *BCL6* gene and BLIMP1 inactivation are mutually exclusive in DLBCL, suggesting that these alterations represent alternative oncogenic mechanisms contributing to lymphomagenesis by blocking post-GC differentiation (Figure 5).

### 4.4.3 PMBCL

*PMBCL* is postulated to derive from thymic B cells in the mediastinum and displays a unique transcriptional profile highly similar to HL [175,176]. A genetic hallmark of both PMBCL and HL is the amplification of chromosomal region 9q24, detected in nearly 50% of patients [69,206]. This relatively large interval encompasses multiple genes of possible pathogenetic significance, including the gene encoding the JAK2 tyrosine kinase and the *PDL1/PDL2* genes, which encode inhibitors of T cell responses [48,69,206,207] and have been linked to impaired antitumor immune responses in several cancers. Other lesions affecting regulators of immune responses in PMBCL include genomic breakpoints and mutations of the MHC class II transactivator gene *CIITA*, which may reduce tumor cell immunogenicity by downregulating surface HLA class II expression [190]. In addition to contributing to immune escape, elevated expression levels of these genes may in part explain the unique features of these lymphoma types, which are characterized by a significant inflammatory infiltrate. PMBCL also shares with HL the presence of genetic lesions affecting the NF-κB pathway and the deregulated expression of receptor tyrosine kinases [59,208–210]. In particular, mutations of the transcription factor *STAT6*, amplifications/overexpression of *JAK2* (which promotes STAT6 activation via IL-3/IL-4), and inactivating mutations of its negative regulator *SOCS1* are highly recurrent in PMBCL. Taken together, these lesions suggest that deregulation of the JAK/STAT signaling pathway is an important contributor to PMBCL pathogenesis.

### 4.4.4 DLBCL Derived from CLL and FL Transformation

Exome sequencing of sequential biopsies of CLL/RS and FL/tFL have provided insights onto the mechanisms underlying these transformation processes. These studies allowed the reconstruction of the evolutionary history of the dominant tumor clone during transformation, revealing that FL and tFL derive from a common mutated precursor clone by divergent evolution involving the disruption of distinct genes and pathways, including *CDKN2A/B* loss, *TP53* loss, *MYC* translocations, ASHM, and *B2M* inactivation [68,163]. Conversely, RS typically derives from the dominant CLL clone through a linear pattern, involving the maintenance of the CLL-associated lesions and the acquisition of new ones, including *NOTCH1* mutations, *CDKN2A/B* loss, *TP53* loss, and *MYC* translocations [70]. Finally, comparison with de novo DLBCL showed that, despite their morphologic resemblance, the genomic landscapes of RS and tFL are largely unique because they are characterized by distinct combinations of alterations otherwise not commonly observed in de novo DLBCL [68,70]. In conclusion, the histologic diagnosis of DLBCL may include at least five genetically distinct diseases: GCB-DLBCL, ABC-DLBCL, PMBCL, tFL, and RS DLBCL. This distinction may have important future implications for the development of targeted therapies.

## 4.5 Chronic Lymphocytic Leukemia

CLL is an indolent malignancy whose precise cellular origin has been remarkably controversial. CLL cases can carry mutated or unmutated *IGHV* genes [211,212] but, based on gene expression profile analysis, they all share a homogeneous signature that is more related to that of CD27$^{+}$ memory and marginal zone B cells [213,214]. The histogenetic heterogeneity of CLL carries prognostic relevance, because cases with mutated *IG* genes associate with a significantly longer survival [215,216]. Intriguingly, 6% of the normal elderly population develops a monoclonal B cell lymphocytosis (MBL) that is considered the precursor to CLL in 1–2% of cases [217]. Analysis of the *IG* gene repertoire in these patients indicates the "stereotypic" use of a recurrent restricted subset of V regions in different CLL cases [218–223], strongly supporting a role for antigen receptor signaling in CLL pathogenesis.

In contrast with other BCLs, and consistent with the derivation from a post-GC or GC-independent B cell in which *IG* gene remodeling mechanisms are no longer active, CLL cases are largely devoid of balanced, reciprocal chromosomal translocations [62,63]. Conversely, CLL is recurrently associated with several numerical abnormalities, including trisomy 12 and monoallelic or biallelic deletion/inactivation of chromosomal regions 17p, 11q, and, the most common, 13q14 [62,63]. The minimal deleted region of 13q14 deletions encompasses a long noncoding RNA (*DLEU2*) and two microRNAs expressed as a cluster, namely miR-15a and miR-16-1 [64–66]. Deletion of this locus in mice leads to the development of clonal lymphoproliferative diseases with features of MBL, CLL, and DLBCL at 25–40% penetrance [224], thus documenting its pathogenetic role in CLL. Trisomy 12 is found in approximately 16% of patients and correlates with poor survival, but no specific genes have been identified [225–227]. Deletions of chromosomal region 11q22–q23 (18% of cases) almost invariably encompass the *ATM* gene and may thus promote genomic instability [228–230]. These lesions can be observed in the patient germ line and may thus account, at least in part, for the familial form of the disease. Deletions of 17p13, which include the *TP53* tumor suppressor and are frequently accompanied by mutation of the second allele [51,231], are observed in ~7% of CLLs at diagnosis but are enriched in cases that underwent transformation to RS, a highly aggressive lymphoma with poor clinical outcome [51,70]. More recently, gain-of-function mutations of *NOTCH1* and mutations of *SF3B1* were discovered in 5–10% of diagnostic CLL samples. These lesions have negative prognostic value because they are preferentially enriched in RS (30% of cases) and chemoresistant cases (25% of cases) [70,232–236].

## 4.6 Hodgkin Lymphoma

HL is characterized by the presence of scattered large atypical cells—the mononucleated Hodgkin cells and the multinucleated Reed-Sternberg (HRS) cells—residing in a complex admixture of inflammatory cells [98,237]. Morphology and phenotype of the tumor cells, as well as the composition of the infiltrate, allow the identification of two major subtypes of HL: nodular lymphocyte-predominant HL (NLPHL) (~5% of cases) and classic HL (cHL), comprising the nodular sclerosis, mixed-cellularity, lymphocyte-depletion, and lymphocyte-rich variants. Both HL types represent clonal populations of B cells, as they harbor clonally rearranged and somatically mutated *IG* genes [238,239]. In about 25% of cHL cases, nonsense mutations disrupt originally in-frame *IGHV* gene rearrangements (crippling mutations), thereby preventing antigen selection and suggesting that HRS cells of cHL have escaped apoptosis through a mechanism not linked to antigen stimulation [239].

Constitutive activation of NF-κB in cHL can be due to heterogeneous lesions, including the frequent amplification of *REL* (50% of cases), associated with increased protein expression levels [240,241], and/or inactivating mutations in genes coding for negative regulators of NF-κB, including *NFKBIA* (20% of cases), *NFKBIE* (15%), and *TNFAIP3* (40%), among others [59,242–244]. Exome sequencing techniques have identified additional lesions including genomic gains of *PDL1* (CD274) and *PDL2* (CD273) and translocations of *CIITA* [48,190] which may lead to alterations in molecules involved in recognition by the immune system. Amplification of *JAK2*, mutations of *STAT6*, and inactivating mutations of *SOCS1*, a negative regulator of the JAK/STAT signaling pathway, are often found in NLPHL [206,210]; in an additional large fraction of cases, constitutive JAK/STAT activity is sustained by autocrine and paracrine signals [237]. *BCL6* translocations have been reported in the lympocytic and histiocytic (L&H) cells of NLPHD, but only rarely in cHL

[245,246], and translocations of *BCL2* or mutations in positive or negative regulators of apoptosis (e.g., *TP53*, *FAS*, *BAD*, and *ATM*) are virtually absent [237]. An important pathogenic cofactor in cHL, but not NLPHL, is monoclonal EBV infection, suggesting that infection precedes clonal expansion [237] and is observed in approximately 40% of cHL and up to 90% of HIV-related HL. Of the viral proteins encoded by the EBV genome, infected HRS cells most commonly express LMP1 and LMP2A, which lead to constitutive BCR and CD40 signaling and may therefore have a role in HL pathogenesis [237].

## REFERENCES


[1] Klein U, Dalla-Favera R. Germinal centres: role in B-cell physiology and malignancy. Nat Rev Immunol 2008;8:22–33.

[2] MacLennan IC. Germinal centers. Annu Rev Immunol 1994;12:117–39.

[3] Rajewsky K. Clonal selection and learning in the antibody system. Nature 1996;381:751–8.

[4] Allen CD, Okada T, Tang HL, Cyster JG. Imaging of germinal center selection events during affinity maturation. Science 2007;315:528–31.

[5] Victora GD, Schwickert TA, Fooksman DR, Kamphorst AO, Meyer-Hermann M, Dustin ML, et al. Germinal center dynamics revealed by multiphoton microscopy with a photoactivatable fluorescent reporter. Cell 2010;143:592–605.

[6] Neuberger MS. Antibody diversification by somatic mutation: from Burnet onwards. Immunol Cell Biol 2008;86:124–32.

[7] Goossens T, Klein U, Küppers R. Frequent occurrence of deletions and duplications during somatic hypermutation: implications for oncogene translocations and heavy chain disease. Proc Natl Acad Sci USA 1998;95:2463–8.

[8] Kuppers R, Zhao M, Hansmann ML, Rajewsky K. Tracing B cell development in human germinal centres by molecular analysis of single cells picked from histological sections. EMBO J 1993;12:4955–67.

[9] Klein U, Tu Y, Stolovitzky GA, Keller JL, Haddad Jr J, Miljkovic V, et al. Transcriptional analysis of the B cell germinal center reaction. Proc Natl Acad Sci USA 2003;100:2639–44.

[10] Cattoretti G, Chang CC, Cechova K, Zhang J, Ye BH, Falini B, et al. BCL-6 protein is expressed in germinal-center B cells. Blood 1995;86:45–53.

[11] Chang CC, Ye BH, Chaganti RS, Dalla-Favera R. BCL-6, a POZ/zinc-finger protein, is a sequence-specific transcriptional repressor. Proc Natl Acad Sci USA 1996;93:6947–52.

[12] Basso K, Saito M, Sumazin P, Margolin AA, Wang K, Lim WK, et al. Integrated biochemical and computational approach identifies BCL6 direct target genes controlling multiple pathways in normal germinal center B cells. Blood 2010;115:975–84.

[13] Niu H, Ye BH, Dalla-Favera R. Antigen receptor signaling induces MAP kinase-mediated phosphorylation and degradation of the BCL-6 transcription factor. Genes Dev 1998;12:1953–61.

[14] Ci W, Polo JM, Cerchietti L, Shaknovich R, Wang L, Yang SN, et al. The BCL6 transcriptional program features repression of multiple oncogenes in primary B cells and is deregulated in DLBCL. Blood 2009;113:5536–48.

[15] Phan RT, Dalla-Favera R. The BCL6 proto-oncogene suppresses p53 expression in germinal-centre B cells. Nature 2004;432:635–9.

[16] Phan RT, Saito M, Basso K, Niu H, Dalla-Favera R. BCL6 interacts with the transcription factor Miz-1 to suppress the cyclin-dependent kinase inhibitor p21 and cell cycle arrest in germinal center B cells. Nat Immunol 2005;6:1054–60.

[17] Ranuncolo SM, Polo JM, Dierov J, Singer M, Kuo T, Greally J, et al. Bcl-6 mediates the germinal center B cell phenotype and lymphomagenesis through transcriptional repression of the DNA-damage sensor ATR. Nat Immunol 2007;8:705–14.

[18] Ranuncolo SM, Polo JM, Melnick A. BCL6 represses CHEK1 and suppresses DNA damage pathways in normal and malignant B-cells. Blood Cells Mol Dis 2008;41:95–9.

[19] Keller AD, Maniatis T. Identification and characterization of a novel repressor of beta-interferon gene expression. Genes Dev 1991;5:868–79.

[20] Turner Jr CA, Mack DH, Davis MM. Blimp-1, a novel zinc finger-containing protein that can drive the maturation of B lymphocytes into immunoglobulin-secreting cells. Cell 1994;77:297–306.

[21] Shaffer AL, Yu X, He Y, Boldrick J, Chan EP, Staudt LM. BCL-6 represses genes that function in lymphocyte differentiation, inflammation, and cell cycle control. Immunity 2000;13:199–212.

[22] Tunyaplin C, Shaffer AL, Angelin-Duclos CD, Yu X, Staudt LM, Calame KL. Direct repression of prdm1 by Bcl-6 inhibits plasmacytic differentiation. J Immunol 2004;173:1158–65.

[23] Saito M, Novak U, Piovan E, Basso K, Sumazin P, Schneider C, et al. BCL6 suppression of BCL2 via Miz1 and its disruption in diffuse large B cell lymphoma. Proc Natl Acad Sci USA 2009;106:11294–9.

[24] Liu YJ, Arpin C, de Bouteiller O, Guret C, Banchereau J, Martinez-Valdez H, et al. Sequential triggering of apoptosis, somatic mutation and isotype switch during germinal center development. Semin Immunol 1996;8:169–77.

[25] Longerich S, Basu U, Alt F, Storb U. AID in somatic hypermutation and class switch recombination. Curr Opin Immunol 2006;18:164–74.

[26] Muramatsu M, Kinoshita K, Fagarasan S, Yamada S, Shinkai Y, Honjo T. Class switch recombination and hypermutation require activation-induced cytidine deaminase (AID), a potential RNA editing enzyme. Cell 2000;102:553–63.

[27] Revy P, Muto T, Levy Y, Geissmann F, Plebani A, Sanal O, et al. Activation-induced cytidine deaminase (AID) deficiency causes the autosomal recessive form of the Hyper-IgM syndrome (HIGM2). Cell 2000;102:565–75.

[28] Kuppers R, Klein U, Hansmann ML, Rajewsky K. Cellular origin of human B-cell lymphomas. N Engl J Med 1999;341:1520–9.

[29] Kuppers R, Dalla-Favera R. Mechanisms of chromosomal translocations in B cell lymphomas. Oncogene 2001;20:5580–94.

[30] Tsujimoto Y, Gorham J, Cossman J, Jaffe E, Croce CM. The t(14;18) chromosome translocations involved in B-cell neoplasms result from mistakes in VDJ joining. Science 1985;229:1390–3.

[31] Tsujimoto Y, Louie E, Bashir MM, Croce CM. The reciprocal partners of both the t(14; 18) and the t(11; 14) translocations involved in B-cell neoplasms are rearranged by the same mechanism. Oncogene 1988;2:347–51.

[32] Ramiro AR, Jankovic M, Eisenreich T, Difilippantonio S, Chen-Kiang S, Muramatsu M, et al. AID is required for c-myc/IgH chromosome translocations in vivo. Cell 2004;118:431–8.

[33] Robbiani DF, Bothmer A, Callen E, Reina-San-Martin B, Dorsett Y, Difilippantonio S, et al. AID is required for the chromosomal breaks in c-myc that lead to c-myc/IgH translocations. Cell 2008;135:1028–38.

[34] Pasqualucci L, Bhagat G, Jankovic M, Compagno M, Smith P, Muramatsu M, et al. AID is required for germinal center-derived lymphomagenesis. Nat Genet 2008;40:108–12.

[35] Takizawa M, Tolarova H, Li Z, Dubois W, Lim S, Callen E, et al. AID expression levels determine the extent of cMyc oncogenic translocations and the incidence of B cell tumor development. J Exp Med 2008;205:1949–57.

[36] Akasaka H, Akasaka T, Kurata M, Ueda C, Shimizu A, Uchiyama T, et al. Molecular anatomy of BCL6 translocations revealed by long-distance polymerase chain reaction-based assays. Cancer Res 2000;60:2335–41.

[37] Chen W, Iida S, Louie DC, Dalla-Favera R, Chaganti RS. Heterologous promoters fused to BCL6 by chromosomal translocations affecting band 3q27 cause its deregulated expression during B-cell differentiation. Blood 1998;91:603–7.

[38] Ye BH, Lista F, Lo Coco F, Knowles DM, Offit K, Chaganti RS, et al. Alterations of a zinc finger-encoding gene, BCL-6, in diffuse large-cell lymphoma. Science 1993;262:747–50.

[39] Ye BH, Rao PH, Chaganti RS, Dalla-Favera R. Cloning of bcl-6, the locus involved in chromosome translocations affecting band 3q27 in B-cell lymphoma. Cancer Res 1993;53:2732–5.

[40] Yoshida S, Kaneita Y, Aoki Y, Seto M, Mori S, Moriyama M. Identification of heterologous translocation partner genes fused to the BCL6 gene in diffuse large B-cell lymphomas: 5′-RACE and LA – PCR analyses of biopsy samples. Oncogene 1999;18:7994–9.

[41] Baron BW, Nucifora G, McCabe N, Espinosa 3rd R, Le Beau MM, McKeithan TW. Identification of the gene associated with the recurring chromosomal translocations t(3;14)(q27;q32) and t(3;22)(q27;q11) in B-cell lymphomas. Proc Natl Acad Sci USA 1993;90:5262–6.

[42] Kerckaert JP, Deweindt C, Tilly H, Quief S, Lecocq G, Bastard C. LAZ3, a novel zinc-finger encoding gene, is disrupted by recurring chromosome 3q27 translocations in human lymphomas. Nat Genet 1993;5:66–70.

[43] Miki T, Kawamata N, Arai A, Ohashi K, Nakamura Y, Kato A, et al. Molecular cloning of the breakpoint for 3q27 translocation in B-cell lymphomas and leukemias. Blood 1994;83:217–22.

[44] Houldsworth J, Mathew S, Rao PH, Dyomina K, Louie DC, Parsa N, et al. REL proto-oncogene is frequently amplified in extranodal diffuse large cell lymphoma. Blood 1996;87:25–9.

[45] Houldsworth J, Olshen AB, Cattoretti G, Donnelly GB, Teruya-Feldstein J, Qin J, et al. Relationship between REL amplification, REL function, and clinical and biologic features in diffuse large B-cell lymphomas. Blood 2004;103:1862–8.

[46] Rao PH, Houldsworth J, Dyomina K, Parsa NZ, Cigudosa JC, Louie DC, et al. Chromosomal and gene amplification in diffuse large B-cell lymphoma. Blood 1998;92:234–40.

[47] Rosenwald A, Wright G, Chan WC, Connors JM, Campo E, Fisher RI, et al. The use of molecular profiling to predict survival after chemotherapy for diffuse large-B-cell lymphoma. N Engl J Med 2002;346:1937–47.

[48] Green MR, Monti S, Rodig SJ, Juszczynski P, Currie T, O'Donnell E, et al. Integrative analysis reveals selective 9p24.1 amplification, increased PD-1 ligand expression, and further induction via JAK2 in nodular sclerosing Hodgkin lymphoma and primary mediastinal large B-cell lymphoma. Blood 2010;116(17):3268–77.

[49] Neri A, Knowles DM, Greco A, McCormick F, Dalla-Favera R. Analysis of RAS oncogene mutations in human lymphoid malignancies. Proc Natl Acad Sci USA 1988;85:9268–72.

[50] Hollstein M, Sidransky D, Vogelstein B, Harris CC. p53 mutations in human cancers. Science 1991;253:49–53.

[51] Gaidano G, Ballerini P, Gong JZ, Inghirami G, Neri A, Newcomb EW, et al. p53 mutations in human lymphoid malignancies: association with Burkitt lymphoma and chronic lymphocytic leukemia. Proc Natl Acad Sci USA 1991;88:5413–7.

[52] Lo Coco F, Gaidano G, Louie DC, Offit K, Chaganti RS, Dalla-Favera R. p53 mutations are associated with histologic transformation of follicular lymphoma. Blood 1993;82:2289–95.

[53] Mandelbaum J, Bhagat G, Tang H, Mo T, Brahmachary M, Shen Q, et al. BLIMP1 is a tumor suppressor gene frequently disrupted in activated B cell-like diffuse large B cell lymphoma. Cancer Cell 2010;18:568–79.

[54] Pasqualucci L, Compagno M, Houldsworth J, Monti S, Grunn A, Nandula SV, et al. Inactivation of the PRDM1/BLIMP1 gene in diffuse large B cell lymphoma. J Exp Med 2006;203:311–7.

[55] Tam W, Gomez M, Chadburn A, Lee JW, Chan WC, Knowles DM. Mutational analysis of PRDM1 indicates a tumor-suppressor role in diffuse large B-cell lymphomas. Blood 2006;107:4090–100.

[56] Compagno M, Lim WK, Grunn A, Nandula SV, Brahmachary M, Shen Q, et al. Mutations of multiple genes cause deregulation of NF-kappaB in diffuse large B-cell lymphoma. Nature 2009;459:717–21.

[57] Kato M, Sanada M, Kato I, Sato Y, Takita J, Takeuchi K, et al. Frequent inactivation of A20 in B-cell lymphomas. Nature 2009;459:712–6.

[58] Novak U, Rinaldi A, Kwee I, Nandula SV, Rancoita PM, Compagno M, et al. The NF-{kappa}B negative regulator TNFAIP3 (A20) is inactivated by somatic mutations and genomic deletions in marginal zone lymphomas. Blood 2009;113:4918–21.

[59] Schmitz R, Hansmann ML, Bohle V, Martin-Subero JI, Hartmann S, Mechtersheimer G, et al. TNFAIP3 (A20) is a tumor suppressor gene in Hodgkin lymphoma and primary mediastinal B cell lymphoma. J Exp Med 2009;206:981–9.

[60] Pasqualucci L, Dominguez-Sola D, Chiarenza A, Fabbri G, Grunn A, Trifonov V, et al. Inactivating mutations of acetyltransferase genes in B-cell lymphoma. Nature 2011;471:189–95.

[61] Morin RD, Mendez-Lago M, Mungall AJ, Goya R, Mungall KL, Corbett RD, et al. Frequent mutation of histone-modifying genes in non-Hodgkin lymphoma. Nature 2011;476:298–303.

[62] Dohner H, Stilgenbauer S, Benner A, Leupolt E, Krober A, Bullinger L, et al. Genomic aberrations and survival in chronic lymphocytic leukemia. N Engl J Med 2000;343:1910–6.

[63] Dohner H, Stilgenbauer S, Dohner K, Bentz M, Lichter P. Chromosome aberrations in B-cell chronic lymphocytic leukemia: reassessment based on molecular cytogenetic analysis. J Mol Med 1999;77:266–81.

[64] Calin GA, Dumitru CD, Shimizu M, Bichi R, Zupo S, Noch E, et al. Frequent deletions and down-regulation of micro- RNA genes miR15 and miR16 at 13q14 in chronic lymphocytic leukemia. Proc Natl Acad Sci USA 2002;99:15524–9.

[65] Kalachikov S, Migliazza A, Cayanis E, Fracchiolla NS, Bonaldo MF, Lawton L, et al. Cloning and gene mapping of the chromosome 13q14 region deleted in chronic lymphocytic leukemia. Genomics 1997;42:369–77.

[66] Migliazza A, Bosch F, Komatsu H, Cayanis E, Martinotti S, Toniato E, et al. Nucleotide sequence, transcription map, and mutation analysis of the 13q14 chromosomal region deleted in B-cell chronic lymphocytic leukemia. Blood 2001;97:2098–104.

[67] Jares P, Colomer D, Campo E. Genetic and molecular pathogenesis of mantle cell lymphoma: perspectives for new targeted therapeutics. Nat Rev Cancer 2007;7:750–62.
[68] Pasqualucci L, Khiabanian H, Fangazio M, Vasishtha M, Messina M, Holmes AB, et al. Genetics of follicular lymphoma transformation. Cell Reports 2014;6:130–40.
[69] Lenz G, Wright GW, Emre NC, Kohlhammer H, Dave SS, Davis RE, et al. Molecular subtypes of diffuse large B-cell lymphoma arise by distinct genetic pathways. Proc Natl Acad Sci USA 2008;105:13520–5.
[70] Fabbri G, Khiabanian H, Holmes AB, Wang J, Messina M, Mullighan CG, et al. Genetic lesions associated with chronic lymphocytic leukemia transformation to Richter syndrome. J Exp Med 2013;210:2273–88.
[71] Martinez-Delgado B, Robledo M, Arranz E, Osorio A, Garcia MJ, Echezarreta G, et al. Hypermethylation of p15/ink4b/MTS2 gene is differentially implicated among non-Hodgkin's lymphomas. Leukemia 1998;12:937–41.
[72] Pasqualucci L, Neumeister P, Goossens T, Nanjangud G, Chaganti RS, Kuppers R, et al. Hypermutation of multiple proto-oncogenes in B-cell diffuse large-cell lymphomas. Nature 2001;412:341–6.
[73] Gordon MS, Kanegai CM, Doerr JR, Wall R. Somatic hypermutation of the B cell receptor genes B29 (Igbeta, CD79b) and mb1 (Igalpha, CD79a). Proc Natl Acad Sci USA 2003;100:4126–31.
[74] Müschen M, Re D, Jungnickel B, Diehl V, Rajewsky K, Küppers R. Somatic mutation of the CD95 gene in human B cells as a side-effect of the germinal center reaction. J Exp Med 2000;192:1833–40.
[75] Pasqualucci L, Migliazza A, Fracchiolla N, William C, Neri A, Baldini L, et al. BCL-6 mutations in normal germinal center B cells: evidence of somatic hypermutation acting outside Ig loci. Proc Natl Acad Sci USA 1998;95:11816–21.
[76] Shen HM, Peters A, Baron B, Zhu X, Storb U. Mutation of BCL-6 gene in normal B cells by the process of somatic hypermutation of Ig genes. Science 1998;280:1750–2.
[77] Cerri M, Capello D, Muti G, Rambaldi A, Paulli M, Gloghini A, et al. Aberrant somatic hypermutation in post-transplant lymphoproliferative disorders. Br J Haematol 2004;127:362–4.
[78] Deutsch AJ, Aigelsreiter A, Staber PB, Beham A, Linkesch W, Guelly C, et al. MALT lymphoma and extranodal diffuse large B-cell lymphoma are targeted by aberrant somatic hypermutation. Blood 2007;109:3500–4.
[79] Gaidano G, Pasqualucci L, Capello D, Berra E, Deambrogi C, Rossi D, et al. Aberrant somatic hypermutation in multiple subtypes of AIDS-associated non-Hodgkin lymphoma. Blood 2003;102(5):1833–41.
[80] Montesinos-Rongen M, Van Roost D, Schaller C, Wiestler OD, Deckert M. Primary diffuse large B-cell lymphomas of the central nervous system are targeted by aberrant somatic hypermutation. Blood 2004;103:1869–75.
[81] Vakiani E, Basso K, Klein U, Mansukhani MM, Narayan G, Smith PM, et al. Genetic and phenotypic analysis of B-cell post-transplant lymphoproliferative disorders provides insights into disease biology. Hematol Oncol 2008;26:199–211.
[82] Hemann MT, Bric A, Teruya-Feldstein J, Herbst A, Nilsson JA, Cordon-Cardo C, et al. Evasion of the p53 tumour surveillance network by tumour-derived MYC mutants. Nature 2005;436: 807–11.
[83] zur Hausen H, Schulte-Holthausen H, Klein G, Henle W, Henle G, Clifford P, et al. EBV DNA in biopsies of Burkitt tumours and anaplastic carcinomas of the nasopharynx. Nature 1970;228:1056–8.
[84] Magrath IT. African Burkitt's lymphoma. History, biology, clinical features, and treatment. Am J Pediatr Hematol Oncol 1991;13:222–46.
[85] Neri A, Barriga F, Inghirami G, Knowles DM, Neequaye J, Magrath IT, et al. Epstein–Barr virus infection precedes clonal expansion in Burkitt's and acquired immunodeficiency syndrome-associated lymphoma. Blood 1991;77:1092–5.
[86] Ballerini P, Gaidano G, Gong JZ, Tassi V, Saglio G, Knowles DM, et al. Multiple genetic lesions in acquired immunodeficiency syndrome-related non-Hodgkin's lymphoma. Blood 1993;81:166–76.
[87] Hamilton-Dutoit SJ, Pallesen G. A survey of Epstein–Barr virus gene expression in sporadic non-Hodgkin's lymphomas. Detection of Epstein–Barr virus in a subset of peripheral T-cell lymphomas. Am J Pathol 1992;140:1315–25.
[88] Horenstein MG, Nador RG, Chadburn A, Hyjek EM, Inghirami G, Knowles DM, et al. Epstein–Barr virus latent gene expression in primary effusion lymphomas containing Kaposi's sarcoma-associated herpesvirus/human herpesvirus-8. Blood 1997;90:1186–91.
[89] Cingolani A, Gastaldi R, Fassone L, Pierconti F, Giancola ML, Martini M, et al. Epstein–Barr virus infection is predictive of CNS involvement in systemic AIDS-related non-Hodgkin's lymphomas. J Clin Oncol 2000;18:3325–30.
[90] Fassone L, Bhatia K, Gutierrez M, Capello D, Gloghini A, Dolcetti R, et al. Molecular profile of Epstein–Barr virus infection in HHV-8-positive primary effusion lymphoma. Leukemia 2000;14:271–7.
[91] Kieff E, Leibowitz D. Oncogenesis by herpesvirus. In: Weinberg RA, editor. Oncogenes and the molecular origin of cancer. Cold Spring Harbor (NY): Cold Spring Harbor Laboratory Press; 1989. p. 259.
[92] Chang Y, Cesarman E, Pessin MS, Lee F, Culpepper J, Knowles DM, et al. Identification of herpesvirus-like DNA sequences in AIDS-associated Kaposi's sarcoma. Science 1994;266:1865–9.
[93] Soulier J, Grollet L, Oksenhendler E, Cacoub P, Cazals-Hatem D, Babinet P, et al. Kaposi's sarcoma-associated herpesvirus-like DNA sequences in multicentric Castleman's disease. Blood 1995;86:1276–80.
[94] Carbone A, Gaidano G. HHV-8-positive body-cavity-based lymphoma: a novel lymphoma entity. Br J Haematol 1997;97:515–22.
[95] Cesarman E, Chang Y, Moore PS, Said JW, Knowles DM. Kaposi's sarcoma-associated herpesvirus-like DNA sequences in AIDS-related body-cavity-based lymphomas. N Engl J Med 1995;332:1186–91.
[96] Gaidano G, Pastore C, Gloghini A, Cusini M, Nomdedeu J, Volpe G, et al. Distribution of human herpesvirus-8 sequences throughout the spectrum of AIDS-related neoplasia. Aids 1996;10:941–9.
[97] Moore PS, Gao SJ, Dominguez G, Cesarman E, Lungu O, Knowles DM, et al. Primary characterization of a herpesvirus agent associated with Kaposi's sarcomae. J Virol 1996;70:549–58.
[98] Swerdlow SH, Campo E, Harris NL, Jaffe ES, Pileri SA, Stein H, et al. WHO classification of tumours of haematopoietic and lymphoid tissues. In: Bosman FT, Jaffe ES, Lakhani SR, Ohgaki H, editors. World health organization of tumours. Lyon: International Agency for Research on Cancer (IARC); 2008.
[99] Tsujimoto Y, Jaffe E, Cossman J, Gorham J, Nowell PC, Croce CM. Clustering of breakpoints on chromosome 11 in human B-cell neoplasms with the t(11;14) chromosome translocation. Nature 1985;315:340–3.
[100] Tsujimoto Y, Yunis J, Onorato-Showe L, Erikson J, Nowell PC, Croce CM. Molecular cloning of the chromosomal breakpoint of B-cell lymphomas and leukemias with the t(11;14) chromosome translocation. Science 1984;224:1403–6.

[101] Erikson J, Finan J, Tsujimoto Y, Nowell PC, Croce CM. The chromosome 14 breakpoint in neoplastic B cells with the t(11;14) translocation involves the immunoglobulin heavy chain locus. Proc Natl Acad Sci USA 1984;81:4144–8.
[102] Motokura T, Bloom T, Kim HG, Juppner H, Ruderman JV, Kronenberg HM, et al. A novel cyclin encoded by a bcl1-linked candidate oncogene. Nature 1991;350:512–5.
[103] Rosenberg CL, Wong E, Petty EM, Bale AE, Tsujimoto Y, Harris NL, et al. PRAD1, a candidate BCL1 oncogene: mapping and expression in centrocytic lymphoma. Proc Natl Acad Sci USA 1991;88:9638–42.
[104] Withers DA, Harvey RC, Faust JB, Melnyk O, Carey K, Meeker TC. Characterization of a candidate bcl-1 gene. Mol Cell Biol 1991;11:4846–53.
[105] Seto M, Yamamoto K, Iida S, Akao Y, Utsumi KR, Kubonishi I, et al. Gene rearrangement and overexpression of PRAD1 in lymphoid malignancy with t(11;14)(q13;q32) translocation. Oncogene 1992;7:1401–6.
[106] Komatsu H, Iida S, Yamamoto K, Mikuni C, Nitta M, Takahashi T, et al. A variant chromosome translocation at 11q13 identifying PRAD1/cyclin D1 as the BCL-1 gene. Blood 1994;84:1226–31.
[107] Wiestner A, Tehrani M, Chiorazzi M, Wright G, Gibellini F, Nakayama K, et al. Point mutations and genomic deletions in CCND1 create stable truncated cyclin D1 mRNAs that are associated with increased proliferation rate and shorter survival. Blood 2007;109:4599–606.
[108] Bea S, Salaverria I, Armengol L, Pinyol M, Fernandez V, Hartmann EM, et al. Uniparental disomies, homozygous deletions, amplifications, and target genes in mantle cell lymphoma revealed by integrative high-resolution whole-genome profiling. Blood 2009;113:3059–69.
[109] Bodrug SE, Warner BJ, Bath ML, Lindeman GJ, Harris AW, Adams JM. Cyclin D1 transgene impedes lymphocyte maturation and collaborates in lymphomagenesis with the myc gene. EMBO J 1994;13:2124–30.
[110] Lovec H, Grzeschiczek A, Kowalski MB, Moroy T. Cyclin D1/bcl-1 cooperates with myc genes in the generation of B-cell lymphoma in transgenic mice. EMBO J 1994;13:3487–95.
[111] Schaffner C, Idler I, Stilgenbauer S, Dohner H, Lichter P. Mantle cell lymphoma is characterized by inactivation of the ATM gene. Proc Natl Acad Sci USA 2000;97:2773–8.
[112] Louie DC, Offit K, Jaslow R, Parsa NZ, Murty VV, Schluger A, et al. p53 overexpression as a marker of poor prognosis in mantle cell lymphomas with t(11;14)(q13;q32). Blood 1995;86:2892–9.
[113] Pinyol M, Hernandez L, Cazorla M, Balbin M, Jares P, Fernandez PL, et al. Deletions and loss of expression of p16INK4a and p21Waf1 genes are associated with aggressive variants of mantle cell lymphomas. Blood 1997;89:272–80.
[114] Bea S, Valdes-Mas R, Navarro A, Salaverria I, Martin-Garcia D, Jares P, et al. Landscape of somatic mutations and clonal evolution in mantle cell lymphoma. Proc Natl Acad Sci USA 2013;110:18250–5.
[115] Kridel R, Meissner B, Rogic S, Boyle M, Telenius A, Woolcock B, et al. Whole transcriptome sequencing reveals recurrent NOTCH1 mutations in mantle cell lymphoma. Blood 2012;119:1963–71.
[116] Bea S, Tort F, Pinyol M, Puig X, Hernandez L, Hernandez S, et al. BMI-1 gene amplification and overexpression in hematological malignancies occur mainly in mantle cell lymphomas. Cancer Res 2001;61:2409–12.
[117] Chapman CJ, Mockridge CI, Rowe M, Rickinson AB, Stevenson FK. Analysis of VH genes used by neoplastic B cells in endemic Burkitt's lymphoma shows somatic hypermutation and intraclonal heterogeneity. Blood 1995;85:2176–81.
[118] Chapman CJ, Zhou JX, Gregory C, Rickinson AB, Stevenson FK. VH and VL gene analysis in sporadic Burkitt's lymphoma shows somatic hypermutation, intraclonal heterogeneity, and a role for antigen selection. Blood 1996;88:3562–8.
[119] Jain R, Roncella S, Hashimoto S, Carbone A, Francia di Celle P, Foa R, et al. A potential role for antigen selection in the clonal evolution of Burkitt's lymphoma. J Immunol 1994;153:45–52.
[120] Tamaru J, Hummel M, Marafioti T, Kalvelage B, Leoncini L, Minacci C, et al. Burkitt's lymphomas express VH genes with a moderate number of antigen-selected somatic mutations. Am J Pathol 1995;147:1398–407.
[121] Dave SS, Fu K, Wright GW, Lam LT, Kluin P, Boerma EJ, et al. Molecular diagnosis of Burkitt's lymphoma. N Engl J Med 2006;354:2431–42.
[122] Hummel M, Bentink S, Berger H, Klapper W, Wessendorf S, Barth TF, et al. A biologic definition of Burkitt's lymphoma from transcriptional and genomic profiling. N Engl J Med 2006;354:2419–30.
[123] Dalla-Favera R, Bregni M, Erikson J, Patterson D, Gallo RC, Croce CM. Human c-myc onc gene is located on the region of chromosome 8 that is translocated in Burkitt lymphoma cells. Proc Natl Acad Sci USA 1982;79:7824–7.
[124] Dalla-Favera R, Martinotti S, Gallo RC, Erikson J, Croce CM. Translocation and rearrangements of the c-myc oncogene locus in human undifferentiated B-cell lymphomas. Science 1983;219:963–7.
[125] Dalla-Favera R. Chromosomal translocations involving the c-myc oncogene in lymphoid neoplasia. In: Kirsch IR, editor. The causes and consequences of chromosomal aberrations. Boca Raton: CRC Press; 1993. p. 312.
[126] Taub R, Kirsch I, Morton C, Lenoir G, Swan D, Tronick S, et al. Translocation of the c-myc gene into the immunoglobulin heavy chain locus in human Burkitt lymphoma and murine plasmacytoma cells. Proc Natl Acad Sci USA 1982;79:7837–41.
[127] Davis M, Malcolm S, Rabbitts TH. Chromosome translocation can occur on either side of the c-myc oncogene in Burkitt lymphoma cells. Nature 1984;308:286–8.
[128] Neri A, Barriga F, Knowles DM, Magrath IT, Dalla-Favera R. Different regions of the immunoglobulin heavy-chain locus are involved in chromosomal translocations in distinct pathogenetic forms of Burkitt lymphoma. Proc Natl Acad Sci USA 1988;85:2748–52.
[129] Pelicci PG, Knowles 2nd DM, Magrath I, Dalla-Favera R. Chromosomal breakpoints and structural alterations of the c-myc locus differ in endemic and sporadic forms of Burkitt lymphoma. Proc Natl Acad Sci USA 1986;83:2984–8.
[130] ar-Rushdi A, Nishikura K, Erikson J, Watt R, Rovera G, Croce CM. Differential expression of the translocated and the untranslocated c-myc oncogene in Burkitt lymphoma. Science 1983;222:390–3.
[131] Hayday AC, Gillies SD, Saito H, Wood C, Wiman K, Hayward WS, et al. Activation of a translocated human c-myc gene by an enhancer in the immunoglobulin heavy-chain locus. Nature 1984;307:334–40.
[132] Rabbitts TH, Forster A, Baer R, Hamlyn PH. Transcription enhancer identified near the human C mu immunoglobulin heavy chain gene is unavailable to the translocated c-myc gene in a Burkitt lymphoma. Nature 1983;306:806–9.

[133] Dominguez-Sola D, Victora GD, Ying CY, Phan RT, Saito M, Nussenzweig MC, et al. The proto-oncogene MYC is required for selection in the germinal center and cyclic reentry. Nat Immunol 2012;13(11):1083–91.
[134] Cesarman E, Dalla-Favera R, Bentley D, Groudine M. Mutations in the first exon are associated with altered transcription of c-myc in Burkitt lymphoma. Science 1987;238:1272–5.
[135] Bhatia K, Huppi K, Spangler G, Siwarski D, Iyer R, Magrath I. Point mutations in the c-Myc transactivation domain are common in Burkitt's lymphoma and mouse plasmacytomas. Nat Genet 1993;5:56–61.
[136] Bhatia K, Spangler G, Gaidano G, Hamdy N, Dalla-Favera R, Magrath I. Mutations in the coding region of c-myc occur frequently in acquired immunodeficiency syndrome-associated lymphomas. Blood 1994;84:883–8.
[137] Gu W, Bhatia K, Magrath IT, Dang CV, Dalla-Favera R. Binding and suppression of the Myc transcriptional activation domain by p107. Science 1994;264:251–4.
[138] Gregory MA, Hann SR. c-Myc proteolysis by the ubiquitin-proteasome pathway: stabilization of c-Myc in Burkitt's lymphoma cells. Mol Cell Biol 2000;20:2423–35.
[139] Dang CV, O'Donnell KA, Zeller KI, Nguyen T, Osthus RC, Li F. The c-Myc target gene network. Semin Cancer Biol 2006;16: 253–64.
[140] Grandori C, Cowley SM, James LP, Eisenman RN. The Myc/Max/ Mad network and the transcriptional control of cell behavior. Annu Rev Cell Dev Biol 2000;16:653–99.
[141] Meyer N, Penn LZ. Reflecting on 25 years with MYC. Nat Rev Cancer 2008;8:976–90.
[142] Felsher DW, Bishop JM. Transient excess of MYC activity can elicit genomic instability and tumorigenesis. Proc Natl Acad Sci USA 1999;96:3940–4.
[143] Adams JM, Harris AW, Pinkert CA, Corcoran LM, Alexander WS, Cory S, et al. The c-myc oncogene driven by immunoglobulin enhancers induces lymphoid malignancy in transgenic mice. Nature 1985;318:533–8.
[144] Kovalchuk AL, Qi CF, Torrey TA, Taddesse-Heath L, Feigenbaum L, Park SS, et al. Burkitt lymphoma in the mouse. J Exp Med 2000;192:1183–90.
[145] Schmitz R, Young RM, Ceribelli M, Jhavar S, Xiao W, Zhang M, et al. Burkitt lymphoma pathogenesis and therapeutic targets from structural and functional genomics. Nature 2012;490:116–20.
[146] Gaidano G, Hauptschein RS, Parsa NZ, Offit K, Rao PH, Lenoir G, et al. Deletions involving two distinct regions of 6q in B-cell non-Hodgkin lymphoma. Blood 1992;80:1781–7.
[147] Lombardi L, Newcomb EW, Dalla-Favera R. Pathogenesis of Burkitt lymphoma: expression of an activated c-myc oncogene causes the tumorigenic conversion of EBV-infected human B lymphoblasts. Cell 1987;49:161–70.
[148] Prevot S, Hamilton-Dutoit S, Audouin J, Walter P, Pallesen G, Diebold J. Analysis of African Burkitt's and high-grade B cell non-Burkitt's lymphoma for Epstein–Barr virus genomes using in situ hybridization. Br J Haematol 1992;80:27–32.
[149] Thorley-Lawson DA, Allday MJ. The curious case of the tumour virus: 50 years of Burkitt's lymphoma. Nat Rev Microbiol 2008;6: 913–24.
[150] Kridel R, Sehn LH, Gascoyne RD. Pathogenesis of follicular lymphoma. J Clin Invest 2012;122:3424–31.
[151] Montoto S, Fitzgibbon J. Transformation of indolent B-cell lymphomas. J Clin Oncol 2011;29:1827–34.
[152] Bakhshi A, Jensen JP, Goldman P, Wright JJ, McBride OW, Epstein AL, et al. Cloning the chromosomal breakpoint of t(14;18) human lymphomas: clustering around JH on chromosome 14 and near a transcriptional unit on 18. Cell 1985;41:899–906.
[153] Cleary ML, Sklar J. Nucleotide sequence of a t(14;18) chromosomal breakpoint in follicular lymphoma and demonstration of a breakpoint-cluster region near a transcriptionally active locus on chromosome 18. Proc Natl Acad Sci USA 1985;82:7439–43.
[154] Cleary ML, Smith SD, Sklar J. Cloning and structural analysis of cDNAs for bcl-2 and a hybrid bcl-2/immunoglobulin transcript resulting from the t(14;18) translocation. Cell 1986;47: 19–28.
[155] Ott G, Katzenberger T, Lohr A, Kindelberger S, Rudiger T, Wilhelm M, et al. Cytomorphologic, immunohistochemical, and cytogenetic profiles of follicular lymphoma: 2 types of follicular lymphoma grade 3. Blood 2002;99:3806–12.
[156] Tsujimoto Y, Finger LR, Yunis J, Nowell PC, Croce CM. Cloning of the chromosome breakpoint of neoplastic B cells with the t(14;18) chromosome translocation. Science 1984;226:1097–9.
[157] Cleary ML, Galili N, Sklar J. Detection of a second t(14;18) breakpoint cluster region in human follicular lymphomas. J Exp Med 1986;164:315–20.
[158] Graninger WB, Seto M, Boutain B, Goldman P, Korsmeyer SJ. Expression of Bcl-2 and Bcl-2-Ig fusion transcripts in normal and neoplastic cells. J Clin Invest 1987;80:1512–5.
[159] Ngan BY, Chen-Levy Z, Weiss LM, Warnke RA, Cleary ML. Expression in non-Hodgkin's lymphoma of the bcl-2 protein associated with the t(14;18) chromosomal translocation. N Engl J Med 1988;318:1638–44.
[160] Petrovic AS, Young RL, Hilgarth B, Ambros P, Korsmeyer SJ, Jaeger U. The Ig heavy chain 3′ end confers a posttranscriptional processing advantage to Bcl-2-IgH fusion RNA in t(14;18) lymphoma. Blood 1998;91:3952–61.
[161] Buchonnet G, Jardin F, Jean N, Bertrand P, Parmentier F, Tison S, et al. Distribution of BCL2 breakpoints in follicular lymphoma and correlation with clinical features: specific subtypes or same disease? Leukemia 2002;16:1852–6.
[162] Morin RD, Johnson NA, Severson TM, Mungall AJ, An J, Goya R, et al. Somatic mutations altering EZH2 (Tyr641) in follicular and diffuse large B-cell lymphomas of germinal-center origin. Nat Genet 2010;42:181–5.
[163] Okosun J, Bödör C, Wang J, Araf S, Yang CY, Pan C, et al. Integrated genomic analysis identifies recurrent mutations and evolution patterns driving the initiation and progression of follicular lymphoma. Nat Genet 2014;46(2):176–81.
[164] Cleary ML, Meeker TC, Levy S, Lee E, Trela M, Sklar J, et al. Clustering of extensive somatic mutations in the variable region of an immunoglobulin heavy chain gene from a human B cell lymphoma. Cell 1986;44:97–106.
[165] Bahler DW, Levy R. Clonal evolution of a follicular lymphoma: evidence for antigen selection. Proc Natl Acad Sci USA 1992; 89:6770–4.
[166] Elenitoba-Johnson KS, Gascoyne RD, Lim MS, Chhanabai M, Jaffe ES, Raffeld M. Homozygous deletions at chromosome 9p21 involving p16 and p15 are associated with histologic progression in follicle center lymphoma. Blood 1998;91:4677–85.
[167] Yano T, Jaffe ES, Longo DL, Raffeld M. MYC rearrangements in histologically progressed follicular lymphomas. Blood 1992; 80:758–67.

[168] Ichikawa A, Hotta T, Takagi N, Tsushita K, Kinoshita T, Nagai H, et al. Mutations of p53 gene and their relation to disease progression in B-cell lymphoma. Blood 1992;79:2701–7.

[169] O'Shea D, O'Riain C, Taylor C, Waters R, Carlotti E, Macdougall F, et al. The presence of TP53 mutation at diagnosis of follicular lymphoma identifies a high-risk group of patients with shortened time to disease progression and poorer overall survival. Blood 2008;112:3126–9.

[170] Sander CA, Yano T, Clark HM, Harris C, Longo DL, Jaffe ES, et al. p53 mutation is associated with progression in follicular lymphomas. Blood 1993;82:1994–2004.

[171] A clinical evaluation of the International Lymphoma Study Group classification of non-Hodgkin's lymphoma. The non-Hodgkin's lymphoma classification Project. Blood 1997;89:3909–18.

[172] Alizadeh AA, Eisen MB, Davis RE, Ma C, Lossos IS, Rosenwald A, et al. Distinct types of diffuse large B-cell lymphoma identified by gene expression profiling. Nature 2000;403:503–11.

[173] Victora GD, Dominguez-Sola D, Holmes AB, Deroubaix S, Dalla-Favera R, Nussenzweig MC. Identification of human germinal center light and dark zone cells and their relationship to human B-cell lymphomas. Blood 2012;120:2240–8.

[174] Wright G, Tan B, Rosenwald A, Hurt EH, Wiestner A, Staudt LM. A gene expression-based method to diagnose clinically distinct subgroups of diffuse large B cell lymphoma. Proc Natl Acad Sci USA 2003;100:9991–6.

[175] Savage KJ, Monti S, Kutok JL, Cattoretti G, Neuberg D, De Leval L, et al. The molecular signature of mediastinal large B-cell lymphoma differs from that of other diffuse large B-cell lymphomas and shares features with classical Hodgkin lymphoma. Blood 2003;102:3871–9.

[176] Rosenwald A, Wright G, Leroy K, Yu X, Gaulard P, Gascoyne RD, et al. Molecular diagnosis of primary mediastinal B cell lymphoma identifies a clinically favorable subgroup of diffuse large B cell lymphoma related to Hodgkin lymphoma. J Exp Med 2003;198:851–62.

[177] Choi WW, Weisenburger DD, Greiner TC, Piris MA, Banham AH, Delabie J, et al. A new immunostain algorithm classifies diffuse large B-cell lymphoma into molecular subtypes with high accuracy. Clin Cancer Res 2009;15:5494–502.

[178] Hans CP, Weisenburger DD, Greiner TC, Gascoyne RD, Delabie J, Ott G, et al. Confirmation of the molecular classification of diffuse large B-cell lymphoma by immunohistochemistry using a tissue microarray. Blood 2004;103:275–82.

[179] Lohr JG, Stojanov P, Lawrence MS, Auclair D, Chapuy B, Sougnez C, et al. Discovery and prioritization of somatic mutations in diffuse large B-cell lymphoma (DLBCL) by whole-exome sequencing. Proc Natl Acad Sci USA 2012;109:3879–84.

[180] Pasqualucci L, Trifonov V, Fabbri G, Ma J, Rossi D, Chiarenza A, et al. Analysis of the coding genome of diffuse large B-cell lymphoma. Nat Genet 2011;43:830–7.

[181] Lo Coco F, Ye BH, Lista F, Corradini P, Offit K, Knowles DM, et al. Rearrangements of the BCL6 gene in diffuse large cell non-Hodgkin's lymphoma. Blood 1994;83:1757–9.

[182] Offit K, Jhanwar S, Ebrahim SA, Filippa D, Clarkson BD, Chaganti RS. t(3;22)(q27;q11): a novel translocation associated with diffuse non-Hodgkin's lymphoma. Blood 1989;74:1876–9.

[183] Offit K, Wong G, Filippa DA, Tao Y, Chaganti RS. Cytogenetic analysis of 434 consecutively ascertained specimens of non-Hodgkin's lymphoma: clinical correlations. Blood 1991;77:1508–15.

[184] Iqbal J, Greiner TC, Patel K, Dave BJ, Smith L, Ji J, et al. Distinctive patterns of BCL6 molecular alterations and their functional consequences in different subgroups of diffuse large B-cell lymphoma. Leukemia 2007;21:2332–43.

[185] Ye BH, Chaganti S, Chang CC, Niu H, Corradini P, Chaganti RS, et al. Chromosomal translocations cause deregulated BCL6 expression by promoter substitution in B cell lymphoma. EMBO J 1995;14:6209–17.

[186] Ying CY, Dominguez-Sola D, Fabi M, Lorenz IC, Hussein S, Bansal M, et al. MEF2B mutations lead to deregulated expression of the oncogene BCL6 in diffuse large B cell lymphoma. Nat Immunol 2013;14:1084–92.

[187] Duan S, Cermak L, Pagan JK, Rossi M, Martinengo C, di Celle PF, et al. FBXO11 targets BCL6 for degradation and is inactivated in diffuse large B-cell lymphomas. Nature 2012;481:90–3.

[188] Cattoretti G, Pasqualucci L, Ballon G, Tam W, Nandula SV, Shen Q, et al. Deregulated BCL6 expression recapitulates the pathogenesis of human diffuse large B cell lymphomas in mice. Cancer Cell 2005;7:445–55.

[189] Challa-Malladi M, Lieu YK, Califano O, Holmes AB, Bhagat G, Murty VV, et al. Combined genetic inactivation of β2-Microglobulin and CD58 reveals frequent escape from immune recognition in diffuse large B cell lymphoma. Cancer Cell 2011;20(6):728–40.

[190] Steidl C, Shah SP, Woolcock BW, Rui L, Kawahara M, Farinha P, et al. MHC class II transactivator CIITA is a recurrent gene fusion partner in lymphoid cancers. Nature 2011;471:377–81.

[191] Iqbal J, Sanger WG, Horsman DE, Rosenwald A, Pickering DL, Dave B, et al. BCL2 translocation defines a unique tumor subset within the germinal center B-cell-like diffuse large B-cell lymphoma. Am J Pathol 2004;165:159–66.

[192] Kawasaki C, Ohshim K, Suzumiya J, Kanda M, Tsuchiya T, Tamura K, et al. Rearrangements of bcl-1, bcl-2, bcl-6, and c-myc in diffuse large B-cell lymphomas. Leuk Lymphoma 2001;42:1099–106.

[193] Ladanyi M, Offit K, Jhanwar SC, Filippa DA, Chaganti RS. MYC rearrangement and translocations involving band 8q24 in diffuse large cell lymphomas. Blood 1991;77:1057–63.

[194] Pfeifer M, Grau M, Lenze D, Wenzel SS, Wolf A, Wollert-Wulf B, et al. PTEN loss defines a PI3K/AKT pathway-dependent germinal center subtype of diffuse large B-cell lymphoma. Proc Natl Acad Sci USA 2013;110:12420–5.

[195] Pasqualucci L, Migliazza A, Basso K, Houldsworth J, Chaganti RS, Dalla-Favera R. Mutations of the BCL6 proto-oncogene disrupt its negative autoregulation in diffuse large B-cell lymphoma. Blood 2003;101:2914–23.

[196] Wang X, Li Z, Naganuma A, Ye BH. Negative autoregulation of BCL-6 is bypassed by genetic alterations in diffuse large B cell lymphomas. Proc Natl Acad Sci USA 2002;99:15018–23.

[197] Capello D, Vitolo U, Pasqualucci L, Quattrone S, Migliaretti G, Fassone L, et al. Distribution and pattern of BCL-6 mutations throughout the spectrum of B-cell neoplasia. Blood 2000;95:651–9.

[198] Migliazza A, Martinotti S, Chen W, Fusco C, Ye BH, Knowles DM, et al. Frequent somatic hypermutation of the 5′ noncoding region of the BCL6 gene in B-cell lymphoma. Proc Natl Acad Sci USA 1995;92:12520–4.

[199] Shen HM, Michael N, Kim N, Storb U. The TATA binding protein, c-Myc and survivin genes are not somatically hypermutated, while Ig and BCL6 genes are hypermutated in human memory B cells. Int Immunol 2000;12:1085–93.

[200] Saito M, Gao J, Basso K, Kitagawa Y, Smith PM, Bhagat G, et al. A signaling pathway mediating downregulation of BCL6 in germinal center B cells is blocked by BCL6 gene alterations in B cell lymphoma. Cancer Cell 2007;12:280–92.
[201] Davis RE, Brown KD, Siebenlist U, Staudt LM. Constitutive nuclear factor kappaB activity is required for survival of activated B cell-like diffuse large B cell lymphoma cells. J Exp Med 2001;194:1861–74.
[202] Lenz G, Davis RE, Ngo VN, Lam L, George TC, Wright GW, et al. Oncogenic CARD11 mutations in human diffuse large B cell lymphoma. Science 2008;319:1676–9.
[203] Ngo VN, Young RM, Schmitz R, Jhavar S, Xiao W, Lim KH, et al. Oncogenically active MYD88 mutations in human lymphoma. Nature 2010;470(7332):115–19.
[204] Davis RE, Ngo VN, Lenz G, Tolar P, Young RM, Romesser PB, et al. Chronic active B-cell-receptor signalling in diffuse large B-cell lymphoma. Nature 2010;463:88–92.
[205] Shapiro-Shelef M, Lin KI, McHeyzer-Williams LJ, Liao J, McHeyzer-Williams MG, Calame K. Blimp-1 is required for the formation of immunoglobulin secreting plasma cells and pre-plasma memory B cells. Immunity 2003;19:607–20.
[206] Joos S, Kupper M, Ohl S, von Bonin F, Mechtersheimer G, Bentz M, et al. Genomic imbalances including amplification of the tyrosine kinase gene JAK2 in CD30+ Hodgkin cells. Cancer Res 2000;60:549–52.
[207] Rui L, Emre NC, Kruhlak MJ, Chung HJ, Steidl C, Slack G, et al. Cooperative epigenetic modulation by cancer amplicon genes. Cancer Cell 2010;18:590–605.
[208] Mestre C, Rubio-Moscardo F, Rosenwald A, Climent J, Dyer MJ, Staudt L, et al. Homozygous deletion of SOCS1 in primary mediastinal B-cell lymphoma detected by CGH to BAC microarrays. Leukemia 2005;19:1082–4.
[209] Melzner I, Bucur AJ, Bruderlein S, Dorsch K, Hasel C, Barth TF, et al. Biallelic mutation of SOCS-1 impairs JAK2 degradation and sustains phospho-JAK2 action in the MedB-1 mediastinal lymphoma line. Blood 2005;105:2535–42.
[210] Weniger MA, Melzner I, Menz CK, Wegener S, Bucur AJ, Dorsch K, et al. Mutations of the tumor suppressor gene SOCS-1 in classical Hodgkin lymphoma are frequent and associated with nuclear phospho-STAT5 accumulation. Oncogene 2006;25:2679–84.
[211] Fais F, Ghiotto F, Hashimoto S, Sellars B, Valetto A, Allen SL, et al. Chronic lymphocytic leukemia B cells express restricted sets of mutated and unmutated antigen receptors. J Clin Invest 1998;102:1515–25.
[212] Oscier DG, Thompsett A, Zhu D, Stevenson FK. Differential rates of somatic hypermutation in V(H) genes among subsets of chronic lymphocytic leukemia defined by chromosomal abnormalities. Blood 1997;89:4153–60.
[213] Klein U, Tu Y, Stolovitzky GA, Mattioli M, Cattoretti G, Husson H, et al. Gene expression profiling of B cell chronic lymphocytic leukemia reveals a homogeneous phenotype related to memory B cells. J Exp Med 2001;194:1625–38.
[214] Rosenwald A, Alizadeh AA, Widhopf G, Simon R, Davis RE, Yu X, et al. Relation of gene expression phenotype to immunoglobulin mutation genotype in B cell chronic lymphocytic leukemia. J Exp Med 2001;194:1639–47.
[215] Damle RN, Wasil T, Fais F, Ghiotto F, Valetto A, Allen SL, et al. Ig V gene mutation status and CD38 expression as novel prognostic indicators in chronic lymphocytic leukemia. Blood 1999;94:1840–7.
[216] Hamblin TJ, Davis Z, Gardiner A, Oscier DG, Stevenson FK. Unmutated Ig V(H) genes are associated with a more aggressive form of chronic lymphocytic leukemia. Blood 1999;94:1848–54.
[217] Rawstron AC, Bennett FL, O'Connor SJ, Kwok M, Fenton JA, Plummer M, et al. Monoclonal B-cell lymphocytosis and chronic lymphocytic leukemia. N Engl J Med 2008;359:575–83.
[218] Widhopf 2nd GF, Rassenti LZ, Toy TL, Gribben JG, Wierda WG, Kipps TJ. Chronic lymphocytic leukemia B cells of more than 1% of patients express virtually identical immunoglobulins. Blood 2004;104:2499–504.
[219] Messmer BT, Albesiano E, Efremov DG, Ghiotto F, Allen SL, Kolitz J, et al. Multiple distinct sets of stereotyped antigen receptors indicate a role for antigen in promoting chronic lymphocytic leukemia. J Exp Med 2004;200:519–25.
[220] Tobin G, Thunberg U, Karlsson K, Murray F, Laurell A, Willander K, et al. Subsets with restricted immunoglobulin gene rearrangement features indicate a role for antigen selection in the development of chronic lymphocytic leukemia. Blood 2004;104:2879–85.
[221] Murray F, Darzentas N, Hadzidimitriou A, Tobin G, Boudjogra M, Scielzo C, et al. Stereotyped patterns of somatic hypermutation in subsets of patients with chronic lymphocytic leukemia: implications for the role of antigen selection in leukemogenesis. Blood 2008;111:1524–33.
[222] Stamatopoulos K, Belessi C, Moreno C, Boudjograh M, Guida G, Smilevska T, et al. Over 20% of patients with chronic lymphocytic leukemia carry stereotyped receptors: pathogenetic implications and clinical correlations. Blood 2007;109:259–70.
[223] Tobin G, Thunberg U, Johnson A, Eriksson I, Soderberg O, Karlsson K, et al. Chronic lymphocytic leukemias utilizing the VH3-21 gene display highly restricted Vlambda2-14 gene use and homologous CDR3s: implicating recognition of a common antigen epitope. Blood 2003;101:4952–7.
[224] Klein U, Lia M, Crespo M, Siegel R, Shen Q, Mo T, et al. The DLEU2/miR-15a/16-1 cluster controls B cell proliferation and its deletion leads to chronic lymphocytic leukemia. Cancer Cell 2010;17:28–40.
[225] Anastasi J, Le Beau MM, Vardiman JW, Fernald AA, Larson RA, Rowley JD. Detection of trisomy 12 in chronic lymphocytic leukemia by fluorescence in situ hybridization to interphase cells: a simple and sensitive method. Blood 1992;79:1796–801.
[226] Hjalmar V, Kimby E, Matutes E, Sundstrom C, Jacobsson B, Arvidsson I, et al. Trisomy 12 and lymphoplasmacytoid lymphocytes in chronic leukemic B-cell disorders. Haematologica 1998;83:602–9.
[227] Juliusson G, Oscier DG, Fitchett M, Ross FM, Stockdill G, Mackie MJ, et al. Prognostic subgroups in B-cell chronic lymphocytic leukemia defined by specific chromosomal abnormalities. N Engl J Med 1990;323:720–4.
[228] Bullrich F, Rasio D, Kitada S, Starostik P, Kipps T, Keating M, et al. ATM mutations in B-cell chronic lymphocytic leukemia. Cancer Res 1999;59:24–7.
[229] Stankovic T, Weber P, Stewart G, Bedenham T, Murray J, Byrd PJ, et al. Inactivation of ataxia telangiectasia mutated gene in B-cell chronic lymphocytic leukaemia. Lancet 1999;353:26–9.
[230] Starostik P, Manshouri T, O'Brien S, Freireich E, Kantarjian H, Haidar M, et al. Deficiency of the ATM protein expression defines an aggressive subgroup of B-cell chronic lymphocytic leukemia. Cancer Res 1998;58:4552–7.

[231] Gaidano G, Newcomb EW, Gong JZ, Tassi V, Neri A, Cortelezzi A, et al. Analysis of alterations of oncogenes and tumor suppressor genes in chronic lymphocytic leukemia. Am J Pathol 1994;144:1312–9.
[232] Fabbri G, Rasi S, Rossi D, Trifonov V, Khiabanian H, Ma J, et al. Analysis of the chronic lymphocytic leukemia coding genome: role of NOTCH1 mutational activation. J Exp Med 2011;208:1389–401.
[233] Puente XS, Pinyol M, Quesada V, Conde L, Ordonez GR, Villamor N, et al. Whole-genome sequencing identifies recurrent mutations in chronic lymphocytic leukaemia. Nature 2011;475:101–5.
[234] Wang L, Lawrence MS, Wan Y, Stojanov P, Sougnez C, Stevenson K, et al. SF3B1 and other novel cancer genes in chronic lymphocytic leukemia. N Engl J Med 2011;365:2497–506.
[235] Quesada V, Conde L, Villamor N, Ordonez GR, Jares P, Bassaganyas L, et al. Exome sequencing identifies recurrent mutations of the splicing factor SF3B1 gene in chronic lymphocytic leukemia. Nat Genet 2012;44:47–52.
[236] Rossi D, Bruscaggin A, Spina V, Rasi S, Khiabanian H, Messina M, et al. Mutations of the SF3B1 splicing factor in chronic lymphocytic leukemia: association with progression and fludarabine-refractoriness. Blood 2011;118:6904–8.
[237] Kuppers R. The biology of Hodgkin's lymphoma. Nat Rev Cancer 2009;9:15–27.
[238] Küppers R, Rajewsky K, Zhao M, Simons G, Laumann R, Fischer R, et al. Hodgkin disease: Hodgkin and Reed-Sternberg cells picked from histological sections show clonal immunoglobulin gene rearrangements and appear to be derived from B cells at various stages of development. Proc Natl Acad Sci USA 1994;91:10962–6.
[239] Kanzler H, Kuppers R, Hansmann ML, Rajewsky K. Hodgkin and Reed-Sternberg cells in Hodgkin's disease represent the outgrowth of a dominant tumor clone derived from (crippled) germinal center B cells. J Exp Med 1996;184:1495–505.
[240] Martin-Subero JI, Gesk S, Harder L, Sonoki T, Tucker PW, Schlegelberger B, et al. Recurrent involvement of the REL and BCL11A loci in classical Hodgkin lymphoma. Blood 2002;99:1474–7.
[241] Barth TF, Martin-Subero JI, Joos S, Menz CK, Hasel C, Mechtersheimer G, et al. Gains of 2p involving the REL locus correlate with nuclear c-Rel protein accumulation in neoplastic cells of classical Hodgkin lymphoma. Blood 2003;101:3681–6.
[242] Jungnickel B, Staratschek-Jox A, Brauninger A, Spieker T, Wolf J, Diehl V, et al. Clonal deleterious mutations in the Ikappa-Balpha gene in the malignant cells in Hodgkin's lymphoma. J Exp Med 2000;191:395–402.
[243] Emmerich F, Meiser M, Hummel M, Demel G, Foss HD, Jundt F, et al. Overexpression of I kappa B alpha without inhibition of NF-kappaB activity and mutations in the I kappa B alpha gene in Reed-Sternberg cells. Blood 1999;94:3129–34.
[244] Schmitz R, Stanelle J, Hansmann ML, Kuppers R. Pathogenesis of classical and lymphocyte-predominant Hodgkin lymphoma. Annu Rev Pathol 2009;4:151–74.
[245] Wlodarska I, Nooyen P, Maes B, Martin-Subero JI, Siebert R, Pauwels P, et al. Frequent occurrence of BCL6 rearrangements in nodular lymphocyte predominance Hodgkin lymphoma but not in classical Hodgkin lymphoma. Blood 2003;101:706–10.
[246] Martin-Subero JI, Klapper W, Sotnikova A, Callet-Bauchu E, Harder L, Bastard C, et al. Chromosomal breakpoints affecting immunoglobulin loci are recurrent in Hodgkin and Reed-Sternberg cells of classical Hodgkin lymphoma. Cancer Res 2006;66:10332–8.

Chapter 23

# B Cells Producing Pathogenic Autoantibodies

**Yong-Rui Zou and Betty Diamond**

*Center for Autoimmune and Musculoskeletal Disease, The Feinstein Institute for Medical Research, Manhasset, NY, USA*

## 1. ORIGIN OF AUTOANTIBODIES

Immunoglobulins (Ig) with specificity for self-antigens can be generated by all subsets of mature B cells. Some autoantibodies belong to the repertoire of natural antibodies, which are mainly derived from B1 cells in the absence of external antigen stimulation. Autoreactive natural antibodies are predominantly of the IgM isotype. Many autoantibodies that are involved in autoimmune diseases are of IgG isotypes produced by B2 cells with T cell help. Although natural IgM autoantibodies are thought to be selected through immune evolution, high-affinity IgG autoantibodies are generated through the breach of immune tolerance at various checkpoints of B2 cell development and activation.

### 1.1 B1 Cells and Natural Autoantibodies

#### 1.1.1 *Natural Antibodies: Their Protective Properties and Autoreactivity*

An appreciation of natural antibodies dates back to the early era of immunology. Landsteiner noted in 1900 that "the serum of healthy humans has an agglutinating effect, not only upon animal blood cells, but frequently upon blood cells from other individuals as well." Circulating antibodies, mostly IgM and, to a lesser extent, IgA and IgG, emerge early in the development of physiologically healthy humans and mice.

Studies conducted in antigen-free and germ-free mice revealed that these mice maintain a normal serum level of IgM with a profile of specificity for endogenous tissue extracts and bacteria that is similar to that of conventional or specific pathogen free mice [1,2]. Human newborns also exhibit a remarkably similar IgM repertoire, even though their mothers have extremely diverse IgM immunoreactivity patterns established through their previous exposures to antigen [3]. Therefore, the repertoire and production of natural antibodies develop largely independent of external antigen load.

Natural antibodies are encoded by germline V gene segments with rare somatic mutations. The repertoire of natural antibodies is relatively restricted and seems to be evolutionarily selected to recognize phylogenetically conserved microbial and self structures, including carbohydrates, phospholipids, and nucleic acids [4]. Because of the ubiquitous presence of these structural motifs, natural antibodies often recognize a broad range of determinants on both microbial and self-antigens and are described as "polyreactive." For example, natural antibodies with T15 idiotype bind phosphorylcholine (PC), a determinant on the cell wall of *Streptococcus pneumoniae* and several other microbes [5–7]. T15 antibodies are highly represented in germ-free mice and dominate anti-PC responses in newborn mice [8]. These pre-existing, naturally occurring T15 antibodies provide an innate-like immune response against virulent pneumococcal infection even before adaptive immunity becomes fully mature or activated [9–11]. At the same time, T15 antibodies also exhibit autoreactivity specific for the PC head group of oxidized phospholipids present on oxidized low-density lipoprotein (OxLDL) and on apoptotic cells [12]. Binding of T15 IgM to apoptotic cells or blebs promotes C1q recruitment and enhances phagocytosis, providing an important pathway for the clearance of cellular debris [13]. In addition, T15 IgM blocks the binding of OxLDL to scavenger receptors on macrophages, thereby inhibiting inflammatory responses mediated by macrophages after activation by oxidized phospholipids [14–16]. Considering that the accumulation of apoptotic cells and oxidized lipoproteins is key to the pathogenesis of atherosclerosis, T15 IgM antibodies are thought to mediate protection against atherogenesis [17,18]. Moreover, engagement of Clq by IgM immune complexes, including T15 immune complexes, leads to activation of LAIR-1, an inhibitory molecule on myeloid cells,

Molecular Biology of B Cells. http://dx.doi.org/10.1016/B978-0-12-397933-9.00023-0

thereby helping to maintain myeloid cells in a quiescent state [19,20].

Autoreactive natural antibodies are largely nonpathogenic. They provide the first line of defense against microbes before adaptive immune responses take effect [21,22]. Multiple mechanisms exist by which natural antibodies mediate protection from infections. Some natural antibodies, such as T15 IgM, directly neutralize bacteria or viruses in the circulation [22,23]. Other natural antibodies enhance phagocytosis of infectious particles by macrophages and polymorphonuclear leukocytes as the first step in their destruction, or eliminate microbes through complement-mediated lysis. In addition, blood-borne pathogens, when they form immune complexes with natural antibodies, are rapidly filtered out by the spleen before they spread to vital organs.

Further protective properties of natural antibodies are conveyed by their specificity for altered self-antigens. A large fraction (~30%) of natural antibodies bind oxidation-specific epitopes [24]. Since oxidative stress is involved in many biological processes such as aging and apoptosis, as well as tissue injury and inflammation, it is not surprising that oxidation-specific epitopes are abundantly distributed in healthy individuals. Natural autoantibodies, through binding to oxidation-specific epitopes, play an important role in regulating tissue homeostasis by clearance of toxic metabolites and cell breakdown products [4,25]. Another set of natural autoantibodies reacts with self-determinants displayed on damaged or senescent red blood cells, such as phosphatidylcholine (PtC) and I/i carbohydrate epitopes [26,27]. It appears that damaged cells of different lineages are tagged with distinct modifications to be recognized by different subsets of natural autoantibodies for clearance. Impaired removal of apoptotic cells often causes inflammation and leads to the autoimmune disease systemic lupus erythematosus (SLE) in murine models [28]. Thus, natural autoantibodies have a potent suppressive effect on inflammatory responses and thereby prevent autoimmunity. For these antibodies to function in an immunosuppression, they must be of the IgM isotype.

Analogous to murine T15 natural antibodies, human anti-PC antibodies arise early in life and are dominant in cord blood [24]. Later in life, lower anti-PC IgM levels are found to be associated with a higher incidence of atherosclerotic cardiovascular diseases [29–31]. Studies of twins discordant for SLE show that the healthy twin has a higher level of anti-PC IgM than the affected twin, suggesting that natural IgM antibodies may reduce risk for SLE in those that are genetically predisposed [32]. Higher levels of IgM anti-PC or anti–double-stranded DNA (dsDNA) antibodies in SLE patients are also correlated with less organ damage [33]. Further support of the protective role of natural antibodies comes from murine studies in which infusion of monoclonal or polyclonal anti-dsDNA IgM delays disease onset and lowers lupus activity in lupus-prone mice [34,35]. Natural antibodies in these settings may provide protection by directly neutralizing autoantigens and thus preventing the formation of immune complexes of pathogenic IgG autoantibodies in tissue which can activate the immune system through their Fc region.

There are certain inflammatory conditions in which natural antibodies exhibit pathogenic properties. This is best represented by IgM natural antibodies in ischemia-reperfusion injury. Reoxygenation of ischemic organs leads to sterile inflammation mediated mainly by neutrophils and causes extensive tissue damage [36]. At the cellular level, ischemia-reperfusion induces mitochondrial dysfunction, production of reactive oxygen species, and the exposure of several neo-epitopes on ischemic cells, including annexin IV and nonmuscle myosin heavy chain II [37,38]. Binding of IgM natural antibodies to these epitopes and subsequent activation of the complement cascade mediates further tissue injury [39–41]. Thus, targeting specific IgM could benefit patients with myocardial infarction and reperfusion, ischemic stroke, or organ transplantation. Similarly, IgM antibodies to blood-borne antigens, erythrocytes, and platelets can opsonize the cell or particle and cause its destruction or elimination by the reticuloendothelial system and, hence, induce the pathology of autoimmune diseases (Figure 1).

### 1.1.2 B1 Cell Development and Selection

In mice, up to 80% of circulating IgM is produced by B1 cells [42]. This subset of B cells was first described in the early 1980s by Herzenberg and colleagues as a population of B cells that preferentially resides in the pleural and peritoneal cavities in mice and expresses CD5 together with a high level of IgM and low level of IgD on the cell surface [43]. B1 cells arise early in ontogeny. In fact, B cell precursors in the fetal liver give rise predominantly to B1 cells [44]. When B lymphopoiesis shifts from the fetal liver to the bone marrow after birth, B1 cell generation drastically declines. In adulthood, conventional B2 cells are continuously generated from bone marrow progenitors, whereas the B1 cell population is largely maintained through self-renewal [45,46].

It is not fully understood what limits B cell development to the B1 lineage in fetal life. The different microenvironment of fetal liver and adult bone marrow may affect the differentiation potential of B cell progenitors. For example, Notch2 signaling, which is activated in the fetal liver, may guide early hematopoietic precursors to preferentially differentiate into B1 cells [47,48]. More importantly, cell-intrinsic properties of fetal and postnatal B cell progenitors may account for the B lineage choice. One of the most distinctive features of fetal B cell progenitors is their lack of expression of terminal deoxynucleotidyl transferase (TdT) [49,50]. TdT, upregulated only after birth, is an enzyme for

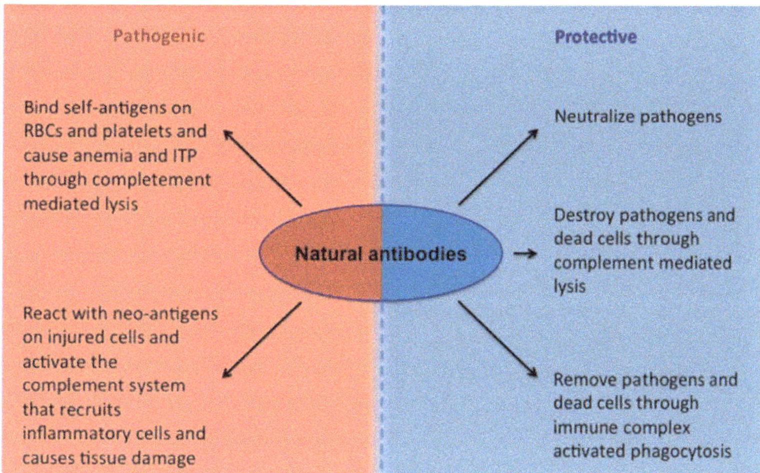


FIGURE 1 **Mechanisms of protective and pathogenic natural antibodies.**

the addition of nontemplated nucleotides at the D-J and V-D junctions. With extremely low TdT activity in fetal B cell progenitors, the length of CDR3 of the Ig heavy chain is shorter and the diversity of VDJ becomes much reduced compared to that of B2 cells [51,52]. In addition, there is biased usage of D proximal $V_H$ and $J_H$ gene segments during fetal B cell development, whereas B2 cells generated in adult bone marrow express a much broader range of $V_H$ genes [53]. For example, $V_H$6-1, the most $J_H$ proximal human $V_H$ gene, is infrequent in the adult B cell repertoire and is overrepresented in human fetal B cells [54]. $V_H$6-1 has been shown to have immunoreactivity to single-stranded DNA and cardiolipin, and is thought to be involved in clearance of DNA-containing cell breakdown products [55]. In mice, $V_H$S107.1, also a $J_H$ proximal $V_H$ gene segment that encodes the PC-specific T15 natural antibodies, is enriched in the B1 cell compartment [56]. Based on these observations, a model has been proposed that the specificity of the B cell receptor (BCR) for endogenous ligands may determine the B1 cell fate.

That self-antigens are required for B1 cell generation was shown first by Hayakawa et al., when they demonstrated that B1 cells specific for CD90/Thy-1 are generated in wildtype but not CD90 knockout mice [57]. Enforced expression of $V_H$12, a $V_H$ gene that is specific for PtC on senescent erythrocytes and typically expressed by B1 cells, leads to the generation of very high numbers of B1 cells in transgenic mice [58]. Interestingly, coexpression of $V_H$12 together with a $V_H$ gene expressed by B2 cells, $V_H$B1-8, drives B cell precursors to generate B2 cells in adult bone marrow. However, loss of the $V_H$B1-8 allele in these B2 cells converts them into B1 type cells in the periphery. It was speculated that $V_H$B1-8 expression diluted the density of self-reactive $V_H$12 on the cell surface, thereby weakening B1-specific BCR signaling for positive selection into the B1 cell compartment [59]. These data support the hypothesis that reactivity to self-antigen drives B1 lineage determination, but they do not address the question of how B1 cells are protected from BCR-mediated negative selection.

Multiple lines of evidence support the hypothesis that the strength of BCR signal plays a determinant role in the bifurcation of B1 and B2 lineages [60]. Mutations in positive regulators of BCR signaling, such as CD19 and Btk, selectively abrogate or impair the development of B cell precursors into B1 cells while leaving B2 maturation largely intact [61–63]. More strikingly, a Btk mutation in $V_H$12 transgenic mice converts the predominant $V_H$12-expressing B1 into cells with a B2 phenotype [64]. In contrast, genetically modified mice with enhanced BCR signaling have an increase in the B1 cell population [65–67]. It is apparent that stronger BCR stimulation selects B cell precursors into the B1 compartment. Hence, self-recognition of phylogenetically conserved immunodeterminants imposes a critical selective pressure on B1 cells generated in fetal life. This also explains how different individuals and even different species share an invariable, conserved B1 cell repertoire. As stated above, a critical unresolved question is why BCR signaling activates tolerance mechanisms in B2 but not B1 cell development.

Mice with enhanced BCR signaling have higher numbers of B1 cells. Moreover, enhanced BCR signaling can also prolong B1 cell survival, as demonstrated in Siglec-G–deficient mice. Siglec-G is a negative regulator that specifically inhibits BCR-mediated calcium signaling in B1 cells [68]. Mice lacking Siglec-G have an enlarged pool of B1 cells that exhibit enhanced expression of NFATc1 and a prolonged lifespan [69]. NFATc1 has been shown to promote resistance to apoptosis [70]. NFATc1-deficient fetal liver cells are unable to repopulate the B1 cell compartment [71]. Interestingly, vitamin A deficiency in mice leads to a decrease in NFATc1 expression and a reduction in B1 cell numbers and in the level of natural antibodies [72]. Thus, both B cell–intrinsic and –extrinsic signals can act synergistically to maintain the B1 cell pool.

It remains unclear where self-renewal of B1 cells takes place and what controls B1 cell homeostatic proliferation. B1 cells in the spleen have a higher turnover rate than they do in body cavities, suggesting that homeostatic proliferation of the B1 cell pool may occur in the spleen [73,74]. Consistent with this notion, the established $CD5^+$ B1 cell compartment in the peritoneal cavity quickly shrinks in adult mice after splenectomy [75]. Surprisingly, splenectomy does not lead to the loss of B1 cells in L2 transgenic mice in which expression of the transgenic λ2 Ig light chain blocks B2 cell development and leads to the generation of uniquely B1 cells. However, analysis of splenectomized L2 transgenic mice showed that the repertoire of $CD5^+$ B1 cells is significantly altered after surgery [76]. These studies reveal that the spleen may provide a selective niche for the expansion of the majority of B1 cells. Nonetheless, other tissues may provide compensatory positive selection in the absence of a spleen, and the expression of distinct self-antigens in different tissues may lead to the expansion of B1 cells with different repertoires, depending on the microenvironment in which they develop. This may have important clinical implications. Overproduction of natural antibodies may contribute to autoimmunity. For example, lupus-prone NZB/W mice have an enlarged B1 cell population, although an expansion of the B1 cell population is not a feature of all lupus-prone mouse strains [43]. Approximately 30% of patients with $CD5^+$ B cell chronic lymphocytic leukemia (B-CLL) develop hemolytic anemia due to an autoreactive BCR on the malignant B cells [77]. Thus, it is possible that overexpansion of autoreactive B1 cells may have a pathogenic outcome.

## 1.2 B2 Cells and Pathogenic Autoantibodies

Although B1 cells can undergo class switch recombination to produce germline-encoded IgG, high-affinity autoantibodies of the IgG isotype are mostly produced by B2 cells with T cell help [78]. Unlike B1 cells that are mainly generated in the fetal liver and preferentially reside in the body cavities, B2 cells continuously differentiate from precursors in the bone marrow and populate peripheral lymphoid organs after birth. Due to TdT upregulation and a broader usage of V genes, the BCR repertoire of B2 cells is much more diverse compared to that of B1 cells. After successful assembly of the BCR complex on the cell surface, newly generated B cells in the bone marrow migrate into the periphery as transitional B cells, and further mature into phenotypically distinct B cells that localize either in the marginal zone or the B cell follicles of the spleen and other peripheral lymphoid tissues.

BCRs recognizing self-antigens are constantly generated through the random process of V-(D)-J DNA recombination in developing B cells in the bone marrow. In fact, up to 75% of newly generated B cells in humans are autoreactive [79]. Thus, a complicated B cell selection system has been evolved to prevent autoimmunity at every stage of B cell development [80,81]. In the bone marrow, central tolerance mechanisms, such as clonal deletion or receptor editing, remove some developing B cells with high-affinity autoreactive BCR before they enter the periphery. Some autoreactive immature B cells are retained in an anergic state. Tolerance mechanisms also operate in the spleen when transitional autoreactive B cells are exposed to peripheral self-antigens.

At the mature stage, B cells may become autoreactive through somatic hypermutation in the Ig V genes during germinal center (GC) reactions after antigen stimulation [82,83]. Most of these GC-derived autoreactive B cells with novel antigen specificities will die due to the lack of positive selection by follicular dendritic cells (FDC) and T helper cells ($T_{FH}$) that are specific for the eliciting antigen. Under special circumstances when autoreactive $T_{FH}$ cells are also present or the newly acquired BCR autospecificity is crossreactive with the eliciting antigen, these autoreactive B cells may gain survival signals from FDCs and $T_{FH}$ cells and may be able to differentiate into memory cells or long-lived plasma cells. There is evidence that RAG1 and RAG2 expression can be induced in activated B cells during and after the GC reaction, and thus autoreactivity acquired by antigen-activated B cells through somatic hypermutation may be eliminated by receptor editing [84,85]. Additional suppression of GC-derived autoreactive B cells comes from T regulatory cells (Treg) in GCs [86–89]. Furthermore, in mice, plasmablasts ($CD138^+$ $B220^+$) that are reactive to the ribonucleoprotein Smith (Sm) are more susceptible to apoptosis than nonautoreactive plasmablasts, and in humans the frequency of autoreactive cells in the bone marrow plasma cell population is lower than in the blood memory cell compartment [90,91]. These observations suggest that there may be tolerance checkpoints at the B cell terminal differentiation stages that prevent autoreactive B cells from entering into the long-lived plasma cell pool.

## 1.3 Mechanisms that Breach B Cell Tolerance

We know more about the immunopathology of autoantibodies than we do about the mechanisms that induce their production. Every autoimmune disorder has a complex etiology that involves both genetic and environmental influences. These cell intrinsic and extrinsic factors may breach B cell tolerance at various checkpoints through modulating B cell activation and responses or regulating survival of autoreactive B cells, culminating in inflammatory responses and eventual tissue damage.

### 1.3.1 Environmental Factors

The importance of environmental or stochastic factors is emphasized by the relatively low concordance for autoimmune disorders between monozygotic twins, generally in the range of 15–30%. The incidence of human autoimmune diseases increases with modern industrial development that affects our diet, medications, environmental pathogens, and stress. Noninfectious environmental factors predisposed to autoantibody-mediated autoimmunity include physical trauma, stress, drugs, and smoking. These factors may contribute to disease by disruption of tissue integrity, thus allowing autoantibodies access to healthy tissues that are not involved in initiating autoimmunity or even to immunologically privileged sites. They may also induce innate and adaptive immune cells to express pro-inflammatory genes. Among all environmental factors associated with autoimmunity, infectious pathogens and gender-related humoral milieu appear to be the two most important mediators for the onset of disease. This section focuses on the mechanisms of infection- and gender-associated B cell autoimmunity.

#### 1.3.1.1 Infection

Many autoimmune diseases have an apparent infectious etiology (Table 1). Rheumatic fever and Guillain-Barré syndrome are classical examples of autoimmune diseases initiated by an acute infection with defined pathogens. Rheumatic fever is triggered by infection with group A *Streptococcus pyogenes.* The streptococcal M protein, the major virulent factor in the cell wall of Streptococci, and the carbohydrate epitope *N*-acetyl-glucosamine (GlcNAc) share striking structural homology with several heart proteins, such as the myosin heavy chain, tropomyosin, and laminin, which all have an α-helical coiled-coil structure [92]. In mice and humans, anti-M protein and anti-GlcNAc antibodies broadly crossreact with these self-antigens and cause damage to multiple organs, including the heart, joints, and kidney. In some individuals, anti-GlcNAc antibodies bind the D2 dopamine receptor and cause symptoms of Sydenham's chorea [93].

**TABLE 1** **Infectious Agents Involved in Autoantibody Mediated Autoimmune Diseases**

| Diseases | Pathogens |
|---|---|
| Rheumatic fever | Group A streptococci |
| Guillain-Barré syndrome | Campylobacter jejuni Cytomegalovirus |
| Myasthenia gravis | Herpes simplex virus |
| Graves' disease | Yersinia enterocolitica |
| Mixed cryoglobulinemia | Hepatitis C virus Mycoplasma pneumonia |

Guillain-Barré syndrome is an autoimmune disease that is characterized by lymphocytic infiltration and demyelination in the peripheral nervous system [94]. The principal infectious microorganism that causes Guillain-Barré syndrome is *Campylobacter jejuni,* which stimulates B cells to produce antibodies against LPS. LPS of some *C. jejuni* strains bears structural similarities to gangliosides on the cell surface membrane of peripheral nerves [95,96]. Thus, antibodies against *C. jejuni* also recognize peripheral nerve antigens and causes motor symptoms. A direct proof of *C. jejuni* as the cause of Guillain-Barré syndrome comes from animal models. Rabbits immunized with *C. jejuni* LPS produce autoantibodies against gangliosides and develop neural pathology analogous to human Guillain-Barré syndrome.

In the case of both rheumatic fever and Guillain-Barré syndrome, autoantibodies are elicited by immunodominant epitopes of infectious agents that share significant sequence or structural homology with self-antigens of the host. B cells activated by the pathogen produce neutralizing antibodies that are also crossreactive to the self and cause tissue injury. This mechanism is termed molecular mimicry. Crossreactivity of autoantibodies with a variety of pathogen-derived molecules has been observed in a number of autoimmune diseases. For example, glycoprotein D of herpes simplex virus (HSV) contains a peptide sequence identical to the α-subunit of the human acetylcholine receptor (AchR) [97]. Antibodies against HSV crossreact with AchR, suggesting that HSV may be associated with the initiation of myasthenia gravis. Many patients with Graves' disease have been infected with *Yersinia enterocolitica*, and autoantibodies against the thyroid-stimulating hormone (TSH) receptor also bind to the bacteria [98]. SLE patients have pathogenic autoantibodies with crossreactivity against structurally distinct immunodominant epitopes on a wide variety of pathogens [99]. Therefore, molecular mimicry represents a mechanism by which infectious agents sidestep B cell tolerance and cause autoimmune diseases.

Epitope spreading is another common strategy by which infectious agents provoke and amplify autoimmunity. In this process, antigen exposure occurs when the pathogen lyses infected host cells or when inflammation induced by the pathogen causes tissue damage. Novel or cryptic epitopes exposed by these processes are presented to B and T cells. This results in an immune response not only against the inciting pathogen but also to self-antigens. In a model of experimental myasthenia gravis, animals that have been immunized with a peptide from the α-subunit of AchR develop autoantibodies not only specific to the immunogen but also to the other subunits of AchR, demonstrating an intramolecular epitope spreading [100]. Epitope spreading also occurs in human SLE, as some patients who initially have anti-Sm autoantibodies later develop antibodies against RNP, another component of the spliceosomal complex [101]. Moreover, the autoreactive B cells function as antigen-presenting cells (APCs), presenting cryptic epitopes of self antigen not previously available to purge the T cell repertoire of autoreactivity; thus, autoreactive T cells can be activated [102].

Pathogen infection often induces T-dependent GC reactions. During this process, autoantibodies can arise from an original antimicrobial humoral response through somatic hypermutation. This was first observed in an in vitro culture system, in which a single base substitution in the $V_H$ gene of PC specific Ig changed its reactivity to DNA [82]. In vivo studies confirmed this model. When B cells of mice immunized with PC coupled to a protein carrier were fused to a Bcl-2–overexpressing cell line to form hybridomas, approximately 40% of the B cells secreting anti-PC antibody produced an antibody crossreactive with DNA [103]. All of the crossreactive antibodies displayed somatic mutation. Similar observations were made in mice immunized with the hapten phenylarsonate coupled to a protein carrier.

Recent studies in transgenic mice show that autoantigens, when abundantly and ubiquitously expressed in the body, delete autoreactive B cells. However, when the target autoantigen is located only in a tissue or organ remote from the GC, B cells capable of reacting against both foreign and self antigens are able to escape the GC and produce autoantibodies [224]. These animal studies provide an additional mechanistic explanation for the induction of autoimmune diseases after an infection.

Beside the aforementioned antigen-specific mechanisms, pathogens can trigger nonspecific activation of autoreactive B cells and promote their expansion. This process is known as bystander activation. Pathogens can activate both innate and adaptive immune cells through pathogen recognition receptors (PRR). It is interesting to note that many experimental autoimmune diseases are induced by immunizing animals with autoantigens emulsified with complete Freund's adjuvant (CFA), which contains killed *Mycobacterium tuberculosis*. For example, administration of the autoantigen AchR together with CFA in C57BL/6 mice induces symptoms similar to those of human myasthenia gravis [104]. It has been found that NK cells activated by CFA in this experimental autoimmune model are required for polarization of autoreactive $T_H1$ cells, production of anti-AchR autoantibodies, and disease development [105]. This observation reveals that NK cells and other innate immune cells, being the first line of defense against infectious agents, can participate in the development of destructive autoimmunity.

After pathogen activation, innate immune cells can augment autoimmunity by enhancing processing of autoantigens to activate autoreactive B cells, and by producing inflammatory cytokines and chemokines to recruit autoreactive B cells and to promote their survival and expansion. These pathogenic roles are exemplified by neutrophils in SLE patients. There is an expanded population of neutrophils in the blood of SLE patients. Unlike those in healthy individuals, circulating neutrophils in SLE patients are of low density and bear an immature phenotype [106]. Activated immature neutrophils undergo a specific form of cell death called NETosis, and form neutrophil extracellular traps (NET) containing self-DNA and nuclear proteins that are the principal autoantigens of SLE [107]. Through NETosis, neutrophils also release inflammatory cytokines, and trigger plasmacytoid dendritic cells (DCs) to produce type-I interferon (IFN) [108]. Upregulation of type-I IFN and its target genes, known as the IFN signature, is well known to contribute to the pathogenesis of several systemic autoimmune diseases, including SLE [109].

There are several families of PRRs through which innate and adaptive immune cells detect damage- or pathogen-associated molecular patterns (PAMPs) [110,111]. Toll-like receptors (TLRs) detect PAMPs on the cell surface or on endosomes and lysosomes. C-type lectin receptors (CLRs) bind β-glucans. The cytosolic PPRs, including NOD-like receptors (NLRs) and RIG-I-like receptors (RLRs), sense intracellular bacterial products and nucleic acids, respectively. Activation of PRRs has been linked with chronic inflammation and autoimmune diseases. In particular, murine studies show that TLR7 is required for the generation of autoantibodies against RNA-containing antigens such as Smith (Sm), and TLR9 for spontaneous production of anti-DNA autoantibodies [112–114]. These studies clearly demonstrate that infection might lead to aberrant activation of innate immune cells and autoreactive lymphocytes and thus contribute to inflammatory responses and autoimmunity.

#### 1.3.1.2 Gender Influences

Most autoimmune disease affects women more commonly than men. There are data that implicate the presence of two X chromosomes in the female predominance of autoimmunity. For example, TLR7 and CD40 ligand (CD40L) are both encoded on the X chromosome and overexpressed on lymphocytes of women with SLE. There are also data demonstrating that hormones contribute to the female predominance. Female castration reduces the incidence and severity of disease in NZB/W lupus-prone mice, whereas male castration increases the incidence of lupus in this strain [115]. Estrogen has been shown to alter the function of numerous cells within the immune system. In particular, it has been shown to allow DNA-reactive transitional B cell to pass an early tolerance checkpoint and to mature into marginal-zone (MZ) B cells [116]. The molecular mechanisms for this are not clear, but estrogen increases expression of Bcl–2, an antiapoptotic molecule, and CD22, an inhibitory coreceptor of the BCR [117]. Estrogen synergizes with type I IFN to increase expression of TLRs and to decrease expression of another inhibitory coreceptor of the BCR, FcγRIIB [118–120]. One study of estrogen receptor α (ERα)–deficient NZB/W mice suggests that the absence of ERα diminishes renal diseases but does not alter autoantibody titer [121]. Other studies show that effects of estrogen on B cells are mediated through ERα, and that an absence of ERα prevents an estrogen-mediated breach of B cell tolerance [122].

### 1.3.2 Genetic Factors

A genetic predisposition to autoimmune diseases is demonstrated by the fact that most of these diseases tend to run in families, and genome-wide association studies (GWAS) have identified some genetic risk factors. Although the exact genetic factors are not known and are very likely to be different for each B cell–mediated autoimmune disorder, animal models and human studies have identified numerous genes that are involved in the pathogenesis of autoimmune diseases. Some of the disease-associated genes act in a cell-autonomous way to breach B cell tolerance, whereas others indirectly modulate B cell function. These genes can be categorized into those that affect B cell activation and those that affect the survival and expansion of autoreactive B cells (Table 2).

**TABLE 2** Genes Involved in Autoantibody Mediated Autoimmune Diseases

| |
|---|
| **Autoantigen Clearance** |
| Complements C1q and C4 (mouse and human) |
| Mfge8 (mouse and human) |
| DNase I (mouse and human) |
| Trex1 (mouse and human) |
| **Antigen Presentation** |
| HLA-DR2, DR3 and DR4 (human) |
| **Signaling** |
| BANK1 (human) |
| CD22 (mouse) |
| Lyn (mouse and human) |
| SHP-1 (mouse) |
| PTPN22 (mouse and human) |
| FcγRIIB (mouse and human) |
| **Growth and Anti-apoptotic Factors** |
| BAFF-R (mouse and human) |
| CD40 (mouse and human) |
| IL-6 (mouse and human) |
| Fas (mouse and human) |

#### 1.3.2.1 Genes that Affect B Cell Activation

Loss of tissue integrity and organ destruction can lead to the exposure of autoantigens and neoantigens to the immune system. This process represents a major trigger of inflammation and autoimmunity. An important mechanism to prevent autoimmune responses is rapid clearance of autoantigens and immune complexes [123]. There is evidence that, in GCs of SLE patients, tingible body macrophages are not able to efficiently engulf apoptotic cells. Studies show that impaired uptake of apoptotic cells by DCs and macrophages in mice deficient for milk fat globule epidermal growth factor 8 (Mfge8) causes spontaneous GC reactions and production of autoantibodies [124]. Efficient removal of immune complexes and opsonized cell debris is dependent on complement components and Fc receptors (FcR) on phagocytic cells. In mice, deletion of C1q and C4 or a point mutation of the complement receptor Cr2 that diminishes C3d binding leads to the loss of tolerance [125]. In humans, C1q and C4 deficiencies, although rare, predispose to SLE. FcγRIIIA, and FcγRIIIB polymorphisms have also been documented to be associated with SLE [126]. DNase I and Trex1 are responsible for the removal of extracellular or cytosolic nucleic acids, respectively. Polymorphisms in these molecules are also linked to SLE [127,128]. Collectively, these data indicate that multiple pathways mediate the rapid clearance of dead cells and prevent the accumulation of autoantigens. Defects in these pathways result in activation of autoreactive B cells and autoantibody production.

A group of disease-associated genes encoding molecules within the BCR signaling pathway directly affect B cell activation. Changes in the BCR activation threshold by these molecules can influence the life and death of a B cell. Among these signaling molecules, BANK1, a positive regulator of the BCR signaling pathway, has been identified by GWAS to be a susceptibility gene for SLE. The disease-associated variants of BANK1 are found to induce sustained BCR signaling and B cell hyperactivity, leading to the proliferation and expansion of autoreactive B cells [129,130]. Several negative regulators of BCR signaling have also been associated with autoimmunity, including PTPN22, Csk, Lyn, CD22, and FcγRIIB. PTPN22 is a phosphatase that associates with Csk to regulate phosphorylation of the Src family kinases in both B and T cells. The function of the disease-associated variant of PTPN22 remains controversial, with the evidence for both gain of function and loss of function. Overall, the data best supports the view that TCR and BCR signaling is enhanced in carriers of the risk allele [131]. Although the mechanisms are not well understood, the PTPN22 risk allele results in an accumulation of autoreactive B cells in the periphery [225]. Csk phosphorylates an inhibitory C-terminal regulatory tyrosine on Lyn. The disease-associated Csk allele leads to enhanced Lyn inactivation [132]. Lyn deficiency represents another compromised negative regulatory feedback pathway that contributes to B cell–mediated autoimmunity. Expression of Lyn is decreased in B cells of SLE patients [133,134]. In mice, Lyn inactivation leads to a severe lupus-like disease. Lyn phosphorylates CD22, which then recruits the tyrosine phosphatase SHP-1 to the CD22/BCR complex and dampens BCR signaling [135]. Lyn also negatively controls BCR signaling by phosphorylating FcγRIIB and recruiting SHIP, an inositol phosphatase. FcγRIIB is the only inhibitory FcR and plays a critical role in maintaining B cell tolerance. Mice carrying a transgenic BCR specific for DNA produce autoreactive IgM but do not develop autoimmune nephritis. Introducing FcγRIIB deficiency in this mouse strain promotes class switch of nonpathogenic autoreactive IgM to pathogenic IgG2a and IgG2b through a TLR9/MyD88-dependent pathway, with resulting kidney damage [136]. This study exemplifies how infections that activate TLRs may trigger a genetically susceptible individual, for instance with FcγRIIB deficiency, to produce pathogenic autoantibodies and to develop full-blown autoimmunity.

#### 1.3.2.2 Genes that Affect Expansion and Survival of Autoreactive B Cells

One of the major genetic mechanisms that abrogates B cell tolerance is the failure of autoreactive lymphocytes to undergo apoptosis. Apoptosis is critical not only for removal of B cells with a high-affinity autoreactive BCR during central and peripheral tolerance but also for maintaining lymphocyte homeostasis. A family of proteins that plays an important role in supporting B cell survival is the TNF family, including BAFF and APRIL [137]. BAFF is essential for the survival of transitional B cells and MZ B cells. It also participates in the positive selection of B cells during GC reactions. APRIL has been shown to be a critical survival factor for plasma cells. High levels of BAFF can rescue autoreactive B cells from apoptosis. Both BAFF and APRIL are expressed by myeloid lineage cells, as well as by nonhematopoietic cells in various tissues. In multiple autoimmune diseases, such as SLE and rheumatoid arthritis (RA), levels of BAFF and APRIL are elevated [138]. The direct evidence that increased BAFF levels lead to the development of autoimmune disease comes from a murine study in which transgenic mice constitutively expressing excessive amounts of BAFF develop a lupus-like disease [139].

Various mutations in genes involved in Fas-mediated apoptotic pathways cause autoimmune lymphoproliferative syndrome (ALPS). Patients with ALPS develop lymphadenopathy and splenomegaly, produce autoantibodies, and often have hemolytic anemia [140]. In contrast, mice with spontaneous mutations in Fas (MLR/lpr) or Fas ligand (MLR/gld) develop lymphoid hyperplasia, anti-DNA antibodies, and glomerulonephritis. Fas is highly expressed in activated and GC B cells. In mice, ablation of Fas specifically in GC B cells leads to fatal lymphoproliferation and accumulation of memory B cells in a T cell–dependent manner. Importantly, deficiency of Fas in B cells leads to activation and hyperproliferation of Fas-sufficient T cells [141]. This study shows that B cell expression of Fas is essential for removal of B cells with suboptimal affinity or autoreactivity in the GCs and for T cell homeostasis. When Fas is nonfunctional, autoreactive GC B cells fail to die and differentiate into plasma cells or memory B cells. In turn, autoreactive memory B cells may act as APC for T cells, thereby causing a lymphoproliferative disorder.

The most predominant genetic factor affecting the susceptibility to almost all autoimmune diseases is HLA class II. HLA class II polymorphisms have been shown to confer high risk for Graves' disease, myasthenia gravis, RA, SLE, and many other autoimmune diseases [142]. In SLE patients, HLA-DR3 haplotypes are related to the production of autoantibodies against ribonucleoprotein complexes (Ro/SSA or La/SSB), HLA-DR2 to the anti-DNA antibodies and nuclear antigen Sm, and HLA-DR4 to antibodies specific for the small nuclear ribonucleoproteins [143,144]. Peptide-binding studies have demonstrated that the polymorphic residues of HLA-DR can determine the repertoire of self-peptides presented by these MHC class II molecules. MHC class II molecules that confer susceptibility to an autoimmune disease may positively select $T_H$ cells with autoreactive TCRs and trigger TCR signaling by presentation of peptides derived from the target antigen. The fact that MHC class II haplotypes shape the specificity of autoantibodies suggests that $T_H$ cells that react with a specific self-peptide drive the development and expansion of autoreactive B cells in a cognate interaction.

T cells can facilitate the production of pathogenic autoantibodies by helping autoreactive B cells to escape tolerance and by promoting them to differentiate into plasma cells that secrete IgG. CD40L is upregulated in activated T cells. The receptor CD40 is expressed on B cells, DCs and other nonhematopoietic cells. Costimulation of the BCR together with CD40 drives B cells to proliferate vigorously, rescues immature and transitional autoreactive B cells from tolerance mechanisms, and prevents spontaneous apoptosis of GC B cells. CD40 signaling also induces B cells and DCs to produce the proinflammatory cytokine IL-6, which in turn promotes class switch recombination and supports the survival of plasma cells [145]. Polymorphisms of CD40 have been shown to be associated with SLE, multiple sclerosis (MS), and RA [146,147]. Not surprisingly, treatment with antibodies blocking the interaction between CD40 and CD40L has been shown to inhibit disease progression in animal models of SLE, MS, and RA [148].

A small population of activated $T_H$ cells gains access to B cell follicles by upregulation of the chemokine receptor CXCR5. These B-cell follicle–residing T cells, termed $T_{FH}$ cells, express high levels of the costimulatory molecules CD40L and ICOS, and produce mitogenic cytokines IL-4, IFNγ, and IL-21 [149,150]. An increase in the number of $T_{FH}$ cells in mice, caused by a mutation in the putative E3 ligase Roquin (Rc3h1), leads to the development of lupus-like disease [151]. That expansion of $T_{FH}$ cells instigates pathogenic GC reactions is also observed in mice with DC-specific ablation of Blimp1 [152]. Blimp1-deficient DCs lose tolerogenicity, producing high amounts of IL-6, which triggers the expansion of $T_{FH}$ cells, spontaneous GC reactions, production of autoantibodies, and development of lupus-like symptoms. Blimp1 polymorphisms have been found to be associated with autoimmune disorders, including SLE [153,154].

In addition, autoimmunity can be suppressed by $CD1d^{hi}$ $CD5^+$ regulatory B cells that produce IL-10 upon TLR and CD40 stimulation (B10 cells) [155]. The development of B10 cells depends on CD40 stimulation and the BCR signal strength. In SLE patients, the number of circulating B10 cells is increased; however, their function is impaired. B10 cells isolated from SLE patients fail to respond to CD40 engagement and are defective in secreting IL-10. $CD19^{-/-}$ NZB/W mice, which lack B10 cells, exhibit accelerated development of nephritis. Adoptive transfer of $CD1d^{hi}$ $CD5^+$ splenic B cells into $CD19^{-/-}$ NZB/W mice prolongs their survival, thus supporting a protective role of B10 cells in lupus [156]. The mechanism of B10 cell protection is likely through IL-10–mediated expansion of Tregs.

## 2. IMMUNODEFICIENCY, B CELL MALIGNANCY, AND AUTOREACTIVITY

### 2.1 Immunodeficiency

Primary immunodeficiency comprises diverse clinical conditions with varied genetic defects causing a partial or severe immunological failure. Since the core functions of the immune system are to discriminate autoantigens from non-self antigens, to rid the host of pathogenic invaders, and to remove cell debris in a nonimmunogenic fashion, persons with defective immune systems have increased susceptibility to infection, and some also have autoimmune disorders. In some but not all immunodeficiencies, individuals possess low numbers of B cells, which can lead to high levels of BAFF and BAFF-mediated rescue of autoreactive B cells. BAFF-mediated maturation of autoreactive B cells emerging from the bone marrow can lead to the production of pathogenic autoantibodies [137].

Hyper IgM syndromes (HIgM) are caused by defective class switch recombination. Multiple genetic defects can cause HIgM, including defects in the CD40 signaling pathway, molecules mediating class switch recombination, and the NFκB pathway that is required for GC formation [157]. Autoimmunity is commonly associated with defects in activation-induced cytidine deaminase (AID), NEMO, and CD40. These patients generally develop polyarthritis, thyroiditis, and autoimmune cytopenia, suggesting that defects in these molecules lead to impaired establishment or maintenance of B cell tolerance. Indeed, the frequency of autoreactive B cells is increased in the periphery of CD40L-deficient patients but not in the bone marrow, indicating that the CD40 signaling pathway, albeit dispensable for central tolerance, is critical for establishing peripheral tolerance [158]. Studies of newly generated B cells and mature B cells from AID-deficient patients show that both B cell compartments have an abnormal Ig repertoire with a high frequency of autoreactivity [159]. Therefore, unlike CD40, AID is required for both central and peripheral B cell tolerance.

Common variable immune deficiency (CVID) is a group of primary immunodeficiency diseases with the hallmark of antibody deficiency and recurrent infections. Autoimmune disorders commonly associated with CVID are hemolytic anemia and thrombocytopenia [160]. The genetic basis of CVID is diverse, and the pathogenesis for CVID and associated autoimmunity is not fully understood. Some CVID patients have altered B cell activation and a loss of inhibitory signals in B cells, which may compromise peripheral B cell tolerance. In other cases, B cell dysfunction in CVID patients is caused by increased BAFF signaling [161]. Approximately 10% of CVID patients have mutations in TACI, the receptor for BAFF and APRIL, and central B cell tolerance is defective in all patients carrying TACI mutations [162].

Some primary immunodeficiency disorders lead to autoantibody production through dysregulated T cells. Autoimmune polyendocrinopathy candidiasis ectodermal dystrophy (APECED) and immune dysregulation, polyendocrinopathy, enteropathy, and X-linked (IPEX) are two rare primary immunodeficiency diseases characterized by the production of autoantibodies to a broad spectrum of self-antigens, as well as by lymphocytic infiltration and damage to endocrine organs. Both diseases are caused by genetic defects affecting T cell development. APECED patients carry mutations in the autoimmune regulator gene (Aire) [163]. Aire functions as a transcriptional regulator that induces the expression of peripheral tissue self-antigens in medullary epithelial cells and DCs in the thymus. In Aire-deficient mice, T cell central tolerance is compromised, because peripheral tissue self-antigens are not expressed in the thymus, and thereby autoreactive T cells escape thymic deletion [164]. APECED shows clearly that autoreactive T cells can drive the production of pathogenic autoantibodies. The importance of T cells for maintaining B cell tolerance is also shown by the Foxp3 mutation that causes IPEX. Foxp3 is the master transcription factor that determines the generation of Tregs [165]. Tregs suppress autoimmunity by inhibiting the activation and proliferation of T effector cells and B cells. Mutations in Foxp3 cause fatal autoimmune disorders in both humans and mice [166]. A small fraction of Tregs, $CD4^+$ or $CD8^+$, express high levels of CXCR5 and migrate into the GCs [86–89]. These Tregs negatively control the number and function of $T_{FH}$ cells. Mouse studies show that $CD8^+$ Tregs express Qa-1–restricted TCRs and interact with Qa-1 on $T_{FH}$ cells. Disruption of this interaction causes a lupus-like disease [89].

Wiskott-Aldrich syndrome (WAS) is a rare X-linked immunodeficiency caused by mutations in the WAS protein (WASp), a key cytoskeletal regulator in all lineages of hematopoietic cells [167]. Around 25% of WAS patients have multiple autoimmune diseases [168]. WASp deficient T cells display abnormal immune synapses and are not able to be fully activated and polarized to become $T_H1$ effector cells. Treg cell number and function are also reduced in WAS patients and mice, which may contribute to the development of autoimmunity. WASp deficient macrophages may also contribute to autoimmunity, as they are unable to engulf cell debris. In addition, pDCs in WAS patients play a role in WAS autoimmunity by producing an increased amount of type-I IFN after TLR9 stimulation. Moreover, WASp deficient B cells become hyperresponsive to BCR and TLR stimulation and may lose B cell tolerance in a cell autonomous way. Indeed, chimeric mice in which only B cells are WASp deficient undergo spontaneous GC reactions, produce autoantibodies and develop fatal nephritis [169]. This is consistent to the observation that WAS patients that have mixed chimerism after stem cell transplantation often develop severe B cell mediated autoimmune diseases [170].

## 2.2 B Cell Malignancy

It has been noted that several B cell mediated autoimmune diseases are associated with B cell malignancy. The co-occurrence of B cell malignancy and autoimmunity is perhaps not surprising, because these two diseases share some common etiologies, such as defective apoptosis. Mutations in *tnfrsf6*, the gene encoding Fas, cause ALPS in early childhood [140]. Like Fas mutated MRL/lpr mice, ALPS patients have multiple immune abnormalities, including autoimmune hemolytic anemia and thrombocytopenia. More than 10% of ALPS patients later develop B cell lymphomas, both Hodgkin's and non-Hodgkin's lymphomas [171].

B-CLL is a B cell malignancy in which malignant clones from different patients exhibit stereotyped BCRs [172]. It has long been thought that engagement of the BCR with specific antigens might drive CCL. Recent studies suggest that the CDR3 region of the Ig heavy chain and an internal epitope of the BCR induce antigen-independent BCR signaling [173]. Small-molecule inhibitory drugs that target BCR signaling molecules, such as PI3K, Btk, and Syk, are clinically effective in treating several B cell malignancies, as well as B cell–mediated autoimmune disease [174]. In addition to BCR activation, cytokine stimulation also plays a critical role in autoimmunity and B cell malignancy. APRIL was initially identified as a cytokine expressed predominantly by tumor tissues. Enhanced APRIL signaling through the PI3K/AKT pathway may represent a crucial pathogenic mechanism of B cell malignancy and autoimmune diseases.

Sustained B cell activation and spontaneous GC reactions in autoimmune disease undoubtedly increase genetic instability that could lead to B cell malignancy. As described above, the GC is a high-risk zone where autoreactive B cells arise through somatic hypermutations. Somatic hypermutations also occur in non-Ig genes such as Fas and Bcl-6; these mutations can contribute to lymphomas. In addition, class switch recombination mediated by AID in GCs can also induce pathogenic chromosomal translocations, such as Bcl-2 and c-Myc translocations, which are responsible for the development of follicular B cell lymphoma and Burkitt's lymphomas, respectively. Thus, uncontrolled GC reactions are causative to both B cell autoimmunity and B cell lymphomagenesis.

Under certain circumstances, malignant B cells secrete autoantibodies that lead to autoimmune disorders. Some B-CLL clones produce autoantibodies specific for erythrocytes and cause autoimmune hemolytic anemia. Paraneoplastic pemphigus (PNP) is another autoimmune disease that is associated with B cell neoplasms, most commonly with non-Hodgkin's lymphoma, B-CLL, and giant follicular hyperplasia (Castleman's disease) [175]. Autoantibodies produced by malignant B cells in PNP patients bind desmogleins 1 and 3 and members of the plakin family of epithelial proteins, leading to a severe blistering disease [176]. The discovery of the interdependent association between B cell malignancy and autoimmunity is of great importance for developing therapies not only to treat autoimmune diseases but also to prevent lymphomagenesis.

# 3. FEATURES OF PATHOGENIC AUTOANTIBODIES

## 3.1 Pathogenicity of Autoantibodies

Antibodies with autoreactive specificity can be detected in virtually all autoimmune diseases. It is important to note that not all autoantibodies isolated from patients are responsible for pathology. For example, patients with type 1 diabetes (T1D) often have a higher frequency of thyroid microsomal autoantibodies, although there may be no apparent thyroid pathology [177,178]. Obviously, an important criterion to define whether the autoantibody is contributory to a particular autoimmune disease is the presence of the autoantigen and the autoantibody in the affected organ. A poorly understood aspect of autoimmune disease, however, is that targeted autoantigens are not always selectively expressed by the target organ. Some ubiquitous autoantigens appear to contribute to organ-specific autoimmune disorders, such as ribonucleoprotein antigens in Sjögren's syndrome and tRNA synthetase in polymyositis [179–181]. This may be due to access of autoantibodies to tissues. For example, in the K/BxN mouse model of arthritis, immune complexes containing autoantibodies specific for the ubiquitous cytoplasmic enzyme glucose-6-phophate isomerase (GPI) cause vascular leakage at the distal joints that leads to joint-localized deposition of GPI immune complexes and arthritis [182]. In other situations, the autoantibody may exhibit a tissue-specific crossreactivity. Identification of disease-inciting autoantigens has been a challenging task and continues to lead to major milestones in autoimmune disease research.

Disease-specific autoantibodies may appear long before manifestation of the clinical symptoms. These autoantibodies thus serve as biomarkers for the prediction and diagnosis of certain autoimmune diseases. In individuals with T1D, autoantibodies recognizing islet cell antigens are present prior to and at the time of diagnosis. In nonobese diabetic (NOD) mice, the early appearance of anti-insulin autoantibodies predicts an early onset of the disease [183]. However, diabetes can develop even in the absence of B cells [184–186]. It is thought that autoantibodies alone in T1D are not pathogenic; instead, they might be by-products of islet cell destruction. Thus, the mere presence of autoantibodies even in the organ that is the target of pathogenic autoreactivity is a necessary but not sufficient condition for their pathogenicity.

An indisputable demonstration of the pathological role of autoantibodies is that transfer of these antibodies can directly induce disease. Autoimmunity arises in neonates from transplacental passage of maternal autoantibodies. This occurs in about 1–2% of infants born to women with SLE who carry autoantibodies against Ro/SSA and La/SSB [187]. Similarly, infants of women with pemphigus vulgaris develop a blistering rash due to transmission of maternal anti–desmoglein-3 antibody through the placenta. Recent studies also link maternal antibodies specific for fetal brain proteins to autism spectrum disorder. These "experiments" of nature demonstrate the causative role of autoantibodies in the pathology of these specific autoimmune diseases [188]. Further proof of pathogenic autoantibodies comes from experimental transfer models in which Ig from the serum of affected patients leads to disease symptoms in recipient animals. For example, newborn mice develop blisters after receiving serum from patients with pemphigus vulgaris [189]. Finally, multiple lines of transgenic mice expressing V genes encoding autoantibodies to erythrocytes or dsDNA, for example, develop symptoms resembling the clinical manifestations of hematolytic anemia or lupus nephritis [190–192].

Autoantibodies that are pathogenic in one genetic background may not be pathogenic in another genetic background. It has been clearly demonstrated in animal studies that there is a genetic component to target organ susceptibility to autoantibody-mediated damage. For example, antimyosin antibodies cause an inflammatory response in the hearts of DBA/2 mice where cardiac myosin is present in the extracellular matrix, but not in BALB/c mice in which there is far less cardiac myosin exposed [193].

## 3.2 Structural Basis of Autoreactivity

There are no features of the Ig variable region that characterize pathogenic autoantibodies. In B1 cells, particular V genes clearly associate with particular autospecificities. For example, the antibodies that are elicited by *Mycoplasma pneumoniae* infection and bind erythrocyte antigens express a stereotypic set of BCRs [194]. In contrast, in B2 cells, multiple $V_H$ genes can encode antibodies with a given autospecificity, and the same V genes are used to encode autoantibodies and protective antibodies, although certain $V_H$ or $V_L$ genes may be favored in antibodies of a particular antigenic specificity. For example, $V_H4$ expressing antibodies are overrepresented in the cerebral spinal fluid of patients with MS. Analysis of antibodies sequenced from B cells of patients with MS has shown a higher than usual frequency of replacement mutations in six codons of $V_H4$ genes, three in framework regions and three in complementarity-determining regions. Several of these antibodies bind to neurons or astrocytes, suggesting that these mutations may contribute to autoreactivity [195]. Some recent data suggest that the anti-desmoglein antibodies that cause blistering in individuals with pemphigus preferentially utilize the $V_H1$-46 gene [196]. There is an extensive literature on rheumatoid factor (RF) antibodies that bind to the Fc region of IgG. Although RF made under physiologic conditions have been shown to exhibit restricted light chain usage and the light chains have been demonstrated to be central to RF specificity, the RF present in individuals with RA display less restricted V gene usage. Now that techniques for the sequencing of large numbers of antibody molecules are available, it will be important to revisit the question of V gene usage in autoantibodies.

Indeed, it is not surprising that there might be no Ig genes specific for the production of autoantibodies. Many autoantibodies have acquired autospecificity through somatic hypermutation; the germline-encoded antibody often fails to display autoreactivity and so must have a different antigenic specificity. The fact that several autoimmune diseases begin after an infection suggests that autoantibodies may arise as part of a dysregulated antimicrobial response in genetically susceptible individuals. An antimicrobial antibody can diversify to have novel antigenic specificities, including autospecificities.

CDR3 regions of autoantibodies are neither shorter nor longer than those present in protective antibodies. Structural studies have not discerned a particular conformation of the antigen-binding site of autoantibodies. Over many decades of study, no general structural features of autoantibodies that apply to multiple autospecificities have emerged. Early studies suggested that high affinity for autoantigen was associated with pathogenicity. Subsequent studies have shown that autoantibodies that differ tenfold in affinity for autoantigen can all be pathogenic. Although there is likely to be an affinity threshold below which there is no pathogenicity, it does not appear that higher affinity necessarily correlates with greater pathogenicity.

The most critical determinant of pathogenicity of autoantibodies lies in the Fc region of Ig, which mediates the different physiological effects of antibodies. IgM antibodies do not penetrate tissue, as they do not bind to FcRn on endothelial cells; therefore, they exhibit pathogenicity only when they target a circulating antigen, such as an erythrocyte or platelet. They opsonize the antigen, and recruit C1q and other complement components that earmark the antigenic target for destruction through the complement membrane attack complex or through ingestion by phagocytic cells. IgM autoantibodies can even be immunosuppressive. Once C1q is bound to the immune complex, the complex may engage LAIR-1 on the surface of monocytes or DCs [20]. LAIR-1 is an inhibitory receptor with two ITIM motifs in its cytoplasmic tail [19]. LAIR-1 engagement prevents monocyte to DC differentiation and activation of DCs through TLR engagement. Tissue damage is initiated after class switch recombination to IgG. If mice producing IgG anti-DNA antibodies are given an infusion of IgM anti-DNA antibodies, disease progression is arrested [34].

IgG antibodies, in contrast to IgM, are able to penetrate tissues. IgG is also the only antibody isotype that can pass through the placenta. Thus, IgG but not IgM antibodies exhibit a criterion for pathogenicity, as they can access the target organs. Moreover, tissue-bound IgG or IgG immune complexes can activate FcR on immune cells and on FcR–bearing, tissue-resident cells. In general, the engagement of these receptors triggers an inflammatory cascade. The FcRγIIB receptor uniquely transduces an inhibitory signal. Sialylation of IgG reduces its binding to Fc receptors and converts the binding to lectin receptors DC-SIGN or SIGN-R1 [197]. This leads to upregulated surface expression of FcRγIIB on inflammatory cells [198]. Thus, it is possible to design IgG antibodies or immune complexes that will suppress immune activation; however, in general, IgG autoantibodies are proinflammatory. Notably, during an inflammatory response, sialylation of IgG is diminished [199–201].

When IgG antibodies engage activating FcR, they trigger downstream pathways resulting in the generation of free radicals, induction of an IFN signature, and release of proinflammatory cytokines. IgG immune complexes also engage the complement cascade leading to destruction of the target antigen and the generation of the C3 and C5 cleavage products, C3a and C5a, respectively. These molecules are chemoattractants that amplify an inflammatory response by recruiting more effector cells into the target organ.

## 4. EFFECTOR MECHANISMS OF PATHOGENIC AUTOANTIBODIES

Antibodies have evolved to protect the host from infections through three main mechanisms: neutralization of the pathogens to prevent viruses and intracellular bacteria from entering cells, opsonization of the pathogen or its toxins to boost phagocytosis, and activation of the complement cascade that directly lyses certain microbes and enhances opsonization. When antibodies target autoantigens, they display no new tricks but utilize the same mechanisms for self-destruction. Opsonization and activation of the complement cascade by autoantibodies can incite inflammation and cause tissue damage. Autoantibodies can also act as either antagonists or agonists of cellular receptors to cause derangement of cellular functions without inducing target organ inflammation. Anti–insulin receptor antibodies function in this manner. Similarly, antibodies to soluble hormones, such as insulin and thyroid hormone, can alter cellular physiology without causing inflammation. An additional pathological property of autoantibodies may relate to their ability to inactivate or deplete regulatory T cells. This section describes the pathogenic mechanisms of autoantibodies in several representative diseases (Table 3).

### 4.1 Noninflammatory Autoimmune Disorders Mediated by Autoantibodies

Autoimmune diseases are often linked to inflammatory conditions that eventually lead to organ damage. However, several autoimmune disorders occur through autoantibody-mediated organ malfunction without inflammation and cytotoxicity.

#### *4.1.1 Autoantibodies that Block Cellular Functions*

A well-established pathogenic mechanism of some autoantibodies is inhibition of cellular activities through neutralizing effector molecules. Patients with pernicious anemia have autoantibodies against gastric intrinsic factor, a glycoprotein produced by gastric parietal cells that is essential for vitamin $B_{12}$ absorption by ileal enterocytes. By neutralizing the $B_{12}$ binding site of intrinsic factor (type I) or blocking interaction between the intrinsic factor–$B_{12}$ complex and its receptor on ileal enterocytes (type II), intrinsic factor–specific autoantibodies prevent vitamin $B_{12}$ uptake and consequently lead to anemia [202].

**TABLE 3** Mechanisms of Pathogenic Autoantibodies

| Mechanisms | Target Antigens | Disease | Manifestations |
|---|---|---|---|
| **Blocking cellular functions anemia** | Intrinsic factor | Pernicious anemia | Defective erythropoiesis, |
| | Voltage-gated calcium channels | Lambert-Eaton myasthenic syndrome | Muscle weakness, paralysis |
| | Acetylcholine receptor | Myasthenia gravis | Muscle weakness, paralysis |
| **Dysregulation of cellular and molecular functions** | Thyroid-stimulating hormone (TSH) receptor | Graves' disease | Hyperthyroidism |
| | PAD3/PAD4 | Rheumatoid arthritis | Arthritis |
| **Antibody-dependent cell-mediated cytotoxicity** | Thyroperoxidase (TPO), thyroglobulin | Hashimoto's thyroiditis | Hypothyroidism |
| | Proteinase 3 (PR3), Myeloperoxidase (MPO) | Granulomatosis with polyangiitis | Vasculitis |
| **Complement-mediated cytotoxicity and phagocytosis** | Fibrinogen receptor (gpIIa/IIIb) | Thrombocytopenic purpura | Hemorrhage |
| | Rh blood group antigens, I antigen | Hemolytic anemia | Hemolysis, anemia |
| **Immune complex-mediated tissue damage** | DNA, nucleoproteins, and others | Systemic lupus erythematosus | Nephritis, vasculitis, rash, and others |

Neutralizing autoantibodies are also found in patients with myasthenia gravis and Lambert-Eaton myasthenic syndrome. In both cases, muscle weakness is due to autoantibodies against proteins in the neuromuscular junction. For a muscle to contract, an action potential induces $Ca^{2+}$ influx through voltage-dependent calcium channels on the presynaptic cell membrane of motor neurons and causes exocytosis of acetylcholine. Acetylcholine binds to its receptor, AchR, on the postsynaptic membrane of skeletal muscle fibers and opens sodium/potassium channels to trigger muscle contraction. Lambert-Eaton myasthenic syndrome is caused by autoantibodies specific to the presynaptic voltage-gated calcium channels and myasthenia gravis by autoantibodies against postsynaptic AchR [203,204]. Binding of autoantibodies to these proteins in the neuromuscular junction antagonizes the ion channels with subsequent disturbance of the proper neuromuscular function. Ultimately, there is also complement mediated destruction of the neuromuscular junction.

### *4.1.2 Autoantibodies that Dysregulate Cellular Functions*

Some symptoms of autoimmune diseases are induced by autoantibodies with agonist activity. This mechanism of action is best exemplified by Graves' disease. Graves' disease is an organ specific autoimmune disorder that specifically affects the thyroid gland. Normally, hormone production by the thyroid is induced by TSH from the anterior pituitary, which itself is negatively regulated by the thyroid hormone levels in a feedback loop. TSH receptor (TSHR), located on the basal surface of thyroid follicular cells, is the prime autoantigen in Graves' disease. In patients with Graves' disease, autoantibodies bind to the TSHR and mimic the action of TSH [205]. Therefore, TSHR specific autoantibodies override the negative feedback mechanism. With uncontrolled stimulation by anti-TSHR autoantibodies, the gland produces an excessive amount of thyroid hormone causing hyperthyroidism. Graves' disease is the most common cause of hyperthyroidism, often affecting women. Mothers with this disease can pass anti-TSHR IgG to the fetus, and the baby will be born with hyperthyroidism.

Autoimmune pathology can also be caused by autoantibody-mediated changes in soluble molecules. A group of patients with RA have autoantibodies against the enzyme peptidylarginine deiminase (PAD) 4 with cross-reactivity with PAD3. These crossreactive anti-PAD antibodies increase calcium binding to PAD4, thus activating PAD4 for protein citrullination at a physiological level of calcium [206]. Citrullinated proteins are one of the major autoantigens in RA. The finding that anti-PAD4 antibodies can modify PAD activity for the production of citrullinated autoantigens illustrates an amplification feedback loop for RA pathogenesis. In RA patients, PAD3/PAD4 cross-reactive antibodies are associated with severity of erosive joint disease [206].

## 4.2 Inflammatory Autoimmune Disorders Mediated by Autoantibodies

Autoantibody-mediated cytotoxicity represents the major known pathogenic mechanism that causes tissue injury.

Like humoral responses against pathogens, autoantibodies engage multiple effector cells and different pathways to mediate lysis of the target cells. In most antibody-mediated autoimmune diseases, several of the following pathological mechanisms operate together, causing inflammation and tissue injury.

### 4.2.1 Antibody-Dependent Cell-Mediated Cytotoxicity

Antibody-dependent cell-mediated cytotoxicity (ADCC) is a lytic mechanism that can be mediated by autoreactive IgG. Here, autoreactive IgG recruits effector cells to the target cells with the Fab portion of the IgG binding to the target cell and the Fc region associating with FcγR on effector cells. Crosslinking FcγR on effector cells activates cells to secrete cytotoxic molecules. Classically, ADCC is mediated by NK cells that express CD16 (FcγRIII). Upon FcγR activation, NK cells release cytotoxic granules containing perforin and granzymes to kill antibody-sensitized target cells. In patients with Hashimoto's thyroiditis, a condition leading to hypothyroidism, anti-thyroperoxidase (TPO) antibodies bind to the basement membrane of the thyroid follicles, after which ADCC destroys the thyroid follicles. Other effector cells can also participate in ADCC, including macrophages and neutrophils. These effector cells may lyse autoantibody opsonized target cells through production of reactive oxygen species (ROS). A rare disease is granulomatosis with polyangiitis (GPA), an autoimmune vasculitis characterized by antineutrophil cytoplasmic antibodies (ANCA) that affect small and medium-size vessels in many organs [207]. Many ANCA bind proteinase 3, which is normally present in the cytoplasm of neutrophils. Activated neutrophils express proteinase 3 on their plasma membrane. Upon ANCA binding, neutrophils are immobilized on vessel walls and trigger inflammation by releasing chemokines that induce further neutrophil influx and, by undergoing a respiratory burst, leading to vascular necrosis. Approximately 70% of GPA patients develop ANCA-associated glomerulonephritis with no evident immune complex deposition (known as pauci-immune), contrary to glomerulonephritis manifested in lupus nephritis, as the antibodies activate neutrophils rather than deposit on glomerular cells.

### 4.2.2 Complement System–Mediated Cytotoxicity and Inflammation

The complement system plays a primary pathogenic role in autoantibody-induced inflammation and cytotoxicity [208]. Both IgM and IgG autoantibodies activate the classic complement pathway by binding to C1q through the Fc portion, which leads to the activation of the autocatalytic enzyme activity of the C1 complex and initiates the complement cascade that eventually leads to the assembly of the membrane–attack complex, an annular structure with a hydrophilic internal channel and a hydrophobic external portion that allows it to insert into the lipid bilayer of a cell membrane. The free passage of electrolytes and water through this transmembrane channel results in colloid osmotic lysis of cells. Complement-mediated cytolysis represents a major pathogenic mechanism underlying autoimmune hemolytic anemia and autoimmune thrombocytopenic purpura. Autoantibodies specific to the cell surface molecules on red blood cells or to glycoprotein IIb/IIIa (the fibrinogen receptor) on platelets activate the classic complement pathway and lead to intravascular lysis of circulating red blood cells and platelets. Compared to IgM, which is a potent activator of the classic complement pathway, IgG is less effective in complement fixation. Thus, complement-mediated blood cell destruction is predominantly a consequence of autoreactive IgM.

In addition to cytolysis, complement also incites inflammation. The autoantibody-activated complement cascade yields the small cleavage products C3a, C4a, and C5a, which are powerful inflammatory mediators that bind receptors on phagocytes and activate them to secrete inflammatory cytokines and chemokines. C5a itself is also a potent chemoattractant that induces homing of macrophages and neutrophils into the target organs. Tissue-bound antibody can activate the complement cascade and generate complement components that recruit and activate phagocytes, which release cytokines and ROS, to cause inflammatory injury. Moreover, C5a binds to and activates endothelial cells. Indeed, C5a binding to brain microvascular endothelial cells potently impairs the integrity of the blood–brain barrier.

### 4.2.3 Phagocytosis-Mediated Cytotoxicity

Both autoreactive IgM and IgG are able to promote phagocytosis of antibody-sensitized cells. In patients with autoimmune hemolytic anemia, IgG antibodies generally recognize protein antigens on erythrocyte surface at physiological temperature and are termed warm agglutinins, whereas IgM antibodies are referred to as cold agglutinins because they react with polysaccharide antigens on erythrocytes at a cold temperature of 0–4 °C [209]. IgG-coated erythrocytes activate phagocytic cells through FcγR crosslinking and trigger phagocytosis. Erythrocytes sensitized with autoreactive IgG are cleared from the circulation by macrophages in the splenic sinusoids. IgM does not bind directly to phagocytes, but activates the complement cascade that produces C3b. C3b and, to a lesser extent, C4b can act as opsonins, which mediate the attachment of IgM-sensitized erythrocytes to phagocytic cells through the complement receptor 1 (CR1). In general, phagocytosis of IgM and C3b opsonized erythrocytes takes place in the liver by Kupffer cells. Thus, IgM-associated erythrocyte elimination occurs both intravascularly through the complement system and extravascularly through hepatic phagocytosis, whereas IgG primarily mediates extravascular hemolysis. Importantly, this phagocytic mechanism has been co-opted for its therapeutic potential. Many therapeutic antibodies used to deplete certain cell subsets do so by opsonizing target cells, thereby priming them for removal by the reticuloendothelial system.

### 4.2.4 Immune Complex-Mediated Inflammation and Tissue Damage

Immune complexes are formed when antibodies bind to soluble autoantigens. Normally, immune complexes do not impose significant harm, as they are rapidly cleared from the bloodstream by phagocytic cells that express both complement receptors and FcγR. Under certain circumstances when either clearance mechanisms fail or an excessive amount of complex is produced, immune complexes persist in the body. Circulating immune complexes may be progressively trapped in tissues, especially synovium and kidney glomeruli, due in part to increased vascular permeability at those sites. Immune complex deposition in tissues activates the complement cascade and FcγR on inflammatory immune cells such as neutrophils, monocytes, macrophages, or DCs. These processes lead to severe inflammatory responses. Immune complexes formed with autoreactive IgG cause damage wherever they precipitate. Arthritis, glomerulonephritis, and vasculitis are often manifestations of autoimmune diseases caused by immune complexes. Interestingly, these complexes can also be made of antibody and microbial antigen as in streptococcal glomerulonephritis [210]. Autoantibodies also bind antigen in tissues, and thus form local noncirculating immune complexes that activate complement in situ and trigger the activation of inflammatory immune cells.

A typical autoimmune disease mediated by tissue-bound immune complexes is SLE. SLE is characterized by the production of antinuclear autoantibodies that lead to multiorgan damage. Although the etiology of this disease is not known, it is evident that abnormal apoptosis and deficient clearance of dead cells occurs in SLE. Nucleoproteins and dsDNA released from dead cells can activate autoreactive B cells. In lupus patients with active disease, an excessive amount of nuclear and intracellular autoantigen can be found in the bloodstream. These autoantigens form immune complexes that contain RNA or DNA and can activate TLRs in both myeloid and lymphoid cells. Thus, with continuous stimulation by autoantigens, innate and adaptive immune cells are activated, resulting in a GC reaction that produces a large number of long-lived plasma cells that produce high-affinity autoreactive IgG. Immune complexes in SLE patients deposit in the skin, the glomerulus, and in many other organs, often as a result of binding to tissue antigens. Immune complexes that form in situ are capable of activating all of the cytotoxic machineries described above. By activating the complement cascade and promoting phagocytosis and ADCC, immune complexes in SLE patients kill cells in tissues where they are deposited. Dead cells, in turn, release more nuclear and intracellular autoantigens continuing to drive the autoimmune response.

### 4.2.5 Autoantibody-Mediated Inactivation of Regulatory T Cells

Tregs play an essential role in suppressing autoimmunity. This heterogeneous population contains a variety of T cells of different lineages with different suppressive functions. The traditional Tregs are $CD25^+Foxp3^+$ $CD4^+$ T cells. These $CD4^+$ Tregs suppress proliferation of both autoreactive T cells and B cells. A much less studied population of Tregs is $CD122^+$ $CD8^+$ T cells. This population of Tregs was found to suppress a number of autoimmune diseases in mice, including experimental autoimmune encephalomyelitis (EAE) and Graves disease [211,212]. $CD122^+$ $CD8^+$ Tregs have been shown to suppress autoreactive GC reactions by inhibiting $T_{FH}$ cell function [89].

A subpopulation of $CD122^+$ $CD8^+$ Tregs expresses CD38. CD38 is a multifunctional molecule, serving as a cell surface receptor and an ectoenzyme, with its ectonucleotidase activity implicated in T cell suppression function [213]. CD38 deficiency has been shown to accelerate development of SLE and autoimmune diabetes in mice. In humans, autoantibody specific to CD38 has been detected in patients with T1D and autoimmune thyroiditis [214,215]. Serum from SLE patients with anti-CD38 autoantibodies induces apoptosis of $CD38^+$ T cells, and the level of anti-CD38 autoantibody in SLE patients correlates with anti-dsDNA autoantibodies and clinical activity [216].

## 5. B CELLS AS THE THERAPEUTIC TARGET IN AUTOIMMUNE DISEASE

Although this chapter is focused on the pathogenic importance of autoantibodies and the B cells that produce them, the contribution of autoreactive B cells to autoimmunity extends beyond antibody production. Even in some autoimmune diseases, such as diabetes and MS, which have traditionally been thought to be mediated by T cells, depletion of B cells has proved to be beneficial. The antibody-independent pathogenic mechanisms of autoreactive B cells include presentation of autoantigens to T cells, production of cytokines and chemokines, and organization of ectopic lymphoid foci. Thus, B cells have become a promising target for treatment of a variety of autoimmune diseases.

As described above, there are three distinct pathologic conditions characterized by the presence of autoreactive B cells: conventional autoimmune diseases, immunodeficiency with autoimmunity, and B cell malignancy with autoreactive BCRs. B cell–directed therapies are being explored in all three conditions.

B cell depletion using antibody targeting CD20 has been explored in several conventional autoimmune diseases: T1D, MS, RA, ANCA vasculitis, and SLE. Interestingly, T1D, MS, and RA are traditionally considered to exhibit T cell–mediated pathology; the therapeutic effect of B cell depletion in these diseases is thought to reflect the importance of B cells as APCs. B cell depletion also leads to an increase in Tregs for reasons that are not entirely understood; thus, it may be that an anti-CD20 antibody-induced increase in Tregs reduces disease activity [217]. Finally, B cell reconstitution leads to a prolonged period with increased number of transitional and naive B cells. Many of these cells produce IL-10, which, in many diseases, is immunosuppressive. Little is known about the generation of B10cells, and B10cell reconstitution has not been well studied in patients treated with B cell depletion.

Moreover, in SLE, IL-10 can exacerbate disease [218,219]; whether IL-10 can exacerbate some other autoimmune disease has not been fully studied. Both ANCA vasculitis and SLE are diseases in which autoantibodies are the major mediators of tissue pathology. Anti-CD20–mediated B cell depletion is effective in ANCA vasculitis but not in SLE. The reasons for the failure of B cell depletion in SLE are not known, although it is clear that BAFF levels increase with B cell depletion and that high BAFF levels correlate with an increased risk of disease flare. A recent study reveals that the liver is the primary site for removing B cells coated with anti-CD20 antibodies [220]. Intravital two-photon imaging in mice of GFP-labeled Kupffer cells show that they arrest circulating B cells in the liver sinusoids and engulf the captives. This process is dependent on FcR, consistent with the observation that FcγR polymorphisms influence the efficiency of anti-CD20 therapy in humans. In addition, these data also show that circulating B cells are more readily depleted than are tissue-residing B cells [220]. These observations may provide an explanation for the inefficient B cell depletion in SLE, as many autoreactive B cells in the lymphoid organs in SLE may not circulate and so may not be susceptible to elimination by Kupffer cells. Depleting CD20 B expressing cells does not deplete plasma cells. In mice, anti-CD20–mediated B cell depletion also fails to eliminate MZ B cells unless the MZ B cells are induced to exit the MZ. In humans, it is thought that MZ B cells circulate and so are vulnerable to antibody-mediated deletion. Indeed, anti-CD20 antibody eliminates MZ B cells in nonhuman primates. Whether the failure to mediate plasma cell depletion or the increase in BAFF or in Tregs differentially affects ANCA vasculitis versus SLE is not currently obvious.

BAFF blockade with an antibody to soluble BAFF has been explored in both RA and SLE. This strategy was not effective in RA but did show efficacy in SLE [221]. Studies of blood B cell subsets have shown that naive and transitional B cells, most dependent on BAFF for survival, are indeed reduced in subjects receiving BAFF blockade. Curiously, memory B cells are initially expanded, perhaps through a homeostatic mechanism. Although BAFF blockade has thus far displayed limited efficacy in autoimmune disease, it warrants continued exploration, as autoreactive B cells have an increased requirement for BAFF and so may be selectively vulnerable to the appropriate use of BAFF blockade. Surprisingly, however, there has been little evidence for a selective decrease in autoantibody titers; total IgM and IgG both decrease continuously over time in individuals on BAFF blockade, as do autoantibody titers. A clinical trial in RA with an antibody to BAFF that is alleged to bind both soluble and membrane BAFF also failed to show efficacy. There have been several studies in SLE of blockade of both BAFF and APRIL with a soluble receptor for both molecules. This agent causes a more profound decrease in IgM and has been associated with infectious complications. Whether it exhibits increased efficacy is not known.

Taking a cue from the therapy of B cell malignancy, immunodeficiency with autoantibodies has been successfully treated with proteasome inhibition, which selectively destroys plasma cells [222]. It would seem that BAFF blockade should be considered in individuals with immunodeficiency-related autoimmunity.

Recent therapeutic strategies for B-CLL include Syk inhibition to diminish BCR signaling and prevent the BCR-mediated signals that promote cell survival [174]. This should diminish viability of the malignant cells and as a consequence autoantibody production as well.

The challenge remains to identify, in each pathologic condition, those B cell subsets that include the autoreactive B cells. It may be that in some diseases the pathogenic B cells are restricted to a single compartment; in other diseases, such as SLE, autoreactive B cells may populate several different compartments depending on the patient or may populate multiple compartments within the same patient.

B cells within any B cell compartment except the plasma cell compartment can function as APCs. It has even been reported that anergic B cells can present antigen [223]. Thus, in conditions in which the antigen presenting function of the B cell is critical, it is important to know the compartment in which pathogenic antigen-presenting B cells reside in order to eliminate them or neutralize their function. Certainly, the approach of blocking costimulation to eliminate B cell antigen presentation function is worth pursuing. Unfortunately, it is difficult to see how this approach could be selective for autoreactive B cells. Indeed, many therapeutic approaches fail to show specificity for autoreactive B cells.

## 6. CONCLUSION

Autoantibodies can mediate both protective and pathologic function. In general, protective autoantibodies are produced by "innate" B1 or MZ B cells, which have a highly restricted antibody repertoire. Pathogenic autoantibodies, in contrast, may be made by any B cell subset and may undergo maturation through a GC response. There are no known unique features of autoantibodies; often their pathogenicity relates to their Fc region and its associated effector function. Thus far, therapeutic approaches have failed to exhibit selectivity for autoantibodies or autoreactive B cells.

## REFERENCES


[1] Haury M, Sundblad A, Grandien A, Barreau C, Coutinho A, Nobrega A. The repertoire of serum IgM in normal mice is largely independent of external antigenic contact. Eur J Immunol 1997;27:1557–63.

[2] Bos NA, Kimura H, Meeuwsen CG, De Visser H, Hazenberg MP, Wostmann BS, et al. Serum immunoglobulin levels and naturally occurring antibodies against carbohydrate antigens in germ-free BALB/c mice fed chemically defined ultrafiltered diet. Eur J Immunol 1989;19:2335–9.

[3] Merbl Y, Zucker-Toledano M, Quintana FJ, Cohen IR. Newborn humans manifest autoantibodies to defined self molecules detected by antigen microarray informatics. J Clin Invest 2007;117:712–8.

[4] Vas J, Gronwall C, Silverman GJ. Fundamental roles of the innate-like repertoire of natural antibodies in immune homeostasis. Front Immunol 2013;4:4.

[5] Briles DE, Forman C, Hudak S, Claflin JL. Anti-phosphorylcholine antibodies of the T15 idiotype are optimally protective against *Streptococcus pneumoniae*. J Exp Med 1982;156:1177–85.

[6] Harnett W, Harnett MM. Phosphorylcholine: friend or foe of the immune system? Immunol Today 1999;20:125–9.

[7] Claflin JL, Cubberley M. Clonal nature of the immune response to phosphocholine. VII. Evidence throughout inbred mice for molecular similarities among antibodies bearing the T15 idiotype. J Immunol 1980;125:551–8.

[8] Sigal NH, Gearhart PJ, Klinman NR. The frequency of phosphorylcholine- specific B cells in conventional and germfree BALB/C mice. J Immunol 1975;114:1354–8.

[9] Yother J, Forman C, Gray BM, Briles DE. Protection of mice from infection with *Streptococcus pneumoniae* by anti-phosphocholine antibody. Infect Immun 1982;36:184–8.

[10] McDaniel LS, Benjamin Jr WH, Forman C, Briles DE. Blood clearance by anti-phosphocholine antibodies as a mechanism of protection in experimental pneumococcal bacteremia. J Immunol 1984;133:3308–12.

[11] Mi QS, Zhou L, Schulze DH, Fischer RT, Lustig A, Rezanka LJ, et al. Highly reduced protection against *Streptococcus pneumoniae* after deletion of a single heavy chain gene in mouse. Proc Natl Acad Sci USA 2000;97:6031–6.

[12] Shaw PX, Horkko S, Chang MK, Curtiss LK, Palinski W, Silverman GJ, et al. Natural antibodies with the T15 idiotype may act in atherosclerosis, apoptotic clearance, and protective immunity. J Clin Invest 2000;105:1731–40.

[13] Chen Y, Park YB, Patel E, Silverman GJ. IgM antibodies to apoptosis-associated determinants recruit C1q and enhance dendritic cell phagocytosis of apoptotic cells. J Immunol 2009;182:6031–43.

[14] Horkko S, Bird DA, Miller E, Itabe H, Leitinger N, Subbanagounder G, et al. Monoclonal autoantibodies specific for oxidized phospholipids or oxidized phospholipid-protein adducts inhibit macrophage uptake of oxidized low-density lipoproteins. J Clin Invest 1999;103:117–28.

[15] Imai Y, Kuba K, Neely GG, Yaghubian-Malhami R, Perkmann T, van Loo G, et al. Identification of oxidative stress and Toll-like receptor 4 signaling as a key pathway of acute lung injury. Cell 2008;133:235–49.

[16] Chang MK, Bergmark C, Laurila A, Horkko S, Han KH, Friedman P, et al. Monoclonal antibodies against oxidized low-density lipoprotein bind to apoptotic cells and inhibit their phagocytosis by elicited macrophages: evidence that oxidation- specific epitopes mediate macrophage recognition. Proc Natl Acad Sci USA 1999;96:6353–8.

[17] Binder CJ, Shaw PX, Chang MK, Boullier A, Hartvigsen K, Horkko S, et al. The role of natural antibodies in atherogenesis. J Lipid Res 2005;46:1353–63.

[18] de Faire U, Su J, Hua X, Frostegard A, Halldin M, Hellenius ML, et al. Low levels of IgM antibodies to phosphorylcholine predict cardiovascular disease in 60-year old men: effects on uptake of oxidized LDL in macrophages as a potential mechanism. J Autoimmun 2010;34:73–9.

[19] Meyaard L, Adema GJ, Chang C, Woollatt E, Sutherland GR, Lanier LL, et al. LAIR-1, a novel inhibitory receptor expressed on human mononuclear leukocytes. Immunity 1997;7:283–90.

[20] Son M, Santiago-Schwarz F, Al-Abed Y, Diamond B. C1q limits dendritic cell differentiation and activation by engaging LAIR-1. Proc Natl Acad Sci USA 2012;109:E3160–7.

[21] Ochsenbein AF, Zinkernagel RM. Natural antibodies and complement link innate and acquired immunity. Immunol Today 2000;21:624–30.

[22] Boes M, Prodeus AP, Schmidt T, Carroll MC, Chen J. A critical role of natural immunoglobulin M in immediate defense against systemic bacterial infection. J Exp Med 1998;188:2381–6.

[23] Ochsenbein AF, Fehr T, Lutz C, Suter M, Brombacher F, Hengartner H, et al. Control of early viral and bacterial distribution and disease by natural antibodies. Science 1999;286:2156–9.

[24] Chou MY, Fogelstrand L, Hartvigsen K, Hansen LF, Woelkers D, Shaw PX, et al. Oxidation-specific epitopes are dominant targets of innate natural antibodies in mice and humans. J Clin Invest 2009;119:1335–49.

[25] Silverman GJ, Gronwall C, Vas J, Chen Y. Natural autoantibodies to apoptotic cell membranes regulate fundamental innate immune functions and suppress inflammation. Discov Med 2009;8:151–6.

[26] Mercolino TJ, Arnold LW, Hawkins LA, Haughton G. Normal mouse peritoneum contains a large population of Ly-1+ (CD5) B cells that recognize phosphatidyl choline. Relationship to cells that secrete hemolytic antibody specific for autologous erythrocytes. J Exp Med 1988;168:687–98.

[27] Hardy RR, Hayakawa K. Development of B cells producing natural autoantibodies to thymocytes and senescent erythrocytes. Springer Semin Immunopathol 2005;26:363–75.

[28] Boes M, Schmidt T, Linkemann K, Beaudette BC, Marshak-Rothstein A, Chen J. Accelerated development of IgG autoantibodies and autoimmune disease in the absence of secreted IgM. Proc Natl Acad Sci USA 2000;97:1184–9.

[29] Gronlund H, Hallmans G, Jansson JH, Boman K, Wikstrom M, de Faire U, et al. Low levels of IgM antibodies against phosphorylcholine predict development of acute myocardial infarction in a population-based cohort from northern Sweden. Eur J Cardiovasc Prev Rehabil 2009;16:382–6.

[30] de Faire U, Frostegard J. Natural antibodies against phosphorylcholine in cardiovascular disease. Ann N Y Acad Sci 2009;1173:292–300.

[31] Frostegard J. Low level natural antibodies against phosphorylcholine: a novel risk marker and potential mechanism in atherosclerosis and cardiovascular disease. Clin Immunol 2010;134:47–54.

[32] Silverman GJ, Srikrishnan R, Germar K, Goodyear CS, Andrews KA, Ginzler EM, et al. Genetic imprinting of autoantibody repertoires in systemic lupus erythematosus patients. Clin Exp Immunol 2008;153:102–16.

[33] Gronwall C, Akhter E, Oh C, Burlingame RW, Petri M, Silverman GJ. IgM autoantibodies to distinct apoptosis-associated antigens correlate with protection from cardiovascular events and renal disease in patients with SLE. Clin Immunol 2012;142:390–8.

[34] Werwitzke S, Trick D, Kamino K, Matthias T, Kniesch K, Schlegelberger B, et al. Inhibition of lupus disease by anti-double-stranded DNA antibodies of the IgM isotype in the (NZB x NZW) F1 mouse. Arthritis Rheum 2005;52:3629–38.

[35] Jiang C, Zhao ML, Scearce RM, Diaz M. Activation-induced deaminase-deficient MRL/lpr mice secrete high levels of protective antibodies against lupus nephritis. Arthritis Rheum 2011;63:1086–96.

[36] Eltzschig HK, Eckle T. Ischemia and reperfusion–from mechanism to translation. Nat Med 2011;17:1391–401.

[37] Li C, Jackson RM. Reactive species mechanisms of cellular hypoxia-reoxygenation injury. Am J Physiol Cell Physiol 2002;282:C227–41.

[38] Zhang M, Alicot EM, Chiu I, Li J, Verna N, Vorup-Jensen T, et al. Identification of the target self-antigens in reperfusion injury. J Exp Med 2006;203:141–52.

[39] Zhang M, Austen Jr WG, Chiu I, Alicot EM, Hung R, Ma M, et al. Identification of a specific self- reactive IgM antibody that initiates intestinal ischemia/reperfusion injury. Proc Natl Acad Sci USA 2004;101:3886–91.

[40] Kulik L, Fleming SD, Moratz C, Reuter JW, Novikov A, Chen K, et al. Pathogenic natural antibodies recognizing annexin IV are required to develop intestinal ischemia-reperfusion injury. J Immunol 2009;182:5363–73.

[41] Haas MS, Alicot EM, Schuerpf F, Chiu I, Li J, Moore FD, et al. Blockade of self-reactive IgM significantly reduces injury in a murine model of acute myocardial infarction. Cardiovasc Res 2010;87:618–27.

[42] Baumgarth N, Herman OC, Jager GC, Brown L, Herzenberg LA, Herzenberg LA. Innate and acquired humoral immunities to influenza virus are mediated by distinct arms of the immune system. Proc Natl Acad Sci USA 1999;96:2250–5.

[43] Hayakawa K, Hardy RR, Parks DR, Herzenberg LA. The "Ly-1 B" cell subpopulation in normal immunodefective, and autoimmune mice. J Exp Med 1983;157:202–18.

[44] Montecino-Rodriguez E, Leathers H, Dorshkind K. Identification of a B-1 B cell- specified progenitor. Nat Immunol 2006;7:293–301.

[45] Hardy RR. B-1 B cell development. J Immunol 2006;177:2749–54.

[46] Baumgarth N. The double life of a B-1 cell: self-reactivity selects for protective effector functions. Nat Rev Immunol 2011;11:34–46.

[47] Walker L, Carlson A, Tan-Pertel HT, Weinmaster G, Gasson J. The notch receptor and its ligands are selectively expressed during hematopoietic development in the mouse. Stem Cells 2001;19:543–52.

[48] Witt CM, Won WJ, Hurez V, Klug CA. Notch2 haploinsufficiency results in diminished B1 B cells and a severe reduction in marginal zone B cells. J Immunol 2003;171:2783–8.

[49] Gregoire KE, Goldschneider I, Barton RW, Bollum FJ. Ontogeny of terminal deoxynucleotidyl transferase-positive cells in lymphohemopoietic tissues of rat and mouse. J Immunol 1979;123:1347–52.

[50] Feeney AJ. Lack of N regions in fetal and neonatal mouse immunoglobulin V-D-J junctional sequences. J Exp Med 1990;172:1377–90.

[51] Souto-Carneiro MM, Sims GP, Girschik H, Lee J, Lipsky PE. Developmental changes in the human heavy chain CDR3. J Immunol 2005;175:7425–36.

[52] Gu H, Forster I, Rajewsky K. Sequence homologies, N sequence insertion and JH gene utilization in VHDJH joining: implications for the joining mechanism and the ontogenetic timing of Ly1 B cell and B-CLL progenitor generation. EMBO J 1990;9:2133–40.

[53] Tornberg UC, Holmberg D. B-1a, B-1b and B-2 B cells display unique VHDJH repertoires formed at different stages of ontogeny and under different selection pressures. EMBO J 1995;14: 1680–9.

[54] Schroeder Jr HW, Mortari F, Shiokawa S, Kirkham PM, Elgavish RA, Bertrand 3rd FE. Developmental regulation of the human antibody repertoire. Ann N Y Acad Sci 1995;764:242–60.

[55] Logtenberg T, Young FM, Van Es JH, Gmelig-Meyling FH, Alt FW. Autoantibodies encoded by the most Jh-proximal human immunoglobulin heavy chain variable region gene. J Exp Med 1989;170:1347–55.

[56] Crews S, Griffin J, Huang H, Calame K, Hood L. A single VH gene segment encodes the immune response to phosphorylcholine: somatic mutation is correlated with the class of the antibody. Cell 1981;25:59–66.

[57] Hayakawa K, Asano M, Shinton SA, Gui M, Allman D, Stewart CL, et al. Positive selection of natural autoreactive B cells. Science 1999;285:113–6.

[58] Arnold LW, Pennell CA, McCray SK, Clarke SH. Development of B-1 cells: segregation of phosphatidyl choline-specific B cells to the B-1 population occurs after immunoglobulin gene expression. J Exp Med 1994;179:1585–95.

[59] Lam KP, Rajewsky K. B cell antigen receptor specificity and surface density together determine B-1 versus B-2 cell development. J Exp Med 1999;190:471–7.

[60] Berland R, Wortis HH. Origins and functions of B-1 cells with notes on the role of CD5. Annu Rev Immunol 2002;20:253–300.

[61] Rickert RC, Rajewsky K, Roes J. Impairment of T-cell-dependent B-cell responses and B-1 cell development in CD19-deficient mice. Nature 1995;376:352–5.

[62] Engel P, Zhou LJ, Ord DC, Sato S, Koller B, Tedder TF. Abnormal B lymphocyte development, activation, and differentiation in mice that lack or overexpress the CD19 signal transduction molecule. Immunity 1995;3:39–50.

[63] Satterthwaite AB, Li Z, Witte ON. Btk function in B cell development and response. Semin Immunol 1998;10:309–16.

[64] Clarke SH, Arnold LW. B-1 cell development: evidence for an uncommitted immunoglobulin (Ig)M+ B cell precursor in B-1 cell differentiation. J Exp Med 1998;187:1325–34.

[65] Sidman CL, Shultz LD, Hardy RR, Hayakawa K, Herzenberg LA. Production of immunoglobulin isotypes by Ly-1+ B cells in viable motheaten and normal mice. Science 1986;232:1423–5.

[66] Hibbs ML, Tarlinton DM, Armes J, Grail D, Hodgson G, Maglitto R, et al. Multiple defects in the immune system of Lyn-deficient mice, culminating in autoimmune disease. Cell 1995;83:301–11.
[67] Nishizumi H, Taniuchi I, Yamanashi Y, Kitamura D, Ilic D, Mori S, et al. Impaired proliferation of peripheral B cells and indication of autoimmune disease in lyn-deficient mice. Immunity 1995;3:549–60.
[68] Hoffmann A, Kerr S, Jellusova J, Zhang J, Weisel F, Wellmann U, et al. Siglec-G is a B1 cell-inhibitory receptor that controls expansion and calcium signaling of the B1 cell population. Nat Immunol 2007;8:695–704.
[69] Jellusova J, Duber S, Guckel E, Binder CJ, Weiss S, Voll R, et al. Siglec-G regulates B1 cell survival and selection. J Immunol 2010;185:3277–84.
[70] Ueffing N, Schuster M, Keil E, Schulze-Osthoff K, Schmitz I. Upregulation of c- FLIP short by NFAT contributes to apoptosis resistance of short-term activated T cells. Blood 2008;112:690–8.
[71] Berland R, Wortis HH. Normal B-1a cell development requires B cell-intrinsic NFATc1 activity. Proc Natl Acad Sci USA 2003;100:13459–64.
[72] Maruya M, Suzuki K, Fujimoto H, Miyajima M, Kanagawa O, Wakayama T, et al. Vitamin A-dependent transcriptional activation of the nuclear factor of activated T cells c1 (NFATc1) is critical for the development and survival of B1 cells. Proc Natl Acad Sci USA 2011;108:722–7.
[73] Deenen GJ, Kroese FG. Murine peritoneal Ly-1 B cells do not turn over rapidly. Ann N Y Acad Sci 1992;651:70–1.
[74] Rothaeusler K, Baumgarth N. Evaluation of intranuclear BrdU detection procedures for use in multicolor flow cytometry. Cytom A 2006;69:249–59.
[75] Wardemann H, Boehm T, Dear N, Carsetti R. B-1a B cells that link the innate and adaptive immune responses are lacking in the absence of the spleen. J Exp Med 2002;195:771–80.
[76] Kretschmer K, Stopkowicz J, Scheffer S, Greten TF, Weiss S. Maintenance of peritoneal B-1a lymphocytes in the absence of the spleen. J Immunol 2004;173:197–204.
[77] Kipps TJ, Carson DA. Autoantibodies in chronic lymphocytic leukemia and related systemic autoimmune diseases. Blood 1993;81:2475–87.
[78] Zou YR, Diamond B. Fate determination of mature autoreactive B cells. Adv Immunol 2013;118:1–36.
[79] Wardemann H, Yurasov S, Schaefer A, Young JW, Meffre E, Nussenzweig MC. Predominant autoantibody production by early human B cell precursors. Science 2003;301:1374–7.
[80] Goodnow CC. Transgenic mice and analysis of B-cell tolerance. Annu Rev Immunol 1992;10:489–518.
[81] Shlomchik MJ. Sites and stages of autoreactive B cell activation and regulation. Immunity 2008;28:18–28.
[82] Diamond B, Scharff MD. Somatic mutation of the T15 heavy chain gives rise to an antibody with autoantibody specificity. Proc Natl Acad Sci USA 1984;81:5841–4.
[83] Guo W, Smith D, Aviszus K, Detanico T, Heiser RA, Wysocki LJ. Somatic hypermutation as a generator of antinuclear antibodies in a murine model of systemic autoimmunity. J Exp Med 2010;207:2225–37.
[84] Ohmori H, Hikida M. Expression and function of recombination activating genes in mature B cells. Crit Rev Immunol 1998;18: 221–35.
[85] Wang YH, Diamond B. B cell receptor revision diminishes the autoreactive B cell response after antigen activation in mice. J Clin Invest 2008;118:2896–907.
[86] Chung Y, Tanaka S, Chu F, Nurieva RI, Martinez GJ, Rawal S, et al. Follicular regulatory T cells expressing Foxp3 and Bcl-6 suppress germinal center reactions. Nat Med 2011;17:983–8.
[87] Linterman MA, Pierson W, Lee SK, Kallies A, Kawamoto S, Rayner TF, et al. Foxp3+ follicular regulatory T cells control the germinal center response. Nat Med 2011;17:975–82.
[88] Wollenberg I, Agua-Doce A, Hernandez A, Almeida C, Oliveira VG, Faro J, et al. Regulation of the germinal center reaction by Foxp3+ follicular regulatory T cells. J Immunol 2011;187:4553–60.
[89] Kim HJ, Verbinnen B, Tang X, Lu L, Cantor H. Inhibition of follicular T-helper cells by CD8(+) regulatory T cells is essential for self tolerance. Nature 2010;467:328–32.
[90] Culton DA, O'Conner BP, Conway KL, Diz R, Rutan J, Vilen BJ, et al. Early preplasma cells define a tolerance checkpoint for autoreactive B cells. J Immunol 2006;176:790–802.
[91] Scheid JF, Mouquet H, Kofer J, Yurasov S, Nussenzweig MC, Wardemann H. Differential regulation of self-reactivity discriminates between IgG+ human circulating memory B cells and bone marrow plasma cells. Proc Natl Acad Sci USA 2011;108:18044–8.
[92] Cunningham MW, Antone SM, Gulizia JM, McManus BM, Fischetti VA, Gauntt CJ. Cytotoxic and viral neutralizing antibodies crossreact with streptococcal M protein, enteroviruses, and human cardiac myosin. Proc Natl Acad Sci USA 1992;89:1320–4.
[93] Diamond B, Honig G, Mader S, Brimberg L, Volpe BT. Brain-reactive antibodies and disease. Annu Rev Immunol 2013;31:345–85.
[94] Hughes RA, Rees JH. Clinical and epidemiologic features of Guillain-Barre syndrome. J Infect Dis 1997;176(Suppl 2):S92–8.
[95] Moran AP. Structure and conserved characteristics of *Campylobacter jejuni* lipopolysaccharides. J Infect Dis 1997;176(Suppl 2):S115–21.
[96] Ang CW, Noordzij PG, de Klerk MA, Endtz HP, van Doorn PA, Laman JD. Ganglioside mimicry of *Campylobacter jejuni* lipopolysaccharides determines antiganglioside specificity in rabbits. Infect Immun 2002;70:5081–5.
[97] Schwimmbeck PL, Dyrberg T, Drachman DB, Oldstone MB. Molecular mimicry and myasthenia gravis. An autoantigenic site of the acetylcholine receptor alpha-subunit that has biologic activity and reacts immunochemically with herpes simplex virus. J Clin Invest 1989;84:1174–80.
[98] Hargreaves CE, Grasso M, Hampe CS, Stenkova A, Atkinson S, Joshua GW, et al. *Yersinia enterocolitica* provides the link between thyroid-stimulating antibodies and their germline counterparts in Graves' disease. J Immunol 2013;190:5373–81.
[99] Doria A, Canova M, Tonon M, Zen M, Rampudda E, Bassi N, et al. Infections as triggers and complications of systemic lupus erythematosus. Autoimmun Rev 2008;8:24–8.
[100] Agius MA, Twaddle GM, Fairclough RH. Epitope spreading in experimental autoimmune myasthenia gravis. Ann N Y Acad Sci 1998;841:365–7.
[101] Deshmukh US, Gaskin F, Lewis JE, Kannapell CC, Fu SM. Mechanisms of autoantibody diversification to SLE-related autoantigens. Ann N Y Acad Sci 2003;987:91–8.
[102] Mamula MJ, Janeway Jr CA. Do B cells drive the diversification of immune responses? Immunol Today 1993;14:151–2. discussion 153–4.
[103] Mandik-Nayak L, Nayak S, Sokol C, Eaton-Bassiri A, Madaio MP, Caton AJ, et al. The origin of anti-nuclear antibodies in bcl-2 transgenic mice. Int Immunol 2000;12:353–64.
[104] Berman PW, Patrick J. Experimental myasthenia gravis. A murine system. J Exp Med 1980;151:204–23.

[105] Shi FD, Wang HB, Li H, Hong S, Taniguchi M, Link H, et al. Natural killer cells determine the outcome of B cell-mediated autoimmunity. Nat Immunol 2000;1:245–51.

[106] Knight JS, Kaplan MJ. Lupus neutrophils: 'NET' gain in understanding lupus pathogenesis. Curr Opin Rheumatol 2012;24:441–50.

[107] Villanueva E, Yalavarthi S, Berthier CC, Hodgin JB, Khandpur R, Lin AM, et al. Netting neutrophils induce endothelial damage, infiltrate tissues, and expose immunostimulatory molecules in systemic lupus erythematosus. J Immunol 2011;187:538–52.

[108] Garcia-Romo GS, Caielli S, Vega B, Connolly J, Allantaz F, Xu Z, et al. Netting neutrophils are major inducers of type I IFN production in pediatric systemic lupus erythematosus. Sci Transl Med 2011;3:73ra20.

[109] Ronnblom L. The type I interferon system in the etiopathogenesis of autoimmune diseases. Ups J Med Sci 2011;116:227–37.

[110] Akira S, Uematsu S, Takeuchi O. Pathogen recognition and innate immunity. Cell 2006;124:783–801.

[111] Mills KH. TLR-dependent T cell activation in autoimmunity. Nat Rev Immunol 2011;11:807–22.

[112] Hwang SH, Lee H, Yamamoto M, Jones LA, Dayalan J, Hopkins R, et al. B cell TLR7 expression drives anti-RNA autoantibody production and exacerbates disease in systemic lupus erythematosus-prone mice. J Immunol 2012;189:5786–96.

[113] Lartigue A, Courville P, Auquit I, Francois A, Arnoult C, Tron F, et al. Role of TLR9 in anti-nucleosome and anti-DNA antibody production in lpr mutation- induced murine lupus. J Immunol 2006;177:1349–54.

[114] Christensen SR, Kashgarian M, Alexopoulou L, Flavell RA, Akira S, Shlomchik MJ. Toll-like receptor 9 controls anti-DNA autoantibody production in murine lupus. J Exp Med 2005;202:321–31.

[115] Roubinian JR, Talal N, Greenspan JS, Goodman JR, Siiteri PK. Effect of castration and sex hormone treatment on survival, anti-nucleic acid antibodies, and glomerulonephritis in NZB/NZW F1 mice. J Exp Med 1978;147:1568–83.

[116] Grimaldi CM, Michael DJ, Diamond B. Cutting edge: expansion and activation of a population of autoreactive marginal zone B cells in a model of estrogen-induced lupus. J Immunol 2001;167:1886–90.

[117] Grimaldi CM, Cleary J, Dagtas AS, Moussai D, Diamond B. Estrogen alters thresholds for B cell apoptosis and activation. J Clin Invest 2002;109:1625–33.

[118] Calippe B, Douin-Echinard V, Delpy L, Laffargue M, Lelu K, Krust A, et al. 17Beta-estradiol promotes TLR4-triggered proinflammatory mediator production through direct estrogen receptor alpha signaling in macrophages in vivo. J Immunol 2010;185:1169–76.

[119] Seillet C, Laffont S, Tremollieres F, Rouquie N, Ribot C, Arnal JF, et al. The TLR-mediated response of plasmacytoid dendritic cells is positively regulated by estradiol in vivo through cell-intrinsic estrogen receptor alpha signaling. Blood 2012;119:454–64.

[120] Panchanathan R, Liu H, Liu H, Fang CM, Erickson LD, Pitha PM, et al. Distinct regulation of murine lupus susceptibility genes by the IRF5/Blimp-1 axis. J Immunol 2012;188:270–8.

[121] Cunningham MA, Naga OS, Eudaly JG, Scott JL, Gilkeson GS. Estrogen receptor alpha modulates Toll-like receptor signaling in murine lupus. Clin Immunol 2012;144:1–12.

[122] Bynote KK, Hackenberg JM, Korach KS, Lubahn DB, Lane PH, Gould KA. Estrogen receptor-alpha deficiency attenuates autoimmune disease in (NZB x NZW)F1 mice. Genes Immun 2008;9:137–52.

[123] Elliott MR, Ravichandran KS. Clearance of apoptotic cells: implications in health and disease. J Cell Biol 2010;189:1059–70.

[124] Hanayama R, Tanaka M, Miyasaka K, Aozasa K, Koike M, Uchiyama Y, et al. Autoimmune disease and impaired uptake of apoptotic cells in MFG-E8-deficient mice. Science 2004;304:1147–50.

[125] Truedsson L, Bengtsson AA, Sturfelt G. Complement deficiencies and systemic lupus erythematosus. Autoimmunity 2007;40:560–6.

[126] Li X, Ptacek TS, Brown EE, Edberg JC. Fcgamma receptors: structure, function and role as genetic risk factors in SLE. Genes Immun 2009;10:380–9.

[127] Yasutomo K, Horiuchi T, Kagami S, Tsukamoto H, Hashimura C, Urushihara M, et al. Mutation of DNASE1 in people with systemic lupus erythematosus. Nat Genet 2001;28:313–4.

[128] Namjou B, Kothari PH, Kelly JA, Glenn SB, Ojwang JO, Adler A, et al. Evaluation of the TREX1 gene in a large multi-ancestral lupus cohort. Genes Immun 2011;12:270–9.

[129] Kozyrev SV, Abelson AK, Wojcik J, Zaghlool A, Linga Reddy MV, Sanchez E, et al. Functional variants in the B-cell gene BANK1 are associated with systemic lupus erythematosus. Nat Genet 2008;40:211–6.

[130] Castillejo-Lopez C, Delgado-Vega AM, Wojcik J, Kozyrev SV, Thavathiru E, Wu YY, et al. Genetic and physical interaction of the B-cell systemic lupus erythematosus-associated genes BANK1 and BLK. Ann Rheum Dis 2012;71:136–42.

[131] Bottini N, Peterson EJ. Tyrosine phosphatase PTPN22: multifunctional regulator of immune signaling, development, and disease. Annu Rev Immunol 2013.

[132] Manjarrez-Orduno N, Marasco E, Chung SA, Katz MS, Kiridly JF, Simpfendorfer KR, et al. CSK regulatory polymorphism is associated with systemic lupus erythematosus and influences B-cell signaling and activation. Nat Genet 2012;44:1227–30.

[133] Flores-Borja F, Kabouridis PS, Jury EC, Isenberg DA, Mageed RA. Decreased Lyn expression and translocation to lipid raft signaling domains in B lymphocytes from patients with systemic lupus erythematosus. Arthritis Rheum 2005;52:3955–65.

[134] Liossis SN, Solomou EE, Dimopoulos MA, Panayiotidis P, Mavrikakis MM, Sfikakis PP. B-cell kinase lyn deficiency in patients with systemic lupus erythematosus. J Investig Med 2001;49:157–65.

[135] Cornall RJ, Cyster JG, Hibbs ML, Dunn AR, Otipoby KL, Clark EA, et al. Polygenic autoimmune traits: lyn, CD22, and SHP-1 are limiting elements of a biochemical pathway regulating BCR signaling and selection. Immunity 1998;8:497–508.

[136] Ehlers M, Fukuyama H, McGaha TL, Aderem A, Ravetch JV. TLR9/MyD88 signaling is required for class switching to pathogenic IgG2a and 2b autoantibodies in SLE. J Exp Med 2006;203:553–61.

[137] Vincent FB, Saulep-Easton D, Figgett WA, Fairfax KA, Mackay F. The BAFF/APRIL system: emerging functions beyond B cell biology and autoimmunity. Cytokine Growth Factor Rev 2013;24:203–15.

[138] Moisini I, Davidson A. BAFF: a local and systemic target in autoimmune diseases. Clin Exp Immunol 2009;158:155–63.

[139] Mackay F, Woodcock SA, Lawton P, Ambrose C, Baetscher M, Schneider P, et al. Mice transgenic for BAFF develop lymphocytic disorders along with autoimmune manifestations. J Exp Med 1999;190:1697–710.

[140] Sneller MC, Wang J, Dale JK, Strober W, Middelton LA, Choi Y, et al. Clincal, immunologic, and genetic features of an autoimmune lymphoproliferative syndrome associated with abnormal lymphocyte apoptosis. Blood 1997;89:1341–8.
[141] Hao Z, Duncan GS, Seagal J, Su YW, Hong C, Haight J, et al. Fas receptor expression in germinal-center B cells is essential for T and B lymphocyte homeostasis. Immunity 2008;29:615–27.
[142] Gough SC, Simmonds MJ. The HLA region and autoimmune disease: associations and mechanisms of action. Curr Genomics 2007;8:453–65.
[143] Hamilton RG, Harley JB, Bias WB, Roebber M, Reichlin M, Hochberg MC, et al. Two Ro (SS-A) autoantibody responses in systemic lupus erythematosus. Correlation of HLA-DR/DQ specificities with quantitative expression of Ro (SS-A) autoantibody. Arthritis Rheum 1988;31:496–505.
[144] Schur PH. Genetics of systemic lupus erythematosus. Lupus 1995;4:425–37.
[145] Grammer AC, Slota R, Fischer R, Gur H, Girschick H, Yarboro C, et al. Abnormal germinal center reactions in systemic lupus erythematosus demonstrated by blockade of CD154-CD40 interactions. J Clin Invest 2003;112:1506–20.
[146] Peters AL, Stunz LL, Bishop GA. CD40 and autoimmunity: the dark side of a great activator. Semin Immunol 2009;21:293–300.
[147] Li G, Diogo D, Wu D, Spoonamore J, Dancik V, Franke L, et al. Human genetics in rheumatoid arthritis guides a high-throughput drug screen of the CD40 signaling pathway. PLoS Genet 2013;9:e36873.
[148] Toubi E, Shoenfeld Y. The role of CD40-CD154 interactions in autoimmunity and the benefit of disrupting this pathway. Autoimmunity 2004;37:457–64.
[149] Crotty S. Follicular helper CD4 T cells (TFH). Annu Rev Immunol 2011;29:621–63.
[150] Tangye SG, Ma CS, Brink R, Deenick EK. The good, the bad and the ugly - TFH cells in human health and disease. Nat Rev Immunol 2013;13:412–26.
[151] Vinuesa CG, Cook MC, Angelucci C, Athanasopoulos V, Rui L, Hill KM, et al. A RING-type ubiquitin ligase family member required to repress follicular helper T cells and autoimmunity. Nature 2005;435:452–8.
[152] Kim SJ, Zou YR, Goldstein J, Reizis B, Diamond B. Tolerogenic function of Blimp-1 in dendritic cells. J Exp Med 2011;208:2193–9.
[153] Han JW, Zheng HF, Cui Y, Sun LD, Ye DQ, Hu Z, et al. Genome-wide association study in a Chinese Han population identifies nine new susceptibility loci for systemic lupus erythematosus. Nat Genet 2009;41:1234–7.
[154] Anderson CA, Boucher G, Lees CW, Franke A, D'Amato M, Taylor KD, et al. Meta-analysis identifies 29 additional ulcerative colitis risk loci, increasing the number of confirmed associations to 47. Nat Genet 2011;43:246–52.
[155] Kalampokis I, Yoshizaki A, Tedder TF. IL-10-producing regulatory B cells (B10 cells) in autoimmune disease. Arthritis Res Ther 2013;15(Suppl 1):S1.
[156] Watanabe R, Ishiura N, Nakashima H, Kuwano Y, Okochi H, Tamaki K, et al. Regulatory B cells (B10 cells) have a suppressive role in murine lupus: CD19 and B10 cell deficiency exacerbates systemic autoimmunity. J Immunol 2010;184:4801–9.
[157] Jesus AA, Duarte AJ, Oliveira JB. Autoimmunity in hyper-IgM syndrome. J Clin Immunol 2008;28(Suppl 1):S62–6.
[158] Herve M, Isnardi I, Ng YS, Bussel JB, Ochs HD, Cunningham-Rundles C, et al. CD40 ligand and MHC class II expression are essential for human peripheral B cell tolerance. J Exp Med 2007;204:1583–93.
[159] Meyers G, Ng YS, Bannock JM, Lavoie A, Walter JE, Notarangelo LD, et al. Activation-induced cytidine deaminase (AID) is required for B-cell tolerance in humans. Proc Natl Acad Sci USA 2011;108:11554–9.
[160] Resnick ES, Cunningham-Rundles C. The many faces of the clinical picture of common variable immune deficiency. Curr Opin Allergy Clin Immunol 2012;12:595–601.
[161] Warnatz K, Voll RE. Pathogenesis of autoimmunity in common variable immunodeficiency. Front Immunol 2012;3:210.
[162] Romberg N, Chamberlain N, Saadoun D, Gentile M, Kinnunen T, Ng YS, et al. CVID-associated TACI mutations affect autoreactive B cell selection and activation. J Clin Invest 2013;123:4283–93.
[163] Kluger N, Ranki A, Krohn K. APECED: is this a model for failure of T cell and B cell tolerance? Front Immunol 2012;3:232.
[164] Anderson MS, Venanzi ES, Klein L, Chen Z, Berzins SP, Turley SJ, et al. Projection of an immunological self shadow within the thymus by the aire protein. Science 2002;298:1395–401.
[165] Josefowicz SZ, Lu LF, Rudensky AY. Regulatory T cells: mechanisms of differentiation and function. Annu Rev Immunol 2012;30:531–64.
[166] Barzaghi F, Passerini L, Bacchetta R. Immune dysregulation, polyendocrinopathy, enteropathy, x-linked syndrome: a paradigm of immunodeficiency with autoimmunity. Front Immunol 2012;3:211.
[167] Takenawa T, Suetsugu S. The WASP-WAVE protein network: connecting the membrane to the cytoskeleton. Nat Rev Mol Cell Biol 2007;8:37–48.
[168] Catucci M, Castiello MC, Pala F, Bosticardo M, Villa A. Autoimmunity in wiskott-Aldrich syndrome: an unsolved enigma. Front Immunol 2012;3:209.
[169] Becker-Herman S, Meyer-Bahlburg A, Schwartz MA, Jackson SW, Hudkins KL, Liu C, et al. WASp-deficient B cells play a critical, cell-intrinsic role in triggering autoimmunity. J Exp Med 2011;208:2033–42.
[170] Ozsahin H, Cavazzana-Calvo M, Notarangelo LD, Schulz A, Thrasher AJ, Mazzolari E, et al. Long- term outcome following hematopoietic stem-cell transplantation in Wiskott-Aldrich syndrome: collaborative study of the European Society for Immunodeficiencies and European Group for Blood and Marrow Transplantation. Blood 2008;111:439–45.
[171] Straus SE, Jaffe ES, Puck JM, Dale JK, Elkon KB, Rosen-Wolff A, et al. The development of lymphomas in families with autoimmune lymphoproliferative syndrome with germline Fas mutations and defective lymphocyte apoptosis. Blood 2001;98:194–200.
[172] Agathangelidis A, Darzentas N, Hadzidimitriou A, Brochet X, Murray F, Yan XJ, et al. Stereotyped B-cell receptors in one-third of chronic lymphocytic leukemia: a molecular classification with implications for targeted therapies. Blood 2012;119:4467–75.
[173] Duhren-von Minden M, Ubelhart R, Schneider D, Wossning T, Bach MP, Buchner M, et al. Chronic lymphocytic leukaemia is driven by antigen-independent cell-autonomous signalling. Nature 2012;489:309–12.
[174] Puri KD, Di Paolo JA, Gold MR. B-cell receptor signaling inhibitors for treatment of autoimmune inflammatory diseases and B-cell malignancies. Int Rev Immunol 2013;32:397–427.

[175] Kaplan I, Hodak E, Ackerman L, Mimouni D, Anhalt GJ, Calderon S. Neoplasms associated with paraneoplastic pemphigus: a review with emphasis on non- hematologic malignancy and oral mucosal manifestations. Oral Oncol 2004;40:553–62.
[176] Anhalt GJ. Paraneoplastic pemphigus. J Investig Dermatol Symp Proc 2004;9:29–33.
[177] Bernard NF, Ertug F, Margolese H. High incidence of thyroiditis and anti-thyroid autoantibodies in NOD mice. Diabetes 1992;41:40–6.
[178] Vexiau P, Helmy-Khalil S, Deschamps I, Homberg JC, Cathelineau G, Woimant G, et al. Organ-specific autoantibodies in HLA genotyped insulin-dependent diabetes mellitus families. J Autoimmun 1988;1:253–61.
[179] Nishikai M, Reichlin M. Heterogeneity of precipitating antibodies in polymyositis and dermatomyositis. Characterization of the Jo-1 antibody system. Arthritis Rheum 1980;23:881–8.
[180] Alspaugh MA, Talal N, Tan EM. Differentiation and characterization of autoantibodies and their antigens in Sjogren's syndrome. Arthritis Rheum 1976;19:216–22.
[181] Martinez-Lavin M, Vaughan JH, Tan EM. Autoantibodies and the spectrum of Sjogren's syndrome. Ann Intern Med 1979;91:185–90.
[182] Matsumoto I, Maccioni M, Lee DM, Maurice M, Simmons B, Brenner M, et al. How antibodies to a ubiquitous cytoplasmic enzyme may provoke joint- specific autoimmune disease. Nat Immunol 2002;3:360–5.
[183] Yu L, Robles DT, Abiru N, Kaur P, Rewers M, Kelemen K, et al. Early expression of antiinsulin autoantibodies of humans and the NOD mouse: evidence for early determination of subsequent diabetes. Proc Natl Acad Sci USA 2000;97:1701–6.
[184] Wong FS, Wen L, Tang M, Ramanathan M, Visintin I, Daugherty J, et al. Investigation of the role of B-cells in type 1 diabetes in the NOD mouse. Diabetes 2004;53:2581–7.
[185] Charlton B, Zhang MD, Slattery RM. B lymphocytes not required for progression from insulitis to diabetes in non-obese diabetic mice. Immunol Cell Biol 2001;79:597–601.
[186] Martin S, Wolf-Eichbaum D, Duinkerken G, Scherbaum WA, Kolb H, Noordzij JG, et al. Development of type 1 diabetes despite severe hereditary B-lymphocyte deficiency. N Engl J Med 2001;345:1036–40.
[187] Tseng CE, Di Donato F, Buyon JP. Stability of immunoblot profile of anti- SSA/Ro-SSB/La antibodies over time in mothers whose children have neonatal lupus. Lupus 1996;5:212–5.
[188] Palmeira P, Quinello C, Silveira-Lessa AL, Zago CA, Carneiro-Sampaio M. IgG placental transfer in healthy and pathological pregnancies. Clin Dev Immunol 2012;2012:985646.
[189] Juhasz I, Lazarus GS, Murphy GF, Shih IM, Herlyn M. Development of pemphigus vulgaris-like lesions in severe combined immunodeficiency disease mice reconstituted with lymphocytes from patients. J Clin Invest 1993;92:2401–7.
[190] Okamoto M, Murakami M, Shimizu A, Ozaki S, Tsubata T, Kumagai S, et al. A transgenic model of autoimmune hemolytic anemia. J Exp Med 1992;175:71–9.
[191] Tsao BP, Ohnishi K, Cheroutre H, Mitchell B, Teitell M, Mixter P, et al. Failed self-tolerance and autoimmunity in IgG anti-DNA transgenic mice. J Immunol 1992;149:350–8.
[192] Dumas G, Pritsch O, Pereira A, Gallart T, Dighiero G. A murine model of human cold agglutinin disease. Br J Haematol 1997;98:589–96.
[193] Liao L, Sindhwani R, Rojkind M, Factor S, Leinwand L, Diamond B. Antibody- mediated autoimmune myocarditis depends on genetically determined target organ sensitivity. J Exp Med 1995;181:1123–31.
[194] Silberstein LE, Jefferies LC, Goldman J, Friedman D, Moore JS, Nowell PC, et al. Variable region gene analysis of pathologic human autoantibodies to the related i and I red blood cell antigens. Blood 1991;78:2372–86.
[195] Cameron EM, Spencer S, Lazarini J, Harp CT, Ward ES, Burgoon M, et al. Potential of a unique antibody gene signature to predict conversion to clinically definite multiple sclerosis. J Neuroimmunol 2009;213:123–30.
[196] Yamagami J, Payne AS, Kacir S, Ishii K, Siegel DL, Stanley JR. Homologous regions of autoantibody heavy chain complementarity-determining region 3 (H-CDR3) in patients with pemphigus cause pathogenicity. J Clin Invest 2010;120:4111–7.
[197] Sondermann P, Pincetic A, Maamary J, Lammens K, Ravetch JV. General mechanism for modulating immunoglobulin effector function. Proc Natl Acad Sci USA 2013;110:9868–72.
[198] Anthony RM, Ravetch JV. A novel role for the IgG Fc glycan: the anti-inflammatory activity of sialylated IgG Fcs. J Clin Immunol 2010;30(Suppl 1):S9–14.
[199] Parekh R, Isenberg D, Rook G, Roitt I, Dwek R, Rademacher T. A comparative analysis of disease-associated changes in the galactosylation of serum IgG. J Autoimmun 1989;2:101–14.
[200] Malhotra R, Wormald MR, Rudd PM, Fischer PB, Dwek RA, Sim RB. Glycosylation changes of IgG associated with rheumatoid arthritis can activate complement via the mannose-binding protein. Nat Med 1995;1:237–43.
[201] Espy C, Morelle W, Kavian N, Grange P, Goulvestre C, Viallon V, et al. Sialylation levels of anti-proteinase 3 antibodies are associated with the activity of granulomatosis with polyangiitis (Wegener's). Arthritis Rheum 2011;63:2105–15.
[202] Schade SG, Abels J, Schilling RF. Studies on antibody to intrinsic factor. J Clin Invest 1967;46:615–20.
[203] Takamori M. Lambert-Eaton myasthenic syndrome: search for alternative autoimmune targets and possible compensatory mechanisms based on presynaptic calcium homeostasis. J Neuroimmunol 2008;201-202:145–52.
[204] Patrick J, Lindstrom J. Autoimmune response to acetylcholine receptor. Science 1973;180:871–2.
[205] Smith BR, Sanders J, Furmaniak J. TSH receptor antibodies. Thyroid 2007;17:923–38.
[206] Darrah E, Giles JT, Ols ML, Bull HG, Andrade F, Rosen A. Erosive rheumatoid arthritis is associated with antibodies that activate PAD4 by increasing calcium sensitivity. Sci Transl Med 2013;5:186ra65.
[207] Falk RJ, Jennette JC. ANCA disease: where is this field heading? J Am Soc Nephrol 2010;21:745–52.
[208] Markiewski MM, Lambris JD. The role of complement in inflammatory diseases from behind the scenes into the spotlight. Am J Pathol 2007;171:715–27.
[209] Gehrs BC, Friedberg RC. Autoimmune hemolytic anemia. Am J Hematol 2002;69:258–71.
[210] Rodriguez-Iturbe B, Carr RI, Garcia R, Rabideau D, Rubio L, McIntosh RM. Circulating immune complexes and serum immunoglobulins in acute poststreptococcal glomerulonephritis. Clin Nephrol 1980;13:1–4.
[211] Saitoh O, Abiru N, Nakahara M, Nagayama Y. CD8+CD122+ T cells, a newly identified regulatory T subset, negatively regulate Graves' hyperthyroidism in a murine model. Endocrinology 2007;148:6040–6.
[212] Lee YH, Ishida Y, Rifa'i M, Shi Z, Isobe K, Suzuki H. Essential role of CD8+CD122+ regulatory T cells in the recovery from experimental autoimmune encephalomyelitis. J Immunol 2008;180: 825–32.

[213] Bahri R, Bollinger A, Bollinger T, Orinska Z, Bulfone-Paus S. Ectonucleotidase CD38 demarcates regulatory, memory-like CD8+ T cells with IFN-gamma-mediated suppressor activities. PLoS One 2012;7:e45234.
[214] Antonelli A, Fallahi P, Nesti C, Pupilli C, Marchetti P, Takasawa S, et al. Anti-CD38 autoimmunity in patients with chronic autoimmune thyroiditis or Graves' disease. Clin Exp Immunol 2001;126:426–31.
[215] Pupilli C, Antonelli A, Iughetti L, D'Annunzio G, Cotellessa M, Vanelli M, et al. Anti-CD38 autoimmunity in children with newly diagnosed type 1 diabetes mellitus. J Pediatr Endocrinol Metab 2005;18:1417–23.
[216] Pavon EJ, Zumaquero E, Rosal-Vela A, Khoo KM, Cerezo-Wallis D, Garcia-Rodriguez S, et al. Increased CD38 expression in T cells and circulating anti-CD38 IgG autoantibodies differentially correlate with distinct cytokine profiles and disease activity in systemic lupus erythematosus patients. Cytokine 2013;62:232–43.
[217] Stasi R, Cooper N, Del Poeta G, Stipa E, Laura Evangelista M, Abruzzese E, et al. Analysis of regulatory T-cell changes in patients with idiopathic thrombocytopenic purpura receiving B cell-depleting therapy with rituximab. Blood 2008;112:1147–50.
[218] Ishida H, Muchamuel T, Sakaguchi S, Andrade S, Menon S, Howard M. Continuous administration of anti-interleukin 10 antibodies delays onset of autoimmunity in NZB/W F1 mice. J Exp Med 1994;179:305–10.
[219] Llorente L, Zou W, Levy Y, Richaud-Patin Y, Wijdenes J, Alcocer-Varela J, et al. Role of interleukin 10 in the B lymphocyte hyperactivity and autoantibody production of human systemic lupus erythematosus. J Exp Med 1995;181:839–44.
[220] Montalvao F, Garcia Z, Celli S, Breart B, Deguine J, Van Rooijen N, et al. The mechanism of anti-CD20-mediated B cell depletion revealed by intravital imaging. J Clin Invest 2013;123:5098–103.
[221] Navarra SV, Guzman RM, Gallacher AE, Hall S, Levy RA, Jimenez RE, et al. Efficacy and safety of belimumab in patients with active systemic lupus erythematosus: a randomised, placebo-controlled, phase 3 trial. Lancet 2011;377:721–31.
[222] Voorhees PM, Dees EC, O'Neil B, Orlowski RZ. The proteasome as a target for cancer therapy. Clin Cancer Res 2003;9:6316–25.
[223] Eris JM, Basten A, Brink R, Doherty K, Kehry MR, Hodgkin PD. Anergic self-reactive B cells present self antigen and respond normally to CD40-dependent T-cell signals but are defective in antigen-receptor-mediated functions. Proc Natl Acad Sci USA 1994;91:4392–6.
[224] Chan TD, Wood K, Hermes JR, Butt D, Jolly CJ, Basten A, et al. Elimination of germinal-center-derived self-reactive B cells is governed by the location and concentration of self-antigen. Immunity 2012;37:893–904.
[225] Menard L, Saadoun D, Isnardi I, Ng YS, Meyers G, Massad C, et al. The PTPN22 allele encoding an R620W variant interferes with the removal of developing autoreactive B cells in humans. J Clin Invest 2011;121:3635–44.

Chapter 24

# The Cellular and Molecular Biology of HIV-1 Broadly Neutralizing Antibodies

Barton F. Haynes[1], Kevin O. Saunders[2], Garnett Kelsoe[3], John R. Mascola[2] and Gary J. Nabel[4]

[1]*Departments of Medicine and Immunology, Duke Human Vaccine Institute, Duke University School of Medicine, Durham, NC, USA;* [2]*Vaccine Research Center, National Institutes of Allergy and Infectious Diseases, National Institutes of Health, Bethesda, MD, USA;* [3]*Department of Immunology, Duke Human Vaccine Institute, Duke University School of Medicine, Durham, NC, USA;* [4]*Sanofi, Cambridge, MA, USA*

## 1. INTRODUCTION

There were 1.7 million acquired immune deficiency syndrome (AIDS)-related deaths, 2.5 million newly human immunodeficiency virus 1 (HIV-1)-infected people, and 34.2 million people living with HIV-1 in 2011 (www.who.int/hiv/data/2012_epi_core_en.png), highlighting the urgent need for an effective vaccine for HIV-1. More than 30 years after the discovery of HIV-1 and the recognition of its causal role in AIDS [1,2], a vaccine for AIDS remains elusive. The challenges for HIV-1 vaccine development have been many; they include a high rate of HIV-1 mutation, glycoprotein masking, and the ability of HIV-1 to integrate into the host genome and establish latency in virus-infected CD4 T cells [3]. Most importantly, high levels of antibodies that can recognize conserved regions on the HIV-1 envelope (Env) glycoprotein and neutralize potently are not commonly made during infection and, to date, have not been induced by vaccination.

An HIV-1 vaccine efficacy trial, called RV144, reported in 2009, showed an estimated vaccine efficacy of 31% and gave hope that a preventive HIV-1 vaccine could be made [4]. This trial was performed in Thailand with a prime–boost vaccine combination including an avian poxvirus vaccine vector carrying HIV-1 genes and Env protein boosts [4]. Analysis of immune correlates of protection generated in the RV144 vaccine trial suggested that antibodies directed to the first and second variable regions (V1V2) of the HIV-1 Env were associated with a decreased risk of transmission [5]. However, no plasma broadly neutralizing antibodies (bnAbs) were induced by the RV144 vaccine [6], and the level of efficacy was insufficient to deploy as a commercial vaccine. Thus, a major goal of the HIV-1 vaccine development field is to improve protective efficacy in human clinical trials. One approach toward this goal is to understand how to induce long-lasting and durable plasma and mucosal levels of bnAbs [7].

Currently licensed vaccines, such as measles, mumps, and polio, typically induce protective neutralizing antibodies, and these vaccines have been made and successfully deployed without a precise understanding of the immunobiology of the control of protective immune responses [8]. In contrast, the failure of several HIV-1 vaccine efficacy trials [9–11] and the minimal efficacy of RV144 [4] raised the critical question, why are bnAbs not readily made during HIV-1 infection or Env vaccination? The answer necessitates an understanding of the pathways required for the induction of bnAbs.

Progress in the development of high-throughput recombinant techniques has enabled the rescue and analysis of human antibodies from infected and vaccinated individuals [12–15]. These recombinant techniques have been coupled with advances in the cloning of human and rhesus macaque memory B cells with culture techniques that provide potent ways to probe the B cell repertoire after infection or vaccination [14,16–24]. In addition, next-generation pyrosequencing has allowed for the analysis of bnAb lineages in chronically infected individuals when they occur [19,24]. Mice have been generated by knock-in of the $V_HDJ_H$ and $V_LJ_L$ genes of bnAbs and used to determine how the mammalian immune system regulates bnAbs [26–31]. In these animals, it has been demonstrated that the development of bnAbs targeted to a gp41 Env neutralizing site that crossreact with host lipids and proteins [32,33] is regulated by B cell tolerance, one of multiple ways that the host can constrain antibody lineages [26–32,34–36]. Thus, bringing new technology to the problem of bnAb induction has begun not only to help elucidate HIV-1 neutralizing antibody responses in particular, but also to bring about a better understanding of host tolerance control mechanisms of antibody induction in general [7,27].

Molecular Biology of B Cells. http://dx.doi.org/10.1016/B978-0-12-397933-9.00024-2

The study of viruses from a large HIV-1 infection cohort enrolled in the earliest stages of acute HIV-1 infection has demonstrated that in ~80% of heterosexually transmitted cases, HIV-1 infection is established by only one transmitted/founder virus [37]. This observation, together with the finding that bnAbs can develop in a minority of chronically infected individuals after years of infection [38–40], has led to the hypothesis that the interplay between virus and induced Env antibodies is required for bnAb induction. Recently, the simultaneous study of bnAb and transmitted/founder virus coevolution in HIV-1-infected individuals from the time of transmission through the development of bnAbs has begun to elucidate the precise steps in the ontogeny of B cells that give rise to bnAbs and the pathway of viral escape mutations in Env that stimulate their bnAb production [19,19a]. Moreover, these new technologies and unique acute HIV-1 infection cohorts followed to bnAb induction have allowed, for the first time, the analysis of the coevolution of an infectious agent and the resulting antibody response at the clonal level.

This chapter discusses the specificity and ontogeny of HIV-1 bnAbs, contrasts their development with that of other HIV-1 antibody specificities, and outlines strategies for induction of bnAbs by vaccination.

## 2. HIGHLY CONSERVED STRUCTURES ON HIV-1 ENV

HIV-1 Env is composed of three identical gp160 protomers. Surface gp160 precursor molecules are cleaved by cellular furin proteases, and the mature Env spike consists of a trimer of noncovalently associated gp120 and gp41 subunits (Figure 1). HIV-1 Env mediates virus–cell fusion through three major steps. First, it engages host cell CD4, the primary HIV-1 receptor. This interaction results in an activated intermediate conformation that engages a secondary chemokine receptor, usually CCR5 and sometimes CXCR4. In the third step, the gp41 subunit rearranges into a six-helix bundle that causes hemifusion of the viral and cell surface membranes (Figure 2) [41]. Antiviral neutralizing antibodies must bind either to the prefusion native viral spike or to an Env intermediate to block the virus–cell interaction.

Numerous bnAbs have been isolated from individuals identified within cohorts with plasma neutralizing activity against diverse strains. BnAbs recognize one of at least four highly conserved yet vulnerable regions of Env: the membrane proximal external region (MPER) of gp41 [18,20,42], a peptide-glycan epitope involving the V1V2 domain (V1V2 glycan) [16,22], a gp120 outer-domain glycan or peptide-glycan region (V3 glycan) [43,44], and the CD4 binding site (CD4bs) region of gp120 [19,23–25,45,46] (Table 1). Recently, a fifth bnAb epitope composed of both the gp41 and gp120 subunits of Env has been defined [46a,46b,46c,46d]. Importantly, advances in cryoelectron microscopy and crystal structures have provided new insights into the native prefusion structure and the conformation of the HIV-1 Env spike [52–54]. Together with the detailed atomic level structures of bnAbs bound to their

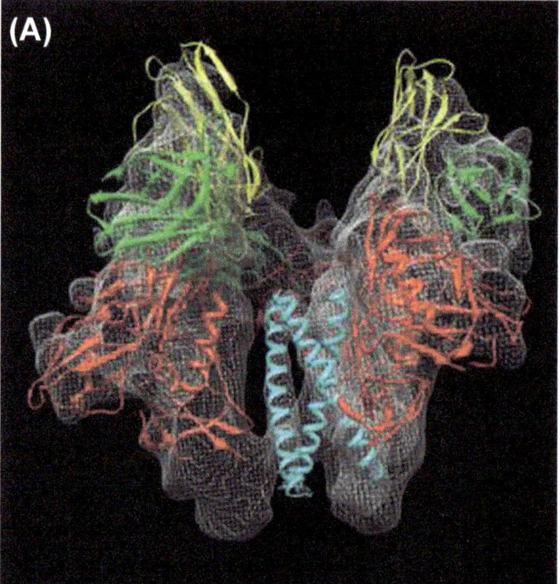


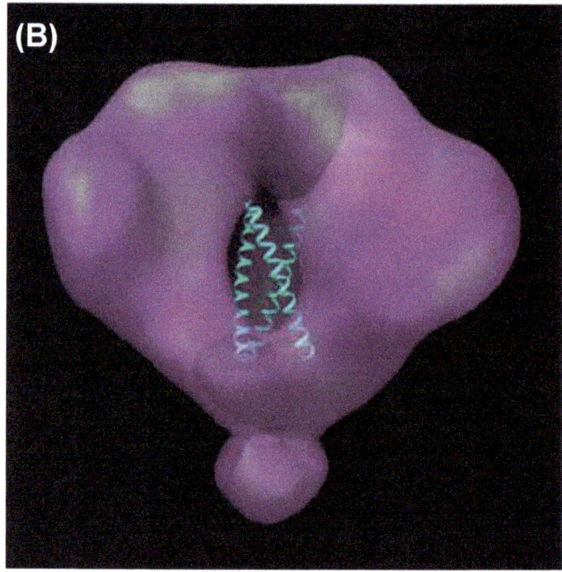


FIGURE 1 **Structure of the open conformation of trimeric envelope (Env) at subnanometer resolution.** (A) Side view of the structure of trimeric Env bound to 17b Fab. The map was fitted with three copies of the X-ray structure for the gp120–17b portion of the 1GC1 coordinates with gp120 (red) and 17b Fv fragments (light chain, yellow; heavy chain, green). One copy of the gp41 N-terminal helix (cyan) of 1AIK coordinates (N34) was fitted individually into each of the three densities, which occupy the central region of the spike that is essentially a cavity in the unliganded state. (B) Side view of the density map from the unliganded native trimer, with the three gp41 N-terminal helices (cyan) superimposed to show that, in the open conformation, they occupy the solvent-filled cavity in the density map of the unliganded state. *From Ref.* [41].

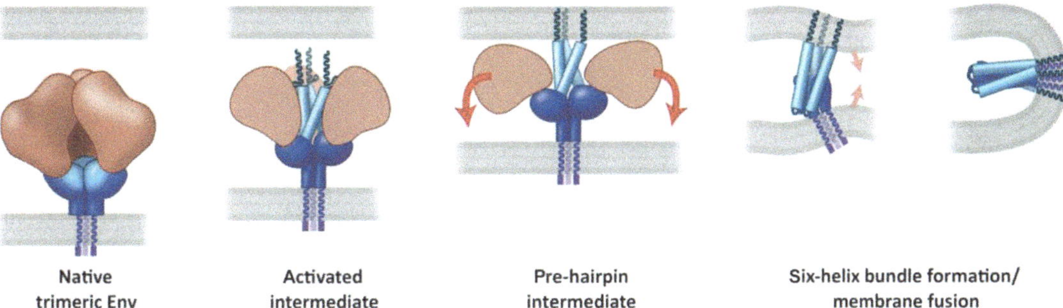


FIGURE 2 **Model for the mechanism of envelope (Env) activation.** The CD4- or coreceptor-triggered activation of the Env spike forms an activated intermediate in which the N-terminal gp41 helices are vulnerable to neutralizing ligands. The prehairpin intermediate is formed upon insertion of the fusion peptide into the target cell membrane and dissociation of gp120, leading to the formation of the six-helix bundle state and subsequent fusion between viral and target cell membranes. *From Ref.* [41].

cognate epitopes, these Env trimeric structures provide tools that facilitate structure-based vaccine design [55–58].

## 3. MECHANISMS OF HIV-1 NEUTRALIZATION BY ANTIBODIES

The target of HIV-1 neutralizing antibodies is the functional Env spike on the virion surface [7,57,59,60], and structural and biochemical studies have provided insight into the molecular mechanisms of HIV-1 neutralization (Figure 3). Antibodies to the CD4bs precisely target this recessed cavity of gp120 [46,51]. The monoclonal antibody (mAb) VRC01 was the first member of a class of CD4bs antibodies isolated from numerous HIV-1-infected donors and that share characteristic genetic, structural and maturation features (Table 1). VRC01 class antibodies contain a heavy chain derived either from the $V_H$1-2 or $V_H$1-46 genes that is heavily mutated during affinity maturation [24,25,51,61,62]. The light chain can arise from either a kappa or a lambda variable gene, but when paired with the $V_H$1-2 gene its CDRL3 is always short, containing five amino acids. These genetic features are critical for the antibody to attain a structure that partially fits into the CD4 binding site on gp120, thus mimicking CD4 itself [61]. By targeting the highly conserved CD4bs binding site region of gp120, VRC01 class antibodies can recognize and neutralize up to 90% of diverse HIV-1 strains.

There are two categories of highly potent bnAbs that bind to the surface of the Env spike and make contact with both glycans and amino acid residues. These antibodies bind either to a V1V2 domain or to an Env third variable loop (V3) domain that is exposed on the surface of the Env viral spike. All known V1V2 bnAbs interact with a glycan at Env position Asn160 and usually with at least one other glycan [63,64]. In addition, V1V2 bnAbs contact amino acid residues within a β-strand of the V2 region. A distinguishing characteristic of this group of antibodies is their preferential binding to the native (quaternary) structure of Env, and as such they bind poorly to soluble gp120 or gp140 proteins. Electron microscopy data indicate that these V1V2-glycan mAbs bind with a stoichiometry of one antibody per Env trimer: the mAb probably interacts with two of the three Env protomers on each trimer, potentially explaining the quaternary preference of these antibodies [65]. The V3-glycan category of antibodies has been defined by their requirement for interaction with an Env glycan at position Asn332 and their usual interaction with at least one other glycan on the outer domain region of Env [43,66,67,67a]. These antibodies also contact Env V3 amino acid residues, and in this respect, they are similar to the V1V2 category of glycopeptide-targeting bnAbs. V3-glycan bnAbs are genetically diverse and appear to adopt various orientations to interact with the V3-glycan site—leading some investigators to term this region a supersite of immune vulnerability on the glycosylated face of HIV-1 Env [58,67a]. Both the V1V2- and the V3-glycan bnAbs utilize a long (extended) heavy chain complementarity determinant region 3 (CDRH3) loop to penetrate the surface glycans and interact with the underlying Env protein surface. It is important to note that there is negative selection against antibodies with long CDRH3s during B cell development in the bone marrow of humans [68].

HIV-1 Env gp41 neutralizing antibodies recognize a specific region of gp41, termed the MPER. The MPER consists of a tryptophan-rich sequence of ~25 sequential amino acids and is the binding site for two of the first bnAbs isolated, 2F5 and 4E10 (Table 1) [47]. BnAb 4E10 displays broad HIV-1 crossreactivity—neutralizing more than 90% of diverse HIV-1 strains, but with rather weak potency. BnAb 2F5 displays modest potency but limited breadth—reacting with ~30% of HIV-1 strains. These antibodies neutralize by interfering with Env conformational changes that occur after CD4 binding and that are necessary for virion–cell fusion [47,69–71]. Antibody 4E10 has been shown to target the prehairpin intermediate conformation of Env instead of the prefusion native Env (Figure 2) [71] and is able to access the transient prehairpin conformation through a two-step model in which it first binds the viral membrane and subsequently binds the HIV Env [72,73]. Because 4E10 does not perturb gp120, it does not impair virus attachment

**TABLE 1** Genetic Characteristics of HIV-1 Broadly Neutralizing Monoclonal Antibodies

| Viral Epitope | Antibody Binding Characteristics[a] | Antibody Clonal Family | Isotype and Subclass[b] | Heavy Chain V-Gene | Light Chain V-Gene (κ or λ) | CDRH3 Length (Kabat AA)[c] | $V_H$ Mutation (nt%)[d] | $V_H$ Mutation (AA%)[d] | Neutralization Breadth[e] | Neutralization Potency[f] | Polyreactive[g] | References |
|---|---|---|---|---|---|---|---|---|---|---|---|---|
| MPER of gp41 | Contiguous sequence | 2F5 | IgG3 | 2–5 | κ1-13 | 22 | 14 | 15 | ++ | ++ | Yes | [47] |
| | | 4E10 | IgG3 | 1–69 | κ3-20 | 18 | 14 | 20 | ++++ | ++ | Yes | [48] |
| | | M66.6 | NR | 5–51 | κ1-39 | 21 | 4.3 | 10 | + | ++ | Yes | [42] |
| | | CAP206-CH12 | IgG1 | 1–69 | κ3-20 | 15 | 12 | 19 | + | ++ | Yes | [20] |
| | | 10E8[l] | IgG3 | 3–15 | λ3-19 | 20 | 21 | 26 | ++++ | +++ | No | [49] |
| V1V2-glycan | Peptidoglycan | PG9, PG16 | NR | 3–33 | λ2-14 | 28 | 12–13 | 17–20 | +++ | +++ | No | [12] |
| | | CH01–04 | IgG1 | 3–20 | κ3-20 | 24 | 14–17 | 23–29 | + | ++ | No[h] | [16] |
| | | PGT 141–145 | NR | 1–8 | κ2-28 | 31–32 | 12–18 | 21–29 | +++ | +++ | NR | [43] |
| Outer domain glycan | Glycan only | 2G12 | IgG1 | 3–21 | κ1-5 | 14 | 21 | 32 | + | ++ | NR | [44] |
| V3-glycan | Peptidoglycan | PGT121–123 | NR | 4–59 | λ3-21 | 24 | 17–21 | 21–27 | ++ | ++++ | NR | [43] |
| | | PGT125–131 | NR | 4–39 | λ2-8 | 19 | 15–23 | 23–33 | ++ | ++++ | NR | [43] |
| | | PGT135–137 | NR | 4–39 | κ3-15 | 18 | 17–20 | 25–29 | + | ++ | NR | [43] |
| CD4 binding site | CDRH3 loop | b12 | NR | 1–3 | κ3-20 | 18 | 13 | 20 | + | ++ | No[i] | [50] |
| | No liganded structure | HJ16 | NR | 3–30 | κ4-1 | 19 | 15 | 37 | + | ++ | NR | [45] |
| | CDRH3 loop | CH103–106 | IgG1 | 4–59 | λ3-1 | 13 | 14–16 | 22 | ++ | ++ | Yes[j] | [19] |

| | | | | | | | | | | | | |
|---|---|---|---|---|---|---|---|---|---|---|---|---|
| | Mimics CD4 via CDRH2 | VRC01–03 | IgG1 | 1–2 | κ3-20[m] | 12–14 | 30–32 | 40–48 | ++++ | +++ | No | [23,51] |
| | | VRC-PG04, 04b | IgG1 | 1–2 | κ3-20 | 14 | 30 | 38–39 | +++ | +++ | No | [24] |
| | | VRC-CH30–34 | IgG1 | 1–2 | κ1-33 | 13 | 23–24 | 36–40 | +++ | +++ | No | [24] |
| | No liganded structure | 3BNC117, 3BNC60[l] | NR | 1–2 | κ1-33 | 10 | 27 | 37–40 | +++ | +++ | Yes[k] | [25] |
| | Mimics CD4 via CDRH2 | NIH45–46[l] | IgG1 | 1–2 | κ3-20 | 16 | 30 | 41 | ++++ | +++ | Yes | [25] |
| | No liganded structure | 12A12, 12A21[l] | NR | 1–2 | κ1-33 | 13 | 23–25 | 38–40 | ++++ | +++ | Yes | [25] |
| | | 8ANC131, 134[l] | NR | 1–46 | κ3-20 | 16 | 27 | 37–38 | +++ | ++ | Yes | [25] |
| | | 1NC9, 1B2530[l] | NR | 1–46 | λ1-47 | 16–19 | 24–26 | 36–38 | + | | | [25] |
| gp120-gp41 | No liganded structure | 35O22 | NR | 1–18 | λ2-14 | 16 | NR | 35 | ++ | +++ | No | [46a] |
| | No liganded structure | PGT151* | NR | 3–30 | κ2D-29 | 28 | 18–22 | 27–31 | ++ | +++ | No | [46b,46c] |
| | Peptidoglycan | 8ANC195* | NR | 1–69 | κ1-5 | 20 | 24 | 47 | ++ | + | Yes | [25,46d] |

*[a]Major binding characteristics indicated by cocrystal structural analysis. mAb 2G12 utilizes unusual domain swap structure—see indicated references. mAb b12 was isolated from a phase display library, and the natural light chain pair was not retained; b12 displays heavy chain only binding in cocrystal structure with gp120.*
*[b]Isotype indicates the natural isotype and subclass of the isolated antibody. In some cases, this was not reported. In many cases, the PCR-amplified heavy and light chain genes are expressed in an IgG1 expression vector even if the natural antibody was a different subclass.*
*[c]Reported CDRH3 lengths are often based on IMGT or Kabat nomenclature. The Kabat definition is generally used for structural studies and is often 2 AA residues shorter than the IMGT definition. For explanation of the differences in nomenclature, see www.imgt.org.*
*[d]The percentage of mutations (nt) in an antibody heavy chain is based on comparison with the inferred germ-line gene. Percentage mutation is sometime also reported based on translated AA sequences (AA).*
*[e]Neutralization breadth is indicated for antibodies tested on large panels of >100 tier 2 Env pseudoviruses and the values shown are the percentage of viruses neutralized at an $IC_{50}$ of <50 μg/ml. ++++ ≥90%; +++ = 75–90%; ++ = 50–74%; + ≤50%. If there are several members of a clonal family, the broadest clonal member was used.*
*[f]Neutralization potency indicated by median or geometric mean $IC_{50}$ (μg/ml) of neutralized viruses (i.e., excluding nonneutralized viruses); ++++, $IC_{50}$ < 0.05; +++, $IC_{50}$ = 0.05–0.49; ++, $IC_{50}$ = 0.5–4.9; +, $IC_{50}$ = 5.0–50.*
*[g]Poly- or self-reactivity as assessed by a combination of common autoimmune assays, including cardiolipin ELISA, Hep2 cell IFA, and Athena assay for antinuclear antibodies.*
*[h]One antibody in this clonal lineage, CH03, is polyreactive.*
*[i]Some have found b12 polyreactive, but studies in b12 $V_HDJ_H + V_LJ_L$ knock-in mice have demonstrated that the degree of polyreactivity for this antibody is not sufficient to predispose these antibodies to tolerance deletion [30].*
*[j]CH103, CH104, CH106 are autoreactive, CH105 is not autoreactive.*
*[k]3BNC117 is reported to be polyreactive, but 3BNC60 is not.*
*[l]Additional members of the clonal family also reported in the primary publication. NIH45-45 is same donor and clonal family as VRC01. For mAbs 1NC9 and 1B2530, neutralization data only available for 1B2530.*
*[m]VRC01 and 02 were initially reported to derive from inferred VK3-11, but additional analysis of the antibody lineages strongly suggests VK3-20. Also, VRC03 was initially inferred to be from different clonal family than VRC01, but more detailed analysis suggests that VRC01, 02 and 03 are clonal relatives (Mascola, J and Kwong, P, unpublished observations).*
*NR, Not reported.*

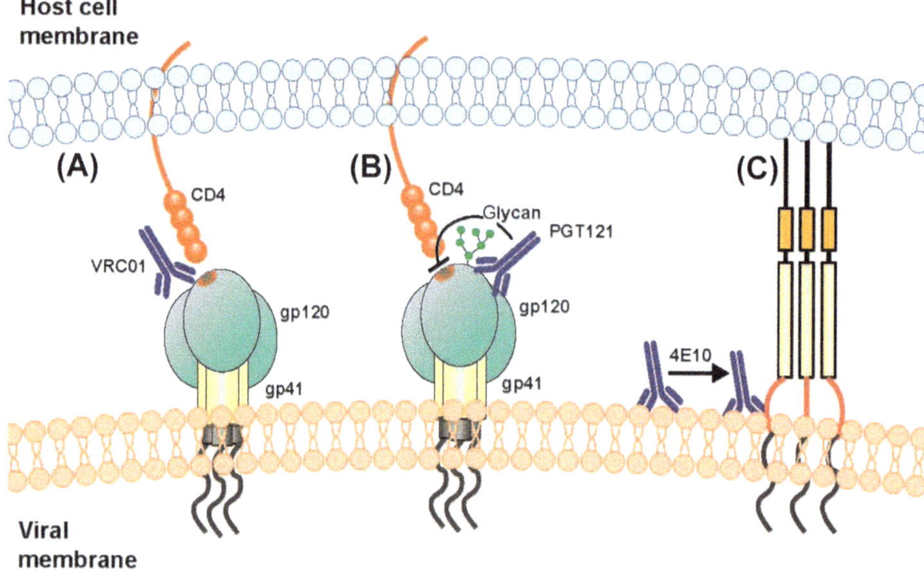


FIGURE 3 **Neutralization mechanisms of human immunodeficiency virus 1 (HIV-1) broadly neutralizing antibodies (bnAbs).** (A) VRC01 inhibition of CD4 binding by direct competition for the CD4 binding site on envelope (Env). Occupancy of the CD4 binding site on gp120 of Env by VRC01 probably precludes CD4 access to the same site, which prevents virus attachment to target cells. (B) Allosteric modulation of CD4 binding to Env by PGT121. The PGT121 family of antibodies interacts with protein and glycan of gp120 near the V3 loop of gp120. The binding of PGT121 to gp120 induces changes to the CD4 binding site that make it no longer accessible by CD4. (C) Inhibition of the fusion machinery of gp41 by an membrane proximal external region (MPER)-specific bnAb. The MPER antibody 4E10 neutralizes in a two-step model in which it first contacts the viral membrane. After CD4 attachment and gp120 shedding 4E10 binds its epitope on the prehairpin conformation of gp41. The 4E10 binding inhibits the conformational changes necessary for virus–cell membrane fusion.

to the cell surface [74]. Instead, bnAb 4E10 inhibits HIV-1 entry by preventing further structural rearrangement of the fusion machinery of HIV Env. This mechanism of neutralization may help prevent transmission of cell-associated virus, because Env–CD4 binding probably occurs rapidly in the junction between uninfected and infected cells, which could resist neutralization by CD4 binding site antibodies [74]. The more recently isolated bnAb 10E8 also recognizes this region, neutralizes with high potency and breadth, and shows minimal autoreactivity, suggesting that bnAbs to this structure could potentially be elicited by vaccination [18].

## 4. ROLE OF NEUTRALIZING ANTIBODIES IN PROTECTION FROM HIV-1 TRANSMISSION

HIV-1-specific neutralizing antibodies confer protection against lentiviral challenge in nonhuman primates models. Specifically, antibodies that neutralize HIV-1 have been shown to prevent infection by a chimeric simian–human immunodeficiency virus (SHIV) containing the Env gene of HIV-1 [75]. Early SHIV challenge studies used intravenous inoculation of the challenge virus. In this setting, high concentrations of neutralizing Ab's were required to block infection, and nonneutralizing antibodies were unable to protect [7,76,77]. Subsequent studies, using a single oral or vaginal inoculation sufficient to infect 100% of control animals, also demonstrated protection by neutralizing Ab's [78–81]. Protection against SHIV infection has been correlated with the neutralization potency of the infused antibodies [82]. Importantly, passive transfer studies have employed low-dose mucosal inoculation of rhesus macaques with chimeric SHIV that requires multiple challenges to infect all control animals [50,83]. This model is the most physiologically relevant to HIV-1 infection in humans, and the data suggest that the level of infused IgG that is delivered at the mucosal surface correlates with the level of protection from retroviral transmission [84].

In one study of the highly potent V3-glycan bnAb PGT121 (Table 1), complete protection against SHIV challenge was achieved at a low infusion dose of 1 mg/kg [85]. Moreover, antibody-mediated suppression of HIV-1 viremia in chronically SHIV-infected rhesus macaques can lead to prolonged absence of plasma virus [86,87]. Thus, data from nonhuman primate challenge models support the concept that potent HIV-1 bnAbs can prevent human HIV-1 infection if they can be sustained by passive transfer or if they can be induced by vaccination.

## 5. BIOLOGY OF BROAD NEUTRALIZING ANTIBODY DEVELOPMENT

A general characteristic of bnAbs is their extraordinary burden of point and insertion/deletion mutations in the V(D)J rearrangements of their H and L chains [7]. It is presumed that these mutations are acquired in germinal centers (GCs) [88,89], where hypermutation and affinity-dependent selection comprise a Darwinian process that refines humoral responses (discussed in [90–92]). Mutations that enhance affinity are less frequent than neutral and/or debilitating mutations, but they confer a survival advantage to GC B cells, perhaps by delivering survival signals by increasing the efficiency of antigen/major histocompatibility complex (MHC) presentation [92,93].

In contrast to the frequencies of V(D)J mutations in H and L chains (6% or 3%, respectively), characteristic of most human IgG memory B cells [94], V(D)J mutations in bnAbs are much higher (14–48%) [7,95]. This elevated frequency of mutations includes nucleotide exchanges in the scaffolding framework regions (FWRs) [95] of bnAbs and normally rare insertion or deletion mutations that enhance broadly neutralizing activity [25,95]. Although it is not surprising that some bnAb FWR mutations result in amino acid residues that contact the HIV-1 ligand [96,97], bnAb FWR mutations that do not make contact with bound antigen maybe necessary for optimal binding affinity and neutralization breadth [95]. This finding has been interpreted to mean that FWR mutations that increase flexibility of the β-sheet scaffold normally preserved in mutated high-affinity antibodies is structurally required for bnAb activity [95].

If the evolution of GC B cells is limited by their interaction with T cells [92,93], affinity-driven selection should have a ceiling that is determined by the off rate of bound antigen [98]; Ab off rates beyond the rate at which bound antigen is internalized and degraded for MHCII-associated presentation cannot be selected [98]. That bnAbs can infrequently originate from HIV-1-infected patients suggests that their characteristic mutational load must arise as a consequence of the lengthy coevolution of virus and GC B cells and selective criteria that are distinct from those that drive clonal evolution to other antigens [19]. At present, the criteria that drive bnAb evolution from ancestral B cells/antibodies that neutralize only autologous virus are not fully understood. Nonetheless, B cells that acquire insertion or deletion and/or FWR mutations must persist over more conventional mutant B cells in the GC, otherwise their clonal lineages are unlikely to persist over the years necessary to acquire bnAb activity [7].

Regardless of whether the high frequency of mutations exhibited by the known bnAbs is fundamentally necessary to generate antibodies that avidly bind to conserved broad neutralization epitopes conserved among many HIV-1 clades [19,24,95] or whether it is needed for other reasons [26,28,32,33,35], all models for GC B cell selection require that bnAb B cells participate in extraordinary numbers of mutation/selection events over the course of several years [19]. A question relevant to the origin of bnAbs is how likely is it that clonal antibodies persist for multiple years in the GC? Are there molecular requirements, or exogenous cues, for B cells to continue along the GC differentiation pathway that are independent of receptor affinity? Processes that disfavor the GC pathway, such as differentiation to plasmacytes or quiescent memory cells, would effectively reduce the probability of any GC B cell lineage acquiring the ~15–30% mutation frequency associated with bnAb activity.

As discussed elsewhere (Chapter 12), after encountering antigen, B cells migrate toward the T-cell-rich areas of secondary lymphoid tissues, whereas antigen-activated T cells migrate toward B cell areas. This mirrored behavior promotes the encounter of antigen-activated B and T cells, their cognate interaction, and joint proliferation. Some fraction of these expanded populations then migrates back into the tissue's B cell areas (follicles) that contain the reticular network of follicular dendritic cells (FDCs) where they establish a nascent GC [88,91–93,99–101]. After a period of proliferation, the GC becomes polarized into light and dark zones (LZ and DZ, respectively) [93].

A general model for how the GC LZ and DZ participate in the generation and selection of higher affinity mutants was initially postulated [90] on the basis of several observations: the rapid labeling of DZ centroblasts by radiolabeled nucleotides and the subsequent transfer of label to LZ centrocytes [102], iterative accumulation of mutations in single GCs that produced clonal lineages [88,99,100], and purifying selection in GCs that acted both inter- and intraclonally [100]. Maclennan proposed that after a single exposure to a nonreplicating protein antigen, the GC would form by the local proliferation of limited numbers of antigen-specific B cells. DZ centroblasts would proliferate to give noncycling centrocytes in the LZ that localized near FDCs that were decorated with complexed antigen. In turn, LZ centrocytes with avid antigen receptors could efficiently endocytose, process, and display antigen fragments in association with MHCII to local T cells. Centrocytes would be selected based on their ability to recover FDC-associated antigen and, thereby, to receive survival signals from LZ T cells. This general model was widely known and generally accepted [103], and its essentials have proved to be correct [93].

The demonstration of iterative accumulations of mutations in a single B cell clone recovered from a single GC strongly indicated a process of clonal evolution that was quintessentially Darwinian: mutation→selection→proliferation→mutation [88,100,103]. Mutations in GC B cells would generate novel antibody phenotypes that would undergo selection, and mutations that provided a survival advantage to GC B cells would be fixed by cell proliferation. Indeed this general model was found mathematically to be an optimal solution for efficiently generating high-affinity B cell mutants [104–107]. Significantly, the mathematicians provided a biological context for their theories: mutated centrocytes that survived selection in the LZ would return to the DZ for rounds of proliferation and mutation; progeny centroblasts that carried selected mutations would subsequently return to the LZ as centrocytes to undergo selection anew. This cyclic-reentry model neatly accounted for the known biology of GCs and for the stepwise nature of affinity maturation [106]. A decade later, initial studies by intravital microscopy [108–110] cast doubt on the cyclic-reentry model, but later, elegant studies utilizing in situ photoactivation confirmed the model [93].

If the high frequency of mutations characteristic of bnAbs is acquired in the GC, they must represent many years of HIV-1-specific B cells migrating between the DZ and the LZ of GCs to acquire and then select mutations that increase their (fitness) affinity for neutralizing epitopes. Indeed, bnAb lineages can now be traced over several years in serial blood samples from individual HIV-1-infected patients to a time that is only weeks after their infection [19] (Figure 4). These observations demonstrate that it is possible for a single, HIV-1-specific B cell clone to expand, diversify, and acquire bnAb activity in the milieu of chronic viral infection and coevolution but not how this is accomplished. A common notion at this time is that rounds of HIV-1 envelope evolution and escape from autologous HIV-1 neutralizing antibodies drive the evolution of bnAb specificities that recognize multiple Env variants [19].

This model is based on at least two requirements that are not easily accomplished. First, it requires that generalist B cells—those recognizing a determinant common to many HIV-1 types but not necessarily optimally expressed by each HIV-1 quasi-species—outcompete specialized GC B cells that recognize epitopes abundant on individual HIV-1 variants. Second, it requires that these generalist B cells enter the memory compartment and be recalled to the GC reaction at some later point by the expansion of HIV-1 variants. The first issue is a classical question in evolutionary biology and has been thoroughly discussed [111]. The second relates to how memory B cell fate is controlled by the balance of transcription factors.

The transcriptional control of terminal B cell fate has been well studied: the ratios of the early B-cell-specific transcription factor Pax5 and Blimp-1 (Prdm1), a regulator of plasma cell differentiation, determine the differentiation of activated B cells into antibody-secreting plasmacytes [112] (see also Chapter 6). In brief, Pax5 ensures the early program of B-cell-specific genes but is repressed by Blimp-1 during the transition of B cells to plasma cells; reciprocally,

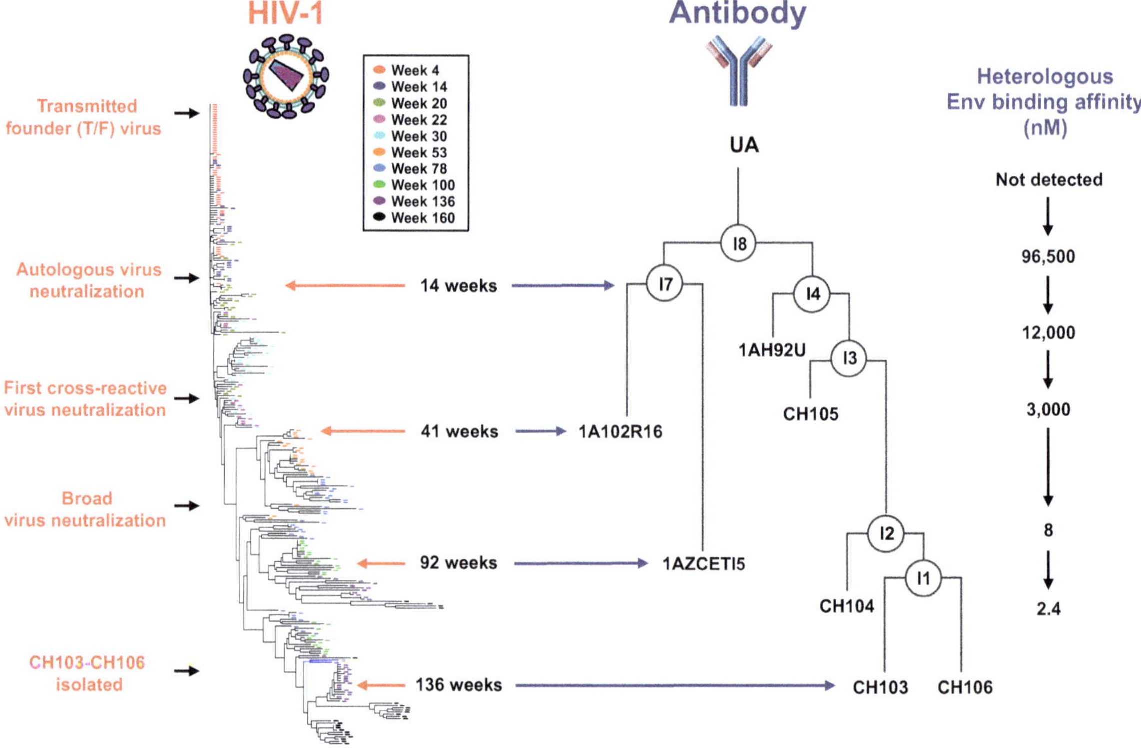


FIGURE 4 **Coevolution of virus and a single antibody lineage in an human immunodeficiency virus 1 (HIV-1) seroconverter.** Mature CD4 binding site antibodies CH103-106 were isolated from circulating memory B cells at week 136. Longitudinal sampling allowed inference and reconstruction of the evolution of the infecting viral sequence and of the specific neutralizing antibody lineage. The heavy chain antibody lineage was augmented with sequences derived from B cell gene sequencing, and bioinformatics analyses were used to infer early intermediates (IA) and the unmutated ancestor (UA). The left part of the figure displays a phylogenetic tree of Env sequences derived from week 4 through week 160. The UA and IA heavy chain sequences of the CH103 antibody lineages are shown alongside viral evolution. This antibody lineage evolved to gain high-affinity Env binding, and virus neutralization evolved from strain-specific autologous virus activity to crossreactive neutralization of heterologous viruses. *From Refs* [7,19].

Blimp-1 represses expression of early B-lineage genes, including Pax5. Another transcription factor, Bach2, first identified as a transcriptional repressor of Blimp-1 [113], acts to control B cell fate in a reciprocal fashion. Expression of Bach2 is elevated in developing B cells in the bone marrow and decreases with their developmental maturation until it is extinguished in plasma cells. By repressing Blimp-1, and thus plasma cell differentiation, Bach2 promotes secondary responses by sustaining their capacity to express activation-induced cytidine deaminase and retain the GC phenotype, but Bach2 levels are diminished in IgG memory B cells [114]. The consequence of this reduction in Bach2 is that class-switched memory B cells are more likely to differentiate into plasmacytes than to return to GCs for additional rounds of mutation and selection [114].

The newly discovered roles for Blimp-1 and Bach2 recall observations that IgM, not IgG, memory B cells are preferentially recruited into secondary GC reactions [115,116]. In fact, Bach2 levels are higher in IgM memory B cells than in their IgG counterparts [114]. Kometani and colleagues note that this different expression of Bach2 could represent: (1) a briefer GC history for IgM compared to IgG memory B cells; (2) the consequence of different signaling qualities of IgM versus IgG antigen receptors; and/or (3) a manifestation of independent memory cell types (IgM versus IgG) that are intrinsically distinct with regard to the GC or plasmacyte fate decision [114].

In the context of generating highly mutated bnAbs, virtually all HIV-1 vaccine immunogens and adjuvants have been selected for their capacity to elicit IgG antibody production and class-switched memory B cells. It is quite possible that this approach is antithetical to generating memory B cells that retain high levels of Bach2 and thus have a high probability of returning to/remaining within the GC over long periods. The notion of IgM "memory stem cells" is equally intriguing as most analysis of HIV-1-driven B cell evolution is based on the recovery of class-switched, IgG or IgA, memory B cells from the tissues of infected patients [19], an approach that minimizes recovery of IgM memory B cell pools that might be a source for continuing bnAb evolution.

## 6. CHARACTERISTICS OF HIV-1 ENV NEUTRALIZING ANTIBODIES

### 6.1 VRC01-Class of CD4 Binding Site bnAbs

The VRC01 class of antibodies is designated by four common characteristics that provide insight into their maturation. First, these antibodies all originate from the $V_H$1-2 gene family [56,61]. This observation is probably due to the presence of key residues at positions 50 and 71 that are conserved during antibody maturation and are postulated to aid in the initial engagement of HIV Env [62]. In contrast, the light-chain variable gene usage is diverse among the members of this class. Second, these antibodies are highly somatically mutated. In the case of VRC01, the $V_H$1-2 gene undergoes nearly 91 nucleotide somatic mutations to facilitate a high-affinity interaction with HIV Env [23,24]. Third and fourth, the VRC01 class antibodies that utilize the VH1-2 gene possess a five-amino-acid CDRL3 and a specifically mutated CDRL1 [46,56,61]. The CDRL1 mutations are generally a two-amino-acid deletion or two glycine substitutions that alleviate a potential clash with loop D of HIV-1 Env [61]. Antibody sequence analysis suggests that the five-amino-acid CDRL3 is a result of germ-line rearrangement and is therefore present at the level of the naive B cell receptor, whereas the CDRL1 mutations are a result of affinity maturation [61]. Overall, these antibodies use different heavy- and light-chain gene pairings, but they all mature to engage the CD4 binding site in a mode remarkably similar to CD4 itself (Figure 5).

### 6.2 V1V2 and Other Glycan-Targeted bnAbs

The V1V2-reactive antibodies and the V3-reactive antibodies comprise the two primary categories of glycan-reactive antibodies. The V1V2 antibodies possess an extraordinarily long CDRH3 (24–32 amino acids) that allows these antibodies to make contact with Env peptide and glycan [16,22,63,64] (Figure 6). The long CDRH3 develops as a consequence of the usage of a long D segment and nucleotide addition during VDJ recombination [19a]. Unlike the VRC01 class of antibodies, these antibodies do not require a specific heavy-chain variable gene and have lower degrees of somatic mutation (11–18% and 9–16% $V_H$ and $V_L$, respectively) [56]. Thus, the long CDRH3 seems to be the critical step in the development of V1V2 and V3 glycan-dependent bnAbs [63,64]. Meffre et al. have demonstrated in humans that antibodies with long CDRH3s are deleted in bone marrow [68]. Thus, one likely explanation for the low frequency of glycan-dependent antibodies is a reduced pool of naive B cells with germ-line $V_HDJ_H$ recombinations of sufficient length and shape to give rise to such antibodies. The CDRH3 of potent V3-glycan-directed antibodies is generally between 18 and 24 amino acids, making it slightly shorter than that of V1V2 antibodies [22,43,56]. The most potent members of this antibody category recognize Env by contacting the V3 peptide in a β-strand confirmation while interacting with glycans present at the base (N332) and tip (N301) of the V3 loop of Env [67]. The crystal structure of PGT128 detailing these interactions suggests that a moderately long CDRH3 is necessary to reach the V3 peptide masked by the N332 and N301 glycans [67]. Whether somatic mutation or gene recombination generates the long CDRH3 is unknown, but interestingly the most potent and broad V3 antibodies originate from the $V_H$4 gene family and through somatic mutation diverge from germ-line sequence up to 30% [43,56]. The importance of the $V_H$4 gene selection is unknown; however, like the VRC01 class of CD4

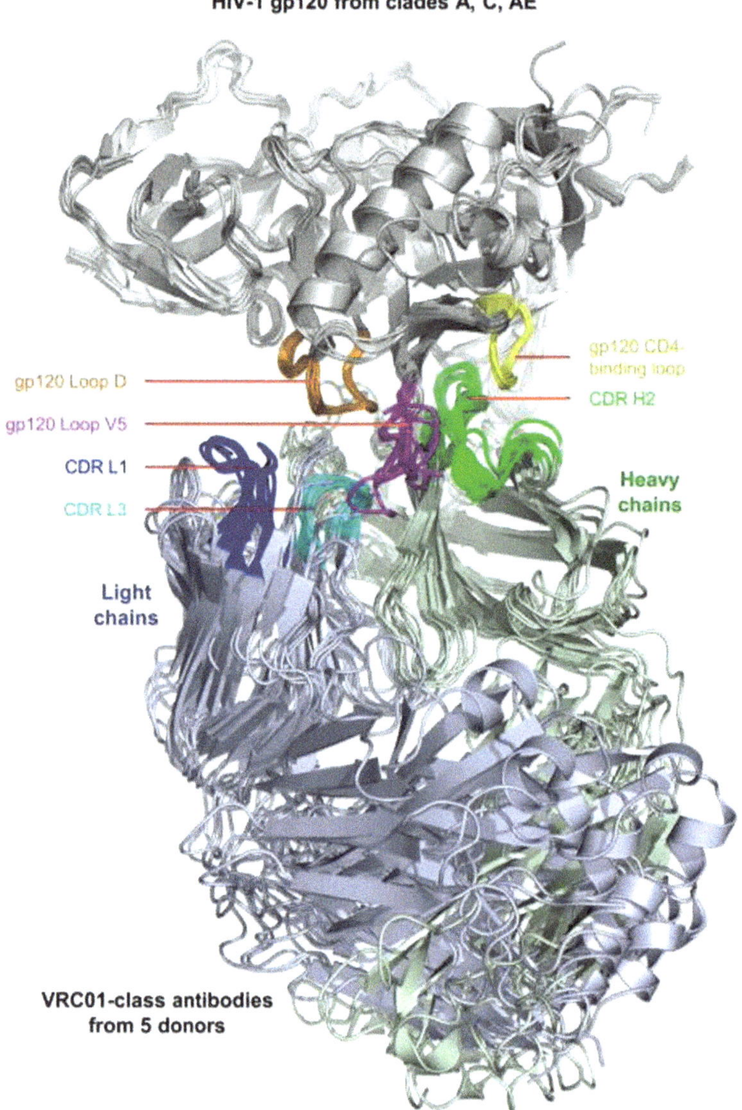


FIGURE 5 **VRC01 class recognition of human immunodeficiency virus 1 (HIV-1) gp120.** Nine cocrystal structures of five VRC01 class antibodies from five donors in complex with HIV-1 gp120 from clades A, C, and AE. Macromolecules are shown in ribbon representation after alignment of gp120, which is shown in gray with the CD4-binding loop (yellow), loop D (orange), and V5 (magenta) highlighted. Antibody light chains are colored light blue, and heavy chains are shown in pale green with CDRH2 (green), CDRL1 (blue), and CDRL3 (cyan) highlighted. *From Ref.* [61]

binding site bnAbs, high levels of somatic mutation of the variable gene are required for neutralization by PGT128 [95]. Future characterization of antibodies at early stages along the pathway of development for these new potent V3 glycan-reactive antibodies will provide more insight into the roles of variable gene selection and somatic mutation.

## 6.3 CD4bs Loop Binding bnAbs

Unlike the VRC01 type of CD4 binding site antibodies, which make predominant contacts via CDRH2, loop-binding antibodies bind to the gp120 CD4 binding site using the CDRH3 loop. This more common CDRH3 mode of recognition allows antibody binding without the requirement for a specific $V_H$ gene [19]. Moreover, loop-binding CD4 binding site antibodies are less unusual than VRC01-like bnAbs, with fewer somatic hypermutated nucleotides, suggesting that loop-binding CD4 binding site antibodies may be more easily induced than VRC01-like antibodies [19].

## 6.4 Env gp41 bnAbs

Among the first bnAbs isolated from chronically HIV-1-infected individuals were two antibodies, 2F5 and 4E10, that were found to be broadly neutralizing and bound to the

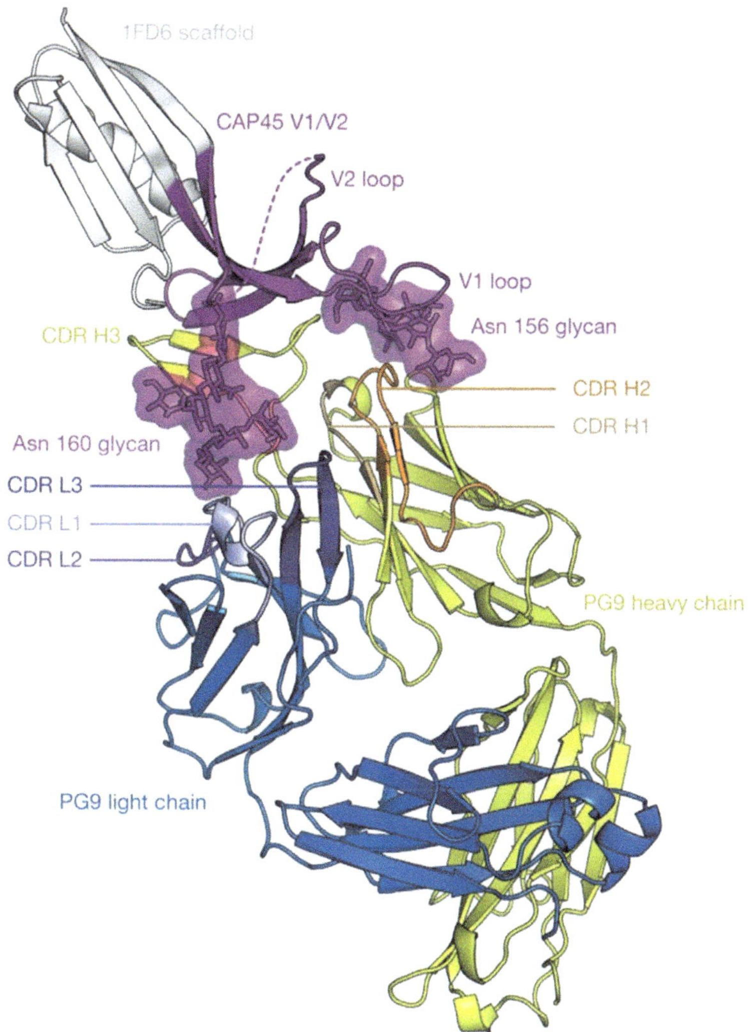


FIGURE 6 **Overall structure of the V1V2 domain of human immunodeficiency virus 1 (HIV-1) gp120 in complex with PG9.** V1V2 from the CAP45 strain of HIV-1 is shown, in magenta ribbons, in complex with the Fab of antibody PG9. The PG9 heavy and light chains are shown as yellow and blue ribbons, respectively, with CDRs in different shades. Although the rest of HIV-1 gp120 has been replaced by the 1FD6 scaffold (shown in white ribbons), the positions of V1V2, PG9, and the scaffold are consistent with the proposal that the viral spike, and hence the viral membrane, is positioned toward the top of the page. The extended CDRH3 of PG9 is able to penetrate the glycan shield that covers the V1V2 cap on the spike and to reach conserved elements of the polypeptide, whereas residues in heavy- and light-chain-combining regions recognize *N*-linked glycans. The disordered region of the V2 loop is represented by a dashed line. *From Ref.* [63].

membrane proximal region of the HIV envelope close to the virion lipid bilayer [47,48,117]. Both 2F5 and 4E10 are autoreactive for lipids and protein antigens, with 2F5 crossreacting with the tryptophan metabolism enzyme kynureninase and 4E10 crossreacting with splicing factor 3B3 [33]. The lipid reactivity confers on 2F5 and 4E10 the ability to bind to the HIV-1 virion in a two-step manner that includes both lipid and membrane proximal envelope, such that these bnAbs neutralize HIV after Env has bound host cell CD4 and CCR5 and the fusion process has begun [72,117,118]. 2F5 and 4E10 have very long CDRH3 regions and are heavily somatically mutated antibodies in addition to being polyreactive. As mentioned above, 2F5 and 4E10 $V_HDJ_H/V_LJ_L$ knock-in mice massively delete these antibodies at the naive B cell stage in the bone marrow, leaving only about 5% of these cells in the periphery in an anergic state [26–29,31,36]. Immunization of 2F5 and 4E10 knock-in mice with gp41 MPER peptide liposomes, however, can break anergy and induce high levels of plasma bnAbs, thus providing hope for inducing these types of antibodies in humans [29]. HIV-1 antibody polyreactivity is most probably due to their derivation from a similar pool of normal polyreactive memory B cells often associated with autoimmune diseases [12,119]. In the case of bnAb 4E10, polyreactivity may be clinically significant. The mAb 4E10 polyreactivity with cardiolipin led to prolongation of the partial thromboplastin time in HIV-1-infected individuals to whom 4E10 was administered, but without any thrombotic side effects [120].

Huang et al. have described a new MPER gp41 bnAb, termed 10E8, which, like 2F5 and 4E10, has a long CDRH3 and is highly somatically mutated [18] (Table 1). However, 10E8 is not particularly polyreactive although it is autoreactive for a nuclear protein, FAM84A (Yang G and Kelsoe G, unpublished observations).Thus, it is possible that this mAb will be affected by tolerance mechanisms.

## 7. HIV-1 ENV ANTIBODIES INDUCED BY CURRENT HIV VACCINE CANDIDATES

To date there have been six HIV vaccine efficacy studies (Table 2) [4,9,10,121–125]. Only one trial, the RV144 poxvirus prime/Env gp120 protein boost vaccine efficacy trial, showed any prevention of HIV-1 transmission, with an estimated 31.2% vaccine efficacy [4]. An immune correlates study of RV144 demonstrated that high levels of antibodies to the first and second variable regions (V1V2) of Env gp120 correlated with decreased transmission risk [5], and a viral genetic analysis of infections in the trial demonstrated that the presence of a lysine (K) at V2 position 169 in infecting viruses that matched the RV144 vaccine V2 sequence led to 48% vaccine efficacy [126]. That is, viruses that infected placebo recipients in the trial were more likely to have a V2 lysine 169 than viruses that infected vaccine recipients—giving rise to the hypothesis that an antibody response targeting K169 was involved in protection in the trial. One dominant antibody response induced in the RV144 trial was targeted at V2 with the epitope centered on K169 [127]. These antibodies do not neutralize primary infecting viruses, but do bind to the surface of primary HIV-1-strain-infected CD4 T cells and mediate natural killer cell antibody-dependent cellular cytotoxicity (ADCC) [127]. This observation led to the hypothesis that the protective effects of antibodies in the

**TABLE 2** HIV-1 Vaccine Efficacy Trials and Immune Correlates

| HIV-1 Efficacy Trial | Vaccine Strategy | Risk Population | Geographic Location | Vaccine Efficacy Outcome[a] | Correlates of HIV-1 Risk/ Incidence | Evidence of Immune Pressure (Virus Sieve) |
|---|---|---|---|---|---|---|
| 1. VAX003 (phase III) | Protein/alum (CRF01_AE/ clade B Env) | Injection drug users | Thailand | No efficacy | No | No |
| 2. VAX004 (phase III) | Protein/alum (clade B Env) | MSM[b]/high risk women | USA | No efficacy | Yes | No |
| 3. Step HVTN502 (phase IIb) | Ad5 vector (clade B Gag/Pol Nef) | MSM/high risk hetero-sexual men and women | North and South America, Australia, Caribbean | No efficacy (efficacy futility determined at first interim analysis after full enrollment) | Yes | Yes |
| 4. Phambili HVTN503 (phase IIb) | Ad5 vector (clade B Gag/Pol/Nef) | Heterosexual men and women | South Africa | No efficacy (vaccinations discontinued: early unblinding due to step results) | *None reported* | *None reported* |
| 5. RV144 (phase III) | ALVAC vector (clade B Gag/ Pro + CRF01_A/E Env) + protein/alum (CRF01_AE/clade B Env) | Community[c] | Thailand | 31.2% efficacy | Yes | Yes |
| 6.HVTN505 (phase IIb) | DNA + Ad5 vector (clade A,B,C *Env, Clade B Gag/Pol*) | MSM and TG[d] (Ad5 seronegative, circumcised) | USA | No efficacy (efficacy futility at first interim analysis after full enrollment) | *Studies in progress* | *Studies in progress* |

Vaccine-induced immune responses were studied to identify immune correlations with infection risk and evidence of virus sieve that can inform the design and evaluation of the next phase of vaccine efficacy trials (shaded boxes indicate reported correlations). Immunizations in two efficacy studies (Phambili/ HVTN 503, HVTN 505) were stopped; however, follow-up of participants continues.
[a]*Vaccine efficacy outcome is noted as "no efficacy" if there was no overall statistically significant vaccine efficacy.*
[b]*MSM: men who have sex with men.*
[c]*Participants meeting enrollment criteria were enrolled from the general population.*
[d]*TG: Transgender.*
From Ref. [121].

RV144 trial were due to a nonneutralizing effector function such as ADCC, virus capture [128], or blocking virion movement across mucosal surfaces [127]. These types of antibodies were easily induced in the RV144 trial, and one strategy to improve on the protection seen is to increase the diversity of the immunogen at the V2 region [121]. However, although the RV144 Env immunogens were reactive with bnAb epitopes, bnAbs have yet to be isolated from vaccine recipients [6,127]. Thus, a considerable effort is now under way to develop strategies for inducing subdominant bnAbs targeted at conserved bnAb sites on HIV-1 Env [7,58,129,130].

## 8. NEW STRATEGIES FOR INDUCTION OF HIV-1 bnAbs

### 8.1 B Cell Lineage Immunogen Design

With currently successful vaccines, protective antibodies are the dominant antibodies induced by the vaccine and there has been no necessity for deciphering B cell lineages induced by the vaccine, such as for rotavirus or polio. In contrast, the dominant antibody responses to HIV-1 are not protective, and the antibody responses to conserved neutralizing epitopes on HIV-1 Env are limited by tolerance or other host immunoregulatory controls [7,27,32,129]. Thus, for HIV-1 vaccines that can induce bnAbs, it has become necessary to define the pathways of bnAbs, when they do arise, and to chart a strategy for immunogen design. Thus, to overcome host constraints on bnAb induction, one strategy that has been proposed is termed B cell lineage immunogen design and is based on the principle that the most potent immunogens have the highest affinity for responding B cell receptors [60,115,129,131,132]. The process has three steps [129]. The first is to isolate clonally related bnAb $V_HDJ_H/V_LJ_L$ pairs by either clonal memory B cell cultures or antigen-specific B cell sorting. Second is to use these sequences and clonally related sequences derived from next-generation $V_HDJ_H/V_LJ_L$ sequencing to reconstruct the clonal lineage by inference from intermediate antibodies and unmutated common ancestor antibodies [19,24,25]. Third is to harness recombinant antibody technology and protein expression to evaluate the clonal lineages for high-affinity binding to antigen or Envs rationally designed for high-affinity binding and sequential immunizations defined based on Env affinity for binding at each clonal lineage step (Figure 7) [129]. A variation on the theme of B cell lineage design is to design Envs that bind to the unmutated common ancestor of bnAbs that are predicted to represent the naive B cell receptors [16,19,129,133].

### 8.2 Envelope Scaffolds

Protein epitope scaffolds can be exploited in immunogen design to focus the immune response to a specific neutralizing epitope on HIV-1 Env (Figure 8). In general, this

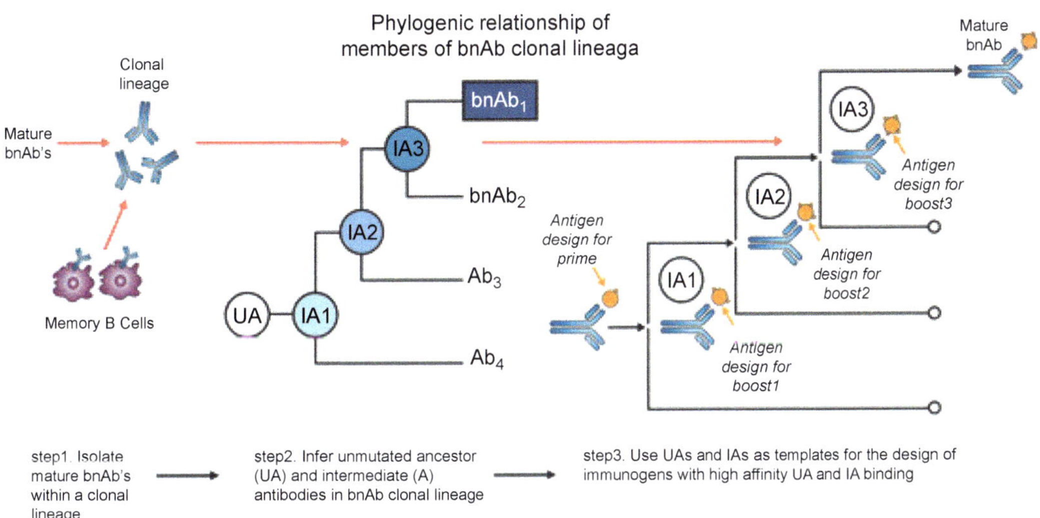


FIGURE 7 **B cell lineage-based approach to vaccine design.** Mature broadly neutralizing antibody (bnAb) responses can be isolated from HIV-1-infected donors using modern methods such as memory B cell culture or sorting of antigen-specific B cells. Based on the known bnAb sequence, next-generation sequencing can be used to find numerous clonal relatives of the mature bnAb. If appropriate longitudinal samples are available, it is possible to infer the full antibody lineage, including the unmutated ancestor (UA) and early intermediates (IA). The expressed UA and IA sequences can then be used as templates for the design of HIV-1 immunogens with high-affinity binding. As the antibody lineage is known to evolve in response to viral evolution, it may be possible to design sequential immunogens with high-affinity binding for the UA and IA, thus guiding the antibody response toward the mature antibody with broad neutralizing activity. HIV-1, human immunodeficiency virus 1. *Adapted from Refs* [7,129].

process involves grafting the neutralizing epitope onto an unrelated protein with a structurally similar motif [136–138]. This strategy has been shown to work for the design of immunogens for the induction of neutralizing antibodies to respiratory syncytial virus [139,140]. The epitope can be transplanted by side-chain grafting in which only the side-chain residues are introduced onto the protein scaffold [141]. The effectiveness of side-chain grafting is constrained by the structural similarity of the epitope and scaffold backbone [142]. Therefore, backbone grafting can be performed where the epitope amino acids replace a similar region of the protein scaffold [141]. Both of these approaches require the grafted epitope to resemble its native conformation in the HIV-1 envelope.

Grafting linear or continuous epitopes in which a single motif is interchanged between the scaffold and the antigen has been accomplished for the epitopes of the MPER-directed antibodies [141,143,144]. Overall, these scaffolds present the epitope in a manner sufficient for 2F5 or 4E10 binding, but have not elicited antibodies that recapitulate their neutralization activity. However, 2F5-like or 4E10-like antibodies were not elicited, probably because the epitope scaffolds lacked a lipid bilayer, a contact required for efficient neutralization [117,144–146]. Thus, the use of linear epitope scaffolds may be better suited to sites of vulnerability that require peptide binding only for neutralization.

Many antibodies, such as the CD4 binding site bnAbs, recognize epitopes that are composed of short segments that are spatially in close proximity after protein folding, but are not adjacent in the primary protein sequence [46,51]. Grafting these types of epitopes is particularly challenging, but computer-aided design coupled with high-throughput in vitro screening yielded a b12 epitope scaffold [142]. The scaffold possessed an extremely defined antigenic specificity for b12 but not other CD4 binding site antibodies.

## 8.3 Envelope Multimers

In vivo recognition of antigen by B cell receptors may occur in microclusters that promote cell signaling [147,148]. Multimerization of immunogens for presentation to B cells may be a means to mimic B cell receptor–antigen interaction (Figure 8). Fusing the desired epitope to proteins that assemble into viral particles or nanoparticles is a method used to generate an ordered array of up to hundreds of copies of a single immunogen [149]. Another multimerization approach uses multiple-antigen peptides, which allow for presentation of two to eight immunogens. The MPER has been applied to each of these approaches to create immunogens that react with 2F5 or 4E10 in binding assays [134,150,151]. Env gp120 nanoparticles have elicited only binding antibodies that failed to protect macaques from SHIV infection [152].

Some bnAb germ-line B cell receptors have low affinity for HIV Env. Immunogens arrayed on nanoparticles may be a productive method for increasing Env recognition by germline precursors of bnAbs. Scheif and colleagues tested this hypothesis by first engineering an HIV Env outer-domain protein, called eOD-GT6, to bind to several bnAb CD4 binding site germ-line B cell receptors [133]. Subsequently they performed *in vitro* B cell signaling assays and observed that the eOD-GT6 nanoparticle induced cell signaling, whereas monomeric protein did not [133]. This result suggests that germ-line immunogen nanoparticles may be a promising vaccine modality for initiating a particular naive B cell to mature toward a specific class of bnAbs.

## 8.4 Minimal Immunogens

Similar to scaffolds, minimal immunogens aim to mimic, in the correct conformation, the epitope present in the native trimer with only short peptide fragments (Figure 8). These designs present the relevant epitope to the immune system

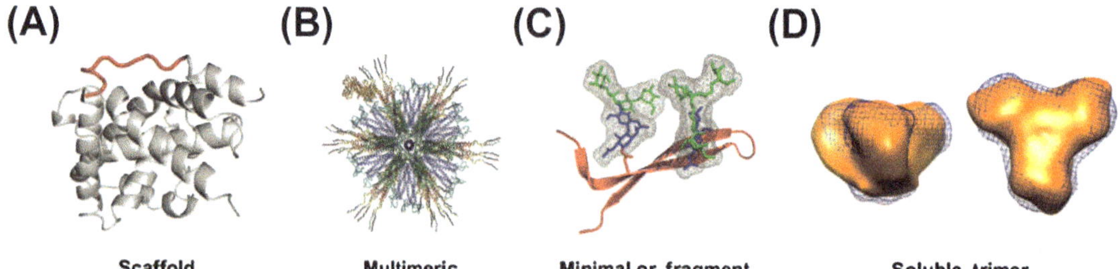


FIGURE 8 **Four novel human immunodeficiency virus 1 (HIV-1) immunogen design approaches.** (A) A rationally designed scaffold protein (gray) incorporating the 2F5 epitope (red) is a representative epitope scaffold immunogen. (B) An MPER nanoparticle derived from oligomerizing peptides fused with the MPER. The self-assembling nanoparticles display multiple copies of the MPER of envelope (Env) that can bind MPER antibodies (gold). *(Image reproduced with permission from Ref.* [134].*)* (C) A V1V2 glycopeptide containing glycans and amino acids sufficient for recognition by PG9/PG16-like antibodies that could be used as a minimal immunogen. *(Image reproduced with permission from Ref.* [135]. *Copyright 2103 American Chemical Society.)* (D) The electron microscopy reconstruction of a thermostable cleaved soluble gp140 SOSIP.664 trimer engineered from the HIV-$1_{BG505}$ Env (orange) compared to the HIV-$1_{KNH1144}$ trimer (blue). This trimer includes novel modifications that allow for the production of highly stable soluble trimers for the native presentation of neutralizing epitopes. *(The image was reproduced with permission from Ref.* [65].*)*

while removing potentially immunodominant epitopes elsewhere in the HIV Env [135]. Comprising this class of immunogens are MPER liposomes [131,132], V1/V2 glycopeptides [135,153,154], and CD4 binding site fragments [155]. The MPER liposomes are promising immunogens because they include the lipid bilayer requisite for 4E10 and 2F5 recognition of the MPER. A fusion-intermediate MPER liposome was able to elicit MPER-specific binding antibodies, but did not induce neutralizing antibodies [131]. V1/V2 glycopeptides have been designed to elicit the PG9 and PG16 class of antibodies [135,153]. These glycopeptides bind tightly to PG9 and PG16, presumably owing to new methods that can specifically glycosylate the key residues on the peptide through chemoenzymatic reactions. Whether these newly described glycopeptides will be immunogenic is yet to be determined. However, they do provide hope for carbohydrate-directed vaccines after the failure of oligomannose immunogens designed to elicit 2G12-like antibodies [156,157]. Finally, two CD4 binding site fragment immunogens called b121a and b122a bind remarkably well to b12, although they include only 70% of its epitope. When used as a prime before a gp120 boost these fragment immunogens elicited neutralizing antibodies in rabbits [155]. Future studies will need to be performed to determine if this finding can be translated to higher animals.

### 8.5 Soluble gp140 Trimers

Antibody binding to the Env trimer on the cell surface correlates with HIV neutralization, providing a rationale for its development as an immunogen [59,158]. The lack of an atomic structure of the HIV trimer had previously encumbered the ability to rationally design immunogenic trimers [159]. Currently, soluble recombinant trimer immunogens are made as gp140 in which the gp41 is truncated before the transmembrane domain. The gp140 can be left uncleaved, which results in increased presentation of nonneutralizing epitopes [160,161], or cleaved to produce a gp120 and a truncated gp41 heterodimer (Figure 8). Unfortunately, the trimerization domain in gp41 is insufficient for stably holding the envelope in the trimeric form, giving rise to gp140 monomers and presumably poor immunogenicity [159]. The current designs for trimer immunogens have improvements that include inclusion of heterologous trimerization motifs at the C-terminus of the gp41 [162,163], the insertion of a disulfide bond between the gp120 and the gp41 [164], the addition of both a disulfide bond between gp120 and gp41 and a proline at position 559 in gp41 [165], optimization of the gp120/gp41 cleavage site [166], and the truncation of gp41 at Env position 664 [158]. As mentioned, four structures, three by cryoelectron microscopy and one by X-ray crystallography, of the HIV-1 trimer have been published [52–54,167]. It is hoped that knowledge of these structures will aid in immunogen design to ensure presentation of the most native structures to the B cell repertoire. However, the design of immunization regimens to also overcome host tolerance and other immunoregulatory controls will be necessary for successful induction of bnAbs.

The immunogenicity profile of these trimers has yet to be determined, but there is optimism for BG505 trimers given the elicitation of neutralizing antibodies in animal models. Immunizations with CZA97012 or gp140-F trimers in guinea pigs and macaques, respectively, have induced primarily tier 1 neutralizing antibodies against multiple HIV-1 clades [168–170]. However, the macaque neutralizing antibodies did not provide significant protection from SHIV infection [169].

## 9. SUMMARY

The major roadblock preventing successful HIV-1 vaccine development is the inability to induce bnAbs. The HIV-1 envelope has conserved sites of vulnerability; however, these sites are not routinely targeted by a dominant antibody response owing to epitope mimicry of host molecules and/or the need for restricted and unusual antibody types. All of the bnAb types appear subdominant in nature, and many are limited in development by immune tolerance controls. Nonetheless, bnAbs do develop in ~20% of HIV-1-infected individuals. Thus, HIV-1 vaccine development differs from other current successful vaccines in that mapping pathways of development is critical to understand how bnAbs develop and, therefore, also is critical to the design of immunogens to induce B cell development pathways similar to those seen in chronically HIV-1-infected people. Once bnAb pathways are defined, immunogens can be designed with characteristics that mimic the Env trimer native conformations that have high affinity for binding to unmutated common ancestors and intermediate antibodies of bnAb lineages (B cell lineage immunogen design). In this manner, it is hoped that host immune controls of bnAbs can be overcome, and bnAbs can be induced as dominant antibody responses by candidate HIV-1 vaccines.

## REFERENCES

[1] Barre-Sinoussi F, Chermann JC, Rey F, Nugeyre MT, Chamaret S, Gruest J, et al. Isolation of a T-lymphotropic retrovirus from a patient at risk for acquired immune deficiency syndrome (AIDS). Science 1983;220:868–71.

[2] Gallo RC, Salahuddin SZ, Popovic M, Shearer GM, Kaplan M, Haynes BF, et al. Frequent detection and isolation of cytopathic retroviruses (HTLV-III) from patients with AIDS and at risk for AIDS. Science 1984;224:500–3.

[3] Eisele E, Siliciano RF. Redefining the viral reservoirs that prevent HIV-1 eradication. Immunity 2012;37:377–88.

[4] Rerks-Ngarm S, Pitisuttithum P, Nitayaphan S, Kaewkungwal J, Chiu J, Paris R, et al. Vaccination with ALVAC and AIDSVAX to prevent HIV-1 infection in Thailand. N Engl J Med 2009;361:2209–20.

[5] Haynes BF, Gilbert PB, McElrath MJ, Zolla-Pazner S, Tomaras GD, Alam SM, et al. Immune-correlates analysis of an HIV-1 vaccine efficacy trial. N Engl J Med 2012;366:1275–86.
[6] Montefiori DC, Karnasuta C, Huang Y, Ahmed H, Gilbert P, de Souza MS, et al. Magnitude and breadth of the neutralizing antibody response in the RV144 and Vax003 HIV-1 vaccine efficacy trials. J Infect Dis 2012;206:431–41.
[7] Mascola JR, Haynes BF. HIV-1 neutralizing antibodies: understanding nature's pathways. Immunol Rev 2013;254:225–44.
[8] Plotkin SA. Complex correlates of protection after vaccination. Clin Infect Dis 2013;56:1458–65.
[9] Hammer SM, Sobieszczyk ME, Janes H, Karuna ST, Mulligan MJ, Grove D, et al. Efficacy trial of a DNA/rAd5 HIV-1 preventive vaccine. N Engl J Med 2013;369:2083–92.
[10] Pitisuttithum P, Gilbert P, Gurwith M, Heyward W, Martin M, van Griensven F, et al. Randomized, double-blind, placebo-controlled efficacy trial of a bivalent recombinant glycoprotein 120 HIV-1 vaccine among injection drug users in Bangkok, Thailand. J Infect Dis 2006;194:1661–71.
[11] Flynn NM, Forthal DN, Harro CD, Judson FN, Mayer KH, Para MF. Placebo-controlled phase 3 trial of a recombinant glycoprotein 120 vaccine to prevent HIV-1 infection. J Infect Dis 2005;191:654–65.
[12] Wardemann H, Yurasov S, Schaefer A, Young JW, Meffre E, Nussenzweig MC. Predominant autoantibody production by early human B cell precursors. Science 2003;301:1374–7.
[13] Tiller T, Meffre E, Yurasov S, Tsuiji M, Nussenzweig MC, Wardemann H. Efficient generation of monoclonal antibodies from single human B cells by single cell RT-PCR and expression vector cloning. J Immunol Methods 2008;329:112–24.
[14] Scheid JF, Mouquet H, Feldhahn N, Walker BD, Pereyra F, Cutrell E, et al. A method for identification of HIV gp140 binding memory B cells in human blood. J Immunol Methods 2009;343:65–7.
[15] Liao HX, Levesque MC, Nagel A, Dixon A, Zhang R, Walter E, et al. High-throughput isolation of immunoglobulin genes from single human B cells and expression as monoclonal antibodies. J Virol Methods 2009;158:171–9.
[16] Bonsignori M, Hwang KK, Chen X, Tsao CY, Morris L, Gray E, et al. Analysis of a clonal lineage of HIV-1 envelope V2/V3 conformational epitope-specific broadly neutralizing antibodies and their inferred unmutated common ancestors. J Virol 2011;85:9998–10009.
[17] Doria-Rose NA, Klein RM, Manion MM, O'Dell S, Phogat A, Chakrabarti B, et al. Frequency and phenotype of human immunodeficiency virus envelope-specific B cells from patients with broadly cross-neutralizing antibodies. J Virol 2009;83:188–99.
[18] Huang J, Ofek G, Laub L, Louder MK, Doria-Rose NA, Longo NS, et al. Broad and potent neutralization of HIV-1 by a gp41-specific human antibody. Nature 2012;491:406–12.
[19] Liao HX, Lynch R, Zhou T, Gao F, Alam M, Boyd SD, et al. Co-evolution of a broadly neutralizing HIV-1 antibody and founder virus. Nature 2013;496:469.
[19a] Doria-Rose NA, Schramm CA, Gorman J, Moore PL, Bhiman JN, DeKosky BJ, et al. Developmental pathway for potent V1V2-directed HIV-neutralizing antibodies. Nature 2014;509:55–62.
[20] Morris L, Chen X, Alam M, Tomaras G, Zhang R, Marshall DJ, et al. Isolation of a human anti-HIV gp41 membrane proximal region neutralizing antibody by antigen-specific single B cell sorting. PLoS One 2011;6:e23532.
[21] Scheid JF, Mouquet H, Feldhahn N, Seaman MS, Velinzon K, Pietzsch J, et al. Broad diversity of neutralizing antibodies isolated from memory B cells in HIV-infected individuals. Nature 2009;458:636–40.
[22] Walker LM, Phogat SK, Chan-Hui PY, Wagner D, Phung P, Goss JL, et al. Broad and potent neutralizing antibodies from an African donor reveal a new HIV-1 vaccine target. Science 2009;326:285–9.
[23] Wu X, Yang ZY, Li Y, Hogerkorp CM, Schief WR, Seaman MS, et al. Rational design of envelope identifies broadly neutralizing human monoclonal antibodies to HIV-1. Science 2010;329:856–61.
[24] Wu X, Zhou T, Zhu J, Zhang B, Georgiev I, Wang C, et al. Focused evolution of HIV-1 neutralizing antibodies revealed by structures and deep sequencing. Science 2011;333:1593–602.
[25] Scheid JF, Mouquet H, Ueberheide B, Diskin R, Klein F, Oliveira TY, et al. Sequence and structural convergence of broad and potent HIV antibodies that mimic CD4 binding. Science 2011;333:1633–7.
[26] Verkoczy L, Diaz M, Holl TM, Ouyang YB, Bouton-Verville H, Alam SM, et al. Autoreactivity in an HIV-1 broadly reactive neutralizing antibody variable region heavy chain induces immunologic tolerance. Proc Natl Acad Sci USA 2010;107:181–6.
[27] Verkoczy L, Kelsoe G, Moody MA, Haynes BF. Role of immune mechanisms in induction of HIV-1 broadly neutralizing antibodies. Curr Opin Immunol 2011;23:383–90.
[28] Verkoczy L, Chen Y, Bouton-Verville H, Zhang J, Diaz M, Hutchinson J, et al. Rescue of HIV-1 broad neutralizing antibody-expressing B cells in 2F5 $V_H \times V_L$ knockin mice reveals multiple tolerance controls. J Immunol 2011;187:3785–97.
[29] Verkoczy L, Chen Y, Zhang J, Bouton-Verville H, Newman A, Lockwood B, et al. Induction of HIV-1 broad neutralizing antibodies in 2F5 knock-in mice: selection against membrane proximal external region-associated autoreactivity limits T-dependent responses. J Immunol 2013;191:2538–50.
[30] Ota T, Doyle-Cooper C, Cooper AB, Doores KJ, Aoki-Ota M, Le K, et al. B cells from knock-in mice expressing broadly neutralizing HIV antibody b12 carry an innocuous B cell receptor responsive to HIV vaccine candidates. J Immunol 2013;191:3179–85.
[31] Doyle-Cooper C, Hudson KE, Cooper AB, Ota T, Skog P, Dawson PE, et al. Immune tolerance negatively regulates B cells in knock-in mice expressing broadly neutralizing HIV antibody 4E10. J Immunol 2013;191:3186–91.
[32] Haynes BF, Fleming J, St Clair EW, Katinger H, Stiegler G, Kunert R, et al. Cardiolipin polyspecific autoreactivity in two broadly neutralizing HIV-1 antibodies. Science 2005;308:1906–8.
[33] Yang G, Holl TM, Liu Y, Li Y, Lu X, Nicely NI, et al. Identification of autoantigens recognized by the 2F5 and 4E10 broadly neutralizing HIV-1 antibodies. J Exp Med 2013;210:241–56.
[34] Haynes BF, Moody MA, Verkoczy L, Kelsoe G, Alam SM. Antibody polyspecificity and neutralization of HIV-1: a hypothesis. Hum Antibodies 2005;14:59–67.
[35] Chen Y, Zhang J, Hwang KK, Bouton-Verville H, Xia SM, Newman A, et al. Common tolerance mechanisms, but distinct cross-reactivities associated with gp41 and lipids, limit production of HIV-1 broad neutralizing antibodies 2F5 and 4E10. J Immunol 2013;191:1260–75.
[36] Finton KA, Larimore K, Larman HB, Friend D, Correnti C, Rupert PB, et al. Autoreactivity and exceptional CDR plasticity (but not unusual polyspecificity) hinder elicitation of the anti-HIV antibody 4E10. PLoS Pathog 2013;9:e1003639.

[37] Keele BF, Giorgi EE, Salazar-Gonzalez JF, Decker JM, Pham KT, Salazar MG, et al. Identification and characterization of transmitted and early founder virus envelopes in primary HIV-1 infection. Proc Natl Acad Sci USA 2008;105:7552–7.

[38] Simek MD, Rida W, Priddy FH, Pung P, Carrow E, Laufer DS, et al. Human immunodeficiency virus type 1 elite neutralizers: individuals with broad and potent neutralizing activity identified by using a high-throughput neutralization assay together with an analytical selection algorithm. J Virol 2009;83:7337–48.

[39] Gray ES, Madiga MC, Hermanus T, Moore PL, Wibmer CK, Tumba NL, et al. The neutralization breadth of HIV-1 develops incrementally over four years and is associated with CD4+ T cell decline and high viral load during acute infection. J Virol 2011;85:4828–40.

[40] Tomaras GD, Binley JM, Gray ES, Crooks ET, Osawa K, Moore PL, et al. Polyclonal B cell responses to conserved neutralization epitopes in a subset of HIV-1-infected individuals. J Virol 2011;85:11502–19.

[41] Tran EE, Borgnia MJ, Kuybeda O, Schauder DM, Bartesaghi A, Frank GA, et al. Structural mechanism of trimeric HIV-1 envelope glycoprotein activation. PLoS Pathog 2012;8:e1002797.

[42] Zhu Z, Qin HR, Chen W, Zhao Q, Shen X, Schutte R, et al. Cross-reactive HIV-1-neutralizing human monoclonal antibodies identified from a patient with 2F5-like antibodies. J Virol 2011;85: 11401–8.

[43] Walker LM, Huber M, Doores KJ, Falkowska E, Pejchal R, Julien JP, et al. Broad neutralization coverage of HIV by multiple highly potent antibodies. Nature 2011;477:466–70.

[44] Trkola A, Purtscher M, Muster T, Ballaun C, Buchacher A, Sullivan N, et al. Human monoclonal antibody 2G12 defines a distinctive neutralization epitope on the gp120 glycoprotein of human immunodeficiency virus type 1. J Virol 1996;70:1100–8.

[45] Corti D, Langedijk JP, Hinz A, Seaman MS, Vanzetta F, Fernandez-Rodriguez BM, et al. Analysis of memory B cell responses and isolation of novel monoclonal antibodies with neutralizing breadth from HIV-1-infected individuals. PLoS One 2010;5:e8805.

[46] Zhou T, Georgiev I, Wu X, Yang ZY, Dai K, Finzi A, et al. Structural basis for broad and potent neutralization of HIV-1 by antibody VRC01. Science 2010;329:811–7.

[46a] Huang J, Kang BH, Pancera M, Lee JH, Tong T, Feng Y, et al. Broad and potent HIV-1 neutralization by a human antibody that binds the gp41-120 interface. Nature; 2014. in press.

[46b] Falkowska E, Le KM, Ramos A, Doores KJ, Lee JH, Blattner C, et al. Broadly neutralizing HIV antibodies define a glycan-dependent epitope on the prefusion conformation of gp41 on cleaved envelope trimers. Immunity 2014;40:657–68.

[46c] Blattner C, Lee JH, Sliepen K, Derking R, Falkowska E, de la Peña AT, et al. Structural delineation of a quaternary, cleavage-dependent epitope at the gp41-gp120 interface on intact HIV-1 Env trimers. Immunity 2014;40:669–80.

[46d] Scharf L, Scheid JF, Lee JH, West Jr AP, Chen C, Gao H, et al. Antibody 8ANC195 reveals a site of broad vulnerability on the HIV-1 envelope spike. Cell Rep 2014;7:785–95.

[47] Muster T, Steindl F, Purtscher M, Trkola A, Klima A, Himmler G, et al. A conserved neutralizing epitope on gp41 of human immunodeficiency virus type 1. J Virol 1993;67:6642–7.

[48] Stiegler G, Kunert R, Purtscher M, Wolbank S, Voglauer R, Steindl F, et al. A potent cross-clade neutralizing human monoclonal antibody against a novel epitope on gp41 of human immunodeficiency virus type 1. AIDS Res Hum Retroviruses 2001;17:1757–65.

[49] Huang YA, Duerr L, Zhang Z, Moodie N, Frahm SC, DeRosa J, et al. Immune-correlates analysis of the step HIV vaccine efficacy trial. AIDS Vaccine 2013;2013.

[50] Hessell AJ, Poignard P, Hunter M, Hangartner L, Tehrani DM, Bleeker WK, et al. Effective, low-titer antibody protection against low-dose repeated mucosal SHIV challenge in macaques. Nat Med 2009;15:951–4.

[51] Zhou T, Xu L, Dey B, Hessell AJ, Van Ryk D, Xiang SH, et al. Structural definition of a conserved neutralization epitope on HIV-1 gp120. Nature 2007;445:732–7.

[52] Julien JP, Cupo A, Sok D, Stanfield RL, Lyumkis D, Deller MC, et al. Crystal structure of a soluble cleaved HIV-1 envelope trimer. Science 2013:1–12.

[53] Lyumkis D, Julien JP, de Val N, Cupo A, Potter CS, Klasse PJ, et al. Cryo-EM structure of a fully glycosylated soluble cleaved HIV-1 envelope trimer. Science 2013;342:1484–90.

[54] Bartesaghi A, Merk A, Borgnia MJ, Milne JL, Subramaniam S. Prefusion structure of trimeric HIV-1 envelope glycoprotein determined by cryo-electron microscopy. Nat Struct Mol Biol 2013;20:1352–7.

[55] Kong L, Sattentau QJ. Antigenicity and immunogenicity in HIV-1 antibody-based vaccine design. J AIDS Clin Res 2012;S8:3.

[56] Kwong PD, Mascola JR. Human antibodies that neutralize HIV-1: identification, structures, and B cell ontogenies. Immunity 2012; 37:412–25.

[57] Kwong PD, Mascola JR, Nabel GJ. Rational design of vaccines to elicit broadly neutralizing antibodies to HIV-1. Cold Spring Harb Perspect Med 2011;1:a007278.

[58] Kwong PD, Mascola JR, Nabel GJ. Broadly neutralizing antibodies and the search for an HIV-1 vaccine: the end of the beginning. Nat Rev Immunol 2013;13:693–701.

[59] Parren PW, Mondor I, Naniche D, Ditzel HJ, Klasse PJ, Burton DR, et al. Neutralization of human immunodeficiency virus type 1 by antibody to gp120 is determined primarily by occupancy of sites on the virion irrespective of epitope specificity. J Virol 1998;72:3512–9.

[60] Corti D, Lanzavecchia A. Broadly neutralizing antiviral antibodies. Annu Rev Immunol 2013;31:705–42.

[61] Zhou T, Zhu J, Wu X, Moquin S, Zhang B, Acharya P, et al. Multidonor analysis reveals structural elements, genetic determinants, and maturation pathway for HIV-1 neutralization by VRC01-class antibodies. Immunity 2013;39:245–58.

[62] West Jr AP, Diskin R, Nussenzweig MC, Bjorkman PJ. Structural basis for germ-line gene usage of a potent class of antibodies targeting the CD4-binding site of HIV-1 gp120. Proc Natl Acad Sci USA 2012;109:E2083–90.

[63] McLellan JS, Pancera M, Carrico C, Gorman J, Julien JP, Khayat R, et al. Structure of HIV-1 gp120 V1/V2 domain with broadly neutralizing antibody PG9. Nature 2011;480:336–43.

[64] Pancera M, McLellan JS, Wu X, Zhu J, Changela A, Schmidt SD, et al. Crystal structure of PG16 and chimeric dissection with somatically related PG9: structure-function analysis of two quaternary-specific antibodies that effectively neutralize HIV-1. J Virol 2010; 84:8098–110.

[65] Julien JP, Lee JH, Cupo A, Murin CD, Derking R, Hoffenberg S, et al. Asymmetric recognition of the HIV-1 trimer by broadly neutralizing antibody PG9. Proc Natl Acad Sci USA 2013;110: 4351–6.

[66] Mouquet H, Scharf L, Euler Z, Liu Y, Eden C, Scheid JF, et al. Complex-type *N*-glycan recognition by potent broadly neutralizing HIV antibodies. Proc Natl Acad Sci USA 2012;109:E3268–77.

[67] Pejchal R, Doores KJ, Walker LM, Khayat R, Huang PS, Wang SK, et al. A potent and broad neutralizing antibody recognizes and penetrates the HIV glycan shield. Science 2011;334:1097–103.

[67a] Sok D, Doores KJ, Briney B, Le KM, Saye-Francisco KL, Ramos A, et al. Promiscuous glycan site recognition by antibodies to the high-mannose patch of gp120 broadens neutralization of HIV. Sci Transl Med 2014;6: 236ra63.

[68] Meffre E, Milili M, Blanco-Betancourt C, Antunes H, Nussenzweig MC, Schiff C. Immunoglobulin heavy chain expression shapes the B cell receptor repertoire in human B cell development. J Clin Invest 2001;108:879–86.

[69] Purtscher M, Trkola A, Gruber G, Buchacher A, Predl R, Steindl F, et al. A broadly neutralizing human monoclonal antibody against gp41 of human immunodeficiency virus type 1. AIDS Res Hum Retroviruses 1994;10:1651–8.

[70] Cardoso RM, Zwick MB, Stanfield RL, Kunert R, Binley JM, Katinger H, et al. Broadly neutralizing anti-HIV antibody 4E10 recognizes a helical conformation of a highly conserved fusion-associated motif in gp41. Immunity 2005;22:163–73.

[71] Frey G, Peng H, Rits-Volloch S, Morelli M, Cheng Y, Chen B. A fusion-intermediate state of HIV-1 gp41 targeted by broadly neutralizing antibodies. Proc Natl Acad Sci USA 2008;105:3739–44.

[72] Alam SM, McAdams M, Boren D, Rak M, Scearce RM, Gao F, et al. The role of antibody polyspecificity and lipid reactivity in binding of broadly neutralizing anti-HIV-1 envelope human monoclonal antibodies 2F5 and 4E10 to glycoprotein 41 membrane proximal envelope epitopes. J Immunol 2007;178:4424–35.

[73] Haynes BF, Nicely NI, Alam SM. HIV-1 autoreactive antibodies: are they good or bad for HIV-1 prevention? Nat Struct Mol Biol 2010;17:543–5.

[74] Abela IA, Berlinger L, Schanz M, Reynell L, Gunthard HF, Rusert P, et al. Cell-cell transmission enables HIV-1 to evade inhibition by potent CD4bs directed antibodies. PLoS Pathog 2012;8:e1002634.

[75] Mascola JR. Passive transfer studies to elucidate the role of antibody-mediated protection against HIV-1. Vaccine 2002;20:1922–5.

[76] Mascola JR, Lewis MG, Stiegler G, Harris D, VanCott TC, Hayes D, et al. Protection of Macaques against pathogenic simian/human immunodeficiency virus 89.6PD by passive transfer of neutralizing antibodies. J Virol 1999;73:4009–18.

[77] Shibata R, Igarashi T, Haigwood N, Buckler-White A, Ogert R, Ross W, et al. Neutralizing antibody directed against the HIV-1 envelope glycoprotein can completely block HIV-1/SIV chimeric virus infections of macaque monkeys. Nat Med 1999;5:204–10.

[78] Baba TW, Liska V, Hofmann-Lehmann R, Vlasak J, Xu W, Ayehunie S, et al. Human neutralizing monoclonal antibodies of the IgG1 subtype protect against mucosal simian-human immunodeficiency virus infection. Nat Med 2000;6:200–6.

[79] Mascola JR, Stiegler G, VanCott TC, Katinger H, Carpenter CB, Hanson CE, et al. Protection of macaques against vaginal transmission of a pathogenic HIV-1/SIV chimeric virus by passive infusion of neutralizing antibodies. Nat Med 2000;6:207–10.

[80] Parren PW, Marx PA, Hessell AJ, Luckay A, Harouse J, Cheng-Mayer C, et al. Antibody protects macaques against vaginal challenge with a pathogenic R5 simian/human immunodeficiency virus at serum levels giving complete neutralization in vitro. J Virol 2001;75:8340–7.

[81] Hofmann-Lehmann R, Vlasak J, Rasmussen RA, Smith BA, Baba TW, Liska V, et al. Postnatal passive immunization of neonatal macaques with a triple combination of human monoclonal antibodies against oral simian-human immunodeficiency virus challenge. J Virol 2001;75:7470–80.

[82] Nishimura Y, Igarashi T, Haigwood N, Sadjadpour R, Plishka RJ, Buckler-White A, et al. Determination of a statistically valid neutralization titer in plasma that confers protection against simian-human immunodeficiency virus challenge following passive transfer of high-titered neutralizing antibodies. J Virol 2002;76:2123–30.

[83] Hessell AJ, Rakasz EG, Poignard P, Hangartner L, Landucci G, Forthal DN, et al. Broadly neutralizing human anti-HIV antibody 2G12 is effective in protection against mucosal SHIV challenge even at low serum neutralizing titers. PLoS Pathog 2009;5:e1000433.

[84] Klein K, Veazey RS, Warrier R, Hraber P, Doyle-Meyers LA, Buffa V, et al. Neutralizing IgG at the portal of infection mediates protection against vaginal simian/human immunodeficiency virus challenge. J Virol 2013;87:11604–16.

[85] Moldt B, Rakasz EG, Schultz N, Chan-Hui PY, Swiderek K, Weisgrau KL, et al. Highly potent HIV-specific antibody neutralization in vitro translates into effective protection against mucosal SHIV challenge in vivo. Proc Natl Acad Sci USA 2012;109:18921–5.

[86] Shingai M, Nishimura Y, Klein F, Mouquet H, Donau OK, Plishka R, et al. Antibody-mediated immunotherapy of macaques chronically infected with SHIV suppresses viraemia. Nature 2013;503:277–80.

[87] Barouch DH, Whitney JB, Moldt B, Klein F, Oliveira TY, Liu J, et al. Therapeutic efficacy of potent neutralizing HIV-1-specific monoclonal antibodies in SHIV-infected rhesus monkeys. Nature 2013;503:224–8.

[88] Jacob J, Kelsoe G, Rajewsky K, Weiss U. Intraclonal generation of antibody mutants in germinal centres. Nature 1991;354:389–92.

[89] William J, Euler C, Christensen S, Shlomchik MJ. Evolution of autoantibody responses via somatic hypermutation outside of germinal centers. Science 2002;297:2066–70.

[90] MacLennan IC. Germinal centers. Annu Rev Immunol 1994;12:117–39.

[91] Kelsoe G. In situ studies of the germinal center reaction. Adv Immunol 1995;60:267–88.

[92] Victora GD, Nussenzweig MC. Germinal centers. Ann Rev Immunol 2012;30:429–57.

[93] Victora GD, Schwickert TA, Fooksman DR, Kamphorst AO, Meyer-Hermann M, Dustin ML, et al. Germinal center dynamics revealed by multiphoton microscopy with a photoactivatable fluorescent reporter. Cell 2010;143:592–605.

[94] Tiller T, Tsuiji M, Yurasov S, Velinzon K, Nussenzweig MC, Wardemann H. Autoreactivity in human IgG+ memory B cells. Immunity 2007;26:205–13.

[95] Klein F, Diskin R, Scheid JF, Gaebler C, Mouquet H, Georgiev IS, et al. Somatic mutations of the immunoglobulin framework are generally required for broad and potent HIV-1 neutralization. Cell 2013;153:126–38.

[96] Sheriff S, Silverton EW, Padlan EA, Cohen GH, Smith-Gill SJ, Finzel BC, et al. Three-dimensional structure of an antibody-antigen complex. Proc Natl Acad Sci USA 1987;84:8075–9.

[97] Padlan EA, Silverton EW, Sheriff S, Cohen GH, Smith-Gill SJ, Davies DR. Structure of an antibody-antigen complex: crystal structure of the HyHEL-10 Fab-lysozyme complex. Proc Natl Acad Sci USA 1989;86:5938–42.

[98] Batista FD, Neuberger MS. Affinity dependence of the B cell response to antigen: a threshold, a ceiling, and the importance of off-rate. Immunity 1998;8:751–9.

[99] Jacob J, Kelsoe G. In situ studies of the primary immune response to (4-hydroxy-3-nitrophenyl)acetyl. II. A common clonal origin for periarteriolar lymphoid sheath-associated foci and germinal centers. J Exp Med 1992;176:679–87.
[100] Jacob J, Przylepa J, Miller C, Kelsoe G. In situ studies of the primary immune response to (4-hydroxy-3-nitrophenyl)acetyl. III. The kinetics of V region mutation and selection in germinal center B cells. J Exp Med 1993;178:1293–307.
[101] Garside P, Ingulli E, Merica RR, Johnson JG, Noelle RJ, Jenkins MK. Visualization of specific B and T lymphocyte interactions in the lymph node. Science 1998;281:96–9.
[102] Hanna Jr MG. An autoradiographic study of the germinal center in spleen white pulp during early intervals of the immune response. Lab Invest 1964;13:95–104.
[103] Kelsoe G. Life and death in germinal centers (redux). Immunity 1996;4:107–11.
[104] Kepler TB, Perelson AS. Cyclic re-entry of germinal center B cells and the efficiency of affinity maturation. Immunol Today 1993;14:412–5.
[105] Kepler TB, Perelson AS. Modeling and optimization of populations subject to time-dependent mutation. Proc Natl Acad Sci USA 1995;92:8219–23.
[106] Kepler TB, Perelson AS. Somatic hypermutation in B cells: an optimal control treatment. J Theor Biol 1993;164:37–64.
[107] Oprea M, Perelson AS. Somatic mutation leads to efficient affinity maturation when centrocytes recycle back to centroblasts. J Immunol 1997;158:5155–62.
[108] Schwickert TA, Lindquist RL, Shakhar G, Livshits G, Skokos D, Kosco-Vilbois MH, et al. In vivo imaging of germinal centres reveals a dynamic open structure. Nature 2007;446:83–7.
[109] Allen CD, Okada T, Cyster JG. Germinal-center organization and cellular dynamics. Immunity 2007;27:190–202.
[110] Hauser AE, Junt T, Mempel TR, Sneddon MW, Kleinstein SH, Henrickson SE, et al. Definition of germinal-center B cell migration in vivo reveals predominant intrazonal circulation patterns. Immunity 2007;26:655–67.
[111] Macarthur R, Levins R. Competition, habitat selection, and character displacement in a patchy environment. Proc Natl Acad Sci USA 1964;51:1207–10.
[112] Shapiro-Shelef M, Calame K. Regulation of plasma-cell development. Nat Rev Immunol 2005;5:230–42.
[113] Muto A, Ochiai K, Kimura Y, Itoh-Nakadai A, Calame KL, Ikebe D, et al. Bach2 represses plasma cell gene regulatory network in B cells to promote antibody class switch. EMBO J 2010;29:4048–61.
[114] Kometani K, Nakagawa R, Shinnakasu R, Kaji T, Rybouchkin A, Moriyama S, et al. Repression of the transcription factor Bach2 contributes to predisposition of IgG1 memory B cells toward plasma cell differentiation. Immunity 2013;39:136–47.
[115] Dogan I, Bertocci B, Vilmont V, Delbos F, Megret J, Storck S, et al. Multiple layers of B cell memory with different effector functions. Nat Immunol 2009;10:1292–9.
[116] Pape KA, Taylor JJ, Maul RW, Gearhart PJ, Jenkins MK. Different B cell populations mediate early and late memory during an endogenous immune response. Science 2011;331:1203–7.
[117] Alam SM, Morelli M, Dennison SM, Liao HX, Zhang R, Xia SM, et al. Role of HIV membrane in neutralization by two broadly neutralizing antibodies. Proc Natl Acad Sci USA 2009;106:20234–9.
[118] Haynes BF, Alam SM. HIV-1 hides an Achilles' heel in virion lipids. Immunity 2008;28:10–2.
[119] Bonsignori M, Wiehe K, Grimm SK, Lynch R, Yang G, Kozink DM, et al. An autoreactive antibody from an SLE/HIV-1 patient broadly neutralizes HIV-1. J Clin Invest 2014;124:1835–43.
[120] Mehandru S, Vcelar B, Wrin T, Stiegler G, Joos B, Mohri H, et al. Adjunctive passive immunotherapy in human immunodeficiency virus type 1-infected individuals treated with antiviral therapy during acute and early infection. J Virol 2007;81:11016–31.
[121] Tomaras G, Haynes BF. Advancing toward HIV-1 vaccine efficacy through the intersections of immune correlates. Vaccine 2014;2:1–14.
[122] Buchbinder SP, Mehrotra DV, Duerr A, Fitzgerald DW, Mogg R, Li D, et al. Efficacy assessment of a cell-mediated immunity HIV-1 vaccine (the step study): a double-blind, randomised, placebo-controlled, test-of-concept trial. Lancet 2008;372:1881–93.
[123] Gilbert PB, Peterson ML, Follmann D, Hudgens MG, Francis D, Gurwith M, et al. Correlation between immunologic responses to a recombinant glycoprotein 120 vaccine and incidence of HIV-1 infection in a phase 3 HIV-1 preventive vaccine trial. J Infect Dis 2005;191:666–77.
[124] Gottardo R, Bailer RT, Korber BT, Gnanakaran S, Phillips J, Shen X, et al. Plasma IgG to linear epitopes in the V2 and V3 regions of HIV-1 gp120 correlate with a reduced risk of infection in the RV144 vaccine efficacy trial. PLoS One 2013;8:e75665.
[125] Gray GE, Allen M, Moodie Z, Churchyard G, Bekker LG, Nchabeleng M, et al. Safety and efficacy of the HVTN 503/Phambili study of a clade-B-based HIV-1 vaccine in South Africa: a double-blind, randomised, placebo-controlled test-of-concept phase 2b study. Lancet Infect Dis 2011;11:507–15.
[126] Rolland M, Edlefsen PT, Larsen BB, Tovanabutra S, Sanders-Buell E, Hertz T, et al. Increased HIV-1 vaccine efficacy against viruses with genetic signatures in Env V2. Nature 2012;490:417–20.
[127] Liao HX, Bonsignori M, Alam SM, McLellan JS, Tomaras GD, Moody MA, et al. Vaccine induction of antibodies against a structurally heterogeneous site of immune pressure within HIV-1 envelope protein variable regions 1 and 2. Immunity 2013;38:176–86.
[128] Liu P, Yates NL, Shen X, Bonsignori M, Moody MA, Liao HX, et al. Infectious virion capture by HIV-1 gp120-specific IgG from RV144 vaccinees. J Virol 2013;87:7828–36.
[129] Haynes BF, Kelsoe G, Harrison SC, Kepler TB. B-cell-lineage immunogen design in vaccine development with HIV-1 as a case study. Nat Biotechnol 2012;30:423–33.
[130] Burton DR, Ahmed R, Barouch DH, Butera ST, Crotty S, Godzik A, et al. A blueprint for HIV vaccine discovery. Cell Host Microbe 2012;12:396–407.
[131] Dennison SM, Sutherland LL, Jaeger FH, Anasti KM, Parks R, Stewart S, et al. Induction of antibodies in rhesus macaques that recognize a fusion-intermediate conformation of HIV-1 gp41. PLoS One 2011;6:e27824.
[132] Dennison SM, Stewart SM, Stempel KC, Liao HX, Haynes BF, Alam SM. Stable docking of neutralizing human immunodeficiency virus type 1 gp41 membrane-proximal external region monoclonal antibodies 2F5 and 4E10 is dependent on the membrane immersion depth of their epitope regions. J Virol 2009;83:10211–23.
[133] Jardine J, Julien JP, Menis S, Ota T, Kalyuzhniy O, McGuire A, et al. Rational HIV immunogen design to target specific germline B cell receptors. Science 2013;340:711–6.

[134] Wahome N, Pfeiffer T, Ambiel I, Yang Y, Keppler OT, Bosch V, et al. Conformation-specific display of 4E10 and 2F5 epitopes on self-assembling protein nanoparticles as a potential HIV vaccine. Chem Biol Drug Des 2012;80:349–57.

[135] Aussedat B, Vohra Y, Park PK, Fernandez-Tejada A, Alam SM, Dennison SM, et al. Chemical synthesis of highly congested gp120 V1V2 N-glycopeptide antigens for potential HIV-1-directed vaccines. J Am Chem Soc 2013;135:13113–20.

[136] Vita C, Roumestand C, Toma F, Menez A. Scorpion toxins as natural scaffolds for protein engineering. Proc Natl Acad Sci USA 1995;92:6404–8.

[137] Vita C, Vizzavona J, Drakopoulou E, Zinn-Justin S, Gilquin B, Menez A. Novel miniproteins engineered by the transfer of active sites to small natural scaffolds. Biopolymers 1998;47:93–100.

[138] Fernandez-Carneado J, Grell D, Durieux P, Hauert J, Kovacsovics T, Tuchscherer G. Surface grafting onto template-assembled synthetic protein scaffolds in molecular recognition. Biopolymers 2000;55:451–8.

[139] McLellan JS, Correia BE, Chen M, Yang Y, Graham BS, Schief WR, et al. Design and characterization of epitope-scaffold immunogens that present the motavizumab epitope from respiratory syncytial virus. J Mol Biol 2011;409:853–66.

[140] McLellan JS, Chen M, Joyce MG, Sastry M, Stewart-Jones GB, Yang Y, et al. Structure-based design of a fusion glycoprotein vaccine for respiratory syncytial virus. Science 2013;342:592–8.

[141] Azoitei ML, Ban YE, Julien JP, Bryson S, Schroeter A, Kalyuzhniy O, et al. Computational design of high-affinity epitope scaffolds by backbone grafting of a linear epitope. J Mol Biol 2012;415: 175–92.

[142] Azoitei ML, Correia BE, Ban YE, Carrico C, Kalyuzhniy O, Chen L, et al. Computation-guided backbone grafting of a discontinuous motif onto a protein scaffold. Science 2011;334:373–6.

[143] Correia BE, Ban YE, Holmes MA, Xu H, Ellingson K, Kraft Z, et al. Computational design of epitope-scaffolds allows induction of antibodies specific for a poorly immunogenic HIV vaccine epitope. Structure 2010;18:1116–26.

[144] Ofek G, Guenaga FJ, Schief WR, Skinner J, Baker D, Wyatt R, et al. Elicitation of structure-specific antibodies by epitope scaffolds. Proc Natl Acad Sci USA 2010;107:17880–7.

[145] Ofek G, Tang M, Sambor A, Katinger H, Mascola JR, Wyatt R, et al. Structure and mechanistic analysis of the anti-human immunodeficiency virus type 1 antibody 2F5 in complex with its gp41 epitope. J Virol 2004;78:10724–37.

[146] Burton DR. Scaffolding to build a rational vaccine design strategy. Proc Natl Acad Sci USA 2010;107:17859–60.

[147] Tolar P, Pierce SK. A conformation-induced oligomerization model for B cell receptor microclustering and signaling. Curr Top Microbiol Immunol 2010;340:155–69.

[148] Tolar P, Sohn HW, Pierce SK. The initiation of antigen-induced B cell antigen receptor signaling viewed in living cells by fluorescence resonance energy transfer. Nat Immunol 2005;6:1168–76.

[149] Raman S, Machaidze G, Lustig A, Aebi U, Burkhard P. Structure-based design of peptides that self-assemble into regular polyhedral nanoparticles. Nanomedicine 2006;2:95–102.

[150] Zhang L, Miao L, Gong X, Zhang H, Yang L, Shi Y, et al. Multiple antigen peptide mimetics containing gp41 membrane-proximal external region elicit broad neutralizing antibodies against human immunodeficiency virus type 1 in guinea pigs. J Pept Sci 2013;19:491–8.

[151] Arnold GF, Velasco PK, Holmes AK, Wrin T, Geisler SC, Phung P, et al. Broad neutralization of human immunodeficiency virus type 1 (HIV-1) elicited from human rhinoviruses that display the HIV-1 gp41 ELDKWA epitope. J Virol 2009;83:5087–100.

[152] Himeno A, Akagi T, Uto T, Wang X, Baba M, Ibuki K, et al. Evaluation of the immune response and protective effects of rhesus macaques vaccinated with biodegradable nanoparticles carrying gp120 of human immunodeficiency virus. Vaccine 2010;28: 5377–85.

[153] Amin MN, McLellan JS, Huang W, Orwenyo J, Burton DR, Koff WC, et al. Synthetic glycopeptides reveal the glycan specificity of HIV-neutralizing antibodies. Nat Chem Biol 2013;9:521–6.

[154] Alam SM, Dennison SM, Aussedat B, Vohra Y, Park PK, Fernandez-Tejada A, et al. Recognition of synthetic glycopeptides by HIV-1 broadly neutralizing antibodies and their unmutated ancestors. Proc Natl Acad Sci USA 2013;110:18214–9.

[155] Bhattacharyya S, Singh P, Rathore U, Purwar M, Wagner D, Arendt H, et al. Design of an *Escherichia coli* expressed HIV-1 gp120 fragment immunogen that binds to b12 and induces broad and potent neutralizing antibodies. J Biol Chem 2013;288:9815–25.

[156] Astronomo RD, Kaltgrad E, Udit AK, Wang SK, Doores KJ, Huang CY, et al. Defining criteria for oligomannose immunogens for HIV using icosahedral virus capsid scaffolds. Chem Biol 2010;17:357–70.

[157] Joyce JG, Krauss IJ, Song HC, Opalka DW, Grimm KM, Nahas DD, et al. An oligosaccharide-based HIV-1 2G12 mimotope vaccine induces carbohydrate-specific antibodies that fail to neutralize HIV-1 virions. Proc Natl Acad Sci USA 2008;105:15684–9.

[158] Sanders RW, Derking R, Cupo A, Julien JP, Yasmeen A, de Val N, et al. A next-generation cleaved, soluble HIV-1 Env Trimer, BG505 SOSIP.664 gp140, expresses multiple epitopes for broadly neutralizing but not non-neutralizing antibodies. PLoS Pathog 2013;9:e1003618.

[159] Koff WC. HIV vaccine development: challenges and opportunities towards solving the HIV vaccine-neutralizing antibody problem. Vaccine 2012;30:4310–5.

[160] Si Z, Phan N, Kiprilov E, Sodroski J. Effects of HIV type 1 envelope glycoprotein proteolytic processing on antigenicity. AIDS Res Hum Retroviruses 2003;19:217–26.

[161] Pancera M, Wyatt R. Selective recognition of oligomeric HIV-1 primary isolate envelope glycoproteins by potently neutralizing ligands requires efficient precursor cleavage. Virology 2005;332:145–56.

[162] Yang X, Florin L, Farzan M, Kolchinsky P, Kwong PD, Sodroski J, et al. Modifications that stabilize human immunodeficiency virus envelope glycoprotein trimers in solution. J Virol 2000;74:4746–54.

[163] Yang X, Lee J, Mahony EM, Kwong PD, Wyatt R, Sodroski J. Highly stable trimers formed by human immunodeficiency virus type 1 envelope glycoproteins fused with the trimeric motif of T4 bacteriophage fibritin. J Virol 2002;76:4634–42.

[164] Binley JM, Sanders RW, Clas B, Schuelke N, Master A, Guo Y, et al. A recombinant human immunodeficiency virus type 1 envelope glycoprotein complex stabilized by an intermolecular disulfide bond between the gp120 and gp41 subunits is an antigenic mimic of the trimeric virion-associated structure. J Virol 2000;74:627–43.

[165] Sanders RW, Vesanen M, Schuelke N, Master A, Schiffner L, Kalyanaraman R, et al. Stabilization of the soluble, cleaved, trimeric form of the envelope glycoprotein complex of human immunodeficiency virus type 1. J Virol 2002;76:8875–89.

[166] Binley JM, Sanders RW, Master A, Cayanan CS, Wiley CL, Schiffner L, et al. Enhancing the proteolytic maturation of human immunodeficiency virus type 1 envelope glycoproteins. J Virol 2002;76:2606–16.

[167] Mao Y, Wang L, Gu C, Herschhorn A, Desormeaux A, Finzi A, et al. Molecular architecture of the uncleaved HIV-1 envelope glycoprotein trimer. Proc Natl Acad Sci USA 2013;110:12438–43.

[168] Kovacs JM, Nkolola JP, Peng H, Cheung A, Perry J, Miller CA, et al. HIV-1 envelope trimer elicits more potent neutralizing antibody responses than monomeric gp120. Proc Natl Acad Sci USA 2012;109:12111–6.

[169] Sundling C, Forsell MN, O'Dell S, Feng Y, Chakrabarti B, Rao SS, et al. Soluble HIV-1 Env trimers in adjuvant elicit potent and diverse functional B cell responses in primates. J Exp Med 2010;207:2003–17.

[170] Chakrabarti BK, Feng Y, Sharma SK, McKee K, Karlsson Hedestam GB, Labranche CC, et al. Robust neutralizing antibodies elicited by HIV-1 JRFL envelope glycoprotein trimers in nonhuman primates. J Virol 2013;87:13239–51.

Chapter 25

# Immune Deficiencies Caused by B Cell Defects

Anne Durandy[1,2,3], Sven Kracker[1,2] and Alain Fischer[1,2,3]

[1]National Institute of Health and Medical Research INSERM U768, Necker Children's Hospital, Paris, France, [2]Descartes-Sorbonne Paris Cité University of Paris, Imagine Institute, France, [3]Department of Immunology and Hematology, Assistance Publique-Hopitaux de Paris, Necker Children's Hospital, Paris, France

Primary antibody deficiencies (PADs) are the most common primary immunodeficiency observed in humans, affecting approximately 1:20,000 individuals. They are all characterized by a peculiar susceptibility to bacterial infections because the production of the different immunoglobulin (Ig) isotypes with high affinity for antigens is essential for efficient protection against infections: IgM constitutes classically the first line of defense against capsulated bacteria in vascular spaces but appears to be also protective against some noncapsulated bacteria being secreted at the mucosal surface like IgA [1]; IgG has a longer half-life and diffuses in the extracellular compartment, IgA is transported to mucosal surfaces (especially in the gut), and IgE is involved in immune responses to helminthes. Moreover, these isotypes differentially bind to their respective Fc receptors and some can activate complement components. These infections predominantly affect the respiratory tracts, leading to the severe complications of sinusitis and bronchiectasis, if left untreated. *Streptococcus pneumoniae*, non-typable *Haemophilus influenzae*, and gram-negative bacteria are the most prevalent microorganisms causing these infections. To a lesser extent, patients with PADs are vulnerable to intestinal tract infections (mainly by *Giardia, Campylobacter jejuni, Salmonella*, or *Helicobacter pylori*) and to bacterial cutaneous infections. The absence of susceptibility to fungal infections indicates that antibodies are dispensable for immunity to fungi. Antibodies seem also to be less important for antiviral responses, with the notable exception of enteroviruses observed in the complete absence of Ig production (agammaglobulinemia).

Autoinflammation and/or autoimmunity is also an important symptom of PADs. The disease mechanisms are not well known and it is not clear whether B cells always have a direct role in the disease pathogenesis. Autoinflammation, typically granuloma formation, is characterized by seemingly unprovoked episodes of inflammation in the absence of detectable autoantibodies or autoreactive T cells. The occurrence of autoimmunity in patients with PADs is frequently associated with the presence of circulating B cells that express low amounts of membrane CD21 molecules ($CD21^{low}$). This B cell population is clearly distinct from other circulating B cells: they barely express the transitional B cell surface marker CD38, show low levels of calcium influx and decreased proliferation downstream of B-cell receptor (BCR) and CD40 stimulation as compared with other B cells, and exhibit polyreactivity and self-reactivity [2,3]. Another complication of PADs relates to the occurrence of lymphomas and other types of cancer. The mechanism underlying the occurrence of B cell lymphomas in PADs remains in most cases undefined, but epithelial carcinomas are thought to result from chronic infections, and possibly ectopic activation-induced cytidine deaminase (AID) expression [4].

Some PADs are secondary to functional impairments of other immune cell lineages: indeed, genetic impairments of T cell differentiation lead to secondary B cell defects: the complete absence of T cells as observed in severe combined immunodeficiency (CID) [5–8], or in thymic aplasia (Di George syndrome) [9] or the defect in T cell activation molecules [10–12] lead to a secondary PAD. More specifically, defects in the generation, maintenance, or activation (STAT3- IL-21-, CD40L-, ICOS-, or SAP-deficiency) of T follicular helper ($T_{FH}$) result also in a PAD [13–17]. As more recently described, PADs can also be caused by functional impairments in innate immune cells as defects in neutrophils or in Toll-like receptor (TLR) pathways result in a partial antibody response defect ([18,19]). However, most of the PADs are caused by a B cell-intrinsic defect that can be either specific for B cells or combined with a defect of other lineages, especially T cells. We herein review the different PADs due to a B cell defect (specific or combined) in

Molecular Biology of B Cells. http://dx.doi.org/10.1016/B978-0-12-397933-9.00025-4

terms of the known mechanisms that are mainly perturbed in the affected patients.

## 1. DEFECTS IN B CELL DEVELOPMENT

### 1.1 Defects in Pre-BCR Expression or Signaling

Genetic defects affecting the expression of pre-B-cell receptor (pre-BCR) or pre-BCR signaling typically result in a B cell-specific defect characterized by the absence of circulating mature B cells and of all Ig isotypes, accompanied by the accumulation of pre-B cells in the bone marrow (Figure 1). This PAD is a rare disease, affecting roughly 1:500,000 births, and is characterized by agammaglobulinemia from 6 months of age, after maternal IgGs have disappeared. Bacterial infections are the most commonly observed infections, but severe enteroviral infections, including meningitis, can occur. No autoimmune manifestations are reported, but inflammatory complications, as arthritis. No cancer susceptibility is described.

As shown in Table 1 and Figure 1, several PADs have been associated with defects that affect the different components of the pre-BCR (the λ5 or μ chains), the pre-BCR and BCR coreceptors Igα and Igβ, and the components of the pre-BCR and BCR signaling pathways, including the p85α subunit of PI3 kinase (PI3K) and the scaffold protein B cell linker (BLNK) [21,22]. These defects all lead to rare, autosomal-recessive forms of agammaglobulinemia (Table 1).

However, the most common PAD that has been associated with defects in B cell development is X-linked agammaglobulinemia (XLA). This condition is caused by mutations in the gene encoding B tyrosine kinase (*BTK*) and accounts for 85% of patients with agammaglobulinemia [23]. BTK is a key molecule in pre-B cell activation and differentiation. Following pre-BCR and BCR engagement later in the development, BTK is phosphorylated by the Src kinase Lyn and interacts with phosphatidylinositol 3,4,5-triphosphate (a PI3K product). Furthermore, activated BTK directly phosphorylates phospholipase Cγ2 (PLCγ2), which in turn generates inositol 1,4,5-triphosphate and diacylglycerol and leads to intracellular $Ca^{2+}$ release [24].

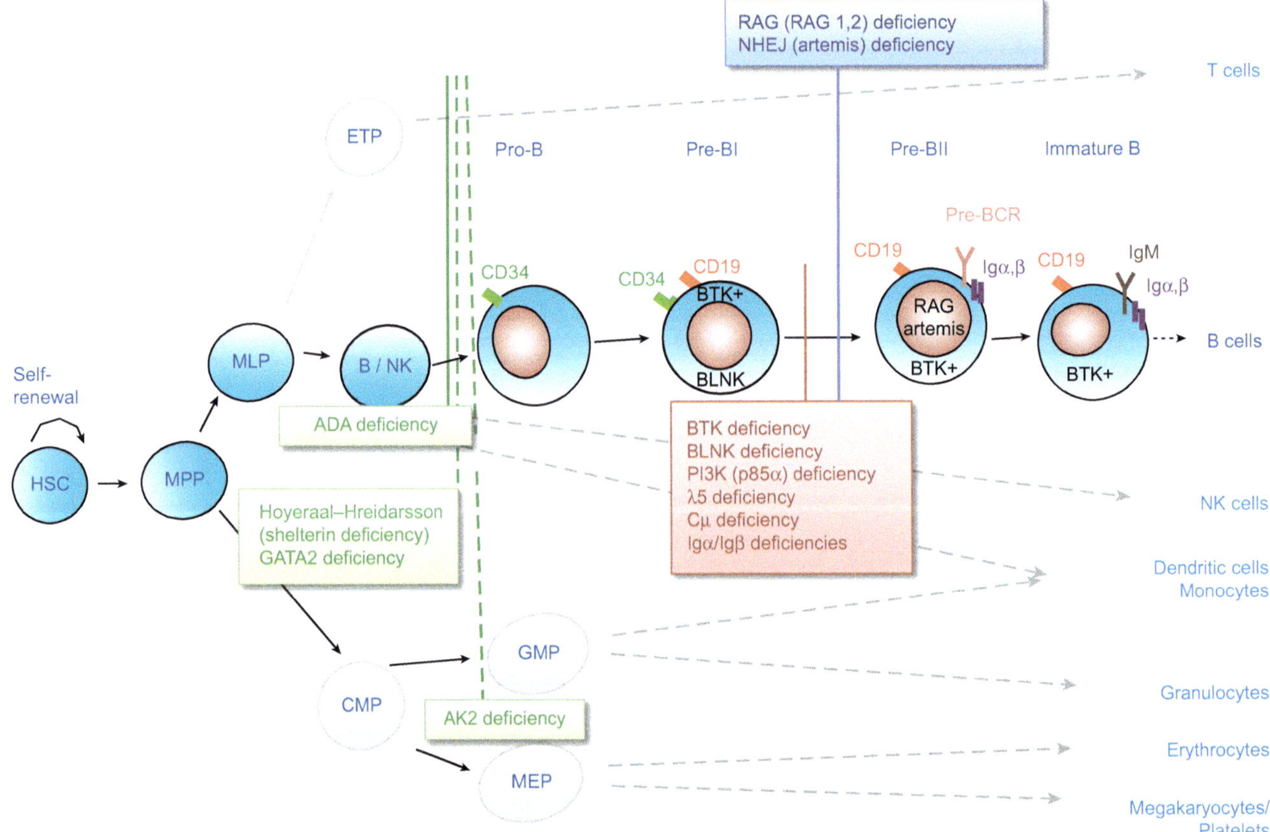


FIGURE 1 **Primary antibody deficiencies (PADs) caused by B cell differentiation defects.** B cell development occurs in bone marrow in an antigen and T-independent manner. Different arrests can occur, either B specific or affecting B and other lineages, leading to PADs. The dotted lines indicate an incomplete block of differentiation. CMP, common myeloid progenitor; ETP, early T cell precursor; GMP, granulocyte/macrophage progenitor; MEP, megakaryocyte/erythrocyte progenitor; MLP, immature lymphoid progenitor; MPP, multipotent progenitor; BLNK, B cell linker; ADA, adenosine deaminase; NK, natural killer; BTK, B tyrosine kinase; PI3K, PI3 kinase; AK, Adenylate kinase. *(Figure adapted from [20]).*

**TABLE 1** Primary Antibody Deficiencies Caused by a Specific B Cell Defect

| Affected Pathway | Affected Protein | Inheritance | Main Clinical Features | Main Biological Features |
|---|---|---|---|---|
| **B cell development** | | | | |
| Agammaglobulinemia | λ5, Cμ, Igα, Igβ, Pi3K(p85α), B cell linker/B tyrosine kinase | AR<br>AR/XL | Bacterial infections (URT)<br>Enterovirus infections | Absence of IgG, IgM and IgA<br>Absence of B cells |
| **B cell survival** | | | | |
| Common variable Immunodeficiency | Transmembrane activator and calcium modulator and cyclophilin ligand interactor<br>BAFFR/TWEAK | AR/AD<br>AR/AD | Bacterial infections, splenomegaly, autoimmunity, granulomas, cancers | Variable pan-hypogammaglobulinemia<br>Decreased frequency of memory CD27$^+$ B cells |
| **B cell activation** | | | | |
| Hypogammaglobulinemia | CD19 complex | AR | Bacterial infections | Variable hypogammaglobulinemia, decreased frequency of memory CD27$^+$ B cells |
| | CD20 | AR | | Defective response to PS |
| | HOIL1 | AR | Autoinflammation | Defective response to PS, decreased frequency of memory CD27$^+$ B cells |
| | CARD11 (gain of function) | AD | | Defective response to PS, increased numbers of transitional B |
| PLAID | PLCγ2 (gain of function) | AD | Allergy, autoimmunity | |
| CSR-D | AID/UNG | AR | Bacterial infections | IgM normal or increased, IgG and IgA low or absent |
| | | (AD)/AR | | Normal frequency of memory CD27$^+$ B cells, absence of switched memory B cells |

PS = polysaccharide antigens, AR = autosomal recessive, AD = autosomal dominant, URT = upper respiratory tract, BAFFR = BAFF receptor, TWEAK = tumor necrosis factor-like weak inducer of apoptosis, PLAID = PLCγ2-associated antibody deficiency and immune dysregulation, AID = activation-induced cytidine deaminase, UNG = Uracil-N glycosylase.

Most *BTK* mutations lead to the lack of protein expression, as witnessed in monocytes (which do not depend on BTK for their development) and a severe block in B cell differentiation. However, the block is less pronounced than that observed in AR-agammaglobulinemia. Moreover, some *BTK* mutations do not affect BTK expression and are generally associated with a milder phenotype (including the presence of low numbers of circulating B cells and residual levels of Igs) [21,25]. This genotype–phenotype correlation is inconstant and the observation of intra-family variations [26] suggests the influence of additional as-yet unknown modifier genes. At one extreme of the phenotypic spectrum, an XLA patient whose only disease feature was defective production of polysaccharide-specific antibodies has been reported [27].

Interestingly $Btk^{-/-}$ mice present with a milder defect as compared with humans, characterized by reduced numbers of mature B cells, especially of B1 cells, serum IgM and IgG3 deficiency, and partially defective antibody production, suggesting a differential requirement for Btk during B cell development in mice versus humans [28]. Compensatory mechanisms likely occur in mice since only the $Btk^{-/-}$ $Tec^{-/-}$ double deficient mouse model mimics the human XLA condition [29].

## 1.2 Defects in B and Other Lineages Development

In addition to the above-mentioned B cell specific defects, some genetic defects affecting not only development of B cells but also other lineages lead to a PAD (Table 2 and Figure 1): Some autosomal recessive (AR) forms of severe CID (SCID), a rare disease accounting for 1:50,000 births, are associated with an early defect in both B cells and T cells and are diagnosed in the first months or years of life. Susceptibility to bacterial and especially opportunistic

**TABLE 2 Primary Antibody Deficiencies Caused by a Non B Cell Specific Defect**

| Affected Pathway | Affected Protein | Inheritance | Other Cells Affected | Main Clinical Features | Main Biological Features |
|---|---|---|---|---|---|
| **B cell development** | | | | | |
| Severe combined immunodeficiency (CID) | ADA, NHEJ<br>RAG | AR<br>AR | T + others<br>T | Bacterial, viral, opportunistic infections | Absence of IgG, IgM, and IgA<br>Absence of B cells. |
| | AK2 | AR | T + PMN + inner ear | + Deafness | Hypogammaglobulinemia |
| Dyskeratosis congenita | Telomerase or shelterin complex | AR or XL | T + cerebellum | + Cutaneous abnormalitis, cancers | Progressive hypogammaglobulinemia, lymphopenia |
| Momac/DCML deficiency | GATA2 | AR | Monocytes, dendritic, NK | | Hypogammaglobulinemia |
| **B cell migration** | | | | | |
| WHIM | CXCR4 (gain of function) | AD | PMN | Bacterial infections, warts | Decreased number of B cells, neutropenia |
| WAS/WAS-like | WASP/WIP | XL/AR | T + platelets | Infections, eczema, hemorragia, autoimmunity, cancers | Decreased IgM levels, decreased response to PS |
| CID | DOCK8<br>MST1 | AR<br>AR | T<br>T | | Hypogammaglobulinemia, decreased numbers of $CD27^+$ B cells<br>Defective Ab production |
| **B cell activation** | | | | | |
| ICF | DNMT3B, ZBTB24 | AR | T | Bacterial infections | Variable hypogammaglobulinemia, absence of memory $CD27^+$ B cells |
| Defects in $Ca^{++}$influx | ORAi1, STIM1 | AR | T + muscle | AED, autoimmunity | Defective Ab production |
| CID | LRBA | AR | T | Infections, autoimmunity, IBD | Variable hypogammaglobulinemia |
| | Polε | AR | T + bone | Infections, short stature | Defective response to PS |
| Defect of NF-κB | CARD11, | AR | T | Bacterial infections | Variable hypogammaglobulinemia<br>Defective Ab production |
| | MALT1 | AR | T | | |
| | IkBβ, IkBα, | AR | T | | |
| | NEMO | XL | T + multi organ | AED | Defective response to PS |
| CSR-D | CD40 | AR | T + monocytes | Bacterial and opportunistic infections | IgM normal or increased, IgG low, IgA low but sometimes normal<br>Decreased frequency of $CD27^+$ B cells |
| DNA repair | ATM, | AR | T + multi organ | Bacterial infections, ataxia | IgM normal or increased, IgG and IgA low, normal or decreased frequency of $CD27^+$ B cells |
| | MRE11, NBS1, RNF128 | AR | T + multi organ | Bacterial infections | |
| | PMS2 | AR | T + multi organ | Early onset cancers, skin spots | Decreased levels of IgA and IgG2.<br>Decreased frequency of $CD27^+$ B cells |
| | Components of NHEJ | AR | T + multiorgan | Bacterial and opportunistic infections | IgM normal, IgG and IgA low, decreased frequency of $CD27^+$ B cells lymphopenia |
| Cytokine signaling | IL21R/ STATE3 | AR/AD | T + multi organ | Bacterial infections | Hyper-IgE, decreased frequency of TFH |
| | JAK3 | AR | T + NK | Bacterial and opportunistic infections | IgM normal or decreased, IgG and IgA low, lymphopenia |

PS = polysaccharide antigens; AR = autosomal recessive $CD27^+$ AD = autosomal dominant $CD27^+$ Ab = antibody $CD27^+$ AED = anhidrotic ectodermal dysplasia $CD27^+$ IBD = inflammatory bowel disease $CD27^+$ ADA = adenosine deaminase $CD27^+$ NHEJ = nonhomologous end joining $CD27^+$ NK = natural killer $CD27^+$ WHIM = Warts, hypogammaglobulinemia, infections, and myelokathexis $CD27^+$ WAS = Wiskott–Aldrich syndrome $CD27^+$ WASP = Wiskott–Aldrich syndrome protein $CD27^+$ WIP = WASP-interacting protein.

infections (*Pneumocystis jirovicii*) is a hallmark of the disease. Several causes underlie these defects: AR adenosine deaminase (ADA) deficiency, which represents 10% of SCID, leads to an accumulation of adenosines and thus death of the lymphocytes, including B lymphocytes. AR adenylate kinase 2 (AK2) deficiency (reticular dysgenesis) is a very rare metabolic defect that mostly affects development of T cells, natural killer (NK) cells, neutrophils, and B cells leading to early-onset hypo or even agammaglobulinemia [30,31]. Defects in V(D)J recombination (as observed in RAG and nonhomologous end joining (NHEJ) deficiencies) typically result in the absence of mature B and T cells [32,33]. The pre-BCR is affected in these cases, leading to PADs, as V(D)J recombination is required for heavy chain rearrangement and pre-BCR expression, thus pre-B cell expansion.

AR or X-linked dyskeratosis congenita, the most severe form being the Hoyeraal–Hreidarsson's syndrome, is caused by mutations in genes encoding components of the telomerase or shelterin complexes [34]; it is characterized by a bone-marrow failure leading to progressive B and T lymphopenia and hypogammaglobulinemia.

Dominant mutations in *GATA-2* (whose product is a transcription factor required for early differentiation of hematopoietic cells in the bone marrow), already identified as a predisposing gene for familial myelodysplastic syndrome/acute myeloid leukemia, also lead to an immune-deficiency characterized by B cell, dendritic cell, monocyte, and NK deficiencies (mono-Mac syndrome) [35].

## 2. DEFECTS IN B CELL MIGRATION

Up to now no specific defect in B cell migration responsible for a PAD has been reported in humans since all defects impairing B cell migration also affect other hematopoietic lineages. In normal conditions, B cells migrate from the bone marrow to the blood and then to the spleen (Figure 2). From there, they recirculate in the peripheral blood and home to secondary lymphoid organs. Abnormalities in this migration pattern lead to PAD pathogenesis (Table 2). Warts, hypogammaglobulinemia, infections, and myelokathexis (WHIM) syndrome is a rare, autosomal dominant (AD) disorder caused by gain-of-function mutations in the C terminal part of the gene encoding the CXC chemokine

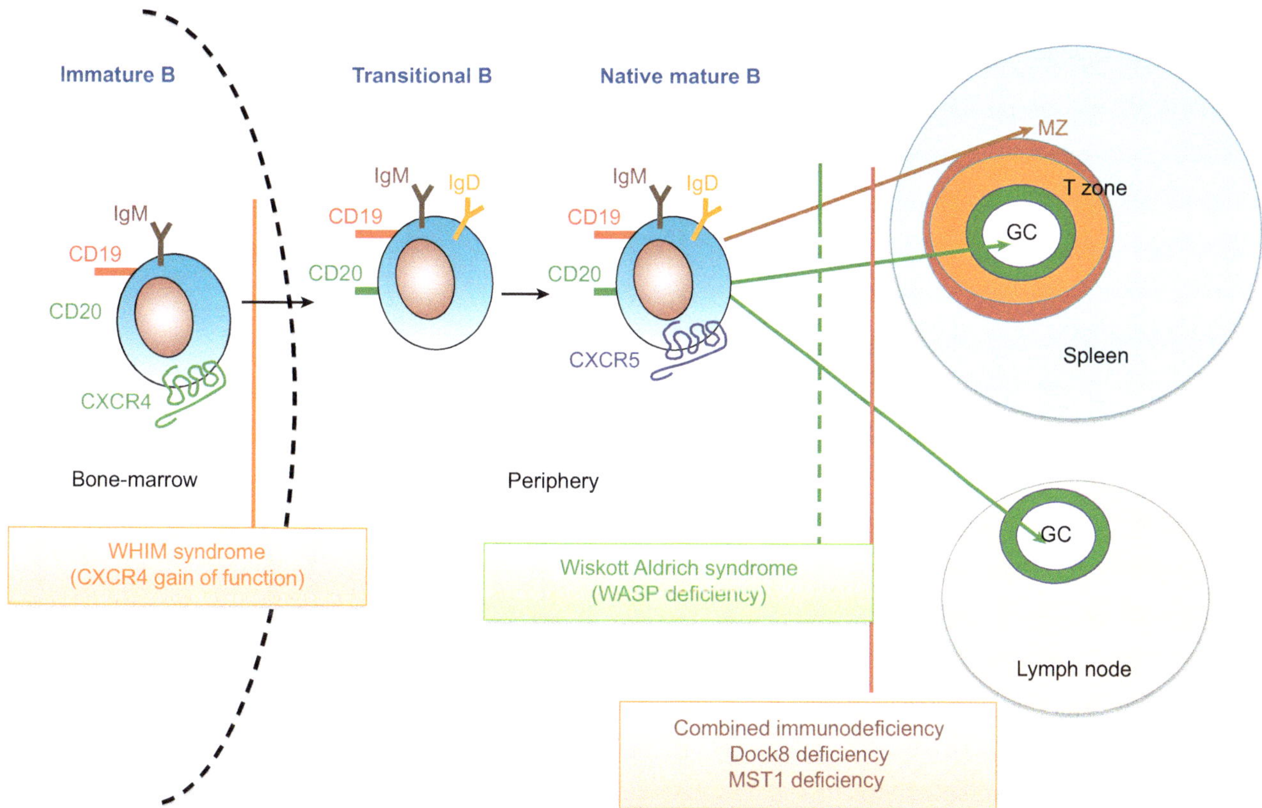


FIGURE 2 **Primary antibody deficiencies (PADs) caused by B cell migration defects.** Immature B cell egresses from the bone marrow after attenuation of CXCR4-CXCL12 interaction and matures in the periphery into a transitional B cell and then a mature naive B cell. Mature naive B cells undergo two different fates upon antigen activation: they either localize in the marginal zone (MZ), where they elicit early T-cell-independent antibody responses, or enter the follicles in response to migration signals (CXCR5-CXCL13) for T cell-dependent antibody response. Defects in migration leads to PADs. The dotted lines indicate an incomplete block in migration. GC, germinal center; WHIM, warts, hypogammaglobulinemia, infections, and myelokathexis; CXCR, CXC chemokine receptor.

receptor 4 (CXCR4). The interaction between CXCR4 and its ligand, the CXC chemokine ligand 12 (CXCL12), retains developing B cells in bone-marrow niches. As a consequence of increased CXCR4-CXCL12 interaction, WHIM syndrome patients have decreased B cell counts and hypogammaglobulinemia [36]. Interestingly, the delineation of the molecular basis of this syndrome led to the use of a CXCR4 inhibitor in patients, a treatment that appears efficient as judged by the increased numbers of circulating leukocytes, including those of B cells.

In addition to the WHIM syndrome, several CID syndromes lead to PAD as a result of defects in B cell migration, at least partly. The X-linked Wiskott–Aldrich syndrome (WAS) is caused by mutations in the *WAS* gene and is characterized by thrombocytopenia, eczema, and immune deficiency, generally diagnosed during the first few years after birth [37]. This disease occurs in about 1:200,000 births. Wiskott–Aldrich syndrome protein (WASP) is required for both adhesion and migration of hematopoietic cells (including B cells) and for actin polarization during the formation of the immunological synapse [38]. Although WASP is dispensable for B cell development, it has a crucial role in peripheral B cell homeostasis, development, and function. As a consequence, patients with *WAS* mutations have low IgM levels, low marginal zone (MZ) B cell counts, and high transitional B cell counts. These features are also observed in *Was*$^{-/-}$ mice as a consequence of defective BCR-mediated integrin signaling in B cells and a defective response to the chemokine CXCL12, which is involved in MZ localization of B cells [39]. The autosomal-recessive deficiency in the WASP-interacting protein (WIP) leads to a similar phenotype [40].

Other CID associated to a defect in B cell migration have been more recently described and their frequency in PAD is not yet exactly known. Deficiency in the cytoskeleton regulator, the dedicator of cytokinesis 8 (DOCK8), an AR syndrome, is characterized by susceptibility to recurrent sinus and pulmonary bacterial infections, viral infections, atopy, early-onset malignancies, and autoimmunity. Reported immunological abnormalities include T, NK, and B compartment defects: the B cell defect involves defective B cell differentiation into MZ B cells and impaired retention of B cells at the germinal centers (GCs) and abnormal BCR affinity maturation. Low numbers of CD27$^{+}$ memory B cells, low IgM (and IgA) serum levels, variable IgG antibody responses, and defective B cell responses to TLR activation are the main features of this PAD. Thus, DOCK8 deficiency results in a CID, and the PAD observed in this syndrome does not likely result only from defects in B cell migration, but also from defects in other B cell and T cell functions. By mutagenesis, N-ethyl-N-nitrosourea-generated mice exhibit a similar phenotype since Dock8-mutant B cells are unable to form MZ B cells or to persist in GCs and undergo affinity maturation, a defect likely linked to defective adhesion during the B cell immunological synapse [41–43].

Apart from proteins involved in cytoskeletal dynamics, mammalian sterile 20-like protein (MST1) has also been implicated in PAD pathogenesis [44]. MST1 has a role in lymphocyte adhesion, migration, and survival. Patients with MST1 deficiency present with a CID characterized by normal Ig levels but defective antibody production [45,46]. This AR B cell immunodeficiency may partially result from the defective localization of B cells in the MZ or the GCs, as MST1 is involved in the HIPPO pathway that regulates FOXO transcription factors and hence chemokine receptor expression. In the murine model, Mst1$^{-/-}$ lymphocytes exhibit an impairment of firm adhesion to endothelia resulting in an inefficient homing capacity. As a consequence, Mst1$^{-/-}$ mice present with defective egress of single positive thymocytes from thymus, hypotrophic peripheral lymphoid tissues, and reduced MZ B cells and dendritic cells in the spleen [47].

Although these B cell defects appear related to abnormal migration (or retention) of B cells, they can also be secondary to defective B cell activation, as virtually all molecules involved in this class of PADs have a role in both types of biological events.

## 3. DEFECTS IN B CELL SURVIVAL

Defects in B cell survival appear as B cell specific, affecting only the B cell compartment. Survival and homeostasis of B cells at discrete stages of their development are regulated by two cytokines—BAFF (B cell-activating factor) and APRIL (a proliferation-inducing ligand)—which are both produced by stromal and hematopoietic cells (Figure 3). BAFF binds to three receptors expressed on B cells (the BAFF receptor (BAFFR), the transmembrane activator and calcium modulator and cyclophilin ligand interactor (TACI) and the B cell maturation antigen (BCMA)), whereas APRIL binds to TACI and BCMA only. To date, only mutations in *TACI* and *BAFFR* have been reported as causing PADs in humans (Table 1). TACI deficiency leads to a subset of the so-called common variable immunodeficiency (CVID), characterized by an hypogammaglobulinemia affecting variably the three main isotypes, and frequently complicated by splenomegaly, autoimmunity, granuloma formation, and cancers. An AD as well as an AR (more rare) TACI deficiency has been reported. However, many of the *TACI* variants described in CVID patients are also observed in healthy controls and may well correspond to CVID predisposition genes rather than disease-causing mutations [48]. Interestingly, *Taci*$^{-/-}$ mice show increased B cell accumulation and marked splenomegaly. Isolated *Taci*$^{-/-}$ B cells hyperproliferate and produce increased amounts of Igs in vitro. In vivo antigen challenge results in enhanced T cell-dependent (TD) antibody production but decreased

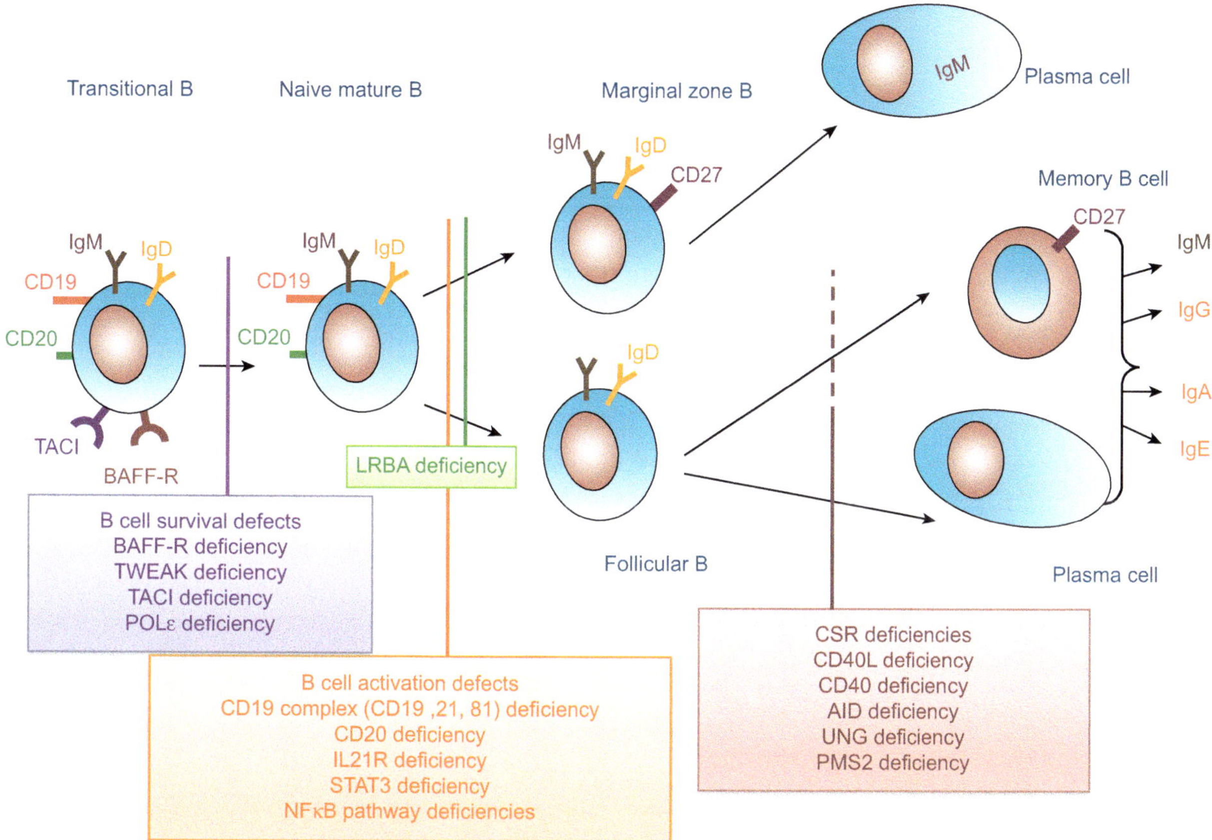


FIGURE 3 **Primary antibody deficiencies (PADs) caused by B cell survival and activation defects.** Survival of B cells in the periphery requires activation of the BCR, costimulatory molecules (the CD19 complex and CD20), and survival signals (via B cell-activating factor (BAFF) and a proliferation-inducing ligand). Upon antigen activation, marginal zone B cells rapidly differentiate into IgM-producing plasma cells and thus act as the first line of defense against systemic, blood-borne antigens. Follicular B cells in GCs undergo class switch recombination (CSR) and somatic hypermutation upon antigen and T cell activation before they differentiate into long-lived memory CD27+ B cells and plasma cells. Blockades on these pathways lead to PADs. The dotted lines indicate an incomplete block in activation. TACI, transmembrane activator and calcium modulator and cyclophilin ligand interactor; LRBA, LPS-responsive beige-like anchor; AID, activation-induced cytidine deaminase; UNG, Uracil-N glycosylase; PMS2, post-meiotic segregation 2; IL21R, IL-21 receptor; TWEAK, tumor necrosis factor-like weak inducer of apoptosis.

T cell-independent (TI) response. Thus, in this model TACI appears rather as an inhibitor for B cell activation that helps maintain immunological homeostasis [49]. Conversely, more recent data using transgenic mice expressing a missense mutation (already described in some patients) on the *Taci*$^{-/-}$ background indicate a phenotype resembling that of humans: mice present with low serum IgA levels and significantly impaired antibody responses to TI antigen. Their B cells are impaired in vitro in their capacity to proliferate and to secrete IgG1 and IgA in response to TACI ligation [50]. These two reports emphasize the controversial role for TACI in antibody responses.

In the few BAFFR-deficient patients reported to date, bi-allelic mutations in *BAFFR* appear to be disease causative, despite some phenotypic variability: patients have B cell lymphopenia, with low numbers of MZ, follicular and memory CD27+ B cells. This emphasizes the role of BAFF in the survival of transitional B cells during their differentiation into mature B cells [51]. In these patients, the TD antibody response to tetanus toxoid is normal but the TI response to polysaccharides is not. IgA levels are normal, indicating that the differentiation of mucosal IgA+ human B cells is independent of BAFFR signaling. However, the related immunopathology has low penetrance, indicating that very likely a combination of defects is required to lead to clinical and immunological consequences. The association of an heterozygous missense mutation in the gene encoding the TNF (tumor necrosis factor)-like weak inducer of apoptosis (TWEAK) with an AD hypogammaglobulinemia (Table 1) further supports the key role for BAFF signaling, as the TWEAK mutant appears to affect B cell survival through its integration with BAFF and the formation of noneffective BAFF complexes [52]. In Baff-R–deficient mice, B cells develop normally up to the stage of transitional B cells but cannot complete maturation in the spleen, as Baff/Baff-R–dependent survival signals are missing. Responses to TD and TI antigens are impaired [53]. As in humans, the mucosal IgA-secreting plasma cells develop normally in the gut.

## 4. DEFECTS IN B CELL ACTIVATION

### 4.1 Defects in BCR Costimulatory Molecules and BCR Signaling

Among the defects in B cell activation that lead to a PAD, some are B cell specific, affecting the CD19 complex that modulates the intensity and threshold of BCR signal transduction. CD19 is expressed together with CD21, CD81, and CD225 on the surface of B cells. CD19 and CD21 are B cell-specific receptors unlike CD81 and CD225, which are also expressed by other immune cells The complement receptor CD21 links the innate and adaptive immune systems by binding complement C3d, thus linking the CD19-complex signaling to the complement pathway [54]. CD19 and CD21 bind each other directly and as CD21 lacks an intracellular domain, it is thought that it signals through CD19, which possesses multiple tyrosine residues. The CD19 complex initiates $Ca^{2+}$ signaling following BCR stimulation. Since antigen recognition by antigen-specific BCR leads to the activation of naive B cells, which in turn differentiate into either plasmablasts or memory B cells, impaired BCR signaling leads to variable pan-hypogammaglobulinemia, although the numbers of circulating B cells remain normal (Table 1 and Figure 3). These AR diseases are typically diagnosed in childhood or early adulthood. Bi-allelic mutations have been reported in *CD19*, *CD21*, and *CD81* genes [55–57] leading to the same phenotype: defective expression of CD19 on B cells, decreased numbers of memory $CD27^+$ B cells, defective antibody response to TD antigens, and defective in vitro BCR-$Ca^{2+}$ signaling. In CD19-deficient mice, the TD responses are severely impaired and the number of $CD5^+$ B1 B cells is found decreased in the peritoneum [58]. Whereas human $CD5^+$ B cells in the blood might not be the functional equivalence of mouse B1 B cells, they are equally affected in CD19-deficient mice and patients.

Interestingly, biallelic mutations in the gene encoding CD20 do not affect BCR-induced $Ca^{2+}$ signaling, but still lead to partial hypogammaglobulinemia and strikingly normal antibody response to TD antigens but impaired response to TI polysaccharides [59], as observed also in $CD20^{-/-}$ mice [60]. The pathophysiological mechanism of CD20 deficiency remains so far unknown.

Although biallelic mutations in the delta subunit of protein kinase C (PKC) have been described as responsible for autoimmunity and lymphoproliferation [61], a recent report suggests that they could also be associated to a PAD: the described patient presents with lymphoid hyperplasia, splenomegaly, severe autoimmunity, and susceptibility to bacterial infections. The B cell defect is characterized by progressive B lymphopenia, defective serum IgG levels, and lack of TD antibody production [62]. PKCδ, a member of the PKC family involved in BCR-mediated signaling downstream of BTK and PLCγ2, is critical for B cell tolerance, and the corresponding knockout mouse shows immune-complex glomerulonephritis, splenomegaly, and lymphadenopathy associated with B cell expansion and defective B cell tolerance to self-antigen, but no typical PAD [63].

Increased intracellular signaling can also lead to a PAD. The AD condition PLAID (PLCγ2-associated antibody deficiency and immune dysregulation) is caused by deletion of the regulatory domain of PLCγ2 and provides an interesting example of how a mast cell- and B cell-associated immunodeficiency can cause both cold urticaria and autoimmunity [64]. In patients with PLAID, mutant PLCγ2 is constitutively active at low temperature in mast cells, hence causing cold urticaria associated with hyper-IgE syndrome. At 37 °C, PLCγ2 function is reduced, likely accounting for the mild B cell immunodeficiency [64]. More recently, a gain of function missense mutation in the autoinhibitory domain of PLCγ2 has been reported as being associated with mild pan-hypogammaglobulinemia and inflammatory manifestations [65].

### 4.2 Defects in T and B Cell Activation

Some immunodeficiencies that involve impaired B cell activation have been associated with genetic defects that affect both B cell- and T cell-signaling (Table 2 and Figure 3). For example, the centromeric instability, facial anomalies syndrome is caused by mutations in the genes that encode DNA methyl transferase DNMT3B [66] and ZBTB24 (a transcription factor possibly involved in DNA methylation [67]). This AR syndrome results in a CID mainly characterized by pan-hypogammaglobulinemia. B cells exhibit a naive phenotype, are prone to apoptosis in vitro, and are potentially self-reactive [68].

Following antigen binding to antigen receptors (such as the BCR), endoplasmic reticulum $Ca^{2+}$ stores are depleted, STIM1 is activated, and ORAI1–CRAC channels open, resulting in store-operated $Ca^{2+}$ entry. This influx results in the activation of the transcription factor NFAT. A defect in this pathway (as observed in the very few patients carrying biallelic mutations in the *STIM1* or *ORAI1* genes) leads to abnormal production of specific antibodies despite the presence of normal Ig levels and to autoimmunity against blood cells.

Mutations in the gene encoding the LPS-responsive beige-like anchor (LRBA) have been recently reported in several patients affected by an AR CID, hypogammaglobulinemia and inflammatory bowel disease [69], or autoimmunity [70]. Their B cells are intrinsically defective, as activation with CD40 ligand (CD40L) and cytokines leads to poor B cell survival, plasma blast generation, and Ig secretion in vitro. They also exhibit an increased susceptibility to apoptosis, which could be a consequence of defective autophagy. LRBA homologs, including LYST—found mutated in the Chediak-Higashi syndrome [71], which is, however, not associated with PAD—are known to function in lysosomal vesicles. The movement of lysosomal vesicles in which these proteins participate might be part of the autophagy process. However, the exact function of LRBA has yet to be defined.

Defects in B cell proliferation also result in a PAD: we recently reported that a homozygous missense mutation in the *POLE1* gene (encoding the catalytic subunit of Polε) caused a new disorder characterized by facial dysmorphism, immunodeficiency, livedo, and short stature (FILS syndrome). Patients suffer from recurrent bacterial infections, low IgM and IgG2 levels, lack of antibodies to TI polysaccharide antigens, and low memory $CD27^+$ B cells counts. In vitro B cells fail to proliferate and exhibit impaired G1- to S-phase progression because of partially defective DNA replication as capacity of DNA synthesis is limited [72]. This is likely to limit all steps of B cell proliferation whatever the trigger, cytokine and/or antigen, is (Table 2). The reason why POLε deficiency mostly results in a B cell deficiency is yet to be characterized. Animal models are not easy to test since $Pole^{-/-}$ mice die in utero whereas those with knockin allele resulting in loss of Polε proofreading but retained polymerase activity, die prematurely of intestinal adenomas and adenocarcinomas [73].

## 4.3 Defects in the NF-κB Pathway

Impairment of the canonical nuclear factor-κB (NF-κB) pathway leads to abnormal B cell activation. The BCR and CD40 trigger the canonical NF-κB pathway via activation of the IKK kinases IKKα, IKKβ, and IKKγ (also known as NEMO). IKK activation leads to phosphorylation of IκBα and the subsequent activation of the p50–p65 NF-κB heterodimer, which then translocates to the nucleus to regulate gene transcription. The scaffolding protein CARD11 links BCR and T cell receptor (TCR) signaling to the canonical NF-κB pathway, so it is not surprising that mutations in *CARD11* result in a type of CID. In detail, patients with *CARD11* mutations exhibit agammaglobulinemia and a striking increase in the numbers of transitional B cells [74] (Table 2, Figure 3). It is noteworthy that dominant gain-of-function mutations in *CARD11* cause a selective deficiency in polysaccharide-specific antibodies and an elevated transitional B cell count. This phenotype may be linked to constitutive NF-κB pathway activation, resulting in polyclonal B cell activation and B cell exhaustion [75]. The previously described $Card11^{-/-}$ mouse exhibits a selective defect in NF-κB activation, leading to diminished antigen receptor–mediated proliferation and defective cytokine production by B (and T) cells [76]. Recently, biallelic mutations affecting the CARD domain of MALT1 have been reported as responsible for a PAD with defective production of specific antibodies to proteinic and polysaccharidic antigens [77]. Although a T cell defect is certainly involved in the phenotype, an associated B cell subset abnormality (not studied) is, however, likely present, as suggested by the decreased number of MZ B cells observed in $malt1^{-/-}$ mice [78]. Hypomorphic mutations in either *NEMO* or *IKBKB* appear in patients with X-linked anhidrotic ectodermal dysplasia with immunodeficiency [79,80] and in a subset of patients with AR CID, respectively. These patients exhibit variable hypogammaglobulinemia as a result of a B cell activation defect. Moreover, AD EDA with immunodeficiency is caused by a gain of function mutation in the gene encoding the inhibitory protein IκBα, resulting in low serum IgG, low or normal IgA, and elevated serum IgM levels [81,82]. Strikingly, specific inactivation of *IKBKB* in follicular dendritic cells causes impairment of ICAM-1 expression and GC formation [83]—suggesting that in AD EDA, both an extrinsic B cell defect (migration) and an intrinsic B cell defect (activation) are intricately linked and result in the observed PAD. *NEMO* nonsense mutations are embryonic lethal in humans as in mice. Heterozygous female carriers can present with dermatopathy and a skewed pattern of X inactivation, especially in case of drastic *NEMO* mutations.

The AR deficiency in HOIL1 (also named RBCK1), a component of the linear ubiquitination chain assembly complex that interacts with NEMO and is essential for NF-κB activation, also results in a defective production of antipolysaccharide-specific antibodies, and hence an increased susceptibility to invasive bacterial infections and reduced numbers of memory B cells, and in some cases has been associated with a severe, autoinflammatory syndrome related to an innate immune defect [84]. Drastic mutations in *HOIL1* have been recently reported as a cause of progressive severe cardiomyopathy, emphasizing the ubiquitous spectrum of the NF-κB pathway [85]. $Hoil1^{-/-}$ deficient mice and cells derived from these mice exhibit a decreased NF-κB signaling induced by proinflammatory cytokines such as IL1 or TNFα [86].

## 4.4 Defects in Ig Class Switch Recombination

Two key steps in the maturation of B cells, namely class switch recombination (CSR) and somatic hypermutation (SHM), are achieved in GCs through close B cell–T cell cooperation involving CD40, which is constitutively expressed on B cells, and CD40L, which is transiently expressed on activated helper T cells and especially on the $T_{FH}$ cell subset. Following the establishment of B cell–T cell contacts, AID, a B cell-specific molecule, creates DNA lesions by deaminating cytosine to uracil into specific regions of the Ig locus (Switch S and V regions). Uracil-N glycosylase (UNG) recognizes and processes the AID-induced uracils in DNA. This step is followed (at least during CSR) by the generation of DNA double strand breaks (DSBs), the process of which is not completely elucidated. Among molecules that could be involved, post-meiotic segregation 2 (PMS2) appears to play a substantial role, likely through its endonuclease activity [87]. DSBs in S regions are sensed by ataxia telangiectasia mutated (ATM) and the MRN complex (MRE11–RAD50–NBS1). Repair is generally performed by the NHEJ enzymes, although an alternative pathway has been recently described [88]. This results in CSR toward IgG, IgA, and IgE. In V regions, the AID-UNG induced DNA lesions are repaired through the mismatch repair (MMR) enzymes and error-prone polymerases, resulting in SHM and further selection of B cells producing high-affinity antibodies [89] (Figure 4).

Defects in the CSR pathway (CSR deficiencies or CSR-D) result in a PAD characterized by normal or elevated serum IgM levels that contrast with very low levels of IgG, IgA, and IgE. These features may or may not (depending on the molecular defect) be associated with defective SHM, and are generally diagnosed during childhood. So far, no PAD related to isolated defective SHM has been reported.

Mutations in *CD40* lead to a rare AR deficiency characterized by a defect in both CSR and SHM [90]. As CD40 is constitutively expressed not only on B cells but on other cells including monocytes and dendritic cells and CD40/CD40L interaction is required for full T cell responses to antigens, patients present with an associated T cell defect (Table 2 and Figure 3). Ig levels are typical of a CSR-D, although a few patients exhibit normal (and sometimes even elevated) IgA levels, suggesting that compensatory mechanisms operate in the intestinal mucosae, possibly involving the APRIL–TACI pathway [91] or activation downstream of TLRs [19]. SHM are reduced on a decreased number of memory CD27+ B cells. However, certain SHM events take place in MZ B cells in some patients with CD40 deficiency [92], suggesting again a CD40-independent maturation pathway. CD40-deficient mice exhibit the very same phenotype [93]. As expected, the phenotype of CD40 deficiency is very similar to that of the most common CSR-D,

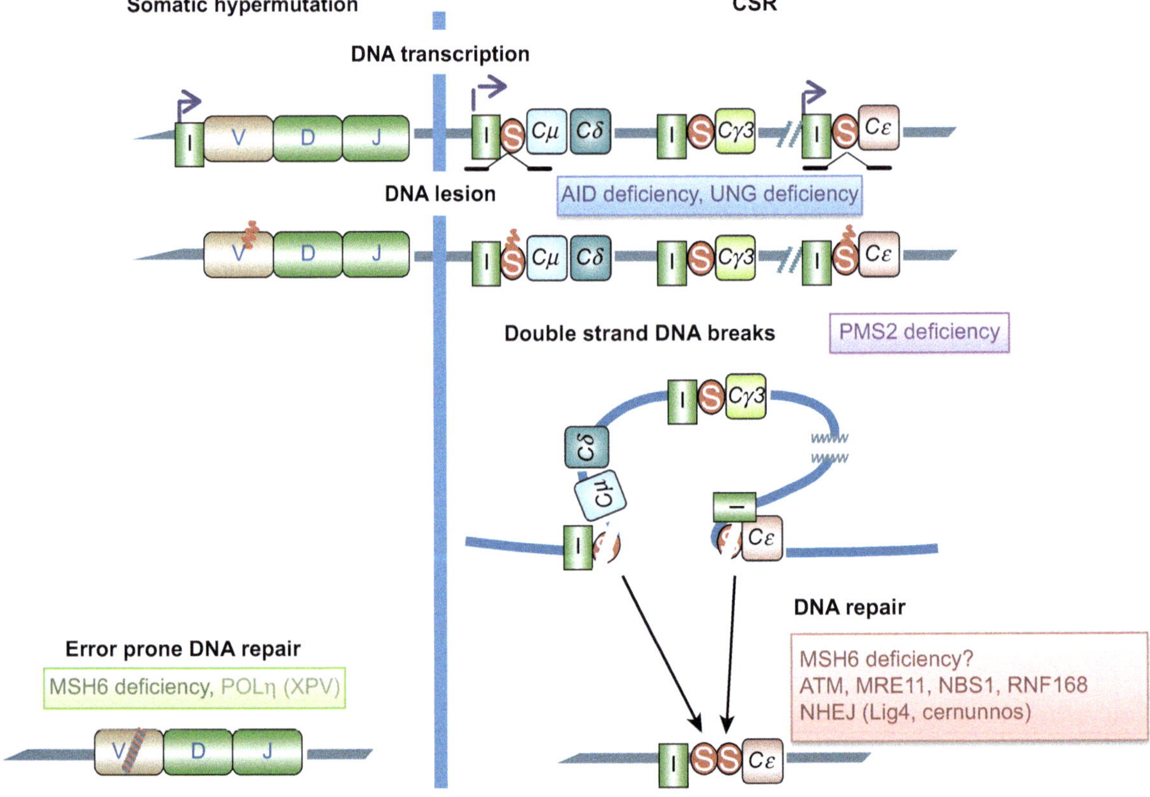


FIGURE 4 **Primary antibody deficiencies (PADs) caused by antibody maturation defects.** Class switch recombination (CSR) involves replacement of the Cμ region by a downstream Cx region from another class of Ig through a DNA recombination process. Somatic hypermutation (SHM) introduces mutations into the Ig V regions. Both processes share the first steps, i.e., transcription of the targeted region (switch region for CSR, V region for SHM) and introduction of a DNA lesion (AID (activation-induced cytidine deaminase)-induced uracils, thereafter uracil excision by uracil-N glycosylase (UNG)). CSR repair is achieved through DNA double strand break generation and repair mostly by the nonhomologous end joining (NHEJ) whereas SHM-repair requires the MMR and error-prone polymerases. The different blockades leading to PADs are shown. So far, no PAD related to isolated defective SHM has been reported. POL eta, error-prone polymerase; XPV, xeroderma pigmentosum variant; PMS2, post-meiotic segregation 2; MSH, Mut S homolog; ATM, ataxia telangiectasia mutated.

the X-linked form due to mutations in the gene encoding *CD40L* (CD154) that accounts for approximately 50% of all CSR-D cases. In this syndrome, B cells are intrinsically normal and are able to undergo CSR upon activation [94], but CD40L-deficient T cells differentiate poorly into $T_{H1}$ and $T_{FH}$ cells. As a consequence, immune cellular responses to opportunistic and viral infectious agents are impaired, as well as B–T cell interactions in GCs, resulting in impairment of GC formation, CSR, and SHM, although residual IgA levels and some SHM can, however, occur (Figure 3).

Other B cell-intrinsic defects can also lead to a CSR-D (Figure 4). For instance, a defect in AID (as observed in the AR AID deficiency or in *aid*$^{-/-}$ mice, which where contemporaneously described) induces a complete lack of both CSR and SHM, emphasizing the key role of AID in the two processes [95,96]. Very high levels of serum IgM with complete lack of switched isotypes is characteristic of this condition. Defective TD antibody IgG production is constant, but IgM antibodies to TI antigen (antipolysaccharide or isohemagglutinins) are normally present. Interestingly, CD27$^+$ expressing B cells are normal in numbers, although devoid of SHM, indicating that CD27 is rather a marker of activated than of memory B cells [97]. A hallmark of AID deficiency is the impressive lymphadenopathies with the presence of typical giant GCs, filled with highly proliferating GC founder cells (CD38$^+$, IgM$^+$ B cells). Autoimmunity is the main complication of this deficiency affecting almost 30% of patients, a complication also observed in aging *aid*$^{-/-}$ mice [98]. Patients present more often with autoimmune hemolytic anemia and thrombocytopenia, but also hepatitis or lupus erythematosus [99]. Both central and peripheral B cell tolerance have been shown to be defective in AID-deficient patients [100]. Such a complication came as a surprise since the development of autoimmune diseases has long been considered to be exclusively associated with hyper-mutated IgG antibodies.

Notably, biallelic mutations located in the carboxy-terminus of AID result in a complete CSR defect but do not affect SHM, which appears normal in frequency and nucleotide substitution pattern. This feature suggests that AID is not only a cytidine deaminase but also has a further role in CSR—possibly as a docking protein that recruits a putative CSR-specific cofactor [101,102]. Another interesting finding is that hemizygous nonsense mutations abrogating the AID nuclear export signal lead to a CSR deficiency [101], with normal SHM, that is transmitted as an AD trait and characterized by a milder phenotype than AR AID deficiency, very close to that observed in CVID [101].

UNG deficiency causes a very rare AR CSR-D associated with an abnormal pattern of SHM [103]. Four patients have been reported so far with susceptibility to bacterial infections, lymphadenopathies, and autoimmunity [103,104]. As UNG belongs to the ubiquitously expressed base excision repair pathway, patients are at risk for cancer complications, even if they have not been reported so far. Interestingly, *ung*$^{-/-}$ mice present a much milder CSR-D as compared with humans, suggesting a compensation by the MMR enzymes that do not occur as well in humans [105]. Nevertheless, they are highly susceptible to lymphomas when aging [106].

Patients with ataxia-telangiectasia resulting from mutations in *ATM* can present with a typical CSR-D with normal SHM, a phenotype comparable to that observed with AID Cter mutations, a similarity that can lead to misdiagnosis. Patients with ataxia-telangiectasia-like disease associated with mutations in *MRE11* or patients with Nijmegen breakage syndrome due to mutations in *NBS1* can develop a variable CSR-D. A CSR-D can also be a predominant abnormality in patients with hypomorphic mutations in genes encoding molecules of the NHEJ pathway, such as DNA ligase 4 or Cernunnos [107], although these mutations are associated with B and T lymphopenia (as NHEJ is required for the expression of the BCR and TCR). A very rare AR CSR-D is caused by mutations in the RNF168 gene (which also underlie the radiosensitivity, immunodeficiency, dysmorphic features and learning difficulties (RIDDLE) syndrome), which demonstrates the role of RNF168-dependent ubiquitination of DNA repair proteins [108]. Defects in the MMR complex (which are also known to lead to early onset occurrence of cancers) are associated with a variable CSR-D that mostly affects IgA and IgG subclasses (as observed in PMS2 and at a lesser degree in MutS homolog 6 (MSH6) deficiencies) [87,109]. Although the in vitro CSR is strongly impaired in B cells from patients with these DNA repair deficiencies, one cannot rule out the possibility that an associated T cell immunodeficiency (affecting the $T_{FH}$ subset) is responsible (at least in part) for the CSR-D observed in vivo.

## 4.5 Defects in Cytokine Signaling

Over recent years, IL-21 has emerged as a major inducer of human B cell proliferation and differentiation. IL-21 binds to the IL-21 receptor (IL21R) on B cells, which then induces JAK3 phosphorylation and STAT3 dimerization. IL21R signaling results in the generation of high-affinity antibodies and survival of memory B cells and plasma cells [13]. An IL21R deficiency (due to biallelic mutations in *IL21R* gene) has recently been described in two different families (Table 2). Although the phenotype of patients with *IL21R* mutations is reminiscent of that of patients with CID (namely, susceptibility to opportunistic infections), unlike CID, a combination of hypo-IgM and hyper-IgE is also noted [14]. IL-21R-deficient mice have normal lymphoid development, but exhibit after immunization a higher production of IgE, but lower IgG1, than wild-type animals [15].

IL-21 signals through STAT3, and a partial STAT3 deficiency underlies an AD hyper-IgE syndrome with increased susceptibility to recurrent cutaneous and mucosal bacterial infections (Table 2). This susceptibility to bacterial infections results partly from poor or absent antibody formation in response to immunization and reduced frequency of circulating memory CD27+ B cells that are commonly observed in these patients [13,110]. The hyper-IgE, which is a hallmark of this disease, might result from the physiological role of IL-21 in the inhibition of IgE synthesis. Given the broad function of STAT3 in many cell types and that STAT3 deficiency is embryonic lethal in mice, mice lacking STAT3 expression in specific cell lineages were generated: mice lacking STAT3 in B cells exhibit normal B cell development, have normal levels of serum IgM, IgG, and IgA, and produce normal levels of specific IgM but no IgG following immunization with T cell-dependent antigens. Interestingly, no hyper-IgE is observed [111]. Although these data point to the role of IL-21 in the delivery of survival and proliferation signals to B cells, IL-21 appears also as a major cytokine for generation and maintenance of $T_{FH}$ cells that are essential for T:B interaction within GC and antibody production.

SCIDs caused by mutations in the genes encoding the common cytokine receptor $\gamma$ chain ($\gamma c$) or JAK3 (Table 2) are characterized by normal circulating B cell counts and absence of T cells. $\gamma c$—and JAK3-deficient B cells fail to respond to IL-21 since IL21R includes the common $\gamma$chain. They are unable to produce high-affinity antibodies, even following the reversal of the T cell deficiency by allogeneic transplantation or gene therapy [112,113]. It is noteworthy that an impairment in IL-7 signaling caused by $\gamma c$ (or IL-7R$\alpha$) deficiency does not impair B cell development [5], pinpointing the distinct function of IL-7 in B cell development in humans and mice.

## 5. PADS WITH UNKNOWN ETIOLOGY

Whereas most of the early blocks in B cell differentiation have been defined in molecular terms, this is not often the case for more distal defects: 25% of CSR deficiencies currently lack a molecular understanding and most of the supposed genetic defects in CVID have yet to be identified. Most of selective Ig deficiencies, although likely caused by an intrinsic B cell defect, have not yet been molecularly defined.

### 5.1 Common Variable Immunodeficiency

CVID, the most frequent PAD, with an incidence of ~1 in 25,000 births, corresponds to a heterogeneous group of disorders, characterized by pan-hypogammaglobulinemia with defective antibody responses and in some cases associated with an increased incidence of granuloma, autoimmunity, and cancer [114,115]. CVID is generally diagnosed in adulthood, although some pediatric cases have been reported. Low numbers of circulating B cells and memory CD27+ B cells and abnormally high numbers of transitional B cells are associated with a bad prognosis. Ten to 20% of diagnosed patients have a family history of CVID or IgA deficiency (IgAD). Although researchers have attempted to determine the genetic variations associated with CVID, over 90% of these cases have not yet been defined on a molecular basis. Sporadic AD and AR forms have been reported. Recently, a genome-wide association approach showed that disease heterogeneity in a large cohort of patients could be related to several variations in the genetic background, including previously reported variations in genes of the MHC locus [116]. Polymorphisms in genes present in this locus (including those encoding TNF, MHC class II, complement factors, and the MMR MutS protein homolog 5 (MSH5)) are preferentially associated with CVID. As discussed above, individuals with some *TACI* variants also have predisposition to CVID [48]. As CVID has a late onset, it is most possibly caused by a combination of several gene variations rather than a defect in a single gene.

### 5.2 Selective Ig Deficiency

Similarly, although selective IgA deficiency (SIgAD) is the most common PAD, affecting near 1/1500 births, its genetic basis has not yet been defined. Most individuals are asymptomatic, suggesting effective compensation by secretory IgM in the mucosae [117], but approximately 25% of symptomatic patients have a family history of either SIgAD or CVID. Polymorphisms in *TACI* [118,119] or in MHC genes [120] have been reported in some cases, although the involvement of *TACI* variations has been controversial.

Selective IgG subclass deficiency is defined as a lack of one or more IgG subclasses with normal overall IgG levels. IgG2 deficiency is most commonly reported and is often associated with IgG4 deficiency and susceptibility to recurrent bacterial infections. IgG2 and IgG4 are barely detectable before the age of 2 years, and IgG4 can be absent in adult healthy females. The etiology of IgG2 deficiency remains ill-defined and it can be transient during childhood. IgG3 deficiency, which is less common, leads to susceptibility to recurrent bacterial infections and is often associated with another IgG subclass deficiency. Deletions of C regions have been reported in few patients [121] but do not account for the majority of cases, as residual amounts of the IgG subclasses are generally detectable.

Selective IgM deficiency (SIgMD) is a rare disorder that leads to recurrent infections, most frequently with encapsulated pathogens, from infancy onward. Its pathogenesis remains unclear. Deletion of the C$\mu$ can be ruled out, as patients' B cells express surface IgM.

Lastly, patients with selective polysaccharide antibody deficiency (SPAD) have normal Ig levels (including IgG2), but lack polysaccharide-specific antibodies, and are therefore prone to bacterial infections, and especially those caused by encapsulated bacteria (Table 1). Given that infants are unable to raise polysaccharide-specific responses before the age of 2, SPAD can be diagnosed only after this time point. MZ B cells are especially reactive to bacterial cell wall components, and splenectomized patients also suffer from SPAD [122]. Although most SPADs are not molecularly defined, NEMO deficiency [79], which impairs NF-κB signaling downstream of several pathways, including the TLR signaling pathways, gain of function mutations in *CARD11* and *HOIL1* deficiency can result in an SPAD, as discussed above. A leaky form of BTK deficiency also reportedly causes SPAD [27].

## 6. THERAPEUTIC APPROACHES

The approaches used for the treatment of PADs depend greatly on the disease severity. Almost all antibody-deficient patients require IgG replacement, with either intravenous or subcutaneous delivery. IgG replacement therapy has considerably limited infectious complications and chronic lung lesions such as bronchectasis. It is not, however, always sufficient, as the other Ig isotypes are not provided. IgM antibodies appear protective against encapsulated bacteria as a first means of defense in spleen MZ and in mucosal surfaces. We could show in a prospective study of a cohort of patients with agammaglobulinemia and CSR-D that received equal IgG replacement therapy that patients with CSR-D had a lower incidence of acute respiratory tract infections and bronchiectasis and a significantly lower risk of non-typable *H. influenzae* carriage [1], a feature that can be linked to the presence of anti-non-typable *H. influenza*-specific IgM antibodies in the serum and saliva. Interestingly, similar protection has been found in AID-deficient patients (who lack SHM) as well as in other patients with CSR-D, in whom SHM is normal. Thus, IgM antibodies that have been actively transported to mucosal surfaces [117] appear microbiologically and clinically protective against some microorganisms, irrespective of whether they have undergone SHM or not. This shows that IgG, which cannot be not transported to mucosal surfaces as it fails to bind the J chain, cannot fully substitute for the immunoglobulins (IgM and IgA) that can be transported at mucosal surfaces. Thus a long-term antimicrobial prophylaxis could be required in agammaglobulinemia in addition to Ig replacement therapy. Administration of antibiotics instead of IgG replacement might also be required, as in the case of patients with SIgMD. Notably, Ig replacement does not prevent complications such as lymphadenopathies, splenomegaly or autoimmunity. However, it could be useful to prevent epithelial cancers by reducing chronic infections.

The morbidity associated with hematopoietic stem cell transplantation (HSCT), the only curative therapy of PAD, means that this therapeutic strategy is restricted to the life-threatening B cell disorders that are generally also associated with a T cell deficiency. All PADs associated to a T cell defect, including CD40 and CD40L deficiencies, can certainly benefit from HSCT. Severe complications such as liver failure observed in these situations require liver transplantation, which should be done before HSCT.

Finally, gene therapy has been successfully performed in SCID related to ADA [123] and γc-deficiencies [113] and in WAS [124] and may also be a promising therapeutic strategy for B cell deficiencies in the future. In $Btk^{-/-}$ and $Tec^{-/-}$ double deficient mice (a mouse model of human XLA), transplantation of stem cells transduced with a lentiviral construct encoding Btk under B-cell specific promoters results in long-term B cells' rescue and antibody responses [125]. However, clinical applications of similar gene therapies will require a very strict assessment of safety issues.

## 7. CONCLUSION

The ongoing genetic definition of PADs, recently facilitated by new genetic screenings such as whole exome or genome sequencing, is essential for accurate diagnosis, including genetic counseling, prenatal diagnosis, and establishment of an appropriate follow-up and treatment. It also allows a better delineation of the complex molecular mechanisms underlying B cell development, migration, survival, and activation, even if all these different steps are not completely understood and obviously overlap. It seems that the genetic analysis of PADs is more complicated than what was previously anticipated: the limited genotype–phenotype correlation observed in many cases suggests that additional genetic and/or environmental factors have a role in PAD pathogenesis. This complexity is further increased by the fact that mutations in the same gene can lead to different phenotypes and even have different patterns of inheritance, depending on their location or on as-yet unknown genetic modifications. Lastly, whereas most primary immunodeficiencies are monogenic, the most prevalent form of PAD, the CVID, appears to have a more complex genetic background [48]. It is very likely that new genetic tools will in the near future help define modifier genes or epigenetic interactions responsible for such diseases.

## REFERENCES


[1] Micol R, Kayal S, Mahlaoui N, Beaute J, Brosselin P, Dudoit Y, et al. Protective effect of IgM against colonization of the respiratory tract by nontypeable *Haemophilus influenzae* in patients with hypogammaglobulinemia. J Allergy Clin Immunol 2012;129:770–7.

[2] Rakhmanov M, Keller B, Gutenberger S, Foerster C, Hoenig M, Driessen G, et al. Circulating $CD21_{low}$ B cells in common variable immunodeficiency resemble tissue homing, innate-like B cells. Proc Natl Acad Sci USA 2009;106:13451–6.

[3] Isnardi I, Ng YS, Menard L, Meyers G, Saadoun D, Srdanovic I, et al. Complement receptor 2/CD21- human naive B cells contain mostly autoreactive unresponsive clones. Blood 2010;115:5026–36.
[4] Matsumoto Y, Marusawa H, Kinoshita K, Endo Y, Kou T, Morisawa T, et al. *Helicobacter pylori* infection triggers aberrant expression of activation-induced cytidine deaminase in gastric epithelium. Nat Med 2007;13:470–6.
[5] Puel A, Ziegler SF, Buckley RH, Leonard WJ. Defective IL7R expression in $T^{-}B^{+}NK^{+}$ severe combined immunodeficiency. Nat Genet 1998;20:394–7.
[6] Dadi HK, Simon AJ, Roifman CM. Effect of CD3delta deficiency on maturation of alpha/beta and gamma/delta T-cell lineages in severe combined immunodeficiency. N Engl J Med 2003;349:1821–8.
[7] de Saint Basile G, Geissmann F, Flori E, Uring-Lambert B, Soudais C, Cavazzana-Calvo M, et al. Severe combined immunodeficiency caused by deficiency in either the delta or the epsilon subunit of CD3. J Clin Invest 2004;114:1512–7.
[8] Rieux-Laucat F, Hivroz C, Lim A, Mateo V, Pellier I, Selz F, et al. Inherited and somatic CD3zeta mutations in a patient with T-cell deficiency. N Engl J Med 2006;354:1913–21.
[9] Greenberg F, Crowder WE, Paschall V, Colon-Linares J, Lubianski B, Ledbetter DH. Familial DiGeorge syndrome and associated partial monosomy of chromosome 22. Hum Genet 1984;65:317–9.
[10] Elder ME, Lin D, Clever J, Chan AC, Hope TJ, Weiss A, et al. Human severe combined immunodeficiency due to a defect in ZAP-70, a T cell tyrosine kinase. Science 1994;264:1596–9.
[11] Huck K, Feyen O, Niehues T, Ruschendorf F, Hubner N, Laws HJ, et al. Girls homozygous for an IL-2-inducible T cell kinase mutation that leads to protein deficiency develop fatal EBV-associated lymphoproliferation. J Clin Invest 2009;119:1350–8.
[12] Hauck F, Randriamampita C, Martin E, Gerart S, Lambert N, Lim A, et al. Primary T-cell immunodeficiency with immunodysregulation caused by autosomal recessive LCK deficiency. J Allergy Clin Immunol 2012;130:1144–52. e11.
[13] Avery DT, Deenick EK, Ma CS, Suryani S, Simpson N, Chew GY, et al. B cell-intrinsic signaling through IL-21 receptor and STAT3 is required for establishing long-lived antibody responses in humans. J Exp Med 2010;207:155–71.
[14] Kotlarz D, Zietara N, Uzel G, Weidemann T, Braun CJ, Diestelhorst J, et al. Loss-of-function mutations in the IL-21 receptor gene cause a primary immunodeficiency syndrome. J Exp Med 2013;210:433–43.
[15] Ozaki K, Spolski R, Feng CG, Qi CF, Cheng J, Sher A, et al. A critical role for IL-21 in regulating immunoglobulin production. Science 2002;298:1630–4.
[16] Bossaller L, Burger J, Draeger R, Grimbacher B, Knoth R, Plebani A, et al. ICOS deficiency is associated with a severe reduction of CXCR5+CD4 germinal center Th cells. J Immunol 2006;177:4927–32.
[17] Qi H, Cannons JL, Klauschen F, Schwartzberg PL, Germain RN. SAP-controlled T-B cell interactions underlie germinal centre formation. Nature 2008;455:764–9.
[18] Puga I, Cols M, Barra CM, He B, Cassis L, Gentile M, et al. B cell-helper neutrophils stimulate the diversification and production of immunoglobulin in the marginal zone of the spleen. Nat Immunol 2011;13:170–80.
[19] Weller S, Bonnet M, Delagreverie H, Israel L, Chrabieh M, Marodi L, et al. $IgM^{+}IgD^{+}CD27^{+}$ B cells are markedly reduced in IRAK-4-, MyD88-, and TIRAP- but not UNC-93B-deficient patients. Blood 2012;120:4992–5001.
[20] Doulatov S, Notta F, Laurenti E, Dick JE. Hematopoiesis: a human perspective. Cell Stem Cell 2012;10:120–36.
[21] Conley ME, Dobbs AK, Farmer DM, Kilic S, Paris K, Grigoriadou S, et al. Primary B cell immunodeficiencies: comparisons and contrasts. Annu Rev Immunol 2009;27:199–227.
[22] Conley ME, Dobbs AK, Quintana AM, Bosompem A, Wang YD, Coustan-Smith E, et al. Agammaglobulinemia and absent B lineage cells in a patient lacking the p85alpha subunit of PI3K. J Exp Med 2012;209:463–70.
[23] Vetrie D, Vorechovsky I, Sideras P, Holland J, Davies A, Flinter F, et al. The gene involved in X-linked agammaglobulinaemia is a member of the src family of protein-tyrosine kinases. Nature 1993;361:226–33.
[24] Fruman DA. Phosphoinositide 3-kinase and its targets in B-cell and T-cell signaling. Curr Opin Immunol 2004;16:314–20.
[25] Lopez-Granados E, Perez de Diego R, Ferreira Cerdan A, Fontan Casariego G, Garcia Rodriguez MC. A genotype-phenotype correlation study in a group of 54 patients with X-linked agammaglobulinemia. J Allergy Clin Immunol 2005;116:690–7.
[26] Bykowsky MJ, Haire RN, Ohta Y, Tang H, Sung SS, Veksler ES, et al. Discordant phenotype in siblings with X-linked agammaglobulinemia. Am J Hum Genet 1996;58:477–83.
[27] Wood PM, Mayne A, Joyce H, Smith CI, Granoff DM, Kumararatne DS. A mutation in Bruton's tyrosine kinase as a cause of selective anti-polysaccharide antibody deficiency. J Pediatr 2001;139:148–51.
[28] Khan WN, Alt FW, Gerstein RM, Malynn BA, Larsson I, Rathbun G, et al. Defective B cell development and function in Btk-deficient mice. Immunity 1995;3:283–99.
[29] Ellmeier W, Jung S, Sunshine MJ, Hatam F, Xu Y, Baltimore D, et al. Severe B cell deficiency in mice lacking the tec kinase family members Tec and Btk. J Exp Med 2000;192:1611–24.
[30] Lagresle-Peyrou C, Six EM, Picard C, Rieux-Laucat F, Michel V, Ditadi A, et al. Human adenylate kinase 2 deficiency causes a profound hematopoietic defect associated with sensorineural deafness. Nat Genet 2009;41:106–11.
[31] Pannicke U, Honig M, Hess I, Friesen C, Holzmann K, Rump EM, et al. Reticular dysgenesis (aleukocytosis) is caused by mutations in the gene encoding mitochondrial adenylate kinase 2. Nat Genet 2009;41:101–5.
[32] Moshous D, Pannetier C, Chasseval Rd R, Deist Fl F, Cavazzana-Calvo M, Romana S, et al. Partial T and B lymphocyte immunodeficiency and predisposition to lymphoma in patients with hypomorphic mutations in Artemis. J Clin Invest 2003;111:381–7.
[33] Schwarz K, Gauss GH, Ludwig L, Pannicke U, Li Z, Lindner D, et al. RAG mutations in human B cell-negative SCID. Science 1996;274:97–9.
[34] Nelson ND, Bertuch AA. Dyskeratosis congenita as a disorder of telomere maintenance. Mutat Res 2012;730:43–51.
[35] Dickinson RE, Griffin H, Bigley V, Reynard LN, Hussain R, Haniffa M, et al. Exome sequencing identifies GATA-2 mutation as the cause of dendritic cell, monocyte, B and NK lymphoid deficiency. Blood 2011;118:2656–8.
[36] Hernandez PA, Gorlin RJ, Lukens JN, Taniuchi S, Bohinjec J, Francois F, et al. Mutations in the chemokine receptor gene CXCR4 are associated with WHIM syndrome, a combined immunodeficiency disease. Nat Genet 2003;34:70–4.
[37] Derry JM, Ochs HD, Francke U. Isolation of a novel gene mutated in Wiskott-Aldrich syndrome. Cell 1994;78:635–44.

[38] Westerberg L, Larsson M, Hardy SJ, Fernandez C, Thrasher AJ, Severinson E. Wiskott-Aldrich syndrome protein deficiency leads to reduced B-cell adhesion, migration, and homing, and a delayed humoral immune response. Blood 2005;105:1144–52.

[39] Westerberg LS, Dahlberg C, Baptista M, Moran CJ, Detre C, Keszei M, et al. Wiskott-Aldrich syndrome protein (WASP) and N-WASP are critical for peripheral B-cell development and function. Blood 2012;119:3966–74.

[40] Lanzi G, Moratto D, Vairo D, Masneri S, Delmonte O, Paganini T, et al. A novel primary human immunodeficiency due to deficiency in the WASP-interacting protein WIP. J Exp Med 2012;209:29–34.

[41] Zhang Q, Davis JC, Lamborn IT, Freeman AF, Jing H, Favreau AJ, et al. Combined immunodeficiency associated with DOCK8 mutations. N Engl J Med 2009;361:2046–55.

[42] Randall KL, Lambe T, Johnson AL, Treanor B, Kucharska E, Domaschenz H, et al. Dock8 mutations cripple B cell immunological synapses, germinal centers and long-lived antibody production. Nat Immunol 2009;10:1283–91.

[43] Jabara HH, McDonald DR, Janssen E, Massaad MJ, Ramesh N, Borzutzky A, et al. DOCK8 functions as an adaptor that links TLR-MyD88 signaling to B cell activation. Nat Immunol 2012;13:612–20.

[44] Pan D. The hippo signaling pathway in development and cancer. Dev Cell 2010;19:491–505.

[45] Nehme NT, Schmid JP, Debeurme F, Andre-Schmutz I, Lim A, Nitschke P, et al. MST1 mutations in autosomal recessive primary immunodeficiency characterized by defective naive T-cell survival. Blood 2012;119:3458–68.

[46] Abdollahpour H, Appaswamy G, Kotlarz D, Diestelhorst J, Beier R, Schaffer AA, et al. The phenotype of human STK4 deficiency. Blood 2012;119:3450–7.

[47] Katagiri K, Katakai T, Ebisuno Y, Ueda Y, Okada T, Kinashi T. Mst1 controls lymphocyte trafficking and interstitial motility within lymph nodes. EMBO J 2009;28:1319–31.

[48] Salzer U, Bacchelli C, Buckridge S, Pan-Hammarstrom Q, Jennings S, Lougaris V, et al. Relevance of biallelic versus monoallelic TNFRSF13B mutations in distinguishing disease-causing from risk-increasing TNFRSF13B variants in antibody deficiency syndromes. Blood 2009;113:1967–76.

[49] Yan M, Wang H, Chan B, Roose-Girma M, Erickson S, Baker T, et al. Activation and accumulation of B cells in TACI-deficient mice. Nat Immunol 2001;2:638–43.

[50] Bacchelli C, Buckland KF, Buckridge S, Salzer U, Schneider P, Thrasher AJ, et al. The C76R transmembrane activator and calcium modulator cyclophilin ligand interactor mutation disrupts antibody production and B-cell homeostasis in heterozygous and homozygous mice. J Allergy Clin Immunol 2011;127:1253–9. e13.

[51] Warnatz K, Salzer U, Rizzi M, Fischer B, Gutenberger S, Bohm J, et al. B-cell activating factor receptor deficiency is associated with an adult-onset antibody deficiency syndrome in humans. Proc Natl Acad Sci USA 2009;106:13945–50.

[52] Wang HY, Ma CA, Zhao Y, Fan X, Zhou Q, Edmonds P, et al. Antibody deficiency associated with an inherited autosomal dominant mutation in TWEAK. Proc Natl Acad Sci USA 2013;110:5127–32.

[53] Thompson JS, Bixler SA, Qian F, Vora K, Scott ML, Cachero TG, et al. BAFF-R, a newly identified TNF receptor that specifically interacts with BAFF. Science 2001;293:2108–11.

[54] Fearon DT, Carroll MC. Regulation of B lymphocyte responses to foreign and self-antigens by the CD19/CD21 complex. Annu Rev Immunol 2000;18:393–422.

[55] van Zelm MC, Reisli I, van der Burg M, Castano D, van Noesel CJ, van Tol MJ, et al. An antibody-deficiency syndrome due to mutations in the CD19 gene. N Engl J Med 2006;354:1901–12.

[56] van Zelm MC, Smet J, Adams B, Mascart F, Schandene L, Janssen F, et al. CD81 gene defect in humans disrupts CD19 complex formation and leads to antibody deficiency. J Clin Invest 2010;120:1265–74.

[57] Thiel J, Kimmig L, Salzer U, Grudzien M, Lebrecht D, Hagena T, et al. Genetic CD21 deficiency is associated with hypogammaglobulinemia. J Allergy Clin Immunol 2012;129:801–10. e6.

[58] Rickert RC, Rajewsky K, Roes J. Impairment of T-cell-dependent B-cell responses and B-1 cell development in CD19-deficient mice. Nature 1995;376:352–5.

[59] Kuijpers TW, Bende RJ, Baars PA, Grummels A, Derks IA, Dolman KM, et al. CD20 deficiency in humans results in impaired T cell-independent antibody responses. J Clin Invest 2010;120:214–22.

[60] Uchida J, Lee Y, Hasegawa M, Liang Y, Bradney A, Oliver JA, et al. Mouse CD20 expression and function. Int Immunol 2004;16:119–29.

[61] Kuehn HS, Niemela JE, Rangel-Santos A, Zhang M, Pittaluga S, Stoddard JL, et al. Loss-of-function of the protein kinase C delta (PKCdelta) causes a B-cell lymphoproliferative syndrome in humans. Blood 2013;121:3117–25.

[62] Salzer E, Santos-Valente E, Klaver S, Ban SA, Emminger W, Prengemann NK, et al. B-cell deficiency and severe autoimmunity caused by deficiency of protein kinase C delta. Blood 2013;121:3112–6.

[63] Miyamoto A, Nakayama K, Imaki H, Hirose S, Jiang Y, Abe M, et al. Increased proliferation of B cells and auto-immunity in mice lacking protein kinase Cdelta. Nature 2002;416:865–9.

[64] Ombrello MJ, Remmers EF, Sun G, Freeman AF, Datta S, Torabi-Parizi P, et al. Cold urticaria, immunodeficiency, and autoimmunity related to PLCG2 deletions. N Engl J Med 2012;366:330–8.

[65] Zhou Q, Lee GS, Brady J, Datta S, Katan M, Sheikh A, et al. A hypermorphic missense mutation in PLCG2, encoding phospholipase Cgamma2, causes a dominantly inherited autoinflammatory disease with immunodeficiency. Am J Hum Genet 2012;91:713–20.

[66] Xu GL, Bestor TH, Bourc'his D, Hsieh CL, Tommerup N, Bugge M, et al. Chromosome instability and immunodeficiency syndrome caused by mutations in a DNA methyltransferase gene. Nature 1999;402:187–91.

[67] de Greef JC, Wang J, Balog J, den Dunnen JT, Frants RR, Straasheijm KR, et al. Mutations in ZBTB24 are associated with immunodeficiency, centromeric instability, and facial anomalies syndrome type 2. Am J Hum Genet 2011;88:796–804.

[68] Blanco-Betancourt CE, Moncla A, Milili M, Jiang YL, Viegas-Pequignot EM, Roquelaure B, et al. Defective B-cell-negative selection and terminal differentiation in the ICF syndrome. Blood 2004;103:2683–90.

[69] Alangari A, Alsultan A, Adly N, Massaad MJ, Kiani IS, Aljebreen A, et al. LPS-responsive beige-like anchor (LRBA) gene mutation in a family with inflammatory bowel disease and combined immunodeficiency. J Allergy Clin Immunol 2012;130:481–8. e2.

[70] Lopez-Herrera G, Tampella G, Pan-Hammarstrom Q, Herholz P, Trujillo-Vargas CM, Phadwal K, et al. Deleterious mutations in LRBA are associated with a syndrome of immune deficiency and autoimmunity. Am J Hum Genet 2012;90:986–1001.

[71] Barbosa MD, Nguyen QA, Tchernev VT, Ashley JA, Detter JC, Blaydes SM, et al. Identification of the homologous beige and Chediak-Higashi syndrome genes. Nature 1996;382:262–5.

[72] Pachlopnik Schmid J, Lemoine R, Nehme N, Cormier-Daire V, Revy P, Debeurme F, et al. Polymerase epsilon1 mutation in a human syndrome with facial dysmorphism, immunodeficiency, livedo, and short stature. FILS syndrome J Exp Med 2012;209:2323–30.

[73] Albertson TM, Ogawa M, Bugni JM, Hays LE, Chen Y, Wang Y, et al. DNA polymerase epsilon and delta proofreading suppress discrete mutator and cancer phenotypes in mice. Proc Natl Acad Sci USA 2009;106:17101–4.

[74] Stepensky P, Keller B, Buchta M, Kienzler AK, Elpeleg O, Somech R, et al. Deficiency of caspase recruitment domain family, member 11 (CARD11), causes profound combined immunodeficiency in human subjects. J Allergy Clin Immunol 2013;131:477–85. e1.

[75] Snow AL, Xiao W, Stinson JR, Lu W, Chaigne-Delalande B, Zheng L, et al. Congenital B cell lymphocytosis explained by novel germline CARD11 mutations. J Exp Med 2012;209:2247–61.

[76] Hara H, Wada T, Bakal C, Kozieradzki I, Suzuki S, Suzuki N, et al. The MAGUK family protein CARD11 is essential for lymphocyte activation. Immunity 2003;18:763–75.

[77] Jabara HH, Ohsumi T, Chou J, Massaad MJ, Benson H, Megarbane A, et al. A homozygous mucosa-associated lymphoid tissue 1 (MALT1) mutation in a family with combined immunodeficiency. J Allergy Clin Immunol 2013;132:151–8.

[78] Ruland J, Duncan GS, Wakeham A, Mak TW. Differential requirement for Malt1 in T and B cell antigen receptor signaling. Immunity 2003;19:749–58.

[79] Zonana J, Elder ME, Schneider LC, Orlow SJ, Moss C, Golabi M, et al. A novel X-linked disorder of immune deficiency and hypohidrotic ectodermal dysplasia is allelic to incontinentia pigmenti and due to mutations in IKK-gamma (NEMO). Am J Hum Genet 2000;67:6.

[80] Hanson EP, Monaco-Shawver L, Solt LA, Madge LA, Banerjee PP, May MJ, et al. Hypomorphic nuclear factor-kappaB essential modulator mutation database and reconstitution system identifies phenotypic and immunologic diversity. J Allergy Clin Immunol 2008;122:1169–77. e16.

[81] Courtois G, Smahi A, Reichenbach J, Doffinger R, Cancrini C, Bonnet M, et al. A hypermorphic IkappaBalpha mutation is associated with autosomal dominant anhidrotic ectodermal dysplasia and T cell immunodeficiency. J Clin Invest 2003;112:1108–15.

[82] Ohnishi H, Miyata R, Suzuki T, Nose T, Kubota K, Kato Z, et al. A rapid screening method to detect autosomal-dominant ectodermal dysplasia with immune deficiency syndrome. J Allergy Clin Immunol 2012;129:578–80.

[83] Victoratos P, Lagnel J, Tzima S, Alimzhanov MB, Rajewsky K, Pasparakis M, et al. FDC-specific functions of p55TNFR and IKK2 in the development of FDC networks and of antibody responses. Immunity 2006;24:65–77.

[84] Boisson B, Laplantine E, Prando C, Giliani S, Israelsson E, Xu Z, et al. Immunodeficiency, autoinflammation and amylopectinosis in humans with inherited HOIL-1 and LUBAC deficiency. Nat Immunol 2012;13:1178–86.

[85] Nilsson J, Schoser B, Laforet P, Kalev O, Lindberg C, Romero NB, et al. Polyglucosan body myopathy caused by defective ubiquitin ligase RBCK1. Ann Neurol 2013;74:914–19.

[86] Ikeda F, Deribe YL, Skanland SS, Stieglitz B, Grabbe C, Franz-Wachtel M, et al. SHARPIN forms a linear ubiquitin ligase complex regulating NF-κB activity and apoptosis. Nature 2011;471:637–41.

[87] Peron S, Metin A, Gardes P, Alyanakian MA, Sheridan E, Kratz CP, et al. Human PMS2 deficiency is associated with impaired immunoglobulin class switch recombination. J Exp Med 2008;205:2465–72.

[88] Yan CT, Boboila C, Souza EK, Franco S, Hickernell TR, Murphy M, et al. IgH class switching and translocations use a robust nonclassical end-joining pathway. Nature 2007;449:478–82.

[89] Durandy A, Revy P, Fischer A. Human models of inherited immunoglobulin class switch recombination and somatic hypermutation defects (hyper-IgM syndromes). Adv Immunol 2004;82:295–330.

[90] Ferrari S, Giliani S, Insalaco A, Al-Ghonaium A, Soresina AR, Loubser M, et al. Mutations of CD40 gene cause an autosomal recessive form of immunodeficiency with hyper IgM. Proc Natl Acad Sci USA 2001;98:12614–9.

[91] Litinskiy MB, Nardelli B, Hilbert DM, He B, Schaffer A, Casali P, et al. DCs induce CD40-independent immunoglobulin class switching through BLyS and APRIL. Nat Immunol 2002;3:822–9.

[92] Weller S, Faili A, Garcia C, Braun MC, Le Deist FF, de Saint Basile GG, et al. CD40-CD40L independent Ig gene hypermutation suggests a second B cell diversification pathway in humans. Proc Natl Acad Sci USA 2001;98:1166–70.

[93] Kawabe T, Naka T, Yoshida K, Tanaka T, Fujiwara H, Suematsu S, et al. The immune responses in CD40-deficient mice: impaired immunoglobulin class switching and germinal center formation. Immunity 1994;1:167–78.

[94] Durandy A, Schiff C, Bonnefoy JY, Forveille M, Rousset F, Mazzei G, et al. Induction by anti-CD40 antibody or soluble CD40 ligand and cytokines of IgG, IgA and IgE production by B cells from patients with X-linked hyper IgM syndrome. Eur J Immunol 1993;23:2294–9.

[95] Revy P, Muto T, Levy Y, Geissmann F, Plebani A, Sanal O, et al. Activation-induced cytidine deaminase (AID) deficiency causes the autosomal recessive form of the Hyper-IgM syndrome (HIGM2). Cell 2000;102:565–75.

[96] Muramatsu M, Kinoshita K, Fagarasan S, Yamada S, Shinkai Y, Honjo T. Class switch recombination and hypermutation require activation-induced cytidine deaminase (AID), a potential RNA editing enzyme. Cell 2000;102:553–63.

[97] Klein U, Rajewsky K, Kuppers R. Human immunoglobulin (Ig) M+IgD+ peripheral blood B cells expressing the CD27 cell surface antigen carry somatically mutated variable region genes: CD27 as a general marker for somatically mutated (memory) B cells. J Exp Med 1998;188:1679–89.

[98] Hase K, Takahashi D, Ebisawa M, Kawano S, Itoh K, Ohno H. Activation-induced cytidine deaminase deficiency causes organ-specific autoimmune disease. PLoS One 2008;3:e3033.

[99] Durandy A, Cantaert T, Kracker S, Meffre E. Potential roles of activation-induced cytidine deaminase in promotion or prevention of autoimmunity in humans. Autoimmunity 2012;46:148–56.

[100] Meyers G, Ng YS, Bannock JM, Lavoie A, Walter JE, Notarangelo LD, et al. Activation-induced cytidine deaminase (AID) is required for B-cell tolerance in humans. Proc Natl Acad Sci USA 2011;108:11554–9.

[101] Imai K, Zhu Y, Revy P, Morio T, Mizutani S, Fischer A, et al. Analysis of class switch recombination and somatic hypermutation in patients affected with autosomal dominant hyper-IgM syndrome type 2. Clin Immunol 2005;115:277–85.

[102] Ta VT, Nagaoka H, Catalan N, Durandy A, Fischer A, Imai K, et al. AID mutant analyses indicate requirement for class-switch-specific cofactors. Nat Immunol 2003;4:843–8.

[103] Imai K, Slupphaug G, Lee WI, Revy P, Nonoyama S, Catalan N, et al. Human uracil-DNA glycosylase deficiency associated with profoundly impaired immunoglobulin class-switch recombination. Nat Immunol 2003;4:1023–8.

[104] aan de Kerk DJ, Jansen MH, ten Berge IJ, van Leeuwen EM, Kuijpers TW. Identification of B cell defects using age-defined reference ranges for in vivo and in vitro B cell differentiation. J Immunol 2013;190:5012–9.
[105] Rada C, Williams GT, Nilsen H, Barnes DE, Lindahl T, Neuberger MS. Immunoglobulin isotype switching is inhibited and somatic hypermutation perturbed in UNG-deficient mice. Curr Biol 2002;12:1748–55.
[106] Nilsen H, Stamp G, Andersen S, Hrivnak G, Krokan HE, Lindahl T, et al. Gene-targeted mice lacking the Ung uracil-DNA glycosylase develop B-cell lymphomas. Oncogene 2003;22:5381–6.
[107] Du L, Peng R, Bjorkman A, Filipe de Miranda N, Rosner C, Kotnis A, et al. Cernunnos influences human immunoglobulin class switch recombination and may be associated with B cell lymphomagenesis. J Exp Med 2012;209:291–305.
[108] Stewart GS, Panier S, Townsend K, Al-Hakim AK, Kolas NK, Miller ES, et al. The RIDDLE syndrome protein mediates a ubiquitin-dependent signaling cascade at sites of DNA damage. Cell 2009;136:420–34.
[109] Gardes P, Forveille M, Alyanakian MA, Aucouturier P, Ilencikova D, Leroux D, et al. Human MSH6 deficiency is associated with impaired antibody maturation. J Immunol 2012;188:2023–9.
[110] Minegishi Y, Saito M, Tsuchiya S, Tsuge I, Takada H, Hara T, et al. Dominant-negative mutations in the DNA-binding domain of STAT3 cause hyper-IgE syndrome. Nature 2007;448:1058–62.
[111] Fornek JL, Tygrett LT, Waldschmidt TJ, Poli V, Rickert RC, Kansas GS. Critical role for Stat3 in T-dependent terminal differentiation of IgG B cells. Blood 2006;107:1085–91.
[112] Recher M, Berglund LJ, Avery DT, Cowan MJ, Gennery AR, Smart J, et al. IL-21 is the primary common gamma chain-binding cytokine required for human B-cell differentiation in vivo. Blood 2011;118:6824–35.
[113] Hacein-Bey-Abina S, Hauer J, Lim A, Picard C, Wang GP, Berry CC, et al. Efficacy of gene therapy for X-linked severe combined immunodeficiency. N Engl J Med 2010;363:355–64.
[114] Chapel H, Lucas M, Patel S, Lee M, Cunningham-Rundles C, Resnick E, et al. Confirmation and improvement of criteria for clinical phenotyping in common variable immunodeficiency disorders in replicate cohorts. J Allergy Clin Immunol 2012;130:1197–8. e9.
[115] Resnick ES, Moshier EL, Godbold JH, Cunningham-Rundles C. Morbidity and mortality in common variable immune deficiency over 4 decades. Blood 2012;119:1650–7.
[116] Orange JS, Glessner JT, Resnick E, Sullivan KE, Lucas M, Ferry B, et al. Genome-wide association identifies diverse causes of common variable immunodeficiency. J Allergy Clin Immunol 2011;127:1360–7. e6.
[117] Brandtzaeg P, Johansen FE. Mucosal B cells: phenotypic characteristics, transcriptional regulation, and homing properties. Immunol Rev 2005;206:32–63.
[118] Salzer U, Chapel HM, Webster AD, Pan-Hammarstrom Q, Schmitt-Graeff A, Schlesier M, et al. Mutations in TNFRSF13B encoding TACI are associated with common variable immunodeficiency in humans. Nat Genet 2005;37:820–8.
[119] Castigli E, Wilson SA, Garibyan L, Rachid R, Bonilla F, Schneider L, et al. TACI is mutant in common variable immunodeficiency and IgA deficiency. Nat Genet 2005;37:829–34.
[120] Rioux JD, Goyette P, Vyse TJ, Hammarstrom L, Fernando MM, Green T, et al. Mapping of multiple susceptibility variants within the MHC region for 7 immune-mediated diseases. Proc Natl Acad Sci USA 2009;106:18680–5.
[121] Lefranc MP, Lefranc G, Rabbitts TH. Inherited deletion of immunoglobulin heavy chain constant region genes in normal human individuals. Nature 1982;300:760–2.
[122] Kruetzmann S, Rosado MM, Weber H, Germing U, Tournilhac O, Peter HH, et al. Human immunoglobulin M memory B cells controlling *Streptococcus pneumoniae* infections are generated in the spleen. J Exp Med 2003;197:939–45.
[123] Aiuti A, Cattaneo F, Galimberti S, Benninghoff U, Cassani B, Callegaro L, et al. Gene therapy for immunodeficiency due to adenosine deaminase deficiency. N Engl J Med 2009;360:447–58.
[124] Boztug K, Schmidt M, Schwarzer A, Banerjee PP, Diez IA, Dewey RA, et al. Stem-cell gene therapy for the Wiskott-Aldrich syndrome. N Engl J Med 2010;363:1918–27.
[125] Kerns HM, Ryu BY, Stirling BV, Sather BD, Astrakhan A, Humblet-Baron S, et al. B cell-specific lentiviral gene therapy leads to sustained B-cell functional recovery in a murine model of X-linked agammaglobulinemia. Blood 2010;115:2146–55.

Chapter 26

# IMGT® Immunoglobulin Repertoire Analysis and Antibody Humanization

Marie-Paule Lefranc

*IMGT®, The International ImMunoGeneTics Information System®, Université Montpellier 2, CNRS, Institut de Génétique Humaine, Montpellier, France*

## 1. IMGT® AND THE BIRTH OF IMMUNOINFORMATICS

IMGT®, the international ImMunoGeneTics information system® (http://www.imgt.org) [1], was created in 1989 by Marie-Paule Lefranc at Montpellier, France (CNRS and Université Montpellier 2). The founding of IMGT® marked the advent of immunoinformatics, a new science, which emerged at the interface between immunogenetics and bioinformatics. For the first time, immunoglobulin (IG) or antibody and T cell receptor (TR) variable (V), diversity (D), joining (J), and constant (C) genes were officially recognized as "genes" as well as the conventional genes [2–5]. This major breakthrough allowed genes and data of the complex and highly diversified adaptive immune responses to be managed in genomic databases and tools.

The adaptive immune response was acquired by jawed vertebrates (or *gnathostomata*) more than 450 million years ago and is found in all extant jawed vertebrate species from fishes to humans. It is characterized by a remarkable immune specificity and memory, which are properties of the B and T cells owing to an extreme diversity of their antigen receptors. The specific antigen receptors comprise the IG or antibodies of the B cells and plasmacytes [2], and the TR [3]. The IG recognize antigens in their native (unprocessed) form, whereas the TR recognize processed antigens, which are presented as peptides by the highly polymorphic major histocompatibility (MH, in humans HLA for human leukocyte antigens) proteins.

The potential antigen receptor repertoire of each individual is estimated to comprise about $2 \times 10^{12}$ different IG and TR, and the limiting factor is only the number of B and T cells that an organism is genetically programmed to produce [2,3]. This huge diversity results from the complex molecular synthesis of the IG and TR chains and, more particularly, of their variable domains (V-DOMAIN) which, at their N-terminal end, recognize and bind the antigens [2,3]. The IG and TR synthesis includes several unique mechanisms that occur at the DNA level: combinatorial rearrangements of the V, D, and J genes that code the V-DOMAIN (the V-(D)-J being spliced to the C gene that encodes the C-REGION in the transcript); exonuclease trimming at the ends of the V, D, and J genes; and random addition of nucleotides by the terminal deoxynucleotidyl transferase (TdT) that creates the junctional N-diversity regions, and later during B cell differentiation, for the IG, somatic hypermutations, and class or subclass switch [2,3].

IMGT® manages the diversity and complexity of the IG and TR and the polymorphism of the MH of humans and other vertebrates. IMGT® is also specialized in the other proteins of the immunoglobulin superfamily (IgSF) and MH superfamily (MhSF) and related proteins of the immune system (RPI) of vertebrates and invertebrates. IMGT® provides a common access to standardized data from genome, proteome, genetics, and two-dimensional (2D) and three-dimensional (3D) structures. IMGT® is the acknowledged high-quality integrated knowledge resource in immunogenetics for exploring immune functional genomics. IMGT® comprises seven databases (for sequences, genes, and 3D structures) [6–11], 17 online tools [12–27], and more than 15,000 pages of web resources (e.g., IMGT Scientific chart, IMGT Repertoire, IMGT Education>Aide-mémoire [28], The IMGT Immunoinformatics page) [1]. IMGT® is the global reference in immunogenetics and immunoinformatics [29–44]. Its standards have been endorsed by the World Health Organization–International Union of Immunological Societies (WHO-IUIS) Nomenclature Committee since 1995 (first IMGT® online access at the 9th International Congress of Immunology, San Francisco, CA, USA) [45,46] and the WHO International Nonproprietary Names (INN) Programme [47,48]. The accuracy and the consistency of the IMGT® data are based on IMGT-ONTOLOGY [49–51], the first, and so far unique, ontology for immunogenetics and immunoinformatics [49–68]. IMGT-ONTOLOGY

Molecular Biology of B Cells. http://dx.doi.org/10.1016/B978-0-12-397933-9.00026-6

manages the immunogenetics knowledge through diverse facets that rely on seven axioms: IDENTIFICATION, DESCRIPTION, CLASSIFICATION, NUMEROTATION, LOCALIZATION, ORIENTATION, and OBTENTION [50,51,55]. The concepts generated from these axioms led to the elaboration of the IMGT® standards that constitute the IMGT Scientific chart: e.g., IMGT® standardized keywords (IDENTIFICATION) [56], IMGT® standardized labels (DESCRIPTION) [57], IMGT® standardized gene and allele nomenclature (CLASSIFICATION) [58], IMGT unique numbering [59–64] and its standardized graphical 2D representation or IMGT Colliers de Perles [65–68] (NUMEROTATION).

This chapter first reviews the fundamental information generated from these IMGT-ONTOLOGY concepts, which led to the IMGT Scientific chart rules. It then describes the major IMGT® tools and databases used for immunoglobulin repertoire analysis and antibody engineering and humanization: IMGT/V-QUEST [12–17] for the analysis of rearranged nucleotide sequence with the results of the integrated IMGT/JunctionAnalysis [18,19]; IMGT/Automat [20,21] and IMGT/Collier-de-Perles tool [26]; IMGT/HighV-QUEST, the high-throughput version for Next-Generation Sequencing (NGS) [22,23]; IMGT/DomainGapAlign [9,24,25] for amino acid sequence analysis; IMGT/3Dstructure-DB for 3D structures [8–10]; and its extension, IMGT/2Dstructure-DB (for antibodies and other proteins for which the 3D structure is not available). IMGT® tools and databases run against IMGT reference directories built from sequences annotated in IMGT/LIGM-DB [6], the IMGT® nucleotide database (170,685 sequences from 335 species in June 2013), and from IMGT/GENE-DB [7], the IMGT® gene database (3107 genes and 4722 alleles from 17 species, of which there were 694 genes and 1420 alleles for *Homo sapiens* and 868 genes and 1318 alleles for *Mus musculus* in June 2013).

An interface, IMGT/mAb-DB [11], has been developed to provide an easy access to therapeutic antibody amino acid sequences (links to IMGT/2Dstructure-DB) and structures (links to IMGT/3Dstructure-DB, if 3D structures are available). IMGT/mAb-DB data include monoclonal antibodies (mAb, INN suffix –mab) (a –mab is defined by the presence of at least an IG variable domain) and fusion proteins for immune applications (FPIA, INN suffix –cept) (a –cept is defined by a receptor fused to a, Fc) from the WHO-INN program [47,48]. This database also includes a few composite proteins for clinical applications (CPCA) (e.g., protein or peptide fused to an Fc for only increasing their half-life, identified by the INN prefix ef–) and some RPI used, unmodified, for clinical applications.

The unified IMGT® approach is of major interest for bridging knowledge from IG repertoire in normal and pathologic situations [69–72], IG allotypes and immunogenicity [73–75], NGS repertoire [22,23], antibody engineering, and humanization [32,39–41,43,76–80].

# 2. FUNDAMENTAL INFORMATION FROM IMGT-ONTOLOGY CONCEPTS

## 2.1 IDENTIFICATION: IMGT® Standardized Keywords

More than 325 IMGT® standardized keywords (189 for sequences and 137 for 3D structures) were precisely defined [56]. They represent the controlled vocabulary assigned during the annotation process and allow standardized search criteria for querying the IMGT® databases and for the extraction of sequences and 3D structures. They have been entered in BioPortal at the National Center for Biomedical Ontology (NCBO) in 2010 (http://bioportal.bioontology.org/ontologies/1491).

Standardized keywords are assigned at each step of the molecular synthesis of an IG. Those assigned to a nucleotide sequence are found in the "DE" (definition) and "KW" (keyword) lines of the IMGT/LIGM-DB files [6]. They characterize, for instance, the gene type, the configuration type, and the functionality type [56]. There are six gene types: variable (V), diversity (D), joining (J), constant (C), conventional-with-leader, and conventional-without-leader. Four of them (V, D, J, and C) identify the IG and TR genes and are specific to immunogenetics. There are four configuration types: germline (for the V, D, and J genes before DNA rearrangement), rearranged (for the V, D, and J genes after DNA rearrangement), partially rearranged (for D gene after only one DNA rearrangement), and undefined (for the C gene and for the conventional genes, which do not rearrange). The functionality type depends on the gene configuration. The functionality type of genes in germline or undefined configuration is functional (F), ORF (for "open reading frame"), or pseudogene (P). The functionality type of genes in rearranged or partially-rearranged configuration is either productive (no stop codon in the V-(D)-J region and in-frame junction) or unproductive (stop codon(s) in the V-(D)-J region, and/or out-of-frame junction).

The 20 usual amino acids (AA) have been classified in 11 IMGT physicochemical classes (IMGT®, http://www.imgt.org, IMGT Education>Aide-mémoire>Amino acids). The amino acid changes are described according to the hydropathy (3 classes), volume (5 classes), and IMGT physicochemical classes (11 classes) [28]. For example Q1>E (+ + −) means that in the amino acid change (Q>E), the two amino acids at codon 1 belong to the same hydropathy (+) and volume (+) classes but to different IMGT physicochemical property ( ) classes [28]. Four types of AA changes are identified in IMGT®: very similar (+ + +), similar (+ + −, + − +), dissimilar (− − +, − + −, + − −), and very dissimilar (− − −).

## 2.2 DESCRIPTION: IMGT® Standardized Labels

More than 560 IMGT® standardized labels (277 for sequences and 285 for 3D structures) were precisely defined

[57]. They are written in capital letters (no plural) to be recognizable without creating new terms. Standardized labels assigned to the description of sequences are found in the "FT" (feature) lines of the IMGT/LIGM-DB files [6]. Querying these labels represents a big plus compared to the generalist databases (GenBank/European Nucleotide Archive (ENA)/DNA Data Bank of Japan (DDBJ)). Thus it is possible to query for the "CDR3-IMGT" of the human rearranged productive sequences of IG-Heavy-Gamma (e.g., 1733 CDR3-IMGT obtained, with their sequences at the nucleotide or amino acid level).

The core labels include V-REGION, D-REGION, J-REGION, and C-REGION, which correspond to the coding region of the V, D, J, and C genes. IMGT structure labels for human IG chains and domains and their correspondence with sequence labels are shown in Table 1. These labels are necessary for a standardized description of the IG sequences and structures in databases and tools.

Highly conserved amino acids at a given position in a V or C domain (Section 2.4) have IMGT labels [57]. Three of them are common to the V and C domains: 1st-CYS (cysteine C at position 23), CONSERVED-TRP (tryptophan W at position 41), and 2nd-CYS (C at position 104) [59–62,64]. Two others are characteristics of the V-DOMAIN and correspond to the first amino acid of the canonical motif F/W-G-X-G (where F is phenylalanine, W tryptophan, G glycine, and X any amino acid) encoded by the J-REGION: J-PHE or J-TRP (F or W at position 118) [59–61,64].

## 2.3 CLASSIFICATION: IMGT® Standardized Genes and Alleles

The IMGT-ONTOLOGY CLASSIFICATION axiom was the trigger of immunoinformatics' birth. Indeed, the IMGT® concepts of classification allowed us, for the first time, to classify the antigen receptor genes (IG and TR) for any

**TABLE 1** Immunoglobulin Receptor, Chain, and Domain Structure Labels and Correspondence with Sequence Labels

| | IG Structure Labels (IMGT/3DStructure-DB) | | | Sequence Labels (IMGT/LIGM-DB) |
|---|---|---|---|---|
| Receptor[a] | Chain[b] | Domain Description Type | Domain[c] | Region |
| IG-GAMMA-1_KAPPA | L-KAPPA | V-DOMAIN | V-KAPPA | V-J-REGION |
| | | C-DOMAIN | C-KAPPA | C-REGION |
| | H-GAMMA-1 | V-DOMAIN | VH | V-D-J-REGION |
| | | C-DOMAIN | CH1 | C-REGION[d] |
| | | C-DOMAIN | CH2 | |
| | | C-DOMAIN | CH3 | |
| IG-MU_LAMBDA | L-LAMBDA | V-DOMAIN | V-LAMBDA | V-J-REGION |
| | | C-DOMAIN | C-LAMBDA-1 | C-REGION |
| | H-MU | V-DOMAIN | VH | V-D-J-REGION |
| | | C-DOMAIN | CH1 | C-REGION[d] |
| | | C-DOMAIN | CH2 | |
| | | C-DOMAIN | CH3 | |
| | | C-DOMAIN | CH4[e] | |

*[a]Labels are shown for two examples of IG (Homo sapiens IgG1-kappa and IgM-lambda). An IG ("Receptor") is made of two identical heavy (H, for IG-HEAVY) chains and two identical light (L, for IG-LIGHT) chains ("Chain") and usually comprises 12 (e.g., IgG1) or 14 (e.g., IgM) domains. Each chain has an N-terminal V-DOMAIN (or V-(D)-J-REGION, encoded by the rearranged V-(D)-J genes), whereas the remaining of the chain is the C-REGION (encoded by a C gene). The IG C-REGION comprises one C-DOMAIN (C-KAPPA or C-LAMBDA) for the L chain, or several C-DOMAIN (CH) for the H chain [2].*

*[b]The kappa (L-KAPPA) or lambda (L-LAMBDA) light chains may associate to any heavy chain isotype (e.g., H-GAMMA-1, H-MU). In humans, there are nine isotypes, H-MU, H-DELTA, H-GAMMA-3, H-GAMMA-1, H-ALPHA1, H-GAMMA2, H-GAMMA-4, H-EPSILON, and H-ALPHA2 (listed in the order 5′–3′ in the IGH locus of the IGHC genes that encode the constant region of the heavy chains [2]) (IMGT® http//www.imgt.org, IMGT Repertoire).*

*[c]The IG V-DOMAIN includes VH (for the IG heavy chain) and VL (for the IG light chain). In higher vertebrates, the VL is V-KAPPA or V-LAMBDA, whereas in fishes, the VL is V-IOTA. The C-DOMAIN includes CH (for the IG heavy chain, the number of CH per chain depending on the isotype [2]) and CL (for the IG light chain). In higher vertebrates, the CL is C-KAPPA or C-LAMBDA, whereas in fishes, the CL is C-IOTA.*

*[d]The heavy chain C-REGION also includes the HINGE-REGION for the H-ALPHA, H-DELTA, and H-GAMMA chains and, for membrane IG (mIG), the CONNECTING-REGION (CO), TRANSMEMBRANE-REGION (TM), and CYTOPLASMIC-REGION (CY); for secreted IG (sIG), the C-REGION includes CHS instead of CO, TM, and CY.*

*[e]For H-MU and H-EPSILON.*

locus (e.g., immunoglobulin heavy (IGH), T cell receptor alpha (TRA)), for any gene configuration (germline, undefined, or rearranged), and for any species (from fishes to humans). In higher vertebrates, there are seven IG and TR major loci (other loci correspond to chromosomal orphon sets, genes of which are orphons, not used in the IG or TR chain synthesis). The IG major loci include the immunoglobulin heavy (IGH), and for the light chains, the immunoglobulin kappa (IGK) and the immunoglobulin lambda (IGL) in higher vertebrates and the immunoglobulin iota (IGI) in fishes (IMGT®, http://www.imgt.org, IMGT Repertoire).

Since the creation of IMGT® in 1989, at New Haven during the 10th Human Genome Mapping Workshop (HGM10), the standardized classification and nomenclature of the IG and TR of humans and other vertebrate species have been under the responsibility of the IMGT Nomenclature Committee (IMGT-NC).

IMGT® gene and allele names are based on the concepts of classification of "Group," "Subgroup," "Gene," and "Allele" [58]. "Group" allows classification of a set of genes that belong to the same multigene family, within the same species or between different species. For example, there are 10 groups for the IG of higher vertebrates: IGHV, IGHD, IGHJ, IGHC, IGKV, IGKJ, IGKC, IGLV, IGLJ, and IGLC. "Subgroup" allows identification of a subset of genes that belong to the same group and that, in a given species, share at least 75% identity at the nucleotide level, e.g., *Homo sapiens* IGHV1 subgroup. Subgroups, genes, and alleles are always associated to a species name. An allele is a polymorphic variant of a gene, which is characterized by the mutations of its sequence at the nucleotide level, identified in its core sequence, and compared to the gene allele reference sequence, designated as allele *01. For example, *Homo sapiens* IGHV1-2*01 is the allele *01 of the *Homo sapiens* IGHV1-2 gene that belongs to the *Homo sapiens* IGHV1 subgroup, which itself belongs to the IGHV group. For the IGH locus, the constant genes are designated by the letter (and eventually number) corresponding to the encoded isotypes (IGHM, IGHD, IGHG3...), instead of using the letter C. IG and TR genes and alleles are not italicized in publications. IMGT-ONTOLOGY concepts of classification have been entered in the NCBO BioPortal.

The IMGT® IG and TR gene names [2–5] were approved by the Human Genome Organisation (HUGO) Nomenclature Committee (HGNC) in 1999 [81,82] and were endorsed by the WHO-IUIS Nomenclature Subcommittee for IG and TR [45,46]. The IMGT® IG and TR gene names are the official international reference and, as such, have been entered in IMGT/GENE-DB [7], in the Genome Database (GDB) [83], in LocusLink at the National Center for Biotechnology Information (NCBI) USA [84], in Entrez Gene (NCBI) when this database (now designated as "Gene") superseded LocusLink [85], in NCBI MapViewer, in Ensembl at the European Bioinformatics Institute (EBI) [86], and in the Vertebrate Genome Annotation (Vega) Browser [87] at the Wellcome Trust Sanger Institute (UK). HGNC, Gene NCBI, Ensembl, and Vega have direct links to IMGT/GENE-DB [7]. IMGT® human IG and TR genes were also integrated in IMGT-ONTOLOGY on the NCBO BioPortal and, on the same site, in the HUGO ontology and in the National Cancer Institute (NCI) Metathesaurus. Amino acid sequences of human IG and TR constant genes (e.g., *Homo sapiens* IGHM, IGHG1, IGHG2) were provided to UniProt in 2008. Since 2007, IMGT® gene and allele names have been used for the description of the therapeutic mAb and FPIA of the WHO-INN program [47,48].

## 2.4 NUMEROTATION: IMGT Unique Numbering and IMGT Collier de Perles

The IMGT-ONTOLOGY NUMEROTATION axiom is acknowledged as the "IMGT® Rosetta stone" that has bridged the biological and computational spheres in bioinformatics [37]. The IMGT® concepts of numerotation comprise the IMGT unique numbering [59–64] and its graphical 2D representation, the IMGT Collier de Perles [65–68]. Developed for and by the "domain," these concepts integrate sequences, structures, and interactions into a standardized domain-centric knowledge for functional genomics. The IMGT unique numbering has been defined for the variable V domain (V-DOMAIN of the IG and TR, and V-LIKE-DOMAIN of IgSF other than IG and TR) [59–61], the constant C domain (C-DOMAIN of the IG and TR, and C-LIKE-DOMAIN of IgSF other than IG and TR) [62], and for information, not as the focus of that chapter, the groove G domain (G-DOMAIN of the MH, and G-LIKE-DOMAIN of MhSF other than MH) [63,88,89]. Thus the IMGT unique numbering and IMGT Collier de Perles provide a definitive and universal system across species, including invertebrates, for the sequences and structures of the V, C, and G domains of IG, TR, MH, IgSF, and MhSF [64,68,90,91].

### *2.4.1 V Domain IMGT® Definitive System*

#### 2.4.1.1 V Domain Definition and Main Characteristics

In the IMGT® definitive system, the V domain includes the V-DOMAIN of the IG and of the TR, which corresponds to the V-J-REGION or V-D-J-REGION encoded by V-(D)-J rearrangements [2,3], and the V-LIKE-DOMAIN of the IgSF other than IG and TR. The V domain description of any receptor, any chain, and any species is based on the IMGT unique numbering for V domain (V-DOMAIN and V-LIKE-DOMAIN) [59–61,64].

A V domain (Figure 1) comprises about 100 amino acids and is made of nine antiparallel beta strands (A, B, C, C′, C″, D, E, F, and G) linked by beta turns (AB, CC′, C″D,

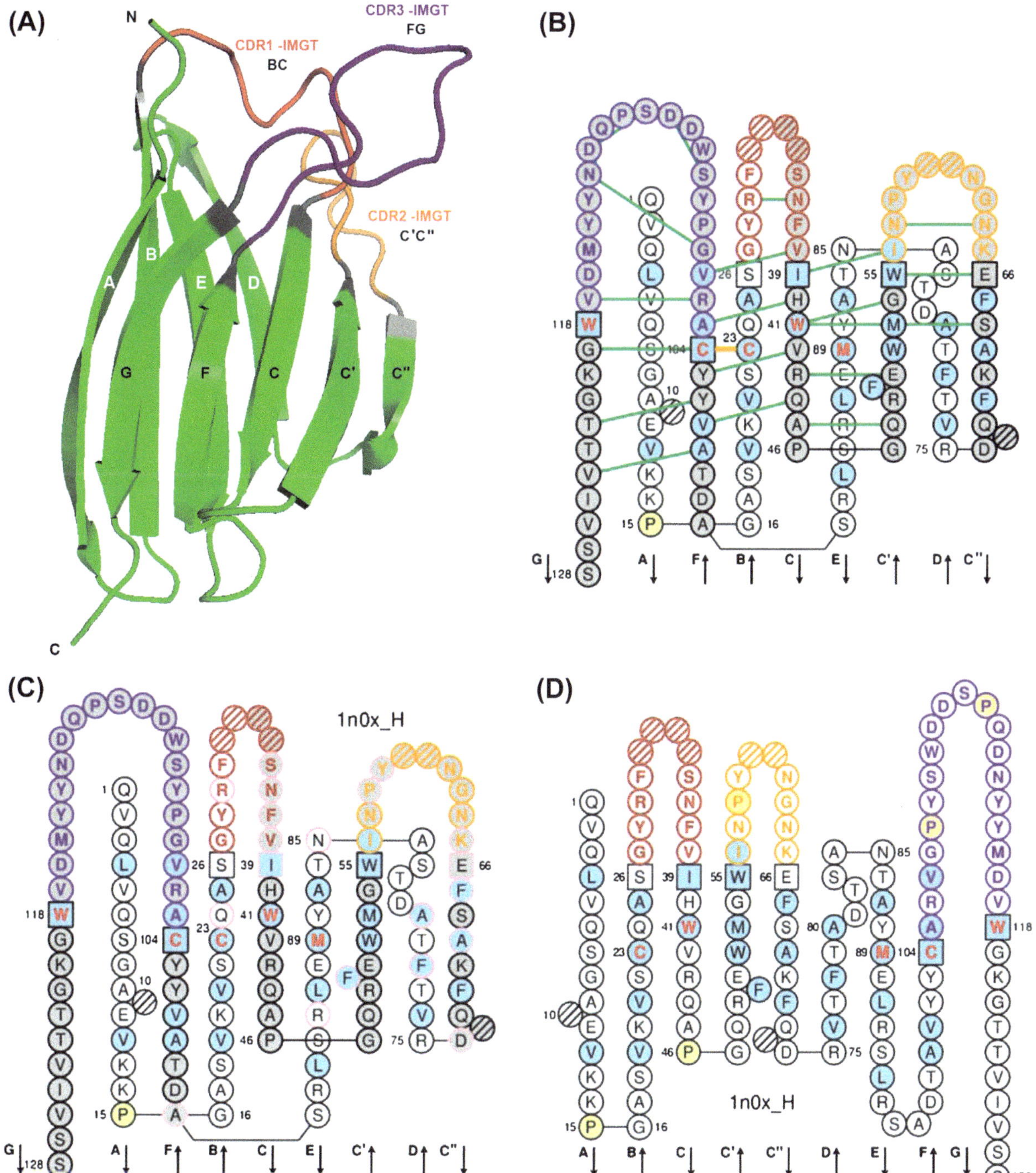


**FIGURE 1 Variable (V) domain. An IG VH (V-DOMAIN) is shown as example.** (A) 3D structure ribbon representation with the IMGT strand and loop delimitations [61]. (B) IMGT Collier de Perles on two layers with hydrogen bonds. The IMGT Collier de Perles on two layers shows, in the forefront, the GFCC′C″ strands (forming the sheet located at the interface VH/VL of the IG) and, in the back, the ABED strands. The IMGT Collier de Perles with hydrogen bonds (green lines online, only shown here for the GFCC′C″ sheet) is generated by the IMGT/Collier-de-Perles tool integrated in IMGT/3Dstructure-DB, from experimental 3D structure data [8–10]. (C) IMGT Collier de Perles on two layers generated from IMGT/DomainGapAlign [9,24,25]. Pink circles (online) indicate amino acid changes compared to the closest genes and alleles from the IMGT reference directory. (D) IMGT Collier de Perles on one layer. Amino acids are shown in the one-letter abbreviation. All proline (P) are shown online in yellow. IMGT anchors are in squares. Hatched circles are IMGT gaps according to the IMGT unique numbering for V domain [61,64]. Positions with bold (online red) letters indicate the four conserved positions that are common to a V domain and to a C domain: 23 (1st-CYS), 41 (CONSERVED-TRP), 89 (hydrophobic), 104 (2nd-CYS) [59–62,64], and the fifth conserved position, 118 (J-TRP or J-PHE), which is specific to a V-DOMAIN and belongs to the motif F/W-G-X-G that characterizes the J-REGION [61,64] (Table 2). The hydrophobic amino acids (hydropathy index with positive value: I, V, L, F, C, M, A) and tryptophan (W) [28] found at a given position in more than 50% of sequences are shown (online with a blue background color). Arrows indicate the direction of the beta strands and their designations in 3D structures. IMGT color menu for the CDR-IMGT of a V-DOMAIN indicates the type of rearrangement, V-D-J (for a VH here, red, orange, and purple) or V-J (for V-KAPPA or V-LAMBDA (not shown), blue, green, and green-blue) [2]. The identifier of the chain to which the VH domain belongs is 1n0x_H (from the *Homo sapiens* b12 Fab) in IMGT/3Dstructure-DB (http://www.imgt.org). The CDR-IMGT lengths of this VH are [8.8.20] and the FR-IMGT are [25.17.38.11]. The 3D ribbon representation was obtained using PyMOL (http://www.pymol.org) and "IMGT numbering comparison" of 1n0x_H (VH) from IMGT/3Dstructure-DB (http://www.imgt.org).

DE, and EF) and three loops (BC, C′C″, and FG), forming a sandwich of two sheets [ABED] [GFCC′C″] [59–61,64]. The sheets are closely packed against each other through hydrophobic interactions, giving a hydrophobic core, and joined together by a disulfide bridge between a first highly conserved cysteine (1st-CYS) in the B strand (in the first sheet) and a second, equally conserved cysteine (2nd-CYS) in the F strand (in the second sheet) [59–61,64].

#### 2.4.1.2 V Domain Strands and Loops (FR-IMGT and CDR-IMGT)

The V domain strands and loops and their delimitations and lengths, based on the IMGT unique numbering for V domain [59–61,64], are shown in Table 2. In the IG and TR V-DOMAIN, the three hypervariable loops BC, C′C″, and FG involved in the ligand recognition (native antigen for IG and pMH for TR) are designated complementarity determining regions (CDR-IMGT), whereas the strands form the framework region (FR-IMGT), which includes FR1-IMGT, FR2-IMGT, FR3-IMGT, and FR4-IMGT (Table 2). In the IMGT® definitive system, the CDR-IMGT have accurate and unambiguous delimitations, in contrast to the CDR described in the literature. Correspondences between the IMGT unique numbering with other numberings, e.g., Kabat [92] or Chothia [93], are available in the IMGT Scientific chart. The correspondences with these previous and heterogenous numberings are useful for the interpretation of previously published data, but nowadays the usage of these numberings has become obsolete in regard to the development of immunoinformatics based on the IMGT® standards [59–68] (IMGT® http://www.imgt.org, IMGT Scientific chart>Numbering>Correspondence between V numberings).

For a V domain, the BC loop (or CDR1-IMGT in a V-DOMAIN) encompasses positions 27–38, the C′C″ loop (or CDR2-IMGT in a V-DOMAIN) positions 56–65, and the FG loop (or CDR3-IMGT) positions 105–117. In a V-DOMAIN, the CDR3-IMGT encompasses the V-(D)-J junction that results from a V-J or V-D-J rearrangement [2,3] and is more variable in sequence and length than the CDR1-IMGT and CDR2-IMGT that are encoded by the V gene region only. For CDR3-IMGT of length >13 AA,

**TABLE 2** V Domain Strands and Loops, IMGT Positions, and Lengths, Based on the IMGT Unique Numbering for V Domain (V-DOMAIN and V-LIKE-DOMAIN)

| V Domain Strands and Loops[a] | IMGT Position[b] | Lengths[c] | Characteristic IMGT Residue@ Position[d] | V-DOMAIN FR-IMGT and CDR-IMGT |
|---|---|---|---|---|
| A-STRAND | 1–15 | 15 (14 if gap at 10) | | FR1-IMGT |
| B-STRAND | 16–26 | 11 | 1st-CYS 23 | |
| BC-LOOP | 27–38 | 12 (or less) | | CDR1-IMGT |
| C-STRAND | 39–46 | 8 | CONSERVED-TRP 41 | FR2-IMGT |
| C′-STRAND | 47–55 | 9 | | |
| C′C″-LOOP | 56–65 | 10 (or less) | | CDR2-IMGT |
| C″-STRAND | 66–74 | 9 (or 8 if gap at 73) | | FR3-IMGT |
| D-STRAND | 75–84 | 10 (or 8 if gaps at 81, 82) | | |
| E-STRAND | 85–96 | 12 | Hydrophobic 89 | |
| F-STRAND | 97–104 | 8 | 2nd-CYS 104 | |
| FG-LOOP | 105–117 | 13 (or less, or more) | | CDR3-IMGT |
| G-STRAND | 118–128 | 11 (or 10) | V-DOMAIN J-PHE 118 or J-TRP 118[e] | FR4-IMGT |

*[a]IMGT® labels (concepts of description) are written in capital letters (no plural) [57]. Beta turns (AB, CC′, C″D, DE, or EF) are individualized only if they have additional amino acids compared to the standard description. If not, they are included in the strands.*
*[b]Based on the IMGT unique numbering for V domain (V-DOMAIN and V-LIKE-DOMAIN) [59–61,64].*
*[c]In number of amino acids (or codons).*
*[d]IMGT Residue@Position is a given residue (usually an amino acid) or a given conserved property amino acid class, at a given position in a domain, based on the IMGT unique numbering [64].*
*[e]In the IG and TR V-DOMAIN, the G-STRAND (or FR4-IMGT) is the C-terminal part of the J-REGION, with J-PHE or J-TRP 118 and the canonical motif F/W-G-X-G at positions 118–121 [2,3]. The JUNCTION refers to the CDR3-IMGT plus the two anchors 2nd-CYS 104 and J-PHE or J-TRP 118 [60,61]. The JUNCTION (positions 104–118) is therefore two amino acids longer than the corresponding CDR3-IMGT (positions 105–117) [60,61].*

additional IMGT positions are added at the top of the loop between 111 and 112 (Table 3).

#### 2.4.1.3 IMGT Colliers de Perles

The loop and strands are visualized in the IMGT Colliers de Perles [65–68], which can be displayed on one layer (closer to the amino acid sequence) or on two layers (closer to the 3D structure) (Figure 1). The three loops, BC, C′C″, and FG (or CDR1-IMGT, CDR2-IMGT, and CDR3-IMGT for a V-DOMAIN) are delimited by the IMGT anchors, which are shown in squares in the IMGT Colliers de Perles. IMGT anchors are positions that belong to strands and represent anchors for the loops of the V domains. IMGT anchors are the key and original concept of IMGT® that definitively solved the ambiguous situation of different CDR lengths and delimitations found in the literature. The six anchors of a V domain are positions 26 and 39 (anchors of the BC loop or CDR1-IMGT in V-DOMAIN), 55 and 66 (anchors of the C′C″ loop or CDR2-IMGT in V-DOMAIN), and 104 and 118 (anchors of the FG loop or CDR3-IMGT in V-DOMAIN). The CDR3-IMGT anchors are highly conserved; they are C104 (2nd-CYS, in F strand) and F118 or W118 (J-PHE or J-TRP, in G strand). The JUNCTION of an IG or TR V-DOMAIN includes the anchors 104 and 118 and is therefore two amino acids longer than the corresponding CDR3-IMGT (positions 105–117).

In biological data, the lengths of the loops and strands are given by the number of occupied positions (unoccupied positions or "IMGT gaps" are represented with hatches in the IMGT Colliers de Perles (Figure 1) or by dots in alignments). The CDR-IMGT lengths are given in number of amino acids (or codons), into brackets and separated by dots: for example [9.6.9] means that the BC, C′C″, and FG loops (or CDR1-IMGT, CDR2-IMGT, and CDR3-IMGT for a V-DOMAIN) have a length of 9, 6, and 9 amino acids (or codons), respectively. Similarly [25.17.38.11] means that the FR1-IMGT, FR2-IMGT, FR3-IMGT, and FR4-IMGT have a length of 25, 17, 38, and 11 amino acids (or codons), respectively. Together, the four FR of a VH domain usually comprise 91 amino acids and the individual FR-IMGT lengths are [25.17.38.11], whereas the four FR of a VL domain usually comprise 89 amino acids and the individual FR-IMGT lengths are [26.17.36.10].

#### 2.4.1.4 Conserved Amino Acids

A V domain has five characteristic amino acids at given positions (positions with bold (online red) letters in the IMGT Colliers de Perles). Four of them are highly conserved and hydrophobic [28] and are common to the C domain: 23 (1st-CYS), 41 (CONSERVED-TRP), 89 (hydrophobic), and 104 (2nd-CYS). These amino acids contribute to the two major features shared by the V and C domain: the disulfide bridge (between the two cysteines 23 and 104) and the internal hydrophobic core of the domain (with the side chains of tryptophan W41 and amino acid 89). The fifth position, 118, is an anchor of the FG loop. It is occupied, in the V domains of IgSF other than IG or TR, by amino acids with diverse physicochemical properties [28]. In contrast, in IG and TR V-DOMAIN, the position 118 is occupied by remarkably conserved amino acids, which consist of a phenylalanine or a tryptophan encoded by the J-REGION and therefore designated J-TRP or J-PHE 118. The bulky aromatic side chains of J-TRP and J-PHE are internally orientated and structurally contribute to the V-DOMAIN hydrophobic core [61].

**TABLE 3** IMGT Additional Positions for CDR3-IMGT

| CDR3-IMGT Lengths | IMGT Additional Positions for CDR3-IMGT Length >13 AA[a] | | | | | | | | | |
|---|---|---|---|---|---|---|---|---|---|---|
| – | | | | | | | | | | |
| 21 | 111 | 111.1 | 111.2 | 111.3 | 111.4 | 112.4 | 112.3 | 112.2 | 112.1 | 112 |
| 20 | 111 | 111.1 | 111.2 | 111.3 | – | 112.4 | 112.3 | 112.2 | 112.1 | 112 |
| 19 | 111 | 111.1 | 111.2 | 111.3 | – | – | 112.3 | 112.2 | 112.1 | 112 |
| 18 | 111 | 111.1 | 111.2 | – | – | – | 112.3 | 112.2 | 112.1 | 112 |
| 17 | 111 | 111.1 | 111.2 | – | – | – | – | 112.2 | 112.1 | 112 |
| 16 | 111 | 111.1 | – | – | – | – | – | 112.2 | 112.1 | 112 |
| 15 | 111 | 111.1 | – | – | – | – | – | – | 112.1 | 112 |
| 14 | 111 | – | – | – | – | – | – | – | 112.1 | 112 |

[a]*For CDR3-IMGT length >13 AA, IMGT additional positions are created between positions 111 and 112 at the top of the CDR3-IMGT loop in the following order: 112.1, 111.1, 112.2, 111.2, 112.3, 111.3, etc. For CDR3-IMGT length <13 AA (not shown), IMGT gaps are created classically from the top of the loop, in the following order: 111, 112, 110, 113, 109, 114, etc. (IMGT®, http://www.imgt.org, IMGT Scientific chart>Numbering).*

#### 2.4.1.5 Genomic Delimitation

A last criterion used in the IMGT® definitive system for the characterization of a V domain is its delimitation, taking into account the exon delimitations, whenever appropriate. The exon rule is not used for the delimitation of the 5′ end of the first N-terminal domain of proteins with a leader (this includes the V-DOMAIN of the IG and TR chains). In those cases, the 5′ end of the first N-terminal domain corresponds to the proteolytic site between the leader (L-REGION) and the coding region of the mature protein. The IG and TR V-DOMAIN is therefore delimited in 5′ by a proteolytic site and in 3′ at the genomic level by the splicing site of the J-REGION. This IMGT® genomic approach integrates the strands A and G, in contrast to structural alignments that usually lack these strands due to their poor structural conservation, and thus bridges the gap between genomic data (exon) and 3D structure (domain).

### *2.4.2 C Domain IMGT® Definitive System*

#### 2.4.2.1 C Domain Definition and Main Characteristics

In the IMGT® definitive system, the C domain includes the C-DOMAIN of the IG and of the TR [2,3] and the C-LIKE-DOMAIN of the IgSF other than IG and TR. The C domain description of any receptor, any chain, and any species is based on the IMGT unique numbering for C domain (C-DOMAIN and C-LIKE-DOMAIN) [62,64].

A C domain (Figure 2) comprises about 90–100 amino acids and is made of seven antiparallel beta strands (A, B, C, D, E, F and G), linked by beta turns (AB, DE and EF), a transversal strand (CD) and two loops (BC and FG), and forming a sandwich of two sheets [ABED] [GFC] [62,64]. A C domain has a topology and a three-dimensional structure similar to that of a V domain but without the C′ and C″ strands and the C′C″ loop, which is replaced by a transversal CD strand [62].

#### 2.4.2.2 C Domain Strands and Loops

The C domain strands, turns, and loops and their delimitations and lengths, based on the IMGT unique numbering for C domain [62,64], are shown in Table 4. Correspondences between the IMGT unique numbering with other numberings (Eu, Kabat) are available in the IMGT Scientific chart. The correspondences with these previous numberings are useful for the interpretation of previously published data but, as for the V domain, the usage of these previous numberings has become obsolete in regard to the development of immunoinformatics based on the IMGT® standards [59–68] (IMGT®, http://www.imgt.org, IMGT Scientific chart > Numbering > Correspondence between C numberings).

#### 2.4.2.3 IMGT Colliers de Perles

The lengths of the strands and loops are visualized in the IMGT Colliers de Perles [66–68], on one layer and two layers (Figure 2). There are six IMGT anchors in a C domain (four of them identical to those of a V domain): positions 26 and 39 (anchors of the BC loop), 45 and 77 (by extension, anchors of the CD strand as there is no C′C″ loop in a C domain [62]), and 104 and 118 (anchors of the FG loop).

#### 2.4.2.4 Conserved Amino Acids

A C domain has five characteristic amino acids at given positions (positions with bold (online red) letters in the IMGT Colliers de Perles). Four of them are highly conserved and hydrophobic [28] and are common to the V domain: 23 (1st-CYS), 41 (CONSERVED-TRP), 89 (hydrophobic), and 104 (2nd-CYS). As mentioned above, these amino acids contribute to the two major features shared by the V and C domain: the disulfide bridge (between the two cysteines 23 and 104) and the internal hydrophobic core of the domain (with the side chains of tryptophan W41 and amino acid 89). The fifth position, 118, is diverse and is characterized as being an FG loop anchor.

#### 2.4.2.5 Genomic Delimitation

In the IMGT® definitive system, the C domains (C-DOMAIN and C-LIKE-DOMAIN) are delimited, taking into account the exon delimitation, whenever appropriate. As for the V domain, this IMGT® genomic approach integrates the strands A and G, which are absent of structural alignments.

### *2.4.3 IMGT/Collier-de-Perles Tool*

The IMGT/Collier-de-Perles tool [26], on the IMGT® Website at http://www.imgt.org, is a generic tool that allows the users to draw IMGT Colliers de Perles [65–68] starting from their own domain amino acid sequences (sequences already gapped according to the IMGT unique numbering, using, for example, IMGT/DomainGapAlign [9,24,25]) (Table 5). IMGT/Collier-de-Perles tool online can be customized to display the IG and TR CDR-IMGT according to the IMGT color menu and the amino acids according to their hydropathy or volume, or to the 11 IMGT physicochemical classes [28].

The IMGT/Collier-de-Perles tool is integrated in IMGT/DomainGapAlign [9,24,25] (users start from V, C, or G amino acid sequences) and in IMGT/V-QUEST [12–17] (users start from IG and TR V-DOMAIN nucleotide sequences) (Table 5). IMGT Colliers de Perles for V, C, and G domains are provided in IMGT/2D structure-DB (for amino acid sequences in the database) and in IMGT/3D structure-DB (on two layers with hydrogen bonds for the V or C domains or with the pMH contact sites for the G domains, for 3D structures in the database) [8–10] (Table 5).

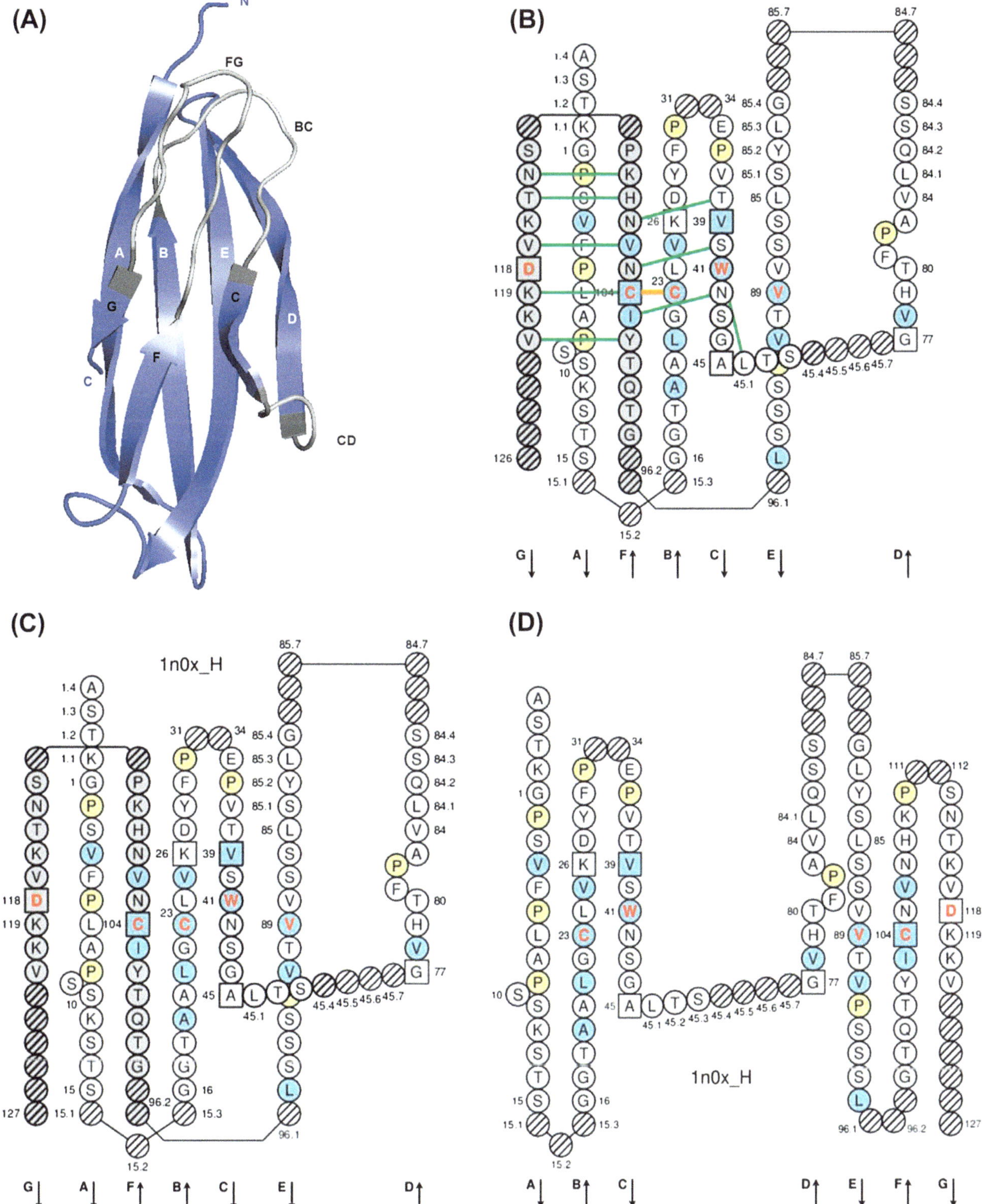


FIGURE 2 **Constant (C) domain. An IG CH (C-DOMAIN) is shown as an example.** (A) 3D structure ribbon representation with the IMGT strand and loop delimitations [62]. (B) IMGT Collier de Perles on two layers with hydrogen bonds. The IMGT Collier de Perles on two layers shows, in the forefront, the GFC strands and, in the back, the ABED strands (located at the interface CH1/CL of the IG), linked by the CD transversal strand. The IMGT Collier de Perles with hydrogen bonds (green lines online, only shown here for the GFC sheet) is generated by the IMGT/Collier-de-Perles tool integrated in IMGT/3Dstructure-DB, from experimental 3D structure data [8–10]. (C) IMGT Collier de Perles on two layers from IMGT/DomainGapAlign [9,24,25]. (D) IMGT Colliers de Perles on one layer. Amino acids are shown in the one-letter abbreviation. All proline (P) are shown online in yellow. IMGT anchors are in squares. Hatched circles are IMGT gaps according to the IMGT unique numbering for C domain [62,64]. Positions with bold (online red) letters indicate the four conserved positions that are common to a V domain and to a C domain: 23 (1st-CYS), 41 (CONSERVED-TRP), 89 (hydrophobic), 104 (2nd-CYS) [59–62,64] (Table 3), and position 118, which is only conserved in V-DOMAIN. The identifier of the chain to which the CH domain belongs is 1n0x_H (from the *Homo sapiens* b12 Fab, in IMGT/3Dstructure-DB, http://www.imgt.org). The 3D ribbon representation was obtained using PyMOL and "IMGT numbering comparison" of 1n0x_H (CH1) from IMGT/3Dstructure-DB (http://www.imgt.org).

**TABLE 4** C Domain Strands, Turns, and Loops, IMGT Positions, and Lengths, Based on the IMGT Unique Numbering for C Domain (C-DOMAIN and C-LIKE-DOMAIN)

| C Domain Strands, Turns, and Loops[a] | IMGT Position[b] | Lengths[c] | Characteristic IMGT Residue@ Position[d] |
|---|---|---|---|
| A-STRAND | 1–15 | 15 (14 if gap at 10) | |
| AB-TURN | 15.1–15.3 | 0–3 | |
| B-STRAND | 16–26 | 11 | 1st-CYS 23 |
| BC-LOOP | 27–31 34–38 | 10 (or less) | |
| C-STRAND | 39–45 | 7 | CONSERVED-TRP 41 |
| CD-STRAND | 45.1–45.9 | 0–9 | |
| D-STRAND | 77–84 | 8 (or 7 if gap at 82) | |
| DE-TURN | 84.1–84.7 85.1–85.7 | 0–14 | |
| E-STRAND | 85–96 | 12 | Hydrophobic 89 |
| EF-TURN | 96.1–96.2 | 0–2 | |
| F-STRAND | 97–104 | 8 | 2nd-CYS 104 |
| FG-LOOP | 105–117 | 13 (or less, or more) | |
| G-STRAND | 118–128 | 11 (or less) | |

[a] *IMGT® labels (concepts of description) are written in capital letters (no plural) [57].*
[b] *Based on the IMGT unique numbering for C domain (C-DOMAIN and C-LIKE-DOMAIN) [62,64].*
[c] *In number of amino acids (or codons).*
[d] *IMGT Residue@Position is a given residue (usually an amino acid) or a given conserved property amino acid class, at a given position in a domain, based on the IMGT unique numbering [64].*

# 3. IMGT® IMMUNOGLOBULIN REPERTOIRE ANALYSIS

## 3.1 IMGT/V-QUEST

IMGT/V-QUEST [12–17] is the IMGT® online tool for the analysis of nucleotide sequences of the IG and TR V-DOMAIN (Table 5). IMGT/V-QUEST identifies the variable (V), diversity (D), and junction (J) genes in rearranged IG and TR sequences and, for the IG, the nucleotide (nt) mutations and AA changes resulting from somatic hypermutations by comparison with the IMGT/V-QUEST reference directory. The tool integrates IMGT/JunctionAnalysis [18,19] for the detailed characterization of the V-D-J or V-J junctions, IMGT/Automat [20,21] for a complete sequence annotation, and IMGT/Collier-de-Perles [26].

The IMGT/V-QUEST most important functionalities include: introduction of "IMGT gaps" in the user nucleotide sequences (and in their translation); alignments and identification of the genes and alleles with the closest germline V, D, and J genes; analysis of the junctions; analysis of somatic hypermutations and amino acid changes; and, if the option "Search for insertions and deletions" was selected, identification of insertions and deletions (indels) and their correction. Customized parameters and results provided by IMGT/V-QUEST and IMGT/JunctionAnalysis have been described elsewhere [12–17]. Here, we briefly highlight major features of the three types of IMGT/V-QUEST outputs ("Detailed view," "Synthesis view," and "Excel file") and recall the content of the IMGT/V-QUEST reference directory.

### *3.1.1 IMGT/V-QUEST Output for "Detailed View"*

#### 3.1.1.1 Sequence and "Result Summary"

The sequence and "Result summary" are shown at the top of each individual result (Figure 3(A)) [14,16]. The IMGT reference directory set against which the sequence was analyzed (e.g., human IG) is indicated. A sequence submitted in antisense orientation will be shown as complementary reverse sequence, that is, in V gene sense orientation. The "Result summary" provides a crucial feature that is the evaluation of the functionality of the V domain rearranged sequence performed automatically by IMGT/V-QUEST: productive (if no stop codon in the V-(D)-J region and in-frame junction) or unproductive (if stop codons in the V-(D)-J region and/or out-of-frame junction). It also summarizes the main characteristics of the analyzed V domain sequence, which include: (1) the names of the closest "V-GENE and allele" (e.g., IGHV1-69*09) and "J-GENE and allele" (e.g., IGHJ4*02) with the alignment score, the percentage of identity, and the ratio of the number of identical nucleotides/number of aligned nucleotides; (2) the name of the closest "D-GENE and allele" (e.g., IGHD2-2*01), determined by IMGT/JunctionAnalysis with the D-REGION reading frame; (3) the FR-IMGT lengths (e.g., [25.17.38.11]), the CDR-IMGT lengths (e.g., [8.8.14]), and the AA JUNCTION sequence, which characterizes a V domain [14,16]. The information shown in Figure 3(A) (online in orange), "Productive IGH rearranged sequence" and "93,75%," is used by clinicians analyzing the level of somatic hypermutations in chronic lymphocytic leukemia (CLL) IGHV rearranged sequences as a prognostic criterion [15,70–72]. IMGT/V-QUEST provides warnings (not shown) that appear as notes in red to alert the user, if potential insertions or deletions are suspected in the V-REGION, or if other possibilities for the J-GENE and allele names are identified [14,16].

**TABLE 5 IMGT® Tools and Databases for Immunoglobulin Repertoire Analysis and Antibody Engineering and Humanization**

| IMGT® Tools and Databases | Results for V or C Domains[a] | Entry Types and Protocol References |
|---|---|---|
| IMGT/Collier-de-Perles [26] | Graphical 2D representation of IMGT Colliers de Perles [65–68] | User "IMGT gapped" V or C domain amino acid sequences (1 sequence per representation) [26] |
| **IMGT Immunoglobulin Repertoire Analysis** | | |
| IMGT/V-QUEST [12–17] | **1.** Introduction of IMGT gaps<br>**2.** Identification of the closest V, D, and J genes and alleles<br>**3.** IMGT/JunctionAnalysis results [18,19]<br>**4.** Description of mutations and amino acid changes<br>**5.** Identification of indels and their correction [16] (option)<br>**6.** IMGT/Automat annotation [20,21]<br>**7.** IMGT Colliers de Perles [65–68] | User nucleotide sequences of V-DOMAIN (1–50 sequences per analysis) [16] |
| IMGT/HighV-QUEST [21–23] | **1.** Introduction of IMGT gaps<br>**2.** Identification of indels and their correction [16] (by default)<br>**3.** Identification of the closest V, D, and J genes and alleles<br>**4.** IMGT/JunctionAnalysis results [18,19]<br>**5.** Description of mutations and amino acid changes<br>**6.** IMGT/Automat annotation [20,21]<br>**7.** Statistical analysis [22]<br>**8.** Characterization of the IMGT clonotypes (AA) [23] | User NGS nucleotide sequences of V-DOMAIN, preferentially long sequences (e.g., from 454) (up to 500,000 sequences per run and statistics on results[b]) [c] [22,23] |
| **IMGT Antibody Engineering and Humanization** | | |
| IMGT/DomainGapAlign [9,24,25] | **1.** Introduction of IMGT gaps<br>**2.** Identification of the closest genes and alleles<br>**3.** Delimitation of the domains<br>**4.** Description of amino acid changes<br>**5.** IMGT Colliers de Perles [65–68] with highlighted AA changes (pink circles online) | User amino acid sequences of V and C domains (1 to several sequences of same domain type) [24,25] |
| IMGT/3Dstructure-DB [8–10] | **1.** Identification of the closest genes and alleles<br>**2.** IMGT/DomainGapAlign results [9,24,25]<br>**3.** IMGT Colliers de Perles [65–68] on two layers with hydrogen bonds (for V and C)<br>**4.** Contact analysis between a pair of domains or between a domain and a ligand<br>**5.** Renumbered IMGT files<br>**6.** IMGT numbering comparison | 2987 structure entries (1973 IG, including 1272 IG/Ag complexes)[d] [10]. |
| IMGT/2Dstructure-DB [10]* | **1.** Identification of the closest genes and alleles<br>**2.** IMGT/DomainGapAlign results [9,24,25]<br>**3.** IMGT Colliers de Perles [65–68]<br>**4.** Renumbered IMGT files | 543 amino acid sequence entries[d] (534 IG, 335 from Kabat and 199 from INN) [10]*. |

An asterisk (*) indicates that parts of the protocol dealing with 3D structures (hydrogen bonds in IMGT Colliers de Perles on two layers, Contact analysis) are not relevant; otherwise all other queries and results are similar to IMGT/3Dstructure-DB.

[a]*V: V domain (includes V-DOMAIN of IG and TR and V-LIKE-DOMAIN of IgSF other than IG and TR) [61]. C: C domain (includes C-DOMAIN of IG and TR and C-LIKE-DOMAIN of IgSF other than IG and TR) [62]. IMGT/DomainGapAlign, IMGT/3Dstructure-DB, and IMGT/2Dstructure-DB also manage the G domain (G-DOMAIN of MH and G-LIKE-DOMAIN of MhSF other than MH) [63] of interest in the context of that chapter when the antibody target belongs to the MH (or more generally to the MhSF), but not detailed here.*

[b]*In November 2013.*

[c]*In May 2014, more than 2.5 billions of sequences analyzed by IMGT/HighV-QUEST by 846 users from 40 countries (62% from USA, 25% from EU, 13% from the remaining world).*

[d]*In May 2014.*

## (A)

### Sequence number 1: SEQ1

Sequence compared with the human IG set from the IMGT reference directory

```
>SEQ1
caggtgcagctggtgcagtctggggctgaggtgaagaagcctggatcctcggtgaaggtc
tcatgtaaggcttctggaggcaccttcaacagctttcctatcaactgggtgcgacaggcc
cctggacaagggcttgagtggatgggaaggatcatccctatcattggtatagcagactac
gcacaggaattccagggcagagtcacgattaccgcggacagatccacgagcacagcctac
atggagctgcgtagtctgatatctgaggacacggccgtgtacttctgtgcgagaccggaa
gccgtaacagtaccagctcctcttgactactggggccagggaaccctggtcaccgtctcc
tcagcctccaccaagggcccatcggtcttccccctggcaccctcctccaagagcacctct
gggggcacagcggccctgggctgcctggtcaaggactacttccccgaaccggtgacggtg
tcgtggaactcaggcgccctgaccagcggcgtgcacaccttcccggctgtcctacagtcc
tcaggactctactccctcagcagcgtggtgaccgtgccctccagcagcttgggcacccag
acctacatctgcaacgtgaatcacaagcccagcaacaccaaggtggacaagacagttgag
cgcaaa
```

<table>
<tr><td>Result summary:</td><td colspan="3">Productive IGH rearranged sequence (no stop codon and in-frame junction)</td></tr>
<tr><td>V-GENE and allele</td><td>Homsap IGHV1-69*09 F</td><td>score = 1273</td><td>identity = 93,75% (270/288 nt)</td></tr>
<tr><td>J-GENE and allele</td><td>Homsap IGHJ4*02 F</td><td>score = 186</td><td>identity = 87,50% (42/48 nt)</td></tr>
<tr><td>D-GENE and allele by IMGT/JunctionAnalysis</td><td>Homsap IGHD2-2*01 F</td><td colspan="2">D-REGION is in reading frame 3</td></tr>
<tr><td>FR-IMGT lengths, CDR-IMGT lengths and AA JUNCTION</td><td>[25.17.38.11]</td><td>[8.8.14]</td><td>CARPEAVTVPAPLDYW</td></tr>
</table>

## (B)

### Sequence number 2: SEQ2

Sequence compared with the human IG set from the IMGT reference directory

```
>SEQ2
atgtctgtctccttcctcatcttcctgcccgtgctgggcctcccatggggtgtcctgtca
caggtacagctgcagcagtcaggtccaggactggtgaagccctcgcagaccctctcactc
acctgtgccatctccggggacagtgtctctAGCagcaacggtgttgcttggaactgggtc
aggcagtccccatcgagaggccttgagtggctgggaaggacatactacaggtccaagtgg
tataatgattatgcagtatctgtgaaaagtcgaataaccatcaacccagacacatccaag
aaccagttctccctgcagctgaactctgtgactcccgaagacacggctgtgtattactgt
gcaagaggccgttggaccgcttttgatttctggggccaagggacaatggtcaccgtctct
tcc
```

<table>
<tr><td rowspan="3">Result summary:</td><td colspan="6">Nucleotide insertions have been detected and automatically removed for this analysis: they are displayed as capital letters in the user submitted sequence above.</td></tr>
<tr><td>localization in V-REGION</td><td>nb of inserted nt</td><td>inserted nt</td><td>causing frameshift</td><td>from V-REGION codon</td><td>from nt position in user submitted sequence</td></tr>
<tr><td>CDR1-IMGT</td><td>3</td><td>AGC</td><td>no</td><td>34</td><td>151</td></tr>
<tr><td colspan="7">IMGT/V-QUEST results after removal of the insertion(s):<br>Potentially productive IGH rearranged sequence: no stop codon and in-frame junction<br>(Check also your sequence with BLAST against IMGT/GENE-DB reference sequences to eventually identify out-of-frame pseudogenes)</td></tr>
<tr><td>V-GENE and allele</td><td colspan="2">Homsap IGHV6-1*01 F</td><td>score = 1444</td><td colspan="3">identity = 98,65% (293/297 nt) [98,32% (292/297 nt)]</td></tr>
<tr><td>J-GENE and allele</td><td colspan="2">Homsap IGHJ3*01 F, or Homsap IGHJ3*02 F</td><td>score = 191</td><td colspan="3">identity = 87,76% (43/49 nt)</td></tr>
<tr><td>D-GENE and allele by IMGT/JunctionAnalysis</td><td colspan="2">Homsap IGHD3-16*02 F</td><td colspan="4">D-REGION is in reading frame 2</td></tr>
<tr><td>FR-IMGT lengths, CDR-IMGT lengths and AA JUNCTION</td><td colspan="2">[25.17.38.11]</td><td>[10.9.10]</td><td colspan="3">CARGRWTAFDFW</td></tr>
</table>

FIGURE 3 **IMGT/V-QUEST results.** (A) Sequence and "Result summary" ("Detailed view"). (B) Sequence and "Result summary" with the option "Search for insertions and deletions" ("Detailed view"). An insertion of 3 nt was detected by IMGT/V-QUEST. The analysis is then performed automatically by the tool after having removed the insertion. (C) "Results of IMGT/JunctionAnalysis" ("Detailed view"). The "List of eligible D" (an option) is shown. "Ngc" is 13/17 as there are 13 "c" or "g" in N1+N2 on a total of 17 nucleotides (aggctggg (N1), gagcgcgtc (N2)). (D) "Results of IMGT/JunctionAnalysis" ("Synthesis view"). The numbering of IMGT additional positions for CDR3-IMGT is according to Table 3. Owing to the figure format, Vmut, Dmut, Jmut, Ngc, molecular mass, and pI are not shown. The SEQ1, SEQ2, SEQ3, SEQ4, SEQ5, SEQ6, and SEQ7 correspond to the AB027435, L21957, AB363166, Z98684, Z98689, Z98734, and Z98715 accession numbers, respectively, from the IMGT/LIGM-DB database [6].

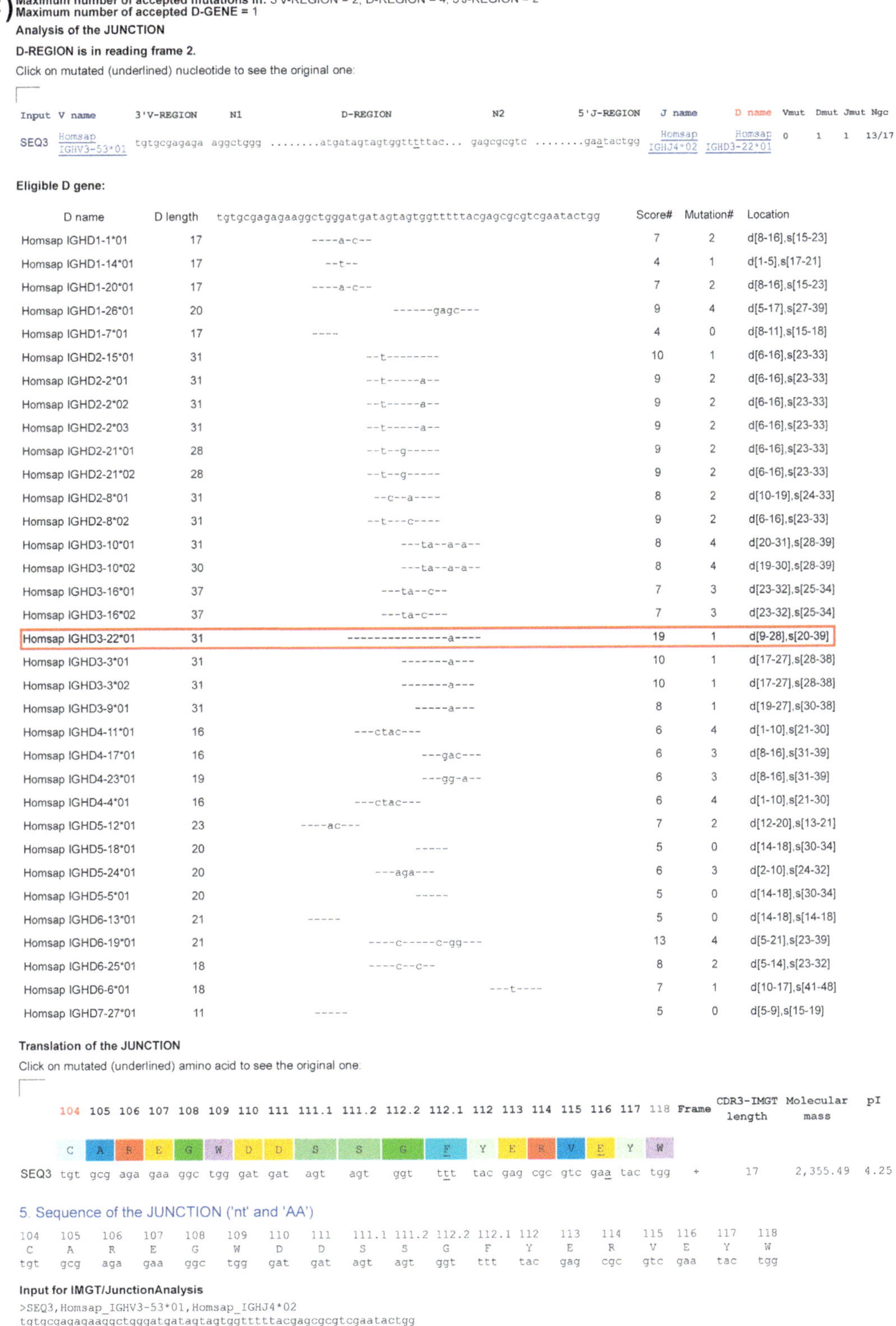


**FIGURE 3** Cont'd

**(D)**

**Results for the IGH junctions**

**Analysis of the JUNCTIONs**

Click on mutated (underlined) nucleotide to see the original one:

| Input | V name | 3'V-REGION | P | N1 | D-REGION | N2 | 5'J-REGION | J name | D name |
|---|---|---|---|---|---|---|---|---|---|
| SEQ1 | Homsap IGHV1-69*09 | tgtgcgaga.. | | ccggaagcc | ........gtaacagtaccagctcct..... | c | ......ttgactactgg | Homsap IGHJ4*02 | Homsap IGHD2-2*01 |
| SEQ7 | Homsap IGHV3-23*01 | tgtgcgaaag. | | | ggtatggt............ | c | ......ttgactgctgg | Homsap IGHJ4*02 | Homsap IGHD1-26*01 |
| SEQ6 | Homsap IGHV3-30*03 | tgtgcgaaaga | tc | ctgcgaa | ......agcagttcgtcc | tgaattgccggt | ................cggtgtggacgtctgg | Homsap IGHJ6*02 | Homsap IGHD6-6*01 |
| SEQ3 | Homsap IGHV3-53*01 | tgtgcgagaga | | aggctggg | ........atgatagtagtggttttac... | gagcgcgtc | ........gaatactgg | Homsap IGHJ4*02 | Homsap IGHD3-22*01 |
| SEQ4 | Homsap IGHV3-73*02 | tgtattaga.. | | | .tattactatgatagcagtggt......... | gacggtgtatc | .............ctgg | Homsap IGHJ4*02 | Homsap IGHD3-22*01 |
| SEQ5 | Homsap IGHV3-73*02 | tgtattaga.. | | | .tattactatgatagcagtggt......... | gacggtgtatc | .............ctgg | Homsap IGHJ4*02 | Homsap IGHD3-22*01 |
| SEQ2 | Homsap IGHV6-1*01 | tgtgcaagag. | | gc | ............................cgttggacc | | ....gcttttgatttctgg | Homsap IGHJ3*01 | Homsap IGHD3-16*02 |

**Translation of the JUNCTIONs**

Click on mutated (underlined) amino acid to see the original one:

| | 104 | 105 | 106 | 107 | 108 | 109 | 110 | 111 | 111.1 | 111.2 | 112.3 | 112.2 | 112.1 | 112 | 113 | 114 | 115 | 116 | 117 | 118 | Frame | CDR3-IMGT length |
|---|---|---|---|---|---|---|---|---|---|---|---|---|---|---|---|---|---|---|---|---|---|---|
| | C | A | R | P | E | A | V | T | | | | | V | P | A | P | L | D | Y | W | | |
| SEQ1 | tgt | gcg | aga | ccg | gaa | gcc | gta | aca | ... | ... | ... | ... | gta | cca | gct | cct | ctt | gac | tac | tgg | + | 14 |
| | C | A | K | G | Y | | | | | | | | | | | G | L | D | C | W | | |
| SEQ7 | tgt | gcg | aaa | ggg | tat | ... | ... | ... | ... | ... | ... | ... | ... | ... | ... | ggt | ctt | gac | tgc | tgg | + | 8 |
| | C | A | K | D | P | A | K | A | V | R | P | E | L | P | V | G | V | D | V | W | | |
| SEQ6 | tgt | gcg | aaa | gat | cct | gcg | aaa | gca | gtt | cgt | cct | gaa | ttg | ccg | | ggt | gtg | gac | gtc | tgg | + | 18 |
| | C | A | R | E | G | W | D | D | S | S | | G | F | Y | E | R | V | E | Y | W | | |
| SEQ3 | tgt | gcg | aga | gaa | ggc | tgg | gat | gat | agt | agt | ... | ggt | ttt | tac | gag | cgc | gtc | gaa | tac | tgg | + | 17 |
| | C | I | R | Y | Y | Y | D | S | | | | | | S | G | D | G | V | S | W | | |
| SEQ4 | tgt | att | aga | tat | tac | tat | gat | agc | ... | ... | ... | ... | ... | agt | ggt | gac | ggt | gta | tcc | tgg | + | 13 |
| | C | I | R | Y | Y | Y | D | S | | | | | | S | G | D | G | V | S | W | | |
| SEQ5 | tgt | att | aga | tat | tac | tat | gat | agc | ... | ... | ... | ... | ... | agt | ggt | gac | ggt | gta | tcc | tgg | + | 13 |
| | C | A | R | G | R | W | | | | | | | | | T | A | F | D | F | W | | |
| SEQ2 | tgt | gca | aga | ggc | cgt | tgg | ... | ... | ... | ... | ... | ... | ... | ... | acc | gct | ttt | gat | ttc | tgg | + | 10 |

FIGURE 3 Cont'd

### 3.1.1.2 Sequence and "Result Summary" with the Option "Search for Insertions and Deletions"

Potential insertions or deletions (indels) are suspected by IMGT/V-QUEST when the V-REGION score is very low (less than 200), and/or the percentage of identity is less than 85%, and/or when the input sequence has different CDR1-IMGT and/or CDR2-IMGT lengths, compared to those of the closest germline V [14,16]. In those cases, the user can go back to the IMGT/V-QUEST Search page and select the option "Search for insertions and deletions" in "Advanced parameters." If, indeed, insertions and/or deletions are detected, they will be described in the "Result summary" row with their localization in FR-IMGT or CDR-IMGT, the number of inserted or deleted nucleotides (and, for insertions, the inserted nucleotides), the presence or absence of frameshift, the V-REGION codon from which the insertion or deletion starts, and the nt position in the user submitted sequence (Figure 3(B)). The insertions are highlighted in capital letters in the user sequence and the tool runs a classical IMGT/V-QUEST search after having removed the insertion(s) from the user sequence (Figure 3(B)). In case of deletions, the tool adds gaps to replace the identified deletions before running a classical IMGT/V-QUEST search. Users should be aware that an insertion or a deletion at the beginning of FR1-IMGT or at the end of the FR3-IMGT may not be detected. The identification and correction of indels by IMGT/V-QUEST is important for calculating the percentage of somatic hypermutations in CLL despite these indels [15,16].

### 3.1.1.3 Alignments for V, D, and J Genes and Alleles

The alignments for V, D, J genes and alleles display the alignments with the five closest V, D, and J gene alleles, respectively, with their alignment score and their identity percentage [12–17]. In case of discrepancy between the alignments for D genes and the results of IMGT/JunctionAnalysis (below), the results of IMGT/JunctionAnalysis are the most accurate (the way of identifying the closest germline D and the evaluation of the alignment score are different between IMGT/V-QUEST and IMGT/JunctionAnalysis). The IMGT/V-QUEST alignment for D gene is still provided (as an option), although less accurate, as it is less stringent and displays several D genes and alleles, and therefore may help solving some ambiguous cases [16].

#### 3.1.1.4 "Results of IMGT/JunctionAnalysis"

The results of IMGT/JunctionAnalysis comprise (Figure 3(C)):

1. "Analysis of the JUNCTION": provides the name of the D gene and allele for IGH sequences and shows the details of the junction at the nucleotide level (nucleotides trimmed at the V, D, and J gene ends and N nucleotides added by the TdT) with an accurate delimitation of the 3′V-REGION (from the 2nd-CYS 104 to the 3′ end of the V-REGION), D-REGION, and 5′J-REGION (from the 5' end of the J-REGION to J-TRP 118 or J-PHE 118). Dots represent the nucleotides that have been trimmed by the exonuclease [19]. The number of mutations in the 3′V-REGION, D-REGION, and 5′J-REGION is indicated under "Vmut," "Dmut," and "Jmut," respectively (Figure 3(C)), and the corresponding mutated nucleotides are underlined in the sequence. Nucleotides of the N-REGION are displayed in N1 and N2. The ratio of the number of g+c nucleotides to the total number of N nucleotides is indicated under "Ngc" (Figure 3(C)) [19].
2. "Eligible D genes": If the option "with full list of eligible D-GENE" was selected in "Display view," its results allow comparing the IMGT/JunctionAnalysis D gene identification with all D genes that match the junction with their corresponding score (Figure 3(C)).
3. "Translation of the JUNCTION": displays the junction with amino acids colored according to the 11 IMGT "Physicochemical" classes [28], the frame (+ and − indicate in-frame and out-of-frame, respectively), the CDR3-IMGT length, the molecular mass, and the isoelectric point (pI). In case of frameshifts, gaps (represented by one or two dots) are inserted to maintain the J-REGION frame. The corresponding codon, which cannot be translated, is represented by "#" [19].

#### 3.1.1.5 "V-REGION Alignment," "V-REGION Translation," and "V-REGION Protein Display"

The results provide three displays of the V-REGION with the FR-IMGT and CDR-IMGT delimitations [14,16]: the "V-REGION alignment" displays the nucleotide sequences; the "V-REGION translation" displays the nucleotide sequence and deduced AA translation of the input sequence, aligned with the closest germline V-REGION; and the "V-REGION protein display" displays the deduced AA translation of the input sequence, aligned with the V-REGION translation of the closest germline V-GENE and, on the third line of the alignment and shown in bold, the AA of the input sequence that are different from the closest germline V-REGION.

#### 3.1.1.6 Analysis of Mutations and Amino Acid Changes

The analysis of the mutations in the V-REGION is performed by comparison of the analyzed sequence with the closest germline V gene and allele [14,16].

The "V-REGION mutation and AA change table" lists the nt mutations and the corresponding AA changes if the mutations are not silent. They are described for each FR-IMGT and CDR-IMGT, with their nt and codon position according to the IMGT unique numbering and for the AA changes according to the IMGT AA classes [28]. For example, c1>g, Q1>E (+ + −) means that the nt mutation (c>g) at nt 1 leads to an AA change (Q>E) at codon 1.

The "V-REGION mutation and AA change statistics" comprises two tables. The first table reports the number of "Nucleotide (nt) mutations," described as silent or nonsilent, transition (a>g, g>a, c>t, t>c) or transversion (a>c, c>a, a>t, t>a, g>c, c>g, g>t, t>g). The second table reports the number of "Amino acid (AA) changes" with the type of each change (e.g., (+ + +) highly similar, (− − +) dissimilar). For both tables, results are given for the V-REGION and for FR-IMGT and CDR-IMGT [14,16].

The positions of the mutations have been correlated with the presence of specific patterns in the germline V gene and allele (IMGT®, http://www.imgt.org, IMGT Education>Tutorials>Immunoglobulins and B cells>Somatic hypermutations). The "V-REGION mutation hot spots" table shows the localization of the hot spot patterns (a/t)**a** (or w**a**) and (a/g)**g**(c/t)(a/t) (or r**g**yw) and their complementary reverse motifs **t**(a/t) (or **t**w) and (a/t)(a/g)**c**(c/t) (or wr**c**y) in the closest germline V gene and allele [14,16].

#### 3.1.1.7 IMGT Collier de Perles

The "IMGT Collier de Perles" [65–68] can be displayed, on one layer or two layers, either as a "link to IMGT/Collier-de-Perles tool" [26] or as a direct "IMGT Collier de Perles (for a number of sequences >5)" representation integrated in IMGT/V-QUEST results, depending on the user selection in the IMGT/V-QUEST Search page [14,16].

### *3.1.2 IMGT/V-QUEST Output for "Synthesis View"*

#### 3.1.2.1 Summary Table

The "Summary table" displays one row for each input sequence with the corresponding results, including: (1) the name of the sequence (Sequence ID); (2) the name of the closest V-GENE and allele; (3) the functionality of the V domain rearranged sequence (productive or unproductive); (4) the V-REGION score; (5) the V-REGION percentage of identity with, between parentheses, the ratio of number of identical nt/number of aligned nt; (6) the name of the closest J-GENE and allele; and (7) provided according to the IMGT/JunctionAnalysis results, the D-GENE and allele name, the D reading frame, the CDR-IMGT lengths, the AA JUNCTION, and the JUNCTION frame. In the absence of results of IMGT/JunctionAnalysis, only the AA JUNCTION defined by IMGT/V-QUEST is displayed [14,16].

#### 3.1.2.2 IMGT/JunctionAnalysis

The results of IMGT/JunctionAnalysis are displayed per locus (e.g., IGH, IGK, or IGL for IG sequences). Results (Figure 3(D)) are similar to those obtained for individual sequences detailed above [18,19].

#### 3.1.2.3 Alignment Displays

The synthesis results provide six different displays (if all were selected): "Alignment for V-GENE," "V-REGION alignment according to the IMGT unique numbering" and "V-REGION translation," and "V-REGION protein displays" in three different formats [14,16].

### *3.1.3 IMGT/V-QUEST Output for Excel File*

"Excel file" allows the users to open and save a spreadsheet including the results of the IMGT/V-QUEST analysis. The file contains 11 sheets (if all were selected in the IMGT/V-QUEST Search page). The "Summary" and "Parameters" are always selected [14,16].

### *3.1.4 IMGT/V-QUEST Reference Directory*

The IMGT/V-QUEST reference directory sets against which the IMGT/V-QUEST is running include IMGT reference sequences from all functional (F) genes and alleles, all ORF, and all in-frame pseudogene (P) alleles. By definition, the IMGT reference directory sets contain one sequence for each allele. By default, the user sequences are compared with all genes and alleles. However, the option "With allele *01 only" is useful for: (1) "Detailed view," if the user sequences need to be compared with different genes, and (2) "Synthesis view," if the user sequences that use the same gene need to be aligned together (independently of the allelic polymorphism) [14,16].

The IMGT/V-QUEST reference directories have been set up for species that have been extensively studied, such as human and mouse. This also holds for the other species or taxons with incomplete IMGT reference directory sets. In those cases, results should be interpreted considering the status of the IMGT reference directory (information on the updates on the IMGT® Website). Links to the IMGT/V-QUEST reference directory sets are available from the IMGT/V-QUEST Welcome page [14,16].

## 3.2 IMGT/HighV-QUEST

IMGT/HighV-QUEST [22], created in October 2010, is the high-throughput version of IMGT/V-QUEST. It is so far the only online tool available on the Web for the direct analysis of complete IG and TR domain sequences from NGS. It analyzes NGS sequences, preferentially long sequences, e.g., obtained from the Roche 454 Life Sciences technology, without the need of computational read assembly [22,23] (Table 5). IMGT/HighV-QUEST analyzes up to 500,000 sequences per run and performs statistical analysis on the results [22,23], with the same degree of resolution and high-quality results as IMGT/V-QUEST [12–17]. IMGT/HighV-QUEST represents a major breakthrough for the analysis and the comparison of the antigen receptor V-DOMAIN repertoires and immunoprofilings of the adaptive immune response [22,23].

The functionalities of IMGT/HighV-QUEST include: the introduction of IMGT gaps; the identification of indels and their correction; the identification of the closest V, D, and J genes and alleles; the IMGT/JunctionAnalysis results; the description of mutations and amino acid changes; the annotation by IMGT/Automat; the NGS statistical analysis; and the characterization of the IMGT clonotypes (AA) [22,23] (Table 5). IMGT/HighV-QUEST provides results in different categories "1 copy" and "More than 1" to avoid redundancy of the analysis, "single allele" and "several alleles (or genes)" (with "single allele" sequences being usually longer than "several alleles") [22]. These categories have been fundamental in the characterization of clonotypes for NGS [23].

As for the other IMGT® databases and tools, IMGT/HighV-QUEST is freely available for academics. However, the IMGT/HighV-QUEST Welcome page requires user identification and provides, for new users, a link to register. User identification has been set to avoid nonrelevant use and overload of the server, and to contact the user if needed. The user identification gives access to the IMGT/HighV-QUEST Search page.

### *3.2.1 MGT/HighV-QUEST Search Page*

The IMGT/HighV-QUEST Search page [22] (Figure 4(A)) is very similar to the classical IMGT/V-QUEST Search page and therefore looks familiar to the biologists. The four additional fields at the top comprise the analysis title (provided by the user), the selection of the species, the selection of the receptor type or locus, and the choice of e-mail notifications. Because of their large number of sequences and use of the tool by multiple users at the same time, the analyses are first queued on the IMGT® server, and they are performed depending on the available resources [22].

The submission consists of the upload of a file (in simple text format) containing the user sequences in FASTA format (from 1 to 500,000 sequences). The customization for the "Detailed View" (13 options), "Files in CSV" (11 files,

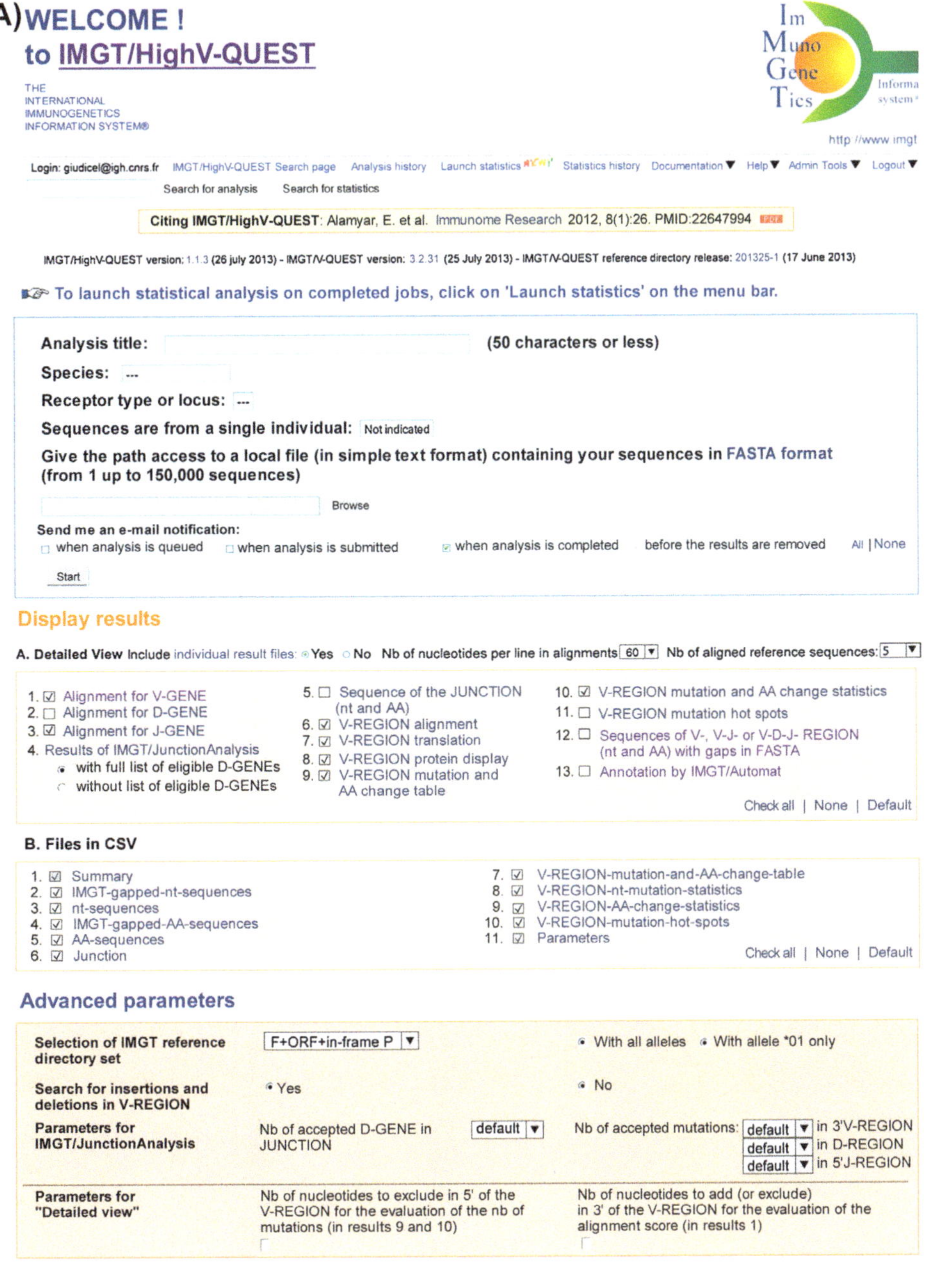


FIGURE 4 **IMGT/HighV-QUEST.** (A) IMGT/HighV-QUEST Search page. (B) IMGT clonotypes (AA). This figure represents a screen shot from IMGT/HighV-QUEST online. "Exp. ID" is the identifier of the "IMGT clonotype (AA)" in the dataset. The IMGT clonotype (AA) definition includes the CDR3-IMGT length (AA), the CDR3-IMGT sequence (AA), the names of the V, D, J genes and alleles, and the anchors 104 and 118 of the junction (for VH, "C, W" for 2nd-CYS 104 and J-TRP 118, respectively) (the order of these columns depends on the query, here on CDR3-IMGT length (AA)). "V %" indicates the percentage identity of the V-REGION of the representative sequence with the closest germline V-REGION; "Sequence length" is the length in nucleotides of the representative sequence; the link in the "Sequence ID" column gives access to the representative sequence in FASTA format; "Nb" indicates the number of sequences assigned to an IMGT clonotype (AA) ("1 copy," "More than 1," and Total). In the "IMGT clonotypes (nt)" column, the link on each "Sequences file" gives access to a file containing the "1 copy" sequences assigned to a given IMGT clonotype (AA), in FASTA format. This figure shows a very small part of the list of the 64,093 unique IMGT clonotypes (AA) identified in this case study. The "IMGT clonotype (AA)" #45309, with an Exp. ID "45309_Xall," is defined by the "IGHV6-1*01 F – IGHD1-26*01 F – IGHJ4*02 F" rearrangement, with a CDR3-IMGT length (AA) of "14 AA" and a CDR3-IMGT sequence (AA) "TRDHDQEGANSFDY," and conserved anchors 104 and 118 "C, W" (recall of the filter). The IMGT clonotype (AA) representative sequence has a V-REGION that is 96.62% identical to that of the germline IGHV6-1*01 and a length of 427 nt. (C) IMGT clonotypes (nt). The nb of different CDR3-IMGT (nt) indicates the nb of IMGT clonotypes (nt) for a given IMGT clonotype (AA) (e.g., two for #41328). The CDR3-IMGT sequence (nt) is shown with the number of different nucleotides ("Nb diff nt") by comparison with that of the representative sequence. "0" indicates that the CDR3-IMGT sequence (nt) is identical to that of the IMGT clonotype (AA) representative sequence. For #41328, there is an IMGT clonotype (nt) with 4 nt differences ("c" instead of "t" at position 9, "c" instead of "a" at position 12, "t" instead of "c" at position 30, and "c" instead of "t" at position 33) compared to the CDR3-IMGT of the representative sequence. #41328 also shows an example of "several alleles" (for V and J) assigned to an IMGT clonotype (AA).

**(B)**

| ID | | IMGT clonotype (AA) definition | | | | | | IMGT clonotype (AA) representative sequence | | | Nb | | | IMGT clonotypes (nt) |
|---|---|---|---|---|---|---|---|---|---|---|---|---|---|---|
| # | Exp. ID | CDR3-IMGT length (AA) | CDR3-IMGT sequence (AA) | V gene and allele | D gene and allele | J gene and allele | Anchors 104,118 | V % | Sequence length | Sequence ID | Total nb of '1 copy' | Total nb of 'More than 1' | Total | Sequences file ('1 copy') |
| 45307 | 45307-Xall | 14 AA | TRDGPVAVHWFFAL | Homsap IGHV6-1*01 F | Homsap IGHD2-8*02 F | Homsap IGHJ2*01 F | C,W | 97.64 | 422 | H6VTD4304JCAGK | 1 | 0 | 1 | Sequences file |
| 45308 | 45308-Xall | 14 AA | TRDGSLTGTKGFDI | Homsap IGHV3-30*01 F | Homsap IGHD1-7*01 F | Homsap IGHJ3*01 F | C,W | 87.85 | 739 | H7CGN0J01CVBLX | 1 | 0 | 1 | Sequences file |
| 45309 | 45309-Xall | 14 AA | TRDHDQEGANSFDY | Homsap IGHV6-1*01 F | Homsap IGHD1-26*01 F | Homsap IGHJ4*02 F | C,W | 96.62 | 427 | H7CGN0J01BQTCY | 24 | 9 | 33 | Sequences file |
| 45310 | 45310-Xall | 14 AA | TRDLDRSGWYGMDV | Homsap IGHV3-49*03 F | Homsap IGHD6-19*01 F | Homsap IGHJ6*02 F | C,W | 95.92 | 424 | H7CGN0J01BQGHH | 1 | 0 | 1 | Sequences file |
| 45311 | 45311-Xall | 14 AA | TRDLGSSGPYYFDY | Homsap IGHV3-49*04 F | Homsap IGHD6-19*01 F | Homsap IGHJ4*02 F | C,W | 86.39 | 424 | H7CGN0J01CBHZL | 1 | 0 | 1 | Sequences file |
| 45312 | 45312-Xall | 14 AA | TRDLWLSGQYTFDY | Homsap IGHV6-1*01 F | Homsap IGHD2-8*02 F | Homsap IGHJ2*01 F | C,W | 98.32 | 427 | H6VTD4302EYLAU | 1 | 0 | 1 | Sequences file |
| 45313 | 45313-Xall | 14 AA | TRDPVAGVPDFFDH | Homsap IGHV3-30*31 F | Homsap IGHD2-15*01 F | Homsap IGHJ4*02 F | C,W | 89.90 | 411 | H7CGN0J01DM7W9 | 2 | 0 | 2 | Sequences file |
| 45314 | 45314-Xall | 14 AA | TRDRALLFGAHFDF | Homsap IGHV1-2*02 F | Homsap IGHD3-10*01 F | Homsap IGHJ4*02 F | C,W | 93.75 | 417 | H7CGN0J01C3729 | 1 | 0 | 1 | Sequences file |
| 45315 | 45315-Xall | 14 AA | TRDRLSGNVWLLED | Homsap IGHV6-1*01 F | Homsap IGHD6-19*01 F | Homsap IGHJ5*02 F | C,W | 96.30 | 424 | H7CGN0J01A5CLN | 1 | 0 | 1 | Sequences file |
| 45316 | 45316-Xall | 14 AA | TRDRLYGDVWLLED | Homsap IGHV6-1*01 F | Homsap IGHD6-19*01 F | Homsap IGHJ5*02 F | C,W | 95.29 | 427 | H7CGN0J01EFJMF | 1 | 0 | 1 | Sequences file |

**(C)**

VH clonotypes (nt) header

| # | CDR3-IMGT length (nt) | Nb diff CDR3-IMGT (nt) | CDR3-IMGT sequence (nt) | Nb diff nt | V gene and allele | D gene and allele | J gene and allele | Anchors 104,118 | V % mean | V-REGION length mean | J % mean | J-REGION length mean | Sequence length mean | Total nb of '1 copy' | Total nb of 'More than 1' | Total |
|---|---|---|---|---|---|---|---|---|---|---|---|---|---|---|---|---|
| 41325 | 41325-Xall | 14 AA | ARDLCDYGLCYFDY | Homsap IGHV1-18*04 F | Homsap IGHD4-17*01 F | Homsap IGHJ4*02 F | C,W | 97.91 | 414 | H7CGN0J01ET06J | 1 | 0 | 1 | Sequences file | | |
| 41325 | 42 | 1 | gcgagagatttatgtgactacggactctgctactttgactac | 0 | Homsap IGHV1-18*04 F | Homsap IGHD4-17*01 F | Homsap IGHJ4*02 F | C,W | 97.91 | 295 | 97.92 | 47 | 414 | 1 | 0 | 1 |
| 41326 | 41326-Xall | 14 AA | ARDLDGSKMFCMDV | Homsap IGHV3-23*04 F | Homsap IGHD5-24*01 F | Homsap IGHJ6*02 F | C,W | 89.25 | 439 | H7CGN0J01BDKK3 | 1 | 0 | 1 | Sequences file | | |
| 41326 | 42 | 1 | gcgagagatctagacggcagcaagatgttctgtatggacgtc | 0 | Homsap IGHV3-23*04 F | Homsap IGHD5-24*01 F | Homsap IGHJ6*02 F | C,W | 89.25 | 287 | 62.9 | 44 | 439 | 1 | 0 | 1 |
| 41327 | 41327-Xall | 14 AA | ARDLDGSSWPIFDY | Homsap IGHV6-1*01 F | Homsap IGHD6-13*01 F | Homsap IGHJ4*02 F | C,W | 100 | 786 | H7CGN0J01ARV9K | 7 | 0 | 7 | Sequences file | | |
| 41327 | 42 | 1 | gcaagagatcttgacggcagcagctggcccatatttgactac | 0 | Homsap IGHV6-1*01 F | Homsap IGHD6-13*01 F | Homsap IGHJ4*02 F | C,W | 96.86 | 304 | 91.67 | 43 | 475 | 7 | 0 | 7 |
| 41328 | 41328-Xall | 14AA | ARDLDSSGWPIFDY | Homsap IGHV6-1*01 F | Homsap IGHD6-19*01 F | Homsap IGHJ4*02 F | C,W | 100 | 427 | H7CGN0J01APFR8 | 11 | 3 | 14 | Sequences file | | |
| | | | gcaagagacctcgatagcagtggctggcctatctttgactac | 4 | Homsap IGHV6-1*01 F | Homsap IGHD6-19*01 F | Homsap IGHJ4*02 F | C,W | 98.22 | 304 | 91.37 | 44 | 424 | 7 | 3 | 10 |
| 41328 | 42 | 2 | gcaagagatctagatagcagtggctggcccatttttgactac | | Homsap IGHV6-1*01 F | Homsap IGHD6-19*01 F | Homsap IGHJ4*02 F | C,W | 99.77 | 305 | 90.97 | 43 | 427 | 3 | 0 | 3 |
| | | | | 0 | Homsap IGHV6-1*01 F or Homsap IGHV6-1*02 F | Homsap IGHD6-19*01 F | Homsap IGHJ4*02 F | C,W | 99.99 | 305 | 91.67 | 43 | 427 | 1 | 0 | 1 |

FIGURE 4 Cont'd

equivalent to the "Excel file" of IMGT/V-QUEST), and "Advanced parameters" (Figure 4(A)) is identical to that of IMGT/V-QUEST (except for two "Detailed View" options, IMGT Collier de Perles and IMGT/Phylogene, which are not available) [22]. The "Search for insertions and deletions in V-REGION," added by default (Figure 4(A)), allows an accurate V-DOMAIN analysis, despite the high frequency of indels due to homopolymer hybridization sequencing errors in NGS 454 sequencing [22,23].

### 3.2.2 *IMGT/HighV-QUEST Results Download*

From their account, users can check the status of their analyses at any time by displaying the "Analysis history" table [22]. When the analysis is completed, a user can "download" the results as a single file in ZIP format. The size of the ZIP file (in Mb) and the number of included files are indicated in the "Analysis history" table. For analyses having less than the expected number of files in the results file, a warning is shown. If the user removes a completed analysis, all related files (sequences and results) are definitively deleted from the server. A user may cancel a queued or a running analysis at any time.

The downloaded ZIP file [22] contains a main folder with 11 files (equivalent to the 11 sheets of the Excel file provided by the classical IMGT/V-QUEST) in CSV format (Table 6), and one subfolder with individual files, in Text, for each sequence (providing "Detailed view" results) [22]. Text and CSV formats have been chosen in order to facilitate statistical studies for further interpretation and knowledge extraction.

### 3.2.3 IMGT/HighV-QUEST Statistical Analysis

Following the IMGT/HighV-QUEST analysis, the results of completed analysis can be submitted for a statistical analysis, which provides a summary with the different results categories (the "1 copy" filtered-in category for the statistical analysis and the other filtered-out categories) and, for the "1 copy," in tables and histograms, the V, D, and J gene and allele usage observed in the "single allele" and "several alleles." For the submission, the user chooses completed results (several results can be chosen, up to 500,000 sequences) [22], a title for the statistical analysis, e-mail notifications, and whether he or she wants a separate copy of the graphical elements (histograms) in the final results. The user can also enter a comment of up to 500 characters and choose to include it in the final PDF reports. All chosen results for a statistical analysis must be compatible (same locus, same value for the options "Search for insertions/deletions" and "IMGT reference directory set") and without warnings. The status of all statistical analyses can be checked using the "Statistics history" table.

The downloaded result file is a single file in ZIP format, comprising of five PDF reports [22]: (1) Summary, (2) Number of "1 copy" "single allele" and "several alleles (or genes)" tables and histograms, (3) V, D, and J genes and alleles tables and histograms for "single allele" and "several alleles (or genes)," (4) CDR3-IMGT tables and histograms, and (5) filtered-out sequences; and another general PDF report containing all these results together. If chosen by the user, the results contain also the graphical elements in a directory named "graphics."

### 3.2.4 IMGT/HighV-QUEST Identification of IMGT Clonotypes (AA)

In the literature, clonotypes are defined differently, depending on the experiment design (functional specificity) or available data. Thus, a clonotype may denote either a complete antigen receptor (e.g., IgG1-kappa), or only one of the two chains of the receptor (e.g., H or L), or one domain (e.g., VH), or the CDR3 sequence of a domain. Moreover, the sequence can be at the AA or nt level, and this is rarely specified. Therefore, the goal of IMGT® was first of all to define clonotypes and their properties, which could be identified and characterized by IMGT/HighV-QUEST, unambiguously.

In IMGT®, the clonotype, designated as "IMGT clonotype (AA)," is defined by a unique V-(D)-J rearrangement (with IMGT gene and allele names determined by IMGT/HighV-QUEST at the nt level) and a unique CDR3-IMGT AA (in-frame) junction sequence [23] (Figure 4(B)). For identifying "IMGT clonotypes (AA)" in a given IMGT/HighV-QUEST dataset, the "1 copy" are filtered to select for sequences with in-frame junction, conserved anchors 104 and 118 ("C" is 2nd-CYS 104, and "F" or "W" is the J-PHE or J-TRP 118) and for V and J functional or ORF, and "single allele" (for V and J) [23].

By essence, an "IMGT clonotype (AA)" is "unique" for a given dataset. For that reason, each "IMGT clonotype (AA)," in a given dataset, has a unique set identifier (column "Exp. ID") (Figure 4(B)) and, importantly, has a unique representative sequence (link in column "Sequence ID") selected by IMGT/HighV-QUEST among the "1 copy" "single allele" (for V and J), based on the highest percent of identity of the V-REGION ("V %") compared to that of the closest germline, and/or on the sequence length (thus the most complete V-REGION) [23].

In a second step, the "1 copy" "single allele" sequences not selected as representative are assigned to a characterized IMGT clonotype (AA). These sequences differ from the representative sequence by a different (usually shorter) length, and/or by sequencing errors in the V-REGION (lower "V %" of identity) or in the J-REGION, and/or by nt differences in the CDR3-IMGT. In these latter cases these sequences are identified as "IMGT clonotypes (nt)," the nt differences resulting from sequencing errors or, if this can be proven experimentally, from molecular convergence. For a given "IMGT clonotype (AA)," the number (nb) of different CDR3-IMGT (nt) or "IMGT clonotypes (nt)," the CDR3-IMGT sequence (nt) and the nb of different nt in the CDR3-IMGT are reported in the results (Figure 4(C)) [23].

Thus, by proceeding stepwise to assign sequences to clonotypes, aggregating to the representative sequence, first the nonrepresentative "1 copy" "single allele," then the other categories ("1 copy" "several alleles [or genes]," and "More than 1" for each "1 copy" [22]), the high quality and specific characterization of the "IMGT clonotype (AA)" remain unaltered [23]. For the first time for NGS antigen receptor data analysis, the IMGT® standardized approach allows a clear distinction and accurate evaluation between the clonal diversity (nb of "IMGT clonotypes (AA)"), and the clonal expression (total nb of sequences assigned, unambiguously, to a given "IMGT clonotype (AA)") [23]. These assignments are clearly described and visualized in detail so the user always has the means of checking clonotypes individually. Indeed, the sequences of each "1 copy"

**TABLE 6 Content of the 11 IMGT/HighV-QUEST Results Files**

| File Number | File Name | Number of Columns | Results Content[a] |
|---|---|---|---|
| #1 | Summary | 25–29 | • Identity percentage with the closest V, D, and J genes and alleles,<br>• FR-IMGT and CDR-IMGT lengths,<br>• Amino acid (AA) JUNCTION,<br>• Description of insertions and deletions if any. |
| #2 | IMGT-gapped-nt-sequences | 18 | • Nucleotide (nt) sequences gapped according to the IMGT unique numbering for the labels V-D-J-REGION, V-J-REGION, V-REGION, FR1-IMGT, CDR1-IMGT, FR2-IMGT, CDR2-IMGT, FR3-IMGT,<br>• nt sequences of CDR3-IMGT, JUNCTION, J-REGION, and FR4-IMGT. |
| #3 | Nt-sequences | 63–78 | • nt sequences of all labels that can be automatically annotated by IMGT/Automat. |
| #4 | IMGT-gapped-AA-sequences | 18 | • AA sequences gapped according to the IMGT unique numbering for the labels V-D-J-REGION, V-J-REGION, V-REGION, FR1-IMGT, CDR1-IMGT, FR2-IMGT, CDR2-IMGT, FR3-IMGT,<br>• AA sequences of CDR3-IMGT, JUNCTION, J-REGION, and FR4-IMGT. |
| #5 | AA-sequences | 18 | Same columns as "IMGT-gapped-AA-sequences" (#4), but sequences of labels are without IMGT gaps. |
| #6 | Junction | 33, 46, 66, or 77 | • Results of IMGT/JunctionAnalysis (JUNCTION, CDR3, N regions, trimmed V(D)J nucleotides, P regions, D regions, and D-REGION reading frame (for IGH, TRB, and TRD sequences), molecular mass, pI. |
| #7 | V-REGION-mutation-and-AA-change table | 11 | • List of mutations (nt mutations, AA changes, AA class identity (+) or change (−)) for V-REGION, FR1-IMGT, CDR1-IMGT, FR2-IMGT, CDR2-IMGT, FR3-IMGT, and germline CDR3-IMGT. |
| #8 | V-REGION-nt-mutation-statistics | 130 | • Number (nb) of nt positions including IMGT gaps, nb of nt, nb of identical nt, total nb of mutations, nb of silent mutations, nb of nonsilent mutations, nb of transitions (a>g, g>a, c>t, t>c), and nb of transversions (a>c, c>a, a>t, t>a, g>c, c>g, g>t, t>g) for V-REGION, FR1-IMGT, CDR1-IMGT, FR2-IMGT, CDR2-IMGT, FR3-IMGT, and germline CDR3-IMGT. |
| #9 | V-REGION-AA-change-statistics | 189 | • nb of AA positions including IMGT gaps, nb of AA, nb of identical AA, total nb of AA changes, nb of AA changes according to AAclassChangeType (+ + +, + + −, + − +, + − −, − + −, − − +, − − −), and nb of AA class changes according to AAclassSimilarityDegree (nb of Very similar, nb of Similar, nb of Dissimilar, nb of Very dissimilar) for V-REGION, FR1-IMGT, CDR1-IMGT, FR2-IMGT, CDR2-IMGT, FR3-IMGT, and germline CDR3-IMGT. |
| #10 | V-REGION-mutation-hotspots | 8 | • Hot spots motifs (a/t)a, t(a/t), (a/g)g(c/t)(a/t), (a/t)(a/g)c(c/t) detected in the closest germline V-REGION with positions in FR-IMGT and CDR-IMGT. |
| #11 | Parameters | | • Date of the analysis,<br>• IMGT/V-QUEST program version, IMGT/V-QUEST reference directory release,<br>• Parameters used for the analysis: species, receptor type or locus, IMGT reference directory set, and Advanced parameters. |

The content of the 11 results files is in CVS format (results equivalent to those of the Excel file from IMGT/V-QUEST online).
[a]*Files 1–10 comprise systematically sequence identification (name, functionality, and names of the closest V, D, and J genes and alleles [2,3]. Files 7–10, which report the analysis of mutations, are used mostly for immunoglobulins (IG).*

assigned to a given "IMGT clonotype (AA)" are available in "Sequences file" (Figure 4(B) and (C)) [23]. The user can easily perform an analysis of these sequences on line with IMGT/V-QUEST (up to 50 sequences, selecting "Synthesis view display" and the option "Search for insertions and deletions"), and/or with IMGT/JunctionAnalysis (up to 5000 junction sequences), which provide a visual representation familiar to the IMGT® users (Figure 3(D)).

Clonal diversity is also visualized in the online results with histograms which represent the number of IMGT clonotypes (AA) per V, D (for IGH), and J genes (in pink) [23]. Clonal expression is visualized with histograms, which represent the number of sequences assigned to IMGT clonotypes (AA) per V (in green), D (in red), and J (in yellow) genes [23]. Values are normalized, respectively, for 10,000 IMGT clonotypes (AA) to represent IG diversity immunoprofiles per V, D (for IGH), and J genes, and for 10,000 sequences assigned to IMGT clonotypes (AA) to represent IG expression immunoprofiles per V, D (for IGH), and J genes [23]. These normalized values allow comparative analysis studies performed with the same IMGT/HighV-QUEST standards [23].

# 4. IMGT® ANTIBODY ENGINEERING AND HUMANIZATION

## 4.1 IMGT/DomainGapAlign

IMGT/DomainGapAlign [9,24,25] is the IMGT® online tool for the analysis of amino acid sequences and 2D structures of domains (V and C for IG) (Table 5). It analyzes domain amino acid sequences by comparison with the IMGT domain reference directory sets (translation of the germline V and J genes and of the C gene domains). IMGT/DomainGapAlign functionalities include: introduction of "IMGT gaps" in the user amino acid sequences, alignments and identification of the genes and alleles by comparison with the closest domain(s), delimitation of the domain(s) (e.g., V or C) in the user sequence (Figure 5), description of the amino acid changes, and IMGT Collier de Perles.

### *4.1.1 Submission to IMGT/DomainGapAlign*

In the IMGT/DomainGapAlign Welcome page, amino acid sequences are submitted in FASTA format (pasted in a text area or uploaded in a file). A precise delimitation of the domain sequences is not required; however, if the sequence contains several domains, the sequence should be split between the different domains. Several domain amino acid sequences can be analyzed simultaneously (up to 50) provided that each sequence has a distinct name and that they all belong to the same domain type (e.g., V or C). If not, the query needs to be launched for each domain type, successively. If the limits and the numbers of domains of an amino acid sequence are unknown, the protein can be analyzed progressively, shortening the sequence once a domain has been identified by the tool (it should be remembered that the first domain identified by the tool is not necessarily the first one in the protein).

### *4.1.2 IMGT/DomainGapAlign Results*

IMGT/DomainGapAlign identifies the closest germline V-REGION and J-REGION alleles (for V domain) and the closest C-DOMAIN alleles (for C domain). IMGT/DomainGapAlign displays the user V-DOMAIN sequences aligned with the closest V and J regions, with IMGT gaps and delimitations of the strands and loops and the FR-IMGT and CDR-IMGT, according to the IMGT unique numbering. In Figure 5(A), the VH sequence (from b12 Fab, 1n0x_H) is identified as having 79.6% and 93.8% identity with the *H. sapiens* IGHV1-3*01 and IGHJ6*03, respectively. The VH CDR-IMGT lengths are [8.8.20] and the FR-IMGT lengths [25.17.38.11]=91 AA. If several closest genes and/or alleles are identified, the user can select the display of each corresponding alignment. Clicking on the user sequence name in the alignment gives access to the IMGT/Collier-de-Perles tool [26], which automatically provides the IMGT Collier de Perles of the analyzed domain [65–68] with highlighted amino acid differences (in pink circles online) with the closest germline sequence (Figure 1(C)).

The user amino acid sequence is also displayed, according to the IMGT color menu, with the delimitations of the V-REGION (in green) and J-REGION (in yellow), and for a VH, of the (N-D)-REGION (in red), identified by the tool (Figure 5(A)). The number of amino acid differences in the FR-IMGT and CDR-IMGT is one of the criteria to evaluate the potential immunogenicity of the V-DOMAIN [76–80]. The characteristics of the AA changes are shown in strands and loops (and, for the V-DOMAIN, in FR-IMGT and CDR-IMGT).

IMGT/DomainGapAlign analyzes the user C-DOMAIN sequences with similar functionalities (Figure 5(B)): alignments and identification of the genes and alleles with the closest C domain; delimitation of the C-DOMAIN in the user sequence; characteristics of the AA changes in strands, turns, and loops; IMGT Collier de Perles of the C-DOMAIN with highlighted amino acid differences (in pink circles online) with the closest reference sequence.

### *4.1.3 IMGT Domain Reference Directory*

The IMGT domain reference directory is the IMGT reference directory for V, C, and G domains. Sequences are from the IMGT Repertoire [1] and from IMGT/GENE-DB [7]. Owing to the particularities of the V-DOMAIN synthesis [2,3], there is no V-DOMAIN in the IMGT reference directory. Instead, the directory comprises the translation of the IG and TR germline V and J genes (V-REGION and J-REGION, respectively). The IMGT domain reference directory provides the IMGT® "gene" and "allele" names.

(A)

FIGURE 5 **IMGT/DomainGapAlign alignment results.** (A) VH. The nine strands (A, B, C, C′, C″, D, E, F, and G) (horizontal arrows) and the three loops (BC, C′C″, and FG) and, on the upper line, the FR-IMGT and CDR-IMGT and their delimitations (start and end positions) are shown according to the IMGT unique numbering for V domain [59–61,64] (Table 2). The closest genes and alleles are identified automatically by IMGT/DomainGapAlign by comparison with the IMGT domain reference directory (V and J genes and alleles for a V-DOMAIN). The analyzed VH sequence is identified as having 79.6% identity with the germline *Homo sapiens* IGHV1-3*01 for its V-REGION and 93.8% identity with the germline *Homo sapiens* IGHJ6*03 for its J-REGION (the VH amino acid sequence is from b12 Fab, 1n0x_H, in IMGT/3Dstructure-DB, http://www.imgt.org, and was submitted ungapped). The VH CDR-IMGT lengths are [8.8.20] and the FR-IMGT lengths [25.17.38.11]=91 AA (results online, below the alignment). The V-REGION of the b12 VH sequence is heavily mutated as shown by the high number (20) of amino acid changes, shown in bold below the alignment, and online (not shown) detailed per strand and per loop online in the IMGT/DomainGapAlign results [9,24,25]. One AA change is also observed in the FR4-IMGT (T125>I). Moving the mouse cursor on a letter in bold displays the AA change according to its IMGT AA classes [28], as shown for "K24>Q (+ − −), dissimilar." Horizontal green and yellow lines identify the V-REGION and J-REGION, respectively. The region localized in the CDR3-IMGT, which results from the V-(D)-J rearrangement and which cannot be identified as being V or J, is the (N-D)-REGION (shown with a horizontal red line). Below the alignment, the domain sequence, delimited by the tool, is also visualized with the identified V-REGION (green), (N-D)-REGION (red), and J-REGION (yellow) (IMGT®, http://www.imgt.org, IMGT Scientific chart>IMGT color menu). (B) CH1 domain. The analyzed CH1 domain is identified as having 100% identity at the amino acid level with the *Homo sapiens* IGHG1*01, IGHG1*02, and IGHG1*04 CH1 (the CH1 amino acid sequence is from b12 Fab, 1n0x_H, in IMGT/3Dstructure-DB, http://www.imgt.org, and was submitted ungapped). IMGT/DomainGapAlign displays the C domain amino acid sequence of the user, with IMGT gaps and delimitations of the strands (shown with arrows, except for the transversal CD strand), turns (AB, DE, and EF), and loops (BC and FG), according to the IMGT unique numbering [62] (Table 4). A horizontal blue line identifies the C-DOMAIN. The lower alignment is by comparison with the IGHG1*03 CH1 to highlight the AA change R120>K, which corresponds to a different Gm allotype [74], as detailed later in the text and in Table 7. The four conserved AA between the V and C domains are in bold and in color online: C23 (pink), W41 (blue), hydrophobic 89 (blue, here M and V, respectively), and C104 (pink) (IMGT®, http://www.imgt.org, IMGT Scientific chart>IMGT color menu).

Data are comprehensive for human and mouse IG (and also TR), whereas for other species and other IgSF and MhSF they are added progressively. The IMGT domain reference directory comprises domain sequences of functional (F), ORF (open reading frame), and in-frame pseudogene (P) genes. As IMGT® alleles are characterized at the nucleotide level, identical sequences at the amino acid level may therefore correspond to different alleles in the IMGT domain reference directory. The sequences can be displayed by querying IMGT/DomainDisplay (http://www.imgt.org).

## 4.2 IMGT/3Dstructure-DB

IMGT/3Dstructure-DB [8–10], the IMGT® structure database, provides IMGT® annotation and contact analysis of IG 3D structures and paratope/epitope description of IG/

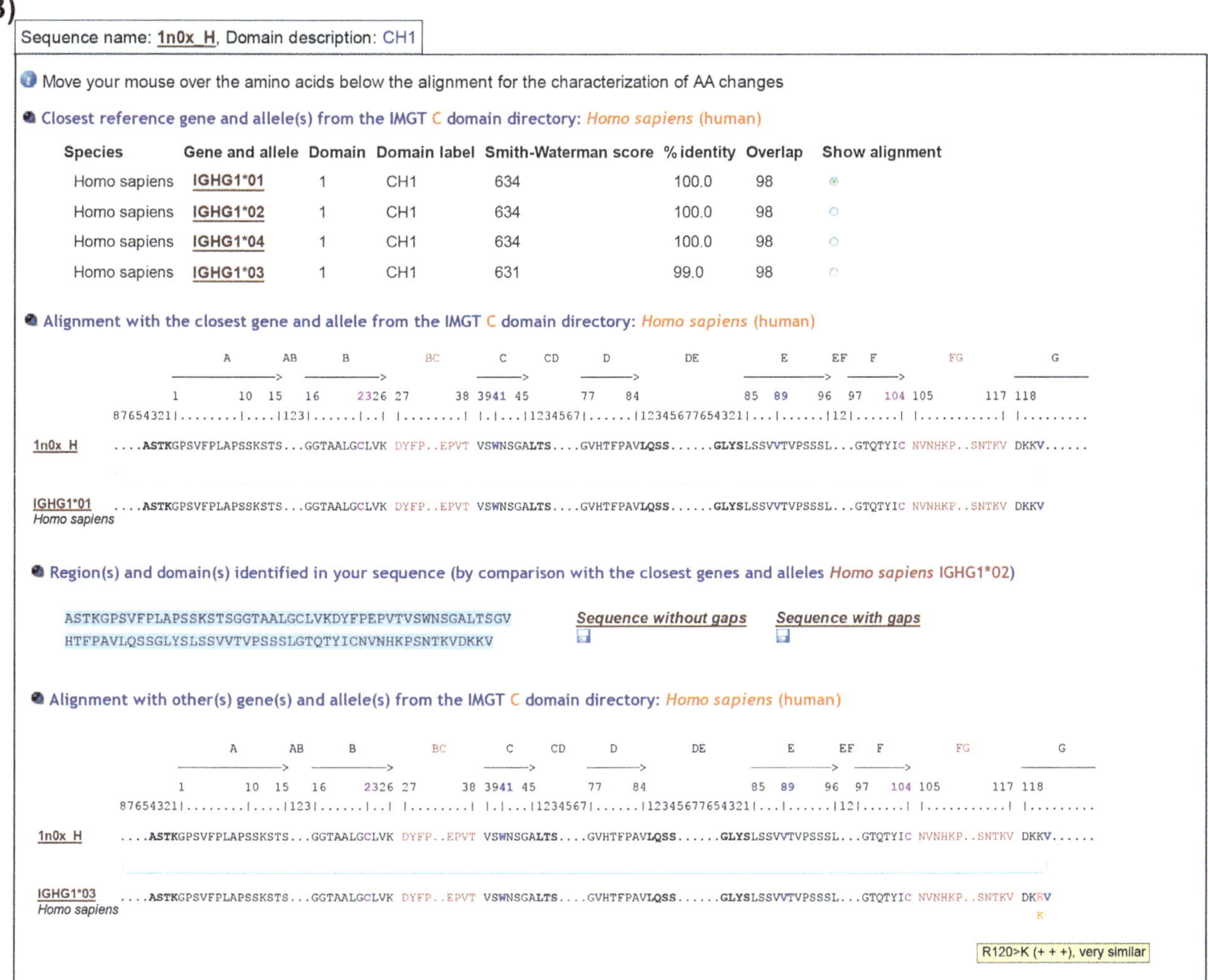


FIGURE 5 Cont'd

antigen complexes (Table 5). There is one "IMGT/3D structure-DB card" per IMGT/3Dstructure-DB entry and this card provides access to all data related to that entry. The "PDB code" (4 letters and/or numbers, e.g., 1n0x) is used as "IMGT entry ID" for the 3D structures obtained from the Research Collaboratory for Structural Bioinformatics (RCSB) Protein Data Bank (PDB) [94]. The IMGT/3D structure-DB card provides eight search/display options: "Chain details," "Contact analysis," "Paratope and epitope," "3D visualization Jmol or QuickPDB," "Renumbered IMGT files," "IMGT numbering comparison," "References and links," and "Printable card" [8–10].

### 4.2.1 IMGT Chain and Domain Annotation

The "Chain details" section comprises information first on the chain itself, then per domain [8–10]. Chain and domain annotation includes the IMGT gene and allele names (CLASSIFICATION), region and domain delimitations (DESCRIPTION), and domain AA positions according to the IMGT unique numbering (NUMEROTATION) [59–64] (Figure 6(A)). The closest IMGT® genes and alleles (found expressed in each domain of a chain) are identified with the integrated IMGT/DomainGapAlign [9,24,25], which aligns the amino acid sequences of the 3D structures with the IMGT domain reference directory.

### 4.2.2 Contact Analysis

"Contact analysis" gives access to a table with the different "Domain pair contacts" of the 3D structure (this table is also accessed from "Chain details" by clicking on "Domain contact (overview)"). "Domain pair contacts" refer to contacts between a pair of domains or between a domain and a ligand. Clicking on "DomPair" gives access to the contacts between amino acids for a given "Domain pair contacts." Contacts between VH and the Ligand (antigen, Ag) and the V-KAPPA and the Ligand (Ag) of an IG/Ag complex are shown in

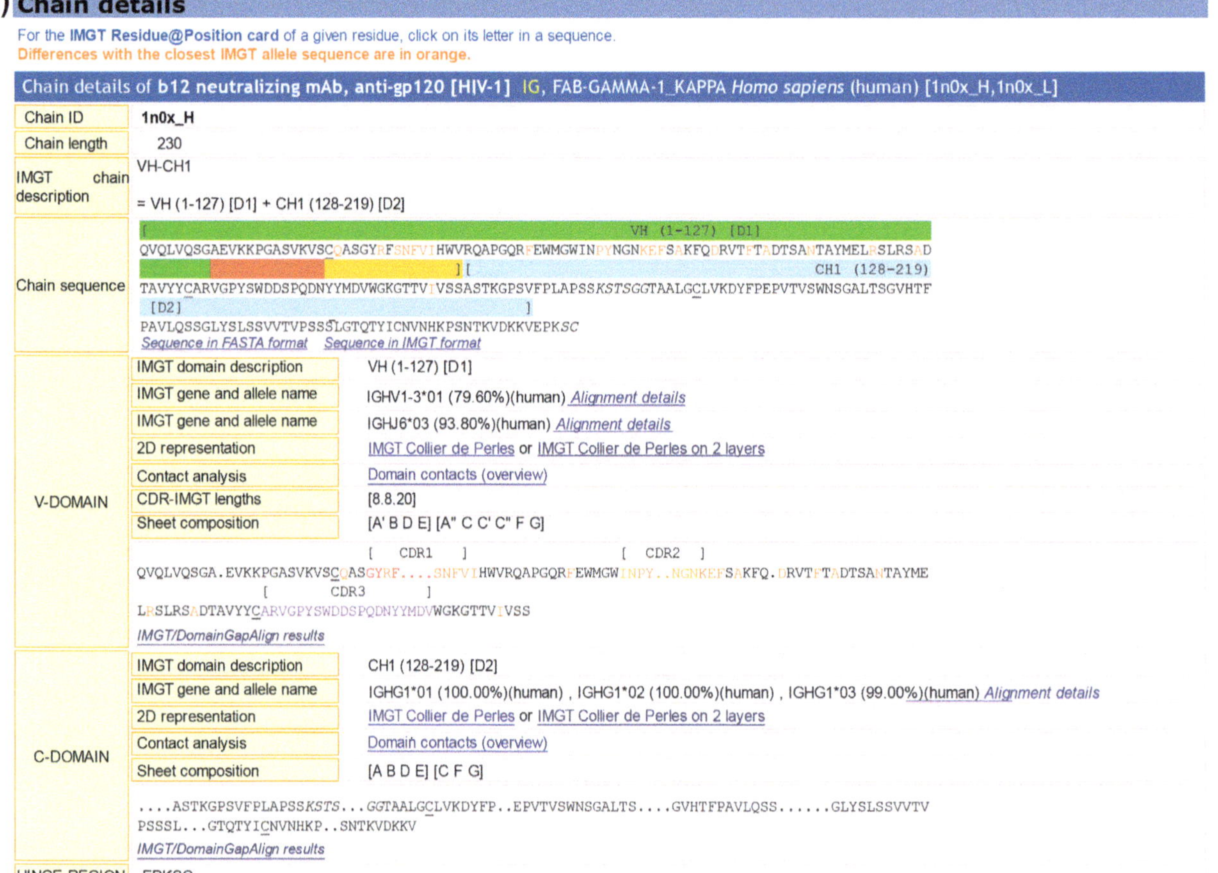


FIGURE 6 **IMGT/3Dstructure-DB.** (A) IMGT/3Dstructure-DB card. The "IMGT/3Dstructure-DB card" is available for each entry of the database. The "Chain details" shows, first, information on the chain (Chain ID, Chain length, IMGT chain description, Chain sequence), then a detailed description of each domain of the chain. The description of the V-DOMAIN (VH) and C-DOMAIN (CH1) of the VH-CH1 chain (1n0x_H) of the b12 Fab is shown. A similar result display interface is provided in IMGT/2Dstructure-DB cards but without "Contact analysis" (and without hydrogen bonds in IMGT Collier de Perles on two layers). (B) IMGT/3Dstructure-DB Domain pair contacts between the "VH" and the "Ligand" (antigen, Ag) of an IG/Ag complex. The VH is from the VH-CH1 chain (1n0x_H) of the b12 Fab and the ligand is a synthetic peptide (1n0x_P). The VH is in contact with the ligand by three AA of the CDR2-IMGT (orange online) (N62, N64, and K65) and two AA of the CDR3-IMGT (purple online) (P112.3 and Q112.2). The two AA that interact with the ligand but do not belong to the CDR-IMGT are the anchors W55 and E66. These contacts are not unexpected given the small size (peptide) of the ligand. (C) IMGT/3Dstructure-DB Domain pair contacts between the "V-KAPPA" and the "Ligand" (Ag) of an IG/Ag complex. The V-KAPPA is from the L-KAPPA chain (1n0x_L) of the b12 Fab and the ligand is the peptide (1n0x_P), as in (B). The V-KAPPA is in contact with the ligand by seven AA of the CDR1-IMGT (blue online) (H27, S28, I29, R30, S36, R37, and R38) and three AA of the CDR3-IMGT (green-blue online) (A109, S114, and S115). "Polar," "Hydrogen bond," and "Nonpolar" are selected by default in "Atom pair contact types" options at the bottom of the page (not shown). The user can also choose to display these contacts by "Atom pair contact categories" (BB), (SS), (BS), and (SB). Clicking on R@P gives access to the IMGT Residue@Position card (not shown). (D) "IMGT paratope and epitope details" of an IG/Ag complex. Each AA that belongs to the IG paratope is characterized by its position in the V-DOMAIN according to the IMGT unique numbering [59–62,64] (AA are listed first for the VH, then for the VL). As examples, "N(62_H)" is the asparagine (N) at position 62 of the VH of the 1n0x_H chain whereas S(28_L) is the serine (S) at position 28 of the V-KAPPA of the 1n0x_L chain. Each AA that belongs to the antigenic determinant (epitope) is characterized by its position in the ligand chain, in the 3D structure. For example, "N(12_P)" means that the asparagine (N) is at position 12 of the 1n0x_P chain. The "IMGT paratope and epitope" analysis of the IG/Ag 3D structure (1n0x) is from IMGT/3Dstructure-DB, http://www.imgt.org.

Figure 6(B) and (C), respectively. These contact analysis representations are important as they demonstrate that most, if not all, contacts with the ligands involved the amino acids of the CDR-IMGT. They definitively confirmed the CDR-IMGT delimitations as the official reference standards [64,68,91].

In IMGT/3Dstructure-DB, all contacts are described as atom pair contacts. Atom pair contacts are obtained by a local program in which atoms are considered to be in contact when no water molecule can take place between them [8,9]. Atom pair contacts are provided by atom contact types (noncovalent, polar, hydrogen bond, nonpolar, covalent, disulfide) and/or atom contact categories (BB, backbone/backbone; SS, side chain/side chain; BS, backbone/side chain; SB, side chain/backbone) [8,9,88,89].

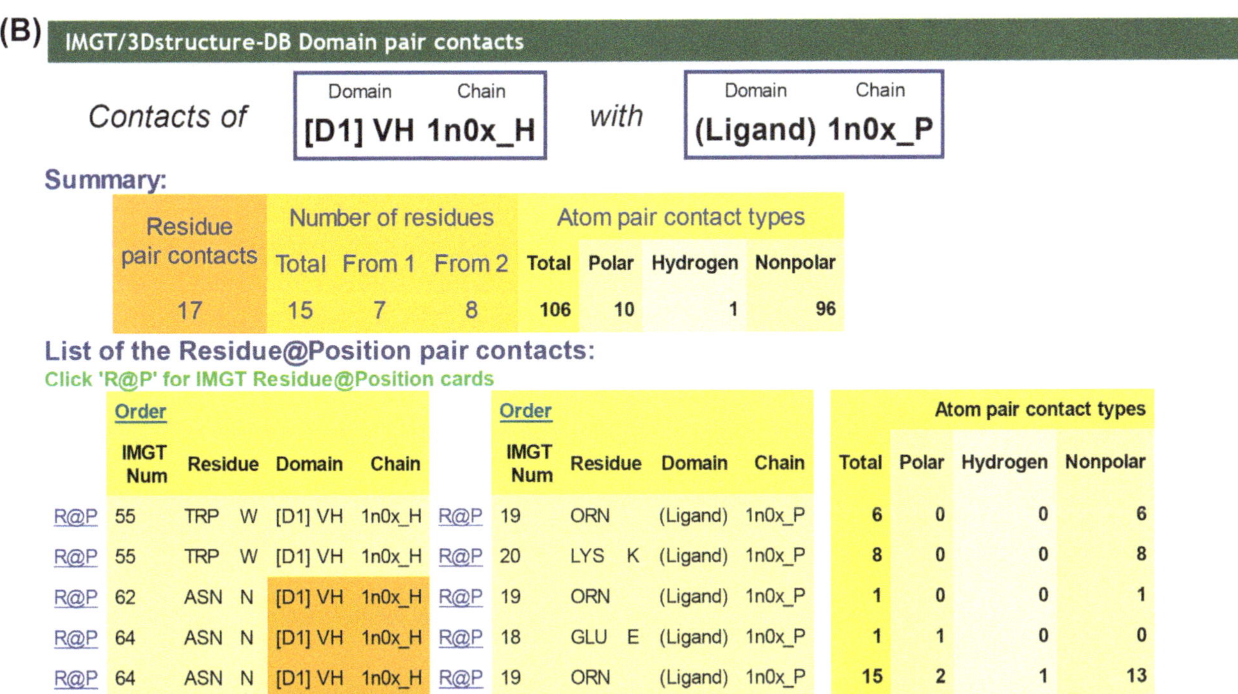

(B) IMGT/3Dstructure-DB Domain pair contacts

Contacts of [D1] VH 1n0x_H (Domain, Chain) with (Ligand) 1n0x_P (Domain, Chain)

Summary:

| Residue pair contacts | Number of residues | | | Atom pair contact types | | | |
|---|---|---|---|---|---|---|---|
| | Total | From 1 | From 2 | Total | Polar | Hydrogen | Nonpolar |
| 17 | 15 | 7 | 8 | 106 | 10 | 1 | 96 |

List of the Residue@Position pair contacts:

Click 'R@P' for IMGT Residue@Position cards

| | Order | | | | | Order | | | | | Atom pair contact types | | | |
|---|---|---|---|---|---|---|---|---|---|---|---|---|---|---|
| | IMGT Num | Residue | Domain | Chain | | IMGT Num | Residue | Domain | Chain | Total | Polar | Hydrogen | Nonpolar |
| R@P | 55 | TRP W | [D1] VH | 1n0x_H | R@P | 19 | ORN | (Ligand) | 1n0x_P | 6 | 0 | 0 | 6 |
| R@P | 55 | TRP W | [D1] VH | 1n0x_H | R@P | 20 | LYS K | (Ligand) | 1n0x_P | 8 | 0 | 0 | 8 |
| R@P | 62 | ASN N | [D1] VH | 1n0x_H | R@P | 19 | ORN | (Ligand) | 1n0x_P | 1 | 0 | 0 | 1 |
| R@P | 64 | ASN N | [D1] VH | 1n0x_H | R@P | 18 | GLU E | (Ligand) | 1n0x_P | 1 | 1 | 0 | 0 |
| R@P | 64 | ASN N | [D1] VH | 1n0x_H | R@P | 19 | ORN | (Ligand) | 1n0x_P | 15 | 2 | 1 | 13 |
| R@P | 64 | ASN N | [D1] VH | 1n0x_H | R@P | 20 | LYS K | (Ligand) | 1n0x_P | 8 | 1 | 0 | 7 |
| R@P | 65 | LYS K | [D1] VH | 1n0x_H | R@P | 19 | ORN | (Ligand) | 1n0x_P | 1 | 1 | 0 | 0 |
| R@P | 65 | LYS K | [D1] VH | 1n0x_H | R@P | 20 | LYS K | (Ligand) | 1n0x_P | 8 | 1 | 0 | 7 |
| R@P | 66 | GLU E | [D1] VH | 1n0x_H | R@P | 20 | LYS K | (Ligand) | 1n0x_P | 9 | 2 | 0 | 7 |
| R@P | 66 | GLU E | [D1] VH | 1n0x_H | R@P | 21 | LYS K | (Ligand) | 1n0x_P | 2 | 0 | 0 | 2 |
| R@P | 112.3 | PRO P | [D1] VH | 1n0x_H | R@P | 4 | SER S | (Ligand) | 1n0x_P | 5 | 0 | 0 | 5 |
| R@P | 112.3 | PRO P | [D1] VH | 1n0x_H | R@P | 5 | TYR Y | (Ligand) | 1n0x_P | 9 | 1 | 0 | 8 |
| R@P | 112.3 | PRO P | [D1] VH | 1n0x_H | R@P | 6 | MET M | (Ligand) | 1n0x_P | 19 | 0 | 0 | 19 |
| R@P | 112.3 | PRO P | [D1] VH | 1n0x_H | R@P | 17 | ALA A | (Ligand) | 1n0x_P | 2 | 0 | 0 | 2 |
| R@P | 112.2 | GLN Q | [D1] VH | 1n0x_H | R@P | 6 | MET M | (Ligand) | 1n0x_P | 10 | 0 | 0 | 10 |
| R@P | 112.2 | GLN Q | [D1] VH | 1n0x_H | R@P | 17 | ALA A | (Ligand) | 1n0x_P | 1 | 0 | 0 | 1 |
| R@P | 112.2 | GLN Q | [D1] VH | 1n0x_H | R@P | 21 | LYS K | (Ligand) | 1n0x_P | 1 | 1 | 0 | 0 |

FIGURE 6 Cont'd

Clicking on "R@P" gives access to the IMGT identity card of a given residue (usually an amino acid) at a given position or Residue@Position. The IMGT R@P card can also be accessed from the amino acid sequences of the IMGT/3Dstructure-DB card or from the IMGT Collier de Perles, by clicking on one amino acid. In an IMGT R@P card, the Residue@Position is defined by the IMGT position numbering in a domain, or if not characterized, in the chain, the amino acid name (3-letter and, between parentheses, 1-letter abbreviations), the IMGT domain description, and the IMGT chain ID, e.g., "103 - TYR (Y) - VH - 1hzh_H" [8–10]. The IMGT R@P card includes: (1) general information (PDB file numbering, IMGT file numbering, residue full name and formula); (2) structural information "IMGT LocalStructure@Position" (secondary structure, Phi and Psi angles (in degrees), and accessible surface area (ASA) (in square angstrom)); and (3) detailed contact analysis with amino acids of other domains [8–10].

### 4.2.3 Paratope and Epitope

In an IG/Ag complex, the amino acids in contact at the interface between the IG and the Ag constitute the paratope on the IG V-DOMAIN surface and the epitope on the Ag surface. For IG/Ag, the paratope and epitope are displayed in Contact analysis (Figure 6(B) and (C)), but for each V domain separately. Clicking on the "Paratope and

(C)

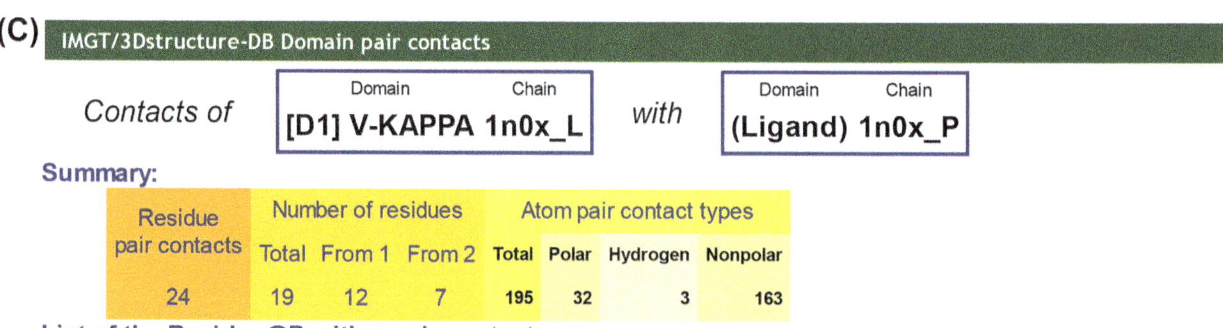


| Residue pair contacts | Number of residues | | | Atom pair contact types | | | |
|---|---|---|---|---|---|---|---|
| | Total | From 1 | From 2 | Total | Polar | Hydrogen | Nonpolar |
| 24 | 19 | 12 | 7 | 195 | 32 | 3 | 163 |

**List of the Residue@Position pair contacts:**

Click 'R@P' for IMGT Residue@Position cards

| | Order IMGT Num | Residue | | Domain | Chain | | Order IMGT Num | Residue | | Domain | Chain | Total | Polar | Hydrogen | Nonpolar |
|---|---|---|---|---|---|---|---|---|---|---|---|---|---|---|---|
| R@P | 1 | GLU | E | [D1] V-KAPPA | 1n0x_L | R@P | 10 | LEU | L | (Ligand) | 1n0x_P | 4 | 0 | 0 | 4 |
| R@P | 2 | ILE | I | [D1] V-KAPPA | 1n0x_L | R@P | 9 | ASP | D | (Ligand) | 1n0x_P | 6 | 0 | 0 | 6 |
| R@P | 2 | ILE | I | [D1] V-KAPPA | 1n0x_L | R@P | 10 | LEU | L | (Ligand) | 1n0x_P | 11 | 0 | 0 | 11 |
| R@P | 27 | HIS | H | [D1] V-KAPPA | 1n0x_L | R@P | 9 | ASP | D | (Ligand) | 1n0x_P | 7 | 1 | 0 | 6 |
| R@P | 27 | HIS | H | [D1] V-KAPPA | 1n0x_L | R@P | 10 | LEU | L | (Ligand) | 1n0x_P | 11 | 1 | 0 | 10 |
| R@P | 28 | SER | S | [D1] V-KAPPA | 1n0x_L | R@P | 9 | ASP | D | (Ligand) | 1n0x_P | 10 | 3 | 0 | 7 |
| R@P | 29 | ILE | I | [D1] V-KAPPA | 1n0x_L | R@P | 9 | ASP | D | (Ligand) | 1n0x_P | 26 | 3 | 0 | 23 |
| R@P | 30 | ARG | R | [D1] V-KAPPA | 1n0x_L | R@P | 7 | PHE | F | (Ligand) | 1n0x_P | 21 | 0 | 0 | 21 |
| R@P | 30 | ARG | R | [D1] V-KAPPA | 1n0x_L | R@P | 9 | ASP | D | (Ligand) | 1n0x_P | 10 | 2 | 0 | 8 |
| R@P | 30 | ARG | R | [D1] V-KAPPA | 1n0x_L | R@P | 12 | ASN | N | (Ligand) | 1n0x_P | 11 | 4 | 0 | 7 |
| R@P | 36 | SER | S | [D1] V-KAPPA | 1n0x_L | R@P | 7 | PHE | F | (Ligand) | 1n0x_P | 8 | 1 | 0 | 7 |
| R@P | 36 | SER | S | [D1] V-KAPPA | 1n0x_L | R@P | 9 | ASP | D | (Ligand) | 1n0x_P | 14 | 4 | 2 | 10 |
| R@P | 37 | ARG | R | [D1] V-KAPPA | 1n0x_L | R@P | 9 | ASP | D | (Ligand) | 1n0x_P | 1 | 1 | 0 | 0 |
| R@P | 38 | ARG | R | [D1] V-KAPPA | 1n0x_L | R@P | 6 | MET | M | (Ligand) | 1n0x_P | 2 | 0 | 0 | 2 |
| R@P | 38 | ARG | R | [D1] V-KAPPA | 1n0x_L | R@P | 7 | PHE | F | (Ligand) | 1n0x_P | 10 | 4 | 1 | 6 |
| R@P | 38 | ARG | R | [D1] V-KAPPA | 1n0x_L | R@P | 8 | SER | S | (Ligand) | 1n0x_P | 3 | 1 | 0 | 2 |
| R@P | 38 | ARG | R | [D1] V-KAPPA | 1n0x_L | R@P | 9 | ASP | D | (Ligand) | 1n0x_P | 9 | 5 | 0 | 4 |
| R@P | 109 | ALA | A | [D1] V-KAPPA | 1n0x_L | R@P | 6 | MET | M | (Ligand) | 1n0x_P | 8 | 0 | 0 | 8 |
| R@P | 109 | ALA | A | [D1] V-KAPPA | 1n0x_L | R@P | 7 | PHE | F | (Ligand) | 1n0x_P | 2 | 0 | 0 | 2 |
| R@P | 109 | ALA | A | [D1] V-KAPPA | 1n0x_L | R@P | 8 | SER | S | (Ligand) | 1n0x_P | 6 | 1 | 0 | 5 |
| R@P | 109 | ALA | A | [D1] V-KAPPA | 1n0x_L | R@P | 10 | LEU | L | (Ligand) | 1n0x_P | 2 | 0 | 0 | 2 |
| R@P | 109 | ALA | A | [D1] V-KAPPA | 1n0x_L | R@P | 15 | ILE | I | (Ligand) | 1n0x_P | 3 | 0 | 0 | 3 |
| R@P | 114 | SER | S | [D1] V-KAPPA | 1n0x_L | R@P | 10 | LEU | L | (Ligand) | 1n0x_P | 2 | 0 | 0 | 2 |
| R@P | 115 | SER | S | [D1] V-KAPPA | 1n0x_L | R@P | 10 | LEU | L | (Ligand) | 1n0x_P | 8 | 1 | 0 | 7 |

FIGURE 6 Cont'd

epitope" tag (displayed in the IMGT/3Dstructure-DB card, only if relevant) gives access to "IMGT paratope and epitope details," which are described in a standardized way (Figure 6(D)). Each amino acid that belongs to the paratope is defined by its position in an IG V-DOMAIN (e.g., VH and V-KAPPA of 1n0x in Figure 6(D)). Each amino acid that belongs to the epitope is defined by its position in the chain in the 3D structure or, if the antigen belongs to an IgSF or MhSF protein and if the epitope is part of a characterized V, C, or G domain, by its position in the domain according to the IMGT unique numbering.

### *4.2.4 Renumbered Flat File and IMGT Numbering Comparison*

"Renumbered IMGT file" allows viewing (or downloading) of an IMGT coordinate file renumbered according to the IMGT unique numbering, and with added IMGT specific

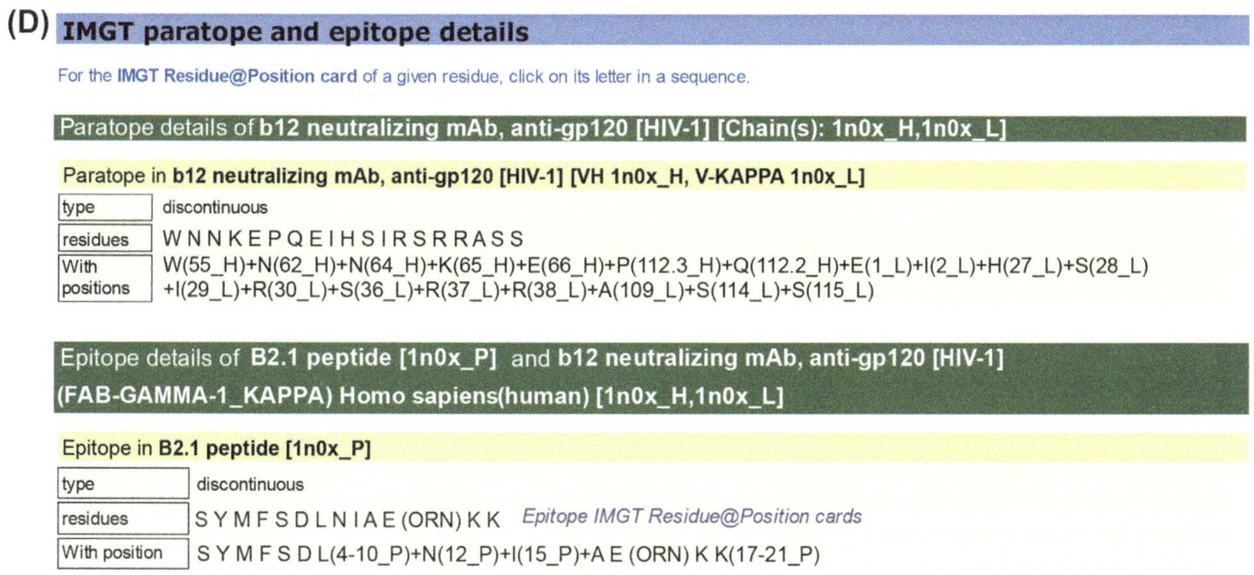

(D) **IMGT paratope and epitope details**

For the **IMGT Residue@Position card** of a given residue, click on its letter in a sequence.

**Paratope details of b12 neutralizing mAb, anti-gp120 [HIV-1] [Chain(s): 1n0x_H,1n0x_L]**

Paratope in **b12 neutralizing mAb, anti-gp120 [HIV-1] [VH 1n0x_H, V-KAPPA 1n0x_L]**

| | |
|---|---|
| type | discontinuous |
| residues | W N N K E P Q E I H S I R S R R A S S |
| With positions | W(55_H)+N(62_H)+N(64_H)+K(65_H)+E(66_H)+P(112.3_H)+Q(112.2_H)+E(1_L)+I(2_L)+H(27_L)+S(28_L) +I(29_L)+R(30_L)+S(36_L)+R(37_L)+R(38_L)+A(109_L)+S(114_L)+S(115_L) |

**Epitope details of B2.1 peptide [1n0x_P] and b12 neutralizing mAb, anti-gp120 [HIV-1] (FAB-GAMMA-1_KAPPA) Homo sapiens(human) [1n0x_H,1n0x_L]**

Epitope in **B2.1 peptide [1n0x_P]**

| | |
|---|---|
| type | discontinuous |
| residues | S Y M F S D L N I A E (ORN) K K *Epitope IMGT Residue@Position cards* |
| With position | S Y M F S D L(4-10_P)+N(12_P)+I(15_P)+A E (ORN) K K(17-21_P) |

FIGURE 6 Cont'd

information on chains and domains (added in the "REMARK 410" lines [blue online] and identical to the "Chain details" annotation).

"IMGT numbering comparison" provides, per domain, the IMGT DOMAIN numbering by comparison with the PDB numbering and the residue (3-letter and 1-letter names), which allows standardized IMGT representations using generic tools (Figures 1(A) and 2(A)).

### 4.2.5 IMGT/3Dstructure-DB Associated Tools

Tools associated with IMGT/3Dstructure-DB include IMGT/StructuralQuery [8] and IMGT/DomainSuperimpose, available online. IMGT/StructuralQuery allows retrieval of the IMGT/3Dstructure-DB entries, based on specific structural characteristics of the intramolecular interactions: phi and psi angles, accessible surface area, type of atom contacts, distance in angstrom between amino acids, IMGT Residue@Position contacts, and, for V-DOMAIN, CDR-IMGT length or pattern [8]. IMGT/DomainSuperimpose allows superimposing of the 3D structures of two domains from IMGT/3Dstructure-DB.

## 4.3 IMGT/2Dstructure-DB

IMGT/2Dstructure-DB was created as an extension of IMGT/3Dstructure-DB [8–10] to describe and analyze amino acid sequences of chains and domains for which no 3D structures were available (Table 5). IMGT/2Dstructure-DB uses the IMGT/3Dstructure-DB informatics frame and interface, which allow one to analyze, manage, and query IG (and also TR and MH, as well as other IgSF and MhSF) and engineered proteins (FPIA, CPCA) as polymeric receptors made of several chains, in contrast to the IMGT/LIGM-DB sequence database that analyzes and manages sequences individually [6]. The amino acid sequences are analyzed with the IMGT® criteria of standardized identification [56], description [57], nomenclature [58], and numerotation [59–64]. The current IMGT/2Dstructure-DB entries include amino acid sequences of antibodies from Kabat [92] (those for which there were no available nucleotide sequences) and amino acid sequences of mAb and FPIA from the WHO-INN program [11,47,48]. Queries can be made on an individual entry using the entry ID or the molecule name. The same query interface is used for IMGT/2Dstructure-DB and IMGT/3Dstructure-DB. Thus a "trastuzumab" query in "Molecule name" allows retrieval of three results: two INN ("trastuzumab" and "trastuzumab emtansine") from IMGT/2Dstructure-DB and one 3D structure ("1nz8") from IMGT/3Dstructure-DB. The IMGT/2Dstructure-DB cards provide standardized IMGT information on chains and domains and IMGT Collier de Perles on one or two layers, identical to that provided for the sequence analysis in IMGT/3Dstructure-DB; however, the information on experimental structural data (hydrogen bonds in IMGT Collier de Perles on two layers, Contact analysis) is only available in the corresponding IMGT/3Dstructure-DB cards if the antibodies have been crystallized.

## 4.4 IMGT® V and C Domain Annotation for Antibody Humanization, Engineering, and Evolution

### 4.4.1 Antibody Humanization

#### 4.4.1.1 CDR-IMGT Delimitation for Grafting

The objective of antibody humanization is to graft at the DNA level the CDR of an antibody V domain, from mouse (or other species) and of a given specificity, onto a human

V domain framework, thus preserving the specificity of the original (murine or other species) antibody while decreasing its immunogenicity [95]. IMGT/DomainGapAlign [9,24,25] is the reference tool for antibody humanization design based on CDR grafting. Indeed, it precisely defines the CDR-IMGT to be grafted and helps in selecting the most appropriate human FR-IMGT by providing the alignment of the amino acid sequences between the mouse (or other species) and the closest human V-DOMAIN.

Analyses performed on humanized therapeutic antibodies underline the importance of a correct delimitation of the CDR and FR. As an example, two amino acid changes were required in the first version of the humanized VH of alemtuzumab, in order to restore the specificity and affinity of the original rat antibody. The positions of these amino acid changes (S28>F and S35>F) are now known to be located in the CDR1-IMGT and should have been directly grafted, but at the time of this mAb humanization they were considered as belonging to the FR according to the Kabat numbering [92]. In contrast, positions 66–74 were, at the same time, considered as belonging to the CDR according to the Kabat numbering, whereas they clearly belong to the FR2-IMGT and the corresponding sequence should have been "human" instead of being grafted from the "rat" sequence (IMGT®, http://www.imgt.org, The IMGT Biotechnology page>Antibody humanization>Alemtuzumab).

#### 4.4.1.2 Amino Acid Interactions between FR-IMGT and CDR-IMGT

IMGT Colliers de Perles from crystallized 3D structures in IMGT/3Dstructure-DB [8–10] highlight two conserved hydrogen bonds between FR-IMGT and CDR-IMGT positions: FR2-IMGT 39 with CDR2-IMGT 56 (or 57) and FR2-IMGT 40 with CDR3-IMGT 105 (Figure 1(B)). Antibody engineering and humanization should therefore preserve these bondings, which stabilize the loops. It is also worthwhile to note that, in VH CDR3, the stem of the CDR3 loop is stabilized by a conserved salt bridge between R106 (arginine contributed by the 3′V-REGION) and D116 (aspartate contributed by the 5′J-REGION of the *Homo sapiens* IGHJ2, IGHJ3, IGHJ4, IGHJ5, or IGHJ6) (IMGT®, http://www.imgt.org, IMGT Repertoire>Proteins and alleles>Alignments of alleles).

### 4.4.2 *IGHG1 Alleles and G1m Allotypes*

Allotypes are polymorphic markers of an IG subclass that correspond to amino acid changes and are detected serologically by antibody reagents [74]. In therapeutic antibodies (human, humanized, or chimeric) [11], allotypes may represent potential immunogenic residues [73], as demonstrated by the presence of antibodies in individuals immunized against these allotypes [74]. The allotypes of the human heavy gamma chains of the IgG are designated as Gm (for gamma marker). The allotypes G1m, G2m, and G3m are carried by the constant region of the gamma1, gamma2, and gamma3 chains, encoded by the IGHG1, IGHG2, and IGHG3 genes, respectively [74]. The gamma1 chains may express four G1m alleles (combinations of G1m allotypes): G1m3, G1m3,1, G1m17,1, and G1m17,1,2 (and in Negroid populations two additional G1m alleles, Gm17,1,28 and Gm17,1,27,28) [74] (Table 7). The C region of the G1m3,1, G1m17,1, and G1m17,1,2 chains differs from that of the G1m3 chains by two, three, and four amino acids, respectively [74]. The correspondence between the G1m alleles and IGHG1 alleles is shown in Table 7. Thus, IGHG1*01 and IGHG1*02 are G1m17,1, IGHG1*03 is G1m3, IGHG1*04 is G1m17,1,2, and IGHG1*05 is G1m3,1. In the IGHG1 CH1, the lysine at position 120 (K120) in strand G corresponds to the G1m17 allotype [74] (Figures 2(D) and 5(B)). The isoleucine I103 (strand F) is specific of the gamma1 chain isotype. If an arginine is expressed at position 120 (R120), the simultaneous presence of R120 and I103 corresponds to the expression of the G1m3 allotype [74]. For the gamma3 and gamma4 isotypes (which also have R120 but T in 103), R120 only corresponds to the expression of the nG1m17 isoallotype (an isoallotype or nGm is detected by antibody reagents that identify this marker as an allotype in one IgG subclass and as an isotype for other subclasses) [74]. In the IGHG1 CH3, the aspartate D12 and leucine L14 (strand A) correspond to G1m1, whereas glutamate E12 and methionine M14 correspond to the nG1m1 isoallotype [74] (Table 7). A glycine at position 110 corresponds to G1m2, whereas an alanine does not correspond to any allotype (G1m2-negative chain) (Table 7). Therapeutic antibodies are most frequently of the IgG1 isotype, and to avoid a potential immunogenicity, the constant region of the gamma1 chains are often engineered to replace the G1m3 allotype by the "less immunogenic" G1m17 (CH1 R120>K) (G1m17 is more extensively found in different populations) [74].

### 4.4.3 *Only-Heavy-Chain Antibodies*

#### 4.4.3.1 Dromedary IgG2 and IgG3 Only-Heavy-Chain Antibodies

Two IgG antibody formats are expressed in the dromedary or Arabian camel (*Camelus dromedarius*) and in Camelidae in general: the conventional IG (with two identical heavy gamma chains associated with two identical light chains) and the "only-heavy-chain" IG (no light chain, and only two identical heavy gamma chains lacking CH1) [96]. The Camdro (for *Camelus dromedarius* in the 6-letter species abbreviation) IGHV1 genes belong to two sets based on four amino acid changes that are characteristic of each set [97]. The first set of IGHV1 genes is expressed in conventional tetrameric IgG1 that constitute

**TABLE 7** Correspondence Between the IGHG1 Alleles and G1m Alleles

| G1m Alleles[a] | | | IMGT Amino Acids[b] and Positions | | | | | |
|---|---|---|---|---|---|---|---|---|
| | | | | CH1 | | CH3 | | |
| | | | 103 | 120 | 12 | 14 | 110 | |
| IGHG1 Alleles | Allotypes | Isoallotypes[c] | | G1m17/ nG1m1 G1m3[d] | | G1m1/ nG1m1 | G1m2/- | Populations [74] |
| IGHG1*01[e] IGHG1*02[e] | G1m17,1 | | I | **K** | **D** | **L** | A | Caucasoid Negroid Mongoloid |
| IGHG1*03 | G1m3 | *nG1m1, nG1m17* | **I** | **R** | E | M | A | Caucasoid |
| IGHG1*07[f] | G1m17,1,2 | | I | **K** | **D** | **L** | **G** | Caucasoid Mongoloid |
| IGHG1*08[f] | G1m3,1 | *nG1m17* | **I** | **R** | **D** | **L** | A | Mongoloid |

[a]*In Negroid populations, the G1m17,1 allele frequently includes G1m27 and G1m28, leading to two additional G1m alleles, G1m17,1,27 and G1m17,1,27,28 [74].*
[b]*Amino acids corresponding to G1m allotypes are shown in bold.*
[c]*The nG1m1 and nG1m17 isoallotypes present on the Gm1-negative and Gm17-negative gamma-1 chains (and on other gamma chains) are shown in italics.*
[d]*The presence of R120 is detected by anti-nG1m17 antibodies, whereas the simultaneous presence of I103 and R120 in the gamma1 chains is detected by anti-Gm3 antibodies [74].*
[e]*The IGHG1*01 and IGHG1*02 alleles only differ at the nucleotide level (codon 85.1 in CH2). The C region of the b12 gamma1 chain (1hzh_H) is encoded by the IGHG1*01 or IGHG1*02 alleles and expresses the G1m17,1 allotypes, based on sequence analysis (CH1 K120 (Figure 2(D)), CH3 D12 and L14) [74]. The presence of an A (Ala) in CH1 121 of 1hzh_H is a PDB file error. It should be a V (Val), as in 1n0x_H (Figure 2(D)) (see IMGT note in IMGT/3D structure-DB, code PDB:1hzh, http://www.imgt.org).*
[f]*IGHG1*07 and IGHG1*08 amino acids are expected [74] but not yet sequenced at the nucleotide level, and therefore the IGHG1*04 and IGHG1*05 alleles are not shown in IMGT Repertoire, Alignments of alleles: Homo sapiens IGHG1 (http://www.imgt.org).*

25% of circulating antibodies. The second set is expressed in the only-heavy-chain antibodies, IgG2 and IgG3, which constitute 75% of the circulating antibodies [96]. The four amino acid changes are located in the FR2-IMGT at positions 42, 49, 50, and 52; the first position, 42, is in the C strand and the three others, 49, 50 and 52, are in the C′ strand (Figure 1, Table 8). They belong to the [GFCC′C″] sheet at the hydrophobic VH-VL interface in conventional antibodies of Camelidae as well as of any vertebrate species, whereas in camelid only-heavy-chain antibodies (no light chains, and therefore no VL), these positions are exposed to the environment with, through evolution, a selection of hydrophilic amino acids.

The respective heavy gamma2 and gamma3 chains are both characterized by the absence of the CH1 domain owing to a splicing defect [98]. It is the absence of CH1 that is responsible for the lack of association of the light chains. Only-heavy-chain antibodies is a feature of the Camelidae IG as they have also been found in the Bactrian camel (*Camelus bactrianus*) of Central Asia and in the llama (*Lama glama*) and alpaca (*Vicugna pacos*) of South America. The genetic event (splicing defect) responsible for the lack of CH1 occurred in their common ancestor before the radiation between the "camelini" and "lamini," dating approximately 11 million years ago.

**TABLE 8** Amino Acid Changes at FR2-IMGT Positions 42, 49, 50, and 52, between the Two Sets of *Camelus dromedarius* IGHV1 Subgroup Genes

| | | *Camelus dromedarius* | |
|---|---|---|---|
| V-REGION | | IGHV1 Set 1 | IGHV1 Set 2 |
| Strand | FR2-IMGT Amino Acid Positions | IGHV1S1 to IGHV1S41[a] | IGHV1S42 to IGHV1S74[a] |
| C | 42 | V | F, Y |
| C′ | 49 | G | E, Q |
| | 50 | L | C, R |
| | 52 | W | F, G, L, W |

It is not excluded that some of these amino acid changes in set 2 may result from somatic mutations. Indeed, sequences may be described as genomic in publications, but most of the time it is not known if they are germline or rearranged (IMGT®, http://www.imgt.org, IMGT Gene tables).
[a]*Temporary nomenclature, waiting for the complete genomic Camelus dromedarius IGH locus.*

The V domains of Camelidae only-heavy-chain antibodies have characteristics for potential pharmaceutical applications (e.g., easy production and selection of single-domain format, extended CDR3 with novel specificities, and binding to protein clefts). They are designated as $VH_H$ when they have to be distinguished from conventional VH (the sequence criteria are based on the four amino acids at positions 42, 49, 50, and 52). The term "nanobody," initially used for describing a single-domain format antibody, is not equivalent to $VH_H$, as it has been used for V domains other than $VH_H$ and for constructs containing more than one V domain (VH and/or $VH_H$) (e.g., caplacizumab, ozoralizumab) (IMGT®, http://www.imgt.org, IMGT Repertoire>Locus and Genes>Gene tables; ibid., IMGT Biotechnology page>Characteristics of the camelidae (camel, llama) antibody synthesis; ibid. IMGT/mAb-DB>caplacizumab; ibid. IMGT/mAb-DB>ozoralizumab).

#### 4.4.3.2 Human Heavy Chain Diseases

The camelidae only-heavy-chain antibodies synthesis is remarkably reminiscent of what is observed in human heavy chain diseases (HCD). These proliferative disorders of B lymphoid cells produce truncated monoclonal immunoglobulin heavy chains, which lack associated light chains. In most HCD, the absence of the heavy chain CH1 domain by deletion or splicing defect may be responsible for the lack of assembly of the light chain [99]. Similar observations have also been reported in mouse variants [99] (IMGT®, http://www.imgt.org, IMGT Education>Pathology of the immune system>Molecular defects in Immunoglobulin Heavy Chain Diseases [HCD]; ibid., IMGT Lexique>Heavy Chain Diseases [HCD]).

#### 4.4.3.3 Nurse Shark IgN

A convergence mechanism in evolution is observed in nurse shark (*Ginglymostoma cirratum*, "Gincir" in the 6-letter species abbreviation) IgN antibodies (previously IgNAR, "immunoglobulin new antigen receptor") [100], which are only-heavy-chain antibodies (homodimeric heavy nu chains without CH1, and no associated light chains). The IGHV genes expressed in the Gincir heavy nu chains belong to the IGHV2 subgroup and are characterized by the absence of the CDR2-IMGT owing to a deletion that encompasses position 54–67. The Gincir IGH genes are organized in duplicated cassettes, and those that express IgN comprise Gincir IGHV2 subgroup genes and an IGHN constant gene (IMGT®, http://www.imgt.org, IMGT Repertoire>Proteins and alleles>Protein display>IGHV>nurse shark (*G. cirratum*); ibid., IMGT Repertoire>Locus and genes>Gene tables>IGHV>Chondrichthyes).

### 4.4.4 IGHG CH Properties and Antibody Engineering

#### 4.4.4.1 N-Linked Glycosylation Site CH2 N84.4

An N-linked glycosylation site is present in the CH2 domain of the constant region of the human IG heavy chains of the four IgG isotypes. The N-linked glycosylation site belongs to the classical N-glycosylation motif N-X-S/T (where N is asparagine, X is any amino acid except proline, S is serine, T is threonine) and is defined as CH2 N84.4. As shown in the IMGT Collier de Perles, this asparagine is localized at the DE turn. The IMGT unique numbering has the advantage of identifying the C domain (here, CH2) and, in the domain, the amino acid and its localization (here, N84.4), which can be visualized in the IMGT Collier de Perles and correlated with the 3D structure [80].

#### 4.4.4.2 Interface Ball-and-Socket-like Joints

The interface between the CH2 and CH3 domains in 3D structures of *Homo sapiens* IGHG2 Fc was recently analyzed and compared with the interface in 3D structures of IGHG1 Fc [101]. This study revealed that in all Fc of gamma chains the movement of the CH2 results from a pivoting around a highly conserved ball-and-socket-like joint [101], in which, using the IMGT numbering, the CH2 L15 side chain (last position of the A strand, next to the AB turn) (the ball) interacts with a pocket (the socket) formed by CH3 M107, H108, E109, and H115 (FG loop) [80]. These amino acids are well conserved between the gamma isotypes and the IGHG genes and alleles except for IGHG3 H115, which shows a polymorphism associated with different G3m allotypes [74]. This ball-and-socket-like joint is a structural feature similar but reversed to that previously described at the VH and CH1 domain interface [102], in which the VH L12, T125, and S127 form the socket, whereas the CH1 F29 and P30 (BC loop) form the ball.

#### 4.4.4.3 Knobs-into-Holes CH3 T22 and Y86

The knobs-into-holes methodology has been proposed for obtaining bispecific antibodies [103]. The aim is to increase interactions between the CH3 domain of two gamma1 chains that belong to antibodies with a different specificity. Two amino acids, CH3 T22 (B strand) and Y86 (E strand), which belong to the [ABED] sheet, at the interface of the two *Homo sapiens* IGHG1 CH3 domains [80], were selected for amino acid changes. Interactions of these two amino acids are described in "Contact analysis" in IMGT/3Dstructure-DB [8–10]. The knobs-into-holes methodology consists of an amino acid change on one CH3 domain (T22>Y) that creates a knob, and another amino acid change on the other CH3 domain (Y86>T) that creates a hole, thus favoring increased interactions between the CH3 of the two gamma1 chains at both positions 22 and 86 [103] (IMGT®, http://www.imgt.org, IMGT Biotechnology page>Knobs-into-holes).

## 5. CONCLUSION

IMGT-ONTOLOGY and the IMGT® information system, which are at the origin of immunoinformatics, have provided the concepts and the knowledge environment and informatics frame for a standardized and integrated analysis of IG, from gene to structure and function [29–44]. IG repertoire analysis and therapeutic antibody engineering and humanization represent two major current fields of immunoinformatics at the forefront of fundamental, clinical and pharmaceutical research owing to major methodological advances and medical implications.

The IMGT® standards for IG are used in clinical applications. Thus, IMGT/V-QUEST is frequently used by clinicians for the analysis of somatic hypermutations in leukemia, lymphoma, and myeloma, and more particularly in chronic lymphocytic leukemia (CLL) [15,70–72], in which the percentage of mutations of the rearranged IGHV gene in the VH of the leukemic clone has a prognostic value for the patients. IMGT/V-QUEST is the recommended standard recommended by the European Research Initiative on CLL (ERIC) for comparative analysis between laboratories [70]. The sequences of the V-(D)-J junctions determined by IMGT/JunctionAnalysis [18,19] are also used in the characterization of stereotypic patterns in CLL [71,72] and for the synthesis of probes specific to the junction for the detection and follow-up of minimal residual diseases (MRD) in leukemias and lymphomas. A new era is opening in hemato-oncology with the use of NGS for analysis of the clonality and MRD identification, making IMGT® standards use more needed than ever. More generally, the IMGT/HighV-QUEST web portal is a paradigm for identification of IMGT clonotype diversity and expression in NGS immune repertoire analysis of the adaptive immune response in infectious diseases, in vaccination, and for next-generation repertoire immunoprofiling [23].

The therapeutic monoclonal antibody engineering field represents the most promising potential in medicine. Standardized genomic and expressed sequence, structure, and interaction analysis of IG is crucial for a better molecular understanding and comparison of the mAb specificity, affinity, half-life, Fc effector properties, and potential immunogenicity. IMGT-ONTOLOGY concepts have become a necessity for IG loci description of newly sequenced genomes [104,105], antibody structure/function characterization, antibody engineering (single chain Fragment variable, phage displays, combinatorial libraries) and antibody humanization (chimeric, humanized, and human antibodies) [32,39,41,43,73–75,80]. IMGT® standardization allows repertoire analysis and antibody humanization studies to move to novel high-throughput methodologies with the same high-quality criteria. The CDR-IMGT lengths are now required for mAb INN applications and are included in the WHO-INN definitions [48], bringing a new level of standardized information to the comparative analysis of therapeutic antibodies.

## 6. AVAILABILITY AND CITATION

Authors who use IMGT® databases and tools are encouraged to cite this article and to quote the IMGT® Home page, http://www.imgt.org. Online access to IMGT® databases and tools are freely available for academics and under licenses and contracts for companies.

## ACKNOWLEDGMENT

We are grateful to Gérard Lefranc for helpful discussion, Véronique Giudicelli and Souphatta Sasorith for help in the figures, Patrice Duroux for upgrade of the IMGT/3Dstructure-DB tools, Eltaf Alamyar for IMGT/HighV-QUEST developments, and the IMGT® team for its expertise and constant motivation. We thank Cold Spring Harbor Protocol Press for the pdf of the IMGT Booklet available in IMGT references. IMGT® is a registered trademark of CNRS. IMGT® is a member of the International Medical Informatics Association (IMIA). IMGT® was funded in part by the BIOMED1 (BIOCT930038), Biotechnology BIOTECH2 (BIO4CT960037), 5th PCRDT Quality of Life and Management of Living Resources (QLG2-2000-01287), and 6th PCRDT Information Science and Technology (ImmunoGrid, FP6 IST-028069) programmes of the European Union (EU). IMGT® is currently supported by the Centre National de la Recherche Scientifique (CNRS), the Ministère de l'Enseignement Supérieur et de la Recherche (MESR), the University Montpellier 2, the Agence Nationale de la Recherche (ANR) Labex MabImprove (ANR-10-LABX-53-01), the Région Languedoc-Roussillon (Grand Plateau Technique pour la Recherche; GPTR). This work was granted access to the HPC resources of CINES under the allocation 036029 (2010–2014) made by GENCI (Grand Equipement National de Calcul Intensif).

## REFERENCES

[1] Lefranc M-P, Giudicelli V, Ginestoux C, Jabado-Michaloud J, Folch G, Bellahcene F, et al. IMGT®, the international ImMunoGeneTics information system®. Nucl Acids Res 2009;37:D1006–12.

[2] Lefranc M-P, Lefranc G. The Immunoglobulin FactsBook. London: Academic Press; 2001.

[3] Lefranc M-P, Lefranc G. The T cell receptor FactsBook. London: Academic Press; 2001.

[4] Lefranc M-P. Nomenclature of the human immunoglobulin genes. In: Coligan JE, Bierer BE, Margulies DE, Shevach EM, Strober W, editors. Current protocols in immunology. Hoboken N.J: John Wiley and Sons; 2000. A.1P.1-A.1P.37.

[5] Lefranc M-P. Nomenclature of the human T cell receptor genes. In: Coligan JE, Bierer BE, Margulies DE, Shevach EM, Strober W, editors. Current protocols in immunology. Hoboken N.J: John Wiley and Sons; 2000. A.1O.1-A.1O.23.

[6] Giudicelli V, Duroux P, Ginestoux C, Folch G, Jabado-Michaloud J, Chaume D, et al. IMGT/LIGM-DB, the IMGT® comprehensive database of immunoglobulin and T cell receptor nucleotide sequences. Nucl Acids Res 2006;34:D781–4.

[7] Giudicelli V, Chaume D, Lefranc M-P. IMGT/GENE-DB: a comprehensive database for human and mouse immunoglobulin and T cell receptor genes. Nucl Acids Res 2005;33:D256–61.

[8] Kaas Q, Ruiz M, Lefranc M-P. IMGT/3Dstructure-DB and IMGT/StructuralQuery, a database and a tool for immunoglobulin, T cell receptor and MHC structural data. Nucl Acids Res 2004;32:D208–10.

[9] Ehrenmann F, Kaas Q, Lefranc M-P. IMGT/3Dstructure-DB and IMGT/DomainGapAlign: a database and a tool for immunoglobulins or antibodies, T cell receptors, MHC, IgSF and MhcSF. Nucl Acids Res 2010;38:D301–7.

[10] Ehrenmann F, Lefranc M-P. IMGT/3Dstructure-DB: querying the IMGT database for 3D structures in immunology and immunoinformatics (IG or Antibodies, TR, MH, RPI, and FPIA). Cold Spring Harb Protoc 2011;6:750–61.

[11] Poiron C, Wu Y, Ginestoux C, Ehrenmann F, Duroux P, Lefranc M-P. IMGT/mAb-DB: the IMGT® database for therapeutic monoclonal antibodies. 11èmes Journées Ouvert Biol Inform Mathématiques (JOBIM), Poster n°101, Montpellier 2010. 7–9 September 2010.

[12] Giudicelli V, Chaume D, Lefranc M-P. IMGT/V-QUEST, an integrated software for immunoglobulin and T cell receptor V-J and V-D-J rearrangement analysis. Nucl Acids Res 2004;32:W435–40.

[13] Giudicelli V, Lefranc M-P. Interactive IMGT on-line tools for the analysis of immunoglobulin and T cell receptor repertoires. In: Veskler BA, editor. New research on immunology. New York: Nova Science Publishers Inc.; 2005. pp. 77–105.

[14] Brochet X, Lefranc M-P, Giudicelli V. IMGT/V-QUEST: the highly customized and integrated system for IG and TR standardized V-J and V-D-J sequence analysis. Nucl Acids Res 2008;36:W503–8.

[15] Giudicelli V, Lefranc M-P. IMGT® standardized analysis of immunoglobulin rearranged sequences. In: Ghia P, Rosenquist R, Davi F, editors. Immunoglobulin gene analysis in chronic lymphocytic leukemia. Italy: Wolters Kluwer Health Italy; 2008. pp. 33–52. [chapter 2].

[16] Giudicelli V, Brochet X, Lefranc M-P. IMGT/V-QUEST: IMGT standardized analysis of the immunoglobulin (IG) and T Cell receptor (TR) nucleotide sequences. Cold Spring Harb Protoc 2011;6:695–715.

[17] Alamyar E, Duroux P, Lefranc M-P, Giudicelli V. IMGT® tools for the nucleotide analysis of immunoglobulin (IG) and T cell receptor (TR) V-(D)-J repertoires, polymorphisms, and IG mutations: IMGT/V-QUEST and IMGT/HighV-QUEST for NGS. In: Christiansen F, Tait B, editors. Immunogenetics. Methods molecular biology, vol. 882. New York: Humana Press, Springer; 2012. pp. 569–604. [chapter 32].

[18] Yousfi Monod M, Giudicelli V, Chaume D, Lefranc M-P. IMGT/JunctionAnalysis: the first tool for the analysis of the immunoglobulin and T cell receptor complex V-J and V-D-J JUNCTIONs. Bioinformatics 2004;20:i379–385.

[19] Giudicelli V, Lefranc M-P. IMGT/JunctionAnalysis: IMGT standardized analysis of the V-J and V-D-J junctions of the rearranged immunoglobulins (IG) and T Cell receptors (TR). Cold Spring Harb Protoc 2011;6:716–25.

[20] Giudicelli V, Protat C, Lefranc M-P. The IMGT strategy for the automatic annotation of IG and TR cDNA sequences: IMGT/Automat. In: "Proceedings of the European conference on computational biology (ECCB 2003)", data and knowledge bases, poster DKB_31. Paris: Institut National de Recherche en Informatique et en Automatique; 2003. pp. 103–4. ECCB 2003, September 27–30, Paris, France.

[21] Giudicelli V, Chaume D, Jabado-Michaloud J, Lefranc M-P. Immunogenetics sequence annotation: the strategy of IMGT based on IMGT-ONTOLOGY. Stud Health Technol Inform 2005;116:3–8.

[22] Alamyar E, Giudicelli V, Shuo L, Duroux P, Lefranc M-P. IMGT/HighV-QUEST: the IMGT® web portal for immunoglobulin (IG) or antibody and T cell receptor (TR) analysis from NGS high throughput and deep sequencing. Immunome Res 2012;8(1):26.

[23] Li S, Lefranc M-P, Miles JJ, Alamyar E, Giudicelli V, Duroux P, et al. IMGT/HighV-QUEST paradigm for T cell receptor IMGT clonotype diversity and next generation repertoire immunoprofiling. Nat Comm 2013;4:2333.

[24] Ehrenmann F, Lefranc M-P. IMGT/DomainGapAlign: IMGT standardized analysis of amino acid sequences of variable, constant, and groove domains (IG, TR, MH, IgSF, MhSF). Cold Spring Harb Protoc 2011;6:737–49.

[25] Ehrenmann F, Lefranc M-P. IMGT/DomainGapAlign: the IMGT® tool for the analysis of IG, TR, MHC, IgSF and MhcSF domain amino acid polymorphism. In: Christiansen F, Tait B, editors. Immunogenetics. Methods molecular biology, vol. 882. New York: Humana Press, Springer; 2012. pp. 605–33. [chapter 33].

[26] Ehrenmann F, Giudicelli V, Duroux P, Lefranc M-P. IMGT/Collier de Perles: IMGT Standardized Representation of Domains (IG, TR, and IgSF Variable and Constant Domains, MH and MhSF Groove Domains). Cold Spring Harb Protoc 2011;6:726–36.

[27] Lane J, Duroux P, Lefranc M-P. From IMGT-ONTOLOGY to IMGT/LIGMotif: the IMGT® standardized approach for immunoglobulin and T cell receptor gene identification and description in large genomic sequences. BMC Bioinformatics 2010;11:223.

[28] Pommié C, Levadoux S, Sabatier R, Lefranc G, Lefranc M-P. IMGT standardized criteria for statistical analysis of immunoglobulin V-REGION amino acid properties. J Mol Recognit 2004;17:17–32.

[29] Lefranc M-P. IMGT, the international ImMunoGeneTics information system. In: Bock G, Goode J, editors. Immunoinformatics: bioinformatic strategies for better understanding of immune function. Novartis foundation symposium, vol. 254. Chichester UK: John Wiley and Sons; 2003. pp. 126–36. discussion 136–142, 216–222, 250–252.

[30] Lefranc M-P, Giudicelli V, Ginestoux C, Chaume D. IMGT, the international ImMunoGeneTics information system: the reference in immunoinformatics. Stud Health Technol Inform 2003;95:74–9.

[31] Lefranc M-P. IMGT databases, web resources and tools for immunoglobulin and T cell receptor sequence analysis. Leukemia 2003;17(1):260–6.

[32] Lefranc M-P. IMGT, the international ImMunoGenetics information system®. In: Lo BKC, editor. Antibody engineering methods and protocols. 2nd ed. Methods molcular biology, vol. 248. Totowa, NJ: Humana Press; 2004. pp. 27–49.

[33] Lefranc M-P. IMGT-ONTOLOGY and IMGT databases, tools and web resources for immunogenetics and immunoinformatics. Mol Immunol 2004;40(10):647–60.

[34] Lefranc M-P. IMGT, the international ImMunoGeneTics information system: a standardized approach for immunogenetics and immunoinformatics. Immunome Res 2005;1:3.

[35] Lefranc M-P. IMGT®, the international ImMunoGeneTics information system® for immunoinformatics. Methods for querying IMGT® databases, tools and Web resources in the context of immunoinformatics. In: Flower DR, editor. Immunoinformatics: Predicting immunogenicity in silico. Methods molecular biology, vol. 409. Totowa N.J.: Humana Press, Springer; 2007. pp. 19–42. [chapter 2].

[36] Lefranc M-P. IMGT-ONTOLOGY, IMGT® databases, tools and web resources for immunoinformatics. In: Schoenbach C, Ranganathan S, Brusic V, editors. Immunoinformatics. Immunomics reviews, series of Springer science and business media LLC, vol. 1. New York: Springer; 2008. pp. 1–18. [chapter 1].

[37] Lefranc M-P, Giudicelli V, Regnier L, Duroux P. IMGT®, a system and an ontology that bridge biological and computational spheres in bioinformatics. Brief Bioinform 2008;9:263–75.

[38] Lefranc M-P. IMGT®, the international ImMunoGeneTics information system® for immunoinformatics. Methods for querying IMGT® databases, tools and web resources in the context of immunoinformatics. Mol Biotechnol 2008;40:101–11.

[39] Lefranc M-P. Antibody databases and tools: the IMGT® experience. In: An Z, editor. Therapeutic monoclonal antibodies: from bench to clinic. Hoboken N.J: John Wiley and Sons; 2009. pp. 91–114. [chapter 4].

[40] Lefranc M-P. Antibody databases: IMGT®, a French platform of world-wide interest [in French]. Bases de données anticorps: IMGT®, une plate-forme française d'intérêt mondial. Médecine/Sciences 2009;25:1020–3.

[41] Ehrenmann F, Duroux P, Giudicelli V, Lefranc M-P. Standardized sequence and structure analysis of antibody using IMGT®. In: Kontermann R, Dübel S, editors. Antibody engineering, vol. 2. Berlin, Heidelberg: Springer-Verlab; 2010. pp. 11–31. [chapter 2].

[42] Lefranc M-P. IMGT, the international ImMunoGeneTics information system. Cold Spring Harb Protoc 2011;6:595–603.

[43] Lefranc M-P, Ehrenmann F, Ginestoux C, Duroux P, Giudicelli V. Use of IMGT® databases and tools for antibody engineering and humanization. In: Chames P, editor. Antibody engineering. Methods molecular biology, vol. 907. New York: Humana Press, Springer; 2012. pp. 3–37. [chapter 1].

[44] Lefranc M-P. IMGT® information system. In: Dubitzky W, Wolkenhauer O, Cho K-H, Yokota H, editors. Encyclopedia of systems biology. Springer Science+Business Media, LLC012; 2013. pp. 959–64.

[45] Lefranc M-P. WHO-IUIS nomenclature subcommittee for immunoglobulins and T cell receptors report. Immunogenetics 2007;59: 899–902.

[46] Lefranc M-P. WHO-IUIS Nomenclature Subcommittee for immunoglobulins and T cell receptors report August 2007, 13th International Congress of Immunology, Rio de Janeiro, Brazil. Dev Comp Immunol 2008;32:461–3.

[47] World Health Organization. International nonproprietary names (INN) for biological and biotechnological substances (a review); 2012. INN Working Document 05.179. Update 2012.

[48] Lefranc M-P. Antibody nomenclature: from IMGT-ONTOLOGY to INN definition. MAbs 2011;3(1):1–2.

[49] Giudicelli V, Lefranc M-P. Ontology for immunogenetics: IMGT-ONTOLOGY. Bioinformatics 1999;15:1047–54.

[50] Giudicelli V, Lefranc M-P. IMGT-ONTOLOGY 2012. Frontiers in bioinformatics and computational biology. Front Genet 2012;3:79.

[51] Giudicelli V, Lefranc M-P. IMGT-ONTOLOGY. In: Dubitzky W, Wolkenhauer O, Cho K-H, Yokota H, editors. Encyclopedia of systems biology. Springer Science+Business Media, LLC012; 2013. pp. 964–72.

[52] Giudicelli V, Lefranc M-P. IMGT-ONTOLOGY: gestion et découverte de connaissances au sein d'IMGT. In: Hacid M-S, Kodratoff Y, Boulanger D, editors. Extraction et gestion des connaissances (EGC'2003); 2003. Actes des troisièmes journées Extraction et Gestion des Connaissances, Lyon, France, janvier 22–24, 2003. Revue des Sciences et Technologies de l'Information, RSTI, série Revue d'Intelligence Artificielle- Extraction des Connaissances et Apprentissage (RIA-ECA), Hermès Science Publications, ISBN 2-7462-0631-5, 17(1–3), pp.13–23. Lavoisier, Cachan, Paris.

[53] Lefranc M-P, Giudicelli V, Ginestoux C, Bosc N, Folch G, Guiraudou D, et al. IMGT-ONTOLOGY for immunogenetics and immunoinformatics. In Silico Biol 2004;4:17–29.

[54] Lefranc M-P, Clément O, Kaas Q, Duprat E, Chastellan P, Coelho I, et al. IMGT-choreography for immunogenetics and immunoinformatics. In Silico Biol 2005;5:45–60.

[55] Duroux P, Kaas Q, Brochet X, Lane J, Ginestoux C, Lefranc M-P, et al. IMGT-Kaleidoscope, the formal IMGT-ONTOLOGY paradigm. Biochimie 2008;90:570–83.

[56] Lefranc M-P. From IMGT-ONTOLOGY IDENTIFICATION axiom to IMGT standardized keywords: for immunoglobulins (IG), T cell receptors (TR), and conventional genes. Cold Spring Harb Protoc 2011;6:604–13.

[57] Lefranc M-P. From IMGT-ONTOLOGY DESCRIPTION axiom to IMGT standardized labels: for immunoglobulin (IG) and T cell receptor (TR) sequences and structures. Cold Spring Harb Protoc 2011;6:614–26.

[58] Lefranc M-P. From IMGT-ONTOLOGY CLASSIFICATION axiom to IMGT standardized gene and allele nomenclature: for immunoglobulins (IG) and T cell receptors (TR). Cold Spring Harb Protoc 2011;6:627–32.

[59] Lefranc M-P. Unique database numbering system for immunogenetic analysis. Immunol Today 1997;18:509.

[60] Lefranc M-P. The IMGT unique numbering for Immunoglobulins, T cell receptors and Ig-like domains. The Immunologist 1999;7: 132–6.

[61] Lefranc M-P, Pommié C, Ruiz M, Giudicelli V, Foulquier E, Truong L, et al. IMGT unique numbering for immunoglobulin and T cell receptor variable domains and Ig superfamily V-like domains. Dev Comp Immunol 2003;27:55–77.

[62] Lefranc M-P, Pommié C, Kaas Q, Duprat E, Bosc N, Guiraudou D, et al. IMGT unique numbering for immunoglobulin and T cell receptor constant domains and Ig superfamily C-like domains. Dev Comp Immunol 2005;29:185–203.

[63] Lefranc M-P, Duprat E, Kaas Q, Tranne M, Thiriot A, Lefranc G. IMGT unique numbering for MHC groove G-DOMAIN and MHC superfamily (MhcSF) G-LIKE-DOMAIN. Dev Comp Immunol 2005;29:917–38.

[64] Lefranc M-P. IMGT unique numbering for the variable (V), constant (C), and groove (G) domains of IG, TR, MH, IgSF, and MhSF. Cold Spring Harb Protoc 2011;6:633–42.

[65] Ruiz M, Lefranc M-P. IMGT gene identification and Colliers de Perles of human immunoglobulins with known 3D structures. Immunogenetics 2002;53:857–83.

[66] Kaas Q, Lefranc M-P. IMGT Colliers de Perles: standardized sequence-structure representations of the IgSF and MhcSF superfamily domains. Curr Bioinforma 2007;2:21–30.

[67] Kaas Q, Ehrenmann F, Lefranc M-P. IG, TR and IgSf, MHC and MhcSF: what do we learn from the IMGT Colliers de Perles? Brief Funct Genomic Proteomic 2007;6:253–64.

[68] Lefranc M-P. IMGT Collier de Perles for the variable (V), constant (C), and groove (G) domains of IG, TR, MH, IgSF, and MhSF. Cold Spring Harb Protoc 2011;6:643–51.

[69] Robert R, Lefranc M-P, Ghochikyan A, Agadjanyan MG, Cribbs DH, Van Nostrand WE, et al. Restricted V gene usage and VH/VL pairing of mouse humoral response against the N-terminal immunodominant epitope of the amyloid β peptide. Mol Immunol 2010;48(1–3):59–72.

[70] Ghia P, Stamatopoulos K, Belessi C, Moreno C, Stilgenbauer S, Stevenson FI, et al. ERIC recommendations on IGHV gene mutational status analysis in chronic lymphocytic leukemia. Leukemia 2007;21:1–3.

[71] Agathangelidis A, Darzentas N, Hadzidimitriou A, Brochet X, Murray F, Yan XJ, et al. Stereotyped B-cell receptors in one third of chronic lymphocytic leukemia: towards a molecular classification with implications for targeted therapeutic interventions. Blood 2012;119(19):4467–75.

[72] Kostareli E, Gounari M, Janus A, Murray F, Brochet X, Giudicelli V, et al. Antigen receptor stereotypy across B-cell lymphoproliferations: the case of IGHV4-59/IGKV3-20 receptors with rheumatoid factor activity. Leukemia 2012;26(5):1127–31.

[73] Jefferis R, Lefranc M-P. Human immunoglobulin allotypes: possible implications for immunogenicity. MAbs 2009;1(4):332–8.

[74] Lefranc M-P, Lefranc G. Human Gm, Km and Am allotypes and their molecular characterization: a remarkable demonstration of polymorphism. In: Christiansen F, Tait B, editors. Immunogenetics. Methods molecular biology, vol. 882. New York: Humana Press, Springer; 2012. pp. 635–80. [chapter 34].

[75] Dechavanne C, Guillonneau F, Chiappetta G, Sago L, Lévy P, Salnot V, et al. Mass spectrometry detection of G3m and IGHG3 alleles and follow-up of differential mother and neonate IgG3. PLoS One 2012;7(9):e46097.

[76] Magdelaine-Beuzelin C, Kaas Q, Wehbi V, Ohresser M, Jefferis R, Lefranc M-P, et al. Structure-function relationships of the variable domains of monoclonal antibodies approved for cancer treatment. Crit Rev Oncol Hematol 2007;64:210–25.

[77] Pelat T, Bedouelle H, Rees AR, Crennell SJ, Lefranc M-P, Thullier P. Germline humanization of a non-human primate antibody that neutralizes the anthrax toxin, by in vitro and in silico engineering. J Mol Biol 2008;384:1400–7.

[78] Pelat T, Hust M, Hale M, Lefranc M-P, Dübel S, Thullier P. Isolation of a human-like antibody fragment (scFv) that neutralizes ricin biological activity. BMC Biotechnol 2009;9:60.

[79] Alamyar M-P, Giudicelli V, Duroux P, Lefranc M-P. Antibody V and C domain sequence, structure and interaction analysis with special reference to IMGT®. In: Ossipow V, Fisher N, editors. Monoclonal antibodies: methods and protocols. 2nd ed. vol. 1131. Springer Science + Business Media, LLC. NewYork: Humana Press; 2014. pp. 337–81.

[80] Lefranc M-P. How to use IMGT® for therapeutic antibody engineering. In: Dübel SK, Reichert J, editors. Handbook of therapeutic antibodies. 2nd ed. Wiley; 2014, [chapter 10] in press.

[81] Wain HM, Bruford EA, Lovering RC, Lush MJ, Wright MW, Povey S. Guidelines for human gene nomenclature. Genomics 2002;79:464–70.

[82] Bruford EA, Lush MJ, Wright MW, Sneddon TP, Povey S, Birney E. The HGNC database in 2008: a resource for the human genome. Nucl Acids Res 2008;36:D445–8.

[83] Letovsky SI, Cottingham RW, Porter CJ, Li PW. GDB: the human genome database. Nucl Acids Res 1998;26(1):94–9.

[84] Maglott DR, Katz KS, Sicotte H, Pruitt KD. NCBI's LocusLink and RefSeq. Nucl Acids Res 2000;28(1):126–8.

[85] Maglott D, Ostell J, Pruitt KD, Tatusova T. Entrez gene: gene-centered information at NCBI. Nucl Acids Res 2007;35:D26–31.

[86] Stabenau A, McVicker G, Melsopp C, Proctor G, Clamp M, Birney E. The Ensembl core software libraries. Genome Res 2004;14:929–33.

[87] Wilming LG, Gilbert JG, Howe K, Trevanion S, Hubbard T, Harrow JL. The vertebrate genome annotation (Vega) database. Nucl Acids Res 2008;36:D753–60.

[88] Kaas Q, Lefranc M-P. T cell receptor/peptide/MHC molecular characterization and standardized pMHC contact sites in IMGT/3D structure-DB. In Silico Biol 2005;5:505–28.

[89] Kaas Q, Duprat E, Tourneur G, Lefranc M-P. IMGT standardization for molecular characterization of the T cell receptor/peptide/MHC complexes. In: Schoenbach C, Ranganathan S, Brusic V, editors. Immunoinformatics, Immunomics reviews, series of Springer science and Business Media LLC, chapter 2. New York: Springer; 2008. pp. 19–49.

[90] Vlachakis D, Feidakis C, Megalooikonomou V, Kossida S. IMGT/Collier-de-Perles: a two-dimensional visualization tool for amino acid domain sequences. Theor Biol Med Model 2013;10:14.

[91] Lefranc M-P. Immunoinformatics of the V, C and G domains: IMGT® definitive system for IG, TR and IgSF, MH and MhSF. In: De RK, Tomar N, editors. Immunoinformatics: from biology to informatics. Methods in molecular biology. 2nd ed. Humana Press; 2014, in press.

[92] Kabat EA, Wu TT, Perry HM, Gottesman KS, Foeller C. Sequences of proteins of immunological interest. Washington, DC: U.S. Department of Health and Human Services (USDHHS), National Institute of Health NIH Publication; 1991. 91–3242.

[93] Chothia C, Lesk AM. Canonical structures for the hypervariable regions of immunoglobulins. Mol Biol 1987;196(4):901–17.

[94] Rose PW, Beran B, Bi C, Bluhm WF, Dimitropoulos D, Goodsell DS, et al. The RCSB Protein Data Bank: redesigned web site and web services. Nucl Acids Res 2011;39:D392–401.

[95] Riechmann L, Clark M, Waldmann H, Winter G. Reshaping human antibodies for therapy. Nature 1988;332:323–7.

[96] Hamers-Casterman C, Atarhouch T, Muyldermans S, Robinson G, Hamers C, Bajyana Songa E, et al. Naturally occurring antibodies devoid of light chains. Nature 1993;363:446–8.

[97] Nguyen VK, Hamers R, Wyns L, Muyldermans S. Camel heavy-chain antibodies: diverse germline VHH and specific mechanisms enlarge the antigen-binding repertoire. EMBO J 2000;19:921–30.

[98] Nguyen VK, Hamers R, Wyns L, Muyldermans S. Loss of splice consensus signal is responsible for the removal of the entire CH1 domain of the functional camel IgG2a heavy chain antibodies. Mol Immunol 1999;36:515–24.

[99] Lefranc M-P, Lefranc G. The constant region genes of the immunoglobulin heavy chains. Mol Genet (Life Sci Adv) 1988;7:39–45.

[100] Greenberg AS, Avila D, Hughes M, Hughes A, McKinney EC, Flajnik MF. A new antigen receptor gene family that undergoes rearrangement and extensive somatic diversification in sharks. Nature 1995;374:168–73.

[101] Teplyakov A, Zhao Y, Malia TJ, Obmolova G, Gilliland GL. IgG2 Fc structure and the dynamic features of the IgG CH2-CH3 interface. Mol Immunol 2013;56:131–9.

[102] Lesk AM, Chothia C. Elbow motion in the immunoglobulin involves a molecular ball-and-socket joint. Nature 1988;335:188–90.

[103] Ridgway JB, Presta LG, Carter P. 'Knobs-into-holes' engineering of antibody CH3 domains for heavy chain heterodimerization. Protein Eng 1996;9:617–21.

[104] Schwartz JC, Lefranc M-P, Murtaugh M-P. Organization, complexity and allelic diversity of the porcine (*Sus scrofa domestica*) immunoglobulin lambda locus. Immunogenetics 2012;64(5):399–407.

[105] Schwartz JC, Lefranc M-P, Murtaugh MP. Evolution of the porcine (*Sus scrofa domestica*) immunoglobulin kappa locus through germline gene conversion. Immunogenetics 2012;64(4):303–11.

# Chapter 27

# Anti-Interleukin-6 Receptor Antibody Therapy Against Autoimmune Inflammatory Diseases

Toshio Tanaka[1,2], Masashi Narazaki[3] and Tadamitsu Kishimoto[4]

[1]*Department of Clinical Application of Biologics, Osaka University Graduate School of Medicine, Osaka University, Osaka, Japan,* [2]*Department of Immunopathology, World Premier International Immunology Frontier Research Center, Osaka University, Osaka, Japan,* [3]*Department of Respiratory Medicine, Allergy and Rheumatic Diseases, Osaka University Graduate School of Medicine, Osaka University, Osaka, Japan,* [4]*Laboratory of Immune Regulation, World Premier International Immunology Frontier Research Center, Osaka University, Osaka, Japan*

## 1. INTERLEUKIN-6 AND ITS RECEPTOR SYSTEM

The growth and differentiation of B-cells into immunoglobulin (Ig)-producing cells require soluble factors from T-cells and macrophages. In 1986, the factor that induces the final differentiation of activated B-cells was cloned as B-cell stimulatory factor-2 (BSF-2) [1,2]. Originally, human BSF-2 was purified from the culture supernatant of a human T-cell leukemia virus type 1 (HTLV-1) transformed T-cell line by monitoring the enhancement of Ig production by Epstein Barr virus (EBV)-transformed B-cell lines. In vivo, B-cells activated with antigen become responsive to BSF-2 and differentiate into Ig-producing cells (Figure 1). B-cell stimulatory factor-2 is produced by activated T-cells as a nonspecific T-cell factor. Almost at the same time, molecules that were later found to be the same as BSF-2 were being studied and were given distinct names based on their biological activity. For example, the factor that induces acute-phase protein synthesis in hepatocytes was called hepatocyte-stimulating factor (HSF), the factor that enhances the growth of fusion cells between plasma cells and myeloma cells was called hybridoma growth factor (HGF), and interferon (IFN)-beta-2 was named for its antiviral activity, although it was later found not to have antiviral activity. However, researchers determined that molecules with different names that were studied by various groups were identical, and the single name interleukin-6 (IL-6) was established [3]. We now know that the function of IL-6 is not restricted to B-cells, but that IL-6 exerts a wide variety of effects on the immune response, inflammation, and hematopoiesis. The human IL-6 gene has been mapped to chromosome 7p21. Human IL-6 consists of 184 amino acids with two N-glycosylation sites and two cysteine–cysteine bridges in the molecule. The core protein is about 20 kDa, but glycosylation accounts for the 21- to 26-kDa size of natural IL-6.

The functional IL-6 receptor (IL-6R) is composed of two chains (Figure 2) [4]. The IL-6R constitutes the IL-6-binding chain [5], whereas gp130 constitutes the signal-transducing chain [6]. After binding of IL-6 to IL-6R, the IL-6/IL-6R complex induces homodimerization of gp130 and activates a downstream signaling cascade [7]. The soluble form of IL-6R (sIL-6R) without the cytoplasmic region is found in healthy human serum at around 76 ng/mL. Interleukin-6 binds to sIL-6R, and the resulting IL-6/sIL-6R complex also induces the IL-6 signaling on gp130-expressing cells, even cells without expression of membrane IL-6R [8]. Thus, the pleiotropic effect of IL-6 is explained by the fact that gp130 is expressed on various types of cells, and serum sIL-6R binds to IL-6 and activates gp130-expressing cells [9]. The activated IL-6R complex forms a hexameric structure composed of two molecules each of IL-6, IL-6R (or sIL-6R), and gp130 [10]. Although IL-6R is a unique receptor for IL-6, the signal-transducing chain, gp130, is shared by members of the IL-6 family of cytokines: leukemia inhibitory factor, oncostatin M, ciliary neurotrophic factor, IL-11, cardiotrophin 1, cardiotrophin-like cytokine, IL-27, and IL-35. Although all of these cytokines bind to their specific receptors, they use the same gp130 for their

Molecular Biology of B Cells. http://dx.doi.org/10.1016/B978-0-12-397933-9.00027-8

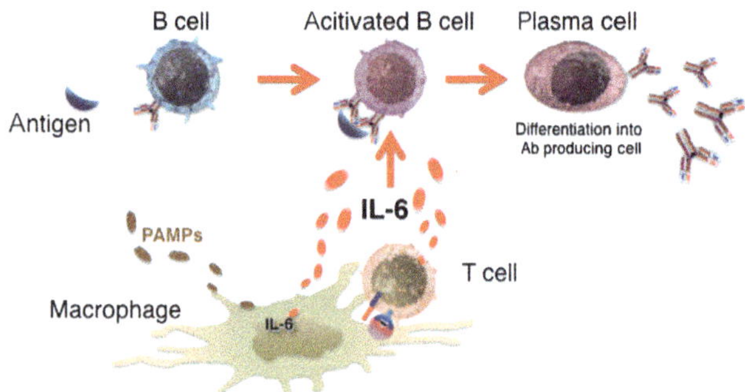


FIGURE 1 **Differentiation of activated B-cells to plasma cells by IL-6.** In the innate immune response, pathogen-associated molecular patterns stimulate macrophage to produce IL-6. In the acquired immune response, the T-cell activated by macrophage in an antigen-specific manner produces IL-6. A specific antigen binds to the B-cell receptor and activates B-cells that differentiate into immunoglobulin-producing plasma cells in the presence of IL-6.

signals [11,12]. The only exception is virus-encoded IL-6, which is the product of Kaposi sarcoma–associated herpes virus (KSHV), also known as human herpes virus 8. Virus-encoded IL-6 can directly bind to and activate gp130 [13]. The mechanism by which members of the IL-6 cytokine family use the common signal-transducer, gp130, clarifies why IL-6 family cytokines show functional redundancy. The molecular elucidation of the IL-6 receptor system finally solved the longstanding mystery of why cytokines are pleiotropic and redundant.

## 2. PLEIOTROPIC ACTIVITY OF IL-6

### 2.1 Activity on Hepatocytes

After IL-6 is produced in the inflammatory region, IL-6 reaches the liver through the bloodstream, followed by the rapid induction of acute-phase proteins such as C-reactive protein (CRP), serum amyloid A (SAA), fibrinogen, haptoglobin, and α1-antichymotrypsin (Figure 3) [14]. On the other hand, IL-6 reduces the production of fibronectin, albumin, and transferrin. Interleukin-6 induces high concentrations of SAA for a long time, leading to the development of amyloid A amyloidosis, which is a serious complication of chronic inflammatory diseases [15]. This results in amyloid fibril deposition in various organs, which causes progressive deterioration of these organs. Interleukin-6 is also involved in the metabolism of serum iron and zinc through regulation of their transporters. Regarding serum iron, IL-6 induces hepcidin production, which blocks the action of the iron transporter ferroportin-1 in the gut and thus reduces serum iron and hemoglobin levels [16]. This means that IL-6 is responsible for the anemia related to chronic inflammation. Interleukin-6 also enhances expression of the zinc importer, ZIP14, on hepatocytes and thus induces the hypozincemia that accompanies inflammation [17].

### 2.2 Activity in the Innate and Acquired Immune Systems

Interleukin-6 is promptly and transiently produced as an SOS signal in response to infections and tissue injury. In the innate immune system, IL-6 is generated in infectious lesions and sends out a signal to the entire body. The signature of exogenous pathogens, known as pathogen-associated molecular patterns, is recognized by immune cells such as monocytes and macrophages via pathogen recognition receptors [18], which include Toll-like receptors (TLRs), retinoic acid-inducible gene-1–like receptors, nucleotide-binding oligomerization domain–like receptors, and DNA receptors. These receptors initiate signaling pathways to activate the transcriptional factor, nuclear factor (NF)-κB, and promote the transcription of mRNA for inflammatory cytokines such as IL-6, tumor necrosis factor (TNF)α, and IL-1β, whereas TNFα and IL-1β also activate transcription factors to produce IL-6.

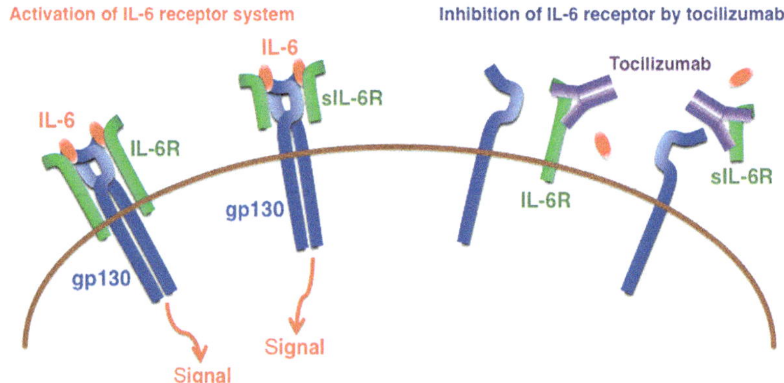


FIGURE 2 **The IL-6 receptor system and a humanized anti-IL-6 receptor antibody, tocilizumab.** Interleukin-6 binds to transmembrane IL-6 receptor (IL-6R) or soluble IL-6R (sIL-6R) and then induces homodimerization of the signal transducer gp130, leading to activation of downstream signaling. A humanized anti-IL-6R antibody, tocilizumab, inhibits IL-6 binding to both types of IL-6R and blocks IL-6–mediated signaling.

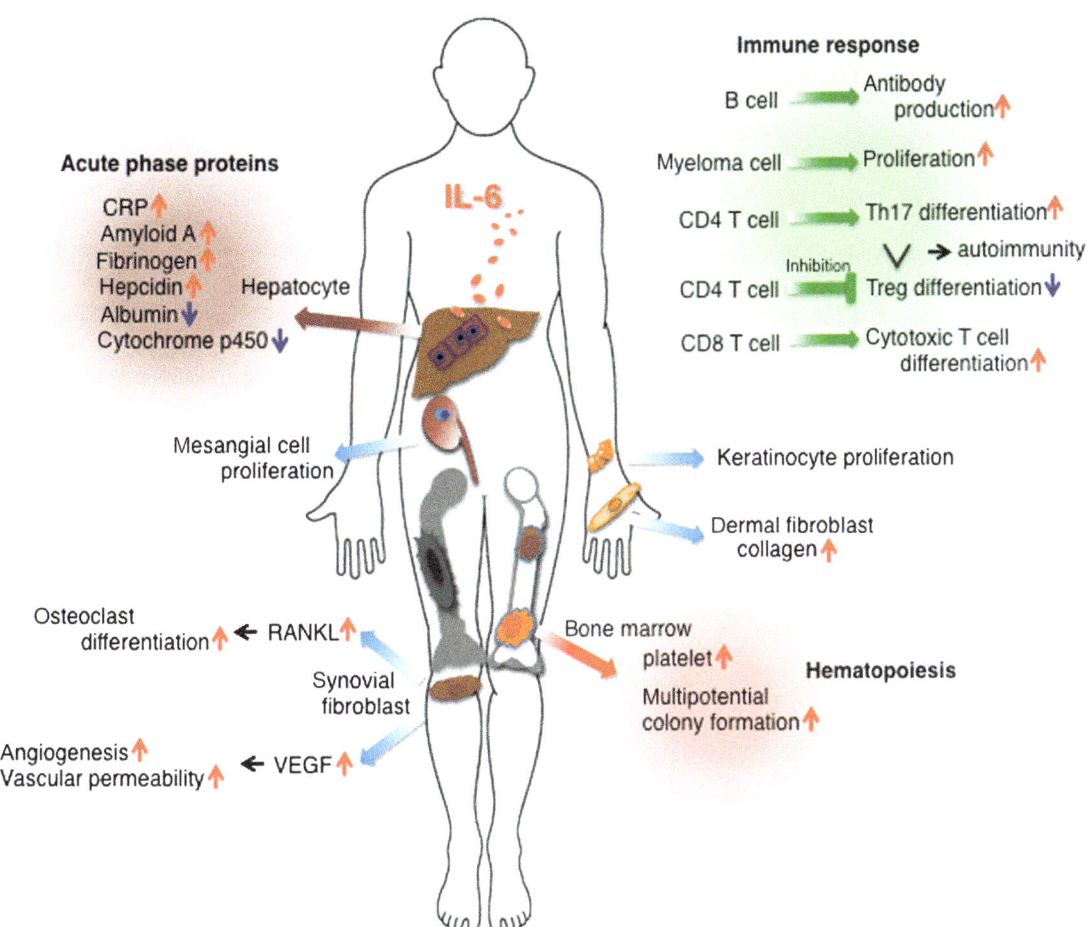


FIGURE 3 **Pleiotropic activity of IL-6 and its pathological implications in disease development.** Interleukin-6 has pleiotropic effects on the liver, immune system, hematopoiesis, and others. Dysregulation of IL-6 production is implicated in the development of various pathological conditions. In hepatocytes, IL-6 induces acute-phase proteins including CRP, serum amyloid A, fibrinogen, and hepcidin, whereas it inhibits production of albumin. Interleukin-6 also has an important role in the immune response by stimulating antibody production by B-cells and inducing development of effector T-cells. Moreover, IL-6 can promote differentiation or proliferation of several non-immune cells. CRP: C-reactive protein; Treg: regulatory T-cell; RANKL: receptor activator of nuclear factor of kappa B (NF-κB) ligand; VEGF; vascular endothelial growth factor.

Interleukin-6 also provides a warning signal in the event of tissue damage. Damage-associated molecular patterns, which are released from damaged or dying cells in noninfectious types of inflammation, such as burns or trauma, directly or indirectly promote inflammation. During sterile surgical operations, the level of serum IL-6 increases before the elevation in body temperature and serum acute-phase protein concentrations [19]. Damage-associated molecular patterns from injured cells contain a variety of molecules such as mitochondrial DNA, high mobility group box 1 (HMGB1), and S100 proteins [20]. Serum mitochondrial DNA levels in trauma patients are reportedly thousands of times higher than in controls, which leads to TLR9 stimulation and NF-κB activation [21], whereas binding of HMGB1 to TLR2, TLR4, or the receptor of advanced glycation end products (RAGE) can promote inflammation. The S100 family of proteins is composed of more than 25 members, some of which also interact with RAGE to evoke sterile inflammation [22].

In the acquired immune response, IL-6 performs an important function. As suggested by its previous name, BSF-2, IL-6 induces the differentiation of activated B-cells into Ig-producing plasma cells, so that continuous overexpression of IL-6 results in hypergammaglobulinemia and autoantibody production. Interleukin-6 is a growth factor for hybridoma cells and myeloma cell lines. The phenotype of IL-6 transgenic mice supports the pathological role of IL-6 in inflammatory symptoms including polyclonal plasmacytosis in spleen, lymph nodes, and thymus, and increased megakaryocytes in bone marrow [23]. The polyclonal plasmacytosis in IL-6 transgenic mice changes to monoclonal plasmacytomas with chromosomal translocation in mice on a certain genetic background such as BALB/c [24]. Injection of pristane or mineral oil induces plasma cell neoplasms in BALB/c mice [25], which can be explained by the production of IL-6 after mineral oil–induced chronic inflammation. Myeloma cells freshly isolated from myeloma patients responded to IL-6 for growth, whereas some myeloma cells themselves can produce IL-6. Thus, IL-6 is an autocrine growth factor in some types of myelomas [26].

Interleukin-6 affects not only B-cells, but also T-cells. Interleukin-6 promotes specific differentiation of naive CD4-positive T-cells. In combination with transforming growth factor (TGF)-β, IL-6 is indispensable for $T_h17$ differentiation from naive CD4-positive T-cells [27], whereas IL-6 inhibits TGF-β–induced regulatory T-cell (Treg) development [27,28]. Disruption of $T_h17$/Treg balance is thought to be responsible for the disruption of immunological tolerance, and is thus pathologically involved in the development of autoimmune and chronic inflammatory diseases [29]. Interleukin-6 also promotes T-follicular helper cell differentiation as well as production of IL-21 [30], which regulates Ig synthesis and IgG4 production in particular. Interleukin-6 also induces the differentiation of CD8-positive T-cells into cytotoxic T-cells (Figure 3) [31].

## 2.3 Activity in Other Cells

As described earlier, because gp130 is expressed in various cells, the complex consisting of IL-6 and soluble IL-6R can activate the gp130-mediated signaling cascade in a variety of cell types. When IL-6 is generated in bone marrow stromal cells, it stimulates the receptor activator of the NF-κB ligand (RANKL) [32], which induces the differentiation and activation of osteoclasts [33], linking chronic inflammation with osteoporosis [34]. Interleukin-6 also induces excess production of vascular endothelial growth factor (VEGF), leading to increased vascular permeability and angiogenesis, which are pathological features of inflammatory lesions that are observed in, for example, synovial tissues of rheumatoid arthritis (RA) or edema of remitting seronegative symmetrical synovitis with pitting edema syndrome [35,36]. Interleukin-6 aids keratinocyte proliferation [37] or the generation of collagen in dermal fibroblasts and their differentiation into myofibroblasts that may account for changes in the skin of patients with systemic sclerosis [38]. Moreover, IL-6 interacts with vascular endothelial cells, the endocrine system including the hypothalamic–pituitary–adrenal axis, neuropsychological systems, and others.

# 3. REGULATION OF IL-6 SYNTHESIS

As mentioned, IL-6 contributes to host defense against environmental stress by stimulating acute-phase responses, hematopoiesis, and immune reactions. The fact that IL-6 issues a warning signal to indicate the occurrence of an emergency accounts for the strict regulation of IL-6 synthesis both transcriptionally and posttranscriptionally. A number of transcription factors have been shown to positively and negatively regulate IL-6 gene transcription. The functional *cis*-regulatory elements in the 5′ flanking region of the human IL-6 gene are binding sites for several proteins including NF-κB, specificity protein-1, nuclear factor IL6 (NF-IL6) (also known as CAAT/enhancer-binding protein beta), activator protein-1, and interferon regulatory factor-1 [39,40]. An interesting finding is that some viral products interact with NF-κB and/or NF-IL6 and enhance their DNA binding activity. These viral products include the transactivator protein (Tax) derived from HTLV-1 [41,42], the transactivator of the transcription (TAT) protein of the human immunodeficiency virus-1 (HIV-1) [43,44], and the human hepatitis-B virus X protein [45,46]. On the other hand, some transcription factors and hormone receptors suppress IL-6 expression. These include fibrates-activated peroxisome proliferator-activated receptor-α [47], estrogen receptor [48], and glucocorticoid receptor [49]. In addition, retinoblastoma protein and p53 repress the IL-6 gene promoter, whereas it is upregulated by mutant p53 [50]. In addition to transcriptional regulation by proteins, some microRNAs directly or indirectly regulate transcription activity. MicroRNA-155 suppresses NF-IL-6 expression by interacting with the 3′-untranslated regions (UTR) of NF-IL6 [51], whereas microRNA-146a/b and −223 indirectly suppress transcription of IL-6 by targeting IL-1 receptor-associated kinase-1 and signal transducer and activator of transcription-3, respectively [52,53].

Regarding posttranscriptional regulation, IL-6 mRNA is controlled through both the 5′- and 3′-UTR [54,55]. The 5′-UTR determines the initiation of mRNA translation, whereas the 3′-UTR controls the stability of the mRNA. A number of RNA binding proteins and microRNAs bind to the 3′-UTRs and regulate the stability of IL-6 mRNA For example, IL-6 mRNA stabilization is promoted by the MAP kinase p38α via 3′-UTRs of IL-6 [56]. Stabilization of both viral and human IL-6 mRNA by the KSHV open reading frame occurs by competing with the binding of microRNA-1293 to viral IL-6 or microRNA-608 to the human IL-6 mRNA [57]. On the other hand, RNA-binding proteins such as tristetraprolin and butyrate response factor-1 and −2 promote IL-6 mRNA degradation [58]. Interleukin-6 mRNA levels are also reduced by microRNAs such as microRNA-365 and -608 through direct interaction with the IL-6 3′-UTR [57,59].

A nuclease known as regulatory RNase-1, Regnase-1, has a role in the destabilization of IL-6 mRNA, and the relevant knockout mice spontaneously develop autoimmune diseases accompanied by splenomegaly and lymphadenopathy [60]. The inhibitor of NF-κB (IκB) kinase (IKK) complex controls IL-6 mRNA stability by phosphorylating Regnase-1 in response to IL-1R/TLR stimulation [61]. Phosphorylated Regnase-1 undergoes ubiquitination and degradation. Regnase-1 that is re-expressed in IL-1R/TLR-activated cells shows delayed kinetics, and Regnase-1 mRNA is negatively regulated by Regnase-1 itself via a

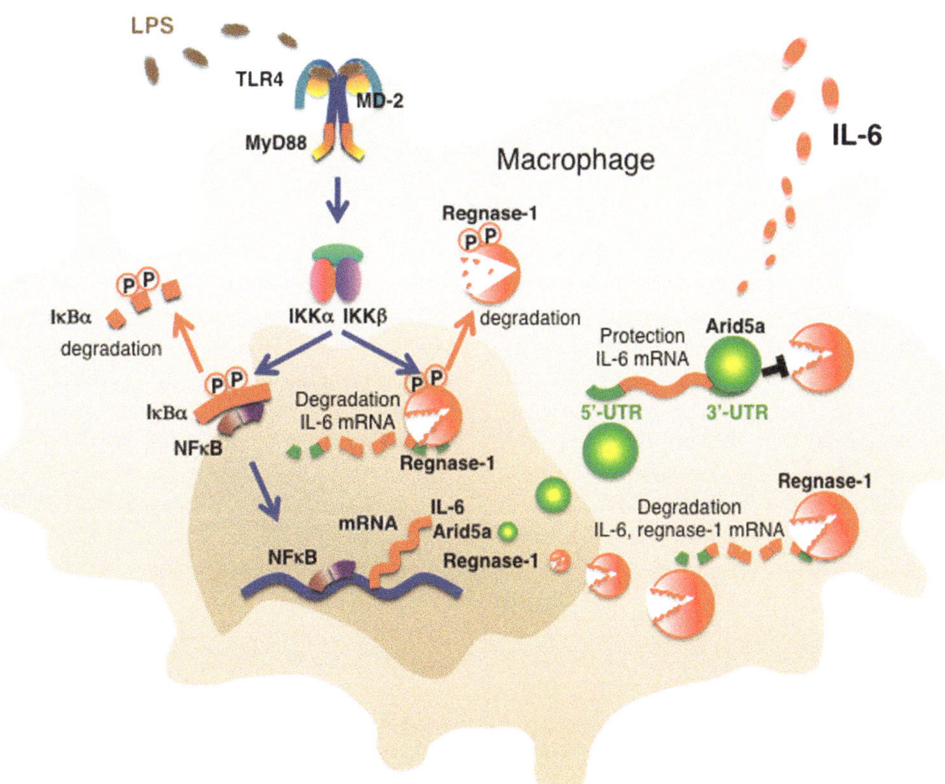


FIGURE 4 **Transcriptional and post-transcriptional regulation of IL-6 synthesis.** In the innate immune response, pathogen-associated molecular patterns are recognized by pathogen recognition receptors to induce proinflammatory cytokines. For example, in this figure, TLR4 recognizes LPS, which is a component of the outer membrane of gram-negative bacteria. MD-2 links TLR4 with LPS. Through MyD88, TLR4 activates the IKKα/β complex and phosphorylates IκBα, leading to the degradation of IκBα and activation of NF-κB, which promotes the transcription of IL-6 mRNA. The activated IKKα/β complex also phosphorylates Regnase-1, leading to the degradation of Regnase-1. Regnase-1 promotes IL-6 mRNA degradation, whereas Arid5a inhibits the destabilizing effect of Regnase-1. The balance between Arid5a and Regnase-1 is important for the regulation of IL-6 mRNA. TLR4: Toll-like receptor 4; LPS: lipopolysaccharide; MD-2: myeloid differentiation protein 2; MyD88: myeloid differentiation primary response 88; UTR: untranslated region; IKK; I-kappa-B kinase; NF-κB: nuclear factor of kappa B; IκBα: inhibitor of NF-κB; Arid5a: AT-rich interactive domain-containing protein 5a.

stem-loop region present in the Regnase-1 3′-UTR. These findings demonstrate that the IKK complex phosphorylates IκBα to activate transcription, but also that Regnase-1 releases the brake on IL-6 mRNA expression. Regnase-1 also regulates the mRNAs of a set of genes, including c-Rel, Ox40, and IL-2 by cleaving their 3′-UTRs in T-cells. T-cell receptor engagement then leads to cleavage of Regnase-1, which frees T-cells from Regnase 1–mediated suppression, indicating that Regnase-1 may have a crucial role not only in IL-6 expression but also in T-cell activation [62].

We recently identified a novel RNA-binding protein, AT-rich interactive domain-containing protein 5a (Arid5a), which also binds to the 3′-UTR of IL-6 mRNA, but results in the selective stabilization of IL-6 but not of TNFα or IL-12 mRNA [63]. Arid5a expression is upregulated in macrophages after stimulation with lipopolysaccharide (LPS), IL-1β, or IL-6, and is induced in T-cells under $T_h17$-polarizing conditions. We also found that Arid5a gene deficiency inhibits elevation of IL-6 levels in LPS-injected mice and preferential $T_h17$ cell development in experimental autoimmune encephalomyelitis. Moreover, Arid5a counteracts the destabilizing function of Regnase-1 on IL-6 mRNA (Figure 4), which indicates that the balance between Arid5a and Regnase-1 has an important role in IL-6 mRNA stability. All of these results suggest that posttranscriptional regulation of IL-6 mRNA by Arid5a and Regnase-1 may have an important role in the expression of IL-6. We hypothesize that the predominance of Arid5a over Regnase-1 causes persistent IL-6 production, leading to the development of autoimmune inflammatory diseases.

## 4. DYSREGULATED PERSISTENT IL-6 SYNTHESIS HAS A PATHOLOGIC ROLE IN THE DEVELOPMENT OF VARIOUS DISEASES

The immediate and transient expression of IL-6 is generated in response to environmental stress factors such as infection with microbial pathogens and tissue damage. When the source of stress is removed from the host, IL-6 production is terminated. However, dysregulated and persistent IL-6 production leads to the development of various diseases. The association of IL-6 with development of a human disease was first demonstrated in a case of cardiac myxoma. Myxoma is a benign tumor in the heart atrium, but one-third of such patients present with inflammatory and autoimmune symptoms such as fever, polyarthritis, thrombocytosis, elevated CRP level, hypergammaglobulinemia, and, occasionally, positivity for antinuclear antibody. Removal of the tumor resolves all of these symptoms. Culturing of the fluid obtained from myxoma tissues produces a large quantity of IL-6, which suggests that IL-6 may contribute to inflammation and autoimmunity [64]. Subsequent studies have demonstrated that elevated levels of IL-6 are found in the synovial fluids of patients with RA [65]. In addition, excess IL-6 production has been observed in the swollen lymph nodes of patients with Castleman disease [66], and in myeloma cells [26] and peripheral blood cells, as well as involved tissues in various other autoimmune and chronic inflammatory diseases, and even malignant cells in cancers [67,68].

Moreover, the pathological role of IL-6 in disease development has been demonstrated in numerous animal models of diseases, as have the findings that IL-6 blockade by gene knockout or administration of anti-IL-6 or anti-IL-6R antibody (Ab) can result in preventive or therapeutic suppression of disease development. For example, IL-6 blockade results in a noticeable reduction in susceptibility to Castleman disease–like symptoms in IL-6 transgenic mice [69]. Similar ameliorative effects are observed in models of RA [70–72], systemic lupus erythematosus [73], systemic sclerosis [74], inflammatory myopathies [75], experimental autoimmune uveoretinitis [76], experimental autoimmune encephalomyelitis [77], and many other diseases.

## 5. A HUMANIZED ANTI-IL-6 RECEPTOR ANTIBODY FOR TREATMENT OF AUTOIMMUNE INFLAMMATORY DISEASES

At the end of the nineteenth century, a German physiologist named Emil Adolf von Behring developed serum therapy against diphtheria toxin. He injected serum obtained from animals immunized with diphtheria toxin into patients with diphtheria and effectively cured the disease. The serum included antitoxin Abs, and the Abs themselves were the active components mediating this effect. For this pioneer achievement in the medical science, he received the first Nobel Prize in Physiology or Medicine in 1901. In 1975, Georges Köhler and César Milstein successfully created a hybridoma by fusing a specific Ab-secreting B cell with a myeloma cell, and thus produced a large amount of specific monoclonal Ab [78]. Because of this prominent discovery, they received the Nobel Prize in 1984. Using hybridoma technology, various mouse antihuman monoclonal Abs have been made that are specific for target molecules responsible for the development of inflammatory and infectious diseases and cancers. Although they possess high sensitivity and specificity for target antigens, the use of mouse monoclonal Abs was initially hampered owing to their low functional activity as well as high immunogenicity in humans. However, advances in genetic engineering have facilitated the development of monoclonal Abs with lower immunogenicity and higher activity, including chimeric, humanized, or fully human Abs [79]. The target molecule can be either a toxin or a specific molecule that has a critical role in the pathogenesis of diseases. One targeted molecule was IL-6 because of its pathological role in various diseases as described previously. Thus, tocilizumab, which is a humanized anti-IL-6R monoclonal Ab of the IgG1 class, was generated by grafting the complementarity determining regions of a mouse antihuman IL-6R Ab onto human IgG1 [80]. Tocilizumab blocks IL-6–mediated signal transduction by inhibiting IL-6 binding to transmembrane and soluble IL-6R (Figure 2). The first clinical trial was performed in a patient with intractable Castleman disease at Osaka University Hospital. The finding that tocilizumab injection ameliorated systemic inflammatory symptoms and serological parameters led to worldwide clinical trials. The outstanding efficacy, tolerability, and safety profile of tocilizumab were verified in numerous clinical trials, resulting in the current approval of this biologic for the treatment of RA in more than 100 countries [81]. Tocilizumab is also approved for systemic juvenile idiopathic arthritis in Japan, the United States, the European Union, and India [82,83] and for Castleman disease in Japan and India [84]. Although other biologics including TNF inhibitors, a T-cell stimulation blocker, a B-cell depletory, and an IL-1R antagonist are available for the treatment of RA, tocilizumab is the only effective biologic as monotherapy for moderate to severe active RA [85] and is now recommended as one of the first-line biologics [86]. Moreover, the prominent efficacy of tocilizumab for systemic juvenile idiopathic arthritis has led to the recognition that a new era has arrived for the treatment of this disease, which had long been considered one of the most intractable juvenile diseases [87].

Furthermore, based on favorable results detailed in numerous recent case reports, series, and pilot studies of off-label use of tocilizumab, this drug may be applicable for the treatment of various autoimmune inflammatory diseases including autoimmune, chronic inflammatory, autoinflammatory, and allergic diseases, as well as type 2 diabetes mellitus and even cancers [88,89]. Clinical trials are in progress to identify additional indications for tocilizumab.

## 6. INTERLEUKIN-6 BLOCKADE AFFECTS B- AND T-CELL FUNCTION IN VIVO; LESSONS FROM TOCILIZUMAB TREATMENT

As described elsewhere, evidence obtained from in vitro studies and animal models indicates that IL-6 is a central cytokine affecting B- and T-cell function. Recent analyses of changes in immunological parameters during tocilizumab treatment appear to confirm its effects on these lymphocytes. During treatment for Castleman disease, a marked reduction in the serum IgG level and a decrease in IgM, IgA, and IgE concentrations were observed [84].

Moreover, IL-6 blockade therapy causes a significant decrease in the serum levels of pathological autoantibodies. In patients with systemic lupus erythematosus, the percentage of $CD38^{high}CD19^{low}IgD^{negative}$ plasma cells in the peripheral blood is increased compared with normal controls (mean, 5.3% versus 1.2%), but 6 weeks of treatment with tocilizumab administered every 2 weeks reduced the percentage from 5.3% to 3.1% in association with a reduction in anti–double stranded DNA Abs [90]. In vitro analyses of plasmablasts in patients with neuromyelitis optica (NMO) showed that IL-6 enhances the survival of the population of $CD19^{int}CD27^{high}CD38^{high}CD180^{negative}$ plasmablasts, which secrete pathological anti-aquaporin 4 (AQP4) Abs in NMO [91]. Furthermore, blockade of IL-6R signaling reduces the survival of the plasmablasts. Tocilizumab treatment resulted in clinical improvement in a patient with NMO and reduced the elevated numbers of plasmablasts and the serum titer of anti-AQP4 autoantibody [92]. These findings indicate that IL-6, which was originally described as BSF-2 or HGF, interferes with the late phase of B-cells. Thus, tocilizumab treatment decreases the elevated serum Ig concentrations, the number of plasma cells, and pathological autoantibodies.

Interleukin-6 inhibition also results in changes within the B-cell compartment. Sixteen patients with RA were treated with tocilizumab, and immunophenotyping was performed. Frequencies of both $CD27^{positive}IgD^{negative}$ postswitch memory B-cells and $CD27^{positive}IgD^{positive}$ preswitch memory B-cells were increased in RA patients compared with healthy controls, whereas the percentages of $CD27^{negative}IgD^{positive}$ naive B-cells were significantly reduced in RA patients. Twenty-four weeks of tocilizumab treatment reduced the frequency of preswitch memory B-cells from a median of 19.6% to 12.3%, and of postswitch memory B-cells from a frequency of 18.6% to 15.0% [93]. $CD19^{positive}IgA^{positive}$ and $CD19^{positive}IgG^{positive}$ B-cells were also significantly decreased. These results indicate that IL-6 blockade suppresses B-cell hyperreactivity. In animal studies, IL-6 produced by immune complex-activated follicular dendritic cells promotes germinal center reactions, IgG responses, and somatic hypermutation [94]. Analysis of Ig receptors (IgR) of single cell-sorted preswitch and postswitch memory B-cells from 11 RA patients demonstrated a reduced mutational frequency in IgR of preswitch memory B-cells during tocilizumab therapy [95], but whether this effect was mediated by direct actions on B-cells or by inhibiting T-helper cell activity remains unknown.

As mentioned earlier, IL-6 is an important differentiation factor for converting naive CD4-positive T-cells into effector T-cell subsets. In particular, persistent production of IL-6 causes the predominance of $T_h17$ over Treg, which is considered a key immunological abnormality in several autoimmune and chronic inflammatory diseases, whereas IL-6 blockade may rectify this imbalance [29,96]. Indeed, two recent reports demonstrated that tocilizumab restores $T_h17$/Treg balance in T cells in the peripheral blood of RA patients [97,98]. Using a mouse model, a recent article showed that the induction of Ab production by IL-6 is indirectly mediated by IL-21 produced by CD4-positive T-cells [99]. Interleukin-6 is necessary to induce IL-21 production by naive and memory CD4-positive T-cells upon T-cell receptor stimulation, whereas IL-21 production by CD4-positive T-cells is required to promote B-cell Ab production in vitro. Furthermore, administration of IL-6 with inactive influenza virus enhances virus-specific Ab in vivo, and this effect is dependent on IL-21. The authors concluded that IL-6 enhances Ab production by promoting the B-cell helper capabilities of CD4-positive T-cells through increased IL-21 production rather than via a direct effect on activated B-cells. In eight RA patients, treatment with tocilizumab reduced IL-21 production by activated/memory CD4-positive T-cells as well as serum levels of IgG4 subclass anti-citrullinated peptide Ab [100]. Further evaluation is required to clarify the effects of tocilizumab in vivo.

## 7. CONCLUDING REMARKS

The first report of the existence of soluble factors for the enhancement of the IgG and IgE Ab response was published by Kishimoto and Ishizaka in 1973 [101], and 13 years elapsed until the successful cloning of the IL-6 gene [1]. Since then, the fundamental research has progressed rapidly, and the entire picture of the IL-6 signaling system was completed in the early 1990s [11], leading to clarification of the molecular basis of the characteristic features of cytokines of redundancy and pleiotropy. In parallel with this development, the pathological involvement of IL-6 in various diseases was also established [19,66,88]. Then, clinical trials with tocilizumab began in the late 1990s, and this humanized monoclonal Ab was approved for the treatment of Castleman disease in 2005 in Japan, nearly 20 years after successful molecular cloning of the IL-6 gene (Figure 5) [102]. In following years, tocilizumab has become one of the first-line biologics and the only one that is effective as monotherapy for the treatment of moderate to severe active

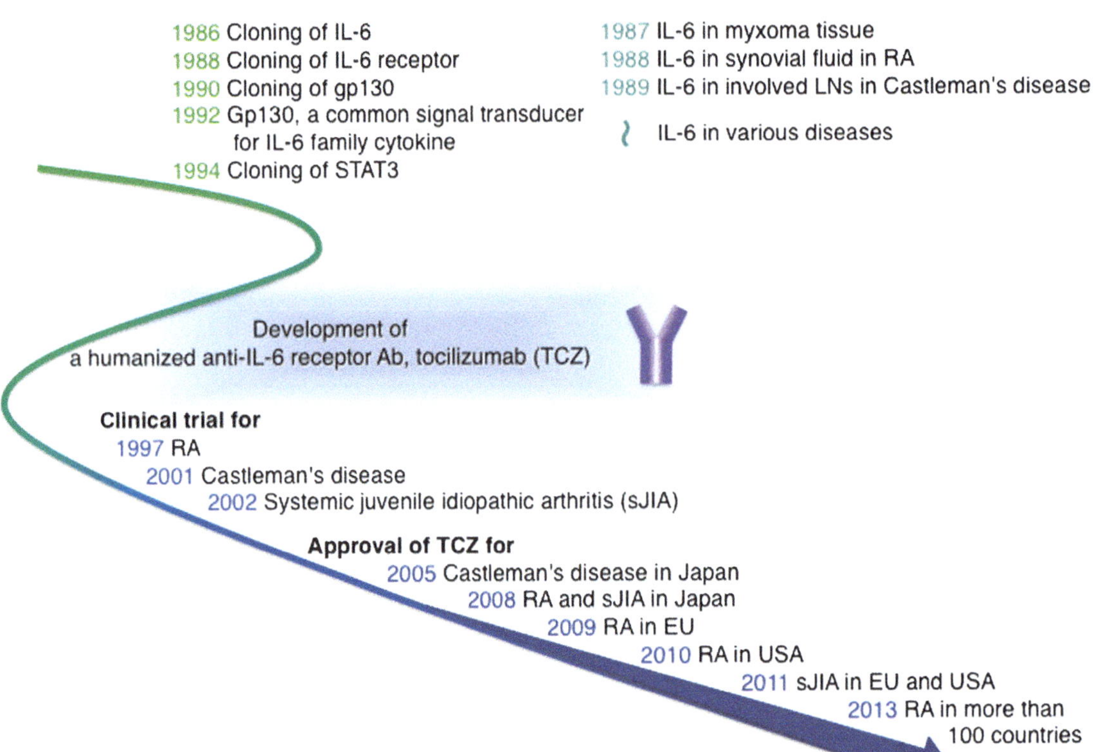


FIGURE 5 **Major discoveries in IL-6–related research and establishment of an IL-6 targeting strategy for autoimmune inflammatory diseases.** Basic research on IL-6 clarified the entire receptor-mediated signaling system and the molecular basis of the characteristics of cytokines, redundancy and pleiotropy. Clinical research revealed the pathological significance of IL-6 in disease development. These findings led to the concept that IL-6 targeting would constitute a novel therapeutic strategy for autoimmune inflammatory diseases, and indeed, tocilizumab, a humanized anti-IL-6R antibody became an innovative biologic for the treatment of refractory diseases such as rheumatoid arthritis (RA), systemic juvenile idiopathic arthritis (sJIA), and Castleman disease in the first decade of the twenty-first century. We expect that this strategy will be widely applicable in other autoimmune and chronic inflammatory diseases. LNs: lymph nodes; STAT3: signal transducer and activator of transcription 3; TCZ: tocilizumab.

RA. Tocilizumab is also approved for the treatment of systemic juvenile idiopathic arthritis. We anticipate that during the next decade, this IL-6 blocker will be widely used for the treatment of various currently intractable diseases and that its use will overcome the refractory nature of such diseases and may replace corticosteroids.

However, to achieve this goal, further clinical trials are essential. Although IL-6–related research has solved several mysteries, the question remains as to why IL-6 is persistently expressed in distinct cell populations in various diseases. Accurate and detailed analyses of proteins such as Arid5a and Regnase-1 and of microRNAs that regulate IL-6 synthesis will be helpful for solving this mystery. Furthermore, clarification of the mechanism(s) involved will certainly facilitate the identification of more specific target molecules and investigations into the pathogenesis of specific diseases.

## 8. CONFLICT OF INTEREST

T. Kishimoto holds a patent for tocilizumab and has received royalties for Actemra. T. Tanaka has received a grant and payment for lectures including service on speaker's bureaus from Chugai Pharmaceutical Co., Ltd. M. Narazaki has received payment for lectures including service on speaker's bureaus from Chugai Pharmaceutical Co., Ltd.

## REFERENCES

[1] Hirano T, Yasukawa K, Harada H, Taga T, Watanabe Y, Matsuda T, et al. Complementary DNA for a novel human interleukin (BSF-2) that induces B lymphocytes to produce immunoglobulin. Nature 1986;324:73–6.
[2] Kishimoto T, Hirano T. Molecular regulation of B lymphocyte response. Annu Rev Immunol 1988;6:485–512.
[3] Kishimoto T. The biology of interleukin-6. Blood 1989;74:1–10.
[4] Kishimoto T, Akira S, Taga T. Interleukin-6 and its receptor: a paradigm for cytokines. Science 1992;258:593–7.
[5] Yamasaki K, Taga T, Hirata Y, Yawata H, Kawanishi Y, Seed B, et al. Cloning and expression of the human interleukin-6 (BSF-2/IFN beta 2) receptor. Science 1988;241:825–8.
[6] Hibi M, Murakami M, Saito M, Hirano T, Taga T, Kishimoto T. Molecular cloning and expression of an IL-6 signal transducer, gp130. Cell 1990;63:1149–57.

[7] Murakami M, Hibi M, Nakagawa N, Nakagawa T, Yasukawa K, Yamanishi K, et al. IL-6-induced homodimerization of gp130 and associated activation of a tyrosine kinase. Science 1993;260:1808–10.

[8] Narazaki M, Yasukawa K, Saito T, Ohsugi Y, Fukui H, Koishihara Y, et al. Soluble forms of the interleukin-6 signal-transducing receptor component gp130 in human serum possessing a potential to inhibit signals through membrane-anchored gp130. Blood 1993;82:1120–6.

[9] Taga T, Kishimoto T. Gp130 and the interleukin-6 family of cytokines. Annu Rev Immunol 1997;15:797–819.

[10] Boulanger MJ, Chow DC, Brevnova EE, Garcia KC. Hexameric structure and assembly of the interleukin-6/IL-6 alpha-receptor/gp130 complex. Science 2003;300:2101–4.

[11] Kishimoto T, Taga T, Akira S. Cytokine signal transduction. Cell 1994;76:253–62.

[12] Kishimoto T, Akira S, Narazaki M, Taga T. Interleukin-6 family of cytokines and gp130. Blood 1995;86:1243–54.

[13] Aoki Y, Narazaki M, Kishimoto T, Tosato G. Receptor engagement by viral interleukin-6 encoded by Kaposi sarcoma-associated herpesvirus. Blood 2001;98:3042–9.

[14] Heinrich PC, Castell JV, Andus T. Interleukin-6 and the acute phase response. Biochem J 1990;265:621–36.

[15] Gillmore JD, Lovat LB, Persey MR, Pepys MB, Hawkins PN. Amyloid load and clinical outcome in AA amyloidosis in relation to circulating concentration of serum amyloid A protein. Lancet 2001;358:24–9.

[16] Nemeth E, Rivera S, Gabayan V, Keller C, Taudorf S, Pedersen BK, et al. IL-6 mediates hypoferremia of inflammation by inducing the synthesis of the iron regulatory hormone hepcidin. J Clin Invest 2004;113:1271–6.

[17] Liuzzi JP, Lichten LA, Rivera S, Blanchard RK, Aydemir TB, Knutson MD, et al. Interleukin-6 regulates the zinc transporter Zip14 in liver and contributes to the hypozincemia of the acute-phase response. Proc Natl Acad Sci USA 2005;102:6843–8.

[18] Kumar H, Kawai T, Akira S. Pathogen recognition by the innate immune system. Int Rev Immunol 2011;30:16–34.

[19] Nishimoto N, Yoshizaki K, Tagoh H, Monden M, Kishimoto S, Hirano T, et al. Elevation of serum interleukin 6 prior to acute phase proteins on the inflammation by surgical operation. Clin Immunol Immunopathol 1989;50:399–401.

[20] Bianchi ME. DAMPs, PAMPs and alarmins: all we need to know about danger. J Leukoc Biol 2007;81:1–5.

[21] Zhang Q, Raoof M, Chen Y, Sumi Y, Sursal T, Junger W, et al. Circulating mitochondrial DAMPs cause inflammatory responses to injury. Nature 2010;464:104–7.

[22] Sims GP, Rowe DC, Rietdijk ST, Herbst R, Coyle AJ. HMGB1 and RAGE in inflammation and cancer. Annu Rev Immunol 2010;28:367–88.

[23] Suematsu S, Matsuda T, Aozasa K, Akira S, Nakano N, Ohno S, et al. IgG1 plasmacytosis in interleukin 6 transgenic mice. Proc Natl Acad Sci USA 1989;86:7547–51.

[24] Suematsu S, Matsusaka T, Matsuda T, Ohno S, Miyazaki J, Yamamura K, et al. Generation of plasmacytomas with the chromosomal translocation t(12;15) in interleukin 6 transgenic mice. Proc Natl Acad Sci USA 1992;89:232–5.

[25] Potter M, Boyce CR. Induction of plasma-cell neoplasms in strain BALB/c mice with mineral oil and mineral oil adjuvants. Nature 1962;193:1086–7.

[26] Kawano M, Hirano T, Matsuda T, Taga T, Horii Y, Iwato K, et al. Autocrine generation and requirement of BSF-2/IL-6 for human multiple myelomas. Nature 1988;332:83–5.

[27] Korn T, Bettelli E, Oukka M, Kuchroo VK. IL-17 and Th17 cells. Annu Rev Immunol 2009;27:485–517.

[28] Bettelli E, Carrier Y, Gao W, Korn T, Strom TB, Oukka M, et al. Reciprocal developmental pathways for the generation of pathogenic effector TH17 and regulatory T cells. Nature 2006;441:235–8.

[29] Kimura A, Kishimoto T. IL-6: regulator of Treg/Th17 balance. Eur J Immunol 2010;40:1830–5.

[30] Ma CS, Avery DT, Chan A, Batten M, Bustamante J, Boisson-Dupuis S, et al. Functional STAT3 deficiency compromises the generation of human T follicular helper cells. Blood 2012;119:3997–4008.

[31] Okada M, Kitahara M, Kishimoto S, Matsuda T, Hirano T, Kishimoto T. IL-6/BSF-2 functions as a killer helper factor in the in vitro induction of cytotoxic T cells. J Immunol 1988;141:1543–9.

[32] Hashizume M, Hayakawa N, Mihara M. IL-6 trans-signalling directly induces RANKL on fibroblast-like synovial cells and is involved in RANKL induction by TNF-alpha and IL-17. Rheumatology (Oxford) 2008;47:1635–40.

[33] Kotake S, Sato K, Kim KJ, Takahashi N, Udagawa N, Nakamura I, et al. Interleukin-6 and soluble interleukin-6 receptors in the synovial fluids from rheumatoid arthritis patients are responsible for osteoclast-like cell formation. J Bone Miner Res 1996;11:88–95.

[34] Poli V, Balena R, Fattori E, Markatos A, Yamamoto M, Tanaka H, et al. Interleukin-6 deficient mice are protected from bone loss caused by estrogen depletion. EMBO J 1994;13:1189–96.

[35] Nakahara H, Song J, Sugimoto M, Hagihara K, Kishimoto T, Yoshizaki K, et al. Anti-interleukin-6 receptor antibody therapy reduces vascular endothelial growth factor production in rheumatoid arthritis. Arthritis Rheum 2003;48:1521–9.

[36] Hashizume M, Hayakawa N, Suzuki M, Mihara M. IL-6/sIL-6R trans-signalling, but not TNF-alpha induced angiogenesis in a HUVEC and synovial cell co-culture system. Rheumatol Int 2009;29:1449–54.

[37] Grossman RM, Krueger J, Yourish D, Granelli-Piperno A, Murphy DP, May LT, et al. Interleukin 6 is expressed in high levels in psoriatic skin and stimulates proliferation of cultured human keratinocytes. Proc Natl Acad Sci USA 1989;86:6367–71.

[38] Duncan MR, Berman B. Stimulation of collagen and glycosaminoglycan production in cultured human adult dermal fibroblasts by recombinant human interleukin 6. J Invest Dermatol 1991;97:686–92.

[39] Libermann TA, Baltimore D. Activation of interleukin-6 gene expression through the NF-kappa B transcription factor. Mol Cell Biol 1990;10:2327–34.

[40] Matsusaka T, Fujikawa K, Nishio Y, Mukaida N, Matsushima K, Kishimoto T, et al. Transcription factors NF-IL6 and NF-kappa B synergistically activate transcription of the inflammatory cytokines, interleukin 6 and interleukin 8. Proc Natl Acad Sci USA 1993;90:10193–7.

[41] Ballard DW, Bohnlein E, Lowenthal JW, Wano Y, Franza BR, Greene WC. HTLV-I tax induces cellular proteins that activate the kappa B element in the IL-2 receptor alpha gene. Science 1988;241:1652–5.

[42] Leung K, Nabel GJ. HTLV-1 transactivator induces interleukin-2 receptor expression through an NF-kappa B-like factor. Nature 1988;333:776–8.

[43] Scala G, Ruocco MR, Ambrosino C, Mallardo M, Giordano V, Baldassarre F, et al. The expression of the interleukin 6 gene is induced by the human immunodeficiency virus 1 TAT protein. J Exp Med 1994;179:961–71.

[44] Ambrosino C, Ruocco MR, Chen X, Mallardo M, Baudi F, Trematerra S, et al. HIV-1 Tat induces the expression of the interleukin-6 (IL6) gene by binding to the IL6 leader RNA and by interacting with CAAT enhancer-binding protein beta (NF-IL6) transcription factors. J Biol Chem 1997;272:14883–92.

[45] Mahe Y, Mukaida N, Kuno K, Akiyama M, Ikeda N, Matsushima K, et al. Hepatitis B virus X protein transactivates human interleukin-8 gene through acting on nuclear factor kB and CCAAT/enhancer-binding protein-like cis-elements. J Biol Chem 1991;266:13759–63.

[46] Ohno H, Kaneko S, Lin Y, Kobayashi K, Murakami S. Human hepatitis B virus X protein augments the DNA binding of nuclear factor for IL-6 through its basic-leucine zipper domain. J Med Virol 1999;58:11–8.

[47] Delerive P, De Bosscher K, Besnard S, Vanden Berghe W, Peters JM, Gonzalez FJ, et al. Peroxisome proliferator-activated receptor alpha negatively regulates the vascular inflammatory gene response by negative cross-talk with transcription factors NF-kappaB and AP-1. J Biol Chem 1999;274:32048–54.

[48] Jilka RL, Hangoc G, Girasole G, Passeri G, Williams DC, Abrams JS, et al. Increased osteoclast development after estrogen loss: mediation by interleukin-6. Science 1992;257:88–91.

[49] Ray A, Prefontaine KE. Physical association and functional antagonism between the p65 subunit of transcription factor NF-kappa B and the glucocorticoid receptor. Proc Natl Acad Sci USA 1994;91:752–6.

[50] Santhanam U, Ray A, Sehgal PB. Repression of the interleukin 6 gene promoter by p53 and the retinoblastoma susceptibility gene product. Proc Natl Acad Sci USA 1991;88:7605–9.

[51] He M, Xu Z, Ding T, Kuang DM, Zheng L. MicroRNA-155 regulates inflammatory cytokine production in tumor-associated macrophages via targeting C/EBPbeta. Cell Mol Immunol 2009;6:343–52.

[52] Zilahi E, Tarr T, Papp G, Griger Z, Sipka S, Zeher M. Increased microRNA-146a/b, TRAF6 gene and decreased IRAK1 gene expressions in the peripheral mononuclear cells of patients with Sjogren's syndrome. Immunol Lett 2012;141:165–8.

[53] Chen Q, Wang H, Liu Y, Song Y, Lai L, Han Q, et al. Inducible microRNA-223 down-regulation promotes TLR-triggered IL-6 and IL-1beta production in macrophages by targeting STAT3. PLoS One 2012;7:e42971.

[54] Chen CY, Shyu AB. AU-rich elements: characterization and importance in mRNA degradation. Trends Biochem Sci 1995;20:465–70.

[55] Anderson P. Post-transcriptional control of cytokine production. Nat Immunol 2008;9:353–9.

[56] Zhao W, Liu M, Kirkwood KL. p38alpha stabilizes interleukin-6 mRNA via multiple AU-rich elements. J Biol Chem 2008;283:1778–85.

[57] Kang JG, Pripuzova N, Majerciak V, Kruhlak M, Le SY, Zheng ZM. Kaposi's sarcoma-associated herpesvirus ORF57 promotes escape of viral and human interleukin-6 from microRNA-mediated suppression. J Virol 2011;85:2620–30.

[58] Palanisamy V, Jakymiw A, Van Tubergen EA, D'Silva NJ, Kirkwood KL. Control of cytokine mRNA expression by RNA-binding proteins and microRNAs. J Dent Res 2012;91:651–8.

[59] Xu Z, Xiao SB, Xu P, Xie Q, Cao L, Wang D, et al. miR-365, a novel negative regulator of interleukin-6 gene expression, is cooperatively regulated by Sp1 and NF-kappaB. J Biol Chem 2011;286:21401–12.

[60] Matsushita K, Takeuchi O, Standley DM, Kumagai Y, Kawagoe T, Miyake T, et al. Zc3h12a is an RNase essential for controlling immune responses by regulating mRNA decay. Nature 2009;458:1185–90.

[61] Iwasaki H, Takeuchi O, Teraguchi S, Matsushita K, Uehata T, Kuniyoshi K, et al. The IkappaB kinase complex regulates the stability of cytokine-encoding mRNA induced by TLR-IL-1R by controlling degradation of regnase-1. Nat Immunol 2011;12:1167–75.

[62] Uehata T, Iwasaki H, Vandenbon A, Matsushita K, Hernandez-Cuellar E, Kuniyoshi K, et al. Malt1-induced cleavage of regnase-1 in CD4+ helper T cells regulates immune activation. Cell 2013;153:1036–49.

[63] Masuda K, Ripley B, Nishimura R, Mino T, Takeuchi O, Shioi G, et al. Arid5a controls IL-6 mRNA stability, which contributes to elevation of IL-6 level in vivo. Proc Natl Acad Sci USA 2013;110:9409–14.

[64] Hirano T, Taga T, Yasukawa K, Nakajima K, Nakano N, Takatsuki F, et al. Human B-cell differentiation factor defined by an anti-peptide antibody and its possible role in autoantibody production. Proc Natl Acad Sci USA 1987;84:228–31.

[65] Hirano T, Matsuda T, Turner M, Miyasaka N, Buchan G, Tang B, et al. Excessive production of interleukin 6/B cell stimulatory factor-2 in rheumatoid arthritis. Eur J Immunol 1988;18:1797–801.

[66] Yoshizaki K, Matsuda T, Nishimoto N, Kuritani T, Taeho L, Aozasa K, et al. Pathogenic significance of interleukin-6 (IL-6/BSF-2) in Castleman's disease. Blood 1989;74:1360–7.

[67] Hirano T, Akira S, Taga T, Kishimoto T. Biological and clinical aspects of interleukin 6. Immunol Today 1990;11:443–9.

[68] Akira S, Taga T, Kishimoto T. Interleukin-6 in biology and medicine. Adv Immunol 1993;54:1–78.

[69] Katsume A, Saito H, Yamada Y, Yorozu K, Ueda O, Akamatsu K, et al. Anti-interleukin 6 (IL-6) receptor antibody suppresses Castleman's disease like symptoms emerged in IL-6 transgenic mice. Cytokine 2002;20:304–11.

[70] Alonzi T, Fattori E, Lazzaro D, Costa P, Probert L, Kollias G, et al. Interleukin 6 is required for the development of collagen-induced arthritis. J Exp Med 1998;187:461–8.

[71] Ohshima S, Saeki Y, Mima T, Sasai M, Nishioka K, Nomura S, et al. Interleukin 6 plays a key role in the development of antigen-induced arthritis. Proc Natl Acad Sci USA 1998;95:8222–6.

[72] Fujimoto M, Serada S, Mihara M, Uchiyama Y, Yoshida H, Koike N, et al. Interleukin-6 blockade suppresses autoimmune arthritis in mice by the inhibition of inflammatory Th17 responses. Arthritis Rheum 2008;58:3710–9.

[73] Mihara M, Takagi N, Takeda Y, Ohsugi Y. IL-6 receptor blockage inhibits the onset of autoimmune kidney disease in NZB/W F1 mice. Clin Exp Immunol 1998;112:397–402.

[74] Kitaba S, Murota H, Terao M, Azukizawa H, Terabe F, Shima Y, et al. Blockade of interleukin-6 receptor alleviates disease in mouse model of scleroderma. Am J Pathol 2012;180:165–76.

[75] Okiyama N, Sugihara T, Iwakura Y, Yokozeki H, Miyasaka N, Kohsaka H. Therapeutic effects of interleukin-6 blockade in a murine model of polymyositis that does not require interleukin-17A. Arthritis Rheum 2009;60:2505–12.

[76] Haruta H, Ohguro N, Fujimoto M, Hohki S, Terabe F, Serada S, et al. Blockade of interleukin-6 signaling suppresses not only Th17 but also interphotoreceptor retinoid binding protein-specific Th1 by promoting regulatory T cells in experimental autoimmune uveoretinitis. Invest Ophthalmol Vis Sci 2011;52:3264–71.

[77] Serada S, Fujimoto M, Mihara M, Koike N, Ohsugi Y, Nomura S, et al. IL-6 blockade inhibits the induction of myelin antigen-specific Th17 cells and Th1 cells in experimental autoimmune encephalomyelitis. Proc Natl Acad Sci USA 2008;105:9041–6.

[78] Kohler G, Milstein C. Continuous cultures of fused cells secreting antibody of predefined specificity. Nature 1975;256:495–7.
[79] Jakobovits A, Amado RG, Yang X, Roskos L, Schwab G. From XenoMouse technology to panitumumab, the first fully human antibody product from transgenic mice. Nat Biotechnol 2007;25:1134–43.
[80] Sato K, Tsuchiya M, Saldanha J, Koishihara Y, Ohsugi Y, Kishimoto T, et al. Reshaping a human antibody to inhibit the interleukin 6-dependent tumor cell growth. Cancer Res 1993;53:851–6.
[81] Tanaka T, Ogata A, Kishimoto T. Targeting of interleukin-6 for the treatment of rheumatoid arthritis: a review and update. Rheumatol Curr Res 2013;73(4):341–56. S4:002.
[82] Yokota S, Imagawa T, Mori M, Miyamae T, Aihara Y, Takei S, et al. Efficacy and safety of tocilizumab in patients with systemic-onset juvenile idiopathic arthritis: a randomised, double-blind, placebo-controlled, withdrawal phase III trial. Lancet 2008;371:998–1006.
[83] De Benedetti F, Brunner HI, Ruperto N, Kenwright A, Wright S, Calvo I, et al. Randomized trial of tocilizumab in systemic juvenile idiopathic arthritis. N Engl J Med 2012;367:2385–95.
[84] Nishimoto N, Kanakura Y, Aozasa K, Johkoh T, Nakamura M, Nakano S, et al. Humanized anti-interleukin-6 receptor antibody treatment of multicentric Castleman disease. Blood 2005;106:2627–32.
[85] Emery P, Sebba A, Huizinga TW. Biologic and oral disease-modifying antirheumatic drug monotherapy in rheumatoid arthritis. Ann Rheum Dis 2013;72:1897–904.
[86] Smolen JS, Landewe R, Breedveld FC, Buch M, Burmester G, Dougados M, et al. EULAR recommendations for the management of rheumatoid arthritis with synthetic and biological disease-modifying antirheumatic drugs: 2013 update. Ann Rheum Dis 2014;73:492–509.
[87] Sandborg C, Mellins ED. A new era in the treatment of systemic juvenile idiopathic arthritis. N Engl J Med 2012;367:2439–40.
[88] Tanaka T, Narazaki M, Kishimoto T. Therapeutic targeting of the interleukin-6 receptor. Annu Rev Pharmacol Toxicol 2012;52:199–219.
[89] Tanaka T, Kishimoto T. Targeting interleukin-6: all the way to treat autoimmune and inflammatory diseases. Int J Biol Sci 2012;8:1227–36.
[90] Illei GG, Shirota Y, Yarboro CH, Daruwalla J, Tackey E, Takada K, et al. Tocilizumab in systemic lupus erythematosus: data on safety, preliminary efficacy, and impact on circulating plasma cells from an open-label phase I dosage-escalation study. Arthritis Rheum 2010;62:542–52.
[91] Chihara N, Aranami T, Sato W, Miyazaki Y, Miyake S, Okamoto T, et al. Interleukin 6 signaling promotes anti-aquaporin 4 autoantibody production from plasmablasts in neuromyelitis optica. Proc Natl Acad Sci USA 2011;108:3701–6.
[92] Araki M, Aranami T, Matsuoka T, Nakamura M, Miyake S, Yamamura T. Clinical improvement in a patient with neuromyelitis optica following therapy with the anti-IL-6 receptor monoclonal antibody tocilizumab. Mod Rheumatol 2012;23:827–31.
[93] Roll P, Muhammad K, Schumann M, Kleinert S, Einsele H, Dorner T, et al. In vivo effects of the anti-interleukin-6 receptor inhibitor tocilizumab on the B cell compartment. Arthritis Rheum 2011;63:1255–64.
[94] Wu Y, El Shikh ME, El Sayed RM, Best AM, et al. IL-6 produced by immune-complex-activated follicular dendritic cells promotes germinal center reactions, IgG responses and somatic hypermutation. Int Immunol 2009;21:745–6.
[95] Muhammad K, Roll P, Seibold T, Kleinert S, Einsele H, Dorner T, et al. Impact of IL-6 receptor inhibition on human memory B cells in vivo: impaired somatic hypermutation in preswitch memory B cells and mutational targeting in memory B cells. Ann Rheum Dis 2011;70:1507–10.
[96] Tanaka T. Can IL-6 blockade rectify imbalance between Tregs and Th17 cells? Immunotherapy 2013;5:695–7.
[97] Samson M, Audia S, Janikashvili N, Ciudad M, Trad M, Fraszczak J, et al. Brief report: inhibition of Interleukin-6 function corrects Th17/Treg cell imbalance in patients with rheumatoid arthritis. Arthritis Rheum 2012;64:2499–503.
[98] Pesce B, Soto L, Sabugo F, Wurmann P, Cuchacovich M, Lopez MN, et al. Effect of interleukin-6 receptor blockade on the balance between regulatory T cells and T helper type 17 cells in rheumatoid arthritis patients. Clin Exp Immunol 2013;171:237–42.
[99] Dienz O, Eaton SM, Bond JP, Neveu W, Moquin D, Noubade R, et al. The induction of antibody production is indirectly mediated by IL-21 produced by CD4+ T cells. J Exp Med 2009;206:69–78.
[100] Carbone G, Wilson A, Diehl SA, Bunn J, Cooper SM, Ricon M. Interleukin-6 receptor blockade selectively reduces IL-21 production by CD4 T cells and IgG4 autoantibodies in rheumatoid arthritis. Int J Biol Sci 2013;9:279–88.
[101] Kishimoto T, Ishizaka K. Regulation of antibody response in vitro. VII. Enhancing soluble factors for IgG and IgE antibody response. J Immunol 1973;111:1194–205.
[102] Kishimoto T. Interleukin-6: from basic science to medicine–40 years in immunology. Annu Rev Immunol 2005;23:1–21.

Chapter 28

# Targeting the IL-17/IL-23 Axis in Chronic Inflammatory Immune-Mediated Diseases

Alexander Tsoukas, Carl S. Goodyear and Iain B. McInnes
*Institute of Infection, Immunity and Inflammation, Glasgow Biomedical Research Centre, University of Glasgow, Glasgow, UK*

## 1. INTRODUCTION

Significant progress has been made over the past 15 years in our understanding of the pathophysiology in autoimmune disease. Earlier studies in animal models led to the detection of the ubiquitous interleukin (IL)-17 cytokine family, which was shown to be vital in the crosstalk between innate and adaptive immunity. Thereafter, the discovery of a specific T-cell subset termed T helper 17 ($T_H17$) cells resulted in a paradigm shift in our perception that autoimmune disease was traditionally caused by T helper 1 ($T_H1$) cells. Several murine and human studies have demonstrated the important role IL-17 cytokines and $T_H17$ cells play in the initiation and propagation of inflammation as well as the pathogenesis of autoimmune processes. This pathway is in turn driven by a group of upstream cytokines, principally IL-1, IL-6, IL-21, transforming growth factor (TGF)-β, and IL-23. The IL-17/IL-23 ($T_H17$) axis has been implicated as a principal effector in several chronic inflammatory disease such as rheumatoid arthritis (RA), psoriasis, psoriatic arthritis (PsA), ankylosing spondylitis (AS), inflammatory bowel disease (IBD), and multiple sclerosis (MS), making it a valuable target for therapeutic manipulation. Rapid progress in the development of immunotherapeutics targeting this axis has given both scientists and clinicians insight into the complex interaction between these cytokines and their receptors in various disease states. Recent clinical trials have given impressive results in the treatment of psoriasis and psoriatic arthritis using therapeutic monoclonal antibodies targeting IL-17A, IL-23p40, and IL-17A receptor. Although still in early development, several clinical trials have shown the potential for immunotherapy of other chronic inflammatory conditions. This chapter will discuss the mechanisms by which immune therapies target the IL-17/IL-23 axis and later focus on clinical trials conducted in several autoimmune and inflammatory diseases.

## 2. THE IL-17 FAMILY

The IL-17 protein was first discovered in 1993 while identifying inducible T-cell products in a murine cytotoxic T-cell hybridoma cDNA library, and was originally named cytotoxic T lymphocyte-associated antigen 8 (CTLA8) [1]. It was subsequently cloned from human T cells, and renamed IL-17 [2]. Early biological assays in synoviocytes from patients with rheumatoid arthritis showed IL-17 to have proinflammatory properties, as it induced prostaglandin E2 ($PGE_2$), granulocyte colony-stimulating factor (G-CSF), IL-6, and IL-8 in a dose-dependent manner, suggesting it may serve in the quick nonspecific immune response to infectious agents [3]. Although it did not increase tumor necrosis factor (TNF)-α, there seemed to be a synergistic effect of TNF-α, IL-1, and interferon (IFN)-γ on IL-17 induced secretion of IL-6. Furthermore, when monoclonal antibodies to IL-17 were added to cultures on synoviocytes obtained from patients suffering from RA, levels of IL-6 fell by 70% [4]. The IL-17 pathway was recognized as a coordinating link between innate and adaptive immunity.

IL-17 was later retermed IL-17A after five additional homologous cytokines were discovered, IL-17B, IL-17C, IL-17D, IL-17E (IL-25), and IL-17F. These glycoprotein molecules are composed of two monomers that are linked by disulfide bridges to form homodimers, with IL-17A also being able to form a heterodimer with IL-17F [5]. IL-17A and IL-17F are syntenic, 50% identical in structure, and bind the same receptor [5]. They are the closest related members within the family; however, there are distinct functional differences. The strength of downstream gene activation by IL-17F is 10–30 times weaker than that of IL-17A, with heterodimer IL-17A/F acting at an intermediate level [6,7]. In animal models, there were distinct pathophysiological effects of each cytokine.

Molecular Biology of B Cells. http://dx.doi.org/10.1016/B978-0-12-397933-9.00028-X

IL-17F knockout mice had reduced colitis after dextran sulfate sodium challenge, compared to IL-17A knockout mice that developed more severe disease [8]. In addition, IL-17A, but not IL-17F, was required for the initiation of experimental autoimmune encephalomyelitis, and mice deficient in IL-17F, but not IL-17A, had defective airway neutrophilia in response to allergen challenge [8]. Differences in response may have to do with differential expression of IL-17 cytokines by effector cells, and diversity of target receptor location and function.

IL-17B–D have lower homology, and IL-17E the least. IL-17C promotes innate defense in epithelial cells, regulates $T_H17$ cell differentiation, and may be implicated in the development of human psoriasiform skin lesions, where lesional biopsies showed levels significantly higher than other IL-17 isoforms and decreased with TNF-a inhibition and improved skin phenotype [9–11]. In experimental colitis, IL-17C knockouts had exacerbated disease, suggesting that IL-17C may play a critical role in maintaining mucosal barrier integrity [12]. IL-17B is less well characterized. It has been shown to have a role in collagen-induced arthritis in mouse models, and its neutralization by a monoclonal antibody significantly suppressed the progression of arthritis and bone destruction [13]. Recently, IL-17B has been shown to be present in human RA synovium at a higher level than other isoforms, likely derives from neutrophils, and acts on fibroblasts to promote transformation from acute to chronic disease [14]. Less studied, IL-17D has also been shown to be proinflammatory, leading to increased expression of IL-6, IL-8, and GM-CSF, and may also have an inhibitory effect on hematopoiesis of myeloid progenitor cells [15]. IL-17E, also known as IL-25, is a $T_H2$ cell-promoting cytokine implicated in allergic responses [16]. It may inhibit the function of $T_H17$ cells, and hence have indirect anti-inflammatory effects [6].

## 3. IL-17 RECEPTOR/PATHWAY

Five single transmembrane domain-containing receptor subunits have been identified, IL-17RA to IL-17RE, which combine to form functional receptor complexes. IL-17RA is the largest subunit and may serve as a common receptor chain since it is used by at least four IL-17 ligands [6]. Oligomerization of IL-17RA with IL-17RC (possibly trimeric combination) is required for cell signaling from both IL-17A and IL-17F [17,18]. IL-17E mediated signaling occurs through an IL-17RA/IL-17RB complex, and IL-17C mediated signaling may occur through an IL-17RA and IL-17RE complex [10,19,20].

IL-17RA receptor mediated signaling differs from other known interleukin pathways as it does not involve Janus kinase (JAK) and signal transducer and activator of transcription (STAT) activation, but rather resembles IL-1R and Toll-like receptors (TLR), resulting in activating of the transcription factor NF-κB and mitogen-activated protein kinases (MAPK) [5,21]. However unlike TLRs, IL-17Rs bind with the novel adaptor protein Act1 through a SEF–IL-17R domain, followed by recruitment of the kinase TAK1 and TNF receptor-associated factor 6 (TRAF6), which mediate activation of nuclear factor kappa-light-chain-enhancer of activated B cells (NF-κB), a chief transcription factor associated with the induction of inflammation [22,23]. IL-17RA pathways upregulate CCAAT/enhancer-binding protein (C/EBP) transcription factors C/EBP-β and C/EBP-δ, which are linked to IL-17A dependent induction of IL-6 and lipocalin-2 expression [24]. In addition to proinflammatory cytokine production, IL-17A stimulation in human fibroblasts and epithelial cells results in increased G-CSF and chemokine CXCR2, CXCL1, CXCL2, and CXCL8 production due to increased mRNA stability by Act1 [25,26]. The ultimate characterization of this response to IL-17 depends on the cellular sources of the cytokines and the diversity of IL-17 receptors on the target cells. Responses may also differ during various time points during the interplay of innate and adaptive immunity.

### 3.1 Type 17 T Cells

The innate immune system produces cytokines that cause the differentiation of naive CD4+ T cells into various T-helper ($T_H$) cell subsets. Classically, the naive T cells divide into $T_H1$ cells, which secrete high levels of IFN-γ and IL-2, activate macrophages, and are required for cell-mediated immunity against intracellular pathogens, and $T_H2$ cells, which predominantly secrete IL-4, IL-5, and IL-13 and are required for clearing extracellular pathogen, particularly parasitic infections [27]. These IL-12-induced proinflammatory $T_H1$ cells were originally postulated to be critical in the pathogenesis of certain autoimmune diseases. However, later studies show that lymphocyte-rich, experimental autoimmune encephalitis (EAE; a murine multiple sclerosis model) can be induced in IFN-γ and IL-12 knockout mice [28,29]. This suggested the presence of a different non-$T_H1$ subset of T cells infiltrating the lesions [28,29]. These activated CD4+ cells were induced by IL-23 and not IFN-γ or IL-4. Upon antigenic stimulation in the presence of IL-23, this subset produced large quantities of IL-17A, IL-17F, and therefore termed $T_H17$ cells, as well as IL-22, IL-6, and TNFα [30]. The passive transfer of these IL-17 producing T cells into other mice, and expanded with IL-23, induced severe EAE (in vitro) [30]. Furthermore, a deficiency in either p40 or p19, subunits that form IL-23, resulted in a decreased quantity of $T_H17$ cells and protection from EAE and collagen-induced arthritis (CIA) [29,31]. Cartilage destruction and bony erosion in a CIA mouse model showed that the IL-23/IL-17 axis, rather than the IL-12-IFN-γ axis, is critical not only for the onset phase, but also for the bone destruction phase of autoimmune arthritis by inducing RANKL [32]. In recent years, the role of the IL-23/IL-17 axis has been expanded to a variety of autoimmune and autoinflammatory disorders in humans as described in further sections.

## 4. $T_H17$ CELL DIFFERENTIATION

Since IL-23 receptor is not naturally present on naive CD4+ T cells, their differentiation into $T_H17$ cells occurs through a distinct pathway to that of $T_H1$ and $T_H2$ cells [33]. In murine models, it was shown that transforming growth factor β (TGF- β) in combination with IL-6 initiated differentiation of naive CD4+ T cells to $T_H17$ cells [34–36]. These cytokines activate the transcription factor retinoid-related orphan receptor (ROR)γt, which in conjunction with RORα and other transcription factors, initiates expression of IL-17A and IL-23R on developing $T_H17$ cells [37,38]. In addition, IL-1β can augment proliferation of $T_H17$ cells and ultimately increase production of IL-17 by $T_H17$ cells and γδ cells [20,39,40]. Developing $T_H17$ cells require further exposure to IL-23 to stabilize their phenotype, help them acquire effector functions, and enhance their production of IL-17A, IL-17F, IL-22, and IL-21, an autocrine cytokine augmenting differentiation [41]. Therefore, loss of any one of cytokines IL-6, IL-21, or IL-23 limits the $T_H17$ response.

Human $T_H17$ cells are similar in that they produce IL-17A, IL-17F, IL-22, IL-23, IL-26, and CCL20 and require ROR transcription factors (RORc in humans) for their differentiation [42–44]. However, while murine $T_H17$ cells originate in the presence of TGF-β and IL-6, human $T_H17$ cells originate in the presence of IL-1β and IL-23 [35,45]. Whether TGF-β is a requirement has been controversial, with studies showing mixed results [44].

$T_H17$ cell differentiation is regulated through reciprocal relationship with CD4+ regulatory T cells ($T_{regs}$) whose main function is to suppress T-cell responses against self and foreign antigens. TGF-β induces forkhead box protein P3 $(Foxp3)^+$ $T_{reg}$ cells from naive CD4+ T cells in the absence of IL-6. This causes suppression of RORγt and leads naive CD4+ T cells in developing into $T_{regs}$ [46,47]. The inverse occurs in the presence of IL-6, where Foxp3 gets downregulated and RORγt gets upregulated [48,49]. Therefore, $T_{regs}$ show plasticity in the presence of inflammatory cytokines and can differentiate into IL-17A producing $T_H17$ cells.

Manipulation of the differentiation pathway between $T_H17$ and $T_{reg}$ cells may allow a shift in balance between proinflammatory and regulatory mechanisms and result in new treatment targets in chronic inflammatory disorders.

## 5. CELLULAR SOURCES AND TARGETS

IL-17A and IL-17F are major contributors to host defense against bacterial and fungal pathogens, primarily through neutrophil recruitment and production of anti-microbial peptides, chemokines, and acute phase reactants. This is particularly important at mucosal surfaces and other environmental interfaces. In quiescent states, IL-17A is expressed in the gut (mainly lamina propria), and to a lesser degree in lymphoid follicles and progenitor cells of the spleen (in mouse models) [50].

In early response to infection or immunization with an adjuvant, IL-17A levels can rapidly increase within hours, suggesting additional sources of its production since differentiation of naive T cells to $T_H17$ cells can take days [20]. This may be due to tissue-based γδ T cells that have been shown to express IL-23R and RORγt and produce IL-17, 21, and 22 in response to IL-1β and IL-23, without T-cell receptor engagement [39]. IL-21 was also shown to amplify subsequent $T_H17$ response [39]. Although CD4+ $T_H17$ cells are the chief source of IL-17 cytokines, they can also be expressed in a subset of natural killer (NK) cells known as NK1.1, myeloid cells, CD8+ T cells, innate lymphoid cells (type 3), and possibly neutrophils and mast cells [20,33,51,52]. IL-23 appears to be an important activating factor in these cell lines; however, mast cells can express IL-17 directly through activation by various TLR-2 ligands as well [52].

Given its involvement at mucosal sites, IL-17A can act on a variety of cells of the immune system such as dendritic cells, macrophages, neutrophils, and lymphocytes, as well as mesenchymal cells such as fibroblasts, osteoblasts, chondrocytes, endothelial, and epithelial cells [20,53]. As an acute proinflammatory cytokine, IL-17A has several dynamic functions. First, IL-17A mediated release of IL-6 and IL-8 (granulocyte chemokine) from local mesenchymal cells and not only initiates an acute phase reaction but also allows naive T cells to differentiate into $T_H17$ cells [20]. IL-17 upregulates production of neutrophilic chemokines (CXCL1, CXCL2, CXCL5) and G-CSF, leading to an influx and maturation of granulocytes, as well as chemokine CCL20, which induces migration of $T_H17$ and dendritic cells to the site of inflammation or microbial invasion. Lastly, acute phase proteins, β-defensins, and S100A members are induced to directly kill microbes [37].

Although IL-17F plays a similar functional role by binding the same receptor, its signaling strength is 10–30 times weaker than IL-17A. Murine gene knockout studies have shown redundancy of IL-17A and IL-17F against mucocutaneous *Staphylococcus aureus*, where knockout of both genes (but not each individual gene) leads to increased disease, whereas it has synergy to colonic *Citrobacter rodentium*, where knockout of either gene leads to similar disease activity as to the knockout of both [54]. This contrasts with murine models of autoimmunity (EAE and autoimmune arthritis). In the same study, IL-17A-deficient mice show reduced autoimmune tissue inflammation compared to

IL-17F-deficient mice. In mice deficient in both IL-17A and IL-17F, there was no additional protection in EAE or arthritis compared to IL-17A-only deficient mice, leading the authors to postulate that while important in infection, the absence of IL-17F is neither protective nor harmful in their autoimmune models [54]. Ongoing studies with immunotherapeutics targeting both IL-17A and IL-17F (e.g., brodalumab through IL-17RA) in humans may provide more insight into these two cytokines' individual roles in autoimmunity and infection.

# 6. THE ROLE OF THE IL-17/23 AXIS IN IMMUNE-MEDIATED INFLAMMATORY DISEASES

## 6.1 Rheumatoid Arthritis

Rheumatoid arthritis (RA) is increasingly recognized to be an autoimmune-mediated disorder in which dysregulated immunity to citrullinated self proteins leads to a perpetual cycle of synovial inflammation in addition to attendant distant inflammatory comorbidity in vascular, bone, and neurologic tissues (Figure 1). Animal arthritis inflammatory models clearly implicate the IL-17 pathway in initiation and chronicity implicit in those models. Synovial tissue expression of IL-17A, IL-17B, and IL-17F is reported. In vitro studies suggest that IL-17A and with less potency IL-17F can activate a variety of relevant cell lineages. Thus, in particular, synovial fibroblasts are induced to release metalloproteinases, chemokines, cytokines, and prostanoids upon addition of IL-17 in vitro, particularly when synergy is sought with, e.g., TNF and IL-1. IL-17 is also implicated in chondrocyte activation toward catabolism and osteoclastogenesis, thus promoting erosion and articular damage over time. Further data also suggest particular proinflammatory properties in the activation and recruitment of neutrophils. Thus IL-17A and its receptor system are attractive targets in RA.

In passing we note that there are rather fewer data supporting the role of IL-23 in RA pathogenesis—it has a potent role demonstrated in several inflammatory arthritis models, but reliable expression in synovial tissues has been rather more difficult to define. For this reason we will focus on clinical trial sets that define a potential role for IL-17A.

### 6.1.1 *Secukinumab*

Secukinumab is a highly selective, fully human monoclonal anti-IL-17A antibody (IgG1κ) that showed preliminary efficacy in proof-of-concept studies. In a phase II double-blind, placebo-controlled study, 237 patients with active RA on stable doses of methotrexate (MTX) were randomized to receive subcutaneous injections of secukinumab (25, 75, 150, and 300 mg), or placebo every 4 weeks for 48 weeks. The primary endpoint was 20% improvement in American College of Rheumatology response criteria (ACR20) at week 16. If patients achieved an ACR20 response, they continued on the same dose, while those who did not were dose-escalated at week 20, so that the 25 and 75 mg groups received 150 mg, and the 150 mg group received 300 mg (the 300 mg remained on the same dose). The placebo group was also reassigned to 150 mg, and all groups were followed through week 52. At week 16, ACR20 responses for the three highest secukinumab dosing groups (75 mg, 47%; 150 mg, 47%; 300 mg, 54%; placebo, 36%) were increased compared to placebo but did not achieve statistical significance. Over 52 weeks, ACR20 non-responders at week 16

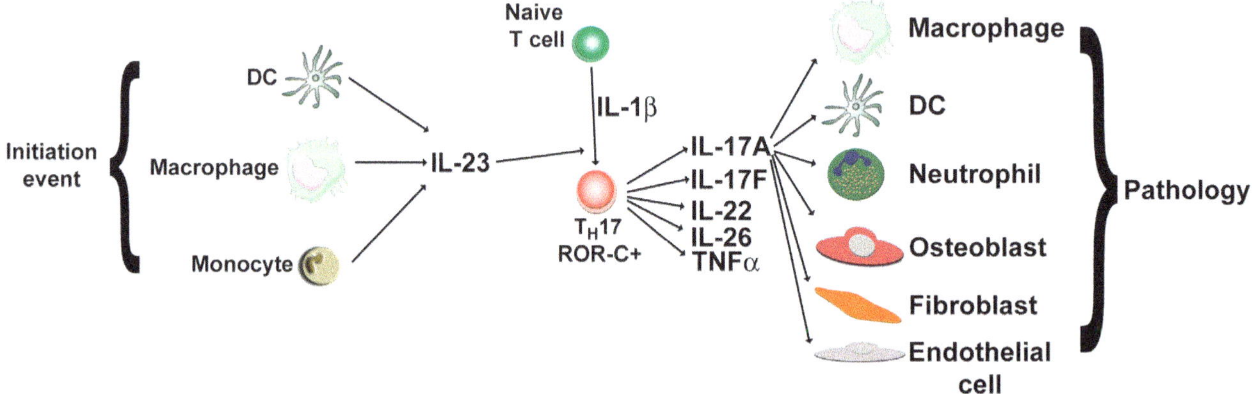


FIGURE 1 The IL-23/IL-17 axis in inflammatory disease. The initiation of inflammatory disease is multifactorial and unique to each individual disease. In various diseases these initiation pathways lead to activation of many compartments included myeloid cells (dendritic cells, macrophages and monocytes). The resulting activation leads to increased production of IL-23 from these cells, which in concert with IL-1β can subsequently lead to $T_H17$ (RORc$^+$) T cell differentiation and maintenance. These T cells can produce numerous cytokines and chemokines including but not limited to IL-17A, IL-17F, IL-22 & IL-26. Each of these cytokines can stimulate multiple signaling cascades in various cell types that leads to disease pathology. IL-17A for example can mediate it effects through the interaction with macrophages, DCs, neutrophils, osteoblasts, fibroblasts and endothelial cells. In the case of fibroblasts this can lead to up-regulation of MMP secretion. In all instances the resulting interaction contributes to disease pathology.

did not gain additional benefit from dose escalation. However significant improvement was seen in the initial week 16 responders on the 150 mg dose ($n=20$), who achieved ACR20 responses of 75% and 90% at weeks 24 and 52, and ACR70 response rates of 20% and 40%, respectively. In addition, the secukinumab 150 mg group achieved European League Against Rheumatism (EULAR) remission rates of 12% at week 16, 30% at week 24, and 40% at week 52. At 52 weeks, there were no safety or tolerability issues, and no dose-dependent adverse effects were noted [55]. Two phase III trials are already in progress, one of which is using abatacept as a comparator arm.

#### 6.1.2 Ixekizumab

Ixekizumab, a highly specific, humanized monoclonal anti-IL-17A antibody (IgG4), underwent a phase I proof-of-concept trial in which RA patients ($n=77$) were randomized into escalating intravenous dosage groups with an efficacy endpoint at 10 weeks [56]. At week 10, response rate was most improved in the highest (2.0 mg/kg) group, with an ACR20 response of 90% (vs. 56% placebo group; $p \leq 0.05$) and an ACR70 response of 24% (vs. 5.6% placebo group; $p \leq 0.05$).

A phase II randomized controlled trial was subsequently conducted in RA patients already on disease-modifying anti-rheumatic drugs (DMARDs) and divided into two groups: those naive to biological therapy ($n=260$) and those resistant to TNF-a inhibitor therapy ($n=188$) [57]. Patients received placebo or subcutaneous ixekizumab 3, 10, 30, 80 or 180 mg at weeks 0, 1, 2, 4, 6, 8, and 10. Ixekizumab produced a statistically significant dose-related response based on logistic regression of ACR20 at week 12 in patients who were biologic naïve ($p=0.03$), achieving their primary endpoint. Concomitant decreases in CRP, DAS28-CRP, and Crohn's disease activity index (CDAI) were also noted. In addition, ACR20 responses were significantly better in the ixekizumab 80 and 180 mg group in TNF-α inhibitor resistant patients compared to placebo. In some dosage groups, responses were seen as early as day 3. There were no major safety or adverse effects in either trial.

#### 6.1.3 Brodalumab

Brodalumab is a highly specific, human monoclonal antibody (IgG2) that binds IL-17RA, and effectively blocks biological activity of IL-17A, IL-17F, and IL-17E. A single phase I randomized controlled study was conducted in patients with moderate to severe RA ($n=40$) who received brodalumab (subcutaneous or intravenous) at ascending dosage groups [58]. At week 13, there was no significant response in any subgroup and a dose–response relationship was not established. This was followed with a phase II randomized control trial of 252 biological naive patients with active RA who received subcutaneous brodalumab at ascending dosage groups [59]. At 12 weeks, there was no significant improvement in ACR response rate or change in DAS28 in any subgroup. There were no major safety signals or adverse effects between groups.

## 7. CROHN'S DISEASE

Crohn's disease is a chronic intestinal inflammatory-mediated disorder where the dysregulation of the intestinal immune response, in the context of a specific commensal environment, results in inflammatory lesions [60]. A critical player in disease pathogenesis is thought to be the $T_H17$ CD4+ T cell, which is expanded in Crohn's disease-associated inflammatory lesions [61]. Interestingly, it has been found that patients have a circulating population of IL-23 responsive gut-homing $T_H17$ CD4+ T cells [62]. In support of the concept that the intestinal environment can drive the IL-17/IL-23 axis, evaluation of lesional intestinal tissue from Crohn's patients has also demonstrated that both IL-17 and IL-23 are raised in areas with active inflammation [63,64]. Furthermore, the expression of IL-23 has been associated with lesional macrophage, whilst the IL-23R has been predominately observed on infiltrating CD4+ and CD8+ T cells and NK cells [65]. In vitro stimulation of mononuclear cells from the lamina propria of patients results in the secretion of increased amounts of IL-17. In addition, studies have demonstrated that the $T_H17$-related effector cytokines have the capacity to induce a proinflammatory cascade (induction of TNF-α, IL-1β, and IL-6), neutrophil recruitment via IL-8, and the secretion of metalloproteinases by intestinal fibroblasts, which emulates disease pathogenesis. It should also be appreciated that in support of the human studies, animal models of intestinal inflammation clearly show that when the IL-17/IL-23 axis is perturbed, disease severity is ameliorated. Thus both IL-17 and IL-23, and their respective receptor systems, are attractive targets in Crohn's disease.

### 7.1 Clinical Trials

#### 7.1.1 Secukinumab

Secukinumab was evaluated in a phase IIa randomized controlled trial of 59 adult patients with moderate to severe Crohn's disease who were on stable doses of azathioprine, MTX, steroids (≤40 mg/dl), or 6-mercaptopurine (excluded if recently on TNFα inhibitor therapy) [66]. Patients were assigned to receive intravenous secukinumab 10 mg/kg versus placebo on days 1 and 22, and frequent safety and efficacy assessments were done up to 18 months. The primary end point was change in mean CDAI at week 6. The difference in mean CDAI change from baseline between secukinumab and placebo at 6 weeks was insignificant as reduction in the placebo group was greater than in the secukinumab group, and the study was prematurely discontinued. Fourteen serious adverse events occurred in 10 patients (seven secukinumab, three placebo); of these, "worsening of disease" occurred in four patients receiving secukinumab versus one on placebo. In addition, 20 infections (four fungal) were reported in the secukinumab group versus none in the placebo group. This trial demonstrates the complexity of targeting the IL-17 ($T_H17$) pathway in IBD, and may even suggest a worsening of disease with secukinumab. Given the higher rates of mucocutaneous candidiasis in this trial (unusual with other biologics), disease exacerbation may be linked to proliferation of *C. albicans*, but this is yet to be proven [67].

### 7.1.2 Brodalumab

Brodalumab was evaluated in a phase II randomized controlled trial in moderate to severe Crohn's disease divided into three ascending dosing groups versus placebo [68]. The primary endpoint was the proportion of subjects achieving CDAI remission at week 6. After 130 of the planned 216 patients were randomized, an independent data review team unblinded the safety data and found a worsening of Crohn's disease in the active treatment groups. Although the study was stopped early, no safety risks beyond worsening of symptoms was noted among the groups.

### 7.1.3 Ustekinumab

Ustekinumab, an anti-IL-12/23 monoclonal antibody, was initially investigated for efficacy in a phase IIa double-blinded, randomized controlled crossover trial of 104 patients with moderate to severe Crohn's disease [69]. Clinical response rates were significantly higher among patients receiving ustekinumab (sc or iv) than placebo (53% vs. 30%; $p=0.02$) at weeks 4 and 6, but then diminished by week 8 (49% vs. 40%, $p=0.34$). Within this study, a subgroup of 49 patients who were previously given infliximab (a monoclonal anti-TNFα antibody) (neither primary or secondary nonresponders) had a clinical response rate to ustekinumab significantly greater than placebo through week 8 ($p<0.05$). This early efficacy led to further investigation of ustekinumab in patients with Crohn's disease, particularly those refractory to anti-TNF therapy.

A phase IIb double-blinded, randomized controlled trial was conducted comparing ustekinumab to placebo in patients with moderate to severe Crohn's disease who had previously failed anti-TNF therapy [70]. This failure was defined by primary or secondary nonresponse, or unacceptable side effects after receiving a TNF antagonist ≥8 weeks before enrollment. Patients were able to continue immunomodulators, oral steroids, or antibiotics during the trial. During the induction phase, 526 patients were randomized to receive 1, 3, or 6 mg/kg intravenous ustekinumab or placebo, and the primary endpoint was clinical response (≥100 point decrease from the baseline CDAI score) at week 6. The proportions of patients who met the primary endpoint were 37%, 34%, and 40% for 1, 3, and 6 mg of ustekinumab, respectively, and 24% for placebo (all $p \leq 0.05$), although there was no statistically significant clinical remission between groups at week 6. In the maintenance phase, 145 patients who had a response to ustekinumab at week 6 were randomly assigned to receive subcutaneous ustekinumab or placebo at weeks 8 (270 mg) and 16 (90 mg). At week 22, those receiving ustekinumab had a significantly increased response rate (70% vs. 43%, $p<0.001$) compared to placebo, as well as increased clinical remission rates (42% vs. 27 %, $p=0.03$).

No serious opportunistic infections, major adverse cardiovascular events, or deaths were reported up to the 36 week safety analysis; however, there was a small increase in serious non-opportunistic infections in the 6 mg/kg ustekinumab group. Phase III trials (UNITI program) are ongoing, which will provide important additional objective efficacy and safety data, and determine whether it is a viable first-line therapy in patients with Crohn's disease.

## 8. PSORIASIS

Although initially though to be driven by hyperproliferation of keratinocytes, it is now appreciated that psoriasis is a chronic immune-mediated inflammatory disorder with the IL-17/IL-23 axis as one of the main components in disease immunopathology. This is supported by genetic, experimental, and clinical data, which clearly demonstrate that the IL-17/IL-23 axis is a key feature of disease initiation and chronicity. Murine models of psoriasis-like disease support the IL-23-driven IL-17A-induced changes in skin [71]. In conjugation, the interrogation of lesional and non-lesional skin has shown that IL-23 and IL-17A are highly expressed in diseased tissue, with activated dendritic cells being one of the main sources of IL-23. The current dogma dictates that early activation of dendritic cells that subsequently express IL-23 drive a $T_H17$ response and the resulting IL-17A, further activating keratinocytes, thereby establishing an inflammatory loop [72]. Thus, IL-17, IL-23, and their respective receptor systems are attractive targets in psoriasis (Table 1).

It should also be noted that in addition to the overt skin disease, psoriasis is associated with an increased risk of comorbidities, including psoriatic arthritis (Section 9), inflammatory bowel disease (Section 7), cardiovascular disease, and metabolic syndrome.

## 8.1 Clinical Trials

### 8.1.1 Secukinumab

In a phase IIa double-blind, proof-of-concept trial, secukinumab (given as a single 3 mg/kg intravenous dose)

**TABLE 1** Main Sources and Effects of IL-23 and IL-17

| Cytokine | Cellular Sources | Receptor | Responder Cells | Activated Signaling Cascades | Functional Outcome |
|---|---|---|---|---|---|
| IL-23 | Dendritic cells<br>Macrophages<br>Monocytes | IL-23R | $T_H17$<br>Entheseal T cells<br>$CD8^+$ T cells<br>NK cells | JAK2<br>STAT3 | IL-6<br>TNFα<br>IL-17 |
| IL-17A | $T_H17$<br>$CD8^+$ T cells<br>Mast cells<br>Monocytes<br>Neutrophils<br>Eosinophils<br>γδ T cells<br>NK cells<br>Innate lymphoid cells | IL-17RA | Dendritic cells<br>Macrophages<br>Neutrophils<br>Osteoblasts<br>Fibroblasts<br>Chondrocytes<br>Endothelial cells<br>Epithelial cells | NF-κB<br>MAPKinase | IL-1β<br>IL-6<br>TNFα<br>PGE2<br>NO<br>Metalloproteinases<br>Chemokines (i.e., IL-8, CXCL1, CXCL2, CXCL5) |

**TABLE 2** Summary of Clinical Trials Targeting IL-23 or IL-17 in Inflammatory Disease

| | | Stage of Clinical Trails | | | | |
|---|---|---|---|---|---|---|
| Drug | Target | Rheumatoid Arthritis | Crohn's Disease | Psoriasis | Psoriatic Arthritis | Ankylosing Spondylitis |
| Secukinumab | IL-17A | Phase III | Phase II (stopped) | Phase III | Phase III | Phase III |
| Ixekizumab | IL-17A | Phase II | - | Phase III | Phase III | - |
| Brodalumab | IL-17RA | Phase II (stopped) | Phase II (stopped) | Phase III | Phase III | - |
| Ustekinumab | IL-12/23 | - | Phase III | Phase III | Phase III | Proof-of-concept |

demonstrated categorical reductions in investigator global assessment score (83% vs. 11%; $p=0.0004$) and significant reductions in mean psoriasis area and severity index (PASI) score compared to placebo at week 4 (58% vs. 4%; $p=0.0001$), both of which were maintained through week 12 [73]. Immunohistological analysis of psoriatic skin samples (week 4) showed significant reduction in dermal IL-17A T cells, and molecular profiling demonstrated selective modulation of a variety of cytokines, including IL-12B, IL-17A, IL-17F, IL-21, IL-22, IL-26, CCL20, and DEFB4 (encodes β-defensin), as well IL-6 and IFN-γ to a lesser degree [37,73](Table 2).

This was followed by a phase IIb dose-ranging (SC) double-blind, randomized controlled trial in which 125 patients with moderate to severe plaque psoriasis were randomized to ascending subcutaneous dosage groups or placebo [74]. Secukinumab 75 and 150 mg (each administered at weeks 0, 4, and 8) were associated with significantly greater rates of PASI 75 response at week 12 compared with placebo (57% and 81% vs. 9%; $p=0.002$ and $p<0.001$), meeting the primary endpoint. The 150 mg dose also significantly improved the PASI 90 rate compared to placebo (52% vs. 5%; $p=0.005$). Patients with prior exposure to anti-TNFα therapies had similar responses, and PASI scores remained improved (albeit diminished) compared to placebo 12 weeks after treatment had ceased.

Given the good results in the highest dose group, a phase IIb regimen-finding study was conducted to assess three different regimens of secukinumab 150 mg in patients with moderate to severe plaque psoriasis [75]. Patients ($n=404$) were randomized to receive secukinumab 150 mg SC weekly (weeks 0, 1, 2, and 4), monthly (weeks 0, 4, and 8) or placebo. The weekly and monthly induction regimens with secukinumab resulted in significantly improved PASI 75 (55% and 42% vs. 2%; both $p<0.001$) and PASI 90 (32% and 17% vs. 2%; both $p<0.001$) response rates than placebo at week 12. Primary responders were re-randomized into maintenance treatment with 150 mg at weeks 12 and 24 versus treating at time of relapse. The fixed-interval maintenance strategy showed significantly higher PASI responses from week 20 to week 28. A third phase IIb regimen-finding study using intravenous secukinumab showed similar results, albeit with six serious adverse events, none of which were deemed specific to treatment [76]. In all phase II trials there appeared to be comparable, low rates of serious adverse events between secukinumab and placebo cohorts, and no signals for increased infection risk. Several phase III trials have commenced.

#### 8.1.2 Ixekizumab

The safety and efficacy of ixekizumab were evaluated in a double-blind, randomized controlled phase IIb trial in moderate to severe psoriasis [77]. Patients ($n = 142$) were randomized to receive subcutaneous injections of 10, 25, 75, or 150 mg of ixekizumab or placebo at 0, 2, 4, 8, 12, and 16 weeks. The primary end point was the proportion of patients with a PASI 75 score at week 12. The three highest doses of ixekizumab were associated with significantly greater rates of PASI 75 response at 12 weeks compared to placebo (150 mg, 82%; 75 mg, 83%; 25 mg, 77%; placebo, 8%; $p < 0.001$ for each comparator). Increased PASI 90 scores were also seen (71%, 59%, and 50% vs. 0%, respectively; $p < 0.001$). Responses were maintained through week 20. Phase III studies are ongoing, with etanercept comparator arms (UNCOVER program). Overall, anti-IL-17A therapies show excellent efficacy in chronic plaque psoriasis, with good tolerability and few documented adverse events.

#### 8.1.3 Brodalumab

Brodalumab was evaluated for safety and efficacy in a double-blind, randomized controlled trial of 198 patients with moderate-to-severe plaque psoriasis [78]. Patients were assigned to receive brodalumab (70, 140, or 210 mg, at day 1 and weeks 1, 2, 4, 6, 8, and 10, or 280 mg monthly) or placebo. The primary endpoint was the percentage improvement from baseline in the PASI score at week 12. At week 12, brodalumab significantly reduced mean PASI scores by 45%, 86%, 86%, and 76%, respectively, compared to 16% in the placebo group ($p < 0.001$ for all comparators), with PASI 75 responses achieved in 77% and 82% of patients (in the 140-mg and 210-mg dose groups), and PASI 90 responses in 72% and 75% of patients, respectively. Quality of life scores were consistently improved. In the brodalumab 210 mg arm two cases of grade 3 neutropenia were reported. PASI 75 responses with brodalumab (82%) compared to secukinumab (81%) and ixekizumab (83%), as previously described.

#### 8.1.4 Ustekinumab

IL-23, involved in $T_H17$ function, seems a potentially interesting pharmacological target in the treatment of psoriasis, and several trials have been conducted since 2005. Ustekinumab was initially shown to be efficacious in a phase II trial that demonstrated PASI 75 responses of up to 81% (90 mg doses × 4 weeks) compared to 2% in the placebo group ($p < 0.001$) [79]. This was followed by two phase III trials. In PHOENIX I, 766 patients were randomly assigned to receive ustekinumab 45 mg or 90 mg at weeks 0 and 4 and then every 12 weeks or placebo, with subsequent crossover to ustekinumab at week 12 [80]. The primary endpoint was the proportion of patients achieving PASI 75 response at week 12, which was achieved in 67% (45 mg), 66% (90 mg), and 3% (placebo) ($p < 0.0001$ for all comparators). Those who were re-randomized to maintenance ustekinumab at week 40 showed better PASI 75 scores at 1 year compared to those taken off therapy. PHOENIX II demonstrated that more frequent administration (8 weeks versus 12 weeks) at higher doses (90 mg) can be beneficial for patients experiencing a partial response [81]. Another trial of 903 patients randomized to receive ustekinumab 45 or 90 mg (0 and 4 weeks) or high-dose etanercept (50 mg twice weekly for 12 weeks) showed superiority of ustekinumab in achieving PASI 75 rates (68–74% versus 57%; $p \leq 0.01$), and safety patterns were similar [82]. After 5 years of exposure to ustekinumab, no dose-related or cumulative toxicity was observed, and malignancy rates were similar to the general population. Adverse events were generally comparable to other biologics [83].

## 9. PSORIATIC ARTHRITIS

Psoriatic arthritis (PsA) is an immune-mediated chronic inflammatory disease with complex, and incompletely understood, immunopathogenesis. The clinical manifestations of both skin and joint disease make it difficult to experimentally model both aspects simultaneously, and thus animal models to define the contribution of IL-17 and IL-23 disease pathology tend to overlap with those studies outlined in the RA and psoriasis (PsO) sections. Such animal models clearly support the IL-17/IL-23 axis in both aspects of disease pathology. More important, the evaluation of clinical samples has clearly demonstrated that elevated IL-17 and IL-23 are present in the psoriatic skin and synovial fluid of PsA patients [84]. However, although more research needs to be done to tease apart the cellular and molecular mechanisms that are specific for this disease, and how they overlap between the skin and joint pathology, sufficient information is available to justify the targeting of the IL-17/IL-23 axis in PsA.

### 9.1 Trials

#### 9.1.1 Secukinumab

The safety and efficacy of secukinumab in PsA were evaluated in a double-blind, randomized controlled phase

IIa proof-of-concept trial in which patients ($n=42$) with moderate to severe PsA were randomized (2:1) to treatment with two doses of secukinumab 10 mg/kg SC, given 3 weeks apart, or placebo [85]. The primary endpoint was ACR20 response rate at week 6. There was no statistically significant difference in ACR20 response rates at 6 weeks between secukinumab and placebo (39% vs. 23%; $p=0.27$). However, there were significant reductions in acute phase reactants (CRP and ESR), and improvement in quality of life measures from baseline to week 6 in the secukinumab group. Given the trend in positive outcomes, a phase III trial is planned.

### 9.1.2 Ixekizumab

No directed trials for Ixekizumab in PsA have been conducted. However, results extrapolated from patients reporting PsA in the phase II psoriasis trial showed statistically significant improvement on the joint pain visual analog scale (VAS) at 12 weeks compared with baseline only in the 150 mg ixekizumab group (8 patients out of 28 in that group) [77]. This prompted a phase III safety and efficacy trial, which is ongoing.

### 9.1.3 Brodalumab

A double-blind, randomized controlled phase II trial was conducted in 168 patients with PsA, with a primary endpoint of ACR20 response rate at 12 weeks [86]. Patients receiving brodalumab 140 mg, 280 mg, or placebo had an ACR20 response rate of 37%, 39%, and 18%, respectively ($p<0.05$ for all comparators), with significant improvement in CRP, DAS28, and ACR50 response rates as well. A phase III trial has commenced.

### 9.1.4 Ustekinumab

The safety and efficacy of ustekinumab were initially evaluated in a phase II double-blind, randomized controlled crossover trial that enrolled 146 patients with active PsA (defined as three or more swollen and tender joints and either CRP ≥15 mg/L or morning stiffness ≤45 min) who had active plaque psoriasis and had been poor responders to NSAIDS, DMARDs, or anti-TNF therapy [87]. The primary endpoint, ACR20 at week 12, was achieved in 42% in the ustekinumab group versus 14% in the placebo group ($p=0.0002$). Secondary outcomes showed superior ACR 50 (25% vs. 7%; $p=0.003$), ACR 70 (11% vs. 0%; $p=0.005$), and improvement in Health Assessment Questionnaire (HAQ) scores.

Subsequently, the P-SUMMIT I trial was a phase III double-blind, randomized controlled trial of 615 patients with biological-naive, active psoriatic arthritis (≥5 tender and ≥5 swollen joints, C-reactive protein ≥3.0 mg/L, and past or present plaque psoriasis) who were randomized to receive ustekinumab 45 mg, 90 mg, or placebo at weeks 0, 4, and every 12 weeks thereafter [88]. At week 16, poor responders (<5% improvement) entered a masked early-escape option and were given 45 mg ustekinumab (if in the placebo group) or 90 mg ustekinumab (if in the 45 mg group). The primary endpoint was proportion with ACR20 response at week 24. At that time, all remaining patients in the placebo group received ustekinumab 45 mg, which they continued thereafter. At week 24, there was a significantly higher proportion in the ustekinumab groups that achieved the primary endpoint (42% in the 45 mg group, 50% in the 90 mg group, 23% in the placebo group; all comparators $p<0.0001$). In addition, improvements were seen in DAS28-CRP, HAQ, BASDAI, and dactylitis, enthesitis, and PASI scores. Improvements persisted through to 52 weeks. The P-SUMMIT II trial was comparable, in which patients ($n=312$) with active PsA had an ACR20 response rate of 44% compared to 20.2% in the placebo group [89]. Among 180 patients previously treated with ≥1 TNF inhibitor (most had received two or more agents with lack of effect), efficacy with ustekinumab was observed (week 24: 36% vs. 15%; $p<0.01$). A pooling of radiographic data from both trials showed statistically significant inhibition of radiographic progression of joint damage at weeks 24 and 52 [90].

## 10. ANKYLOSING SPONDYLITIS

Our understanding of the spinal and peripheral joint pathogenesis of ankylosing spondylitis (AS) is limited. It has been known for several decades that HLA-B27 is associated with disease, but it is only within the last decade that genome-wide association studies have identified new disease-associated loci. One of the most interesting comes in the form of the IL-23 receptor, which importantly has been associated with other disorders, including Crohn's disease and psoriasis. This is not the only line of investigation that has implicated the IL-23/IL-17 axis in ankylosing spondylitis, as it is known that the unfolded protein response (HLA-B27 misfolding) enhances IL-23 production. Furthermore, circulating CD4$^{+}$ T$_{H}$17 T cells are increased in AS and a significant infiltration of IL-17 producing innate immune cells is observed in inflamed tissue. These clinical findings are supported by numerous animal studies, which have, for example, illustrated that the overexpression of IL-23 can drive entheseal inflammation and bone remodeling, reminiscent of spondyloarthritis-like disease. The current consensus on the pathology of AS is that of a complicated relationship between the expression of HLA-B27 (unfolded protein response), biomechanical stress, and the gut microbiome. This drives increased IL-23 expression, which could enhance both CD4+ and CD4−/CD8$^{-}$ T cells to produce IL-17, thus driving inflammation and bone loss. In essence the existing data support the need for IL-17/IL-23 targeting clinical trials in AS.

## 10.1 Clinical Trials

### 10.1.1 Secukinumab

The safety and efficacy of secukinumab in ankylosing spondylitis were evaluated in a double-blind, randomized controlled trial in which 30 patients with AS (defined by 1984 modified New York criteria) received secukinumab (10 mg/kg IV at weeks 0 and 3) or placebo [91]. The primary endpoint was the rate of Assessment of SpondyloArthritis International Society-20 (ASAS-20) response at week 6. At that time, 59% on secukinumab versus 24% on placebo achieved an ASAS-20 response (99.8% probability that secukinumab is superior to placebo). An MRI sub-study was done in 27 patients with AS who had a spinal MRI at baseline [92]. After two doses of secukinumab (10 mg/kg IV at weeks 0 and 3), spinal inflammation as measured through MRI scores (modified Berlin score) improved as early as week 6, and was maintained to week 28.

### 10.1.2 Ustekinumab

Ustekinumab was evaluated in a recent open-label, single-arm, proof-of-concept clinical trial (TOPAS) in patients with active AS, defined by the modified New York criteria and a Bath Ankylosing Spondylitis Disease Activity Index (BASDAI) score ≥4 [93]. Patients ($n=20$) received ustekinumab 90 mg SC at baseline, week 4, and week 16 and were followed through to week 28. Primary end point was a 40% improvement in disease activity (ASAS-40) at week 24, which was reached by 65% of the patients. ASAS-20 response was reached in 75%, ASDAS (AS Disease Activity Score) inactive disease in 35%, and a ≥50% improvement of the BASDAI (BASDAI-50) was seen in 55% of the patients. There was a significant improvement in quality of life indicators, MRI scores, and decline in acute phase reactants. There was no issue in safety or tolerability. These excellent results make it a promising future therapeutic option.

## 10.2 Other Conditions

In addition to the foregoing, there are many studies either ongoing or being considered in many other conditions, including, for example, asthma, multiple sclerosis, uveitis, sarcoid, and Sjogren's. The potential therefore for these pathways offering benefit beyond the detailed discussions above is well recognized; however, these remain to be determined.

## 10.3 Adverse Effects/Safety/Potential Risks

The IL-17 family is implicated in host defense, particularly against mucosal site infections and fungal pathogens. Individuals with genetic deficiencies in IL-17A and IL-17F are prone to chronic mucocutaneous candidiasis and possibly *Staphylococcus aureus* [94]. However, in phase II trials, anti-IL-17 therapies demonstrated infection rates (particularly candidal) only slightly higher than those seen with placebo. Larger trials with longer duration are necessary. As previously mentioned, worsening of bowel inflammation in those treated with secukinumab and brodalumab may represent superimposed *Candida* infection; however, this has yet to be proven [67]. In addition, IL-17 cytokines affect neutrophil biology. Several cases of neutropenia were noted with all IL-17 inhibitors (albeit asymptomatic and unrelated to infection) and may be due to defective neutrophil trafficking in blood; however, this needs to be further characterized.

As they are on the same axis, there is also a theoretical risk of infection with anti-IL-12/IL-23 therapy, particularly in those who are genetically deficient in IL-12/IL-23p40 and IL-12Rβ1. This group has an inherent susceptibility to weakly virulent mycobacterial and *Salmonella* infections [95,96]. The safety of ustekinumab has been evaluated in several phase II and III trials in psoriasis, PsA, and Crohn's disease. While the serious infection risk is similar to other biologics, there does not appear to be an increased risk of mycobacterial diseases, disseminated salmonellosis, or systemic fungal infections. In addition, no particular infections related to IL-12/IL-23 deficiency were identified [83], and in pooled data from five psoriasis trials, no cases of latent tuberculosis infection (LTBI) reactivation were observed in patients receiving concomitant INH prophylaxis for LTBI [97]. It is possible that IL-12/IL-23 inhibition is incomplete and does not hinder host defense toward these opportunistic infections [83]. A 4 year safety analysis of 3117 patients with psoriasis who received at least one dose of ustekinumab showed no increased incidence of serious infections, malignancies (other than non-melanoma skin cancer), and major adverse cardiovascular events [98]. Their incidences were consistent with expected levels based on population-matched rates. An extension of this to 5 years showed consistency [83]. Safety profiles in PsA and Crohn's disease are similar to patients with psoriasis. However, follow-up for long-term complications such as cardiac events and malignancy is necessary before conclusions can be drawn.

# 11. SUMMARY

Table 3 summarizes all trials and their phase for the various medications and antibody types and targets.

**TABLE 3** Antibody Types and Targets by Drug in Ongoing Trials

| | Antibody Type | Target | Route | Disease | Phase | Ongoing Trials |
|---|---|---|---|---|---|---|
| Secukinumab | Human IgG1-κ | IL-17A | SC | RA | III | NCT01377012 |
| | | | IV | Psoriasis | III | NCT01406938 |
| | | | | Psoriatic Arthritis | III | NCT01392326 |
| | | | | AS | III | NCT02008916 |
| Ixekizumab | Humanized IgG4 | IL-17A | SC | RA | II | NCT00966875 |
| | | | | Psoriasis | III | NCT01646177 |
| | | | | AS | III | NCT01695239 |
| Brodalumab | Human IgG2 | IL-17RA | SC | Psoriasis | III | NCT01708590 |
| | | (receptor of IL-17A, C, E, F) | | Psoriatic Arthritis | III | NCT02024646 |
| Ustekinumab | Human IgG1-κ | p40 subunit of IL-12 & 23 | SC | Psoriasis | III | NCT00267969 |
| | | | IV | Psoriatic Arthritis | III | NCT01009086 |
| | | | | IBD | III | NCT01369342 |
| | | | | AS | II | NCT01330901 |

## REFERENCES


[1] Rouvier E, Luciani MF, Mattei MG, Denizot F, Golstein P. J Immunol 1993;150:5445–56.

[2] Yao Z, Painter SL, Fanslow WC, Ulrich D, Macduff BM, Spriggs MK, et al. J Immunol 1995;155:5483–6.

[3] Fossiez F, Djossou O, Chomarat P, Flores-Romo L, Ait-Yahia S, Maat C, et al. J Exp Med 1996;183:2593–603.

[4] Chabaud M, Durand JM, Buchs N, Fossiez F, Page G, Frappart L, et al. Arthritis Rheum 1999;42:963–70.

[5] Wright JF, Guo Y, Quazi A, Luxenberg DP, Bennett F, Ross JF, et al. J Biol Chem 2007;282:13447–55.

[6] Gaffen SL. Nat Rev Immunol 2009;9:556–67.

[7] Chang SH, Dong C. Cell Res 2007;17:435–40.

[8] Yang XO, Chang SH, Park H, Nurieva R, Shah B, Acero L, et al. J Exp Med 2008;205:1063–75.

[9] Pfeifer P, Voss M, Wonnenberg B, Hellberg J, Seiler F, Lepper PM, et al. Am J Respir Cell Mol Biol 2013;48:415–21.

[10] Chang SH, Reynolds JM, Pappu BP, Chen G, Martinez GJ, Dong C. Immunity 2011;35:611–21.

[11] Johnston A, Fritz Y, Dawes SM, Diaconu D, Al-Attar PM, Guzman AM, et al. J Immunol 2013;190:2252–62.

[12] Reynolds JM, Martinez GJ, Nallaparaju KC, Chang SH, Wang YH, Dong C. J Immunol 2012;189:4226–30.

[13] Yamaguchi Y, Fujio K, Shoda H, Okamoto A, Tsuno NH, Takahashi K, et al. J Immunol 2007;179:7128–36.

[14] Kouri VP, Olkkonen J, Ainola M, Li TF, Bjorkman L, Konttinen YT, et al. Rheumatol Oxf 2014;53:39–47.

[15] Starnes T, Broxmeyer HE, Robertson MJ, Hromas R. J Immunol 2002;169:642–6.

[16] Dong C. Immunol Rev 2008;226:80–6.

[17] Toy D, Kugler D, Wolfson M, Vanden Bos T, Gurgel J, Derry J, et al. J Immunol 2006;177:36–9.

[18] Ely LK, Fischer S, Garcia KC. Nat Immunol 2009;10:1245–51.

[19] Song X, Zhu S, Shi P, Liu Y, Shi Y, Levin SD, et al. Nat Immunol 2011;12:1151–8.

[20] Miossec P, Kolls JK. Nat Rev Drug Discov 2012;11:763–76.

[21] Shibata K, Yamada H, Hara H, Kishihara K, Yoshikai Y. J Immunol 2007;178:4466–72.

[22] Qian Y, Liu C, Hartupee J, Altuntas CZ, Gulen MF, Jane-Wit D, et al. Nat Immunol 2007;8:247–56.

[23] Chang SH, Park H, Dong C. J Biol Chem 2006;281:35603–7.

[24] Ruddy MJ, Wong GC, Liu XK, Yamamoto H, Kasayama S, Kirkwood KL, et al. J Biol Chem 2004;279:2559–67.

[25] Hartupee J, Liu C, Novotny M, Li X, Hamilton T. J Immunol 2007;179:4135–41.

[26] Henness S, Johnson CK, Ge Q, Armour CL, Hughes JM, Ammit AJ. J Allergy Clin Immunol 2004;114:958–64.

[27] Mosmann TR, Coffman RL. Annu Rev Immunol 1989;7:145–73.

[28] Ferber IA, Brocke S, Taylor-Edwards C, Ridgway W, Dinisco C, Steinman L, et al. J Immunol 1996;156:5–7.

[29] Cua DJ, Sherlock J, Chen Y, Murphy CA, Joyce B, Seymour B, et al. Nature 2003;421:744–8.

[30] Langrish CL, Chen Y, Blumenschein WM, Mattson J, Basham B, Sedgwick JD, et al. J Exp Med 2005;201:233–40.

[31] Becher B, Durell BG, Noelle RJ. J Clin Invest 2002;110:493–7.

[32] Sato K, Suematsu A, Okamoto K, Yamaguchi A, Morishita Y, Kadono Y, et al. J Exp Med 2006;203:2673–82.

[33] Awasthi A, Kuchroo VK. Int Immunol 2009;21:489–98.

[34] Bettelli E, Carrier Y, Gao W, Korn T, Strom TB, Oukka M, et al. Nature 2006;441:235–8.

[35] Veldhoen M, Hocking RJ, Atkins CJ, Locksley RM, Stockinger B. Immunity 2006;24:179–89.
[36] Mangan PR, Harrington LE, O'Quinn DB, Helms WS, Bullard DC, Elson CO, et al. Nature 2006;441:231–4.
[37] van den Berg WB, McInnes IB. Semin Arthritis Rheum 2013;43: 158–70.
[38] Dong C. Exp Mol Med 2011;43:1–6.
[39] Sutton CE, Lalor SJ, Sweeney CM, Brereton CF, Lavelle EC, Mills KH. Immunity 2009;31:331–41.
[40] Chung Y, Chang SH, Martinez GJ, Yang XO, Nurieva R, Kang HS, et al. Immunity 2009;30:576–87.
[41] Korn T, Bettelli E, Oukka M, Kuchroo VK. Annu Rev Immunol 2009;27:485–517.
[42] Acosta-Rodriguez EV, Rivino L, Geginat J, Jarrossay D, Gattorno M, Lanzavecchia A, et al. Nat Immunol 2007;8:639–46.
[43] Chen Z, Tato CM, Muul L, Laurence A, O'Shea JJ. Arthritis Rheum 2007;56:2936–46.
[44] Annunziato F, Romagnani S. Arthritis Res Ther 2009;11:257.
[45] Bettelli E, Korn T, Oukka M, Kuchroo VK. Nature 2008;453: 1051–7.
[46] Hori S, Nomura T, Sakaguchi S. Science 2003;299:1057–61.
[47] Fontenot JD, Gavin MA, Rudensky AY. Nat Immunol 2003;4:330–6.
[48] Ichiyama K, Yoshida H, Wakabayashi Y, Chinen T, Saeki K, Nakaya M, et al. J Biol Chem 2008;283:17003–8.
[49] Pasare C, Medzhitov R. Science 2003;299:1033–6.
[50] Ivanov II, McKenzie BS, Zhou L, Tadokoro CE, Lepelley A, Lafaille JJ, et al. Cell 2006;126:1121–33.
[51] Weaver CT, Hatton RD, Mangan PR, Harrington LE. Annu Rev Immunol 2007;25:821–52.
[52] Mrabet-Dahbi S, Metz M, Dudeck A, Zuberbier T, Maurer M. Exp Dermatol 2009;18:437–44.
[53] Iwakura Y, Nakae S, Saijo S, Ishigame H. Immunol Rev 2008;226:57–79.
[54] Ishigame H, Kakuta S, Nagai T, Kadoki M, Nambu A, Komiyama Y, et al. Immunity 2009;30:108–19.
[55] Genovese MC, Durez P, Richards HB, Supronik J, Dokoupilova E, Mazurov V, et al. Ann Rheum Dis 2013;72:863–9.
[56] Genovese MC, Van den Bosch F, Roberson SA, Bojin S, Biagini IM, Ryan P, et al. Arthritis Rheum 2010;62:929–39.
[57] Genovese MC, Greenwald M, Cho CS, Berman A, Jin L, Cameron GS, et al. Arthritis Rheum 2014. doi: 10.1002/art.38617 (accepted online March 2014).
[58] Martin DA, Churchill M, Flores-Suarez L, Cardiel MH, Wallace D, Martin R, et al. Arthritis Res Ther 2013;15:R164.
[59] Pavelka Karel, Chon Yun, Newmark Richard, Erondu Ngozi, Lin Shao Lee. A randomized, double-blind, placebo-controlled, multiple-dose study to evaluate the safety, tolerability, and efficacy of brodalumab (AMG 827) in subjects with rheumatoid arthritis and an inadequate response to methotrexate. Arthritis Rheum 2012;64(Suppl. 10):831. 2014.
[60] Xavier RJ, Podolsky DK. Nature 2007;448:427–34.
[61] Maynard CL, Weaver CT. Immunity 2009;31:389–400.
[62] Kleinschek MA, Boniface K, Sadekova S, Grein J, Murphy EE, Turner SP, et al. J Exp Med 2009;206:525–34.
[63] Fujino S, Andoh A, Bamba S, Ogawa A, Hata K, Araki Y, et al. Gut 2003;52:65–70.
[64] Schmidt C, Giese T, Ludwig B, Mueller-Molaian I, Marth T, Zeuzem S, et al. Inflamm Bowel Dis 2005;11:16–23.
[65] Liu Z, Yadav PK, Xu X, Su J, Chen C, Tang M, et al. J Leukoc Biol 2011;89:597–606.
[66] Hueber W, Sands BE, Lewitzky S, Vandemeulebroecke M, Reinisch W, Higgins PD, et al. Gut 2012;61:1693–700.
[67] Colombel JF, Sendid B, Jouault T, Poulain D. Gut 2013;62: 800–1.
[68] Targan SR, Feagan BG, Vermeire S, Panaccione R, Melmed GY, Blosch Consuelo, et al. Mo2083 A Randomized, double-blind, placebo-controlled study to Evaluate the safety, tolerability, and efficacy of AMG 827 in subjects with moderate to severe Crohn's disease. Gastroenterology 2012;143(3):e26. 2014.
[69] Sandborn WJ, Feagan BG, Fedorak RN, Scherl E, Fleisher MR, Katz S, et al. Gastroenterology 2008;135:1130–41.
[70] Sandborn WJ, Gasink C, Gao LL, Blank MA, Johanns J, Guzzo C, et al. N Engl J Med 2012;367:1519–28.
[71] Rizzo HL, Kagami S, Phillips KG, Kurtz SE, Jacques SL, Blauvelt A. J Immunol 2011;186:1495–502.
[72] Lynde CW, Poulin Y, Vender R, Bourcier M, Khalil S. J Am Acad Dermatol 2014, in press. http://www.ncbi.nlm.nih.gov/pubmed/24655820 (accepted Online March 2014).
[73] Hueber W, Patel DD, Dryja T, Wright AM, Koroleva I, Bruin G, et al. Sci Transl Med 2010;2:52ra72.
[74] Papp KA, Langley RG, Sigurgeirsson B, Abe M, Baker DR, Konno P, et al. Br J Dermatol 2013;168:412–21.
[75] Rich P, Sigurgeirsson B, Thaci D, Ortonne JP, Paul C, Schopf RE, et al. Br J Dermatol 2013;168:402–11.
[76] Papp KA, Matheson RT, Tu JH. Secukinumab, a novel fully human antibody to interleukin-17A, in the treatment of moderate-to-severe plaque psoriasis: efficacy and safety interim results from a phase II intravenous induction dose-ranging study; Presented at: 20th congress of the European Academy of Dermatology and Venereology; 20–24 October 2011, Lisbon, Portugal. Abstract 0630.
[77] Leonardi C, Matheson R, Zachariae C, Cameron G, Li L, Edson-Heredia E, et al. N Engl J Med 2012;366:1190–9.
[78] Papp KA, Leonardi C, Menter A, Ortonne JP, Krueger JG, Kricorian G, et al. N Engl J Med 2012;366:1181–9.
[79] Krueger GG, Langley RG, Leonardi C, Yeilding N, Guzzo C, Wang Y, et al. N Engl J Med 2007;356:580–92.
[80] Leonardi CL, Kimball AB, Papp KA, Yeilding N, Guzzo C, Wang Y, et al. Lancet 2008;371:1665–74.
[81] Papp KA, Langley RG, Lebwohl M, Krueger GG, Szapary P, Yeilding N, et al. Lancet 2008;371:1675–84.
[82] Griffiths CE, Strober BE, van de Kerkhof P, Ho V, Fidelus-Gort R, Yeilding N, et al. N Engl J Med 2010;362:118–28.
[83] Papp KA, Griffiths CE, Gordon K, Lebwohl M, Szapary PO, Wasfi Y, et al. Br J Dermatol 2013;168:844–54.
[84] Suzuki E, Mellins ED, Gershwin ME, Nestle FO, Adamopoulos IE. Autoimmun Rev 2014;13:496–502.
[85] McInnes IB, Sieper J, Braun J, Emery P, van der Heijde D, Isaacs JD, et al. Ann Rheum Dis 2014;73:349–56.
[86] Mease PJ. Efficacy of brodalumab, an anti-IL-17R antibody, in subjects with psoriatic arthritis; 2014.
[87] Gottlieb A, Menter A, Mendelsohn A, Shen YK, Li S, Guzzo C, et al. Lancet 2009;373:633–40.
[88] McInnes IB, Kavanaugh A, Gottlieb AB, Puig L, Rahman P, Ritchlin C, et al. Lancet 2013;382:780–9.
[89] Ritchlin C, Rahman P, Kavanaugh A, McInnes IB, Puig L, Li S, et al. Ann Rheum Dis 2014;73(6):990–9. doi: 10.1136/annrheumdis-2013-204655. Epub 2014 Jan 30.

[90] Kavanaugh A, Ritchlin C, Rahman P, Puig L, Gottlieb AB, Li S, et al. Ann Rheum Dis 2014;73(6):1000–6. doi: 10.1136/annrheumdis-2013-204741. Epub 2014 Feb 19.
[91] Baeten D, Baraliakos X, Braun J, Sieper J, Emery P, van der Heijde D, et al. Lancet 2013;382:1705–13.
[92] Baraliakos J, Braun DD, Laurent D, Baeten D, van der Heijde J, Sieper P, et al. 2014.
[93] Poddubnyy D, Hermann KG, Callhoff J, Listing J, Sieper J. Ann Rheum Dis 2014;73(5):817–23. doi: 10.1136/annrheumdis-2013-204248. Epub 2014 Jan 3.
[94] Puel A, Cypowyj S, Bustamante J, Wright JF, Liu L, Lim HK, et al. Science 2011;332:65–8.
[95] van de Vosse E, Ottenhoff TH. Microbes. Infect 2006;8:1167–73.
[96] Filipe-Santos O, Bustamante J, Chapgier A, Vogt G, de BL, Feinberg J, et al. Semin Immunol 2006;18:347–61.
[97] Tsai TF, Ho V, Song M, Szapary P, Kato T, Wasfi Y, et al. Br J Dermatol 2012;167:1145–52.
[98] Reich K, Papp KA, Griffiths CE, Szapary PO, Yeilding N, Wasfi Y, et al. J Drugs Dermatol 2012;11:300–12.

Chapter 29

# Discovery and Development of Anti-TNF Therapy: Pillar of a Therapeutic Revolution

Marc Feldmann

*Nuffield Department of Orthopaedics, Rheumatology, and Musculoskeletal Sciences, Kennedy Institute of Rheumatology, University of Oxford, Headington, Oxford, UK*

## GLOSSARY

**CD** Cluster of differentiation
**CRP** C-reactive protein
**CsA** Cyclosporin A
**CTLA4** Cytotoxic T-lymphocyte-associated protein 4
**GM-CSF** Granulocyte–macrophage colony-stimulating factor
**HIV** Human immunodeficiency virus
**Ig** Immunoglobulin
**IL** Interleukin
**MHC** Major histocompatibility complex
**MTX** Methotrexate
**OA** Osteoarthritis
**R** Receptor
**RA** Rheumatoid arthritis
**TGF** Transforming growth factor
**TNF** Tumor necrosis factor

## 1. INTRODUCTION

The need for healing and therapy is ingrained in the human race, probably predating civilization. In the past few centuries traditional medicines, mostly herbal, have been supplemented by the products of the chemical industry, which in the past century have become increasingly sophisticated, but also hugely costly to develop. With the beginning of the twenty-first century, we have seen the birth of a new era of "biologicals," chiefly monoclonal antibodies and their analogs and derivatives.

It is beyond the scope of this chapter to discuss all the possible aspects of this biologicals therapeutic stream. We concentrate here on how our contribution, defining tumor necrosis factor (TNF) as a target and developing anti-TNF treatment through to therapeutic success, has helped drive this therapeutic revolution.

The 1970s were an exciting time to be a scientist. New technologies were emerging that promised to revolutionize scientific progress in medicine. Recombinant DNA technology enabled the cloning of cDNA and subsequently exploring the nature of genes and the expression of recombinant proteins. Molecular biology has become the foundation of a new biology, providing ever-expanding molecular insights, extending to genome mapping and whole genome sequencing, and sequencing of all the expressed RNAs.

Immunology was revolutionized by the generation of monoclonal antibodies, first in mice by Kohler and Milstein [1]. This enabled the generation of unlimited quantities of antibodies to any given structure and specificity and enabled the detailed characterization of the proteins on the surface of cells, among other successes. But these mouse monoclonal antibodies were highly immunogenic and so only one mouse monoclonal antibody was commercialized as a therapeutic, Orthoclone OKT3 [2], reacting with CD3 and useful in patients with graft rejection.

With the involvement of molecular biology in monoclonal antibody technology, monoclonal antibodies evolved from research tools into potential therapeutics. First the backbone of the mouse antibodies was replaced, generating chimeric antibodies with only the binding site, the Fv region, remaining murine. This was the work of Morrison, Oi, and Herzenberg [3]. This then evolved to fully human antibodies generated in a variety of ways, of which Greg Winter's [4,5] contribution of phage display has been highly significant, and now human anti-TNF tops the charts of best-selling medicines (Adalimumab, Humira).

## 2. HOW WAS TNF DEFINED AS A THERAPEUTIC TARGET?

After my colleagues and I proposed a new hypothesis [6] for how autoimmune diseases might be triggered by upregulated antigen presentation and cytokine production, testing

**Molecular Biology of B Cells. http://dx.doi.org/10.1016/B978-0-12-397933-9.00029-1**

this concept with human Graves's disease [7], it was logical to pursue a major disease with significant unmet medical need, with local features. Rheumatoid arthritis (RA) is a very common (~1%) and, if not appropriately treated, crippling and life-shortening disease [8]. Thus in the 1980s numerous patients were in wheelchairs or walking frames. RA was a suitable disease for study, as the disease tissues are readily accessible, by biopsy or by surgery, and were thus amenable to detailed cellular and molecular analysis, using well-characterized patients in a collaboration with Ravinder Maini.

The only known upregulators of antigen presentation in the mid-1980s were cytokines, such as the interferons defined by Steeg and Oppenheim [9]. Both Ravinder Maini, from his work with Dudley Dumonde on mitogenic supernatants of lymphoid cells [10], and I, from my work in soluble mediators of cell collaboration with Tony Basten [11], were deeply interested in soluble mediators of activation. So we decided to explore which cytokines might be of importance in RA, using newly developed tools of molecular biology, but also of immune cell culture.

The key to understanding the molecular mechanism of a chiefly local disease is to know what is being produced at the disease site. We thus chose to monitor cytokine synthesis, monitoring mRNA expression in the disease site using cDNA probes. These were generous gifts from academics (Kishimoto, Taniguchi) [12,13] as well as biotechnology companies (Genentech, etc.) [14,15]. From biopsy or surgical samples of RA synovium (the lining layer of the joints), mRNAs for all the cytokines for which we had appropriate probes were detected in essentially all samples [16–18]. This initial observation was of great interest. Normally cytokines are transiently expressed in response to stimuli, so it would be expected that not all samples would express any given cytokine mRNA. But we found all the cytokines in all the samples, which indicated that in disease tissue, cytokine synthesis was continuous: clearly this is highly consistent with a chronic disease and augured well for potentially finding a cytokine therapeutic target [16–18].

The reasons for this cytokine overexpression were explored; mRNA stability was not altered, and so it was due to continuous induction of mRNA in the cell milieu of the disease tissue. Culturing synovial cells from RA joints provided confirmation of this. The technique used for this work was different from those of prior experiments reported. Most "synovial" cultures comprised just the fibroblast-like synovial cells, representing about 10% of the synovium. The rest, the great majority, are immune and inflammatory cells. Prior culture techniques discarded the inflammatory/immune cells and kept the fibroblast-like cells. To an immunologist like myself this made no sense. Our analyses were of the supernatant products of this complex cell mixture, comprising ~20–40% lymphocytes and 40% macrophages, with smaller numbers of plasma cells, dendritic cells, and endothelial cells [19]. We found that there was long-term production into culture supernatants of proinflammatory mediators such as IL-1, GM-CSF, and IL-6, consistent with mRNA analysis [20,21].

Progressively we were cataloging the expression of multiple proinflammatory cytokines, e.g., TNFα, IL-1, IL-6, interferon-γ, GM-CSF, and IL-2, as well as several antiinflammatory cytokines, e.g., IL-10, TGFβ [22]. Other groups using different techniques were also cataloging cytokine expression in joints, e.g., Jean-Michel Dayer, Gordon Duff, Gary Firestein, etc., and yielding similar conclusions [23–25].

The expression of multiple cytokines raised a challenge. The key question was which, if any of the proinflammatory cytokines might be therapeutic targets. Based on the evidence that injected or in vitro recombinant proinflammatory cytokines, e.g., IL-1, TNF, GM-CSF, and IL-6, have markedly overlapping biologies, a property known as "cytokine redundancy" [26], many groups abandoned the possibility that cytokines might be good therapeutic targets for RA, as clearly the new products of biotechnology, engineered monoclonal antibodies [27] or receptor Ig fusion proteins [28], would target only a single mediator. The remaining cytokines would be able to drive disease.

Our approach was to delve deeper into cytokine regulation in the joints, using our mixed synovial cultures, to investigate the molecules driving the production of proinflammatory mediators. As immunologists, the tools we used were antibodies to potential activators, including cytokines [20], but also immune complexes [29].

Because the synovial culture production of IL-1 was prolonged in our cultures, and many experts believed based chiefly on a rabbit joint injection model that IL-1 was the major driver of RA, we explored the regulation of synovial IL-1 production.

The key experiment that opened up the therapeutic path was performed in 1988 by Fionula Brennan, a postdoctoral student of mine at the time, who subsequently became a professor and a lead scientist at the Kennedy Institute. The results were striking [20].

Anti-TNF antibody rapidly switched off the production of IL-1, reducing it in most cases to baseline. This was not the case in osteoarthritis synovial cultures, and so there was the first clue that TNF might be a therapeutic target, as it regulated the production of other proinflammatory cytokines, IL-1, and subsequently GM-CSF and IL-6, in human disease tissue culture, used as a model of what might be happening in vivo [22]. With other antibodies tested, including antibodies to TNF's close sibling, lymphotoxin, which binds to the same cell surface receptor as TNF, the results were negative. So blocking TNF was special.

In the late 1980s animal models of RA were not believed to be highly predictive. That of course has not changed.

Collagen-induced arthritis (CIA) has a genetic predisposition in the major histocompatibility complex and many histological similarities to RA [30]. But T cell depletion with anti-CD4 monoclonal antibodies, which was highly effective in this mouse model [31], was not successful in human patients [32]. These findings emerged at the time we were considering testing our hypothesis that TNF was a therapeutic target in an in vivo situation.

Hamster antimouse TNFα antibody was generously given to us by Bob Schreiber [33] and this was very effective if administered after the disease onset, reducing inflammation leukocyte infiltration and preventing damage to cartilage and bone in the joints of the treated mice [34]. Concurrently other groups also showed that TNF blockade ameliorated CIA [35] and also that human TNF-overexpressing transgenic mice developed severe erosive arthritis [36].

Thus we had generated a scientific rationale for anti-TNF therapy for RA based on three aspects: first, TNF and TNF receptor are upregulated in the joint [37]; second, TNF is dysregulated and at the apex of a cytokine cascade driving IL-1, IL-6, and GM-CSF [20,21,38]; and third, anti-TNF therapy postonset was efficacious in an animal model of arthritis [34,35].

The next and major challenge was to obtain anti-TNF monoclonal antibody for testing our hypothesis by treating active RA patients. In the 1980s Tony Cerami and his colleagues Bruce Beutler and Kevin Tracey had established that animal models of bacterial septic shock were driven by TNF, with successful anti-TNF therapy administered very early after the bacteria [39,40]. The major unmet medical need of septic shock (300,000 deaths per year) led many biotechnology and pharmaceutical companies to generate TNF-inhibitory monoclonals or Ig–TNF receptor fusion proteins. Thus there were multiple companies that had generated anti-TNF biologicals that were being used in patients with sepsis.

There was thus no need to generate one's own anti-TNF for the specific purpose of treating RA. This would have delayed us by several years. However, a key challenge was to convince these companies to work with us to test our hypothesis, which at the time appeared "heretical." Many more thought IL-1 was the better target, but probably the majority thought that it was unlikely that blocking a single mediator would have major clinical benefit in a complex multigenic disease with many proinflammatory mediators [41].

We were fortunate when an ex-colleague, Dr James Woody, joined one of the companies (Centocor) that had made an anti-TNF monoclonal antibody. This was a chimeric antibody with a human backbone [42]. He was readily convinced to work with us while his own rheumatologists at Centocor continued to work on anti-CD4 antibodies, which was a much more favored approach.

## 3. ESTABLISHING THE CLINICAL UTILITY OF ANTI-TNF THERAPY

Despite the fact that anti-TNF had been used in hundreds, if not thousands, of patients with very, very severe infection, bacterial sepsis, we were unable to go straight to a placebo-controlled clinical trial, because of the potential infectious risk of blocking an important host defense molecule, TNF, with anti-TNF.

So we had to start (in May 1992) with an open-label trial (all patients treated with active agent) to evaluate what happened, with Maini and Feldmann as the principal investigators. Because of the potential risk, as patients with severe RA have a compromised immune system, the infusions of anti-TNF were performed on the subjects as inpatients (in hospital), and our nurse stayed with the patients overnight to ensure that any problems or discomfort could be promptly dealt with.

The results were dramatic. Patients who had active disease for 10 or more years, after having failed all available treatments, now rapidly started to feel better: there was less fatigue within hours, joint pain was reduced in a few days, and the patients were very pleased with the response. For many weeks some thought they may have been cured, but all relapsed between 12 and 18 weeks after the onset of treatment (20 mg/kg over 2 weeks) [43]. But as all patients had responded, although to varying degrees, and serum inflammatory mediators (CRP, IL-6, etc.) rapidly diminished, we were convinced this was not a placebo response, but a genuine therapeutic effect, and we wanted to scale up as quickly as possible.

Another 10 patients were then successfully treated, and based on the degree of clinical responses, patient enthusiasm, and apparent safety, permission was obtained to re-treat some of these patients for several courses after they had clinically relapsed. Because they had responded well to 10 mg/kg, this was the dose used for re-treatment. Again there was a very good clinical and serological response, although the degree of response and its duration varied [44].

With the half-dose, and the very small number of patients retreated (seven patients) it was difficult to know whether an apparently briefer duration of clinical response might be due to the immunogenicity of the therapeutic, which is chimeric, ¼ mouse, ¾ human, or whether it was the lower dose. But an important fact had been established, that patients who had stopped responding to anti-TNF as the drug was metabolized could nevertheless respond to up to three further courses. This appeared to validate the concept of a TNF-dependent cytokine cascade.

So the next challenges were to formally prove that anti-TNF was effective: many open clinical trials in RA, e.g., anti-CD4 and anti-CD5, which were encouraging as open studies, failed as proper placebo-controlled randomized clinical trials were performed. If that hurdle could be passed, the net challenge was to define a strategy for long-term use of anti-TNF and to minimize potential risks and immunogenicity.

## 4. PROOF OF EFFICACY

Based on the rapid effectiveness of 10 mg/kg anti-TNF, the placebo-controlled randomized trial was performed in 1993 with two doses of anti-TNF, 10 mg/kg and 1 mg/kg, and a placebo control that was human serum albumin. Albumin was used, as it seemed unethical to potentially immunize patients to mouse immunoglobulin; that could be deleterious in the future.

This was a multicenter European trial, with the Kennedy Institute working with Professor Joachim Kalden in Erlangen, Professor Ferdinand Breedveld in Leiden, and Professor Josef Smolen in Vienna [45].

This trial was performed after a minimum 4-week washout of other antirheumatic drugs, although low-dose steroids and nonsteroidal antiinflammatory drugs, both at a stable doses, were permitted.

The results, analyzed at 4 weeks using the Paulus composite criteria (the precursor of the ACR criteria response), were very clear. Seventy-nine percent responded at 10 mg/kg, 44% to 1 mg/kg, and 7% to placebo. Clinical trials, especially small ones like this (73 patients), do not come clearer than this [45].

## 5. OPTIMIZING LONG-TERM USE

Because of the risk of immunogenicity to anti-TNF used for long-term therapy in a chronic disease, we went back to the laboratory to investigate approaches to enhance the therapeutic benefit and reduce immunogenicity. The CIA model was used and Richard Williams showed the dramatic effect of combining anti-TNF, at therapeutic or subtherapeutic doses, which were not efficacious if used after disease onset, with anti-CD4 [46]. This enhanced therapeutic effect was also noted with other anti-T cell agents, the immunosuppressant drug cyclosporin A (CsA) or CTLA4–Ig [47]. Most notably the duration of response was augmented, and the dose of anti-TNF could be reduced.

Despite the fact that Centocor had a chimeric anti-CD4, which it had used in RA clinical trials, it was not possible, for regulatory (and safety) reasons, to use two experimental drugs together in a clinical trial. So with the complexity of titrating doses of CsA and its potential toxicity excluding it from serious consideration, it was a challenge to decide what therapeutics to combine with anti-TNF to enhance its efficacy. Many candidates were considered, and we settled on methotrexate (MTX), which reduces interferon-γ production and hence has some anti-T cell effects. With very low dose MTX being increasingly used in RA, testing anti-TNF in patients not doing well in MTX was the longer term trial that we eventually performed, and it was focused on an unmet medical need, what to do with patients not responding well to MTX [48].

This clinical trial, as we were concerned about the risks of infection with combination, was hampered by Centocor's parlous financial state, so groups of only 15 patients were used for exploring the interaction of ultra-low-dose MTX (7.5 mg/week) with various doses of anti-TNF (1, 3, 10 mg/kg). This European clinical trial began in 1995 and finished in 1996.

The results of five doses given over 12 weeks, followed subsequently to week 26, were striking. There was a greater and more durable response in the presence of "almost homeopathic" very low dose MTX. The dose at 7.5 mg/week was much lower than most patients had "failed" on, but was used for safety reasons as we were concerned about the risks of infection with combination.

Since then all anti-TNFs have been tested in combination with MTX [48,49] and the majority of RA patients are on the combination. This combination is a pointer for the future: combination treatment is the way to get augmented benefits, as very clearly shown in HIV [50,51].

Despite the positive effect of the MTX combination in a small number of patients with Crohn disease, this combination is not used, which may explain the less predictable results for Crohn disease [52].

## 6. PHASE III CLINICAL TRIALS

Because of the enhanced clinical effect with low-dose MTX, phase III was performed (3 mg/kg and 10 mg/kg) on top of existing MTX therapy given every 4 or 8 weeks. The duration of treatment was prolonged and the data were analyzed at 1 and 2 years.

This was performed in both Europe and the United States and amply confirmed the result of the small phase II trial. Based on the relatively good efficacy of 3 mg/kg every 8 weeks after a loading regimen this became the approved starting dose for cost and supply reasons. Because of the prolonged treatment it was possible to also analyze the effects of therapy on joint protection. There was a marked degree of both bone and cartilage protection at all anti-TNF doses compared to MTX alone [53,54].

## 7. CONCLUSIONS

The development of anti-TNF therapy was atypical at the time. The therapeutic target was discovered by academics, and as is usual for academics, but not for industry, it was put into the public domain.

The clinical trials were performed by industry, but were academically led first by Ravinder Maini and then joined by Peter Lipsky for the phase III trial.

Once the pilot of 10 patients, the clinical proof of principle, was performed in April/May of 1992, permission was requested to present the data at a translational research conference, "From the Laboratory to the Clinic," which we have run since 1984. In 1992, it was held in Arad, Israel, by the Dead Sea, in September.

Centocor gave us permission to present the data. Were they aware of its potential importance? Rapid disclosure of what seemed to be an effective target resulted in many other companies, e.g., Roche and Immunex, that had already generated anti-TNF therapeutics for sepsis, changing their strategy to initiate RA clinical studies.

The rest of the story is well known in rheumatology. Anti-TNF therapy is now the standard of care, once patients have not responded well enough to MTX. What had been a very pleasant surprise has become the degree of efficacy and relative safety.

Prior to anti-TNF the existing therapies did not arrest joint destruction. Anti-TNF did, and it is interesting that bone damage protection occurs even in patients who, by conventional criteria, which assess chiefly inflammation, are "nonresponders" to anti-TNF. Cartilage damage is also reduced, but to a lesser degree. So it is of considerable interest that blocking a single molecule can lead to such wide-ranging effects, ameliorating all aspects of the disease.

The mechanisms of action and of the clinical benefit have been extensively studied and are partially understood [55,56]. Less well understood is why certain patients, about 20–30%, respond very little, and perhaps 50% respond insufficiently. It is clearly not genetic, as nonresponders can respond to some or another anti-TNF a few months later. Understanding the nature of low responsiveness to anti-TNF is a major challenge, which if solved will have significant clinical benefit [57].

For companies involved in TNF blockade the results are dramatic, as the sales of anti-TNFs have now reached $25 billion per year. So anti-TNF is the best-selling drug class since some time in 2012. For many big pharmaceutical companies, Abbott, J&J, Amgen, their anti-TNF is their best-selling drug, and thus the credibility of antibodies as therapeutics is beyond challenge. Not so many years ago, the chemists working in many pharmaceutical companies considered antibodies as "tools," which would be rapidly replaced by small chemicals.

But antibodies, the evolutionary products of millions of years of selection for effective multipurpose defense molecules, are very effective and very specific.

The safety of antibodies, including anti-TNF, has been very good. If properly selected and effectively manufactured, there are no off-target effects, which so bedevil small-molecule therapeutics.

Antibodies have come of age as a therapeutic revolution. Apart from "naked" antibodies, there is in clinical development a horde of potential improvements, cut-down antibodies, single-chain-domain antibodies, and bifunctional antibodies, which can allow antibodies to be made more cheaply.

But there are still challenges. Anti-TNF therapy ameliorates RA and many other diseases. Approved indications apart from RA also include Crohn disease, ulcerative colitis, psoriasis, psoriatic arthritis, ankylosing spondylitis, and juvenile RA [58–61].

And anti-TNF could be very beneficial in other diseases, e.g., fibrosis of the hand and postoperative cognitive dysfunction, for which my colleagues and I have abundant preclinical data [62–64].

Anti-TNF is not a cure for RA, however. Only a very few patients, treated very early, are close to a drug-free cure. The real challenge is how to get closer to a cure. I believe this will, like the now very effective therapy for HIV and some leukemias, depend on combination therapy. Defining which targets to block to safely get closer to a cure will be a difficult task. But with the tools and knowledge available, it is not an unreachable goal.

## ACKNOWLEDGMENTS

The work described could not have been done without the help of many friends and colleagues. Sir Ravinder Maini and I have worked since 1984 on this project and it has been a lot of fun to exchange ideas and solve challenges together. A key early postdoctoral student was Glenn Buchan, who helped adapt many of the molecular biology techniques to work on small human samples. Fionula Brennan took this over when he left and was a key contributor over a more than 20 years time span. Regrettably, both died in their 50s of cancer, both still had much to contribute to science and medicine beyond their invaluable contributions to this work. Richard Williams has been critical in validating the effect on anti-TNF in mice and exploring potential interactions with other therapeutics. Ravinder Maini, Peter Taylor, Peter Lipsky, Josef Smolen, Ferdinand Breedveld, and Joachim Kalden led the clinical trials. Peter Charles, Ewa Paleolog, and Fionula Brennan contributed to exploring the mechanism of action. But there are many more. A pivotal colleague was James N. Woody. He was the Chief Scientist at Centocor when we needed a suitable anti-TNF to test our ideas in humans. He worked with us when others had decided that our ideas were too risky and not worth testing. This work was funded by Arthritis Research UK, the MRC, The Wellcome Trust, the Nuffield Foundation, and increasingly by the Kennedy Trust for Rheumatological Research. The clinical trials were funded by Centocor, who also funded research.

## REFERENCES

[1] Kohler G, Milstein C. Continuous cultures of fused cells secreting antibody of predefined specificity. Nature 1975;256:495–7.

[2] Cosimi AB, Colvin RB, Burton RC, Rubin RH, Goldstein G, Kung PC, et al. Use of monoclonal antibodies to T-cell subsets for immunologic monitoring and treatment in recipients of renal allografts. N Engl J Med 1981;305:308–14.

[3] Morrison SL, Johnson MJ, Herzenberg LA, Oi VT. Chimeric human antibody molecules: mouse antigen-binding domains with human constant region domains. Proc Natl Acad Sci USA 1984;81:6851–5.

[4] Winter G, Harris WJ. Humanized antibodies. Immunol Today 1993;14:243–6.

[5] McCafferty J, Griffiths AD, Winter G, Chiswell DJ. Phage antibodies: filamentous phage displaying antibody variable domains. Nature 1990;348:552–4.

[6] Bottazzo GF, Pujol-Borrell R, Hanafusa T, Feldmann M. Role of aberrant HLA-DR expression and antigen presentation in induction of endocrine autoimmunity. Lancet 1983;2:1115–9.

[7] Londei M, Bottazzo GF, Feldmann M. Human T-cell clones from autoimmune thyroid glands: specific recognition of autologous thyroid cells. Science 1985;228:85–9.

[8] Pincus T, Callahan LF. What is the natural history of rheumatoid arthritis? Rheum Dis Clin North Am 1993;19:123–51.

[9] Steeg PS, Moore RN, Johnson HM, Oppenheim JJ. Regulation of murine macrophage Ia antigen expression by a lymphokine with immune interferon activity. J Exp Med 1982;156:1780–93.

[10] Maini RN, Bryceson AD, Wolstencroft RA, Dumonde DC. Lymphocyte mitogenic factor in man. Nature 1969;224:43–4.

[11] Feldmann M, Basten A. Specific collaboration between T and B lymphocytes across a cell impermeable membrane in vitro. Nat New Biol 1972;237:13–5.

[12] Hirano T, Yasukawa K, Harada H, Taga T, Watanabe Y, Matsuda T, et al. Complementary DNA for a novel human interleukin (BSF-2) that induces B lymphocytes to produce immunoglobulin. Nature 1986;324:73–6.

[13] Taniguchi T, Matsui H, Fujita T, Takaoka C, Kashima N, Yoshimoto R, et al. Structure and expression of a cloned cDNA for human interleukin-2. Nature 1983;302:305–10.

[14] Pennica D, Nedwin GE, Hayflick JS, Seeburg PH, Derynck R, Palladino MA, et al. Human tumour necrosis factor: precursor structure, expression and homology to lymphotoxin. Nature 1984;312:724–9.

[15] Gray PW, Aggarwal BB, Benton CV, Bringman TS, Henzel WJ, Jarrett JA, et al. Cloning and expression of cDNA for human lymphotoxin, a lymphokine with tumour necrosis activity. Nature 1984;312:721–4.

[16] Hirano T, Matsuda T, Turner M, Miyasaka N, Buchan G, Tang B, et al. Excessive production of interleukin 6/B cell stimulatory factor-2 in rheumatoid arthritis. Eur J Immunol 1988;18:1797–801.

[17] Buchan G, Barrett K, Fujita T, Taniguchi T, Maini R, Feldmann M. Detection of activated T cell products in the rheumatoid joint using cDNA probes to Interleukin-2 (IL-2) IL-2 receptor and IFN-gamma. Clin Exp Immunol 1988;71:295–301.

[18] Buchan G, Barrett K, Turner M, Chantry D, Maini RN, Feldmann M. Interleukin-1 and tumour necrosis factor mRNA expression in rheumatoid arthritis: prolonged production of IL-1 alpha. Clin Exp Immunol 1988;73:449–55.

[19] Brennan FM, Chantry D, Jackson AM, Maini RN, Feldmann M. Cytokine production in culture by cells isolated from the synovial membrane. J Autoimmun 1989;(2 Suppl.):177–86.

[20] Brennan FM, Chantry D, Jackson A, Maini R, Feldmann M. Inhibitory effect of TNF alpha antibodies on synovial cell interleukin-1 production in rheumatoid arthritis. Lancet 1989;2:244–7.

[21] Haworth C, Brennan FM, Chantry D, Turner M, Maini RN, Feldmann M. Expression of granulocyte-macrophage colony-stimulating factor in rheumatoid arthritis: regulation by tumor necrosis factor-alpha. Eur J Immunol 1991;21:2575–9.

[22] Feldmann M, Brennan FM, Maini RN. Role of cytokines in rheumatoid arthritis. Annu Rev Immunol 1996;14:397–440.

[23] Rooney M, Symons JA, Duff GW. Interleukin 1 beta in synovial fluid is related to local disease activity in rheumatoid arthritis. Rheumatol Int 1990;10:217–9.

[24] Roux-Lombard P, Modoux C, Vischer T, Grassi J, Dayer JM. Inhibitors of interleukin 1 activity in synovial fluids and in cultured synovial fluid mononuclear cells. J Rheumatol 1992;19:517–23.

[25] Xu WD, Firestein GS, Taetle R, Kaushansky K, Zvaifler NJ. Cytokines in chronic inflammatory arthritis. II. Granulocyte-macrophage colony-stimulating factor in rheumatoid synovial effusions. J Clin Invest 1989;83:876–82.

[26] Paul WE. Pleiotropy and redundancy: T cell-derived lymphokines in the immune response. Cell 1989;57:521–4.

[27] Morrison SL, Wims L, Wallick S, Tan L, Oi VT. Genetically engineered antibody molecules and their application. Ann N Y Acad Sci 1987;507:187–98.

[28] Mohler KM, Torrance DS, Smith CA, Goodwin RG, Stremler KE, Fung VP, et al. Soluble tumor necrosis factor (TNF) receptors are effective therapeutic agents in lethal endotoxemia and function simultaneously as both TNF carriers and TNF antagonists. J Immunol 1993;151:1548–61.

[29] Chantry D, Winearls CG, Maini RN, Feldmann M. Mechanism of immune complex-mediated damage: induction of interleukin 1 by immune complexes and synergy with interferon-gamma and tumor necrosis factor-alpha. Eur J Immunol 1989;19:189–92.

[30] Courtenay JS, Dallman MJ, Dayan AD, Martin A, Mosedale B. Immunisation against heterologous type II collagen induces arthritis in mice. Nature 1980;283:666–8.

[31] Ranges GE, Sriram S, Cooper SM. Prevention of type II collagen-induced arthritis by in vivo treatment with anti-L3T4. J Exp Med 1985;162:1105–10.

[32] van der Lubbe PA, Dijkmans BA, Markusse HM, Nassander U, Breedveld FC. A randomized, double-blind, placebo-controlled study of CD4 monoclonal antibody therapy in early rheumatoid arthritis. Arthritis Rheum 1995;38:1097–106.

[33] Sheehan KC, Ruddle NH, Schreiber RD. Generation and characterization of hamster monoclonal antibodies that neutralize murine tumor necrosis factors. J Immunol 1989;142:3884–93.

[34] Williams RO, Feldmann M, Maini RN. Anti-tumor necrosis factor ameliorates joint disease in murine collagen-induced arthritis. Proc Natl Acad Sci USA 1992;89:9784–8.

[35] Thorbecke GJ, Shah R, Leu CH, Kuruvilla AP, Hardison AM, Palladino MA. Involvement of endogenous tumor necrosis factor alpha and transforming growth factor beta during induction of collagen type II arthritis in mice. Proc Natl Acad Sci USA 1992;89:7375–9.

[36] Keffer J, Probert L, Cazlaris H, Georgopoulos S, Kaslaris E, Kioussis D, et al. Transgenic mice expressing human tumour necrosis factor: a predictive genetic model of arthritis. EMBO J 1991;10:4025–31.

[37] Brennan FM, Gibbons DL, Mitchell T, Cope AP, Maini RN, Feldmann M. Enhanced expression of tumor necrosis factor receptor mRNA and protein in mononuclear cells isolated from rheumatoid arthritis synovial joints. Eur J Immunol 1992;22:1907–12.

[38] Brennan FM, Maini RN, Feldmann M. TNF alpha – a pivotal role in rheumatoid arthritis? Br J Rheumatol 1992;31:293–8.

[39] Beutler B, Milsark IW, Cerami AC. Passive immunization against cachectin/tumor necrosis factor protects mice from lethal effect of endotoxin. Science 1985;229:869–71.

[40] Tracey KJ, Fong Y, Hesse DG, Manogue KR, Lee AT, Kuo GC, et al. Anti-cachectin/TNF monoclonal antibodies prevent septic shock during lethal bacteraemia. Nature 1987;330:662–4.

[41] Campion GV. The prospect for cytokine based therapeutic strategies in rheumatoid arthritis. Ann Rheum Dis 1994;53:485–7.

[42] Knight DM, Trinh H, Le JM, Siegel S, Shealy D, McDonough M, et al. Construction and initial characterization of a mouse-human chimeric anti-TNF antibody. Mol Immunol 1993;30:1443–53.

[43] Elliott MJ, Maini RN, Feldmann M, Long-Fox A, Charles P, Katsikis P, et al. Treatment of rheumatoid arthritis with chimeric monoclonal antibodies to tumor necrosis factor α. Arthritis Rheum 1993;36:1681–90.

[44] Elliott MJ, Maini RN, Feldmann M, Long-Fox A, Charles P, Bijl H, et al. Repeated therapy with monoclonal antibody to tumour necrosis factor alpha (cA2) in patients with rheumatoid arthritis. Lancet 1994;344:1125–7.

[45] Elliott MJ, Maini RN, Feldmann M, Kalden JR, Antoni C, Smolen JS, et al. Randomised double-blind comparison of chimeric monoclonal antibody to tumour necrosis factor α (cA2) versus placebo in rheumatoid arthritis. Lancet 1994;344:1105–10.

[46] Williams RO, Mason LJ, Feldmann M, Maini RN. Synergy between anti-CD4 and anti-tumor necrosis factor in the amelioration of established collagen-induced arthritis. Proc Natl Acad Sci USA 1994;91:2762–6.

[47] Williams RO, Mauri C, Mason LJ, Marinova-Mutafchieva L, Ross SE, Feldmann M, et al. Therapeutic actions of cyclosporine and anti-tumor necrosis factor alpha in collagen-induced arthritis and the effect of combination therapy. Arthritis Rheum 1998;41:1806–12.

[48] Maini RN, Breedveld FC, Kalden JR, Smolen JS, Davis D, Macfarlane JD, et al. Therapeutic efficacy of multiple intravenous infusions of anti-tumor necrosis factor alpha monoclonal antibody combined with low-dose weekly methotrexate in rheumatoid arthritis. Arthritis Rheum 1998;41:1552–63.

[49] Breedveld FC, Weisman MH, Kavanaugh AF, Cohen SB, Pavelka K, van Vollenhoven R, et al. The PREMIER study: a multicenter, randomized, double-blind clinical trial of combination therapy with adalimumab plus methotrexate versus methotrexate alone or adalimumab alone in patients with early, aggressive rheumatoid arthritis who had not had previous methotrexate treatment. Arthritis Rheum 2006;54:26–37.

[50] Weinblatt ME, Kremer JM, Bankhurst AD, Bulpitt KJ, Fleischmann RM, Fox RI, et al. A trial of etanercept, a recombinant tumor necrosis factor receptor: Fc fusion protein, in patients with rheumatoid arthritis receiving methotrexate. N Engl J Med 1999;340:253–9.

[51] Hammer SM, Squires KE, Hughes MD, Grimes JM, Demeter LM, Currier JS, et al. A controlled trial of two nucleoside analogues plus indinavir in persons with human immunodeficiency virus infection and CD4 cell counts of 200 per cubic millimeter or less. AIDS Clinical Trials Group 320 Study Team. N Engl J Med 1997;337:725–33.

[52] Schroder O, Blumenstein I, Stein J. Combining infliximab with methotrexate for the induction and maintenance of remission in refractory Crohn's disease: a controlled pilot study. Eur J Gastroenterol Hepatol 2006;18:11–6.

[53] Lipsky PE, van der Heijde DM, St Clair EW, Furst DE, Breedveld FC, Kalden JR, et al. Infliximab and methotrexate in the treatment of rheumatoid arthritis. Anti-Tumor Necrosis Factor Trial in Rheumatoid Arthritis with Concomitant Therapy Study Group. N Engl J Med 2000;343:1594–602.

[54] Maini R, St Clair EW, Breedveld F, Furst D, Kalden J, Weisman M, et al. Infliximab (chimeric anti-tumour necrosis factor alpha monoclonal antibody) versus placebo in rheumatoid arthritis patients receiving concomitant methotrexate: a randomised phase III trial. ATTRACT Study Group. Lancet 1999;354:1932–9.

[55] Paleolog EM, Hunt M, Elliott MJ, Feldmann M, Maini RN, Woody JN. Deactivation of vascular endothelium by monoclonal anti-tumor necrosis factor alpha antibody in rheumatoid arthritis. Arthritis Rheum 1996;39:1082–91.

[56] Taylor PC, Peters AM, Paleolog E, Chapman PT, Elliott MJ, McCloskey R, et al. Reduction of chemokine levels and leukocyte traffic to joints by tumor necrosis factor alpha blockade in patients with rheumatoid arthritis. Arthritis Rheum 2000;43:38–47.

[57] Ulfgren AK, Andersson U, Engstrom M, Klareskog L, Maini RN, Taylor PC. Systemic anti-tumor necrosis factor alpha therapy in rheumatoid arthritis down-regulates synovial tumor necrosis factor alpha synthesis. Arthritis Rheum 2000;43:2391–6.

[58] Smolen JS, Weinblatt ME. When patients with rheumatoid arthritis fail tumour necrosis factor inhibitors: what is the next step? Ann Rheum Dis 2008;67:1497–8.

[59] van Dullemen HM, van Deventer SJ, Hommes DW, Bijl HA, Jansen J, Tytgat GN, et al. Treatment of Crohn's disease with anti-tumor necrosis factor chimeric monoclonal antibody (cA2). Gastroenterology 1995;109:129–35.

[60] Rutgeerts P, Sandborn WJ, Feagan BG, Reinisch W, Olson A, Johanns J, et al. Infliximab for induction and maintenance therapy for ulcerative colitis. N Engl J Med 2005;353:2462–76.

[61] Antoni CE, Kavanaugh A, Kirkham B, Tutuncu Z, Burmester GR, Schneider U, et al. Sustained benefits of infliximab therapy for dermatologic and articular manifestations of psoriatic arthritis: results from the infliximab multinational psoriatic arthritis controlled trial (IMPACT). Arthritis Rheum 2005;52:1227–36.

[62] van der Heijde D, Kivitz A, Schiff MH, Sieper J, Dijkmans BA, Braun J, et al. Efficacy and safety of adalimumab in patients with ankylosing spondylitis: results of a multicenter, randomized, double-blind, placebo-controlled trial. Arthritis Rheum 2006;54:2136–46.

[63] Terrando N, Monaco C, Ma D, Foxwell BM, Feldmann M, Maze M. Tumor necrosis factor-alpha triggers a cytokine cascade yielding postoperative cognitive decline. Proc Natl Acad Sci USA 2010;107:20518–22.

[64] Verjee LS, Verhoekx JS, Chan JK, Krausgruber T, Nicolaidou V, Izadi D, et al. Unraveling the signaling pathways promoting fibrosis in Dupuytren's disease reveals TNF as a therapeutic target. Proc Natl Acad Sci USA 2013;110:E928–37.

# Index

*Note*: Page numbers followed by "f" and "t" indicate figures and tables respectively.

## A

## B

## C

## D



## E



## F

## G



## H

## I

## N

## T



## U



## V

## W



## X



## Y



## Z